Parallel and Distributed
Computing Handbook

Computer Engineering Series

CHEN • *Computer Engineering Handbook,* 0-07-010924-9

CHEN • *Fuzzy Logic and Neural Network Handbook,* 0-07-073020-2

DEVADAS, DEUTZER, GHASH • *Logic Synthesis,* 0-07-016500-9

HOWLAND • *Computer Hardware Diagnostics for Engineers,*
   0-07-030561-7

LEISS • *Parallel and Vector Computing,* 0-07-037692-1

PERRY • *VHDL,* Second Edition, 0-07-049434-7

ROSENSTARK• *Transmission Lines in Computer Engineering,*
   0-07-053953-7

ZOMAYA • *Parallel and Distributed Computing Handbook,* 0-07-049906-3

Related Titles of Interest:

KIELKOWSKI • *SPICE Practical Parameter Modeling,* 0-07-911524-1

KIELOWSKI • *Inside SPICE,* 0-07-911525-X

MASSABRIO, ANTOGNETTI • *Semiconductor Device Modeling with SPICE,* Second Edition, 0-07-002469-3

To order or receive additional information on these or any other
McGraw-Hill titles, please call 1-800-822-8158 in the United States.
In other countries, contact your local McGraw-Hill representative.   **BC15XXA**

Parallel and Distributed Computing Handbook

Albert Y. Zomaya, Editor

McGraw-Hill

New York   San Francisco   Washington, D.C.   Auckland   Bogotá
Caracas   Lisbon   London   Madrid   Mexico City   Milan
Montreal   New Delhi   San Juan   Singapore
Sydney   Tokyo   Toronto

Library of Congress Cataloging-in-Publication Data

Parallel and distributed computing handbook / Albert Y. Zomaya, editor
     p.   cm. — (Computer engineering series)
  Includes index.
  ISBN 0-07-073020-2
  1. Parallel processing (Electronic computers). 2. Electronic data
processing—Distributed processing. I. Zomaya, Albert Y.
II. Series.
QA76.58.P3635   1996
004'.35—dc20   95-32594
                                     CIP

McGraw-Hill

A Division of The McGraw·Hill Companies

  2 3 4 5 6 7 8 9 0   BKP/BKP   9 0 0 9 8 7 6

ISBN 0-07-073020-2

The sponsoring editor for this book was Stephen S. Chapman and the production supervisor was Pamela Pelton. It was set in Times New Roman by J. K. Eckert & Company, Inc.

Printed and bound by Quebecor Book Press.

This book is printed on acid-free paper.

For the kindred spirits who contributed to the growth of parallel and distributed computing

"To accomplish great things, we must not only act but also dream, not only plan but also believe."

Anatole France

Contents

Foreword xix
Preface xxi
Acknowledgments xxiii
List of contributors xxv

Part I   Theory

Foundations

Chapter 1. Parallel and Distributed Computing: The Scene, the Props, the Players 5

   Albert Y. Zomaya

1.1	**A Perspective**	
1.2	**Parallel Processing Paradigms**	7
1.3	**Modeling and Characterizing Parallel Algorithms**	11
1.4	**Cost vs. Performance Evaluation**	13
1.5	**Software and General-Purpose PDC**	15
1.6	**A Brief Outline of the Handbook**	16
1.7	**Recommended Reading**	19
1.8	**References**	21

Chapter 2. Semantics of Concurrent Programming 24

   J. Desharnais, A. Mili, R. Mili, J. Mullins, and Y. Slimani

2.1	**Models of Concurrent Programming**	25
2.2	**Semantic Definitions**	27

2.3	**Axiomatic Semantic Definitions**	30
2.4	**Denotational Semantic Definitions**	36
2.5	**Operational Semantic Definitions**	54
2.6	**Summary and Prospects**	57
2.7	**References**	57

Chapter 3. Formal Methods: A Petri Nets Based Approach   59

Giorgio De Michelis, Lucia Pomello,
Eugenio Battiston, Fiorella De Cindio, and Carla Simone

3.1	**Process Algebras**	61
3.2	**PETRI Nets**	68
3.3	**High-Level Net Models**	78
3.4	**Conclusions**	84
3.5	**References**	86

Chapter 4. Complexity Issues in Parallel and Distributed Computing   89

E. V. Krishnamurthy

4.1	**Introduction**	89
4.2	**Turing Machine as the Basis, and Consequences**	93
4.3	**Complexity Measures for Parallelism**	101
4.4	**Parallel Complexity Models and Resulting Classes**	103
4.5	**VLSI Computational Complexity**	121
4.6	**Complexity Measures for Distributed Systems**	121
4.7	**Neural Networks and Complexity Issues**	121
4.8	**Other Complexity Theories**	123
4.9	**Concluding Remarks**	124
4.10	**References**	125

Chapter 5. Distributed Computing Theory   127

Hagit Attiya

5.1	**The Computation Model**	129
5.2	**A Simple Example**	131
5.3	**Leader Election**	132
5.4	**Sparse Network Covers and Their Applications**	138
5.5	**Ordering of Events**	142
5.6	**Resource Allocation**	146
5.7	**Tolerating Processor Failures in Synchronous Systems**	146
5.8	**Tolerating Processor Failures in Asynchronous Systems**	151
5.9	**Other Types of Failures**	154
5.10	**Wait-Free Implementations of Shared Objects**	156
5.11	**Final Comments**	157
5.12	**References**	158

Models

Chapter 6. PRAM MODELS   163

Lydia I. Kronsjö

6.1   Introduction   163
6.2   Techniques for the Design of Parallel Algorithms   165
6.3   The PRAM Model   168
6.4   Optimality and Efficiency of Parallel Algorithms   171
6.5   Basic PRAM Algorithms   175
6.6   The NC-Class   180
6.7   P-Completeness: Hardly Parallelizable Problems   180
6.8   Randomized Algorithms and Parallelism   181
6.9   List Ranking Revisited: Optimal $O(\log n)$ Deterministic List Ranking   184
6.10 Taxonomy of Parallel Algorithms   186
6.11 Deficiencies of the PRAM Model   186
6.12 Summary   189
6.13 References   189

**Chapter 7. Broadcasting with Selective Reduction: A Powerful Model of
              Parallel Computation**   192

Selim G. Akl and Ivan Stojmenović

7.1   Introduction   192
7.2   A Generalized BSR Model   197
7.3   One Criterion BSR Algorithms   200
7.4   Two Criteria BSR Algorithms   212
7.5   Three Criteria BSR Algorithms   215
7.6   Multiple Criteria BSR Algorithms   218
7.7   Conclusions and future Work   220
7.8   References   221

Chapter 8. Dataflow Models   223

R. Jagannathan

8.1   Kinds of Dataflow   224
8.2   Data-Driven Dataflow Computing Models   225
8.3   Demand-driven Dataflow Computing Models   230
8.4   Unifying Data-Driven and Demand-Driven   234
8.5   Lessons Learned and Future Trends   235
8.6   Summary   236
8.7   References   237

Chapter 9. Partitioning and Scheduling   239

Hesham El-Rewini

9.1   Program Partitioning   241
9.2   Task Scheduling   243

9.3   Scheduling System Model   244
9.4   Communication Models   248
9.5   Optimal Scheduling Algorithms   249
9.6   Scheduling Heuristic Algorithms   255
9.7   Scheduling Nondeterministic Task Graphs   262
9.8   Scheduling Tools   266
9.9   Task Allocation   267
9.10   Heterogeneous Environments   268
9.11   Summary and Concluding Remarks   270
9.12   References   272

Chapter 10. Checkpointing in Parallel and Distributed Systems   274

   Avi Ziv and Jehoshua Bruck

10.1   Introduction   274
10.2   Checkpointing Using Task Duplication   276
10.3   Techniques for Consistent Checkpointing   288
10.4   Conclusions and Future Directions   299
10.5   References   300

Chapter 11. Architecture for Open Distributed Software Systems   303

   Kazi Farooqui and Luigi Logrippo

11.1   Introduction to Open Distributed Systems Architecture   303
11.2   Computational Model   307
11.3   Engineering Model   315
11.4   ODP Application   324
11.5   Conclusion and Directions for Future Research   327
11.6   References   328

Algorithms

Chapter 12. Fundamentals of Parallel Algorithms   333

   Joseph F. Jájá

12.1   Introduction   333
12.2   Models of Parallel Computation   334
12.3   Balanced Trees   339
12.4   Divide and Conquer   342
12.5   Partitioning   345
12.6   Combining   350
12.7   Conclusions and Future Trends   353
12.8   Acknowledgment   354
12.9   References   354

Chapter 13. Parallel Graph Algorithms   355

    Stephan Olariu

  13.1 **Graph-Theoretic Concepts and Notation**   356
  13.2 **Tree Algorithms**   358
  13.3 **Algorithms for General Graphs**   372
  13.4 **Algorithms for Particular Classes of Graphs**   384
  13.5 **Concluding Remarks**   401
  13.6 **References**   402

Chapter 14. Parallel Computational Geometry   404

    Mikhail J. Atallah

  14.1 **Parallel CG: Why New Techniques Are Needed**   405
  14.2 **Basic Subproblems**   407
  14.3 **CG on the PRAM**   409
  14.4 **CG on the Mesh**   416
  14.5 **CG on the Hypercube**   419
  14.6 **Other Parallel Models**   420
  14.7 **Conclusions and Future Work**   422
  14.8 **References**   423

Chapter 15. Data Structures for Parallel Processing   429

    Sajal K. Das and Kwang-Bae Min

  15.1 **Arrays and Balanced Binary Trees**   430
  15.2 **Linked Lists**   432
  15.3 **Trees and Euler Tour**   434
  15.4 **General Trees and Binarized Trees**   435
  15.5 **Euler Tour vs. Parentheses String**   436
  15.6 **Stacks**   440
  15.7 **Queues**   445
  15.8 **Priority Queues (Heaps)**   448
  15.9 **Search Trees/Dictionaries**   455
  15.10 **Conclusions**   463
  15.11 **References**   464

Chapter 16. Data Parallel Algorithms   466

    Howard Jay Siegel, Lee Wang, John John E. So, and Muthucumaru Maheswaran

  16.1 **Chapter Overview**   466
  16.2 **Machine Model**   467
  16.3 **Impact of Data Distribution**   469
  16.4 **CU/PE Overlap**   476
  16.5 **Parallel Reduction Operations**   480
  16.6 **Matrix and Vector Operations**   487

16.7   Mapping Algorithms onto Partitionable Machines   489
16.8   Achieving Scalability Using a Set of Algorithms   492
16.9   Conclusions and Future Directions   494
16.10   References   497

Chapter 17. Systolic and VLSI Processor Arrays for Matrix Algorithms   500

            D. J. Evans and M. Gusev

17.1   Processor Array Implementations   500
17.2   VLSI Processor Arrays   501
17.3   Systolic Array Algorithms   508
17.4   Mathematical Methods in DSP   510
17.5   Implementation of Systolic Algorithms in DSP   517
17.6   Conjugate Gradient Method   530
17.7   Summary   535
17.8   References   536

Chapter 18. Direct Interconnection Networks   537

            Ivan Stojmenović

18.1   Topological Properties of Interconnection Networks   537
18.2   Hypercubic Networks   547
18.3   Routing and Broadcasting   555
18.4   Conclusions   563
18.5   References   564
18.6   Suggested Readings   565

Chapter 19. Parallel and Communication Algorithms on Hypercube
            Multiprocessors   568

            Afonso Ferreira

19.1   Topological Aspects   569
19.2   Communication Issues   573
19.3   Useful Algorithmic Tools   577
19.4   Solving Problems   581
19.5   Conclusions and Future Directions   587
19.6   References   588

Part II   Architectures and Technologies

Architectures

Chapter 20. RISC Architectures   595

            Manolis Katevenis

20.1   What is RISC?   596
20.2   Pipelining and Bypassing   599

20.3 Dependences and Parallelism in CISC and in RISC   604
20.4 Instruction Alignment, Size, and Format   609
20.5 Implementation Disadvantages of CISC   615
20.6 History, Perspective, and Conclusions   617
20.7 References   619

Chapter 21. Superscalar and VLIW Processors   621

        Thomas M. Conte

21.1 Superscalar Processors   622
21.2 VLIW Processors   634
21.3 Superscalar vs. VLIW: Which Is Better?   645
21.4 Bibliography   647

Chapter 22. SIMD-Processing: Concepts and Systems   649

        Michael Jurczyk and Thomas Schwederski

22.1 Basic Concepts   649
22.2 SIMD Machine Components   654
22.3 Associative Processing   660
22.4 Case Studies of SIMD Systems   662
22.5 Applications and Algorithms   669
22.6 Languages and Programming   673
22.7 Conclusions   677
22.8 References   677

Chapter 23. MIMD Architectures: Shared and Distributed Memory Designs   680

        Ralph Duncan

23.1 Proliferation of MIMD Designs   681
23.2 Shared Memory Architectures   682
23.3 Distributed Memory Architectures   689
23.4 Hybrid Shared/Distributed Memory Architectures   695
23.5 Conclusion   696
23.6 References   697

Chapter 24. Memory Models   699

        Leonidas I. Kontothanassis and Michael L. Scott

24.1 Memory Hardware Technology   700
24.2 Memory System Architecture   702
24.3 User-Level Memory Models   707
24.4 Memory Consistency Models   711
24.5 Implementation and Performance of Memory Consistency Models   714
24.6 Conclusions and Trends   718
24.7 References   719

Technologies

Chapter 25. Heterogeneous Computing   725

> *Howard Jay Siegel, John K. Antonio, Richard C. Metzger, Min Tan, and Yan Alexander Li*

25.1   Introduction   725
25.2   Mixed-Mode Systems   727
25.3   Examples of Existing Mixed-Machine HC Systems   733
25.4   Examples of Software Tools for Mixed-Machine HC Systems   735
25.5   A Conceptual Model for Automatic Mixed-Machine HC   739
25.6   Task Profiling and Analytical Benchmarking   741
25.7   Matching and Scheduling for Mixed-Machine HC Systems   747
25.8   Conclusions and Future Directions   756
25.9   References   758

Chapter 26. Cluster Computing   762

> *Louis Turcotte*

26.1   Technological Evolution   763
26.2   Overview of Clustering   767
26.3   Distinct Uses of Clusters   771
26.4   Open Issues   777
26.5   References   779

Chapter 27. Massively Parallel Processing with Optical Interconnections   780

> *Eugen Schenfeld*

27.1   Parallel Processing Motivations   782
27.2   General-Purpose Parallel Computers   785
27.3   How Much Interconnection?   793
27.4   Considerations in Choosing the Interconnection Topology   795
27.5   Optical Communication: Free-Space Interconnection   796
27.6   Conclusions and Future Work   808
27.7   References   808

Chapter 28. ATM-Based Parallel and Distributed Computing   811

> *Salim Hariri and Bei Lu*

28.1   Introduction   811
28.2   Broadband Integrated Service Data Network (B-ISDN)   812
28.3   ATM Protocols   814
28.4   ATM Switches   824
28.5   Host-to-Network Interfaces   831
28.6   Parallel and Distributed Computing Environment Over ATM   835
28.7   Conclusions and Future Directions   836
28.8   References   837

Part III Tools and Applications

Development Tools

Chapter 29. Parallel Languages   **843**

    R. H. Perrott

  29.1   Introduction   843
  29.2   Language Categories   844
  29.3   Programming Languages   846
  29.4   Summary   861
  29.5   References   862

Chapter 30. Tools for Portable High-Performance Parallel Computing   **865**

    Doreen Y. Cheng

  30.1   Introduction   865
  30.2   Criteria for Evaluating Portability Support   867
  30.3   Portable Message-Passing Libraries   871
  30.4   Language-Centered Tools   878
  30.5   Parallelizing Compilers and Preprocessors   886
  30.6   Conclusion   892
  30.7   References   893

Chapter 31. Visualization of Parallel and Distributed Systems   **897**

    Michael T. Heath

  31.1   Performance Monitoring   898
  31.2   Performance Visualization   899
  31.3   Example   910
  31.4   Future Directions   913
  31.5   References   915

**Chapter 32. Constructing Numerical Software Libraries for High-Performance
Computer Environments**   **917**

    Jack J. Dongarra and David W. Walker

  32.1   Introduction   917
  32.2   The BLAS as the Key to Portability   924
  32.3   Block Algorithms and Their Derivation   925
  32.4   LU Factorization   930
  32.5   Data Distribution   932
  32.6   Parallel Implementation   935
  32.7   Optimization, Tuning, and Trade-Offs   941
  32.8   Conclusions and Future Research Directions   948
  32.9   References   951

Chapter 33. Testing of Distributed Programs 955

K. C. Tai and Richard H. Carver

33.1   SYN-Sequences of Distributed Programs   956
33.2   Definitions of Correctness and Faults for Distributed Programs   961
33.3   Approaches to Testing Distributed Programs   963
33.4   Test Generation for Distributed Programs   968
33.5   Analysis and Replay of Program Executions   973
33.6   Building Testing Tools for Distributed Programs   975
33.7   Conclusions and Future Work   976
33.8   References   977

Applications

Chapter 34. Scientific Computation 981

Timothy G. Mattson

34.1   Programming Models for Parallel Computing   982
34.2   Algorithms for Parallel Scientific Computing   983
34.3   Case Studies: Molecular Modeling   989
34.4   Trends   997
34.5   Further Reading   999
34.6   Conclusion   1000
34.7   References   1001

Chapter 35. Parallel and Distributed Simulation of Discrete Event Systems 1003

Alois Ferscha

35.1   Simulation Principles   1003
35.2   "Classical" LP Simulation Protocols   1009
35.3   Conservative vs. Optimistic Protocols   1037
35.4   Conclusions and Outlook   1037
35.5   References   1039

Chapter 36. Parallelism for Image Understanding 1042

Viktor K. Prasanna and Cho-Li Wang

36.1   Vision Tasks   1046
36.2   A Model of CM-5   1050
36.3   Scalable Parallel Algorithms   1051
36.4   Implementation Details and Experimental Results   1060
36.5   Concluding Remarks   1068
36.6   References   1069

**Chapter 37.   Parallel Computation in Biomedicine: Genetic and
          Protein Sequence Analysis**   1071

          Tieng K. Yap, Ophir Frieder, and Robert L. Martino

    37.1   **The Origin of Genetic and Protein Sequence Data**   1072
    37.2   **An Example Database: GenBank**   1074
    37.3   **Residue Substitution Scoring Matrices**   1077
    37.4   **Sequence Comparison Algorithms**   1080
    37.5   **Parallel Techniques for Sequence Similarity Searching**   1083
    37.6   **Performance**   1089
    37.7   **Discussion and Conclusions**   1093
    37.8   **Future Work**   1095
    37.9   **References**   1095

**Chapter 38.   Parallel Algorithms for Solving Stochastic Linear
          Programs**   1097

          Amal De Silva and David Abramson

    38.1   **Stochastic Linear Programming**   1098
    38.2   **Techniques for Solving Stochastic Linear Programs**   1104
    38.3   **Comparison of Methods**   1113
    38.4   **Conclusion and Future Directions**   1115
    38.5   **References**   1115

Chapter 39.   Parallel Genetic Algorithms   1118

          Andrew Chipperfield and Peter Fleming

    39.1   **What Are Genetic Algorithms?**   1118
    39.2   **Major Elements of the Genetic Algorithm**   1121
    39.3   **Parallel GAs**   1130
    39.4   **Conclusions and Future Trends**   1140
    39.5   **References**   1141

Chapter 40.   Parallel Processing for Robotic Computations: A Review   1144

          Tarek M. Nabhan and Albert Y. Zomaya

    40.1   **Overview of Robotic Systems**   1144
    40.2   **The Task Planner**   1145
    40.3   **Sensing**   1147
    40.4   **Robot Control**   1149
    40.5   **Applications of Advanced Architectures for Robot Kinematics
           and Dynamics**   1150
    40.6   **Summary, Conclusions, and Future Directions**   1154
    40.7   **References**   1155

Chapter 41. Distributed Flight Simulation: A Challenge for Software
Architecture   **1160**

Rick Kazman

41.1   **The Challenge of Distributed Flight Simulation**   **1160**

41.2   **A Generic Flight Simulator**   **1162**

41.3   **Introduction to Software Architecture**   **1164**

41.4   **Structural Modeling**   **1166**

41.5   **Motivations for Structural Modeling**   **1167**

41.6   **Flight Simulator Software Architecture: Overview**   **1168**

41.7   **Flight Simulator Software Architecture: Base Types**   **1169**

41.8   **A Simplified Software Structure**   **1173**

41.9   **Requirements of Flight Simulation**   **1173**

41.10 **Lessons Learned/Future Directions**   **1176**

41.11 **Summary**   **1176**

41.12 **References**   **1176**

Index   1179

Foreword

It is now 50 years since the introduction of the first electronic digital computer. Even as early as the decade following that introduction, researchers were busy suggesting various kinds and forms of parallel processors which could be used to speed up the execution of a program. By 1966, multiprocessors, vector processors, array processors, and dataflow-driven pipelined processors already had been built or proposed. It seemed that the future of computers was *parallel*. I recall 1966 rather well because at that time I wrote a paper describing the state of the art of parallel processors and suggested the use of the stream concept (SIMD, MIMD, etc.) to categorize the various emerging forms of parallelism. I expected that the immediate future would be "the time of the parallel processor."

Things don't always work out as predicted, however. The decades that followed saw unexpected and extraordinary advances in uniprocessor technology. Memory costs on a per-bit basis dropped by a factor of a million. Uniprocessor price-to-performance scaled in a similar fashion. More significantly, the realization of efficient parallel processors proved to be a *hard* problem. Indeed, even as we enter the sixth decade following the introduction of the electronic computer, the realization of large parallel processor systems remains a hard problem.

There are two manifestations of this difficulty:

- finding large degrees of parallelism
- efficiently managing parallelism to achieve rapid program execution

The first problem is referred to as either the *algorithmic* or the *partitioning* problem. Given an application, we must either find a new parallel algorithm or a tool to decompose an existing algorithm into many small tasks which can be concurrently executed. Even when we can achieve high degrees of parallelism, we have no guarantee of rapid program execution.

The parallel execution may not proceed in a uniformly rapid execution. This brings us to the second problem, realizing program *speed-up*. This has many aspects, except for isolated cases: scheduling tasks so that required data is available at the appropriate times, communicating results through complex networks, and ensuring that computations are properly synchronized and that the resultant data is stored in a way that achieves a result consistent with the result of a uniprocessor.

It is my feeling that there is a fundamental problem that lies beyond these manifestations. The problem is simply that we cannot easily represent complex applications in ways that allow efficient parallel processing. Parallel processes do not resemble the conscious reasoning of the human mind. Hence, the representations which have evolved to suit our reasoning needs are necessarily cumbersome when mapped over to the regime of the parallel processor. It is only when we can look at applications in a fresh way, recognizing the requirements of the parallel processor, that we may be able to derive important new representations and a new methodology that would pose a fundamentally new computational model for describing applications.

This book is an important statement of where we are: a state of the art of parallel processor research in the decade of the 1990s. It is easy to recognize the enormous effort of the many able authors and especially the effort of the editors in assembling this comprehensive statement.

I am very pleased to see the broad scope (theory, algorithms, technology) of this work and especially its emphasis on tools and applications. Advances in theory and algorithms provide a basis for understanding an emerging computational model more suited to parallel computation. Technology and architecture provide a current parameterization of physical limitations of parallel processor implementations. These parameters aid in refining and directing the parallel computational model. Applications and tools for parallelizing applications provide a testbed for the parallel processor and parallel model. This book addresses each of these areas with an in-depth discussion of the significant ideas which define the various aspects of the parallel processor field. It is in the integration of these basic ideas of theory, algorithms, language, and architecture in the service of applications that we can expect the advances of this decade. In all of this are the seeds of the new understanding required for the next decade to finally see the easy, ready, and efficient use of parallel processors.

Michael J. Flynn
Stanford, California
September, 1995

Preface

Parallel and distributed computing is considered to be one of the most exciting technologies to achieve prominence since the invention of computers in the 1940s. It is expected that this decade will witness a proliferation in the use of parallel and distributed computing systems. It is also strongly believed that this paradigm of computing will influence the design of all future computers. This unrivalled development has been facilitated by the rapid advances in electronics and integrated circuit technologies.

Nowadays, in many academic departments, such as electrical engineering, computer engineering and computer science, courses related to parallel and distributed computing are becoming widely available not only at the graduate level but also at the undergraduate level. These new directions are motivated by the existence of many commercial parallel machines and hardware components that can be used to build inexpensive prototype systems. This is making parallel and distributed computing facilities accessible, at a reasonable cost, to a wide range of potential users.

Moreover, parallel and distributed computing systems will play an ever increasing role in the so-called "information superhighway." This means that today's practicing scientists or engineers are constantly being exposed to many new ideas and applications of this exciting technology. As time goes by, it is becoming a more difficult task to be familiar with the trends in theoretical work, hardware architectures, software tools, and many other issues that embody the discipline of parallel and distributed computing.

However, researchers and practitioners must be kept aware of the essential ideas and techniques of the field, and the role of this handbook is to promote this understanding. The handbook explores the current and future developments of this area of computing by a range of chapters which cover the mature work and the developing domains. These chapters were identified as key areas in parallel and distributed computing by the editor and the editorial board.

One of the main objectives of this handbook is to extract from the broad literature on parallel and distributed computing the central ideas and essential methods for analysis, design, and application. The handbook also aims at directing the reader to more reference material for the details beyond initial inquiry. The bibliography that accompanies each chapter should be considered to be of a selective rather than an exhaustive nature.

The primary reader of the handbook is a professional scientist or engineer engaged in the study, design, and operation of parallel and distributed computing environments, technologies, and applications. The main philosophy that was followed in organizing this handbook was to "let the experts write." It was realized from the beginning that it is impossible for one person to write a book on parallel and distributed computing that covers all the important aspects of the field and maintain the same depth and consistency of coverage. Therefore, the idea was to identify some of the key topics in this area, realizing that we cannot cover everything in one volume, and invite many people to contribute. All the chapters are written by well recognized experts in the field.

We tried to avoid the "hasty" approach of an encyclopedia of trying to cover everything and failing to convey any useful information to the reader. At the same time, we were keen on confining the handbook to one volume to make it more manageable and accessible to as many people as possible. Hence, every chapter in the handbook provides the reader with a comprehensive overview of the main results, development, and future directions in a certain area of this rapidly evolving field.

All in all, it is hoped that the chapters featured in this handbook will contribute to progress in this exciting field. Computer professionals will be able to use the handbook as an informative technical introduction to areas of interest. The contents of the handbook should provide the reader with the needed background to pursue any desired further information in the references listed. With this understanding, the reader can then employ parallel and distributed systems, as well as expand the knowledge base about their design, analysis, and use.

Acknowledgments

This handbook grew out of months of preparation and consultation with the publisher and the editorial board. The success of the editor of a technical handbook is made possible only through the cooperation efforts of a great team. I would like to express my sincere gratitude and appreciation to the members of the editorial board, whose names are listed on page xxiii, for their time, help, and support throughout.

I am deeply grateful to the contributors to this handbook, who lent me their efforts and patience in preparing this timely account of their experiences. This handbook would have not been possible to produce without their hard work, dedication, and punctuality.

Many thanks also to those friends and colleagues who helped me each in his/her own way, in particular: Arvind (MIT), Gary Bundell (University of Western Australia), Ajoy K. Datta (University of Nevada, Las Vegas), Doug DeGroot (Texas Instruments), Narsingh Deo (University of Central Florida), Alan Fekete (Sydney University), Geoffrey Fox (Syracuse University), Lee Giles (NEC Research Institute), Garry Greenwood (University of Western Michigan), Ajay Gupta (University of Western Michigan), Kai Hwang (University of Southern California), Oscar Ibarra (UC, Santa Barbara), Thomas LeBlanc (University of Rochester), Nancy Lynch (MIT), Rami Melhem (University of Pittsburgh), Venu Murthy (Queensland University of Technology), Lionel Ni (Michigan State University), David Nicol (The College of William and Mary), John Rice (Purdue University), John Rosenberg (Sydney University), Sartaj Sahni (University of Florida), Janet M. Siegel (Purdue University), Per Stenstrom (Lund University), and Daniel Tabak (George Mason University). Without their help and the help of others the enormity of the task would have surpassed my capabilities. I am also grateful to Mike Flynn (Stanford University) for his enthusiasm and encouragement and for agreeing to write the foreword to the handbook.

Special thanks must go to Mr. Steve Chapman, editor, for his skills and energy, a tribute to McGraw-Hill. He first got me to put the initial outline on paper, and also had the vision and courage to support the evolving ideas. I am greatly indebted to Mr. Jeff Eckert and Mr. Orlo Day at J. K. Eckert & Company, Inc., for taking on the enormous task of producing the handbook. Jeff should be credited for his professionalism and patience in dealing with many of the problems that a such project is expected to encounter. Thanks are also due to the reviewers for their reparté on the initial outline.

Finally, although all the chapters were reviewed by at least two reviewers (and sometimes more), one cannot guarantee the existence of no errors with a project of such diverse outline. I am sure that this was beyond the control of the editor, the editorial board, and the contributors. However, I would greatly appreciate it if the readers would bring to my attention any errors or omissions that were overlooked.

Albert Y. Zomaya
Perth, Western Australia
zomaya@ee.uwa.edu.au

Contributors

Editor in Chief

Albert Y. Zomaya
The University of Western Australia, Australia

Editorial Advisory Board

Selim G. Akl
Queen's University, Canada

Mikhail J. Atallah
Purdue University, U.S.A.

Jack J. Dongarra
University of Tennessee, U.S.A.

David J. Evans
Loughborough University of Technology, U.K.

Ophir Frieder
George Mason University, U.S.A.

Allan Gottlieb
New York University, U.S.A.

Rick Kazman
University of Waterloo, Canada

E. V. Krishnamurthy
Australian National University, Australia

Ted G. Lewis
Naval Postgraduate School, U.S.A.

Yoichi Muraoka
Waseda University, Japan

Heiko Schroeder
University of Newcastle, Australia

Howard Jay Siegel
Purdue University, U.S.A.

Ivan Stojmenović
University of Ottawa, Canada

Authors/Contributors

David Abramson
School of Computing and Information
    Technology
Griffith University
Nathan, Queensland, 4111
Australia

Selim G. Akl
Department of Computing and Information
    Science
Queen's University
Goodwin Hall
Kingston, Ontario K7L 3N6
Canada

John K. Antonio
Computer Science Department
Texas Tech University
Box 43104
Lubbock, TX 79409-3104
U.S.A.

Mikhail J. Atallah
Department of Computer Science
Purdue University
West Lafayette, IN 47907
U.S.A.

Hagit Attiya
Department of Computer Science
The Technion, Haifa 32000
Israel

Eugenio Battiston
Department of Information Sciences
University of Milano
Via Comelico 39
20135 Milano
Italy

Jehoshua Bruck
California Institute of Technology
Mail Code 116-81
Pasadena, CA 91125
U.S.A.

Richard H. Carver
Department of Computer Science
George Mason University
Fairfax, VA 22030-4444
U.S.A.

Doreen Y. Cheng
Philips Research Palo Alto
Philips Electronics North American Corp.
4005 Miranda Ave, Suite 175
Palo Alto, CA 94304
U.S.A.

Andrew Chipperfield
Department of Automatic Control and Systems
   Engineering
University of Sheffield
P.O. Box 600, Mappin Street
Sheffield S1 4DU
U.K.

Thomas M. Conte
Department of Electrical and Computer
   Engineering
North Carolina State University
Raleigh, NC 27695-7911
U.S.A.

Sajal K. Das
Center for Research in Parallel and Distributed
   Computing
Department of Computer Sciences
University of North Texas
Denton, TX 76203-3886
U.S.A.

Fiorella De Cindio
Department of Information Sciences
University of Milano
Via Comelico 39
20135 Milano
Italy

Giorgio De Michelis
Department of Information Sciences
University of Milano
Via Comelico 39
20135 Milano
Italy

Jules Desharnais
Département d'Informatique
Université Laval
Quebec, PQ G1K 7P4
Canada

Amal De Silva
The Preston Group
488 Victoria Street
Richmond, Victoria 3121
Australia

Jack J. Dongarra
Computer Science Department
University of Tennessee
Knoxville TN 37996-1301
U.S.A.

Ralph V. Duncan
Keltic Computing Services
P.O. Box 451227
Atlanta, GA 31145
U.S.A.

Hesham El-Rewini
Department of Computer Science
University of Nebraska at Omaha
Omaha, NE 68182-0243
U.S.A.

David J. Evans
Parallel Algorithms Research Centre
Department of Computer Studies
Loughborough University of Technology
Loughborough, Leicestershire LE11 3TU
U.K.

Kazi Farooqui
Department of Computer Science
University of Ottawa
Ottawa, Ontario K1N 9B4
Canada

Afonso Ferreira
Laboratoire de l'Informatique du Parallelisme
Ecole Superieure de Lyon
46 Alee de l'Italie
69364 Lyon
Cedex 07
France

Alois Ferscha
Department of Applied Computer Science
University of Vienna
Lenaugasse 2/8
A-1080 Vienna
Austria

Peter Fleming
Department of Automatic Control and Systems
    Engineering
University of Sheffield
P.O. Box 600, Mappin Street
Sheffield S1 4DU
U.K.

M. Gusev
"Kiril i Metodij" i Skopje
PMF Inst. za Informatika
Skopje
Macedonia

Salim Hariri
Department of Electrical and Computer
    Engineering
Syracuse University
111 College Place
Syracuse, New York 13244-4100
U.S.A.

Michael T. Heath
Department of Computer Science
University of Illinois
1304 West Springfield Avenue
Urbana, IL 61801-2987
U.S.A.

R. Jagannathan
Computer Science Laboratory
SRI International, M/S EL-280
333 Ravenswood Avenue
Menlo Park, California 94025
U.S.A.

Joseph JáJá
Institute for Advanced Computer Studies
University of Maryland
College Park, MD 20742
U.S.A.

Michael Jurczyk
Institute for Microelectronics Stuttgart
Allmandring 30a
W-7000 Stuttgart 80 (Vaihingen)
Germany

Manolis G. H. Katevenis
Science & Technology Park of Crete
Vassilika Vouton
P. O. Box 1385
GR 711 10 Heraklion, Crete
Greece

Rick Kazman
Department of Computer Science
University of Waterloo
Waterloo, Ontario N2L 3G1
Canada

Leonidas I. Kontothanassis
Department of Computer Science
University of Rochester
Rochester, NY 14627-0226
U.S.A.

E.V. Krishnamurthy
Computer Sciences Laboratory
Research School of Physical Sciences
Australian National University, ACT 2601
Australia

Lydia Kronsjö
Department of Computing Science
The University of Birmingham
Edgbaston, B15 2TT
U.K.

Yan Alexander Li
Parallel Processing Laboratory
School of Electrical and Computer Engineering
Purdue University
West Lafayette, IN 47907-1285
U.S.A.

Luigi Logrippo
Department of Computer Science
University of Ottawa
Ottawa, Ontario K1N 9B4
Canada

Bei Lu
Department of Electrical and Computer
    Engineering
121 Link Hall
Syracuse University
Syracuse, New York 13244-4100
U.S.A.

Muthucumaru Maheswaran
Parallel Processing Laboratory
School of Electrical and Computer Engineering
Purdue University
West Lafayette, IN 47907-1285
U.S.A.

Robert L. Martino
National Institutes of Health
DCRT/CBEL, Building 12A, Room 2033
12 South Dr MSC 5624
Bethesda, MD 20892-5624
U.S.A.

Tim Mattson
Intel Supercomputers Systems Division
Cornel Oaks 6, Zone 8
14924 N.W. Green Brier Pkwy
Beaverton OR 97006
U.S.A.

Richard C. Metzger
Software Engineering Branch (C3CB)
Rome Laboratory
Griffiss AFB, NY 13441-5700
U.S.A.

Ali Mili
Department of Computer Science
University of Ottawa
Ottawa, Ontario K1N 9B4
Canada

R. Mili
Department of Computer Science
University of Ottawa
Ottawa, Ontario K1N 9B4
Canada

Kwang-Bae Min
Center for Research in Parallel and Distributed
    Computing
Department of Computer Sciences
University of North Texas
Denton, TX 76203-3886
U.S.A.

J. Mullins
Department of Computer Science
University of Ottawa
Ottawa, Ontario K1N 9B4
Canada

Tarek M. Nabhan
ERG Group of Companies
247 Balcatta Road
Balcatta, Western Australia 6021
Australia

Stephan Olariu
Department of Computer Science
Old Dominion University
Norfolk, VA 23529
U.S.A.

Ron H. Perrott
Department of Computer Science
Queen's University
Belfast BT7 1NN
U.K.

Lucia Pomello
Department of Information Sciences
University of Milano
Via Comelico 39
20135 Milano
Italy

Viktor K. Prasanna
Department of Electrival Engineering—
    Systems
EEB-244
University of Southern California
Los Angeles, CA 90089-2562
U.S.A.

Eugen Schenfeld
NEC Research Institute
4 Independence Way
Princeton, NJ 08540
U.S.A.

Thomas Schwederski
Institute for Microelectronics Stuttgart
Allmandring 30a
W-7000 Stuttgart 80 (Vaihingen)
Germany

Michael L. Scott
Department of Computer Science
University of Rochester
Rochester, NY 14627-0226
U.S.A.

Howard Jay Siegel
Parallel Processing Laboratory
School of Electrical and Computer Engineering
Purdue University
West Lafayette, IN 47907-1285
U.S.A.

Yahia Slimani
Departement d'Informatique
Faculte des Sciences
Université de Tunis II
Belvedere 1002
Tunisia

Carla Simone
Department of Information Sciences
University of Milano
Via Comelico 39
20135 Milano
Italy

John John E. So
Parallel Processing Laboratory
School of Electrical and Computer Engineering
Purdue University
West Lafayette, IN 47907-1285
U.S.A.

Ivan Stojmenović
Department of Computer Science
University of Ottawa
Ottawa, Ontario K1N 9B4
Canada

K.C. Tai
Department of Computer Science
North Carolina State University
Box 8206
Raleigh, NC 27695-8206
U.S.A.

Min Tan
Parallel Processing Laboratory
School of Electrical and Computer Engineering
Purdue University
West Lafayette, IN 47907-1285
U.S.A.

Louis H. Turcotte
NSF Engineering Research Center for
    Computational Field Simulation
P.O. Box 6176
Mississippi State, MS 39762
U.S.A.

David W. Walker
Oak Ridge National Laboratory
P.O. Box 2008, Building 6012
Mathematical Science Section
Oak Ridge, TN 37821-6367
U.S.A.

Cho-Li Wang
Department of Electrical Engineering—
    Systems
EEB-244
University of Southern California
Los Angeles, CA 90089-2562
U.S.A.

Lee Wang
Parallel Processing Laboratory
School of Electrical and Computer Engineering
Purdue University
West Lafayette, IN 47907-1285
U.S.A.

Tieng K. Yap
National Institutes of Health
DCRT/CBEL, Building 12A, Room 2033
12 South Dr., MSC 5624
Bethesda, MD 20892-5624
U.S.A.

Avi Ziv
Information Systems Laboratory
Stanford University
Stanford, CA 94305-4055
U.S.A.

Albert Y. Zomaya
Parallel Computing Research Laboratory
Department of Electrical and Electronic
   Engineering
The University of Western Australia
Nedlands, Perth, Western Australia 6907
Australia

Parallel and Distributed
Computing Handbook

I

Theory

Foundations

1

Parallel and Distributed Computing: The Scene, the Props, the Players

Albert Y. Zomaya

The past four decades have witnessed startling advances in computing technology that were stimulated by the availability of faster, more reliable, and cheaper electronic components. These resolute developments enabled the solution of a wide range of computationally intensive problems [54, 65]. However, better devices, albeit essential, are not the sole factor contributing to high performance. This fact made it evident that the performance of uniprocessor computers is limited, because device technology places an upper bound on the speed of any single processor.

The evolutionary transition from sequential to parallel and distributed computing (PDC) offers the promise of a quantum leap in computing power that can be brought to bear on many important problems. Although this technology has risen to prominence during the last 20 years, there remain many unresolved issues. The field is in a state of rapid flux where advances are still being made on several fronts. PDC systems are recognized today as an important vehicle for the solution of many problems, especially those known as the "Grand Challenges" [48]. These include climate modeling, human genome mapping, semiconductor and superconductor modeling, pollution dispersion, and pharmaceutical design. These and many other applications require computing power greater than is obtainable with many conventional computers [19, 68].

The notion that parallel and distributed computing are two separate fields is now beginning to fade because technological advances have been bridging the gap, for the last decade or so, between these two disciplines. Hence, it is with this prevailing opinion that this handbook presents the underlying theories, designs, and applications of PDC. The collection of ideas presented in the different chapters of this volume are considered to be the necessary ingredients for providing the high-performance computing (HPC) capabilities required to solve computationally intensive problems in science and engineering [17, 18]. The remainder of this chapter provides a brief introduction to the field and an overview of the handbook.

1.1   A Perspective

The overwhelming majority of today's computers are conceptually very similar. Their architectures and modes of operation follow, more or less, the same basic design principles formulated in the late 1940s by John von Neumann and coworkers [22]. The von Neumann approach centered around the simple idea of a control unit that fetches an instruction and its operands from a memory unit and sends them to a processing unit. In the processing unit the instruction is executed, and the result is sent back to memory (Fig. 1.1).

A von Neumann computer, in principle, consists of three components:

1. a central processing unit (CPU)
2. a main-memory system
3. an input-output (I/0) system

In such a computer, the instructions are executed sequentially, where the CPU executes, or appears to execute, one operation at a time. Conventional von Neumann machines provide a single pathway for addresses and a second pathway for data and instructions [8]. Thus, the speed by which the memory is accessed or the speed of input and output devices can slow the computation rate. This problem is better known as the *CPU-to memory bottleneck,* which mainly results from the disparity between high CPU speeds and much lower memory speeds. Other variations of von Neumann machines exist, such as Harvard architectures, which provide independent pathways for data, data addresses, instructions, and instruction addresses [8].

Several techniques were introduced to improve these early designs, such as *cache memories* and *pipelining* which led to the conception of the early *supercomputers* [50, 61]. However, this trend is, unfortunately, coming to an end because of the fundamental limitations of fast sequential computers and the excessive cost associated with such designs.

In the case of PDC systems, the approach followed in solving a given problem is entirely different. Also, the time needed to solve the problem is significantly reduced by using PDC systems as opposed to traditional uniprocessor (or sequential) computers. A parallel or distributed architecture is one that consists of a collection of processing units (processors or computers) that cooperate to solve a given problem by working simulta-

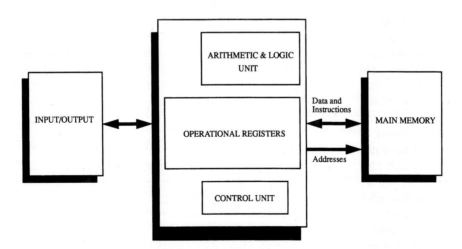

Figure 1.1 The von Neumann computer

neously on different parts of that problem. Clearly, there is no limit, at least in principle, to the number of actions that can be executed in parallel; hence, the concept offers an arbitrary degree of improvement in computing speed.

The research in PDC is motivated by a variety of factors; for example, advances in very large scale integration (VLSI) technology facilitated the production of cheaper electronic components that have efficient power consumption, higher speeds, and superior reliability. However, sheer performance is the raison d'être behind the research in this area, and only by solving computationally intensive problems better than other architectural approaches, can the notion of parallelism fully realize its full potential.

1.2 Parallel Processing Paradigms

Computers operate simply by executing instructions on data. A stream of instructions inform the computer of what to do at each step. Flynn [16] classified the architecture of a computer on the basis of how the machine relates its instructions to the data being processed (Fig. 1.2). The multiplicity of instruction streams and data streams in the system was used to produce four classifications.

- single instruction stream, single data stream (SISD)
- single instruction stream, multiple data stream (SIMD)
- multiple instruction stream, single data stream (MISD)
- multiple instruction stream, multiple data stream (MIMD)

All von Neumann machines belong to the SISD class. An algorithm running on a computer in this class is known as a *sequential algorithm.*

A SIMD machine consists of N processors, N memories, an interconnection network, and a control unit. All the processing elements are supervised by the same control unit, and the processors operate on different data sets from distinct data streams. For example, for a SIMD computer with N processors, each processor will be executing the same instruction at the same time, but each on its own data. The processors in this case operate synchronously. SIMD machines can be classified into two categories: *shared-memory* (SM) and *local-memory* (LM) [45]. In SM-SIMD machines, the processors do not have any local memory, while in LM-SIMD machines, each one of the processors has its own local memory.

High performance on SIMD machines requires the rewriting of conventional algorithms to manipulate many data simultaneously by sending instructions to all processors. The programming of such machines is different from that of sequential machines; however, in an ideal situation, a sequential algorithm can be converted into a SIMD algorithm by replacing each inner loop with a single instruction that implements the complete loop. A large class of problems fits this model extremely well and has provided the incentive for the design and construction of these machines (e.g., target tracking, image processing, and database operations).

In a MISD computer, each of the processors has its own control unit and shares a common memory unit where data reside. Therefore, parallelism is realized by enabling each processor to perform a different operation on the same datum at the same time. Systolic arrays are known to belong to this class of architectures [30, 36].

Many problems do not lend themselves to efficient execution in SIMD environments. The operations required for such problems cannot be easily organized into repetitive operations on uniformly structured data. MIMD machines are the most general and powerful in the paradigm of parallel computing. In this case, there are N processors, N streams of in-

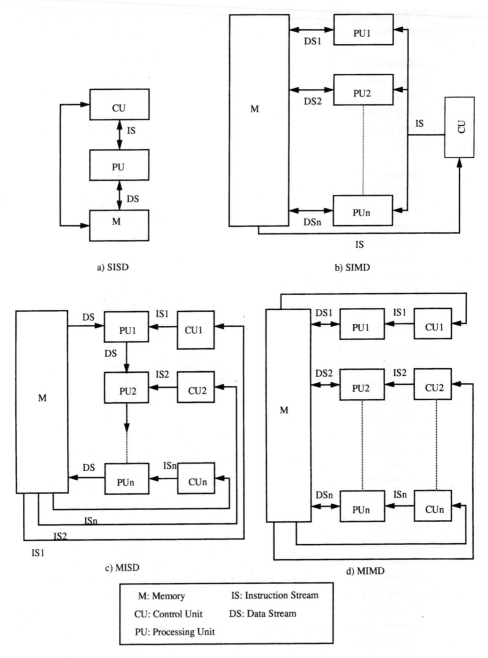

a) SISD

b) SIMD

c) MISD

d) MIMD

M: Memory	IS: Instruction Stream
CU: Control Unit	DS: Data Stream
PU: Processing Unit	

Figure 1.2 Flynn's taxonomy

structions, and N streams of data. The processors are of the same type used in a SISD computer; that is, each processor has its own control unit in addition to its local memory and arithmetic and logic unit (ALU). Hence, these processors are considered to be more powerful than the ones used for SIMD computers. A MIMD computer is tightly coupled if the degree of interactions among processors is high. Otherwise, it is considered loosely coupled. Of the four approaches, the MIMD model is the most popular for building commercial parallel machines, followed by the SIMD model, while the MISD is the least popular [3].

As an addition to Flynn's taxonomy, another class known as SPMD (single program, multiple data) is used to describe cases where many programs that have the same process type (or same code) are executed on different data sets, synchronously (as in SIMD) or asynchronously (as a special case of MIMD). This is discussed further in the next section. The reader can find other classification schemes for computer architectures in the literature [29, 35, 57].

1.2.1 Organization of PDC systems: architectural features

Theoretically, any parallel algorithm can be executed efficiently on the MIMD model. Thus, this model can be used to build parallel computers for a wide variety of applications. Such computers are said to have a general-purpose architecture. In practice, it is quite reasonable in many applications to assemble several processors in a configuration specifically designed for the problem under consideration. The result is a computer that solves a particular problem efficiently but that, in general, cannot be used for any other purpose. Such a computer is said to have a special-purpose architecture.

Although there is a variety of paradigms for building parallel computers, they all fall into several general categories:

- *Pipelining.* This is a classical way to exploit parallelism and concurrency in computer architectures. This situation arises when many data items need to be operated upon at the same time. In a pipeline architecture the different processing elements are connected to each other in a chain and data items are passed sequentially, so that each processing element will operate on each datum as it passes through [52]. An important class of computers that employ pipelining is the reduced instruction set computer (RISC) [59]. Other classes that benefit from pipelining are superscalar processors [50].
- *Array processors.* These are systems that consist of a number of processing elements with nearest-neighbor connection that operate in parallel under the direction of a single control unit. Each of the processing elements performs the same operation at the same time, on different data elements [42]. Array processors can be viewed as a subclass of SIMD computers.
- *Vector processors.* most vector machines have a pipelined structure and specially designed register sets that can hold vectors. When one pipeline is not enough to achieve high performance, usually several pipelines are provided. In such processors, high performance is not only achieved through using pipelining but also by providing the capability for the parallel operation of several pipelines on independent streams of data [4, 27].
- *Multiprocessors.* These are small-scale MIMD machines. In this case, the system is built of a number of independent processors and facilities for controlling their interactions and cooperation. Multiprocessors are usually classified into shared-memory or distributed memory architectures [15, 60].
- *Dataflow machines.* In a dataflow architecture, instructions are automatically executed when data are available. Hence, there is no control flow mechanism or program counter to determine when instructions are executed [7, 64].
- *Neural networks.* Like dataflow machines, they represent another radical departure from the von Neumann approach, and instead are loosely based on models of biological systems. Neural networks resemble dataflow models in that they do not execute conventional programs. Neural computers learn or animate the behaviors of complex systems

through the use of a large number of processors and interconnections. The interconnections tend to play an important role, since the learning process is usually achieved by increasing or decreasing the strength of a given link between two processors [26].

Another popular classification of PDC systems is by the degree of coupling between the different processing components in the system [39]. Processing elements can be wired directly together, communicate via buses or cables, or be remotely located and communicate by radio or microwave signals. Hence, the type and mechanism for coupling often can be used to distinguish between distributed and parallel processing. Because it was agreed from the beginning to treat PDC as one discipline, it is important to introduce another approach for organizing processing resources:

- *Clustering.* This is a paradigm that enables multiple computers to cooperate simultaneously to solve a computationally intensive problem. Of course, the application in this case must be parallelizable. This form of computing is dependent on advances in networking technology [53].

Multicomputers can be connected using various forms of LANs (local area networks) and WANs (wide area networks) to provide high performance that approaches supercomputer levels [38]. Also, the recent emergence of asynchronous transfer mode (ATM), which is considered to be the ultimate in networking technology, is another step in achieving supercomputing capabilities by connecting several computers together [31].

Of course, the architect looks for the least expensive set of interconnections that achieve good performance over a large class of applications. Several factors affect the suitability of a given network for a class of application under study, for example,

- *Communication distance.* This is the physical distance or the number of switches that a message has to traverse before reaching its destination. For long distances, the performance may degrade due to network contention.
- *Expandability.* This refers to the amount of additional hardware needed to include an extra component (e.g., processors, memory modules). Of course, the cost should be minimal.
- *Fault tolerance.* The existence of multiple paths between each source and each destination [55] provides multiple paths that a message can follow from its source to its destination.
- *Bandwidth.* This is the amount of data that can be sent over the network per unit time.

Note that PDC paradigms and topologies are interrelated. An important design issue in parallel processor and multicomputer systems is the type and topology of the *interconnection network*. A network is used to connect processors or computers together or processors to memories. The availability of more efficient and reliable networks is essential for achieving high performance. While some algorithms can make excellent use of a certain network topology, they might not fit well into another. Many network topologies have been proposed in the literature, for example, bus architectures, ring networks, crossbars, meshes, shuffle exchanges networks, and hypercube ensembles [55].

There are other important factors that characterize a parallel algorithm in relation to the host parallel architecture. For example, the number of processors can vary from a small number (as few as four) to thousands. Also, the capability of these processors could range from the very simple, capable of performing a single floating-point operation at a time, to the very powerful, which can be considered to be a computer in its own right.

Another factor that characterizes a parallel architecture is the mode of control. In most computers, the control mode is command driven, which means that the different events are driven by the sequence of instructions. Other computers employ a data-driven approach (e.g., dataflow machines). In this case, the control-flow mechanism is triggered by the availability of data. Another mode of control is demand driven, whereby computations take place only if their results are requested by other events. Alternate modes of control are based on combinations of these approaches. Other factors that can influence the choice of an architecture for the implementation of a given algorithm are the required communication structure, the size of memory, and the presence or absence of synchronization mechanisms [10, 45].

1.3   Modeling and Characterizing Parallel Algorithms

A von Neumann type of architecture can be studied using an abstract model which is called the random access machine (RAM) (Fig. 1.3). The RAM model can be used to theoretically study sequential algorithms and evaluate their complexities [2].

The parallel counterpart to the RAM is the parallel random access machine (PRAM) [34], which is a theoretical model that plays a central role in studying parallel algorithms (see Fig. 1.4).

A PRAM is basically a set of synchronous processors connected to a shared memory. A main feature in a PRAM is the capability for different processors to simultaneously make

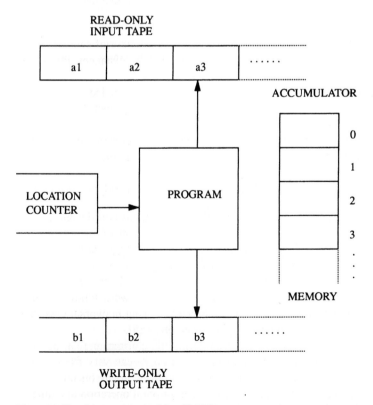

Figure 1.3 Random access machine (RAM)

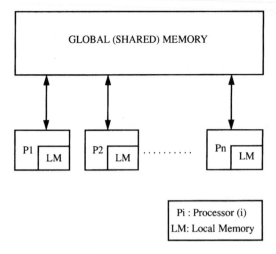

Figure 1.4 Parallel random access machine (PRAM)

references to distinct memory cells. Several variations of PRAM have been introduced in the literature to allow simultaneous "reads" and/or "writes" to the same memory location [23, 32, 34]. PRAMs exhibit two limitations. The first is that although the different processors can execute different programs, it is assumed that all the instructions take the same amount of time to execute. The second limitation is the lack of an interconnection network. The first limitation makes the PRAM a synchronous shared-memory machine, while the second imposes a zero communication time, which means that the communication time is not accounted for when studying a given parallel algorithm. Nevertheless, PRAMs are useful for studying parallel algorithms and evaluating their behavior and properties. This is because, if an algorithm does not perform well on a PRAM, it will be pointless to try to implement it on a realistic parallel architecture.

A parallel computation may involve millions of small operations. The mind obviously cannot comprehend such a multitude of simultaneous events in detail. To impose some order on the complexity, one must describe it in terms of a number of general features. Some of these features are highlighted in this section.

Parallelism in computations or programs can be exploited at several levels:

- job level, among jobs or phases of jobs
- task or procedure level, among procedures and within loops (usually included as a feature of the language)
- instruction level, among phases of an instruction cycle (e.g., fetch, decode, and execute)
- arithmetic and bit level, among bits within arithmetic circuits (This is the lowest level of parallelism, and it is fairly standard in computers today.)

Processes also can be divided into several levels in terms of their *granularity*. In reality, granularity is the single most distinguishing factor that can be used to characterize PDC systems. Granularity can be measured as the ratio of the compute time needed to execute a given task to the communication overhead incurred during the computation of the same task [61]. Accordingly, granularity can be classified as *very coarse, coarse, medium,* and *fine*. For example, fine-grain parallelism might refer to processing at the bit level, while coarse-grain parallelism describes a process that could consist of millions of operations. In most systems, mixed levels of granularity are usually employed [56].

1.4   Cost vs. Performance Evaluation

In either sequential or PDC systems, the architecture is characterized by functional components, the communication topology and facilities, and control structures and mechanisms. However, there are several issues related to parallelization that do not arise in sequential programming [10].

The first issue is *task allocation,* which is the breakdown of the total workload into smaller tasks assigned to different processors, and the proper sequencing of the tasks when some of them are interdependent and cannot be executed simultaneously. To achieve the highest level of performance, it is important to ensure that each processor is properly utilized. This process is called *load balancing* or *scheduling,* and it is considered to be extremely "formidable" to solve. The scheduling problem belongs to a class of problems called *nondeterministic-polynomial-time-complete* (NP-complete) [12, 20, 41, 62].

The second issue that was previously mentioned is the communication time between processors and the choice of the interconnection network. The problem of communication becomes more serious when the number of processors increases to hundreds or thousands. The interconnection network plays an important role in achieving high performance. Interconnection networks are also used to enable PDC systems to be more fault-tolerant [14, 47, 55]. Thus, the quest for better technologies that can be employed to build better interconnection networks, especially for massively parallel processor (MPP) systems, is still at its earnest. A good example of such a technology is based on optical interconnections [24].

Once a new algorithm for a given problem has been designed, it is usually evaluated. The evaluation helps in quantifying how well an architecture matches a particular algorithm. Some of the measurements used for performance evaluation are discussed below.

1.4.1   Execution time

Speeding up computations appears to be one of the main reasons behind the interest in building PDC systems, hence the reader should always bear in mind that the only complete and reliable measure of performance is time, which is defined as *the time taken by the algorithm to solve a problem on a parallel computer,* or *the time elapsed from the moment the algorithm starts to the moment it terminates.*

If the various processors do not all begin and end their operation simultaneously, then the execution time is equal to the time elapsed between the moment the first processor begins computing and the moment the last processor ends computing [3].

Performance can also be measured in terms of the execution rate of the different instructions. One of the most common measures is the millions of instructions per second (MIPS) a computer can execute. Other popular measures are millions of floating-point operations per second (MFLOPS) and billions (or *giga*) of floating-point operations per second (GFLOPS). These are especially useful for supercomputers and scientific and engineering applications [17, 50, 61].

1.4.2   Speedup

It is quite natural to evaluate a parallel algorithm in terms of the best available sequential algorithm for that problem. Thus, a good indication of the quality of a parallel algorithm is the speedup it produces. This is defined as:

$$S = \frac{\text{Running time of the best available sequential algorithm}}{\text{Running time of the parallel algorithm}} \quad (1.1)$$

Obviously, the larger the speedup, the better the parallel algorithm. Ideally, one hopes to achieve the maximum speedup of N when solving such a problem using N processors operating in parallel. In practice, such speedup cannot be achieved for every problem, because

1. It is not always possible to decompose a problem into N tasks, each requiring $1/N$th of the time taken by one processor to solve the original problem, and
2. In most cases the structure of the parallel computer used to solve a problem usually imposes restrictions that render the desired running time unattainable (e.g., synchronization overhead).

Amdahl [5] has pointed out that the speedup is limited by the amount of parallelism inherent in the algorithm which can be characterized by a parameter f, the fraction of the computation that must be done sequentially. He reasons that the maximum speedup of an N-processor system in executing an algorithm as a function of f is given by

$$S_N \leq \frac{1}{f + (1-f)/N} \leq \frac{1}{f} \quad \text{for every } N \tag{1.2}$$

Note that $(S_{max} = 1)$ (no speedup) where $(f = 1)$ (all computations must be sequential), and $(S_{max} = N)$ where $(f = 0)$ (all computations must be in parallel) [33].

Moreover, in many algorithms, the interprocessor communication time significantly contributes to the overall execution time. Hence, the effect of communication (or communication penalty) can be characterized by the following ratio,

$$cp_i = \frac{E_i}{C_i} \tag{1.3}$$

where E_i is the total execution time spent by processor i to run the algorithm, and C_i is the corresponding time attributed only to communications. Note that the large cp_i are obtained in coarse-grain parallelism because the time intervals between interprocessor data transfers are high, and the overhead can be amortized over many computational cycles. On the other hand, low values of cp_i are usually observed in fine-grain parallelism, when there is more frequent interprocessor communication. Also, small values of cp_i indicate relatively high communication overhead. In most cases, maximum performance is achieved by balancing parallelism against the communication overhead [41].

1.4.2.1 Number of processors
Another important criterion in evaluating a parallel algorithm is the number of processors it requires to solve the problem. It costs money to purchase, maintain, and run computers. When several processors are present, the problem of maintenance in particular is compounded, and the price paid to guarantee a high degree of reliability and fidelity rises sharply. Therefore, the larger the number of processors an algorithm uses to solve a problem, the more expensive the solution becomes. A trade-off needs to be made between the solution of a given problem and the cost involved.

1.4.2.2 Efficiency
To describe the speed advantage of the parallel algorithm, the following ratio is used:

$$\phi = \frac{S}{N} \tag{1.4}$$

where N is the number of processors and $\phi \leq 1$. In the ideal case $\phi = 1$, or 100 percent. For this to happen, the parallel algorithm should be such that no processor ever remains idle or does any unnecessary work. This ideal situation is unattainable.

An important issue pertaining to efficiency is that of scalability [25, 30]. The idea is that when the size of a problem is scaled up sufficiently, then any required efficiency can be achieved on any number of processors. Nevertheless, as sequential computers' performance is bound by the laws of physics [9], parallel computers also have their limitations [6, 45]. In general, one can expect speedup improvements that result from increasing the number of processors to be limited by the number of tasks.

1.4.2.3 Utilization

A popular metric for evaluating the design efficiency is the resource utilization factor U, defined as the ration between the total usage time over the total available time of the measured resource. It depends on how the resource is defined (e.g., processor, memory). Simply stated, utilization measures the percentage of resources kept busy during the running of a parallel application,

$$U = \frac{O(N)}{NT(N)} \tag{1.5}$$

where $O(N)$ is the "actual" total number of unit operations performed by an N-processor machine, while $NT(N)$ represents the number of operations that could have been performed with N processors in $T(N)$ time units. In general, the goal is to utilize all the processors and avoid having poorly utilized resources in the network.

Other criteria for measuring performance are *throughput* and *responsiveness*. The former is dependent on the rate of input and output of data, while the latter pertains to applications that require rapid responses (e.g., real-time processing [67]). In some applications, other criteria need to be considered such as *weight, power consumption,* and *ease of programming* [3].

1.5   Software and General-Purpose PDC

One of the most important of the problems that face PDC today is the construction of standard parallel languages and software portability [1, 43]. In reality, these are the same factors that made sequential computing very popular. Some researchers even claim that it is worth sacrificing some performance if one is to solve the software problem.

The process of writing parallel programs must be made more systematic. In other words, programmers should be able to write a parallel program, translate it into a workable version, and port it to a variety of PDC platforms [1]. To write a parallel program in a machine-independent fashion requires the development of abstract models for parallel computations without sacrificing the implementability of these programs. Of these paradigms, one can list *communicating sequential processes* (CSP), *calculus of communicating systems* (CCS), and *Petri nets* [28, 37, 44, 46]. More concerted work is required to produce models that combine the benefits of programming in a high level of abstraction and at the same time can be used to study and prove certain properties for some practical systems.

When a program is executed on a computer, two of the most important issues are *how long it will take to execute* and *how much memory it will require* [40, 49]. Turing machines are the simplest and most widely used theoretical model to study sequential computation [49]. Although Turing machines are very slow, they tend to capture the essence of the computing process. The importance of a theoretical model, such as the Turing machine, stems from the fact that such a model enables one to highlight the strengths and limitations of a given computational model. Moreover, Turing machines are also capable of computing any function that is computable by any other theoretical model (e.g., finite automata, pushdown automata, linear bounded automata) [13].

A similar universal model of computation is required for parallel computers. As it was noted earlier, the PRAM model is widely used for modeling ideal parallel computers. Other models exist that provide a reasonable level of abstraction for the design and evaluation of parallel programs [58, 63]. However, the need for a universally acceptable model is paramount.

The development of parallel languages is another important factor that influences the implementation of parallel algorithms. There are several directions to specify and support parallelism in PDC systems: sequential language extensions, parallelizing compliers, and new parallel languages and coordination languages [1, 21, 51, 66]. When a sequential language is extended to support parallelism, suitable machine constructs are used to enhance performance. These constructs vary from one machine to another, which adversely affects portability. The introduction of new machine-independent parallel languages will enhance portability.

Furthermore, there is a great need for extensive sets of portable libraries and tools that can be used for debugging and handling the huge amounts of data involved in most PDC applications. These tools should be easily integrated in a variety of environments [11].

In summary, many outstanding research issues related to the development of software for PDC systems need to be addressed before a successful transition from sequential to parallel computing can be fully realized.

1.6   A Brief Outline of the Handbook

Here we give a short rundown of the material presented in each of the chapters of this volume. The purpose is to identify the contents and also to aid dissimilar readers in assessing just what chapters are pertinent to their pursuits and desires.

The chapters are categorized into three main parts: Part I—Theory, Part II—Architectures and Technologies, and Part III—Tools and Applications. Each chapter should provide the reader with the equivalent of consulting an expert in a given discipline by summarizing the state of the art, interpreting trends, and providing pointers to further reading.

PART I: THEORY

Foundations

In Chapters 2 and 3, the authors consider certain paradigms that can be applied to describe the behavior of computational models. These paradigms can be used to produce computational models that are machine independent. However, that does not mean that these paradigms cannot be used to study and prove certain properties for some given practical or realistic systems.

Chapter 4 presents a framework for studying the complexity of computational problems in an implementation-independent manner. The chapter also investigates complexity issues pertaining a variety of computational models (e.g., parallel, distributed, neural networks). In Chapter 5, the author gives an overview of distributed computing theory and presents formal methods that can be used to reason about such systems. These methods and ideas can be used to solve problems in many applications, such as operating systems, transaction processing, communication networks, and parallel processors.

Models

In Chapter 6, the author presents a more detailed discussion of the PRAM model, which is a powerful tool for studying and designing parallel algorithms. Chapter 7 introduces an alternate model which is called *broadcasting with selective reduction* (BSR). The authors show that the BSR is a stronger model for studying parallel algorithm and this superiority does not incur any increase in resources. The method is applied to a variety of computational problems.

The dataflow model, which is the topic of Chapter 8, represents a radical departure from the von Neumann approach. This model is nonprocedural, that is, the programmer specifies an algorithm by giving its dataflow graph, and not by writing a program like the case with conventional programming techniques. The execution of the different operations is governed by the availability of data and computational resources.

Chapter 9 reviews the problems of scheduling and program partitioning in PDC systems. The solution of these problems is important if parallel systems are to produce the high performance capabilities expected of them. What makes scheduling problems interesting and challenging is the fact that they are known to be NP complete, and in most practical cases, heuristics must be used to solve them.

The authors of Chapter 10 describe a method for fault-tolerant computing. Fault-tolerance techniques are important because they often provide the capacity to detect the presence of faults, and also the ability of to reconfigure by replacing a faulty component (software or hardware) with an operational one. Chapter 11 presents a framework for designing distributed systems. The approach can be used to standardize the design process of such systems.

Algorithms

Several chapters in this section present and review a variety of parallel algorithms. In Chapter 12, a general review of the available paradigms and models for the design of efficient parallel algorithms is presented. This is followed by a description of several parallel graph and geometric algorithms in Chapters 13 and 14, respectively. These problems arise in many real-world applications, and high-performance computing capabilities are necessary for the efficient execution of this type of algorithms. Chapter 15 examines the use of efficient parallel data structures to facilitate the development of parallel algorithms.

In Chapters 16 and 17, the authors investigate algorithms that take advantage of certain data regularities in the data or the operations. The data-parallel model of computation is discussed in Chapter 16, along with the associated implementation issues on SIMD architectures. Then, Chapter 17 explores the use of systolic and VLSI-based processor arrays in solving several matrix oriented problems.

An important aspect of parallel algorithms design is the communication requirements and topology. Interconnection networks are used to connect the different components of a system together (e.g., processors and memories). A wide range and topology types of interconnection networks are reviewed in Chapter 18, while in Chapter 19 the author studies hypercube networks.

PART II: ARCHITECTURES AND TECHNOLOGIES

Architectures

Chapter 20 reviews the basic design principles of RISC architectures, and Chapter 21 highlights the underlaying design concepts of superpipelined and VLIW processors. These powerful processors are used as the main building blocks in many high-performance computers.

Chapters 22 and 23 describe in detail the design of SIMD and MIMD computer architectures, respectively. These two paradigms have been used to build many commercial parallel machines. As is the case with sequential computers, efficient memory architectures are essential for getting the best performance out of any parallel or distributed computing system. Memory systems performance and design issues, which are far more complicated in parallel computers, are explored in Chapter 24.

Technologies

In Chapter 25, the authors address the *heterogeneous computing paradigm,* which is a very powerful approach to exploit parallelism in many applications. An application can be divided into several subtasks that have different granularities and parallel requirements. To maximize performance each of the subtasks is executed on the most suitable computing platform. Chapter 26 reviews a new approach that provides HPC capabilities at a viable cost. Clusters of workstations (hence the term *cluster computing*) can be arranged together to provide a powerful platform for solving computationally intensive problems. The computing paradigms presented in Chapters 25 and 26 are dependent on networking technology, which is the topic of the next two chapters.

The use of optics technology in the construction of massively parallel processor systems is motivated by the need for faster interconnection networks. For systems with hundreds or thousands of processors, the efficiency and speed in which information is transferred from one processor to another can compromise the overall performance of the system. These and other issues are discussed in Chapter 27. Chapter 28 highlights the concepts and applications of ATM technology, which is considered to be an ideal medium for building HPC systems.

PART III: TOOLS AND APPLICATIONS

Development tools

The collected chapters in this section review the state of the art in software and tools for parallel programming. The success and popularity of PDC in the future will rely very much on the development of software and supporting tools. Chapter 29 reviews some of the language categories used to program parallel machines. This is followed by Chapter 30 in which the author presents a survey of portable tools that support parallel programming. Portability is an important issue that will enable PDC systems to achieve the generality of sequential computing.

Chapter 31 reviews the use of visualization techniques in PDC systems. Visualization is essential for managing the huge amounts of parallel execution data that are usually generated by many applications. In Chapter 32, the authors investigate the development of linear algebra software libraries for implementation on MIMD machines. Chapter 33 deals with the testing of programs in PDC environments. The testing problem involves the execution of a program with some benchmark inputs to check for faults. The testing problem is more complicated in PDC systems than in sequential computers.

Applications

PDC systems are having an unprecedented impact on many application areas. In fact, several volumes can be written on the many applications of PDC systems in science and engineering. In this section, a sample of these applications are reviewed.

Chapter 34 is devoted to reviewing what today is called *scientific computing*. This is a new discipline that emerged because of advances in computing in general, and parallel computing in particular. Chapter 35 outlines the development of more efficient simulation tools. Simulation is an important tool that can be used to study and evaluate a wide variety of systems, and simulation as a discipline can benefit from the computational power that PDC systems can offer. In Chapter 36, the authors deal with computer vision and image understanding problems. Their work spans the different levels of vision (low, intermediate, and high) accompanied by implementations. In Chapter 37, the authors investigate an area of research that is becoming more reliant on parallel processing to provide the necessary computing power: inter- and intra-sequence homology searching for sequence pattern analysis activities.

The solution of optimization problems arises in many disciplines. However, most of these problems require massive computing power. This makes PDC systems the ideal candidates to provide the necessary computing power. The next two chapters in this section are devoted to providing solutions to this problem. Chapter 38 is concerned with solving large linear stochastic optimization problems, while Chapter 39 deals with the use of parallel genetic algorithms for solving stochastic optimization problems.

Chapter 40 presents a taxonomy of parallel processing applications to several robotics computational tasks. The use of parallel processing for robotics computations is motivated by the need for more computing power. Most, if not all, robotics algorithms must be computed in situ, thus placing a great demand for more powerful computing systems. Finally, in Chapter 41, the author reviews another problem that requires real-time computing capabilities: flight simulation. This is computationally very demanding, and it must be "realistic" enough to be useful, thus the use of a distributed computing environment seems to be the natural choice. The author reviews the software development problem, which is not only important for distributed flight simulators but also crucial for a wide range of distributed applications.

Acknowledgments

The author would like to acknowledge the support of the Australian Research Council. Thanks are also due to Professor H. J. Siegel and Professor E.V. Krishnamurthy for their critique of the initial draft of this chapter.

1.7   Recommended Reading

Each chapter in the handbook is accompanied by its own bibliography. However, it is necessary for the reader to refer to journals and conference proceedings to keep up with the recent developments in the field.

Some of the important journals in the field are:

- *IEEE Transaction on Parallel and Distributed Systems* (IEEE Computer Society)
- *IEEE Parallel and Distributed Technology* (IEEE Computer Society)
- *Journal of Parallel and Distributed Computing* (Academic Press)
- *Journal of Distributed Computing* (Springer Verlag)
- *Journal of Parallel Algorithms and Applications* (Gordon and Breach)
- *International Journal of High Speed Computing* (World Scientific)
- *Parallel Computing* (North Holland)
- *Parallel Processing Letters* (World Scientific)
- *Concurrency: Practice and Experience* (Wiley)

A sample of some other journals that regularly publish topics related to PDC are:

- *Proceedings of the IEEE*
- *IEEE Transactions on Computers*
- *IEEE Transactions on Software Engineering*
- *IEEE Transactions on Systems, Man and Cybernetics*
- *IEEE Transactions on Computer Aided Design*
- *IEEE Transactions on Pattern Analysis and Machine Intelligence*
- *IEEE Computer*
- *IEEE Micro*
- *IEEE Software*
- *IEEE Spectrum*
- *Journal of the ACM*
- *Communications of the ACM*
- *ACM Transactions on Computer Systems*
- *ACM Computing Surveys*
- *ACM Letters on Programming Languages and Systems*
- *ACM Transactions on Database Systems*
- *SIAM Journal on Computing*
- *SIAM Reviews*
- *The Computer Journal* (British Computer Society)
- *International Journal of Real-Time Systems* (Kluwer Academic Publishers)
- *IEE Proceedings on Computers and Digital Techniques (Part-E) (Institution of Electrical Engineers, U.K.)*
- *International Journal in Computer Simulation* (Ablex Publishing)
- *Information Processing Letters* (Elsevier)
- *Journal of Neurocomputing* (Elsevier)
- *Journal of Evolutionary Computing* (MIT Press)
- *Neural Networks Journal* (Pergamon Press)
- *Information and Computation Journal* (Academic Press)
- *Journal of Complexity* (Academic Press)
- *Journal of Micro and Mini Computers* (Acta Press)
- *International Journal of Supercomputer Applications* (MIT Press)
- *Journal of Microprocessors and Microsystems* (Butterworth-Heinemann)

The reader is also encouraged to refer to the proceedings of some of the main events and conferences in the area, for example,

- International Conference on Parallel Processing (Pennsylvania State University)
- International Parallel Processing Symposium (IEEE Computer Society)

- International Conference on Supercomputing (ACM)
- International Symposium on Computer Architecture (IEEE Computer Society/ACM)
- Symposium on the Frontiers of Massively Parallel Computation (IEEE Computer Society)
- IEEE International Symposium on High Performance Distributed Computing (IEEE Computer Society)
- IEEE Conference on Foundations of Computer Science (IEEE Computer Society)
- ACM Symposium on Parallel Algorithms and Architectures (ACM)
- Annual ACM Symposium on Principles of Distributed Computing (ACM)
- International Conference on Parallel and Distributed Systems
- International Conference on Massively Parallel Processing Using Optical Interconnections (IEEE Computer Society)
- ACM Symposium on Theory of Computation
- CONPAR: Conference on Algorithms and Hardware for Parallel Processing

1.8   References

1. Aggarwal, J. K. and Chillakanti, P. 1995. Chapter 12, Software for parallel computing—A perspective, in *Parallel Computing: Paradigms and Applications,* ed. A. Y. Zomaya. London: International Thomson Computer Press, pp. 355–373.
2. Aho, A., Hopcroft, J., and Ullman, J. 1974. *The Design and Analysis of Computer Algorithms.* Reading, Mass.: Addison-Wesley.
3. Akl, S. G. 1989. *The Design and Analysis of Parallel Algorithms.* Englewood Cliffs, N.J.: Prentice Hall.
4. Almasi, G. S. and Gottlieb, A. 1989. *Highly Parallel Computers.* Menlo Park, Calif.: Benjamin/Cummings Publishing Co.
5. Amdahl, G. M. 1967. Validity of the single-processor approach to achieving large scale computing capabilities, in *Proc. AFIPS,* vol. 30, pp. 483–485. Washington, D.C.: Thompson.
6. Amdahl, G. M. 1988. Limits of expectation. *The International Journal of Supercomputer Applications,* vol. 2, no. 1, pp. 88–97.
7. Arvind, and Culler, D. E. 1986. Dataflow architectures. *Annual Review of Computer Science,* vol. 1, pp. 225–253.
8. Baron, R. J. and Higbie, L. 1992. *Computer Architecture.* New York: Addison-Wesley.
9. Bennett, C. H. and Landauer, R. 1985. The Fundamental Physical Limits of Computation. *Scientific American,* pp. 38–46.
10. Bertsekas, B. P. and Tsitsiklis, J. N. 1989. *Parallel and Distributed Computation: Numerical Methods.* Englewood Cliffs, N.J.: Prentice-Hall.
11. Cheng, D. Y., 1993. *A Survey of Parallel Languages and Tools,* Report RND 93-005, NAS Systems Development Branch, NAS Systems Division, NASA Ames Research Center.
12. Coffman, E. G. 1976. *Computer and Job-Shop Scheduling Theory.* New York: John Wiley & Sons.
13. Cohen, D. A. 1986. *Introduction to Computer Theory.* New York: John Wiley & Sons.
14. Cristian, F. 1991. Understanding fault-tolerant distributed systems. *Communications of the ACM,* vol. 34, no. 2, pp. 56–78.
15. Duncan, R. 1995. Chapter 1, Developments in scalable MIMD architectures, in *Parallel Computing: Paradigms and Applications,* ed. A. Y. Zomaya. London: International Thomson Computer Press, 3–24.
16. Flynn, M. J. 1966. Very high-speed computing systems, in *Proc. IEEE,* vol. 54, pp. 1901–1909.
17. Fox, G. 1990. Applications of parallel supercomputers: Scientific results and computer science lessons, in *Natural and Artificial Parallel Computation,* ed. M. A. Arbib and J. A. Robinson, pp. 47–90. Cambridge, Mass.: MIT Press.
18. Fox, G. and Mills, K. 1995. Opportunities for HPCC use in industry: Chapter 17, Opportunities for a new software industry in HPCC, in *Parallel Computing: Paradigms and Applications,* ed. A. Y. Zomaya. London: International Thomson Computer Press, 453–463.
19. Fox, G. C., Williams, R. D., and Messina, P. C., eds. 1994. *Parallel Computing Works!* San Mateo, Calif.: Morgan Kaufmann Publishers.
20. Garey, M. R. and Johnson, D. S. 1979. *Computers and Interactability: A Guide to the Theory of NP Completeness.* San Francisco: W. H. Freeman.

21. Gelernter, D., Nicolau, A., and Padua, D. 1990. *Languages and Compilers for Parallel Computing.* Cambridge, Mass.: MIT Press.

22. Godfrey, M. D. and Hendry, D. F. 1993. The computer as von Neumann planned it. *IEEE Annals of the History of Computing,* vol. 15, no. 1, pp. 11–21.

23. Goldschlager, L. 1982. A unified approach to models of synchronous parallel machines. *J. Assoc. Comput. Mach.,* vol. 29, pp. 1073–1086.

24. Guha, A., Bristow, J., Sullivan, C., and Husain, A. 1990. Optical interconnections for massively parallel architectures. *Applied Optics,* vol. 29, no. 8, pp. 1077–1093.

25. Gustafson, J. L. 1988. Reevaluating Amdahl's law. *Communications of the ACM,* vol. 21, no. 5, pp. 532–533.

26. Hecht-Nielsen, R. 1990. *Neurocomputing.* Reading, Mass.: Addison-Wesley.

27. Hillis, W. D. 1986. *The Connection Machine.* Cambridge, Mass.: MIT Press.

28. Hoare, C. A. R. 1978. Communicating sequential processes. *Communications of the ACM,* vol. 21, no. 8, pp. 666–677.

29. Hockney, R. W. 1981. A structural taxonomy of computers, in *Workshop on Taxonomy in Computer Architecture,* ed. G. A. Blaauw and W. Handler, pp. 77–92, Nurnberg: Friedrich Alexander Universitat Erlangen-Nurnberg.

30. Hwang, K. 1993. *Advanced Computer Architecture: Parallelism, Scalability, and Programmability.* New York: McGraw-Hill.

31. *IEEE Spectrum,* vol. 31, no. 6 (1994).

32. Jájá, J. 1992. *An Introduction to Parallel Algorithms.* Reading, Mass.: Addison-Wesley.

33. Karp, A. H. and Flat, H. P. 1990. Measuring parallel processor performance. *Communications of the ACM,* vol. 33, no. 5, pp. 539–543.

34. Karp, R. M. and Ramachandran, V. 1990. Parallel algorithms for shared-memory machines, in *Handbook of Theoretical Computer Science,* ed. J. Van Leeuwen, pp. 869–941. Amsterdam: Elsevier Science Publishers.

35. Kavi, K. M. and Gragon, H. G. 1983. A conceptual framework for the description and classification of computer architecture. *Proceedings of the IEEE International Workshop on Computer Systems,* pp. 10–19. Los Angeles: IEEE Computer Society Press.

36. Kung, H. T. 1982. Why systolic architectures? *IEEE Computer,* pp. 37–46.

37. Lamport, L. and Lynch, N. 1990. Distributed computing: models and methods, in *Handbook of Theoretical Computer Science,* ed. J. Van Leeuwen, pp. 1157–1199. Amsterdam: Elsevier Science Publishers.

38. Langholz, G., Francioni, J., and Kandel, A. 1989. *Elements of Computer Organization.* Englewood Cliffs, N.J.: Prentice-Hall.

39. Lawson, H. W. 1992. *Parallel Processing in Industrial Real-Time Applications.* Englewood Cliffs, N.J.: Prentice Hall.

40. Lewis, H. R. and Papammltriou, C. H. 1981. *Elements of the Theory of Computation.* Englewood Cliffs, N.J.: Prentice Hall.

41. Lewis, T. G. and El-Rewini, H. 1992. *Introduction to Parallel Computing.* Englewood Cliffs, N.J.: Prentice Hall.

42. Louie, T. 1981. Array processors: a selected bibliography. *IEEE Computer,* vol. 14, no. 9, pp. 53–57.

43. May, D. 1990. Towards general-purpose parallel computers, in *Natural and Artificial Parallel Computation,* ed. M. A. Arbib and J. A. Robinson, pp. 91–121. Cambridge, Mass.: MIT Press.

44. Milner, R. 1980. *A Calculus of Communicating Systems (Lecture Notes in Computer Science).* New York: Springer-Verlag.

45. Moldovan, D. I. 1993. *Parallel Processing: From Applications to Systems.* San Mateo, Calif.: Morgan Kaufmann Publishers.

46. Murata, T. 1989. Petri-nets: properties, analysis and applications. *Proceedings of the IEEE,* vol. 77, no. 4, pp. 544–580.

47. Nelson, V. P. 1990. Fault-tolerant computing: fundamental concepts. *IEEE Computer,* vol. 23, no. 7, pp. 19–25.

48. NSF. 1992. *Grand Challenge: High-Performance Computing and Communications,* Report, Committee on Physical, Mathematical, and Engineering Sciences. Washington, D.C.: U.S. Office of Science and Technology Policy, National Science Foundation.

49. Papadimitriou, C. H. 1994. *Computational Complexity.* Reading, Mass.: Addison-Wesley.

50. Patterson, D. A. and Hennessy, J. L. 1994. *Computer Organization and Design: The Hardware/Software Interface.* San Mateo, Calif.: Morgan Kaufmann Publishers.

51. Perrott, R. H. 1987. *Parallel Programming.* Reading, Mass.: Addison-Wesley.

52. Ramamoorthy, C. V. and Li, H. F. 1977. Pipeline architecture. *ACM Computing Surveys,* vol. 9, no. 1, pp. 61–102.

53. Reed, D. A. and Grunwald, D. C. 1987. The performance of multicomputer interconnection network. *IEEE Computer,* vol. 20, no. 6, pp. 63–73.

54. Shurkin, J. 1984. *Engines of the Mind: A History of the Computer.* New York: W. W. Norton & Company.

55. Siegel, H. J. 1990. *Interconnection Networks for Large-Scale Parallel Processing: Theory and Case Studies,* ed. 2. New York: McGraw-Hill.

56. Siegel, H. J., Schwederski, T., Nation, W. G., Armstrong, J. B., Wang, L., Kuehn, J. T., Gupta, R., Allemang, M. D., Meyer, D. G., and Watson, D. G. 1995. Chapter 3, The design and prototyping of the PASM reconfigurable parallel processing system, in *Parallel Computing: Paradigms and Applications,* ed. A. Y. Zomaya. London: International Thomson Computer Press, pp. 78–114.

57. Skillicorn, D. B. 1988. A taxonomy for computer architectures. *IEEE Computer,* vol. 21, no. 11, pp. 46–57.

58. Skillicorn, D. B. 1991. Models for practical parallel computation. *International Journal of Parallel Programming,* vol. 20, no. 2, pp. 133–158.

59. Stallings, W. 1990. *Reduced Instruction Set Computer (RISC).* 2d ed. Washington, D.C.: IEEE Computer Society Press.

60. Stenstrom, P. 1995. Chapter 2, Shared-memory multiprocessors—a cost-effective approach to high-performance parallel computing, in *Parallel Computing: Paradigms and Applications,* ed. A. Y. Zomaya. London: International Thomson Computer Press, pp. 25–77.

61. Stone, H. S. 1990. *High-Performance Computer Architectures.* 2d ed. Reading, Mass.: Addison-Wesley.

62. Ullman, J. 1975. NP-complete scheduling problems. *Journal of Computer and System Sciences,* vol. 10, pp. 384–393.

63. Valiant, L. G. 1990. General purpose parallel architectures, in *Handbook of Theoretical Computer Science,* ed. J. Van Leeuwen, pp. 944–971. Amsterdam: Elsevier Science Publishers.

64. Veen, A. H. 1986. Dataflow machine architecture. *ACM Computing Surveys,* vol. 18, no. 1, pp. 365–396.

65. Wilkes, M. V. 1985. *Memoirs of a Computer Pioneer.* Cambridge, Mass.: MIT Press.

66. Zima, H. and Chapman, B. 1990. *Supercompiler for Parallel and Vector Computers.* Reading, Mass.: Addison-Wesley.

67. Zomaya, A. Y. 1992. *Modelling and Simulation of Robot Manipulators: A Parallel Processing Approach.* Singapore: World Scientific Publishing.

68. Zomaya, A. Y., ed. 1995. *Parallel Computing: Paradigms and Application.* London: International Thomson Computer Press.

2

Semantics of Concurrent Programming

J. Desharnais, A. Mili, R. Mili, J. Mullins, and Y. Slimani

In sequential programming, a program in execution has a single thread of control, so that when it is executing, a single instruction is active at any one time. *Concurrent programming* refers to programs whose execution may have more than one thread of control, i.e., more than one active instruction at a time. Due to its scientific interest and its practical application potential, concurrent programming has attracted a great deal of attention since the beginning of computing.

The study of programming language semantics deals with assigning meanings to programs or program parts written in a specific programming language. A number of concurrent programming languages are known nowadays: Some, such as Brinch Hansen's *Concurrent Pascal* [8], are designed for educational and research purposes. Others, such as *Ada*, are designed to be used for applications development. Ada, e.g. has a large community of users as well as a wide range of compilers running on a wide range of platforms. By and large, the semantic definitions of programming languages can be divided into three broad categories: *axiomatic semantic definitions*, which define the semantics of a language by making statements (under the form of axioms) about each construct of the language; *operational semantic definitions*, which define the semantics of a language by defining the effect of each construct of the language on the underlying machine (or, equivalently, how the underlying machine proceeds to execute each statement of the language); and *denotational semantic definitions*, which define the semantics of a language by mapping each one of its constructs into a mathematical object that represents its *meaning*.

A number of proposals have been presented in the past for the semantic definition of concurrent programming languages. These semantic definitions can be characterized by three features, namely: (1) the concurrent programming language, (2) the computation model that the language supports, and (3) the method that is selected for the semantic definition. The semantic definition method, in turn, is characterized by means of three features: (1) the meaning of a program or program part (this issue involves such questions as what is the meaning of a program, what aspects of a program's behavior do we want to capture, and what properties do we want to formulate and prove about programs, and so on); (2) how this meaning is defined for trivial programs; and (3) how this meaning is inductively defined for complex programs. In Section 2.1 we discuss the various models of concurrent programming, as well as the languages that support them, and in Section 2.2 we review the various methods of semantic definition as well as their mathematics. In Sections 2.3, 2.4, and 2.5 we use the insights gained in the previous two sections to discuss some of the existing proposals of semantic definitions of concurrent languages. Finally, in Section 2. 6 we discuss the main tenets of the field of semantics of concurrency, outline the current state of the art in this field, and then sketch prospects for future research in the field and direct the interested reader to more detailed sources.

2.1 Models of Concurrent Programming

The interest of the field of concurrent programming stems as much from the technical challenge of the problems that it poses as it stems from the diversity of interpretations to which it lends itself. Nowhere is this diversity more apparent than in defining a *model* of concurrent programming. We have identified a number of *attributes* of concurrency models; each model can be defined by the values that it takes for these attributes.

Level of granularity. We consider two agents who are accessing a centralized database simultaneously (e.g. two bank tellers or two travel agents). If we observe their activities at the transaction level, then we find naturally that their activities are concurrent, given that they are performing their transactions at the same time. If, on the other hand, we observe their activities at the level of the CPU cycle, then we find that their activities are sequential, because the CPU takes turns serving their queries. The activities of these agents can be considered as concurrent or sequential depending on the level of granularity at which we observe them. One of the key features of any concurrent programming model is the level of granularity (also called *level of atomicity*) at which actions are considered.

Sharing the clock. When several processes operate concurrently and must eventually interact (e.g. to exchange information), it is important to determine whether they share the same clock. The options they have for the exchange of information vary considerably, depending on whether they share the same clock.

Sharing memory. When several processes operate concurrently and must eventually interact, it is important to determine whether they share some memory space. If they do, this has several implications: first, this space may be used for sharing information; second, access to this space may have to be mutually exclusive, hence affecting the operation of the processes; third, the level of granularity of the processes's activities is then typically larger than or equal to a memory access.

Pattern of interaction. There exist two forms of interaction between processes: *synchronization* and *communication*. Synchronization defines a chronological order between events taking place within different processes. Communication defines a transfer of information from one process to another. Even though the definitions of these patterns appear quite distinct, they in fact have similar functional properties: it is possible to achieve communication by means of synchronization constructs (using these constructs to grab the attention of the communicating processes); also, it is possible to achieve synchronization by means of communication constructs (using communication constructs to transfer information about the occurrence of events at different processes).

Pattern of synchronization. There exist two patterns of synchronization between processes: *mutual exclusion* and *mutual admission*. Mutual exclusion is defined by an encapsulated sequence of actions whose execution is indivisible; once a process starts executing this sequence, it may not be interrupted until the sequence is completed. Mutual admission is defined by an encapsulated sequence of actions which must be executed by two processes simultaneously; if one process is ready and the other is not, it must wait until both are ready to give the sequence their undivided attention. An example of mutual admission is Ada's *rendez-vous* mechanism. An important distinction between these patterns is that, in mutual exclusion, the processes that interact may ignore each other totally; each process may ignore the number, the names, and the functions of the other processes. In mutual admission, however, the participating processes know each other to some extent; this pattern is typically used in the context where one process offers a service to a community of user processes. Typically the service users know the exact identity of the service provider, whereas the service provider has a general knowledge of the user community.

Pattern of communication. There exist two modes of communication between processes: *synchronous communication* and *asynchronous communication*. In synchronous communication, the transfer of information is achieved while both communicating processes are giving their undivided attention to the transfer. In asynchronous communication, the transfer is achieved in two steps: first, the sender deposits the message in a mailbox area; second the receiver picks it up from the mailbox. The implementation of the transfer depends to a large extent on the characteristics of the mailbox: whether the mailbox has finite capacity or infinite capacity; whether messages are ordered by their time of arrival or placed in an arbitrary order; and whether the mailbox is public or private. In the case where the mailbox is private, the implementation also depends on whether the mailbox is labeled with the sender's identification or the receiver's identification.

Specifying concurrency. It is legitimate to distinguish between two types of concurrent programming: *application concurrency* and *implementation concurrency*. In application concurrency, the problem we wish to solve or the situation we wish to simulate is intrinsically concurrent; concurrency constructs are then required to represent the problem/situation at hand in a more faithful/ more natural manner. Such is the case, for example, of an Ada program that simulates the *sleeping barber* problem or the *cigarette smokers'* problem [36]. In implementation concurrency, the problem we wish to solve is intrinsically sequential; concurrency is introduced at design time or implementation time for the sake of efficiency (e.g., to make use of a multiprocessor architecture). Such is the case, for example, of a Fortran program where we can identify statements that can be executed in parallel because they have no functional dependencies. An important difference between these two types of concurrency is that, in the first type, concurrency appears at the specification of the problem, whereas in the second type the specification makes no mention of concurrency (because concurrency is a feature of the design/implementation rather than a feature of the specification).

Implementation of concurrency. Given a problem that is intrinsically concurrent, we may want to implement it in one of two ways: either by *effective concurrency* or by *simulated concurrency*. In effective concurrency, the concurrent processes are executed on distinct processors and interact in real time. In contrast, in simulated concurrency, the concurrent processes are executed sequentially on a single processor, and concurrency is simulated by the scheduling mechanism of the shared processor.

Interaction protocol. There exist two interaction protocols between concurrent processes: *cooperation* and *competition*. In a cooperation protocol, the interacting processes know each other mutually and fulfill complementary functions to serve a common objective or interest. In a competition protocol, the interacting processes do not necessarily know each other; they fulfill distinct functions and typically (although not necessarily) serve different interests. It is tempting to believe that mutual exclusion is used to implement competition, and mutual admission is used to implement admission, but this is by no means required; cooperation (e.g. in a producer-consumer arrangement) involves mutual exclusion (e.g. mutually exclusive access to the transfer buffer).

Given that a model of concurrency depends on so many features, it is no wonder there are so many models nowadays that seem to be talking about different things. It is useful to add, however, that these features are not strictly orthogonal but are interdependent: for example, if we choose *synchronization* as our pattern of interaction (versus communication), then it makes no sense to discuss possible patterns of communication.

2.2   Semantic Definitions

A programming language is defined by two features: its *syntax*, which defines the set of legal sentences in the language; and its *semantics,* which assigns meaning to legal sentences of the language. Three broad techniques are known nowadays for the definition of programming language semantics: *axiomatic definition; operational definition*; and *denotational definition*. We review these briefly in this section as they pertain to sequential languages. In Sections 2.3 through 2.5, we see how these techniques have been applied to the semantic definition of concurrent languages.

2.2.1   Axiomatic semantics

The axiomatic semantic definition of a programming language defines the meaning of language constructs by making statements (in the form of axioms or inference rules) about the effect of such constructs on the state of the program. The effect of the programming construct on the state of the program is defined by a *precondition* and a *postcondition*. The precondition describes the state of the program before execution, and the postcondition describes the state after the execution. This method of semantic definition, which comes about as a by-product of Hoare's program verification method [18], is based on the following notation:

$$\{p\} \, S \, \{q\}$$

where p and q are state predicates and S is a program or a statement. The formula is interpreted as follows: If statement S is executed in a state where p holds, then it terminates, and q holds upon termination of S.

The semantic definition of a language is constructed inductively on the structural complexity of a program: the semantics of complex constructs is derived from the semantics of their components; at the bottom of this inductive process is the semantics of assignment statements, which is defined by means of an axiom.

For the sake of illustration, we briefly show the semantic definition of a simple Pascal-like language which has assignment, sequence, and alternation statements.

Assignment statement axiom. The semantics of assignment statements is defined by means of the following axiom:

$$\{q \, (E \, (s)) \, \} \, s := E \, (s) \, \{q \, (s) \, \}$$

where s is the program state (e.g. a variable, if the program has a single variable), E is an expression of the language, and q is a predicate defined on the space of the program. This axiom provides that whatever was true of $E(s)$ (i.e. $q(E(s))$) before execution of the assignment statement $s := E(s)$ becomes true of s (i.e., $q(s)$) once the statement has executed and terminated.

Sequence statement rule. The semantics of sequence statements is defined by means of the following rule:

$$\frac{\{p\}\,S_1\,\{int\} \qquad \{int\}\,S_2\,\{q\}}{\{p\}\,S_1\,;\,S_2\,\{q\}}$$

To prove the conclusion of this rule, it suffices to find predicate int such that both premises hold.

Alternation statement rule. The axiomatic semantics of alternation statements is defined by means of the following inference rule:

$$\frac{\{p \wedge t\}\,S\,\{q\} \qquad \{p \wedge \neg t\}\,T\,\{q\}}{\{p\}\ \text{if } t \text{ then } S \text{ else } T\,\{q\}}$$

To establish the conclusion of this rule, it suffices to prove that the premises hold about statements S and T; this rule defines the meaning of an alternation statement.

2.2.2 Operational semantics

While axiomatic semantics capture the meaning of programming language constructs by focusing on the effect of these constructs on the program state, operational semantics focuses on *how* the state of the program is affected. In operational semantics, each statement of the language is defined by describing the process that a computer goes through to execute that statement; typically, the process is described in terms of a lower-level language whose semantics is predefined. The most archetypical kind of operational semantic definition is the compiler: a compiler defines the semantics of the source language by mapping it into a target language, typically an assembly-level language.

As an example, if we use the assembly language of the M68000 family [15], we can define the semantics of Pascal's alternation statement

$$\text{if } t \text{ then } S \text{ else } T$$

by the following assembly language program:

```
              <assign CCR bits>      ; used for conditional branch
              Bcc elselabel          ; branch on carry clear
              <code for S>           ; then clause
              Bra next               ; branch to next statement
    elselabel: <code for T>          ; else clause
    next:      <next statement>      ; code for next statement
```

A very common operational semantic definition which is in widespread use nowadays is *syntax directed translation*. In this method, the semantic definition is built on top of the language's syntactic definition: to each BNF rule of the language (which we assume context free), we associate a sequence of actions that the compiler must undertake upon reducing the input string by means of the rule. As an illustrative example, we show the syntax directed translation of simple arithmetic expressions (terminated by an end of line symbol) whose meaning is taken to be their numeric value.

```
line        : expr '\n'              {printf("%d\n",$1);}
            ;
expr        : expr '+' term          {$$ = $1 + $3;}
            ;
term        : term '*' factor        {$$ = $1 * $3;}
            ;
factor      : '(' expr ')'           {$$ = $2;}
            | DIGIT
            ;
```

The double dollar sign refers to the meaning associated with the left-hand side of the rule, whereas the numbered dollar signs refer to the grammar symbols on the right-hand side of the rule. Such semantic definitions are submitted to parser generators (the most common is Yacc), which generate parsers accordingly.

2.2.3   Denotational semantics

In denotational semantics, each construct of the language is mapped into a mathematical object that defines its meaning. These objects vary according to the language at hand, the computation model underlying the language, and the properties that we want the semantic definition to capture.

A common form of denotational semantic definition maps programming language constructs into functions on the space of the program, where the space of the program is defined by the values that program variables may take [21, 27, 26]. As an illustration, we give below a denotational semantic definition of the three Pascal-like statements whose axiomatic definition is given above (in Section 2.2.1); to each statement we assign a function, i.e. a set of pairs, that represents the set of initial state/final state pairs that the statement defines. The function that we assign to a statement S, which we call the *denotation* of the statement, is represented by $[S]$.

Assignment statement. Let x be the name of a variable in the program at hand, s be the name of a common state of the program (defined by all the variables of the program, say x, $y_1, y_2, \ldots, y_k$); and E be an expression that can be evaluated in state s. Then we pose:

$$[x := E(s)] = \{ (s, s') \mid x(s') = E(s) \wedge y_1(s') = y_1(s) \wedge$$

$$y_2(s') = y_2(s) \ldots \wedge y_k(s') = y_k(s) \}$$

where $y_1, y_2, \ldots, y_k$ are the names of the other variables of the program.

Sequence statement. We pose, by definition,

$$[S_1;S_2] \; = \; [S_1] \circ [S_2]$$

where o represents the relative product of relations.

Alternation statement. We pose, by definition,

$$[\text{if } t \text{ then } S \text{ else } T] \; = \; \{\, (s, s') \,|\, t(s) \wedge (s, s') \in [S] \vee \neg t(s) \wedge (s, s') \in [T] \,\}$$

The semantic definition presented above focuses on the function of a program, irrespective of how simple or how complex the program is. In Ref. [35], Schmidt presents a denotation for programs which reflects not only the functional properties of a program, but also its structural properties. In Schmidt's semantic definition, a program is represented not by a function, but by a matrix of functions, where each entry of the matrix represents a node in the flowgraph of the program at hand. As an illustration, we give below the flowgraph and the matrix representation of the following program:

$$\text{while } x > 5 \text{ do } x := x - 1$$

Note that node 1 represents the entry point of the program while node 3 represents the exit point. Each arc of the graph is labeled with the function computed along that arc. Note that, despite our intuition, even a condition defines a function, normally a subset of the identity (hence e.g. the while condition $x > 5$ defines the function $\{\, (x, x') \,|\, x > 5 \wedge x' = x \,\}$).

$$
\begin{array}{c}
\begin{array}{ccc}
1 & 2 & 3
\end{array} \\[4pt]
\begin{array}{c}
1 \\ 2 \\ 3
\end{array}
\left(
\begin{array}{ccc}
\emptyset & x > 5 \wedge x' = x & x \leq 5 \wedge x' = x \\
x' = x - 1 & \emptyset & \emptyset \\
\emptyset & \emptyset & \emptyset
\end{array}
\right)
\end{array}
$$

Flowgraph: node 1 → node 2 labeled $x > 5 \wedge x' = x$; node 2 → node 1 labeled $x' = x - 1$; node 1 → node 3 labeled $x \leq 5 \wedge x' = x$.

These denotations for programs prove to be very useful in practice, and very interesting. In particular they enable us to manipulate programs by performing operations on their matrices, which are quite similar to traditional matrix operations, and can be interpreted quite naturally in programming terms.

Using the insights gained in Sections 2.1 and 2.2, we wish to discuss some semantic definitions of concurrent programming languages and models. Our intent is not so much to do an exhaustive survey of past proposals as it is to discuss some representative solutions.

2.3   Axiomatic Semantic Definitions

2.3.1   A semantic definition of await-then

In Ref. [30], Owicki and Gries define a concurrent programming language, and provide an axiomatic proof technique for the language; this work is extended by Owicki and Lamport [31] to prove liveness properties of concurrent programs. The language of Owicki and Gries is derived from Pascal by adding two constructs: a *cobegin-coend* construct, which serves to activate procedures concurrently, and an *await-then* construct, which serves to synchronize two concurrent processes. The general format of a cobegin-coend block is the following (where we assume, for simplicity, that only two processes are involved):

```
cobegin
S1, S2
coend.
```

When the control flow reaches cobegin, both S_1 and S_2 start executing concurrently; the coend keyword is reached when both processes have completed their execution.

The general format of an await-then statement (which appears in the body of a process that is called in a cobegin-coend statement) is the following:

```
await t then b,
```

where t is a condition and b is a block of code. Condition t potentially involves foreign variables (i.e. variables that are under the control of other processes); if condition t holds then block b is executed in an indivisible fashion, else the process is suspended until such time as t becomes true.

The language that Owicki and Gries define uses synchronization (by means of await-then) as a means of interaction; the processes involved in the synchronization share the same clock and may share variables in memory. The await-then construct may be used to program mutual exclusion as well as mutual admission; also concurrent processes in a cobegin-coend block may interact in mutual cooperation (e.g. if they collaborate to compute a function) or in mutual competition (e.g. if they simulate a resource allocation situation). Owicki and Gries have provided an axiomatic proof system for this language, which constitutes in effect an axiomatic semantic definition thereof. The sequential parts of the language are defined by means of traditional Pascal-like axioms and rules [23]; the concurrent features of the language are defined by means of the following rules.

Await-then rule. The semantics of the await-then statement is defined by means of the following inference rule:

$$\frac{\{p \wedge t\}\, b\, \{q\}}{\{p\}\ \text{await } t \text{ then } b\, \{q\}}$$

To appreciate the meaning of this statement, and how this rule reflects the meaning, it is instructive to compare this rule to the *if-then* rule for sequential Pascal-like programs, which we give below:

$$\frac{\{p \wedge t\}\, b\, \{q\} \qquad p \wedge \neg t \Rightarrow q}{\{p\}\ \text{if } t \text{ then } b\, \{q\}}$$

The second clause of the if-then rule provides for the case when condition t is not true; for the await-then statement, we make no provision for this case since the statement is enabled only when condition t becomes true. In the sequential case, if t is not true when the if-then statement is reached then it makes no sense to wait until it is true, since there is a single thread of control, namely the thread we are following; in the concurrent case, it makes sense to wait for t to be true since other processes may eventually switch t to true.

Cobegin-coend rule. The semantics of a cobegin-coend statement is defined by the following inference rule:

$$\{p_1\} S_1 \{q_1\}$$

$$\{p_2\} S_2 \{q_2\}$$

$$\overline{\{p_1 \wedge p_2\} \text{ cobegin } S_1, S_2 \text{ coend } \{q_1 \wedge q_2\}}$$

provided the proofs of S_1 and S_2 are free from mutual interference. Now, we say that these proofs are free from interference if each can be established without reference to the other. In practice, the proofs of S_1 and S_2 can be made interference-free by weakening their intermediate assertions sufficiently so that the assertions of each do not refer to the other.

2.3.2 An axiomatic proof theory

In Ref. [2], Apt and Olderog discuss the semantics of concurrent programs in the context of a more general discussion on program verification. They introduce a guarded commands language, which they then extend with various constructs to discuss the semantics of concurrent programs.

2.3.2.1 Disjoint parallel programs. The first family of programs that they consider is the set of *disjoint parallel programs*; two programs S_1 and S_2 are said to *disjoint* if none of them can change the variables accessed by the other. For example, the programs

```
S1:     x:=z;
S2:     y:=z;
```

are disjoint. Disjoint programs can be combined by means of the *disjoint parallel composition*, which we denote as shown below:

```
S:      [S1 || S2].
```

The semantics of this statement is defined by the following inference rule.

$$\{p_1\} S_1 \{q_1\}$$

$$\{p_2\} S_2 \{q_2\}$$

$$\overline{\{p_1 \wedge p_2\} [S_1 \| S_2] \{q_1 \wedge q_2\}}$$

where S_1 and S_2 are disjoint and for each S_i, $i=1,2$, the set of free variables of p_i and q_i includes no variable that S_i changes.

2.3.2.2 Parallelism with shared variables. Apt and Olderog enrich their model by introducing shared variables as a means for parallel programs to interact. Consequently, they also introduce the notion of *atomic action* (denoted by $\langle S \rangle$, and defined as a statement whose execution is indivisible) and the notion of *proof outline* (denoted by S^*, and defined as the program S which has been annotated by appropriate intermediate assertions) and redefine the property of interference freedom as follows.

Definition A. Let S be a component program. Consider a standard proof outline $\{p\} S^* \{q\}$ for total correctness and a statement R with the precondition *pre(R)*. We say that *R does not interfere with* $\{p\} S^* \{q\}$ if the following conditions hold:

1. For all assertions r in the proof outline $\{p\} S^* \{q\}$, the formula

$$\{r \wedge pre\,(R)\} R \{r\}$$

holds in the sense of total correctness.
2. For all bound functions t in $\{p\} S^* \{q\}$, the formula

$$\{t = z \wedge pre\,(R)\} R \{t \le z\}$$

holds in the sense of total correctness where z is some fresh variable not occurring in t, R, and *pre(R)*. ❑

Definition B. Let $[S_1 \| S_2]$ be a parallel program. Standard proof outlines

$$\{p_i\} S_i^* \{q_i\}$$

for total correctness, $i = 1,2$, are called *interference free* if no normal assignment or atomic region of a component S_i interferes with the proof outline of another component. ❑

By means of this altered definition of interference freedom, Apt and Olderog obtain a rule to the effect that if the standard proof outlines $\{p_i\} S_i^* \{q_i\}$, $i = 1,2$ are interference free, then the formula

$$\{p_i \wedge p_2\} [S_1 \| S_2] \{q_1 \wedge q_2\}$$

is a theorem.

2.3.2.3 Synchronization and deadlock freedom. Apt and Olderog further enrich their model by introducing synchronization, which they implement by means of Owicki and Gries's *await-then* construct, and obtain the same proof rules as above (Section 2.3.1). With the introduction of synchronization comes the potential for deadlocks between interacting processes; to deal with this contingency, Apt and Olderog define the notion of *weak correctness*, whereby a formula

$$\{p\} S \{q\}$$

holds in the sense of weak total correctness if every execution of S starting in a state satisfying p is finite and either terminates in a state satisfying q or gets blocked.

Definition C. Consider a parallel program S: $[S_1 \| S_2]$. A pair $\langle R_1, R_2 \rangle$ of statements is called *potential deadlock* of S if the following holds:

1. Each $\overset{\cdot}{R_i}$ is either an await-statement in the component S_i or the symbol E, which stands for the empty statement and represents termination of S_i.
2. At least one R_i, $i = 1,2$, is an *await*-statement in S_i. ❏

Definition D. Consider a parallel program S: $[S_1 \| S_2]$. Given interference-free standard proof outlines $\{p_i\} S_i^* \{q_i\}$ for weak total correctness, we associate with every potential deadlock of S a pair $\langle R_1, R_2 \rangle$ of assertions, where

1. $r_i = pre(R_i) \wedge \neg B$ if R_i is *await B then S end*, and
2. $r_i = q_i$ if $R_i = E$. ❏

Using these definitions, we introduce the proof rule for *parallelism with deadlock freedom*:

> the standard proof outlines $\{p_i\} S_i^* \{q_i\}$
> for weak total correctness are interference free,
> for every potential deadlock $\langle R_1, R_2 \rangle$ of $[S_1 \| S_2]$
> the corresponding tuple of assertions $\langle r_1, r_2 \rangle$ satisfies $\neg(r_1 \wedge r_2)$
> ───
> $\{p_i \wedge p_2\} [S_1 \| S_2] \{q_1 \wedge q_2\}$

2.3.3   A deductive logic of concurrent programming

In Ref. [10], Chandy and Misra introduce a concurrent programming language under the name *UNITY* and define its semantics by means of a *programming logic*. A *UNITY* program contains four sections, namely a *declare* section, where program variables are declared, an *always* section, where relationships between variables are declared (for documentation purposes), an *initially* section, which defines initial values for the program variables, and finally an *assign* section, which defines assignments to the variables. The language supports simultaneous assignments and conditional expressions, as well as quantified simultaneous assignments (e.g. assigning expressions to arrays cells, where each cell receives a different expression as a function of its index).

To define the semantics of the *UNITY* language, Chandy and Misra use the notation

$$\{p\} S \{q\}$$

but depart significantly from its original interpretation [18, 23, 30]. First, they interpret such a formula to mean that if the *UNITY* program S executes on some state verifying predicate p *and* if it terminates *then* it produces a state that verifies predicate q. Second, and more importantly, such assertions do not make statements about specific positions in the control flow of a program (as they do in their traditional interpretation), because *UNITY* has no notion of control flow; rather they make statements about properties of a *UNITY* program as a whole, such as 'predicate *inv* is always true' or 'if p ever becomes true, then so will q, eventually'. This separation of programs and proofs is valuable in the following sense: in order to combine programs, it is not necessary to have the detailed code of the programs; rather it suffices to have a description of their relevant properties, using the proposed notation. Third, for most continuous processes (such as operating systems or process control systems) there are no preconditions and postconditions in the traditional sense [18, 23]; what is important to capture, however, is a characterization of the states of the program during its execution. The interpretation of formulas such as $\{p\} S \{q\}$ is adjusted to accommodate this emphasis. Fourth, this notation is enriched by introducing quantification on its middle term, namely the statement S; hence it is possible to assert that a formula of the form given above holds for all statements S in a *UNITY* program, or that it holds for some statement. Interestingly, the universal quantification of such formulas helps us prove *safety* properties of *UNITY* programs, whereas the existential quantification helps us prove *progress* properties. Finally, to make the notation more powerful, Chandy and Misra introduce a number of operators for composing assertions about programs: *unless, stable, invariant, ensures, leads-to,* and *fixed-point*.

For the sake of illustration, we give below the definitions of some of these operators:

Unless. For a given *UNITY* program F, the assertion p *unless* q is defined as follows:

$$p \text{ } unless \text{ } q \stackrel{\text{def}}{=} (\forall S{:}S \in F{:} \{p \wedge \neg q\} S \{p \vee q\})$$

If p holds at any point during the execution of F, then either q never holds and p continues to hold forever, or q holds eventually and p continues to hold at least until q holds.

Stable. The assertion *stable(p)* is defined as follows:

$$stable\,(p) \stackrel{\text{def}}{=} p \text{ } unless \text{ } false$$

Whenever p becomes true during execution of F, it will remain true forever. If p and $\neg p$ are both stable, we say that p is *constant*.

Invariant. The assertion *invariant(p)* is defined as follows:

$$invariant\,(q) \stackrel{\text{def}}{=} (initial \text{ } condition \Rightarrow q) \wedge stable\,(q)$$

A predicate is invariant if it is true throughout the execution of the program (true initially, and stable).

Ensures. The assertion p *ensures* q is defined as follows:

$$p \text{ } ensures \text{ } q = (p \text{ } unless \text{ } q \wedge (\exists S \in F{:} \{p \wedge \neg q\} S \{q\}))$$

If p is true at some point in the execution of F, p remains true as long as q is false, and eventually q becomes true.

Also, for the sake of illustration of Chandy and Misra's programming logic, we give below an example of inference rule of this logic.

p unless q
invariant (¬q)
stable (p)

Finally, we give illustrative examples of theorems about this logic.

Reflexivity of Unless
p unless p

Antireflexivity of Unless
p unless ¬p

Cancellation of Unless

$$\frac{\begin{array}{c} p \ unless \ q \\ q \ unless \ r \end{array}}{(p \lor q) \ unless \ r}$$

Impossibility of Ensure
$$\frac{p \ ensures \ false}{\neg p}$$

2.4   Denotational Semantic Definitions

2.4.1   Communicating sequential processes

In Ref. [19], Hoare introduces a logic of concurrent programming under the title *Communicating Sequential Processes* (or *CSP*, for short). In this logic, concurrent systems are made up of *processes*, where each process is defined by a sequence of events, and communicates with the external environment by means of *send* and *receive* operations over virtual channels.

2.4.1.1 Processes. A process P can be represented by the notation $x \rightarrow Q$, where x is an event and Q is a process; such a representation (read as: x then Q) defines P as the process which first engages in the event x then behaves exactly as described by Q. Each process is defined with respect to a particular alphabet (namely the set of events in which it may engage), and the expression $x \rightarrow Q$ is defined only if event x is an element of the alphabet of Q; the alphabet of process Q is denoted by $\alpha(Q)$. A simple vending machine that serves a single customer then breaks down can be represented as the following process on alphabet {*coin, candy, stop*}:

$$VM_0 \stackrel{\text{def}}{=} (coin \rightarrow (candy \rightarrow stop))$$

Recursive definitions can be used to define arbitrarily more complex processes; hence for example the vending machine that serves an indefinite number of customers can be represented as the following process on alphabet $\{coin, candy\}$:

$$VM_1 \overset{\text{def}}{=} (coin \rightarrow (candy \rightarrow VM_1))$$

Using the construct of *choice*, it is possible in CSP to specify several alternative behaviors depending on the stimulus of the external environment; hence for example, the vending machine that serves a candy for a coin deposit and a toffee for a notebill deposit can be represented as the following process on alphabet $\{coin, candy, notebill, toffee\}$:

$$VM_2 \overset{\text{def}}{=} (coin \rightarrow (candy \rightarrow VM_2) \mid notebill \rightarrow (toffee \rightarrow VM_2))$$

Also, it is possible to specify nondeterministic choice between two behaviors, where the selection among the two alternative choices is made arbitrarily; hence a vending machine that, for each coin deposit, may either return a candy or return the coin then stop can be represented as the following process on alphabet $\{coin, candy, coinreturn, stop\}$:

$$VM_3 \overset{\text{def}}{=} (coin \rightarrow (candy \rightarrow VM_3)) \sqcap (coin \rightarrow (returncoin \rightarrow stop))$$

Mutual recursion is also possible; hence a vending machine that accepts a single kind of coin and will return alternatively either a candy or a chocolate bar can be represented as the following process on alphabet $\{coin, candy, bar\}$:

$$VM_C \overset{\text{def}}{=} (coin \rightarrow (candy \rightarrow VM_B))$$

$$VM_B \overset{\text{def}}{=} (coin \rightarrow (chocolate \rightarrow VM_C))$$

Depending on whether we want our vending machine to start with delivering a candy or a chocolate bar, we let it be defined as VM_C or VM_B.

2.4.1.2 Concurrency. Processes can be combined with the parallel construct, whether they share a common alphabet or not. The parallel construct satisfies a number of identities, some of which we present below:

$$P \parallel Q = P \parallel P$$

$$(P \parallel Q) \parallel R = P \parallel (Q \parallel R)$$

$$P \parallel stop = stop$$

$$(c \rightarrow P) \parallel (c \rightarrow Q) = c \rightarrow (P \parallel Q)$$

$$(c \rightarrow P) \parallel (d \rightarrow Q) = stop, \text{ if } c \neq d$$

$$(a \rightarrow P) \parallel (c \rightarrow Q) = a \rightarrow (P \parallel (c \rightarrow Q))$$

2.4.1.3 Communication. In CSP, interaction between concurrent processes takes place by means of message passing via named virtual channels. A communication is an event of the form $c.v$ where c is the name of a channel on which the communication takes place and v is the value of the message which passes. A process which first outputs (sends) value v on the channel c and then behaves like P is represented as:

$$(c!v \rightarrow P)$$

The only event in which this process is initially prepared to engage in is the communication event $c!v$.

A process that is initially prepared to input any value communicable on the channel c and then behave like $P(x)$ is represented as:

$$(c?x \rightarrow P(x))$$

The only event in which this process is initially prepared to engage in is the communication event $c?v$.

The communication primitives satisfy a number of laws, such as, e.g.,

$$(c!v \rightarrow P) \| (c?x \rightarrow Q(x)) = c!v \rightarrow (P \| Q(v)).$$

2.4.1.4 Denotations of CSP processes. For the sake of our survey, we content ourselves with covering the denotational semantic definition of deterministic processes. Under this restriction, each process is assigned a denotation (meaning) under the form of a pair (A,S), where A is the alphabet of the process, and S is its set of traces (where the trace of a process is sequence of events that the process may engage in); by abuse of notation, we make no distinction between a process and its denotation (under the form (A,S)).

We now address the question of how to determine the denotation (meaning) of a process X from the formula that defines this process; without loss of generality (the examples given above bear witness to this generality) we assume that the definition of process X takes the form $X = F(X)$ for some function F. An ordering relation is defined between processes by the following formula:

$$(A, S) \sqsubseteq (B, T) \stackrel{\text{def}}{=} A = B \wedge S \subseteq T$$

This ordering means that each trace that the first process can follow can be followed by the second process as well. Given a process X defined by the recursive equation $X = F(X)$, the denotation of X is defined as the minimal (with respect to the ordering relation $\sqsubseteq$) element, if it exists, that satisfies this equation. This is called the *least fixpoint* of F and denoted by:

$$\mu XF(X)$$

A *chain* in a partial ordering is an infinite sequence of elements $(P_0, P_1, P_2,...)$ such that for all $i \geq 0$, $P_i \sqsubseteq P_{i+1}$. Given a chain, (P_i), we define the *limit* (least upper bound) of this chain as the process denoted by $\bigsqcup_{(i \geq 0)} P_i$ and defined by:

$$\bigsqcup_{i \geq 0} P_i = (\alpha(P_0), \bigcup_{i \geq 0} traces(P_1))$$

A partial ordering is said to be *complete* if and only if it has a minimal element, and all chains have a least upper bound. The ordering defined above among processes with the same alphabet proves to be complete (with *stop* as the minimal element).

Typically, function F that describes recursive definitions of processes preserves the alphabet of its argument and increases its set of traces. Therefore, if we consider consecutive applications of F to some argument, we find a chain with respect to $\sqsubseteq$. Fixpoint theory provides that under these conditions the least fixpoint of F can be written as:

$$\mu X.F(X) = \bigsqcup_{i \geq 0} F^i(stop)$$

As an illustration of this formula, we find that the least fixpoint of the recursive definition

$$clock \stackrel{def}{=} (tick \to clock)$$

is the limit (least upper bound) of the following chain:

$$F^0(stop) = (\{tick\}, \{\langle\ \rangle\})$$

$$F^1(stop) = (\{tick\}, \{\langle\ \rangle, \langle tick\rangle\})$$

$$F^2(stop) = (\{tick\}, \{\langle\ \rangle, \langle tick\rangle, \langle tick, tick\rangle\})$$

$$F^3(stop) = (\{tick\}, \{\langle\ \rangle, \langle tick\rangle, \langle tick, tick\rangle, \langle tick, tick, tick\rangle\})$$

2.4.2 The temporal logic of actions

There is an intimate relationship between the semantics of a programming language and the methods used to prove properties about programs in that language: whether a property holds or does not hold depends critically on the meaning of the program. In Ref. [20], Lamport recognizes that, traditionally, whenever we want to verify that a program satisfies a property, we introduce three separate concepts: the *program*, the *property* (or specification), and the *satisfaction* relationship (or correctness formula). Lamport then proposes a solution that unifies these three concepts into one single concept, namely the *temporal logic of actions* (TLA). TLA is the combination of two logics: a logic of actions and a temporal logic.

We consider a set of variables, and assume that each variable may takes *values* in a specified range. We let a *state* be an assignment of values to variables; hence if s is a state and x is a variable, then $s(x)$ is the value that state s associates to variable x. A *state function* is an expression built from variables and values; a variable is a special case of state function, one whose expression is made up of the mere name of the variable. A *predicate* is a state function whose value is boolean. An *action* is any boolean-valued expression formed from variables, primed variables (representing subsequent values of these variables) and values; for example, if x, y and z are variables, then $x' + 1 = y$ and $x - 1 \notin z'$ are examples of actions.

A temporal logic is built from elementary formulae using boolean operators and the unary operator $\square$ (read: *always*); from this operator we deduce the $\Diamond$ operator (read: *eventually*) by the following formula:

$$\Diamond F = \neg \Box \neg F$$

The semantics of temporal logic is based on behaviors, where a behavior is an infinite sequence of states. A temporal logic formula is interpreted as an assertion about behaviors. The temporal notation allows us to express such statements as: some property F will be true infinitely often ($\Box \Diamond F$) ; property F will eventually be always true ($\Diamond \Box F$) ; property F leads to property G, i.e. if F ever becomes true then G will be true then or will become true later ($\Box (F \Rightarrow \Diamond G)$) . The *raw temporal logic of actions*, or *RTLA*, is obtained by letting elementary temporal formulae (i.e. formulae from which complex temporal expressions are built) be actions.

RTLA can be used both to write programs and to specify program properties. Also, a program is deemed to satisfy a property if the RTLA formula that represents the property is a logical consequence (in the RTLA logic) of the formula that represents the program. This achieves Lamport's aim of unifying programs, specifications, and correctness relationships in a single notation. This logic is used to prove functional properties of programs (pertaining to results delivered by such programs) but also to prove operational properties of such programs, such as liveness and fairness.

Lamport concludes with a genesis of the TLA/RTLA method and discusses related methods of defining the semantics of concurrent programs. As an illustration of Lamport's notation, we show a simple program written in a traditional programming notation then show its RTLA equivalent. Note how RTLA's temporal operators are used to capture the notion of execution. Given that the program is represented as a formula in RTLA (designated by Φ), we can now prove any RTLA-formulated property (say, Ψ) of this program by establishing $(\Phi \Rightarrow \Psi)$ in the RTLA logic. First we give the program:

```
var natural x, y=0;
do
      true --> x:=x+1
      []
      true --> y:=y+1
      od.
```

We give below a RTLA formula that represents the program above:

$$Init_\Phi \stackrel{\text{def}}{=} (x = 0) \wedge (y = 0)$$

$$M_1 \stackrel{\text{def}}{=} (x' = x + 1) \wedge (y' = y)$$

$$M_2 \stackrel{\text{def}}{=} (y' = y + 1) \wedge (x' = x)$$

$$M_3 \stackrel{\text{def}}{=} M_1 \vee M_2$$

$$\Phi \stackrel{\text{def}}{=} Init_\Phi \wedge \Box M$$

The idea of iteration is captured in the *always* operator (denoted by $\Box$) which is applied to formula M.

2.4.3 Algebraic semantic definition of concurrency

In Ref. [14], Desharnais et al. introduce a semantic definition of a concurrent programing notation based on the seminal work of Schmidt [35]. In this work, Schmidt represents programs by flowgraphs and captures their semantics by means of matrices whose entries are relations. Furthermore, he shows that traditional composition of sequential programs can be captured by simple operations on the matrices that represent these programs. Our aim here is to investigate the operations that capture the concurrent composition of programs; to this effect, we use Tarski's algebra of relations [39].

2.4.3.1 Relational algebras. The algebra of relations originated in the last century with the works of De Morgan, Pierce, Dedekind and Schröder. Their work has been followed up by Chin and Tarski [11, 39], who provide an axiomatization of homogeneous relational algebras. The following definition is attributable to Ref. [35].

Definition A. A *homogeneous relational algebra* is a structure of the form $(\mathfrak{R}, \cup, \cap, ^-, \mathrm{o}, \hat{\ })$ on some non-empty set $\mathfrak{R}$ whose elements are called *relations*. The following conditions are satisfied:

1. $(\mathfrak{R}, \cup, \cap, ^-)$ is a complete boolean algebra, with ϕ for zero element and L for universal element. Set $\mathfrak{R}$ is provided with a partial ordering, which we denote by $\subseteq$.
2. Relational composition is associative and has I for neutral element:

$$P \mathrm{o} (Q \mathrm{o} R) = (P \mathrm{o} Q) \mathrm{o} R \text{ and } I \mathrm{o} R = R \mathrm{o} I = R$$

3. $P \mathrm{o} Q \subseteq R \Leftrightarrow \hat{P} \mathrm{o} \bar{R} \subseteq \bar{Q} \Leftrightarrow \bar{R} \mathrm{o} \hat{Q} \subseteq \bar{P}$ (Schröder's rule).
4. If $R \neq \phi$, then $L \mathrm{o} R \mathrm{o} L = L$ (Tarski's rule). ❏

The usual model for these axioms is that of binary relations on some set. In this model, the union ($\cup$), intersection ($\cap$) and complement ($^-$) are the usual set-theoretic operations, the inverse of a relation R is defined by

$$\hat{R} = \{ (x, y) \mid (y, x) \in R \}$$

and relational composition is defined by the following formula:

$$Q \mathrm{o} R = \{ (x, z) \mid \exists y : (x, y) \in Q \wedge (y, z) \in R \}$$

In the sequel, we may want to use a heterogeneous algebra—one whose model is that of relations from one set to another (rather from one set to itself). The definition of heterogeneous relational algebra is quite similar to that given in Section 2.4.1 for homogeneous algebras [35]; nevertheless, we must introduce the notion of type and that of partial operations. Hence, if $Q, R \subseteq S \times T$, operators $Q \cup R$ and $Q \cap R$ are defined. Likewise, if $Q \subseteq S \times T$ and $R \subseteq T \times U$, then the composition $Q \mathrm{o} R$ is defined. When several sets are involved, there may be more than one universal relation, more than one identity, and more than one zero relation (e.g., $I \subseteq S \times S$ and $I \subseteq T \times T$); for the sake of simplicity, we will denote all of them with $L, \varnothing, I$, respectively. The precedence of relational operators is defined as follows, from highest to lowest precedence: $^-$ and $\hat{\ }$, have the same priority, they are followed by o, then $\cap$ and finally by $\cup$. In the sequel, we omit the composition symbol (o) and merely write QR to represent $Q\circ R$. In addition, we write $\widehat{(R)}$ rather than $\widehat{(R)}$ whenever we are dealing with lengthy parenthesized expressions.

From this definition we derive the usual rules of the calculus of relations (see, for example, Refs. [11] and [35]). We assume that these rules are known to the reader and content ourselves with presenting some of them, including some boolean laws.

Definition B. Let P, Q, R be relations. Then

1. $\overline{Q \cup R} = \overline{Q} \cap \overline{R}$

2. $\overline{Q \cap R} = \overline{Q} \cup \overline{R}$

3. $\overline{\overline{R}} = R$

4. $P \cap Q \subseteq R \Leftrightarrow P \subseteq \overline{Q} \cup R$

5. $Q \subset R \Leftrightarrow \overline{R} \subset \overline{Q}$

6. $P(Q \cap R) \subset P(Q \cap PR)$

7. $(P \cap Q)R \subset PR \cap QR$

8. $P(Q \cup R) = PQ \cup PR$

9. $(P \cup Q)R = PR \cup QR$

10. $Q \subset R \Rightarrow PQ \subset PR$

11. $P \subset Q \Rightarrow PR \subset QR$

12. $Q \subset R \Leftrightarrow \hat{Q} \subset \widehat{R}$

13. $(Q \cup R)\hat{} = \hat{Q} \cup \hat{R}$

14. $(Q \cap R)\hat{} = \hat{Q} \cap \hat{R}$

15. $(QR)\hat{} = \hat{R}\hat{Q}$

16. $\hat{\hat{R}} = R$

17. $\overline{\hat{R}} = \hat{\overline{R}}$ ❏

Definition C. Relation R is said to be *deterministic* if and only if $\hat{R}R \subseteq I$; a relation R is said to be *total* if and only if $L = RL$ (or equivalently, $I \subseteq R\hat{R}$); a relation R is said to be *injective* if and only if $\hat{R}$ is deterministic (i.e., $R\hat{R} \subseteq I$); a relation R is said to be *surjective* if and only if $\hat{R}$ is total (i.e., $LR = L$, or $I \subseteq \hat{R}R$). ❏

Definition D. Let P, Q, and R be relations. Then,

1. P deterministic $\Rightarrow P(Q \cap R) = PQ \cap R$

    P injective $\Rightarrow (Q \cap R)P = QP \cap RP$

2. P deterministic $\Rightarrow (Q \cap R\hat{P})P = Q(P \cap R)$

    P injective $\Rightarrow P(\hat{P}Q \cap R) = Q \cap PR$ ❑

Definition E. The vector of relations $(\pi_1, \ldots, \pi_n)$ is said to be a *direct product* if and only if

$$\hat{\pi}_i \pi_i = I \quad (i = 1 \ldots n),$$

$$\pi_i L = L \quad (i = 1 \ldots n),$$

$$\bigcap_{i=1}^{n} \pi_i \hat{\pi}_i = I$$

We say that this direct product is *full* if

$$i \neq j \Rightarrow \hat{\pi}_i \pi_j = L \quad (i, j = 1 \ldots n)$$

Relations π_i are called *projections*. ❑

This definition implies, among other things, that a projection is a total, deterministic, and surjective relation. We consider, for example, sets S, T and U. We check easily that projections

$$\pi_1 : S \times T \times U \rightarrow S \times T$$

and

$$\pi_2 : S \times T \times U \rightarrow T \times U$$

define a product. Note that

$$\pi_1 = \{((s, t, u), (s, t)) \mid s \in S \wedge t \in T \wedge u \in U\}$$

and

$$\pi_2 = \{((s, t, u), (t, u)) \mid s \in S \wedge t \in T \wedge u \in U\}$$

Likewise, projections

$$\pi_s : S \times T \times U \rightarrow S, \ \pi_T : S \times T \times U \rightarrow T, \ \text{and} \ \pi_U : S \times T \times U \rightarrow U$$

define a full direct product. The definition of full direct product given above is called *direct product* in Ref. [35]; we owe the distinction between direct product and full direct product to Maddux [22].

Proposition. Let $(\pi_1,\ldots,\pi_n)$ be a direct product, and let R_i $(i = 1\ldots n)$ be n arbitrary non-empty relations. These relations satisfy the identity:

$$\hat{\pi}_i \left(\cap_{j=1}^{n} \pi_j R_j \hat{\pi}_j \right) \pi_i = R_i \qquad (i = 1\ldots n).$$

If we interpret R_i's as concrete relations on some set, then this proposition is easy to prove if we note that:

$$\pi_1 R_1 \hat{\pi}_1 \cap \pi_2 R_2 \hat{\pi}_2 = \{ ((x_1, x_2), (y_1, y_2)) \mid (x_1, y_1) \in R_1 \wedge (x_2, y_2) \in R_2\}$$

2.4.3.2 Program flowgraphs and parallel composition. In this section, we introduce formally the notion of the *program flowgraph* (attributable to Schmidt [35]), then we define an operator of parallel composition which allows us to combine two flowgraphs into a flowgraph that represents the parallel execution of the programs that are represented by the component flowgraphs.

Definition F. We consider sets S (the set of *states*), and V (the set of control points). A *program flowgraph* on S and V is a relation P on $S \times V$, i.e.,

$$P \subseteq (S \times V) \times (S \times V)$$

An an illustration of this definition we refer the reader to the flowgraph and matrix that are given in Section 2.2.3, which represent the following program on variable x of type **natural**.

```
1: while x>5 do
     2: x:=x-1;
3:
```

This flowgraph can be represented by the matrix that is given in Section 2.2.3 or, equivalently, by the following relation on $\mathbf{N} \times \{1, 2, 3\}$:

$$\{ ((x, 1), (x', 2)) \mid x > 5 \wedge x' = x\} \cup \{ ((x, 1), (x', 3)) \mid (x \leq 5 \wedge x') = x\}$$

$$\cup \{ ((x, 2), (x', 1)) \mid x' = x - 1\}$$

Note that with the matrix representation as well as with the set-theoretic representation, we can no longer determine which is the entry node and which is the exit node; given that we do not need this information in our subsequent study, we do not concern ourselves with this loss. In the sequel, we adopt a semantic viewpoint, hence we may confuse the program with its associated program flowchart; also, we will in general omit to represent the column and row numbers on our matrices (as identifications of the control points). For a complete axiomatic definition of program flowgraphs the reader is referred to Refs. [34] and [35].

A set of matrices of appropriate dimension defines a relational algebra (in the sense of Ref. [35]), provided we define the following operators (where R_{ij} designates the i,j entry of matrix R):

$$(Q \cup R)_{ij} = (Q_{ij} \cup R_{ij})$$

$$(Q \cap R)_{ij} = (Q_{ij} \cap R_{ij})$$

$$(QR)_{ij} = \bigcup_{k} Q_{ik} R_{kj}$$

$$(\bar{R})_{ij} = \overline{R_{ij}}$$

$$(\hat{R})_{ij} = (R_{ij})^\wedge.$$

For example, we consider the composition of two matrices:

$$
\begin{pmatrix} R_{11} & R_{12} \\ R_{21} & R_{22} \end{pmatrix}
\begin{pmatrix} R_{11} & R_{12} \\ R_{21} & R_{22} \end{pmatrix}
=
\begin{pmatrix} R_{11}R_{11} \cup R_{12}R_{21} & R_{11}R_{12} \cup R_{12}R_{22} \\ R_{21}R_{11} \cup R_{22}R_{21} & R_{21}R_{12} \cup R_{22}R_{22} \end{pmatrix}
$$

Note how the product of matrices, as it is defined here, captures at the same time the traditional composition of graphs and the composition of relations on the paths defined by the compound graph. Hence, e.g., there exist two possible paths from node 1 to node 2 in the compound graph above:

- From node 1 to node 1 by the first matrix then from node 1 to node 2 by the second matrix. The composition of relations on this path is $R_{11}R_{12}$.
- From node 1 to node 2 by the first matrix then from node 2 to node 2 by the second matrix. The composition of relations on this path is $R_{12}R_{22}$.

This explains the contents of entry (1,2) of the compound matrix.

Finally, note that for matrices of size 2×2 for example, the zero, universal, and identity relations are given respectively by:

$$
\begin{pmatrix} \varnothing & \varnothing \\ \varnothing & \varnothing \end{pmatrix},
\begin{pmatrix} L & L \\ L & L \end{pmatrix}, \text{ and }
\begin{pmatrix} I & \varnothing \\ \varnothing & I \end{pmatrix}
$$

This, by virtue of the fact that constant relations are denoted by the same symbol irrespective of their dimension, yields expressions such as:

$$
L = \begin{pmatrix} L & L \\ L & L \end{pmatrix}
$$

We are now ready to define the parallel composition of programs represented by relational matrices.

Definition G. We consider programs P_1 and P_2, such that P_1 is a relation on $T \times S_1 \times V_1$, and P_2 is a relation on $T \times S_2 \times V_2$ (the set of states of P_i is $T \times S_i$, and the set of its control points is V_i, for $i = 1,2$). The components S_1 and S_2 are P_1 and P_2's own components, respectively, whereas T is a shared component. Let (π_1, π_2) be a direct product, where π_1 et π_2 are the projections

$$\pi_1: T \times S_1 \times S_2 \times V_1 \times V_2 \rightarrow T \times S_1 \times V_1$$

$$\pi_2: T \times S_1 \times S_2 \times V_1 \times V_2 \rightarrow T \times S_2 \times V_2$$

The *parallel composition* of P_1 and P_2, denoted by $P_1 \| P_2$, is the relation on $T \times S_1 \times S_2 \times V_1 \times V_2$ defined by

$$P_1 \| P_2 \stackrel{\text{def}}{=} \pi_1 P_1 \hat{\pi}_1 \cap \pi_2 P_2 \hat{\pi}_2 \cup \pi_1 \hat{\pi}_1 \cap \pi_2 P_2 \hat{\pi}_2 \cup \pi_1 P_1 \hat{\pi}_1 \cap \pi_2 \hat{\pi}_2 \quad \square$$

To illustrate the above definition, let P_1 and P_2 be defined by the following predicates p_1 and p_2:

$$P_1 = \{ ((t, s_1, v_1), (t', s'_1, v'_1)) \mid p_1 (t, s_1, v_1, t', s'_1, v'_1) \}$$

$$P_2 = \{ ((t, s_2, v_2), (t', s'_2, v'_2)) \mid p_2 (t, s_2, v_2, t', s'_2, v'_2) \}$$

It is easy to verify that

$$\pi_1 P_1 \hat{\pi}_1 \cap \pi_2 P_2 \hat{\pi}_2 = \{ ((t, s_1, s_2, v_1, v_2), (t', s'_1, s'_2, v'_1, v'_2)) \mid$$

$$p_1 (t, s_1, v_1, t', s'_1, v'_1) \wedge (p_2 (t, s_2, v_2, t', s'_2, v'_2)) \}$$

The first term of the definition of $P_1 \| P_2$ represents the simultaneous transitions of P_1 and P_2. These transitions are possible if programs P_1 et P_2 have the same effect on the shared space T; indeed, the conditions imposed on component T are defined by the conjunction of p_1 and p_2. Also, we can prove that

$$\pi_1 \hat{\pi}_1 \cap \pi_2 P_2 \hat{\pi}_2$$

$$= \{ ((t, s_1, s_2, v_1, v_2), (t', s'_1, s'_2, v'_1, v'_2)) \mid t' = t \wedge s' = s_1 \wedge$$

$$v'_1 = v_1 \wedge p_2 (t, s_2, v_2, t', s'_2, v'_2) \}$$

The second term of $P_1 \| P_2$ is then the set of transitions carried out by P_2 while P_1 is inactive. These transitions are possible provided program P_2 does not modify the shared component T. The third term of the definition of $P_1 \parallel P_2$ has symmetric interpretation. The identities 4.2(8,9) allow us to give an alternative expression for $P_1 \parallel P_2$:

$$P_1 \parallel P_2 = \pi_1\,(P_1 \cup I)\,\hat{\pi}_1 \cap \pi_2 P_2 \hat{\pi}_2 \cup \pi_1 P_1 \hat{\pi}_1 \cap \pi_2 \hat{\pi}_2$$

This is the formula we will use in the sequel.

Note that Definition G can trivially be applied in the case when one of the components T, S_1, S_2 is absent. We give below a concrete example of parallel product. We show in particular how projections that we have used in the definition of parallel combination can be represented by relational matrices, just like programs.

2.4.3.3 Parallel composition of transition systems.

We consider the transition systems P_1 and P_2 that are defined as follows (where we omit to identify an initial node and a final node):

We let a, b, and c be relations on some spaces A, B, and C, so that P_1 is a relation on $A \times B \times V_1$, and P_2 is a relation on $A \times C \times V_2$, with $V_1 = V_2 = \{1,2\}$. Hence, P_1 and P_2 share a common component, namely A, and each has its own component (respectively, B and C). We impose the following constraints on relations a, b, and c:

$$a \cap I = \varnothing,\ b \cap I = \varnothing,\ c \cap I = \varnothing,\ a \neq \varnothing,\ b \neq \varnothing,\ \text{and } c \neq \varnothing$$

For example, we could let $A = B = C = \{0,1\}$ and $a = b = c = I$.

Relations P_1 and P_2 are given as follows:

$$P_1 = \begin{pmatrix} \varnothing & a\underline{I} \\ \underline{I}b & \varnothing \end{pmatrix}_{AB} \qquad\qquad P_2 = \begin{pmatrix} \varnothing & \underline{I}c \\ a\underline{I} & \varnothing \end{pmatrix}_{AC}$$

The subscripts of the matrices serve as indicators that the entries of P_1 are relations on $A \times B$, whereas the entries of P_2 are relations on $A \times C$. We consider the following projections:

$$\pi_{A_1} : A \times B \to A \ , \text{and}\ \ \pi_{B_1} : A \times B \to B$$

We use the notation $\underline{xy}$ to reduce the size of matrices; this abbreviation is defined by

$$\underline{xy} \stackrel{\text{def}}{=} \pi_{A_1} x \hat{\pi}_{A_1} \cap \pi_{B_1} y \hat{\pi}_{B_1}$$

For example, the entry $a\underline{I}$ in matrix P_1 is the relation on $A \times B$ whose A component is defined by a and whose $\overline{B}$ component is defined by the identity relation I. Relations $a\underline{I}$

and Ib of matrix P_1 (for example) satisfy the following identities (see the Proposition on p. 44):

$$\hat{\pi}_{A_1}\,aI\pi_{A_1} = a \qquad \hat{\pi}_{A_1}\,Ib\pi_{A_1} = I \qquad \hat{\pi}_{B_1}\,aI\pi_{B_1} = I \qquad \hat{\pi}_{B_1}\,Ib\pi_{B_1} = b$$

$$\pi_{A_1}\,a\hat{\pi}_{A_1} = \underline{aL} \qquad \pi_{A_1}\,I\hat{\pi}_{A_1} = \underline{IL} \qquad \pi_{B_1}\,I\hat{\pi}_{B_1} = \underline{LI} \qquad \pi_{B_1}\,b\hat{\pi}_{B_1} = \underline{Lb}$$

This can trivially be generalized to the case where the number of components of matrices is different from 2.

For the example discussed here, relations π_1 and π_2 of Definition I have the following signatures:

$$\pi_1: A \times B \times C \times V_1 \times V_2 \to A \times B \times V_1$$

$$\pi_2: A \times B \times C \times V_1 \times V_2 \to A \times C \times V_2$$

These projections can be represented by matrices, just like program flowgraphs. For π_1, π_2 we propose the following definitions:

$$\pi_1 = \begin{array}{c} \\ 11 \\ 12 \\ 21 \\ 22 \end{array} \begin{array}{cc} 1 & 2 \\ \left(\begin{array}{cc} \omega_1 & \varnothing \\ \omega_1 & \varnothing \\ \varnothing & \omega_1 \\ \varnothing & \omega_1 \end{array}\right) \end{array} \qquad \pi_2 = \begin{array}{c} \\ 11 \\ 12 \\ 21 \\ 22 \end{array} \begin{array}{cc} 1 & 2 \\ \left(\begin{array}{cc} \omega_2 & \varnothing \\ \varnothing & \omega_2 \\ \omega_2 & \varnothing \\ \varnothing & \omega_2 \end{array}\right) \end{array}$$

where $\omega_1 : A \times B \times C \to A \times B$ and $\omega_2 : A \times B \times C \to A \times C$ are projections which involve exclusively the sets of states (versus control points). These matrices are determined to within a permutation, since the numbering of control points is arbitrary. The structure of matrices determines the projections of $V_1 \times V_2$ on V_1 or V_2; for example, π_1 projects the pair (1,2) on 1, whereas π_2 projects it on 2. The projections that pertain to the sets of states are carried out by ω_1 and ω_2. The pair (π_1,π_2) is a direct product (refer to Definition E). We will check, e.g. that this pair satisfies $\hat{\pi}_1\pi_1 = I$; the other conditions can be checked in a similar manner.

$$\hat{\pi}_1\pi_1 = \left(\begin{array}{cccc} \hat{\omega}_1 & \hat{\omega}_1 & \varnothing & \varnothing \\ \varnothing & \varnothing & \hat{\omega}_1 & \hat{\omega}_1 \end{array}\right) \left(\begin{array}{cc} \omega_1 & \varnothing \\ \omega_1 & \varnothing \\ \varnothing & \omega_1 \\ \varnothing & \omega_1 \end{array}\right) = \left(\begin{array}{cc} \hat{\omega}_1\omega_1 & \varnothing \\ \varnothing & \hat{\omega}_1\omega_1 \end{array}\right) = \left(\begin{array}{cc} I & \varnothing \\ \varnothing & I \end{array}\right) = I$$

where we have used the values in Definition G as well as the fact that projection ω_1 satisfies $\hat{\omega}_1 \omega_1$ (refer to Definition E).

A projection such as π_1, which cancels two components, may be expressed as the product of two projections where each cancels a single component. Indeed,

Definition H.

$$
\pi_1 = \begin{pmatrix} \omega_1 & \varnothing \\ \omega_1 & \varnothing \\ \varnothing & \omega_1 \\ \varnothing & \omega_1 \end{pmatrix} = \begin{pmatrix} \omega_1 & \varnothing & \varnothing & \varnothing \\ \varnothing & \omega_1 & \varnothing & \varnothing \\ \varnothing & \varnothing & \omega_1 & \varnothing \\ \varnothing & \varnothing & \varnothing & \omega_1 \end{pmatrix} \begin{pmatrix} I & \varnothing \\ I & \varnothing \\ \varnothing & I \\ \varnothing & I \end{pmatrix} = \Omega_1 \rho_1 = \begin{pmatrix} I & \varnothing \\ I & \varnothing \\ \varnothing & I \\ \varnothing & I \end{pmatrix} \begin{pmatrix} \omega_1 & \varnothing \\ \varnothing & \omega_1 \end{pmatrix} = \rho'_1 \Omega'_1
$$

(We check easily that the compositions of these matrices do yield π_1.) The signatures of $\Omega_1, \rho_1, \Omega'_1,$ and ρ'_1 are given below.

$$
\Omega_1 : A \times B \times C \times V_1 \times V_2 \to A \times B \times V_1 \times V_2
$$

$$
\Omega'_1 : A \times B \times C \times V_1 \to A \times B \times V_1
$$

$$
\rho_1 : A \times B \times V_1 \times V_2 \to A \times B \times V_1
$$

$$
\rho'_1 : A \times B \times C \times V_1 \times V_2 \to A \times B \times C \times V_1
$$

We propose to compute the four terms of the definition of $P_1 \parallel P_2$ (Definition I). First, by virtue of Definitions G and J,

$$
\pi_1 (P_1 \cup I) \hat{\pi}_1 = \begin{pmatrix} \omega_1 & \varnothing \\ \omega_1 & \varnothing \\ \varnothing & \omega_1 \\ \varnothing & \omega_1 \end{pmatrix} \begin{pmatrix} \underline{II} & a\underline{I} \\ \underline{Ib} & \underline{II} \end{pmatrix}_{AB} \begin{pmatrix} \hat{\omega}_1 & \hat{\omega}_1 & \varnothing & \varnothing \\ \varnothing & \varnothing & \hat{\omega}_1 & \hat{\omega}_1 \end{pmatrix}
$$

$$
= \begin{pmatrix} \omega_1 \underline{II} \hat{\omega}_1 & \omega_1 \underline{II} \hat{\omega}_1 & \omega_1 a\underline{I} \hat{\omega}_1 & \omega_1 a\underline{I} \hat{\omega}_1 \\ \omega_1 \underline{II} \hat{\omega}_1 & \omega_1 \underline{II} \hat{\omega}_1 & \omega_1 a\underline{I} \hat{\omega}_1 & \omega_1 a\underline{I} \hat{\omega}_1 \\ \omega_1 \underline{Ib} \hat{\omega}_1 & \omega_1 \underline{Ib} \hat{\omega}_1 & \omega_1 \underline{II} \hat{\omega}_1 & \omega_1 \underline{II} \hat{\omega}_1 \\ \omega_1 \underline{Ib} \hat{\omega}_1 & \omega_1 \underline{Ib} \hat{\omega}_1 & \omega_1 \underline{II} \hat{\omega}_1 & \omega_1 \underline{II} \hat{\omega}_1 \end{pmatrix}_{ABC} = \begin{pmatrix} \underline{IIL} & \underline{IIL} & a\underline{IL} & a\underline{IL} \\ \underline{IIL} & \underline{IIL} & a\underline{IL} & a\underline{IL} \\ \underline{IbL} & \underline{IbL} & \underline{IIL} & \underline{IIL} \\ \underline{IbL} & \underline{IbL} & \underline{IIL} & \underline{IIL} \end{pmatrix}_{ABC}
$$

(Note how this matrix is similar to that of $P_1 \cup I$; it is a sort of expansion thereof, whose entries indicate that the relation on component C is universal.) Likewise, we find

$$\pi_2 P_2 \hat{\pi}_2 = \begin{pmatrix} \omega_2 & \varnothing \\ \varnothing & \omega_2 \\ \omega_2 & \varnothing \\ \varnothing & \omega_2 \end{pmatrix} \begin{pmatrix} \varnothing & \underline{Ic} \\ \underline{aI} & \varnothing \end{pmatrix}_{AC} \begin{pmatrix} \hat{\omega}_2 & \varnothing & \hat{\omega}_2 & \varnothing \\ \varnothing & \hat{\omega}_2 & \varnothing & \hat{\omega}_2 \end{pmatrix} = \begin{pmatrix} \varnothing & \underline{ILc} & \varnothing & \underline{ILc} \\ \underline{aLI} & \varnothing & \underline{aLI} & \varnothing \\ \varnothing & \underline{ILc} & \varnothing & \underline{ILc} \\ \underline{aLI} & \varnothing & \underline{aLI} & \varnothing \end{pmatrix}_{ABC}$$

$$\pi_1 P_1 \hat{\pi}_1 = \begin{pmatrix} \omega_1 & \varnothing \\ \omega_1 & \varnothing \\ \varnothing & \omega_1 \\ \varnothing & \omega_1 \end{pmatrix} \begin{pmatrix} \varnothing & \underline{aI} \\ \underline{Ib} & \varnothing \end{pmatrix}_{AB} \begin{pmatrix} \hat{\omega}_1 & \hat{\omega}_1 & \varnothing & \varnothing \\ \varnothing & \varnothing & \hat{\omega}_1 & \hat{\omega}_1 \end{pmatrix} = \begin{pmatrix} \varnothing & \varnothing & \underline{aIL} & \underline{aIL} \\ \varnothing & \varnothing & \underline{aIL} & \underline{aIL} \\ \underline{IbL} & \underline{IbL} & \varnothing & \varnothing \\ \underline{IbL} & \underline{IbL} & \varnothing & \varnothing \end{pmatrix}_{ABC}$$

$$\pi_2 \hat{\pi}_2 = \begin{pmatrix} \omega_2 & \varnothing \\ \varnothing & \omega_2 \\ \omega_2 & \varnothing \\ \varnothing & \omega_2 \end{pmatrix} \begin{pmatrix} \hat{\omega}_2 & \varnothing & \hat{\omega}_2 & \varnothing \\ \varnothing & \hat{\omega}_2 & \varnothing & \hat{\omega}_2 \end{pmatrix} = \begin{pmatrix} \underline{ILI} & \varnothing & \underline{ILI} & \varnothing \\ \varnothing & \underline{ILI} & \varnothing & \underline{ILI} \\ \underline{ILI} & \varnothing & \underline{ILI} & \varnothing \\ \varnothing & \underline{ILI} & \varnothing & \underline{ILI} \end{pmatrix}_{ABC}$$

We must now add the various terms of our expression

$$\pi_1 (P_1 \cup I) \hat{\pi}_1 \cap \pi_2 P_2 \hat{\pi}_2 \cup \pi_1 P_1 \hat{\pi}_1 \cap \pi_2 \hat{\pi}_2$$

We find

$$P_1 \| P_2 = \begin{pmatrix} \varnothing & \underline{IIc} & \varnothing & \varnothing \\ \varnothing & \varnothing & \underline{aII} & \varnothing \\ \hline \underline{IbI} & \underline{Ibc} & \varnothing & \underline{IIc} \\ \varnothing & \underline{IbI} & \varnothing & \varnothing \end{pmatrix}_{ABC}$$

$P_1 \| P_2$ is a relation on $A \times B \times C \times V_1 \times V_2$. Matrix $P_1 \| P_2$ is divided into four sub-matrices. The position of each sub-matrix corresponds to a transition of P_1, whereas the position inside a sub-matrix corresponds to a transition in P_2. For example, the entry $\underline{aII}$, which appears in the sub-matrix at position (1,2), corresponds to a transition of P_1 from node 1 to node 2 (with label a); the same entry corresponds to a transition of P_2 from node 2 to node 1 (which is also labeled with a). This corresponds to a simultaneous transition of P_1 and P_2 on their shared component A. The entries $\underline{IIc}$ correspond to a transition of P_2 on its own component C while P_1 does not perform a transition. The entry $\underline{Ibc}$ corresponds to a simultaneous transition of P_1 on its own component B and a transition of P_2 on its own component C. We observe that the parallel composition obtained by our definition corresponds to the product of transitions, which is synchronized on their shared transitions (performed on their shared spaces).

2.4.4 A Relational logic for await-then programs

In Ref. [37], Slimani proposes a relational logic for concurrent programs as an extension of Mills' logic of sequential Algol-like programs [21, 26, 27]. To this effect, Slimani uses a block structured language whose scoping rules are inspired from Ada, and whose synchronization constructs are Owicki and Gries' **await-then**.

2.4.4.1 Syntax. The overall structure of an *await program* is given by the following BNF rules:

```
<awaitprogram> ::= program <identifier> <block>.

<block> ::=          begin
                     use <usepart>;
                     var <varpart>;
                     exec <statement>
                     end

<statement> ::=      <atomicstatement> |
                     <structuredstatement>

<atomicstatement> ::= <skipstatement> |
                     <readstatement> |
                     <writestatement> |
                     <awaitstatement> |
                     <assignstatement>
<compoundstatement> ::= <block> |
                     <compoundstatement> |
                     <sequencestatement> |
                     <conditionalstatement> |
                     <alternationstatement> |
                     <whilestatement> |
                     <parallelstatement>
```

Blocks are a key feature of this language; they are similar to Algol's block in the sense that they fulfill two functions at once, namely that of encapsulating statements and that of defining a scope. Unlike Algol, however, these blocks do not automatically inherit the space of the enclosing block. Rather the space of a block is defined by two components: the *var* section, where internal variables are declared, and the *use* section, where external variables are *explicitly* imported. The outermost block of the program refers systematically to variables *input* and *output* in its *use* section; these variables are supposed to be known outside of the program (e.g. by the operating system).

The semantic definition of this language proceeds in two steps: first we define the composition of spaces (defined by declarations), then we define the composition of relations (defined by statements).

2.4.4.2 Composing spaces. A number of operations are introduced on spaces: The *cartesian sum* of two spaces S (defined, say, by variables a and b) and S' (defined, say, by variables c and d) is the space denoted by $S + S'$ and defined by the union of the variables of S and S' (in this case, a, b, c, and d); the *cartesian intersection* of two spaces S (defined by, say, *a* and *b*) and *S'* (defined by, say, *b* and *c*) is the space denoted by *S* • *S'* and defined by the intersection of the variables of *S* and *S'* (in this case, *b*). If *S* and *S'* have no variables in common then their cartesian intersection is a special space, which we denote by Φ; this space is the neutral element of the cartesian sum.

To illustrate how spaces and operations on paces are used in this programming language, we consider the following annotated program.

```
program sample;
{space: input+output}

begin
        {space: input+output}
        use input, output;
        var v, w: integer;
        exec
                {space: input+output+v+w}
                w:=0;
                begin
                        {space: input+v}
                        use input, v;
                        var x: integer;
                        exec
                                {space: input+v+x}
                                read(x);
                                v:= x;
                end
                ||
                begin
                        {space: output+w}
                        use output, w;
                        var y: integer;
                        exec
                                y:= w;
                                write(y);
                end;
                w:= 1;
        end.
```

This example can be explained by two simple premises: the space of a block is defined by the *use* part of the block; the space of each statement is defined by the sum of the *use* part and the *var* part of the innermost block in which the statement appears. Also, space definitions must satisfy the following constraints: if two blocks are arranged sequentially, then they must have the same space; if two blocks are arranged in parallel, then the sum of their spaces must be the same as the space of the enclosing block.

2.4.4.3 Composing relations. In Ref. [37], Slimani proposes a denotational semantic definition of the proposed language, which proceeds by bottom-up evaluation of the source program and works its way up through the hierarchical structure of the program. The bottom up evaluation concludes by mapping the overall program into a relation on the space *input + output*.

The details of this semantic definition are beyond the scope of our survey; we will content ourselves with discussing some features thereof. The premises on which this semantic definition is based are the following:

- The level of atomicity of this language is defined, and each atomic statement is captured by means of a relation on the space of the *exec* section where the statement appears.
- A sequential combination of atomic statements is captured by means of a list of relations on the same space.
- Because of the possibility that several control paths may exist in a given block (because of conditional statements, alternation statements and while statements) the *exec* section of a block is captured by means of sets of lists of relations on the overall space of the block (*use* part plus *var* part).
- A block is captured by means of a set of lists of relations on the *use* space of the block.
- The parallel combination of two blocks is captured by a single relation on the overall space of the enclosing block.
- The overall program is captured by a single relation on the space *input + output* (typically making no reference to the initial value of variable *output*).

It is worthwhile to discuss how the semantics of a parallel combination of two blocks, say A and B, is computed from the semantics of the individual blocks. Each block is captured by a set of lists of relations; the relation that represents the parallel combination is obtained as the union all the terms that can be obtained by combining a list of block A with a list of block B. The combination of two lists is captured by the union of terms, where each term represents a possible interleaving of atomic actions of A with atomic actions of B. The relation that captures an await statement has a limited domain (limited to those states that satisfy the await condition); on the other hand, atomic actions that make the condition of the await statement false have a range which is disjoint from the domain of the relation of await. The interplay between these two conditions will cancel all the terms that represent an illegal sequencing of atomic actions.

2.4.4.4 Capturing deadlock properties. Let A and B be two blocks that are combined in parallel in a program, and let λ_A and λ_B be arbitrary labels in A and B. We denote by λ the compound label $\langle \lambda_A, \lambda_B \rangle$, and we define the past function and the future function at label λ as follows.

Definition I. The *past function* at label λ is obtained as the parallel combination of the code preceding λ_A in block A with the code preceding λ_B in block B.

Definition J. The *future function* at label λ is obtained as the parallel combination of the code following λ_A in block A with the code following λ_B in block B.

Using these notions, we now introduce three deadlock-related definitions.

Definition K. The parallel combination of blocks A and B is said to be *deadlock-free* at label λ if and only if the range of the past function at label λ is a subset of the domain of the future function at label λ.

Definition L. The parallel combination of blocks A and B is said to be *deadlock-prone* at label λ if and only if the range of the past function at label λ is not a subset of the domain of the future function at label λ.

Definition M. The parallel combination of blocks A and B is said to be *deadlock-doomed* at label λ if and only if the range of the past function at label λ and the domain of the future function at label λ are disjoint.

2.5   Operational Semantic Definitions

2.5.1   Process equivalences

This section is again concerned with concurrent processes as terms of an algebraic language (see Section 2.4.1.1) but under the operational point of view.

The behaviors of processes are described using transitions. Families of transitions can be arranged as labeled graphs, concrete representation of the behavior processes. However, this operational description is not sufficient to entirely describe the meaning of processes. Indeed, complete meaning means in particular to get a criterion to decide when two processes may be considered to be the same.

This problem has been particularly studied in the process algebra framework such as the Milner's *Calculus for communicating systems* (or *CCS*, for short) [28]. The point of view adopted by Milner is close to the approach used by Landin to define the semantics of λ-calculus by the mean of a stack machine. He begins by defining a small language whose constructions reflect simple operational ideas. The meaning of these constructions is presented by means of *structured operational semantics*, a method first introduced by Plotkin [33].

2.5.2   Hennessy-Milner logic

We introduce first an abstract approach to examine process behaviors in terms of behavioral properties. In this approach, processes are equivalent if they have the same properties. To express these properties one needs a specification language. Such a language describing local capabilities of processes, has been introduced by Hennessy-Milner [16, 17]. Formulas of the Hennessy-Milner logic are built up from

- the constants **tt** and **ff** (read: *true* and *false*)
- boolean connectives $\vee$ and $\wedge$
- modal operators $[a]$ (read: *box a*) and $\langle a \rangle$ (read: *diamond a*) for each event a.

The following syntax definition specifies these formulas:

- **tt** and **ff** are formulas
- if Φ and Ψ are formulas then so are $\Phi \vee \Psi$ and $\Phi \wedge \Psi$
- if a is an event and Φ is a formula, then $[a]\Phi$ and $<a>\Phi$ are formulas

The meanings of modal formulas refer to the transition behavior of a process: a process P has the property

- $[a]\Phi$ if after every execution of the event a the resulting process has the property Φ
- $<a>\Phi$ if there exits an event a such that, after its execution, the resulting process has the property Φ

The simple modal formula $\langle coin \rangle$ **tt** expresses the capability for performing the event *coin*. The vending machine VM_0 from Section 2.4.1.1 has this property. In contrast, $\langle candy \rangle$ **ff** expresses the inability to perform the event candy. VM_0 has that property too. Indeed, the only capability of process VM_0 is to perform the event *coin* and then the resulting process has the capability to perform the event *candy*. Such basic properties can be embedded within modal operators and between boolean connectives.

For instance the formula $\Phi = \; [coin] < candy > \mathbf{tt} \, (\, [coin] \, \mathbf{ff})$ expresses the property that after any coin event, it is possible to perform *candy* but not to perform *coin* again. VM_0 satisfies Φ.

2.5.3   Behavioral equivalence

The consequence of the existence of a formal language like the Hennessy-Milner logic to describe properties of processes is the introduction of a precise definition of equivalence between processes. Two processes can then be considered as equivalent if they satisfy the same formulas. Since the Hennessy-Milner logic is intended to describe process behaviors, this means that two processes are equivalent if they have the same behavior or equivalently if placed in the same context the resulting processes have the same properties. As an example consider the following three new versions of vending machines (see Section 2.4.1.1):

$$VM_4 \stackrel{\text{def}}{=} coin \rightarrow (coin \rightarrow ((candy \rightarrow VM_4) \mid (bar \rightarrow VM_4)))$$

$$VM_5 \stackrel{\text{def}}{=} coin \rightarrow ((coin \rightarrow (candy \rightarrow VM_5)) \mid (coin \rightarrow (bar \rightarrow VM_5)))$$

$$VM_6 \stackrel{\text{def}}{=} (coin \rightarrow (coin \rightarrow (candy \rightarrow VM_6))) \mid (coin \rightarrow (coin \rightarrow (bar \rightarrow VM_6)))$$

which have the same observable traces but Hennessy-Milner logic is sufficiently specific to discriminate between them: that is, for the three vending machines $VM_I, 4 \geq i \geq 6$ there exits a formula Φ_j such that VM_j has the property Φ_j unlike VM_i for $i \neq j$:

$$\Phi_4 \equiv [coin] \, [coin] \, (\langle candy \rangle \wedge \langle bar \rangle \mathbf{tt})$$

$$\Phi_5 \equiv [coin] \, (\langle coin \rangle \langle candy \rangle \mathbf{tt} \wedge \langle coin \rangle \langle bar \rangle \mathbf{tt} \wedge \langle coin \rangle \, (\langle bar \rangle \mathbf{ff} \vee \langle candy \rangle \mathbf{ff}))$$

$$\Phi_6 \equiv [coin] \, (\langle coin \rangle \langle candy \rangle \mathbf{ff} \vee \langle candy \rangle \langle bar \rangle \mathbf{ff})$$

2.5.4   Observation and bisimulation

Equivalences for CCS processes depends on *bisimulation* relations [29, 32] based on the idea that an observer can repeatedly interact with a process and that two processes are bisimular if they have the capability to execute the same testing sequences of events. We formalize first the notion of observations [38] using background ideas from *game theory*.

An *observation sequence* from a pair of processes (P_0, Q_0) is a finite or infinite sequence of pairs of processes of the form

$$(P_0, Q_0) \dots (P_i, Q_i) \dots$$

For each j the pair (P_{j+1}, Q_{j+1}) is determined from $P_{j+1} \rightarrow P_j, Q_j$ in the following iterative algorithm.

1. Non-deterministically, choose between
    (a) P_{j+1} if there is an event a such that $(P_j) \stackrel{\text{def}}{=} (a \rightarrow P_{j+1})$

    (b) Q_{j+1} if there is an event a such that $Q_j \stackrel{\text{def}}{=} a \rightarrow Q_{j+1}$

2. Try to complete the pair in finding a corresponding transition on the event chosen in step 1.

An observation sequence is successful if the previous iterative algorithm:

1. eventually get stuck at step 1 or
2. fails to terminate.

Definition N. Two processes P and Q are *observation equivalent* if every observation sequence from (P, Q) is successful.

This means that there is no way to detect a difference between processes. As an example, VM_5 and VM_6 are not interactive equivalent. Indeed, it is possible to build an unsuccessful observation sequence from VM_5, VM_6 as follows.

1. Step 1: choose

$$coin \rightarrow (coin \rightarrow (candy \rightarrow VM_6))$$

   Step 2: forces

$$coin \rightarrow ((coin \rightarrow (candy \rightarrow VM_5)) \mid (coin \rightarrow (bar \rightarrow VM_5)))$$

2. Step 1: choose

$$coin \rightarrow (bar \rightarrow VM_5)$$

   Step 2: forces

$$coin \rightarrow (candy \rightarrow VM_6)$$

3. Step 1: choose

$$bar \rightarrow VM_5$$

   Step 2: get stuck!

Definition O. A binary relation $\Re$ between processes is a *bisimulation* if whenever $(P, Q) \in \Re$, then for all events a,

1. if $P \overset{\text{def}}{=} a \rightarrow P'$ then $\exists_{Q'} Q \overset{\text{def}}{=} a \rightarrow Q'$ and $(P', Q') \in \Re$
2. if $Q \overset{\text{def}}{=} a \rightarrow Q'$ then $\exists_{P'} Q \overset{\text{def}}{=} a \rightarrow P'$ and $(P', Q') \in \Re$

Note that the conditions 1 and 2 are hereditary.

Definition P. Two process are *bisimulation equivalent* (or *bisimular*) if there is a bisimulation which relates them.

A bisimulation relation may be viewed as a schedule for choice at step 2 making sure that the iterative algorithm never gets stuck at that step so it is easily seen that bisimulation equivalence is the same as equivalence associated with Hennessy-Milner logic.[*]

Another important result known as *Hennessy-Milner theorem* provides that this operational equivalence is the same as the logical equivalence associated with Hennessy-Milner logic.

2.6   Summary and Prospects

In this chapter we have given a brief outline of the state of the art in the field of *semantics of concurrency*. We have observed in particular that a semantic definition depends on a number of parameters, which include: the programming notation; the computational model supported by the notation; the method of semantic definition. Further, the latter can in turn be defined by means of three features, namely: the meaning associated to each construct of the language; how this meaning is defined for trivial constructs; and how this meaning is inductively defined for complex constructs. The wide range of computational models, notations, and semantic definition methods has spawned, in the past, a profusion of proposals for semantic definitions. In the short survey we have conducted in this chapter, we have barely started to cover the vast domain of *semantics of concurrency*; we do expect, however, that this survey gives the reader some idea about the issues involved in defining the semantics of a concurrent language, as well as some flavor for the solutions that have been proposed in the past. There is a sense in the research community that the field of semantic definitions of concurrency is saturated, and that it is perhaps more advantageous to enrich and extend existing proposals than to devise new solutions; in keeping with this premise, a large volume of research effort seems to be concentrated on *communicating sequential processes* and on the *calculus for communicating systems*. Several current conference series seem to devote some time to the study of the semantics of concurrency; these include the *CONCUR* series (*International Conference on Concurrency Theory* [6]), the *PARLE* series (*International Conference on Parallel Architecture and Language—Europe*} [7]), and the workshop series titled *Languages and Compilers for Parallel Computing* [4].

2.7   References

1. Andrews, G.R. and F.B. Schneider. Concepts and Notations for Concurrent Programming. *ACM Computing Surveys* 15:1, 3–43.
2. Apt, K.R., and E.R. Olderog. Introduction to Program Verification. In: *Formal Description of Programming Concepts*, ed. E.J. Neuhold and M. Paul. Berlin: Springer Verlag, 1991.
3. A. Arnold. *Systemes de transitions finis et sémantique de processus communicants.* Paris: Masson, 1992.
4. Banerjee, U., D. Gelernter, A. Nicolau, and D. Padua, eds. *Proceedings, Sixth International Workshop on Languages and Compilers for Parallel Computing.* Portland, Oregon, August 1993. Lecture Notes in Computer Science, Vol 768. Springer Verlag.
5. A. Bergeron. A unified approach to control problems in discrete event processes. *RAIRO Informatique Théorique et Applications* 27:6, 1993.
6. Best, E., ed. *Proceedings, CONCUR'93: 4th International Conference on Concurrency Theory.* Hildesheim, Germany, August 1993. Lecture Notes in Computer Science, Vol 715. Springer Verlag.
7. Boda, A., M. Reeve, and G. Wolf, eds. *Proceedings, PARLE'93: 5th International Conference on Parallel Architectures and Languages Europe.* Munich, Germany, June 1993. Lecture Notes in Computer Science, volume 694. Springer Verlag.
8. Brinch Hansen, P. *The Architecture of Concurrent Programming.* Englewood Cliffs, NJ: Prentice Hall, 1971 .
9. K.L. Calvert and S.S. Lam. Formal methods for protocol conversion. *IEEE J. on Selected Areas in Communicans* 8:1 1990, 127–142.

*Under a constraint of image finiteness.

10. Chandy, K.M. and J. Misra. *Parallel Program Design: A Foundation.* Reading, Mass.: Addison Wesley, 1988.
11. Chin, L.H., and A. Tarski. Distributive and modular laws in the arithmetic of relation algebras. *University of California Publications* 1, 1951, 341–348.
12. Desharnais, J., A. Jaoua, F. Mili, N. Boudriga, and A. Mili. A relational decision operator: The conjugate kernel. *Theoret. Comput. Sci.* 114, 1993, 247–272.
13. Desharnais, J., A. Mili, and F. Mili. On the mathematics of sequential decompositions. *Sci. Comput. Program.* 20, 1993, 253–289.
14. Desharnais, J. et al. *Une Approache Relationnelle a la Decomposition Parallele.* Quebec City, Canada: Laval University, 1994.
15. Ford, W. and W. Topp. *Assembly Language and Systems Programming for the M68000 Family.* Lexington, Mass.: D.C. Heath, 1992.
16. Hennessy, M., and R. Milner. On observing nondeterminism and concurrency. *Lecture notes in Comp. Sci.* 85, 1980, 295–309.
17. Hennessy, M., and R. Milner. Algebraic laws for nondeterminism and concurrency. *J. Ass. of Comp. Sci.* 85, 1985, 137–162.
18. Hoare, C.A.R. An axiomatic basis for computer programming. *Communications of the ACM* 12:10, 576–583.
19. Hoare, C.A.R. *Communicating Sequential Processes.* Englewood Cliffs, N.J.: Prentice Hall, 1985.
20. Lamport, L. *The Temporal Logic of Actions.* Digital Equipment Corporation, Systems Research Center, Report 79, Dec. 1991.
21. Linger, R.C., H.D. Mills, and B.I. Witt. *Structured Programming: Theory and Practice.* Reading, Mass.: Addison Wesley, 1979.
22. Maddux, R.D. *On the derivation of identities involving projection functions.* Ames, Iowa: Department of Mathematics, Iowa State University, 1993.
23. Manna, Z. *Mathematical Theory of Computation.* New York: McGraw Hill, 1974.
24. Mili, A., J. Desharnais, and F. Mili. Relational heuristics for the design of deterministic programs. *Acta Inform.* 24:3, 1987, 239–276.
25. Mili, F. and A. Mili. Heuristics for the construction of while loops. *Sci. Comput. Program.* 18, 1992, 67–106.
26. Mili, A., J. Desharnais, and F. Mili. *Computer Program Construction.* New York: Oxford University Press, 1994.
27. Mills, H.D., V. Basili, J.D. Gannon, and R.G. Hamlet. *Principles of Computer Programming: A Mathematical Approach.* Boston: Allyn and Bacon, 1986.
28. Milner, R. *A calculus for communicating systems.* Lect. Notes in Comp. Science 272, Springer, 1980.
29. Milner, R. *Communication and Concurrency.* Englewood Cliffs, N.J.: Prentice Hall, 1989.
30. Owicki, S. and D. Gries. An axiomatic proof technique for parallel programs. Acta Informatica 6 (a976), 319–340.
31. Owicki., S. and L. Lamport. Proving liveness properties of concurrent programs. *ACM TOPLAS* 4:3 (1982), 455–495.
32. Park, D. Concurrency and automata on infinite sequences. Lect. Notes in Comput. Sci. 154, Springer, 1981, 561–572.
33. Plotkin, G. *A structural approach to operational semantics.* Report DAIMI FN-19, Computer Science Dept., Aarhus University, 1981.
34. Schmidt, G. Programs as partial graphs I: Flow equivalence and correctness. *Theoret. Comput. Sci.* 15, 1981, 1–25.
35. Schmidt, G. and T. Ströhlein. *Relations and Graphs.* EATCS Monographs in Computer Science. Berlin: Springer-Verlag, 1993.
36. Silberschatz, A., J. Peterson, and P. Galvin. *Operating Systems Concepts.* Reading, Mass.: Addison Wesley, 1991.
37. Slimani, Y. *Une Logique Relationnelle de la Programmation parallele.* Doctorat es-Sciences d'Etat, University of Oran, Algeria, 1994.
38. Stirling, C. Modal and temporal logic. In Handbook of Logic in Computer Science, Vol. 2, ed. S. Abramsky, D. Gabbay, and T. Maibaum. New York: Oxford University Press, 369–383.
39. Tarski, A. On the calculus of relations. J. Symb. Log. 6:3 (1941) 73–89.

3

Formal Methods: A Petri Nets Based Approach

Giorgio De Michelis, Lucia Pomello,
Eugenio Battiston, Fiorella De Cindio, and Carla Simone

Distributed systems are basically systems constituted by a set of (sequential) components synchronizing one another on the basis of a well defined policy. Various synchronization mechanisms have been proposed for distributed systems: among them we can recall semaphores, critical regions, shared resources, etc. In the late 1970s, almost independently, Tony Hoare [1] and Robin Milner [2] introduced synchronous communication as a means of specification for distributed systems, showing that, on the one hand, almost every synchronization mechanism can be defined in its terms and, on the other, it has an elegant and effective formal (algebraic) representation.

The latter characteristic has been the basis for a wide and rapid research development, bringing forth various formal methods for specifying distributed systems through algebraic calculi called *process algebras* (for an introduction see Refs. [3] and [4]). Process algebras in tandem with Petri nets [5], introduced in the 1960s by Carl Adam Petri [6] as an extension of sequential automata characterizing systems through local states and transitions, have allowed one to overcome the limitations that formal methods defined for sequential systems exhibit with respect to distributed systems. Distributed systems have been characterized by their concurrency, their communication protocols, the invariant properties of their never-ending components; and their behavior has been characterized through new classes of properties such as safeness [5], liveness [5], fairness [7], etc.

Formal methods based on both Petri nets and process algebras characterize distributed systems at two levels: as a means for understanding and, more recently, as a tool for specifying distributed systems within well defined design paradigms and methods. At the elementary level they allow the clear and precise formulation of behavioral and structural properties of distributed systems and their mutual relations.

Elementary net systems [8, 9] have been developed as a theoretical model whose main characteristic is to assume concurrency (i.e., causal independency) as a fundamental property [10] that is exhibited by both system and behavior models. Despite the fact that net systems lack, until now, an algebraic compositional definition, net theory has an elegant linear algebra representation offering interesting methods for characterizing and evaluating both safeness and liveness properties [5].

Thanks in particular to the fundamental contribution of Robin Milner [2, 11, 12], pure process algebras (without value passing) characterize distributed system behavior from the viewpoint of an observer experimenting with it in a sequential manner: she cannot distinguish the concurrent occurrence of two actions from their occurrence in arbitrary sequence.

The main points in favor of process algebras are, on the one hand, their being characterized in compositional terms and, on the other, their defining suitable equivalence notions that are congruences with respect to the operators of the algebras (see Refs. [13–17]). Process algebras are an evolving subject: quite recently, new algebraic calculi for distributed systems have been defined [18], allowing one to specify systems dynamically changing their mutual links and/or their components.

In the 1980s, the distance between process algebras and Petri nets was reduced [4, 19, 20] through various studies defining translation algorithms between (subclasses of) them and, on the one hand, distinguishing the concurrency of a term of a process algebra and, on the other, introducing into net systems process algebra operations. These comparative studies have been very important, because even when they pointed out limitations to any of the two approaches, they contributed to deepening our understanding of concurrency.

Within the theory of Petri nets, also classes of higher level models can be defined, defining the tokens moving in the net as individuals carrying on some information: *predicate/transition net systems* by Hartmann Genrich and Kurt Lautenbach [21] and *colored nets* by Kurt Jensen [22, 23] are the most popular examples of *high-level Petri nets* introduced in the literature. High-level Petri nets have been quite successful in various applicative domains to specify complex systems such as protocols, production systems, office processes, etc., and to verify their properties [19]. New developments are going on to define *high-level net systems* whose tokens are objects, matching the object oriented design paradigm with the well defined discipline of synchronization of Petri nets [24–26].

Process algebras, in turn, have also been studied as an applicative formal method [27]. In particular, they are the basis for the definition of a language for specifying distributed systems, LOTOS, whose aim consists in providing a rich set of formal methods for the design of real complex systems [28]. Also CSP, the programming language proposed by Hoare in the late 1970s, has been developed as a specification language for distributed systems [29, 30].

We must also recall that temporal logic has been developed as a theoretical framework for specifying distributed system behavior, with interesting applications to specification languages—in particular UNITY, the language by K. M. Chandy and J. Misra, has been quite successful [31].

Space limitations prevent us from introducing all the proposals we have listed above, even in synthetical terms; we will therefore restrict our presentation to some features of Petri nets and process algebras, addressing the interested reader to the literature listed at the end of this chapter for the rest. Also in regard to the approaches that are presented, our ambition is not to give the reader complete information but only to provide a taste of them, provoking curiosity and inspiring continuing study by reading some of the texts we propose in the reference list. For this reason, the subsection "Additional readings" in each section is very important.

3.1 Process Algebras

3.1.1 Some examples

Let us consider a very simple machine, a two-places buffer. The buffer is operated through two actions *in* and *out*, respectively putting one item in the buffer and taking one item from it. It is impossible to put one item into a full buffer (when the buffer already contains two things); it is impossible to take one item from the empty buffer. Let us assume that at the beginning the buffer is empty; the buffer behaves as follows: at first only the *in* operation is possible, after *in* is performed (the buffer contains one item), it is possible to perform both *in* and *out*. If *in* is performed, afterwards the buffer is full and only *out* is possible; if *out* is performed, afterward the buffer is empty and only *in* is possible.

We can represent the two-places buffer in graphical terms (see Fig. 3.1), as a *transition system*. Let us now build a two-places buffer through the composition of two one-place buffers.

The two-places buffer can be built identifying the *out* operation of the first one-place buffer with the *in* operation of the second one-place buffer, giving rise to an internal (unobservable) action of the system transferring one item from the first buffer to the second. Let us assume that at the beginning both one-place buffers are empty and that the system behaves as follows: at first only the *in* operation on the first buffer is possible, after *in* is performed (the first buffer contains one item), only the unobservable transfer of the item from the first buffer to the second is possible. Once the transfer is performed, the first buffer is again empty, while the second one is full; it is possible to perform both *in* on the first buffer and *out* on the second. If *in* is performed, afterwards both buffers are full and only *out* (on the second buffer) is possible; if *out* is performed, afterwards both the buffers are empty and only *in* (on the first buffer) is possible. In graphical terms (where the unobservable transfer of the item from the first to the second buffer is represented by a τ) the system can be represented by the transition system of Fig. 3.2.

As is apparent in Fig. 3.2, the transition system representation of the system correctly shows the sequences of operations it may engage. But it neither shows that the system is composed of two components, nor does it explain that in this system, when the first buffer is empty and the second one is full, the *in* and *out* operations (respectively, on the first and second buffers) are independent and may be performed concurrently. To emphasize the components of the system, below we will introduce process algebras that provide, with full compositionality, the models of concurrent systems. In the next section, we will introduce Petri net models of concurrent systems, where their degree of concurrency is clearly

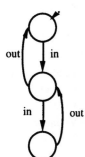

Figure 3.1 The transition system representing a two-place buffer

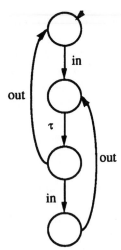

Figure 3.2 The transition system representation of a two-place buffer constructed through two one-place buffers

represented. Before moving on to those topics, let us introduce the equivalence notions making transition systems adequate semantics models for concurrent systems.

3.1.2 Transition systems and observation bisimilarity

In what follows, we will always consider systems that are able to perform both observable and unobservable actions. The observable actions will be represented by the symbols of an alphabet A, while any unobservable action (i.e., considered as an internal action the observer cannot see) will be represented by τ. Whenever we say that the alphabet of a system is A, we are considering the possibility that the system performs an action from $A \cup \{\tau\}$. We use the letters $a, b, c, \ldots$ to indicate elements of A, while we use $u, v, \ldots$ to indicate elements of $A \cup \{\tau\}$. Moreover we will use w to indicate elements of $(A \cup \{\tau\})^*$ and ow to indicate elements of A^*.

Concurrent systems may be represented as machines. Since the early 1950s, machines have been modeled as transition systems (automata).

Definition A. Transition system
A *transition system* is a quadruple $\mathbb{A} = (A, S, \text{----}>, s_0)$, where

1. A is an alphabet.
2. S is a possibly infinite set of states.
3. $\text{----}> \in S \times (A \cup \{\tau\}) \times S$ is the transition relation.
4. s_0 is the initial state.

5. Let s_1, s_2 range over S. Then $(s_1, u, s_2) \in \text{----}>$ will be also denoted $s_1 \text{--}^u\text{--}> s_2$. With

$\text{----}>^*$ we will denote the transitive closure of $\text{----}>$: let $w = u_1.u_2 \ldots u_n \in (A \cup \{\tau\})^*$, then $s\text{--}^w\text{--}>^* s'$ if there exist $s_1, s_2, \ldots, s_n, s_{n+1} \in S$ such that

$$s = s_1 \text{--}^{u_1}\text{--}> s_2 \text{--}^{u_2}\text{--}> \ldots s_n\text{--}^{u_n}\text{--}> s_{n+1} = s'$$

Let $\mathbb{A} = (A, S, \text{----}>, s_0)$ be a transition system. Then $s \in S$ is a reachable state of $\mathbb{A}$ if there exists $w = u_1.u_2 \ldots u_n \in (A \cup \{\tau\})^*$ such that $s_0 \text{--}^w\text{--}>^* s$.
The set of all the reachable states of $\mathbb{A}$ is denoted by reach($\mathbb{A}$).

If we restrict our attention to observable actions only, then we need some more notations.

We say that there is a visible move, a, from s to s', $s==^a==>s'$, if there exist $s_1, s_2, s_3, s_4 \in S$ and $u_1, u_2 \in \{\tau\}^*$ such that $s = s_1--^{u_1}-->^*s_2 --^a-->s_3--^{u_2}-->^*s_4 = s'$.

With $====>^*$ we will denote the transitive closure of $====>$. We say that there is a visible path $ow = a_1.a_2\ldots a_k \in A^*$ from s to s', $s==^{ow}==>^* s'$, if there exist $s_1, s_2, \ldots, s_{2k+2} \in S$ and $u_1, u_2, \ldots u_{k+1} \in \{\tau\}^*$ such that

$$s = s_1--^{u_1}-->^*s_2 --^{a_1}-->s_3--^{u_2}-->^*s_4 --^{a_2}-->s_5. \ldots s_{2k}--^{a_k}->s_{2k+1} --^{u_{k+1}}-->^*s_{2k+2} = s'$$

From the above definition it becomes apparent that for any $w \in \{\tau\}^*$ and for any couple of states s, s' of S, if $s--^w-->^*s'$ then $s = ^\varepsilon=>^*s'$.

Transition systems have a nice graphic representation through oriented graphs. States are represented by nodes of the graph, and transitions are represented by arrows linking nodes with the label corresponding to their action names. The initial state is indicated by a small arrow entering in it. The examples given above are all examples of transition systems.

To use the class of transition systems as a semantic model for concurrent systems, we need to define when two transition systems are equivalent. As we have said above, in automata theory, the main equivalence notion is based on the comparison of the observable paths from the initial state. With respect to concurrent systems, we need a greater distinctive power.

In fact, while the equivalence of the two two-place buffers we have introduced above can be proved showing that for both of them the set of observable paths from the initial state is $\{\varepsilon, in, in.in, in.out, in.in.out, in.out.in, in\ in.out.out, ..\}$, there are some cases in which the comparison of observable paths from the initial state is not enough. Let us discuss an example to make this point clear. Let us have the following two transition systems (Fig. 3.3).

From the viewpoint of the set of observable paths from the initial state they are equivalent, since for both it is $\{\varepsilon, a, ab, ac\}$. But while the former, after the execution of a, is ready to execute both b and c, the latter is either ready to execute b (but not c) or ready to execute c (but not b). The second transition system is not deterministic, in the sense that after the execution of a it can reach two different states, opening different sets of possibilities. As Robin Milner has pointed out [2], this type of nondeterminism is highly relevant

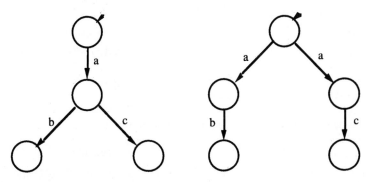

Figure 3.3 Two non-observation bisimilar transition systems

for concurrent systems because its presence affects the way a system interacts with another system giving rise to a larger system.

Let us use the two above machines to build two new machines, in a similar way as we did building a two-places buffer through two one-place buffers.

First we build a machine by composing the two above systems, with the second one having its actions *a, b, c* renamed *e, b, c*. The new system is the result of identifying each couple of actions with the same name in both the systems giving rise to an observable action maintaining that name. It is easy to see that the new system can execute (in any order) *a* and *e*, and afterwards either *b* or *c*.

If we build a new system by composing two copies of the second transition system of Fig. 3.3, renaming actions *a, b, c* of the second copy with *e, b, c*, and again identifying each couple of actions with the same name in both the systems giving rise to an observable action, maintaining its name, we obtain a system that may deadlock after the execution of *a* and *e*.

To capture the different ways of interacting of the two systems above, we need to introduce a new equivalence notion: observation bisimilarity, i.e. an equivalence capturing the fact that each of the two systems is capable of simulating the other.

Definition B. Observation bisimilarity

Let $A_i = (A_i, S_i, ---->_i, s_{0i})$, i = 1, 2, be two transition systems. A_1 and A_2 are *observation bisimilar*, $A_1 \approx A_2$, if $A_1 = A_2$ and there is a relation $B \subseteq S_1 \times S_2$ such that:

1. $(s_{01}, s_{02}) \in B$
2. if $(s_1, s_2) \in B$ and s_1 --^{a}--> s_1', then there is s_2' such that $(s_1', s_2') \in B$ and s_2 --^{a}--> s_2'
3. if $(s_1, s_2) \in B$ and s_2 --^{a}--> s_2', then there is s_1' such that $(s_1', s_2') \in B$ and s_1 --^{a}--> s_1'

It is easy to see that while the two two-places buffers in Fig. 3.1 and Fig. 3.2 are also observation bisimilar, the two transition systems of Fig. 3.3 are not observation bisimilar, because the state reached by the first one after *a* is not in the B relation with any of the two states reached after *a* by the second one.

Observation bisimilarity is therefore a good equivalence notion with respect to concurrent systems.

3.1.3 A process algebra

Let us now introduce a second view on concurrency, whereby concurrent systems are recursive terms over certain operator symbols, i.e. they form the smallest class of terms generated by the operator symbols and closed under recursion (without parameters). Systems of this type are generally called *process algebras.* In the following, we will introduce an algebra based on the one proposed by E. R. Olderog [4] because it is a good representative of the many introduced in the literature. The minimal changes we have introduced are due to the necessity to compress the presentation.

Definition C. Process terms

Let *Act* be defined as $A \cup \{\tau\}$, where *A* is a finite alphabet of (observable) actions, whose elements will be denoted by *a, b,* and τ is the symbol of the unobservable action, not belonging to *A*. The set of *process terms, Proc*, whose elements will be denoted by *P, Q, R*, is defined as follows:

$P ::= stop{:}A$ (deadlock) $| \ div{:}A$ (divergence) $| \ a.P$ (prefix) $| \ P + Q$ (choice) $| \ P \| Q$ (parallelism) $| \ P_{[a/b]}$ (renaming) $| \ P_{\backslash\{b\}}$ (hiding) $| \ X$ (identifier) $| \ \mu X.P$ (recursion).

The signature of *Proc* consists, for each action alphabet, A, of two nullary operator symbols, respectively *stop:A* and *div:A*, of a unary prefix symbol a. for each action symbol in A, of two unary postfix symbols $[a/b]$ and $\setminus\{b\}$ respectively for each couple of action symbols a, b and for each action symbol b, and, finally, of two binary infix symbols $+$ and $\|$.

An occurrence of an indentifier X within a process term P is said to be *bound* if it occurs within a subterm of the form $\mu X.Q$. In all the other cases it is said to be *free*. An identifier X is said to be *free* in P if all its occurrences within P are free. We will write $P\{Q/X\}$ to denote the result of substituting Q for every free occurrence of X in P. A process term is said *closed* if it does not contain free occurrences of identifiers. The class of closed process terms is called *CProc*.

In what follows, we will pay attention only to closed process terms.

We can associate to each process term $P \in CProc$ an alphabet $A(P)$ as follows:

1. Every maximal sequential subterm of P, Q, has as alphabet $A(Q)$ the union of the action symbols occurring in it; every subterm of Q has as alphabet the alphabet of Q, $A(Q)$.
2. Every process term of the form $P \| Q$, has as alphabet $A(P) \cup A(Q)$; $A(P) \cap A(Q)$ is its synchronization alphabet.
3. Every recursive process term $\mu X.P$ has as alphabet the alphabet of P, $A(P)$; the alphabet of X, $A(X)$, is also the alphabet of P, $A(P)$.

The application of operators and recursion is regulated by priorities $(+, \| < a. , \mu X. < [a/b] ,\setminus\{b\})$; $+$ and $\|$ are commutative and associative.

$P_{[a/b]}$ denotes the term we obtain from P substituting in it every occurrence of a with b; $P_{\setminus\{b\}}$ denotes the term we obtain from P substituting in it every occurrence of b with τ.

Process terms are *action guarded*: i.e., for any term P in *CProc*, every recursive subterm in P, $\mu X.Q$, is such that every free occurrence of X in Q occurs within a subterm of the form $a.R$.

The meaning of the process terms can be explained in intuitive terms as follows:

- *stop:A* denotes a process which doesn't perform any action.
- *div:A* denotes a process which performs an infinite sequence of unobservable actions τ.
- *a.P* denotes a process that performs a and then behaves like P.
- $P + Q$ denotes a process that behaves either as P or as Q.
- $P \| Q$ denotes a process whose behavior is the result of the concurrent behavior of P and Q, taking into account that they must synchronize on the actions of their synchronization alphabet (a more precise description will be given here below within the interleaving semantics of *Proc*).
- $P_{[a/b]}$ denotes a process that behaves like P, except for the fact it executes b whenever P executes a.
- $P_{\setminus\{b\}}$ denotes a process that behaves like P, except for the fact it executes the unobservable action τ whenever P executes b.
- $\mu X.P$ denotes a process that behaves like P, except for the fact it executes again $\mu X.P$ whenever it had to execute X.

Let us now define the terms representing the systems we have introduced above as transition systems.

The sequential two-places buffer (Fig. 3.1) is the following term with a double recursion:

$$2B1 = \mu X.in.\mu Y.(in.out.X + out.Y)$$

The two-places buffer built up through two one-place buffers (Fig. 3.2), is defined by the following equations:

$$1B = \mu X.in.out.X$$

$$2B2 = (1B[link\ /out]\ //\ 1B[link\ /in])\backslash\{link\}$$

where the first equation defines the one-place buffer, and the second equation defines the two-places buffer as parallel composition of two instances of the one-place buffer, the first one of which has the *out* action renamed as *link*, the second one the *in* action also renamed *link*, and finally with *link* action, constituting the synchronization alphabet, made unobservable by hiding.

The first system of Fig. 3.3 is given by the following term:

$$C1 = a.(b.stop\{a,\ b,\ c\} + c.stop\{a,\ b,\ c\})$$

whereas the second system of the same figure is given by the following term:

$$C2 = (a.b.stop\{a,\ b,\ c\} + a.c.stop\{a,\ b,\ c\})$$

The parallel composition of *C1* and *C2* is given, evidently, by

$$A1 = C1 \parallel C2$$

while the parallel composition of two instances of *C2* is given by

$$A2 = C2 \parallel C2$$

To move from the informal semantics given above to a formal one, let us explain the behavioral rules of process terms by means of a transition relation.

The transition relation of process terms is defined by the following axioms and rules:

1. $a.P \;\text{--}^{a}\text{---}> P$
   (prefix axiom)

2. $div{:}A \;\text{--}^{\tau}\text{---}> div{:}A$
   (divergence axiom)

3.
$$\frac{P \;\text{--}^{u}\text{--}> P'}{P \parallel Q \;\text{--}^{u}\text{--}> P' \parallel Q,\ Q \parallel P \;\text{--}^{u}\text{--}> Q \parallel P'}$$
   $(u \notin A(P) \cap A(Q)$—parallel composition without synchronization rule)

4.
$$\frac{P \;\text{--}^{a}\text{--}> P',\ Q \;\text{--}^{a}\text{--}> Q'}{P \parallel Q \;\text{--}^{a}\text{--}> P' \parallel Q'}$$
   (parallel composition with synchronization rule)

5.
$$\frac{P\ \text{--}^u\text{--}>\ P'}{P+Q\ \text{--}^u\text{--}>\ P'+Q,\ Q+P\ \text{--}^u\text{--}>\ Q+P'}$$
(choice rule)

6.
$$\frac{P\ \text{--}^u\text{--}>\ Q}{P\,[a\,/\,b\,]\ \text{--}^u\text{--}>\ Q\,[a\,/\,b\,],\ P\backslash\{b\}\ \text{--}^u\text{--}>Q\,\{b\}}$$
(renaming and hiding rule)

7.
$$\frac{P\{\mu X.P\,/X\}\ \text{--}^u\text{--}>\ Q}{\mu X.P\ \text{--}^u\text{--}>\ Q}$$
(recursion rule)

The transition relation given above is the main building block to associate to each process term its semantics as a transition system. In order to grant the maximal generality of our construction process, we introduce an abstract version of transition systems, disregarding unreachable states as well as the names given to reachable states.

Definition D. Weak isomorphism

Let $\mathbb{A}_i = (A_i, S_i, \text{----}>_i, s_{0i})$, $i = 1, 2$, be two transition systems. $\mathbb{A}_1$ and $\mathbb{A}_2$ are weakly isomorphic, $\mathbb{A}_1 =_{\text{isom}} \mathbb{A}_2$, if $A_1 = A_2$ and there is a bijection
β: $reach(\mathbb{A}_1) \text{----}> reach(\mathbb{A}_2)$ such that $\beta(s_{01}) = s_{02}$ and, for all $s_1, s_2 \in reach(\mathbb{A}_1)$ and for all $u \in A_1 \cup \{\tau\}$, $s_1 \text{--}^u\text{--}> s_2$ iff $\beta(s_1) \text{--}^u\text{--}> \beta(s_2)$.
The bijection β is said a weak isomorphism between $\mathbb{A}_1$ and $\mathbb{A}_2$.

Weakly isomorphic transition systems are equal but for the renaming of the states. The equivalence classes of transition systems with respect to weak isomorphism are called *abstract transition systems*. The class of abstract transition systems is called *Trans*.

The semantics we are now introducing does not take into account the degree of parallelism of a concurrent system: from its point of view a true concurrent behavior and its interleaved sequential simulation are not distinguishable. For this reason we call this semantics, interleaving semantics.

The interleaving semantics for process terms is a mapping $\mathbb{A}\,[[\,.\,]]$: *CProc* ----> *Trans* which assigns to every process term P the abstract transition system $\mathbb{A}\,[[\,P\,]] = (A(P), Proc,$ $\text{----}>|_{A(P)}, P)$ where ----> is the transition relation given above and $\text{----}>|_{A(P)}$ is its restriction with respect to $A(P)$: $\text{----}>|_{A(P)} = \{P\text{--}^u\text{--}>Q \mid u \in \{\tau\} \cup A(P)\}$.

It is easy to see that the semantic function $\mathbb{A}$ associates to each term listed above the abstract transition system containing, as isomorphism class, the transition system to which it relates. We can therefore derive from the bisimilarity relation defined for transition systems a bisimilarity relation for process terms.

3.1.4   Additional readings

Process Algebras has been a very active field of research for fifteen years. After the *Calculus of Communicating Systems* (CCS) by Robin Milner [2] and the algebraic model of the *Communicating Sequential Processes* (CSP) [1] called *Theoretical CSP* (TCSP) by Tony Hoare and co-workers [32], many proposals of algebraic calculi as well as of equivalence notions have been proposed [3, 4, 12–14]. References [4] and [12] can be considered good introductory papers for the interested reader. Applications of process algebras can be found in Refs. [27] and [28].

3.2 PETRI Nets

3.2.1 An example

Figure 3.4a shows a net-based system modeling the mutual exclusion of a reader and a writer in the access to an area in main memory. The circles, called S-elements, places or conditions, represent the local states, while the boxes represent the local atomic transitions. The black dots, called *tokens*, indicate in which local states are the involved processes; both writer and reader are doing "other processing," and the "access" to the memory is "free." The behavior of the system can be simulated by moving the tokens inside the places following the directions of the arcs and the following *transition rule*: a transition may occur when each of the places connected to it by an arc ingoing to the transition contains a token, and its occurrence takes away the tokens from those places and puts a token on each place connected to it by an arc outgoing from the transition. For example, after the occurrence of transition t_1, the process writer is "ready to write" and, since the condition "access free" holds, i.e., contains a token, transition t_2 may occur and its occurrence takes away the tokens from "ready to write" and "access free" and puts a token into "writing". Now if the reader becomes, after the occurrence of t_4, "ready to read" it can start "reading" only when the condition "access free" holds again. This happens after the occurrence of t_3, when the writer has finished writing.

3.2.2 Elementary net systems: basic features of concurrency

The static structure of a distributed system is described by a net.

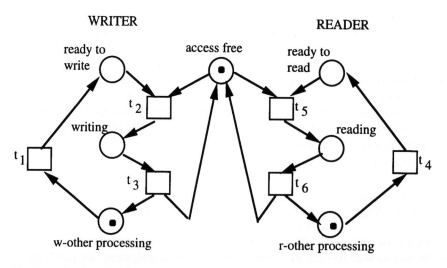

Figure 3.4a A net-based system

Definition E. Net

A *Net* is a triple $N = (S, T, F)$, where:

1. $S \cap T = \emptyset$ and $S \cup T \neq \emptyset$
2. $F \subseteq (S \times T) \cup (T \times S)$ is such that: $\mathrm{dom}(F) \cup \mathrm{ran}(F) = S \cup T$

Thus a *net* is a bipartite directed graph. S is the set of *S-elements*, the *local* atomic states, T is the set of *T-elements*, the *local* atomic transitions, and F is the *flow relation*, which captures the neighborhood relationship between local states and local transitions.

Also, $X = S \cup T$ is the set of elements of N. Condition 1 means that S and T form a partition of X and that X, and thus the Net, cannot be empty. Condition 2 means that the net has no isolated elements. In graphical representations, the S-elements are drawn as circles, the T-elements as boxes and the members of the flow relation are indicated through appropriate directed arcs. For $x \in X$, $\bullet x = \{y \in X : (y, x) \in F\}$ is the set of *pre-elements* of x, and $x \bullet = \{x \in X : (x, y) \in F\}$ is the set of *post-elements* of x.

There is a variety of subclasses of Nets depending on restrictions on the Net structure, we present two of them.

Definition F. Pure and simple nets

Let $N = (S, T, F)$ be a Net:

1. N is *Pure* if $\forall\, x \in X$: $\bullet x \cap x \bullet = \emptyset$.
2. N is *Simple* if $\forall\, x, y \in X$: $[(\,\bullet x = \bullet y$ and $x \bullet = y \bullet\,) \Rightarrow x = y]$.

In the case of not pure nets, the places s such that $\bullet s \cap s \bullet \neq \emptyset$ are called *side conditions* and the transitions t such that $\bullet t \cap t \bullet \neq \emptyset$ are called *impure transitions*. The net shown in Fig. 3.4a is pure and simple.

Depending on the applications, a wide range of interpretations can be attached to S- and T-elements. In the case of *elementary net systems* [8, 9], S-elements represent *conditions*, whereas T-elements represent *events* and are usually denoted by B and E, respectively.

Definition G. Elementary net system

An *elementary net system*, abbreviated *EN system,* is a 4-tuple $\Sigma = (B, E, F, c_{in})$, where (B, E, F) is a net, called the underlying net of Σ, and $c_{in} \subseteq B$ is the initial case of Σ.

An EN system is *finite* if $B \cup E$ is a finite set.

The initial case describes the initial state of the system in terms of the set of conditions which initially holds. In graphical representations, a case $c \subseteq B$ is shown by placing a *token* on those conditions that are members of c.

The system given in Fig. 3.4a is an example EN system. The evolution of the system, hence its behavior, is defined through the *transition rule* specifying the conditions under which an event can occur, and how the event occurrence modifies the holding of conditions (see Fig. 3.4b).

Definition H. Transition rule for EN systems

Let $N = (B, E, F)$ be a Net, $e \in E$ and $c \, \hat{\underline{E}} \, B$.

1. e is said to be *enabled* at c, denoted $c[e>$, if $\bullet e \subseteq c$ and $e \bullet \cap c = \emptyset$.
2. If e is enabled at c, then the *occurrence* of e leads from c to c', denoted $c\,[e> c'$, if $c' = (c - \bullet e) \cup e \bullet$.

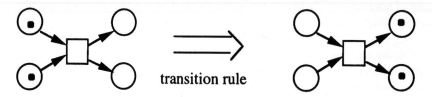

Figure 3.4b Event occurrence modifies holding of conditions

The first part of the transition rule is the *enabling rule*, or rule of concession: an event e can occur if and only if its preconditions hold and its postconditions do not hold; the second part of the transition rule is the *occurrence rule*: the preconditions of e cease to hold and the postconditions of e begin to hold; the remaining part of the case is unaffected by the event occurrence. The change of state is therefore *local*: it is confined to the immediate neighborhood of the occurring event.

Concurrency is a basic notion in net theory. A set u of events can *concurrently* occur at a given case c (u is a *step*) if the events in the set can individually occur at c without interfering each others (i.e., they are *independent*).

Definition I. Steps
Let $N = (B, E, F)$ be a net and $\emptyset \neq U \subseteq E$.

1. U is *independent* if $\forall e_1, e_2 \in U: [\, e_1 \neq e_2 \Rightarrow (\bullet e_1 \cup e_1 \bullet) \cap (\bullet e_2 \cup e_2 \bullet) = \emptyset \,]$.
2. Let $c \subseteq B$. Then U is a *step enabled* at c, $c[U>$, if U is independent and $\forall e \in U$: $c[e>$.
3. Let $c_1, c_2 \subseteq B$. Then U is a *step* leading from c_1 to c_2, $c_1[U> c_2$, if $c_1 [U>$ and $c_2 = (c_1 - \bullet U) \cup U^\bullet$.

Obviously, when a step U is a singleton $\{e\}$, then $c_1[U> c_2$ corresponds to the previously introduced occurrence of the single event e.

In the EN system of Fig. 3.4a, $\{t_1, t_4\}$ is independent and is a step enabled at

$$c_{in} = \{\text{w-other processing, access free, r-other processing}\}$$

that is,

$$c_{in}[\{t_1, t_4\}>\{\text{ready to write, access free, ready to read}\}$$

Definition J. Set of cases, set of steps
Let $\Sigma = (B, E, F, c_{in})$ be an EN system. The *set of cases* of Σ, denoted C_Σ, is the smallest subset of 2^B such that:

1. $c_{in} \in C_\Sigma$
2. $\forall c \in C_\Sigma, \forall c' \subseteq B, \forall e \in E: [c \,[e> c' \Rightarrow c' \in C_\Sigma]$

The *set of steps* of Σ, denoted U_Σ, is the set $\{U \subseteq E : \exists c \in C_\Sigma \, c \,[U> \}$.

Therefore, the behavior of an elementary net system can be described both in a sequential and in a nonsequential way. In the case of sequential behavior, we have an *interleaving semantics* given by (finite or infinite) *occurrence sequences* $c_{in}[e_1> c_1 \ldots c_{n-1}[e_n> c_n$ or by (finite or infinite) *event sequences* $c_{in}[e_1 e_2 \ldots e_n> c_n$.

In the case of nonsequential behavior, we have a so called *step semantics* and a *partial-order semantics*. Step semantics is given by (finite or infinite) *step occurrence sequences* $c_{in}[U_1>c_1...c_{n-1}[U_n>c_n$ or by (finite or infinite) step sequences $c_{in}[U_1U_2...U_n>c_n$.

The partial order semantics of an EN system is given by the non sequential processes (called simply *processes*) [10, 33] describing the system behavior in terms of a partial order of event occurrences and of condition holding. For space constraints we don't present this notion here.

An overview of the behavior of an EN system is given through its state space, its *case graph*, i.e., through the collection of all the global states together with the transitions between them.

Definition K. Case graph

Let $\Sigma = (B, E, F, c_{in})$ be an EN system. The *case graph* of Σ, denoted $CG(\Sigma)$, is the edge-labeled graph with an initial node $(C_\Sigma, U_\Sigma, A, c_{in})$, where: C_Σ is the set of nodes, U_Σ is the alphabet; A is the set of labeled edges $\{(c, a, c')| c, c' \in C_\Sigma, a \in U_\Sigma$, and $c[a>c'\}$, and c_{in} is the initial node.

In Fig. 3.5a, an EN system Σ with its case graph $CG(\Sigma)$ is presented.

Let us now consider again the enabling rule of an event e at a case c. It has two components: an input component, $\bullet e \subseteq c$; and an output component, $e\bullet \cap c = \emptyset$. The property of *contact-freeness* says that it is sufficient that $\bullet e \subseteq c$ for the event e to be enabled.

Definition L. Contact-freeness

The EN system $\Sigma = (B, E, F, c_{in})$ is *contact-free* if $\forall e \in E, \forall c \in C_\Sigma$:

$$[\bullet e \subseteq c \Rightarrow e\bullet \cap c = \emptyset]$$

There is an algorithm for transforming an EN system Σ into a contact-free EN system Σ' such that Σ and Σ' are behaviorally equivalent in a strong sense [9]. In what follows, we will very often restrict our attention to contact-free EN system without any loss of generality.

Let $\Sigma = (B, E, F, c_{in})$ be a contact-free EN system. At a case c, two events e_1 and e_2 can be related to each other in the following ways:

- *sequence*: e_1 and e_2 are in sequence at c if $c[e_1>$ and $\neg(c[e_2>)$ and $c'[e_2>$, where $c[e_1>c'$.

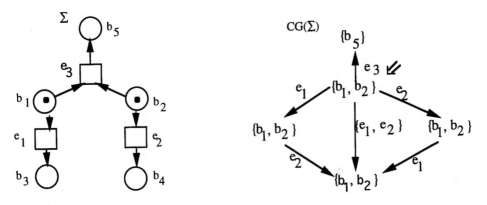

Figure 3.5a An EN system Σ with its case graph $CG(\Sigma)$

- *choice (conflict)*: e_1 and e_2 are in conflict at c if $c[e_1>$ and $c[e_2>$ and $\neg(c[\{e_1, e_2\}>)$. Whether e_1 will occur or e_2 will occur is left unspecified, the system exhibits "non-determinism".

- *concurrency*: e_1 and e_2 can occur concurrently at c if $c[\{e_1, e_2\}>$.

  No order is specified in regard to their occurrences. The occurrences of events and the holdings of conditions can result *partially ordered*. The system can exhibit "non-sequential behavior".

- *confusion*: it is a combination of choice and concurrency. (c, e_1, e_2) is a confusion at c if $c[\{e_1, e_2\}>$ and $cfl(e_1,c) \neq cfl(e_1, c_2)$, where $c[e_2>c_2$ and $cfl(e', c') = \{e \in E : e' \neq e$ and $c'[e>$ and $\neg c[\{e, e'\}>\}$.

Let us consider the EN system Σ_1 of Fig. 3.5b. For $c = \{b_1, b_2, b_3\}$, (c, e_1, e_2) is a confusion. In fact, there could be disagreement over whether or not a conflict was resolved going from case c to case $c' = \{b_4, b_5\}$ via step $\{e_1, e_2\}$. There are two possible interpretations: (a) e_1 occurred first without being in conflict with other events; then e_2 occurred, or (b) e_2 occurred first, and then e_1, e_3 got in conflict; this conflict was resolved in favor of e_1. There is an overlapping of concurrency and conflict. Following the given definition, (c, e_1, e_2) is a confusion since $cfl(e_1,c) = \emptyset \neq \{e_3\} = cfl(e_1, c_2)$, where $c_2 = \{b_1, b_3, b_4\}$.

Systems exhibiting confusion are difficult to analyze, while the class of net-based systems, like the so called *free-choice systems,* in which choice and concurrency are combined in a confusion-free manner, admits a nice theory. See, for example Ref. [34].

3.2.3 Place/transition systems: behavioral properties and structural methods

In this section we will present the fundamental notions of *place/transition* (P/T) systems. EN systems will turn out as a special case in this framework [35].

Let us consider again the problem of the access right to a memory area (see Fig. 3.4a). Now we want to model a configuration of two processes with write access and four processes with read access, in which at most three reader processes may overlap in their access to the memory; when the memory is being changed by some writer process no other process may have access.

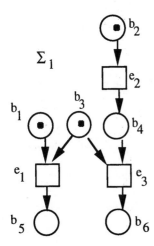

Figure 3.5b A confusion in an EN system

Figure 3.6a shows this system as a place/transition system. The S-elements are called *places* and may contain more than one token. Two of the arcs are labeled by 3. In this case when the appropriate transition occurs, the token count on the place s is reduced or increased by 3 instead of 1 [5]. The place/transition system of Fig. 3.6 is based on a *place/ transition net* with places with unlimited capacity. This one will be the model presented here. In Ref. [5] *place/transition nets and systems,* where places have a limited capacity (i.e., may carry a limited number of tokens), are also considered.

Definition M. Place/transition net
A *place/transition net*, also called P/T net, is a 5-tuple N = (S, T, F, W), in which:

- (S, T, F) is a net, where the S-elements are called places and the T-elements are called transitions.
- W: F → $\mathbb{N}^+$ is a *weight* function.

In the graphical representation of P/T nets, the arcs f ∈ F are labeled by W(f) whenever W(f)>1.

Definition N. Place/transition system
A *place/transition system*, (P/T system), is a 6-tuple Σ = (S, T, F, W, M_0), where:

- (S, T, F, W) is a P/T net, the underlying net of Σ.
- M_0 : S → $\mathbb{N}$ is the *initial marking* function.

A P/T system is *finite* if S ∪ T is a finite set.
In the graphical representation, the marking M_0 is indicated by $M_0(s)$ tokens on each place s. Again, the behavior is defined through the *transition rule* (see Fig. 3.6b).

Definition O. Transition rule for P/T system
Let Σ = (S, T, F, W, M_0) be a P/T system. Let M be a marking, i.e. a function M: S → N.

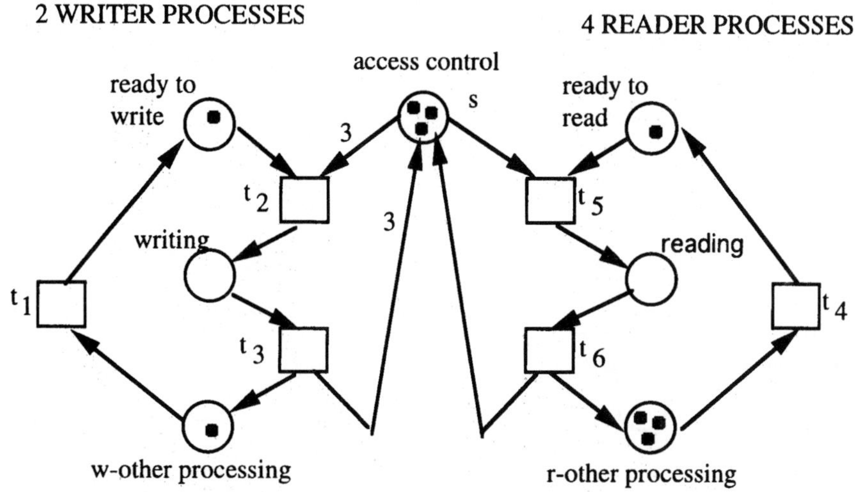

Figure 3.6a A place/transition system

transition rule for P/T systems

Figure 3.6b Event occurrence modifies holding of conditions

- A transition $t \in T$ is *enabled* at M, $M[t>$, if $\forall s \in S$: $W(s,t) \leq M(s)$.
- A transition t enabled at M *may occur* (*may fire*) yielding a new marking M', $M[t>M'$, such that $\forall s \in S$: $M'(s) = M(s) - W(s,t) + W(t,s)$.
- A multiset of transitions $U: T \to N$ is a *step enabled* at M, $M[U>$, if $\forall s \in S$:

$$\sum_{t \in T} U(t)W(s,t) \leq M(s)$$

- A step U enabled at M *may occur* yielding a new marking M', $M[U>M'$, such that $\forall s \in S$: $M'(s) = M(s) - \sum_{t \in T} U(t) W(s,t) + \sum_{t \in T} U(t)W(t,s)$.
- The *set of (reachable) markings* of Σ, $[M_0>$, is the smallest subset of N^S such that:
  1. $M_0 \in [M_0>$
  2. if $M \in [M_0>$ and $\exists t \in T$: $M[t>M'$ then $M' \in [M_0>$

Similarly to EN systems, the behavior of P/T systems can be described by *interleaving semantics*, by means of (finite or infinite) *firing sequences*; by *step semantics*, by means of (finite or infinite) *step sequences*; and by *partial-order semantics*, by means of *processes*.

Also in P/T systems, at a given marking two transitions can be related to each other in different ways as exemplified in the following.

Let us consider the P/T systems Σ_1 and Σ_2 given in Fig. 3.7. Transitions t_1 and t_2 of Σ_1 are in *conflict* at M_0. t_1 is enabled at M_0, it is even *concurrently enabled* with itself: $2t_1$ is a step at M_0. Also t_2 is enabled at M_0, while t_1 and t_2 are *not* concurrently enabled at M_0. In Σ_2 the multiset $(2t_1 + t_2)$ is a step at M_0.

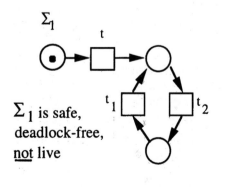

Σ_1 is safe, deadlock-free, not live

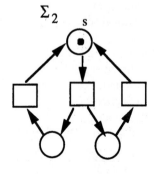

Σ_2 is live, not safe

Figure 3.7 P/T systems Σ_1 and Σ_2

While in finite EN systems the set of reachable cases is finite, in finite P/T systems with infinite place capacity the set of reachable markings is in general infinite. A finite way of representing the state space of P/T systems is given by *coverability trees* [35].

EN systems can be seen as a subclass of P/T systems [37]. Another subclass of P/T system, called either *marked nets* or *Petri nets*, is of particular interest since most publications on P/T Systems are based on this model.

Definition P.   Marked nets

A P/T system $\Sigma = (S, T, F, W, M_0)$ is a *marked net* iff $\forall t \in T$: $[W(s, t) \le 1$ and $W(t, s) \le 1]$. We denote marked nets as (S, T, F, M_0), since W is now redundant.

By definition, marked nets are contact-free, so their transition rule is the one for contact-free P/T systems.

Definition Q.   Safe marked nets

Let $n \in \mathbb{N}$. A marked net (S, T, F, M_0) is *n-safe (bound)* iff $\forall M \in [M_0>$, $\forall s \in S$: M(s) $\le n$. It is *safe* iff $\forall M \in [M_0>$, $\forall s \in S$: M(s) ≤ 1.

In marked nets, impure transitions can be enabled. Therefore, the subclass of safe marked nets does not correspond to EN Systems, where impure events are never enabled. If we consider only pure nets, than we get the correspondence, i.e., the class of pure contact-free EN systems corresponds to the class of pure safe marked nets.

Definition R.   Behavioral properties

Let $\Sigma = (S, T, F, W, M_0)$ be a P/T system.

1. An transition t is *dead* in Σ if $\nexists M \in [M_0>$: M[t>.
2. Σ is *deadlock-free* if $\forall M \in [M_0>$, $\exists t \in T$: M[t>.
3. Σ is *live* if $\forall M \in [M_0>$, $\forall t \in T$, $\exists M' \in [M >$: M'[t>.

From the definition above, it follows that liveness is a stronger property than deadlock-freeness. Both systems given in Figs. 3.4a and 3.6a are live, and then also deadlock-free. The system of Fig. 3.4a, being an EN system, can be seen also as a safe marked net. The P/T system of Fig. 3.6a is bound, it is 4-safe. Let us now consider the P/T systems given in Fig. 3.7. Σ_1 is safe, deadlock-free but not live; after its first occurrence, transition t becomes dead, and it will never be enabled again. Σ_2 is live, therefore also deadlock-free, but it is not safe; there is no limit to the number of tokens contained in place s.

3.2.4   An operational net semantics for process terms

Before concluding this section, let us recall that net systems have been widely used to give the operational semantics of process algebras. Without entering into a full presentation of the issues raised by the Petri net semantics of process algebras (see Ref. [4] for a good introduction to this topic), we will show how a particular class of *labeled net systems* (where the labeling function assigns (action-) names to transitions), namely 1-safe *labeled marked nets,* can be used to define an operational semantics for the process terms introduced in Section 1, such that it captures and makes explicit its concurrency.

Let us first introduce the definition of 1-safe labeled marked nets and then characterize the operational net semantics for process terms.

Definition S.   1-safe labeled marked nets

Let (S, T, F, M_0) be a 1-safe marked net. Then $((S, T, F, M_0), A, \alpha)$, where A is a finite alphabet (of action names) and α:T--->A $\cup \{\tau\}$ is a labeling function, is called a *1-safe labeled marked net.*

A 1-safe labeled marked net, $((S, T, F, M_0), A, \alpha)$, can also be represented as follows, $(A, S, ---\!>|_A, M_0)$, where $---\!>|_A = \{(M, u, M') \mid M, M' \in [M_0> \text{ and } \exists\, t \in T: (M[t>M' \text{ and } \alpha(t) = u)\}$.

The definition of the semantic function requires two steps. First, we need to decompose a process term into sequential components; second, we need a transition rule for sets of sequential components.

The set of sequential components, *Sequ*, consists of all the terms generated by the following production rules:

$$C ::= stop{:}A \mid div{:}A \mid a.P \mid C+D \mid C\,\|_A\mid A\|D \mid C_{[a\,/\,b\,]} \mid C\backslash\{b\})$$

The operators $+$, $\|_A$, $_A\|$, $[a\,/\,b\,]$, $\backslash\{b\}$ can be applied also to sets of sequential components, represented by **P, Q, R**, and they have to be understood as the element by element extensions of the operators for single sequential components.

To transform a process term into a set of sequential components, we introduce a function, *dex*: *CProc* $---\!> 2^{Sequ}$, defined as follows:

$$dex(stop{:}A) = \{stop{:}A\}$$

$$dex(div{:}A) = \{div{:}A\}$$

$$dex(a.P) = \{a.P\}$$

$$dex(P + Q) = dex(P) + dex(Q)$$

$$dex(P\,\|Q) = dex(P)\|_A \cup A\|dex(Q), \text{ where } A = \alpha(P) \cap \alpha(Q)$$

$$dex(P_{[a/\,b\,]}) = dex(P)_{[a/\,b\,]}$$

$$dex(P\backslash\{b\}) = dex(P)\backslash\{b\}$$

$$dex(\mu X.P) = dex(P\{\mu X.P\,/X\})$$

It is easy to see that *dex* is well defined.

The Petri net transition relation of sets of sequential components is defined by the following axioms and rules:

1. $\{a.P\} --^a---\!> dex\,(P) - (\text{prefix axiom})$

2. $\{div{:}A\} --^\tau---\!> \{div{:}A\} - (\text{divergence axiom})$

3.
$$\frac{\mathbf{P} --^u--\!> \mathbf{P'}}{\mathbf{P}\|_A --^u--\!> \mathbf{P'}\|_A, \; _A\|\mathbf{P} --^u--\!> {}_A\|\mathbf{P'}}$$
$(u \notin A - \text{parallel composition without synchronization rule})$

4.
$$\frac{\mathbf{P} --^a--\!> \mathbf{P'} , \; \mathbf{Q} --^a--\!> \mathbf{Q'}}{\mathbf{P}\|_A \cup A\|\mathbf{Q} --^a--\!> \mathbf{P'}\|_A \cup A\|\mathbf{Q'}}$$
$(a \in A - \text{parallel composition with synchronization rule})$

5.
$$\frac{\mathbf{P_1} \cup \mathbf{P_2} --^u--\!> \mathbf{P'}}{\mathbf{P_1} \cup (\mathbf{P_2} + \mathbf{Q}) --^u--\!> \mathbf{P'} , \; \mathbf{P_1} \cup (\mathbf{Q} + \mathbf{P_2})--^u--\!> \mathbf{P'}}$$
$(\mathbf{P_1} \cup \mathbf{P_2} = \varnothing \text{ and } \mathbf{Q} \text{ is complete} - \text{choice rule})$

6.

$$\frac{P \text{ --}^u\text{--> } Q}{P[a/b] \text{ --}^u\text{--> } Q[a/b], P\{b\} \text{ --}^u\text{-->} Q\{b\}}$$

(renaming and hiding rules)

The *dex* function and the transition relation given above are the main building blocks to associate to each process term its operational semantics as a net system (1-safe marked net). To grant the maximal generality of our construction process, we introduce an abstract version of 1-safe marked nets, disregarding the names given to places.

Definition T. Weak isomorphism
Let $N_i = (A_i, S_i, \text{ --->}_i, M_{0i})$, i = 1,2, be two 1-safe labeled marked nets.
N_1 and N_2 are *weakly isomorphic*, $N_1 ==_{isom} N_2$, if $A_1 = A_2$ and there is a bijection β: $S_1 \text{ ---> } S_2$ such that $β(M_{01}) = M_{02}$, and, for all $M_1, M_2 \in [M_{0i}>$ and for all $u \in A_1 \cup \{τ\}$, $M_1 \text{ --}^u\text{--> } M_2$ if $β(M_1) \text{ --}^u\text{--> } β(M_2)$.
The bijection β is said a *weak isomorphism* between N_1 and N_2.

Weakly isomorphic 1-safe labeled marked nets are equal but for the renaming of the *places*. The equivalence classes of transition systems with respect to weak isomorphism are called *abstract 1-safe labeled marked nets*. The class of abstract 1-safe labeled marked nets is called *Anet*.

The operational net semantics for process terms is a mapping $N[[.]] : CProc \text{ ---->}$ *Anet* which assigns to every process term P the abstract 1-safe labeled marked net $N[[P]] = (A(P), Sequ, \text{ ---->}|_{A(P)}, dex(P))$ where ----> is the transition relation given above, and $\text{ ---->}|_{A(P)}$ is its restriction with respect to $A(P) : \text{ ---->}|_{A(P)} = \{P\text{--}^u\text{-->}Q \mid u \in \{τ\} \cup A(P)\}$.

It can be proved that $N[[.]]$ is well defined and has all the properties we require for semantic functions [4]. We will not enter into these details. To conclude this section let us show two examples of Petri net semantics of process terms.

The operational Petri net semantics of the process term is term, a.b.c.stop:{a,b,c}‖ d.b.e.stop:{d,b,e}, is given by the 1-safe labeled marked net of Fig. 3.8a, whereas the operational Petri net semantics of the recursive process term, μX.a.b.X, is given by the 1-safe labeled marked net of Fig. 3.8b, where the reader can see how recursive terms are manipulated by the *dex* function.

3.2.5   Additional readings

The literature on net theory and its application is very large. A good introductory text is Ref. [5]. A review of the history and application areas of Petri nets with suggestions for further reading referring to as many as 315 publications is contained in the tutorial paper listed as Ref. [36], which also presents properties and analysis methods, an introduction both to stochastic nets with application to performance modeling and to high-level nets with application to logic programming.

The proceedings of the second advanced course held in 1986 in Bad Honnef [37, 19], contain the state of the art of Petri nets theory, the presentation of tools supporting the design and analysis of Petri nets, a variety of applications and [19] also covers the relationship of Petri nets with other models of concurrency.

Two series of international conferences are held yearly. One is on application and theory of Petri nets; its proceedings are published by Springer-Verlag within the *Lecture Notes in Computer Science* series. The other one places emphasis on timed and stochastic nets and their applications to performance evaluation; its proceedings are published by the IEEE Computer Society Press.

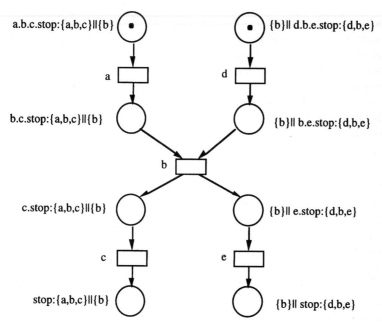

Figure 3.8a The operational net semantics of a.b.c.stop:{a,b,c}‖ d.b.e.stop: {d,b,e}

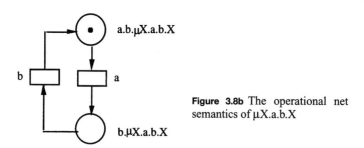

Figure 3.8b The operational net semantics of μX.a.b.X

Moreover, the subseries of the Lecture Notes in Computer Science, *Advances on Petri Nets*, periodically presents the most significant recent results in the application and theory of Petri Nets, as well as surveys of various topics (e.g., Refs. [20, 38]).

3.3 High-Level Net Models

The languages presented in the previous sections are suitable for representing concurrent systems where the control flow is—or is represented as—independent of data manipulation, i.e., fully hidden in action labels. This limited expressive power is not adequate for specifying real-world systems whose description requires the explicit data representation. In fact, modeling each value of a data structure via a distinguished place would make the size of the net unacceptable. To fulfil this requirement, combinations of basic models of concurrency with various approaches to data representation have been proposed (e.g. LOTOS [28] based on CCS [2] and ACT-ONE [39]). In the framework of Petri nets, at the beginning of 1980, *high-level* Petri nets were introduced. The main idea is to attach information to the tokens, which therefore are no longer "black dots." To specify the information carried by such *structured* tokens, *predicate/transition* (Pr/T) nets [21, 40] use first-order predicate logic and *colored Petri nets* [22, 23] a functional language such as Standard ML. More recently, algebraic specifications have been considered to achieve a more abstract specification of the structured tokens [25]. While Pr/T and colored nets are presently more used in practical applications (see Ref. [41] for a collection), the second one is gaining attention due to its capability to combine data abstraction with the control abstraction provided by the hosting theory of concurrency. In the following, the main features of high-level nets are presented through an example, specified according to a class of modular Petri nets [24, 42].

3.3.1   Some examples

Let us consider a set of 3 senders and 2 receivers: each receiver has a mailbox, consisting of a one-cell buffer, where senders asynchronously put messages. Their 1-safe net models (see Definition Q) are given in Fig. 3.9.

The model of the whole system is obtained by merging 3 + 2 copies of the nets given in Fig. 3.9, through the superposition of the transitions having the same name. Since each one of the N = 3 senders may choose in a nondeterministic way one of the M = 2 mailboxes to send it a message, then the resulting set, shown in Fig. 3.10, contains N*M = 6 copies of the transition *Send*. It would be quite impractical to draw the net model of a system with N = 30 senders and M = 20 receivers.

To reduce the size of the system model, in high-level nets it is possible to attach information to the tokens. In particular, tokens may have a distinguished individuality: in this way the subnets modeling different instances of the same component may be folded, as shown in Fig. 3.11. Let us notice that now arcs are labeled: S may assume a natural number value from 1 to 3, to identify the different senders; M and R a value from 1 to 2, to identify the possible receivers. It is easy to see that in this case it is trivial to extend these ranges of values to model systems with more nodes. The predicate (M = R), inscribed in the transition *Read* specifies that a receiver can read only its own mailbox. Places are implicitly

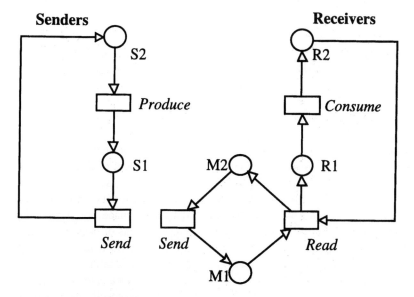

Figure 3.9 1-safe net models

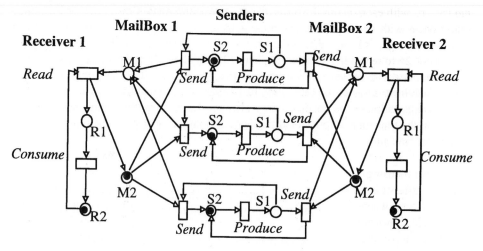

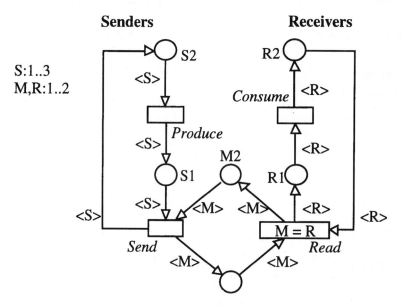

Figure 3.10 Resulting net

Figure 3.11 Subnets modeling different instances of the same component

typed by the type of the tokens occurring in them. In a more realistic specification, we would want senders to produce messages directed to a precise destination. This is modeled in High-Level Nets by associating further information with each token in Fig. 3.11. The tokens representing senders become triples <S,msg,dest>: the Sender S aims to send a message msg to the Receiver dest. Tokens representing Mailboxes and Receivers become couples <id,msg>. In the initial marking the three Senders <1,-,->+<2,-,->+<3,-,-> are in S2, the two Receivers <1,->+<2,-> are in R2 and their Mailboxes <1,->+<2,-> in M2. The dashes represent "don't care" values, i.e., the empty message and the unknown destination. Initially, transition *Produce* is enabled to occur according to three different *occurrence modes*, i.e., with each one of the three tokens in its input place. Moreover, it could concurrently occur with itself because the three occurrence modes are independent. Its occurrence with, e.g., <3,-,->, removes the token from S2 and adds to S1 the token <3,"Hello",2> to represent that Sender 3 wants to say "Hello" to Receiver 2. Now the transition *Produce* with senders 1 and 2 is still enabled. Moreover the transition *Send* is enabled involving the Sender token <3,"Hello",2> and the Mailbox token <2,->. In fact, these two tokens constitute an occurrence mode because they satisfy the predicate (M=dest) inscribed in the transition. The occurrence of *Send* removes <3,"Hello",2> from S1 and <2,-> from M2 and adds <3,-,-> to S2 and <2,"Hello"> to M1.

This quite informal description of the information carried by the different tokens can be formalized by using whatever data type specification. However, usually mailboxes are not one-cell buffers but lists of messages. This can be modeled in a straightforward way using algebraic nets, where the information associated with tokens is algebraically specified. Figure 3.12 shows the algebraic net of our example with the associated algebraic specification, given according to the OBJ3 syntax [43]. A list of messages is an abstract data type provided with the constant value empty list (*eL*) and with the usual operators *head, tail,* and *append* (@). The reader familiar with algebraic specifications will notice that this specification is simpler than usual: thanks to the associated net, the algebraic specification does not have to care about partial operations (head and tail). In fact, place M2 contains empty mailboxes, while place M1 is marked by tokens representing mailboxes with at least

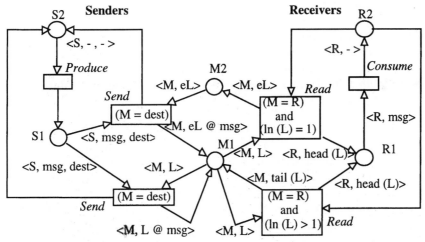

obj LISTOFMESSAGES is sort ListOfMessages.
    protecting MESSAGE NAT.
    op el : ->ListOfMessages.
    op _@_ : ListOfMessages Message -> ListOfMessages.
    op tail : ListOfMessages -> ListOfMessages.
    op head: ListOfMessages->Message.
    op len : ListOfMessages -> Nat.
    var m : Message. var n : Nat. var Lm: ListOfMessages.
    eq head (Lm@m) = if (Lm == eL) then m else head (Lm) fi.
    eq tail (Lm@m) = if (Lm == eL) the eL else (tail (Lm)) @ m fi.
    eq ln (eL) = 0.
    eq ln(Lm @ m) = 1 + ln (Lm).
endo.

Figure 3.12 An algebraic net

one message. Tokens "flow" from M2 to M1 by the occurrence of transition *Send* which appends the message "Hello" produced by the sender 3 to the right destination mailbox (<2,eL@"Hello">). At this point mailbox 2 may receive another message, let us say "Good Bye": the token <2,eL@"Hello"> "stays" in M1 changing according to the arc label into <2,(eL@"Hello")@"Good Bye">. The list length is now equal to 2. Therefore a first occurrence of transition *Read* leaves the token in M1 rewriting it as <2,eL@"Good Bye"> according to the algebraic specification of the operator *tail* : tail((eL@"Hello")@"Good Bye") -----> (tail(eL@"Hello"))@"Good Bye" -----> eL@"Good Bye".

3.3.2 Algebraic nets

In the following we give the definition of the net systems used in the previous examples: namely we introduce SPEC-inscribed nets, i.e., the more general class of algebraic high-level nets, and OBJSA nets. The following definition is a shortened version of the one given in [25].

Definition U. SPEC-inscribed net
Given an algebraic specification SPEC=(S,OP,E), a *SPEC-inscribed net* is a triple (N, ins, E) with an initial marking M_0, where:

- N=(P,T,F) is the underlying net.
- ins=(φ, λ,η) is a SPEC-inscription of N where:
  - φ is a sort assignment which "types" places by sorts in S, and
    for each transition t, λ is a φ-respecting arc labeling, i.e. it labels the transition surrounding arcs with a formal sum of terms of the corresponding sort;
  - η inscribes each transition with a predicate, i.e. a boolean operator whose arguments are the variables occurring in the transition labeling.
- M_0 associates with each place a multiset of tokens, i.e., of ground terms of the corresponding sort.

OBJSA Nets are a subclass of SPEC-inscribed nets stressing the possibility of building the system model through composition of its (sequential nondeterministic) components and encourages the incremental development of the specification and its reusability. To this aim OBJSA nets take *superposed automata nets* [42] as the underlying net model and adopt OBJ [43] as the algebraic specification language. All the nets presented in this chapter are OBJSA nets. Places are partitioned into disjoint classes and typed accordingly, to represent the different components of the modeled system. The tokens have a fixed, algebraically specified structure. Each token is structured in a name part which cannot be modified by transition occurrence, and a data part which can be modified according to arc and transition labeling.

In the following definition, given a specification SPEC=(S,OP,E), we extend it to allow the coupling of sorts, by introducing for each couple of sorts N, D∈ S, a sort $2Tuple_{N,D}$, a constructor operator $Make2Tuple_{N,D}$ and two projectors, $1^*_{N,D}$ and $2^*_{N,D}$, respectively, for the first and second elements of the couple. In OBJ this corresponds to instantiate the predefined parameterized OBJect 2TUPLE with suitable actual parameters. I

Definition V. OBJSA component
Given an algebraic specification SPEC=(S,OP,E), *an OBJSA component* is a SPEC-inscribed net (N, ins, E) with an initial marking M_0, where:

- $N=(P,T,F,W,\Pi)$ is the underlying net, in particular it is an *Extended SA* net where: places and transitions are partitioned in two disjoint sets denoting open and closed places (OP and CP) and transitions (OT and CT); Π is a partition of P into disjoint classes $\Pi_1, \ldots,$ Π_m; each arc in F is weighted by a function $W : F \rightarrow$ Nat such that $\forall\, i\ (1 \le i \le m)$:

$$\Pi_i \subseteq OP \text{ exor } \Pi_i \subseteq CP$$

and

$$\forall\, t \in T: \sum_{p \in (\Pi_i \cap\, \bullet t)} W(p,t) = \sum_{p \in (\Pi_i \cap\, t\bullet)} W(t,p)$$

the nets generated by the classes of Π are called *elementary subnets* of N;
- $ins=(\varphi, \lambda, \eta)$ is a SPEC-inscription of N where:
  - φ is a sort assignment which "types" places by sorts in S; in particular φ is Π-respecting and $\varphi(p)$ is a sort $2Tuple_{N,D}$ where N and D are sorts in S which specify respectively the token name and the token data part;
  - for each transition t, such that $^\circ t = \{p_1, \ldots, p_a\}$ and $t^\circ = \{p'_1, \ldots, p'_b\}$, λ is a φ-respecting arc labeling, i.e. it labels the transition surrounding arcs: $f=(p_i,t)$ with a formal sum $y_{i,1}<+>\ldots<+>y_{i,W(f)}$ of as many variables of the sort attached to p_i, as the weight of the arc is; let X_t denote the tuple of variables labeling the transition input arcs; $f=(t,p_k)$ with a formal sum $y'_{k,1}(X_t)<+>\ldots<+>y'_{k,W(f)}(X_t)$ of as many terms of sort (p_k) as the weight of the arc is; for each $y_{i,j}\ \exists!\ y'_{k,w}(X_t)$ such that $name(y_{i,j}) == name(y'_{k,w}(X_t))$.
  - η inscribes each transition with a predicate $\eta(t)(X_t)$.
- M_0 associates with each place a multiset *(without multiplicity)* of tokens, i.e., of ground terms of the corresponding sort, *under the condition that if a token name occurs in the marking of a place, it must not appear in the marking of any other place of the same elementary component)*.

Definition W. OBJSA net

- An OBJSA component is said to be *closed* (resp., *open*) if the underlying net N is closed, i.e., if $OT = \varnothing$ (resp., open, i.e., if $OT \ne \varnothing$).
- An *OBJSA net system*, or *OBJSA net* for short, is an OBJSA *closed* component.

OBJSA nets are obtained by composing OBJSA open components, through transition fusion. Details on the composition are omitted for sake of space; the interested reader finds the definition of the composition in Ref. [24]. As usual for high-level Petri nets in general, and for SPEC-inscribed nets and OBJSA nets in particular, when (1) the input places of a transition contain at least as many tokens as specified by the labeling of the transition input arcs; and (2) there exists a binding of these tokens to the free variables occurring in the transition input arcs such that the boolean condition (predicate) associated with the transition is satisfied, then the transition is enabled in the *occurrence mode* given by the tuple of tokens. Its occurrence removes them from the input places and adds to output places the tokens obtained according to the labeling of the transition output arcs, under the same binding. Transitions can occur concurrently when they are enabled in disjoint (i.e., nont involving the same individual token more than once) occurrence modes.

The development of OBJSA components, their composition and simulation are supported by the environment ONE *(OBJSA Nets Environment)* which has been developed using DesignML [44] and includes the OBJ3 interpreter [43].

3.3.3   Further readings

Reference [41] collects the more significant papers on high-level nets. The reader can find there the presentation of predicate/transitions nets, colored nets, SPEC-inscribed nets, the first definition of OBJSA nets and, moreover, two further algebraic net formalisms: Segras [45] and many-sorted high-level nets [46]. Moreover, Ref. [41] collects papers on high-level nets analysis techniques and applications developed using high-level nets. An up-to-date list of the tools for high-level nets is available at petri@crim.ca. An updated reference for colored Petri nets is Ref. [47].

After their definition in Ref. [24], OBJSA nets had a number of improvements. The current state of the art can be found in technical papers available via ftp anonymous at ftp.ghost.dsi.unimi.it. In particular [48] presents OBJSA closed and open components and their composition, and sketches the main functionalities of the environment. References [48] and [49] present OBJSA nets in an object-oriented perspective. The environment ONE can be obtained upon request to the authors.

3.4   Conclusions

3.4.1   Research directions

Formal methods for concurrency is still a hot topic in theoretical computer science. Despite the relevant results that have been already achieved—this chapter offers a largely incomplete picture of them—there are several open problems giving rise to interesting research directions. It can be useful to recall those we consider the most relevant to suggest to the interested reader research areas to look at in the next years.

3.4.1.1 Mobile processes. The process algebras to which we have made reference in Section 3.1 are such that processes have a static structure. It can be interesting to extend them in such a way that processes may have a changing structure, i.e., that they are able to exchange information modifying their mutual links. The idea is to offer effective ways to model dynamically evolving scalable concurrent systems, i.e., concurrent systems that are in some sense mobile. In the mid 1970s, Carl Hewitt already had introduced mobility into his actor model [51], but only after a long maturation period some algebraic calculi with mobility appeared. Among them we must mention the π-calculus developed by Robin Milner, Joachim Parrow, and David Walker [52], which is now the reference point for all those studying the issues raised by mobility.

3.4.1.2 The categorical framework. Category theory is going to become one of the most widely adopted algebraic frameworks for studying concurrent systems. It offers in fact a nice and elegant language for establishing the correspondence between a system specification and its semantics, avoiding cumbersome proofs for the equivalence and/or congruence between (a canonical representative of) the former and latter.

Moreover, it offers an adequate abstraction level to directly compare interleaving and concurrent models of distributed systems and/or different theoretical frameworks for dealing with concurrent systems, from process algebras to Petri nets, to various types of logics (temporal logics, observational logics, etc.). Good starting points for this research area can be Refs. [53] and [54].

Category theory has been also used by J. Meseguer and U. Montanari to characterize Petri nets in algebraic terms [55].

3.4.1.3 The synthesis problem. In Section 3.1, we introduced a (sequential) semantic function associating to each process term a transition system, and in Section 3.2 we introduced a (concurrent) semantic function associating to each process term a labeled net system, respecting retrievability, i.e., such that its sequential semantics is isomorphic to the sequential semantics of the process term.

We are therefore able to associate in a straightforward and consistent manner a transition system to any concurrent system specified by means of both a process term or a labeled net system.

Quite recently, G. Rozenberg, with Ehrenfeucht [56] and M. Nielsen and P.S. Thiagarajan [57], has proposed a new approach allowing one not only to associate to each net system a transition system describing its sequential behavior, but also to associate to each (elementary) transition system a net system whose sequential behavior is described by that transition system, i.e., to synthesize a net system from a description of its sequential behavior. The work of Rozenberg and co-workers, which is limited to the basic class of net systems, namely the *elementary set systems* (see Section 3.2.2) and to the associated basic class of elementary transition systems, is based on the notion of Region. A region is a subset of the state-space of a transition system uniformly traversed by transitions with the same action name. The regions of an elementary transition system can be in fact considered the places of the (saturated) elementary net system whose behavior is described by that elementary transition system.

A synthesis result for asynchronous net systems has been proved by Zielonka in a different way [58]. The comparison between the two approaches will allow a deeper understanding of the synthesis problem as well as enlarge its applicability.

The work of Rozenberg and co-workers has excited great interest, giving rise to various promising research lines. On the one hand, how the synthesis procedure can be improved in order to make it more efficient in terms of time consumption and of dimensions of the synthesized net system [59] is being investigated. On the other, the region concept is being studied in order to extend it so that it can give rise to synthesis procedures for larger classes of net systems. The extensions of the region concept are generally oriented to move from a traversing to a gradient property for transitions with the same name; while Mukund has shown how with an extended region notion it is possible to synthesize P/T net systems [60], in Ref. [61] generalized regions are analyzed as a means of evaluating synchronic distances on transition systems. The generalized regions of a transition system have nice algebraic properties, whose investigation will open new possibilities in terms of efficient computation of concurrent system properties.

3.4.1.4 *Concurrency and object-orientation.* Concurrent object-oriented programming (COOP) is today one of the hottest topics in the development of adequate programming environments for MIS, telecommunications, and other key segments of the research in computer science. Problems of interest arise in the areas of object distribution, concurrency mechanisms, and methodological approaches, where formal methods for concurrent system design have challenging testbeds of their expressive power and usability. *Communications of the ACM* has dedicated a large space to COOP on different occasions [62, 63].

3.4.2 Summary

At the end of this chapter, having introduced the reader to many approaches, concepts and formalisms without offering the chance for a deeper understanding of any of them, it can be useful to summarize the most relevant issues we have dealt with and omitted.

Formal methods for concurrency are still a research area, where different approaches are under development, in competition with one another.

We have dedicated Section 1 to process algebras, which after CCS [2] represent the main algebraic framework for concurrent systems specification and analysis.

We have dedicated Sections 2 and 3 to Petri nets [5], which, also thanks to their graphical representation, are the most widely used theoretical framework for concurrency.

We have omitted, for space reasons, any introduction of the various attempts to define logics for concurrent systems specification. The interested reader can read Refs. [31] and [64] to get some insight on this topic.

As mentioned in the introductory pages, the whole chapter is a rather superficial survey of a rich and interesting research area. The references concluding it can help the reader to address it in greater depth.

3.5 References

1. C.A.R. Hoare. 1978. Communicating sequential processes. *Communications of the ACM,* Vol. 21, No. 8, 666–677.
2. R. Milner 1980. *A Calculus of Communicating Systems*, Lecture Notes in Computer Science 92. Berlin: Springer-Verlag.
3. J. C. M. Baeten and P. Weijland. 1991. *Process Algebra*. Cambridge, U.K.: Cambridge University Press.
4. E. R. Olderog. 1991. *Nets, Terms and Formulas*. Cambridge, U.K.: Cambridge University Press.
5. W. Reisig. 1985. *Petri Nets: An Introduction*. Berlin: Springer-Verlag.
6. C. A. Petri. 1962. Kommunikation mit Automaten. *Schrift No. 2*, Rheinisch-Westfaelisches Institut fuer Instrumentelle Mathematik and der Universitaet Bonn. (Also: 1966. Coommunication with Automata. *Technical Report RADC-TR-65-377*, Vol. 1, Suppl. 1. New York: Griffiss Air Force Base.)
7. N. Francez. 1986. *Fairness*. Berlin: Springer-Verlag.
8. G. Rozenberg and P.S. Thiagarajan. 1986. Petri Nets: basic notions, structure, behaviour. In *Current Trends in Concurrency*, J.W. de Bakker, W.P. de Rover and G. Rozenberg, eds., Lecture Notes in Computer Science 224. Berlin: Springer-Verlag, 585–668.
9. P.S. Thiagarajan. Elementary net systems. In Ref. [37], 26–59.
10. C.A. Petri. 1977. Non-sequential processes. *ISF-Bericht ISF-77-5*. Bonn: ISF-GMD..
11. R. Milner. 1989. *Communication and Concurrency*. Englewood Cliffs, N.J.: Prentice Hall.
12. R. Milner. 1990. Operational and Algebraic Semantics of Concurrent Processes. In *Handbook of Theoretical Computer Science, Volume B: Formal Models and Semantics*, J. Van Leeuwen, ed. Amsterdam: Elsevier, 1201–1242.
13. R. De Nicola and M. Hennessy. 1984. Testing equivalences for processes. *Theoretical Computer Science*, Vol. 34, No. 1, 83–134.
14. S.D. Brookes and A.W. Roscoe. 1985. An improved failure model for communicating sequential processes. In *Proceedings of the NSF-SERC Seminar on Concurrency*, Lecture Notes in Computer Science 197. Berlin: Springer-Verlag, 281–305.
15. P. Degano and U. Montanari.1987. Concurrent histories: a basis for observing distributed systems. *J. of Computer and System Science*, Vol. 34, 442–461.
16. E. R. Olderog and C. A. R. Hoare. 1986. Specification-oriented semantics for communicating processes. *Acta Informatica*, Vol. 23, 9–66.
17. M. Hennessy. 1988. *Algebraic Theory of Processes*. Cambridge Mass.: MIT Press.
18. R. Milner, J. Parrow, D. Walker. 1992. A calculus of mobile processes, I & II. *Information and Computation*, Vol. 100, 1–77.
19. W. Brauer, W. Reisig, G. Rozenberg, eds. 1987. *Petri Nets: Applications and Relationships to Other Models of Concurrency*. Lecture Notes in Computer Science 255. Berlin: Springer-Verlag.

20. L. Pomello, G. Rozenberg, C. Simone. 1992. A survey of equivalence notions for net based systems. In *Advances in Petri Nets 1992*, G. Rozenberg, ed. Lecture Notes in Computer Science 609, Berlin: Springer-Verlag 410–472.

21. H. J. Genrich and K. Lautenbach. 1991. System modelling with high–level Petri nets. *Theoretical Computer Science*, Vol. 13, 109–136. (Also in Ref. [41].)

22. K. Jensen. 1981. Coloured Petri nets and the invariant method. *Theoretical Computer Science* Vol. 14, 317–336. (Also in Ref. [41].)

23. K. Jensen. 1990. Coloured Petri nets: A high level language for system design and analysis. In *Advances in Petri Nets*, G. Rozenberg, ed. Lecture Notes in Computer Science 483. Berlin: Springer-Verlag, 342–416, (Also in Ref. [41]).

24. E.Battiston, F. De Cindio, G. Mauri. 1988. OBJSA nets: a class of high level nets having objects as domains. In *Advances in Petri Nets 88*, G. Rozenberg, ed. Lecture Notes in Computer Science 340. Springer-Verlag, 20–43. (Also in Ref. [41].)

25. W. Reisig. 1991. Petri Nets and Algebraic Specifications. *Theoretical Computer Science*, Vol. 80, 1–34. (Also in Ref. [41].)

26. J. Vautherin. 1987. Parallel Systems specification with coloured Petri nets and algebraic specification. In *Advances in Petri Nets 87*, G. Rozenberg, ed. Lecture Notes in Computer Science 266. Berlin: Springer-Verlag, 293–308.

27. J. C. M. Baeten, ed. 1990. *Applications of Process Algebras*. Cambridge, U.K.: Cambridge University Press.

28. P. H. J. Van Eijk, C. A. Vissers, M. Diaz, eds.. 1989. *The Formal Description Technique LOTOS*. Amsterdam: North-Holland.

29. C.A.R. Hoare. 1985. *Communicating Sequential Processes*. Series in Computer Science. Englewood Cliffs, N.J.: Prentice Hall.

30. C.A.R. Hoare, ed. 1990. *Developments in Concurrency and Communication*. Reading, Mass.: Addison-Wesley.

31. K. M. Chandy and J. Misra. 1988. *Parallel Program Design—A Foundation*, Reading, Mass.: Addison-Wesley.

32. S.D. Brookes, C.A.R. Hoare, A.W. Roscoe. 1984. A theory of communicating sequential processes. *Journal of the ACM* , Vol. 31 No. 3, 560–599.

33. E. Best and C. Fernandez. 1988. *Nonsequential Processes, a Petri Net View*. Berlin: Springer-Verlag.

34. E. Best. Structure theory of petri nets: the free choice hiatus. In Ref. [37], 168–206.

35. W. Reisig. Place/transition systems. In Ref. [37], 117–141.

36. T. Murata. 1989. Petri nets: Properties, analysis and applications. *Proceedings of the IEEE*, Vol.77, No. 4, 541–580.

37. W. Brauer, W. Reisig, G. Rozenberg, eds. 1987. *Petri Nets: Central Models and Their Properties*. Lecture Notes in Computer Science 254. Berlin: Springer-Verlag.

38. L. Bernardinello and F. De Cindio. 1992. A survey of basic net models and modular net classes. In *Advances in Petri Nets 1992*, G. Rozenberg, ed. Lecture Notes in Computer Science 609. Berlin: Springer-Verlag 304–351.

39. H. Ehrig, W. Fey, H. Hansen. 1983. ACT-ONE: An algebraic specification Language with two level of semantics. *Technical Report TR 83-01*. Berlin: Tech. Univ. Berlin.

40. H. Genrich. Predicate/transition nets. In Ref. [37], 207–247.

41. K. Jensen, G. Rozenberg, eds.. 1991. *High-level Petri Nets. Theory and Application*. Berlin: Springer-Verlag.

42. F. De Cindio, G. De Michelis, L. Pomello, C. Simone. 1982. Superposed Automata Nets. In *Application and Theory of Petri Nets*, C.Girault, W.Reisig, eds.. IFB 52, Berlin: Springer-Verlag, 269–279.

43. J. A. Goguen, T. Winkler. 1988. Introducing OBJ3. *Technical Report SRI–CSL–88–9*, SRI International, Computer Science Lab.

44. J. Malhotra. 1990. *DesignML: Reference Manual*. Cambridge, Mass.: Meta Software Co.

45. B. Kramer and H. W. Schmidt. 1987. Types and modules for net specifications. in *Concurrency and Nets*, K.Voss, H.Genrich, G.Rozenberg, eds. Berlin: Springer-Verlag, 269–286. (Also in Ref. [41].)

46. J. Billington. Many–sorted high–level nets. In *Proc. Third Int. Workshop on Petri Nets and Performance Models*, Kyoto, Japan. New York: IEEE Computer Society Press,166–179. (Also in Ref. [41].)

47. K. Jensen. 1992. *Coloured Petri Nets: Basic Concepts, Analysis Methods and Practical Use. Vol. 1: Basic Concepts*. Berlin: Springer-Verlag.

48. E. Battiston, F. 1994. De Cindio, G. Mauri. 1994. A Class of Modular Algebraic Nets and its Support Environment. *Technical Report i/4/105*, CNR, Progetto Finalizzato, Sistemi Informatici e Calcolo Parallelo.

49. E. Battiston, F. De Cindio. 1993. Class orientation and inheritance in modular algebraic nets. In *Proceedings of 1993 International Conference on Systems, Man and Cybernetics*, Vol.2, Le Touquet, France.

50. C.A.R. Hoare. 1991. The transputer and Cccam: a personal story. *Concurrency: Practice and Experience,* Vol. 3, No. 4, 249–264.

51. C. E. Hewitt. 1977. Viewing control structures as patterns of message passing. *Journal of Artificial Intelligence,* Vol. 8, 323–364.

52. R. Milner, J. Parrow, D. Walker. 1992. A Calculus for mobile processes. I, II. *Information and Computation,* Vol. 100, 1–77.

53. M. Nielsen and G. Winskel.1993. Categories of models for concurrency. In *Handbook of Logic in Computer Science.* New York: Oxford.

54. G. L. Ferrari and U. Montanari. 1990. Towards the Unification of Models for Concurrency. In CAAP '90, Lecture Notes in Computer Science 431. Berlin: Springer-Verlag, 162–176.

55. J. Meseguer and U. Montanari. 1990. Petri nets are monoids. *Information and Computation,* Vol. 88, 105–155.

56. A. Ehrenfeucht and G. Rozenberg. 1990. Partial (set) 2-structures. I, II. *Acta Informatica,* Vol. 27, No. 4, 315–368.

57. M. Nielsen, G. Rozenberg, P. S. Thiagarajan. 1992. Elementary transition systems. *Theoretical Computer Science,* Vol. 96, No. 1, 3–33.

58. W. Zielonka. 1987. Notes on Finite Asynchronous Automata. *RAIRO,* Vol. 21, No. 2, 99–135.

59. L. Bernardinello. Synthesis of net systems. In *Application and Theory of Petri Nets.* Lecture Notes in Computer Science 691. Berlin: Springer-Verlag, 89–105.

60. M. Mukund. 1992. Petri nets and step transition systems. *International Journal of Foundations of Computer Science,* Vol. 3, No. 3.

61. L. Bernardinello, G. De Michelis, K. Petruni, S. Vigna. 1994. On Synchronic Structure of Transition Systems. *DSI Internal Report.* Milano: University of Milano.

62. G. Agha. 1990. Concurrent object-oriented programming. *Communications of the ACM,* Vol. 33, No. 9, 125–141.

63. 1993. Special Issue: Concurrent Object-Oriented Programming—Introduction. *Communications of the ACM,* Vol. 36, No. 9, 34–138.

64. A. Pnueli.1989. Temporal logic. In *Proceedings of the University of Texas Year of Programming, I, Concurrent Programming,* C. A. R. Hoare, ed. Reading, Mass.: Addison-Wesley.

4

Complexity Issues in Parallel and Distributed Computing

E. V. Krishnamurthy

4.1 Introduction

The *complexity theory* estimates the amount of computational resources (such as the computing time and the storage space) needed to solve a problem, ignoring the details of implementation of the algorithm in a specific computing machine. To be more precise, this theory provides a functional relationship between the input size of a *solvable problem* and the amount of resources needed for solving that problem on a universal computational model; such a functional relationship is expressed using the model-based measures for the storage space and computing time. Using the model-based measures, the complexity theory then constructs a spectral classification chart for the resource bounds needed for solving different classes of problems.

In this spectral classification chart, the far end of the spectrum is the class of intractable problems whose solutions require exponentially growing computational time and space resources; the intermediate zone is the class of problems solvable by consumption of a feasible amount of time and space resources that grows like a polynomial function; and the near end of the spectrum is the class of problems that can be solved ultrafast using a large number of processors, consuming sublinear or logarithmically bounded time and space resources.

4.1.1 Complexity theory—its basic aims

The *three* basic aims of complexity theory are:

1. *Introducing a notation to specify complexity.* The first aim is to introduce a mathematical notation to specify the functional relationship between the input size of the problem and the consumption of computational resources, e.g., computational time and memory space.
2. *Choice of a machine model to standardize the measures.* The second aim is to specify an underlying machine model to prescribe an associated set of measures for the consumption of resources. These measures are standardized so that these are invariant

for possible variations in algorithms, methods of problem formulations, specific problem instances, variations in the choice of models, and the nature of resources used. The basic model chosen is the classic *Turing machine* (TM) and the prescribed abstract space and time measures are respectively the number of cells used in the memory tape and the number of control moves; accordingly, the measures based on this model will be called *Turing measures*. The Turing measures are applicable to algorithms that have been traditionally used in sequential machines with a practically feasible amount of consumption of time and space resources (mathematically, this is represented by a polynomial function). Why a Turing machine is chosen as the basis and what are its consequences will be discussed in Section 4.2.

3. *Refinement of the measures for parallel computation.* Having obtained the Turing measures, the third aim is to understand how fast we can solve certain problems when a large number of processors are put together to work in parallel. To answer this question, based on common sense, we must somehow manage to relate the space (or time) measure in the Turing machine to time (or space) measure under parallelism; that is, the time and space complexities must somehow be traded for one another under transformation from sequential to parallel computation.

In Sections 4.3 and 4.4, we study these aspects and also refine the Turing measures for superfast algorithms that consume sublinear space and time resources. The refinement of the measures require the choice of newer more practical models—such as the *parallel random access machines* (PRAMs), *bulk-synchronous processor model* (BSP) that are better suited for the description and programming of parallel algorithms, and the circuit models with bounded fan-in (CKT) and unbounded fan-in (UCKT) that are better suited for the hardware realization of parallel algorithms.

4.1.2 Complexity notation

The first aim is to provide a good notation to specify the functional nature of the requirement of resources and also introduce the asymptotic measures to specify upper, lower and exact bounds on the complexity.

4.1.2.1 Resource-bounding functions. Let n and k belong to the set of nonnegative integers N and f(n) be a function from $(N \times N)$ to N representing a suitable function of the input size n (≥ 0) and k (≥ 0) a constant. We use the following resource-bounding functions:

constant: $f(n) \leq k$

log n: $f(n) \leq k \log n$

lin n: $f(n) \leq kn$

poly n: $f(n) \leq k.n^k + k$

Explin n: $f(n) \leq k.2^{k.n}$

Exp n: $f(n) \leq k.2^{n^k}$

Note that poly contains lin and constant. Also Explin is contained within Exp if the second exponential k is chosen to be 1. Also we may use a combination of the functions such as:

polylog n: $f(n) \leq k(\log n)^k$; also denoted by $k(\log^k n)$

doubleexp n = 2exp: $f(\exp n) \leq 2^{2^n}$

tripleexp n = 3exp: $2^{2^{2^n}}$

expoly n = $2^{\text{poly } n}$, **expolylog** n = $2^{(\log n)^k}$

4.1.2.2 Asymptotic measures. To specify the complexity measures (such as linear, logarithmic, exponential) the following mathematical notation is used. This notation is helpful for specifying the asymptotic order—namely, the upper bound O (worst case), the lower bound Ω (best case) or the exact bound Θ—of the complexities.

Let f and g be two functions defined on the set of natural numbers N; then:

1. $f(n) = O[g(n)]$: (called "*Big oh*" notation) means $f(n)$ is order at most $g(n)$ if and only if there are constants c > 0 and d > 0 such that, for all $n \geq d$, $f(n) \leq c\, g(n)$. That is O gives the upper bound ignoring constant factor; that is the *worst case* complexity. Mathematically, this means that for limit n tending to infinity, $f(n)/g(n) = c$. That is, f grows no faster than g.

    For example, if $g(n) = n^3/2$ and $f(n) = 37\, n^2 + 120n + 89$, then $f = O(g)$; but g is not $O(f)$.

2. $f(n) = \Omega[g(n)]$: (called "*Omega*" notation) means $f(n)$ is order at least $g(n)$ if and only if there are constants c > 0 and d > 0 such that, for $n \geq d$, $f(n) \geq c\, g(n)$. That is, Ω gives the lower bound ignoring constant factor or the *best case* complexity; this means f grows at least as fast as g.

3. $f(n) = \Theta[g(n)]$: (called "*Theta*" notation) means $f(n)$ is order exactly $g(n)$ if and only if $f(n) = O[g(n)]$ and $f(n) = \Omega[g(n)]$. That is, Θ gives the exact bound ignoring constant factor; that is the *exact bound*. That is f grows as fast as g.

For further mathematical properties of these functions, see Refs. [5] and [57].

4.1.2.3 Why choose a polynomial as the standard function? Among the different resource bounding functions, the polynomial is chosen as the standard function for the following reasons.

1. *Practicality*: The polynomial function n^m (mth degree polynomial of the input size n) grows relatively slowly compared to any exponential function of n. When the machine hardware speeds up by a constant factor k, for a given amount of processing time, a polynomial algorithm of degree m can handle another instance of the same problem with an input size larger by a multiplicative factor of mth root of k. If the input size n of another instance of the same problem is p-fold larger, the number of computational steps needed (or the processing time) increases by a multiplicative factor p^m.

    However, when the machine hardware speeds up by a factor of k, for a given amount of processing time, the exponential function m^n (m a positive integer) can permit only a marginal increase in the size of the input for another instance of the same problem by a small additive factor $\log_m k$. Also, if the input size of another instance of the same problem is p-fold larger, the number of computational steps needed (or the processing time) grows exponentially with an exponent p. Accordingly, the notion of polynomial time solvability has been taken as an equivalent to efficient solvability [13, 15, 29].

As an example, consider the polynomial function n^m; for $m = 2$, when the machine speed increases by a multiplicative factor of $k = 10$, we can take a new input size that is larger by a multiplicative factor equal to the square root of 10. However, for an exponential function m^n, for $m = 2$, when the machine speed increases by a multiplicative factor of $k = 10$, we can take a new input size that is only larger by an additive factor $\log_2 10$.

2. *Model interchangeability.* Because an increase in the machine speed by a constant results in a constant muliplicative factor, a problem solvable in polynomial time in one model is also solvable in another model in polynomial time of the same order O. This hypothesis is widely accepted and is called "extended Church-Turing thesis." This thesis forms the basis for defining complexity classes independent of specific models.

3. *Closure.* Unlike the exponential functions, the polynomial functions have a practically useful set of mathematical properties: they are closed under addition, multiplication and composition. These mathematical properties when translated into practice mean the following: if a composite program consists of many component programs, where each component program runs in polynomial time, then the composite program that uses a constant number of compositions or a constant number of recursive or iterative calls of its component programs can still run in polynomial time. Accordingly, within the specification of a polynomial function we include polynomials of polynomials and polynomials of logarithmic functions.

4. *Reducibility.* The polynomial function also provides a useful method to relate the complexities of any two different problems by a constructive transformation that maps an instance of the first problem into an equivalent instance of the second. (By an instance of a problem, we mean a particular case of the generic problem obtained by substituting values for the variables in the generic problem). Such a method is called *reduction,* and it provides a means for converting any algorithm that solves the first problem into an algorithm to solve the second problem. If we show that there exists a deterministic polynomial time algorithm that can be simulated in a TM to reduce one problem to another, we say that the two problems are polynomial-time reducible. If one of the problems has a polynomial time algorithm, then obviously due to the composition property of the polynomials the other problem is also polynomial-time solvable. Therefore, the relation polynomial reducibility is transitive. This transitivity relation of the polynomial reduction is a key concept used in complexity theory.

We observe that this transitivity of reduction also holds for transformations computable in logarithmic work space, because these are also closed under polynomial-time reductions.

4.1.3 Invariance of model and associated measures

To guarantee invariance, the machine model and the complexity measures should satisfy the following five principles, called the *invariance thesis:*

1. *Invariance across algorithms.* To achieve invariance across algorithms in solving a given problem we measure the problem complexity.
2. *Invariance across problems formulations.* To achieve invariance across problem formulations all problems are converted to decision problems—that is, to problems with

only yes or no answers. That is each problem is expressed as a boolean formula with arbitrary number of variables, which can only be true or false on substitution of the truth values to the boolean variables. The conversion of any solvable problem into a decision problem is achievable in polynomial time so that the polynomial reduction holds.

3. *Invariance across different instances.* To achieve invariance across different instances of a problem, we assign different possible truth values to variables in the problem, and encode these instances as a binary string with a defined length n. This length n is used as the input size and the complexity is measured as a function of this length.

4. *Invariance across resources* (e.g., energy, time). The invariance across resources is achieved by showing that the solution to the problem would demand the consumption of any abstract resource as described by a suitable growth function of the input size.

5. *Invariance across several possible computers.* The invariance across different brands of computers is achieved by choosing the standard computational model—the deterministic Turing machine (DTM) (and its nondeterministic or probabilistic versions or their generalizations) and the polynomial growth function measure. While any other model provided with features of a modern computer—a processing unit, read-only input tape, write-only input tape, a program and a memory unit—may appear attractive as a better alternative to a TM, it turns out that these are no more powerful than a TM to adequately describe the complexity classes under the polynomial growth measurement of resources. Furthermore, these models can be simulated by a TM or its variants at worst, requiring only a practically feasible amount of (polynomially bounded) resources. This result follows from the polynomial *model interchangeability* (Section 4.1.2.3), and it is the most important among the five basic principles of *invariance thesis* [56].

4.2   Turing Machine as the Basis, and Consequences

We now consider the second aim of the complexity theory, namely, the choice of an underlying machine model and the prescription of an associated set of measures for the consumption of resources. I turns out that the Turing machine serves as a very good basic machine for the following reasons.

4.2.1   Simplicity of abstraction

The Turing machine is the simplest model of an abstract machine. It consists of three basic components: a finite state machine, a read-write head, and an infinite tape memory marked-off into square cells, where each cell can hold only one symbol of the alphabet. The computation consists in initializing the scanning head to a given tape cell and a given state of the finite state machine. Then, depending on the symbol in the cell and the current state of the finite state machine, the tape head prints a new symbol or does nothing, and the machine changes to a new (possibly the same) state and moves the head left or right. This sequence of actions is called a *move*. If this move is unique for a current state of the finite state machine and the input symbol currently scanned by the head, then the TM is called a *deterministic* Turing machine (DTM); if, under these conditions, several alternative moves are possible, then the TM is called a *nondeterministic* Turing machine (NDTM). Turing showed that the DTM, though simple in its actions, can formalize the notion of an effective procedure. That is, the TM serves as the basic machine model to classify problems into two classes—the solvable (or the decidable) class, and the unsolvable (or undecidable) class.

The extra power of a TM for solving higher-level problems over other lower-power machines (such as finite state machines or push-down stack machines) arises due to its infinite tape memory. Hence, one can restrict the tape length or restrict the direction of the movement of the head to limit the power to simulate a lower order machine.

For example, if we restrict the head movement to one direction on a finite tape, the TM behaves like a finite-state machine. Thus, the tape is the fundamental resource on which the complexity study firmly rests by finding how often the access to the tape is needed, how frequent are the changes in the direction (or reversal) of head movement, and how many cells in the tape are scanned while solving a problem containing parameters of a given size.

4.2.2 Dual nature of computation

A Turing machine can simulate any special-purpose algorithm; this is the speciality property. Also it can as well simulate any other Turing machine; this is the generality or universality property. These dual features, namely, speciality and universality, make it a standard model for studying the theory of computation.

4.2.3 Inclusion of nondeterminism

As mentioned in Section 4.2.1, it is possible to add nondeterminism in a TM by providing several alternative moves and allowing the TM to choose one of these alternatives nondeterministically. This modified TM is called a *nondeterministic* TM (NDTM). The essential difference between the NDTM and DTM is the manner in which the strings are accepted or recognized. A DTM computation accepts a string if and only if there is a straight-line sequence of configuration to a final state. However, the NDTM computation can only be modeled by a branching tree. A computation halts if and only if there is a sequence that leads to an accepting state. The NDTM rejects an input if and only if there is no possible sequence of moves that leads to an acceptance. Thus, the NDTM carries out nondeterministic moves and performs a search through the tree of possibilities, called an OR-tree; in other words, it simulates the action of an existential quantifier, by guessing and verifying the guesses whether it is an accepting state. This ability of a NDTM to choose a path through the tree that leads to a correct solution may be thought of as a guess made by the NDTM. Accordingly, we say that an NDTM solves a problem if there are some sequences of guesses or a path in the tree of possibilities to reach the solution. While nondeterminism does not enlarge the class of solvable problems, it can help speed up computation of a solvable problem through guess work which provides for unbounded parallelism.

4.2.4 Realization of unbounded parallelism

The NDTM can be simulated using DTMs by allowing unbounded parallelism in computation. Each time a choice is made, the DTM makes several copies of itself. One copy is made for each possible choice. Thus, many copies of the DTM are executing at the same time at different nodes in the tree. The copy of DTM that reaches a solution early halts all other DTMs; if a copy fails, it only terminates. Hence, to simulate a NDTM, we need to use k^d copies of DTM, where k is the number of branches of the OR-tree at each level, and d is the depth of the tree. This means the NDTM is equivalent to an exponential number of DTMs working in parallel. That is, we need an exponential amount of time with a single DTM to do this job. Since the NDTM is assumed to have the magical gift, it can always guess the right choice, carry out singly all the work done by an exponential number of DTM on its own, and that, too, in polynomial time (without requiring exponential time). If we have any doubt and challenge the NDTM, it can provide us a convincing argument in polynomial time to verify the truth of the solution it has obtained. Thus, the NDTM is a guessing machine that, after having solved the problem, can provide us with a deterministic verification of the solution in polynomial time.

As an example, consider the problem of determining whether a positive integer is composite or not. If the NDTM can guess a factor, we can verify the truth in polynomial time using a DTM in polynomial time.

4.2.5   Provision of an easy abstract measure for resources

The measurement of abstract resources in a DTM can be carried out by counting each move as one time unit and each tape cell scanned as one space unit. For example, if for every input of length n bits, in solving a given instance of a problem the machine makes at most T(n) moves before halting on the solution, it takes a time bound T(n). Similarly, for every input of length, the head scans at most S(n) cells it has a space bound S(n). In the case of NDTM, the time complexity is the smallest number of moves required for an accepting sequence of moves, since there can be more than one accepting sequence; if there is no accepting sequence it is undefined. The time used is the depth of the tree and the space is the maximum (over all configurations in the tree) of the number of cells in use.

4.2.6   Describing the classes P and NP

The DTM classifies the two fundamental complexity classes of problems: those solvable in polynomial time, and those that are not. Those problems for which the DTM takes a time T(n) that is a polynomial in n is called *polynomial time computable problems* or *tractable problems.* This defines the class of problems called P and is denoted by

$$P = DTIME \, (poly(n))$$

When a decision problem cannot be answered "yes" by a DTM in polynomial time but can be answered "yes" by a NDTM in polynomial time, we say the problems belong to the NP-class; this definition is one-sided in the sense we are only concerned with the "yes" answer. Once the solutions are obtained by a NDTM using guesswork, their truth is verifiable by a DTM in polynomial time. The NP class is denoted by NP = NTIME(poly(n)). It is believed that NP contains P properly; but so far this has not been proved and it remains an open problem! In fact, whether the set difference NP-P is empty or not is an open problem and the relationships that arise between deterministic time/space and nondeterministic time/ space are not fully understood. Some known relationships are the containments:

$$DTIME \, (poly \, (n)) \subseteq NTIME \, (poly \, (n))$$

$$NTIME \, (poly \, (n)) \subseteq DTIME \, (Expoly \, (n))$$

4.2.7   Description of the relatives of NP

We can define the class co-NP that consists of those problems that are complement of the problems in NP. That is, co-NP is the class of decision problems for which a "no" answer can be given in polynomial time. Again this definition is one-sided but complementary to the NP class. It is not known whether $NP = co\text{-}NP$, nor is it known whether the problems in the intersecting class $NP \cap co\text{-}NP = P$.

An example in co-NP is the decision problem "whether a given number is composite," and an example of a decision problem in NP is "whether a given number is a prime." It so happens that both these decision problems are in $NP \cap co\text{-}NP$. Such problems are called *open* because they may be in P. It is believed that P is properly contained in this intersection $NP \cap co\text{-}NP$ (see Fig. 4.1).

4.2.8 Suggestion of the existence of NP-complete problems

The NP-class contains another class of problems called *NP-hard*. A problem X is called NP-hard, if every problem in NP can be polynomially reduced to X. If X itself is NP, then X is *NP-complete.* Thus NP-complete problems are contained within NP-hard problems. That is NP-hard problems are at least as hard as NP-complete problems; but they do not form an equivalence class. The NP-complete problems are mutually polynomially reduc-

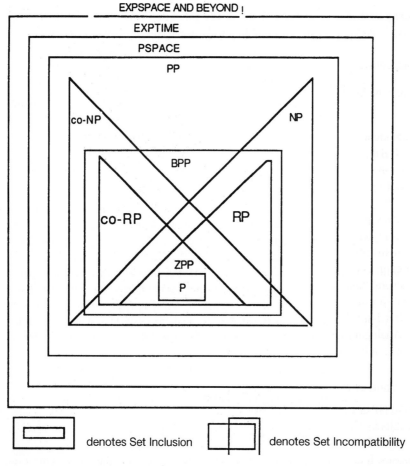

Figure 4.1 Higher-level Turing complexity classes

ible forming an equivalence class. Accordingly, if any one NP-complete problem is solved in polynomial time by a DTM, then all other problems in NP can be solved in polynomial time by a DTM and we will have NP = P!

4.2.9   Provision of a basis to prove NP-completeness

The TM model provides a basis to prove NP-completeness of a problem X using the following *four* essential steps:

Step 1   Convert X to a decision problem with an yes or no answer.
Step 2   Choose a known NP complete problem Y(NPC).
Step 3   Transform Y(NPC) to X, ensuring that the yes (or no) answer for an instance of X is an yes (or no) answer to the transformed instance.
Step 4   Show that the transformation in step 3 is polynomial bounded in time.

When the NP completeness proof was first provided by Cook in 1971, there were no NP complete problems and consequently the first proof was tedious. Cook established that the problem of satisfiability, to be defined below is NP complete.

The satisfiability problem. Let E be a Boolean expression in the conjunctive normal form (CNF); that is, it is an expression consisting of "**and**" (or conjunct) of several "**or**" clauses.

For example, consider E given by:

(x **or** y **or** z) **and** (x **or not** y **or not** w) **and** (**not** z **or** w) **and** (**not** x)

is in CNF.

A truth assignment for E is called *satisfiable* if there is some assignment of truth values T (true) and F (false) to the variables in E such that the resulting value of E is T. This E is satisfiable under the assignment x = z = w = F and y = T. On the other hand, the expression (x **or** y) **and** (**not** x) **and** (**not** y) is not satisfiable for any assignment of T and F to x and y, as it always results in the expression assuming the value F.

The satisfiability problem is to determine whether a given Boolean CNF expression is satisfiable. Its complementary problem, namely whether the given CNF is *unsatisfiable* is called the unsatisfiability problem. This problem is the same as the satisfiability problem of a Boolean expression in the disjunctive normal form (DNF), which consists of "**or**" (disjunct) of several "**and**" clauses. If there are n literals per conjunct in the satisfiability problem, we call it *n-satisfiability problem*. If n = 2, we call it *2-satisfiability problem;* and if n = 3, we call it *3-satisfiability problem.* Cook showed that the n-satisfiability for n ≥ 3 is NP-complete. Analogously, the 3-unsatisfiablity is co-NP complete. However, the 2-satisfiability problem is in P; see also Section 4.4.1.3.

4.2.10   Suggestion of the existence of the class P-SPACE

The problems in NP or co-NP, however, can be solved in polynomial amount of space by a DTM. This results in a larger class of problems called P-Space, which is the class of problems solvable by a DTM in polynomial amount of space. The P-Space contains the NP and co-NP classes. It is believed that this containment is strict; that is, there may be some problems belonging to P-Space but that are not in NP or co-NP. The class P-Space contains another class called *P-Space complete*, which are the problems such that if any one of them is in P-Space, then all problems in P-Space can be reduced to it. We can define the class P-Space-hard problem analogous to the NP-hard problem; that is, a problem X is P-Space hard if every problem in P-Space can be reduced to X. Accordingly, if X itself is in P-Space, then X is P-Space complete.

A practical problem that is P-Space complete is the problem of *quantified satisfiability*. Here is given a Boolean expression of the form:

$$\forall x_1 \exists x_2 \forall x_3 \exists x_4 \ldots \forall x_{(2n-1)} \exists x_{2n} B(x_1, x_2, \ldots, x_{2n})$$

where B is in conjunctive normal form ("and" of a set clauses made up of "or" of literals) is given. We are required to find whether the expression is satisfiable; that is, whether for all choices of truth values for x_1, there is a choice of truth value for x_2, such that for all choices for x_3, etc., and finally there is a choice for x_{2n} which leaves all the clauses true at least for one literal. The above problem is P-Space complete for 3 literals or more [49].

The existence of P-Space complete problems demonstrates that there are problems that can be solved in reasonable amount (polynomial) of sequential space, but not in reasonable sequential time even in nondeterministic TM. The P-Space complete problems stand as a strong evidence for computational intractability. It is highly likely that P-Space complete problems are harder than NP-Complete problems. It has been conjectured that the answers to P-Space problems are not likely to be verified in polynomial time using a polynomial length guess.

It turns out that all polynomial time problems can be solved in polynomial space in DTM, but not conversely. This means space as a resource seems to have a greater potential than time—this property could probably be attributed due to the reusability of space unlike time.

4.2.11 Generalized TM relating time and space measures

To define the class PSPACE and also relate the time and space measures, we can generalize the NDTM to a more powerful TM called an *alternating Turing machine* (ATM). The ATM is very similar to NDTM except for the rules used for acceptance of the input strings. The ATM time complexity classes closely correspond to TM and NDTM space complexity classes. Also, the ATM computations directly correspond to first-order formulas in prenex form with alternating quantifiers.

An ATM consists of an infinite memory tape and a finite state control as in a TM. The states of ATM are, however, partitioned into normal, existential, and universal states. The existential and universal states have two successor states. Normal states can be accepting, rejecting, or undefined. Given an initial state of the finite control and the input tape position for the head, the state of the machine is completely determined.

The NDTM rules allow a single machine move to reach several possible configurations; if there exists at least one configuration that is an acceptance state, the NDTM halts. The ATM behaves differently. In a universal state, a configuration can reach several configurations as in an NDTM, but now there is a difference; it leads to an acceptance (a rejection), if and only if **all of its** (if **one of its**) immediate branching descendents lead to an acceptance (a rejection). This corresponds to an AND tree or is equivalent to the simulating a universal quantifier.

In an existential state, a computation is accepting (rejecting) if and only if at least **one of its** (**all its**) immediate descendents is accepting (are rejecting). This corresponds to the simulation of an OR-tree or the existential quantifier. In a normal state, the computation leads to an acceptance (rejection) if and only if its unique descendent is accepting (rejecting). If none of the above rules are applicable, it can lead to an undefined state that corresponds to an infinite computation.

Thus, an ATM halting computation consists of a finite subtree of the infinite tree of computation containing the root, existential branches with one of the daughter subtrees, and universal branches containing all their daughter subtrees. We can measure the time (space) consumed by an ATM as the maximal time (space) consumed along any branch in this finite computation tree. As a result, the set of problems solvable in polynomial space in an ATM is a much larger class than that solved by an NDTM and corresponds to the problems that are solvable by a DTM in exponential time. The verification of the truth of the result cannot, however, be done in polynomial time as it could be done for the NDTM.

In summary, an NDTM is a weaker ATM in which there are no universal branches, while a DTM is a weaker NDTM having only one possible move at each step. Thus, the computation tree of a DTM is a straight-line decision tree, whereas that of an NDTM is an OR-tree and that of an ATM is an AND-OR tree. Note, however, that the ATM by definition is a deterministic device. Adding nondeterminism to an ATM does nothing to it. That is, NP-SPACE is the same as PSPACE!

It turns out that the class of problems solvable by an ATM in polynomial time is the class PSPACE solvable by a DTM in a polynomial amount of space; that is:

$$\text{ATIME[poly}(n)] = \text{DSPACE (poly}(n)) = \text{PSPACE}$$

Also,

$$\text{ALOGSPACE} = \text{P; APSPACE} = \text{DTM(EXPTIME); AEXPTIME} = \text{DTM(EXPSPACE)}$$

4.2.12   Describing the classes beyond P-Space

Beyond P-SPACE we have the class of Exppoly time, the class of problems solvable by DTM in time $O(c^{f(n)})$ for any constant c and polynomial f(n). The class of problems solvable in polynomial amount of space by an ATM is the class Expoly time in DTM; or, ASPACE(poly(n)) = DTIME(Expoly(n)). Thus, the class of problems solvable by a DTM in Exppoly time contains both NP and PSPACE.

Beyond the class of Expoly time we have the class of Expoly space; the class of problems solvable by an ATM in Expoly time is the class Expoly space problems in DTM; or ATIME(Expoly(n)) = DSPACE(Expoly(n)). For details see Ref. [20], and also see Fig. 4.1.

4.2.13   Describing the probabilistic classes

There are certain problems for which nondeterministic polynomial time algorithms are known, but deterministic polynomial time algorithms are not yet available. It is tempting to explore whether such problems can be solved in polynomial time by replacing the concept of nondeterminism by the probabilistic concept and obtain answers with an arbitrarily small probability of error as desired. The fact that nondeterminism has a higher degree of uncertainty (or a lesser degree of knowledge) than a probabilistic distribution (which has a higher degree of knowledge) forms the philosophical basis for exploring the probabilistic approaches.

A direct application of the probabilistic approach is for the detection of primality of an integer, which is in NP [20]. A probabilistic algorithm can detect primality in k trials within a probability of error is less than $1/2^k$ (see Ref. [17]). Essentially, in this approach, we weaken the reliability of the answer in order to gain speed. Accordingly, it is interesting to ask whether a TM provided with the probabilistic choices could speed up the computation and describe the new probabilistic complexity classes.

The probabilistic choices can be introduced in a TM in three different ways:

1. by replacing the nondeterministic acceptance in NDTM by a probabilistic acceptance
2. by replacing the nondeterministic rejection in NDTM by a probabilistic rejection
3. by replacing the universal quantification, namely the "and" acceptance in an ATM, by a majority logic acceptance

The first two approaches result in the *random Turing machine* (RTM), and the *probabilistic Turing machine* (PTM), respectively. The third approach results in the *stochastic Turing machine* (STM).

We will now describe these three classes of machines, along with the complexity classes they represent.

4.2.13.1 Random Turing machine. A random Turing machine is an NDTM such that for each possible input string there are no accepting computations, or else *at least half* of all the computations are accepting. Thus the probability of getting an erroneous output "no" when the true answer is "yes" is 1/2 or less. The probability that we get an erroneous output "yes" when the true answer is "no" is zero. This is because it is still an NDTM, which certainly rejects the input for the true answer "no." The *one-sidedness in error probability* permits us to iterate the RTM computations independently k times to get the error probability in the "yes" answer arbitrarily close to zero. The class RP of problems solvable in polynomial time by an RTM satisfies $P \subseteq RP \subseteq NP$ (see Fig. 4.1).

We can also define the complement of RP (co-RP), for which the answer is "no" when there are no accepting computations or else *at least half* of the computations are accepting. Thus, the probability of getting an erroneous output "yes" in an RTM when the true answer is "no" is 1/2 or less. The probability that we get an erroneous output "no" when the true answer is "yes" is zero. This is because it is still an NDTM which certainly rejects the input for the true answer "yes." Again, due to the one-sidedness of definition, we can reduce the error probability in "no" answer arbitrarily close to zero by independent iterations. The methods used for solving the classes RP and co-RP correspond to the *Monte carlo methods,* and these have one-sided errors.

Using RP and co-RP, we can define the intersecting class $ZPP = RP \cap co\text{-}RP$ of problems that can be solved by RTM and always yields the correct answer in expected polynomial time (see Fig. 4.1). The class ZPP is also the class solvable by *Las Vegas algorithms* that produce correct answers in polynomial time. To realize this algorithm, we need to run both the RP and co-RP algorithms repeatedly with a pair of independent random numbers at each step.

4.2.13.2 Probabilistic Turing machine. A probabilistic Turing machine is an NDTM such that for each possible input string the answer is "yes" if more than half of the computations terminate in "yes" answers; and the answer is "no" if more than half the computations terminate in "no" answers. If the number of yes and no answers are equal to exactly half the answer is "don't-know."

The class of decision problems solvable by a PTM in polynomial time in which the answers have probability *greater than 1/2* is denoted by PP. If, in addition, we demand that the probability of the answer being correct is at least $1/2 + d$, for some fixed $d > 0$, then this class of problems solvable by a PTM in polynomial time is denoted by BPP. The classes P, ZPP, RP, BPP, PP and PSPACE satisfy the following relationship (see Fig. 4.1):

$$P \subseteq ZPP \subseteq RP \subseteq BPP \subseteq PP \subseteq PSPACE$$

4.2.13.3 Stochastic Turing machine. A stochastic Turing machine) is a probabilistic ATM in which the universal quantifier (or AND acceptance) is replaced by a majority acceptance; that is, if more than half of the AND-tree branches are accepting, then the answer is "yes." This definition implies that it can be no more powerful than ATM. It turns out that the class of problems solvable by STM in polynomial time (denoted by PPSPACE) is identical to the class PSPACE and hence $PP \subseteq PPSPACE$ (see Fig. 4.1).The STM forms a basis for devising zero-knowledge verification protocols [42] used widely in user authentication.

4.2.14 TM can simulation of a random access machine

The classical hypothetical TM consists of a sequential linear memory tape which needs to wrap around to access the information in cells. It is possible to modify this TM by using another TM that can simulate the random-access memory operations. This composite machine is called a *random access TM* (RACTM). Alternatively, one can think of a hypothetical TM which replaces the standard tape by a random access tape and a read-only register; this permits the head to address any cell in the tape using a index register and to receive the contents of the cell in one time step. This random-access memory can be added to ATM or NDTM. The random access variants of TM are useful for defining the fine structure complexity classes within P.

4.3   Complexity Measures for Parallelism

We now consider the third aim of the complexity theory—namely, to explore how fast we can solve certain problems when a large number of processors work in parallel. As mentioned earlier, the answer to this question is based on how we could relate the space and the time complexity measures. The pioneering attempt to relate space and time in computation was made by Charles Babbage. In the famous book, *Passages from the Life of a Philosopher,* published by Longman in 1864, Babbage asserts,

> It is impossible to construct machinery occupying unlimited space; but it is possible to construct finite machinery, and to use it through unlimited time. It is this substitution of the infinity of time for the infinity of space which I have made use of to limit the size of the engine and yet retain its unlimited power.

This thesis (called Babbage's thesis) states that time and space complexities are related and can be traded for one another. In other words, the product of computation time and the space occupied (number of processors) could serve as a guideline to measure the workload!

4.3.1   Parallel computation thesis

Relating TM space resource to parallel time resource—uses a variant of the Babbage's thesis called *the parallel computation thesis*. In simple terms, this thesis states that "whatever can be solved in polynomially bounded space in a TM using unlimited time can be solved in polynomially bounded time on a parallel machine using unlimited number of processors (or space) and conversely." More formally, we can say that a problem can be solved in time polynomial T(n) by a parallel machine (with an unlimited number of processors) if and only if it can be solved by a sequential machine in an amount of space which is a polynomial in T(n) (in unlimited time). The statement of this thesis is written formally thus:

$$\text{Sequential bounded (poly) space} = \text{Parallel bounded (poly) time}$$

Equivalently, this can be rewritten as:

$$\text{Sequential polylog space} = \text{Parallel polylog time}$$

This thesis, like the Church-Turing thesis, has not been proven. Using the parallel computation thesis, the question "whether there are intractable problems that can become tractable under parallelism" can be reformulated into an equivalent question, "whether complexity class PSPACE contains intractable problems."

The parallel computation thesis, however, is merely a guideline. To deal with matters of practical interest, we need to consider how real parallel machines are organized to satisfy technological and engineering constraints. We will discuss these aspects in the next subsection.

4.3.2   Introduction of new measures

In the basic TM model, we considered only the complexity of consumption of the abstract resources, and these were expressed in terms of time and space complexities and their relationships. In building real parallel machines with a large number of processors, we need to consider the complexity of solving a problem not only at the system or macro level but also at the basic subsystem level—including the component or micro level. This would permit us to exploit the full benefit of parallelism and achieve speedup at all possible levels. At the macro level, for example, we need to consider the message communication time and routing through the links for transfer of data among processors in a network, memory contention, and read-write conflicts. At the micro level, we need to consider factors such as energy dissipation, communication time through the links for transfer of data among processors, wiring space required, delay due to capacitive and inductive phenomena, switching energy dissipated as heat, and the geometry of layout used in the design of VLSI circuits (or chips). Therefore, to carry out the complexity analysis, it is necessary to know the manner in which the processors are organized—namely, whether they form a parallel system or a distributed system, and at what level the complexity reduction is to be achieved for a speedup.

In a parallel system, the processors are identical or homogeneous and communicate using a shared memory or a fixed interconnection network, and they work simultaneously in synchrony. In a distributed system, the processors are dissimilar and heterogeneous, and they work competitively or cooperatively and communicate through message passing. In the former case the speedup at micro level is more appropriate, while in the latter case the speedup at macro level is more relevant. The complexity factors involved and measures to be used for these two systems are widely different and need to satisfy engineering and technological constraints. The refinement of Turing measures and introduction of new measures at both macro and micro levels have been very active areas of research because of their commercial importance. In Section 4.4, we will describe the important parallel complexity models, such as the *parallel random access memory machines* (PRAMs) and the Boolean circuit model. Section 4.5 briefly considers the complexity measures for VLSI based on space (chip area) and computation time. In Section 4.6, we will consider the aspects of complexity in a distributed system at the macro level. The neural network paradigm and connectionist model are considered by many as important paradigms for fast parallel computation. Section 4.7 examines the complexity aspects of these new paradigms. Section 4.8 summarizes the newer developments in complexity theories that are machine-model independent. Finally, in Section 4.9, we provide some concluding remarks.

4.4 Parallel Complexity Models and Resulting Classes

In a synchronous parallel system (where all the identical processors work in synchrony under the control of a centralized clock), it is easy to measure the time in terms of clock cycles. In such a system, the communication time among the processes and processors is much shorter than the computation time, since the interprocessor communication links are predetermined and directly hardwired. Synchronous parallel computation can be used to solve many numerical and scientific problems; it is also the basis for different special-purpose computing techniques, such as systolic computing, pipelining, and data-parallel computing. Hence, significant research efforts have gone in speeding up synchronous computation both at high (or macro) and micro levels. These efforts have provided considerable insight on the fine structure within the class P and also the superfine structure of a subset of P called the *polynomial logarithmic class* (also known as *Nick's class* or NC). This class NC permits hardware realizable superfast parallel algorithms for many problems of practical importance—for example, sorting, weighted averaging, basic polynomial-matrix arithmetic, graph connectivity, and the evaluation of straight-line programs. We will now describe the fine structure of complexity classes within P-, which we shall call *parallel complexity classes* because they provide a basis for parallel speedup.

4.4.1 Fine-structure complexity classes within P

Earlier, we were concerned with problems in P class and problems outside the P. We now look at problems that are solvable in time and space whose growth functions are much slower than polynomials. That is the class of problems solvable by sublinear algorithms that can provide superfast solutions with space and time complexities described by poly-log functions. We will first classify them using DTM and ATM using sublinear space and time bounds.

4.4.1.1 Sublinear Turing complexity classes inside P. We now define classes that are much easier than P in the sense they require sublinear space and time resources in a Turing machine (see Figs. 4.2 and 4.3).

1. DLOGSPACE (denoted by L): L is the class of decision problems solvable by a DTM in a workspace bounded by O(log n).
2. POLYLOGSPACE (denoted by POLYLOG): POLYLOG are the classes of decision problems solvable, respectively, by a DTM in a workspace bounded by $O((\log n)^k)$ for k, any positive integer.

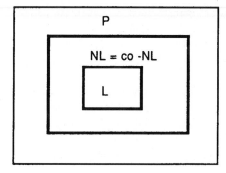

Figure 4.2 Fine structure inside P

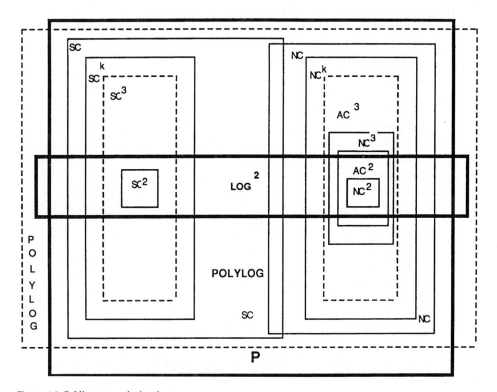

Figure 4.3 Sublinear complexity classes

3. SC (Steve's class): SC is the class of all decision problems solvable by a DTM with simultaneous polylog space and polynomial time bounds; this is denoted by an ordered pair of space-time resource requirements: $SC(\log n^{O(1)}, n^{O(1)})$; this class is also known as Steve's class in honor of Steve Cook, a complexity theorist. Also, SC^k is the class of decision problems solvable in simultaneous $(\log n)^k$ space and polynomial time in a DTM. Note that SC^k is contained in SC and $SC^1 = DLOG = L$.

4. NLOGSPACE (denoted by NL): NL is the class of decision problems solvable by a NDTM in a space bound $O(\log n)$.

5. Co-NLOGSPACE (denoted by co-NL): co-NL is the class of problems complementary to NL; this is identical to NL.

4.4.1.2 Complete problems under NL, P, NP, and PSPACE. The four important complexity classes L, NL, P, NP and PSPACE satisfy the following set inclusions:

$$L = DLOGSPACE \subseteq NL = NLOGSPACE = co\ NL \subseteq P \subseteq NP \subseteq PSPACE$$

It is not known whether these inclusions are proper (see Fig. 4.2).

All five classes are closed under polynomial-time reduction. The classes L and NL need to use the stricter logspace reduction. We can generalize the notion of hardness and completeness in each of the above classes of problems as follows:

Let x and y be two problems, and x **reduces** to y in polynomial time or logspace reduction; we denote this by: x **Red**(poly or log) y.

If, in addition, C denotes one of the five classes above and C **Red**(poly or log) x, then we say x is C hard. Also, we say x is C complete with respect to **Red**(poly or log) if C **Red**(poly or log) x and x is a member of C.

According to this definition, we can have NLOG-complete, P-complete, NP-complete, and P-space complete problems. Recall that in Section 4.2.9, we gave an example of NP-Complete problem.

4.4.1.3 NLOG completeness. The 2-unsatisfiability problem mentioned earlier in Section 4.2.9 is in P. In fact, the 2-unsatisfiability is co-NLOG complete, which is the same as NLOG complete.

Also, the graph accessibility problem, to determine whether a given directed graph has a directed path between any two specified nodes, is NLOG complete.

4.4.1.4 P-completeness. The first P-complete problem was identified by Cook in 1974. These are the problems that are highly resistant to parallelization. According to our earlier definition, a decision problem is P-complete if every problem in P is logspace reducible to it. As mentioned earlier, the logspace reducibility is transitive. One of the well known problems that has been shown P-complete is the *circuit value problem* (CVP). This problem determines the value of an output from a given combinational circuit with certain specified inputs. Formally, such a problem can be defined as a sequence of equations among m Boolean variables X1, X2,..., Xm. Each equation is of the form:

$$Xi = 1, X1 = 0 \text{ or } Xi = Xj * Xk$$

for some j, k, i where * is any one of the sixteen Boolean connectives (such as and, or, exclusive-or), and each variable appears exactly once on the left-hand side of each equation. Variables can occur any number of times in the right-hand side of the equations.

Given such equations, we may assign truth values to all the variables in order X1, X2,..., Xm. The condition j, k < i ensures that there are no cyclic dependencies among the variables. The CVP consists in finding the set of sequences of equations such that Xm =1. Ladner [31] proved that the CVP is P-complete.

4.4.2   Other refined models

We now introduce other refined models of computation popularly known as the PRAM (parallel random access machine) model and the circuit model. There are at least four basic reasons for using these refined models to study the parallel complexity issues. These reasons are given below:

1. *TM is idealistic, but not realistic.*

   The Turing machine model proposed in the 1930s is idealistic, too abstract, and is far removed from the present-day realistic design of computers. It was designed to demonstrate the notion of computability rather than the efficient programmability of algorithms using a general purpose high-level programming language for different data structures and transportability of the resulting software. Therefore, to study the complexity and performance of the present day computers, it is desirable to devise a refined model:

   (a) with the same power as a Turing machine, yet more closely resembling the logical structure of a modern parallel computer at the high level (where processors are sharing memory or are connected by a fixed interconnection network),

   (b) which is useful as a vehicle for the description, as well as programming of parallel algorithms which use different data structures such as arrays, lists, trees and graphs, and

   (c) which also provides a basis for studying the abstract measure for the running time and algorithmic performance hiding away the details of memory management and communication issues from the programmer.

2. *Invariance under sublinear forms*

   Within the sublinear functional forms, the model invariance assumption among the different models may not be valid. As a consequence, the resulting classes of problems may be incomparable with respect to their set-theoretic containments.

3. *Parsimony in problem reduction*

   Also for problems with sublinear algorithms we must replace the concept of polynomial reducibility by a stricter or parsimonious reduction scheme in order to define equivalence among problems, since the additional overhead of resources needed for the problem reduction should not be larger than the essential resource bound that defines the complexity class. As we earlier observed, the logarithmic work space reduction, which is transitive can be used as the basis, since logspace $\subseteq$ polynomial time , and hence is closed under functional composition.

4. *VLSI cost effectiveness*

   We want to obtain refined complexity measures at the micro level or at the basic hardware level that can predict the size and cost reduction, assured physical realizability and desired high speed. At the hardware level parallelism is employed at the expense of extra space for processing elements and extra wiring space. The Turing measures we described earlier cannot be used as effective cost measures for the circuits. The circuit model can describe the micro level (where gates and wires are packed into planar layers) more effectively and can serve as a basis for the physical realizability of an algorithm in a VLSI chip or *"efficient algorithms on silicon "*.

We will briefly describe the RAM, PRAM, and circuit models, their interrelationships, and their significance in the following subsections. For a detailed study on PRAM models, see Chapter 1, by L. Kronsjo, in this book.

4.4.3 The RAM and PRAM models

The *random access memory machine* is popularly known as RAM. A RAM model is a one accumulator computer in which instructions are not permitted to modify themselves. A RAM consists of a read-only tape, a single processor, a write-only output tape, a program, and a memory unit. The memory unit consists of an infinite number of registers, R0, R1,... each of which can hold an integer of arbitrary size. The program is a finite sequence of labeled instructions not stored in the memory but read from the input tape. The basic instruction set has the following form, where * denotes the standard binary operations, such as add, subtract, multiply, modulo, etc., on integers:

$R_i \leftarrow$ constant (load register with a constant)

$R_i \leftarrow R_i * R_k$ (binary operation)

$R_i \leftarrow R_{R_j}$ (indirect load)

$R_{R_i} \leftarrow R_j$ (indirect store)

Halt; go to label m if $R_i \geq 0$

A RAM program can simulate both the DTM and the NDTM. Also, the RAM model can be used for the measurement of complexity for sequential programs based on the assumption that each instruction takes unit time, memory references take zero time, and the space used is the maximum number of memory cells needed for any computational step. Because the loops ("for" and "while") essentially contribute to the execution time, the sequential time complexity is easy to compute using the RAM model.

The parallel version of a RAM is called PRAM model. It consists of N *deterministic and synchronous* RAM processors all operating and communicating via a shared memory. The number of processors is usually a function of the input size of a problem to be solved. Each processor has an infinite number of registers as in a RAM and a unique read-only memory partitioned into cells; the number of cells in the shared memory is assumed to be infinite.

A PRAM has a finite set of instructions, where each instruction is of the form:

1. perform a local computation
2. transfer of control
3. read a value from shared memory
4. write a value in the shared memory

The instructions of type (1) and (2) are local to each RAM, whereas types (3) and (4) instructions are communication commands among different RAMS. At each step, each processor is in some state; the actions and the next state depend upon its current state, the values read from its own memory and the shared memory cells. If we allow several processors to access the same shared memory cell at the same time, we must take care of the conflicts that may arise.

Although we can perform the read action by any processor and the write action by the same processor one after another at each time step, we need to resolve the conflicts that may arise between different processors for read or write action involving the same shared memory cell. The various variants of PRAM model differ in the manner in which the conflicts among the different processors are resolved for accessing a given shared memory cell for either read or write actions by different processors.

4.4.3.1 Variants of PRAMs

Variant 1: Exclusive read—exclusive write (EREW) PRAM

In a EREW model, each processor can either read or write in a given shared memory cell under mutual exclusion from other processors; that is no two processors can simultaneously access a given cell for a read or a write action. This model is also called PRAC(for Parallel random access computer) and is considered a realistic model.

Variant 2: Concurrent read-exclusive write (CREW) PRAM

In a CREW model, each processor can only have an exclusive access to a given cell for a write action; but two or more processors can perform read actions from a given memory cell at the same time. The design of the algorithm should use these features effectively.

Variant 3: Concurrent read-concurrent write (CRCW) PRAM

In a CRCW model, as in CREW model, the different processors can have simultaneous access to a given cell for read actions; however, to perform a write action in a given memory cell, the competition among the different processors have to be resolved so that they can write a unique value for the object in that cell. The conflict resolution can be done using three different approaches, and each approach defines a variant of the CRCW model:

Variant 3a: COMMON PRAM (identical-value write)

In this variant, each processor must write an identical value in the same cell of the shared memory; otherwise access is denied to all processors. This notion of identical-value write can be given different semantic interpretations. For example, it can be interpreted as the sum of all the values, or any other associative binary operation among the values (such as a prefix sum, weighted averages, logical operations, a majority or threshold logic decision). For a given interpretation, the algorithm is to be designed so that the identical—value write takes place as one indivisible or atomic action and is completed without any interruption; otherwise, it is illegal to use this PRAM model.

Variant 3b: ARBITRARY PRAM (nondeterministic write selection)

In this variant, any one of the processors can succeed to have a non-deterministic access to the given cell for exclusive write action; the other processors that do not succeed cannot have an access; this corresponds to a nondeterministic selection of one of the data values computed by different processors. Such a selection should ensure the serializability of the resulting program when different processes are interleaved. The design of the algorithm for this PRAM should meet this requirement.

Variant 3c: Processor-precedence for write (PRIORITY) PRAM

In this variant, the processors are assigned priority numbers (for example a timestamp when each processor starts a process) that establishes some total ordering among the processors; then the processor that has the maximum priority wins in accessing the given cell for a write action; the other processors that have a lower priority cannot access. Such a precedence should ensure the serializability of the resulting program when different processes are interleaved. The design of the algorithm for this PRAM should meet this requirement.

The weakest of these variants in terms of speed of performance is the EREW model, while the strongest is the priority CRCW model; the CREW model has a speed of performance in between the EREW and the CRCW models. It must, however, be noted that the most powerful priority model CRCW can be simulated by a EREW PRAM in O(log p) time using the same number of processors p.

4.4.3.2 Nondeterministic, probabilistic and random PRAMs. The nondeterministic and probabilistic or random computations are organized in two phases:

Phase 1: guessing/tossing (random number) to obtain an answer
Phase 2: deterministic verification of the validity of the answer

These two phases work interactively and are repeated many times; thus the nondeterministic and probabilistic computations are no more than guess-check or toss-check repetitive trials. Thus, in both these computations, two fundamental questions arise:

1. How much computational resource *must we use* if we eliminate nondeterminism or probabilistic chance?
2. How much computational resource *can we save* if we are granted nondeterminism or probabilistic chance?

The PRAMs, by design, are synchronous and deterministic, although a form of nondeterminism is introduced in the write actions in the ARBITRARY PRAM model. Based on this notion, we can include nondeterminism at the process level. Also, the probabilistic or random versions can be easily realized using coin flipping devices or suitable random number generators in read or write statements in any PRAM model. The required random numbers are appended along with the input strings to select a possible answer in the solution space and then check its validity, thereby implementing the two required phases in the solution of the problem. These can realize one-sided error or zero-error probabilistic parallel algorithms as a counterpart to a PTM or a RTM. For some problems, it turns out to be beneficial to use the random variant of the PRAM and save computational resources.

4.4.3.3 Relative power of different PRAMs. It has been observed by many researchers that there are problems for which the CRCW performs much better than the best possible algorithm in EREW. This is not surprising, since it is the most powerful PRAM. But then the question arises as to how much is CRCW more powerful than EREW? The answer to this question can be obtained by simulation. As already mentioned in Section 4.4.3.1, a CRCW PRAM with p processors can be simulated by an EREW or CREW PRAM in $O(\log p)$ time.

The inclusion of randomness in PRAMs seems to speed up the solution of many problems. However, it still remains an open question as to whether they indeed create conceptually new complexity classes of problems that are not solvable by purely deterministic means.

4.4.4   PRAM-based complexity classes

We now describe the important PRAM based complexity classes within P.

4.4.4.1 Nick's class. The class of decision problems solvable by a PRAM in simultaneous polynomial bound on number of processors (space) and polylog time bound is called the Nick's class (NC) in honor of Nicholas Pippenger, a complexity theorist. The class NC is denoted by the ordered pair: $NC(n^{O(1)}, (\log n)^{O(1)})$. The NC class can be thought of as a dual under space-time interchange of the Steve's class $SC((\log n)^{O(1)}, n^{O(1)})$ which has simultaneous polylog space bound and polynomial time bound. Pippenger showed that the class NC is solvable in polynomial time by a deterministic TM whose tape head makes at most polynomial log n reversals. This class of problems is also solved by ATM in a log n space bound and a polynomial log n time bound; that is NC = ATM (space log n, time polylog n).

4.4.4.2 NC reduction. We can think of an NC reduction of a problem X to another problem Y (analogous to the polynomial time reduction using a TM) using PRAM in polylog time using at most a polynomial number of processors. Also we can say two problems are NC-equivalent if they are mutually NC-reducible.

4.4.4.3 FNC (search NC). Some complexity theorists believe that the class of search problems should be distinguished as a separate class, since it requires an additional overhead of time and space for reduction. The class FNC is the class of all those decision problems reduced from search problems that are solvable by a PRAM with simultaneous poly n bound on the number of processors and polylog bound on time. Thus $NC \subseteq FRNC$ by definition.

4.4.4.4 RNC (randomized decision NC). This class is a counterpart to NC, as RP is for P. This class permits random parallel solution to a decision problem. A decision problem (of input of size n) is in RNC, if it is solvable with one-sided error (like a RTM) by a PRAM with simultaneous poly(n) bound on number of processors and polylog(n) bound on time, when random numbers of size poly(n) are appended along with the inputs. It is **not known** whether NC = RNC.

4.4.4.5 FRN C (randomized search NC). FRNC is the class of those decision problems reduced from search problems that are solvable with one-sided error by a PRAM with simultaneous poly(n) bound on number of processors and polylog n bound on time, when random numbers of size poly n are appended along with the inputs. Thus $NC \subseteq FRNC$ by definition. It is **not known** whether FNC = FRNC.

4.4.4.6 ZPNC (zero error randomized algorithm). This class is the counterpart of $ZPP = RP \cap co\text{-}RP$; it is defined by $ZPNC = RNC \cap co\text{-}RNC$; that is, the class of problems solvable with zero error in answer by a PRAM with simultaneous poly n bound on number of processors and polylog n bound on time, when a pair of random numbers of size poly n are appended along with the inputs.

   The problem called "perfect matching" in a graph belongs to this class, although we do not know whether it is in NC. A perfect matching in a graph G is a set of edges S such that each node of G is incident with exactly one edge of S. Based on the determinant evaluation of an integer matrix with enries in the set $\{-n, \ldots, 1, \ldots n\}$, where n is the number of nodes in G, this problem is in RNC [26, 36].

4.4.4.7 NC and SC. In Section 4.4.1.1, we defined the class SC using simultaneous polynomial time and polylog space bounds on a DTM. Although, NC^2 and SC^2 are contained in LOG^2, so far it has not been proven that SC = NC; it is likely that they are incomparable, and the set SC-NC contains NLOG complete problems such as the graph accessibility (Section 4.4.1.3). For the set-theoretic containment relationship of these classes see Fig. 4.3.

4.4.4.8 P-completeness and NC. In the P class, there are problems that have not been shown to be in NC; that is, for these problems no parallel algorithms are known to solve them in polylog time with polynomial number of processors. If any one of these problems can be solved in polylog time, then so can every member of this class. Hence, the P completeness of problems is also known as log-space completeness for P [31]. Proving that a problem is P-Complete is viewed as an evidence that there may not be a parallel algorithm to solve that problem in polylog n time using a polynomial number of processors. However, this statement must not be interpreted as saying that classes in NC are definitely solved more efficiently than those which are P-complete. For example, there are two algorithms for parsing context-free languages: one that uses $O(n^6)$ processors in $O(\log n)$ time, and another that uses $O(n^2)$ processors in time $O(n)$. Obviously, the former algorithm is less efficient than the latter. Also consider a sequential algorithm with running time $n^{1/2}$. This is certainly superior to $(\log n)^3$ for a fairly large n (up to 10^6).Thus solving a problem in $O(\log n)$ time using a polynomial number of processors need not be any more efficient than solving it in polynomial time using a polynomial number of processors. Thus NC-class does not imply high efficiency in the popular sense that if we use N processors in parallel, a speedup by a factor of N is achievable. This is not implied in the definition of NC. In fact it can turn out to be resource-wasteful compared to a sequential algorithm [30].

The existence of the circuit value problem described in Section 4.4.1.4 implies that there may be problems that are essentially sequential and not amenable for fast parallel solution, even if a large number of processors are used. In summary, the class NC has the following relationship among the complexity classes:

$$L = DLOGSPACE \subseteq NL = NLOGSPACE \subseteq NC \subseteq P \subseteq NP \subseteq PSPACE$$

It is believed that these inclusions are strict (see Figs. 4.2, 4.3, and 4.4).

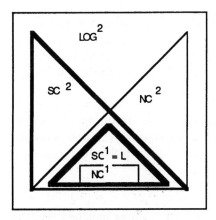
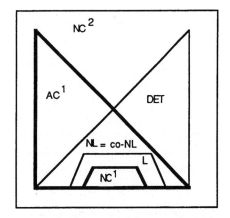

Figure 4.4 Inside LOG^2 and NC^2

While it is known that NLOGSPACE-complete and P-complete problems exist, it is not yet known whether there are NC-complete problems. However, there seem to be complete problems in between NC and P as, for example, a variant of the circuit value problem called *comparator circuit value problem*. Here, all the elements of the circuit are comparators which are two-input "**and**" and "**or**" gates.

4.4.5 Arithmetic and other macro-operator PRAM models

To deal with algebraic or polynomial and arithmetic computations, it is quite feasible to extend the PRAM to model macro operations over the suitably defined elements in unit time. This extension is valid as long as such macro operators are convertible to Boolean circuits of polynomial size in time log n, where n is the length of the binary representation of the elements, to satisfy the model interchangeability under NC. Some authors prefer to differentiate the arithmetic NC class from NC.

For example, the Boolean matrix powering is in NC class, but the problem of integer matrix powering is in arithmetic NC class.

4.4.6 PRAMs—relationship to real parallel machines

We now study the relationship of PRAM to practical parallel machines that are in use. Two different classification or taxonomical schemes are used for classifying these parallel architectures:

1. Flynn's classification scheme [28]
2. Schwartz's classification scheme [44]

4.4.6.1 Flynn's parallel machine classes. The most widely used classification of parallel computers is due to Flynn. Here, the parallel machines are classified into four classes:

1. SISD (single instruction, single data stream), which describes a classical sequential machine.
2. SIMD (single instruction, multiple data stream), which describes machines that execute a given type of instruction simultaneously on different sets of data, e.g., vector processing operations.
3. MISD (multiple instruction, single data stream), which describes a machine that executes different instructions on the same data moving as a stream, e.g., pipeline computing.
4. MIMD (multiple instruction, multiple data stream), which describes machines that can execute different types of instructions on different sets of data. The actions can be simultaneous through sychronization or can be asynchronous as in a distributed computing system.

All the above classes of machines can be simulated by a PRAM. In fact, the PRAM acts like a synchronous MIMD machine with an unbounded number of processors and a shared memory which allows simultaneous reads from the same memory cell, but disallows simultaneous write on the same cell.

4.4.6.2 Schwartz's parallel machine classes. Flynn's classification does not take into account the manner in which the information transfer takes place between different processors. To incorporate this feature, Schwartz introduced the notion of paracomputers and ultracomputers. The paracomputers have access to shared memory and, hence, the transfer of information between different processors take place in constant time. Due to the sharing of memory, it is possible to have read-write conflicts. The PRAM model is in fact a paracomputer [30].

In the ultracomputer, each processor has its own memory, and the communication takes place among processors through a fixed interconnection network. The pairwise completely connected network of processors, mesh connected network of processors, hypercube processor network, and binary trees of processors are examples of ultracomputers. The ultracomputer can be abstracted as a directed or undirected graph in which the nodes are the processors, and the directed or undirected edges correspond to interconnections. Accordingly, the maximum degree of the nodes and the maximum path length among the nodes play an important role in determining the parallelism and its time complexity. The maximum degree of the nodes determines the total number of messages that can be sent or received by processors concurrently in one unit of time; hence it is called the *node degree* of the network. The maximum among the shortest path lengths from one processor to another processor determines the maximum or worst possible time taken by a message to get through; hence, this length is called the *diameter* of the network.

Examples

1. The completely connected network of n nodes has a node degree (n–1) and diameter 1. The two-dimensional mesh connected square network with $n = m^2$ processors has a node degree 4 and diameter 2(m–1).
2. For a ring connected network of n nodes, the node degree is 2 and the diameter is the lower integral part of n/2.
3. The hypercube with $n = 2^d$ processors has a node degree d and diameter d.

4.4.6.3 Module parallel computer. A theoretical model that is useful to study ultracomputers and relate them to the PRAMs was proposed by Alt et al. [4]. This is called a "*module parallel computer*" (MPC), and it uses a completely connected network of processors. The MPC can simulate a PRAM with n processors and a shared memory of size m in probabilistic time O(log n) and deterministic time O(log m). One step of MPC can be simulated by a bounded degree network of n processors in O(log n) steps. Thus the PRAM with n processors and size m shared memory can be simulated by a bounded degree network of n processors in time $O((\log n)^2)$ probabilistic time or O(logm.logn) deterministic time.

4.4.6.4 Is PRAM the right high-level parallel model? There are some important points of criticism against the PRAM models:

1. Although the PRAM model abstracts tightly-coupled machines, it fails to account for the latency of accessing shared memory in real machines.
2. The CRCW version is impractical because concurrent read and write are not practical, and the faster running time of CRCW over EREW is not achievable.
3. The PRAM model does not meet the needs of the programmer, given that it requires a complete description of the partitioning of code among the processors and of the way in which memory is arranged to provide communication. Also, it does not help

the programmer in the management of parallel threads from the communication and synchronization point of view.

4. PRAM is a good model for SIMD synchronous algorithms, but it does not take into account the communication time and, hence, is not directly applicable to parallel computers based on interconnection network. Note, however, that a PRAM can be simulated by a bounded degree network of processors as mentioned in the previous subsection.

5. The invariance notion among different computers based on the polynomial interchangeability and the associated order notation are too coarse to provide a precise quantitative performance and cost measurement. They can only provide qualitative class divisions.

While criticisms (1) and (2) can be satisfied by introducing switches to realize concurrent read and write, the other criticisms cannot be easily met. In spite of these points of criticism, the PRAM model is useful for the design of parallel algorithms with widely different data structures and for the analysis of algorithmic performance independent of the real machines.

4.4.6.5 XPRAM (BSP), LogP and transactional models. Valiant [54] has described a model called XPRAM or bulk synchronous PRAM (also known as BSP model) that is both realistic and idealistic. It improves the PRAM model by adding global communication features in PRAMs. Also, the XPRAM model attempts to bring in universality among the general-purpose parallel computers, much in the same way the Turing machine does for the sequential machines. While the PRAM model ignores the communication and storage management, XPRAM addresses both these issues. In fact, it addresses the general problem of universality in distributed computing by considering networks that contain only nodes with memory, processing or switching capabilities.

The XPRAM model consists of a set of sequential processors, each with a local memory and connected by a sparse message passing communication network to a global memory. This permits XPRAM to perform computations using frequent local references and some less frequent global references. XPRAM can simulate many special-purpose network architectures and the different types of PRAMs efficiently. The communication time is of the order of log p, where p is the number of processors.

Also, the XPRAM model seems to address many of the issues related to computation and communication in both the distributed and centralized systems. The dual properties, i.e., the speciality and universality, possessed by a TM for sequential computation are also possessed by the XPRAM. Thus some believe that the XPRAM may turn out to be an appropriate model for parallel architecture much as the von Neumann is in the sequential case. The XPRAM is conceptually simple and uses message passing synchronization in bulk, and it can be efficiently implemented in both the VLSI and optical technologies. Also, it serves as a useful host on to which higher-level communication and storage management functions can be compiled. However, to use this model for mapping an algorithm on to a target architecture, some mechanism is needed to recast the computations into partially ordered sequences of varying lengths to minimize global memory traffic and maximize local memory traffic. Other modified models that meet the points of criticism against the PRAMs have been proposed [1, 3, 39, 48]. Skillicorn [48] summarizes and compares a number of different parallel computational models that can provide the following characteristics:

1. *architectural independence* that is general enough to model several architectures
2. *congruence* that reflects the cost of execution at the model level
3. *descriptive simplicity* that provides abstractions for decomposition for parallelism, communication, and synchronization

Recently, another model called the LogP model has been proposed [11]. The LogP model generalizes the idea of message passing used in BSP so that the processors may communicate point to point. Another model, Campbell's lenient unified model of parallel systems (CLUMPS) generalizes the LogP model [9].

The transactional model introduced in Murthy and Krishnamurthy [37] provides an objective approach to represent and extract parallelism in solving a problem. This model is based on the notion of processes and data objects. Each basic process is identified with a transaction that provides atomicity, isolation, and consistency. The set of processes that participate in a computation is called a *task*. The computation problem is then represented as a directed acyclic graph with its nodes corresponding to transactions and arcs corresponding to communication of values using either message passing or shared memory approach. A computation schedule consists of a sequence of computation and communication steps that can be executed in a partial order, ensuring serializability of the resulting program. This abstract partial order can be obtained using a topological sort on the given transactions, based on purely syntactic details such as read and write actions or send and receive primitives.

Such an abstract partial order, however, only provides a skeletal structure that results in a serializable total program. For practical scheduling one needs to refine this partial order by considering the space and time-limitations arising from register and processor allocation and their reusability, and the processor-register-memory traffic. These are achieved through critical path scheduling of the real time transactions.

The transactional (or process) model can be set in correspondence with the different PRAM architectural models (Section 4.4.3.1) thus:

- EREW: Transactions under locks (or semaphores and barrier synchronization) correspond to the EREW PRAM model.
- CREW: Transactions under shared reads-lock and Exclusive write lock correspond to the CREW PRAM model.
- CRCW: Transactions can also realize all the variants of CRCW PRAM model and ensure serializability of the resulting program; their nondeterministic nature permits the realization of arbitrary CRCW PRAM; the use of time stamps permit the realization of Priority PRAM; and the common write PRAM can be easily realized using a transaction that reads other transactions and writes a suitable common value.

The transactional model can be represented by a finite state machine model. Two types of events can occur at a transaction triggering its transition function:

1. an internal computation event that can result in sending messages to the network, in addition to a local state change
2. the delivery of a pending message

Based on these events, the transactional model fits well in the context of both the local memory and global communication. Hence, this model can simulate the BSP and LogP model. If the transactions are permitted to invoke subtransactions (resulting in a nested transactions), the extended model is useful for real-time scheduling [14].

A logical foundation for the transactional model can be provided based on the *transactional logic,* a form of temporal logic, developed by Bonner and Kifer [8]. The transaction logic is both declarative and procedural; that is, the user can specify what kind of actions are to be done and also how to do them. Hence, the transactional model of concurrent and parallel computation turns out to be a very useful aid for the design, synthesis and analysis of parallel programs.

4.4.6.6 Parallel models and programming costs. The ultimate purpose of any parallel complexity model is to provide a cost system for the parallel programming of algorithms. For the sequential programming, the RAM model provides a cost measurement based on the assumption that each instruction takes unit time, memory references take zero time, and the space used is the maximum number of memory cells needed for any computational step. In the sequential programming case, the loops ("for" and "while") essentially contribute to the execution time and, hence, the execution time is easy to compute.

In the PRAM model, the execution of basic operations and the access to a shared memory by a processor is assumed to take unit time. Communication between processors take place through a write action of one processor in a given shared memory cell which is eventually read by another processor. This communication time is taken to be a constant. The cost of computation is then determined by the number of processors used and the total number of parallel time steps taken. The assumption of constant communication time is not realizable in practice.

The BSP model assumes uniform communication, by the use of random techniques that bound the message delivery times even if there are competing messages. This model is more realistic in providing the communication costs. A BSP program uses both local computational steps and occasional global super computational steps; the global supersteps involves the read and write actions to shared global memory, but after that, mainly local computations take place with a higher frequency. The performance of a BSP program depends on the problem size, the duration of the supersteps, and the architectural details such as the number of processors used, the ratio of computation to communication time and the latency of communication across the machine. Although the BSP provides a good abstraction of communication and computation costs, it does not provide guidance to a programmer as to how to compute costs during the development of the program.

The transactional model can provide a suitable guidance for computing costs during the program development based on the abstract partial order and time and space resource limited partial order.

In a practical programming situation, we still need to consider the three key issues: cost of data movement between different levels of memory hierarchies (main memory/cache/registers), the ratio of active to idle number of processors at a given time step and the communication overhead. This means we must be able to balance the load among different processors and provide an optimal solution to memory-processor traffic. These issues are algorithm and architecture dependent and so it is not possible for a general parallel model to provide detailed answers to all these specialized issues.

4.4.7   The Boolean circuit model

Boolean circuits (named after the logician Boole) are made up of AND, OR, and NOT gates. Every solvable problem can be solved by an infinite collection of Boolean circuits $\{B_1, B_2, \ldots B_n, \ldots\}$, where B_i denotes a circuit with i inputs and producing the desired answers as outputs. We call a circuit *uniform* if its description can be generated automatically by an appropriately resource-bounded deterministic Turing machine.

The Boolean circuit serves as a good mathematical and practical model because it represents a decision problem directly and tells us about its constructibility as a VLSI circuit. Also, it provides a classification of the fine structure of complexity classes inside NC as well as other superfine structures within. The circuit model uses a bounded input (usually 2-input fan-in) combinational Boolean circuit. It is a directed acyclic graph in which the nodes are labeled input, constant, AND, OR, NOT, and output. For practical reasons, it is assumed that the circuit has a two-dimensional structure with a depth, and a size. We need to use the "invariance thesis" for standardizing the following measures:

1. *Depth*. The maximum of lengths among different possible paths from an input node to an output node is the depth of the circuit; it is a measure of parallel time, since it tells us how long does it takes to obtain the result. We use d (a function of the input size n) to denote the depth.
2. *Size*. The circuit size represents the number of basic operations carried out and therefore can be measured in terms of the number of circuit elements (the number of nodes when the inputs are bounded fan-in, or the number of edges when the inputs are unbounded fan-in). Let s (a function of the input size n) denote the size.

That the CREW PRAM model with p processors can simulate a bounded fan-in circuit model of size s and depth d in $O[d + (s/p)]$ parallel time steps. Hence, if we require that the number of parallel time steps is to be of the order of d, then we need to use $O(s)$ processors. A similar result holds for the EREW PRAM model.

Also, if a p processor PRAM takes time t for an algorithm, then a PRAM with p', where $p' < p$, can simulate the same algorithm in time t', where $t' = O(pt/p')$. In other words, $(p'.t') = O(p.t)$. This product is called the *work*, and is nearly an invariant (Babbage thesis) and corresponds to the sequential time (see Chapter 6 in this book).

Because a given circuit can compute only a fixed-size input problem, we must use a family F(n) of fixed size circuits for each n and compute the complexity of constructing F(n) for any arbitrary n. For this purpose we need to make valid assumptions for the construction and classification of F(n). Four different constructive methods are used; all these methods assume that some kind of inductive modular construction or uniformity condition exists, as defined below:

1. *P-uniform family*. Given the circuit interconnections of the family F(n–1), a DTM can build or describe the circuit interconnections for the family F(n) in polynomial time.
2. *Log space uniform family*. Given the circuit interconnections of the family F(n–1), a DTM can build or describe the circuit interconnections for the family for F(n) using space O(log n).
3. *Ruzzo-uniform family* [43] *or ALOGTIME-uniform family*. Using a random access ATM (RACATM) (see Section 4.2.14) as the basis, the ALOGTIME-uniformity of circuits is defined thus: Given the circuit interconnections of the family F(n–1), a RACATM can recognize this language and describe the interconnections for the family F(n) in O(log n) time. This uniformity is also denoted by U_{E*} uniformity. We define ALOGTIME to be the class of all those decision problems by a random access ATM in O(log n) bounded time where n is the length of the input.

4. *DLOGTIME-uniform family.* Given a description of the interconnections of the family F(n–1), a random access DTM can recognize this language and describe the interconnections for the family F(n) in O(log n) time. We define DLOGTIME to be the class of all those decision problems by a random access DTM in O(log n) bounded time, where n is the length of the input. The concept of uniformity provides us an inductive technique to generate solution for any input size. Also this concept helps us to define complexity classes which have interesting relationships with the earlier classes.

If the families do not belong to any of the above, they are called *nonuniform families.*

We may also remove the restriction on *bounded fan-in* and allow arbitrarily large or *unbounded fan-in*. A simple example of this circuit is the programmable logic array (PLA) in which a bus is used to distribute data. The unbounded fan-in circuit model of size p(n) and depth d(n), denoted by UCKT(p(n), d(n)) can be realized by a bounded fan-in circuit model of size of O(p(n) and depth O(d(n).log p(n)). Hence, the unbounded fan-in model is depth-wise (and hence parallel-time-wise!) more efficient than the bounded fan-in model. Such an unbounded fan-in model is the equivalent of the CRCW PRAM model. In the unbounded fan-in, the number of edges in the circuit rather than the number of nodes is used as the size of the circuit. To prove this equivalence, one establishes that the CRCW can simulate the unbounded fan-in circuit model of size s (number of edges) and depth d, in exactly d steps using s processors.

Since the above families of circuits have special language recognition capabilities, they serve as a basis for classifying the complexity classes of languages.

4.4.8   Circuit complexity classes

We use the notation CKT(s,d) for specifying bounded fan-in circuit model and UCKT(s,d) for denoting the unbounded fan-in circuit model, each of size s (number of nodes in CKT or number of edges in UCKT) and depth d. Note that s is a space measure, and d is a time measure.

4.4.8.1 NC^k. This is the class of languages recognized by the logspace uniform family of circuits of polynomial size and depth $O(\log n)^k$; that is $NC^k = CKT(poly(n), (\log n)^k)$, $k \geq 1$.

It has been shown that $NC^k = ATM(\text{space } \log n, \text{time } (\log n)^k)$. Hence,

$$NC^1 = ATM(\text{space } \log n, \text{time } \log n)$$

Furthermore, if we define $\cup_{k \geq 1} NC^k$, then as said earlier, we have

$$NC = ATM(\text{space } \log n, \text{time polylog } n)$$

4.4.8.2 AC^k. This is the class of languages recognized by the logspace uniform family of circuits of polynomial size and depth $O(\log n)^k$ with *unbounded fan-in*; that is $AC_k = UCKT(poly(n), (\log n)^k)$. Also, it has been shown that for $k \geq 0$:

$$NC^k \subseteq AC^k = UCK(Tpoly(n), (\log n)^k) \subseteq NC^{k+1} \subseteq AC^{k+1},$$

Hence,

$$\cup_{k \geq 0} AC^k = AC = NC$$

For a set-theoretic containment relationship of these classes, see Fig. 4.3.
    Also, we have the relations:

$$CKT\,(poly\,(n),\,(\log n)^k) \subseteq UCKT\,(poly\,(n),\,(\log n)^k/\log\log n)$$

$$UCKT\,(poly\,(n),\,(\log n)^k/\log\log n) \subseteq CRCW\,(poly\,(n),\,(\log n)^k/\log\log n)$$

Furthermore, a PRIORITY CRCW PRAM can be simulated by UCKT or

$$PRIORITYCRCW\,(poly\,(n),\,(\log n)^k/\log\log n) \subseteq UCKT\,(poly\,(n),\,(\log n)^k/\log\log n)$$

4.4.8.3 DET. The class DET consists of all those problems that are logspace reducible to the problem of computing the determinant of an integer matrix. This class is contained in NC^2; also DET contains the probabilistic classes that require simultaneous logspace and expected polynomial time bound. The following problems have been shown to be complete for DET by Cook [1985]: Matrix powering, iterated matrix product, and matrix inverse.

4.4.8.4 AC^1 and DET. AC^1 is contained in NC^2, DET is contained in arithmetic NC^2; but AC^1 and DET are incomparable. The problem of transitive closure of Boolean matrices is in AC^1 (see Fig. 4.4).

4.4.8.5 AC^0_k for $k \geq 0$. This is the class of problems solvable by DLOGTIME uniform, depth k, polynomial size, unbounded fan-in circuits. We define: $AC^0 = \cup_{k>1} AC^0_k$. The class of decision problems DLOGTIME solvable by a random access DTM in $O(\log n)$ time contains AC^0_0 and is contained in AC^0_2 and incomparable with AC^0_1.

4.4.8.6 BW^0_k for $k > 1$. This is the class of problems solvable by polynomial size, bounded fan-in circuits of width k or less. If we define

$$BW^0 = \cup_{k>1} BW^0_k$$

then it has been shown that [20]

$$BW^0 = BW^0_4 = ALOGTIME = U_{E*} \text{ uniform } NC^1$$

Also,

$$AC^0_k \subseteq BW^0_k \text{ for } k \leq 3$$

4.4.8.7 TC^0. This is the class of problems solvable by polynomial size, bounded depth, unbounded fan-in Boolean circuits augmented by threshold gates that have unbounded fan-in and produces an output 1 when a majority of its inputs are 1. This class lies in between AC^0 and $ALOGTIME = BW^0$ but is incomparable to BW^0_3.
    For a set-theoretic containment relationship of these classes, see Fig. 4.5.

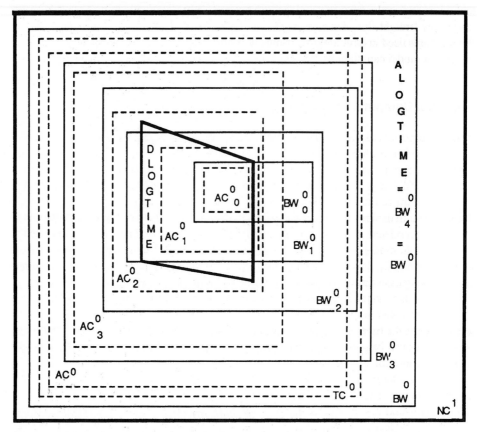

Figure 4.5 Set-theoretic containment relationship

4.4.8.8 Examples of problems inside NC and NC^1

1. The addition/subtraction operation is in AC^0.
2. The multiplication operation is in NC^1 and not in AC^0.
3. The division operation is not in NC^1 and is in NC^2; also the operations reciprocal, powering and iterated multiplication operations are not in NC^1.
4. Matrix powering, iterated matrix product, and matrix inverse are in DET.
5. The problem of Boolean matrix multiplication, successive Boolean matrix squaring, and the transitive closure of Boolean matrices, are in AC^1.

It is an open problem as to whether the greatest common divisor of two positive integers is in NC. (See Ref. [25] for additional examples.)

4.4.8.9 Nondeterminism and randomness in circuit model.
It is possible to use nondeterminism and randomness to realize nondeterministic algorithms and Las Vegas algorithms in VLSI circuits, since random number generators in VLSI technology seems in the realm of the possible [33].

4.4.8.10 Is the circuit model a good parallel micro-level model?
All the basic computations in computers are performed by microelectronic circuits. These consist of logic gates

and wires and evaluate some Boolean expressions. In the language of programming they are simply straight-line programs which can be modeled by acyclic directed graphs. Since the measures used in circuit model can be related to related to Turing and PRAM measures the circuit model can describe the superfine structure of the complexity classes inside NC.

4.5 VLSI Computational Complexity

Another important aspect is the derivation of complexity bounds for VLSI circuit implementation of algorithms. The practical uses of many algorithms, such as systolic algorithms, depend upon the design of VLSI chips that can perform computation with maximum efficiency. This means the time for information transfer and the signal propagation are minimized in the VLSI realization. To study the efficiency, different models are used. The main difference among these models is the manner in which the signal propagation time is modeled. Most circuit designers evaluate a VLSI algorithm under the constant or logarithmic model where the signal propagation time is proportional to the logarithm of the wire length. Some designers, however, prefer the use of a linear model, where the signal propagation time is proportional to the wire length.

Two fundamental parameters decide the efficiency of a VLSI chip: the area of the circuit and the time taken to produce an output for a given input. The area gives the space complexity, while the time taken gives the time complexity. Intuitively, because time and space complexities can be traded for one another, the product of area and time is used as a complexity measure for VLSI circuits.

Switching energy of a VLSI chip can be used as a measure of its complexity. Specific results in this direction are available in Leighton [32]. Also time complexity studies for synchronous and asynchronous schemes for VLSI systems are available in Afghahi and Svensson [2]. Detailed complexity aspects and performance studies of circuits, interconnections, and packaging for VLSI are available in Bakoglu [6].

4.6 Complexity Measures for Distributed Systems

Measuring time and space complexities for asynchronous distributed computations is difficult due to the intrinsic differences between the parallel and distributed computing systems. In the asynchronous distributed system, the communication time can be much larger in comparison to the magnitude of computation time because the communication links are usually longer by several orders of magnitude. Furthermore, in the distributed systems, there are multiple processes, and processors that communicate and cooperate with each other. Accordingly, the complexity issues that arise are entirely of a different nature, and these pertain to the efficiency and reliability in cooperation and communication, avoidance of deadlocks, providing fairness to users and processes, program verification, and termination. In this sense, the problems involved are in the category of performance analysis, although some algorithms and protocols within these problems can be subjected to complexity analysis. For a detailed study, see Bodlaender [7] and Raynal [40, 41].

4.7 Neural Networks and Complexity Issues

In recent years, the theory and practice of neural networks have turned out to be a very vast area of research [18, 24, 47, 52, 58]. It is not known whether this nontraditional way of computing can ameliorate the hardness encountered in solving a problem and, as a consequence, we can break the intractability and the unsolvability barrier. Based on some problem instances, some authors have reported unusual efficiency of neural computing methods and have even attributed to them super-Turing capabilities [16, 23]. None of these claims are well founded and conclusive.

Recently, Cybenko [12], Judd [22], Orponen [38], and Wiedermann [59] have studied the complexity theory to neural networks, and the results show that whereas most known discrete neural network models are formally capable of solving any conventional computational problem, they can neither enlarge the class of problems solvable by a Turing machine nor ameliorate the complexity of resource consumption, Schwarz [45]. It has been proven by Wiedermann [59] that the Hopfield network of size $(n \times n)$ can simulate a Boolean circuit of size $S(n)$ and depth $D(n)$ in parallel time $O(D(n))$. Thus, Hopfield network has polynomial circuit complexity and will have the same power as the UCKT or CKT models and can solve problems in NC.

Also, the discrete version of the Botzmann machine can be modeled by a symmetric neural network and a probabilistic mechanism that assigns a temperature to neuron. As seen earlier, if the circuit size is polynomial bounded this network can solve in polylog time the problems in class RNC. Since it is not known whether RNC = NC, no assertion could be made about the additional power derived using probabilistic decisions.

4.7.1   Relating neural and conventional computing

Neural computing can be related to conventional computing under the following correspondence: In conventional computing, there is a finite state control that executes the algorithm; also, the computation needs a finite amount of memory space. The complexity of solving a problem can be separated into three (not necessarily disjoint, but interrelated) complexities:

1. The number of repetitions of the control loops in the algorithm gives us an estimate of the *time complexity.*
2. The memory size used to store and retrieve intermediate results gives us an estimate of the *space complexity.*
3. The length of the algorithm (or the number of production rules used in the Markov algorithm) can be taken as a measure of the *information or Kolmogorov complexity* (see Section 4.8). While solving a problem using a neural network, the number of iterations in a neural net gives the time complexity, the number of neurons gives the space complexity, and the algorithmic information contained in the synaptic interconnections reflects the program or Kolmogorov complexity. It is not known how these three complexities are interrelated or what kind of a trade-off exists among them.

An analog neural network consists of an interconnected arrangement of analog processors which evolve cooperatively toward a stable configuration. The system has no clock and follows a trajectory determined by certain differential equations and converges to the solution of the problem. Because there are no discrete steps involved in the convergence to solution (unlike the manner in which the solution is obtained for difference equations in the digital setup), we cannot easily work out the complexity. It is widely believed that the analog neural network provides a speedy solution to the NP-complete problems [19, 21, 52]. Takefuji [52] experimentally observes that in many problems (such as searching, sorting, graph-coloring, tiling and routing) the size of the problems has no influence on the time for solution, and essentially the networks converge in a constant time. This observation suggests that computational complexity theory has to be enlarged to include analog computing area.

Some neural networks update the state of their neurons continuously and asynchronously in time. These are difficult to analyze from complexity viewpoint for the following reasons:

1. In a continuous time real valued competitive or cooperative dynamic neural network, we cannot isolate the computation into disjoint actions involving capabilities of switching, processing, or memory [45]. The basic space-time trade-off questions cannot be considered without this isolation, because all our arguments are based on counting of time steps or memory cells, or well defined finite arithmetic or Boolean or threshold operations. Furthermore, in such competitive dynamical networks, the problem of memory read-write conflicts do not seem have any role, given that partial and intermediate results can get mixed up. This situation is difficult to understand from the computer technology point of view.

2. There seems to be no associated set of formal axioms for a dynamical neural network model that permit us to apply the invariance principles across problems, algorithms, or models. Also, the concept of polynomial machine interchangeability or reducibility seems to be inapplicable to equilibrium dynamics based on physical principles in which there are real or complex valued inputs and outputs. In fact, we cannot simply measure the complexity in terms of physical size or time or energy used for computation.

3. Furthermore, the notion of uniformity, which is the basis for scalability in parallel complexity theory, cannot be applied, because problem instances need not use a standardized input of binary strings.

4. Most neural networks turn out to be special-purpose problem solvers for particular instances; that is, they have only the speciality property, but not the universality property. The hardware structure does not remain invariant with respect to problem instances. While some of these issues can be tackled by using successive iterations, the convergence issues are difficult to understand.

4.8   Other Complexity Theories

Several other approaches have been used as the basis for the measurement of resources that can adequately describe the complexity factors in solving a problem. These are:

1. machine independent complexity theory
2. Kolmogorov complexity theory
3. algebraic complexity theory
4. information based complexity theory
5. complexity theory of real functions

The *machine independent complexity theory* is concerned with the program transformation as a computable function within certain resource bounds [46].

The *Kolmogorov complexity theory* has its roots in probability theory, information theory based on entropy formalism, combinatorics, randomness, and the theory of algorithms. It was developed by Kolmogorov, Chaitin, and Solomonoff. The complexity notion is formulated in terms of specifying a minimal length program that can specify an object or a string in any domain using any programming method (a formal notation or a programming language). In other words, this theory relates the two fundamental concepts namely, the algorithm and the entropy or information. For a detailed account of Kolmogorov complexity, see Li and Vitanyi [34, 35] and Storer [50].

The *algebraic complexity theory* is concerned with the complexity of numerical and algebraic problems—for example, symbolic computation, univariate and multivariate polynomial factorization, function approximation, algebraic function manipulation, and group theoretic problems [51].

The *complexity theory of real functions* attempts to extend the notion of computability and complexity to real numbers. This theory can be applied to problems in maximization, integration, differentiation, roots, polynomial approximation, differential equations, and optimization in control theory. The computational complexity of these problems forms a hierarchy analogous to the hierarchy in discrete complexity theory. For a detailed treatment of this subject, see Ref. [27].

The *information-based numerical complexity theory* has been developed by Traub, Wasilokowski, and Wozniakowski [53]. Here, the complexity is defined as the minimal cost of computing an approximation with errors not exceeding a prespecified threshold. The model of computation assumes basic costs of fundamental operations such as add, multiply, comparison of real functions, and evaluation of certain elementary functions. Accordingly, its application is limited to essentially numerical problems.

4.9   Concluding Remarks

The Turing machine-based complexity theory provides a classification hierarchy for the complexity of resource requirements for solvable problems and suggests polynomial-based time and space measures for solving these problems using sequential computers. This classification and the resulting measures turn out to be inadequate to identify classes of problems that are very much easier than the polynomial time class and are amenable to fast parallel solution using many computers in parallel. Therefore, it becomes necessary to introduce new models such as the PRAM and circuit models. These new models give rise to new parallel complexity classes inside the polynomial time class P. However, these models seem to be inadequate to provide practical quantitative performance measures for the diversity of available architectures. Also, the complexity issues for distributed systems are difficult to study using these models. New models—such as Valiant's XPRAM or bulk synchronous model (BSP), LogP model, and transaction model—address some of these issues which are related to computation and communication in both the distributed and centralized systems.

Much work remains to be done in understanding the complexity issues of other non-conventional computing methods such as the dynamic and analog neural networks. So far, there are no formal approaches to understand the complexity of these machines and relate the digital complexity classes with the analog complexity classes. This direction of study will be valuable from the point of view of emulating the brains by using machines.

Acknowledgments
Thanks to Drs. A. Burkitt, L.Kronsjo, V. K. Murthy, and H. Schroder for discussions on several aspects relating to this chapter. Also, the author expresses gratitude to the reviewers and editor-in-chief, Dr. A. Zomaya, for suggesting improvements to earlier versions.

4.10   References

1. Abolhassen, A., R. Drefenstedt, and J. Keller. 1993. Physical design of PRAMs. *Computer Journal,* Vol.36, 756–762.

2. Afghahi, M., and C. Svensson. 1992. Performance of synchronous and asynchronous schemes for VLSI systems. *IEEE Transactions on Computers,* Vol. C41, 858–872.

3. Aggarwal, A., A. K. Chandra, and M. Snir. 1990. Communication complexity of PRAMs. *Theoretical Computer Science,* Vol. 71, 3–28.

4. Alt, H., T. Hagerup, K. Melhorn, and F. P. Preparata. 1987. Deterministic simulation of idealized parallel computers on more realistic ones. *SIAM J. Computing,* Vol. 16, 808–835.

5. Baase, S. 1988. *Computer Algorithms.* Reading, Mass.: Addison Wesley.

6. Bakoglu, H. B. 1990. *Circuits, Interconnections and Packaging for VLSI.* Reading, Mass: Addison Wesley.

7. Bodlaender, H. L. 1987. *Distributed computing: Structure and complexity.* Amsterdam: Centre for Mathematics and Computer Science. CWI Tract 43.

8. Bonner, A. J., and M. Kifer. 1994. Applications of transaction logic to knowledge representation. In *Temporal Logic, Lecture notes in Computer Science,* Vol. 827. New York: Springer-Verlag, 67–81.

9. Campbell, D. K. G., and S. J. Turner. 1994. CLUMPS: A model of efficient general purpose parallel computation. *Proceedings of the IEEE TENCON Conference,* August 1994, Singapore, Vol. 2., 723–727.

10. Cook, S. A. 1985. A taxonomy of problems with fast parallel algorithms. *Information and Control,* Vol. 64, 2–22.

11. Culler, D., R. Karp, D. Patterson, A. Sahay, K. E. Schauser, E. Santos, R. Subromanian, and T. von Eicken. 1993. LogP: Towards a realistic model of parallel computation. In *Fourth ACM SIGPLAN Symposium on Principles and Practice of Parallel Programming,* May 1993, 1–12.

12. Cybenko, G. 1990. Complexity theory of neural networks and classification problems. In *Neural Networks, Lecture Notes in Computer Science,* Vol. 412, 26–44. New York: Springer Verlag.

13. Edmonds, J. 1965. Paths, trees and flowers. *Canadian J. Math.,* Vol. 17, 449–465.

14. Elmagarmid, A. K. 1992. *Database Transaction Models for Advanced Applications.* San Mateo, Calif: Morgan Kauffman.

15. Garey, M.R., and D. S Johnson. 1979. *Computers and Intractability.* San Francisco: W. H. Freeman.

16. Garzon, M. 1990. Cellular automata and discrete neural networks. *Physica,* Vol. 45 D, 431–440.

17. Harel, D. 1992. *Algorithmics.* Reading, Mass.: Addison Wesley.

18. Hecht-Nielsen, R. 1991. *Neurocomputing.* Reading, Mass.: Addison Wesley.

19. Hopfield, J. J. 1994. Neurons, Dynamics and Computation. *Physics Today,* Vol. 47, 40–47.

20. Johnson, D. S. 1990. A catalog of complexity classes. In *Handbook of Theoretical Computer Science, Vol. A,* J. van Leeeuwen, ed. Amsterdam: North Holland, Chapter 2, 66–161

21. Johnson, J. L. 1989. A neural network approach to the 3-satisfiability problem. *J. Parallel and Distributed Computing,* Vol. 6, 435–449.

22. Judd, J. S. 1990. *Neural Network Design and the Complexity of Learning.* Cambridge, Mass.: M.I.T. Press.

23. Judd, K. T., and K. Aihara. 1993. Pulse propagation networks: A neural network model that uses temporal coding by action potentials. *Neural Networks,* Vol. 6, 203–215.

24. Karayiannis, N. B., and A. K. Venetsanopoulos. 1993. *Artificial Neural Networks.* Norwell, Mass.: Kluwer Academic Publishers.

25. Karp, R. M., and V. Ramachandran. 1990. Parallel Algorithms for shared-memory machines. In *Handbook of Theoretical Computer Science, Vol. A,* J. van Leeeuwen, ed. Amsterdam: North Holland, 869–943.

26. Karp, R. M., E. Upfal, and A. Wigderson. 1986. Constructing perfect matching is in random NC. *Combinatorica,* Vol. 6, 35–48.

27. Ko, K-I. 1991. *Complexity Theory of Real Functions.* Boston: Birkhauser.

28. Krishnamurthy, E. V. 1989. *Parallel Processing.* Reading, Mass.: Addison Wesley.

29. Kronsjo L. 1987. Computational Complexity of Sequential and Parallel Algorithms. New York: John Wiley & Sons.

30. Kruskal, C. P., L. Rudolph, and M. Snir. 1990. A complexity theory of efficient parallel algorithms. *Theoretical Computer Science,* Vol. 71, 95–132.

31. Ladner, R. E. 1975. The circuit value problem is logspace complete for P. *SIGACT News,* Vol. 7(2), 18–20.

32. Leighton, F. T. 1983. *Complexity issues in VLSI.* Cambridge, Mass.: M.I.T. Press.

33. Lengauer, T. 1990. VLSI theory. In *Handbook of Theoretical Computer Science, Vol. A.,* Chapter 16, J. van Leeeuwen, ed. Amsterdam: North Holland, 835–868.

34. Li, M., and P. M. B. Vitanyi. 1990. Kolmogorov Complexity and its Applications. In *Handbook of Theoretical Computer Science, Vol. A.,* Chapter 4, J. van Leeeuwen, ed. Amsterdam: North Holland, 187–254.

35. Li, M. and P. B. Vitanyi. 1994. *An Introduction to Kolmogorov Complexity and its Applications.* New York: Springer Verlag.

36. Mulmuley, K., and E. V. Krishnamurthy. 1995. Transactional paradigm: applications to distributed programming. *Proc. of the ICA³P Conference* Vol. 2 (April 1995, Brisbane), 554–558.

37. Murthy, V. K., and E. V. Krishnamurthy. Transactional paradigm for programmer-assisted parallelization of programs. (Unpublished work.)

38. Orponen. P. 1993. Neural networks and complexity theory. In *Mathematical Foundations of Computer Science, Vol. 629,* Lecture Notes in Computer Science, I. M. Havel and V. Koubek, eds. New York: Springer Verlag, 51–61.

39. Papadimitriou, C. H., and M. Yannakakis. 1990. Towards an architecture independent analysis of parallel algorithms. *SIAM J. Computing,* Vol. 19, 322–328.

40. Raynal, M. 1988. *Networks and Distributed Computation.* Cambridge, Mass.: M.I.T. Press.

41. Raynal, M. 1987. *Distributed Algorithms and Protocols.* London: Wiley.

42. Rivest, R. L. 1990. Cryptography. In *Handbook of Theoretical Computer Science, Vol. A.,* Chapter 13, J. van Leeeuwen, ed. Amsterdam: North Holland, 717–755.

43. Ruzzo, W. L. 1981. On uniform circuit complexity. *J. Computer and System Sciences,* Vol. 22, 365–383.

44. Schwartz, J. T. 1980. Ultracomputers. *ACM Transactions on Programming Languages,* Vol. 2, 4, 484–521.

45. Schwarz, G. 1992. Connectionism, processing and memory. *Connection Science,* Vol. 4, 207–227.

46. Seiferas, J. I. 1990. Machine independent complexity theory. In *Handbook of Theoretical Computer Science, Vol. A,* Chapter 3, J. van Leeeuwen, ed. Amsterdam: North Holland, 165–186.

47. Simpson, P. K. 1990. *Artificial Neural Systems.* New York: Pergamon Press.

48. Skillicorn, D.B. 1991. Models for practical parallel computation. *International Journal of Parallel Programming,* Vol. 20, 133–158.

49. Stockmeyer, L. 1987. Classifying the computational complexity of problems. *J. Symbolic Logic,* Vol. 52, 1–43.

50. Storer, J.A. 1989. *Data Compression.* Rockfall, Md.: Computer Science Press.

51. Strassen, V. 1990. Algebraic complexity theory. In *Handbook of Theoretical Computer Science, Vol. A,* Chapter 11, J. van Leeeuwen, ed. Amsterdam: North Holland, 632–672.

52. Takefuji, T. 1992. *Neural Network Parallel Computing.* Boston: Kluwer Academic Publishers.

53. Traub, J. F., G. W. Wasilkowski, and H. Wozniakowski. 1988. *Information Based Complexity.* New York: Academic Press, New York.

54. Valiant, L. G. 1990. General purpose parallel architectures. In *Handbook of Theoretical Computer Science, Vol. A,* Chapter 18, J. van Leeeuwen, ed. Amsterdam: North Holland, 943–973.

55. Valiant, L. G. 1990. A bridging model for parallel computation. *Comm. ACM,* Vol. 33, 103–111.

56. van Emde Boas, P. 1990. Machine models and simulation. In *Handbook of Theoretical Computer Science, Vol. A,* Chapter 1, J. van Leeeuwen, ed. Amsterdam: North Holland, 1–66.

57. Wagner, K. and G. Wechsung. 1986. *Computational Complexity.* Boston: D. Reidel Publishing.

58. Wasserman, P. D. 1989. *Neural Computing-Theory and Practice.* New York: Van Nostrand Reinhold.

59. Wiedermann, J. 1989. On the computational efficiency of symmetric neural networks. *Lecture Notes on Computer Science, Vol 379,* New York: Springer Verlag, 545–552.

5

Distributed Computing Theory

Hagit Attiya

Distributed computer systems have become very common as organizations employ networking to share resources, enhance communication, and increase performance. Examples of distributed systems range from the Internet, to workstations on a local area network (LAN) within a building, to processors within a single multiprocessor. They are characterized by the presence of independent activities, loosely coupled parallelism, heterogeneous software and hardware. The major obstacle in distributed computing is *uncertainty* due to differing processor speeds, varying communication delays, occasional failures of components, and interactive behavior. Consequently, it is difficult but necessary to reason about distributed systems in order to assure correct operation.

Taking a formal approach to studying distributed systems can make it easier to reason about them, as has been successfully done for sequential systems in classical algorithm analysis. The theoretical paradigm begins by identifying and abstracting out fundamental problems and stating them precisely. Then, algorithms to solve these problems can be designed, proven correct, and analyzed for their costs. Finally, impossibility results and lower bounds can be shown for the problems.

The potential payoffs are many. First, carefully specifying the problem to be solved and its environment will clarify the solution process. Just as software engineering methods provide checks to ensure nothing gets left out, so formal models of distributed systems lead to a more complete definition of the requirements. Second, rigorous description and analysis increase confidence in the correctness of a proposed solution. Third, lower bound and impossibility results indicate inherent limitations in solving the problem in a particular environment and may suggest other directions, such as modifying the requirements or providing an environment with stronger guarantees.

Some application areas that have provided classic problems are operating systems, distributed database systems, process control, communication networks, and multiprocessor architectures. These varied settings lead to a variety of models for distributed computing. In this chapter, we discuss the two major models studied in the theory of distributed computing—*massage passing* and *shared memory.* The message-passing model is helpful in investigating computation in communication networks, while the shared memory model is more suited for describing multiprocessors. Historically, certain problems were studied more in the message-passing model, while others were studied more in the shared-memory model; this was motivated by intended applications. Recently, several transformations between the models were presented; these allow us to carry results proven for one model to the other.

Message passing systems are described by a *communication graph,* where the nodes of the graph represent the processors, and (undirected) edges represent two-way communication links between processors. Each processor is an independent processing unit equipped with local memory, and each runs a local program. The local programs contain internal operations, sending messages, and receiving messages. A distributed algorithm is a collection of local programs for the different processors. Executions of the algorithms are produced by running the local programs independently (under some restrictions).

An important characteristic of the system is the degree of synchrony. At one extreme, the system can be *synchronous,* where the computation is performed in rounds. At the beginning of a round, each processor sends messages, and waits to receive messages that were sent by its neighbors in this round. Upon receiving these messages, the processor performs some internal operations and then decides what messages to send in the next round. At the other extreme, in an *asynchronous* system, messages incur an unbounded and unpredictable (but finite) delay, and processors take steps at arbitrary rates. There are also intermediate models of partially synchronous systems, where there are various limitations on message delivery time and processors' step time.

An important (and often confusing) feature of distributed systems is that seemingly minor changes in the assumptions on the system may have drastic implications. For example, the capabilities of synchronous and asynchronous systems are very different. There are problems that may be solved efficiently in the synchronous model, whereas in the asynchronous model many resources are required for solving them. Moreover, there are problems that can be solved in the synchronous model but not in the asynchronous model.

In a *shared memory* system, processors communicate via a common memory area that contains a set of shared variables (also called *registers* or *objects*). In this model, the local programs contain internal operations and operations that access the shared memory. Here, we only consider asynchronous shared memory systems, in which processors take steps at arbitrary rates.*

Several types of shared variables can be employed. The most common type is the *read/write* register, in which the atomic operations are reads and writes. Read/write registers are further characterized according to their access patterns; that is, how many processors can access a specific variable. Other types of shared variables support more powerful atomic operations like *read-modify-write, test&set* or *compare & swap.* Not surprisingly, the type of shared variables used for communication influences the possibility and the complexity of solving a given problem.

After presenting a formal model of a distributed system (Section 5.1), the first part of this chapter is dedicated to reliable systems where no failures occur (Sections 5.2 through 5.6). The second part of the chapter addresses various aspects of tolerating failures in message passing and shared memory systems (Sections 5.7 through 5.10). We conclude, in Section 5.11, with some final comments.

---

*Synchronous shared memory systems were studied in the PRAM model of parallel computation, discussed elsewhere in this book.

Disclaimer: It is impossible to present in this limited space all the beautiful topics studied in the theory of distributed computing. I am aware of several important results that I do not mention, or discuss only briefly, and I am sure there are many others that I am forgetting. Obviously, the material in this chapter is slanted toward the areas that I am interested in, although I have attempted to minimize this bias.

5.1   The Computation Model

Here we outline the basic elements of our formal model of a distributed system. With these simplifications, the model shares many ideas with other models, e.g., see Refs. [53, 67]. In particular, it is a simplified version of the I/O Automaton model of Lynch and Tuttle [73]; our model does not incorporate composition of automata and does not address general issues of fairness in the composed system. We first describe message passing systems and then how to model shared memory systems. The model presented here is more detailed than needed for the rest of the chapter; it is included to provide the reader with an example of what a formal model of distributed systems looks like.

We represent processors as automata. Two types of events can occur at a processor, triggering its transition function: (1) an internal computation step, which can result in sending messages in addition to a local state change, and (2) the delivery of a pending message.

In more detail, a system consists of n processors $p_1, \ldots, p_n$ and the network *net*. Each processor p_i is modeled as a (possibly infinite) state machine with state set Q_i. The state set Q_j contains a distinguished initial state, $q_{0,i}$. We assume the state of processor p_i contains a special component, $buff_i$, in which incoming messages are buffered. Processor p_i's transition function takes as input a state of p_i and produces as output another state of p_i and a set of messages. Each message consists of the body of the message and indicates the sender and recipient. The network topology, in particular p_i's neighbors in the communication graph, is indicated by restrictions on which processors may be the recipients of p_i's messages. The collection of n state machines (local programs) constitutes the *algorithm*. The network *net* is modeled by the set of messages that are in transit.

A *configuration* is a vector $C = (q_1, \ldots, q_{n,M})$ where q_i is the local state of p_i, and M is the state of the network, i.e., the set of messages in transit. The initial configuration is the vector $(q_{0,1}, \ldots, q_{0,n}, \phi)$; each processor is in its initial state and the network is empty.

We model an execution of the system as a sequence of configurations alternating with events. Each event is either a *computation event,* representing a computation step of a single processor, or a *delivery event,* representing the delivery of a message to a processor.

A computation event is specified by *comp(i),* where i is the index of the processor taking the step. The result of the computation step is that p_i changes state and sends some set of messages. Each delivery event has the form *del(i, j, M),* where i and j are processor indices and M is a message body. In the delivery step associated with event *del(i, j, M)* the message M from p_i is removed from the network and added to $buff_j$.

An *execution* α of a system is a (finite or infinite) sequence of the following form:

$$C_0, \phi_0, C_1, \phi_1, C_2, \phi_2, \ldots$$

where C_k are configurations, and ϕ_k are events. If α is finite, then it must end in a configuration. Furthermore, the following conditions must be satisfied:

- $C\phi$ is the initial configuration.
- If $\phi_k = del(i, j, M)$, then M is in the network in C_k, M is not in the network in C_{k+1}, and M is added to $buff_j$ in C_{k+1}.

■ If $\phi_k = comp(i)$, then the only changes in going from C_k to C_{k+1} are that p_i changes state according to its transition function, and a set of messages is added to the network according to p_i's transition function. These messages are said to be *sent* at this event.

With each execution, we associate a *schedule* which is the sequence of events in the execution; that is, $\phi_0, \phi_1, \phi_2, \dots$ .Notice that if the local programs are deterministic, then the execution is uniquely determined by the initial configuration and the schedule.

In modeling particular types of systems, we sometimes put further requirements on executions (see the synchronous model below for an example). In addition to the safety properties embodied by the definition of an execution, we will also require various liveness properties. They are captured by the notion of *admissibility*. The term can also be applied to a schedule.

In the asynchronous model, an execution is *admissible* if each processor has an infinite number of computation events, and every message sent is eventually delivered. The requirement for an infinite number of computation events models the fact that processors do not fail. It does not imply that the processors local program must contain an infinite loop; the informal notion of termination of an algorithm can be accommodated by having the transition function not change the processor's state after a certain point.

In the asynchronous model, we typically assume that processor p_i has a computation event immediately after each delivery event of the form $del(i, j, M)$. In this case, we merge the message delivery event and the computation event and refer to the computation taken by the processor upon receiving the message.

In the synchronous model, processors execute in lock-step. The definition of an execution is further constrained as follows. The computation events appear in *rounds*. We assume that each processor has exactly one computation event in each round and that computation events of round r appear after all computation events of round $r - 1$. Furthermore, we assume all messages sent in round r are delivered before the computation events of round $r + 1$. An execution is admissible if it is infinite. Because of the round structure, this implies that every processor takes an infinite number of computation steps, and every message sent is delivered.

We now describe the formal model of shared memory systems. As in the case of message passing systems, we model processors as state machines and model executions as alternating sequences of configurations and events. The only difference is the nature of configurations and events. Below we discuss in detail the new features of the model and only briefly mention those that are similar to the message passing model.

We assume the system contains n processors, $p_1, \dots, p_n$, and m registers, $R_1, \dots, R_n$. A configuration in the shared memory model is a vector $C = (q_1, \dots, q_n, r_1, \dots, r_m)$ where q_i is the local state of p_i, and r_j is the value of register R_j. In the initial configuration, all processors are in their (local) initial states and all registers contain some initial value. The *events* in a shared memory system are computation steps by the processors and are denoted by the index of the processor.

An *execution* of the algorithm is a (finite or infinite) sequence of the following form:

$$C_0, \phi_0, C_1, \phi_1, C_2, \phi_2 \dots$$

where C_k are configurations, C_0 is the initial configuration, and ϕ_k are events. Furthermore, the application of ϕ_k to C_k results in C_{k+1}, in the natural way. That is, if $\phi_k = i$, then C_{k+1} is the result of applying p_i's transition function to p_i's state in C_k, and applying p_i's memory access operations to the registers in C_k, in the obvious manner (and there are no other changes).

Sometimes there are further restrictions depending on the type of memory accesses we allow. For example, if we only have read/write registers than each transition either writes to a single register or reads from a single register. A *read-modify-write* operation accesses a single register, reads the current value, applies some function to it, and writes back the new value, all in a single atomic operation; the operation returns the old value of the register. A *test&set* register is a restricted read-modify-write register which supports two operations, *reset* and *test & set*. A *test & set* register contains a binary value; the reset operation sets the register to 0, while the test & set operation sets the register to 1; both operations return the previous value of the register.

Schedules are defined as in the message passing model. Note that a schedule in the shared memory model is just a sequence of processors' indices. An occurrence of index i in a schedule is referred to as a step of processor p_i in the schedule. We only consider asynchronous executions; an execution is admissible if each processor has an infinite number of computation events.

5.2 A Simple Example

In this section, we discuss a very simple example of a message passing system where the communication network is a tree. We show an algorithm to conduct simple computations in this system, and discuss how to measure its performance. In addition to providing a basic example of distributed algorithms, tree computation is an important building block in several distributed algorithms.

Assume the communication network is a tree; a single processor is distinguished as a root, and every processor knows which of its edges lead to its children and which lead to its parent (i.e., toward the root). Each processor p_i starts with some input value x_i, and processors want to collect these inputs at the root in order to compute some function of them. The following simple procedure, called *broadcast* and *converge-cast,* collects the inputs:

1. The root sends an `initiate` message to all its children.
2. Processor p_i receives an `initiate` and forwards it to its children.
3. If p_i is a leaf, then it sends `report(x_i)` message to its parent.
4. An internal node p_i of the tree waits until it receives `report` messages from all its children; p_i concatenates the inputs from its children with its own input to a string $\vec{x}$, and sends `report(`$\vec{x}$`)` to its parent.
5. The root collects all the `report` messages and reconstructs the set of inputs.

It is simple to see that this procedure collects all inputs at the root. Steps 1–2 are the *broadcast* phase, while Steps 3–5 are the *converge-cast* phase of the procedure. Note that for certain functions of the inputs, e.g., sum, there is no need to send the whole set of inputs in a `report` message—only the partial sums. Further note that if all processors need to know the inputs (or some function of them), then the root can initiate another broadcast phase with the appropriate information.

We would like to quantitatively measure the efficiency of this algorithm using the two major complexity measures used in the theory of distributed computing: number of messages and time.

The *message complexity* of the algorithm is the number of messages sent in executions of the algorithm. We usually consider the worst-case message complexity, which is the maximal number of messages sent in an admissible execution. Sometimes, especially for randomized algorithms, we study the average message complexity, which is taken over all admissible executions.

The broadcast and converge-cast algorithm described above requires the same number of messages, regardless of the execution. It is simple to see that exactly one `initiate` message and one report message are sent over each edge. Therefore, the total number of messages sent is $2(n-1)$, since in a tree with n nodes there are $n-1$ edges.

Measuring the *time complexity* of an algorithm in synchronous executions is fairly simple. It is defined as the number of rounds until termination of the algorithm. Once again, we can consider either the worst-case or the average-case time complexity. Defining the time complexity of an algorithm in asynchronous executions is less trivial. The best way is to assign occurrence times to the events in the execution under the restriction that the time between the sending of a message and its delivery is *at most* one. Note that there can be several ways to assign occurrence times to events in an execution. The time complexity of the execution is the maximal difference between the time assigned to the last event of the algorithm, and the time assigned to the beginning of the algorithm. Again, we can the take either the worst case or the average case over all admissible executions of the algorithm.

Consider a node p_i at distance ℓ from the root. It is simple to see that the time assigned to the delivery of the initiate message at p_i is at most ℓ (in both the synchronous and the asynchronous models). Let D be the depth of the tree (that is, the length of the longest path from the root to a leaf). By time D, all nodes receive `initiate` messages (under any time assignment to the events). At this point, report messages are sent from the leaves. By similar reasoning, we can see that by time $2D$ the root receives report messages from all its children.

We have seen that the existence of a predefined structure on the communication network allows the system to perform computations efficiently. Later, we show how to induce a tree structure on the network.

5.3 Leader Election

In this section, we study the leader election problem, in which the processors must "choose" one of them as a leader. The existence of a leader can simplify coordination among processors and is helpful in achieving fault-tolerance and saving resources. Furthermore, the techniques developed for the leader election problem are useful for other problems as well.

Informally, the problem is for each processor to eventually decide whether it is the leader or not, subject to the constraint that exactly one processor decides that it is the leader. In terms of our formal model, an algorithm is said to solve the leader election problem if it satisfies the following conditions:

- Each processor has a subset of *elected* states and a disjoint subset of *not-elected* states. Once a processor enters an elected (resp., not-elected) state, its transition function will only move it to another (or the same) elected (resp., not-elected) state.
- In every admissible execution, exactly one processor (the *leader*) enters an elected state, and all the remaining processors enter a not-elected state.

5.3.1 Leader election in rings

The *ring* is a very convenient structure for message passing systems, which corresponds to physical communication systems, e.g., token rings. We assume that the ring is *oriented;* that is, processors distinguish between the links to their left and right neighbors. Furthermore, if p_i is p_j's left neighbor, then p_j is p_i's right neighbor. A discussion of orientation appears in Ref. [14].

5.3.1.1 Anonymous rings. In the model we have defined, the transition function may depend on the processor's index; this can also be captured by assuming that each processor has a distinct integer identifier (ID). In contrast, a system is *anonymous* if the local programs of all processors are identical and the processors do not have access to their IDs. In this case, there is no deterministic leader election algorithm. Intuitively, in an anonymous ring, processors start in identical states, and this symmetry cannot be broken throughout the computation. For simplicity, we discuss this result in synchronous rings; this immediately implies the same result for asynchronous rings.

In any algorithm for an anonymous ring, all processors are identical and execute the same program (i.e., they have the same state machine). In a synchronous system, an algorithm proceeds in rounds. In the first round, a processor sends some initial set of messages. In the second round, the processor receives the messages sent in the first round, and executes some conditional statement that determines what messages should be sent in the second round. This continues until, at some round, the processor decides to enter an elected or a not-elected state. Assume that all processors in the anonymous ring start in the same state. Since they are identical, all processors send exactly the same messages in every round; thus, they all receive the same messages in every round. Consequently, if one of the processors terminates its program in an elected state, then so do all of the processors. This implies:

Theorem 5.3.1 (from Angluin [7]). *There is no algorithm for leader election in anonymous rings.*

This result can be extended to hold for any system in which the communication graph is very regular, e.g., a clique. One way to overcome this impossibility result is by randomization, i.e., by employing algorithms that flip coins; the reader is referred to Refs. [58, 60] for further information. Another way is to compute specific functions that do not require symmetry breaking (see Ref. [14]). For the rest of this chapter however, we assume that processors have distinct IDs.

5.3.1.2 Asynchronous rings. When the ring is asynchronous, there exists a very simple leader election algorithm that requires $O(n^2)$ messages [70]. The algorithm elects the processor with the smallest ID. In the algorithm, each processor sends a message with its ID to its left neighbor and then waits for messages from its right neighbor. When it receives a message, it checks the ID in this message. If the ID is smaller than its own ID, it forwards the message to the left; otherwise, it "swallows" the message and does not forward it. If a processor receives a message with its own ID, it sends a termination message to its left neighbor and terminates in an elected state. A processor that receives a termination message forwards it to the left and terminates in a non-elected state. Note that only the message of the processor with the minimal ID is never swallowed. Therefore, only the processor with the minimal ID receives a message with its own ID and terminates in an elected state. All other processors receive termination messages and terminate in non-elected states, implying the correctness of the algorithm.

The above algorithm has an execution in which $\Theta(n^2)$ messages are sent (but no more than that). There is a better algorithm that elects a leader in $O(n \log n)$ messages [57]. The algorithm presented here assumes that communication links are bidirectional; algorithms with the same message complexity exist also for the case where links are unidirectional [40, 77].

To describe the algorithm, we define the *l-neighborhood* of a processor p_i in the ring to be the processors that are at most l away from p_i in the ring (there are $2l + 1$ such processors, including p_i itself). The algorithm operates in phases; in phase ℓ, a processor checks if it is has the minimum ID in its 2^ℓ-neighborhood. Only processors that succeed in phase ℓ continue to phase $\ell + 1$. Therefore, only a constant fraction of the processors proceed to higher phases. Eventually, only one processor remains and it is the elected leader.

In more detail, in phase 0, each processor sends a message containing its ID to its 1-neighborhood, i.e., to each of its two neighbors. If the ID of the neighbor receiving the message is smaller than the one in the message, it swallows the message; otherwise, it returns the message. If the processor's messages return from both its neighbors, then it has the minimum ID in its 1-neighborhood, and continues to phase 1. In phase ℓ, if p_i is has the minimum ID in its $2^{\ell-1}$-neighborhood, then it sends a message with its ID in each direction. Each such message traverses 2^ℓ processors one by one. A message is swallowed by a processor if it contains an ID that is larger than its own ID. If the message arrives at the last processor in the 2^ℓ-neighborhood without being swallowed, then that last processor returns the message to p_i. If p_i's messages return from both directions, it has the minimum ID in its 2^ℓ-neighborhood, and it starts executing phase $\ell + 1$. A processor that receives on its left edge a message that it sent on its right edge (or vice versa) terminates the algorithm in an elected state.

It is simple to see why the algorithm is correct. To measure its message complexity, note that for any $\ell > 1$, the number of processors that send a message at phase ℓ is at most

$$\frac{n}{2^{\ell-1}}$$

Since there are n processors, there are at most $\lceil \log n \rceil$ phases. Furthermore, each temporary leader in phase e is responsible for at most $4 \cdot 2^\ell$ messages. Thus, the total number of messages sent in each phase is at most $8n$. We have:

Theorem 5.3.2 (from Hirschberg and Sinclair [57]). *There exists a leader election algorithm for asynchronous rings whose worst-cast message complexity is* $O(n \log n)$.

The best message complexity to date for this problem is $1.271 n \log n + O(n)$ [59]. This bound is optimal (within constant multiplicative factor). That is, any algorithm for electing a leader in an asynchronous ring sends at least $\Omega(n \log n)$ messages [28]. We outline here the ideas of this lower bound proof for the slightly simpler case where the elected leader must be the processor with the minimal ID in the ring and all the processors must know who is the elected leader. The result for the more general problem follows by reduction.

Assume A is an algorithm that solves the above variant of the leader election problem. It can be shown that there exists an execution of A in which $\Omega(n \log n)$ messages are sent. This is done by building a "wasteful" execution of the algorithm for rings of size $n/2$, in which many messages are sent. Then, we "paste together" two different rings of size $n/2$ to form a ring of size n, in such a way that we can combine the wasteful executions of the smaller rings and force $\Theta(n)$ additional messages to be sent.

A schedule can be "pasted" if it is *open;* that is, there exists an edge on which no message is delivered. Note that an open schedule need not be admissible; it can be finite, and processors may have not terminated yet. The argument assumes that the algorithm works in the same manner for every ring size. Intuitively, since the processors do not know the size of the ring, we can paste together two open schedules of two small rings to form an open schedule of a larger ring.

Assume that n is an integral power of 2. The proof shows by induction that there exists a ring of size n that has an open schedule in which at least $M(n)$ messages are sent, where $M(2) = 1$ and

$$M(n) = 2M\left(\frac{n}{2}\right) + \frac{1}{2}\left(\frac{n}{2} - 1\right) \text{ for } n > 2$$

Since $M(n) = \Theta(n \log n)$, this implies the desired lower bound.

For the base case, consider a ring of two processors, p_1 and p_2, where the ID of p_1 is smaller than the ID of p_2. Since p_2 must know p_1's ID, it must receive some message. The execution in which only one message is sent (but is not received) is clearly open.

As mentioned before, the inductive step of the proof takes two open schedules, pastes them together and forces extra messages to be sent. Intuitively, one can see why two open schedules can be pasted together and still behave the same. The key step, however, is forcing the additional messages to be sent. The idea is that after the two smaller rings are pasted together, at least one half must learn about the leader of the other half (where the minimum is located). We unblock the messages delayed on the connecting edges, continue the schedule, arguing that many messages must be sent. Our main problem is how to do that in a way that yields an open schedule on the bigger ring (so that the argument can be applied inductively). If we pick in advance which of the two edges connecting the two parts to unblock, then the algorithm can choose to wait for messages on the other edge. To avoid this problem, we first create a "test" schedule, learning which of the two edges, when unblocked, causes the larger number of messages to be sent. We then go back to our original pasted schedule and only unblock that edge. Omitting many details, these ideas can be used to prove:

Theorem 5.3.3 (from Burns [28]). *Any algorithm that elects a leader in an asynchronous ring has worst-case message complexity $\Omega(n \log n)$.*

5.3.1.3 Synchronous rings. We now turn to synchronous rings. The proof of the $\Omega(n \log n)$ lower bound for leader election in an asynchronous ring heavily depends on delaying messages arbitrarily long. It is natural to wonder whether better results can be achieved in the synchronous model, where message delay is fixed. In the synchronous model, information can be obtained not only by receiving a message but also by *not* receiving a message in a certain round. This is exemplified by the following simple algorithm that elects the processor with the minimal ID to be the leader. It works in phases, each consisting of n rounds. At the beginning of the ith phase, if a processor's ID is i, and it has not terminated yet, the processor sends a message around the ring and terminates in an elected state. Since a phase includes n rounds, the message has a chance for reaching every processor on the ring. If the processor's ID is not i and it receives a message in phase i, it forwards the message and terminates the algorithm in a non-elected state. Notice that if no message is received until the ith phase, all processors know that the minimal ID in the ring is at least i. Clearly, the algorithm elects the unique processor with minimal ID as a leader. Moreover, exactly n messages are sent in the algorithm because only the processor with minimal ID sends a message. The algorithm depends on knowing n and synchronized start of the processors; there exist more complicated algorithms with the same message complexity that do not make these assumptions [49, 88].

Obviously, $O(n)$ is optimal message complexity for leader election. However, the above mentioned leader election algorithms with linear message complexity have two undesirable properties: They use the IDs in a nonstandard manner (to decide when to send a message) and the number of rounds in each execution depends on the IDs of processors. It can be shown that any leader election algorithm that uses IDs for comparison only, or whose running time does not depend on the IDs, requires $\Omega(n \log n)$ messages.

Roughly speaking, an algorithm is *comparison based* if it behaves the same on rings with ID assignments that are order-equivalent; that is, processors' IDs in matching locations in the rings satisfy the same order relationships. An algorithm behaves the same on two rings if messages are sent at the same rounds and the same decisions are made.

The lower-bound proof considers an ID assignment that is highly symmetric in its order patterns; that is, there are many neighborhoods with the same order relations between IDs in matching locations. Very informally, as long as two processors see neighborhoods with the same order pattern, they behave the same under A. We derive the lower bound by executing A on a highly symmetric ring and arguing that, if a processor sends a message in a certain round, then all processors with order equivalent neighborhoods also send a message in that round. Constructions of highly symmetric ID assignments appear in Refs. [14, 49].

The result for time bounded algorithms is derived by using the finite version of Ramsey's theorem. The proof shows that any time bounded algorithm behaves as a comparison based algorithm on a subset of the inputs, provided that the input set is sufficiently large.

Theorem 5.3.4 (from Frederickson and Lynch [49]). *If a leader election algorithm is either comparison based or time bounded, then it requires $\Omega(n \log n)$ messages.*

5.3.2   Other special topologies

Another special topology is the *clique,* that is, a system with a complete communication graph. In such systems, every processor is connected by $n - 1$ bidirectional edges, numbered $1, \ldots, n - 1$, to all other processors. There exist leader election algorithms for a complete asynchronous network with message complexity $O(n \log n)$ messages [65, 78]. Later work improved the time complexity of these algorithms [3]. This bound is tight, and there exists an $\Omega(n \log n)$ lower bound for electing a leader even in a synchronous complete network, assuming that processors do not start executing the algorithm simultaneously [3, 65].

Unlike the lower bound proved for leader election in synchronous rings, the lower bound proof for cliques does not put any restrictions on the algorithm (e.g., that it be comparison based or time bounded). Note that this implies that the message cost of electing a leader in a complete network is essentially the same in synchronous and asynchronous systems. The upper bound has another interesting implication. It shows that it is not necessary to explore each and every edge in the system. That is, although the number of edges in a complete graph is $O(n^2)$, a leader can be elected with as few as $O(n \log n)$ messages.

In general, it was shown that a leader election algorithm with low message complexity exists if there is a method to traverse the communication network with a small number of messages [10, 51, 64].

5.3.3   General networks

We now consider asynchronous systems with an arbitrary (connected) communication graph. We study the problem of finding a *minimum spanning tree* (MST) of the communication graph, assuming that edges have unique (positive) weights. As we have seen in Section 5.2, having a spanning tree for the network simplifies many control problems in the network. Given a spanning tree a leader can be elected simply by picking the root of the tree. When edges' weights represent the cost of sending messages along the corresponding link, using an MST allows messages to be sent along cheapest routes. Below, we discuss an algorithm for finding the MST of a system with communication graph $G(V, E)$; the total message complexity of the algorithm is $O(n \log n + m)$, where n is the number of nodes, and m is the number of edges [52].

5.3.3.1 Preliminaries.

Consider a system with an arbitrary undirected, connected communication graph $G(V, E)$ with n nodes and m edges. With each edge, we associate a weight that is a real number. The problem is to find the spanning tree of G with the minimum sum of edges' weights, called the *minimum spanning tree* (MST). At the end of the algorithm, each processor should know which of its adjacent edges belongs to the MST (and which of them is oriented toward the root). Notice that by Theorem 5.3.1, it is necessary to assume that either edges have distinct weights or nodes have unique IDs. It can be shown that in this case, G has a unique MST.

A *fragment* is a connected subgraph of the MST. An outgoing edge of a fragment is an edge with one adjacent node in the fragment and another outside the fragment. We have:

Lemma 5.3.5 *Let $G(V, E)$ be a connected graph with distinct weights, and let $T(V, E')$ be its unique MST. For any fragment F of T, the outgoing edge of F with minimum weight is in T.*

This lemma implies that if an algorithm starts with fragments that consist of single nodes and then proceeds by combining fragments using their minimum outgoing edges until only one fragment remains, then this fragment is the MST. This is the basis of the well known sequential algorithms (Prim-Dijkstra and Kruskal) for finding an MST for a graph. We next describe how to employ these ideas in a distributed MST algorithm.

5.3.3.2 The distributed MST algorithm.

The distributed algorithm has the same general structure as the sequential algorithms. At the beginning, each node is a separate fragment. In each stage of the algorithm, each fragment finds its minimum outgoing edge and attempts to combine with the fragment at the other end of the edge. Lemma 5.3.5 implies that this combination yields a new fragment of the MST. The algorithm ends when there is only one fragment, which is the MST. The distributed algorithm differs from the sequential algorithms in the parallelism of the fragments' combinations. There are two major difficulties in applying the sequential ideas in the distributed setting.

One difficulty is the need that all nodes of a fragment coordinate their actions. For example, they have to cooperate in order to find the fragment's minimum outgoing edge. In addition, they have to know that they are in the same fragment. We can associate with every fragment an ID that is known to all nodes in the fragment. In this case, two neighboring nodes can compare their IDs and find out whether they are in the same fragment or not. Since the fragmentation of the graph is dynamic, care should be taken to avoid the following synchronization problem: two nodes can be in the same fragment but not be aware of this fact yet.

Another difficulty is that there are many ways to combine fragments in order to end up with one big fragment. By maintaining an MST for each fragment and using broadcast and converge-case (Section 5.2), it is possible to find the outgoing edge with minimum weight with message complexity proportional to the fragment's size. Therefore, combining fragments of (approximately) the same size balances the number of messages sent and keeps it small. In contrast, if we allow one big fragment to combine with a single node in each phase of the algorithm until it spans the whole MST, then the total number of messages is $O(n^2)$.

In general, the structure of the algorithm is as follows: Each processor starts the algorithm as an individual fragment. Each fragment tries to join with other fragments. At each stage of the algorithm, each fragment has an MST with a unique edge of the MST called the *core* of the fragment. The two nodes adjacent to the core of a fragment coordinate the activity of the fragment, and all nodes of the fragment's MST are oriented toward the core. After a new fragment is created, its first action is to choose its minimum outgoing edge. First, every processor in the fragment finds the minimum outgoing edge that is adjacent to it. Then the minimum edge is chosen among the edges that were found by the nodes. Finally, the fragment tries to join with the fragment on the other end of the chosen edge by sending a connection request.

A fragment is identified by the weight of its *core* and its *level*. There are two ways to join two fragments. The first way is by *combination* of fragments with the same level and the same minimum outgoing edge. In this case, a new fragment with a new core and level is created. The second way is by *absorption,* which happens when a fragment with small level sends a connection request to a fragment with bigger level.

A fragment containing only a single node is defined to be at level 0. When two fragments of level $L - 1$ are combined, a new fragment of level L is formed. The edge on which the last combination took place is the core of the new fragment. When a fragment is absorbed into another fragment, it takes on the identity of the other fragment. After a new fragment is created, as a result of combination or absorption, the ID of this new fragment is sent to all its nodes, and a new phase begins.

5.3.3.3 Correctness and complexity. Due to its complexity we do not discuss the proof of correctness for this algorithm. There are several attempts to prove the correctness of this algorithm formally and precisely (see Refs. [36, 84, 89]).

To bound the number of messages sent during an execution of the algorithm, note that a fragment of level L contains at least $2L$ nodes. Therefore, a fragment can go through, at most, $\lceil \log n \rceil$ levels.

Some types of messages are sent at most once in each direction of a link; the total number of messages of these types is $O(m)$. All other messages are sent by a node at most once each time the level of a fragment increases; the total number of messages of these types is $O(n \log n)$. Therefore, the total message complexity is $O(n \log n + m)$. The time complexity of the algorithm is $O(n \log n)$. Further work improved the time complexity of finding an MST, finally yielding an $O(n)$ time (and $O(n \log n + m)$ messages) algorithm [16].

5.4   Sparse Network Covers and Their Applications

In this section, we discuss *sparse network covers,* a graph theoretic concept with several applications in distributed algorithms. Very roughly, we take a collection of subgraphs that cover all the nodes of the graph, and conduct computations inside each subgraph, and between neighboring subgraphs. If the cover is sparse (i.e., nodes are not covered by too many subgraphs) this method can be tuned to save a lot on communication, time, and memory. In this section, we present two applications of this method: to message routing and for reducing message complexity of algorithms. Another application appears later, in Section 5.5.3.

5.4.1 Sparse covers

We start with some graph theoretic notions and terminology. Consider a graph $G(V, E)$; as before, we denote $n = |V|$ and $m = |E|$. Let $dist(v,u)$ be the *distance* between v and u, that is, the length of the shortest path in G between the nodes v and u. The diameter of the graph is the maximum distance between two nodes, that is,

$$diam\,(G)\ =\ max_{v,\,u\,\in\,V}\ dist\,(v, u)$$

A *cluster* of G is a set of nodes $S \subseteq V$ that induces a connected subgraph; the diameter of S, $diam(S)$, is calculated in the graph induced by S. Let $\mathbf{S} = \{S_1,\ldots,S_k\}$ be a collection of clusters and denote $|\mathbf{S}| = k$ the number of clusters in $\mathbf{S}$. $\mathbf{S}$ is a *cover* of G if

$$\bigcup_{i=1}^{k} S_i\ =\ V$$

Denote $diam(\mathbf{S}) = max_{i=1\ldots k}\ diam(S_i)$. The *volume* of $\mathbf{S}$ is

$$vol\,(\mathbf{S})\ =\ \sum_{i=1}^{k} |S_i|$$

If $\mathbf{S}$ covers G, then $vol(\mathbf{S}) > n$; in this case, the volume captures how many "repetitions" there are in $\mathbf{S}$. In particular, $vol(\mathbf{S})/n$ is the average number of occurrences of any node in clusters $S_i \in \mathbf{S}$. Given two collections of clusters $\mathbf{S} = \{S_1,\ldots,S_k\}$ and $\mathcal{T} = \{T_1,\ldots,T_r\}$, we say that $\mathcal{T}$ is a *coarsening* of $\mathbf{S}$ if for every $S_i \in \mathbf{S}$ there exists $T_j \in \mathcal{T}$ such that $S_i \subseteq T_j$.

Given a cover $\mathbf{S}$ of G, we would like to obtain a cover of G with a smaller volume to reduce the average number of clusters a node belongs to. A simple way to do so is by unifying clusters from $\mathbf{S}$; however, this tends to increase the diameter of the cover. The next theorem shows how to trade the volume of the cover with its diameter.

Theorem 5.4.1 (from Awerbuch and Peleg [21]). *For any graph $G(V, E)$, a cover $\mathbf{S}$ of G, and an integer $k \geq 1$, there exists a coarsening $\mathcal{T}$ of $\mathbf{S}$ such that*

1. $diam(\mathcal{T}) \leq 4k \cdot diam(\mathbf{S})$, *and*

2. $vol(\mathcal{T}) \leq |\mathbf{S}|^{1 + \frac{1}{k}}$

The proof of this theorem is constructive and provides a sequential algorithm for finding the coarsening. There exist distributed algorithms for constructing the coarsening; we do not discuss them here and refer the reader to a recent paper on this topic [17].

5.4.2 Routing messages

We now discuss the problem of routing messages in a communication network and how to employ sparse covers for solving it efficiently, following Ref. [21].

5.4.2.1 The problem. Assume some processor p_i wants to send a message to another processor p_j. A routing algorithm specifies the route on which the message will be sent in the network by telling each intermediate node on the route on which outgoing edge the message should be sent, depending on the destination. We are interested in the number of messages that are required for transferring the message and the amount of memory dedicated in each processor to the routing information. There is a trade-off between these two measures, as demonstrated by the following two solutions.

One solution is to route messages on the shortest path between the originator of the message and its destination. Specifically, each processor p_i maintains a routing table which contains the identity of p_i's neighbor that is on the shortest path to p_j, for every processor p_j. When p_i has to send (or forward) a message addressed to p_j, it uses the table to find on which link to send the message. Clearly, this guarantees optimal routes. However, it has high memory requirements since the routing table maintained by each processor requires $O(n \log n)$ bits, yielding a total of $O(n^2 \log n)$ bits in the whole network.

Another solution is to send the message by *flooding* without using any routing information. That is, if p_i wants to send a message M to p_j, it sends M to all its neighbors, which forward it to all their neighbors, and so on. If there is a path from p_i to p_j, then the message will eventually arrive at p_j. This solution requires no memory. However, it has high message complexity, as $O(m)$ messages are sent even if the source and the destination are very close.

Below, we describe a scheme which yields a better trade-off between memory and messages, using sparse covers. To describe it, we need to define our cost measure more precisely. For a fixed network G and algorithm A, let $cost(v, u)$ be the number of messages sent by A to deliver a message from v to u. The best we could hope for is that $cost(v, u)$ is $dist(v, u)$. Therefore, a good measure for the quality of the algorithm's behavior is how close $cost(v, u)$ is to $dist(v, u)$, that is, how much overhead the algorithm introduces. Denote

$$stretch = max_{v, u} \frac{cost(v, u)}{dist(v, u)}$$

and note that a small stretch value implies that the algorithm uses routes that are not much longer then the best routes available. For example, the stretch of the first solution mentioned above is 1, while the stretch of the second solution is m.

5.4.2.2 The general routing scheme. For every k, we present an algorithm with stretch factor $O(k)$, which uses

$$O\left(n^{1 + \frac{1}{k}} \log^2 n \right)$$

total memory bits. Note that this scheme provides a trade-off in the form of a reduction in the stretch factor at the cost of an increase in the memory requirements, and vice versa.

The main idea of the algorithm is to construct a hierarchy of schemes. Each scheme is a *regional* (C, l)-*routing scheme,* with the following property: for any pair of nodes v and u, the scheme delivers a message sent from v to u if $dist(v, u) \leq l$ and notifies v otherwise; furthermore, $O(C \cdot l)$ messages are sent in both cases.

We use regional (C, l)-routing schemes, for $l = 2, 4, ..., 2^{\lceil \log D \rceil}$, where $D = diam(G)$. Let R_i be the regional $(C, 2^i)$-routing scheme. To send a message from v to u, routing is attempted in $R_1, ..., R_{\lceil \log D \rceil}$ until some regional scheme succeeds. The properties of the regional routing schemes imply that routing succeeds in $R_{\lceil \log d \rceil}$ at the latest, where

$d = dist(v, u)$. This implies the correctness of the general scheme and also implies that the total number of messages sent is at most

$$\sum_{i=0}^{\lceil \log d \rceil} C \cdot 2^i = O(C \cdot d)$$

That is, the stretch factor of the scheme is C.

Intuitively, the scheme maintains detailed information on close-by nodes and less information on nodes that are far away. When the distance is great, even an inefficient solution has a good stretch factor.

5.4.2.3 The regional routing scheme. We now describe how to use sparse covers to construct a regional (k, l)-routing scheme for a given parameter k. The regional scheme requires a total of

$$O\left(n^{1 + \frac{1}{k}} \log n \right)$$

memory bits in the nodes.

Recall that the l-*neighborhood* of a node v is the set of processors whose distance from v is at most l; clearly, it is a cluster. Let $\mathcal{L}$ be the set containing the l-neighborhoods of all nodes in the graph. Note that $\mathcal{L}$ covers the graph, $|\mathcal{L}| = n$ and $diam(\mathcal{L}) = 2l$. By applying Theorem 5.4.1 to $\mathcal{L}$ with some $k \geq 1$, we obtain a cover $\mathcal{T}$ such that the l-neighborhood of every node is a subset of some cluster in $\mathcal{T}$, $diam(\mathcal{T}) < 4k \bullet 2l$, and

$$vol(\mathcal{T}) \leq n^{1 + \frac{1}{k}}$$

For each node, we distinguish one of the clusters that contain its l-neighborhood as its *home cluster*. In each cluster, we choose some node as a root, and construct a *cluster tree*, which is the shortest-path tree from the root to all nodes in the cluster.

A simple scheme for routing inside the cluster is as follows. Number the nodes of each cluster by running a depth first search (DFS) on the cluster tree, and denote the DFS number of a node v by $dfs(v)$. Every node saves its own DFS number, as well as the DFS numbers of its children in the cluster tree. The root maintains a table with the DFS number of each node v in the cluster. The routing of a message from node v to another node u is done through the root of the cluster: v sends the message up the tree to the root, and the root sends the message to u by propagating it down the tree. At node w with children $w_1, \ldots, w_i$ (ordered by increasing DFS numbers) the message is sent to the child w_i such that $dfs(w_i) \leq dfs(u) < dfs(w_{i+1})$. (Setting $dfs(w_0) = -\infty$ and $dfs(w_{l+1}) = \infty$.)

Clearly, at most $2 \bullet diam(\mathcal{T})$ messages are sent, since in the worst case we use the longest path in the cluster twice. This implies that this is a regional (k, l)-routing scheme. Note that the amount of information maintained in the root is proportional to the number of nodes in the cluster times $\log n$ (in bits). This also dominates the total amount of information maintained at other nodes in the cluster. Therefore, the total amount of information maintained in all nodes is proportional to the volume of $\mathcal{T}$ times $\log n$, which is

$$O\left(n^{1+\frac{1}{k}} \log n\right) \text{ bits}$$

Therefore the total memory overhead of the routing tables, using log n regional routing schemes, is

$$O\left(n^{1+\frac{1}{k}} \log^2 n\right) \text{ bits}$$

This solution optimizes the total memory requirements of the routing scheme. However, some nodes (e.g., the cluster roots or nodes that belong to many clusters) maintain much more information than others. It is possible to optimize also the maximal memory requirements; see Ref. [21] for further details.

5.4.3   Reducing message complexity of distributed algorithms

Another application of sparse covers, which we only discuss briefly, is to reduce the message complexity of distributed algorithms. There is a transformation that takes a distributed algorithm whose message complexity is $O(c \cdot m)$ and produces an algorithm to solve the same problem with message complexity $O(c \cdot n \log n + m \log n)$ [5]. An interesting application of this transformation is to find all the shortest paths in the network with $O(n^2 \log n)$ messages.

Roughly, the transformation uses a sparse cover; in each cluster there is a central node and there are simple paths connecting the centers of two clusters that share a node. To run the algorithm, each central node executes the algorithm for all nodes in its cluster. Messages sent within the cluster need not be sent at all. Messages between nodes in different clusters are sent between the corresponding central nodes. The cost of sending messages from a specific node to all its neighbors is bounded by the number of clusters in which the node has neighbors times the distance between the corresponding central nodes. We bound these quantities by applying Theorem 5.4.1. For example, if we apply it to the cover consisting of the *1*-neighborhood of every node and $k = \log n$, we obtain a cover in which each node is (on the average) in $O(n^{1/\log n})$ clusters, and the distance between central nodes of two neighboring clusters is $O(\log n)$. This suffices for the all pairs shortest paths problem. The details can be found in Ref. [5].

5.5   Ordering of Events

For many distributed applications, it is convenient to obtain an ordering between the events in the system; for example, an ordering between events by different processors that request permission to use some shared resource can be used to grant the resource in a fair manner. In the synchronous model, there is a very clear ordering between events at different processors, as they either happen at the same round or at strictly ordered rounds. In contrast, events in asynchronous systems are completely unordered. In this section, we discuss various mechanisms for ordering events in an asynchronous system.

5.5.1   Logical clocks

We first discuss the notion of *logical clocks,* which assigns logical time stamps to events based on the causal ordering among them [66]. The intuitive idea is to provide a *partial* ordering between events that affect each other. Specifically, in the message passing model, an event ϕ_1 directly influences another event ϕ_2 if one of the following conditions holds:

1. ϕ_1 and ϕ_2 occur at the same processor and ϕ_1 precedes ϕ_2 in the execution
2. ϕ_1 causes a message to be sent and ϕ_2 is the delivery of that message

The *influence*[*] relation is the transitive closure of the directly influence relation. Clearly, it is a partial order. Intuitively, if one event influences another then there is information flow between them, either by explicit messages or by consecutive events at the same processor. Two events are *concurrent* if they are not ordered by the influence relation.

We are interested in assigning *logical time stamps* to events to capture the influence relation. That is, with each event ϕ we associate a time stamp $LT(\phi)$, such that if ϕ_1 influences ϕ_2, then $LT(\phi_1) < LT(\phi_2)$, for any two events ϕ_1 and ϕ_2. There is a simple algorithm for assigning logical time stamps: Each processor p_i keeps a counter L_i that is incremented with each event that occurs at p_i. To each message sent, p_i attaches the current value of L_i. When receiving a message with time stamp T, p_i increases L_i, if necessary, to be strictly larger than T. The time stamp of an event ϕ at p_i, $LT(\phi)$, is the value of L_i after ϕ occurs.

We leave to the reader to verify that this indeed provides logical time stamps. Note that two concurrent events, which are not related by the influence relation, may have the same logical time stamp (e.g., if no messages are sent at all). However, these events must be at different processors. Therefore, we can extend the partial order induced by logical time stamps into a total order, using processor IDs to break ties.

A similar notion can be defined for shared memory systems. In that case, an event directly influences an event of another processor by writing a value that is read by the other processor. We do not discuss the details of this definition here.

5.5.2   Distributed snapshots

The influence relation allows us to define the important notion of distributed snapshots [32]. A snapshot provides a global view of the system's state, which can be used to check the status of the algorithm. Distributed snapshots have immense practical importance, since when running a distributed application we sometime need to know global properties; e.g., whether the application is deadlocked, the computation has terminated, etc.

Intuitively, a snapshot is a global state of the system at a specific time, just like in photography. In an asynchronous system, however, there is no notion of time, and it is not possible to record the states of processors at the "same" time. Instead, we take the states of processors at concurrent times.

Order processors' states by the influence relation according to the influence relation on the preceding events. States are *concurrent* if they are not ordered by the influence relation. A snapshot is a set of n states, one for each processor, that are concurrent. We assume the states contain a list of messages sent and received on each adjacent communication link, so it is possible to reconstruct the set of messages in transit on every link. A *distributed snapshot* algorithm runs in a system that is already performing some computation, and generates a current snapshot, without interfering with the ongoing computation. If

---

[*]This notion is called *happens before* in Ref. [66].

communication links are FIFO (that is, messages are delivered in the order in which they are sent), then there exists a simple distributed snapshot algorithm (see Ref. [32]).

In Section 5.1, we defined configurations, which are global states of the system. A natural question to ask is, "What is the relation between the snapshot returned by a distributed snapshot algorithm and the configurations that appear in the execution?" It is possible to present an execution in which the global state returned by the distributed snapshot algorithm does not correspond to any configuration. However, in this case, there is another execution, which looks exactly the same to the individual processors, in which this global state appears. Therefore, as far as processors can tell, the state returned by the algorithm could have happened. The reader is referred to Ref. [32] for a precise formalization of this notion.

A distributed snapshot can be used to determine *persistent* properties of the algorithm by checking the global state of the algorithm. A property is persistent if, once it holds, it continues to hold in all subsequent states. Therefore, if a persistent property holds in the global state returned by the distributed snapshot, then it will hold throughout the rest of the execution. An example is the problem of termination detection [47]. This problem also has other, more direct solutions (see, e.g., Ref. [75]).

5.5.3 Executing synchronous algorithms in asynchronous systems

Many problems are much easier to solve in synchronous systems, given that the possible behaviors of a synchronous system are more restricted than the possible behaviors of an asynchronous system. For example, if the system is synchronous, then obtaining a snapshot is simple: collect the states of the processors at the beginning of a fixed round. We now discuss how to run algorithms designed for synchronous systems in an asynchronous system, using a general simulation technique called a *synchronizer* [15].

The general idea of a synchronizer is to generate a sequence of pulses, numbered $0, 1, \ldots$, at every node of the network. When the kth pulse is generated, a processor sends messages it would have sent in the kth round of the simulated synchronous algorithm. The pulses should satisfy the following property:

> When pulse number $k + 1$ is generated at node v, v has already received all messages sent to it by its neighbors at their kth pulse, for $k \geq 0$.[*]

The main difficulty in implementing a synchronizer is that a node usually does not know which of its neighbors has sent a message to it (until the message is received). Since there is no bound on the delay a message may incur, simply waiting long enough before generating the next pulse is not sufficient; additional messages have to be sent to achieve synchronization.

To generate a pulse at a node, the node must locally detect the end of the previous pulse. That is, a node u must detect whether every message sent to it in the previous pulse has arrived. The idea is to have the neighbors of u check whether all their messages were received and have them notify u. It is simple for a node to know whether all its message were received if we require each node to send an acknowledgment on every message (of the original synchronous algorithm) received. If all messages sent by a node at pulse k have

---

[*]Note that, according to our definition, the synchronous model has an additional property: Each processor performs its kth step only after all processors have performed at least $(k - 1)$ steps. Although this is a very useful property, there are important synchronous algorithms in which it is not used (see Ref. [15]).

been acknowledged, then the node is *safe* in pulse k. Observe that the acknowledgments only double the number of messages sent by the original algorithm, thus not increasing its asymptotic message complexity. Also observe that each node detects that it is safe in pulse k a constant time after generating pulse k. A node can generate its next pulse once all its neighbors are safe. There are several ways the nodes can notify their neighbors that they are safe, and we discuss two of them.

The simplest synchronizer is called α. As described before, a node detects it is safe using the acknowledgments mechanism. A safe node directly informs all its neighbors. When all the neighbors of node u are safe in pulse k, u knows that all messages sent to it in pulse k have arrived, and it generates pulse $k + 1$. Clearly, each pulse is generated $O(1)$ time after the previous one, at the cost of $O(m)$ messages per pulse (regardless of the number of original messages sent during this pulse).

The second synchronizer, called η, provides an interesting trade-off between time and messages [82]; it uses sparse covers (Section 5.4). For a given trade-off parameter k, η generates the next pulse within $O(\log_k n)$ time after the previous pulse and sends $O(k \cdot n)$ messages per pulse.

Synchronizer η requires a preprocessing stage in which Theorem 5.4.1 is applied to the cover containing the 1-neighborhoods of all the nodes in the network, with parameter $\log k$. The theorem yields a cover $\mathcal{T}$ such that the *1*-neighborhood of every node is a subset of some cluster in $\mathcal{T}$, $diam(\mathcal{T}) = O(\log_k n)$, and $vol(\mathcal{T}) = O(k \cdot n)$. For each node u, we distinguish one of the clusters that contain its *l*-neighborhood as its home cluster. In each cluster, we choose some node as a root, and construct a cluster tree, which is the shortest-path tree from the root to all nodes in the cluster.

As in α, a node detects it is safe using the acknowledgments mechanism. A convergecast mechanism (Section 5.2) is used to collect the safety information up each cluster tree. That is, for each cluster tree it belongs to, a node performs the following procedure: If it is a leaf in the tree, it sends a `safe` message to its parent; if it is an internal node in the tree, it waits for `safe` messages from all its children in this tree, and then sends a `safe` message to its parent.[*] When the root of a cluster tree receives a `safe` message from all its children, it broadcasts a pulse message down the tree. When a node receives a `pulse` message from its parent in its home cluster's tree, it generates the next pulse.

Clearly, each node generates each pulse exactly once. Furthermore, when a node v generates a pulse, it knows that all the nodes in the node's home cluster are safe. Since v's home cluster includes its 1-neighborhood, this implies that all v's neighbors are safe. Therefore, all the messages sent to v at the previous pulse were received, as needed. Note that the time between two consecutive pulses is proportional to the diameter of the cover, i.e., $O(\log_k n)$. The number of messages sent per pulse is proportional to the total size of the cluster trees, which in turn, is proportional to the volume of the cover, i.e., $O(k \cdot n)$.

5.5.4   Clock synchronization

The most powerful way to relate events at different processors is to assume that the processors have access to *clocks*. Various assumptions can be made about the quality of the local clocks, in terms of their closeness to real time and closeness to each other. A *clock synchronization* algorithm attempts to improve the quality of local clocks by exchanging information between the processors and applying appropriate adaptations to the clocks. The quality of the clock synchronization achievable in the system depends on various

---

[*]Messages should be labeled with the cluster's ID.

assumptions on message delay, failures, and the local clocks. We refer the reader to Ref. [83] for a survey of this topic.

5.6   Resource Allocation

In this section, we briefly discuss problems relating to sharing of resources in distributed systems. Such problems arise naturally in the context of operating systems and distributed database systems where processors need to share resources (e.g., CPU time, disk space, printers, or shared data) that should be accessed in an exclusive manner. Problems in this class are determined, largely, by the type of exclusion properties that should be provided and the manner in which processors communicate with each other.

In all versions of the problem, the program is partitioned into four regions:

1. *trying*—where the processor tries to acquire the resources
2. *critical*—where the resources are used
3. *exit*— where some cleanup is performed (optional)
4. *remainder*—the rest of the code

When the processor is not interested in the shared resources, it is in the remainder; to gain access to them it executes the trying region, eventually entering the critical region to use them; afterwards, the processor may announce that it is done, in the exit region. The critical and remainder regions are part of the application and are not available to the resource allocation algorithm; which supplies the trying and exit regions. The algorithm should guarantee *safety* properties (i.e., that the resources are accessed correctly) and *progress* properties (i.e., that access to the shared resources is eventually granted).

The simplest resource allocation problem is *mutual exclusion,* which naturally arises in operating systems that support multiple processes. In this problem, processors wish to obtain exclusive access to a single resource. The problem was studied in the message passing and the shared memory models, and many results were obtained, including algorithms with different progress properties and lower bounds on the amount of communication required for solving it. The reader is referred to a book dedicated to this topic [791.

The *drinking philosophers* problem is the most general resource allocation problem on a communication network [30]. In this problem, a processor requests exclusive access to some set of resources each time it enters the trying region. These requests induce a dynamic *conflict graph,* whose nodes are processors, and there is an edge between two nodes if the two corresponding processors are requesting the same resource. We assume that, at any time, the conflict graph is a subgraph of the communication graph. That is, there is an edge between any pair of processors that may compete on a resource.

There are several algorithms that allow processors to coordinate access to shared resources in a way that guarantees safety; a recent paper allows processors to acquire the resources relatively fast [22].

5.7   Tolerating Processor Failures in Synchronous Systems

So far, we have assumed that the system is completely reliable. In real systems, however, the various components do not operate correctly all the time. In the rest of this chapter we deal with problems that arise when system components fail. We start with the restricted case of tolerating processor failures in synchronous message passing systems. It is assumed that communication links are reliable, but processors are not reliable. We concentrate on *coordination problems,* which require processors to agree on outputs, based on their inputs. Such problems are typically very easy to solve in reliable systems of the kind we have seen so far (especially if we are not concerned with efficiency).

A good representative of coordination problems is the fundamental problem of reaching consensus among processors [69, 76], which we study in the rest of this section. Consensus is a canonical problem in distributed computing because it captures the inherent difficulties of coordination in the presence of failures. In fact, a consensus algorithm can be used to implement many distributed services [81].

Throughout this section and the next one, we assume a complete communication graph, i.e., processors are located at the nodes of a clique. The consensus problem in other network topologies is discussed in [38].

5.7.1 Crash failures

We first consider a mild form of failure, where processors halt during execution. Specifically, a processor *crashes* in round r if it operates correctly in round $1, \ldots, r-1$, and it does not take a step in rounds $r+1$ and on; some messages it sends in round r arrive at their destination, others do not. A processor that crashes is called *faulty*; other processors are called *nonfaulty.*

In the *consensus* problem, each processor p_i starts with a binary input value x_i,[*] and it should irreversibly decide on an output value y_i such that the following conditions are satisfied:

- *Agreement:* $y_i = y_j$ for any nonfaulty processors p_i and p_j. That is, all nonfaulty processors decide on the same value.
- *Validity:* $y_i \in \{x_1, \ldots, x_n\}$ for any nonfaulty processor p_i. That is, the output of a nonfaulty processor is one of the inputs.

Implicit in the definition is the requirement that nonfaulty processors decide within a finite number of rounds; once a processor crashes, it is of no interest to the algorithm, and no requirements are made on its decision. Note that the validity condition implies that if all processors have the same input v, then all nonfaulty processors should decide on v.

Below we outline matching upper and lower bounds of $f+1$ on the number of rounds required for solving this problem, where f is a known upper bound on the number of processors that can fail in any execution.

5.7.1.1 A simple algorithm. The following algorithm is a simplified version of an algorithm that uses authentication to handle more severe failures [42]. In the algorithm, each processor maintains a set of the values it knows to exist in the system; initially, this set contains only its own input. In later rounds, a processor updates its set by joining it with the sets received from other processors, and broadcasts the new set to all processors. After $f+1$ rounds, the processor decides on the smallest value in the set it holds. Clearly, the validity condition is maintained since the decision value is an input of some processor. The crucial point is that all nonfaulty processors have the same set of values after $f+1$ rounds. Intuitively, this holds because among the $f+1$ rounds of the algorithm there is at least one

---

[*]See Ref. [86] about extending this problem to the nonbinary case.

round in which no processor crashes; in this round all processors receive exactly the same set of messages. Thus we have:

Theorem 7.1. *There exists an algorithm that solves the consensus problem in the presence of f crash failures within f + 1 rounds.*

5.7.1.2 Lower bound on the number of rounds.

We now describe the lower bound of $f + 1$ on the number of rounds required for achieving consensus. This result holds even if processors fail in the most benign manner, i.e., by crashing.

The lower bound holds because, when executions are too short, processors cannot distinguish between executions in which they should make different decisions. This intuition can be formalized by the notion of *similarity;* we are only concerned if a *nonfaulty* processor cannot distinguish between the executions. Notice that if one execution looks the same to another for some nonfaulty processor p_i, then p_i decides on the same value in both executions. By the agreement condition, all nonfaulty processors decide on the same value in both executions.

To provide intuition on the structure of the proof, we consider the special case of $f = 1$. Assume, by way of contradiction, that there exists an algorithm that solves the problem in strictly less than two rounds. That is, each execution of the algorithm contains at most one round.

The general proof strategy is to go through a chain of executions which are similar to each other. (See Fig. 5.1, for the case $n = 3$. In the figure, solid arrows represent messages, while dashed arrows represent messages that are not delivered.) We start with the execution in which all processors are nonfaulty and have input 0; hence processors should decide 0. We fail processor p_1 gradually, going through a sequence of n executions, each of which contains one message less from p_1 (executions (a) through (c)). It can be seen that consecutive executions in this sequence look the same to some nonfaulty processor.[*] This way, we can cleanly fail p_1 in the first round.[†] Then, we change p_1's input value from 0 to 1 (execution (d)). In this case, all nonfaulty processors do not see a difference (because no message is received from p_1). At this point, we start correcting p_1 by restoring its first-round messages, one at a time (executions (d) through (f)). We end with an execution in which all processors are nonfaulty, p_1 has input 1, and all other processors have input 0. We repeat this procedure for p_2, p_3, and so on until we change the input values of all processors to be 1. Note that in this chain of executions, each pair of consecutive executions look the same to some nonfaulty processor. By transitivity, this implies that the same deci-

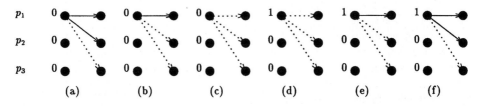

p_1

p_2

p_3

    (a)   (b)   (c)   (d)   (e)   (f)

Figure 5.1 Executions used in lower bound proof; $f = 1$, $n = 3$

---

[*]We assume that there are at least three processors.
[†]Processor p_i's crash in round r is *clean* if none of the messages it sends in this round is delivered.

sion is made in the execution where all processors have input 0 and in the execution where all processors have input 1, which contradicts the validity requirement.

In executions with one round, it is simple to omit a message sent to a nonfaulty processor, and argue that other nonfaulty processors do not distinguish between executions. This is because a nonfaulty processor does not get a chance to communicate the change in its view to other nonfaulty processors. When $f > 1$ and we consider executions with more than one round, this does not hold, since a nonfaulty processor can notify other nonfaulty processors that some message was omitted, say, in the first round. To overcome this problem, we do things more gradually. First, we cause a nonfaulty processor to fail in a clean manner in the later round, then we omit the message sent to it, and finally we "correct" the processor. Failing the nonfaulty processor in the later round is done inductively using a similar sequence of executions. Omitting many details, this shows:

Theorem 5.7.2. *No algorithm solves the consensus problem in strictly less than $f + 1$ rounds in the presence of f crash failures, if $n \geq f + 2$.*

This lower bound was first proved by Fischer and Lynch [44] assuming severe failures of the type discussed in Section 7.2 below. Later it was extended to crash failures by Dolev and Strong [42].

5.7.2 Byzantine failures

We now consider synchronous systems with malicious failures, where failed processors can behave arbitrarily. This is the most severe type of processor failure considered in the theory of distributed computing; clearly, an algorithm that tolerates these failures can tolerate any kind of failure. Faulty processors are often called *Byzantine* because of the following metaphorical description of the consensus problem.

Several divisions of the Byzantine army are camped outside an enemy city. Each division is commanded by a general, and the generals should decide on a common plan of action—that is, whether to attack the city. The generals can communicate with each other only by messengers, which are reliable. The new wrinkle is that some of the generals may be traitors (that's why they are in the Byzantine army) and may try to prevent the loyal generals from agreeing. To do so, the traitors send conflicting messages to different generals, falsely report on what they heard from the other generals, and even conspire and form a coalition.

Paraphrasing the terminology of distributed computing, we consider a system where processor p_i has a binary input x_i. A faulty processor can behave arbitrarily and even maliciously; e.g., it can send different messages to different processors (or not send messages at all) when it is supposed to send the same message. Moreover, the faulty processors may coordinate their actions.

The maximum number of faulty (Byzantine) processors is f.[*] The requirements of the problem are similar to those of the benign failure model discussed in Section 7.1, slightly modified not to consider the decisions of Byzantine processors.

In the case of benign failures, there are algorithms that tolerate any number of failures. In contrast, when failures are Byzantine, we have:

Theorem 5.7.3. *In a system with n processors, f of which may be Byzantine, there is no algorithm that solves the consensus problem if $n \leq 3f$.*

---

[*]The upper bound on the number of Byzantine processors is often denoted t, for *traitors*.

This result is shown for the special case of a system with three processors, one of which might be byzantine; the general result is derived by reduction to this special case. This theorem was originally proved in Ref. [76]; a simpler presentation appears in Ref. [45].

Many algorithms have been suggested for solving this problem, each exhibiting a different trade-off between the number of rounds for solving the problem, the size of the messages and the number of faults that are tolerated. The best algorithm to date solves the problem in the presence of f Byzantine failures, using $f + 1$ rounds and polynomial size messages, for any n and f such that $n > 3f$ [50]. The reader is referred to this paper for detailed overview of these algorithms. Here we discuss two algorithms that represent two extreme ends of the trade-off. The first algorithm takes $f + 1$ rounds and only requires that $n > 3f$, thus matching the lower bounds of Theorems 5.7.2 and Theorem 5.7.3. Unfortunately, the messages sent by the algorithm have exponential size. The second algorithm uses messages of constant size, but doubles the number of rounds and requires that $n > 4f$.

5.7.2.1 An exponential algorithm.
The following algorithm uses messages of exponential size, takes $f + 1$ rounds, and only assumes that $n > 3f$ [76]. The algorithm has two stages. In the first stage, information is collected by communication among the processors, while in the second stage, each processor uses this information to compute locally its decision value.

It is convenient to describe the information collected by each processor as a tree with $f + 1$ levels. Intuitively, p_i stores in a vertex of depth r the value that "p_{i_r} said that p_{i_2} said that...p_{i_1} said," for any sequence of r distinct processors. The tree is filled in $f + 1$ rounds; in each round, each processor broadcasts the values stored in the level of its tree that was filled in the last round. After $f + 1$ rounds, the decision value is computed by applying to each node a recursive majority vote, starting from the leaves. For a good description of this algorithm, see Ref. [23].

Theorem 5.7.4. *There exists an algorithm that solves the consensus problem in the presence of Byzantine failures within $f + 1$ rounds.*

In each round, every processor sends a message to every processor. Therefore, the total message complexity of the algorithm is $O(n^2 f)$. Unfortunately, in each round, every processor broadcasts a whole level of its tree, and thus the message size is exponential.

5.7.2.2 An algorithm with constant-size messages.
The following simple algorithm uses messages of constant size, takes $2(f + 1)$ rounds, and assumes that $n > 4f$ [24]. The algorithm contains $f + 1$ phases, each taking two rounds. Each processor has a preference for each phase, initially its input. At the first round of each phase, all processors send their preference to each other. Let $v_i \in \{0, 1\}$ be the majority value in the set of values received by processor p_i. In the second round of the phase, a distinct processor, called the *king* of the phase, sends its majority value to all processors. (The king of the phase is determined in advance at the beginning of the algorithm.) If p_i receives more than $n/2 + f$ copies of v_i (in the first round of the phase), then it sets its preference for the next phase to be v_i; otherwise, it sets its preference to be the phase king's preference (as sent in the second round of the phase). After $f + 1$ phases, the processor decides on its preference.

The first property of the algorithm to notice is *persistence of agreement:* if all nonfaulty processors prefer v at the beginning of phase r, then they all prefer v at phase $r + 1$. This is because each processor receives $n - f$ copies of v in the first round of phase r. Since $n > 4f$, $n - f > n/2 + f$, which implies that all nonfaulty processors will prefer v in phase $r + 1$.

This immediately implies the validity property: If all nonfaulty processors start with the same input v, they continue to prefer v throughout the phases; finally, they decide on v at the end of phase $f + 1$.

If each phase has a different king, then at least one phase's king is nonfaulty. Let g be the smallest phase whose king is nonfaulty. We claim that all nonfaulty processors finish phase g with the same preference. Clearly, the claim holds if all nonfaulty processors pick the king's preference to be their preference in the next round. Otherwise, some nonfaulty processor p_j sees its majority value u at least $n/2 + f$ times. But this implies that every non-faulty processor sees v at least $n/2$ times and, in particular, the phase's king majority value is v. Therefore, the claim holds in this case, too. Therefore, at phase $g + 1$ all processors have the same preference, and the persistence of agreement implies that they will decide on the same value at the end of the algorithm.

5.8 Tolerating Processor Failures in Asynchronous Systems

In this section, we discuss processor failures in asynchronous systems. In this model, the canonical coordination problem we studied in the previous section, consensus, cannot be solved deterministically. We discuss how to overcome this impossibility result either by using randomization or by solving weaker coordination problems.

5.8.1 Impossibility of solving consensus deterministically

We have seen that the consensus problem can be solved in synchronous systems in the presence of failures, both benign (crash) and malicious (Byzantine). If the system is completely asynchronous, then there is no deterministic consensus algorithm, even in the presence of a single processor failure. Here, we consider only the case where processors fail in the most benign manner, i.e., by crashing, and communication is completely reliable.

Crucial to this proof is the fact that processing is completely asynchronous; that is, we make no assumptions on the relative speeds of processors or the delay of messages. We also assume that processors do not have access to clocks or any other time-measurement device. A processor cannot distinguish between a crashed processor and a slow processor.

The problem is defined exactly as in the previous section, except for a change in the termination condition, which is defined only for infinite executions. A processor is *nonfaulty* in an execution if it has infinitely many computation events in the execution; otherwise, it is *faulty*. We require that a nonfaulty processor decides within a finite number of its own steps, provided at least $n - 1$ processors are nonfaulty in the execution.

The impossibility result holds regardless of the way processors communicate; that is, it holds both for shared memory and message-passing systems. In fact, the proof in the two models has exactly the same structure, with only few technical changes which are caused by the different communication modes.

The proof is by contradiction: We assume the existence of a deterministic algorithm that solves the consensus problem in the presence of one failure and construct an infinite execution in which processors remain indecisive forever. First, we argue that there is some initial configuration in which the decision is not already predetermined (this uses the validity condition). Then, we take an indecisive configuration and a processor, and show that either the processor can take a step (perhaps after other processors have taken steps) and the configuration remains indecisive, or there is a processor that is a "decider" in the configuration. Furthermore, if there is a decider, then there exists an execution in which processors reach conflicting decisions, contradicting the agreement condition. Finally, these elements are used to construct an infinite execution in which a decision is not reached although all processors are nonfaulty.

The crux of the impossibility proof is to show the existence of a decider. Intuitively, a *decider* (for a given configuration) is a processor which can determine the decision of the whole system, but the other processors cannot tell which decision was made. Clearly, since other processors do not know which decision was made, they cannot decide if the decider fails. These ideas form the basis for proving the following important theorems:

Theorem 5.8.1. *In the message-passing model, no consensus algorithm is correct in the presence of one failure.*

Theorem 5.8.2. *In the shared memory model with only read/write operations, no consensus algorithm is correct in the presence of one failure.*

When binary test&set registers are allowed, there is a simple consensus algorithm for two processors. A shared test&set variable l serves as a lock, with initial value 0. Each processor writes its input into a separate variable, and then tries to grab the lock by performing a test&set operation on l. If it succeeds, it decides on its input; otherwise, it decides on the other processor's input. The reader can check that this algorithm is indeed correct and tolerates one failure. However, ideas similar to those discussed above show:

Theorem 5.8.3. *In the shared memory model with test&set operations (as well as read/write operations), no consensus algorithm is correct in the presence of two failures.*

The impossibility of achieving consensus in an asynchronous system was first proved by Fischer, Lynch, and Paterson [46]. Their proof dealt only with message-passing systems. Later, the impossibility result was extended to the shared memory model in Ref. [71] and (implicitly) in Ref. [39]. The results for test&set registers appear in Ref. [71]. The notion of a decider is from Ref. [27].

5.8.2   Randomized algorithms for consensus

We now briefly describe a randomized asynchronous algorithm for reaching consensus, whose time complexity is $O(1)$. Randomization helps us overcome the impossibility result proved in the previous section, since the algorithm has executions that do not terminate, albeit with zero probability. Furthermore, on the average, the randomized algorithm is faster than the lower bound on the number of rounds discussed in Section 5.7.1.2, but some of its executions are very long.

The best algorithm to date solves this problem in the presence of f Byzantine failures, in constant expected time, assuming $n > 3f$ [29]. For further details, we refer the reader to this paper and to an earlier survey of this topic [34]. Here we discuss a weaker, but conceptually much simpler, asynchronous randomized consensus algorithm that tolerates f crash failures, assuming that $n > 4f$. It has the same general structure as the more sophisticated algorithms.

The algorithm shares many ideas with the algorithm presented in Section 5.7.2.2. However, it uses a global coin to break ties between zeroes and ones, instead of a predetermined king. Informally, a global coin simulates the public tossing of a biased coin such that all processors see the coin landing on side v with probability at least p, for every

$v \in \{0, 1\}$; it is possible that processors will not see the coin landing on the same value, but this happens with smaller probability. Specifically, a coin procedure has no input and produces a binary output. The procedure is an *f-resilient global coin with bias p* if in any execution in which at least $n - f$ processors are nonfaulty, nonfaulty processors output v with probability at least p, for any value $v \in \{0, 1\}$.

The consensus algorithm proceeds in a sequence of asynchronous phases. Each processor has a preference for each phase, initially its input. At each phase, all processors send their preferences to each other. A processor waits to hear at least $n - f$ preferences for its current round. (Note that it cannot wait for more values since f processors might crash.) Let $v_i \in \{0, 1\}$ be the majority value in the set of values received by processor p_i. If p_i receives more than $n/2 + f$ copies of v_i, then it decides on v_i;[*] if p_i receives more than $n/2$ copies of v_i, then it sets its preference to be v_i and continues to the next phase; otherwise, p_i sets it preference to be the value of the global coin for this phase and continues to the next phase.

Notice that if all processors prefer v in phase r, then all nonfaulty processors decide on v. This is because each nonfaulty processor receives $n - f$ copies of v in phase r. Since $n > 4f$, $n - f > n/2 + f$, which implies that all nonfaulty processors decide on v. This immediately implies the validity property.

If some processor decides on v, then all nonfaulty processors see at least a majority of v in the same phase and will either decide on v or prefer v at the next phase. In either case, they decide on v (either in the same phase or in the next phase). This implies the agreement property.

It only remains to argue termination (with high probability), where the global coin plays the main role. The crux is to show that in every phase, nonfaulty processors prefer the same value with probability at least p, where p is the bias of the global coin. Note that this implies that nonfaulty processors decide in each phase with probability at least p. If all processors obtain their preference for the next phase from the global coin then the probability that they all obtain the same value is at least $2p$. Also, if some processor prefers v then every other processor either prefers v or obtains its preference from the global coin (but no processor prefers $\bar{v}$). With probability p, the processors obtaining their preference from the global coin also prefer v, as needed. Thus, the expected number of rounds until termination is $O(1/p)$, where p is the bias of the global coin. Therefore, the expected time complexity of the algorithm is $O(T/p)$, where T is the time complexity of the global coin. There exist global coin implementations with constant T and p (e.g., see Ref. [29]) which imply a randomized consensus algorithm with constant expected time complexity.

There also exist randomized consensus algorithms in the shared memory model that use read/write registers and tolerate any number of crash failures. Many of these algorithms have the same overall structure as discussed above, although the communication mechanisms, as well as the implementation of the global coin, are different. We refer the reader to a recent paper on this topic [9] and references therein.

5.8.3 Other problems

Another way to sidestep the impossibility results of Section 5.8.1 is to solve problems with weaker requirements than consensus. Here, we briefly describe three problems that can be solved in the presence of failures in an asynchronous system, approximate agreement, k-consensus, and renaming.

---

[*] A deciding processor continues to send its decision value in the next phase.

The approximate agreement problem (see Ref. [41]) is the real-valued version of the consensus problem. Each processor p_i starts with an input x_i (real number) and has to decide on an output y_i (real number) that is valid, i.e., in the range of the inputs. Furthermore, all outputs have to be within ε of each other for a parameter ε; that is:

ε-*Agreement:* $|y_i - y_j| < \varepsilon$, for any i and j

The approximate agreement problem can be solved for any value of ε [41]. However, as ε gets smaller, the algorithm takes more time.

The *k-consensus* problem (see Ref. [33]) is a discrete version of the approximate agreement problem, and is defined as follows. Each processor starts with an input value x_i taken from some input range (not necessarily binary), and has to decide on an output value y_i. The decision values have to satisfy validity, as defined for the consensus problem, as well as:

k-Agreement: There are at most k different values in the set $\{y_1, \ldots, y_n\}$

The algorithm should tolerate some number f of crash failures. That is, in every execution in which at least $n - f$ processes are nonfaulty, every nonfaulty processor decides. There is a trade-off between k and f; that is, is the smallest k we can achieve depends on the number f of crash failures tolerated. Clearly, the impossibility of solving consensus implies that there is no l-consensus algorithm, for any $f \geq 1$; that is, $k > 1$, regardless of f. If $k > f$, then there is a simple k-consensus algorithm [33]. Recently, it was shown that if $k < f$, then there is no deterministic k-consensus algorithm in the presence of f crash failures [26, 55, 80].

While k-consensus and approximate agreement generalize the consensus problem and require processors to converge on a small set of common values, the renaming problem requires processors to decide on distinct values. Specifically, in the *M-renaming* problem (see Ref. [13]), each processor starts with a distinct name, from an unbounded totally ordered domain, and should decide on a new name $n_i \in [1 \ldots M]$ such that:

Uniqueness: No two processors decide on the same name, i.e., $n_i \neq n_j$, for every $i \neq j$.

Once again, there is a trade-off between minimizing M, the size of the new names space, and maximizing f, the number of failures to be tolerated. Clearly, $M > n$; furthermore, the impossibility result of Section 5.8.1 can be modified to prove $M > n + 1$ for $f > 1$. There is an algorithm for renaming with $M = n + f + 1$, for any $f > 1$ [13]. Recently, it was shown that this bound is tight, that is, $M > n + f$, for any $f > 1$ [55].

Herlihy and Shavit [55] provide a characterization, using algebraic topology, of the coordination problems that can be solved using read/write registers. These results can be translated to the message passing model, using results of Ref. [12].

5.9   Other Types of Failures

In this section, we briefly discuss other types of failures that were studied in the theory of distributed computing. The reader is encouraged to follow the references we mention for further details.

5.9.1   Dynamic networks

The algorithms discussed so far rely on the assumption that the communication network is *static*, i.e., its structure does not change during the execution of the algorithm. However, in real communication networks, e.g., the Internet, communication links are not reliable; they fail and recover quite often. This implies that the topological structure of the network keeps changing *dynamically*, complicating the execution of applications in the system, especially if it is asynchronous. We now mention work that addresses this problem.

If communication link failures can not be detected, then Theorem 5.8.1 implies that there is no algorithm for solving most non-trivial problems, since the network is asynchronous. Therefore, we assume an appropriate notification when a link fails and recovers. This assumption is reasonable for link protocols in most of today's network architectures.

Clearly, if link failures are too numerous (that is, the network's topology changes too often, or nodes are disconnected from each other), then no computation can be performed. Two assumptions can be made to enable algorithms to run in dynamic networks. The first is *eventual stability* [2], which assumes that changes in the network eventually subside, and the network topology stabilizes for a period long enough to allow the application to run from start to completion. The second is *eventual connectivity* [87], which assumes that all nodes remain connected throughout the execution, although the paths connecting them might change over time.

For eventual stability, the most prevalent approach is to reset the algorithm periodically such that, eventually, there is an execution of the application which starts after the network stabilizes. Several such algorithms were suggested—the most recent one (see Ref. [19]) employs the notion of synchronizer, discussed in Section 5.5.3, to yield an efficient method for running synchronous algorithms under this assumption.

For eventual connectivity, the most prevalent approach is to provide a reliable end-to-end primitive that allows two nodes to communicate reliably despite topological changes, as long as they are connected. Several algorithms were suggested in this case too—most recently, a simple and efficient one in Ref. [4].

5.9.2   Self-stabilization

So far, we have assumed that executions start with the system in an initial configuration, satisfying certain conditions. Most of the algorithms we discussed rely on the fact that processors' states and shared variables are initialized correctly and that communication links are initially empty. A *self-stabilizing* algorithm tries to cope with systems that start at some arbitrary state, which does not necessarily satisfy the initialization conditions. It is required that the algorithm will eventually converge to an acceptable state or correct behavior, regardless of the state it starts from. This property was originally suggested by Dijkstra [37] for token-passing algorithms that use read-modify-write operations on shared memory.

A self-stabilizing algorithm provides a new type of fault tolerance. It allows the system to overcome a transient failure that affects the states of the processors, the shared memory or the communication network. The system automatically reorganizes itself and returns to good behavior, without outside intervention or any other special mechanism.

In recent years, this topic has attracted much attention and many results were obtained; we mention only a few of them. Early work in this area extended Dijkstra's original work by considering different types of communication mechanisms, various atomicity assumptions and diverse topologies. See Ref. [43] for a recent overview of this topic. More recent work provides automatic transformations for taking an arbitrary protocol and making it self-stabilizing. This approach was suggested in Ref. [62]; later work improved the efficiency of the transformation, most recently Ref. [20].

5.10   Wait-Free Implementations of Shared Objects

We have mentioned several types of shared variables that can be used by processors for communication and synchronization, e.g., read/write, test&set, etc. An *object* is a variable of a particular type, supporting a certain set of operations. An interesting and important question is how to implement an object of one type from objects of other types. Such implementations allow to run algorithms that use certain objects in systems that do not support them. In addition, stronger objects (even if they do not exist in hardware) provide useful abstractions and simplify the task of programming multiprocessor systems.

Informally, an *implementation* is a set of procedures, one for each operation of the *high-level* (implemented) object. The procedures use only operations supported by the *low-level* (implementing) objects. The operations generated by the procedures should look as if they were applied directly to the high-level object. Specifically, it should be possible to linearize the executions generated by the procedures of the implementation [56]. Consider an execution α of low-level operations that were generated by the implementation. If all low-level operations in the execution of a high-level operation op, appear before all low-level operations of another high-level operation op_2, then op_1 precedes op_2 in α, denoted

$$op_1 \xrightarrow{\alpha} op_2$$

An execution α can be linearized if there is a permutation of all high-level operations in the execution that extends the $\xrightarrow{\alpha}$ relation. That is, there is a sequence that contains all the high-level operations in α and preserves the real-time order of non-overlapping operations.

In the past, the prevalent method for implementing shared objects relied on mutual exclusion: The shared object is accessed only inside a critical section, so that only one processor is modifying the low-level objects at each time. This method is very sensitive to processor failures and delays. If a processor inside the critical section fails or is very slow then all processors attempting to access this object are delayed. *Wait-free* implementations guarantee that each processor completes each procedure (i.e., implementation of a high-level operation) within a finite number of low-level operations, regardless of the behavior of other processors.[*] A slightly weaker requirement is for the implementation to be *non-blocking,* that is, if several high-level operations are in progress, at least one completes within finite number of low-level operations.

A large body of search was dedicated to wait-free implementations between different types of read/write registers. Much of this research deals with minimizing the number of low level operations invoked to perform a high-level operation, and the size of the memory used by the implementation. We refer the reader to a survey of these algorithms [63]. The most sophisticated object that can be implemented from read/write registers seems to be the atomic *snapshot* object [1, 6]. An atomic snapshot object is a shared data structure partitioned into segments; a processor can either *update* an individual segment, or instantaneously scan all segments of the object. It is similar to, but not exactly the same as, distributed snapshots discussed in Section 5.5.2.

---

[*]Another version of this notion requires the number of operations to be *bounded.*

It is clear that wait-freedom is an extreme case of fault-tolerance. Specifically, it can be seen that a wait-free implementation tolerates the (crash) failure of $n - 1$ processors. We borrow the term and say that an algorithm is *wait-free* if it tolerates the failure of $n - 1$ processors. Recall that there is a wait-free algorithm for consensus using test&set registers (in addition to read/write registers) for two-processor systems. It is simple to see that a wait-free implementation of test&set registers from read/write registers would imply a wait-free algorithm for consensus using read/write registers, for two-processor systems. However, Theorem 5.8.2 implies that there is no wait-free algorithm for consensus using read/write registers, for any system of two or more processors. This implies:

Theorem 5.10.1 (Herlihy [54]). *There is no wait-free implementation of test&set registers from read/write registers.*

The same argument can be used to show that there is no wait-free implementation of read-modify-write registers from test&set registers, in a system with three processors or more, as well as other results of this type [54].

Interestingly, consensus can also be used to provide wait-free implementations of one object from another. If there is a wait-free algorithm for consensus in a system with n processors using objects of type X, then there is a wait-free implementation of any object from objects of type X (see Ref. [54]). Because there are randomized algorithms for consensus using read/write registers, this yields a randomized implementation of any object from read/write registers.

These facts motivated an attempt to classify objects into different levels of a hierarchy according to their *consensus number,* that is, the maximal number of processors for which there exists a wait-free consensus algorithm using objects of this type (as well as read/write registers). For additional discussion of this hierarchy, see Refs. [54, 61].

5.11   Final Comments

This chapter concentrated on a computability and complexity theory of distributed computing, emphasizing clean formulations of problems and environments. We discussed which problems can and cannot be solved and the cost of solving them, usually in terms of communication and time. We did not address the important topic of specifying and verifying distributed programs; the reader is referred to several recent books on this subject, e.g., Refs. [8, 31, 48, 74].

In addition to the specific references mentioned in this chapter, the reader is encouraged to consult several more general references, e.g., Refs. [11, 25, 68, 72, 85]. The best place to follow current research on the theory of distributed computing is the proceedings of the annual ACM conference on Principles of Distributed Computing (PODC).

Acknowledgments. I would like to thank Yehuda Afek, Shlomi Dolev, Nissim Francez, Shmuel Katz, David Peleg, and Jennifer Welch for discussions on the contents of this chapter; Eyal Dagan, Eyal Kushilevitz, Rinat Rappoport, Sergio Rajsbaum, and Jennifer Welch for helpful comments on a draft of the chapter; and all the people who provided updated references to their work.

The work of the author is partially supported by grant No. 92-0233 from the United States Israel Binational Science Foundation (BSF), Jerusalem, Israel; Technion V.P.R.— Argentinian Research Fund; and the fund for the promotion of research in the Technion.

5.12   References

1. Afek, Y., H. Attiya, D. Dolev, E. Gafni, M. Merritt, and N. Shavit. 1993. Atomic snapshots of shared memory. *Journal of the ACM,* Vol. 40, No. 4, 873–890.

2. Afek, Y., B. Awerbuch, and E. Gafni. 1987. Applying static network protocols to dynamic networks. In *Proceedings of the 28th Annual IEEE Symposium on Foundations of Computer Science,* 358–370.

3. Afek, Y., and E. Gafni. 1991. Time and message bounds for election in synchronous and asynchronous complete networks. *SIAM J. on Computing,* Vol. 20, No. 2, 376–394.

4. Afek, Y., E. Gafni, and A. Rosen. 1992. The slide mechanism with applications in dynamic networks. In *Proceedings of the 11th Annual ACM Symp. on Principles of Distributed Computing,* 35–46.

5. Afek, Y., and M. Ricklin. 1993. Sparser: A Paradigm for Running Distributed Algorithms. *Journal of Algorithms,* Vol. 14, 316–328.

6. Anderson, J. Composite registers. *Distributed Computing,* Vol. 6, No. 3, 141–154.

7. Angluin, D. 1980. Local and global properties in networks of processors. In *Proceedings of the 12th ACM Symposium on Theory of Computing,* 82–93.

8. Apt, K.,and E.-R. Olderog. 1991. *Verification of Sequential and Concurrent Programs.* New York: Springer Verlag.

9. Aspnes, J., and O. Waarts. 1992. Randomized consensus in expected $O(n \log^2 n)$ operations per processor. In *Proceedings of the 33rd Annual IEEE Symposium on Foundations of Computer Science,* 137–146.

10. Attiya, H.1987. Constructing efficient election algorithms from efficient traversal algorithms. In *Proceedings of the 2nd International Workshop on Distributed Algorithms,* 337–344. Lecture Notes in Computer Science #312, Springer-Verlag.

11. Attiya, H. 1994 (January). Lecture Notes for Course #236357: Distributed Algorithms, Department for Computer Science, The Technion.

12. Attiya, H., A. Bar-Noy, and D. Dolev. 1995. Sharing memory robustly in message-passing systems. *Journal of the ACM.*

13. Attiya, H., A. Bar-Noy, D. Dolev, D. Peleg, and R. Reischuk. 1990. Renaming in an asynchronous environment. *Journal of the ACM,* Vol. 37, No. 3.

14. Attiya, H., M. Snir, and M. Warmuth. 1988. Computing in an anonymous ring. *Journal of the ACM,* Vol. 35, No. 4, 845–876.

15. Awerbuch, B. 1985. Complexity of network synchronization. *Journal of the ACM,* Vol. 32, No. 4, 804–823.

16. Awerbuch, B. 1987. Optimal distributed algorithms for minimum weight spanning tree, counting, leader election and related problems. 1987. In *Proceedings of the 19th ACM Symposium on Theory of Computing,* 230–240.

17. Awerbuch, B., B. Berger, L. Cowen, and D. Peleg. 1993. Near-linear cost sequential and distributed constructions of sparse neighborhood covers. In *Proceedings of the 34th Annual IEEE Symposium on Foundations of Computer Science,* 638–647.

18. Awerbuch, B., and S. Even. 1986. Reliable broadcast protocols in unreliable networks. *Networks,* Vol. 16, 381–396.

19. Awerbuch, B., B. Patt-Shamir, D. Peleg, and M. Saks. 1992. Adapting to asynchronous dynamic networks. In *Proceedings of the 24th ACM Symposium on Theory of Computing,* 557–570.

20. Awerbuch, B., B. Patt-Shamir, and G. Varghese. 1991. Self-stabilization by local checking and correction. In *Proceedings of the 32nd lEEE Symposium on Foundations of Computer Science,* 268–277.

21. Awerbuch, B., and D. Peleg. 1992. Routing with polynomial communication-space trade-off. *SIAM J. on Discrete Mathematics,* Vol. 5, 151–162.

22. Awerbuch, B., and M. Saks. 1990. A dining philosophers algorithm with polynomial response time. In *Proceedings of the 31st IEEE Symposium on Foundations of Computer Science,* 65–75.

23. Bar-Noy, A., D. Dolev, C. Dwork, and R. Strong. 1987. Shifting gears: changing algorithms on the fly to expedite byzantine agreement. In *Proceedings of the 6th Annual ACM Symposium on Principles of Distributed Computing,* 42–51.

24. Berman, P., and J. Garay. 1989. Asymptotically optimal distributed consensus. In *Proceedings of the Int. Colloquium on Automata, Languages and Programming,* 80–94. Lecture Notes in Computer Science #372, Springer-Verlag.

25. Bodlaender, H. L. 1987. *Distributed Computing: Structure and Complexity.* Amsterdam: CWI tract 43.

26. Borowsky, E., and E. Gafni. 1993. Generalized FLP Impossibility result for *t*-resilient asynchronous computation. In *Proceedings of the 25th ACM Symposium on Theory of Computing,* 91–100.

27. Bridgland, M., and R. Watro. 1987. Fault-tolerant decision making in totally asynchronous distributed systems. In *Proceedings of the 6th Annual ACM Symposium on Principles of Distributed Computing,* 52–63.

28. Burns, J. A 1980. *Formal Model for Message Passing Systems.* Technical Report TR-91. Bloomington, Ind.: Computer Science Dept., Indiana University.

29. Canetti, R., and T. Rabin. 1993. Fast asynchronous byzantine agreement with optimal resilience. In *Proceedings of the 25th ACM Symposium on Theory of Computing*, 42–51.

30. Chandy, K., and J. Misra. 1984. The drinking philosophers problem. *ACM Transactions on Programming Languages and Systems*, Vol. 6, No. 4, 632–646.

31. Chandy, M., and J. Misra. 1988. *Parallel Program Design.* Reading, Mass.: Addison-Wesley.

32. Chandy, M., and L. Lamport. 1985. Distributed snapshots: Determining global states of distributed systems. *ACM Transactions on Computer Systems*, Vol. 3, No. 1, 63–75.

33. Chaudhuri, S. 1993. More choices allow more faults: set consensus problems in totally asynchronous systems. *Information and Computation*, Vol. 105, No. 1, 132–158.

34. Chor, B., and C. Dwork. 1989. Randomization in Byzantine agreement. *Advances in Computing Research,* Vol. 5, 443–497.

35. Chor, B., A. Israeli, and M. Li. 1987. On processor coordination using asynchronous hardware. In *Proceedings of the 6th Annual ACM Symposium on Principles of Distributed Computing*, 86–97.

36. Chou, C.-T., and E. Gafni. 1988. Understanding and verifying distributed algorithms using stratified decomposition. In *Proceedings of the 7th Annual ACM Symposium on Principles of Distributed Computing*, 44–55.

37. Dijkstra, E. 1974. Self-stabilizing systems in spite of distributed control. *Communications of the ACM*, Vol. 17, 643–644.

38. Dolev, D. 1982. The Byzantine generals strike again. *Journal of Algorithms*, Vol. 3, No. 1, 14–30.

39. Dolev, D., C. Dwork, and L. Stockmeyer. 1987. On the minimal synchronism needed for distributed consensus. *Journal of the ACM*, Vol. 34, No. 1, 77–97.

40. Dolev, D., M. Klawe, and M. Rodeh. 1982. An $O(n \log n)$ unidirectional distributed algorithm for extrema finding in a circle. *Journal of Algorithms*, Vol. 3, No. 3, 245–260.

41. Dolev, D., N. Lynch, S. Pinter, E. Stark, and W. Weihl. 1986. Reaching approximate agreement in the presence of faults. *Journal of the ACM*, Vol. 33, No. 3, 449–516.

42. Dolev, D., and H. R. Strong. 1983. Authenticated algorithms for Byzantine agreement. *SIAM Journal on Computing*, Vol. 12, No. 4, 656–666.

43. Dolev, S., A. Israeli, and S. Moran. 1993. Self-Stabilization of Dynamic Systems Assuming Only Read/Write Atomicity. *Distributed Computing*, Vol. 7, No. 1, 3–16.

44. Fischer, M., and N. Lynch. 1982. A lower bound for the time to assure interactive consistency. *Information Processing Letters*, Vol. 14, No. 4, 183–186.

45. Fischer, M., N. Lynch, and M. Merritt. 1986. Easy impossibility proofs for distributed consensus problems. *Distributed Computing*, Vol. 1, 26–39.

46. Fischer, M., N. Lynch, and M. Paterson. 1985. Impossibility of distributed consensus with one faulty process. *Journal of the ACM*, Vol. 32, No. 2, 374–382.

47. Francez, N. 1980. Distributed termination. *ACM Transactions on Programming Languages and Systems*, Vol. 2, No. 1, 42–55.

48. Francez, N. 1988. *Program Verification.* Reading, Mass.: Addison-Wesley.

49. Frederickson, G., and N. Lynch. 1987. Electing a leader in a synchronous ring. *Journal of the ACM*, Vol. 34, No. 1, 98–115.

50. Garay, J., and Y. Moses. 1993. Fully polynomial Byzantine agreement in $t + 1$ rounds. In *Proceedings of the 25th ACM Symposium on Theory of Computing*, 31–41.

51. Gafni, E., and Y. Afek. 1984. Election and traversal in unidirectional networks. In *Proceedings of the 3rd Annual ACM Symp. on Principles of Distributed Computing*, 190–198.

52. Gallager, R., P. Humblet, and P. Spira. 1983. A distributed algorithm for minimum-weight spanning trees. *ACM Transactions on Programming Languages and Systems*, Vol. 5, No. 1, 66–77.

53. Harel, D. 1987. Statecharts: A visual formalism for complex systems. *Science of Computer Programming,* Vol. 8, 231–274.

54. Herlihy, M. 1991. Wait-free synchronization. *ACM Transactions on Programming Languages and Systems,* Vol. 13, No. 1, 124–149.

55. Herlihy, M., and N. Shavit. 1993. The asynchronous computability theorem for t-resilient tasks. In *Proceedings of the 25th ACM Symposium on Theory of Computing*, 111–120.

56. Herlihy, M., and J. Wing. 1990. Linearizability: A correctness condition for concurrent objects. *ACM Transactions on Programming Languages and Systems*, Vol. 12, No. 3, 463–492.

57. Hirschberg, D., and J. Sinclair. 1980. Decentralized extrema-finding in circular configurations of processes. *Communications of the ACM*, Vol. 23, 627–628.

58. Higham, L. 1988. *Randomized Distributed Computing on Rings.* Ph.D. thesis. Vancouver, B.C.: University of British Columbia.

59. Higham, L., and T. Przytycka. 1993. A simple, efficient algorithm for maximum finding on a ring. In *Proceedings of the 7th International Workshop on Distributed Algorithms*, 249–263. New York: Springer-Verlag. Lecture Notes in Computer Science #725.

60. Itai, A., and M. Rodeh. 1990. Symmetry breaking in distributed networks. *Information and Computation,* Vol. 88, No. 1, 60–87.

61. Jayanti, P. 1993. On the robustness of Herlihy's hierarchy. In *Proceedings of the 12th Annual ACAI Symp. on Principles of Distributed Computing,* 145–157.

62. Katz, S., and K. Perry. 1993. Self-stabilizing extensions for message-passing systems. *Distributed Computing,* Vol. 7, No. 1, 17–26.

63. Kirousis, L., and E. Kranakis. 1989. A brief survey of concurrent readers and writers. *CWI Quarterly,* Vol. 2, No. 4.

64. Korach, E., S. Kutten, and S. Moran. 1990. A modular technique for the design of efficient leader finding algorithms. *ACM Transactions on Programming Languages and Systems,* Vol. 12, No. 1, 84–101.

65. Korach, E., S. Moran, and S. Zaks. 1984. Tight lower and upper bounds for some distributed algorithms for a complete network of processors. In *Proceedings of the 3rd Annual ACM Symp. on Principles of Distributed Computing,* 199–207.

66. Lamport, L. 1978. Time, clocks and the ordering of events in a distributed system. *Communications of the ACM,* Vol. 21, No. 7, 558–565.

67. Lamport, L. 1983. Specifying concurrent program modules. *ACM Transactions on Programming Languages and Systems,* Vol. 5, No. 2, 190–222.

68. Lamport, L., and N. Lynch. 1990. Distributed computing: models and methods. In *Handbook of Theoretical Computer Science,* J. van Leeuwen, ed. New York: Elsevier Publishers.

69. Lamport, L., R. Shostak, and M. Pease. 1982. The Byzantine generals problem. *ACM Transactions on Programming Languages and Systems,* Vol. 4, No. 3, 382–401.

70. LeLann, G. 1977. Distributed systems: towards a formal approach. In *IFIP Congress,* 155–160.

71. Loui, M., and H. Abu-Amara. 1987. Memory requirements for agreement among unreliable asynchronous processes. *Advances in Computing Research,* Vol. 4, 163–183.

72. Lynch, N., and B. Patt-Shamir. 1993. *Distributed Algorithms: Lecture Notes for 6.852.* MIT/LCS/RSS-20, Cambridge, Mass.: Massachusetts Institute of Technology.

73. Lynch, N., and M. Tuttle. 1989. An introduction to input/output automata. *CWI-Quarterly,* Vol. 2, No. 3.

74. Manna, Z., and A. Pnueli. 1992. *The Temporal Logic of Reactive and Concurrent Systems.* New York: Springer-Verlag.

75. Mattern, F. 1987. Algorithms for distributed termination detection. *Distributed Computing,* Vol. 2, No. 3, 161–175.

76. Pease, M., R. Shostak, and L. Lamport. 1980. Reaching Agreement in the Presence of Faults. *Journal of the ACM,* Vol. 27, No. 2, 228–234.

77. Peterson, G. 1982. An $O(n \log n)$ unidirectional distributed algorithm for the circular extrema problem. *ACM Transactions on Programming Languages and Systems,* Vol 4, 758–762.

78. Peterson, G. 1984. *Efficient Algorithms for Elections in Meshes and Complete Networks,* TR 140. Rochester, N.Y.: University of Rochester.

79. Raynal, M. 1986. *Algorithms for Mutual Exclusion.* Boston: MIT Press.

80. Saks, M., and F. Zaharoglou. 1993. Wait-free k-set Agreement is impossible: The topology of public knowledge. In *Proceedings of the 25th ACA Symposium on Theory of Computing,* 101–111.

81. Schneider, F. 1990. Implementing Fault-tolerant services using the state-machine approach: A tutorial. *ACM Computing Surveys,* Vol. 22, No. 4, 299–319.

82. Shabtay, L., and A. Segall. Low complexity network synchronization. To appear in *Proc. of the 8th International Workshop on Distributed Algorithms.*

83. Simons, B., J. L. Welch, and N. Lynch. 1990. An overview of clock synchronization. In Lecture Notes in Computer Science #448, B. Simons, A. Spector, eds. New York: Springer-Verlag, 84–96.

84. Stomp, F., and W. de Roever. 1989. Designing distributed algorithm by means of formal sequentially phased reasoning. In *Proceedings of the 3rd International Workshop on Distributed Algorithms,* 242–253. New York: Springer-Verlag. Lecture Notes in Computer Science #392.

85. Tel, G. 1994. *Introduction to Distributed Algorithms.* Cambridge, Mass.: Cambridge University Press.

86. Turpin, R., and B. Coan. 1984. Extending binary Byzantine agreement to multivalued Byzantine agreement. *Information Processing Letters,* Vol 18, No. 2, 73–76.

87. Vishkin, U. 1983 (June). A distributed orientation algorithm. *IEEE Transactions on Information Theory.*

88. Vitanyi, P. 1984. Distributed elections in an Archimedean ring of processors. In *Proceedings of the 16th ACM Symposium on Theory of Computing,* 542–547.

89. Welch, J., L. Lamport, and N. Lynch. 1988. A lattice-structured proof technique applied to a minimum spanning-tree algorithm. In *Proceedings of the 7th Annual ACM Symp. on Principles of Distributed Computing,* 28–37.

Models

6

PRAM Models

Lydia I. Kronsjö

6.1   Introduction

Parallel processing can be defined as a technique for increasing the computation speed for a task, by dividing the algorithm into several subtasks and allocating multiple processors to execute multiple subtasks simultaneously.

The main objective of exploration of parallel processing is to achieve higher efficiency of computation as compared to sequential computation. Ideally, when using p processors, we want to solve a problem p times faster. Most of the efficient serial algorithms become rather inefficient parallel algorithms. Experience suggests that the design of parallel algorithms requires new paradigms and techniques as illustrated by the following example.

Example   Find the Smallest Value in the Set of n Values. The following simple example illustrates how the run time of a parallel algorithm is affected by the number of processors available for a parallel computation.

Consider the problem of finding the smallest value in a given set $L = L[1:n]$ of n elements. Three different parallel algorithms are presented, each requiring a different number of processors for the computation.

Variant one solution. This variant assumes that the number of processors available to solve the problem is such that all possible comparisons of the pairs of elements from L are carried out simultaneously, each processor executing one operation of comparison. Under this assumption, the number of processors is interpreted as unlimited; that is, there are as many processors as needed to accommodate the fastest possible computation. (The actual number of processors required here is $p = n(n-1)/2$.) The parallel computation then runs in a small fixed number of steps.

Let each pair (i_1, i_2) of indices, such that $1 \leq i_1 < i_2 \leq n$, correspond to a different i, such that $1 \leq i \leq n(n-1)/2$. We assume that each processor P_i is able to determine on which element pair $(L[i_1], L[i_2])$ of the set L it is to perform the operation of comparison (see, for example, Note 1). Each processor reads the relevant element pair and compares the two elements. After the operation of comparison is concluded more than one processor may attempt to write to the shared memory.

Algorithm 1
```
Do in parallel
     Pᵢ derives the pair (i₁,i₂) of indices that corresponds to a different
          i (see Note 1).
     Pᵢ reads value L[i₁] and reads value L[i₂].
     If L[i₁] ≥ L[i₂]
     then Pᵢ sends NO-outcome to Pᵢ₁
     else Pᵢ sends NO-outcome to Pᵢ₂
```

<<Comment. At this stage the only active processor is P_j, $1 \le j \le n$, which did not receive a negative outcome.>>

P_j reads the value L[j] and writes it into the output cell

Note 1. One possible ordering i on the pairs (i_1, i_2), such that $1 \le i_1 < i_2$, is given as follows: $i = 1 + 2 + \ldots + (i_2 - 2) + i_1 = (i_2 - 2)(i_2 - 1)/2 + i_1$. So, element pair (L_1, L_6) is processed by processor P_{11}, element pair (L_4, L_7) by processor P_{19}, and so on.

Variant two solution. This variant assumes that $\lceil n/2 \rceil$ processors are available for the computation. The parallel computation then requires $O(\log n)$ steps.

Algorithm 2
```
Do in parallel
     Pi reads value L[2i-1] and reads value L[2i].
     If L[2i-1] < L[2i]
     then Pᵢ sends value L[2i-1] to P⌈i/2⌉
     else Pᵢ sends value L[2i] to P⌈i/2⌉
```

<<Comment. At the end of the first step the processors $P_1, \ldots, P_{\lceil i/2 \rceil}$ hold the elements of L that have not been eliminated.>>
```
Repeat
     Do in parallel
     Pi determines itself active if and only if it has been sent some values
     of L in the previous stage.
          If Pᵢ is active then
               If two values are present in Pᵢ then
                    If s1 < s2
                    then Pᵢ sends s1 to P⌈i/2⌉
                    else Pᵢ sends s2 to P⌈i/2⌉
               else Pᵢ sends the value to P⌈i/2⌉
Until no more active Pᵢ
```

<< Comment. After O(log n) stages only P_1 is active, and it holds a single value of L.>>
P₁ returns the value to the output.

The parallel computation process is illustrated in Fig. 6.1.

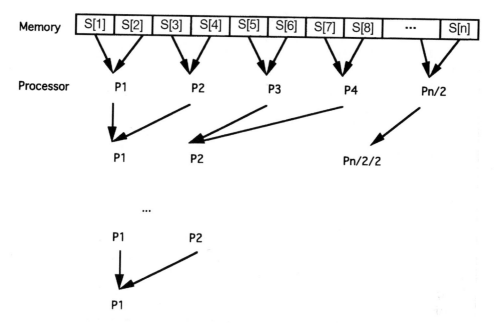

Figure 6.1 Information flow in Algorithm 2

Variant three solution. This variant assumes $p < \lceil n/2 \rceil$ processors. The parallel computation then requires $O(n/p + \log p)$ steps.

Algorithm 3
```
Do in parallel
      Pᵢ computes(reads) p
      Pᵢ finds in O(n/p) steps the smallest value in
      {L[⌈n/p⌉(i-1)+1],...,L[⌈n/p⌉i]}
Determine in O(log p) steps the smallest value among the p values
as in Algorithm 2.
```

6.2   Techniques for the Design of Parallel Algorithms

A number of basic techniques are employed in the design of parallel algorithms. Typically,
they appear many times as subtasks within more complex application and scientific algorithms. In all these techniques, the binary tree structure appears in many guises: as a data
structure, as the structure of computation, or as a structure of networked processors. We
describe these techniques below.

6.2.1   The divide-and-conquer technique

This is the most common technique. A given problem is divided into a number of independent subproblems that are dealt with recursively.

The solution for a problem at one level of the recursion has to be composable from the
solutions of its subproblems.

Example   Evaluate the polynomial P(x) of degree 7 at x = x_0:

$$a + bx + cx^2 + dx^3 \quad + \quad ex^4 + fx^5 + gx^6 + hx^7$$

The polynomial is decomposed into two sections as follows.

$$[a + bx + cx^2 + dx^3] \quad + \quad x^4[e + fx + gx^2 + hx^3]$$

The computation is then organized, in a top-down fashion, as a binary tree where the leaves hold the contstants and the variable's value. Figure 6.2 illustrates the divide-and-conquer technique.

6.2.2   The balanced binary tree method

Here each internal node corresponds to the computation of a subproblem with the root corresponding to the overall problem.The problems are solved in bottom-up order, with those at the same depth in the tree being computed in parallel.

Example   Find the maximum of n elements. A bottom-up approach to solve this problem using the balanced binary tree method is shown in Fig. 6.3.

6.2.3   Compression (collapsing) technique

Again, this technique is particularly suitable for managing the computations involving arrays or lists.

Let $A[1:n]$ be an array of size n. Recursively, for each odd value of i in parallel, the two entries $A[i]$ and $A[i + 1]$ are compressed into a single entry with the value of maximum of $\{A[i], A[i + 1]\}$.

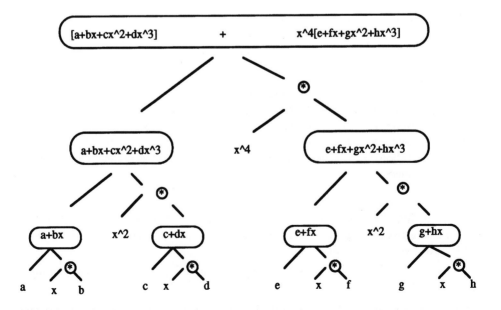

Figure 6.2 Top-down, divide-and-conquer approach to evaluate a polynomial

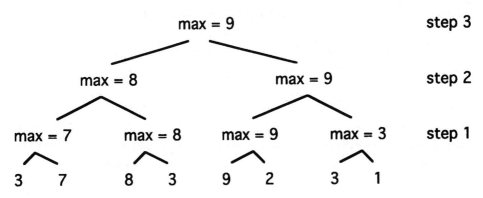

Figure 6.3 A bottom-up approach to solve the "Find-the-maximum" problem

Example

$$[3\ 7\ 8\ 3\ 9\ 2\ 3\ 1] \rightarrow [7\ 8\ 9\ 3] \rightarrow [8\ 9] \rightarrow [9]$$

Here, the compression technique amounts to the balanced binary tree method. This approach is widely used in general applications, particularly in graph algorithms (i.e. graph compression).

6.2.4 Brent's scheduling principle

A general theorem of Brent relates the number of processors *(computational network)* to the time required to compute something in parallel. In this presentation we follow Smith [48].

A computational network can be defined as a directed acyclic graph whose vertices are subdivided into three sets:

Input vertices:	these vertices have no incoming edges.
Output vertices:	these vertices have no outgoing edges.
Interior vertices:	these vertices have incoming edges and a single outgoing edge.

Fan-in of a vertex is the number of its incoming edges, and *fan-out* is the number of its outgoing edges. The maxima of these two quantities over the entire graph is called, respectively, the fan-in and fan-out of the graph. Each vertex in a graph is labelled with an elementary operation. The depth of the computational network is the depth of the longest path from any vertex to any output vertex.

The computation on a computation network, on a given set of inputs implies the following set of actions:

- Apply the input data to the input vertices.
- Transmit data along directed edges. Whenever an interior vertex is encountered, wait until data arrives along all its incoming edges, then perform the indicated elementary computation.
- Transmit the result of the computation along all of the outgoing edges.
- Terminate the procedure when there is no data at interior vertices.

Brent's theorem leads to a general principle in the design of parallel algorithms known as the *Brent Scheduling Principle*. This principle often makes it possible to reduce the number of processors used in parallel algorithms without increasing the asymptotic execution time. In general, the execution time increases somewhat when the number of processors is reduced, but not by an amount that increases the asymptotic time. If an algorithm has an execution time of $O(\log^k n)$, then the execution time might increase by a constant factor.

The Brent Scheduling Principle. Let A be a given algorithm with a parallel time of t units. Suppose that A involves a total number of m computational operations, that is $w(A) = m$. Then A can be implemented using p processors in $O(m/p + t)$ parallel time, where $m = max\{m_i\}$ and m_i is the number of operations performed in parallel in step i.

Proof
Let m_i be the number of computational operations performed in parallel in step i of A. With p processors, where $p < n/2$, this work can be achieved in time at most $m_i/p + 1$. By summing over all i, $1 \le i \le t$, we have the result since $m = m_1 + m_2 + \ldots + m_t$.

In the above simulation it is assumed that processor allocation is not a problem. In fact, it may often be difficult to partition m_i into p pieces.

In our earlier problem of finding the smallest in a set of n elements it is easy to allocate processors and, in fact, the method presented in Algorithm 3 is standard and is used for an optimal solution of many parallel algorithms. Assume that each processor initially receives n/r elements of L; it sequentially compares elements and writes the local smallest number into a shared memory cell. After that, the reduced set in the shared memory contains only r elements. Hence, the remaining execution will be implemented using r processors as in Algorithms 1 or 2, depending on additional constraints on the number of processors.

Examples of applications of the Brent Scheduling Principle are given in sequel.

6.3   The PRAM Model

The question of how to model parallel computation is subtle and has significant impact on both the design of parallel systems and the design of parallel algorithms. One way to uncover the strength of efficient parallel processing is to concentrate on the computation and to ignore other practical considerations of parallel processing such as, for example, communication between the processors. One simply assumes that communication does not cost anything, and the only time consuming operations are those of the actual computation.

A computation model that embodies these assumptions is the *parallel random access machine* (PRAM), first proposed by Fortune and Wyllie in 1978 as a general model for parallel computation. At about the same time, another researcher, Schwartz, was also advocating the use of a PRAM for studying the limits of parallel computation. Since its inception, the PRAM has played a central role in studies of how inherent parallelism within problems can be exploited for efficient computation. At first glance, the PRAM model might not appear to be suitable as a general model for designing efficient parallel algorithms due to some unrealistic features. Yet, it has been used by algorithm designers with remarkable success. Many very efficient PRAM algorithms have been designed that allow efficient implementation on real parallel architectures.

The PRAM model of computation is an idealization that draws its power from three facts and consequences [57]:

- The model ignores algorithmic complexity concerning machine connectivity and communication contention, data locality, synchronization, and reliability. It allows the algorithm designer to focus on the fundamental computational difficulties of the problem at hand. As a result, a substantial number of efficient algorithms have been designed in this model, as well as a number of design paradigms and utilities for designing such algorithms.
- Many of the design paradigms have turned out to be strikingly robust, having as a result found applications in models outside the PRAM domain, including VLSI, where each wire element and gate is carefully accounted in the complexity cost.
- Recent advances have shown PRAM algorithms to be formally emulatable on high-interconnect machines. Formal machine designs that support a large number of virtual processes can, in fact, give a speedup that approaches the number of processors for some efficiently large problems. Some new machine designs are aimed at realizing idealizations that support pipelined, virtual unit time access PRAMs.

Many computer scientists believe that the future success of general-purpose parallel computation depends on parallel architecture designers supporting in their design a virtual PRAM or a model of parallel computation that is very close to the PRAM. For the last decade or so, along with the work on design paradigms and efficient algorithms, theoretical computer scientists have been developing a body of theory centered around parallel algorithms and parallel architectures with the PRAM model of computation in the center of the studies. The theory of parallel algorithms and complexity focuses on the trade-off between the time for a parallel computation and the number of processors required. Of particular interest are the so-called *efficient* algorithms; this name is given to algorithms with *polylogarithmic* parallel run time, i.e., the parallel computation time bounded by a fixed power of the logarithm in the size of the problem input, and with the time × processor product exceeding the number of steps in an optimal sequential algorithm by at most a *polylogarithmic* factor. (A function $f(n) = c \log(p(n))$, where $p(n)$ is a polynomial in n and c is constant, is said to be polylogarithmic.) Many problems with polynomial time sequential algorithms can be solved by parallel algorithms with polylogarithmic run times.

The problems that can be solved by algorithms with polylogarithmic parallel time using a polynomial number of processors are said to form the **NC**-class. The theory of the **NC**-algorithms has helped to classify the problems in terms of their complexity. The fundamental result of this theory is the observation that parallel computation does not help to speed up computation of the so-called hard problems. The problems that take a notoriously long time to compute sequentially remain difficult to compute using parallel framework. More recent studies suggest that new exciting opportunities for efficient solving of notoriously difficult problems are offered by parallel randomized algorithms.

6.3.1 Description of the PRAM

The PRAM of Fortune and Wyllie can be viewed as the parallel analog of the sequential RAM, an abstraction of the von Neumann machine.

A PRAM consists of p general purpose sequential processors, $P_1, P_2, \ldots, P_p$, all of which are connected to a large shared random access memory, M. Each processor has a private, or local, memory for its own computation. All communication among the processors is done via the shared memory. In one step, each processor can access, either reading from it or writing to it, one memory location or execute a single RAM operation. The processors synchronously execute the same program through the central main control. The processors generally operate on different data but performing the same instructions. Hence, the model is sometimes called *shared-memory single instruction stream, multiple data stream* (SM SIMD) machine (see Fig. 6.4).

A parallel time step is defined as consisting of three phases:

1. the read phase, in which the processor may read from a shared memory cell
2. the computation phase, in which the processor does some calculations
3. the write phase, in which the processor may write to a shared memory cell

PRAMs are classified according to restrictions on shared memory access. The read and write conflicts are resolved by allowing one of the following mechanisms:

Exclusive-read, exclusive-write PRAM model (EREW PRAM)
Concurrent-read, exclusive-write PRAM model (CREW PRAM)
Concurrent-read, concurrent-write PRAM model (CRCW PRAM)
Exclusive-read, concurrent-write PRAM model (ERCW PRAM)

The most common model, EREW PRAM, does not allow simultaneous access by more than one processor to the same memory location for read or write purposes. This model is followed by a CREW PRAM, which allows concurrent access for reads but not for writes, and a CRCW PRAM, which allows concurrent access by more than one processor to the same memory location for read or write purposes. In a concurrent write model, it is normally assumed that the write conflicts are resolved in favor of the processor that has the least index among those processors involved in the conflict, also known as PRIORITY conflict resolution. Finally, the variant COMMON for treating read/write conflicts assumes that all processors, which simultaneously write to the same memory cell, write the same value.

The PRAM model is remarkably simple while providing a clean medium for expressing algorithms. The PRAM may be viewed as a virtual design space for a parallel machine and not as a parallel machine as such; for example, the PRAM could serve as a design model for the development of very fast computational paradigms as well as to a core of problems that can be computed fast on PRAM. An improvement in parallel running time of a PRAM algorithm could then benefit the designers in reducing the actual running time on real parallel architectures.

An important, more recent concept in PRAMs is *slackness* in processors [53, 34]. Suppose that a PRAM algorithm, that is efficient for up to p_2 PRAM processors, is simulated on a real parallel machine with p_1 processors. Processor slackness is then defined as the ratio of p_2 to p_1. It may be argued then that even if p_1 is fixed, having a larger p_2 (and

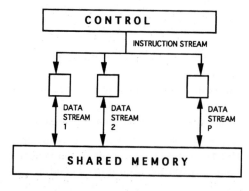

Figure 6.4 PRAM model of parallel computation

therefore larger processor slackness) leads to a more efficient simulation by the real machine. The primary intellectual challenge is to obtain the fastest possible time by a processor-efficient algorithm. The techniques that will be developed to do this are likely to have practical importance [57].

6.3.2   Broadcasting a datum

Sometimes it is useful to disallow the simultaneous read and write operations, which of course means that the EREW PRAM model is used. It is possible to simulate the concurrent read and write operations at a price of $O(\log n)$ time. The standard algorithm used in these situations is known as broadcasting. The algorithm is as follows.

Broadcasting a datum. Let D be a location in shared memory holding a datum that all n processors need at a given moment during the execution of an algorithm.

We shall assume that an array L of length n is present in memory. The array is initially empty and is used by the algorithm as a working space to distribute the contents of D to the processors. Its ith position is denoted by $L[i]$.

Algorithm: Broadcast

```
Processor P1
     reads the value in D
     stores it in its local memory
     writes it in L[1]
for i ← 0 to log(n-1) do
     for j ← 2^i+1 to 2^(i+1) do in parallel
          Processor Pj
          reads the value in L[j-2^i]
          stores it in its local memory
          writes it in L[j]
     end
end
```

When the algorithm terminates, all processors have stored the value of D in their local memories for later use. Since the number of processors having read D doubles in each iteration, the procedure terminates in $O(\log n)$ time.

The memory requirement of Broadcast is an array of length n. In fact, an array of half that length will be sufficient because, in the last iteration of the algorithm, all the processors have received the value in D and need not write it back in L. Hence, an array L of length $n/2$ only is needed.

6.4   Optimality and Efficiency of Parallel Algorithms

For sequential algorithms, computational complexity involves analysis of the amount of computing resources needed to solve a problem as a function of the problem's size. The most obvious and most important resource is time.

Time function remains of fundamental importance for parallel algorithms. In addition, the number of processors, p, is an important parameter, since the availability of many processors to carry out the computation simultaneously is a fundamental assumption in parallel processing.

In an ideal model, one can assume that the number of processors is infinite and, hence, an ideal optimal computation time. However, in practice, one will always reach a stage at which the size of the problem exceeds the number of processors available. Hence, realistic algorithms will inevitably run on an architecture with a very finite number of processors. We need to determine the optimal values involved.

Let $\prod$ be a problem instance of size n. Assume that $\prod$ can be solved on a PRAM by a parallel algorithm A in time $t(n)$ and by employing $p(n)$ processors.

The work of a parallel algorithm A is the product of its time, $t(n)$, and the number of processors, $p(n)$:

$$w(n) = t(n) \times p(n)$$

Of the two parallel algorithms, we say that one algorithm is more efficient than the other if its work $w(n)$ is smaller, and its parallel time $t(n)$ is smaller.

Any PRAM algorithm that requires work $w(n)$ can be converted into a sequential algorithm of time $w(n)$. This can be achieved by letting the sequential processor to simulate each parallel step of PRAM in $p(n)$ time units.

6.4.1   Optimality

A parallel algorithm is said to be optimal if its $w(n)$ is asymptotically equal to the sequential complexity of the problem. An optimal parallel algorithm corresponds to an optimal (often linear) time sequential algorithm for the problem.

A fully parallel algorithm is a parallel algorithm that runs in constant time using an optimal number of processors. The notion of a fully-parallel algorithm represents an ultimate theoretical goal for designers of parallel algorithms.

The concept of slowing down a PRAM algorithm by reducing the number of processors while retaining the same time is also useful. Any PRAM algorithm A with work $w(n) = t(n) \times p(n)$ can be converted into a PRAM algorithm B for the same problem of work $O(w(n))$ and any number of processors $p \leq p(n)$, whose running time is

$$t(n) \times \left\lceil \frac{p(n)}{p} \right\rceil$$

Each processor of B is simply made to do the computation of several processors of A sequentially, thus increasing the time of each step of A to $\lceil p(n)/p \rceil$ at most. (See Brent's Scheduling Principle, discussed earlier.)

6.4.2   Speedup and efficiency

The two most important criteria for evaluation of parallel algorithms are speedup and efficiency of a parallel algorithm.

The speedup achieved by a parallel algorithm running on $p(n)$ processors is the ratio between the sequential time, $T(n)$, taken by that parallel computer executing the fastest serial algorithm and the parallel time, $t(n)$, taken by the same parallel computer executing the parallel algorithm using $p(n)$ processors:

$$\text{speedup} = \frac{T(n)}{t(n)}$$

(The speedup is said to be *unbounded* if $T(n)/t(n)$ tends to ∞ when n increases. The speedup measures an improvement in the running time of an algorithm due to parallelism.

Efficiency measures the work reduction achieved as a result of using several processors to run a parallel algorithm as opposed to a single processor to run a serial algorithm.

Efficiency of a parallel algorithm running on $p(n)$ processors is the speedup divided by $p(n)$, which is, of course, the ratio between the work $W(n)$ of a fastest serial algorithm executed on a parallel computer and the work $w(n)$ of the parallel algorithm, running on $p(n)$ processors, executed on the same parallel computer:

$$efficiency = \frac{speedup}{p(n)} = \frac{T(n)}{w(n)}$$

Example   A parallel matrix multiplication. Given matrices $\mathbf{A}$ of size $q \times r$ and $\mathbf{B}$ of size $r \times s$, one defines the product matrix $\mathbf{C}$ as the matrix of size $q \times s$ whose elements are given by

$$C_{ik} = \Sigma_{j=1,r} A_{ij} \times B_{jk}$$

To appreciate the degree of parallelism inherent in this problem, we note that the elements of the product matrix are completely independent of each other. Thus, given qrs processors and an appropriate distribution of the input data (e.g., each of the q processors holds one row of elements of A, and each of the s processors holds one row of elements of B), we could conceivably achieve an immediate speedup

$$speedup = \frac{qsr}{r} = qs$$

Furthermore if the inner products were performed in a parallel divide-and-conquer manner (see Section 6.1.1), we could achieve an additional speedup of $r/\log r$.

This is, of course, in theoretical terms only. In reality, communication costs and other factors greatly affect the achievable speedup. These other factors account for the many variations of the parallel matrix multiplication's algorithm.

6.4.3   Rates of Improvement and Inefficiency.

Two important rates of improvement in the run time of a parallel algorithm over that of a sequential algorithm that make the parallel algorithm worthwhile, are polynomial and polylogarithmic rates.

A parallel algorithm is said to be polynomially fast if its time, $t(n)$, is a polynomial function of the best sequential time, $T(n)$.

$$t(n) = (T(n))^{\varepsilon}, \varepsilon < 1$$

A parallel algorithm is said to be *polylogarithmically fast* if its time, $t(n)$, is a power of $O(1)$ of the logarithmic function of the time of the best sequential time $T(n)$.

$$t(n) = \log^{O(1)} T(n)$$

The upper bounds on the time are determined by the fastest known algorithms.

The lower bound determines the minimum amount of time needed to solve the problem by an arbitrary parallel algorithm. This is the complexity of the problem studied; particularly important is the question of whether an algorithm with minimal execution time is known—that is, whether the lower bound of the execution time has been achieved.

The *worst-case time,* or the (time) *complexity,* of a parallel algorithm is a time function $t(n)$ that is the maximum, over all inputs of size n, of the time elapsed from when the first processor begins execution of the algorithm until the last processor terminates algorithm execution. The *computational complexity* C_n of the parallel computation of a problem Π of size n is the smallest number of parallel steps required to obtain a solution with arbitrary input data. *Asymptotic complexity* is the behavior of C_n as n tends to ∞.

Another useful measure in the analysis of algorithm performance is *inefficiency,* which is defined as the ratio inverse to that of efficiency:

$$\text{inefficiency} = \frac{w(n)}{W(n)}$$

An algorithm is said to be of a *constant inefficiency,* of a *polylogarithmically bounded inefficiency* or of a *polynomially bounded inefficiency* if its cost, respectively, is as follows:

$$w(n) = t(n) \times p(n) = \begin{cases} O(T(n)) \\ T(n) \times \log^{O(1)}(T(n)) \\ T(n)^{O(1)} \end{cases}$$

Similar polynomial bounds can be defined on the number of available processors.

Let a parallel algorithm solve a problem Π of size n on $p(n)$ processors. If for a certain polynomial K and for all n, $p(n) \le K(n)$, then the number of processors is considered *polynomially bounded,* and in other cases *polynomially unbounded.*

The interest in algorithms for unbounded parallelism in terms of the number of processors is mainly of theoretical nature since these algorithms give the limits of parallel computations and allow a deeper understanding of the intrinsic structure of algorithms.

However, of real importance are parallel algorithms of bounded parallelism, where the number of processors is finite and much smaller than the problem size. In practice, even if one could afford to have as many processors as data for a particular problem instance size, it may not be desirable to design an algorithm based on that assumption: a larger problem instance would render the algorithm totally useless.

Generally, we would expect that an efficient parallel algorithm should exhibit the following properties:

- *The running time of the parallel algorithm is considerably smaller than that of the comparable sequential algorithm.* The primary motive for building parallel computers is to speed up the computational process. To be useful, a parallel algorithm should be significantly faster than the best sequential algorithm for the problem at hand.
- *The running time is decreasing with the number of parallel processors increasing.* Ideally, one hopes to have an algorithm whose running time decreases as more processors are used. In practice, it is usually the case that a limit is eventually reached beyond which no speedup is possible regardless of the number of processors used.
- *The number of parallel processors required by a parallel algorithm is considerably fewer than the input of the problem instance.* It is unrealistic when designing a parallel algorithm to assume that we have at our disposal more, or even as many, processors as

there are items of data. This is particularly true when n is very large. It is important that $p(n) = n^k, 0 < k < 1$.

- *The number of parallel processors is decreasing with the size of the input data increasing*. The availability of additional computing power always means that larger and more complex problems will be attacked than was possible before. Algorithms using a number of processors $p(n)$, such as $\log n$ or $\sqrt{n}$, are really not acceptable either due to their fixed, inflexible dependence on the number of processors. One would like to develop algorithms that possess the intelligence to adapt to the actual number of processors available on the computer being used.

6.5 Basic PRAM Algorithms

A substantial body of PRAM algorithms, including many for fundamental problems, have been obtained over the past decade or so. The PRAM model has also been extensively used to uncover fundamental paradigms and design techniques, and these are of use in many models of physically available parallel architectures. These techniques have led to efficient fast parallel algorithms in a variety of areas, including comparison problems, graph problems, computational geometry, pattern matching, matrix manipulation, and solution of linear equations.

An elaborate overview of the logical structure of parallel computation and of a taxonomy of PRAM parallel algorithms is given in Refs. [31, 12].

For PRAMs, one would like to have simple algorithms that are easy to specify and code. A few key methods have emerged as fundamental subroutines in the design of efficient and optimal parallel algorithms. We shall now review these basic techniques and algorithms.

6.5.1 Computing prefix sums

The *prefix sums* problem plays a major role in parallel algorithms. Given an array $a[1: n]$, it is often useful to compute the sum $S_i = a_1 + a_2 + \ldots + a_i$, $1 \leq i \leq n$. An example is the problem of compacting a sparse array: for an array of n elements, many of which are zero, one wishes to generate a new array containing the non-zero elements in their original order. The position of each non-zero element in the new array can be computed by assigning value 1 to the non-zero elements and computing prefix sums.

Assuming that each processor P_i holds in its local memory a number a_i, $1 \leq i \leq n$, a PRAM algorithm computes all sums $a_1 + a_2 + \ldots + a_i$, $1 \leq i \leq n$, in $O(\log n)$ time with n processors.

Algorithm: Parallel prefix sums

```
Input: an array [a₁, a₂, ...,aₙ] of numbers
if n = 1 then S₁ ← a[1] else
    for j ← 0 to log n -1 do
        for i ← 2ʲ + 1 to n do in parallel
            Processor Pᵢ
                (i) obtains a[i-2ʲ] from Pᵢ₋₂ⱼ via shared memory and
                (ii) replaces a[i] with a[i-2ʲ] + a[i]
```

The algorithm runs on EREW PRAM, because there are no conflicts in the memory access, and requires $O(\log n)$ time, because the number of processors that have finished their computation doubles at each stage. It also requires work $w(n) = O(n)$, which follows from the observation that $w(n) = w(n/2) + O(n)$, with $w(1) = 0$. By invoking Brent's Scheduling principle, we see that this is an optimal EREW PRAM algorithm for $p(n) = O(n/\log n)$. On a CRCW PRAM, the algorithm can be modified to run optimally in $O(\log n/\log \log n)$ time when the a_i are $O(\log n)$-bit numbers [47].

Applying Brent's Theorem yields an optimal prefix sum algorithm with $p(n) = n/\log n$. It was noted earlier that in the general case allocation of processors may be quite difficult, however in this case it is easy to allocate processors. Therefore, using the standard method an optimal prefix sums algorithm is derived as follows.

6.5.2   Optimal prefix sums

Input: An array a[1:n] of elements. Element a_i is in location M_i of shared memory. An EREW PRAM with $p \le n/\log n$ processors is employed. Let $q = \lceil n/p \rceil$.

Assign processor P_i to memory locations $(i-1)q + 1, (i-1)q + 2, \ldots, iq$, $1 \le i \le p$. Processor P_i stores these values in its local memory and then adds them and puts the result in M_i. This requires time $O(n/p)$.

The algorithm then runs simple prefix sums on the new reduced array of p locations with p processors. This takes $O(\log n)$ time.

Finally, in an additional $O(n/p)$ time the P_ith processor computes prefix sums for its local array with S_{i-1} the first element.

The method's parallel time is $O(n/p + \log n)$. For $p = n/\log n$, we get $t(n) = O(\log n)$ and $w(n) = O(n)$; that is, optimal work. On a CRCW PRAM the algorithm can be modified to optimally run in time $O(\log n/\log \log n)$.

The general prefix sums algorithm solves any problem where the addition operation is replaced by any other (associative) binary operation, denoted by *, i.e.,

$$S_i = a_1 * a_2 * \ldots * a_i$$

Examples of such operations on numbers are finding the larger or smaller of two numbers, and multiplication and recognition of any regular language whose input size is restricted to n. Other operations that apply to a pair of logical quantities (or a pair of bits) are **and**, **or**, and **xor**.

6.5.3   List ranking

A variant of the prefix sums problem is the list ranking problem: Given a linked list of n elements, compute the suffix sums of the last i elements of the list, $i = 1, \ldots, n$. Thus, here the ordered sequence of elements is given in a form of list rather than an array, and the sums are computed from the end rather than from the beginning. One special case of this problem is to obtain for each element the number of elements ahead of it in the list; that is, for each element to obtain its rank. This special case has given the name to the problem.

The parallel suffix on a list is more difficult than parallel prefix on arrays because, on the list, there is no knowledge of any global structure; say, assuming that the list is stored in the shared memory locations M_1 to M_n, by looking at M_i, only local information can be obtained. The algorithm is as follows.

Let the linked list be represented in two arrays, the contents array c[1:n] and the successor array s[1:n]. We will need an operation of replacing a pointer s[i] by its pointer s[s[i]], i.e., $s[i] \leftarrow s[s[i]]$, which we shall call *pointer jumping*. In fact this operation is a fundamental technique in parallel algorithm design. The algorithm is as follows:

Algorithm: List ranking by pointer jumping
```
iteration ← 1
repeat
      for i ← 1 to n do in parallel
      c[i] ← c[i] + c[s[i]]
      s[i] ← s[s[i]]
until iteration = ⌈log n⌉
```

If the rank $r[i]$ of the element in location i is the distance of this element from the end of the input linked list, then in the algorithm at the end of the computation $c[i]$ is the rank $r[i]$ of the element in location i. To show this, one needs to show that the following inductive hypothesis is preserved: At the start of each step, $c[i]$ equals the sum of elements in the input list with ranks $r[i]$, $r[i] - 1, \ldots, r[s[i]] + 1$ for the current $s[i]$. After $\lceil \log n \rceil$ iterations, $s[i]$ becomes nil for all i.

By assigning a processor to each location i, we obtain an n-processor parallel algorithm of time $O(\log n)$. However this gives $p(n) = n$ and the work $w(n) = O(n \log n)$, and so the algorithm is not optimal given that the sequential time to rank a list is $T(n) = n$. A straightforward extension of the technique used to optimize the prefix sums is not applicable here because there is no obvious way to know in advance the locations of the elements of the reduced list as is the case with the array elements. Some possible strategies for optimal list ranking are discussed in Refs. [30, 47].

6.5.4   Sorting by comparison and merging

In the sorting problem, the input is an array of n elements from a linearly ordered set. The task is to rearrange the elements into nondecreasing order. In the merging problem, there are two input arrays, each of which is known to be in nondecreasing order. The task is to merge the two arrays into one array of nondecreasing order. In comparison algorithms, the only means of gathering information about the elements is via pairwise comparisons; that is, tests of the form: "Is x less than y?" where x and y are elements of the array. In this discussion, for convenience, we shall restrict the problems to the instances where the n elements are distinct.

The elements to be sorted have keys governing the sorting or merging process. Assume that n keys are stored in the array $k[1:n]$. Further assume that the keys are from a totally ordered domain.

One of the earliest and celebrated parallel sorting algorithms is the sorting network due to Batcher. The algorithm is easily implemented on an EREW PRAM. The algorithm is conveniently presented using the balanced binary tree as shown for a simple example of eight elements in Fig. 6.6. The following discussion is based on Spirakis and Gibbons [49].

The leaves of the tree contain the n keys of the array. The operation at non-leaf nodes is a merge of two sorted subarrays, and the result in each non-leaf node is a sorted sequence of all the keys that are stored in leaf descendants of that node. Thus, the root node contains the sorted sequence of all the keys.

Let $m(n)$ be the time to merge two sorted sequences of n keys each. The depth of the binary tree is $O(\log n)$. If all merges of the same depth are done in parallel, then a parallel sorting time is $t(n) = O(m(n) \times \log n)$. The Batcher method ensures that $m(n) = O(\log n)$.

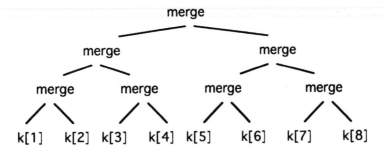

Figure 6.5 Example of a merge sort tree

A model of *parallel comparison* algorithms proposed by Valiant defines the so-called *degree of parallelism, d,* which is a number of comparisons performed at each parallel step. In a general setting it is not required that the d pairs of elements compared at a given step be disjoint. The choice of the d comparisons to be performed at a given step can depend in an arbitrary manner on n, the number of elements, and on the outcome of previous comparisons. The algorithm terminates when it has acquired enough information about the input to specify the answer, for example, on the permutation required to put the elements in nondecreasing order. The execution time of an algorithm is the number of steps performed.

The key idea of Batcher's bitonic merge is the notion of a bitonic sequence. Let S_1 and S_2 be two sorted sequences each of length $n/2$. Let S'_2 be the sequence S_2 in reverse order. Let $S = S_1 \circ S'_2$, where $\circ$ stands for concatenation. The sequence S is called *unimodal.* A sequence is called *bitonic* if it is cyclic shift of a unimodal sequence.

Batcher's Basic Lemma
Let $A = (a_1, a_2, \ldots, a_{2N})$ be a bitonic sequence of even length, of distinct elements in a linearly ordered set. Let the sequences $L(A)$ and $R(A)$ be

$$L(A) = (\min\{a_1, a_{N+1}\}, \min\{a_2, a_{N+2}\}, \ldots, \min\{a_N, a_{2N}\}) \text{ and}$$
$$R(A) = (\max\{a_1, a_{N+1}\}, \max\{a_2, a_{N+2}\}, \ldots, \max\{a_N, a_{2N}\})$$

Then $L(A)$ and $R(A)$ are both bitonic, and each element of $L(A)$ is less than each element of $R(A)$.

A direct translation from Batcher's Basic Lemma yields the following recursive bitonic merge algorithm.

Algorithm: Recursive bitonic merge
```
Input: a bitonic sequence S
if S is of length 1 then stop
else
      form L(S) and R(S)
      do in parallel
          L(S) ← Recursive Bitonic Merge (L(S))
          R(S) ← Recursive Bitonic Merge (R(S))
      concatenate L(S))`R(S))
```

Optimal sorting has been studied very intensively [3, 47]. A pipelined version of merge sort on EREW PRAM has been given by Cole [8]; the algorithm is optimal with $O(\log n)$

time and $O(n)$ processors. Cole's algorithm easily follows if a merge of two nondecreasing sequences of lengths n and m can be done in time $O(\log \log n)$ by using $n + m$ processors. Then, by starting with $O(n)$ processors, the merge tree can be climbed with at most $\log \log n$ time per level. This gives a parallel sorting technique of time $O(\log n \log \log n)$ with $O(n)$ processors. The algorithm yields a straightforward implementation on a CREW PRAM.

6.5.5   The Euler tour technique

One of the most commonly occurring structures in computations of all kinds is the tree. The tree structure is particularly suitable for many graph problems: a rooted spanning tree is the starting point for many graph algorithms, which involve the computation of simple tree functions such as pre- and post order numbering of vertices in the tree, the level and height of each vertex in the tree, and the number of descendants of each vertex in the tree. These operation can be efficiently computed using a recent novel algorithmic technique devised by Tarjan and Vishkin [51]. The technique is called the *Euler tour technique on trees*. The method works by reducing the computation of the tree functions to list ranking. This discussion follows closely Spirakis and Gibbons [49].

An Eulerian circuit (the so-called *Euler tour*) is a circuit in a graph which traverses every edge precisely once. A simple theorem proves that a directed connected graph contains an Eulerian circuit if and only if for every vertex v, the number of edges with v as an endpoint and directed toward v (*in-degree(v)*) is equal to the number of edges which have v as an end-point and which are directed away from v (*out-degree(v)*).

Given an undirected tree, the Euler tour technique proceeds by first replacing each edge of the tree by two anti-parallel edges and then by finding an Eulerian circuit of the resulting directed graph. Suppose an undirected graph is stored as an adjacency list of lists. Then, the first step is to make each adjacency list circular by setting up a pointer from the final item to the first. This is achieved by applying pointer jumping operation, as in the list ranking algorithms, so as to cause a pointer to be directed from the first item of each list to the final item. This pointer can then be employed to pass the address of the first item to the final item.

Given the circular lists, the Euler tour is constructed by determining for each edge (i, j), the edge which follows (i, j) on the Euler tour. The effect is always to produce a circular list of edges which contains every edge. The Euler tour can be constructed by an optimal $O(\log n)$ time algorithm using pointer jumping as described.

A so-called *traversal list* for the tree is obtained by breaking the Euler tour at an arbitrary tree node. Traversal lists are employed for many basic computations, including the construction of a directed out- or in-tree rooted at an arbitrary vertex u given an undirected tree, the computation of the number of descendants for each tree node of a rooted tree, the determining a preorder numbering of the tree nodes and the establishment for all pairs of nodes (i, j) of a rooted tree whether i is an ancestor, a descendent or neither of j. The optimal list ranking algorithm implies optimal $O(\log n)$ time EREW PRAM algorithms for all these problems.

The design of efficient parallel algorithms for graph problems has presented a challenge since the most useful techniques of depth-first and breadth-first search for designing sequential algorithms have resisted attempts to be parallelized efficiently. Techniques called *ear decomposition search* (EDS) and a special type of ear decomposition called *open ear decomposition* were suggested as a replacement for DFS in the context of efficient and fast parallel algorithms [38]. These decompositions can be computed efficiently in parallel and lead to efficient parallel algorithms for several fundamental graph problems, particularly for undirected graphs. However, as reported by Ramachandran [46], no such success has yet been found for directed graphs due to a certain phenomenon known as the *transitive closure bottleneck*. The reasons are as follows. Currently the most efficient highly parallel algorithm known for the basic problem of testing if one vertex is reachable from another in a directed graph computes the transitive closure of the adjacency matrix of the input graph and reads out the result from this matrix. This computation requires well in excess of a number of processors quadratic in the number of vertices in the input graph. Since this problem can be computed in linear time sequentially, the parallel algorithm is not efficient. As a result, most problems on general directed graphs do not have efficient parallel algorithms. The phenomenon is often called the transitive closure bottleneck. Further details can be found in Ref. [46].

6.6   The NC-Class

It has now been universally agreed that an efficient parallel algorithm means an algorithm that runs fast and uses a reasonable number of processors. In the theoretical community, this concept has been consolidated in the notion of the complexity class **NC**, the class of algorithms running in polylogarithmic time, in the input size, with polynomially many processors.

The **NC**-class is defined as follows. An algorithm that takes polylogarithmic time using a polynomial number of processors is deemed to be an efficient parallel algorithm, and the problems that can be solved within these constraints are universally regarded as having efficient parallel solutions. These problems are said to belong to the class **NC**. (The class name is an abbreviation for Nick (Pippenger)'s Class.)

Among important problems that lie in the class **NC** are the basic arithmetic operations, transitive closure and Boolean matrix multiplication, the computation of the determinant, the rank or the inverse of a matrix, the evaluation of certain classes of straight-line programs, and the construction of a maximal independent set of vertices in a graph. So, if **P** is the class of problems solvable in polynomial sequential time, and **NC** is the class of problems solvable in polylogarithmic parallel time using a polynomial number of processors, then probably the most important question about parallel computations today is, "does **P** = **NC**?"

A general belief, but not a proof, is that **P** ≠ **NC**, and hence there are problems in **NC**. Thus, **NC** is the class of well parallelizable problems. For a more detailed exposition see [33] and Chapter 4 in this book.

6.7   P-Completeness: Hardly Parallelizable Problems

The study of parallel complexity within the PRAM model has led to some important negative results. Efficiently parallelizable problems that we know of are in **NC**. In contrast, there have been identified certain problems which do not seem to admit efficient parallelization readily: called "hardly parallelizable," they form the class of the so called **P**-*complete* problems.

A problem **L** ∈ **P** is said to be **P**-complete if every other problem in **P** can be transformed to **L** in polylogarithmic parallel time using a polynomial number of processors (such a transformation is said to be an **NC**-reduction). In other words, **L** is a hardest problem in **P** from the point of view of finding efficient parallel computations. Clearly, if we could prove (an unlikely event) that **L** ∈ **NC**, then it would follow that **P** = **NC**.

Using a theory of reducibility analogous to the theory of **NP**-completeness, it has been possible to show that **P**-complete problems, though solvable sequentially in polynomial time, do not lie in the class **NC** unless every problem solvable in sequentially polynomial time lies in **NC**. The **P**-complete problems evidently are inherently resistant to ultrafast parallel solution. If an efficient parallel solution for any **P**-complete problem could be found, then a similar solution would exist for any other **P**-complete problem.

A **P**-complete problem is the problem that

- is solvable sequentially in polynomial time, but
- does not lie in the class **NC** unless every problem solvable in sequential polynomial time lies in **NC**

The classic definition of **P**-completeness is in terms of logarithmic space reducibility and can be found in Goldschlager [19].

Among examples of **P**-complete problems are the maximum-flow problem and the problems of evaluating the output of a monotone Boolean circuit when all the inputs are fixed at constant values [16]. A large list of **P**-complete problems have been compiled by Greenlaw, Hoover, and Ruzzo [18]. The list is periodically updated to incorporate newly discovered **P**-complete problems.

The border between well-parallelizable and hardly parallelizable problems lies somewhere between **NC** and the class of so-called **P**-complete problems.

6.8   Randomized Algorithms and Parallelism

Today, it is recognized that, in a wide range of applications, randomization is an extremely important tool for the design of algorithms. Randomized algorithms are the algorithms where execution is controlled at one or more points by randomly made choices; an early example of a randomized algorithm is the work done on primality testing by Rabin.

The presence of randomization in an algorithm may have the effect that the execution time, the result, or both, are randomly distributed on successive runs with the same input. Under these conditions, substantial speedup may be obtained by simply running many instances of the algorithm simultaneously and selecting the best or the first result. Such implementations are extremely easy to develop and may yield results equal to or better than those of much more complex implementations.

The randomness in randomized algorithms is a property of the algorithms themselves, and not solely of the input. This is different from the random behavior of a deterministic algorithm applied to a random selection of inputs. Perhaps the first probabilistic algorithm ever developed is that of Buffon proposed in 1733 [48]. The algorithm requires a needle of a precisely-known length and a floor marked with parallel lines with the property that the distance between every pair of neighboring lines is exactly double the length of the needle. It turns out that if the needle is dropped on this floor, the probability that it will touch one of the lines is equal to $1/\pi$. It follows that if a person randomly drops this needle onto the floor and keeps a record of the number of times it hits one of the lines, the observer can calculate an approximation of π by trials. The algorithm converges very slowly and therefore is not very practical, particularly considering that today there are much better ways to compute π.

Random algorithms provide intuition about deterministic, harder solutions; the input to a randomized algorithm is assumed to be the worst case, and so we always have the guaranteed run time. Recently, randomized algorithms have been used in a number of optimization and approximation problems [23, 25, 40].

6.8.1   Types of randomized algorithms

A disadvantage of randomized algorithms is that they return a good solution only most of the time. There are several varieties of randomized algorithms.

Numerical algorithms. These algorithms involve performing a large number of independent trials and produce an answer that converges to the correct answer as the number of trials increase. They are ideally suited to parallelization, since the trials are completely independent, no communication between processors is required.

Monte Carlo algorithms. These are the algorithms that make random choices that cause the algorithm to either produce the correct answer, or a completely wrong answer. In this case, the wrong answers are not approximations to the correct answer. Monte Carlo algorithms are equipped with procedures for comparing answers so that the probability of having a recognizably correct answer increases with the number of trials. However, there is always a finite probability that the answer produced by a Monte Carlo algorithm is wrong; the problem is that the algorithm does not "tell" us whether the answer returned is correct or wrong. This is a strong disadvantage of Monte Carlo algorithms. Monte Carlo algorithms are implementable on SIMD machines. An example of a Monte Carlo algorithm is a randomized algorithm for testing whether an integer is composite [29].

Las Vegas algorithms. These are algorithms that, unlike Monte Carlo algorithms, never produce incorrect answers. The disadvantage of Las Vegas algorithms is that they make random choices that sometimes prevent them from producing an answer at all. The random choices that Las Vegas algorithms make alter the flow of control, so they need independent processes, and therefore they generally do not lend themselves to SIMD implementation.

An example of a Las Vegas algorithm is a randomized algorithm for pattern matching in strings: determine whether or not a certain pattern occurs in a long text [29]. The most naive method for solving this problem moves the short pattern across the entire text, making brute-force comparisons in every position, character-by-character, between the symbols in the pattern and the corresponding symbols in the text. The method has the worst-case run time proportional to mn, where m is the length of the pattern, and n is the length of the text. More sophisticated approaches using structures lead to deterministic methods that run in time $O(m + n)$.

An efficient randomized method for pattern matching of Karp and Rabin [30] follows the brute-force approach of sliding the pattern $X = x_1x_2 \ldots x_m$ across the text $Y = y_1y_2 \ldots y_n$, but instead of comparing the pattern with each block $Y(i) = y_iy_{i+1} \ldots y_{i+m-1}$ of the text, the algorithm compares the *fingerprint* $H_p(X)$ of the pattern with the fingerprints $H_p(Y[i])$ of the blocks of text. [A *fingerprint* of x is a (shorter) string that might be obtained from the given string x by applying some standard hashing or check sum technique.] The fingerprints in the pattern matching are easy to compute by observing that when a shift is made from one block of text to the next, the fingerprint of the new block $Y[i + 1]$ can be computed from the fingerprint of $Y(i)$ using the following formula:

$$H_p(Y[i+1]) = H_p(Y[i]) + H_p(Y[i]) - 2^n y_i + y_{i+n} \pmod p$$

where p is a prime number.

A (sequential) algorithm: randomized pattern matching

```
Input: A pattern X of length m and a text string Y of length n
Choose p at random from {q\ 1 ≤ q ≤ n²m, q prime}
Set match ← false
Set i ← 1
While match = false and 1 ≤ i ≤ n − m + 1 do
    if Hp(X) = Hp(Y[i])
    then match ← true
    else i ← i + 1
    compute Y[i + 1]
```

A false match is only reported if for some i there will be $X \neq Y[i]$ but $H_p(X) = H_p(Y[i])$. The probability of a false match does not exceed $2/n$.

6.8.2   PRAM classification of randomized algorithms

Following Karp and Ramachandran [31], assume that a given problem Π is defined in terms of a binary input-output relation $S(X, Y)$. On input X, the task is to find an output Y satisfying $S(X, Y)$, if such a Y exists. A randomized algorithm will return one of the following three answers:

a suitable value for Y
a report that no such Y exists
a report "failure"—that is, the inability to find whether such a Y exists

We now distinguish between *zero-error* algorithms (these are Las Vegas algorithms and denoted a class **Z** algorithms) and *one-sided error* algorithms (these are Monte Carlo algorithms and denoted as class **m-R** algorithms). If, on input X, there exists a Y: $S(X, Y)$, then both types are alike; each type produces a suitable Y with probability greater than $1/2$, and otherwise reports failure. On input X such that there is no Y satisfying $S(X, Y)$, the two types of algorithms behave differently. A zero-error algorithm reports "no suitable Y exists" with probability greater than $1/2$, and otherwise reports failure, while a one-sided error algorithm always reports failure.With regard to various PRAM implementations, randomized algorithms are classified in terms of the same parallel complexity classes as for the deterministic algorithms. Thus, for example, a problem is in ZCREW(k) if it is solvable by a zero-error randomized algorithm that runs in time $O(\log^k n)$ using poly(n) processors on a CREW PRAM. To be reliably implemented randomized algorithms require the existence of a source of adequately random bits. This is particularly relevant for parallel randomized algorithms because they require huge number of random bits to be generated very quickly. These random bits may be assumed to come out of the support of special hardware. For randomized algorithms on PRAMs, each processor is assumed to have the ability to generate random $(\log n)$-bit numbers.

6.8.3   The classes RNC and m-RNC

The class **RNC** is defined as a class of problems that are solved by the random algorithms that always work and always produce correct answers, but their running time is indeterminate as a result of random choices they make. For these algorithms, the probable running time may be very short, but the worst-case run time, due to an unlucky choice, may be extremely long. More formally, the class **RNC** is the class of problems that can be solved by randomized algorithms in probable polylogarithmic time with polynomial number of processors. This type of algorithms is of particular interest in parallel processing.

Definition A.   A randomized parallel algorithm A possesses the following properties:

- The time distribution of this algorithm is a function $d(t)$ such that the probability of the execution-time of the algorithm being between t_0 and t_1 is given by

$$\int_{t_0}^{t_1} d(t)\,dt$$

- The expected execution-time of the algorithm is

$$\mu(A) \;=\; \int_0^{\infty} td(t)\,dt$$

- A problem instance of size n is in the class **RNC** if there exists a randomized parallel algorithm for it that uses a number of processors that is a polynomial in n and which has expected execution time of $O(\log^k n)$ for some value of k.

The class **m-RNC** is similarly defined as a class of problems for which there are randomized one-sided error algorithms (Monte Carlo algorithms). These algorithms are of less interest in practice due to the fact that they do not necessarily produce the correct answer in expected polylogarithmic time; they only produce an answer with some degree of confidence. Such an algorithm can be repeated many times to increase the level of confidence, but the algorithm will still be in **m-RNC**.

6.9   List Ranking Revisited: Optimal O(log n) Deterministic List Ranking

We shall now recall an earlier discussion where it was pointed out that it seems to be impossible to design an optimal deterministic algorithm for list ranking given that an extension of the technique used to optimize the prefix sums is not applicable for list ranking because there is no obvious way to know in advance the locations of the elements of the reduced list. We shall now discuss an efficient randomized algorithm for list ranking. A randomized algorithm for this problem was discovered by Vishkin in 1984, and then Anderson and Miller [4] developed a simplified version of the algorithm. This RNC algorithm for computing the suffix sums has the expected execution time $O(\log n)$ using $O(n/\log n)$ processors on an EREW PRAM computer.

Therefore, the problem is: given a linked list L of n elements, compute the suffix sums of the last i elements of the list, $i = 1, \ldots, n$. The main outline of the algorithm is as follows.

Step 1   Use contraction, done probabilistically, until a contracted list of size $\leq n/\log n$ is obtained.

Step 2   On the contracted list, use the list ranking by pointer jumping.

Step 3   Expand to produce ranks of all elements.

If we note that Step 2 of the algorithm can be done in $O(\log n)$ time with $w(n) = O(n)$ for $n/\log n$ processors, then the major point of interest is Step 1, the contraction phase, as Step 3 is at most bounded by the time and work of Step 1.

The contraction phase is a recursive algorithm. It begins by assigning $O(\log n)$ elements of the linked list to each processor in an arbitrary way. The process proceeds in phases. Each phase selects certain elements of the list, deletes them, and splices the pieces of the list together. After carrying out these two steps, the contraction algorithm is called recursively on the new, shorter list. After an entry from the linked list has been deleted, the following entry's value is modified so that it is equal to its original value plus the sum of the values of the entries that have been deleted. It can be shown that $O(\log \log n)$ contractions are enough to shrink the list down to $O(n/\log n)$ size. The total probability of error is at most $O(\log \log n \times \exp(-n/64))$ and thus is negligible.

A kind of deterministic optimal algorithm which is based on the above ideas was proposed by Cole and Vishkin [9]. The algorithm assumes that the operation of element deletions must satisfy the condition that no two adjacent elements of the list are ever deleted in the same step. In addition, parallelism is maximized if the condition is observed that at most one of the elements of the linked list that are assigned to a given processor is deleted in any phase of the algorithm. This condition also makes it easier to implement the algorithm. These conditions pave a way to the so called *deterministic coin tossing* technique of Cole and Vishkin.

Definition B.   Given a list of size n, a subset S of elements is called an *r-ruling* set if no pair in S is adjacent on the list, and every element e not in S is at a distance no more than $\varepsilon \times r$ on the list from an element in S, where ε is a constant.

Let $\log^{(k)} n = \log \log \ldots \log n$, where log is iterated k times and put $r = \log^{(k)} n$. The algorithm Ruling of Cole and Vishkin finds an r-ruling set of an n-element list in $O(k)$ time with n processors.

Algorithm: Ruling

Input: A list of n elements with successor pointers s[i] and predecessor pointers c[i].

```
For i from 1 to n do c[i] ← i
For k iterations do
    For each i do in parallel
        Find the rightmost bit position q such that the qth bit of c[i]
        differs from the qth bit of c[s[i]].
        Let b be the qth bit of c[i].
        Put c[i] ← b concatenated with the binary representation of q
For each i do in parallel
    If {c[p[i]] ≤ c[i] and c[s[i]] ≤ c[i]}
    then i is into the ruling set
```

At the end of algorithm Ruling, any element in the list is within distance $O(\log^{(k)} n)$ of an element in the ruling set. For example, if $r = \log^{(k)} n$ is set to 3, then in $O(\log \log \log n)$ time with n processors, we get a 3-ruling set algorithm. Each element in the ruling set can then, in constant time, find its successor by following list pointers. Thus, a contracted list will supervene. A complete algorithm is thus a deterministic optimal algorithm for list ranking with time $O(\log n)$ on EREW PRAM.

6.10   Taxonomy of Parallel Algorithms

Efficient parallel PRAM algorithms have been developed for a number of important problems. They include string matching, set manipulation, computational geometry and graph algorithms. Extensive surveys of efficient parallel algorithms are given in Eppstein and Galil [12] and in Gibbons and Rytter [16]. An excellent bird's-eye view of the state-of-art in parallel algorithmics can be found in Vishkin [55].

6.11   Deficiencies of the PRAM Model

The major drawback of the PRAM model is that it has no mechanism for representing communications between the processors. Its storage management and communication issues are hidden from the algorithm designer. At the same time, in parallel computation, communication is often much slower than computation. Excessive communication may create a bottleneck to the speed of an algorithm. The "communication complexity" of the problem is a major consideration in the design of efficient parallel algorithms as, in practice, communication resources are often limited.

Major challenges for computing science and for the computing industry in the 1990s is to descry an architecture independent programming model for scalable parallel computing and to determine frontiers of the general parallel computing. Researchers are pursuing the concept of a bridging model on which a comprehensive framework for the real-life general-purpose parallel computing systems can be built. The argument for the necessity of such a model is based on an analogy with the von Neumann model of sequential computation. The von Neumann machine serves equally well for hardware and software technologies. Hardware designers can share the common goal of realizing efficient von Neumann machines using ever rapidly changing technology and architectural ideas, without having to be too concerned about the software that is going to be executed. Similarly, the software industry in all its diversity can aim to write programs that can be executed efficiently on this model, without explicit consideration of the hardware.

No single model of parallel computation has yet come to dominate developments in parallel computing in the way the von Neumann model has dominated sequential computing. Instead, a variety of models such as VLSI systems, systolic arrays, and distributed memory multicomputers are gradually shaping the framework of the parallel processing. Careful exploitation of network locality is crucial for algorithmic efficiency.

6.11.1   Bulk-synchronous parallel model

A more recent model for parallel computation has been put forward by Valiant [53]. The *bulk-synchronous parallel* (BSP) model is suggested as candidate model of a unifying model for parallel computation. The BSP model is intended to act as a standard of what is meant by and expected from parallel processing, a standard on which people can agree. Over the last few years, it has been demonstrated that the BSP model can be efficiently realized on a wide variety of parallel architectures. A number of interesting new BSP algorithms for important computational problems have been developed [15]. The BSP arguably provides a robust model on which to base the future development of general-purpose parallel computing systems.

A BSP computer consists of three major components:

- a set of processor-memory pairs, each performing processing and/or memory functions
- a communication network that delivers messages point-to-point between processor-memory pairs
- a mechanism for the efficient barrier synchronization of all, or a subset, of the processors at regular intervals

There are no specialized broadcasting, duplicating or combining facilities.

Assuming that a time step is defined to be the time required for a single local operation, such as addition or multiplication, on locally held data values, the performance of a BSP computer is characterized by the following four parameters:

1. p, the number of processors
2. s, the processor speed (i.e., number of time steps per second)
3. L, synchronization periodicity (i.e., minimal number of time steps between successive synchronization operations)
4. G, the ratio of the total number of local operations performed by all processors in one second to the total number of words delivered by the communications network in one second

A computation consists of a sequence of parallel *supersteps*. During a superstep, each processor is allocated a task consisting of some combination of local computation steps, message transmissions and (implicitly) message arrivals from other processors.

Each superstep is followed by a global check (barrier synchronization) to determine whether the superstep has been completed by all the components. If it has, the machine proceeds to the next superstep. Otherwise, the next period of fixed number of units is allocated to the unfinished superstep. Thus, in the model, the tasks of computation and communication can be separated.

A major feature of the BSP model is that it allows some parallel slackness. Each processor has its own physically local memory model. All other memory is non-local and is accessible in a uniformly efficient way; that is, the time taken for a processor to read from or write to a non-local memory element in another processor-memory pair is independent of which physical memory module the value is held in. This allows the algorithm designer not to worry about such details as memory management, communication assignment, and realization of low-level synchronization, which the model handles itself. Instead, performance of the communication network is estimated only in terms of its global properties, such as the maximum time required to perform a non-local memory operation and the maximum number of non-local memory operations that can simultaneously exist in the network at any time.

The strongest feature of the BSP model is that it lifts considerations of network performance from the local level to the global level. It is no longer relevant whether the network is implemented on a 2D array, a butterfly, a hypercube, or indeed in VLSI. In the design and implementation of a BSP computer, the values of L and G that can be achieved depend on the capabilities of the available technology. With the growth of computational performance of computers, captured by the parameters p and s, the communication hardware will increasingly demand higher and higher percentage of the total cost of the machine for the L and G to remain low.

6.11.2   Algorithms and complexity in BSP

In this presentation, we shall closely follow McColl [37]. Let the work w be the maximum number of local computation steps executed by any BSP processor during a period D. Let m_s and m_r be the maximum numbers of messages sent and received, respectively, by any processor during D. The cost of period D is given as max $\{L, w, Gm_s, Gm_r\}$ time steps. For $G = 1$, the BSP computer corresponds closely to a PRAM, with L determining the degree of parallel slackness required to achieve optimal efficiency. For a BSP computer with low G value, an efficient memory management can be achieved using hashing. Hash functions for this parallel context have been proposed and analyzed by Mehlhorn and Vishkin [39], Carter and Wegman [8] and Karlin and Upfal [26]. For $L = G = 1$, we have the idealized PRAM where no parallel slackness required.

In designing algorithms for a BSP computer with high G value, one must ensure that, for every request for non-local memory access, approximately G operations are performed on local data. This is the *communication slackness*. In other words, the architecture independent parallel algorithms in the BSP model must be parameterized not only in the size of the problem, n, and the number of processors, p, but also in L and G. The resulting algorithms can then be efficiently implemented on a range of BSP architectures with widely differing L and G values.

Very recent work [37, 54] has been exploring a new framework for the design and analysis of BSP computations. Let $U(n)$ denote an upper bound on the computational complexity of a problem. The BSP algorithm design aims to produce a parallel algorithm with an upper bound of $U(n)/p$ for the widest possible range of values for L and G. The following is a simple example from [49] that illustrates the approach.

Consider the problem of multiplication of two $n \times n$ matrices A and B on $p \leq n^2$ processors. The standard sequential algorithm of $O(n^3)$ is adapted to run on p processors as follows. Each processor computes an $(n/\sqrt{p}) \times (n/\sqrt{p})$ submatrix of $C = A \times B$. To do so, each processor will require $n^2/\sqrt{p}$ elements from A and the same number from B. Therefore, for each processor, there is a computational requirement of $O(n^3/p)$ operations, since each inner product requires $O(n)$ operations, and a communications requirement of $O(n^3/p)$ for the number of non-local reads, since $p \leq n^2$. Assuming that both A and B are distributed uniformly among the p processors, with each processor receiving $O(n^2/p)$ of the elements from each matrix, the processors can simply replicate and send the appropriate elements from A and B to the $2/\sqrt{p}$ processors requiring them. Therefore, a communication requirement for messages sent is approximately $n^2/\sqrt{p} = O(n^3/p)$. Hence, a total parallel time complexity is $O(n^3/p)$, provided $L = O(n^3/p)$ and $G = O(n/(\sqrt{p}))$.

6.11.3   Other refinements and improvements of parallel processing models

Several other models related to and expanding on the PRAM have been proposed [1, 17, 24, 36, 41]. The distribution random access machine, DRAM model of Leiserson and Maggs includes a communication network that conveys information between processors and memory banks. The DRAM consists of a set of p processors. All memory in the DRAM is local to the processors, with each processor holding a small number of $O(\log p)$-bit registers. A processor can read, write and perform arithmetic and logical functions on values stored in its local memory. A processor can also read from and write to memory in other processors.

The DRAM is a restricted version of PRAM that models communication costs. In a DRAM, the processors are interconnected as a graph and routing of messages is performed by the processors. The communication cost is measured in terms of the number of messages that must cross a *cut* of the network. A *cut* $S = (A, \hat{A})$ of a network is a partition of the network into two sets of processors A and $\hat{A}$. The capacity $cap(S)$ is the number of wires connecting processors in A with processors in $\hat{A}$; that is, the bandwidth of communication between A and $\hat{A}$. For a set M of messages, the load of M on a $cut S = (A, \hat{A})$ is defined as the number of messages in M between a processor in A and a processor in $\hat{A}$. A lower bound on the time required to deliver a set of messages is provided by the load factor of M on S, which is defined as the ratio of $load(M, S)$ and $cap(M)$, denoted $L(M, S) = load(M, S)/cap(S)$. The load factor of M on the entire network is $L(M) = max \{L(M, S)\}$ for all S.

The phase PRAM of Gibbons incorporates barrier synchronization in a similar way to BSP model but uses a shared memory.

The local-memory PRAM, or LRAM, of Aggarwal et al [1] is a CREW PRAM in which each processor is provided with an unlimited amount of local random access memory. In the LPRAM, an algorithm is represented by an acyclic directed graph (*dag*) in which the nodes correspond to operations and there is an arc from node v to node w if the result of v is required by w. In PRAM models, it is the depth of the *dag* that dominates parallel complexity, whereas in real parallel architectures, the communication delay is significant. The LPRAM model addresses the notion of delay by making each time step to comprise a communication step and a computation step. An LPRAM processor is restricted to just one communication request outstanding at any time.

Yet another model of PRAM with communications facilities uses a parameter *delay* defined as the ratio of the message delivery time to the instruction cycle time [41].

6.12 Summary

We introduced the parallel random access machine (PRAM) models that are commonly used for the design of efficient parallel algorithms. The primary motive for building parallel computers is to speed up the computational process, and the PRAM approach often serves as a useful guideline for achieving this purpose. Many real-life parallel architectures approximate theoretical parallel models.

Efficiency in parallel processing comes from both the performance of computers themselves and from communication between parallel processors. Theoretical models that do not take into account the communication complexity of parallel processing can play only a limited role in the quest for the ultimate parallel architecture. Research is ongoing to extend PRAM models to include the communication costs. One of the most promising models that includes a measure for communication costs is the *bulk-synchronous parallel* (BSP) model of Valiant. These recent advances in modeling of the parallel processing indicate that the next few years will see a rapid convergence in the field of parallel computer systems. It is very likely that various classes of parallel computer that currently exist (distributed memory architectures, shared-memory multiprocessors, networks of workstations) will become more and more alike.

6.13 References

1. Aggarwal, A., A. Chandra, and M. Snir. 1990. Communication complexity of PRAMs. *Theory of Comp. Sci.*, 71, 3–28.
2. Akl, S. G. 1989. *The Design and Analysis of Parallel Algorithms*. Englewood Cliffs, N.J.: Prentice Hall International Editions.
3. Anderson, R., and E. Mayr. Parallelism and Greedy Algorithms. *Advances in Computing Research*, Vol. 24, JAI Press, 1987, 17–38.

4. Anderson, R. J., and G. L. Miller. 1988. A simple randomized parallel algorithm for list ranking, *unpublished* (as in Smith, 1993).

5. Batcher, K. 1968. Sorting networks and their applications. *Proceedings of AFIPS Spring Joint Summer Computer Conf.*, Vol. 32, 307–314.

6. Borodin, A., and J. E. Hopcroft. 1985. Routing, merging and sorting on parallel models of computation. *Journal of Comp. Sys. Sci.*, Vol. 30, 130–145.

7. Brent, R. P. 1974. The parallel evaluation of general arithmetic expressions. *J. of ACM*, Vol. 21, 201–206.

8. Carter, J. L., and M.N. Wegman. 1979. Universal classes of hash functions. *J. Comput. Syst. Sci.*, Vol. 18, 143–154.

9. Cole, R., and U. Vishkin. 1986. Deterministic coin tossing with applications to optimal parallel list ranking. *Inform. and Control*, Vol. 70, 32–53.

10. Cook, S.A., and R.A. Reckhow. 1973. Time-bounded random access machines. *J. Comput. System Sci.*, Vol. 7, 354–375.

11. Edmonds, J. Paths, trees and flowers. *Canad. J. Math.*, Vol. 17, 1965, 449–467.

12. Eppstein, D., and Z. Galil. 1988. Parallel algorithmic techniques for combinatorial computation. *Ann. Rev. Comput. Sci.*, Vol. 3, 233–283.

13. Fich, F. E. 1983. New bounds for parallel prefix circuits. *Proc. 15th Ann. ACM Symp. on Theory of Computing*, 27–36.

14. Fortune, S, and J. Wyllie. 1978. Parallelism in Random Access Machines. *Proc. of the 10th Annual ACM Symp. on Theory of Computing*, 114–118.

15. Gerbessiotis, A.V., and L.G. Valiant. 1992. Direct bulk-synchronous parallel algorithms. *Technical report TR-10-92* (Extended version), Aiken Computation Laboratory, Harvard University. Shorter version appears in *Proc. 3rd Scandinavian Workshop on Algorithm Theory*, July 8–10, LNCS Vol. 621, 1–18, Springer Verlag, 1992.

16. Gibbons, A., and W. Rytter, 1988. *Efficient Parallel Algorithms*. Cambridge, Mass.: Cambridge University Press.

17. Gibbons, P. B. 1989. A more practical PRAM model. *Proc. of the 1989 ACM Symp. on Parallel Algs. and Architectures*, 158–168.

18. Greenlaw, R., H. J. Hoover, and W. L. Ruzzo. 1991. A compendium of problems complete for P. (file name tr91-11.dvi.Z), aval. via anonymous ftp://thorhild.cs.ualberta.ca.

19. Goldschlager, L. 1977. The monotone and planar circuit value problems are log-space complete for **P**. *SIGACT News* Vol. 9, No. 2, 25–29.

20. Gurari, E. M. 1989. *An Introduction to Theory of Computation*. Comp. Sci. Press.

21. Harel, D. 1992. *Algorithmics: The Spirit of Computing,* 2d ed., Reading, Mass.: Addison-Wesley.

22. Hartmanis, J. 1990. New developments in structural theory. *Theoretical Computer Science,* Vol. 71, 79–93.

23. Janakiram, V. K., D.P. Agrawal, and R. Mehrota. 1988. A randomized parallel branch-and-bound algorithm. *Proc. ICPP*, 69–75.

24. Jesshope, C. R., and M. Abdul Qadar. 1993. BSPC and the N-computer. *Workshop of the PPSG BCS on General Purpose Parallel Computing*, University of Westminster, 22 December.

25. Julien-Laferriere, P., F. H. Lee, G. S. Stiles, A. Raghuram and T. W. Morgan.1989. Stochastic optimisation of distributed database networks. *Proc. IEEE Phoenix Conf. on Comp. and Communications*, 452–460.

26. Karlin, A., and E. Upfal. 1988. Parallel hashing—an efficient implementation of shared memory. *J. ACM*, Vol. 35, No. 4, 876–89.

27. Karloff, H.J. 1986. A Las Vegas RNC algorithm for maximum matching. *Combinatorica*, Vol. 6, 387–392.

28. Karp, R. M. 1986. Combinatorics, complexity and randomness. *Comm. of ACM*, Vol. 29, 98–111.

29. Karp, R. M. 1991. An introduction to randomized algorithms. *Discrete Applied Mathematics*, Vol. 34, 165–201.

30. Karp, R. M., and M. Rabin. 1987. Efficient randomized pattern-matching algorithms. *IBM J. Res. Develop.*, Vol. 31, 249–260.

31. Karp, R. M., and V. Ramachandran. 1990. Parallel algorithms for shared-memory machines. *Handbook of Theoretical Computer Science*, J. van Leeuwen, ed., 870–941.

32. Karp, R., E. Upfal, and A. Widerson. 1986. Constructing a perfect matching is in RNC. *Combinatorica*, Vol. 6, 35–48.

33. Krishnamurthy, E.V. 1989. *Parallel Processing: Principles and Practice*. Reading, Mass.: Addison-Wesley.

34. Kruskal, C. P., L. Rudolph, and M. Snir. 1988. A complexity theory of efficient parallel algorithms. *Proc. of 15th ICALP*, Springer LNCS 317, 333–346.

35. Ladner, R. E., and M. J. Fischer. 1980. Parallel prefix computation. *J. ACM*, Vol. 27, 831–838.

36. Leiserson, C., and B. Maggs. 1986. Communication-efficient parallel graph algorithms. *International Parallel Processing Conf.* New York: IEEE.

37. McColl, W.F. 1993. An architecture independent programming model for scalable parallel computing. *Workshop of the PPSG BCS on General Purpose Parallel Computing*, University of Westminster, 22nd December.

38. Maon, Y., B. Schieber, and U. Vishkin. 1986. Parallel ear-decomposition search (EDS) and st-numbering in graphs. *Theoretical Computer Science*, Vol. 47, 277–298.

39. Mehlhorn, K., and U. Vishkin. 1984. Randomized and deterministic simulations of PRAMs by parallel machines with restricted granularity of parallel memories. *Acta Informatica*, Vol. 21, 339–374.

40. Mehrota, R., and E. F. Gehringer. 1985. Superlinear speedup through randomized algorithms. *Proc. ICPP*, 291–300.

41. Papadimitriou, C. H., and M. Yannakakis. 1988. Towards an Architecture-independent Analysis of Parallel Algorithms. *Proc. of the 20th ACM Symp. on Theory of Computing*, 510–513.

42. Pippenger, N. 1979. On simultaneous resource bounds. *Proc. of the 20th IEEE Annual Symp. on Foundation of Computer Science*, 307–311.

43. Rabin, M.O. 1976. Probabilistic algorithms. In *Algorithms and Complexity*, J.F. Traub, ed. New York: Academic Press, 21–39.

44. Rabin, M.O., and Y. Aumann. 1992. Clock construction in fully asynchronous parallel systems and PRAM simulation. *25th Int. Seminar on the Teaching of Computing Science*, University of Newcastle, U.K.

45. Raghavan, P. 1988. Probabilistic construction of deterministic algorithms: approximating packaging integer programs. *J Computer and System Sciences*, Vol. 37, No. 4, 130–143.

46. Ramachandran, V. 1993. Efficient parallel graph algorithms. In *Lectures on Parallel Computation*, A. Gibbons and P. Spirakis, eds. Cambridge University Press.

47. Reif, J. H. 1985. An Optimal Parallel Algorithm for Integer Sorting. *Proc. 26th Ann. IEEE Symp. on Foundations of Computer Science*, 496–504.

48. Smith, J. R. 1993. *The Design and Analysis of Parallel Algorithms*. New York: Oxford University Press.

49. Spirakis, P. G., and A. Gibbons. 1993. PRAM models and fundamental parallel algorithmic techniques. In *Lectures on Parallel Computation*, A. Gibbons and P. Spirakis, eds. Cambridge University Press.

50. Schwartz, J. T. 1980. Ultracomputers. *ACM Transactions on Programming Languages and Systems*, Vol. 2, No.4, 484–521.

51. Tarjan, R.E., and U. Vishkin. 1985. An Efficient Parallel Biconnectivity Algorithm. *SIAM J. of Computing*, Vol. 14, 862–874, also in FOCS 1984, 12–20.

52. Valiant, L.G. 1975. Parallelism in comparison problems. *SIAM J. Comput.*, Vol. 4, 348–355.

53. Valiant, L.G. 1990. A bridging model for parallel computation. *Comm of ACM*, Vol. 33, No. 8, 103–111.

54. Valiant, L.G. 1990. General purpose parallel architectures. In *Handbook of Theoretical Computer Science*, J. van Leeuwen, ed. Amsterdam: North Holland, 945–969.

55. Vishkin, U. 1983. Synchronous parallel computation—a survey. *Technical Report TR 71*. New York: Dept. of Computer Science, Courant Institute, New York University.

56. Vishkin, U. 1984. Randomized speedups in parallel computation. *Proc. 16th Annual ACM Symp. on the Theory of Computing*, 230–238.

57. Vishkin, U. 1993. Structural parallel algorithmics. *Lectures on Parallel Computation,* A. Gibbons and P. Spirakis, eds. Cambridge University Press.

7

Broadcasting with Selective Reduction: A Powerful Model of Parallel Computation

Selim G. Akl and Ivan Stojmenović

Broadcasting with selective seduction (BSR) is a model of parallel computation in which N processors share M memory locations. During the execution of an algorithm, several processors may read from or write to the same memory location simultaneously, such that each processor gains access to at most one location. An additional type of memory access is also permitted in BSR by means of which all processors may gain access to all memory locations at the same time for the purpose of writing. At each memory location, a subset of the incoming broadcast data is selected (according to one or more appropriate selection criteria) and reduced to one value (using an appropriate reduction operator). This value is finally stored in the memory location. It has been shown that BSR, while more powerful than the strongest variant of the PRAM, namely the COMBINING CRCW PRAM, demands no more resources to be implemented than its weakest variant, namely the EREW PRAM. This chapter reviews several existing implementations of BSR. It also presents BSR solutions to a variety of computational problems. These solutions allow us to illustrate effectively the power and elegance of BSR as shown by the conciseness and simplicity of the algorithms it affords. For all the problems addressed, the BSR algorithms run in constant time and use a number of processors linear in the size of the input. To our knowledge, BSR is the only existing model of parallel computation capable of achieving these results.

7.1 Introduction

Since the early days of the field of parallel computation, a very attractive model of a parallel computer, to theoreticians and practitioners alike, has been one where the processors share a common memory from which they can all read at the same time, and to which they can all write at the same time. Because all data exchanges take place through the shared memory, the algorithm designer can focus on the parallelism inherent in the problem to be solved, rather than being concerned with the details of communications among processors.

Of all the theoretical shared memory models that were proposed over the last 15 years, the parallel random access machine (PRAM) is without doubt the most widely accepted and used [25]. Several variants of the PRAM exist, and they differ from one another according to the way that the processors are allowed simultaneous access to memory. In its weakest manifestation, the *exclusive read exclusive write* (EREW) PRAM, all processors may gain access to memory at the same time (either for reading or for writing), provided that each processor gains access to a different memory location. Thus, the model disallows reading from or writing to the same memory location by more than one processor at the same time. In its strongest manifestation, the *concurrent read concurrent write* (CRCW) PRAM, the model allows several processors to gain access to the same memory location for the purpose of either reading or writing. When two or more processors attempt to write into the same memory location, the model uses one of several conflict resolution rules to determine the value that ends up in the memory location. The most powerful of these rules is the COMBINING rule, whereby the values that the processors wish to write in a given memory location are combined (using, for example, addition, minimization, logical AND, etc.) into a single value that is stored in that memory location [1, 17, 36].

Another shared memory model (that was first proposed in Ref. [2]) is *broadcasting with selective reduction* (BSR). It consists of N processors numbered $1, 2, \ldots, N$ sharing M memory locations $u_1, u_2, \ldots, u_M$. The model possesses the same features as the CRCW PRAM. In particular, during the execution of an algorithm, several processors may read from or write into the same memory location simultaneously, such that each processor gains access to at most one location. An additional type of memory access is permitted in BSR, namely a BROADCAST instruction, by means of which all processors may gain access to all memory locations at the same time for the purpose of writing. At each memory location, a subset of the incoming broadcast data is selected and reduced to one value (using an appropriate reduction operator, e.g., minimization); this value is finally stored in the memory location. The selection process is carried out as follows. Along with each datum d_i, a tag t_i is also broadcast, while each memory location u_j is associated with a limit l_j. A selection rule σ (e.g., $<$) is used to test the condition $t_i \, \sigma \, l_j$: if the latter is true, then d_i is selected for reduction, otherwise, d_j is rejected. This is done simultaneously for all broadcast data d_i, $1 \le i \le N$, and all memory locations u_j, $1 \le j \le M$, as illustrated in Fig. 7.1.

It is instructive to compare BSR with the PRAM. In particular, consider the PRAM's weakest variant, namely the EREW PRAM, and its strongest variant, namely the COMBINING CRCW PRAM. It is shown in Ref. [3] that BSR, while more powerful than the COMBINING CRCW PRAM, demands no more resources than the EREW PRAM. This result is obtained through the following two observations:

1. The BROADCAST instruction, which is equivalent to M concurrent writes, can be executed in the same amount of time required by a single concurrent write instruction on the COMBINING CRCW PRAM.
2. The BROADCAST instruction can be implemented by a combinational circuit (connecting the processors to the shared memory), which is no larger asymptotically than that required to implement the EREW PRAM.

In fact, a number of implementations of BSR exist. They include:

1. *Using memory buses* [2]. As shown in Fig. 7.2, each processor possesses a bus (for broadcasting) to which all memory locations are connected through switches (to perform the selection test) and a binary tree (to implement the reduction operation).

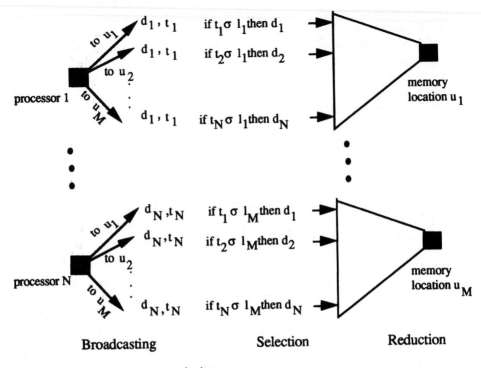

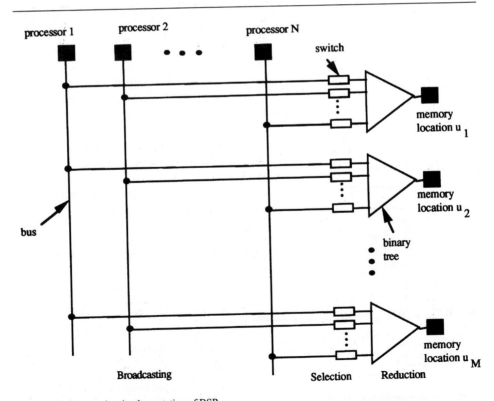

Figure 7.1 Broadcasting with selective reduction

Figure 7.2 Memory bus implementation of BSR

2. *Using a mesh of trees* [3]. As shown in Fig. 7.3, each of the N processors is connected to the root of a binary tree of M leaves (used for broadcasting), and each of the M memory locations is connected to the root of a binary tree of N leaves (used for reduction). The leaves of the memory trees are linked to those of the processor trees through switches (used for selection).

3. *Using circuits for sorting and prefix computation* [21]. As shown in Fig. 7.4, the combinational circuit connecting the N processors to the M memory locations consists of six "boxes" labeled A through F. Boxes A, C, D, and F are sorting circuits, while boxes B and E are (slightly modified) circuits for parallel prefix computation. Reduction is done in box B (using the output of box A), selection in box D (using the outputs of boxes A and C), and broadcasting in boxes E and F. For $N = O(M)$, all

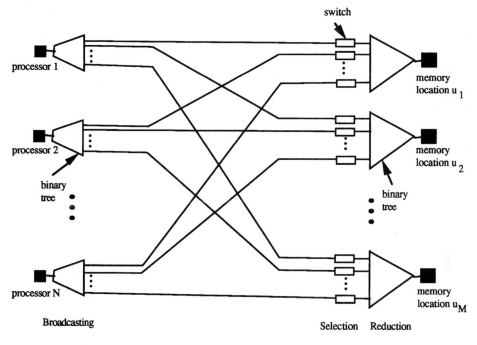

Figure 7.3 Mesh of trees implementation of BSR

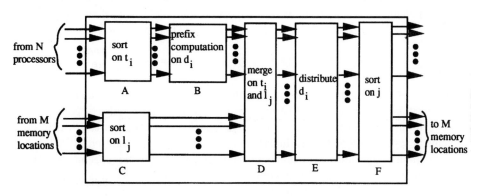

Figure 7.4 Implementation of BSR using circuits for sorting and prefix computation

boxes have depth $O(\log M)$. Thus, the entire circuit has depth $O(\log M)$ and size $O(M \log M)$, which is optimal.

In all of the above implementations, the BROADCAST instruction requires constant time, since a single concurrent write in the COMBINING CRCW PRAM is usually assumed to require constant time. It is also important to note, as observed in Ref. [21], that a combinational circuit implementing the BROADCAST instruction allows the execution of all (less powerful) forms of memory access permitted on BSR and the different variants of the PRAM (such as exclusive read, concurrent read, exclusive write, and concurrent write).

The power and elegance of BSR were demonstrated by using it to obtain constant time solutions to a variety of problems. These problems include prefix sums, element uniqueness, sorting, sieve of Erathostenes, maximal subsequences in one and two dimensions, maximal vectors, and convex hull [2, 3, 4, 5, 21]. In the solutions, the number of processors needed is equal to the input size, except for the convex hull of n points which requires n^2 processors. Recently, BSR solutions have been proposed for several new problems, such as digital geometry problems [30], parenthesis matching and problems on trees [40], and a number of problems in computational geometry [9]. To illustrate the nature of BSR algorithms, we provide a simple example. Consider the (first quadrant) maximal vectors problem, where a set of n points in the plane are given by their (positive) cartesian coordinates (x_i, y_i), and it is required to identify those points, called *maximal*, both of whose coordinates are not smaller than the corresponding coordinates of any other point. This problem can be solved on BSR in one constant time step using n processors. An additional variable m_i is used, initially set equal to y_i; at termination (x_i, y_i) is maximal if and only if $m_i \le y_i$. The BSR algorithm, performed simultaneously for all $1 \le i, j \le n$, is simply as follows: set m_j to the largest y_i such that $x_i > x_j$. Here, the datum and tag broadcast by processor i are y_i and x_i, respectively, the limit associated with memory location m_j is x_j, the selection rule is >, and the reduction operator is maximization.

While the mathematical definition of BSR [3] allows each of t_i, σ, and l_j, to be vectors, all implementations of BSR, and all problems solved on it, have assumed them to be scalars. It is suggested in Refs. [7] and [21] that in some problems it may be necessary for the datum d_i to satisfy more than one criterion before being selected by memory location u_j. In other words, in order for a datum d_i received by u_j to participate in the reduction process, it must first pass several tests of the form $t_i \, \sigma \, l_j$.

For example, consider the following problem introduced in Ref. [37], where it is called *general prefix computation* (GPC). Let $f(1), f(2), \dots, f(n)$, and $y(1), y(2), \dots, y(n)$ be two sequences of elements with a binary associative operator "*" defined on the f-elements, and a linear order "<"defined on the y-elements. It is required to compute the sequence:

$$D(m) = f(j_1) * f(j_2) * \dots * f(j_k), \text{ for } m = 1, 2, \dots, n$$

where $j_1 < j_2 < \dots < j_k$, and $\{j_1, j_2, \dots, j_k\}$ is the set of indices $j < m$ for which $y(j) < y(m)$. This problem is a generalization of basic prefix computation and can also be considered as a special formulation of two-dimensional range searching. It can be used to solve other problems in several different areas. In particular, an efficient algorithm for GPC allows us to obtain efficient algorithms for a number of problems in computational geometry such as ECDF searching, two-set dominance counting, and maximal vectors. It is pointed out in Ref. [7] that in order to solve GPC in constant time on the BSR model, "double selection" is required: the first to select $j_i < m$, and the second to select $y(j_i) < y(m)$. The implementation of BSR described in Ref. [21] does not allow this operation. However, an implementation of BSR is given in Ref. [14] which allows data to be tested for *two* criteria.

The objective of this chapter is to expose the various attractive features of BSR by showing it to be, in its full generality, a feasible and effective model of parallel computation. A detailed description of a BSR implementation that allows the data to be tested for their satisfaction of k criteria, where $k \geq 1$, is first presented. This implementation differs from the ones described above in that the model's memory is not directly shared (and cannot be used as a bulletin board by the processors). Instead, the memory locations are distributed among the processors, thus forming a local memory for each processor (these local memories being, of course, accessible indirectly by other processors). It is then shown how a number of computational problems can be solved on BSR in constant time. It should be noted that the constant time solutions presented are, to our knowledge, the only such solutions in existence for the problems addressed that use a number of processors linear in the size of the input. Furthermore, these solutions allow us to illustrate effectively the power and elegance of BSR as shown by the conciseness and simplicity of the algorithms it affords. The remainder of the chapter is organized as follows. Section 7.2 presents our implementation of a multiple criteria BSR. In Sections 7.3 to 7.6, we describe how problems involving one, two, three, and an arbitrary number of criteria, respectively, can be solved on the new implementation. These problems include, for example, reconstruction of a binary tree from its traversal (one criterion), counting inversions in a permutation (two criteria), vertical segment visibility (three criteria), and d-dimensional maximal elements ($d - 1$ criteria). Conclusions and open problems are in Section 7.7. For ease of presentation, we assume henceforth that on BSR the number of processors is equal to the number of memory locations, that is, $N = M = n$.

7.2   A Generalized BSR Model

In this section, a generalization of the BSR model is presented where each broadcast datum can be tested for its satisfaction of k criteria, for some $k \geq 1$. The mathematical notation used to express the multiple criteria BSR instruction is first introduced. A detailed description is then provided of an implementation of this generalized model.

7.2.1   Notation

Let us define the following symbols:

n = number of processors
d_i = datum broadcast by processor i, $1 \leq i \leq n$
σ_h = selection operation, $1 \leq h \leq k$, taken from the following set $\{<, \leq, =, \geq, >, \neq\}$
$t(i, h)$ = tag broadcast by processor i for criterion σ_h, $1 \leq h \leq k, 1 \leq i \leq n$
$l(j, h)$ = limit value broadcast by processor j for criterion σ_h, $1 \leq h \leq k, 1 \leq j \leq n$
$\Re$ = binary associative reduction operation, where $\Re \in \{\Sigma, \Pi, \wedge, \vee, \otimes, \cap, \cup\}$, denoting sum, product, AND, OR, XOR, max, and min, respectively
u_j = result of a multiple criteria BROADCAST instruction, $1 \leq j \leq n$

The k criteria BSR BROADCAST instruction is denoted by

$$u_j = \overset{\Re d_i}{\underset{1 \leq i \leq n}{}} \mid \overset{\wedge}{\underset{1 \leq h \leq k}{}} t(i, h)\, \sigma_h\, l(j, h)\, , \text{ for } 1 \leq j \leq n$$

When the ranges of the variables are understood, this can be abbreviated as

$$u_j = \Re d_i \big| \wedge t(i, h) \, \sigma_h \, l(j, h)$$

The above notation can be interpreted as follows. If $t(i, h) \, \sigma_h \, l(j, h)$ is satisfied for each $h, 1 \leq h \leq k$, then d_i is "accepted" by location u_j. The set of all data accepted by u_j is reduced to a single value by means of the binary associative operation $\Re$, and stored in u_j. If no data are accepted by a given memory location u_j then u_j receives the value of the neutral element for operation $\Re$. If only one datum is accepted, then u_j is assigned the value of that datum. The operation is performed for all $j, 1 \leq j \leq n$, in parallel.

7.2.2   A multiple criteria BSR implementation

We describe an implementation of BSR which allows k selection criteria, and is a generalization of the CRCW PRAM and the one criterion BSR implementation of Ref. [2]. Consider n^2 switches $s(i, j)$, $1 \leq i, j \leq n$, organized as a mesh (see Fig. 7.5). Switch $s(i, j)$ is connected to processors i and j. Switches $s(i, j)$ for fixed j and $1 \leq i \leq n$ are located at the leaves of a binary tree that we call (after Ref. [2]) a concurrent access tree CAT_j. The latter performs a reduction: it computes from the data, received through the switches, one value to be stored in its associated memory location u_j. The concurrent access trees can be implemented using combinational circuits, i.e., circuits without feedback lines or memory elements (as described in Ref. [2] for the one criterion BSR). Each "switch" $s(i, j)$ con-

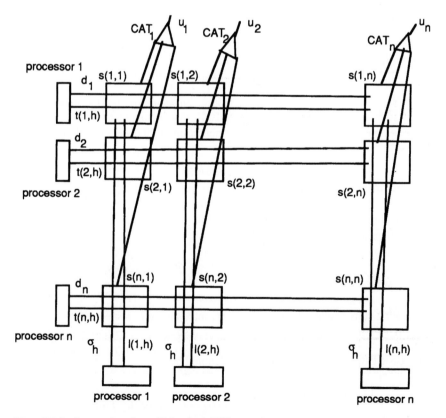

Figure 7.5 Implementation of a multiple criteria BSR

sists of k gates (see Fig. 7.6) whose k output lines go into an AND gate. The output of the AND gate goes into a leaf of CAT_j. Each of the k gates tests a different "tag, selection, limit" condition as required by the algorithm and issues a true or false value. If the AND gate receives k true values, then the datum d_i under consideration is allowed to go through, otherwise it is stopped, and the neutral element $e(\Re)$ of the reduction operation $\Re$ goes through (the neutral elements of sum, product, AND, OR, XOR, max, and min are 0, 1, true, false, false, $-\infty$ and $+\infty$, respectively). Note that the k tags may all be different. This works similarly for the limits and for the selection operations.

The above implementation may lead, in some rare instances, to ambiguities. For example, there exists in the implementation the problem of how to distinguish between: (1) a successful AND reduction leading to "true," and (2) a "true" caused by no data being accepted. (Similar problems exist for other operations, besides AND, as discussed below). There are at least two solutions to the problem. One is to use a "failure" symbol Φ instead of the neutral element, such that $\Phi \Re x = x \Re \Phi = x$ for any argument x and reduction operation $\Re$. Conceptually, this solution is very simple; however, it may require the use of additional hardware. In particular, CAT_j must implement a three-valued logic in case of the logical operations AND, OR, and XOR. Also, arithmetic reduction operations are complicated to implement with this approach. A second solution is to use the neutral element as described above, with minor modifications at the programming level. Thus, for example, we can distinguish between the two cases of the AND operation by replacing the instruction:

$$u_j := \wedge\, d_i \mid \text{criteria}$$

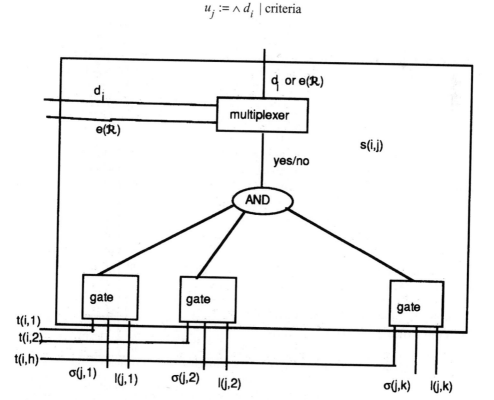

Figure 7.6 Design of a switch using multiple criteria BSR implementation

with the instruction:

$$u_j := \Sigma 1 \mid \text{criteria}$$

(That is, true/false data values d_i are replaced with 1/0 and AND reduction with sum without changing the selection criteria). A zero sum then clearly indicates all failures. In fact, this instruction may be used instead of the problematic AND reduction, since a successful AND is recognized easily by the value of the obtained sum. As a second example, consider the case of the instruction:

$$u_j := \Sigma \, d_i \mid \text{criteria}$$

where the d_i are allowed to take the value 0. Then, $u_j = 0$ appears as the result of reduction in two cases: no d_i was accepted, and some d_i, all equal to 0, were accepted. These two cases can be separated by introducing an additional BROADCAST instruction, namely:

$$v_j := \vee \text{ true} \mid \text{criteria}$$

which uses the same selection criteria but replaces the sum operation with OR and d_i with the value true. A false value for v_j indicates all failures. Other ambiguities are handled similarly. This approach, while requiring additional programming steps, has the advantage of not affecting the hardware.

To function properly, switches should receive all data (tags, selection criteria, limits and datum) at once. One way to accomplish this is to allow each switch to have a memory of size $O(k)$ to store the data. However, this leads to an $O(kn^2)$ memory requirement. A better alternative is to assign $O(k)$ memory to each processor while switches and CATs remain purely combinational circuits. Processors will store $O(k)$ data in memory registers that are directly linked to switches via buses. There are n horizontal and n vertical buses. Processor i is responsible for sending data on the ith horizontal and ith vertical buses, $1 \le i \le n$. Each bus contains $O(k)$ separate lines (subbuses), one per memory register. Processor i supplies its registers with data (one by one in $O(k)$ time) and then sends an enable signal that releases all data from all registers and makes them simultaneously available to all switches for the selection and reduction processes. CAT_j is constructed on top of the jth vertical bus. The design is illustrated in Fig. 7.5. For simplicity, processor i is drawn twice (in row i and column i). Also, the root of CAT_j leads to u_j; the latter should be assumed to be part of the memory of processor j.

The network has therefore $O(n^2)$ switches, each with an $O(k)$ size combinational circuit and a local memory of size $O(k)$ per processor; this implies an overall memory size of $O(kn)$ and an $O(kn^2)$ circuit size.

7.3   One Criterion BSR Algorithms

This section offers a survey of BSR algorithms requiring each broadcast datum to pass a single test before being allowed to participate in the reduction process. The algorithms range from fundamental computations, such as prefix sums and sorting, to computations arising in geometry and graph theory.

7.3.1   Fundamental algorithms

Given n numbers $x_1, x_2, \ldots, x_n$, the prefix sums problem is to compute $y_j := x_1 + x_2 + \ldots + x_j$ for $j = 1, 2, \ldots, n$. This computation can be performed by one BROADCAST instruction as follows [2]:

$$y_j := \Sigma \, x_i \, | \, i \leq j$$

We note the prefix sums computation can be performed in constant time on the scan model of computation [12], which is weaker than BSR. However, to our knowledge, none of the problems surveyed in what follows can be solved in constant time (with a number of processors linear in the size of the input) on any model other than BSR.[*]

Sorting

The problem of sorting was solved in the first paper on BSR [2]. Another sorting algorithm on BSR is given in Ref. [40] as follows. The algorithm consists of three BROAD-CAST instructions. The first instruction gives the rank r_j of element x_j in the array x_1, $x_2, \ldots, x_n$, i.e. the number of elements in the array that are smaller than or equal to x_j, and is expressed as follows:

$$r_j := \Sigma \, 1 \, | \, x_i \leq x_j$$

When all elements in the array are distinct, an exclusive write operation suffices to place every datum x_j in its position r_j in the sorted array. In case of duplicates, however, all equal elements will receive the same rank after the first BROADCAST instruction. Since it is not known a priori whether there are duplicates, the second BROADCAST instruction is as follows:

$$s_j := \Sigma \, 1 \, | \, t_i \leq l_j \text{ where } t_i := r_i - \frac{1}{i} \text{ and } l_j := r_j - \frac{1}{j}$$

Note that different r_j values, being integers, differ by at least 1. If $r_i < r_j$ then clearly

$$r_i - \frac{1}{i} < r_j - \frac{1}{j}$$

---

[*]The scan model introduced by G. E. Blelloch is a single instruction multiple data (SIMD) vector model of computation. The primitive operations of the model work on vectors (one dimensional arrays) of values, with three types of primitive operations: elementwise arithmetic and logical operations, permutation operations, and scan operations. A scan operation is a type of prefix computation; it takes a binary associative operator $\oplus$ and a vector $[a_1, \ldots, a_n]$ and returns the vector $[a_1, a_1 \oplus a_2, \ldots, a_1 \oplus a_2 \oplus \ldots \oplus a_n]$. Although many problems have been solved on the scan model in faster time than on any PRAM model, there are few known constant time algorithms for the model. In Djokic [94], it is shown that many important characteristics of binary images can be determined in constant time on the scan model of parallel computation. These include computing the convex hull, diameter, width, smallest enclosing box, perimeter, area, digital convexity, and parallel and point visibility of an image, computing the smallest, largest, and Hausdorff distances between two images, linear separability of two images, and the recognition of digital lines, rectangles and arcs. All these problems can be solved on BSR in constant time, and with algorithms that in most cases are considerably simpler than the corresponding ones for the scan model.

On the other hand, if $r_i = r_j$ and $i < j$, then $r_i - \dfrac{1}{i} < r_j - \dfrac{1}{j}$

Therefore, all elements $r_i - (1/i)$ are distinct and are ordered in the same fashion as array x_i. Thus, the second BROADCAST instruction correctly finds the rank s_j of x_j for each j, $1 \leq j \leq n$. The elements can be forwarded to their sorted positions by a BROADCAST instruction that corresponds to an exclusive write on PRAM and is as follows [21]:

$$y_j := \Sigma \; x_i \big| s_i = j$$

Parenthesis matching

The problem is to find the pairs of matching parentheses in a given "legal" sequence $l_1, l_2, \ldots, l_n$ of parentheses. By "legal" it is meant that every parenthesis has its matching parenthesis in the sequence. For example, for the input sequence ()((()))(()), the output is an array giving the indices of the matching parenthesis for each parenthesis: 2, 1, 8, 7, 6, 5, 4, 3, 12, 11, 10, 9. The constant time parenthesis matching solution on the BSR model with n processors goes as follows [40]. First, parentheses are numbered as follows: assign 1 to each left parenthesis and -1 to each right one. Let the sequence of assignments be $b_1, b_2, \ldots, b_n$. Then, if the sth element is a left parenthesis, its "ordinal number" is equal to $p_s = b_1 + b_2 + \ldots + b_s$. Otherwise, it is $p_s = b_1 + b_2 + \ldots + b_s + 1$. The ordinal numbers can be assigned by the prefix sums algorithm. If there is a negative ordinal number, or the last element is not 1, then the sequence is not legal. Each right parenthesis s will match a left parenthesis s' so that $p_s = p_{s'}$, $s' < s$, and s' is the greatest index that satisfies this condition. As stated, there are two criteria for selecting the matching parenthesis. To reduce the number of criteria to one only, consider, as in the case of integer sorting, the modified array $p_s - (1/s)$. The modified array satisfies the following properties: if $p_i < p_j$ then

$$p_i - \frac{1}{i} < p_j - \frac{1}{j}$$

If $p_i = p_j$ and $i < j$, then

$$p_i - \frac{1}{i} < p_j - \frac{1}{j}$$

Consider all elements $p_i - (1/i)$ for which $p_i = p_s$: according to the choice of s', $p_{s'} - (1/s')$ is the largest such element smaller than $p_s - (1/s)$. Moreover, there is no element k in the whole array such that

$$p_{s'} - \frac{1}{s'} < p_k - \frac{1}{k} < p_s - \frac{1}{s}$$

Therefore the matching parenthesis for the right parenthesis numbered s is determined by the maximal index j such that

$$p_j - \frac{1}{j} < p_s - \frac{1}{s}$$

The above description leads to the following algorithm for solving the parenthesis matching problem:

```
For each processor j do in parallel
     if l_j = '(' then b_j := 1 else b_j := -1
     p_j := Σ b_i | i ≤ j
     if b_j = -1 then p_j := p_j + 1
     p'_j := p_j - (1/j)
     q_j := -1, t_j := 0, r_j :=0
     q_j := ∩ p'_i | p'_i < p'_j
     t_j := ∩ i | p'_i = q_j
     r_j := ∪ i | t_i = j
     if l_j = ')' then m_j := t_j else m_j := r_j
```

In the above algorithm, m_j is the index of the parenthesis matching the jth parenthesis. It is determined after the second BROADCAST instruction for right parentheses, and after the last BROADCAST instruction for left ones. Note that the above instructions do not check whether the sequence of parentheses is "legal." This can be done with the following "no criterion" BSR instructions:

```
For each processor j do in parallel
     y_j := Σ b_i
     x_j := ∪ p_i
     if y_j ≠ 0 or x_j < 1 then "parenthesis sequence is not legal"
```

The computation tree form of an arithmetic expression is its binary tree representation which defines the order of operations. In Ref. [11], a parallel algorithm for finding the computation tree form of an arithmetic expression on a *concurrent read exclusive write* (CREW) PRAM model is given. It consists of few steps, the most expensive being the parenthesis matching. The remaining steps run in constant time on a CREW PRAM with n processors, where n is the length of the input. Since the CREW PRAM is weaker than BSR, and the parenthesis matching algorithm above runs in constant time, it follows that the computation tree form of an algebraic expression can be computed in constant time on BSR with n processors.

7.3.2 Maximal sum subsegment

The maximal sum subsegment problem is defined as follows. Given an array of numbers $d_1, d_2, \ldots, d_n$, it is required to find a contiguous subarray of maximal sum. The BSR algorithm of Ref. [5] solves the problem in constant time using n processors. First, the prefix sums are computed. Then, at each element in the sequence, the largest prefix sum to its right is found (this is done in constant time using BROADCAST). That sum is then broadcast in order to find the point at which such a sum ends. It is now a simple matter to compute the sum and endpoint of a maximal sum subsegment starting at each element. A maximizing concurrent write is used to find a maximal sum subsegment overall. Finally, another BROADCAST is used to find where such a subsegment actually begins. The algorithm is as follows:

```
s_j := Σ d_i | i ≤ j          {s now contains the prefix sums}
m_j := ∩ s_i | i ≥ j          {m_j is the largest sum to the right of d_j}
e_j := ∩ i | s_i = m_j        {e_j is the index of m_j}
m_j := m_j - s_j + d_j        {compute the sum of a maximal sum subsegment start-
                               ing at d_j}
```

```
t := ∩ mᵢ              {find the sum of the overall maximal sum subsegment}
x := ∩ i | mᵢ = t      {find its starting point}
y := eₓ                {find its ending point}
```

The maximal sum subrectangle problem (in two-dimensional arrays) is also solved in Ref. [5] in constant time but with $O(n^3)$ BSR processors.

7.3.3 Computational geometry

This section describes solutions to various problems in computational geometry, including some problems on intervals, reporting the intersection of two convex polygons and implementing the merging slopes technique with applications. A constant time Voronoi diagram algorithm that uses $O(n^3)$ processors is given in Ref. [9].

Geometry of intervals. Assume that we are given a set I of n intervals $g_i = (a_i, b_i)$ on a straight line. We consider several problems defined on such intervals. We start with the following question: Do any intervals overlap? More generally, for each interval it is required to determine whether it is overlapped by any other interval. In Ref. [33], the problem is reduced to the maximal element problem, which is defined as follows:

Given a set $S = \{p_1, p_2, \dots, p_n\}$ of n points in plane, a point p_i dominates a point p_j iff $x_i \geq x_j$ and $y_i \geq y_j$, where $p_i = (x_i, y_i)$, $1 \leq i \leq n$ is given by its x and y coordinates. The maximal elements problem is to determine those points of S which are dominated by no other point. (Note that this problem differs from the maximal vectors problem mentioned in Section 7.1 and whose definition depends on the quadrant containing the given points [2].)

Interval g_i is contained in g_j when $a_j \leq a_i \leq b_i \leq b_j$, which is equivalent to $-a_i \leq -a_j$ and $b_i \leq b_j$. The latter is true iff the point $(-a_i, b_i)$ is dominated by point $(-a_j, b_j)$. The maximal element points correspond exactly to intervals not overlapped by any other interval. Therefore, the intervals not overlapped by any other interval can be found in constant time by using either the BSR maximal vectors solution [2] (with minor modifications) or the algorithm developed in Section 7.6 for $d = 2$.

The problem of computing the measure of the union of a set of intervals is defined as follows. Given a set I of n intervals $g_i = (a_i, b_i)$ on a real line, find the size of the interval resulting from their union. In Ref. [27], an $O(n \log n)$ time sequential algorithm is given to solve this problem. The following parallel algorithm runs in constant time on BSR with n processors [9]:

1. Sort the $2n$ endpoints. Let the sorted list be $(x_i, flag_i)$ where x_i is the x-coordinate of the endpoint and $flag_i$ has the value 1 or -1, depending on whether the point is the left or right endpoint of its corresponding interval, respectively.
2. Apply a parallel prefix sum computation on the $2n$ flags. The obtained result is $p_i = flag_1 + flag_2 + \dots + flag_i$. The interval between $(x_i, flag_i)$ and $(x_i + 1, flag_i + 1)$ belongs to an interval from I iff $p_i > 0$.
3. Let $m_i = x_{i+1} - x_i$ iff $p_i > 0$ and $m_i = 0$, otherwise. Find the sum $M = m_1 + m_2 + \dots + m_{2n}$. M is the measure of the union of intervals.

The ε-closeness problem is defined as follows. Given $n + 1$ real numbers $x_1, x_2, \dots, x_n$ and $\varepsilon > 0$, determine whether any two x_i and x_j $(i \neq j)$ are at distance less than ε from each other. It can be solved in constant time on BSR by applying sorting, finding differences between neighboring elements in the sorted list and then determining the minimum such difference. The maximum gap problem is to determine the maximum possible positive difference between any two neighboring numbers (after sorting) out of n numbers x_1, $x_2, \ldots, x_n$. An optimal solution for a mesh connected computer is given in [38]. The constant time solution on BSR can be obtained as follows [9]:

1. Sort the numbers (assume now that $x_1, x_2, \ldots, x_n$ is sorted).
2. Find differences $d_i = x_i + 1 - x_i$.
3. Find the maximal d_i, which is the maximum gap D.

Reporting the intersection of two convex polygons. To give a parallel BSR algorithm for finding the intersection of two convex polygons P and Q with a total of $O(n)$ vertices, [9] modified the slab method of [35] and the parallel algorithm on the hypercube model [39].

By drawing a vertical line through each vertex of P and Q we divide P and Q into slabs. The leftmost and the rightmost vertices of P and Q (which can be found easily in constant time on BSR) divide both P and Q into two chains: the upper and the lower chain of vertices. The intersections Al, A2, and A3 of a vertical line passing through a vertex A of one chain with the remaining three chains can be obtained in parallel (one processor per vertex A) by computing the nearest points A′ and A″ of A to the left and to the right, respectively, in the chain under consideration (this can be done by merging the chains) and finding the intersection of A′A″ with the vertical line through A. Let $P \cup Q$ denote the list of vertices of P and Q sorted together by x-coordinate (it can be constructed in constant time by merging the upper and lower chains of P and Q). Consider each slab defined by two neighboring points A and B. On the basis of the coordinates of points A, Al, A2, A3, B, Bl, B2, and B3 (translation by 1 can be used to exchange the data), one can decide in constant time in parallel whether P and Q intersect within the slab and determine the (at most two) points of $P \cap Q$ which are located in the slab (these are either intersections of edges of P and Q or vertices of P or Q which are located inside the other polygon). So far, we have detected all vertices of the intersection of P and Q in constant time. However, we should order them to obtain their convex hull. For each vertex of the intersection, we decide whether it is a vertex of the upper or lower chain of $P \cap Q$ (this can be done in constant time). Also, we assign the vertex to the left vertex of the corresponding slab defined by points of $P \cup Q$. Thus, each vertex of P or Q will have assigned to it zero or one vertex of $P \cap Q$ from the upper convex hull chain (and similarly for the lower convex hull chain). Now, the upper chain of $P \cap Q$ can be obtained by sorting its vertices (similarly, we find the lower convex hull chain).

This algorithm also solves the problem of detecting the intersection of two convex polygons in parallel (linear separability). Clearly, P and Q intersect if at least one vertex of $P \cap Q$ is found. The time complexity is still constant.

The merging slopes technique with applications. The extremal search problem $ES(P, Q, \alpha)$ is to find, for each edge e of P, a vertex v of Q which is closest to it, after Q is rotated in the counterclockwise direction by an angle α. The distances are measured (positive or negative) from vertices to straight lines containing edges. The problem can be solved by the merging slopes technique [39]. If edges of P are followed in the counterclockwise order, their corresponding vertices in Q also follow a counterclockwise order. Moreover, the slopes of two edges from Q that are incident to the corresponding vertex v

are exactly the first smaller and first greater slopes than the slope of e among (rotated) edges of Q. Thus the problem can be solved by merging slopes of edges of P with those of (rotated) edges of Q.

We now describe a constant time solution [23] for the extremal search problem on a BSR with n processors. Let s_i ($1 \leq i \leq n$) be the slopes of edges e_i of P. Suppose that the ith ($1 \leq i \leq m$) edge of Q has slope q_i and its endpoints are (a_i, b_i), ordered clockwise with respect to Q. First, the slopes of rotated edges of P are expressed in the interval $[0, 2\pi)$ (the same is assumed for edges of Q). Let p_j be the slope of rotated edge e_j. The associated vertex of e_j of P is the vertex a_i of Q such that the slope q_i of edge (a_i, b_i) has the minimal possible slope which is larger than p_j. If no such slope is found, then the result is vertex a_1. The algorithm is expressed as follows:

```
for each processor j, 1 ≤ j ≤ m do in parallel
    pj = sj + α mod 2π
    vj := ∩ ai | qi ≤ pj
    if all slopes were rejected then vj := a1
```

Clearly, the broadcasting operation can be repeated to obtain, one by one, all necessary data about the associated vertex (the index, x and y coordinates, etc.). The algorithm runs in constant time on BSR with n processors. The merging slopes technique can be applied to solve some other problems, as discussed in Ref. [39] for hypercubes and in Ref. [8] for the star and pancake models. Without giving details or definitions, we note here that the BSR solution to the merging slopes technique leads to constant time solutions with a number of processors equal to the input size for the following problems [23]: finding the external watchman route of a convex polygon (i.e., the shortest closed path such that each point in the exterior of a polygon is visible to some point along the path), computing the Minkowski (or vector) sum, finding critical support lines and determining the separability of two convex polygons, computing the maximum distance between two convex polygons, and computing the smallest enclosing box, diameter, and width of a convex polygon.

7.3.4   Tree algorithms

This section illustrates the application of BSR to the solution of graph theoretic problems defined on trees.

Decoding binary trees. Two binary trees are considered identical if they have the same shape (once the information they carry is disregarded). More precisely, two binary trees are identical if they are both empty, or else their left and right subtrees are pairwise identical as binary trees. In computer science applications, the binary tree is a data structure consisting of an array of records, where each record corresponds to a tree node and contains the data it carries, and pointers to the parent and its two children. In some applications the binary tree is encoded as a tree sequence of integers, and a variety of such encodings are proposed in the literature (see the reference list in Ref. [6]). The reverse process (i.e., reconstructing the tree data structure from its tree sequence) is referred to as *decoding*.

One of the most commonly used encodings is the bitstring code proposed by Zaks [41]. All nodes of binary tree T (see Fig. 7.7) are labeled 1, and all missing children are replaced by nodes labeled 0. The preorder traversal of such an extended tree $e(T)$ (see Fig. 7.8) de-

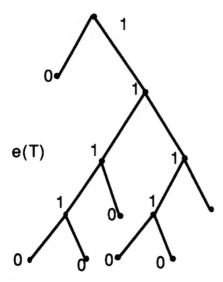

Figure 7.7 Extended binary tree

fines the bitstring tree sequence. The extended tree is a regular tree, meaning that each node has 0 or 2 children. Usually the last 0 is omitted, and thus the string has $2n$ elements. For example, the binary tree of Figs. 7.7 and 7.8 is encoded 101110001100.

The decoding algorithm of Ref. [31] uses the parenthesis matching technique, for which a constant time solution on BSR is given in Section 7.3.1. It produces the corresponding "decoded" tree with nodes numbered $1,2,\ldots,n$ following the preorder traversal of the tree (see Fig. 7.8). Given a $2n$ bitstring sequence (where n is the number of nodes in the tree), the algorithm finds for every 1 its unique matching 0, where 1 is treated as '(' and 0 as ')'. Let z_i be the position of the ith 1 in the bitstring sequence (called the z-sequence in Ref. [41]), and let q_i be the position of its matching 0. For example, in 101110001100 (which corresponds to the parenthesis sequence in the example of Section 7.3.1) we have $z_1 = 1$, $z_2 = 3$, $z_3 = 4$, $z_4 = 5$, $z_5 = 9$, $z_6 = 10$, $q_1 = 2$, $q_2 = 8$, $q_3 = 7$, $q_4 = 6$, $q_5 = 12$, $q_6 = 11$. The z sequence satisfies the following property: $z_1 = 1$, $z_{i-1} < z_i \leq 2i - 1$, $2 \leq i \leq n$. If $q_i = z_i + 1$, then the node has no left child; otherwise, the $(i + 1)$th 1 is its left child. Next, if the $(q_i + 1)$th element in the bitstring is a 0, then the ith node has no right child; otherwise, node j such that $z_j = q_i + 1$ is the right child of the ith node.

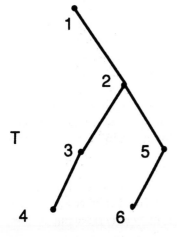

Figure 7.8 Preorder traversal of binary tree in Fig. 7.7

The decoding algorithm can be formally described as follows [40]. The first step is an application of the parenthesis matching algorithm.

```
for each processor j, 1 ≤ j ≤ 2n do in parallel
    m_j := index of the bit matching the jth bit
for each processor j, 1 ≤ j ≤ n do in parallel
    z'_j := Σ b_i | i ≤ j;                              {1 ≤ i ≤ 2n}
    z_j := ∪ i | z'_i = j                               {1 ≤ i ≤ n}
    q_j := ∪ i | m_i = z_d                              {1 ≤ i ≤ 2n}
    if q_j = z_j + 1 then leftch_j := 0 else leftch_j := j + 1
    r_j := ∪ b_i | i = q_j + 1
    if r_j = 0 then rightch_j := 0 else rightch_j := ∪ i | z_i := q_j + 1
```

Generating binary trees. We now consider the problem of listing all binary tree shapes, where the content in the node is disregarded. Given a regular tree P, let r_P be the number of children of the root of P; i.e., $r_P = 2$ or $r_P = 0$. Let P_L and P_R denote the left and right subtrees of P. Two regular trees P and Q are in B-order, $P < Q$, if:

1. $r_P < r_Q$,
2. $r_P = r_Q$ and $P_L < Q_L$, or
3. $r_P = r_Q$, and $P_L = Q_L$, and $P_R < Q_R$

Two binary trees T' and T'' are in B-order, $T' < T''$ if $e(T') < e(T'')$.

Thus, B-order corresponds to the lexicographic order of bitstring sequences. It is considered in the literature to be the natural order of binary trees.

A sequential algorithm is described in [41] for generating binary trees in B-order. The following simple BSR algorithm for generating all z-sequences in constant time per sequence (i.e., with constant delay between pairs of consecutive sequences) is given in Ref. [40]:

```
for each processor j, 1 ≤ j ≤ n do in parallel
    z_j := j
    repeat
        report z-sequence
        if z_j < 2j - 1 then x_j := j else x_j := 0
        a_j := ∩ x_j
        c_j := ∩ z_i | i = a_j
        if j ≥ a_j then z_j = c_j + 1 + j - a_j
    until z_1 = 2
```

Note that one of the two BROADCAST instructions here ($a_j := ∩ x_j$) is used with no criteria. (However, an artificial criterion, always satisfied, can be introduced if needed.) The above algorithm also demonstrates that BSR is suitable for implementing various backtracking algorithms.

Reconstruction of a binary tree from its traversals. The problem of constructing a binary tree from its preorder and inorder traversals is well studied in the literature. Two variants of the problem are considered. In the first variant, nodes are labeled arbitrarily. In the second (and more restricted) variant node labels come from the set $\{1,2,\ldots,n\}$, where n is the number of nodes in the binary tree. Most parallel algorithms assume the latter labeling to overcome the bottleneck that a sorting step would create. However, the linear time algorithms described in Refs. [10] and [29] for the general version show that sorting is not inherent to the problem, and that the real challenge is to solve the reconstruction problem with nodes labeled arbitrarily. A constant time BSR solution to the problem is given in Ref. [40], using the ideas given in Ref. [22]. The input consists of two sequences $I(i)$ and $P(i)$, $1 \le i \le n$, denoting the inorder and preorder traversals, respectively. As stated, the elements of the sequences are arbitrary distinct data (real numbers, integers, letters, words, etc.). Since they refer to the same tree, naturally $\{I(1), I(2),\ldots,I(n)\} = \{P(1), P(2),\ldots,P(n)\}$.

The output consists of pointers to the children, leftch(i) and rightch(i), and to the parent, parent(i), for each node i, $1 \le i \le n$, following their preorder traversal (i.e., node i has $P(i)$ in the corresponding data field). The pi sequence of a binary tree is an integer sequence produced by the following algorithm [22]:

1. Label the nodes of the tree as accessed in preorder $1,2,\ldots,n$ (see Fig. 7.9).
2. Output these numerical labels as the nodes are accessed in inorder.

For example, in Fig. 7.9, the preorder and inorder traversals and pi-sequence are:

index	1	2	3	4	5	6	7	8	9
P	Z	P	V	W	Y	T	X	R	U
I	P	W	V	Y	Z	R	X	T	U
pi	2	4	3	5	1	8	7	6	9

In other words, if the ith node in preorder (which is labeled P(i)) is the same as the jth node in inorder (which is labeled I(j)) then the jth element in the pi sequence is equal to i. Therefore if $P(i) = I(j)$ then $pi(j) = i$. For instance, in Fig. 9, $P(4) = I(2) = W$, and thus

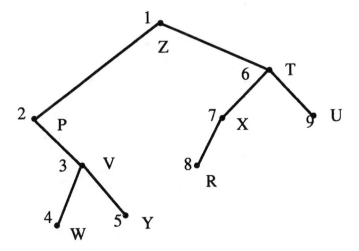

Figure 7.9 Binary tree with nodes numbered according to their preorder traversal

$pi(2) = 4$. The computation of the *pi*-sequence can be achieved by a single BROADCAST operation on BSR as follows:

$$pi(j) := \cup i | P(i) = I(j)$$

All pairs (parent, left child) are easily determined by pairs $(pi(i), pi(i) + 1)$ such that $pi(i) + 1$ precedes $pi(i)$ in the array *pi*. They can be determined on BSR as follows:

```
r_j := ∪ i | pi(i) = pi(j) + 1
if r_j < j then lc(j) := true else lc(j) := false
r_j := ∨ lc(i) | pi(i) = j
if r_j = true then leftch(j) := j + 1 else leftch(j) := 0
```

In the preceding instructions, $lc(j)$ is true if and only if $pi(j)$ has a left child. The left to right maxima (LR maxima for short) of a sequence are those elements larger than all elements preceding them. The set of all (parent, right child) pairs in the tree is determined by pairs $(pi(LRM(j + 1) - 1, m_j + 1)$ where $LRM(j)$ is the index in the *pi* sequence of the *j*th LR maximum in the sequence, and m_j is the value of this maximum [22]. The BSR implementation of this statement is as follows:

```
r_j := ∩ p(i) | i ≤ j
if r_j = pi(j) then mlr(j) := 1 else mlr(j) := 0
r_j := Σ mlr(i) | i ≤ j
LRM(j) := ∪ i | r_i = j
pr(j) := LRM(j + 1) - 1
m_j := ∪ pi(i) | i = LRM(j)
cr(j) := m_j + 1
r_j := ∪ i | pr(i) = j
d_j := ∪ cr(i) | pr(i) = j
if r_j > 0 then rightch(j) := d_j else rightch(j) := 0
```

In the preceding instructions, $pr(j)$ is the parent of right child $cr(j)$. To determine parent pointers, the following BSR procedure finalizes the reconstruction algorithm:

```
parr(j) := ∪ i | rightch(i) = j
parl(j) := ∪ i | leftch(i) = j
if parl(j) > 0 and parl(j) ≤ n then parent(j) := parl(j) else
     parent(j) := parr(j)
```

7.3.5   Digital geometry

We conclude this section on one criterion BSR algorithms by presenting constant time solutions to two problems defined on digital images.

Histogramming. Suppose that an $m \times m$ image is given by its gray level values I_{ij}, $0 \le I_{ij} < g, 1 \le i, j \le m$. Thus, each pixel has one of g possible integer values, denoting its gray level. The histogram of an image is a list of all gray values paired with their counts, i.e., the number of pixels having such gray value. Assume that the image is stored one pixel per processor on a one criterion BSR with $n = m^2$ processors. A constant time algorithm for computing the histogram on the model is presented in Ref. [30] as follows. For simplicity, assume that pixel indices are in the range from 1 to n (in rowmajor order, say). The number s_j of pixels having the same gray value as pixel j is obtained by a single BROADCAST instruction:

$$s_j \leftarrow \Sigma \; 1 \,|\, g_i = g_j$$

To produce a list of all different gray level values with their counts, the list (g_j, s_j) is sorted on g_j, with duplicates eliminated.

Distance transform. A binary image is usually represented by an $m \times m$ matrix I such that $I_{ij} = 0$ if the pixel in the ith row and jth column is white, and $I_{ij} = 1$ if the pixel is black, $1 \le i, j \le m$. The distance transform problem is to find the nearest black pixel to each white pixel in the image. Important applications of distance transforms are expanding and shrinking objects (by thresholding the transform), constructing shortest paths and computing shape factors, reconstructing objects from parts of the boundary, and skeletization.

Various metrics were used in the literature to find the distance transforms, mostly belonging to the family of L_k metrics. The distance transform for the L_1 metric can be found sequentially in $O(n)$ time, where $n = m^2$. For the mesh of trees model, Ref. [34] gave an $O(\log n)$ time algorithm for computing the distance transform under the L_1 metric.

Since the image has n pixels, we assume a BSR model with n processors, one per pixel. A solution to the distance transform problem is given in Ref. [30]. The algorithm for finding the distances to the nearest black pixel is based on the following simple property of binary images. Suppose that b' is the closest black pixel to a white pixel a'. Let w be the pixel that lies in the same row as a' and same column as b'. Then, obviously, b' is the closest black pixel to w among black pixels from the same column to which w belongs. This property follows from the monotonicity of distance functions with respect to absolute values of differences in x- or y-coordinates of two pixels.

This property suggests a simple algorithm for finding the distance transform. Let the processors be numbered by their row and column indices (corresponding to the pixel indices), from 1 to m. The algorithm consists of a columnwise scan followed by a rowwise one. In the columnwise scan, each (black or white) pixel (j, k) finds the distance rjk to the closest black pixel in the same column as itself, if it exists. The closest black pixel (i, k) to pixel (j, k) is at distance $|i-j|$. For each column k ($1 \le k \le m$) a one criterion BSR model with n processors is applied to compute r_{jk} as follows. Indices j and i are used as in the BSR definition, while k is considered a constant for a given column. The BSR instructions are as follows:

```
if I_ik = 1 (i.e. (i, k) is a black pixel) then b_ik ← i else b_ik ← m + 1
r'_jk ← ∪ b_ik | i ≥ j
if I_ik = 0 then b_ik ← 0
r"_ik ← ∩ b_ik | i ≤ j
r_jk ← 3n
if r'_jk < m + 1 then r_jk ← r'_jk - j
if r"_jk > 0 and j - r"_jk < r_jk then r_jk ← j - r"_jk
```

The index in_{jk} of the selected black pixel which is at distance r_{jk} from pixel (j, k) is determined as either $j + r'_{jk}$ or $j - r''_{jk}$, depending on whether the selected pixel is above or below (j, k), respectively. The information contained in the selected black pixel (like a label) can then be broadcast. In the rowwise scan, each pixel (k, j) finds the minimum value $|i - j| + r_{ki}$ which is the distance to its closest black pixel. For each row k ($1 \leq k \leq m$), a one criterion BSR with n processors is used as follows:

```
c'ki ← rki + i for each pixel (k, i)
s'kj ← ∪ c'ki | i ≥ j
c"ki ← -i + rki for each pixel (k, i)
s"kj ← ∪ c"ki | i ≤ j
skj ← ∪ (s'ki - j, j + s"ki) *****
```

To find the index in_{kj} of the pixel that "supplied" indirectly the nearest black pixel, two candidates are found by the following BROADCAST instructions:

```
rinkj ← ∩ i | c'ki = s'kj
linkj ← ∪ i | c"ki = s"kj
```

and one of them, corresponding to the choice for s_{kj}, is taken to be in_{kj}. The information from the nearest black pixel (its indices or component label) is then available via processor (k, j), since it can be collected in the previous columnwise step.

The algorithm for computing the distance transform can be adapted in a straightforward way to calculate the farthest black pixel for each pixel in the image. It should be noted that in the above solution we used columnwise and rowwise scans and considered each column or row to be a separate BSR model with n processors. However, we assumed merely that we have a single BSR with n processors, such that a BROADCAST instruction selects data from all processors to extract a datum for a particular processor. There are two ways to resolve the problem. One is to add additional criteria, requesting data from a particular row or column only (where row and/or column indices are given for the purpose of selection); this solution increases the number of needed criteria, which may not be desirable. A better solution is to sort all data in either column major or row major order (depending on the type of scan) and modify the algorithm accordingly. For example, in the columnwise scan for the distance transform algorithm, pixels are assigned indices from 1 to n, and each pixel finds the nearest black pixel with a greater and a smaller index; if the obtained indices exceed the appropriate bounds for a given column, the corresponding result is ignored and information that no pixel is found is recorded. For the rowwise scan, we can introduce one more "trick": because all meaningful distances after the first step are between 1 and n, it is possible to increase all r_{ki} values by km ($1 \leq k \leq m$). In this way, there is a considerable gap between neighboring rows, which does not allow one to select data from a pixel in a different row. Similar modifications can be made for other algorithms on binary images presented in this chapter.

7.4   Two Criteria BSR Algorithms

In this section, we use the problem of counting inversions in a permutation, as well as several computational geometric problems, to illustrate BSR algorithms where broadcast data are tested for their satisfaction of two criteria before being allowed to take part in the reduction process.

7.4.1   Counting inversions in a permutation

Given a permutation $\pi(1)$, $\pi(2),\dots,\pi(n)$ of the numbers 1, 2,...,n, the problem is to count the number of inversions, i.e., the number of pairs (i, j) which are in disorder: $i < j$ but $\pi(i) > \pi(j)$.

This problem can be solved on BSR as follows:

```
y_j := Σ 1 | i < j ∧ π(i) > π(j)
x_j := Σ y_i
```

7.4.2   Computational geometry

As was shown in Section 7.3.3, BSR lends itself naturally to the efficient solution of computational geometric problems. Further evidence of this is presented below for problems as varied as geometric dominance, intersection, and proximity.

7.4.2.1 ECDF searching and 2-set dominance counting in the plane. The ECDF searching problem consists of computing for each $p \in S$, the number $E(p, S)$ of points of S dominated by p. Given two point sets A and B, the 2-set dominance counting problem is to determine for each point p from B the number $D(p, A)$ of points from A that p dominates.

For points in the plane (2-dimensional space), these two problems can be solved as special cases of the GPC technique, as follows [9]:

ECDF searching:

$$E\left(p_j, S\right) := \Sigma\, 1 \mid p_i[1] \leq p_j[1] \wedge p_i[2] \leq p_j[2]$$

2-set dominance counting:

$$D\left(p_j, A\right) := \Sigma\, w_i \mid p_i[1] \leq p_j[1] \wedge p_i[2] \leq p_j[2]$$

where $w_i = 1$ if the ith point belongs to A and $w_i = 0$ otherwise.

Both algorithms require a two criteria BSR, and a number of processors equal to the number of points, in order to be executed in constant time.

Intersection of isothetic line segments. The problem is to determine, given a set S of n horizontal and/or vertical segments in the plane, for each segment s, the number of segments in S that intersect s (see Fig. 7.10). An $O(\log n)$ time CREW PRAM solution using n processors is given in Ref. [25], where the problem is solved by applying the 2-set dominance problem in the plane. Let L and R be the sets of left and right endpoints of horizontal segments, respectively. Then, the number of horizontal segments crossing a vertical segment pq is given by $(D(p, L) - D(p, R)) - (D(q, L) - D(q, R))$ (see Ref. [25, p. 299]).

A direct solution of the problem (without the use of 2-set dominance) is possible in constant time but with 4 criteria BSR. On the other hand, applying 2-set dominance reduces the number of criteria to two. The solution takes constant time on BSR with n processors.

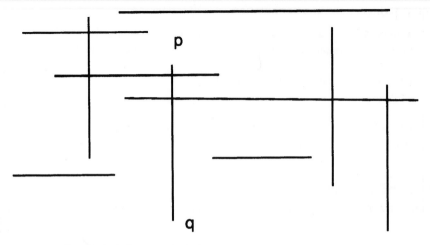

Figure 7.10 Set of isothetic line segments in the plane

All nearest and farthest neighbors in the L₁ metric. Given a set of n points $p_i = (x_i, y_i)$, the all nearest neighbors problem is to find for each point the point that is nearest to it among the $n - 1$ remaining points. A constant time BSR solution to this problem that uses n processors is described in Ref. [30].

The definition of BSR does not allow the use of absolute value functions. Without the function, the distance of p_i to p_j depends on whether p_i is above or below, and to the left or to the right of p_j. Therefore, there are four different definitions of the distance, depending on the respective positions of two points. A BSR operation with two criteria can be applied for each of these positions, and the minimum nn_j of the obtained values is then selected. The distance of a point p_i that is above and to the right of p_j is $d(p_j, p_i) = x_i + y_i - x_i - y_i$. Thus, the closest point in the quadrant is at distance $ar_j - s_j$ where $s_j = x_j + y_j$; and ar_j is the minimal of values $s_i = x_i + y_i$ for points in the quadrant. The closest points in other quadrants are found in a similar way. The algorithm can be written as follows:

```
sⱼ ← xⱼ + yⱼ
tⱼ ← xⱼ - yⱼ
arⱼ ← ∪ sᵢ | xᵢ ≥ xⱼ ∧ yᵢ > yⱼ
blⱼ ← ∩ sᵢ | xᵢ ≤ xⱼ ∧ yᵢ < yⱼ
alⱼ ← ∩ tᵢ | xᵢ < xⱼ ∧ yᵢ ≥ yⱼ
brⱼ ← ∪ tᵢ | xᵢ > xⱼ ∧ yᵢ ≤ yⱼ
nnⱼ ← ∪ (sⱼ - blⱼ, tⱼ - alⱼ, -tⱼ + brⱼ, -sⱼ + arⱼ)
```

The algorithm finds the distance to the nearest neighbor point. The index in_j of the nearest neighbor point of point p_j can be found by using additional BSR instructions. (If there are several points at the same nearest neighbor distance, only one of them is selected.) The points are assumed to be sorted by their x-coordinates in increasing order. The additional instructions are:

```
iarⱼ ← ∩ i | yᵢ > yⱼ ∧ sᵢ = arⱼ
iblⱼ ← ∪ i | yᵢ < yⱼ ∧ sᵢ = blⱼ
```

```
ial_j ← ∪ i | y_i ≥ y_j ∧ s_i = al_j
ibr_j ← ∩ i | y_i ≤ y_j ∧ s_i = abr_j
if nn_j = -s_j + ar_j then in_j ← iar_j
if nn_j = s_j - bl_j then in_j ← ibl_j
if nn_j = t_j - al_j then in_j ← ial_j
if nn_j = -t_j + br_j then in_j ← ibr_j
```

In the above solution we note that one criterion, comparison along x-coordinates, is omitted. This is possible because if a nearest neighbor is in a given quadrant then the choice of max or min guarantees that a point from the quadrant is selected. Finally, if the information co_i contained in the nearest neighbor point i for point j is desired, then the following simple BROADCAST instruction can be used to obtain it (obviously min can be replaced with max or sum, since there is exactly one successful candidate):

```
inf_j ← ∪ co_i | i = in_j
```

The all farthest pairs problem is to find the farthest point for each point in a given set. The latter can be obtained in a similar way as the nearest neighbor. More precisely, it suffices to interchange all references to min and max. The closest pair of points in a set is defined by two points having the minimal possible distance. It can be obtained by minimizing all nearest neighbor distances. Similarly, the diameter is the distance between two farthest points, and is derived by maximizing all farthest pair distances. In Ref [30], a construction of a discrete Voronoi diagram of labeled images in constant time on a two criteria BSR with n processors is also given.

7.5 Three Criteria BSR Algorithms

All the examples in this section, illustrating the use of BSR to solve problems in constant time with three selection criteria, originate in computational geometry.

7.5.1 Vertical segment visibility

We first consider the horizontal and vertical adjacency maps. Because of similarity, we define only the horizontal adjacency map. Consider a set V of n vertical segments. Through each endpoint p of each member of V, we trace a horizontal half-line to the right and one to the left; each of these half-lines either terminates on the vertical segment closest to p (in the given direction) or, if no such intercept exists, the half-line continues to infinity. In this manner the plane is partitioned into regions, of which two are half-planes and all others are rectangles, possibly unbounded in one direction. Each region consists of points in the plane that have the same closest segments in both horizontal directions (in other words, the same segments are visible in both horizontal directions). The number of regions is $3|V|+1$ (see Ref. [32, p. 357]). In Ref. [25], Section 6.3.2, an $O(\log n)$ time CREW PRAM solution is given to the problem of computing adjacency maps, and it uses $O(n)$ processors and a sophisticated plane sweep tree method.

A 3 criteria BSR solution using n processors (one per segment) can be given as follows [9]. Let (x_i, t_i) and (x_i, b_i) be the top and bottom endpoints of the ith vertical segment, respectively. The indices lt_j, lb_j, rt_j, rb_j of the closest segment to the left/right of the top/bottom endpoints of the jth vertical segment are obtained by the last four BROADCAST instructions in the following algorithm (see Fig. 7.11):

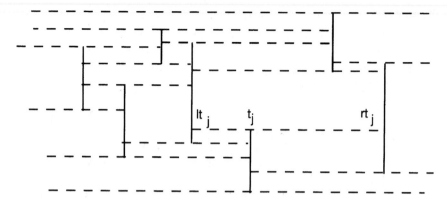

Figure 7.11 Horizontal adjacency map of a set of vertical line segments in the plane

```
sort the vertical segments by x-coordinate such that x₁ ≤ x₂ ≤ ... ≤ xₙ
ltⱼ := ∩ i | xᵢ < xⱼ ∧ tᵢ ≥ tⱼ ∧ bᵢ ≤ tⱼ
lbⱼ := ∩ i | xᵢ < xⱼ ∧ tᵢ ≥ bⱼ ∧ bᵢ ≤ bⱼ
rtⱼ := ∪ i | xᵢ > xⱼ ∧ tᵢ ≥ tⱼ ∧ bᵢ ≤ tⱼ
rbⱼ := ∪ i | xᵢ > xⱼ ∧ tᵢ ≥ bⱼ ∧ bᵢ ≤ bⱼ
```

The vertical segment visibility problem is defined as follows. Given a set of vertical disjoint segments in the plane, determine and report all visible pairs of segments. (A pair is visible if there is a horizontal segment between them which does not cross other segments.) This problem is studied in Ref. [28] where an $O(\log n)$ time solution on a mesh of trees of size n^2 is presented. The same problem is also investigated in Ref. [15], and an $O(\log n)$ time CREW PRAM solution using $O(n)$ processors is given. Here we present a solution that is different from the two just mentioned in the sense that it uses the horizontal adjacency map. Furthermore, we believe that using the horizontal adjacency map may simplify the algorithms of Refs. [28] and [15] on their respective models.

We observe that if the pth and qth segments are visible, then there exists a j such that both segments are visible from one of the endpoints of the jth segment. The choice of j may not be unique, and both endpoints of the jth segment may satisfy the condition. Also, j can be equal to either p or q, or different from both. On the other hand, a given endpoint of a given segment may determine at most one pair of visible vertical segments, because it has at most one visible segment in each of the two horizontal directions. Thus, there are at most $2n$ visible pairs of segments. They can be obtained by the following algorithm:

1. Construct the horizontal adjacency map as shown above.
2. Save the pair (lt_j, rt_j) if $lt_j > 0$ and $rt_j > 0$, otherwise eliminate it.
3. Save the pair (lb_j, rb_j) if $rb_j > 0$ and $lb_j > 0$, otherwise eliminate it.
4. Save the pair $(lt_j, {}_j)$ if $lt_j > 0$, otherwise eliminate it.
5. Save the pair (j, rt_j) if $rt_j > 0$, otherwise eliminate it.
6. Save the pair (lb_j, j) if $lb_j > 0$, otherwise eliminate it.
7. Save the pair (j, rb_j) if $rb_j > 0$, otherwise eliminate it.
8. Sort all saved pairs to eliminate duplications.

Clearly BSR with n processors and constant time suffices to solve the vertical segment visibility problem.

7.5.2 All nearest and farthest foreign neighbors

In Ref. [24], the following modifications of the closest point and all nearest neighbor problems is considered. Let S be a set of n colored points. In the closest foreign pair problem one has to find a closest foreign pair, i.e., a bichromatic pair of points that are closest. In the all nearest foreign neighbors problem one has to find for each point in the configuration S a nearest neighbor with different color. Plane sweep algorithms are presented in Ref. [24] for solving the closest foreign pair and all nearest foreign neighbor problems in L_1 and L_∞ metrics in optimal $O(n \log n)$ sequential time. Using the city block metric, we may solve both problems on BSR with n processors in constant time. The solution is obtained by a modification to the closest pair and all nearest neighbor algorithms (which are special cases of the problems with all points colored in different colors) by adding one more criterion to all BSR instructions so that points with the same color are ignored. For example, the instruction

$$ar_j \leftarrow \cup s_i \big|\; x_i \geq x_j \wedge y_i > y_j$$

becomes

$$ar_j \leftarrow \cup s_i \big|\; x_i \geq x_j \wedge y_i > y_j \wedge c_i \neq c_j$$

where c_i is the color of pixel i. Similarly one can define the all farthest foreign pair and the foreign diameter problems and appropriate solutions.

7.5.3 Medial axis transform

The medial axis transform (MAT) is an image representation scheme proposed in Ref. [13]. The essential idea in the MAT is to find a minimal set of upright squares whose union corresponds exactly to the regions in the image I that have value 1 (i.e., cover black pixels). The sequential solution in Refs. [16] and [26] is followed in Ref. [30] to develop a constant time algorithm using n processors for a three criteria BSR. Assume that the top left corner of the top left pixel is indexed $(1,1)$, and the bottom right corner of the bottom right pixel is indexed $(m + 1, m + 1)$. Let $M[i, j]$ be the height of the largest square with top left corner $[i, j]$, all of whose image values are 1. Obviously $M[m + 1, j] = M[j, m + 1] = 0$, $1 \leq j \leq m$. In fact, to simplify we assume that additional white pixels $(m + 1, j)$ and $(j, m + 1)$ are added, $1 \leq j \leq m$. The top left corners of the squares in the MAT are marked by the value $T[i, j] = \text{true}$ while other pixels have $T[i, j] = \text{false}$; and $T[i, j] = \text{true}$ iff $\max(M[i, j - 1], M[i - 1, j], M[i - 1, j - 1]) \leq M[i, j]$. This assumes that $M[0, j] = M[j, 0] = 0$, $1 \leq j \leq m$. The sequential and parallel algorithms in Refs. [16] and [26] have more definitions and computations which we do not need to introduce for solving the MAT problem on a BSR model. Thus, we have two steps in our algorithm:

Step 1   Compute $M[i, j]$
Step 2   Compute $T[i, j]$

Here we use a different approach to compute $M[i, j]$ on a three criteria BSR model. We use the L_∞ metric, defined as $L_\infty((x_1, y_1), (x_2, y_2)) = \max(|x_1 - x_2|, |y_1 - y_2|)$. Let us observe that all pixels which are to the right and below pixel (i, j) and are at L_∞ distance less than or equal to $M[i, j]$ must be black (otherwise the square of size $M[i, j]$ with top left corner (i, j) contains a white pixel, contrary to the definition of $M[i, j]$). The BSR algorithm in Ref. [30] is based on finding the white pixel to the right and below that is closest to the pixel (i, j), and its distance to (i, j) will determine $M[i, j]$.

Suppose that a white pixel (k, l) is the closest pixel to the pixel (i, j) among pixels to the right of and below pixel (i, j). Then $k \geq i$ (to the right), $l \geq j$ (below) and the distance is $M[i, j] + 1 = \max(k - i, l - j)$. The latter distance is $k - i$ if $k - i \geq l - j$; i.e., $k - l \geq i - j$, and $l - j$ otherwise (see Fig. 12). On a three criteria BSR, $M[i, j]$ and $T[i, j]$ can be found as follows (the ranges of k and l are between 1 and $m + 1$):

```
if I_kl = 0 then {k'←k; l'←l; k1←k - l} else {k'←n+1; l'←m + 1; k1←3m}
ij←i - j
c'ij←∪ k' | k ≥ i ∧ l ≥ j ∧ k1 ≥ ij
c"ij←∪ l' | k ≥ i ∧ l ≥ j ∧ k1 ≤ ij
M[i, j]←∪ (c'ij - i, c"ij - j) - 1
T[i,j] ← true if and only iff max (M[i,j-1], M[i-1j,], M[i-1,j-1]) ≤ M[i,j]
```

7.6   Multiple Criteria BSR Algorithms

We conclude our presentation of BSR algorithms by describing constant time solutions to problems where broadcast data must pass an arbitrary number of tests.

7.6.1   Maximal elements, ECDF searching and 2-set dominance

Given a set $S = \{p_1, p_2, \ldots, p_n\}$ of n points in d-dimensional space, a point p_i dominates a point p_j iff $p_i[k] \geq p_j[k]$ for $k = 1, 2, \ldots, d$, where $p[k]$ denotes the kth coordinate of a point p.

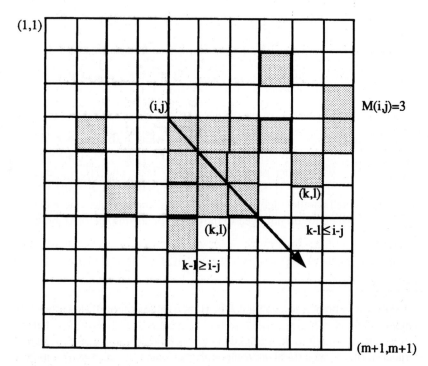

Figure 7.12 Computation of the medial axis transform

The maximal elements problem is to determine those points of a set which are dominated by no other point. The d dimensional maximal elements problem can be solved on a d-1 criteria BSR as follows.

$$f_j := \cap p_i[d] \mid p_i[1] \geq p_j[1] \wedge p_i[2] \geq p_j[2] \wedge \ldots \wedge p_i[d-1] \geq p_j[d-1]$$

The jth point is maximal iff $p_j[d] \geq f_j$.

Next, ECDF searching in d dimensions can be solved on a d criteria BSR as follows:

$$D(p_j, S) := \Sigma\ 1 \mid p_i[1] \leq p_j[1] \wedge p_i[2] \leq p_j[2] \wedge \ldots \wedge p_i[d] \leq p_j[d]$$

Finally, 2-set dominance counting in d-dimensions can be solved on a d criteria BSR as follows:

$$D(p_j, A) := \Sigma\ w_i \mid p_i[1] \leq p_j[1] \wedge p_i[2] \leq p_j[2] \wedge \ldots \wedge p_i[d] \leq p_j[d]$$

where $w_i = 1$ if the ith point belongs to A, and $w_i = 0$ otherwise.

7.6.2   Rectangle containment in d-dimensional space

Given a set A of m points in R^d and a set R of r d-dimensional rectangles whose sides are parallel to the coordinate axes (also called isothetic rectangles), the rectangle containment problem is to determine, for each rectangle, the number of points of A that lie inside R.

Let B be the set of all vertices of R. Apply the 2-set dominance counting procedure in d dimensions. The number of points of A that are inside a given rectangle with vertices a_i, $i = 0, 1, \ldots, 2^d - 1$ [where indices are arranged as in a d-dimensional hypercube, with indices 0 and $2^d - 1$ corresponding to two points with minimal (maximal, respectively) coordinates in all dimensions] is equal to the sum $\Sigma \pm D(a_i, A)$ where the sign + or − for each member is determined as $(-1)^{d+t}$, with t being the sum of the corresponding cube indices. For $d = 2$, the formula is well known (see Ref. [25]) and reduces to $D(a, A) + D(c, A) - D(b, A) - D(e, A)$, where a, b, c, and e are vertices of a rectangle. (Vertices a and c are the bottom left and upper right corners.)

Consider an example for $d = 3$. Let the corners of the cube be $a_{000}, a_{001}, a_{010}, a_{011}, a_{100}, a_{101}, a_{110}, a_{111}$ (or a_0 to a_7). Vertex a_{000} is the one with all minimal coordinates, and a_{111} is the vertex with all maximal coordinates. The number of points inside the cube is $D(a_{111}, A) + D(a_{100}, A) + D(a_{010}, A) + D(a_{001}, A) - D(a_{110}, A) - D(a_{101}, A) - D(a_{011}, A) - D(a_{000}, A)$.

The formula can be proved as follows [9]. Suppose that the vertices of the given rectangle have coordinates $(x_1, x_2, \ldots, x_d)$ where each $x_i = \min_i$ or $x_i = \max_i$, with $\min_i < \max_i$. Each point from A has coordinates $a = (a_1, a_2, \ldots, a_d)$. If $a_i > \max_i$ for any i then no rectangle vertex dominates a. The formula then gives all 0s in the sum, which is the correct value. Suppose that this is not the case. Then either $a_i < \min_i$ or $\min_i \leq a_i \leq \max_i$ for any i. Without loss of generality, suppose the former case happens for $i = 1, 2, \ldots, q$ and the latter happens for $i = q + 1, \ldots, d$. Then the vertices of the rectangle that dominate the point are exactly $(x_1, x_2, \ldots, x_q, \max_{q+1}, \ldots, \max_d)$, i.e., 2^q vertices. According to the formula, if $q > 0$, then 2^{q-1} summands have sign + and 2^{q-1} summands have sign − in the sum; thus the point is counted a total of zero times in the sum, which is the correct value since the point is outside the rectangle. If $q = 0$, then the point is inside the rectangle, and it will be counted once only, corresponding to points dominated by vertex $(\max_1, \max_2, \ldots, \max_d)$. Therefore, a d criteria BSR with n processors solves the rectangle containment problem in d dimensions in constant time.

7.6.3   Rectangle enclosure counting in R^d

Given a set S of n isothetic rectangles in R^d, the rectangle enclosure problem is to find for each rectangle the number of rectangles enclosed by it. Let $R_i = [\min_i[1], \min_i[2],\dots,\min_i[d]] \times [\max_i[1], \max_i[2],\dots,\max_i[d]]$ be a given rectangle, i.e., the set of all points $x = (x_1,x_2,\dots,x_d)$ in R^d such that $\min_i[j] \le x_i[j] \le \max_i[j]$ for all $j = 1,2,\dots,d$.

It is well known that rectangle R_p encloses rectangle R_q if the following conditions are satisfied: $\min_p[j] \le \min_q[j]$ and $\max_p[j] \ge \max_q[j]$ for each $j = 1,2,\dots,d$, which is equivalent to saying that the point $(\max_p[1], \max_p[2],\dots,\max_p[d], -\min_p[1], -\min_p[2],\dots, -\min_p[d])$ dominates the point $(\max_q[1], \max_q[2],\dots, \max_q[d], -\min_q[1], -\min_q[2],\dots, -\min_q[d])$ in R^{2d}. The correspondence is found in Ref. [20] for arbitrary d; it is used in Ref. [18] to solve the problem in optimal time on a mesh connected computer for arbitrary d. Therefore, the problem can be solved by applying 2-set dominance counting in 2^d dimensional space. The problem can thus be solved on a 2^d criteria BSR in constant time [9].

7.6.4   Rectangle intersection counting in R^d

Given a set S of n isothetic rectangles in R^d, the rectangle intersection counting problem is to find for each rectangle the number of rectangles intersected by it.

To solve the problem, the relationship established in Ref. [20] between this problem and 2-set dominance counting in R^{2d} is used in Ref. [9]. Map each rectangle R_i into two points $r'i = (-\min_i[1], -\min_i[2],\dots, -\min_i[d], \max_i[1], \max_i[2],\dots, \max_i[d])$ and $r''i = (-\max_i[1], -\max_i[2],\dots,-\max_i[d], \min_i[1], \min_i[2],\dots, \min_i[d])$ in R^{2d}. Two rectangles R_p and R_q in R^d intersect iff r'_p dominates r''_q in R^{2d}. Therefore, the problem can be solved by applying 2-set dominance counting in $2d$ dimensional space. The same relationship was used in Ref. [18] to solve the problem in optimal time on a mesh connected computer for arbitrary d.

7.7   Conclusions and Future Work

Broadcasting with selective reduction is a model of parallel computation combining simplicity, power, and feasibility. These seemingly contradictory properties are achieved by exploiting the simultaneous access by processors to a shared memory, whether this memory is central or distributed. In particular, the model takes full advantage of the network connecting processors to memory locations: this network is asked to perform a number of important computations on data while routing them from the processors to their destinations in memory. As a result, BSR allows the derivation of algorithms that are mathematically elegant and computationally efficient for a host of important problems.

The feature that distinguishes BSR from other models is the so-called BROADCAST instruction which is executed in three constant-time phases:

1. the *broadcast* phase, where each processor broadcasts a datum
2. the *selection* phase, where a subset of the broadcast data is selected for each memory location if they satisfy a certain number of criteria
3. the *reduction* phase, where the subset of data, selected for each memory location, is reduced to one value and stored in that location

One direction for future research would be to identify other areas in which the multiple criteria BSR implementation presented here might be applied, and exhibit algorithms for problems in these areas. Graph theory and numerical analysis are examples of such areas that are yet to be explored.

Another problem that has so far eluded an answer concerns the existence of a multiple criteria n processor BSR implementation that requires a number of switches asymptotically smaller than n^2. In this regard, the implementation described in Section 7.2.2, like the one criterion implementations of Refs. [2] and [3], uses $O(n^2)$ switches. On the other hand, the "optimal" implementation of a one criterion n processor BSR presented in Refs. [3] and [21] requires $O(n \log n)$ switches. Note, however, that the latter implementation is based on sorting circuits and is therefore restricted to solving problems whose inputs obey a linear order.

Finally, two questions, originally posed in Ref. [2], still remain open. The first seeks a general characterization of problems for which BSR provides a fast solution. This would allow a classification of problems according to whether they are amenable to constant time solution by BSR. The second question calls for relaxing the constant time requirement imposed (implicitly) on all BSR algorithms studied to date. An investigation of this possibility may reveal problems for which a BSR algorithm that does not run in constant time is nevertheless faster than all previously known CRCW PRAM solutions.

7.8 References

1. Akl, S. G. 1989. *The Design and Analysis of Parallel Algorithms.* Englewood Cliffs, N.J.: Prentice Hall.
2. Akl, S .G., and G. R. Guenther. 1989. Broadcasting with selective reduction, *Proceedings of the 11th IFIP Congress,* San Francisco, 515–520.
3. Akl, S. G., L. Fava Lindon, and G.R. Guenther. 1991. Broadcasting with selective reduction on an optimal PRAM circuit. *Technique et Science Informatiques,* 10, 4, 261–268.
4. Akl, S. G. 1991. *Memory access in models of parallel computation: From folklore to synergy and beyond, in: Algorithms and Data Structures.* F. Dehne, J.-R. Sack, and S. Santoro, eds. Berlin: Springer-Verlag, 92 – 104.
5. Akl, S. G., and G. R. Guenther. 1991. Application of BSR to the maximal sum subsegment problem. *International J. of High Speed Computing,* 3, 2, 107–119.
6. Akl, S. G., and I. Stojmenović. 1992. Generating binary trees in parallel. *Proceedings of the Allerton Conference on Communication, Control and Computing,* Monticello, Illinois, 225–233.
7. Akl, S. G., and K. A. Lyons. 1993. *Parallel Computational Geometry.* Englewood Cliffs, N.J.: Prentice Hall.
8. S. G. Akl, Qiu K., and Stojmenović, I. 1993. Fundamental algorithms for the star and pancake interconnection networks with applications to computational geometry. *Networks,* 23, 215–225.
9. Akl, S. G., and I. Stojmenović. 1994. Multiple criteria BSR: An implementation and applications to computational geometry problems. *Proceedings of the 27th Hawaii International Conference of System Sciences,* Maui, Hawaii, 1994, Vol. II, 159–168.
10. Anderson, A., and S. Carlsson. 1990. Construction of a tree from its traversals in optimal time and space. *Information Processing Letters* 34, 1, 21–25.
11. Bar-On, I., and U. Vishkin,. 1985. Optimal parallel generation of a computation tree form. *ACM Trans. on Programming Languages and Systems,* 7,2, 348–357.
12. Blelloch, G. E. 1989. Scans as primitive parallel operations. *IEEE Trans. on Computers,* 38, 11, 1526–1608.
13. Blum, H. 1967. *Models for Perception of Speech and Visual Form.* Cambridge, Mass.: MIT Press, 362–380.
14. Burke, R. F. 1994. Implementations of broadcasting with selective reduction. M.Sc. thesis, Department of Computing and Information Science, Queen's University, Kingston, Ontario, Canada.
15. Chan, I. W. , and D. K. Friesen. 1991. An optimal parallel algorithm for the vertical segment visibility reporting problem. *ICCI '91,* Lecture Notes in Computer Science Vol. 497, Berlin: Springer-Verlag, 323–334.
16. Chandran, S., S. Kim, and D. Mount. 1992. Parallel computational geometry of rectangles. *Algorithmica,* 7, 25–49.
17. Cormen, T. H. , C.E. Leiserson, and R.L. Rivest. 1990. *Introduction to Algorithms.* Cambridge, Mass.: MIT Press.
18. Dehne, F., and I. Stojmenović. 1988. An $\sqrt{n}$ time algorithm for the ECDF searching problem for arbitrary dimensions on a mesh-of-processors. *Information Processing Letters,* 28, 67–70.

19. Djokić, B., Ruppert J., and Stojmenović, I. 1994. Constant time digital geometry algorithms on the scan model of parallel computation. *International J. of High Speed Computing,* 6,4,501–517.
20. Edelsbrunner H., and M. H. Overmars. 1982. On the equivalence of some rectangle problems. *Information Processing Letters,* 14, 3, 124–127.
21. L. Fava Lindon, and S. G. Akl. 1993. An optimal implementation of broadcasting with selective reduction. *IEEE Trans. on Parallel and Distributed Systems,* 4, 3, 256–269.
22. Gabrani, N., and P. Shankar. 1992. A note on the reconstruction of a binary tree from its traversals. *Information Processing Letters,* 42, 117–119.
23. Gewali, L. P., and Stojmenović. 1994. Computing external watchman routes on PRAM, BSR, and interconnection network models of parallel computation. *Parallel Processing Letters ,* 4, 1, 83–93.
24. Graf, T., and K. Hinrichs. 1993. Algorithms for proximity problems on colored point sets. *Proceedings of the 5th Canadian Conference on Computational Geometry,* Waterloo, Ontario, 420–425.
25. JáJá, J. 1992. *An Introduction to Parallel Algorithms.* Reading, Mass.: Addison-Wesley.
26. Jenq, J., and S. Sahni. 1992. Serial and parallel algorithms for the medial axis transform. *IEEE Trans. on Pattern Analysis and Machine Intelligence,* 14, 12, 1218–1224.
27. Klee, V. 1977. Can the measure of union of intervals be computed in less than O(n log n) steps? *American Mathematical Monthly,* 84,4, 284–285.
28. Lodi, E., and L. Pagli. 1986. A VLSI solution to the vertical segment visibility problem. *IEEE Transactions on Computers,* C-35, 10, 923–928.
29. Makinen, E. 1989. Constructing a binary tree from its traversals. *BIT 29,* 572–575.
30. Melter, R. A., and Stojmenović. 1993. Solving city block metric and digital geometry problems on the BSR model of parallel computation. *Proceedings of the Vision Geometry II,* SPIE Vol. 2060, Boston, 39–48; *Journal of Mathemancal Imaging and Vision,* 5 (1995), 119–127.
31. Olariu, S., J. L. Schwing, and J. Zhang. 1992. Optimal parallel encoding and decoding for trees. *International J. of Foundations of Computer Science,* 3, 1–10.
32. Preparata, F. P., and M.I. Shamos. 1985. *Computational Geometry, An introduction.* New York: Springer-Verlag.
33. Sarkar, D., and I. Stojmenović. 1989. An optimal parallel circle-cover algorithm. *Information Processing Letters,* 32, 3–6.
34. Schwarzkopf, O. 1991. Parallel computation of discrete transforms. *Algorithmica* 6, 685–697.
35. Shamos, M. I., and D. Hoey. 1976. Geometric intersection problems. *Proceedings of the 17th IEEE Symposium on Foundations of Computer Science,* Houston, Texas, 208–215.
36. Smith, J. R. 1993. *The Design and Analysis of Parallel Algorithms.* New York: Oxford University Press.
37. Springsteel, F., and I. Stojmenović. 1989. Parallel general prefix computations with geometric, algebraic, and other applications. *International Journal of Parallel Programming,* 18, 6, 485–503.
38. Stojmenović, I. 1988. Computational geometry on a mesh-connected computer. *Univ. u Novom Sadu Zb. Rad. Prirod.-Mat. Fak. Ser. Mat.,* Novi Sad, 18, 2, 127–136.
39. Stojmenović, I. 1988. Computational geometry on a hypercube. *Proceedings of the International Conference on Parallel Processing,* III, St. Charles, Illinois, 100 103.
40. Stojmenović, I. 1993. Constant time BSR solutions to parenthesis matching, tree decoding and tree reconstruction from its traversals. *TR-93-16, Computer Science,* University of Ottawa, Ottawa, Ontario.
41. Zaks, S. 1980. Lexicographic generation of ordered trees. *Theoretical Computer Science,* 10, 63–82.

8

Dataflow Models

R. Jagannathan

Dataflow has different meanings depending on the context in which it is used. Perhaps the most popular usage of dataflow (or data flow) is in software engineering where it refers to the flow of information between data processing entities. The earliest use of the word "dataflow" in the context of parallel computing dates back to 1966 when Karp and Miller [21] studied theoretical properties of a dataflow-like model of computation. The earliest descriptions of dataflow models of computation were by Adams in 1968 [1] and Rodriguez in 1969 [24]. Dataflow got its theoretical underpinnings when Kahn [20] in 1974 conjectured correctly that a dataflow network can be given mathematical meaning by thinking of it as a set of recursion equations and deriving its least fix point. At the same time, Dennis proposed the first architecture to embody the dataflow computing model [12, 15]. It is this model, or more accurately, this family of *computing models* that is the principal focus of this chapter. (What we will not discuss in this chapter is the considerable theoretical work since Kahn's seminal paper on nondeterministic dataflow computing models.)

A computing model is a method that describes how a program is to be evaluated. It is tied to a particular model of programming that has primacy. It is abstract in the sense that it assumes an idealized implementation. The most common computing model is the "von Neumann" control-flow computing model. This model assumes that a program is a series of addressable instructions, each of which either specifies an operation along with memory locations of the operands or specifies transfer of control to another instruction unconditionally or when some condition holds. The method for executing this program is to start at the first instruction of the program, execute it, and proceed to the next instruction unless the executed instruction requires transfer of control. In the latter case, instruction execution continues from the instruction to which control is transferred. For example, consider a program to evaluate the following expression:

```
c = if n = = 0 then a + b else a - b fi
```

When using a control-flow computing model, the program will be translated into a series of instructions starting with an instruction comparing n to 0 and either transferring control to an instruction adding a to b or to an instruction subtracting b from a and, in both cases, executing an instruction that stores the result in c. What a control-flow computing model essentially specifies is the next instruction to be executed, depending on what happened when executing the current instruction. A dataflow computing model, on the other hand, is not based on flow of control; instead, it is based on flow of data. Unlike the control-flow computing model, it assumes that a program is a data-dependency graph (or dataflow graph) whose nodes denote operations and whose edges denote dependencies between operations. The dataflow graph for the expression given above is shown in Fig. 8.1.

When this graph is evaluated using a dataflow computing model, it executes any operation denoted by a node as soon as its incoming edges have the necessary operands. In particular, if n is available, the operation $==$ can be applied to its operands n and constant 0. Similarly, if a and b are available, both of the operations $+$ and $-$ can be applied to a and b, even though only one of these results is needed. In fact, these two operations could be executed even before the comparison between n and 0 has been completed. The essential difference between control-flow and dataflow computing models is that in control-flow, program execution corresponds to instructions in motion operating on data at rest, whereas in dataflow, program execution corresponds to data in motion being processed by instructions (more accurately, operations) at rest.

8.1 Kinds of Dataflow

The dataflow computing model that we have described above can be more accurately characterized as data-driven dataflow, although the "data-driven" qualification is often implied. Data-driven dataflow was invented as an architectural abstraction, based on which parallel computers, known as *dataflow computers,* could be built. The model of programming, dataflow program graphs, was subsequently defined for which data-driven

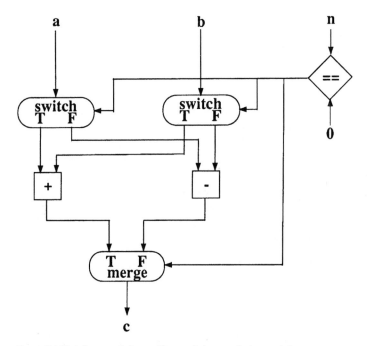

Figure 8.1 Dataflow graph for $c =$ if $n == 0$ then $a + b$ else $a - b$ fi

evaluation is an appropriate computing model. A less well-known branch of the family of dataflow computing models is *demand-driven* dataflow. It was invented as a computing model to correctly and efficiently evaluate programs in non-strict dataflow languages. To some, the phrases "demand-driven" and "dataflow" refer to diametrically opposite methods of evaluating dataflow program graphs. We adopt a broader view in this chapter—we think of dataflow not as a particular kind of computing model but as a general principle in which program execution is driven by flow of data instead of flow of control. Therefore, we discuss both data-driven and demand-driven dataflow. The chapter is organized as follows. We start off by discussing data-driven dataflow computing models and follow this with a discussion of demand-driven dataflow computing models. We then discuss two frameworks for unifying these two families of dataflow computing models. We conclude the chapter by summarizing the "state-of-art" in dataflow models and by considering its future trends.

8.2 Data-Driven Dataflow Computing Models

The model of programming for data-driven dataflow computing models is based on dataflow program graphs (or dataflow graphs for short). Although several variants of the graph language have been proposed, what we describe next is a common yet complete subset.

8.2.1 Dataflow graphs

A dataflow graph is a directed graph whose *nodes* (also known as *actors*) denote operations and whose *edges* (also known as *arcs*) denote dependencies between operations denoted by nodes. Values produced and consumed by nodes are carried in tokens which flow along the edges from tails of edges to their heads.

We saw earlier that it is natural to represent expressions as dataflow graphs. In addition to the usual complement of arithmetic, relational, and logical operation nodes, special nodes (known as *control nodes*) are required for the programming model to be able to express more than just expressions such as conditionals and loops.

The first control node is the *switch* node, which routes the token from the first input depending on the boolean-value of the token along the second input. It has two outputs, one for when the boolean is *true* and one for when the boolean is *false*.

The second control node is the *merge* node, which is used to select the token from one of its first two inputs, depending on the boolean value of the token on the third input. The selected token is reproduced on the output edge.

The third control node is the *apply* node, which is used to invoke user-defined functions. Tokens denoting "function graphs" flow on one of the edges, and the other input edges carry tokens that denote arguments to the function. The output edge of the *apply* node carries tokens that denote the results of function applications. We give an example dataflow program graph (Fig. 8.2) for the formula given below that illustrates the use of these control nodes.

$$s = \sum_{i=1}^{n} f(x_i)$$

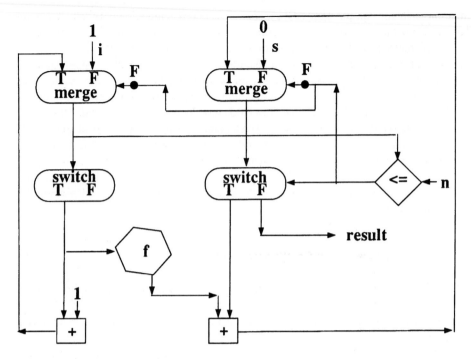

Figure 8.2 Example dataflow program graph

The kind of nodes that we have considered so far deal with tokens that denote simple values. What about complex data structures? While, in principle, we can think of tokens as denoting complex data structures, and we can define nodes that consume and produce data structures, this is practically unfeasible. Thus, there is no standard way of dealing with data structures that is prescribed by dataflow graphs. Instead, different data-driven computing models advocate their different approaches. We will discuss each approach when discussing the computing models.

8.2.2   Classic static model

The classic static dataflow computing model [12, 13, 15] is based on the following simple rule to decide when an operator is ready for execution (or executable):

> *The operator denoted by a node is executable when tokens (values) are present on each of the input edges.*

Execution of a node causes the tokens on the input edges to be removed and the output token to be produced on the output edge. The execution rule has two caveats:

1. Each edge can hold only one token at a time. This means that an executable operator can actually execute only when its output edge has no token on it. This is handled by using acknowledgment signals on a per-node basis.
2. For certain operators, the operator can execute even when some input tokens are missing. For example, with the *merge* operator, if the boolean input token denotes *true*, only the token on the first value-edge is needed for *merge* to execute. The token on the second value-edge can be discarded when it arrives.

The classic static computing model can exploit structural parallelism (as in different un-related operators executing at the same time) and pipeline parallelism (as in different parts of the graph consuming different tokens of a stream of tokens at the same time).

Because each edge can hold only one token at a time, only one iteration or one function invocation can be active. Thus, it cannot exploit dynamic forms of parallelism such as loop parallelism (from simultaneously executing different unrelated iterations of a loop body) or recursive parallelism (from simultaneously evaluating multiple recursive function calls). It also cannot deal with intermittent streams, i.e., streams in which some of tokens are undefined. This is because the model provides no way for an operator to "look past" the current set of tokens on its input edges. The model is thus well suited for applications with regular numerical computational structures such as signal processing and image processing applications [14] that do not make heavy use of iterative or recursive program structures.

8.2.2.1 Data structures as DAGs. We now turn to how the static dataflow model deals with complex data structures. Each data structure is represented as a finite *directed acyclic graph* (DAG). For example, an N by N array can be represented as a two-level tree where each interior vertex has N children and the leaf vertices store the values of the array. Associated with each vertex is a reference count that indicates how many other vertices (and tokens) point to the vertex.

Copying a data structure, e.g., an array, is straightforward using a COPY node which produces a token that points to the same root vertex of the DAG as the input token. The reference count of the root vertex is increased by one to reflect the additional pointer to it. Consider what happens when an element of the array represented by a DAG is modified by a node. The root vertex is duplicated, and the reference counts are adjusted down by 1. In addition, each of the vertices in the path from the root vertex to the new leaf vertex (with the modified value) is duplicated.

The main problem with this approach is that, while excessive copying is avoided by reference-count-based sharing, the DAG corresponding to a data structure can only be modified one at a time. Thus, for example, modifying two independent elements of an array cannot occur concurrently but, rather, has to occur sequentially.

8.2.2.2 Evolution of the classic static model. The classic static model has evolved in two different directions—one dictated by concern over architectural inefficiency, and the other by concern over the model's inability to exploit dynamic parallelism.

The main source of inefficiency in the classic static architecture is due to the use of tokens not just as data carriers but as implicit signal carriers [14]. This has been addressed by refining the classic model in the following manner. The edges of a graph no longer represent flow of actual data in tokens; instead, they are simply paths for signals that indicate availability of data values. When a node has all its signals present, it causes the corresponding instruction to execute in a conventional sense—by retrieving operands from memory and storing the result in memory.

The static model's inability to exploit dynamic parallelism has been dealt with in two simple ways. One has been to pipeline data streams through a graph for a loop body, which is effective when the graph is sufficiently deep. The other has been to replicate graphs (in-line substitution) to effectively deal with small loop bodies and nonrecursive functions.

8.2.3   Classic dynamic model

While there are several variants of the dynamic dataflow computing model, we describe the classic model. (In doing so, we use much of the constructs and notation used by Arvind and Gostelow [5, 6].) The model assumes that associated with each token is a "tag" that uniquely identifies the conceptual position of the token in the stream of tokens that flow on a given edge. Recall that in the static dataflow computing model, the conceptual position of a token is implicit, since it corresponds to its physical position with respect to other tokens of the stream.

The tag associated with each token is a four-tuple $<c, i, b, a>$ where c is the invocation ID, i is the iteration ID, b is the code block address, and a is the instruction address within the code block. The invocation ID distinguishes between tokens of different function invocations, which are alike in other respects. The iteration ID distinguishes tokens belonging to different iterations, which are alike in other respects. The code block address along with the instruction address of a token identify its destination.

The execution rule for the basic dynamic dataflow computing model is as follows:

> *The operator associated with a node is executable when each of the input edges contains tokens whose tags are identical.*

Each edge can contain several tagged tokens at the same time, and there is no bound on the number of such tokens that can exist at any given time. When a node is executed, the matching tokens are removed from the input edges, and a token with the appropriate tag is produced on the output edge.

In this model, three pairs of special nodes, D and D^{-1}, L and L^{-1}, and A and A^{-1} are used for tag manipulation. Node of type D increments the iteration ID, whereas node of type D^{-1} resets the iteration ID. Node of type L creates a new invocation ID (when entering an inner loop), whereas a node of type L^{-1} retrieves the previous invocation ID (when leaving an inner loop). Node of type A creates a new invocation ID (when entering a function) whereas node of type A^{-1} retrieves the previous invocation ID (when "returning" from a function invocation).

This computing model is called "dynamic" because tagging enables both loop parallelism and recursive parallelism that arise (dynamically) at run time to be exploited. The price for exploiting such parallelism is that any architecture embodying the model must support the "matching function"—a token that has just been generated can find another token that already has been produced such that the two tokens have the same tag. Such matching, for which an associative store is needed, has proven to be the Achilles' heel of the implementations.

8.2.3.1 Incremental data structures.
Let us consider how the dynamic dataflow model deals with manipulating complex data structures. The most common approach is to use what are known as I-structures (for incremental structures) [7]. The basic idea is to associate with each atomic unit of a data structure with status bits and a queue of deferred reads. The status of the atomic unit can be one of PRESENT, which means that it can be read but not written, ABSENT, which means that read has to be deferred but it can be written, and WAITING, which means that at least one read has been deferred and nothing has been written. When an atomic unit of the data structure is defined (which it can be precisely once), all deferred reads are immediately satisfied. Thus, it is possible to use a data structure before it is fully defined, and it is not necessary for a data structure to be fully defined

before a subsequent data structure is incrementally defined to deal with a modification to the original.

8.2.3.2 Evolution of the classic dynamic model.

The main problem with the classic dynamic dataflow model is that it results in excessive exploitation of parallelism. In most implementations of this model, it leads to severe resource shortages, resulting in deadlock [4, 11, 25] or processor underutilization [26]. Four approaches have been proposed to sensibly constrain program-inherent parallelism to facilitate its effective execution. We discuss these next.

Loop bounding. Loop bounding is a technique proposed by Culler and Arvind to deal with unbridled unraveling of loops resulting in unmanageable amounts of parallelism. The basic idea is to bound the number of concurrent iterations of a loop to some run-time-determined constant k. This is accomplished by using "triggers" such that an iteration can proceed only if it is triggered. (Initially, there are k triggers.) A termination tree determines when an iteration has been completed and supplies a new trigger to start another iteration.

Loop bounding is essentially a program transformation technique to constrain exploitation of parallelism in loop schemas assuming a dynamic dataflow model. It appears unlikely that the technique is applicable to programs with general recursion. The technique can also be applied to deal with other resource management issues such as reclamation and reuse of I-structure storage and recycling of tags to reasonably limit tag size.

From a computing model standpoint, loop bounding can be seen as a way of limiting the number of tokens that can be present on an edge. Recall that the static model limited it to one, whereas the dynamic model did not impose any limit. By expressing loop bounding through program augmentation, additional implementation complexity (such as acknowledgment signals to maintain bounded edges) is avoided.

Coarse-grain throttling. Coarse-grain throttling is a technique proposed by Ruggeiro and Sargeant for constraining excessive exploitation of parallelism in the Manchester Dataflow Machine. The basic idea is to defer execution of processes (where each process denotes a loop body or a recursive function call) when resources of the underlying machine are in short supply and only allow nodes to execute one at a time until resources are more abundant. The choice of the process to activate when resources are in short supply is determined by the order of processes in a depth-first left-to-right traversal of the process execution tree constructed at run time.

For example, the process execution tree for a doubly nested parallel loop consists of a three-level tree with the second level denoting iterations of the outer loop and the third level denoting iterations of the inner loop per outer loop iteration. The choice of processes would be as if the loop were executed sequentially. Similarly, the process execution tree of a recursive function consists of a vertex for each function invocation with independent invocations being represented as sibling vertices. Again, the choice of processes would be as if the function were invoked sequentially.

From a computing model standpoint, two modifications to the dynamic model are being suggested. The first is to "coarsify" the dataflow graph into processes that are really macronodes denoting a loop body or invocation of a non-trivial recursive function. The second is to use breadth-first execution as implied by data-driven execution when machine resources are available, and depth-first execution when machine resources are scarce. The first modification enables the throttling to be more efficient as it is applied at a coarse level, and the second modification enables computational progress to be made even if resources are scarce without getting into deadlock.

Strongly-connected arc model. The strongly connected arc model forms the basis for the EM-4 dataflow machine under construction at Electrotechnical Laboratory in Japan [26]. It is an extension to the classic dataflow model, which is principally motivated by the need to reduce the overhead associated with firing each operation and by the need to reduce the number of tokens that have to be "matched."

In this model, the edges of a dataflow graph are classified as either a *normal* edge or a *strongly connected* edge. The set of nodes that are connected by strongly connected edges is called a *strongly connected block*. The standard firing rule is that a node is executable when all input edges have matching tokens. The additional firing rule is that a strongly connected block is executable if its "source" nodes are firable and the execution of nodes within the block are conducted as a unit, without requiring the standard firing rule for each node execution. The model uses the standard dataflow firing rule to invoke strict functions but uses an adaptation of demand-driven execution to invoke non-strict functions.

Multithreaded models. One of the key realizations made by the data-driven dataflow community is that pure dataflow and pure control-flow are not orthogonal to each other but are at two ends of a continuum. Iannucci [17] proposes a "multithreaded" model of computation for executing dataflow graphs. The unusual feature of this model is the absence of a data-driven firing rule; instead, each thread (known as scheduling quantum) can decide which other thread to pass control to.

Papadopoulos and Traub [22] suggest a revisionist view of dataflow—replacing consumption of tokens and firing of enabled operations with a dynamic set of multiple, interacting sequential threads, where thread creation and destruction is very efficient. The model tries to improve the efficiency of computation (through use of sequential threads) while being able to perform very rapid synchronization (as in dataflow).

Subcompact process model. Bic [10] proposes a model that is similar in spirit to multithreaded models except that scheduling is done using the standard data-driven principle. In this model, a dataflow graph is divided into "sequential code segments" (SCS) which are conceptually similar to Iannucci's scheduling quantum. Each SCS executes sequentially and multiple SCSs can execute simultaneously. An SCS is active if the first operation has all its inputs are available, it is blocked when a subsequent operation has some inputs missing, it is ready when an operation of the SCS has all input available, and it is running when it is actually executing the operation. Care must be taken that SCSs do not have too small a granularity. When this is the case, multiple SCSs can be coalesced at the expense of parallelism. Care must also be taken to deal with conditionals that manifest as *switch* and *merge* operators. The model can also be extended to deal with loops by having multiple copies of the loop body (as an SCS) and reusing these copies for different iterations. The effect is the same as loop bounding.

8.3 Demand-Driven Dataflow Computing Models

We now turn our attention to dataflow computing models that are based on demand-driven execution. The basic idea behind demand-driven execution is that an operation of a node will be performed only if there are tokens on all the input edges and there is a demand for the result of applying the operation. The main motivation for using demand-driven execution is to deal efficiently with non-strictness, something that data-driven execution cannot efficiently deal with in general.

There are two programming models that demand-driven execution is used to evaluate: operator nets [9], which are quite similar to dataflow graphs, and functional programs [2], which are based on lambda calculus. We describe operator nets and models for their demand-driven execution first and then briefly consider demand-driven evaluation of functional programs.

8.3.1   Operator nets

The language of operator nets is a simple, graphical programming model whose programs by themselves have no meaning but can be given mathematical semantics under a particular data algebra and particular sets of inputs. Although operator nets appear similar to dataflow graphs, there are significant differences between the two programming models. The first difference is that operator nets do not make any assumptions about how they will be evaluated. For example, it is possible for them to be either data-driven or demand-driven. They can be given denotational semantics, which has primacy over operationally-derived meanings using different models of computation. The second difference is that an operators net always denotes a function—an operator net denotes a function that is a composition of other functions corresponding to subnets.

The syntax of operator nets is as follows. An operator net is a "main" network of vertices, together with zero or more "subsidiary" networks (that are used as definitions of functions). Each network is just a directed graph that has zero or more input vertices and one output vertex. These form interfaces with the "outside world" (i.e., with the user) in the case of the main network, and with the "calling environment" in the case of function networks. Input vertices have no incoming edges and one outgoing edge, whereas output vertices have one incoming edge and no outgoing edges.

In addition to input and output vertices, there are two other types of vertices: split vertices and non-split vertices. A split vertex has exactly one incoming edge and two or more outgoing edges. A non-split vertex is labeled with an operator symbol, or a user-defined function name, or a subcomputation symbol. The operator symbols that are used to label vertices are defined in the underlying data algebra, whereas the user-defined function names labeling vertices are associated with the subsidiary networks (that define the functions), mentioned earlier. These non-split vertices are called *operator vertices* and *function vertices,* respectively. They have one or more incoming edges and exactly one outgoing edge. The number of incoming edges for an operator vertex must match the "arity" of the corresponding operator, and the number of incoming edges for a function vertex must match the number of input vertices of the subsidiary network that is the function's definition.[*] The vertices can be connected together in any way, provided that these conventions about numbers of incoming and outgoing edges are adhered to. Each subcomputation symbol labeling a vertex can occur only once in the net. Corresponding to a subcomputation symbol is a subsidiary structured net such that the number of incoming edges of the vertex matches the number of input vertices of the subsidiary net. Figure 8.3 gives an operator net to evaluate the formula mentioned above.

---

[*]Thus, all vertices labeled with a particular user-defined function name must have the same number of incoming edges.

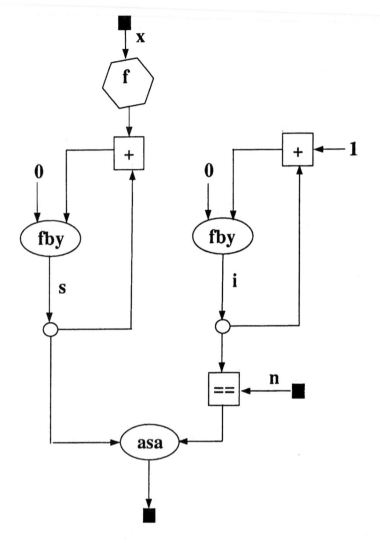

Figure 8.3 Example operator net

In operational terms, an edge of an operator net denotes a sequence of "datons," where a daton is a value-holder that can be uniquely identified among all datons on that edge. The identity of a daton is referred to as its "tag," which consists of the name of the edge and the conceptual position of the daton in the sequence denoted by that edge. Each vertex of an operator net denotes a function that is applied on datons from the incident incoming edges, producing datons on the incident outgoing edge.

We think of a computation to be the execution of an operator net given a set of sequences of input daton values and a sequence of needs for values of output datons of the net. We view a computing model as evaluating a net in steps. At the initial step, none of the values of datons except possibly the values of input datons are defined. At each subsequent step, some of the undefined datons are computed and eventually, the values of the needed output datons are computed.

Next, we consider two demand-driven computing models for evaluating operator nets.

8.3.2 Eduction model

The word *eduction* is defined in the *Oxford English Dictionary* as follows:

> The action of drawing forth, eliciting, or developing from a state of latent, rudimentary, or potential existence; the action of educing (principles, results of calculations) from the data.

The basic idea behind eduction is as follows.

> *An operator denoted by a node can be evaluated only when a particular token on its output edge is demanded and the necessary tokens to produce the demanded output token are available on the input edges.*

Tokens are produced on an edge only after they have been explicitly demanded. Therefore, when an output token is demanded, it in turn causes the necessary input tokens to be demanded, and these demands propagate further until they are satisfied. There are two situations in which a demand for a token can be satisfied:

1. when the token is demanded from a node which is either an input node or a constant node
2. when the token is demanded from a split node, and the value of the token has already been produced as a result of an earlier demand

One of the commonly stated disadvantages of demand-driven execution is that propagation of demands must be performed before any tokens are actually produced. Therefore, on the surface, it would appear that it takes twice as long to evaluate a program using demand-driven execution than with data-driven execution. There are two ways in which this inefficiency can be addressed:

1. It is possible to pipeline multiple demands so that demand propagation for different outputs overlap. This causes demanded outputs to be produced at the same rate as when using data-driven execution, but the time taken to produce a given token is twice as long as with data-driven execution [8].
2. It is possible to eliminate demands for tokens produced by strict operations by executing them in a data-driven way. Demands are needed only when tokens are produced by non-strict operations or split nodes [18].

It is worth reiterating that data-driven execution is difficult to use to evaluate certain non-strict operations, whereas demand-driven execution can be used naturally. For example, consider the non-strict operation *whenever*, which produces those tokens on its first input edge for which the associated token on the second input edge is *true*. Thus, when the *i*th token is demanded, the tag of the token to be demanded from the first input edge is determined by the number of tokens that are *true* on the second input edge. Although data-driven execution can select the appropriate tokens from the first input edge, it would be cumbersome to determine their tags.

The crucial advantage of eduction is that none of the tokens produced are superfluous—unlike the classic dynamic data-driven model. Of course, resources could still be scarce in an eductive implementation, but the nature of demand propagation helps distinguish tokens that are more critical from tokens that are not. The main issue that needs to be addressed with eduction is when to discard values that are stored in a split node. Although, in some cases, it is possible to precisely predict at compile-time how many times a value will be needed, in many other cases it is simply not possible. Therefore, it is either necessary to adjust the usage count at run time or use a heuristic for "retiring" values that are unlikely to be demanded again.

8.3.2.1 Intensional Data Structures. One of the unusual and powerful aspects of the eductive view of computing is that all data structures are distributed. That is, just like a stream is viewed in terms of its elements and their position in the stream, a 2D array is viewed in terms of elements of a stream that varies in two orthogonal spatial dimensions. Similarly, a tree is viewed in terms of elements of a stream that varies in a "branching time" dimension. Thus, each element not only has a value associated with it, but it completely identifies its position not only temporally but spatially. The way operator nets manipulate "data structures" or objects that vary in spatial dimensions is simply by augmenting each operator with a suffix that indicates which dimension it is to be applied to. Such data structures are referred to as *intensional data structures* [8a].

8.3.3   Eazyflow model

The eazyflow model [18] is a hybrid of eduction and eager demand-driven evaluation. The motivation for eazyflow is to produce a bounded number of tokens before they are even demanded, thereby speeding up operator net execution.

The eazyflow model assumes that edges of an operator net are classified as *eager* or *lazy*. An eager edge has the property that each token depends on its predecessor, and it takes a bounded amount of computation to produce each token. All edges that are not eager are classified as lazy. When a token of a lazy edge is demanded, it is produced exactly as in the eduction model. When a token of an eager edge is demanded, a fixed number of its immediate successor tokens are also demanded in anticipation of their eventual use. Thus, if these tokens were to be demanded in the future, they are likely to have been already computed. If these tokens were not to be demanded, then only a bounded number of useless tokens of that edge would have been produced.

It can be shown that, although eazyflow produces a few useless tokens, for the most part only useful tokens are produced. It can also be shown that, for a given operator net, the eazyflow model is at least as "fast" as eduction. That is, with eazyflow, demands for output tokens result in tokens being produced no later than when using eduction.

8.3.4   Demand-driven execution of functional programs

The basic idea is to use a data-driven architecture to implement reduction, which is an abstract computational model for evaluating functional programs [2]. It is suggested that demand-driven execution can be used to naturally realize lazy outermost evaluation of a function. It is pointed out that demand-driven execution is expensive to implement and is slower than data-driven execution.

8.4   Unifying Data-Driven and Demand-Driven

A contentious question is whether data-driven execution can subsume demand-driven execution, or whether demand-driven execution can subsume data-driven execution. The data-driven camp claim that data-driven subsumes demand-driven by showing that demand driven can be efficiently "simulated" on a data-driven machine. The demand-driven camp claim that demand-driven subsumes data-driven by arguing that data-driven is simply demand-driven with implicit always-present demands. We present the two claims next.

8.4.1   Data-driven subsumes demand-driven

It is suggested that demand-driven execution of a dataflow graph can be encoded into a transformed graph which, when data driven, produces the same results that would have been produced if the original program graph were to be demand driven [23]. Thus, it would be possible to "simulate" demand-driven execution on a data-driven machine. The basic idea is to explicitly introduce a demand-token to model demand for each token on an edge and to explicitly augment the original graph with "demand-propagation" operators to mimic demand propagation. Although the transformation is given for a particular initial language, which for example does not permit recursive functions or loop unfolding, it is claimed that the technique can be used for more expressive languages.

8.4.2   Demand-driven subsumes data-driven

It is suggested that data-driven execution is simply all "optimized" version of demand-driven execution [18]. Therefore, it is possible to simulate any of the data-driven models using demand-driven execution, whereas the reverse is not possible without transforming the program itself. Each dataflow computing model simply consists of demanding specific tokens at specific stages of the computation and applying the operators when the necessary tokens are available.

The following is a description of the classic static dataflow computing model (with a bound of one token per edge) in terms of demand-driven execution:

- At the initial stage, demand the initial tokens of all edges.
- At each successive stage, demand a token on an edge only if the previous token, if any, has been consumed.

The crucial point is that, although the classic static dataflow model can be thought of in terms of explicit demands, the demands are implicit, making it appear data driven.

The following is a description of the classic dynamic dataflow model in terms of demand-driven execution:

- At the initial stage, demand all tokens on all edges.

If one were to think of demands as implicit, the effect is exactly that of dynamic data-driven dataflow—tokens are produced as soon as possible, regardless of whether they are needed.

The framework further suggests using data-driven execution when dealing with strict operators by having demands "leapfrog" over such operators and using demand-driven execution with all other operators. One way to think of dataflow program graphs is as consisting only of strict operators except for conditionals, which are dealt using the special control nodes, *merge* and *switch*. Operator nets, on the other hand, allow both strict and non-strict operators. They do not use control nodes, because they do not assume that operators will only be data driven.

It is worth noting that demand-driven execution is used in a restricted sense in data-driven architectures—I-structures in the Monsoon machine [3] and non-strict function invocation in the EM-4 [26].

8.5   Lessons Learned and Future Trends

The static data-driven computing models have evolved on account of the following observed considerations:

- Keeping data and control together (as was done in the original model using "activity templates") causes the data values to have extra movements that can be avoided by separating data and control.
- Although controversial, the significant overhead of "fine-grain" scheduling appears to dampen efficient exploitation of parallelism. This has led to allowing coarser grains of sequentially executed instructions (or threads), which works well when control and data are separated [16].
- The model is adequate for exploiting parallelism in regular numeric computations, but it is less than adequate for general-purpose scientific computing.

The dynamic data-driven dataflow computing models have also evolved. This evolution has been significantly shaped by the following considerations observed by various research groups:

- Matching of tokens is a time-consuming activity that invariably is the bottleneck in the circular dataflow pipeline. Therefore, it is necessary to use the matching store at a "coarser" level.
- Dynamic data-driven execution not only exploits maximal useful parallelism but also significant useless parallelism. Furthermore, it is quite difficult to separate useful parallelism from useless parallelism. Therefore, it is important to constrain parallelism so that machine resources are uniformly utilized over time.
- Pure data-driven execution is unsuited for manipulating complex data structures or invoking non-strict functions. It is necessary in such cases to use demand-driven approaches.

The evolution of demand-driven computing models has been fueled by pragmatic considerations.

- Demand-driven execution is at least as hard to implement in hardware as data-driven execution. A major issue is the management of the store that retains values that are used multiple times.
- The apparent inefficiency of demand propagation can be effectively obviated by raising the granularity of parallelism [19] or by speculative computations [18].
- Demand-driven execution may be replaced by data-driven execution when possible and when useful, but the former is a much more general execution mechanism.

Although we can only speculate on future trends of dataflow computing models, it is safe to say that the days of "pure" dataflow, at least at the implementation level, are behind us. The main trend appears to be toward some form of a hybrid between data-driven or demand-driven and control-driven models. The data-driven/control-driven hybrid has given rise to a new class of architectures known as *multithreaded* architectures, while the demand driven/control-driven hybrid will enable effective coarse-grain programming of conventional large-scale parallel computers.

8.6   Summary

In this chapter, we have described varieties of dataflow computing models from their inception to their current state. We described both static and dynamic data-driven models that more or less assume dataflow graphs as their programming model. We discussed how both kinds of data-driven models have evolved to address their shortcomings. We then described demand-driven models that are based on operator nets. We considered two models, eduction and eazyflow, and also considered how data structures are dealt with when using demand-driven execution. We then summarized two attempts to relate data-driven execution to demand-driven execution. Finally, we reviewed the lessons learned and considered future trends.

8.7   References

1. Adams, D. A. 1968. *A computation model with dataflow sequencing.* Technical Report CS 117, Computer Science Department, Stanford University.

2. Amamiya, M., and R. Hasegawa. 1984. Dataflow computing and eager and lazy evaluations. *New Generation Computing,* 12:105–129.

3. Arvind, L. Bic, and T. Ungerer. 1991. Evolution of data-flow computers. In *Advanced Topics in Data-Flow Computing.* Englewood Cliffs, N.J.: Prentice Hall, 3–33.

4. Arvind, and D.E. Culler. 1986. Managing resources in a parallel machine. In *Fifth Generation Computer Architectures,* J. V. Woods, ed. Amsterdam: North-Holland, 103–121.

5. Arvind, and K.P. Gostelow. 1977. Some relationships between asynchronous interpreters of a dataflow language. In *Formal Description of Programming Languages,* E. J. Neuhold, ed. Amsterdam: North-Holland.

6. Arvind, and K.P. Gostelow. 1982. The U-interpreter. *COMPUTER,* 15(2).

7. Arvind, and R.E. Thomas. 1981. *I-structures: An efficient data type for functional languages.* Technical Report MIT/LCS/TM-210, Laboratory of Computer Science, M.I.T.

8. Ashcroft, E. A., A.A. Faustini, and B. Huey. 1985. Eduction—a model of parallel computation and the programming language Lucid. In *Proceedings of the Fourth Annual Phoenix Conference on Computers and Communications.* New York: IEEE, 9–15.

8a. Ashcroft, E.A., Faustini, A. A., Jagannathan, R., and Wadge, W. W. 1995. *Multidimensional Programming.* Cambridge, Mass.: Oxford University press.

9. Ashcroft, E. A., and R. Jagannathan. 1986. Operator nets.In *Fifth Generation Computer Architectures,* J. V. Woods, ed. Amsterdam: North-Holland, 177–202.

10. Bic, L. 1990. A process-oriented model for efficient execution of dataflow programs. *Journal of Parallel and Distributed Computing,* 8:42–51.

11. Culler, D. E., and Arvind. 1988. Resource requirements of dataflow programs. In *Proceedings 15th Annual International Symposium on Computer Architecture.* New York: IEEE Computer Society and ACM SIGARCH, 141–150.

12. Dennis, J. B. 1974. First version of a data flow procedure language. In *Lecture Notes in Computer Science,* Vol. 19. New York: Springer-Verlag, 362–376.

13. Dennis, J. B. 1980. Dataflow supercomputers. *COMPUTER,* 13(11): 48–56.

14. Dennis, J. B. 1991. Evolution of "static" data-flow architecture. In *Advanced Topics in Data Flow Computing.* Englewood Cliffs, N.J.: Prentice Hall, 35–91.

15. Dennis, J. B., and D.P. Misunas. 1975. A preliminary architecture for a basic data-flow processor. In *Proceedings of the Second Annual Symposium on Computer Architecture.* ACM, 126–132.

16. Gao, G.R. 1991. A flexible architecture model for hybrid data-flow and control-flow evaluation. In *Advanced Topics in Data-Flow Computing.* Englewood Cliffs, N.J.: Prentice Hall, 327–346.

17. Iannucci, R.A. 1988. Toward a dataflow/von Neumann hybrid architecture. In *Proceedings 19th Annual International Symposium on Computer Architecture.* ACM Press, 131–140.

18. Jagannathan, R. 1988. *A descriptive and prescriptive model for dataflow semantics.* Technical Report SRI-CSL-88-S, SRI International, Computer Science Laboratory, Menlo Park, Calif.

19. Jagannathan, R. and A.A. Faustini. 1991. GLU: A system for scalable and resilient large grain parallel processing. In *Proceedings of 24th Hawaii International Conference on System Sciences,* Kauai, Hawaii.

20. Kahn, G. 1974. A semantics of a simple language for parallel processing. In *Proceedings IFIP Congress 1974.* Amsterdam: Elsevier North Holland, 471–475.

21. Karp, R.M., and R.E. Miller. 1966. Properties of a model for parallel computations: Determinacy, termination, queueing. *SIAM Journal of Applied Mathematics,* 14(6):1390–1411.

22. Papadopoulos, G.M., and K.R. Traub. 1991. Multithreading: A revisionist view of dataflow architectures. In *Proceedings of 18th Annual International Symposium on Computer Architecture,* Toronto, Canada. New York: IEEE Computer Society and ACM SIGARCH, 342–351.

23. Pingali, K.K., and Arvind. 1985. Efficient demand-driven evaluation, Part 1. *ACM Transactions on Programming Languages and Systems,* 7(2):311–333.

24. Rodriguez, J. E. 1960. *A graph model for parallel computation.* Technical Report TR 64, Project MAC, Massachusetts Institute of Technology, Cambridge, Massachusetts.

25. Ruggiero, C.A., and J. Sargeant. 1987. Control of parallelism in the manchester dataflow ma chine. In *Proceedings of the 1987 Conference on Functional Programming and Computer Architecture.* ACM.

26. Yamaguchi, Y., S. Sakai, K. Hiraki, Y. Kodama, and T. Yuba. 1989. An architectural design of a highly parallel machine. In *Proceedings Information Processing 89* (IFIP 89), G.X. Ritter, ed. New York: Elsevier Science Publishers.

9

Partitioning and Scheduling

Hesham El-Rewini

Program partitioning and *task scheduling* are two distinguishing features of parallel versus sequential programming. Partitioning and scheduling techniques are essential to high-performance computing on both homogeneous and heterogeneous systems. The partitioning problem deals with how to detect parallelism and determine the best trade-off between parallelism and overhead. The program is first analyzed to determine the *ideal parallelism* revealed by the control and data dependences. Two operations potentially can be executed in parallel if they are not related by control or data dependences. This kind of parallelism is usually referred to as *fine grain parallelism* because the unit of parallelism is normally an operation or a single statement. Several operations or statements may be combined into a larger grain to reduce communication overhead. The problem is to find the best grain size that maximizes parallelism while reducing overhead. The estimated parallel execution time reflects a trade-off between parallelism and overhead and is minimized at an optimal intermediate granularity of parallelism. The best choice of grain size depends on operation execution times, synchronization and communication overheads, and resource limits.

The general scheduling problem emerges whenever there is a choice as to the order in which a number of tasks can be performed, and the assignment of tasks to servers for processing. In the context of parallel and distributed computing, the scheduling problem arises because the concurrent parts of a parallel program must be arranged in time and space so that the overall execution time of the program is minimized. The problem of scheduling program tasks on multiprocessor systems is known to be NP-complete in general as well as in several restricted cases. There are few known polynomial-time scheduling algorithms, even when severe restrictions are placed on the program and the machine models. The intractability of the scheduling problem has led to a large number of heuristics, each of which may work under different circumstances.

Figure 9.1 illustrates how partitioning and scheduling fit in the implicit and explicit approaches to using parallel and distributed systems. In the implicit approach, the underlying computing environment is entirely concealed from the programmer. Compilers are re-

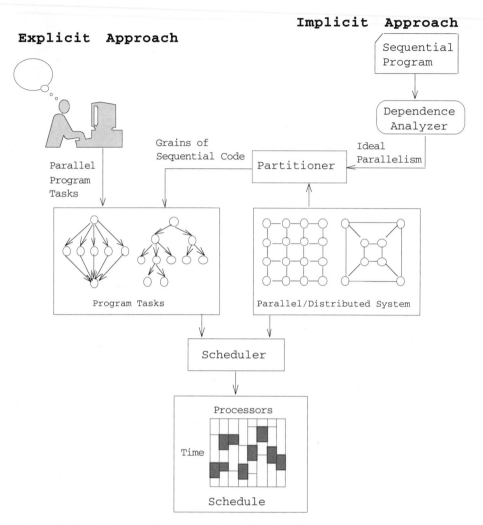

Figure 9.1 Partitioning and scheduling in implicit and explicit parallelism

quired to analyze the application to explore embedded parallelism and perform partition-ing. This approach is highly desirable because it removes the burden of dealing with the increased complexity from the shoulders of programmers. However, the burden is shifted to the compiler writer. On the other hand, in the explicit approach, existing languages are extended, or entirely new languages are introduced, to express parallelism directly. This re-quires that programmers learn new kinds of constructs. It is the programmer's responsibil-ity to identify parallelism within the application. These two approaches are sometimes referred to as compiler-directed and programmer-directed. Regardless of the approach used, efficient strategies for program partitioning and task scheduling are needed.

This chapter addresses the problem of program partitioning and task scheduling in all of its variations and is organized as follows. In Section 9.1, we study the problem of parti-tioning sequential programs into parallel tasks. An introduction to the scheduling problem and its taxonomy is given in Section 9.2. Modeling scheduling systems and the different forms of communication are presented in Sections 9.3 and 9.4, respectively. In Section 9.5, we present several optimal algorithms to solve the scheduling problem and review the

NP-complete results. Sections 9.6 covers a number of scheduling heuristics. Scheduling nondeterministic programs dynamically and statically is summarized in Section 9.7. In section 8, we survey two scheduling CASE tools that represent the recent work in this area. In Section 9.9, we study a special case of the scheduling problem, that is, the task allocation problem. A generalization of partitioning and scheduling in heterogeneous environments is covered in Section 9.10. We give our summary and concluding remarks in Section 9.11.

9.1 Program Partitioning

9.1.1 Data and function partitioning

There are two basic approaches to the partitioning problem: (1) *data partitioning* and (2) *function partitioning*. In data partitioning, the program consists of a number of tasks that perform the same computation on different sets of data concurrently. Data partitioning is appropriate for applications that perform the same operations repeatedly on large collections of data. Algorithms such as matrix multiplication or Fourier transformations and applications such as ray tracing or signal processing adapt well to data partitioning. On the other hand, in function partitioning, the program is decomposed into a number of concurrent tasks that perform different computations. Function partitioning is suitable for applications that perform many different operations on the same data. Application such as flight simulation and process control adapt well to function partitioning. Several applications lend themselves to a combination of these two methods.

9.1.2 Grain size and data locality

A *grain* is defined as one or more sequential instructions packed together to make a module that is sequentially executed on a single processor. The size of a grain is altered by adding or removing instructions. A grain can be as small as a single operation or as large as a whole procedure. If a grain is too large, parallelism is reduced because potentially concurrent tasks are grouped together and executed sequentially by one processor. On the other hand, when the grain size is too fine, more overhead in the form of context switching, scheduling time, and communication delay is added to the overall execution time. With the increasing success of building the shared-memory model on top of a physically distributed memory, it has become very important to partition the program so that the data used by a grain is kept local to the greatest degree possible. The idea is to reduce data movement between the grains running on different processors. Even at high data transfer rates, there is still latency that is many times longer than the local memory latency. There is a trade-off between maximizing locality and maximizing parallelism. At one extreme, running a program sequentially on one processor maximizes locality but does not exploit the parallelism in the program. The parallel execution time of a program can be minimized at an optimal intermediate grain size in which locality is maximized and potential parallelism is also exploited.

9.1.3 Program Dependence Graph

A dependence between two program statements can be either data or control dependence. A *control dependence* is a consequence of the flow of control in a program. For example, the execution of a statement in one path under an `if` test is contingent on the `if` test taking that path. Thus, the task under control of the `if` is control dependent upon the `if` test. *Data dependence* is a consequence of the flow of data in a program. A statement that uses a variable in an expression is data dependent upon the statement that computes the value of the variable. Dependence relations between statements forming a program can be viewed as precedence relations. Two statements can potentially be executed in parallel if they are not related by control or data dependences. The control and data dependences constraints can be also used to determine whether iterations of a given loop can be executed in parallel. If there are any *loop-carried* control or data dependences, the loop must be executed sequentially. A premature exit from within a loop is an example of a loop-carried control dependence. Loop-carried data dependences occur when different loop iterations may perform conflicting read/write accesses on a shared variable.

A *program dependence graph* (PDG) consists of a set of nodes that represents program grains and a set of arcs that represents data and control dependences. The control dependence arcs and the PDG nodes form the *control dependence subgraph* of the PDG. Similarly, the data dependence arcs and the PDG nodes together form the *data dependence subgraph* of the PDG. The arcs and nodes of the PDG are usually annotated with cost and average execution times. Building the PDG from a sequential program is a well known procedure in the field of compiler construction. The details of how to build the PDG can be found in Ref. [5].

9.1.4   Task tree definition

The task tree of a sequential program can be constructed from the PDG. Its structure is a fan-out tree that is based on the control dependence subgraph of the PDG. There are two types of nodes in the task tree: (1) nodes that correspond to statements in the sequential program and (2) other nodes that have special meanings such as loop *preheader* and *postexit* nodes. Associated with the nodes that represent a loop entry (*preheader*) is a Boolean variable *parloop*; *parloop(ph)* = *true* indicates that there is no loop-carried dependence, and it is legal to execute the loop with preheader node *ph* in parallel.

The arcs that correspond to control dependences define the parallel constructs present in the program. Associated with an arc is the control condition (u, l), where u is a statement in the source node and $l \in \{T, F, U\}$. The labels T and F represent *true* and *false* conditional branches, respectively. Label U represents an unconditional branch. Precedence relations due to data dependence are also insured using synchronization arcs.

9.1.5   Partitioning sequential programs

In this section, we present an algorithm introduced by Sarker for partitioning a FORTRAN program dependence graph into parallel tasks [26]. The algorithm starts with the initial *task tree,* which can be obtained from the program dependence graph. The task tree reveals the maximum possible parallelism that can be exploited in the input program. This parallelism is restricted by the constraints imposed by the control and data dependence and by the structure of the parallel constructs supported by the target parallel machine. The partitioning algorithm iteratively merges adjacent tasks in the task tree on the basis of critical path (longest path) length values and overhead values until the task tree is reduced to a single task. Among all task trees in this iterative sequence, the one with the smallest parallel execution time value is selected as the optimized task partition for the current program. The details of this method is shown in Algorithm 1.

Algorithm 1

1. Start with the initial task tree
2. Repeat steps 3-5 until no further merging is possible. Keep track of the best *parallel execution time* value obtained among all partitions generated during the following iterations.
3. Pick the task with the largest average decrease in overhead. Call the selected task T_a. The average decrease in overhead for a task is calculated by summing the decrease in total overhead obtained over all possible merging choices for the task and then dividing by the number of merging choices.
4. Evaluate the parent, sibling, and child tasks of T_a as candidates for merging with T_a. Of these tasks, pick the one that yields the smallest value of the critical path length of the entire task tree, when it is merged with T_a. Call the selected task T_b.
5. Merge tasks T_a and T_b.
6. When no further merging is possible, reconstruct the partition with the best *parallel execution time* value by reinitializing the task tree and repeating steps 3, 4, and 5, until the partition with the best *parallel execution time* value is obtained.

The overhead is calculated based on the following target machine parameters:

$T_{start-up}$	the start-up overhead of a task
$T_{fork-join}$	the total fork-join overhead incurred in a parent task for creating and terminating its child tasks
T_{signal} and T_{wait}	the synchronization overhead

The parallel execution time is calculated based on the number of processors, the overhead, and the average execution frequencies.

Consider the FORTRAN program shown in Fig. 9.2. The task tree of this program is shown in Fig. 9.3. It can be seen that the *true parloop* mapping is shown as annotation to the preheader node *PH*1.The synchronization arcs are shown as dotted arrows. Assume that the target machine parameters are:

$$\text{Number of processors} = 4$$
$$T_{start-up} = 10 \text{ cycles}$$
$$T_{fork-join} = 60 + 300k \text{ cycles, where } k = \text{number of child tasks created}$$
$$T_{signal} = T_{wait} = 0$$

Figure 9.4 shows the final task tree resulting from applying Algorithm 1 to the initial task tree in Fig. 9.3.

9.2 Task Scheduling

After program partitioning, tasks must be optimally scheduled on the processors such that the program execution time is minimized. The scheduling problem has been studied in a number of different ways in different fields. The classical problem of job sequencing in production management has influenced most of the work in this problem. Task scheduling is one of the most challenging problems in parallel and distributed computing and known to be NP-complete in its general form as well as several restricted cases. Several researchers have studied restricted forms of the problem by constraining the models representing the parallel program or the target machine in order to obtain optimal algorithms. Obviously, the special cases that can be solved optimally do not fully represent real-world sys-

```
PARAMETER(ISIZE = 800000)
REAL*4 X(ISIZE), Y(ISIZE), Z(ISIZE)
1:    ENTRY
2:    Y(1) = 0
3:    Z(1) = 0
4:    PI = 3.14159265
5:    READ(5,*) N
6:    IF (N .LE. ISIZE) THEN
7:        IF (N .GE. 1) THEN
8:            DO I = 1,N
9:                X(I) = SIN(PI*I/FLOAT(N))*PI/FLOAT(N)
10:           END DO
11:           DO J = 2,N
12:               Y(J) = X(J)/2.0 + Y(J-1)
13:           END DO
14:           DO K = 2,N
15:               Z(K) = X(K)/3.0 - Z(K-1)
15:           END DO
17:           WRITE(6,*) Y(N),Z(N)
              STOP
          END IF
      END IF
18:   WRITE(6,*) 'Bad value of N'
      END
```

Figure 9.2 Example FORTRAN Program

tems. In an attempt to solve the problem in the general case, several fast heuristics have been introduced. These heuristics do not guarantee an optimal solution to the problem, but they try to find near optimal solutions most of the time.

Scheduling techniques can be classified based on the availability of program task information as deterministic and nondeterministic. In *deterministic scheduling,* all the information about tasks to be scheduled and their relations to one another is entirely known prior to execution time. In contrast, in *nondeterministic scheduling,* some information may not be known before the program executes. Conditional branches and loops are two program constructs that may cause non-determinism.

Scheduling nondeterministic programs can be achieved using static or dynamic methods. The distinction indicates the time at which the scheduling decisions are made. With static scheduling, information regarding the task graph representing the program must be estimated prior to execution, whereas in dynamic scheduling, the parallel processor system schedules tasks on the fly. This is usually implemented as some kind of load balancing heuristic. The disadvantage of dynamic scheduling is the overhead incurred to determine the schedule while the program is running. A combination of static and dynamic methods is referred to as a *hybrid* method. In deterministic (static) scheduling, each task in the program has a static assignment to a particular processor, and each time that task is submitted for execution, it is assigned to that processor. Many algorithms have been proposed in the literature, especially for static scheduling. In the following sections, we present most of the significant results in the task scheduling area.

9.3 Scheduling System Model

A scheduling system consists of parallel program, target machine, schedule, and performance criterion. In what follows, we study each one of these four components and show

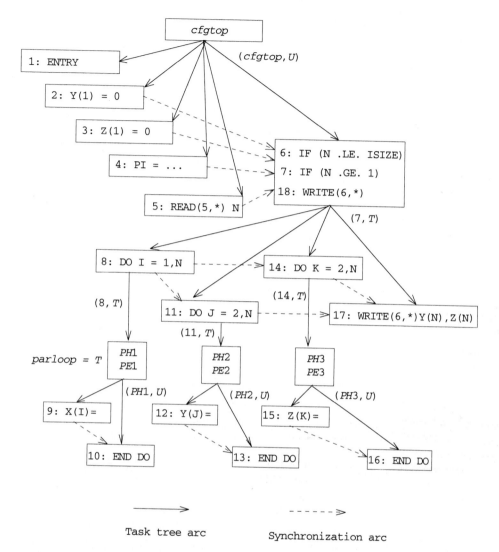

Figure 9.3 Initial task tree of program of Fig. 9.2

how the program and target machine parameters can be used to estimate execution times and communication delays.

9.3.1 Parallel program tasks

The characteristics of a parallel program can be defined as the system $(T, <, [D_{ij}], [A_i])$ as follows:

1. $T = \{t_1, ..., t_n\}$ is a set of tasks to be executed.
2. $<$ is a partial order defined on T which specifies operational precedence constraints. That is $t_i < t_j$ means that t_i must be completed before t_j can start execution.
3. $[D_{ij}]$ is an $n \times n$ matrix of communication data, where $D_{ij} \geq 0$ is the amount of data required to be transmitted from task t_i to task t_j, $1 \leq i, j \leq n$.

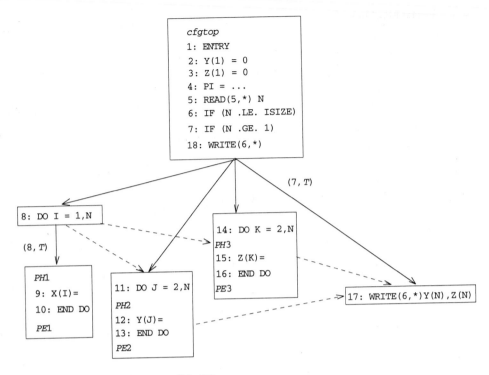

Figure 9.4 Final task tree of program of Fig. 9.2

4. $[A_i]$ is an n vector of the amount of computations, i.e., $A_i > 0$ is the number of instructions required to execute t_i, $1 \le i \le n$.

The partial order $<$ is conveniently represented as a directed acyclic graph called a *task graph*. An arc (i, j) between two tasks t_i and t_j specifies that t_i must be completed before t_j begins. Figure 9.5 shows an example of a task graph consisting of 7 nodes ($n = 7$), where each node represents a task. The number shown in the upper portion of each node is the node number, the number in the lower portion of a node i represents the parameter A_i (the amount of computation needed by task t_i), and the number next to an edge (i, j) represents the parameter D_{ij}. For example $A_2 = 10$, $D_{13} = 2$. Note that tasks are referred to simply by their indices (e.g., 1 is used rather than t_1).

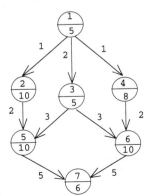

Figure 9.5 A task graph

9.3.2 Target Machine

The *target machine* is assumed to be made up of m heterogeneous processing elements connected using an arbitrary interconnection network. Each processing element can run one task at a time and all tasks can be processed by any processing element. Formally, the target machine characteristics can be described as a system $(P, [P_{ij}], [S_i], [I_i], [B_i], [R_{ij}])$ as follows:

1. $P = \{P_1, ..., P_m\}$ is a set of processors forming the parallel architecture.
2. $[P_{ij}]$ is an $m \times m$ interconnection topology matrix.
3. S_i, $1 \leq i \leq m$, specifies the speed of processor p_i.
4. I_i, $1 \leq i \leq m$, specifies the startup cost of initiating a message on processor p_i.
5. B_i, $1 \leq i \leq m$, specifies the startup cost of initiating a process on processor p_i.
6. R_{ij} is the transmission rate over the link connecting two adjacent processors p_i and p_j.

The connectivity of the processing elements can be represented using an undirected graph called the target machine graph. Figure 9.6 shows an example of a target machine consisting of 16 processors ($m = 16$) forming a mesh. Processors are referred to simply by their indices (e.g., 1 is used rather than P_1).

The execution time of task i when executed on processor j (T_{ij}) can be computed as follows.

$$T_{ij} = \frac{A_i}{S_j} + B$$

The communication delay (over a free link) between tasks i_1 and i_2 when they are executed on adjacent processing elements j_1 and j_2 can be computed as follows.

$$C(i_1, i_2, j_1, j_2) = \frac{D_{i_1 i_2}}{R_{j_1 j_2}} + I_{j_1}$$

9.3.3 The schedule

A schedule of the task graph $G = (V, A)$ on a target machine that is made up of m processors is a function f that maps each task to a processor and a starting time. Formally, $f : V \to \{1, 2, ..., m\} \times (0, \infty)$. If $f(v) = (i, t)$ for some $v \in V$, we say that task v is scheduled to be

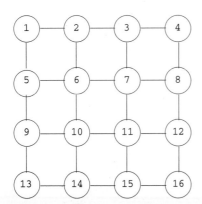

Figure 9.6 Mesh target machine graph

processed by processor p_i starting at time t. A schedule f is a feasible schedule if it preserves all precedence relations and communication restrictions. The function f can be illustrated as a Gantt chart where the start and finish times for all tasks can be easily shown. Figure 9.7 shows an example of a Gantt chart. For example, task 1 starts on processor P_2 at time 0 and finishes at time 10, while task 5 starts on processor P_3 at time 0 and finishes at time 30. The shaded areas illustrates communication delay. For example, the shaded area between times 10 and 11 on processor P_2 shows the communication delay as a result of a message sent from task 2 on P_1 to task 4 on P_2.

9.3.4 Performance measures

Our scheduling goal is to minimize the total completion time of a parallel program. This performance measure is known as the *schedule length* or *maximum finishing time*. Schedule length can be described as follows. Given a task graph $G = (V,A)$ and its schedule on m processors f, the length of schedule f of the G is the maximum finishing time of any task in G. Formally, $length(f) = t_{max}$, where $t_{max} = $ maximum $\{t + T_{ij}\}$ where $f(i) = (j, t)$ $\forall i \in V$ and $1 \le j \le m$. Recall that T_{ij} is the execution time of task i on processor j.

9.4 Communication Models

There are two key components that contribute to the total completion cost: (1) execution time and (2) communication delay. The communication delay can be computed using several different models. The following are three models that can be used to compute the cost of executing a parallel program on a set of processing elements. Communication delay is the key element that differentiate these models. Given a task graph $G = (V, A)$ and its schedule on m processors f, we define $proc(v)$ to be the processor assigned to task v in f, for every $v \in V$.

Model-A. In this model, the Gantt chart that represents the schedule f does not reflect the communication delay, but it shows that the precedence relations between tasks are preserved. Program completion cost can be computed as:

$$\text{total cost} = \text{communication cost} + \text{execution cost}$$

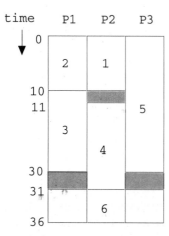

Figure 9.7 A Gantt chart

where

execution cost = schedule length

communication cost = number of messages * cost per message

number of messages = the number of node pairs (u, v) such that $(u, v) \in A$ and $proc(u) \neq proc(v)$

Model-B. This model is similar to Model-A, but it uses a more practical definition of number of messages. By counting each arc (u, v) such that $proc(u) \neq proc(v)$ as communication using Model-A, we may be passing the same information between a given pair of processors several times. For instance, if tasks u, v, and $w \in V$ such that (u, v), $(u, w) \in A$ and $proc(v) = proc(w) \neq proc(u)$, then the result of computing u must be communicated to the processor computing v and w. This communication is counted twice in model-A, and it is counted only once in model-B, as follows:

$$\text{total cost} = \text{communication cost} + \text{execution cost}$$

where

execution cost = schedule length

communication cost = number of messages * cost per message

number of messages = the number of processor-task pairs (P, v) such that processor P does not compute task v but computes at least one direct successor of v.

Model-C. This model will be our basic model to compute communication cost through this chapter. We assume the existence of an I/O processor associated with every processor in the system. The term *processing element* is used to imply the existence of an I/O processor. Observe that a processing element can execute a task and communicate with another processing element at the same time. Communication time between two tasks allocated to the same processing element is assumed to be zero time units. For any two tasks $i, j \in V$, if $i < j$ and $f(i) = (k, t)$, then j should be scheduled on **either**:

- processor k on time t_1 and $t_1 \geq t + T_{ik}$ **or**
- processor l, $l \neq k$ on time t_2 and $t_2 \geq t + T_{ik} + C(i, j, k, l)$

In some cases, we will assume that $T_{ik} = C(i, j, k, l) = 1$ for all tasks i, j and all processors k, l. In this case, the task j should be scheduled on **either**:

- processor k on time t_1 and $t_1 \geq t + 1$ **or**
- processor l, $l \neq k$ on time t_2 and $t_2 \geq t + 2$

Using this model, the communication delay can be easily shown on the Gantt chart representing the schedule. Note that a task can be scheduled in the communication holes in a Gantt chart. In other words, a task can be assigned to a processing element for execution while this processing element is communicating with another processing element. The program completion time is computed as total cost = schedule length.

9.5 Optimal Scheduling Algorithms

The scheduling problem is known to be NP-complete in its general form and in many restricted cases. There are few known polynomial-time scheduling algorithms even when severe restrictions are forced on the task graph representing the program and the parallel processor models. Table 9.1 summarizes the NP-complete results, and Table 9.2 shows all the known optimal algorithms to solve the scheduling problem.

In this section, we present some efficient optimal algorithms that have polynomial time complexity. When communication cost is ignored, we present optimal algorithms in the following three cases: 1) when the task graph is a tree, 2) when the task graph is an interval order and 3) when there are only two processors available. In the three cases all tasks are assumed to have the same execution time. There is no loss of generality if we assume that all the tasks have unit execution times. ($T_{ik} = 1$ for all tasks i on all processing elements k). When communication cost is considered, we present two algorithms to schedule interval orders and tree structured task graphs. In these two cases, we assume that $T_{ik} = C(i, j, k, l) = 1$ for all tasks i,j and all processors k, l.

9.5.1   Scheduling tree structured task graphs

9.5.1.1 Scheduling in-forests/out-forests without communication. We present one of the first polynomial time algorithms to solve the scheduling problem when the task graph is either an *in-forest* (i.e., each node has at most one immediate successor) or an *out-forest* (i.e., each node has at most one immediate predecessor). Algorithm 2, which was introduced by Hu, finds an optimal schedule when all tasks have the same execution time and communication delay is ignored [18]. The time complexity of the algorithm is $O(n)$. The

TABLE 9.1 NP-Complete Results (assuming fully corrected homogeneous target machine)

Task Graph	# Processors	Exec. Time	Comm. Time	Comm. Model
Arbitrary	m	1	0	—
Arbitrary	2	1 or 2	0	—
Opposing Forest	m	1	0	—
Interval Order	2	Arbitrary	0	—
Tree	m	1	1	Model-A
Tree	m	1	1	Model-B
Arbitrary	2	1	1	Model B
Arbitrary	Unlimited	1	>1	Model C

TABLE 9.2 Scheduling Optimal Algorithms (assuming fully connected homogeneous target machine, unit task time, n: number of tasks, e: number of arcs in the task graph, and communication Model-C)

Task Graph	# Processors	Comm. Time	By	Complexity
Arbitrary	2	0	Coffman and Graham, 1972	$O(n^2)$
Tree	m	0	Hu, 1961	$O(n)$
Interval Order	m	0	Papadimitriou and Yanakakis, 1979	$O(n + e)$
Arbitrary	Unlimited	Fixed τ	Jung et al., 1989	$O(n^{\tau+1})$
Interval Order	m	1	Ali and El-Rewini, 1993	$O(e + nm)$
Tree	2	1	El-Rewini and Ali, 1994	$O(n^2)$

general strategy used in the algorithm is the *highest level first,* where the *level* of a node *x* in a task graph is the maximum number of nodes (including *x*) on any path from *x* to a terminal node.

In this section we assume that the task graph is an in-forest of *n* tasks. We refer to the task that has no predecessors or all its predecessors have already been executed as *ready task.*

Algorithm 2

1. The level of each node in the task graph is calculated and used as each node's priority.
2. Whenever a processor becomes available, assign it the unexecuted ready task with the highest priority. Ties are broken arbitrarily.

The above algorithm can be used in the out-forest case with simple modification. However, this algorithm will not produce an optimal solution in the case of an opposing forest. An *opposing forest* is the disjoint union of an in-forest and an out-forest. Scheduling an opposing forest is proven to be NP-complete. Figure 9.8 shows an in-forest and its schedule on three processors using Algorithm 2.

9.5.1.2 Scheduling in-forests/out-forests with communication. The algorithm presented here was introduced by El-Rewini and Ali in 1994 [10]. The algorithm is based on the idea of adding new precedence relations to the task graph in order to compensate for communication. The task graph after adding the new precedence relations is called the *augmented task graph.* Scheduling the augmented task graph without considering communication is equivalent to scheduling the original task graph with communication. Algorithm 3 produces an optimal schedule when the task graph is an in-forest and the target machine has two processors. It can be used in the out-forest case with simple modification. We first define *node depth* and *operation swapall.*

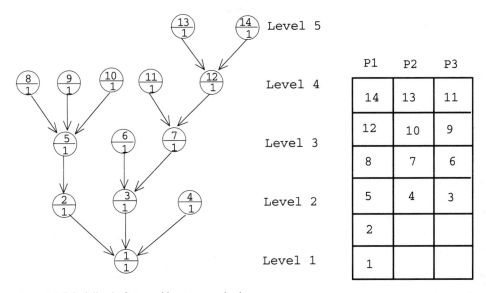

Figure 9.8 Scheduling in-forests without communication

Definition A. *Node depth*. The depth of a node is defined as the length of the longest path from any node with depth zero to that node. A node with no predecessors has a depth of zero. In other words, $depth(u) = 1 + max\{depth(v)\}$, $\forall\ v \in predecessors(u)$; and $depth(u) = 0\ \forall\ u, predecessors(u) = \phi$.

Definition B. *Operation swapall*. Given a schedule f, we define the operation *Swapall(f,x,y)*, where x and y are two tasks in f scheduled to start at time t on processors i and j, respectively. The effect of this operation is to swap all the task pairs scheduled on processors i and j in the schedule f at time t_1, $\forall\ t_1, t_1 \geq t$.

Algorithm 3

```
1.   Given an in-forest G = (V, A), identify the sets of siblings:
     S₁, S₂, . . . , Sₖ, where Sᵢ is the set of all nodes in V with a common child
     child(Sᵢ).
2.   A1 ← A
3.   For every set Sᵢ
     • Pick node u ∈ Sᵢ with the maximum depth
     • A1 ← A1 - (v, child(Sᵢ)) ∀ v ∈ Sᵢ and v ≠ u
     • A1 ← A1 ∪ (v, u) ∀ v∈ Sᵢ and v ≠ u
4.   Obtain the schedule f by applying Algorithm 2 on the augmented in-forest
     F = (V, A1)
5.   For every set Sᵢ in the original in-forest G

     if node u (with the maximum depth) is scheduled in f in the time slot
     immediately before child(Sᵢ) but on a different processor, then apply
     the operation swapall(child(Sᵢ),x, f), where x is the task scheduled in
     the time slot immediately after u on the same processor.
```

Algorithm 3 selects the node u that has the maximum depth from every set of siblings S_i and places it after the other members of S_i, but before the common child of S_i, $child(S_i)$. In other words, Algorithm 3 adds an arc from every node v, $v \in S_i$ and $v \neq u$ to the node u. Note that the augmented task graph constructed is also an in-forest. These added arcs compensate for communication delay. Thus, Algorithm 2 is applied to this augmented in-forest to obtain a schedule where communication delays are considered. The operation swapall is applied when communication restrictions are violated in the output schedule. The time complexity of the algorithm is $O(n^2)$. This algorithm is optimal in the two processor case and it generates optimal algorithm in some cases when the number of processors equals $m > 2$. The time complexity of this heuristic is $O(n^2 + nm)$. Figure 9.9 shows an in-forest and its augmented graph. The final schedule on 3 processors is shown in Fig. 9.10.

9.5.2 Scheduling interval ordered tasks

In this section, we deal with a special class of system tasks called interval ordered tasks. The term *interval ordered tasks* is used to indicate that the task graph which describes the precedence relations among the system tasks is an interval order. A task graph is an *interval order* when its elements can be mapped into intervals on the real line and two elements are related if and only if the corresponding intervals do not overlap. The interval order has a special structure that is established by the following property. For any interval ordered pair of tasks u and v, either the successors of u are also successors of v or the successors of v are also successors of u. Letting $G = (V, A)$ be an interval order and v be any node in V, we define the following:

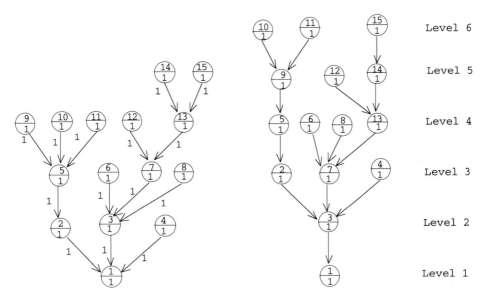

Figure 9.9 Original in-forest with communication (G) and its augmented graph (F)

$N^+(v)$ = set of out-neighbors (all children) of node v
$\quad\quad = \{u \in V: v < u \text{ in } G\}$
$n_2(v) = |N^+(v)|$
$\quad\quad$ = out-degree of node v

9.5.2.1 Scheduling interval orders without communication. The following algorithm was introduced by Papadimitriou and Yannakakis to solve the unit execution time scheduling problem for interval orders when communication is not considered. This algorithm can produce an optimal algorithm for an arbitrary number of processors. The time complexity of the algorithm is $O(n + e)$ where n is the number of tasks and e is the number of arcs in the interval order [25].

P1	P2	P3
15	11	10
14	12	9
13	8	6
7	5	4
	2	
	3	
	1	

Figure 9.10 Scheduling in-forests G with communication

Algorithm 4
```
1.   Use n₂(v) as the priority of task v and ties are broken arbitrary.
2.   Whenever a processor becomes available, assign it the unexecuted ready
     task with the highest priority.
```

   Figure 9.11 shows an interval order and its schedule without communication on three processors using Algorithm 4. Note that the underlined number next to each node is its priority.

9.5.2.2 Scheduling interval orders with communication. We introduce an optimal algorithm to schedule interval orders on an arbitrary number of processors when communication delay is considered. This algorithm was introduced by Ali and El-Rewini in 1993 to solve the problem when execution time is the same for all tasks and is identical to communication delay [3]. We first define the following.

   Definition C. $start\text{-}time(v, i, f)$: the earliest time at which task v can start execution on processor p_i in schedule f.

   Definition D. $task(i, t, f)$: the task scheduled on processor p_i at time t in schedule f. If there is no task scheduled on processor p_i at time t in schedule f, then $task(i, t, f)$ returns the empty task ϕ. Note that $n_2(\phi) < n_2(v)$, for any task $v \in V$.

Algorithm 5
```
1.   Use n₂(v) as the priority of task v and ties are broken arbitrary.
2.   Nodes with highest priority are scheduled first.
3.   Each task v is assigned to processor pᵢ with the minimum start time.
4.   If start-time(v, i, f) = start-time(v, j, f), 1 ≤ i, j ≤ m, task v is
     assigned to processor pᵢ if task(i, start-time(v, i, f) - 1, f) has the
     minimum n₂.
```
   The time complexity of the algorithm is $O(e + nm)$ where n is the number of tasks, e is the number of arcs in the interval order, and m is the number of processors. Figure 9.12 shows an interval order and its schedule with communication on three processors using Algorithm 5. Note that the underlined number next to each node is its priority.

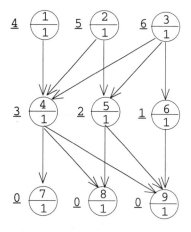

Time	P1	P2	P3
0	3	2	1
1	4	5	6
2	7	8	9
3			

Figure 9.11 Scheduling an interval order without communication on three processors

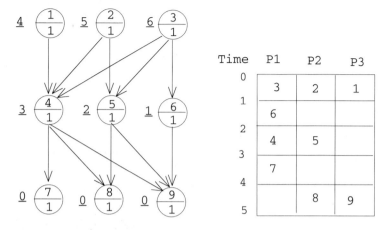

Figure 9.12 Scheduling an interval order with communication on three processors

9.5.3 Two processor scheduling

The first polynomial time algorithm for scheduling unit time task graphs on two processors, based on matching techniques, was introduced by Fujii, et al. [15]. The time complexity of Fujii's algorithm is $O(n^{2.5})$. Improved algorithms have been obtained by Coffman and Graham, Sethi, and Gabow [8, 16, 27]. The time complexity of these three algorithms are $O(n^2)$, $O(min(en, n^{2.61}))$, and $O(e + n\alpha(n))$, respectively, where n is the number of nodes and e is the number of arcs in the task graph. In this section, the algorithm introduced by Coffman and Graham is presented. The idea is similar to that for in-forests and interval orders. The only difference is the way node priorities are assigned.

Algorithm 6

```
1.   Assign 1 to one of the terminal tasks.
2.   Let labels 1,2,...,j -1 have been assigned. Let S be the set of unas-
     signed tasks with no unlabeled successors. We next select an element of
     S to be assigned label j. For each node x in S define l(x) as follows:
     Let y₁, y₂,...,yₖ be the immediate successors of x. Then l(x) is the
     decreasing sequence of integers formed by ordering the set {L(y₁),
     L(y₂),..., L(yₖ)}. Let x be an element of S such that for all x' in S,
     l(x) ≤ l(x') (lexicographically). Define L(x) to be j.
3.   Use L(v) as the priority of task v and ties are broken arbitrary.
4.   Whenever a processor becomes available, assign it the unexecuted ready
     task with the highest priority. Ties are broken arbitrarily.
```

Since each task executes for one unit of time, processors 1 and 2 both become available at the same time. We assume that processor 1 is scheduled before processor 2. Figure 9.13 shows a task graph and its schedule without communication on two processors using Algorithm 6. Note that the underlined number next to each node is its priority.

9.6   Scheduling Heuristic Algorithms

Optimal schedules can be obtained in restricted cases that do not necessarily represent real world situations. In order to provide solutions to real world scheduling problems,

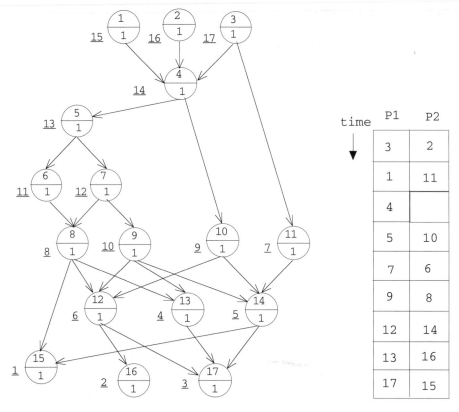

Figure 9.13 Scheduling an interval order with communication on three processors

restrictions on the parallel program and target machine representations must be relaxed. Recent research in this area has emphasized heuristic approaches. A heuristic produces an answer in less than exponential time, but it does not guarantee an optimal solution. Intuition is usually used to come up with heuristics that make use of special parameters that affect the system in an indirect way. A heuristic is said to be better than another heuristic if solutions fall closer to optimality more often, or if the time taken to obtain a near-optimal solution is less. The effectiveness of these scheduling heuristics depends on several parameters of the parallel program and the target machine. A heuristic that can optimally schedule a particular task graph on a certain target machine may not produce optimal schedules for other task graphs on other machines. As a result, a number of heuristics have been proposed, each of which may work under different circumstances. In what follows, we study a number of the scheduling heuristics. More details on scheduling heuristic algorithms can be found in Refs. [1, 9, 17, 19, 24, and 30].

9.6.1 List scheduling

One class of scheduling heuristics, to which many schedulers are classified is *list scheduling*. In list scheduling each task is assigned a priority, then a list of tasks is constructed in decreasing priority order. A ready task with the highest priority is scheduled on its "best" available processor. The schedulers in this class differ in the way they assign priorities to tasks and the criteria used to select the "best" processor to run the task. Algorithm 7 shows the general list scheduling algorithm.

Algorithm 7

1. Each node in the task graph is assigned a priority. A priority queue is
    initialized for ready tasks by inserting every task that has no imme-
    diate predecessors. Tasks are sorted in decreasing order of task pri-
    orities.
2. As long as the priority queue is not empty do the following:
    a. A task is obtained from the front of the queue.
    b. An idle processor is selected to run the task.
    c. When all the immediate predecessors of a particular task are exe-
        cuted, that successor is now ready and can be inserted into the
        priority queue.

Priority assignment results in different schedules because nodes are selected in a different order. The *level* and *co-level* of a task are two examples of task priority. We define the
following terms:

Definition E. *Path length.* The length of a path in a task graph is the summation of the
weights of all nodes and arcs along the path including the initial and final nodes.

Definition F. *Level.* The level of a node in a task graph is defined as the length of the
longest path from the node to an exit node. (An exit node is the one with no successors)

Definition G. *Co-level.* The co-level of a node in a task graph is defined in the same
way as a level except that lengths are measured from the starting points of the task graph
rather than from the exit node.

Adam et al. conducted an empirical performance study of list scheduling heuristics and
showed that among all priority schedulers, level priority are the best at getting close to the
optimal schedule [1]. Highest level first (HLF), which is also known as critical path (CP),
was shown to be superior to others because it provided schedules that are within 5 percent
of the optimal in 90 percent of random cases. This type of heuristic, when communication
is not considered, is most appropriate for a shared-memory parallel processor where messages are passed at memory cycle speeds. But in MIMD distributed-memory parallel
computers, the communication rate is much higher than that of shared-memory systems.
The question now is how the presence of communication may affect the list scheduling
heuristics.

Heuristics that use level numbers or critical path length face the problem that the level
numbers do not remain constant when communication delays are considered. This is because the level of each node changes as the length of the path leading to the exit node
changes. The path length varies depending on communication delay, and the communication delay changes depending on task allocation. Communication delay is zero if tasks
are allocated to the same processor and non-zero if tasks are allocated to different processors. Also, the number of hops between processors makes a difference in computing
the communication delay portion of the level. This is the level number problem for parallel processor scheduling. Some heuristics assume identical processors and compute a
node's level as the summation of the node amount of computations along the path to the
exit node, excluding the communication delay. A better approximation of level number
can be obtained by iteration: schedule, then calculate node level, schedule, and so on.
The time complexity would be increased, and the resulting level number would remain
only an approximation.

9.6.2 Mapping heuristic

The *mapping heuristic* (MH) introduced by El-Rewini and Lewis is a modified list sched-uling technique. MH considers several real-world parameters such as: interconnection topology, processor speed, link transfer rate, and contention delays. MH uses the general model described in Section 9.3 to model the parallel program and the target machine. It is assumed that the transmission rate over a link connecting any two adjacent processors is the same and equals R. The time to initiate message passing is the same for all processors and is equal to I. MH uses the given system parameters to compute the execution time and the communication delay as follows:

- $$T_{ij} = \frac{A_i}{S_j} + B_j$$

- $$C(i_1, i_2, j_1, j_2) = \left(\frac{D_{i_1 i_2}}{R} + I \right) * H_{j_1 j_2} + CD_{j_1 j_2}$$

where

$H_{j_1 j_2}$ = the number of hops between processors j_1 and j_2

$CD_{j_1 j_2}$ = the contention delay on the route from j_1 to j_2

MH constructs and maintains routing tables to hold estimate values of these two variables. For each processor in the system, MH maintains a routing table that has contention infor-mation indexed by, and containing one entry for, each other processor. This entry contains three parts: the number of hops, the preferred outgoing line to use for that destination, and the communication delay due to contention.

MH schedules the task with the highest level first and ties are broken in favor of the task with the largest number of immediate successors in the task graph. When a task is ready (i.e., all its predecessors have been scheduled), it is scheduled on the processor with the earliest finish time. The finish time of a task is determined by considering the following:

1. processor speed
2. link transfer rate
3. message passing route
4. number of hops
5. delay due to contention

The details of MH can be found in [12].

Figure 9.15 shows the Gantt chart constructed by MH when used to schedule the task graph of Fig. 9.14 on an eight-processor hypercube. The labeled dark segments correspond to task execution; the lighter segments correspond to communication delays; and the clear segments mean the processor is idle. The number in each communication segment refers to the source task that is sending a message to the destination task. (When a destination task is waiting for more than one source task, the largest message is the only one shown.) In the example, it should be pointed out that the tasks 7, 8, and 9 are delayed in their exe-cution because they are placed on processors that are two hops away from processor 1. This introduces additional communication delay. Tasks on the same processor enjoy zero delays in communication.

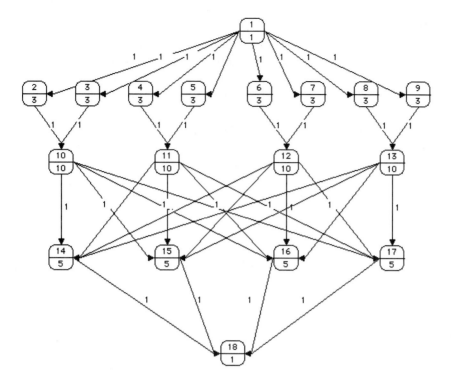

Figure 9.14 A program task graph

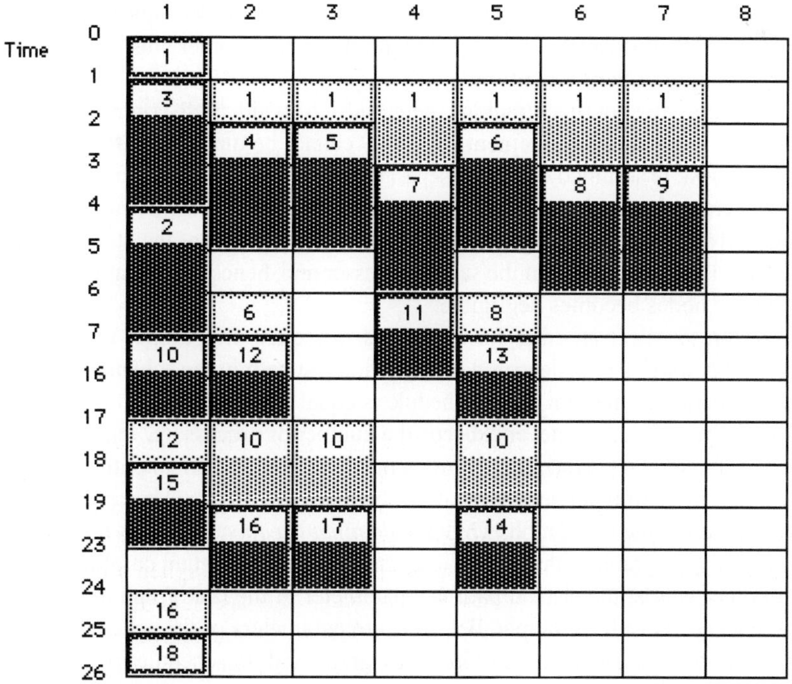

Figure 9.15 Scheduling the task graph of Fig. 9.14 on an eight-processor hypercube

9.6.3   Scheduling using clustering heuristics

A different type of scheduling heuristics tries to partition the scheduling process into two phases: (1) processor assignment, which is the process of allocating tasks to the system processors; and (2) task ordering, which is the process of scheduling the tasks allocated on each processor. Clustering of task graphs can be used as an intermediate phase to solve the allocation problem of the scheduling process. The following is a scheduling heuristic based on clustering that was introduced by Gerasoulis and Yang.

Algorithm 8

```
1.   Cluster the tasks assuming an unlimited number of fully-connected pro-
     cessors. Two tasks in the same cluster are scheduled in the same pro-
     cessor.
2.   Map the clusters and their tasks onto the given number of processors
     (m). In this step, the following optimizations are performed:
     a.   Cluster merging. If the number of clusters is greater than the num-
          ber of available processors, the clusters are merged into m clus-
          ters.
     b.   Physical mapping. The actual architecture is not fully connected.
          A mapping must be determined such that overall communication be-
          tween clusters is minimized.
     c.   Task execution ordering. After the processor assignment of tasks
          is fixed, the execution ordering is determined to assure the cor-
          rect dependence order between tasks.
```

9.6.3.1 Clustering. Clustering algorithms start with an initial clustering and then perform a sequence of clustering refinements to achieve a specific objective. A cluster is a set of tasks that will execute on the same processor. Clusters are not tasks, since tasks that belong to a cluster are permitted to communicate with the tasks of other clusters immediately after the completion of their execution.

Clustering heuristics are non-backtracking heuristics to avoid high complexity, i.e., once the clusters are merged in a refinement step, they cannot be unmerged afterward. At the initial step, each task is assumed to be in a separate cluster. A typical refinement step is to merge two clusters and zero the edge that connects them. *Zeroing* the communication cost on the edge between the two merged clusters is due to the fact that the start and end nodes of this edge will be scheduled on the same processor and, hence, the communication cost between the two nodes becomes negligible.

A typical criterion to select an edge for zeroing is to reduce the parallel time of the schedule. The *parallel time* (PT) is determined by the longest path in the scheduled graph. In other words, the parallel time of a given schedule is equal to its completion time, if we assume that the number of clusters never exceeds the number of processors. There are two other important parameters in performing the refinement steps—the critical path of a clustered task graph and the dominant sequence of a scheduled task graph. The *critical path* (CP) is the longest path in the task graph, while the *dominant sequence* (DS) is the longest path of the scheduled task graph or the path whose length equals the actual parallel time of the schedule. In other words, the critical path is a parameter of the task graph only, while the dominant sequence, as well as the parallel time, are parameters of the schedule of the task graph. For example, consider the clustered task graph and its schedule in Fig. 9.16. The tasks 1, 2, and 7 form the CP with length 9, while a DS of the schedule consists of the tasks 1, 3, 4, 5, 6, and 7 with length 10.

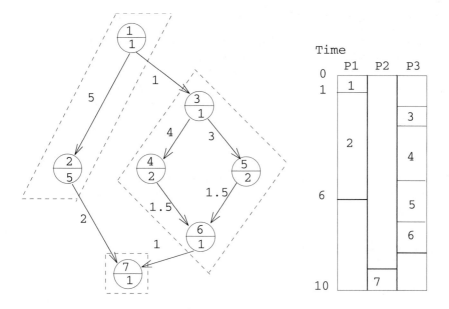

Figure 9.16 Clustered task graph and its schedule on three processors

We present two clustering heuristics. The first heuristic was introduced by Sarker (Algorithm 9), and the other one was introduced by Yang and Gerasoulis (Algorithm 10) [17].

Algorithm 9

1. Initially all edges are marked unexamined and each task forms a separate cluster.
2. Sort all the edges of task graph in a descending order according to their communication costs.
3. **Repeat**
   a. Zero the highest unexamined edge in the sorted list if the parallel time does not increase.
   b. Mark the edge examined.
   c. When two clusters are merged, the tasks are ordered according to the highest level first rule.
   until all edges are marked examined

Algorithm 10

1. Initially each task is in a separate cluster and mark all edges unexamined
2. Initialize $r = 0$
3. Compute the initial dominant sequence DS_0
4. Initialize the list of free node to contain the nodes that have no predecessors
5. **While** there is an edge marked unexamined **do**
   a. Let (t_k, t_j) be the top unexamined edge in DS_r
   b. Mark this edge examined
   c. Suspend zeroing the edge until its head node t_j becomes free
   d. Choose a free node t_i which belongs to the longest path going through any of the free nodes in the scheduled task graph at this

point
e.   Zero the incoming edges of t_i that minimize the co-level(t_i)
f.   Schedule the node t_i after the last scheduled node in its cluster
     in the scheduled task graph at this point
g.   Add any node to the free list if node t_i was its only unscheduled
     predecessor
h.   If all edges in DS_r are examined then increment $r = r + 1$ and find
     a new DS_r
endwhile

The time complexity of Algorithm 9 is $O(e(n + e))$, while the time complexity of Algorithm 10 is $O((e + n)\log n)$, where n and e are the number of tasks and the number of arcs in the task graph, respectively. Figure 9.17 shows the final clustering of a task graph using Algorithms 9 and 10.

9.7   Scheduling Nondeterministic Task Graphs

The scheduling techniques discussed in the earlier sections can be classified under deterministic scheduling in which all the information about program tasks and their relations to one another is entirely known prior to execution time. When some information may not be known before the program starts its execution, we have to deal with nondeterminism. The problem of scheduling nondeterministic task graphs arises in several situations in parallel programs, particularly in the cases of loops and conditional branching. Nondeterminism arises in loops because the number of loop iterations may not be known before the execution of the program. Since loops form a restricted class of conditional branching, there is a higher degree of nondeterminism associated with scheduling conditional branching. In this case, the direction of every branch remains unknown before run time. Consequently, entire subprograms may or may not be executed, which in turn increases

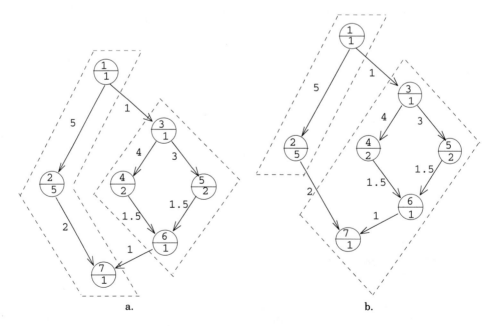

Figure 9.17 Final clustering (a) using Algorithm 9 and (b) using Algorithm 10

the amount of nondeterminism and complicates the scheduling process. Also, having conditional branching within a node in the task graph may cause variable task execution time and communication delay.

Scheduling nondeterministic programs can be achieved dynamically on the fly. However, dynamic scheduling consumes time and resources, and these lead to overhead during program execution. The overhead of extra communication delays, additional memory, and time for the scheduler itself to work detract from dynamic scheduling. In addition, dynamic scheduling can lead to *task thrashing* where a task is moved back and forth between processors, consuming yet more time. Therefore, we must be extremely conservative when applying dynamic scheduling techniques. To eliminate (or reduce) the overhead involved with dynamic scheduling, static or hybrid methods can be applied. In static scheduling, information regarding the task graph representing the program must be estimated prior to execution. This approximation may affect the quality of the produced schedule.

9.7.1 Dynamic scheduling

Dynamic scheduling is any run time technique for placing tasks onto processors and deciding when best to execute them such that the parallel program finishes in the earliest possible time. The most elementary approach to dynamic scheduling attempts to balance processor load across m processors using very local information. In its simplest form, $m+1$ processors are used: one processor runs a scheduler that dispatches tasks on a first-in-first-out (FIFO) basis to all other m processors (see Fig. 9.18). Each of the m processors maintains a private list of waiting tasks called the *waiting queue*. This FIFO list holds all tasks assigned to the processor. As soon as one task is finished, the processor takes the next waiting task from its queue and processes it to completion. As tasks are processed, they may require the services of other tasks. Thus, requests for new tasks are made by the m processors as they do their work. These requests are placed on the *scheduler queue,* maintained by the scheduler processor.

The scheduler processor also dispatches tasks in first-come-first-served order. But what rule is used to decide which waiting queue to select? Many selection criteria have been suggested in the literature. The simplest heuristic attempts to balance the load on all processors using a variant of the following FIFO heuristic: (1) The processor with the cur-

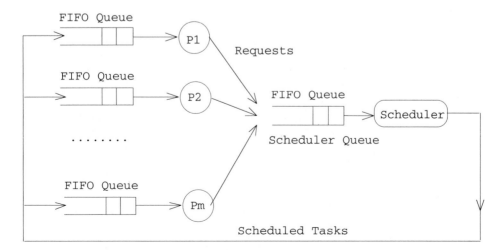

Figure 9.18 Dynamic scheduling

rently shortest Waiting Queue is selected, (2) the processor with the currently smallest waiting queue time is selected. (Queue time is the sum of all of the task execution times waiting in the queue.) When it is not possible to obtain accurate estimates of waiting queue times, the first heuristic above is used. But, clearly, it is often non-optimal, especially when the size of tasks varies greatly.

9.7.2   Static scheduling

Static scheduling of nondeterministic programs requires the use of a different program model that distinguishes between conditional branching and precedence relations among parallel program tasks. This model consists of two directed graphs: branch graph and precedence graph. Before we explain the difference between branch and precedence graphs, we first present an informal definition of an execution instance of a parallel program. An execution instance of a program is defined as the set of tasks that are selected for execution at one time for some input. The branch graph represents the control dependences and the different execution instances of the program, while the precedence graph represents the data dependence among tasks. Associated with each branch arc (u, v) is the probability of having v in the same execution instance with u. The summation of the probabilities associated with the arcs, leaving a node in the branch graph, is always one. These probabilities are obtained by running the program several times for different sets of input. We solve this problem by using the probabilities of executing program tasks to construct a schedule that favors the tasks that are likely to be selected for execution most of the time. As shown in Fig. 9.19, we present a multi-phase approach to provide a static schedule for parallel programs that contain conditional branches [11].

In the first phase, an attempt is made to reduce the amount of nondeterminism associated with conditional branching. This is accomplished by studying some graph theoretic properties of the two directed graphs of the model. The idea is to reduce the degree of nondeterminism in the task graph and construct a reduced graph model from the original parallel program model. The tasks that form the different branches of a certain predicate may have many features in common. These common features include the way in which the alternative tasks communicate with other tasks in the program. Also, the amount of computation of the alternative tasks might be comparable. If we can find a set of tasks with common properties, we should be able to represent this set by a single task. Thus, we can reduce, or hopefully remove, the nondeterminism associated with conditional branching.

There are three different types of possible dissimilarities among the alternative tasks of a branching statement: (1) the set of successors and predecessors in the precedence graph, (2) the cost of communication between the tasks and any common successor or predecessor in the precedence graph, and (3) the amount of computation needed for execution. The first phase is controlled by three parameters that correspond to these three types of possible dissimilarities. These parameters are referred to as tolerance parameters α, β, and γ. When the amount of dissimilarities for a set of alternatives are bounded by the values of α, β, γ, these alternatives are treated as if they are identical. The output of this process is two graphs: reduced branch graph and reduced precedence graph. In the second phase, a number of execution instances is generated. If the number of all possible execution instances is bounded, all possible instances are generated. But if the number of all possible instances is exponential, then only some of the execution instances are considered. These are the ones that are among the most likely to happen and cover all the program tasks; i.e., each task will be included in at least one execution instance.

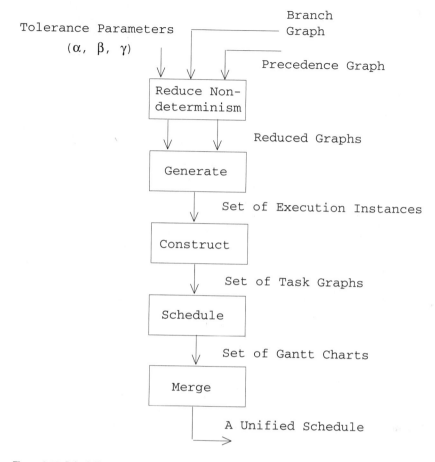

Figure 9.19 Scheduling conditional branching

In the third phase, we construct a precedence task graph for each execution instance generated in the second phase. Each task graph consists of the nodes given in the corresponding execution instance and the precedence relations among them. It also shows the amount of computation needed at each node as well as the size of the data messages passed among the nodes. In the next phase, each task graph is scheduled independently using one of the techniques used in scheduling branch-free task graphs. Given a task graph and a target machine description, a schedule in the form of a Gantt chart is fed to the merge phase.

In the final phase, a number of Gantt charts are combined into a unified schedule. The schedule is given in the form of processor allocation and execution order of the tasks allocated to the same processor. The allocation of each task is obtained by considering all Gantt charts. The summation of the probabilities as well as the number of times a task is assigned to the same processor are used to determine the allocation of that task. The execution order of each pair of nodes assigned to the same processor is obtained by considering all the Gantt charts in which both tasks are assigned to the same processor. In the event that there are two different orders for the two tasks, a weighted majority function is used to determine the order in the unified schedule. The weighted majority function is defined as the summation of the probability of occurrence.

9.8 Scheduling Tools

To develop a parallel application, there are a number of scheduling related questions that a programmer needs help to answer. What is the best grain size of the program tasks? What is the best scheduling heuristic to use? How can the performance be improved? How many processors should be used? Where should synchronization primitives be inserted in the code? Answers to these questions and many others can be figured out in cooperation between a software tool and a human program developer. In this section, we study scheduling software tools that can be used at two different phases of the software development life cycle: (1) design and (2) code generation. We describe Parallax as an example of design tools, and PYRROS, which is a code generation tool.

9.8.1 Parallax

Parallax is a software tool that aids in parallel program design by automating a number of scheduling heuristics and performance analysis tools. Parallax which was introduced by Lewis and El-Rewini produces: (1) schedules in the form of Gantt Charts for a number of scheduling algorithms, (2) performance charts in the form of line and bar graphs, and (3) critical path analysis. With Parallax, a user can: (1) model a parallel program as a task graph, (2) choose a method of optimization from several scheduling heuristics which will automatically produces a schedule, (3) choose the topology of the desired target machine (or design an arbitrary topology for the parallel processor of interest), and (4) observe anticipated scheduling and performance estimates obtained from scheduling the task graph onto the target machine. Parallax supports most of the scheduling heuristics introduced in the literature.

Parallax is a tool for investigating scheduling heuristics prior to the actual execution of a parallel program. That is, Parallax is a design tool as opposed to a programming tool. It can also be used to refine an existing parallel program after performance data has been collected from one or more runs. The basic idea of Parallax is to aid a human user to create a parallel program design as a task graph, enter the target machine as a graph, and then perform a number of "what if..." analyses. Figure 9.20 shows some of Parallax displays. More details about Parallax can be found in Refs. [20] and [21].

9.8.2 PYRROS

PYRROS is a compile-time scheduling and code generation tool that was introduced by Yang and Gerasoulis in 1992 [31]. The input of PYRROS is a task graph and the associated sequential C code. The output is a static schedule and a parallel C code for a given architecture. PYRROS has the following components: (1) task graph language with an interface to C, (2) scheduling system, (3) graphic displayer, and (4) code generator. The task graph language allows users to define partitioned programs and data. The scheduling system is used for clustering the graph, load balancing and physical mapping, and computation/communication ordering. The graphic displayer is used for displaying task graphs and scheduling results. The code generator inserts synchronization primitives and performs code optimization for a variety of parallel machines.

A user first edits the program using the task graph language to specify the dependence information between partitioned program segments, the associated C code, weights, and the maximum number of processors available. PYRROS can display the input dependence graph to help the user verify the correctness of the flow dependence between tasks and also generate the schedule and the code. The user can check the scheduling result by let-

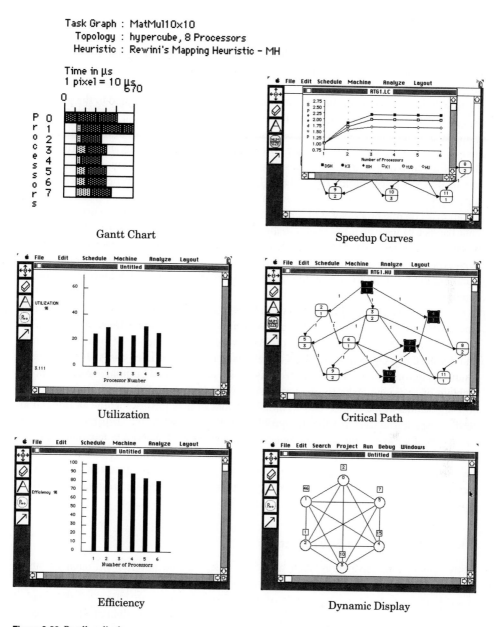

Task Graph : MatMul10x10
Topology : hypercube, 8 Processors
Heuristic : Rewini's Mapping Heuristic - MH

Gantt Chart

Speedup Curves

Utilization

Critical Path

Efficiency

Dynamic Display

Figure 9.20 Parallax displays

ting PYRROS display the schedule Gantt chart in the graph window and the statistics in information in another text window.

9.9 Task Allocation

The term *task allocation* has been used interchangeably with the term *task scheduling* in several places in the literature. Although the allocation of tasks in distributed systems may be considered a special case of task scheduling, failing to distinguish between these two terms causes some misunderstanding. In this section, we briefly review the problem of task allocation and give references to the main results in the literature. The problem of task allocation that arises when specifying the order of executing the system tasks is not required. In other words, system tasks might interact or communicate without imposed precedence relations. In a distributed computing systems made up of several processors, the interacting tasks constituting a distributed program must be assigned to the processors so as to make use of the system resources efficiently.

In the assignment of tasks to processors there are two types of cost: the cost of execution of a task on a processor, and the cost of interprocessor communication. To improve the performance of a distributed system, two goals need to be met: (1) interprocessor communication has to be minimized and (2) the execution cost needs to be balanced among different processors. These two goals seem to conflict with one another. On one hand, having all tasks on one processor will remove interprocessor communication cost but result in poor balance of the execution load. On the other hand, an even distribution of tasks among processors will maximize the processor utilization but might also increase interprocessor communication. Thus, the purpose of a task allocation technique is to find some task assignment in which the total cost due to interprocessor communication and task execution is minimized.

The task allocation problem is known to be NP-complete. A formal proof that the problem is NP-hard even in the restricted case when there are only two values of communication cost between tasks allocated on different processing elements: zero and one can be found in Ref. [4]. Optimal algorithms are obtained in very restricted cases. The first major result was introduced by Harold Stone in 1977. He suggested an optimal algorithm for the problem of assigning tasks to two processors by making use of the well known network flow algorithms in the related two-terminal network graphs [28]. Recently, an optimal algorithm was introduced to solve the problem in the case when the distributed system is composed of a linear array of any number of processors [22]. There are few other cases in which optimal solutions can be found using mathematical programming techniques, but in these cases, severe restrictions are imposed on the problem. Again, the intractability of the problem has led to the introduction of many heuristics. Several related results and heuristic algorithms can be found in Refs. [2, 6, 9, and 23].

9.10   Heterogeneous Environments

Many applications have more than one type of embedded parallelism, such as single instruction multiple data (SIMD) and multiple instructions multiple data (MIMD). Homogeneous systems use one mode of parallelism in a given machine and thus cannot adequately meet the requirements of applications that require more than one type of parallelism. As a result, a machine may spend its time executing code for which it is poorly suited. Heterogeneous computing offers a cost-effective approach to this problem by using existing systems in an integrated environment. Heterogeneous computing is defined as the well orchestrated and coordinated effective use of a suite of diverse high-performance machines to provide superspeed processing for computationally demanding tasks with diverse computing needs. Heterogeneous processing may use a variety of parallel, vector, and special architectures together as a suite of machines. The problem of partitioning and scheduling in homogeneous environments can be considered a special case of the problem when the target computer is a suite of heterogeneous machines. For example, code classification is another objective of program partitioning in a heterogeneous environment. The code needs to be classified based on the type of the embedded parallelism such as SIMD and MIMD. Matching the code type to the machine type will also add more constraints to the scheduling problem. Scheduling in heterogeneous environments can be done at two levels. At the system level, each task is assigned to one or more machines in the system so that the parallelism embedded in the task matches the machine type. At the machine level, portions of the task are assigned to individual processors in the machine.

9.10.1 Heterogeneous problem model

The parallel task T is divided into subtasks t_i, $1 \le i \le N$. Each subtask t_i is further divided into code segments t_{ij}, $1 \le j \le S$, which can be executed concurrently. Each code segment within a subtask can belong to a different type of parallelism (i.e., SIMD, MIMD, vector, and so forth) and should be mapped onto a machine with a matching type of parallelism. Each code segment may further be decomposed into several concurrent code blocks with the same type of parallelism. These code blocks t_{ijk}, $1 \le k \le B$, are suited for parallel execution on machines having the same type of parallelism. This decomposition of the task into subtasks, code segments, and code blocks is shown in Fig. 9.21.

9.10.2 Mapping applications onto heterogeneous systems

In this section, we present a methodology, introduced by Chen et al. in 1993, for mapping algorithms onto heterogeneous systems [7]. First, both the application and the heterogeneous systems must be represented using the *hierarchical cluster-M model* (HCM) as shown below.

The HCM system representation is obtained as follows:

1. **System clustering layer**. Each computer is in a cluster by itself. Each clustering level is constructed by merging clusters from the lower level that are completely connected. This is continued until no more clustering is possible.
2. Each resulting cluster is labeled according to the type of parallelism present in the cluster.

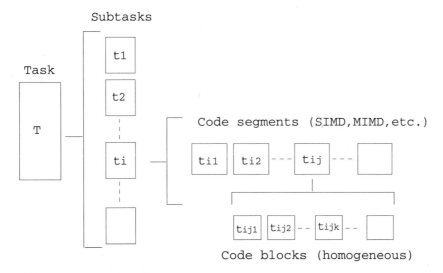

Figure 9.21 Heterogeneous application model

3. **Machine clustering layer**. Each processor in a computer is in a cluster by itself. All completely connected clusters are merged to form the next level of clustering. The highest level of clustering consists of one cluster containing all processors in the computer.

The HCM application representation is obtained as follows:

1. **Subtask clustering layer**. Each subtask t_i is represented by single cluster at level i.
2. **Code segment clustering layer**. Each code segment t_{ij} of subtask t_i is represented by a cluster. Each cluster is labeled with the parallelism type of its corresponding code segment. Code segment clusters, in the same subtask clustering at level i, are connected if results from the clusters are used by a single cluster of subtask clustering at level $i + 1$.
3. **Code block clustering layer**. Each cluster in this layer corresponds to a code block t_{ijk}. Each code block cluster is labeled with the type of parallelism present in the block.
4. **Instruction clustering layer**. This step yields the lowest layer of HCM clustering. It exploits fine grain parallelism at the individual instruction level.

Once both the HCM problem specification and system representation are obtained, the mapping process at several levels is carried out as follows.

1. **Code segment cluster mapping**
    a. For each cluster in a code segment clustering level of the task specification, find a system-level cluster that matches the type of parallelism in the segment. Assign code segment clusters to the appropriate system clusters.
    b. If a system with a matching type of parallelism is not found, then a cluster with the next best type pf parallelism is selected. This selection is based on information collected from analytical benchmarking.
    c. Code segment clusters that are connected in the task specification are mapped onto connected system level clusters. If appropriate connected clusters are not found, then map each two connected segment clusters onto suitable system clusters with a minimum communication cost.
    d. The above steps are repeated for all code segment clustering levels.
2. **Code block cluster mapping**
    Following the completion of code segment cluster mapping for all levels, code block clusters contained in each code segment cluster are mapped onto several subclusters contained in the corresponding system cluster.
3. **Instruction level cluster mapping**
    For each instruction clustering level, find all system clustering levels with the closest matching number of clusters. These levels are possible mapping level candidates.

9.11   Summary and Concluding Remarks

We discussed one of the important problems facing parallel computing; that is, the problem of optimally partitioning applications into modules and then scheduling these modules onto parallel or distributed environments. Partitioning includes parallelism detection and grain packing. The code is analyzed to determine potential parallelism in the program. Several operations may be combined into a larger grain to reduce communication overhead. The basic idea of grain packing is to find the best grain size that maximizes parallelism while reducing communication overhead. Program partitioning for heterogeneous environments also involves code classification based on the type of parallelism. The goal of scheduling is to determine an assignment of tasks to processors and an order in which tasks are executed to optimize some performance measure. Scheduling tasks on heterogeneous environments also involves matching the code type to the machine type. The program is assigned to one or more machines in the system so that the parallelism embedded in each task matches its machine type. Partitioning and scheduling is a computationally intensive problem and known to be NP-complete. Due to the intractability of the problem, recent research has emphasized heuristic approaches. Optimal algorithms can be obtained only in some restricted cases. In this chapter, we surveyed the field and explored solutions that have been proposed in the literature. We presented coverage of essential topics including: program and system models, optimal algorithms, heuristic algorithms, deterministic versus nondeterministic programs, software tools, and homogeneous versus heterogeneous environments.

Models for the parallel program, target machine, and different communication models were provided. The NP-complete results were summarized. When communication cost is ignored, optimal algorithms were presented to schedule trees and interval orders on a given number of processors, and arbitrary task graphs on two processors. We also presented two algorithms to schedule interval orders and tree structured task graphs on a given number of processors when communication cost is considered. Scheduling heuristics were introduced to solve the general form of the problem. We discussed heuristics that are based on list scheduling in which each task is assigned a priority, and the ready task with the highest priority is scheduled first. We also covered another type of heuristics that is based on the idea of task clustering. In addition, the problem of scheduling nondeterministic task graphs was discussed. This problem arises in several situations in parallel programs, particularly in the cases of loops and conditional branching. We showed that scheduling nondeterministic programs can be achieved dynamically, on the fly, with considerable amount of overhead. To reduce the overhead involved with dynamic scheduling, we indicated that static or hybrid methods can be applied. We also covered two scheduling software tools: Parallax, as an example of design tools, and PYRROS, which is a code generation tool. The task allocation problem was summarized and several references that cover the most important work in that area were provided. Finally, we briefly discussed the general partitioning and scheduling problem when the target computer is a suite of heterogeneous machines.

The topics discussed in this chapter covers a wide range of issues and research points that can be grouped in the following categories: optimal algorithms, heuristic construction, and software tools. Even though several instances of the problem have been proven to be NP-complete, there are still several open problems. For example, scheduling task graphs when communication is not considered and all tasks take the same amount of time on fixed $m \geq 3$ processors is still an open problem. More research is needed to obtain optimal algorithms when certain restrictions are relaxed in the cases that have already been solved. For example, communication delay is a major parameter in parallel and distributed systems that should be considered. Since optimal schedules can be obtained in restricted cases that may not necessarily represent real-world situations, a simplified suboptimal approach to the general form of the problem is needed. Recent research in this area has emphasized heuristic construction, evaluation, and application. The challenge is to incorporate real-world parameters in our solutions to the problem. Another research direction is the development of scheduling software tools to help design parallel programs, automatically generate parallel code, and estimate the performance. The following section provides a number of references for further reading in the field.

9.12 References

1. Adam, T. L., K. M. Chandy, and J. R. Dickson. 1974. A comparison of list schedules for parallel processing systems. *Comm. ACM.*, Vol. 17, 685–690.
2. Ali, H., and H. El-Rewini. 1993. Task allocation in distributed systems: a split graph model. *Journal of Combinatorial Mathematics and Combinatorial Computing*, Vol.14, 15–32.
3. Ali, H., and H. El-Rewini. 1993. The time complexity of scheduling interval orders with communication is polynomial. *Parallel Processing Letters*, Vol. 3, No. 1, 53–58.
4. Ali, H., and H. El-Rewini. 1994. On the intractability of task allocation in distributed systems. *Parallel Processing Letters*, Vol. 4, Nos. 1 & 2, 149–157.
5. Aho, A., R. Sethi, and J. Ullman. 1986. *Compilers: Principles, Techniques, and Tools.* Reading, Mass.: Addison-Wesley.
6. Bokhari, S. 1981. A shortest tree algorithm for optimal assignments across space and time in distributed processor system. *IEEE Transaction on Software Engineering*, vol. SE-7, no. 6.
7. Chen, S., et al. 1993. A selection theory and methodology for heterogeneous supercomputing. *Proc. Workshop on Heterogeneous Processing.* Los Alamitos, Calif.: IEEE CS Press, Order No. 3532–02.
8. Coffman, E. G. 1976. *Computer and Job-Shop Scheduling Theory.* New York: John Wiley & Sons.
9. El-Rewini, H., T. Lewis, and H. Ali. 1994. *Task Scheduling in Parallel and Distributed Systems.* Englewood Cliffs, N.J.: Prentice Hall.
10. El-Rewini, H., and H. Ali. 1994. On considering communication in scheduling task graphs on parallel processors. *J. of Parallel Algorithms and Applications,* Vol. 3, 177–191.
11. El-Rewini, H., and H. Ali. 1995. Static Scheduling of containing conditional branching in parallel programs. *Journal of Parallel and Distributed Computing* (January), 42–54.
12. El-Rewini, H., and T. Lewis. 1990. Scheduling parallel program tasks onto arbitrary target machines. *Journal of Parallel and Distributed Computing* (June), 138–153.
13. El-Rewini, H., and T. Lewis. 1991. Schedule-driven loop unrolling for parallel processors. *Proc. 24th Hawaii International Conference on System Sciences*, Kauai, Hawaii, 458–467.
14. Freund, R., and H. Siegel. 1993. Heterogeneous processing. *IEEE Computer*, Vol. 26 (June), 13–17.
15. Fujii, M., T. Kasami, and K. Ninomiya. 1969. Optimal sequencing of two equivalent processors. *SIAM Journal of Appl. Math.*, Vol. 17, No. 4.
16. Gabow, H. 1982. An almost linear algorithm for two-processor scheduling. *J. ACM*, Vol. 29, No. 3, 766–780.
17. Gerasoulis, A., and T. Yang. 1992. A comparison of clustering heuristics for scheduling DAGs on multiprocessors. *J. of Parallel and Distributed Computing*, Vol. 16, No. 4, 276–291.
18. Hu, T. C. 1961. Parallel sequencing and assembly line problems. *Operations Research*, Vol. 9, No. 6, 841–848.
19. Lewis, T., and H. El-Rewini. 1992. *Introduction To Parallel Computing.* Englewood Cliffs, N.J.: Prentice Hall.
20. Lewis, T., and H. El-Rewini. 1993. Parallax: a tool for parallel program scheduling. *IEEE Parallel and Distributed Technology: Systems and Applications*, Vol. 1, No. 2, 62–72.
21. Lewis, T., H. El-Rewini, P. Fortner, J. Chu, and W. Su. 1990. Task Grapher: a tool for scheduling parallel program tasks. *Proceedings of the 5th Distributed Memory Computing Conference*, 1171–1178.
22. Lee, C., D. Lee, and M. Kim. 1992. Optimal task assignment in linear array networks. *IEEE Trans. on Computers*, Vol 41, No. 7, 877–880.
23. Lo, V. 1988. Heuristic algorithms for task assignment in distributed systems. *IEEE Trans. on Computers*, Vol 37, No. 11, 1384–1397.
24. McCreary, C., and H. Gill. 1989. Automatic determination of grain size for efficient parallel processing. *Comm ACM*, 245–251.
25. Papadimitiou, C. H., and M. Yannakakis. 1979. Scheduling interval-ordered tasks. *SIAM Journal of Computing*, Vol. 8, 405–409.
26. Sarkar, V. 1991. Automatic partitioning of a program dependence graph into parallel tasks. *IBM J. Res. Develop.*, Vol 35, No. 5/6.
27. Sethi, R. 1976. Scheduling graphs on two processors. *SIAM J. Comput.*, Vol. 5, No. 1, 73–82.

28. Stone, H. 1977. Multiprocessor scheduling with the aid of network flow algorithms. *IEEE Trans. Software Eng.*, 85–93.

29. Ullman, J. 1975. NP-complete scheduling problems. *Journal of Computer and System Sciences*, Vol. 10, 384–393.

30. Wu, M., and D. Gajski. 1990. Hypertool: a programming aid for message-passing systems. *IEEE Trans. Parallel and Distributed Systems*, Vol. 1, No. 3, 101–119.

31. Yang, T., and A. Gerasoulis. 1992. PYRROS: static task scheduling and code generation for message passing multiprocessors. *Proc. 6th ACM International Conference on Supercomputing*, 428–443.

10

Checkpointing in Parallel and Distributed Systems

Avi Ziv [*] *and Jehoshua Bruck* [†]

Fault tolerance techniques enable systems to perform tasks in the presence of faults. The likelihood of faults grows as systems are becoming more complex and applications are requiring more resources. This chapter provides an overview of the issues related to achieving fault tolerance at the application/system software layer in parallel and distributed systems. The chapter is focused on the description of checkpointing schemes that are based on task duplication and on techniques for achieving state consistency in checkpointing in parallel and distributed systems.

10.1 Introduction

Fault tolerance techniques enable systems to perform tasks in the presence of faults. The likelihood of faults grows as systems are becoming more complex and applications are requiring more resources, including execution speed, storage capacity and communication bandwidth.

Reliability and resilience are critical issues in parallel and distributed systems. These systems combine between a plurality of computing, communication, and storage resources. Fault tolerance is associated with various levels of parallel and distributed systems. The first level is the application/system software level. Reliability is typically addressed at this level by utilizing checkpointing and recovery schemes. The second level is the architecture level, here reliability is achieved by designing architectures that employ redundancy and facilitate computing in the presence of faults. The third level is the actual hardware level. At this level the focus is on schemes that ensure the reliability of particular functional hardware and low level software entities. In many cases, the focus at this level is on detection mechanisms.

[*]Partially supported by the IBM Almaden Research Center, San Jose, California,
[†]Partially supported by the NSF Young Investigator Award CCR-9457811.

The goal of this chapter is to provide an overview of the issues related to achieving fault tolerance at the application/system software layer. The focus of the presentation will be to describe the techniques of checkpointing in resilient parallel and distributed systems.

10.1.1   Terminology

We start by providing some terminology. The common approach in achieving fault tolerance consists of detection and location of a fault in the system followed by reconfiguration of the system to restore its operational condition [15, 34]. Fault-tolerant schemes typically consist of the following steps:

Step 1   *Fault detection.* This is the process of recognizing that an error has occurred in the system.

Step 2   *Fault location.* After the error in the system is detected, the location of the part in the system that caused the error is identified.

Step 3   *Fault containment.* After the fault is located, the faulty part is isolated to prevent a possible propagation of the fault to the rest of the system.

Step 4   *Fault recovery or resilience.* This is the process of restoring the operational status of the system to a previous fault-free consistent state. This step can include reconfiguration of the system.

There are a number fault sources in systems, including physical failure of components, environmental interference, software errors, security violations, and operator errors. Faults can be classified into two types: permanent and transient faults [18]. Permanent faults are faults that cause a permanent damage to some part of the system. Recovery from permanent faults must include containment of the damaged part and reconfiguration of the system. Transient faults are faults that appear for a short period of time and do not cause permanent damage. Recovery from transient faults is easier than recovery from permanent faults, because reconfiguration of the system is not needed. However, detection of transient faults might be more difficult, because they might disappear without a detectable effect on the system.

Early studies showed that the rate of transient faults is 10 to 30 times higher than the rate of permanent faults [35]. As semiconductor technology advances, we will see systems operating with higher clock rates and lower power. This trend increases the sensitivity of systems to environmental effects and may result in a higher rate of transient faults.

10.1.2   Why checkpointing?

In scientific and commercial applications, the execution of the program that has been interrupted by the fault has to be restarted from the beginning. As a result, the applications are completed only if a long enough fault-free interval of time exists in the system. It has been shown (for example, in Ref. [11]) that the average execution time of a program in the presence of faults grows exponentially with the length of the program. Checkpointing is commonly used to avoid losing all the useful processing done before a fault has occurred. Checkpointing consists of intermittently saving the state of the program in a reliable storage medium and, upon detection of a fault, completing the fault recovery process and then restoring the previous consistent state. Checkpointing enables the execution of the program to be resumed from a previous consistent state rather than its beginning; hence, the amount of useful processing lost because of the fault is much less. With checkpointing, the average execution time of a program grows only linearly with the length of the program [11].

Checkpoints can be set at fixed time intervals or at distinct points in the program code. Checkpoints can be set by the programmer [8], the compiler [25], or the system hardware [5]. The interval between checkpoints should be selected carefully. On one hand, placing checkpoints at small intervals introduces a large overhead for the storage of the program state. On the other hand, if the interval between checkpoints is long, the amount of useful processing lost due to the fault is increased. The length of the optimal interval between checkpoints has been extensively studied (for example, see Refs. [13, 22, 27]).

10.1.3   Organization of this chapter

Parallel and distributed systems enable efficient task duplication that is useful in creating checkpointing schemes; however, the distributed nature of the execution requires that special attention be given to achieving state consistency at checkpoints. Specifically, the two key aspects that we will address are checkpointing schemes that are based on task duplication and techniques for achieving state consistency in checkpointing in parallel and distributed systems.

The chapter is organized as follows. In the next section we will describe checkpointing schemes that are based on task duplication; those schemes enable detection as well as recovery. In addition, we will describe analytical techniques that help in performance evaluation of those schemes. In Section 10.3, we will describe checkpointing schemes for parallel/distributed environments in which application programs consist of several processes that communicate with each other. We will focus the presentation on the issue of efficient checkpointing of a consistent state and present coordinated as well as independent checkpointing schemes.

Because of space limitations, we will not discuss the aspects of recovery in database management systems under faults and security violations. Good references to this vast area are the books by Bernstein, Hadzilacos, and Goodman [4], Krishnamurthy and Murthy [20], and a paper by Weihl [41]; for recovery when security violations occur, see the paper by Murthy and Krishnamurthy [26].

10.2   Checkpointing Using Task Duplication

The dominant cause for errors in computer systems are transient faults [35]. Parallel and distributed computing systems provide hardware and software redundancy that helps to reduce the probability of an undetected transient fault, using task duplication [1]. Task duplication is analogous to shadowing techniques used for recovery in database transaction processing systems, where all the information is duplicated (see Ref. [20]). The detection of a transient faults is enabled by comparing the states of the same task that was executed in parallel on different processors. This approach provides a mechanism for increasing the reliability of those systems [31].

The checkpoints serve two purposes: (1) detecting faults that occurred during the execution of a task and (2) reducing the time spent in recovering from faults. Fault detection is achieved by duplicating the task into two or more processors and comparing the states of the processors at the checkpoints. We assume that the probability of two faults resulting in identical states is very small; hence, two matching states is an indication of a correct execution. By saving the state of the task at each checkpoint, we avoid the need to restart the task after each fault. Instead, the task can be rolled back to the last correct checkpoint and execution resumed from there to shorten the fault recovery process.

In this section we describe several checkpointing techniques that use task duplication. We also present analytical methods that are useful in the performance evaluation of those schemes.

10.2.1   The system model

Implementation of a system which uses checkpointing can be done in various ways. Figure 10.1 illustrates a logical view of the system architecture. The system consists of a set of processing elements (PEs). Each PE has a processor and a private memory—either its own memory in a distributed memory system or part of a shared memory system. All the PEs in the system are identical. The system has a stable storage that can be accessed by all processing elements, and it is used to store the states of the PEs at checkpoints. A special processor called the *checkpoint processor* is used to coordinate the execution of tasks on the PEs. The checkpoint processor assigns tasks to processors, detects failures in the PEs by comparing the states of the processors at checkpoints, and coordinates the recovery according to the recovery scheme used.

Figure 10.1 illustrates an example of the architecture of the system. Notice that the checkpoint processor can be implemented as part of the processing elements or as a separate processor. The stable storage can be part of a shared memory, a separate part of a distributed memory, or stored in offline memory (disk). Different implementations of the architecture provide different execution times for operations in the fault recovery schemes and affect performance.

In checkpointing schemes, the task is divided into n intervals. At the end of each interval, a checkpoint is placed either by the programmer [8], the compiler [25], or the system hardware[*] [5]. The execution of a task is done in steps. Each step consists of a series of operations. The first operation in each step is executing a single interval of the task by all the processors that are assigned to it. Note that it is not necessary that all the processors execute the same interval at the same step. After the execution of each interval is completed, the checkpoint processor performs those operations necessary to achieve fault detection and recovery; i.e., it stores the states of the processors at the checkpoint in the stable storage and compares those states. Based on the result of the comparison and the scheme used, the checkpoint processor decides what further action should be taken.

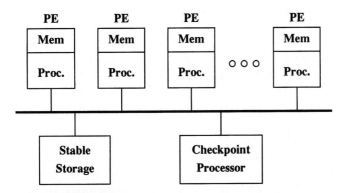

Figure 10.1 Logical system model

---

[*]Note that if the checkpoints are placed by the system hardware, it has to ensure that each checkpoint is placed at the same point in all processors.

Next, we describe three checkpointing techniques that illustrate the various ideas related to checkpointing with task duplication.

10.2.2 Checkpointing with simple rollback

Here we describe the most basic scheme with task duplication. In this technique, the task is duplicated and executed in parallel on two processors. When a checkpoint is reached, the state of the processors that execute the task are compared. If the states match, a correct execution of the interval before the checkpoint is assumed. In this case, the checkpoint processor saves the correct state of the processors, and the processors proceed to the execution of the next interval. If the states do not match, a fault has occurred in (at least) one of the processors. In this case, the checkpoint processor rolls back both processors to the last saved checkpoint state, and the execution of the last interval is repeated. Figure 10.2 presents the flow-chart for this scheme.

It is clear that program execution time can be reduced if more than two processors execute the task in parallel. For example, if three processors are used in parallel, rollback is not necessary if only a single fault in one of the processors has occurred. In this case, the correct state at the checkpoint can be copied from one of the non-faulty processors to the faulty processor, and the execution of the task can be resumed. This scheme with three processors is called the *triple modular redundancy with look forward* (TMR-F) scheme in Refs. [24] and [44]. The use of more than two processors reduces the execution time of a task, but it has two disadvantages: it uses more of system resources, and it is less reliable than the simple rollback scheme with two processors. (The probability that two copies of the a task will result in the same faulty state is higher.)

10.2.3 Checkpointing with lookback and rollback

One of the drawbacks of the foregoing simple rollback technique is that a correct execution of the interval has to occur at both processors at the same time. In lookback schemes [24], the task is executed in parallel by two processors and, when a fault occurs, both processors are rolled back and execute the same interval again. The difference between this scheme and the simple rollback scheme is that all the unverified checkpoints are stored

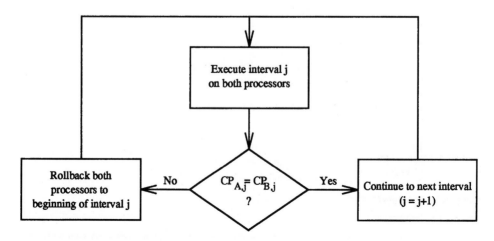

Figure 10.2 Flowchart for simple rollback scheme

and compared, not just the checkpoints of the last step. Hence, two steps with a single non-faulty execution are sufficient for verifying a correct execution.

For example, consider the case when one of the processors that executes a task always generates a random state. In this case, the simple rollback scheme will never be able to match the states of the processors, and the execution of the task will never end. On the other hand, when lookback is used, two correct executions on the other processor will match, and the execution of the task will progress and eventually terminate. The flowchart of a lookback scheme called *double modular redundancy with two lookback processors* (DMR-B-2) [24] is given in Figure 10.3.

Using a lookback scheme instead of a simple rollback scheme might reduce the reliability of the system, because of the increased probability of two matching states being produced by two faulty processors. For example, in the case described above, if the faulty processor generates the same random state every time, the checkpoint processor might decide that this is the correct state and will use it to continue the execution. As in the case of the simple rollback schemes, the task execution time can be reduced when more than two processors are used in parallel to execute the task.

10.2.4   Roll-forward checkpointing scheme

In the schemes described above (i.e., the simple rollback and the lookback schemes), the checkpoint processor rolls back both processors each time a mismatch in the states is detected. Pradhan [29] has described a recovery technique that tries to avoid rolling back the processors by using undedicated spare processors and roll-forward recovery. Unlike the TMR-F scheme where rollback is avoided by continuous utilization of a third processor, in roll-forward schemes, the task is executed by two processors, and spare processors are used for a short period of time in the case of faults.

In this section we describe the *roll-forward checkpointing scheme* (RFCS) [30]. Other schemes that use the roll-forward approach are the DMR-F-1 and DMR-F-2 (for *double modular redundancy with roll-forward* and one or two recovery processors) described in Ref. [24]. A roll-forward scheme that uses three processors without non-dedicated spares is described in Ref. [33]. In a roll-forward checkpointing scheme (RFCS), the task is executed by two processors during normal execution. When a fault is detected, a third spare

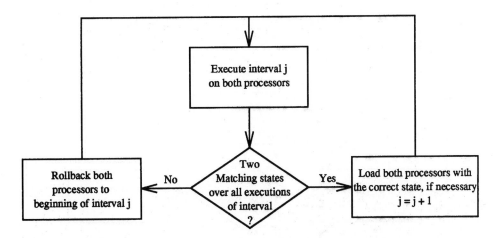

Figure 10.3 Flowchart for lookback scheme

processor is added to identify a regular processor with a correct execution. Figure 10.4 describes the RFCS scheme.

The RFCS scheme consists of three phases.

1. *Normal execution phase.* During normal execution, the task is executed on two processors in parallel. At the end of each interval, the states of the processors are compared. If the states match, no fault is assumed, and the execution continues to the next interval. If the states do not match, the fault recovery process begins.
2. *Recovery, phase I.* In the first phase of fault recovery, the spare processor is loaded with the last verified checkpoint and it tries to verify the next checkpoint, while the two regular processors continue with the normal execution of the next interval. If the state of the spare processor does not match either of the checkpointed states associated with the two regular processors, the task is rolled back to the last verified check-

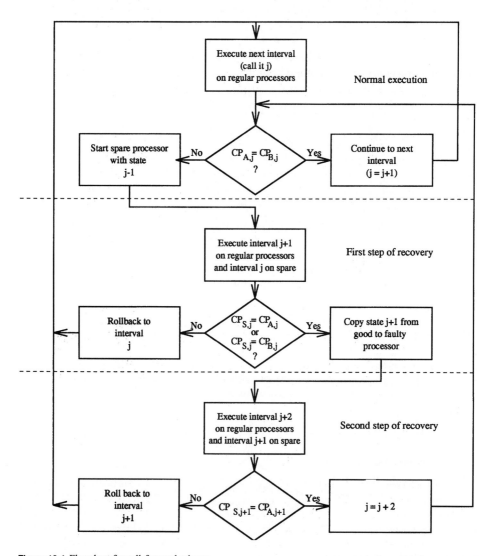

Figure 10.4 Flowchart for roll-forward scheme

point, and normal execution is resumed from that point. If the spare processor succeeds in verifying the first checkpoint, the state of the correct processor is copied to the faulty processor, and the second phase of fault recovery takes place.

3. *Recovery, phase II.* In the second phase of fault recovery, the spare processor tries to verify the checkpoint that corresponds to the first phase of recovery (which has only one potential correct execution). At the same time, the regular processors continue the execution of the next interval. If the spare processor is unable to verify the checkpoint state of the non-faulty processor, the task is rolled back to the last verified checkpoint, and normal execution is resumed from that point. If the verification succeeds, then the state of the two regular processors is compared, and action is taken as in normal execution.

In RFCS, as in other roll-forward schemes, the fault recovery process tries to avoid rolling back the task by using special processors that repeat the execution of the interval with the fault in it, while other processors continue to execute the task. If the recovery process succeeds, then the time to re-execute the faulty interval is saved. On the other hand, if the recovery process fails, the recovery attempt time is wasted. Hence, roll-forward schemes perform well when there is no correlation between faults in different processors, and the probability of successful roll-forward recovery is high. On the other hand, when there is a high correlation between faults in different processors, the probability of successful roll-forward recovery is low, and rollback schemes (with and without lookback) perform better [43].

10.2.5 Self-detection of faults

So far, we have assumed that the only way to detect transient faults is by duplication of the task and comparison of the task states. In most real-life systems, processors have some ability to detect faults by themselves (for example, using parity bits). Self-detected faults can be used to improve the performance of checkpointing schemes. One way to use self-detected faults to improve the performance of a checkpointing scheme is by starting the fault recovery process as soon as the a fault is detected (there is no need to wait for the state comparison at the checkpoint to detect that a fault has occurred).

Another way where self detection of faults can be used to improve the system performance is by exploiting the trade-off between execution time and reliability. In Ref. [31], two versions of the simple rollback and RFCS schemes are presented. The two versions of each scheme differ in the way they handle the case where exactly one processor has a self-detected fault. In the first version, this scenario is treated in the same way as if the comparison at the checkpoint failed, causing rollback or beginning of the fault recovery process.[*] In the second version, when this scenario occurs, the processor without the self-detected fault is assumed to be non-faulty. So, using this version of the schemes, the state of the processor without the self-detected fault is used as the correct state without verifying it. The results given in Ref. [31] show that using the second version of the scheme reduces the execution time of a task, with only a small reduction in the system reliability.

10.2.6 Performance analysis

Performance evaluation of checkpointing schemes is important when selecting a scheme that fits the user's requirements and for optimizing the performance of a given scheme in

---

[*]Actually, in RFCS, the system can immediately jump to the second step of the recovery process because the system knows that the processor with the self-detected fault cannot have the correct state; hence, the state of the other processor can be copied to it immediately.

a specific system. The characteristics of the system on which the scheme is implemented and the environment in which the scheme operates have a major effect on the absolute and relative performance of the schemes [44, 43]. In this section, we briefly describe three techniques that are used to evaluate the performance of checkpointing schemes. We also describe three important performance aspects of checkpointing schemes; namely, the execution time of a task, the processors work needed to execute a task, and the reliability of a scheme. We compare the performance of the three types of schemes described in this section.

Considerable theoretical work has been devoted to the analysis of checkpointing schemes in uniprocessor systems and to determining the optimal checkpoint intervals. Gelenbe [13] showed that to maximize the availability in transactions systems, checkpoint intervals should be deterministic and of the same length. L'Ecuyer and Malenfant [22] derived a numerical approach for availability in dynamic checkpointing strategies when the fault rate is not constant. Nicola and van Spanje [27] compared analysis and optimization of several checkpointing models that differ in the checkpoints' placement and fault occurrence in transaction systems. Kulkarni, Nicola, and Trivedi [21] investigated the effects of checkpointing on the execution time of a program in queueing systems. Coffman and Gilbert [9] described optimal strategies for placement of checkpoints in a single program. However, in the work described above, it is assumed that fault detection is done internally by the processors, and the only role of checkpoints is to shorten the recovery time from an error.

In the checkpointing schemes described in this section, checkpoints are used for both fault detection and speedy recovery. Performance evaluation of the multiprocessors checkpointing schemes is more complicated than performance evaluation of checkpointing schemes in uniprocessor systems, because the recovery process in multiprocessors systems is more complicated. In multiprocessors schemes, the recovery process can include activation of spare processors (in the roll-forward schemes [29]), using checkpoint states from the past (e.g., the lookback schemes [24]), and so on, but in uniprocessor schemes the recovery process includes only rolling the task back to the last checkpoint.

10.2.6.1 Performance analysis techniques. One of the techniques to evaluate the performance of a checkpointing scheme is to use a simulation program. In Refs. [24] and [33], simulation programs were used to evaluate the average execution time of a task using several checkpointing schemes. The advantage of using simulation programs to evaluate the performance of checkpointing schemes is that simulation can be used in cases where the analytical methods are complex—for example, when the model of one of the components of the evaluation (e.g., the system or the environment) is too complicated. One of the disadvantages of simulation programs is that they take a long time to execute and, as a result, only a limited number of cases can be examined.

In Ref. [44], a method for analysis of checkpointing schemes was presented. The analysis of checkpointing scheme is based on modeling the scheme and the behavior of the environment as a *Markov reward model* (MRM) [14]. The analysis of the scheme is done in three steps. In the first step, the scheme is modeled as a Markovian state-machine; that is, a state-machine in which a transition depends only on the current state, not how it was reached. In the second step of the analysis, rewards that represent the behavior of the system and quantities related to the measures of interest are assigned to the edges of the state machine. In this step, the transition probabilities that depend on the error probabilities of the processors are assigned to the edges. In the third and final step, the MRM that was created in the first two steps is analyzed to derive the values of the measures of interest.

As an example for the analysis technique, we describe the average execution time analysis of the DMR-B-2 scheme [24] that was described in Section 10.2.3. We assume that the length of the task is single time unit and that the task is divided into n equal length intervals, each of length $t_i = 1/n$. We also assume that the faults in the processors are independent of each other and that the probability that a fault will occur in a processor during one step of execution is F.

The state machine that describes the operation of the scheme is given in Fig. 10.5. The state machine has two states: state 0 and state 1. The state number indicates the number of correct executions of the current interval that have taken place so far. Specifically, the execution of a new interval begins when the state-machine is at state 0. If, during the execution of the interval, no fault occurred in either processor, the task moves to the next interval, and the state-machine remains at state 0 (transition via edge 0). If both processors had faults, then no correct execution took place, and the state-machine stays at state 0, but the task remains in the same interval. If only one of the processors had a fault, then one correct execution took place, and the state-machine moves to state 1. When the state-machine is at state 1, the execution of the interval can be completed if at least one of the processors did not have a fault. In this case the state-machine moves to state 0 to begin the execution of the next interval.

Note that more than one transition is possible between two states. For example, the state-machine can move from state 0 to state 0 via edges 0 and 2. Parallel edges differ in some of their properties; for example, edge 0 is used when the execution of the task continues in the next interval, whereas edge 2 is used when the current interval is repeated. The events that cause transition in the different edges in the state-machine and their probabilities are given in Table 1.

We associate each edge with two properties (or rewards) that describe the time it takes to execute a step and the progress in the execution. The time to execute a step includes the execution time of the interval (t_I) and the time to store and compare the checkpoint states (t_{ck}). If at least one of the processors had a fault, then additional time is needed to perform fault recovery operations. If a second correct execution was not found, then both processors have to be rolled back to the beginning of the interval (t_r). If a second correct execution is found, but one of the processors had a fault, then the state of correct processor is copied to the faulty one (t_{cp}). The values of the time-to-execute property are given in the 4th column of Table 10.1. The progress in the execution of task indicates if the execution of the task continues to the next interval (value 1), or the execution of the same interval is repeated (value 0). The progress property is given in the last column of the table in Table 10.1.

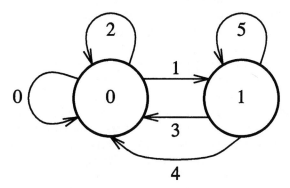

Figure 10.5 State machine for the DMR-B-2 scheme

TABLE 10.1 Transition Description for the DMR-B-2 Extended State Machine

Edge. No.	Event	Transition Probability (p_i)	Time to Execute (t_i)	Progress (v_i)
0	No faults	$(1-F)^2$	$t_I + t_{ck}$	1
1	Single fault	$2F(1-F)$	$t_I + t_{ck} + t_r$	0
2	Double fault	F^2	$t_I + t_{ck} + t_r$	0
3	No fault	$(1-F)^2$	$t_I + t_{ck}$	1
4	Single fault	$2F(1-F)$	$t_I + t_{ck} + t_{cp}$	1
5	Double fault	F^2	$t_I + t_{ck} + t_r$	0

Using the transition probabilities, the transition matrix of the Markov chain that describes the behavior of the scheme is

$$P = \begin{bmatrix} p_0 + p_2 & p_1 \\ p_3 + p_4 & p_5 \end{bmatrix} = \begin{bmatrix} (1-F)^2 + F^2 & 2F(1-F)) \\ 1-F^2 & F^2 \end{bmatrix}$$

with steady-state probabilities of the states

$$\pi = \{\pi_0, \pi_1\} = \left\{ \frac{1+F}{1+3F}, \frac{2F}{1+3F} \right\}$$

Using these steady-state probabilities, the steady-state probability of transition via edge i from state m to n is

$$e_i = \pi_m \cdot p_i$$

The average number of steps needed to complete one interval is

$$S = \frac{1}{\displaystyle\sum_{i=0}^{5} e_i v_i} = \frac{1+3F}{(1-F)(1+F)^2}$$

and the average time to execute a step is

$$T_s = \sum_{i=0}^{5} e_i t_i = t_I + t_{ck} + \frac{4F^2(1-F)t_{cp} + (2F + F^2 + F^3)t_r}{(1+3F)}$$

The average time to execute a single interval is

$$T_1 = ST_s = \frac{(1+3F)(t_I + t_{ck}) + 4F^2(1-F)t_{cp} + (2F + F^2 + F^3)t_r}{(1-F)(1+F)^2}$$

and the average time to execute the whole task is

$$T_n = nT_1 = \frac{(1+3F)(1+nt_{ck}) + 4F^2(1-F)nt_{cp} + (2F+F^2+F^3)nt_r}{(1-F)(1+F)^2}$$

Analysis of checkpointing schemes provides a way to study the performance and enables the comparison between the various schemes and the selection of optimal values for scheme parameters—for example, the selection of the optimal number of checkpoints [43]. Using analysis tools enables the design of a scheme that is the best over all known schemes given the fault rate and the dependencies between faults in different processors.

The performance of the schemes is affected by the fault model, and some schemes perform better than others in different fault models. For example, the roll-forward scheme uses the roll-forward technique to avoid rollback and reduce the execution time of the task. This technique is useful when there is little correlation between faults in different processors. On the other hand, when faults in different processors are correlated, the roll-forward techniques do not perform well.

Scheme performance is also affected by the architecture of the system in which the scheme is implemented, and the exact details of the implementation. For example, when workstations connected by a LAN are used to implement the schemes, operations that involve more than as single workstation and need the LAN take a longer time to execute than operations that can be done locally. In this case, schemes that mostly rely on local operations (e.g., the lookback schemes) perform better than schemes that use the network extensively [44]. To further reduce the use of the network, examples of a new type of checkpointing schemes were presented in Ref. [45]. In these schemes, the states of the processors are not compared at every checkpoint. Instead, each processor saves its own state locally at each checkpoint, and the states are compared every small number of checkpoints. Analytical and experimental results show that this approach for checkpointing can significantly reduce the overhead time caused by the checkpointing.

10.2.6.2 Comparison of checkpointing schemes. Next, we discuss some of the important aspects of the performance of checkpointing schemes and use results from Refs. [31] and [44] to compare among the various schemes that were presented earlier in this section. The performance aspects of a checkpointing scheme that we consider here are the execution time of a task, the amount of processor work that is required to complete the execution of a task, and the reliability of the scheme.

Execution time. The execution time of a task is defined as the total time elapsing from the beginning of the execution of the task until the last checkpoint at the end of the task is verified. The execution time includes the time spent on actually executing the task code, as well as the time required by the system to perform all the operations required by the scheme. The execution time of a task is important in many applications, such as interactive systems where a fast response is desired, and real-time control systems that face hard deadlines.

The average execution time of a task using several schemes is compared in Ref. [44]. The comparison is done for the case when faults in different processors are independent of each other. Figure 10.6 shows the average execution time of three schemes that represent the three techniques that were described earlier in this section. The three schemes that are compared are as enumerated below.

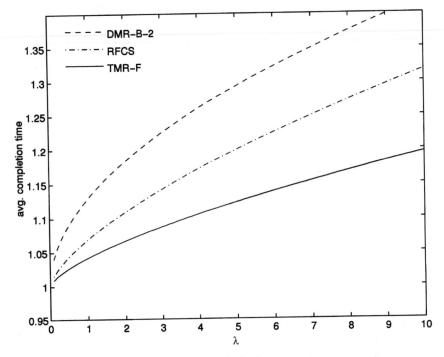

Figure 10.6 Average execution time with optimal checkpoints

- TMR-F [24]. This is the simple rollback scheme with three processors that was described in Section 10.2.2.
- DMR-B-2 [24]. This is the look-back scheme that was described in Section 10.2.3.
- RFCS [30]. This is the roll-forward scheme that was described in Section 10.2.4.

Figure 10.6 shows the average execution time of the above three schemes for a task of length l, when overhead of 0.002 occurs at each checkpoint. The figure shows the average execution time as a function of the fault rate, λ.

The figure shows that the TMR-F scheme, despite being the simplest of the three schemes, has the shortest execution time. The TMR-F scheme has better execution time because it uses more processors than the other schemes and thus has a much lower probability of failing to find two identical states at a checkpoint. The DMR-B-2 scheme is the worst because it uses only two processors and does not use spare processors to try to overcome faults. RFCS uses spare processors during fault recovery and thus has better performance than DMR-B-2.

Processor work. Another aspect of a scheme performance is the processor work that is required to complete the execution of a task. The total processor work to complete the execution is defined as the sum of the execution times over all the processors participating in the scheme. The processor work to complete a task depends not only on the time to complete the task but also on the number of processors used. It is an important performance aspect in transaction systems, where high availability of the system is required. Thus, the system should use as few resources as possible. In systems of this type, reducing the total work to complete a task means increasing the total throughput of the system.

In Ref. [44], the processor work of several checkpointing schemes are compared. The results there show that schemes with low execution times are not work efficient. It also indicates that the lowest amount of work is done using schemes that use a small number of processors and have higher execution time. Figure 10.7 shows the average processor work for the same three schemes, namely, the TMR-F scheme, the DMR-B-2 scheme, and the RFCS scheme. The comparison is performed for a task of length 1 with overhead time of $t_{oh} = 0.002$ at each checkpoint. As can be seen from the figure, the results here are the reverse of the results in the average execution time. The best scheme here is the DMR-B-2, which always uses only two processors. The RFCS, which uses two processors during normal execution and adds spare processors during fault recovery, requires more work. The TMR-F scheme, which uses three processors, is the worst scheme.

Scheme reliability. An important property of a checkpointing scheme its reliability. The reliability of a scheme is defined as the probability that a task is completed correctly. So far, we assumed that two faults cannot produce identical states. Under this assumption, the schemes described in this section are totally reliable; that is, tasks are always completed correctly. While the probability that two faulty processors produce identical states is small, this probability is not zero and should be considered in systems where extremely high reliability is required.

There is a trade-off between the reliability of a scheme and its performance. For example, instead of deciding that the occurrence of two identical states corresponds to a correct execution, one can require that three or more identical states correspond to a correct execution. This requirement increases both the execution time of a task and the number of processors that are needed to execute the task; however, this requirement reduces the probability that a wrong decision will be made; i.e., it increases the task reliability.

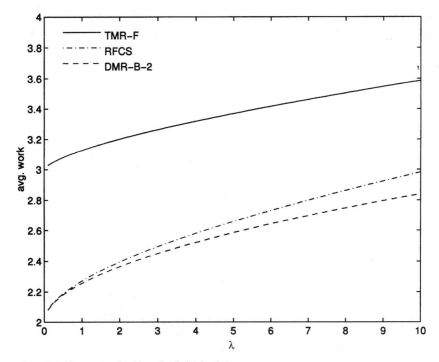

Figure 10.7 Average work with optimal checkpoints

The reliability of a task depends on the number of possible states that can exist at each checkpoint [31]. If instead of comparing the whole state of the processors at a checkpoint, only a small signature is used for comparison, the probability that two faulty states are represented by the same signature increases, and the reliability of the task decreases. On the other hand, using signatures for comparisons reduces the comparison time and improves the execution time. In [31], Pradhan and Vaidya showed that by using self-detected faults the execution time of a task can be reduced while having a reduction in the scheme reliability. The results in Ref. [31] also show that, in most cases, simple rollback schemes are more reliable than roll-forward schemes.

10.3   Techniques for Consistent Checkpointing

In the previous section we have described how task duplication can help in detection and recovery from transient faults in multiprocessor systems. In this section we show how checkpointing can be used to achieve fault-tolerant computation for parallel and distributed programs that are executed on distributed memory message passing systems. Distributed memory multiprocessors consist of a set of processors connected by some interconnection network. Each processor has its own local memory, and processors communicate using messages. In this section, we focus on the fault recovery process while assuming that some fault detection mechanism exists. (For example, the task duplication techniques that were described in Section 10.2 can be used.)

Fault recovery is more complicated in multiprocess programs than in sequential programs. In a multiprocess program, it is not enough to roll back the process that was interrupted by the fault to its last saved checkpoint. Instead, the whole system has to be rolled back to a consistent global state [7]. The state of processor p at checkpoint t consists of all the events that happened in that processor prior to t. An event can be either a local computation or sending/receiving of a message. A global state of the system consists of the states of all its processes. A consistent global state of a system can be defined in two different ways.

- A consistent global state is a global state in which the send events corresponding to all the receive events in the global state are in the global state [19]. That is, if process p received a message from process q, and the event of receiving the message is in the global state, then the event corresponding to sending the message from q is also in the global state.
- A consistent global state is a global state in which the send events corresponding to all the receive events in the global state are in the global state and, in addition, the receive events corresponding to all the send events in the global state are in the global state [37].

Note that every consistent global state according to the second definition is a consistent state according to the first definition as well. We can consider the global state to be a cut in the time diagram of the execution, connecting the local states of the processes. In this time diagram, the event of sending or receiving a message is described as an arrow that starts at the point where the message is sent and ends at the point where the message is received. According to the first definition, a consistent global state is a state in which no arrow starts at the right side of the cut and ends at the left side of it. According to the second definition, a consistent global state is a state in which no arrow crosses the cut. Figure 10.8 shows three global states: state (a) is consistent according to both definitions, state (b) is consistent only according to the first definition, and state (c) is inconsistent.

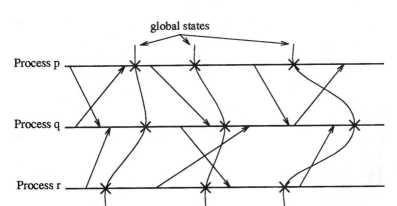

Figure 10.8 Consistent and inconsistent cuts

When a fault is detected in one of the processes, this process is rolled back to its last saved checkpoint. In order to maintain a consistent global state, some other processes that communicated with this process had to be rolled back as well, which might cause other processes to rollback, and so on, until the program might be rolled back to its beginning. This effect is called the *domino effect* [32]. (It is also called *cascade-abort* in transaction processing systems [4, 20]). Figure 10.9 presents an example of the domino effect. In this example, a fault occurred in process q, so it has to be rolled back to checkpoint c_2. After this rollback, the state of the system is inconsistent because message m_4 that was received by process p, has not been sent yet. So, process p is rolled back to c_1. This cause another inconsistent state because of m_3 that forces r to be rolled back to its beginning, and so on, until all three processors are rolled back to the beginning of the program.

A number of checkpointing schemes for multiprocess programs on distributed memory systems have been proposed. The main goal of those schemes is to create a consistent global state and avoid the domino effect. The known checkpointing schemes can be classified to three main groups according to the method by which the consistent global state is achieved. The first method is based on checkpoints that are inserted by a programmer or a

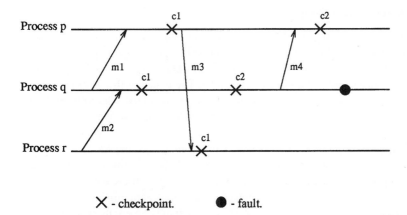

Figure 10.9 Consistent and inconsistent cuts

compiler. In the other two techniques, the checkpointing is done automatically by the system and is transparent to the user. The second group of checkpointing schemes consists of schemes in which the processes save their states in a coordinated way. In the third group, the checkpoints are taken independently by each process, and the system keeps track of its state and performs the necessary operations to avoid an inconsistent state, such as message logging. Next, we describe in more detail the ideas involved in the three methods.

10.3.1 User-driven checkpointing

The first approach for maintaining a consistent global state is to let the user (the application programmer or the compiler) specify exactly where checkpoints should be placed and the cuts that describe consistent global states. Using this approach, the programmer determines the place where checkpoints are placed and specifies the consistent global state to which they belong. After a fault has been detected in a process, the system rolls the process to the last saved state, and all the other processes are rolled back to the consistent state to which this state belongs. User-driven checkpointing is one of the checkpointing techniques that are used in the FTMPS project [5].

User-driven checkpointing puts an extra burden on the application programmer because the programmer has to identify the locations of the checkpoints and the consistent states to which they belong. In many cases, such as iterative calculations, these positions are easy to find—for example, at the end of loops. While this approach puts an extra burden on the programmer, it has many advantages [10]:

- *Smaller storage requirements.* Checkpoints are placed by the programmer in known positions. Hence, checkpoints do not have to include all the variables used by the program, but only those that are needed later.
- *Lower fault-free overhead.* The operation of the process is suspended only when the checkpoint is taken and no extra work is required from the system (e.g., logging messages).
- *Simple recovery.* The system does not have to search for the last consistent global state— this state is supplied by the programmer.
- *Performance tuning.* The programmer can fine-tune the interval between checkpoints to achieve the desired performance requirements.

10.3.2 Coordinated checkpointing

One way to achieve a consistent global state is to checkpoint all the processes at the same time. This approach ensures that every set of checkpoints is a consistent global state. In coordinated checkpointing, after the decision to perform a checkpoint is made, the system is frozen until the states of all the processes are saved. The decision to perform a checkpoint is done either by the system [5] or by one of the processes [19]. The rollback to the last consistent checkpoint, in coordinated checkpointing, is simple because every set of checkpoints represents a consistent global state.

There are two main approaches for coordinated checkpointing: system-level checkpointing and interacting processes checkpointing. In system-level checkpointing, the state of the whole system is saved at the same time. The checkpoint initiator sends a signal to all the processes in the system, indicating the beginning of a checkpoint. This signal causes all application processes to freeze until the state of the whole system is saved. In interacting processes checkpointing, only processes that interact with each other are saved together.

10.3.2.1 Global coordinated checkpointing. In Ref. [38], a system-level coordinated checkpointing scheme is described. In this scheme, the checkpointing process is started by a checkpoint coordinator. The checkpoint coordinator starts the checkpointing process when it receives a hardware interrupt, caused by a checkpoint timer. The checkpointing processes is done in two phases. In the first phase, all the processors in the system propagate the checkpoint request to all their neighbors and, after receiving an acknowledgment from the neighbors, save their states. After all the processors in the system have finished the save operation, the second phase, in which the systems commits to the newly saved checkpoint and resumes normal operation, can begin. While the system is in the first phase of the checkpointing process and a fault is detected, it is rolled back to the old checkpoint that was previously saved. The two phases of the checkpointing process are required to ensure a consistent state in case of a fault during the checkpointing process, when some of the processors have already saved their state while others have not started the checkpointing process. Koo and Toueg [19] proved that two sets of checkpoints and two phases are necessary in coordinated checkpointing.

Checkpointing. To make the system's state at the checkpoint consistent and to avoid losing messages, all the messages that are in transient when the checkpointing process begins have to be accounted for by either the sender or receiver. In Ref. [38], the messages are accounted for by the receiver in the following way: When a processor receives a *begin checkpoint* message from one of its neighbors for the first time, it stops sending normal messages, and it forwards the *begin checkpoint* message to all its neighbors. After the processor sends the *begin checkpoint* message, it continues to receive and handle normal messages from all the neighbors that have not sent it a *begin checkpoint* message. After it has received the *begin checkpoint* message from a neighbor, it knows that no new normal messages are going to arrive from that neighbor until the checkpointing process is completed. Therefore, a processor that has received the *begin checkpoint* message from all its neighbors cannot receive new normal messages, and thus it can save its state. Algorithms 1 and 2 summarize the checkpointing process for the checkpoint coordinator and a checkpoint participant.

Algorithm 1: The checkpoint process for global coordinated checkpointing for the checkpoint coordinator

1. Receive checkpoint interrupt from hardware.
2. Stop working on all application processes, and stop sending normal messages to neighbors.
3. Send **begin checkpoint** message to all neighbors.
4. Wait for **begin checkpoint** message from all neighbors.
5. If a normal message arrives before the **begin checkpoint** message, include it in the processor's state that must be saved.
6. After **begin checkpoint** message was received from all neighbors, send the whole processor state to the disk, so it can be saved.
7. After the state is saved, send **state saved** message to coordinator (itself).
8. After **state saved** message was received from all processors, increase the value of **version** and send **resume** message to all neighbors and resume normal operation.

Algorithm 2: The checkpoint process for global coordinated checkpointing for the checkpoint participant

1. Receive **begin checkpoint** message from one of neighbors.
2. Stop working on all application processes, and stop sending normal messages to neighbors.

3. Send **begin checkpoint** message to all neighbors.
4. Wait for **begin checkpoint** message from all neighbors, except the one from which the message was received.
5. If a normal message arrives before the **begin checkpoint** message, include it in the processor's state that must be saved.
6. After **begin checkpoint** message was received from all neighbors, send the whole processor state to the disk so it can be saved, and set the value of **version** to **unknown**.
7. After the state is saved, send **state saved** message to coordinator.
8. Wait for **resume** message from one of the neighbors.
9. When the **resume** message is received, restore old value of **version** and increase it by 1, and resume normal operation.

Recovery. The recovery process is simple unless the fault occurred in the middle of the checkpointing process. If the fault occurs outside the checkpoint process, the recovery request is sent to the checkpoint coordinator, which sends a *begin recovery* message to all the processors in the system. As in the checkpointing process, a processor stops normal operation and propagates the *begin recovery* message to all its neighbors as soon as it receives the *begin recovery* message, but the recovery itself (i.e., the restore of the state from the disk) is done only after the begin recovery message has been received from all neighbors. Algorithm 3 shows the recovery process for a participant processor.

When a fault occurs in the middle of the checkpoint process, the states of the processors have to be rolled back to the same checkpoint—either the old or the new state. The two phases of the checkpointing process ensure that the processors can easily agree on the checkpoint to roll back to. Because the second phase of the checkpointing process cannot begin before all the processors finished the first phase, the system can be in one of two states:

- Some of the processors (perhaps all of them) have finished the first phase, while others are in the middle of that phase or have not started the checkpointing yet. In this case, the system should be rolled back to the old state.
- All the processors have finished the first phase, and their new state is saved on the disk. Some of the processors finished the second phase and resumed normal operation, while others are still waiting for the *resume* message. In this case, the system should be rolled back to the new state because some of the processors have already committed to it.

To make sure that all the processors roll back to the same checkpoint, the following simple scheme is used: Each processor holds a variable named *version* that holds the number of the checkpoint to which the system should be rolled back. In the first phase of the checkpointing process, after the processor saves it state, it sets the value of this variable to a special value *unknown*, while saving the old value in a temporary variable. This special value indicates that this processor does not know, and does not care, to which state the system is rolled back; it saved the new state but has not yet committed to it. When a processor receives the *resume* message during the second phase of the checkpointing process, it increases the old value of *version* by one and uses it as the new value for it. Because of the two-phase checkpointing process, there are three possible cases for values of the *version* variable

- *The fault occurred outside the checkpointing process.* In this case all the processors agree on the value of *version*, and the system is rolled back to that state.

- The fault occurred while in the first phase of the checkpointing process. In this case some of the processors have the old value in *version*, while others have the unknown value.
- The fault occurred while in the second phase of the checkpointing process. In this case some of the processors have the new value in *version*, while others have the unknown value.

Overall, it is impossible that some processors have the old version number, while other have the new version number, and processors that do not have the *unknown* value have the correct version of the checkpoint to which to roll back. So, to make a consistent recovery, each processor that has a valid version number propagates it to its neighbors, and each processor that has the *unknown* value waits for a valid version before its begins it recovery. To avoid the case where all the processors have the *unknown* value, the checkpoint coordinator cannot set its version number to *unknown*. Instead, it increase its version number as soon as it starts the second phase and keeps the old value while waiting for the *state saved* messages from the other processors. Algorithm 3 shows how the *version* variable is handled during the checkpointing and recovery processes.

Global coordinated checkpointing, similar to the one described here, is used in the FT-MPS system [5] and the DAMP system [3].

Algorithm 3: The recovery process for global coordinated checkpointing for the Checkpoint participant

1. Receive **begin recovery** (**version**) message from one of neighbors.
2. If version number is unknown, update it from message
3. Stop working on all application processes, and stop sending normal messages to neighbors.
4. Send **begin recovery** (**version**) message to all neighbors.
5. Wait for **begin recovery message** from all neighbors, except the one from which the message was received.
6. If a normal message arrives before the **begin recovery** message, ignore it.
7. After **begin recovery** message has been received from all neighbors, restore the processor state.
8. After the state is restored, send **state restored** message to coordinator.
9. Wait for **resume** message from one of the neighbors.
10. When the **resume** message is received resume normal operation.

10.3.2.2 Process-level coordinated checkpointing. The global coordinated checkpointing technique has two disadvantages. The first disadvantage is that it might take a long time to perform a checkpoint, because the checkpointing has to be performed by all the processors in the system before the operation of the system can be resumed. The second disadvantage is that it forces all the applications that are executed to be saved together at the same intervals. To overcome these disadvantages and still benefit from the big advantage of coordinated checkpointing, namely, the simple recovery process, a few schemes that perform coordinated checkpointing at the application or process level were developed. The main idea behind these schemes is that processes that do not interact do not have to be saved together, because any set of states of these processes can be a part of a consistent global state. This idea is used in several schemes, such as in Refs. [2, 19, 23, and 37]. Process-level coordinated checkpointing shortens the time to checkpoint and enables each application to checkpoint at its own pace without making the recovery process more complicated.

In the process-level checkpointing schemes, each process can initiate a checkpoint. This checkpoint should include all the processes that belong to its interacting set. We define the interact relation in the following way: Process p interacts with process q if p sent or received a message from q after its last checkpoint. The interacting set of processor p is the transitive closure of the interact relation. Note that because we checkpoint all the processes in the interacting set of p, p interacts with q means that q interacts with p. Therefore, at any given time, the interacting sets are disjoint. When a process starts a checkpoint, all the processes that belong to its interacting set at that time should be identified. This can be done by either piggy-backing information about the interacting sets on messages [19] or keeping internal data in each process about processes it interacted with [2, 37].

In Ref. [37], a checkpointing scheme is proposed that is based on checkpointing consistent states of interacting processes. When an error is detected, all the processes that could have been affected by it are identified, and the members of the interacting sets of these processors are determined. All the processes that belonged to the interacting sets of those processes at their last checkpoint are rolled back.

The checkpointing and recovery sessions require coordination, which is achieved by using a checkpoint or recovery coordinator. Unlike the global coordinated checkpointing, the coordinator is selected dynamically. The first thing the coordinator does is to determine the interacting set of itself and create a spanning tree that includes all the members of its interacting set, with the coordinator as the tree root. To identify the members of the interacting set, the system keeps for each process the list of all the processes it has communicated with since its last checkpoint in the *direct communication* list. This list is used by the processes to create the spanning tree by identifying potential children for each process. It is possible that a few processes that belong to the same interacting set will initiate a checkpointing or recovery sessions simultaneously. In this case, a priority mechanism, based on the identification number of the processes, is used to create a single tree with the process with the highest priority among all processes that initiated the sessions as the root.

To ensure that the state that is saved by the checkpointing session is consistent and avoid losing messages in transit the following precautions are used:

- When a process is added to the spanning tree, it stops sending new messages.
- A process in the spanning tree continues to receive messages from processes in the interacting set that are not in the tree yet. Those messages are queued and saved together with the process state.
- Messages from processes that are not in the interacting set are not delivered to the processes in the spanning tree but are sent back to their sender.
- The state of the processes is saved only after the spanning tree is completed, and no new messages can be received.

The checkpointing session is similar to the checkpointing process for the global coordinated checkpointing described earlier, with a small addition that is used to prevent a process from joining a new interacting set and being involved in another checkpointing session before all the processes in its current interacting set commit to the new checkpoint. The overall checkpointing session is described in Algorithm 4.

The recovery session is identical to the checkpointing session, and the only difference is that the *begin commit* message is replaced by a *resume* message, and the processes that receive it resume normal operation without waiting for an acknowledgment.

Algorithm 4: The checkpointing session for process-level coordinated checkpointing

1. Build the spanning tree of the interacting set, and propagate **begin checkpoint** message along the tree. Each process, upon receiving the **begin checkpoint** message, stops its normal operation and stops sending new messages.

2. Propagate **tree completed** message from the leaves of the tree to the root. A process sends the **tree completed** message to its parent only after it received it from all its children.

3. The checkpoint coordinator, upon receiving the **tree completed**, saves the process state and propagates a **begin sav**e message toward the leaves.

4. Each process, upon receiving the **begin save** message from its parent, saves its state and forwards the message to its children.

5. After saving their states, the leaves of the tree forward a **save done** message toward the root. Each process, after receiving the message from all its children and saving its state, sends the message to its parent.

6. When the root receives the **save done** message from all its children, it notifies the system to commit to its new checkpoint and sends a **begin commit** message down the tree.

7. Each node, after receiving the **begin commit** message from its parent, notifies the system to commit to the checkpoint and forward the message to its children.

8. After the leaves of the tree receives the **begin commit** message, they send a **resume** message to their parents and resume normal operation. A process that received this message from all its children forwards it to its parent and resumes normal operation.

Similarly to the global schemes, two sets of checkpoints are saved during the checkpoint session, and each process keeps a *version* variable that keeps track in what phase of the session it is, so that all the process can be rolled back to the same consistent checkpoint.

10.3.3   Independent checkpointing

In coordinated checkpointing, the state of the whole system or the state of some subset of processes is saved at the same time to create a consistent global state. This approach ensures that every set of checkpoints is consistent; hence, fault recovery by means of rollback can be performed quickly. On the other hand, the system or the processes involved in the checkpointing are frozen during the whole checkpointing process. This can cause the checkpointing to take a long time. The fact that many processes try to save their state at the same time might overload the I/O system and be a source of faults by itself. A different approach to checkpointing in distributed systems is to let each process checkpoint its own state independently of other processes. To maintain a consistent state in this approach, the system keeps track of the interprocess communications and uses this data either during the checkpointing process [42] or during the recovery process [17].

In this part we describe two methods to maintain a consistent global state with in dependent checkpointing. In the first method, the consistent state is achieved by logging interprocess messages [28]. In this method, the system logs all the interprocess messages. Whenever a fault is detected and a process is rolled back, the system plays back to that process all the messages it had received since its last checkpoint. The system also intercepts the messages this process sends to other processes, if these messages were already recorded as sent by the system.

In the second independent checkpointing method, the consistent state is achieved by forcing the processes to checkpoint at dangerous points, just before events that might cause an inconsistent state [42]. In this method, the system does not have to keep record of the interprocess messages to maintain a consistent state.

10.3.3.1 Message logging. In message logging, the consistent global state of the system is ensured by logging all messages between the processes and replaying them after recovery. That is, instead of making sure that no messages cross the global state cut, message logging creates a consistent state by reproducing the messages that cross the cut. When message logging is used, each process can save its state independently of the other processes. For message logging to perform correctly, the following conditions must be satisfied:

- *The execution of a process is deterministic.* Message logging schemes recover from faults by replaying messages that were received by the failed process and discarding messages that the faulty process sent before the fault. If the process is not deterministic, it may produce different messages after the recovery and create an inconsistent state. For example, suppose process *p* sent message *m* before its failure. This message is recorded by the system and received by the destination process *q*. If, after the fault and rollback of *p*, it sends a different message *m'*, then *m'* is discarded by the system, *q* continues to operate as if *m* had been sent, and *p* considers m' to be the message that was sent. This cause an inconsistent system state.
- Detection of faults must be immediate, or at least a fault must be detected before the faulty process sends its next message. If the faulty process sent a message that should not be sent due to the fault, there is no way that the system can cause the receiver of the message to discard it without rolling it back to a checkpoint prior to the reception of the message. As the checkpoints of the processes are independent of each other, this rollback might cause the domino effect that the message logging tried to avoid. Note that this problem does not exist in the coordinated checkpointing, because in coordinated checkpointing all the processes that interacted with the faulty process are rolled back automatically to a consistent global state; hence, the faulty message is discarded.

Synchronous message logging. The Auragen system [6] provides consistent global state by message logging. The system achieves fault tolerance by keeping an inactive copy of each process. This copy of a process saves the state of the process at the last checkpoint, the list of all the messages that the process received after the checkpoint, and the number of messages that it sent after the last checkpoint. To keep this record, each message that is sent by a process has three destinations:

1. the original destination
2. the backup process of the original destination, where the message is logged
3. the backup process of the source, where the counter of number of sent messages is increased

When a fault is detected, the backup process is started from the last save checkpoint state. All the messages that were received and logged since that checkpoint are placed in the receive queue, and the messages that were sent prior to the fault are discarded; that is, if the sent messages counter has the value of *n* when the fault occurred, the first n_1 messages that the backup wants to send are discarded. The ordering of received messages in a process and its backup must be kept; that is, if process *p* received messages m_1 and m_2, then the backup process must record the messages in that order. This requires a complete synchronization of events between the two processes.

The *publishing* system [28] is another message logging checkpointing system. In this system, which is implemented on a broadcast communication system, a passive listener is added. This listener logs all the messages in the system, as well as the process states at checkpoints. During fault recovery, the passive listener sends the faulty process state to one of the processors in the system, along with all the logged messages and instructions and how many messages were sent from that process prior to its failure, so that these messages can be discarded. The *publishing* system also requires synchronization between the receiver and the passive listener; that is, a message is recorded at the same time it is delivered to the receiver.

Asynchronous message logging. The synchronization requirement of the two systems described above can create a significant overhead in the execution time of an application and degrade the performance of the system. In Ref. [17], a scheme is introduced that does not require synchronization in the logged messages. The scheme assumes that only a single process can fail at a time. In this scheme, each processor keeps a log of the messages that it sent. To each logged message a *receiver sequential number* (RSN) is added that indicates the order in which messages are received in the receiver. When a failed process is recovered, it broadcasts the last sequence number of message it is aware of, and the other processes retransmit all the messages that have higher RSNs to that process. Each message includes a *sender sequence number* (SSN), that enables the processes in the system to discard duplicated messages that arrive from the recovered process.

In Ref. [36], another technique for asynchronous recovery is described. This technique optimizes the technique in Ref. [17] in two ways:

1. The technique uses the fact that the sender can reconstruct all the messages it sent to reduce the amount of data that is stored on the disk. When reconstruction of messages is used, the receiver only needs to save the SSN, RSN pair for each message to ensure correct recovery. The fact that the sender can be reconstructed allows the system to tolerate more then a single fault.
2. The technique reduces the number of message sequences that are stored in the stable storage by not recording expected messages (that is, messages that the receiver has the highest probability of receiving). The receiver uses a prediction function to predict the next message it is going to receive, based on its current state. If the received message is indeed the predicted one, its reception is not recorded. For example, consider the case when process p sent two messages to two other processes q and r, requesting some kind of service. Process p predicts that q is going to supply the service first, and if the message from q arrives before the message from r, it is not recorded.

The problem with message reconstruction is that, to reconstruct its messages, the sender must receive all the messages that caused this message, which in turn causes other processes to rollback, and so on. To avoid the domino effect, the system keeps track of the dependency between messages and remove such dependencies by storing messages in the stable storage. The system also keeps track of checkpoints that might be needed to reconstruct messages. When no one depends on a checkpoint anymore, it can be discarded from the stable storage. Even with this storage of messages, the system might need to roll back farther than the most recent set of checkpoints to recover from a fault.

One of the problems in asynchronous message logging is that it may take a long time before the system can commit to a checkpoint (i.e., ensure that the system will never roll back behind that checkpoint). Fast commitment to a checkpoint is important because it cannot release the output to the outside world before the it makes sure that this output will not be generated again [12, 16]. In the Manetho system [12], a checkpointing technique is described that allows fast commitment to checkpoints. The technique is based on tracking dependencies between nondeterministic events in processes (i.e., reception of messages and asynchronous operations of the operating systems). In Ref. [16], another approach for fast commitment is introduced. In this approach, each process performs checkpointing independently and logs its messages. When a process wants to commit its checkpoint, it starts a commit algorithm that forces all the processes with whom it communicated to commit their states as well. This committing is similar to the coordinated checkpointing that was presented earlier.

Another asynchronous checkpointing scheme is presented in Ref. [39]. In this scheme, the process checkpoint is itself independent of the other processes, but each process keeps data about dependency between checkpoints. When a fault is detected, a process is rolled back, and the process performs a simple search algorithm that finds the most recent consistent state. To avoid the domino effect, the messages are logged in the system. The algorithm does not assume deterministic execution, so if a sender is rolled back after its message is received, the receiver has to be rolled back as well. To avoid losing messages that were received but not recorded, extra dependencies are added between checkpoints before unrecorded messages in the receiver and the checkpoint after they were sent in the sender. This ensures that any unrecorded message will be sent again after rollback.

10.3.3.2 Adaptive independent checkpointing.

A recent method to achieve a consistent global state while still checkpointing a single process at a time is described in Ref. [42]. In the checkpointing scheme described there, a process detects an event that might cause an inconsistent state and performs a checkpoint just before this event happens. By performing the checkpoint just before the dangerous point that causes an inconsistent state, the process does not have to coordinate with the other processes to ensure that the checkpoint belongs to a consistent global checkpoint. This scheme is a hybrid between the coordinated checkpointing scheme and the message logging schemes. In this scheme, like the message logging schemes, the checkpoint is performed in each process independently of the other processes, not in a coordinated fashion. On the other hand, in this scheme there is no need to log the interprocess messages, and hence the execution of a task does not have to be deterministic. Also, in this scheme, like the coordinated checkpointing schemes, the recovery process includes rolling back a set of the processes in the system, not just the faulty one.

Xu and Netzer [42] have identified a necessary and sufficient condition for an inconsistent state, based on *zigzag paths*. A zigzag path is defined as follows.

Definition A. Let $C_{p,i}$ be the ith checkpoint of process p, and let $I_{p,i}$ be the execution interval between $C_{p,i}$ and $C_{p,i+1}$. A zigzag path from $C_{p,i}$ to $C_{q,j}$ exists if and only if there are messages $m_1, m_2, \ldots, m_n$ such that

1. Message m_1 is sent from process p after $C_{p,i}$.
2. If m_k ($1 < k < n$) is received by process r in interval $I_{r,l}$, then m_{k+1} is sent by r after $C_{r,l}$. Note that m_{k+1} may be sent before m_k is received.
3. Message m_n is received by process q before $C_{q,j}$.

Figure 10.10 shows a zigzag path from $C_{p,0}$ to $C_{p,1}$ with the messages m_1, m_2, m_3. Xu and Netzer proved that a checkpoint is useful (i.e., belongs to a consistent global state) if and only if it belongs to no zigzag cycle.

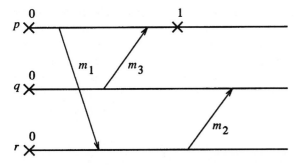

Figure 10.10 Zigzag path

The algorithm: The algorithm presented in Ref. [42] tries to detect zigzag cycles and prevent their occurrence by inserting checkpoints before the reception of the message that causes the zigzag cycle. Note that it is not easy to detect zigzag paths. For example, in the path in Fig. 10.10, the zigzag path from $C_{p,0}$ to $C_{p,1}$ is created only after m_2 is sent and received, which can happen at any time in the future. The algorithm in Ref. [42] finds only the zigzag cycles that are created by a casual path. A *casual path* is a zigzag path in which each message m_{k+1} is sent after m_k is received. The algorithm keeps track of all the casual paths in the system by keeping a vector called *DV* in each process. Entry i in the *DV* vector of process p contains latest checkpoint of process number i that have a casual path to that process. The vector is piggy-backed to each of the outgoing messages of the process, and is used by the other processes to update their vector. When processor p performs a checkpoint, it copies its *DV* vector to another vector *ZV.* Each time p sends a message to g, it adds the entry corresponding to q in the *ZV* vector in the message. If the number of the latest checkpoint in q equals to the number just received from p, a zigzag cycle that involves the last checkpoint of p is about to be created. To prevent this from happening, q inserts a checkpoint before processing the message.

The algorithm described above can be added to any existing checkpointing strategy—for example, periodic checkpointing of each process. Experimental results show that the number of checkpoints that are added to prevent zigzag cycles is small (less than 4 percent). On the other hand, the number of intervals that must be rolled back to maintain consistent state is reduced by up to 80 percent.

A similar approach to the one presented in Ref. [42] appears in Ref. [40]. In this paper, Wang and Fuchs describe an induced checkpoints algorithm. In this algorithm, a process detects a dangerous point where inconsistent state might occur, whereupon the process takes an induced checkpoint. Unlike the Xu and Netzer algorithm, in this algorithm the processes hold a *lazy counter* and take an induced checkpoint only when the distance to the last safe checkpoint reaches the lazy counter. Results given in Ref. [40] show that even for a high lazy counter of 5, only two checkpoints on average have to be rolled back.

10.4   Conclusions and Future Directions

Checkpointing is a fault recovery technique that enables us to reduce the total execution time of a task in the presence of faults. In checkpointing schemes, the state of the task is saved periodically on a stable storage, so when faults occur, the execution of the task does not have to be rolled back to the beginning. Instead, the state of the last saved checkpoint is restored, and the execution of the task is resumed from that point.

In this chapter, we focused on describing two main issues that are related to checkpointing in parallel and distributed systems. The first issue is the utilization of parallel and distributed systems in fault detection and recovery by task duplication. Parallel and distributed computing systems provide hardware redundancy that helps to achieve fault-detection by duplicating a task into a multiple number of processors and comparing the states of the processors at checkpoints. We described several techniques to achieve fault detection and fast recovery in parallel and distributed systems. All the techniques use task duplication for fault recovery but differ in the way they recover from faults. We presented several performance evaluation techniques for checkpointing schemes with task duplication and compared various performance aspects of the schemes.

The second topic we addressed in this chapter is related to techniques for maintaining consistent global states at checkpoints. In parallel and distributed systems, the processors exchange information by means of message passing. Therefore, a change in the state of one processor, when it is rolled back, can affect the states of the other processors and create inconsistency in the global state of the system. To avoid the inconsistency, other processors might be forced to roll back as well, and that might lead to a domino effect. We presented several checkpointing schemes that maintain a consistent state in the system while avoiding a major rollback. We described three types of checkpointing schemes: User-driven schemes, where the user provides the points at which to checkpoint the state of the processor; coordinated checkpointing, where the states of all the processors (or a subset of them) are checkpointed at the same time to keep a consistent state; and independent checkpointing, where each processor checkpoints its state independently, and consistency is maintained by either keeping track of the messages in the system or insertion of extra checkpoints to avoid inconsistent states.

Because of space limitations, we did not cover the vast area of recovery in database systems.

A challenging research direction is to create schemes that simultaneously address both relevant aspects of checkpointing in parallel and distributed systems—namely, the design of an efficient checkpointing scheme that achieves a consistent global state and uses task duplication for fault detection. The combined checkpointing schemes have to address the complexity associated with both the detection of faults and maintaining of a consistent state. For example, if two copies of the same process receive messages in a different order, their states are going to be different, and a comparison of the states will fail even if no fault has occurred. Also, when task duplication is used for fault detection, it may cause a delayed detection of faults, as faults are detected only when comparison is performed. Specifically, faults in one process can spread to other processes when this process sends messages to other processes. In the checkpointing schemes described in Section 10.3, we assumed that a faulty processor halts immediately when the fault occurs. The spread of the faults outside the faulty processor can cause a misdetection of faults or create the domino effect.

10.5 References

1. Agrawal, P. 1988. Fault tolerance in multiprocessor systems without dedicated redundancy. *IEEE Transactions on Computers,* 37:358–362.
2. Barigazzi, G., and L. Strigini. 1983. Application-transparent setting of recovery points. In *Proceedings of the 13th IEEE International Symposium on Fault-Tolerant Computing.* New York: IEEE, 48–55.
3. Bauch, A, B. Bieker, and E. Maehle. 1992. Backward error recovery in the dynamical reconfigurable multiprocessor system DAMP. In *Proceedings of Workshop on Fault-Tolerant Parallel and Distributed Systems,* 36–43.

4. Bernstein, P. A., V. Hadzilacos, and N. Goodman. 1989. *Concurrency Control and Recovery in Database Systems.* Reading, Mass.: Addison Wesley.

5. Bieker, D., G. Deconinck, E. Maehle, and J. Vounckx. 1994. Reconfiguration and checkpointing in massively parallel systems. *EDCC-1.*

6. Borg, A., J. Baumbach, and S. Galzer. 1983. A message system supporting fault-tolerance. In *Proceedings of the 9th ACM Symposium on Operating Systems Principles,* 90–99.

7. Chandy, K. M., and L. Lamport. 1987. Distributed snapshots: Determining global states of distributed system. *ACM Transactions on Computer Systems,* 3:63–75.

8. Chandy, K. M., and C. V. Ramamoorthy. 1972. Rollback and recovery strategies for computer programs. *IEEE Transactions on Computers,* 21:546–556.

9. Coffman, E. G., and E. N. Gilbert. 1990. Optimal strategies for scheduling checkpoints and preventive maintenance. *IEEE Transactions on Reliability,* 39:9–18.

10. Deconinck, G., J. Vounckx, R. Lauwereins, and J. A. Peperstraete. 1993. Survey of backward error recovery techniques for multicomputers based on checkpointing and rollback. In *IASTED Int. Conf. on Modelling and Simulation,* 262–265.

11. Duda, A. 1983. The effects of checkpointing on program execution time. *Information Processing Letters,* 16:221–229.

12. Elnozahy, E. N., and W. Zwaenepoel. 1992. Manetho: Transparent rollback recovery with low overhead, limited rollback and fast output commit. *IEEE Transactions on Computers,* 41:526–531.

13. Gelenbe, E. 1979. On the optimum checkpoint interval. *Journal of the ACM,* 26:259–270.

14. Howard, R. A. 1971. *Dynamic Probabilistic Systems Vol II: Semi Markov and Decision Processes.* New York: John Wiley & Sons.

15. Johnson, B. W. 1989. *Design and Analysis of Fault Tolerant Digital Systems.* Reading, Mass.: Addison-Wesley.

16. Johnson, D. B. 1993. Efficient transparent optimistic rollback recovery for distributed application programs. In *Proceedings of the 12th symposium on Reliable Distributed Systems,* 86–95.

17. Johnson, D. B., and W. Zwaenepoel. 1987. Sender-based message logging. In *Proceedings of the 17th IEEE International symposium on Fault-Tolerant Computing,* 14–19.

18. Kamal, S. 1975. An approach to the diagnosis of intermittent faults. *IEEE Transactions on Computers,* 24:461–467.

19. Koo, R., and S. Toueg. 1987. Checkpointing and rollback-recovery for distributed systems. *IEEE Transactions on Software Engineering,* 13:23–31.

20. Krishnamurthy, E. V., and V. K. Murthy. 1991. *Transaction Processing Systems.* Englewood Cliffs, N.J.: Prentice Hall.

21. Kulkarni, V. G., V. F. Nicola, and K. S. Trivedi. 1990. Effects of checkpointing and queueing on program performance. *Communications of Statistics—Stochastic Models,* 6:615–648.

22. L'Ecuyer, P., and J. Malenfant. 1988. Computing optimal checkpointing for rollback and recovery systems. *IEEE Transactions on Computers,* 37:491–496.

23. Leu, P. Y., and B. Bhargava. 1988. Concurrent robust checkpointing and recovery in distributed systems. In *Proceedings of the 4th IEEE International Symposium on Data Engineering,* 154–163.

24. Long, J., W. K. Fuchs, and J. A. Abraham.1990. Forward recovery using checkpointing in parallel systems. In *Proceedings of the 19th International Conference on Parallel Processing,* 272–275.

25. Long, J., W. K. Fuchs, and J. A. Abraham. 1992. Compiler-assisted static checkpoint insertion. In *Proceedings of the 22nd IEEE International Symposium on Fault-Tolerant Computing,* 58–65.

26. Murthy, V. K., and E. V. Krishnamurthy. 1994. Knowledge-based security control for on-line database transaction processing systems. *SIGSAC Review,* 12:7–14.

27. Nicola, V. F., and J. M. van Spanje. 1990. Comparative analysis of different models of checkpointing and recovery. *IEEE Transactions on Software Engineering,* 16:807–821.

28. Powell, M. L., and D. L. Prestto. 1983. PUBLISHING: A reliable broadcast communication mechanism. In *Proceedings of the 9th ACM Symposium on Operating Systems Principles,* 100–109.

29. Pradhan, D. K. 1989. Redundancy schemes for recovery. *TR89-cse-16,* ECE Department, University of Massachusetts, Amherst.

30. Pradhan, D. K., and N. H. Vaidya. 1992. Roll-forward checkpointing scheme: Concurrent retry with nondedicated spares. In *Proceedings of the IEEE Workshop on Fault-Tolerant Parallel and Distributed Systems,* 166–174.

31. Pradhan, D. K., and N. H. Vaidya. 1994. Roll-forward and rollback recovery: performance reliability tradeoff. In *Proceedings of the 24th IEEE International Symposium on Fault-Tolerant Computing.*

32. Randell, B. 1975. System structure for software fault tolerance. *IEEE Transactions on Software Engineering,* 1:220–232.

33. Sharma, D. D., and D. K. Pradhan. 1993. A static roll-forward checkpointing scheme using three processors. *TR-93-061,* Computer Science Department, Texas A&M University.

34. Shrivastava, S. K. 1985. *Reliable Computer Systems.* New York: Springer-Verlag.

35. Siewiorek, D. P., and R. S. Swarz. 1982. *The Theory and Practice of Reliable System Design.* Digital Press.

36. Storm, R. E., and S. A. Yemini. 1984. Optimistic recovery: An asynchronous approach to fault-tolerance in distributed systems. In *Proceedings of the 14th IEEE International symposium on Fault-Tolerant Computing,* 374–379.

37. Tamir, Y. and T. M. Frazier. 1989. Application-transparent process level error-recovery for multicomputers. In *Proceedings of the Hawaii International Conference on System Sciences,* 296–305.

38. Tamir, Y. and C. H. Sequin. 1984. Error recovery in multicomputers using global checkpoints. In *Proceedings of the 13th International Conference on Parallel Processing,* 32–41.

39. Wang, Y.-M., and K. Fuchs. 1992. Optimistic message logging for independent checkpointing in message-passing systems. In *Proceedings of the 11th symposium on Reliable Distributed Systems,* 147–154.

40. Wang, Y.-M., and K. Fuchs. 1993. Lazy checkpoint coordination for bounding rollback propagation. In *Proceedings of the 12th symposium on Reliable Distributed Systems,* 78–85.

41. Weihl, W. E. 1989. The impact of recovery on concurrency control. In *Proceedings of the 8th ACM symposium on Principles of Database Systems,* 259–269.

42. Xu, J., and R. H. B. Netzer. 1993. Adaptive independent checkpointing for reducing rollback propagation. In *Proceedings of the 5th IEEE symposium on Parallel and Distributed Processing,* 754–761.

43. Ziv, A. 1995. *Analysis and performance optimization of checkpointing schemes with task duplication.* Ph.D. thesis, Stanford University.

44. Ziv, A., and J. Bruck. 1994. Analysis of checkpointing schemes for multiprocessor systems. In *Proceedings of the 13th symposium on Reliable Distributed Systems,* 52–61.

45. Ziv, A., and J. Bruck. 1994. Efficient checkpointing schemes over local area networks. In *Proceedings of the 1994 IEEE Workshop on Fault-Tolerant Parallel and Distributed Systems.*

11

Architecture for Open Distributed Software Systems

Kazi Farooqui and Luigi Logrippo

This chapter describes a common conceptual framework for the design of distributed systems that is gaining a wide degree of acceptance within the distributed systems research community. In an ideal open distributed system, it should be possible for distributed applications developed in different environments to interact. This can be achieved if distributed environments conform to a common conceptual model or architecture. This is the basis for the standardization of the *Reference Model for Open Distributed Processing* (RM-ODP).

The chapter is divided into five major sections. Section 11.1 is an introduction to the architecture for open distributed systems. A set of five abstraction levels—*enterprise viewpoint, information viewpoint, computational viewpoint, engineering viewpoint,* and *technology viewpoint*—are identified in RM-ODP for the specification of distributed software systems. While all the viewpoints are relevant to the design of distributed systems, the computation and engineering models are the ones that bear most directly on the design and implementation of distributed systems. From a distributed software engineering point of view, the computational and engineering viewpoints are the most important; they reflect the software structure of the distributed application most closely. In this chapter, we concentrate on the computational and engineering viewpoints in Section 11.2 and Section 11.3, respectively. Section 11.4 is an illustration of the application of ODP architectural concepts in a simple client-server scenario. Conclusions are drawn in Section 11.5.

11.1 Introduction to Open Distributed Systems Architecture

11.1.1 Motivation

Today's distributed systems are complex structures, composed of many types of hardware and software components. In some systems, components are developed separately by different implementors and then combined, resulting in a heterogeneous system. Well known examples of such systems, which fall within the scope of the distributed systems architecture addressed in this chapter, include telecommunication systems (advanced intelligent networks), computer communication networks (Internet), automated manufacturing systems, office automation systems, client-server systems (banking and airline reservation systems), and so on.

To reason about such systems, it is necessary to develop appropriate concepts. These concepts may vary according to the point of view from which the system is being considered. For example, from the end user's point of view, the system will be described in terms of user objectives and requirements. From the application designer's point of view, it will be described in terms of components communicating with each other in some way, with each component performing some function. The system designer instead is concerned with the communication protocols, etc., required to accomplish the communication between application components. Finally, the technical personnel in charge of putting together the system will see the software and hardware products connected in some way.

Different conceptual frameworks have been devised over the years by implementors and researchers. However, such frameworks are usually adapted to specific vendor's architectures, and fail for heterogeneous systems. This situation has been very damaging in practice, because such frameworks are essential in order to design and maintain heterogeneous systems. Therefore, it should not be surprising that much of the existing work on these unifying concepts has been done within international standardization bodies, mainly the International Organization for Standardization (ISO) and the International Telecommunication Union (ITU, formerly CCITT).

At the time of this writing, a set of documents being assembled by the committees of the ISO and ITU, called the Reference Model for Open Distributed Processing (RM-ODP) [1–4] constitutes what many researchers consider to be the most complete and authoritative statement of the state of research in this area. RM-ODP is based largely on preexisting research work in the field of distributed systems, especially the work done in U.K. on the Advanced Networked Systems Architecture project (ANSA) [5].

RM-ODP builds on other previously established ISO and ITU standards dealing with heterogeneous systems, such as the standards for Open Systems Interconnection (OSI) [6] and Distributed Applications Framework (DAF) [7], Integrated Services Digital Network, and Common Channel Signalling System.

In contrast to OSI, ODP is not restricted to communication between heterogeneous systems. It deals also with the provision of various distribution transparencies within systems, and with application portability across systems. RM-ODP deals with the application *interaction* problems rather than the pure *interconnection* problems addressed in the OSI model. In this sense, ODP encompasses and extends OSI. OSI becomes a (communication) enabling technology for ODP applications; i.e., OSI and other related standards (ISDN, CSS#7) provide the communication protocols that are required to support the communication between distributed applications.

11.1.2   Introduction to ODP

RM-ODP is an architectural framework for the integrated support of distribution, interworking, interoperability and portability of distributed applications. It provides an object-oriented reference model for building open distributed systems. It defines an architecture for distributed systems which enables multivendor, multidomain, heterogeneous, networked computing.

RM-ODP prescribes a methodology for the design of distributed systems by describing different abstraction levels called *viewpoints*. The ODP framework of viewpoints is quite generic. A set of concepts, structures, and rules is given for each viewpoint, providing a *language* for specifying ODP systems in that viewpoint.

11.1.3   ODP Framework of viewpoints

For any given information processing system, there are a number of user categories (or, more accurately, a number of *roles*) that have an interest in the system. Examples include the members of the enterprise who use the system, the architects who design it, the programmers who implement it, and the technicians who install it. Each role is interested in the same system, but their relative views of the system are different; they see different issues, they have different requirements, and they use different vocabularies (or languages) when describing the system [9].

Rather than attempting to deal with the full complexity of distributed systems, RM-ODP attempts to recognize these various interests by considering the system from different viewpoints or projections, each of which is chosen to reflect one set of design concerns. Each viewpoint represents a different abstraction of the original distributed system, without the need to create one large model describing the whole of it [8].

The ODP framework of viewpoints partitions the concerns to be addressed in the design of distributed systems. A viewpoint leads to a representation of the system with emphasis on a specific set of concerns, and the resulting representation is an abstraction of the system; that is, a description which recognizes some distinctions (those relevant to the concern) and ignores others (those not relevant to the concern). Different viewpoints address different concerns of the software engineering process, but there is a common ground between them.

RM-ODP defines the following five viewpoints. Together they provide the complete description of the system: *enterprise* viewpoint, *information* viewpoint, *computational* viewpoint, *engineering* viewpoint, and *technology* viewpoint. Each viewpoint has its own conceptual model to represent the relevant aspects of the system.

11.1.3.1 Enterprise viewpoint. The enterprise viewpoint is directed to describing the *needs* of the *users* of an information system. It provides the members of an enterprise in which information systems are to operate with a view of how and where a system is placed and used within the enterprise [10].

An enterprise view covers the enterprise *objectives* of an information system. It focuses on the *requirements* that an organization places on a distributed system and the *role* of the distributed system within the organization.

The enterprise viewpoint is the most abstract of the ODP framework of viewpoints, stating high-level enterprise requirements, system management policies, and organization structures. In terms of software engineering, this viewpoint is related to requirements capture and transformation and to the early design of distributed systems [17]. The design decisions made using the enterprise viewpoint concern *what* a system is to do and *who* it is doing it for.

11.1.3.2 Information viewpoint. The information viewpoint focuses on the information content of the enterprise. It defines the *information semantics* of the distributed system, i.e., the meaning that a human would ascribe to the data stored or exchanged between components of a distributed system. From this viewpoint, the information processing facilities are seen as black boxes. The parts of the information processing facilities that are to be automated are not differentiated from those to be performed manually [10]. In this model, the distribution of processing is not visible, although the natural distribution of the enterprise itself may, of course, need to be modeled. The model deals with the information, information processing, and information exchange aspects of a distributed system.

The information model is expressed in terms of abstract objects which represent the information elements manipulated by the enterprise. The information modeling activity consists of identifying: *information structures* (or elements) of the system, *constraints* and *manipulations* that may be performed on these information structures, and *information flows* (both the information sources and sinks within the system). These definitions are entirely implementation independent; no restrictions are placed on how the information is represented in a real system, or the means by which it is manipulated.

The information specification of an ODP application could be expressed using a variety of methods, e.g., entity-relationship models, conceptual schemas, Z language, and so forth. RM-ODP gives descriptive terminology and tools for information modeling.

11.1.3.3 Computational viewpoint. The computational viewpoint represents the distributed system as seen by application designers and programmers. It deals with the logical partitioning of a distributed application, breaking it up on the basis of flows of invocation and provision of service. It is here that the idea of particular sets of application components being related by their roles as client or server in an interaction becomes important.

The computational viewpoint regards distributed processing in terms of application components and their interactions, *independent* of any specific distributed environment (operating system, communication system) on which they run. It *hides* from the application designer the details of the realization of the underlying abstract machine (engineering model) that supports it. A computational model thus may be characterized as focusing on applications rather than on the mechanisms used to distribute or, more generally, support them in the system [11].

The computational model describes the coarse-grained structure of a distributed application, i.e., application components and their interactions (in terms of operation invocations) at an abstract, system-independent level. Each coarse-grained entity of a distributed application is represented by an object, called *computational object,* with a set of well defined interfaces, called *computational interfaces.* Computational objects may run concurrently and exhibit internal parallelism.

The computational viewpoint provides a service-oriented view of the distributed application. Computational specifications are expressed declaratively; i.e., they state *what* (kind of environment) is required (to support distributed application components and their interactions) and not *how* it is to be provided [9]. Hence, the computational view of a distributed application is expressed in terms of computational objects, computational interfaces, and distribution-transparent interactions between these interfaces.

Computational viewpoint can be specified using an array of programming languages, interface definition languages, and formal description techniques such as LOTOS [12], SDL [13], and Estelle [14].

11.1.3.4 Engineering viewpoint. The engineering viewpoint addresses the issue of system support for distributed applications. It deals with aspects resulting from physical distribution of applications. It provides an infrastructure or a distributed platform for the support of the computational model. It provides generic services and mechanisms capable of supporting distributed applications and their interactions, specified in the computational model.

The engineering viewpoint is centered around the ways the application may be *engineered* into the (distributed) system. It is concerned with concrete application configurations, component placement and distribution, remote object communication and usage of underlying transport protocols, provision of distribution transparency mechanisms, and application-specific support services.

The ODP engineering model is not a detailed description of how to implement a particular environment. Rather, it identifies the functionality of basic system components that must be present, in one form or another, to support the computational environment described in the computational view. It identifies specific interfaces to identified components, allowing implementation freedom. There may be several engineering models for a particular computational environment, reflecting the use of different system components (mechanisms) and their configurations to realize the same computational environment.

The mechanisms visible in the engineering model are processing and storage resources, distribution transparency mechanisms (access transparency, location transparency, concurrency transparency, migration transparency, replication transparency, resource transparency, failure transparency, federation transparency, etc.), communication support mechanisms (communication protocols), and other application-specific support mechanisms.

11.1.3.5 Technology viewpoint. The technology viewpoint concentrates on the realized technical components, or the real-world artifacts, from which the distributed processing system is built. The technology model identifies the possible technical solution to the engineering mechanisms. The technical artifacts (realized components) from which the network of information system is built include the hardware and software comprising the local operating system, communication system components, and so on. The technology model shows how these technical artifacts are mapped to the (technology independent) designs identified in the engineering viewpoint.

An example of technology specification is the prescription that the communication support will be provided by OSI stack, that internode communication will employ X.25, or the protocol used to convey file data will be FTAM. A more detailed technology view of a specific environment would also specify: specific support environment technologies, e.g., Sun running Unix, Vax machines running VMS, etc.

11.1.4 Summary of viewpoints

The purpose of the RM-ODP framework of viewpoints is to position services relative to one another, to guide the selection of appropriate models of services, and to help in the placement of boundaries upon ODP. The framework of viewpoints is used to partition the concerns to be addressed when describing all facets of an ODP system, so that the task is made simpler. A summary of ODP viewpoints is given in Table 11.1.

11.2 Computational Model

11.2.1 What is a computational model?

The ODP computational model is a framework for describing the structure, specification and execution of the (components of the) distributed application on the distributed com-

TABLE 11.1 Summary of ODP Viewpoints

Viewpoint	Enterprise	Information	Computation	Engineering	Technology
Areas of concern	Enterprise needs of IS Objectives and roles of IS in the organization	Information models Information structures Information flows Information manipulation	Logical partitioning of application components, component interfaces, component interactions Service-oriented view of distributed application	Distributed platform infrastructure Distribution transparency, communication support, and other distributed enabling, regulating, and hiding generic mechanisms System-oriented view of distributed application	Technological artifacts required for realizing engineering mechanisms
Main concepts	Agents, artifacts, communities, roles, etc.	Schemas, relations, integrity roles, etc.	Computational object, computational interface, environmental constraints, computational interactions, etc.	Basic engineering objects, transparency objects, stubs, binders, protocol object, nucleus, etc.	Technological solutions corresponding to engineering mechanisms and structures
Whom it concerns	System procurers, corporate managers	Information analysts, system analysts, information engineers	Application designers and programmers	Operating system designers, communication system designers, system designers	System integrators, system vendors
Language/notation	Requirement description languages	Entity-relationship models, conceptual schemas, etc.	Application programming environments, tools, programming languages, etc.	Distributed platforms, engineering support environments, etc.	Technology mappings, identification of technical artifacts, etc.
Role in software engineering	Requirement capture and early design of distributed system	Conceptual design and information modeling	Software design and development	System design and development	Technology identification, procurement, installation

puting platform. It is the abstract model to express the concepts of the computational viewpoint.

The computational model provides a set of basic (albeit abstract) concepts and elements for the construction of a distributed programming (specification) language for which the model does not provide any syntax. Using the computational model, one can specify (program) a distributed application without worrying about the details of the underlying distributed execution platform (the engineering model). The design principle of the computational model is to minimize the amount of engineering detail that the application programmer is required to know, yet at the same time allowing the programmer to exploit the benefits of distributed computing.

The computational model focuses on the organization of applications into distributable components, identification of interactions between application components, and the identification of the distribution requirements (from the underlying distributed execution environment) for the support of interactions between application components.

The *computational specification* of a distributed application consists of the composition of *computational objects* (which represent application components) interacting, by operation invocations, at their interfaces. It identifies the activities that occur within the computational objects, and the interactions that occur at their interfaces.

11.2.2 Computational model: an object-oriented view of distributed application

The computational model is based on a *distributed-object model*. It prescribes an object oriented view [33] of the distributed application. Applications are collections of interacting objects. In this model, objects are the units of distribution, encapsulation, and failure.

The computational model is an "object world" populated with concurrent (computational) objects interacting with each other, in a *distribution-transparent* abstraction, by invoking operations at their interfaces [9]. An object can have multiple interfaces, and these interfaces define the interactions that are possible with the object.

A distributed computation progresses by operation invocations at object interfaces. The *activity* in an object (invoking object) can pass into another object (invoked object) by invoking *operations* in the interface of the invoked object. Activities carry the state of their computations with them; i.e., when an activity passes into an operation, it carries the parameters for that invocation, and returns carrying results. In the computational model, concurrency within an object and communication between objects are separate concerns. While concurrency is modeled by the concept of activity, communication between objects is modeled as (remote) invocation of an operation [15].

11.2.3 Distribution issues in the computational model

The computational model places few constraints on the extent to which application programs can be distributed. Most of the constraints on distribution of application components stem from discussion in other projections, such as enterprise viewpoint or information viewpoint.

Computational specifications are intended to be distribution-transparent, i.e., written without regard to the specifics of a physically distributed, heterogeneous environment. However, the expression of *environment constraints* in the computational interface template provides a hint of the application requirements from the distributed platform (e.g., distribution transparencies, security mechanisms, specific resource requirements, etc.).

At the computational level, user applications are unaware of how the underlying distributed platform is structured or how the distribution enabling and regulating mechanisms are realized.

11.2.4   Elements of the computational model

The design philosophy of the computational model has been to find the smallest number of concepts (elements) needed to describe distributed computations and to propose a *declarative* approach to the formulation of each concept [16].

The basic elements of the computational model are: *computational object, computational interface, operation* invocation at computational interface, *activities* that occur within a computational object, *environment constraints* on operation invocation, and so on.

11.2.4.1 Activity. Activity is agency by which computations make progress [15]. It is the unit of concurrency of the computational object. A computational object may have multiple activities threading through it, of which one or more may actually be executing on a processor at any one instant, depending on the number of processors available. An activity may pass from one object to another by the first invoking an *operation* on the interface of the second.

11.2.4.2 Computational operation. Computational objects may support multiple interfaces as *service* provision points. A *service* is an association between object state (some data) and the programs that operate upon them [15]. The ways that a user can interact with a service are completely defined by the set of *operations* that the service supports. Operations affect the state of the object. An operation is a service primitive. Each operation has two parts: the *operation signature,* which defines how the operation is invoked by a use of the service (*client*), and the *operation body,* which is the piece of program code executed by the provider of the service (*server*) when that operation is invoked.

An *operation signature* template has three parts [15]:

1. The *operation name* is an intrinsic part of the operation. When a client wishes to invoke an operation in a particular server interface, it identifies it by its name within that interface. To ensure that there is no ambiguity, no two operations in the same service may have the same name.
2. The *parameter part* of an operation specifies the number and types of the parameters and the order in which they are passed to the operation when it is invoked.
3. The *result part* of an operation specifies the number and types of result for each possible outcome from the operation.

Operations have distinct outcomes, each of which can convey different numbers and types of results. An operation's possible outcomes are called *terminations* and are distinguished by their names.

In the computational model, the (engineering) infrastructure failures in invoking an operation on a (remote) interface are reported (to the clients) by the infrastructure objects through the use of termination mechanisms. This permits the detection of invocation failures that occur in the infrastructure.

11.2.4.3 Computational interface. While computational objects are the units of structure and encapsulation of (application-specific) services, interfaces are the units of provision of services; they are the places at which objects can interact and obtain services. The specification of interaction between application components and of their requirements of distribution are captured in *computational interface templates* (see Figure 11.1).

The computational interfaces model different interaction concerns of computational objects. An application component acting as a client may request a number of other components to perform operations and thus needs a different interface with each of these. Similarly, the application component acting as a server may perform actions requested by a number of client components. There is no reason to restrict a server to provide interfaces with identical specifications to each of its clients. Allowing a server to provide multiple interfaces with distinct specifications enables a computational specification to directly reflect the different roles identified in the enterprise specification, especially with regard to access control. Multiple interfaces also enable the knowledge of the services provided by the object to be more tightly scoped [16]. For example, a client that is allowed access to an interface of server object (and may not have access to other interfaces of server object) can observe only a subset of the service provided by the object.

A computational object may support multiple computational interfaces that need not be of the same type. Interfaces of the same type may be provided by objects that are not of the same type. Each object may provide interfaces that are unlike those provided by the other object.

In the ODP computational model, interactions are specified in terms of either operational or stream interfaces.

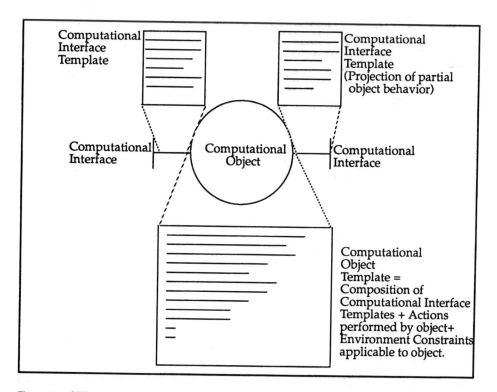

Figure 11.1 ODP computational model concepts

Operational interface. The specification of an operational interface template consists of the *operation specification, property specification, behavior specification,* and *role indication,* as described in the following text.

Operation specification. This is the definition of operations that are supported by the interface. Operation specification includes:

- *Operation name.* Each operation has a local name within an interface template. No two operations, within the interface, may have the same name.

Data specification includes:

- The number, sequence, and type of arguments that may be passed in each operation.
- The number, sequence, and type of results that may be returned in each termination.

The operation name, along with the type of argument and result parameters, constitutes the *operation signature.* Both the operation names and the arguments can be represented as abstract data types.

Most interface specifications, to date, have concentrated on the syntactic requirements of the interface, such as the operation signature. Aspects other than pure syntax are also important in facilitating the interaction between a pair of objects. This additional semantic information falls into two categories [5]:

1. *Information affecting how the infrastructure supports the interactions.* This information constrains the type of distribution transparencies, choice of communication protocols, etc., that must be placed in the interaction path between the interacting objects.
2. *The behavior (or the semantics) of the service offered at the interface.* An interface is viewed as a projection of an object's behavior, seen only in terms of a specified set of observable actions.

As a result, signature compatibility is less discriminating than interface compatibility. Interface compatibility includes not only the signature compatibility of the operations (supported by the interface) but also the behavior observable at the interface and the check on the property specification (see below).

Property specification. The property specification in the computational interface template defines the following attributes:

1. distribution transparency requirement on operation invocation (e.g., migration transparency, transaction transparency, etc.)
2. quality of service (including communication quality of service) attributes associated with the operations
3. temporal constraints on operations (e.g., deadlines)
4. dependability constraints (e.g., availability, reliability, fault tolerance, security, etc.)
5. location constraints on interfaces and, hence, their supporting objects (e.g., an object be located on a fault-tolerant computer system)

These attributes may be associated with individual operations or the entire interface.

Behavior specification. This defines the behavior exhibited at the interface. All possible ordering of operation invocations at or from this interface can be specified. This includes ordering and concurrency constraints between operations as well as sequential and parallel operation invocations. The behavior constitutes the protocol part of the interface.

Role indication. Often, an object assumes the role of either *client* or *server*. All interactions of an object, both as a client and as a server, between it and its environment, occur at object interfaces. It is convenient to partition client-role interaction concerns from server role interaction concerns in different interfaces.

Stream interface. The computational objects may both perform the information processing task and act as containers of information. There is a need to model both the interfaces that provide "service" and those interfaces that model "continuous" information flow. Such interfaces are modeled, in the computational model, as *stream interfaces.*

The stream interface is a set of information flows whose behavior is described by a single action that continues throughout the lifetime of the interface. Information media such as voice and video inherently consist of a continuous sequence of symbols. Such media are described as *continuous,* and the term *stream* is used to refer to the sequence of symbols constituting such a medium [18].

Examples include the flow of audio or video information in a multimedia application or the continuous flow of periodic sensor readings in a process control application. The computational description does not need to be concerned with detailed mechanisms; the fact that the flow is established and continues during the relevant period is enough.

The template for a stream interface consists of the following:

Stream signature. A specification of the type of each information flow contained in a stream interface and, for each flow, the direction in which the flow takes place.

Environment constraints. Continuous media have strict timing and synchronization requirements. The environment constraints that are relevant to stream interfaces include synchronization and clocking properties, time constraints, priority constraints, throughput, jitter, delay, media-specific communication quality requirements, etc., in addition to the properties applicable to operational interfaces.

Role. A role exists for each information flow, e.g., a producer object or a consumer object.

11.2.4.4 Computational object. The components of a distributed application are represented as computational objects in the computational model. The computational objects are the units of (application) structure and distribution. A computational object is an encapsulation of (application-specific) state and mechanisms which are not directly accessible to any other object. The computational objects model both the application components that perform information processing and those components that store the information. Objects can create interfaces or stop them during their lifetime.

As shown in Figure 11.1, *a computational object template* consists of the following:

1. a set of computational interface templates (both operational and stream) that the object can instantiate
2. an action template for initializing the state of new instances of the object
3. a specification of environment constraints applicable to the object as a whole

11.2.5   Multiple views on the computational model

There are several ways in which the general computational model can be described. This section identifies its major aspects. The computational model can be viewed as

1. an *interaction model*—an environment for interaction between computational objects
2. a *construction model*—construction of the configuration of computational objects
3. *a programming model*—an application programming environment

11.2.5.1 Interaction model. One view of the computational model is as an environment that supports the existence of and the interaction between computational objects. Computational objects interact by invoking operations at their interfaces. The interaction model defines an *invocation scheme* and a *type scheme* [19].

The invocation scheme describes the permitted forms of interaction, i.e., how clients may use the interfaces provided by the server. It defines the mechanisms for parameter passing between interfaces.

The type scheme provides a set of types into which computational interfaces can be classified. It defines a conformance relation over interface types which are a set of matching rules between interfaces which must be satisfied before a binding between interfaces can be established.

The interaction model (invocation scheme) is simple and uniform. It is based on the concept of *operation* invocation. The interaction between computational interfaces is via operation invocations that carry input argument parameters. The result of operation execution is returned to the invoker of the operation via *termination*.

The interaction model (invocation scheme) supports two styles of interactions between computational objects (or, more precisely, between computational interfaces). These are *interrogations* and *announcements*, used to model interactions with and without the reply, respectively.

Interrogation is a synchronous *request-response* invocation style; the activity that invoked the interrogation passes (via *operation*) to the object that provides the invoked operation and subsequently returns (via *termination*) to the object from which the invocation was made. There is no change in the degree of concurrency of the system using an interrogation type of invocation.

Announcement is an asynchronous *request-only* invocation style; a new *activity* is created in the object that provides the invoked operation, and the invoking activity continues in the object from which it made the invocation. Invoking an announcement increases the concurrency in the system, and the completion of the evaluation of the body of an announcement decreases the concurrency in the system. The object that invoked the announcement is informed neither of the completion of evaluation (of body of operation) nor of the results delivered (if any). The concept of announcement on the ODP computational model supports the idea of spawning concurrent and independent activities.

11.2.5.2 Construction model. The construction model is concerned with the construction of the configuration of computational objects, and it supports the creation of complex networks of interacting objects, giving the rules that govern object composition and decomposition. The computational objects can be connected in various ways, and networks of such objects can be treated as a single computational object. Similarly, a single computational object can be decomposed into networks of computational objects [5].

11.2.5.3 Programming model. The computational model provides an abstract, distribution transparent, language-independent *specification* and *programming model* for distributed applications, and a model of their execution and interaction semantics. Concerns in this viewpoint essentially include specification/programming language and programming system interface issues. The computational model expresses the programmability of the distributed platform [11].

From this viewpoint, an ODP system appears as a large programming environment capable of building and executing applications on the supporting engineering infrastructure. The distributed programming model, provided by the computational model, abstracts away, in an integrated framework, the generic set of distributed services provided by the engineering model from distributed applications designers and programmers. The ODP engineering model that describes the structure and organization of these distribution enabling and regulating services constitutes a *virtual machine* model for executing distributed programs conforming to the ODP computational model [20].

Hence, the computational model provides the equivalent of a programming language, for use on top of the *abstract machine* realized by the engineering infrastructure. Such a computational model will contain programming language features that are commonly found in advanced object-based distributed platforms. As such, the computational model can be seen as some form of implementation language for building applications on top of ODP systems [21].

The computational model hides the actual degree of distribution of an application from its programmer, thereby ensuring that application programs contain no deep-seated assumptions about which of their components are co-located and which are separated. Because of this, the configuration and degree of distribution of the underlying platform on which ODP applications are run can easily be altered without having a major impact on the applications software [22]. This desirable characteristic is called *distribution transparency.*

By conforming to the computational model, application programmers are given a guarantee that their programs will operate in a variety of different quality environments without modification of the source. The engineering model offers standardized system programming interfaces for supporting the computational programming environment [23].

11.3   Engineering Model

11.3.1   What is an engineering model?

The engineering model is an abstract model designed to express the concepts of the engineering viewpoint. It contains concepts such as operating systems, distribution transparency mechanisms, communication systems (i.e., protocols and networks), processors, storage, and so forth. As the notions of processor, memory, transport network play a more indirect role in a distributed system, the term *engineering model* is used here in a more general way to describe a framework oriented toward the organization of the underlying distributed infrastructure and targeted to the application support. It mostly focuses on what services may be provided to applications and what mechanisms should be used to obtain these services. The term *platform* is used to refer to the (configuration of) services offered to applications by the infrastructure [11].

The engineering model is still an abstraction of the distributed system, but it is a different abstraction from that of the computational model. Distribution is no longer transparent, but we still need not concern ourselves with real computers or with the implementations (technology) of mechanisms or services identified in the engineering model [24]. The engineering model provides a machine-independent execution environment for distributed applications.

11.3.2   Engineering model: an object-based distributed platform

The ODP engineering model is an architectural framework for the provision of an object-based distributed platform. The basic services and mechanisms, identified in the engineering model, are modeled as a collection of interacting objects which together provide support for the *realization* of interactions between distributed application components.

The engineering model can be considered as an extended operating system spanning a network of interconnected computers. In the *networked-operating system*[*] view of the model, the linked computers preserve much of their autonomy and are managed by their local operating systems, which are enhanced with mechanisms to enable, regulate, and (if desired) hide distribution.

11.3.3   Engineering model: animation of computational model

The interest of the computational model is directly related to the existence of a mapping, enabling it to relate to engineering concerns [15]. This means, for instance, being able to map computational concepts onto the engineering structures.

The engineering model provides an infrastructure or a distributed platform for the support of the computational model. The model provides generic services and mechanisms capable of supporting distributed applications specified in the computational model. The model is concerned with *how* an application, as specified in the computational model, may be *engineered* onto the distributed platform. The selection of distribution transparency and communication (protocol) objects, among many other support mechanisms, tailored to application needs, forms an important task [17].

The engineering model identifies the *functionality* of basic system components that must be present, in one form or another, to support the computational model. Hypothetically, there may be several engineering models for a particular computational environment, reflecting the use of different system components and mechanisms to achieve the same end. The issue in the computational model is *what* (interactions, distribution requirements); the engineering model prescribes solution as to *how* to realize these interactions, satisfying the stated requirements.

11.3.4   Structure of the engineering model

The engineering model reveals the structure of the distributed platform, i.e., the ODP infrastructure which supports the computational model. The services or mechanisms that enable, regulate, and hide distribution in the ODP infrastructure are modeled as objects, called *engineering objects,* which may support multiple interfaces.

There are different kinds of engineering objects in the engineering model, corresponding to different distribution (enabling, regulating, hiding) functions required in distributed environment. This is illustrated in Fig. 11.2. Some engineering objects correspond to the

---

[*]In contrast, in the distributed operating systems view, the system management would be global and individual computers have little autonomy.

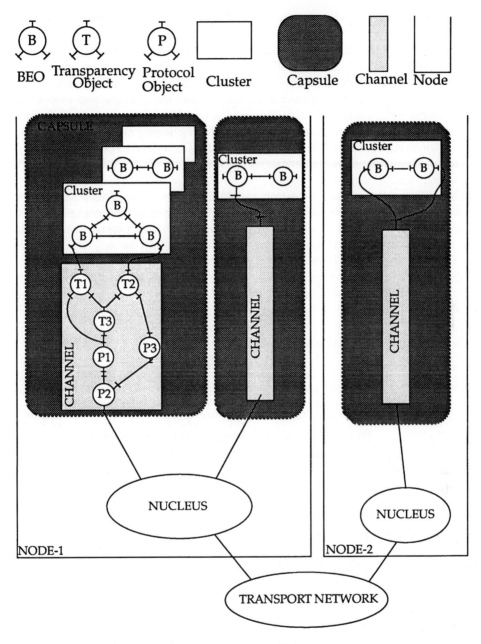

Figure 11.2 ODP engineering model: organization of distributed infrastructure

application functionality and they are referred to as *basic engineering objects* whereas those that provide distribution functions are classified as *transparency objects, protocol objects, supporter objects,* etc. At a given host, the basic engineering objects belonging to an application may be grouped into *clusters*. A host may support multiple clusters in its addressing domain, known as *capsule*. A capsule consists of clusters of basic engineering objects, a set of transparency objects, protocol objects, and other local operating system facilities.

From an engineering viewpoint, the ODP infrastructure consists of interconnected autonomous computer systems (hosts), which are called *nodes*. Each node supports a *nucleus object* and multiple capsules. The nucleus encapsulates computing, storage, and communication resources at a node. All the objects in the node share common processing, storage, and communication resources encapsulated in the nucleus object of the node.

As mentioned before, the engineering model *animates* the computational model. The computational-level interactions between a pair of computational objects (or their interfaces) are supported through *channel* structures in the engineering model. A channel binds basic engineering objects in different clusters, capsules, or nodes. The channel is a configuration of transparency objects, protocol objects, and so forth that provide distribution support.

The services and mechanisms currently identified in the engineering model are generic in nature and can support distribution requirements of applications in a broad range of enterprise domains (telecoms, office information systems, computer integrated manufacturing, etc.). However, domain-specific supporting functions will be defined in domain-specific engineering models (which are the specialization of ODP engineering model).

The following is a brief description of the engineering objects and structures currently identified in the ODP engineering model. Table 11.2 gives a relationship between the engineering objects and the real-world system.

TABLE 11.2   System Abstractions in the Engineering Model

Engineering Object	System Representation
Node	Single computer system, network of workstations managed by a distributed operating system, any autonomous information processing system with independent nucleus resources and failure characteristics
Nucleus	Processing, storage, and communication resources of a *node*
Capsule	The concept of address space in operating systems
Cluster	The concept of "linked" modules to form an executable program image
BEO	The program module that may not be executed in isolation
Channel	The run-time "binding" between distributed BEOs
Transparency object	Special-purpose modules that enhance the operating system environment of the *node* and can be dynamically linked into the distributed application program

11.3.4.1 Basic engineering object. Basic engineering objects (BEOs) are the run time representation of computational objects (obtained through compilation, interpretation, or through some other transformation of computational objects) which encapsulate application functionality. A basic engineering object is the corresponding computational object (computationally) enriched with extra interfaces to interact with objects in the channel. In general, a computational object can be mapped onto a single basic engineering object, but (because of refinement, decomposition, and replication) a computational object will often map to several basic engineering objects.

11.3.4.2 Cluster. A cluster is a configuration of basic engineering objects. Clusters are used to express related objects (which belong to the same application) that should be local to one another, i.e., those groups of objects that should always be on the *same* node at all times.

A cluster is a collection of BEOs in a capsule such that members of the cluster have no interfaces bound directly to interfaces of objects in other clusters. Objects within a cluster communicate directly, whereas objects in different clusters interact through *channels*.

11.3.4.3 Cluster Manager. A cluster is associated with a cluster manager, which coordinates the management of the cluster. The cluster manager performs the operations of activating a cluster, passivating a cluster, checkpointing a cluster, migrating a cluster, and other policy-specific operations.

11.3.4.4 Capsule. A capsule consists of clusters of basic engineering objects, *transparency objects,* and *protocol objects* bound to a common nucleus in a distinct address space from any other capsule. A capsule consists of active clusters, cluster manager objects, (one for each cluster in the capsule), *transparency stub, transparency binder,* and *protocol objects* for each *channel* bound to an interface of a basic engineering object within any of the active clusters, and a capsule manager. A cluster is always contained within a single capsule. A capsule is always contained within a single *node*.

11.3.4.5 Nucleus. A nucleus is an object that provides access to basic processing, storage, and communication functions of a *node* for use by basic engineering objects, *transparency objects,* and *protocol objects* bound together into capsules. A nucleus may support more than one capsule. A nucleus has the capability of interacting with other nuclei (through its communication function), providing the basis for intercapsule and internode communication.

11.3.4.6 Node. A node consists of one nucleus object, a node manager, and a set of capsules. All of the objects in a node share common processing, storage, and communications resources.

11.3.4.7 Node manager. The node manager performs the bootstrapping of the node. It initializes the services on the node. It is a repository of capsule templates.

11.3.4.8 Channel. A channel object is a configuration of transparency objects, protocol objects, application specific supporting objects, etc. providing a binding between a set of interfaces to basic engineering objects, through which interaction can occur. The structure of the channel is dependent on the distribution function requirements of the interaction between basic engineering objects. A general structure of the channel is described in the next section.

11.3.4.9 Supporting object. A supporting object is an object, outside of a channel, which cooperates with objects within the channel for the provision of distribution transparency. The supporting objects are shown in Fig. 11.3. The supporting objects are the repositories of information required by the *transparency objects* and *protocol objects* within the channel. For example, the *location transparency binder* object registers and retrieves object locations via a supporting object known as the *relocator*.

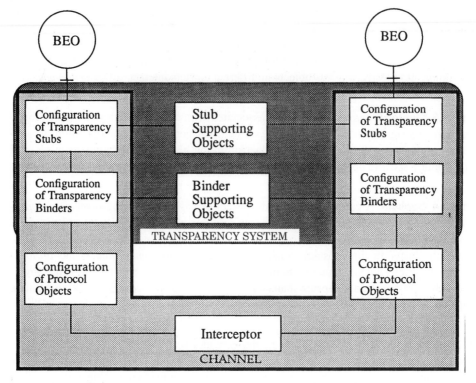

Figure 11.3 Simplified generic channel structure

11.3.5 Structure of a Channel

This section describes the generic structure of the channel, which provides the binding between basic engineering objects. A channel supports *distribution transparent* interaction between a pair of (interfaces to) basic engineering objects located in different clusters.

As shown in Fig. 11.3, a *channel* is a configuration of *transparency objects, protocol objects, application-specific supporting objects,* and *interceptor objects.* It is parameterized by a set of communication interfaces. The configuration of the channel can be dynamically negotiated when establishing the binding between basic engineering objects. The configuration of objects in the channel provides the medium through which (remote) interactions among basic engineering objects pass.

The channel is composed of a variety of *transparency objects.* The transparency objects that make up the channel are classified as either *stub objects* or *binder objects.* Both stub objects and binder objects contribute to the provision of distribution transparency between interacting basic engineering objects, but they differ in that the stub objects actually modify the information exchanged across the channel, while binder objects control various aspects of the binding between the interfaces of remote basic engineering objects.

Figure 11.3 is a simplified view of the channel[*] that illustrates the object types used in the structure. In practice, a channel may be much more complex than this and may contain

---

[*]In spite of the resemblance of Fig. 11.3 to the layered communication models, such as OSI, there is not necessarily a peer-to-peer relationship between the objects in the two halves of the channel. There may exist peer-to-peer relationship between the protocol objects. In contrast, binder objects may obtain the required information from support objects outside the channel.

several different types of stub objects, binder objects, and so on, depending on the transparency properties required [24].

11.3.5.1 Stub object. A stub is an object that acts to a basic engineering object as a *representative* of another basic engineering object located in different clusters, thus contributing toward distribution transparency. Stub objects are bound to the basic engineering objects for the purpose of hiding certain aspects resulting from distribution (or heterogeneity).

Stub objects have direct access to the basic engineering objects. The operation invocations on the interfaces of basic engineering objects are *intercepted* by stub objects to hide some aspects of distribution such as concurrency in the system or to modify the information exchanged between basic engineering objects, thus masking the heterogeneity in the distributed system.

Stub objects add further interactions and/or information to interactions between interacting basic engineering objects to support distribution transparency. As an example, a stub object may provide adaptation functions such as marshalling and unmarshalling of operation parameters to enable *access transparent* interactions between interfaces of basic engineering objects.

Examples of stub objects include *access transparency objects* and *concurrency transparency objects* discussed in the next section.

11.3.5.2 Binder object. A binder is an object that *controls* and *maintains* the binding between interacting basic engineering objects, contributing toward the provision of distribution transparency.

Binder objects maintain the binding between basic engineering objects, even if they are migrated, reactivated at new location, or replicated. Examples of binder objects include *location transparency objects, migration transparency objects, replication transparency objects, failure transparency objects,* and *resource transparency objects.* Transparency support through the combination of stubs and binders is discussed in Section 11.3.6.1.

Binder objects interact with each other to maintain the integrity of the binding between the interacting basic engineering objects. Binder objects in the channel can interact with each other using other objects in the channel or via interaction with supporting objects outside the channel. Binder objects are interconnected by protocol objects.

11.3.5.3 Protocol object. A protocol object encapsulates communication protocol functionality for supporting communication between basic engineering objects. A channel may be composed of a number of protocol objects corresponding to different communication support requirements of interactions between basic engineering objects. Protocol objects interact with other protocol objects to support interaction between basic engineering objects. When protocol objects are in different (administrative) domains they interact via an *interceptor.* When they are in same domain, they interact directly.

11.3.5.4 Interceptor object. An interceptor is an object which masks administrative and technology domain boundaries by performing transformation functions such as protocol conversion, type conversion etc. It enables interactions to cross administrative and communication domains, thus contributing toward *federation transparency.*

11.3.6   Transparency system

Distributed systems exhibit a number of properties, inherent in distribution, not found in centralized systems. Consequently, an application designed to work on a distributed system must take these additional properties into account. However, the application designer does not have to deal explicitly with these properties, if they are made transparent. The complexities of distributed systems may be hidden through the notion of distribution transparencies defined by ODP.

The concept of *transparency* is related to the notion of *abstraction,* where irrelevant details are ignored. Transparency is the property of *hiding* from the user (in the computational environment) some aspects of the potential behavior of the underlying ODP infrastructure [25].

This section describes the distribution transparency system (See Fig. 11.3) that binds a pair of basic engineering objects within the *channel* of the engineering model. Currently, the engineering model identifies a set of transparency mechanisms, which are by no means exhaustive. There is scope for the definition of more generic distribution transparencies in the engineering model. The distribution transparencies, currently identified, can be used in a broad range of enterprise domains.

11.3.6.1 Transparency support through stubs and binders. The transparency objects cooperate to perform the *transparency function* by bringing uniformity to some aspect of the distribution of the engineering objects they support. Some forms of transparency require supporting services; for example, if basic engineering objects can move from one location to another, a means of recording and discovering the current location of the object is required. Supporting functions are modeled as engineering objects so that the architecture provides a maximum degree of configuration flexibility. As shown in Fig. 11.3, the *transparency system* is composed of stub objects and binder objects in the channel and supporting objects outside the channel.

As mentioned in the previous section, the engineering model classifies transparency objects as either stub objects or binder objects. While stub objects address masking of some aspects of distribution (those arising due to the presence of *heterogeneity* and *concurrency* in the distributed system), the binder objects address aspects of distribution resulting from change of location of objects. The migration of the object may be required for any of the following reasons:

1. *Load balancing, reduction of access time, etc.* This aspect of distribution is masked by *location transparency binder* and *migration transparency binder.*
2. *Failure of object at one location and its reactivation at another location.* This aspect of distribution is masked by *failure transparency binder.*
3. *Unavailability of (nucleus) resources at one location and its (re)activation at another location.* This aspect of distribution is masked by *resource transparency binder.*
4. *Replication of objects at different locations.* For example, if the server object is replicated, then it is required to maintain the binding between the client and the set of replicated server objects. Changes to the membership of the replica group, such as addition of a server object, would require establishing the binding with the new member.

In all of these cases, the binding between the basic engineering objects is susceptible to being broken down, resulting in a disruption of the service to the client. The binder objects attempt to maintain the integrity of the binding between basic engineering objects. Hence, they are called transparency *binder* objects. The location transparency binder provides the basic service. All other binders require the support of the location transparency binder. ODP permits distribution transparency to be selectively enabled in any binding between basic engineering objects and specifies channel configuration rules to achieve or avoid specific transparencies.

Transparency mechanisms provide an enhanced environment positioned on top of the low-level operating systems and communications facilities of the distributed platform, for the support of the distribution transparent programming environment offered by the computational model. The technique for providing transparency services is based on the principle of replacing an original service by a new service that combines the original service with the transparency service and permits clients to interact with it as if it were the original service. The clients need not be aware of how these combined services are achieved [26].

Because the interaction between the objects occur at their interfaces, these transparencies are applicable to individual interfaces or to specific operations of the interfaces. An interface may have a set of transparency requirements that may be different from those of other interfaces of the same object. A summary of transparency mechanisms is presented in Table 11.3.

TABLE 11.3   ODP Distribution Transparencies

Transparency	Central Issue	Result of Transparency
Access	Method of access to objects (invocation mechanism and data representation)	Client is unaware of access mechanisms at the server interface.
Concurrency	Concurrent access to objects in the distributed system	Clients are masked from the effects of concurrent access to the server interface.
Location	Location of object in the distributed system	Clients are unaware of the physical location of the server
Migration	Dynamic relocation of objects during the "bind session"	Clients are unaware of the dynamic migration of the server
Replication	Multiple invocations of replicated objects, multiple responses, and consistency of replicated data	Client invokes a replicated server group as if it were a single server. Distribution of requests, collation of responses, consistency of data, and membership changes are hidden.
Resource	Resource management policies of the node (deactivation and reactivation of objects)	Client is unaware of the deactivation and reactivation of the server.

11.3.6.2 Access transparency. Access transparency hides from a client object the details of the access mechanisms for a given server object, including details of data representation and invocation mechanisms (and vice versa). It hides the difference between local and remote provision of the service, and it enables interworking across heterogeneous computer architectures, operating systems, and programming languages.

11.3.6.3 Concurrency transparency. Concurrency transparency hides from the client the existence of concurrent accesses being made to the server. Concurrency transparency hides the *effects* due to the existence of concurrent users of a service from individual users of the service.

11.3.6.4 Location transparency. Location transparency hides from a user (client) the location of the object (server) being accessed.

11.3.6.5 Migration transparency. Migration transparency hides from the user of the service (client) the effects of the provider of the service moving from one location to another during the provision of the service (between successive operation invocations). Migration transparency is the dynamic case that arises if the server interface can move while the client object is interacting with it, without disturbing those interactions.

11.3.6.6 Replication transparency. Replication transparency hides the presence of multiple copies of services and the maintenance of the consistency of multiple copies of data from the users of the services. It enables a set of objects (their interfaces) organized as a *replica group* to be coordinated so as to appear to interacting objects (or their interfaces) as if they were a single object (interface).

There are two main aspects of replication transparency. The first hides the difference between a replicated and a nonreplicated provider of a service from users of that service, and the second hides the difference between replicated and nonreplicated users of a service from providers of that service. Users are unaware of multiple providers of a service and need not be concerned about making multiple operation invocations or dealing with multiple responses.

11.3.6.7 Resource transparency. Resource transparency hides from a user (client) the mechanisms that manage allocation of resources by activating or passivating (server) objects as demand varies. With resource transparency in place, clients can invoke operations on the server irrespective of whether the server is currently active or passive.

11.4   ODP Application

11.4.1   Application of ODP architectural concepts: an illustration

This section illustrates the application of ODP architectural concepts discussed in Sections 11.2 and 11.3 through a simple client-server example. The distributed application consists of a set of distributed file servers and file users (clients). The clients and servers are located on different *nodes* of a distributed platform. The distributed platform conforms to the ODP engineering model.

In the following we illustrate the *computational* and *engineering* modeling of this client-server distributed application. We focus on a single file user (F-CLIENT) and a file server (F-SERVER) interaction to illustrate the issues involved in the computational and engineering modeling of distributed systems. Henceforth the modeling activity is restricted to a pair of client and (possibly replicated) server(s) and interactions between them.

11.4.1.1 Computational modeling. From the computational viewpoint, the application consists of a file user and a file server interacting in a distribution-transparent abstraction. The computational modeling activity consists of the identification of the following:

1. *Computational objects.* The computational objects in this example are F-USER and F-SERVER. The activities performed by these objects are specified in the corresponding *computational object templates.*
2. *Computational interfaces.* In this example, each computational object consist of a single computational interface. (More realistically, the file-server object may possess multiple interfaces to model its interactions with different clients).
3. *Computational operations.* The operations that are supported by (server) interfaces and invoked from (client) interfaces are identified for each computational interface. They are F-Open, F-Read, F-Write, F-Close, and so on.
4. *Environment constraints.* The environment constraints associated with the computational object and their interfaces are specified.

    The environment constraint associated with the F-SERVER object may include, for example, certain *security* requirements (with respect to the *node* in which that object is placed). The environment constraints associated with the F-CLIENT and F-SERVER interfaces include distribution transparency requirements such as access transparency, concurrency transparency, location transparency, replication transparency, communication protocol requirements such as (a specific file transfer protocol) FTAM [27], and connection-oriented session and transport protocols of an OSI stack. The specification of environment constraints in the computational interface template has a direct relationship to realized engineering structures and mechanisms.
5. *Computational interactions.* The F-USER and F-SERVER objects interact by exchanging operations at their interfaces. The rules for operation exchange constitute the behavior (exhibited by the computational object) at the interface.

Items 3, 4, and 5 are specified as part of the *computational interface template.* The F-CLIENT and F-SERVER interfaces are bound on the basis of the matching of their computational interface templates, performed by a special ODP system object called the *trader* [28, 29]. The computational view of this application is shown in the top part of Fig. 11.4.

11.4.1.2 Engineering modeling. In passing from the computational viewpoint to the engineering viewpoint, concerns shift from the specification of computational structures (e.g., computational objects, computational interfaces, etc.) and statements of necessary *properties of* interactions between object interfaces (e.g., distribution transparency requirements) to engineering mechanisms capable of realizing these properties.

At the heart of the separation between the ODP computational and engineering models is the idea of a tool-driven transformation between the abstract computational description of a distributed application and its mechanization in terms of the engineering model. The engineering model animates the computational model [30].

The engineering modeling activity consists of identification of *basic engineering objects* corresponding to computational objects, realization of interactions between computational interfaces through the configuration and instantiation of appropriate *channel* structures between the (corresponding) interfaces of basic engineering objects, and the placement of engineering objects in the appropriate environments (*nodes*). The composition of objects in the channel must satisfy the environment constraints specified in the computational interface template.

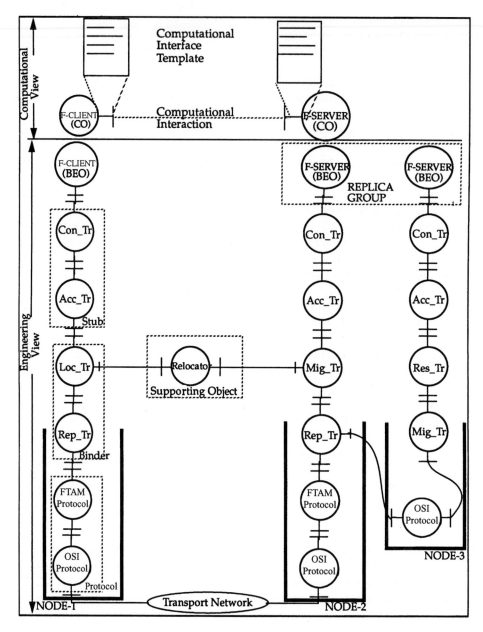

Figure 11.4 Client-server interaction in an ODP environment

1. *Basic engineering objects.* The F-CLIENT and F-SERVER basic engineering objects are obtained through a compilation or any other transformation of the corresponding computational objects, and may result in the decomposition of computational objects and/or identification of additional interfaces to BEOs to interact with objects in the channel. The computational interfaces are not decomposed.
2. *Channel.* The channel structure between the basic engineering objects carries the operations invoked by the interfaces of the BEO. In this example, the channel between the F-USER and F-SERVER BEOs is composed of the following engineer-

ing objects, which- satisfy the distribution transparency requirements of the computational interface.

(a) *Access transparency object.* The F-CLIENT and F-SERVER objects may be coded in different programming languages and compiled on different machines. An access transparency object is interposed to enable them to talk to each other. Each of the access transparency objects on the client-and-server half of the channel convert the operations into messages in the network format, and vice versa. For example, the client may be coded in Smalltalk and the server in C. The client uses the Smalltalk *methods* to invoke the server. However, the server (written in C) cannot understand Smalltalk *methods*. It can only respond to C *function calls*. The access transparency objects on both sides of the channel convert the local invocation and data parameters into messages in the network format and vice versa.

(b) *Concurrency transparency object.* The F-SERVER may have multiple clients invoking operations on its interface. The concurrency transparency object hides the effect of concurrency existing at the F-SERVER from the F-CLIENT [31].

(c) *Location transparency object.* The F-SERVER object may be migrated on a different node (due to load balancing). If an operation for a migrated F-SERVER is received, the location/migration transparency object on the server side of the channel informs its counterpart on the client side of the channel. The client location transparency object obtains the current F-SERVER address from the *relocator* (a supporting object) and redirects the client operation invocation.

(d) *Replication transparency object.* The example shows that there exist two replicas of F-SERVER. If the F-CLIENT wishes that all its operations on the F-SERVER be transparently performed on a server replica (instead of making separate invocations), then the replication transparency objects on both ends of the channel perform the functions of distribution of client requests and server responses, collation, and ordering to perform consistent replication [32].

(e) *Resource transparency object.* If the F-CLIENT wishes that the deactivation and reactivation of F-SERVER be transparent between operation invocations, the resource transparency object on the server side performs the reactivation of a passivated F-SERVER when an operation invocation is received for it.

Note that, as discussed before, transparencies are selective. Only those transparencies that are specified in the computational interface template are included in the channel. Some transparencies require peer objects (on each side of the channel) because they require peer-to-peer protocols to achieve the transparency.

The objects satisfying the communication requirements of the computational interface are as follows.

In this example, the *file transfer access and management* (FTAM) protocol is configured in the channel along with the appropriate subset of OSI session and transport protocols to support the communication of operation exchanges between F-USER and F-SERVER over the unreliable transport network.

11.5   Conclusion and Directions for Future Research

Using the five ODP viewpoints to examine system issues encourages a clear separation of concerns, which in turn leads to a better understanding of the problems being addressed, namely:

- describing the role of the enterprise (enterprise viewpoint) independently of the way in which that role is automated
- describing the information content of the system (information viewpoint) independently of the way in which the information is stored or manipulated
- describing the application programming environment (computation viewpoint) independently of the way in which that environment is supported
- describing the components, mechanisms used to build systems independently of the machines on which they run
- describing the basic system hardware and software (technology viewpoint) independently of the role it plays in the enterprise.

The field of open distributed processing offers numerous research opportunities related to both theoretical and practical aspects of viewpoint models. One of the main research problems is ensuring consistency between viewpoint specifications. This includes identifying the requirements and constraints that originate in the viewpoint models, identifying the consistency constraints between the models, and defining consistency-preserving transformations between the models. Similarly, it is required to ensure consistency between the system design and the viewpoint specifications.

The computational and engineering viewpoints are the most important from the point of view of distributed software engineering. They offer uniform and consistent abstraction levels for the specification of the system and its engineering on the distributed infrastructure. They offer powerful design concepts for the development of application programming environments and distributed platforms.

The ODP model is very generic. It can be applied in various application domains. Currently, it is being used in the field of advanced intelligent networks, distributed network management, and others. Its application in other domains is an area of active interest.

11.6 References

1. Draft Recommendation ITU-T X.901 / ISO 10746-1: Basic Reference Model of Open Distributed Processing—Part-1: Overview.
2. Draft International Standard ITU-T X.902 / ISO 10746-2: Basic Reference Model of Open Distributed Processing—Part-2: Descriptive Model, 1994.
3. Draft International Standard ITU-T X.903 / ISO 10746-3: Basic Reference Model of Open Distributed Processing—Part-3: Prescriptive Model, 1994.
4. Draft Recommendation ITU-T X.904 / ISO 10746-4: Basic Reference Model of Open Distributed Processing—Part-4: Architectural Semantics.
5. *ANSA Reference Manual,* Volume A, B, C., Release 01.01, Advanced Projects Management Limited, Cambridge, U.K., July 1989.
6. International Standard ISO 7498: Information Processing Systems–Open System Interconnection–Basic Reference Model, 1984.
7. CCITT Com. 7 Rxx, Study Group 7, Q19/7 Distributed Applications Framework (DAF) Report, February 1990.
8. Linington, P. F. 1992. Introduction to open distributed processing basic reference model. *Proceedings of the IFIP TC6/WG6.4 International Workshop on Open Distributed Processing* (October 1991), Amsterdam: North Holland 3–14.
9. Proctor, S. 1992. An ODP analysis of OSI systems management. *Proceedings of the Third Telecommunication Information Networking Architecture Workshop,* (TINA 92), Narita, Japan, 23.2.1–23.2.22.
10. van Griethuysen, J. J. 1992. Enterprise modelling, a necessary basis for modern information systems. *Proceedings of the IFIP TC6/WG6.4 International Workshop on Open Distributed Processing* (October 1991), Amsterdam: North Holland 29–68.
11. Bregant, G. 1992. Platform Modelling Requirements from the ROSA Project. *Proceedings of the Third Telecommunication Information Networking Architecture Workshop,* (TINA 92), Narita, Japan, 11.1.1–11.1.15.

12. International Standard ISO 8807: Information Processing Systems-Open System Interconnection-LOTOS-A Formal Description Technique based on temporal ordering of observational behavior, 1988.

13. CCITT Specification and Description Language (SDL), Z.100, Geneva, March 1988.

14. International Standard ISO 9074: Information Processing Systems–Open System Interconnection–Estelle–A Formal Description Technique based on Extended State Transition Model, 1989.

15. *ANSA: An Application Programmer's Introduction to the Architecture, TR.017.00.* Cambridge, U.K.: Advanced Projects Management Limited, November 1991.

16. *ANSA: An Engineer's Introduction to the Architecture, TR.03.02.* Cambridge, U.K.: Advanced Projects Management Limited, November 1989.

17. Schill, A., and M. Zitterbart. 1992. A systems framework for open distributed processing. *Proceedings of the International Workshop on Distributed Systems: Operations and Management,* 1992.

18. *ANSA Technical Report: Integrating Multimedia into ANSA Architecture, TR.028.00.* Cambridge, U.K.: Advanced Projects Management Limited, February 1993.

19. *ANSA Computational Model, AR.001.01.* Cambridge, U.K.: Advanced Projects Management Limited, February 1993.

20. Stefani, J. B., and E. Najm. A Formal Semantics for the ODP Computational Model. *CN&ISDN* special issue on Open Distributed Processing (in publication).

21. Stefani, J. B. 1990. Open distributed processing: the next target for the application of formal description techniques. *Proceedings of the IFIP TC6/WG6.1 Third International Conference on Formal Description Techniques for Distributed Systems and Communication Protocols,* FORTE'90, 427–442.

22. *ANSA Technical Report: DPL Programmers Manual, TR.031.00.* Cambridge, U.K.: Advanced Projects Management Limited, February 1993.

23. Stefani, J. B. 1991. Towards a reflexive architecture for intelligent networks. *Proceedings of the Second Telecommunication Information Networking Architecture Workshop,* (TINA 91), Chantilly, France, 1–13.

24. Taylor, C.J., 1993. Object-oriented concepts in distributed systems. *Computer Standards and Interfaces,* Vol. 15, No. 2/3.

25. 5th Deliverable, The ROSA Architecture, Release Two, RACE Project R1093, May 1992.

26. *ANSA: A System Designer's Introduction to the Architecture, RC253.00.* Cambridge, U.K.: Advanced Projects Management Limited, April 1991.

27. International Standard ISO 8571: Information Processing Systems–Open System Interconnection–File Transfer Access and Management, Parts 1–4, 1988.

28. Indulska, J., K. Raymond, and M.Bearman. 1993. A type management system for an ODP trader. *Proceedings of International Conference on Open Distributed Processing,* ICODP-93, Berlin, 169–180.

29. Goscinski, A., and Y. Ni. 1993. Object trading in open systems. *Proceedings of International Conference on Open Distributed Processing,* ICODP-93, Berlin, 145–156.

30. Watson, A. 1992. *ISA Project Report: Types and Projections, Ref: APM/RC/.258.03.* Cambridge, U.K.: Advanced Projects Management Limited, April 1992.

31. Warne, J.P., and O. Rees. 1993. *ANSA Atomic Activity Model and Infrastructure, AR.004.01.* Cambridge, U.K.: Advanced Projects Management Limited, February 1993.

32. *ANSA: A Model for Interface Groups, AR.002.01.* Cambridge, U.K.: Advanced Projects Management Limited, February 1993.

33. Rumbaugh, J., M.Blaha, W. Premerlani, F. Eddy, and W. Lorensen. 1993. *Object-Oriented Modelling and Design.* Englewood Cliffs, N.J.: Prentice Hall.

Algorithms

12

Fundamentals of Parallel Algorithms

Joseph F. JáJá

We describe in this chapter a number of general paradigms for designing efficient parallel algorithms. Since the performance measure of a parallel algorithm depends on the model of parallel computation used, we introduce some of the major current models with a special emphasis on two models: the data parallel model and a distributed memory model with a shared address space. It turns out that a strategy applied to solve a problem can lead to substantially different efficient algorithms, depending on the underlying model. The paradigms discussed are classified as balanced trees, divide and conquer, partitioning, or combining. We examine detailed case studies that include computing the sum of *n* numbers, computing the discrete Fourier transform, sorting, and list ranking. In all these cases, we describe algorithms that are the best possible under the data parallel model and the distributed memory with shared address space model.

12.1 Introduction

The rapid successful expansion of the software base for microprocessors is due in part to the existing fundamental techniques and data structures for designing efficient sequential algorithms. Such a foundation allows the fast development of application software that runs efficiently on various types of microprocessors. In spite of many efforts to extend these techniques to the parallel domain, no solid framework exists for developing parallel software that is portable and efficient across various multiple processors. The lack of such techniques is mostly due to the absence of a universal model of parallel computation, an absence largely caused by the diversity of the existing architectures and parallel programming models. As we will mention in Section 12.2, there is currently a definite trend toward a common architecture framework that can be integrated with software environments that support the three main programming models: the *data parallel,* the *shared memory,* and the *message passing* models. Such a trend is leading to the introduction of new models of parallel computation that are more likely than earlier models to accurately predict performance on the emerging multiprocessors.

The purpose of this chapter is to introduce a number of strategies for designing efficient parallel algorithms on almost any of the existing parallel models. We illustrate the use of these strategies on a number of specific computations and show how they can be used to derive efficient algorithms on two distinct parallel models: the *data parallel model* and a *distributed memory model* with a shared address space. It turns out that a strategy to solve a given problem can lead to substantially different efficient algorithms on different models. The examples discussed include computing the sum of *n* numbers, computing the Discrete Fourier transform, sorting, and list ranking.

The rest of this chapter is organized as follows. We introduce in Section 12.2 the three major parallel models that have been used in the literature for the development of techniques for parallel algorithms. These models are the *parallel random access model* (PRAM), the *shared memory model,* and the *network model.* Two recent models, the LogP [5] and the *block distributed memory* [8] models, are also introduced. In Section 12.3, we describe the balanced tree strategy and illustrate its use on the problem of computing the sum of *n* numbers. The divide-and-conquer strategy is discussed in Section 12.4 and applied to the problem of computing the discrete Fourier transform. The resulting algorithms for the data parallel model and the distributed memory model are quite different. Section 12.5 is devoted to the partitioning strategy and its application to the development of an efficient randomized sorting algorithm. Finally, we outline in Section 12.6 the strategy of combining a number of different algorithms to solve a given problem, each of which is efficient when the input size falls within a certain range. This strategy is then applied to the list ranking problem.

12.2 Models of Parallel Computation

The purpose of a parallel computation model is to give a simple framework for describing and analyzing parallel algorithms, provided that the analysis captures in a significant way the actual performance of these algorithms on real machines. The model also serves as a link between the software model, which is at a higher level, and the hardware model, which is at a lower level. Thus far, there is no single computation model that is universally accepted as the model for parallel computation. This is mostly due to the existing diversity of multiprocessor architectures and of parallel programming models. This situation has started to change due to the impact of recent developments in microprocessor and communication technologies. It is now widely accepted that a common hardware organization of multiprocessors will consist of a collection of standard powerful processors connected by a fast and robust communication network. In addition, a common set of software and hardware primitives can be used to support the three main programming models (i.e., the data parallel, the shared memory, and the message passing). However, it is expected that most applications software will be developed using a mix of the data parallel and the shared memory tools. These trends are likely to rekindle the efforts to identify a universal model for parallel computation that is as widely accepted as the *random access model* (RAM) has been for sequential computations.

We briefly introduce the major models used in the literature for designing and analyzing parallel algorithms. We also describe two recent models, the LogP model introduced in Ref. [5] and the *block distributed memory* (BDM) model introduced in Ref. [8]. The two latter models build on the recent trends outlined above.

12.2.1 The PRAM model and the related data parallel model

The PRAM is a natural extension of the sequential model of computation. It assumes the presence of an arbitrary number of processors that can access synchronously a large shared memory. Each processor has its own local memory and can execute its own local program. All processors operate synchronously in lock step and can use a processor number to access shared memory locations. There are a number of versions of the PRAM model, depending on the assumptions regarding the handling of the simultaneous access of several processors to the same location of the shared memory. The *exclusive read exclusive write* (EREW) PRAM does not allow any simultaneous read or write into a single memory location. The *concurrent read exclusive write* (CREW) PRAM allows concurrent read accesses only. Concurrent read or write instructions are allowed on the *concurrent read concurrent write* (CRCW) PRAM.

There are three principal varieties of CRCW PRAMs corresponding to the method used for resolving the concurrent write conflicts. The COMMON CRCW PRAM assumes that all concurrent writes into the same location involve the same value. For each set of concurrent writes into the same location, the ARBITRARY CRCW PRAM allows an arbitrary processor to succeed, and the PRIORITY CRCW PRAM assumes that the processor with the minimum index succeeds. Other variations of the CRCW PRAM model exist. It turns out that these models do not differ substantially in their computational powers, and that the CRCW is strictly more powerful than the CREW, which is strictly more powerful than the EREW.

The PRAM model is perhaps the most popular parallel model, and the model for which a rich set of efficient techniques and paradigms exist [7]. The main complexity measure used for evaluating the performance of a PRAM algorithm is the running time $T_p(n)$, where p is the number of processors available and n is the input length. The algorithm achieves an optimal speedup if

$$T_p(n) = \Theta\left(\frac{T^*(n)}{p}\right)$$

where $T^*(n)$ is the best sequential time. That is, a PRAM algorithm is optimal if the work, defined as the product $T_p(n) \times p$, is $O(T^*(n))$.

Another related model, developed independently of the PRAM model, is the *data parallel model* initially introduced in Ref. [10]. A data parallel algorithm consists of a sequence of parallel steps, where each parallel step involves concurrent primitive operations applied across large data sets. As such, the model does not use the number p of processors as a parameter but assumes that there are enough "virtual" processors to execute all the concurrent operations of a single parallel step in $O(1)$ time. As in the case of the PRAM, the model makes two essential assumptions. The first is that all variables are accessible by all the virtual processors and, hence, the shared memory model (a better term would be a *single* or *shared address space*) is assumed. The second is that all concurrent operations in a parallel step are executed synchronously and, hence, the next parallel step can safely use the results produced by the preceding step. However, the PRAM is a lower-level model in the sense that a number p of processors is given, and we are supposed to write a program for each processor. The allocation and the scheduling of the data parallel operations among the processors is referred to as the *processor allocation problem*. Hence, a data parallel algorithm and a solution to the corresponding processor allocation problem yield a PRAM algorithm for solving the same problem.

A data parallel algorithm can be evaluated in terms of two performance measures: the parallel time $T(n)$ and the work $W(n)$. The parallel time is computed assuming no upper bound on the number of processors, whereas the work $W(n)$ refers to the total number of operations used by the algorithm. A data parallel algorithm is *optimal* if $W(n) = O(T^*(n))$, where, as before, $T^*(n)$ is the best sequential time. It is *strongly optimal* if it is optimal and $T(n)$ is as small as possible.

It is clear that an arbitrary PRAM algorithm running in time $T_p(n)$ and using p processors can be translated into a data parallel algorithm executing $T_p(n)$ parallel steps and performing $W(n) = p \times T_p(n)$ work. On the other hand, we can use Brent's Scheduling Principle [4] to show that a data parallel algorithm with $T(n)$ parallel steps and using $W(n)$ work almost always can be translated into a PRAM algorithm with p processors running in

$$\leq \left\lfloor \frac{W(n)}{p} \right\rfloor + T(n)$$

parallel steps. A formal model, called the *vector RAM* (VRAM), for studying data parallel algorithms has been introduced by Blelloch [2]. The VRAM is a sequential RAM processor with the addition of (1) a *vector memory,* each location of which can hold an arbitrarily long vector of atomic values, and (2) a *vector processor* that can operate on a fixed number of vectors from the vector memory. Blelloch introduced a set of primitive vector operations that include elementwise operations, scalar vector operations, and scan operations. The relationship between the VRAM and the PRAM is explored in [2].

12.2.2   The shared memory model

An essential feature of PRAM and data parallel algorithms is the lock-step synchronous execution of instructions. Such a feature simplifies considerably the programming of parallel algorithms but can cause a substantial loss in efficiency if it is strictly enforced on current multiprocessors. The shared memory model retains the single address space of the PRAM but assumes that the processors execute their operations asynchronously. Each processor has its own local memory and control unit and executes its program asynchronously. On the shared memory model, it is the programmer's responsibility to set appropriate synchronization points and to protect shared data. The programming style used is typically the *single program multiple data* (SPMD) model, which results in the same program being run on all the processors that operate asynchronously on different portions of the problem (resulting from either data or function decomposition).

As in the case of the PRAM model, we are interested in designing algorithms that achieve the best running time $T_p(n)$ for an input of size n on a p processor machine. We can charge for communication and synchronization costs as illustrated, for example in Refs. [1] and [6]. The *block distributed model* to be described shortly assumes the shared memory programming model but accounts for communication costs based on a distributed memory organization.

12.2.3   The network model

The models introduced earlier do not make any assumption about the structure and the control of the interconnection network between the processors. The network model takes the topology of the interconnection network as the basis for modeling the communication between different processors. More precisely, a network can be viewed as a graph $G = (V, E)$, where each node $i \in V$ represents a processor, and each edge $(i, j) \in E$ represents a two-way communication link between processors i and j. Each processor is assumed to have its own local memory, and no shared memory is available. Adjacent processors can communicate in constant time, while distant processors must communicate via a connecting path, and so the communication cost is higher for distant processors. Also, we have to ensure the absence of congestion on any link during the execution of the algorithm.

In describing algorithms for the network model, it is usual to use the *message passing model* in which processors communicate explicitly be sending and receiving messages. A typical communication cost is the length of the path that connects the two communicating processors.

Some of the most popular networks studied in the literature include meshes, hypercubes, rings, and trees. For an extensive coverage of network algorithms, the reader can consult Ref. [12].

12.2.4   The LogP model

The LogP model captures performance of distributed-memory multiprocessors in which processors communicate by point-to-point messages. The parameters of the model are L, o, g, and P, defined as follows. The parameter L is an upper bound on the latency incurred in communicating a message from a source processor to a destination processor, whereas the parameter o is defined to be the overhead during which a processor is busy in the transmission or reception of a message. The parameter g refers to the minimal time interval between consecutive message transmissions or receptions, whereas the parameter P is the number of processors available.

A number of optimal algorithms have been designed for the LogP model [11]. These algorithms are substantially different from those developed for the PRAM model. For example, the single item broadcasting problem can be optimally solved in the LogP model by using a Fibonacci-type broadcasting tree. Most of the described algorithms are quite involved and require low-level handling of messages.

12.2.5   The block distributed memory model

As indicated before, the hardware organizations of multiprocessor systems seem to be converging toward a collection of powerful processors connected by a fast and robust communication network. The structure of the communication network for *massively parallel systems* (MPPs) is hierarchical in which the communication between each cluster of "nearby" processors can be modeled by a complete graph, subject to the restrictions imposed by the latency and the bandwidth properties of the cluster. Communication between processors that belong to different clusters can be modeled in the same way, but with the latency depending on the level of the hierarchy through which the processors have to communicate. In what follows, we concentrate on the model for each cluster.

The *block distributed memory* (BDM) [8] is defined in terms of four parameters: p, r, σ, and m. As we will see later, the parameter σ can be dropped without loss of generality. The parameter p refers to the number of processors; each such processor is viewed as a unit cost random access machine (RAM). In addition, each processor has an interface unit to the interconnection network that handles communication among the different processors. Data are communicated between processors via point-to-point messages; each message consists of a packet that holds m words from *consecutive* locations of a local processor memory. Since we will assume the *shared memory programming model,* each request to a remote location involves the preparation of a request packet, the injection of the packet into the network, the reception of the packet at the destination processor and, finally, the sending of a packet containing the contents of m consecutive locations, including the requested value, back to the requesting processor. We will model the cost of handling the request to a remote location (read or write) by the formula $\tau + m\sigma$, where τ is the maximum latency time it takes for a requesting processor to receive the appropriate packet, and σ is the rate at which a processor can inject (receive) a word into (from) the network. Moreover, no processor can send or receive more than one packet at a time. As a result, we note the following two observations. First, if π is any permutation on p elements, then a remote memory request issued by processor P_i and destined for processor $P_{\pi(i)}$ can be completed in $\tau + m\sigma$ time for all processors P_i, $0 \leq i \leq p - 1$, simultaneously. Second, k remote access requests issued by k distinct processors and destined to the same processor may take up to $k(\tau + m\sigma)$ time to be completed; in addition, we do not make any assumption on the relative order in which these requests will be completed.

On the BDM model, we allow the use of the *prefetching* technique to hide memory latency. In particular, k prefetch read operations issued by a processor can be completed in $\tau + km\sigma$ time.

Unlike the LogP model, the BDM model does not allow low-level handling of message passing primitives except implicitly through data accesses. In particular, an algorithm written in the BDM model can specify the initial data placement among the local memories of the p processors, can use the processor ID to refer to specific data items, and can use synchronization barriers to synchronize the activities of various processors whenever necessary. Remote data accesses are charged according to the communication model specified above. As for synchronization barriers, we make the assumption that, on the BDM model, they are provided as primitive operations. There are two main reasons for making this assumption. The first is that barriers can be implemented in hardware efficiently at a relatively small cost. The second is that we can make the latency parameter τ large enough to account for synchronization costs. The resulting communication costs will be on the conservative side, but that should not affect the overall structure of the resulting algorithms.

The complexity of a parallel algorithm on the BDM model will be evaluated in terms of two measures: the computation time T_{comp} and the communication time T_{comm}. The measure T_{comp} refers to the maximum of the local computation performed on any processor as measured on the standard sequential RAM model. The communication time T_{comm} refers to the total amount of communication time spent by the overall algorithm in accessing remote data. Our main goal is the design of parallel algorithms that achieve optimal or near-optimal computational speedups in such a way that the total communication time T_{comm} is minimized.

Since T_{comm} is treated separately from T_{comp}, we can normalize this measure by dividing it by σ. Hence, an access operation to a remote location takes $\tau + m$ time, and k prefetch read operations can be executed in $\tau + km$ time. Note that the parameter τ should now be viewed as an upper bound on the maximum number of words in transit from or to a processor. In our estimates of the bounds on the communication time, we make the simplifying (and reasonable) assumption that τ is an integral multiple of m.

In the remaining sections of this chapter, we describe four major strategies for designing parallel algorithms and use these strategies to solve specific problems on the data parallel and the BDM models.

12.3   Balanced Trees

A common strategy for designing parallel algorithms is to build a balanced tree on the input elements and traverse the tree forward and backward while computing and storing some information in each internal node. In fact, many of the parallel algorithms for the PRAM model use such a strategy either implicitly or explicitly [7]. We will illustrate this technique on the simple example of computing the sum of n elements. As we will see, the algorithms developed for the data parallel model and the BDM model differ significantly in spite of the fact they are each based on a balanced tree scheme.

A data parallel algorithm

Consider the problem of computing the sum S of $n = 2^k$ elements stored in an array A, i.e.,

$$S = \sum_{i=0}^{n-1} A[i]$$

The following is an optimal data parallel algorithm for computing S based on a balanced binary tree.

Algorithm 3.1 (Data Parallel Sum)

Input: n = 2^k numbers stored in an array A.

Output: The sum $S = \sum_{i=0}^{n-1} A[i]$

```
begin
for h = 1 to k do
      for 0 ≤ i ≤ n/2^h - 1 do in parallel
          A[i] := A[2i] + A[2i + 1]
S := A[0]
end
```

It is clear that the parallel complexity of algorithm Data Parallel Sum is $T(n) = O(\log n)$. The number of operations executed during the hth iteration of the loop is $n/2h$, so the total number of operations is given by

$$W(n) = \sum_{h=1}^{\log n} (n/2^h) + 1 = O(n)$$

Therefore this algorithm is optimal on the data parallel model.

A distributed memory algorithm

We now describe a sum algorithm on the BDM model [8]. It turns out that the best algorithm in this model is based on a balanced r-ary tree, where $r = (\tau/m) + 1$, which is not necessarily a binary tree. To make the presentation simpler, we begin by addressing the related problem of broadcasting a single item to p processors.

Let A be a p-dimensional array such that $A[i]$ is held in processor P_i, $0 \leq i \leq p - 1$. The single item broadcasting problem is to copy the element $A[0]$ into the remaining entries of

A. This can be viewed as a concurrent read operation from location $A[0]$ executed by processors $P_1, P_2, \ldots, P_{p-1}$. The next lemma provides a simple algorithm to solve this problem; we later use this algorithm to derive an optimal broadcasting algorithm.

Lemma 3.1 Given a p-processor BDM and an array $A[0 : p - 1]$ where $A[j]$ resides in processor P_j, the element $A[0]$ can be copied into the remaining entries of A in $\tau + m(p - 1)$ communication time.

Proof: A simple algorithm consists of $p - 1$ rounds that can be pipelined using prefetch read instructions. During the rth round, each processor P_j reads $A[(j + r) \bmod p]$, for $1 \le r \le p - 1$; however, only $A[0]$ is copied into $A[j]$. Since these rounds can be realized with $p - 1$ pipelined prefetch read operations, the resulting communication complexity is $r + m(p - 1)$. $\square$

We are now ready for the following theorem that essentially establishes the fact that a k-ary balanced tree broadcasting algorithm is the best possible for $k = (\tau/m) + 1$ (recall that earlier we made the assumption that τ is an integral multiple of m).

Theorem 3.1 Given a p-processor BDM, an item in a processor can be broadcast to the remaining processors in

$$\le 2\tau \left\lceil \frac{\log p}{\log\left(\dfrac{r}{m} + 1\right)} \right\rceil$$

communication time. On the other hand, any broadcasting algorithm that only uses read, write, and synchronization barrier instructions requires at least

$$\tau \frac{\log p}{\log\left(\dfrac{\tau}{m} + 2\right)} + m \log p$$

communication complexity.

Proof: We start by describing the algorithm. Let k be an integer to be determined later. The algorithm can be viewed as a k-ary tree rooted at location $A[0]$; there are $\lceil \log_k p \rceil$ rounds. During the first round, $A[0]$ is broadcast to locations $A[1], A[2], \ldots, A[k - 1]$, using the algorithm described in Lemma 3.1, followed by a synchronization barrier. Then, during the second round, each element in locations $A[1], A[2], \ldots, A[k - 1]$ is broadcast to a distinct set of $k - 1$ locations, and so on. The communication cost incurred during each round is given by $\tau + (k - 1)m$ (Lemma 3.1). Therefore, the total communication cost is

$$T_{comm} \le (\tau + (k - 1)\, m) \lceil \log_k p \rceil$$

If we set $k = (\tau/m) + 1$, then

$$T_{comm} \le \left(\tau + \frac{\tau}{m} m\right) \left\lceil \frac{\log p}{\log\left(\dfrac{\tau}{m} + 1\right)} \right\rceil = 2\tau \left\lceil \frac{\log p}{\log\left(\dfrac{\tau}{m} + 1\right)} \right\rceil$$

We next establish the lower bound stated in the theorem. Any broadcasting algorithm using only read, write, and synchronization barrier instructions can be viewed as operating in phases, where each phase ends with a synchronization barrier (whenever there are more than a single phase). Suppose there are s phases. The amount of communication to execute phase i is at least $\tau + k_i m$, where k_i is the maximum number of copies read from any processor during phase i. Hence, the total amount of communication required is at least

$$\sum_{i=1}^{s} (\tau + k_i m)$$

Note that by the end of phase i, the desired item has reached, at most

$$(k_1 + 1)(k_2 + 1) \cdots (k_i + 1)$$

remote locations. It follows that if, by the end of phase s, the desired item has reached all the processors, we must have $p \leq (k_1 + 1)(k_2 + 1) \cdots (k_i + 1)$. The communication time

$$\sum_{i=1}^{s} (\tau + k_i m)$$

is minimized when $k_1 = k_2 = \cdots = k_s = k$, and hence $(k + 1)^s \geq p$. Therefore,

$$s \geq \frac{\log p}{\log (k + 1)}$$

and the communication time is at least

$$s (\tau + km) \geq (\tau + km) \frac{\log p}{\log (k + 1)}$$

We complete the proof of this theorem by proving the following claim.

Claim: $\dfrac{\tau + km}{\log (k + 1)} \geq \dfrac{\tau}{\log\left(\dfrac{\tau}{m} + 2\right)} + m$, for any $k \geq 1$.

Proof: Let

$$r = \frac{\tau}{m}, f_1(k) = \frac{\tau}{\log (k1)}, f_2(k) = \frac{km}{\log (k + 1)}, \text{ and } f(k) = f_1(k) + f_2(k) = \frac{\tau + km}{\log (k + 1)}$$

Then,

$$f'(k) = \frac{m (k + 1) \log (k + 1) - (\log e) (\tau + km)}{(k + 1) \log^2 (k + 1)}$$

(Case 1) $(1 \leq k \leq r + 1)$. Since $f_1(k)$ is decreasing and $f_2(k)$ is increasing in this range, the claim follows easily by noting that $f_1(k) \geq f_1(r + 1) = \tau/[\log(\tau/m) + 2]$ and $f_2(k) \geq f_2(1) = m$.

(Case 2) $(k > r + 1)$. We show that $f(k)$ is increasing when $k > r + 1$ by showing that $f'(k) > 0$ for all integers $k \geq r + 1$. Note that since $k \geq (\tau/m) + 1$, we have that

$$m(k+1)\log(k+1) - (\log e)(\tau + km)$$

is at least as large as

$$M[(k+1)\log(k+1) - (\log e)(2k-1)]$$

which is positive for all nonzero integer values of k. Hence $f(k) \geq f[(\tau/m) + 1)$, and the claim follows. $\square$

The *sum* of p elements on a p-processor BDM can be computed in at most

$$2\tau \left\lceil \frac{\log p}{\log\left(\dfrac{\tau}{m} + 1\right)} \right\rceil$$

communication time by using a similar strategy. Based on this observation, it is easy to show the following theorem.

Theorem 3.2 Given n numbers distributed equally among the p processors of a BDM, we can compute their sum in

$$O\left(\frac{n}{p} + \frac{\tau \log p}{m \log \dfrac{\tau}{m}}\right)$$

computation time, and at most

$$2\tau \left\lceil \frac{\log p}{\log\left(\dfrac{\tau}{m} + 1\right)} \right\rceil$$

communication time. The computation time reduces to $O(n/p)$ whenever $p \log p \leq n/\tau$. $\square$

12.4 Divide and Conquer

The general divide-and-conquer strategy consists of (1) partitioning the input into several partitions of almost equal sizes, (2) solving recursively and in parallel the subproblems induced by the first step, and (3) merging the solutions of the different subproblems into a solution of the original problem. The success of this strategy depends on whether we can perform the first and the third steps efficiently. This strategy is used extensively for designing sequential and parallel algorithms. We illustrate this technique on the problem of computing the discrete Fourier transform (DFT).

The DFT of an n-dimensional complex vector x is defined by $y = W_n x$, where

$$W_n(j, k) = w_n^{jk}$$

for $0 \leq j,\ k \leq n-1$, and $w_n = e^{i(2\pi/n)} = \cos(2\pi/n) + i \sin(2\pi/n)$, and $i = \sqrt{-1}$. The well known fast Fourier transform (FFT) algorithm uses the divide-and-conquer strategy to compute the DFT in $O(n \log n)$ operations. We first give a brief overview of this algorithm and later sketch two parallel implementations, one tailored for the data parallel model and the other for the distributed memory model.

A data parallel algorithm
Assume that n is a power of 2. One can show that the vector

$$z^{(1)} = [y_0, y_2, \ldots, y_{n-2}]^T$$

is the DFT of the vector

$$[x_0 + x_{(n/2)},\, x_1 + x_{(n/2)+1}, \ldots, x_{(n/2)-1} + x_{n-1}]^T$$

One can also show that the vector

$$z^{(2)} = [y_1, y_3, \ldots, y_{n-1}]^T$$

is the DFT of the vector

$$[x_0 + x_{(n/2)},\, \omega(x_1 - x_{(n/2)+1}), \ldots, \omega^{(n/2)-1}(x_{(n/2)-1} - x_{n-1})]^T$$

Therefore, a straightforward data parallel implementation can be accomplished as follows.

Algorithm 4.1 (Data Parallel FFT)
Input: An n-dimensional vector **x** over the complex numbers, and $\omega = e^{i(2\pi/n)}$, where n is assumed to be a power of 2.
Output: The discrete Fourier transform of **x**.
begin
1.   **if** n = 2 **then** {y_0 := x_0 + x_1; y_1 := x_0 - x_1; **exit**}
2.   **for** $0 \leq 1 \geq (n/2) - 1$ **do in parallel**
      {u_1 := x_1 + $x_{(n/2)+i}$;
      v_i := $\omega^1(x_1 - x_{(n/2)+1}$}
3.   Recursively and in parallel, compute the DFTs
            $z^{(1)} = [z_0^{(1)}, z_1^{(1)}, \ldots, z_{(n/2)-1}]^T$ and
            $z^{(2)} = [z_0^{(2)}, z_1^{(2)}, \ldots, z_{(n/2)-1}]^T$
            of the vectors $[u_0, u_1, \ldots, u_{(n/2)-1}]$ and $[v_0, v_1, \ldots, v_{(n/2)-1}]$
            respectively.
4.   **for** $0 \leq j \leq n-1$ **do in parallel**
            {j even: y_j := $z_{j/2}^{(1)}$;
            j odd: y_j := $z_{(j-1)/2}^{(2)}$}
end

Theorem 4.1 The FFT Algorithm (Algorithm 4.1) computes the discrete Fourier transform of an *n*-dimensional vector in $O(\log n)$ time using a total of $O(n \log n)$ arithmetic operations.

A distributed memory algorithm

We now describe an optimal BDM implementation of the FFT algorithm [8]. The details are quite different from those given in Algorithm Data Parallel FFT.

Our implementation on the BDM model is based on the following well known fact. Let the n-dimensional vector x be stored in the $(n/p) \times p$ matrix X in row-major order form, where p is an arbitrary integer that divides n. Then the DFT of the vector x is given by

$$W_p \left[\overline{W}_n * W_{\frac{n}{p}} X \right]^T \tag{12.1}$$

where $\overline{W}_n$ is the submatrix of W_n consisting of the first n/p rows and the first p columns (twiddle-factor scaling), and $*$ is elementwise multiplication. Notice that the resulting output is a $p \times n/p$ matrix holding the vector $y = W_n x$ in column major order form. Eq. (12.1) can be interpreted as computing DFT(n/p) on each column of X, followed by a twiddle-factor scaling, and finally computing DFT(p) on each row of the resulting matrix.

Let the BDM machine have p processors such that p divides n and $n \geq p^2$. The initial data layout corresponds to the row major order form of the data; i.e., the local memory of processor P_i will hold $x_i, x_{i+p}, x_{i+2p}, \ldots, 0 \geq i \geq p - 1$. Then the algorithm suggested by Eq. (12.1) can be performed in the following three stages. The first stage involves a local computation of a DFT of size n/p in each processor, followed by the twiddle-factor scaling (elementwise multiplication by $\overline{W}_n$). The second stage is a communication step that involves a matrix transposition. Finally, n/p^2 local FFTs each of size p are sufficient to complete the overall FFT computation on n points. A matrix transposition algorithm suitable for implementation on a distributed memory machine is given next.

The matrix transposition problem can be defined as follows. Let A be a $q \geq p$ matrix such that processor P_i holds the ith column of A, and p divides q evenly without loss of generality. The data held in A is supposed to be rearranged into the $q \times p$ array A' so that the first column of A' contains the first q/p consecutive rows of A laid out in row major order form, the second column of A' contains the second set of q/p consecutive rows of A, and so on. Clearly, if $q = p$, this corresponds to the usual notion of matrix transpose.

An efficient algorithm to perform matrix transposition on the BDM model is similar to the algorithm reported in Ref. [5]. There are $p - 1$ rounds that can be fully pipelined by using prefetch read operations. During the first round, the appropriate block of q/p elements in the ith column of A is read by processor $P_{(i+1) \bmod p}$ into the appropriate locations, for $0 \leq i \leq p - 1$. During the second round, the appropriate block of data in column i is read by processor $P_{(i+2) \bmod p}$, and so on. The resulting total communication time is given by

$$T_{comm} = \tau + (p - 1) m \left\lceil \frac{1}{pm} \right\rceil \leq \tau + \left(q - \frac{q}{p} \right) + (p - 1) m$$

and the amount of local computation is $O(q)$. Clearly, this algorithm is optimal whenever pm divides q. Hence we have the following lemma.

Lemma 4.1 A $q \times p$ matrix transposition can be performed on a p-processor BDM in

$$\tau + (p - 1) m \left\lceil \frac{q}{pm} \right\rceil$$

This bound reduces to $t + [q - (q/p)]$ whenever pm divides q. $\square$

Using the matrix transposition algorithm just described and the observations made earlier about the two-dimensional mapping of the FFT algorithm, we obtain the following theorem.

Theorem 4.2 Computing an n-point FFT can be done in $O[(n \log n)/p]$ computation time and

$$\tau + (p-1) \left\lceil \frac{n}{p^2 m} \right\rceil m$$

communication time if $n \geq p^2$. When mp^2 divides n, the communication time reduces to

$$\tau + \left(\frac{n}{p} - \frac{n}{p^2} \right)$$

$\square$

12.5 Partitioning

The partitioning strategy is perhaps the most natural strategy for designing parallel algorithms. It consists of preprocessing the input so that the initial problem can be broken into a set of completely independent subproblems. A degenerate case of this strategy is when the original problem consists of a set of independent computations in which case the initial preprocessing is not required. Contrast this strategy with the divide-and-conquer strategy where the input is partitioned into equal pieces, the induced subproblems are recursively solved in parallel, and a method is specified to merge the solutions of the subproblems into a solution of the original problem. The crux of the divide-and-conquer strategy is usually in the merging procedure. We illustrate the partitioning strategy by developing an efficient randomized sorting algorithm. As before, we begin with a data parallel algorithm and later develop another algorithm for the distributed memory model. Both algorithms are based on the partitioning strategy, but the detailed implementations are quite different.

A data parallel algorithm

The *quicksort* algorithm is a very important practical sorting algorithm whose performance can be captured by a probabilistic analysis. The basic algorithm can be described as follows. The elements of the input array are partitioned into *buckets* $\{B_j\}$, where each element of bucket B_j is smaller than any element in the next bucket B_{j+1} (unless B_j is the last bucket). Once the input is partitioned in this fashion, the problem of sorting the given set of elements reduces to sorting the different buckets separately, and thus the buckets can be sorted concurrently. The main difficulty in implementing this approach lies in developing an efficient scheme for partitioning the input in such a way that the buckets are of approximately equal sizes.

Let A be the array containing the n distinct elements to be sorted, and let k be a positive integer to be determined later. The k-way partitioning strategy consists of picking k *splitters* that will cause the elements of A to be divided into $k+1$ buckets $\{B_j\}$ of almost equal sizes. The elements in bucket B_j have ranks that are between the ranks of the $(j-1)$st and the jth splitters when the splitters are in sorted order. For the data parallel algorithm, we choose $k = \sqrt{n}$ keys randomly from the input A, sort the resulting $\sqrt{n}$ splitters, and then rearrange the elements of A into $\sqrt{n} + 1$ buckets such that each element in B_j is smaller than any element in B_{j+1}. The algorithm is then applied recursively to each bucket until the size of each bucket is small. The details are given in the following algorithm.

Algorithm 5.1 (Data Parallel Sample Sort)

`Input:` n elements stored in an array A.
`Output:` The array A in sorted order.
`begin`

1.   If the number of elements to be sorted is ≤ 30, then use the bitonic sorting algorithm. Else do Steps 2-5.

2.   Draw a set S of $\sqrt{n}$ random samples from the input A. These elements define the splitters.

3.   Compare all pairs of splitters and store the outcomes in a $\sqrt{n} \times \sqrt{n}$ table T. The rank of each element of S can then be easily computed from T.

4.   Rearrange the elements of A into buckets $\{B_i\}_{i=1}^{\sqrt{n}+1}$ such that the elements of B_i are those of A that are between the (i − 1)st smallest and the ith smallest splitters. The 0th splitter is defined to be −∞, and the ($\sqrt{n}$ + 1)st splitter is defined to be +∞.

5.   For all buckets in parallel, sort the elements in each bucket recursively.

`end`

The proof of the following theorem can be found in Ref. [7].

Theorem 5.1 With high probability, Algorithm 5.1 terminates in $O(\log n)$ time using $O(n \log n)$ operations and, hence, the algorithm is optimal.

A distributed memory algorithm

We describe a version of the sample sort algorithm that sorts on the BDM model in at most

$$3\tau + (p-1)\left\lceil \frac{5\ln n}{m} \right\rceil m + 6\frac{n}{p}$$

communication time and $O[(n \log n)/p]$ computation time whenever $p^2 < n/(6 \ln n)$. The complexity bounds are guaranteed with high probability [8]. Before describing our algorithm, we need an efficient algorithm to handle a general routing problem described next.

Let A be an $n/p \times p$ array of n elements initially stored one column per processor in a p-processor BDM machine. Each element of A consists of a pair (data, i), where i is the index of the processor to which the data has to be relocated. We assume that at most $\alpha(n/p)$ elements have to be routed to any single processor for some constant $\alpha \geq 1$. We describe in what follows a randomized algorithm that completes the routing in $2[\tau + c(n/p)]$ communication time and $O(n/p)$ computation time, where c is any constant larger than

$$\max\left\{ 1 + \frac{1}{\sqrt{2}}, \alpha + \frac{\sqrt{\alpha}}{2} \right\}$$

The complexity bounds are guaranteed to hold with high probability, that is, with probability $\geq 1 - n^{-\varepsilon}$, for some positive constant ε, as long as $p^2 < n/(6 \ln n)$ where ln is the logarithm to the base e.

The overall idea of the algorithm has been used in various randomized routing algorithms on the mesh. Here, we follow more closely the scheme described in Ref. [13] for randomized routing on the mesh with bounded queue size.

Before describing our algorithm, we introduce some terminology. We use an auxiliary array A' of size $(cn/p) \times p$ for manipulating the data during the intermediate stages and for holding the final output, where

$$c > \max \left\{ 1 + \frac{1}{\sqrt{2}}, \alpha + \frac{\sqrt{\alpha}}{2} \right\}$$

Each column of A' will be held in a processor. The array A' can be divided into p equal size slices, each slice consisting of (cn/p^2) consecutive rows of A'. Hence, a slice contains a set of (cn/p^2) consecutive elements from each column, and such a set is referred to as a *slot*. We are ready to describe our algorithm (introduced in Ref. [8]).

Algorithm 5.2 (Randomized Routing)

Input: An input array A[0 : (n/p) - 1, 0: p - 1] such that each element of A consists of a pair (data, i), where i is the processor index to which the data has to be routed. No processor is the destination of more than α(n/p) elements for some constant α.

Output: An output array A'[0 : (cn/p) - 1, 0 : p - 1] holding the routed data, where c is any constant larger than max{1 + (1/ $\sqrt{2}$), α + ($\sqrt{\alpha}$ /2)}.

begin
1.   Each processor P_j distributes randomly its n/p elements into the p slots
      of the jth column of A'.
2.   Transpose A' so that the jth slice will be stored in the jth processor,
      for 0 ≤ j ≤ p - 1.
3.   Each processor P_j distributes locally its ≤ cn/p elements such that every
      element of the form (*, i) resides in slot i, for 0 ≤ i ≤ p - 1.
4.   Perform a matrix transposition on A' (hence, the jth slice of the layout
      generated at the end of Step 3 now resides in P_j).
end

The next two facts will allow us to derive the complexity bounds for our randomized routing algorithm. For the analysis, we assume that $p^2 < n/(6 \ln n)$.

Lemma 5.1 At the completion of Step 1, the number of elements in each slot is no more than cn/p^2 with high probability, for any $c > 1 + (1/\sqrt{2})$.

Proof: The procedure performed by each processor is similar to the experiment of throwing n/p balls into p bins. Hence, the probability that exactly cn/p^2 balls are placed in any particular bin is given by the binomial distribution

$$b(k; N, q) = \binom{N}{k} q^k (1 - q)^{N-k}$$

where $k = cn/p^2$, $N = n/p$, and $q = 1/p$. Using the following Chernoff bound for estimating the tail of the binomial distribution

$$\sum_{(j \geq (1+\varepsilon))Nq} b(j; N, q) \leq e^{-\varepsilon^2 N(q/3)}$$

we obtain that the probability that a particular bin has more than cn/p^2 balls is upper bounded by

$$e^{-(c-1)^2 \frac{n}{3p^2}}$$

Therefore, the probability that any of the bins has more than cn/p^2 balls is bounded by

$$p^2 e^{-\varepsilon^2 N(q/3)}$$

and the lemma follows. $\square$

Lemma 5.2 At the completion of Step 3, the number of elements in any processor that are destined to the same processor is at most cn/p^2 with high probability, for any $c > \alpha + (\sqrt{\alpha}/2)$.

Proof: The probability that an element is assigned to the jth slice by the end of Step 1 is $1/p$. Hence the probability that cn/p^2 elements destined for a single processor fall in the jth slice is bounded by

$$b\left(\frac{cn}{p^2}; \frac{\alpha n}{p}, \frac{1}{p}\right)$$

since no processor is the destination of more than $\alpha n/p$ elements. Since there are p slices, the probability that more than cn/p^2 elements in any processor are destined for the same processor is bounded by

$$pe^{-\left(\frac{c}{\alpha} - 1\right)^2 \frac{\alpha n}{3p^2}}$$

and so the lemma follows. $\square$

From the previous two lemmas, it is easy to show the following theorem.

Theorem 5.2 The routing of n elements stored initially in an $n/p \times p$ array A of a p-processor BDM such that at most $\alpha(n/p)$ elements are destined to the same processor can be completed with high probability in $2[\tau + c(n/p)]$ communication time and $O(n/p)$ computation time, where c is any constant larger than

$$\max\left\{1 + \frac{1}{\sqrt{2}}, \alpha + \frac{\sqrt{a}}{2}\right\}$$

and $p^2 < n/(6 \ln n)$. $\square$

Remark: Since we are assuming that $p^2 < n/(6 \ln n)$, the effect of the parameter m is dominated by the bound $c(n/p)$ (as $p(n/p) > 6p \ln n \geq mp$, assuming $m \leq 6 \ln n$). $\square$

We now return to our sorting problem. The overall idea of the BDM sorting algorithm has been used in various sample sort algorithms. Our algorithm described below follows more closely the scheme described in Ref. [3] for sorting on the connection machine CM-2; however the first three steps are different.

Algorithm 5.3 (BDM Sample Sort)

```
Input:     n elements distributed evenly over a p-processor BDM such that
           p² < n/6 ln n.
Output:    The n elements sorted in column major order.
begin
1.    Each processor Pᵢ randomly picks a list of 5 ln n elements from its local
      memory.
2.    Each processor Pᵢ reads all the samples from all the other processors;
      hence each processor will have 5p ln n samples after the execution of
      this step.
3.    Each processor Pᵢ sorts the list of 5p ln n samples and picks
      (5 ln n + 1)st, (10 ln n + 1)st,..., samples as the p- 1 pivots.
4.    Each processor Pᵢ partitions its n/p elements into p sets, Sᵢ,₀,
      Sᵢ,₁,...,Sᵢ,ₚ₋₁, such that the elements in set Sᵢ,ⱼ belong to the inter-
      val between jth pivot and (j + 1)st pivot, where 0th pivot is —∞, pth
      pivot is +∞, and 0 < j < p - 1.
5.    Each processor Pᵢ reads all the elements in the p sets, S₀,ᵢ,
      S₁,ᵢ,...,Sₚ₋₁,ᵢ, by using Algorithm Randomized Routing.
6.  Each processor Pᵢ sorts the elements in its local memory.
end
```

The following lemma can be immediately deduced from the results of Ref. [3].

Lemma 5.3 For any $\alpha > 0$, the probability that any processor contains more than a $\alpha(n/p)$ elements after Step 5 is at most

$$ne^{-\left(1-\frac{1}{\alpha}\right)^2 \frac{5\alpha \ln n}{2}}$$

$\square$

Next, we show the following theorem.

Theorem 5.3 Algorithm Sample Sort can be implemented on the p-processor BDM in $O[(n \log n)/p]$ computation time and in at most

$$\left(3\tau + (p-1)\left\lceil\frac{5\ln n}{m}\right\rceil m + 6\frac{n}{p}\right)$$

communication time with high probability, if $p^2 \leq n/(6 \ln n)$.

Proof: Step 2 can be done in $\tau + (p-1) \lceil (5\ln n)/m \rceil m$ communication time by using a technique similar to that used to prove Lemma 3.2. By Lemma 5.1, the total number of elements that each processor reads at Step 5 is at most $2(n/p)$ elements with high probability. Hence, Step 5 can be implemented in $2\tau + 6(n/p)$ communication time with high probability using Theorem 3.4. The computation time for all the steps is clearly $O[(n \log n)/p]$ with high probability if $p^2 < n/(6 \ln n)$, and the theorem follows. $\square$

12.6   Combining

To obtain an optimal algorithm, it is sometimes necessary to use two or more algorithms, each of which is suitable for handling an input size that falls within a certain range. We illustrate this strategy by developing optimal parallel algorithms for the list ranking on the data parallel and the distributed memory models.

A data parallel algorithm

Consider a linked list L of n nodes whose order is specified by an array S such that $S(i)$ contains a pointer to the node following node i on L, for $1 \le i \le n$. We assume that $S(i) = 0$ when i is the end of the list. The *list ranking problem* is to determine the distance of each node i from the end of the list. That is, we want to compute an array R such that $R(i)$ is equal to the distance of node i from the end of L. Initially we set $R(i) = 1$ for all nodes i, except for the last node whose R value is set to 0.

We start by stating a data parallel algorithm based on a technique called *pointer jumping*. This technique involves the repeated substitution of the successor of each node by the successor of the successor. The details are given in the following algorithm.

Algorithm 6.1 (List Ranking Using Pointer Jumping)
Input: A linked list of n nodes such that (1) the successor of each node i is given by S(i), and (2) the S value of the last node on the list is 0.
Output: For each 1 ≤ i ≤ n, the distance R(i) of node i from the end of the list.
begin
```
1.    for 1 ≤ i ≤ n do in parallel
            if (S(i) ≠ 0) then R(i) := 1;
                          else R(i) := 0;
2.    for 1 ≤ i ≤ n do in parallel
      while (S(i) ≠ 0 and S(S(i)) ≠ 0) do
            {R(i) := R(i) + R(S(i));
             S(i) := S(S(i)); }
```
end

Lemma 6.1 Given a linked list of n nodes, Algorithm 6.1 generates the rank $R(i)$ of each node i in $O(\log n)$ time wing $O(n \log n)$ operations.

Clearly, this algorithm is not optimal since a simple sequential algorithm computes all the ranks in $O(n)$ time. A strategy for solving the list ranking problem optimally is outlined next. First, we shrink the linked list L until only $O(n/\log n)$ nodes remain. Then, we apply the pointer jumping algorithm on the short list of the remaining nodes and, finally, we restore the original list and rank all the nodes removed during the first step.

We have just examined how the algorithm is to execute the second phase. Since the number of elements is $O(n/\log n)$, Algorithm 6.1 takes $O(\log n)$ time using $O(n)$ operations. The operations required to execute the last phase correspond to the reverse process executed in the first phase. The main difficulty lies in performing the shrinking phase in

$O(\log n)$ time using a linear number of operations. To do that, we need to introduce the notion of independent sets.

A set I of nodes is *independent* if, whenever $i \in I$, $S(i) \notin I$. We can remove each node $i \in I$ by adjusting the successor pointer of the predecessor of i. Since I is independent, this process can be applied concurrently to all the nodes in I. The information concerning the removed nodes should be stored somewhere so that later the original list can be restored and the nodes in I can be ranked properly.

The problem of finding an independent set can be handled by coloring the nodes of the list L. A k-coloring of L is a mapping from the set of nodes in L into $\{0,1,\ldots,k-1\}$ such that no adjacent vertices are assigned the same color. A k-coloring of L can be used to determine an independent set of L as follows. A node u is a *local minimum* (*maximum*) with respect to this coloring if the color of u is smaller (larger) than the colors of its predecessor and its successor. The following lemma can be easily shown (see for example Ref. [7]).

Lemma 6.2 Given a k-coloring of the nodes of a list L of size n, the set I of local minima (or maxima) is an independent set of size $\Omega(n/k)$. The set I can be identified in $O(1)$ time using a linear number of operations.

A large independent set can be obtained by using an optimal algorithm to 3-color the vertices of a cycle given in Ref. [7]. Using Lemma 6.2, we see that the corresponding independent set is of size greater than or equal to cn, for some constant $0 < c < 1$ (more precisely, we can choose $c = 1/5$). Contracting this independent set reduces the size of the list by a constant factor, and hence this process can be repeated $\alpha \lceil \log \log n \rceil$ times to produce a list of size less than or equal to $n/\log n$, for some $\alpha > 0$.

We are ready to give a description of simple optimal data parallel algorithm for ranking the nodes of a list.

Algorithm 6.2 (Data Parallel List Ranking)

```
Input: A linked list with n nodes such that the successor of each node i is
given by S(i).
Output: For each node i, the distance of i from the end of the list.
begin
1.    n_0 := n; k := 0;
2.    while n_k > n/ log n do
      a.   k := k + 1;
      b.   Color the nodes of the list with three colors, and identify the
           set I of local minima.
      c.   Remove the nodes in I, and store the appropriate information re-
           garding the removed nodes.
      d.   Let n_k be the size of the remaining list. Compact the list into
           consecutive memory locations.
3.    Apply the pointer jumping technique to the resulting list.
4.    Restore the original list and rank all the removed nodes by reversing
      the process performed in Step 2.
end
```

The proof of the following theorem can be found in Ref. [7].

Theorem 6.1 Algorithm 6.2 ranks a linked list L with n nodes in $O(\log n \log \log n)$ time using a linear number of operations and, hence, the algorithm is optimal in the data parallel model.

A distributed memory algorithm

We now develop a randomized list ranking algorithm on the BDM model that uses a linear number of operations [9]. Our overall strategy consists of repeating the process of identifying a large independent set and removing the nodes in this set until the length of the list is small enough. For the remaining short list, we use an appropriately modified version of the pointer jumping algorithm in such a way that each node has, at most, one predecessor during the execution of the algorithm. This latter fact is necessary to avoid the case where many elements need to be routed to a single processor to get the information concerning their successors, something that would have made the algorithm very inefficient on the BDM model. This unique predecessor property can be achieved as follows. Initially mark the last node, and leave all the other nodes, unmarked; in successive iterations, let only the unmarked nodes access their successors and mark themselves if they access marked nodes as their successors.

Algorithm B.3 (List Ranking)

Input: A linked list of n nodes represented by a set of pairs $\langle u, S(u) \rangle$ distributed evenly over p processors. Each node u has a fixed home processor throughout the algorithm.
Output: The distance D(u) from each node u to the end of the list.
begin
1.   Mark all nodes live.
2.   **repeat** Steps a - e until the number of live nodes is at most n/log n.
      a.   Each processor P_i assigns a random label l(u) of 0 or 1 to each
           live node u in its local memory.
      b.   Route $\langle u, S(u) \rangle$) together with l(u) to the processor containing
           $\langle S(u), S(S(u)) \rangle$ using the randomized h-relation algorithm.
      c.   Each processor replaces $\langle u, S(u) \rangle$ by $\langle u, S(S(u)) \rangle$ for all nodes such
           that l(u) = 1 and l(S(u)) = 0. The node S(u) is marked not live.
      d.   Route back each $\langle u, S(S(u)) \rangle$ to the home location of node u by using
           the randomized h-relation algorithm.
      e.   Each processor creates a list of its local live nodes without
           changing their actual locations.
3.   Use the modified version of the pointer jumping algorithm on the re-
      maining list.
4.   Restore the linked list and compute the ranks of all the nodes.
end

The correctness of the algorithm can be established using several well known techniques. Concerning its complexity bounds, we need to establish a few facts. Throughout the remainder of this section, we use the notation $\alpha = 15/16$. Let n_i be the number of live nodes after performing iteration i. Then, it is easy to show that $n_i + 1 \leq n_i \alpha$ with high probability when $n_i \geq n/\log n$ by using the Chernoff bound:

$$Pr\{X \geq (1+\varepsilon)\,sn\} \leq e^{-\frac{\varepsilon^2 ns}{3}}$$

where X is a random variable denoting the number of successes during n independent Bernoulli trials that have a success probability of s and a failure probability of $r = 1 - s$.

It follows that the total number of iterations of the repeat loop is $O(\log \log n)$ with high probability. We also need to show that the number of live nodes in each processor is decreasing by a constant factor in each iteration, since this fact is essential for the algorithm to be performed optimally. Let $n_{i,j}$, for $j = 0, 1, \ldots, p - 1$, be the number of live nodes in processor P_j after performing iteration i.

Lemma 6.3 The value of $n_{i,j} \leq (n\alpha)/p$ with high probability, for $j = 0, 1, \ldots, p - 1$, whenever $p^2 < n/(6 \ln n \log n)$.

Proof: The probability that each live node is eliminated from the list in Step 3 is 1/4, and the expected number of these vertices is $n/4$. If we consider every other node on a list, then their probabilities are independent, and the probability that a node is live after the first iteration is at most 7/8. Hence, the probability that the number of live nodes in P_j is $\geq (n\alpha)/p = (n\alpha/p) = (7n/8p)(1 + \varepsilon)$ is at most

$$e^{-\varepsilon^2 \frac{7n}{24p}}$$

(where $\varepsilon \approx 0.07$) by the Chernoff bound. The probability that the number of live nodes in a processor is $\geq (n\alpha)/p$ is at most

$$p \cdot e^{-\varepsilon^2 \frac{7n}{24p}}$$

and the lemma follows. $\square$

We can similarly show that $n_{i,j} \leq (n\alpha^i)/p$ with high probability, for $j = 0, 1, \ldots, p - 1$. This implies that iteration $(i + 1)$ of the loop can be performed in $O[\tau + (n\alpha^i)/p]$ communication time and $O[(n\alpha^i)/p)]$ computation time with high probability, whenever $p^2 < [n/(6 \ln n \log n)]$, if we maintain all the live nodes in each processor in another linked list. Hence, the total complexity for the **repeat** loop is $O[\tau \log \log n + (n/p)]$ communication time and $O(n/p)$ computation time with high probability. Clearly, Step 6 of the algorithm can be done in $O[\tau \log n + (n/p)]$ communication time and $O(n/p)$ computation time with high probability. Thus, we have the following theorem [9].

Theorem 6.2 Algorithm list ranking determines the ranks of all the nodes of a linked list of length n in $O[\tau \log n + (n/p)]$ communication time and $O(n/p)$ computation time. The communication bound is guaranteed with high probability whenever $p^2 < n/(6 \ln n \log n)$.

$\square$

12.7 Conclusions and Future Trends

A number of methodologies for designing efficient parallel algorithms were described in detail within the context of two parallel models: the data parallel and the distributed memory model with shared address space. These methodologies were then applied to a number of case studies, including computing the sum of n numbers, computing the discrete Fourier transform, sorting, and list ranking. The resulting algorithms are the best possible under the two parallel models mentioned above. With the current trend toward clusters of workstations connected by a fast switch, it is necessary to extend the existing models to capture the performance of parallel algorithms on such platforms. Also, it is important to incorporate a hierarchical structure into the current models (memory and interconnect) in which memory or communication latency depends on the levels of the hierarchy through which data have to go through. The paradigms covered in this chapter should be applicable to these newer models as well, but the resulting algorithmic techniques will probably be different. Another crucial direction that should be explored is the development of benchmarks for testing the efficiency of these methodologies on various architectures.

12.8 Acknowledgment

The work reported in this chapter has been partially supported by the National Science Foundation, Grant No. CCR9103135, and by the National Science Foundation Engineering Research Center Program NSFD CD 8803012. All the results related to the distributed memory model are joint work with Kwan Woo Ryu.

12.9 References

1. Aggarwal, A., A. Chandra, and M. Snir. 1989. On communication latency in PRAM computations. *Proceedings of the 1st ACM Symposium on Parallel Algorithms and Architectures,* 11–21.
2. Blelloch, G. E. 1990. *Vector Models for Data-Parallel Computing.* Cambridge, Mass.: MIT Press.
3. Blelloch, G. E., et al. 1991. A comparison of sorting algorithms for the connection machine CM-2. *Proceedings of the 3th Symposium on Parallel Algorithms and Architectures,* 3–16.
4. Brent, R. P. 1974. The parallel evaluation of general arithmetic expressions. *Journal of the ACM,* 21(2), 201–208.
5. Culler, D., R. Karp, D. Patterson, A. Sahay, K. E. Schauser, E. Santos, R. Subramonian, and T. V. Eicken. 1993. LogP: toward a realistic model of parallel computation. *Proceedings of the 4th ACMPPOPP,* 1–12.
6. Gibbons, P. B. 1990. Asynchronous PRAM algorithms. In *Synthesis of Parallel Algorithms,* J. H. Reif, ed. San Mateo, Calif.: Morgan-Kaufman.
7. JáJá, J. 1992. *An Introduction to Parallel Algorithms.* Reading, Mass.: Addison Wesley.
8. JáJá, J, and K.W. Ryu. 1994. The Block Distributed Memory Model for Shared Memory Multiprocessors. *Proceedings of the International Parallel Processing Symposium,* 752–756.
9. JáJá, J. and K.W. Ryu. Randomized Parallel Algorithms on the Block Distributed Memory Model. UMIACS Technical Report 94-43, University of Maryland, College Park, Md. 20742.
10. Hillis, W. D., and G. L. Steele. 1986. Data parallel algorithms. *Communications of the ACM,* 29(12), 1170–1183.
11. Karp, R. M., A. Sahay, E. E. Santos, and K. E. Schauser. 1993. Optimal Broadcast and Summation in the LogP Model. *Proceedings of the 5th Symposium on Parallel Algorithms and Architectures,* 142–153.
12. F. T. Leighton, F. T. 1992. *Introduction of Parallel Algorithms and Architectures: Arrays-Trees-Hypercubes.* San Mateo, Calif.: Morgan Kaufmann.
13. Rajasekaran, S., and T. Tsantilas. 1992. Optimal routing algorithms for mesh connected processor arrays. *Algorithmica,* (8), 21–38.

13

Parallel Graph Algorithms

Stephan Olariu

In the area of computing, the only thing that seems to increase faster than processing speed is the need for ever more processing speed. The last few decades have seen cycle times move from milliseconds to nanoseconds, yet applications ranging from medical research to graphics to management of large networks to the design of the processing chips themselves have resulted in a demand for faster processing than is available. It has been known for some time that this phenomenal growth cannot continue forever; the speed of light and the laws of quantum mechanics act as formidable barriers that cannot be breached. Economic constraints are equally significant; in an attempt to make them faster, processor, bus, and memory technologies steadily have become more and more exotic and expensive.

Long before electronic processing speeds began to approach limitations imposed by the laws of physics, experts expressed the opinion that the future of computer science was in parallel processing. The technological difficulties of this fundamental change to the von Neumann model of computing were considerable but, recently, the many years of research have finally begun to pay off, and the technology needed for parallel processing is finally a reality.

It is an established paradigm in computer science to express real-life problems in terms of graphs, with the solution taking the form of a graph computation. In turn, due to their wide spectrum of applications, graph problems require fast solutions. But parallel machines by themselves will not necessarily bring about a dramatic improvement in the running time of graph algorithms unless more efficient solutions to graph problems are obtained. Indeed, in many contexts, inefficient solutions to graph problems constitute a bottleneck that hampers any increase in computational power from translating into increased performance of the same order of magnitude.

Some computations are readily adapted to parallel processing, but many others, including algorithms in VLSI design, network topologies, and distributed data base processing, require fundamental changes in the approach to the problem. Graph theory is no exception. The commercial availability of massively parallel machines, combined with the maturation of the field of parallel computation theory, has motivated a natural shift in the research goals of the graph algorithms community. The main research impetus has gradually been redirected from efforts to parallelize existing sequential algorithms to the task of finding new paradigms that would lead to efficient parallel algorithms that run as fast as possible. Research in parallel graph algorithms has been conducted essentially in two different but not necessarily disjoint directions: (1) devising parallel algorithms for fundamental problems on general graphs, and (2) devising parallel algorithms for several restricted classes of graphs arising in computer science and engineering.

This two-pronged approach is motivated by the well-known fact that many interesting problems in graph theory are NP-complete on general graphs [18]. In practical applications, however, one rarely has to contend with general graphs: typically, a careful analysis of the problem at hand reveals sufficient structure to limit the graphs under investigation to a restricted class. As a rule, two broad problems are of interest when dealing with restricted classes of graphs: devising fast and processor-efficient parallel algorithms to recognize graphs belonging to the given class, and devising fast parallel algorithms for particular problems in the given class. These particular problems are application specific but often include coloring, computing a minimum path cover, identifying a hamiltonian cycle (resp. path) if one exists, and computing a maximum matching, to name just a few.

The principal goal of this chapter is to introduce the reader to some of the basic techniques used to design parallel graph algorithms, both for general graphs and for special graph classes. We do not purport to present an exhaustive picture of the state of the art, nor do we claim to provide a complete summary of the multitude of trends that one witnesses in parallel graph algorithmics today. Within the limited space available for this chapter, such an exhaustive picture cannot be accurately presented and, in fact, no particular topic can be covered in the detail it deserves.

Given all these limitations and constraints, we believe that the best approach is to brush a succinct picture of the basic algorithmic tools that underlie the area of the design of parallel graph algorithms and then to illustrate by presenting a number of sample graph algorithms. In addition, we shall provide the reader with a list of sources for further reading.

This chapter has been specifically written with the non-specialist in mind. The reader is assumed to be familiar with basic graph-theoretic terminology at the level of introductory textbooks and to have some understanding of issues involved in the design and analysis of computer algorithms.

The remainder of this chapter is organized as follows. Section 13.1 reviews fundamental concepts and terminology—both graph-theoretic and algorithmic; Section 13.2 introduces the model of parallel computation assumed throughout the chapter; Section 13.3 presents, at the elementary level, basic algorithmic tools that are key ingredients of parallel algorithms; Section 13.4 is devoted to a number of fundamental tree algorithms; Section 13.5 presents a sample of parallel algorithms for classic problems involving general graphs; and Section 13.6 discusses a number of sample parallel algorithms for particular classes of graphs that are of import in various practical applications. Finally, Section 13.7 offers concluding remarks.

13.1   Graph-Theoretic Concepts and Notation

To make this chapter self-contained we shall now review basic algorithmic and graph-theoretic concepts and terminology and establish notation. For more information, the reader is referred to Refs. [2, 9, 16] and to other relevant chapters in this handbook.

An undirected graph is an ordered pair (V, E) consisting of a set V of vertices and a set E of edges, each edge being an unordered pair of distinct vertices. For example, the edge involving vertices x and y will be denoted by $\{x, y\}$. When no confusion is possible, we will use xy as a shorthand for $\{x, y\}$. Vertices x and y are said to be adjacent if xy is an edge. We also say that the edge xy is incident with x and y. As usual, we shall write $G = (V, E)$ to denote a graph G with vertex-set V and edge-set E. A subgraph of G is a graph $G' = (V', E')$ such that $V' \in V$ and $E' \in E$. Given a set A of vertices of G, the subgraph induced by A is denoted by $G(A) = (A, E(A))$ where $E(A) = \{xy \mid x \in A \text{ and } y \in A\}$. For a vertex x in G, the degree $d(x)$ stands for the number of edges incident with x. In Fig. 13.1, for example, the degree of vertex 6 is 3.

A path is a sequence $v_0, v_1, \ldots, v_p$ of distinct vertices of G with $v_{i-1}v_i \in E$ for all i $(1 \leq i \leq p)$. A graph is connected if there is a path between any pair of its vertices. The connected components of G are the maximal connected subgraphs of G. For example, the graph G in Fig. 13.1 has two connected components, induced by $\{1, 3, 5, 6, 8\}$ and $\{2, 4, 7\}$, respectively. A cycle of length $p + 1$ is a sequence $v_0, v_1, \ldots, v_p$ of distinct vertices of G such that $v_{i-1}v_i \in E$ for all $i (1 \leq i \leq p)$ and $v_p v_0 \in E$.

A graph $G = (V, E)$ is biconnected if the graph obtained from G by removing an arbitrary vertex along with all edges incident to it is still connected. For example, the subgraph induced by $\{2, 4, 7\}$ in Fig. 13.1 is biconnected. A vertex v of G is said to be a *cut vertex* if the graph induced by $V - \{v\}$ is not connected. For example, vertex 6 is a cut vertex in the graph induced by $\{1, 3, 5, 6, 8\}$. An edge $\{x, y\}$ of G is said to be a *bridge* if the graph obtained from G by removing the edge $\{x, y\}$ is not connected.

A graph that contains no cycle is termed *acyclic*. A (legal) coloring of a graph is an assignment of integers to the vertices of the graph in such a way that adjacent vertices always receive distinct integers (i.e., colors).

A clique in a graph is a set of pairwise adjacent vertices. For example, $\{2, 4, 7\}$ is a clique in the graph featured in Fig. 13.1. A set of pairwise nonadjacent vertices is said to be *independent*. Referring again to Fig. 13.1, the set $\{2, 3, 5, 8\}$ is an independent set. A clique cover for a graph G is a set of cliques that contain all the vertices of the graphs. A clique cover is minimal when it contains the fewest possible cliques that together cover all the vertices. For example, $\{6, 8\}$, $\{6, 3\}$, $\{5, 1\}$, and $\{2, 4, 7\}$ constitute a clique cover. A set D of vertices of a graph is said to be *dominating* if every vertex outside D is adjacent to a vertex in D. For example $\{1, 2, 5, 6\}$ is a dominating set in the graph of Fig. 13.1.

A matching in a graph is a set of edges sharing no common vertex. For example, the set $\{\{6,8\},\{1,5\},\{2,7\}\}$ is a matching in the graph of Fig. 13.1. Consider a graph G along with a matching M in G. It is customary to call the edges in M matched and the edges outside M free. Vertices not incident with matched edges are referred to as exposed. A path P:

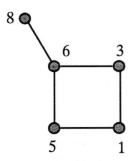 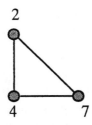

Figure 13.1 A graph G

$x_1, x_2, \ldots, x_p$ is called *alternating* if the edge $\{x_i, x_{i+1}\}$ is free for all odd values of i and matched for all even values of i. If both x_1 and x_p are exposed vertices, then the path P is termed *augmenting*. A fundamental theorem of Berge [9] states that a matching M is maximum if and only if there is no augmenting path with respect to M.

A vertex cover for a graph is a subset C of its vertices such that every edge is incident with a vertex in C. The vertex cover problem is one of the fundamental problems of algorithmic graph theory [9, 18].

A tree is a connected acyclic graph. When dealing with trees, it is customary to refer to vertices as *nodes*. In a tree, every two nodes are connected by a unique path. When a distinguished node termed the root of the tree is selected, the tree is said to be *rooted*. For a node u of a tree, its ancestors are all the nodes on the unique path to the root; the descendents of u are the nodes in the subtree of the original tree rooted at u. A leaf in a tree is a node of degree 1. If u is a node of a rooted tree T, then T_u represents the subtree of T rooted at u and containing all the descendents of u in T. Given any two nodes x and y of a rooted tree T, the lowest common ancestor of x and y, denoted lca(x, y), is the unique node z of T that is an ancestor of both x and y and such that no descendent of z is an ancestor of both x and y.

An ordered tree is either empty or it consists of a distinguished node called the *root* and a (possibly empty) list $T_1, T_2, \ldots, T_k$ of subtrees enumerated from left to right. A binary tree is either empty or else it consists of a root r and two binary subtrees (the left subtree and the right subtree) whose roots are the children of r. A binary tree in which the degree of every non-leaf is precisely 2 is usually referred to as *full*.

13.2   Tree Algorithms

Trees are in some sense the "simplest" of graphs; they are connected, and they contain no cycles. Given this appealing simplicity and the fact that trees find applications to such diverse areas of computer science as VLSI design, natural language processing, database design, and knowledge base design, it comes as no surprise that there is a vast amount of literature on tree algorithms, both sequential and parallel. The purpose of this section is to present a sample of parallel algorithms for a number of fundamental tree problems.

13.2.1   Computing tree functions

13.2.1.1 Computing a postorder numbering. A preorder traversal of an ordered tree T involves enumerating the root and then performing a preorder traversal of the subtrees T_1, $T_2, \ldots, T_k$ in left to right order. A postorder traversal of T amounts to enumerating in postorder the subtrees $T_1, T_2, \ldots, T_k$ followed by the root of T. A postorder (resp. preorder) numbering of a tree is a numbering of the nodes of the tree in the order in which they occur in the postorder (resp. preorder) traversal of the tree.

Given a rooted ordered tree T along with its Euler tour, the task of computing a postorder (resp. preorder) numbering of the nodes of T can be performed as follows. To begin, for every node v of T assign a weight of 0 to $v_1, v_2, \ldots, v^{d(v)}$ and a weight of 1 to $v^{d(v)+1}$, as illustrated in Fig. 13.2. Next, compute the weighted rank of every node in the Euler tour $L(T)$. For every node $v^{d(v)+1}$, the corresponding weighted rank is the postorder number of v. Refer to Fig. 13.3 for the final result of the postorder numbering.) It is easy to see that the entire computation runs in $O(\log n)$ time using an optimal number of processors in the EREW-PRAM.

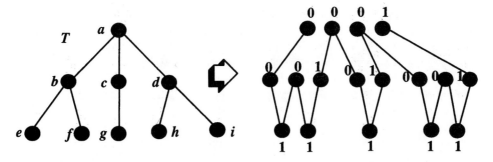

Figure 13.2 Illustrating the weight assignment

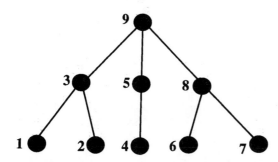

Figure 13.3 Postorder numbering of the tree in Fig. 13.2

13.2.1.2 Computing the number of descendents. Consider an arbitrary rooted tree T. In many applications it is important to compute for every node u in T the number of descendents of u or, equivalently, the number of nodes in the subtree T_u rooted at u. It is easy to devise an $O(n)$ time sequential algorithm to solves the problem [25]. As we are about to show, the Euler tour technique provides an elegant cost-optimal algorithm.

Indeed, once the Euler tour $L(T)$ is available, we assign a weight of 1 to u^1 and a weight of 0 to $u^2,\ldots,u^{d(v)+1}$. Now compute the prefix sums of the resulting sequence (equivalently, compute the weighted rank of every node in $L(T)$). Let $r(u^1)$ and $r(u^{d(u)+1})$ be the corresponding values of the prefix sum at u^1 and $u^{d(u)+1}$, respectively. The number of descendents of u is exactly $r(u^{d(u)+1}) - r(u^1) + 1$. Refer to Fig. 13.4, where the above algorithm is used to compute the number of descendents of every node in the corresponding tree. The top row features the tree and the weight assignment to the Euler tour. The bottom row shows the values of the prefix sums and the number of descendents.

The reader should have no difficulty confirming that the entire computation runs in $O(\log n)$ time using an optimal number of processors in the EREW-PRAM.

13.2.1.3 Level computation. Consider an arbitrary rooted tree T. The level of a node v of T is defined as the number of edges on the path joining v to the root. The task of computing, in parallel, the level of every node in a rooted tree can be solved elegantly using the Euler tour technique discussed previously. Consider again a generic node v of T. After having computed the Euler tour $L(T)$, we assign a weight of 1 to v^1 and a weight of -1 to every v^j with $j \neq 1$. Next, compute the weighted rank of every node v^1 in $L(T)$. It is easy to confirm that the resulting value is precisely the level of v in T. As an illustration, refer to Fig. 13.5,

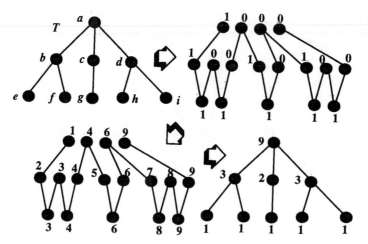

Figure 13.4 Computing the number of descendents

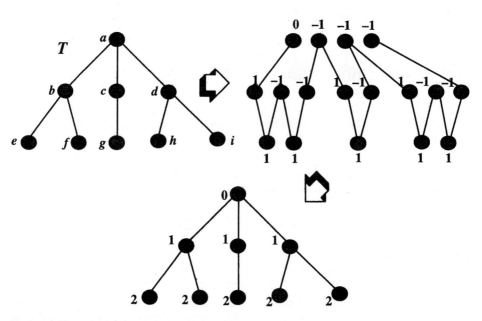

Figure 13.5 Illustrating level computation

where the above algorithm is used to compute the level of every node in the corresponding tree.

The reader should have no difficulty confirming that the entire computation runs in $O(\log n)$ time using an optimal number of processors in the EREW-PRAM.

13.2.1.4 Path identification. Given a rooted tree T and nodes u and v in T, we are interested in identifying the unique path from u to v. We begin by solving a slightly simpler problem, specifically, that of identifying the unique path between u and v in case v is an ancestor of u. Once this problem is solved the original problem is solved as follows. If v is an ancestor of u we are done. Otherwise, we identify the unique path from u to the root r of T and the unique path from v to r. We then identify the (unique) node w in T that belongs to the two paths and has the lowest level. Finally, one retains the subpath from u to w and the subpath from v to w, out of which the desired solution is obtained immediately.

We propose a cost-optimal parallel algorithm to solve path identification problem assuming that the tree T is represented in the standard form. To begin, we compute an Euler-tour $L(T)$ of T and the level of every node in T. Furthermore, assign a weight of $+1$ to $u^{d(u)+1}$ and 0 to all other nodes in $L(T)$ and, after having compacted $L(T)$ in an array, compute the prefix sums of the resulting sequence. The net effect is to isolate the sublist of P starting at u and ending at the smallest copy v^i of v which has received a value of 1 in this prefix sum. Once P is known, perform a prefix minimum on the level of every node in P starting at u, marking the first occurrence of every distinct value in the result. Obviously, what results is the unique path from u to v. To argue for the complexity of this simple algorithm, we note that every step takes $O(\log n)$ time and at most $(n/\log n)$ processors. Therefore, our algorithm is cost-optimal.

13.2.2 Lowest common ancestor computation

Because of its practical applications and theoretical interest, the lowest common ancestor problem has received considerable attention in the literature. In particular, Tsin and Chin [40] proposed a parallel algorithm that runs in $O(n/k \log n)$ time using nk processors ($k > 0$) in the CREW-PRAM model. Later, Tsin [39] proposed two algorithms to solve the same problem running in $O[(n2/p) \log n]$ and $O[(n2/p) + \log n]$ time and using $p (p > 0)$ processors in the CREW-PRAM model.

More recently, Schieber and Vishkin [35] proposed an algorithm that allows k off-line lca queries to be answered in $O(\log n)$ time using $(n + k)/\log n$ processors, provided that read conflicts are allowed. After $O(\log n)$ EREW preprocessing time with $n/\log n$ processors, they can answer n^2 on-line lca queries in constant CREW time using n^2 processors. By using the standard simulation of a CREW-PRAM by an EREW-PRAM, the complexity of their algorithm is $O(\log n)$ with $O(n^2)$ EREW processors. We note that Schieber and Vishkin [35] do not specifically address the issue of finding the lcas for all pairs of nodes in the tree.

Quite recently, Lin and Olariu [28] proposed a very simple optimal parallel algorithm that computes the lowest common ancestor information for all pairs of nodes in a rooted, ordered tree. For the reader's benefit, we now briefly summarize the main differences between the algorithm of Ref. [28] and the algorithm in Ref. [35]. In the preprocessing stage of the algorithm of Ref. [35], a mapping from the nodes of the input tree T to a complete binary tree B is computed. This mapping is such that all the nodes of T mapped into the same node of B form a path in T and, furthermore, for each node w of T, the descendents of w in T are mapped into descendents of the image of w in B. This mapping, along with additional information, makes it possible to answer an lca query in constant time.

The approach of Ref. [28] is quite different: they show that computing lowest common ancestor information for a pair of nodes in the tree can be done efficiently if the preordering of the nodes in the original tree T is known. Therefore, the problem is reduced to that of computing such a preordering; naturally, this can be done efficiently by the well known Euler-tour technique discussed previously. The algorithm of Ref. [28] runs in $O(\log n)$ time using $O(n^2/\log n)$ processors in the EREW-PRAM model. Since there are $O(n^2)$ distinct pairs of nodes in an n-node tree, it is easy to see that this algorithm is cost optimal.

Let T be a rooted, ordered tree with n nodes. For convenience, we assume that the tree is stored in contiguous memory cells of an unordered array, with every node storing a pointer to its parent in the tree. To obtain a preorder numbering of the nodes in T, it is useful to have T represented in a slightly different form; i.e., along with the parent pointer we need for every node in the tree a doubly linked list of children. To achieve this, we only need group all the nodes having the same parent pointer in a doubly linked list. One way to do this is by sorting: if we use Cole's algorithm [11], the entire process takes $O(\log n)$ time using $O(n)$ processors in the EREW-PRAM. Note that as a by-product of sorting, we can compute for every node u of T the number $d(u)$ of its children. As a further by-product of sorting the nodes of T, we can identify for every child of a node u in T its rank in the doubly linked list of children of u.

For later reference, every node u in T records the interval $[l(u), r(u)]$ with $l(u)$ and $r(u)$ standing for the weighted ranks of u^1 and $u^{d(u)+1}$, respectively. For simplicity, we shall assume, without loss of generality, that the nodes of T are denoted by their preorder number. In this notation, it is immediate that $[l(u), r(u)]$ denotes all the nodes contained in the subtree T_u of T rooted at u. To simplify the notation, we shall let lca(x, y) stand for the lowest common ancestor of nodes x and y.

Let u be an arbitrary node in T with children $w_1, w_2, \ldots, w_{d(u)}$. Central to the approach in Ref. [28] is the following result, specifying necessary and sufficient conditions for a node u of T to be the lowest common ancestor of nodes x and y. In other words, we specify how to compute lowest common ancestor information in the presence of a preorder numbering of the nodes of T.

Lemma 1. The lowest common ancestor of nodes x and y is u if, and only if, x (resp. y) is equal to u, and y (resp. x) belongs to T_u, or $x \in [l(w_i), r(w_i)]$ and $y \in [l(w_j), r(w_j)]$ for some i and j such that $1 \le i \ne j \le d(u)$. $\square$

To store the lowest common ancestor information for pairs of nodes in T, a matrix $A[1..n, 1..n]$, initialized to 0 is used; this can be done in $O(\log n)$ time using $O(n^2/\log n)$ processors in the obvious way. Note that Lemma 1 suggests the following simple algorithm to compute lca(x, y) for all pairs x, y of nodes in T.

Procedure Find-lca(T)

```
{Input: a rooted, ordered tree T;
Output: the matrix A[1..n, 1..n] containing lowest common ancestor informa-
tion for all nodes in T}
1.    begin
2.    compute a preorder numbering of the nodes in T;
3.    for all nodes u of T pardo
4.    compute [l(u),r(u)];
5.    store [l(wi), r(wi)] for all children wi of u;
6.    for all v ∈ [l(u), r(u)] pardo
7.         A(u, v) ← u;
8.    endfor
9.    for all x ∈ [l(wi), r(wi)] and y ∈ [l(wj), r(wj)] (1 ≤ i < j ≤ d(u)) pardo
10.        A(x, y) ← u;
11.        endfor
12.   endfor
13.   return(A)
14.   end; {Find-lca}
```

Note that lines 4 and 5 can be performed altogether in $O(\log n)$ time using $O(n/\log n)$ processors by using the Euler-tour technique. The processors are assigned to the nodes of T in two stages. Initially, we assign to every node u with $d(u)$ children $\lceil d(u)/\log n \rceil$ processors. The assignment of processors to nodes of T is done as follows. Let $u_1, u_2, \ldots, u_n$ be the nodes of T enumerated in preorder. Now, the first $(n/\log n)$ processors compute $a_i = \lceil d(u_i)/\log n \rceil$ for every i $(1 \leq i \leq n)$ in $O(\log n)$ time. Furthermore, the same processors will compute the prefix sums of the sequence $a_1, a_2, \ldots, a_n$ and then write i in $B[a_1 + a_2 + \ldots + a_i]$; here, B is an array of size n. Now, a simple broadcast operation sends i to all $B[j]$ with $a_1 + a_2 + \ldots + a_{i-1} \leq j \leq a_1 + a_2 + \ldots + a_i$. Finally, every processor P_i with $(1 \leq i \leq a_1 + a_2 + \ldots + a_n)$ is assigned to the node u_i of T. Note that, all together, the number of processors allocated to all the nodes of T is

$$\sum_{u \in T} \left\lceil \frac{d(u)}{\log n} \right\rceil \leq \sum_{u \in T} \left(\frac{d(u)}{\log n} + 1 \right) \leq \left\lceil \frac{\sum_{u \in T} d(u)}{\log n} \right\rceil + n = \left\lceil \frac{n-1}{\log n} \right\rceil + n \in O(n)$$

Next, we define

$$\sigma_i(u) = \sum_{i < j \leq d(u)} [r(w_i) - l(w_i) + 1] * [r(w_j) - l(w_j) + 1]$$

At this point it is worth noting that for each pair of nodes there is a unique node u such that the nodes fall into distinct subtrees of T_u. To understand the motivation for introducing this notation, we note that for every node u of T,

$$\sum_{i-1}^{d(u)-1} \sigma_i(u)$$

counts the number of pairs of nodes in distinct subtrees of T_u. Consequently, the quantity

$$\sum_{u \in T} \sum_{i=1}^{d(u)-1} \sigma_i(u)$$

is bounded above by the number of pairs of nodes in T which is at most $n(n-1)/2$.

Furthermore, it is easy to confirm that for all $i(1 \leq i \leq d(u) - 1)$, $l(w_i + 1) = 1 + r(w_i)$. Consequently, we can write

$$\sigma_i(u) = [r(w_i) - l(w_i) + 1] * \sum_{i < j \leq d(u)} [r(w_j) - l(w_j) + 1]$$

$$= [r(w_i) - l(w_i) + 1] * [r(w_{d(u)}) - l(w_{i+1}) + 1]$$

To compute all the values $\sigma_i(u)$ $(1 \leq i \leq d(u) - 1)$, we use the $\lceil d(u_i)/\log n \rceil$ processors $P_1(u), P_2(u), \ldots, P_{\lceil d(u)/\log n \rceil}(u)$ allocated to node u. First, we broadcast $r(w_d(u))$ to these processors in $O(\log \lceil d(u_i)/\log n \rceil) = O(\log n)$ time. Next, every processor is responsible for computing $\log n$ of the $\sigma_i(u)$ values. Therefore, the total running time is $O(\log n)$. Once this computation has been performed, every node of T releases all its processors.

In the second stage, assign to each node u of T exactly

$$\sum_{i=1}^{d(u)-1} \left\lceil \frac{\sigma_i(u)}{\log n} \right\rceil$$

processors. By a previous observation, the total number of processors assigned to all the nodes of T in this second stage is

$$\sum_{u \in T} \sum_{i=1}^{d(u)-1} \left\lceil \frac{\sigma_i(u)}{\log n} \right\rceil \leq \sum_{u \in T} \sum_{i=1}^{d(u)-1} \left(\frac{\sigma_i(u)}{\log n} + 1 \right) \leq \frac{n(n-1)}{2\log n} + (n-1) \in O\left(\frac{n^2}{\log n} \right)$$

With this processor assignment, lines 6 and 7 take $O(\log n)$ time using the first

$$\left\lceil \frac{r(v)-(v)}{\log n} \right\rceil$$

of the processors allocated to every node u of T.

To perform the computation specified in lines 8 and 9, we proceed as follows. For all $i(1 \leq i \leq d(u)-1)$ $\lceil \sigma_i(u)/\log n \rceil$ processors are responsible for writing a "u" in the $\sigma_i(u) = [r(w_i) - l(w_i) + 1] * [r(w_{d(u)}) - l(w_{i+1}) + 1]$ entries of the rectangular submatrix $L[l(w_i) .. r(w_i), l(w_{i+1}) .. r(w_{d(u)})]$. Let us refer to each such submatrix as a group. By performing prefix sums we compute in the obvious way the starting position of every group. By performing a broadcast operation among the processor responsible for each group we can distribute the starting position as well as the value to be written to all these processors. This is easily done in $O(\log n)$ time using $O(n^2/\log n)$ processors. Since no read/write conflicts arise, the computation can be carried out in the EREW-PRAM model. Thus, we state the following result of Ref. [28].

Theorem 1. [28] With a rooted ordered tree T with n nodes as input, procedure Find-lca correctly computes the lowest common ancestor of every pair of nodes in T in $O(\log n)$ time using $O(n^2/\log n)$ processors in the EREW-PRAM model. $\square$

13.2.3   Reconstructing binary trees from traversals

In this subsection, we consider the classic problem of reconstructing a binary tree $T = (V, E)$ with vertices $[1, 2, \ldots, n\}$ given its inorder traversal and either its preorder or its postorder traversal. It is well known that a binary tree can be reconstructed from its inorder traversal along with either its preorder or its postorder traversals [25]. Recently, sequential and parallel solutions to this problem have reported in the literature. The algorithm in Ref. [37] runs in $O(\log n)$ time using $O(n)$ processors in the CREW PRAM; the solution in Ref. [7] takes $O(\log \log n)$ time using $O(n/\log \log n)$ processors in the CRCW-PRAM.

Recently, Olariu et al. [31] presented a new parallel algorithm for this problem. The main idea of this algorithm is to reduce the reconstruction process to merging two sorted sequences. With the best results for parallel merging [23], their algorithm can be implemented in $O(\log \log n)$ time using $O(n/\log \log n)$ processors in the CREW-PRAM or in $O(\log n)$ time using $O(n/\log n)$ processors in the EREW-PRAM. Thus, the algorithm in Ref. [31] improves on the previous results, either in the time complexity or in the model of computation.

As pointed out in Ref. [7], optimal parallel algorithms that run in $O(\log \log n)$ time usually need to be implemented in the CRCW-PRAM. The only known doubly logarithmic optimal CREW-PRAM algorithm is the parallel merging algorithm [23], which runs in $O(\log \log n)$ time using $O(n/\log \log n)$ processors in the CREW-PRAM. Thus, the tree reconstruction algorithm in Ref. [31] provides another example of a doubly logarithmic time parallel algorithm in the CREW-PRAM.

Let T be a binary tree. Every node v of T is split into three copies—v_1, v_2, and v_3—all having the same node label as v. For each of the resulting nodes, we define a *next* field as follows: If v has no left child, then $v_1.next = v_2$. If v has no right child, then $v_2.next = v_3$. If w is the left child of v, then $v_1.next = w_1$, and $w_3.next = v_2$. If w is the right child of v, then $v_2.next = w_1$ and $w_3.next = v_3$. What results is a linked list, called an Euler path, which starts at $root_1$ and ends at $root_3$ and traverses each edge of T exactly once in each direction. Letting T be a binary tree with left subtree T_1 and right subtree T_2, the Euler path $\Lambda(T)$ of T can be expressed as $root_1\Lambda(T_1)root_2\Lambda(T_2)root_3$.

When no confusion is possible, we let the Euler path denote the sequence of node labels contained in the corresponding linked list. Obviously, the Euler path of a binary tree contains three copies of each node label in the tree. An interesting property of the Euler path of a binary tree T is that keeping only the first copy of each label results in a preorder traversal of T; keeping only the second copy of each label gives an inorder traversal of T; and keeping only the third copy of each label yields a postorder traversal of T [38].

For convenience, we define a *preorder-inorder path* to be a sequence of labels obtained by deleting the third copy of each label in an Euler path. Similarly, we can define an *inorder-postorder path* as a sequence of labels obtained by deleting the first copy of each label in an Euler path. Since a binary tree can be reconstructed from its inorder traversal along with either its preorder or postorder traversal, it follows that a binary tree is completely determined by its preorder-inorder path or its inorder-postorder path. For example, Fig. 13.6 features a binary tree along with the associate preorder-inorder path, preorder traversal, and inorder traversal.

Lemma 2. [31] A sequence of labels $b_1, b_2, \ldots, b_{2n}$ represents the preorder-inorder path of an n-node binary tree T if and only if the following two conditions hold: (1) there are exactly two copies of each label in the sequence, and (2) there exist no integers i, j, k, m with $1 \le i < j < k < m \le 2n$ such that $b_i = b_k$ and $b_j = b_m$. $\square$

Intuitively, Lemma 2 states that any preorder-inorder path containing distinct labels u and v must be of the form "$\ldots u \ldots v \ldots v \ldots u \ldots$" or "$\ldots u \ldots u \ldots v \ldots v \ldots$"; it is impossi-

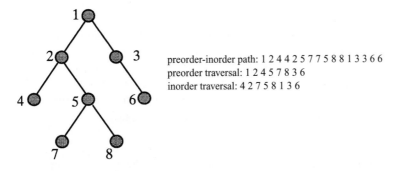

preorder-inorder path: 1 2 4 4 2 5 7 7 5 8 8 1 3 3 6 6
preorder traversal: 1 2 4 5 7 8 3 6
inorder traversal: 4 2 7 5 8 1 3 6

Figure 13.6 Binary tree with associated paths

ble for a preorder-inorder path to be of the form "...u...v...u...v..." which we refer to as *overlapping* of labels u and v.

Let $b_1, b_2, \ldots, b_{2n}$ be a preorder-inorder path with $2n$ labels. For any pair of duplicate labels b_i and b_j ($1 \leq i < j \leq n$), call b_i the first copy of its duplicate label and b_j the second copy of its duplicate label.

The reconstruction algorithm of Ref. [31] has two steps. In the first step they construct the preorder inorder path of the binary tree determined by the traversals. Once the preorder-inorder path is available, it is straightforward to reconstruct the corresponding binary tree. The details of the second step are spelled out as follows:

Procedure Convert-Path-to-Tree(T)

```
{Input: a preorder-inorder path b₁,b₂, . . . ,b₂ₙ, in which every label remembers
the position of its duplicate.
Output: A binary tree with root r and nodes 1,2, . . . ,n;
1.   begin
2.     r ← b₁;
3.     for each label bᵢ (1 ≤ i ≤ 2n) do
4.          if (bᵢ is the first copy of its duplicate) then
5.               if (bᵢ₊₁ is the first copy of its duplicate label) then
6.                    leftchild(bᵢ) ← bᵢ₊₁;
7.          else {bᵢ is the second copy of its duplicate}
8.               if (bᵢ₊₁ is the first copy of its duplicate label) then
9.                    rightchild(bᵢ) ← bᵢ₊₁;
10. end; {Convert-to-Tree}
```

The following result is easily seen.

Lemma 3. Given a preorder-inorder path with $2n$ labels in which each label remembers the position of its duplicate, procedure Convert-to-Tree correctly reconstructs the corresponding binary tree in $O(1)$ time using $O(n)$ processors in the EREW-PRAM. □

We now turn to the first step of the tree reconstruction algorithm, that is, the construction of the preorder-inorder path from the traversals. The main idea of this step is suggested by Lemma 2, which says that overlapping of labels cannot occur in a preorder-inorder path. The idea of Ref. [31] is to "merge" the given preorder and inorder traversals into a sequence that satisfies the above properties. To give the reader some intuition about this idea, we show that this merging process can actually be carried out with the help of a stack, as in the following procedure.

Procedure Construct-Preorder-Inorder-Path(T)

```
{Input: c₁,c₂, . . . ,cₙ and d₁,d₂, . . . ,dₙ, the preorder and inorder traversals
of T;
Output: the preorder-inorder path b₁,b₂, . . . ,b₂ₙ};
1.   begin
2.     Stack ← Φ; j ← k ← 1;
3.     for i ← 1 to 2n do
4.          if (dₖ matches the label on the top of Stack) then
5.               bᵢ ← dₖ;
6.               α ← pop Stack;
7.               α and dₖ remember each other's position in the sequence
                    b₁,b₂, . . . ,b₂ₙ;
```

```
8.                   k ← k+1
9.          else
10.                b_i ← c_j;
11.                push c_j onto Stack;
12.                j ← j + 1;
13.   return b_1,b_2, . . . ,b_2n;
14. end; {Construct-Preorder-Inorder-Path}
```

The reader should have no difficulty confirming that the above procedure, given the preorder and the inorder traversals of an n-node binary tree T, correctly constructs in $O(n)$ time the preorder inorder path of T such that every label remembers the position of its duplicate in the path. Consequently, an n-node binary tree can be reconstructed from its preorder and inorder traversal in $O(n)$ time with $O(n)$ extra space.

Although it is easy to implement procedure Construct-Preorder-Inorder-Path sequentially, it is not obvious how the procedure can be implemented in parallel. The contribution of Ref. [31] was to show that the idea behind the *Stack* being used in the procedure can lead to the definition of a linear order. With this linear order, both the preorder and the inorder traversals are sorted sequences; and the construction of the preorder-inorder path can be done by merging these two sorted sequences according to the linear order.

Let $c_1, c_2, \ldots c_n$ and $d_1, d_2, \ldots, d_n$ be the preorder and inorder traversals of T. For simplicity, we assume that $c_1, c_2, \ldots c_n$ is $1, 2, \ldots, n$. The case where $c_1, c_2, \ldots c_n$ is a permutation of $1, 2, \ldots, n$ can be reduced to this case easily. Construct two sequences of triples: a sequence $(1, j_1, c_1), (1, j_2, c_2), \ldots, (1, j_n, c_n)$ such that $d_{j_i} = c_i$, $(i = 1, 2, \ldots, n)$ (i.e., j_i is the position of c_i in sequence $(d_1, d_2, \ldots, d_n)$), and a sequence $(2, 1, d_1), (2, 2, d_2), \ldots, (2, n, d_n)$.

Write $\Pi = \{(1, j_1, c_1), (1, j_2, c_2), \ldots, (1, j_n, c_n), (2, 1, d_1), (2, 2, d_2), \ldots, (2, n, d_n)\}$ and define a binary relation $\ll$ on Π as follows: for arbitrary triples (α, β, γ) and $(\alpha', \beta', \gamma')$ in Π we have:

Rule 1. $((\alpha = 1) \wedge (\alpha' = 1)) \rightarrow (((\alpha, \beta, \gamma) \ll (\alpha', \beta', \gamma')) \leftrightarrow (\gamma < \gamma'))$

Rule 2. $((\alpha = 2) \wedge (\alpha' = 2)) \rightarrow (((\alpha, \beta, \gamma) \ll (\alpha', \beta', \gamma')) \leftrightarrow (\beta < \beta'))$

Rule 3. $((\alpha = 1) \wedge (\alpha' = 2)) \rightarrow (((\alpha, \beta, \gamma) \ll (\alpha', \beta', \gamma')) \leftrightarrow ((\beta < \beta') \vee (\gamma < \gamma')))$

The motivation behind Rules 1 and 2 is the fact that keeping only the first (second) copies of the labels in the preorder-inorder Euler path gives the preorder (inorder) traversal of the binary tree. Therefore, when combining the traversals, the original relative order between labels from the same traversal should be maintained. (For the case of Rule 1, notice the assumption that $c_1, c_2, \ldots c_n$ is $1, 2, \ldots, n$.)

The motivation for Rule 3 is the following: For any c_i and d_j, their relative order in the merged sequence should be determined in a way consistent with the fact that overlapping of labels is not allowed in the preorder-inorder path (see Lemma 2), and that the copy of a label from the preorder traversal should always come before the copy of the same label from the inorder traversal.

As it turns out, the binary relation $\ll$ is total on Π and, in addition, it is also transitive (the reader is referred to Ref. [31] for details). Consequently, $\ll$ is a linear order on Π.

Rules 1 and 2 guarantee that $\ll$ is consistent with both sequences $(1, j_1, c_1), (1, j_2, c_2), \ldots, (1, j_n, c_n)$ and $(2, 1, d_1), (2, 2, d_2), \ldots, (2, n, d_n)$. By merging these two sequences according to $\ll$ we obtain a sorted sequence of $2n$ triples, $(\alpha_1, \beta_1, g_1), (\alpha_1, \beta_1, g_1), \ldots, (\alpha_{2n}, \beta_{2n}, \gamma_{2n})$. It is shown in Ref. [31] that is $\gamma_1, \gamma_2, \ldots, \gamma_{2n}$ is the preorder-inorder path of the tree determined by the traversals.

Up to this point, we have successfully reduced computing the preorder-inorder path to parallel merging. We now discuss the complexity of this reduction. First, consider the complexity to construct Π. For this purpose, let us see how to construct from the given traversals, sequences $(1, j_1, c_1), (1, j_2, c_2), \ldots, (1, j_n, c_n)$ and $(2, 1, d_1), (2, 2, d_2), \ldots, (2, n, d_n)$ such that $c_i = d_{j_i}$ $(i = 1, 2, \ldots, n)$. We note that this can be done easily with an auxiliary array $A[1 \ldots n]$. Since $c_1, c_2, \ldots c_n$ is $1, 2, \ldots, n$, and $d_1, d_2, \ldots d_n$ is a permutation of $1, 2, \ldots, n$, we can compute the entries of $A[1 \ldots n]$ as follows: $A[d_i] = i$ $(1, 2, \ldots, n)$ in $O(1)$ time in the EREW-PRAM using n processors. To determine the subscript j_i satisfying $c_i = d_{j_i}$ $(i = 1, 2, \ldots, n)$, we simply take $j_i = A[c_i]$ $(1, 2, \ldots, n)$. This again can be computed in $O(1)$ time in the EREW-PRAM using n processors. Consequently, Π can be constructed in $O(1)$ time using n processors in the EREW-PRAM.

Next, we consider the complexity to merge $(1, j_1, c_1), (1, j_2, c_2), \ldots, (1, j_n, c_n)$ and $(2, 1, d_1), (2, 2, d_2), \ldots, (2, n, d_n)$ according $\ll$. Optimal parallel algorithms are proposed in Refs. [3, 23]. Thus, one can state the following result.

Theorem 2. [31] A binary tree $T = (V, E)$ ($V = \{1, 2, \ldots, n\}$), can be reconstructed from its preorder and inorder traversals in $O(\log \log n)$ time using $O(n/\log \log n)$ processors in the CREW PRAM, or in $O(\log n)$ time using $O(n/\log n)$ processors in the EREW-PRAM, using $O(n)$ extra space. $\square$

13.2.4 A breadth-first algorithm for ordered trees

Let T be an ordered tree. For convenience, we assume that T is maintained in an unordered array. For every node u of T we keep the following information:

- $p(u)$—the parent of u in T
- $rs(u)$—the right sibling of u in T
- $fc(u)$—the first child of u in left to right order

We begin by converting T to a linked list $L(T)$ similar to the Euler tour previously discussed. Specifically, we split every node u of T into two nodes u_1 and u_2. We then proceed to construct the linked list $L(T)$ by establishing the links between various nodes. To simplify the notation, we refer to u_1 and u_2 as the first and second occurrence of u, respectively. The details are spelled out by the following simple procedure.

Procedure Construct-List(T)

```
{Input: an n-node ordered tree T rooted at r;
Output: the linked list L(T) as illustrated in Fig. 13.7}
1.    begin
2.    if T is empty then return(φ);
3.    for all nodes u of T pardo
4.         replace u by u₁ and u₂;
5.         if rs(u) ≠ nil then
6.              link(u²) ← rs(u)¹
7.         else
8.              link(u²) ← p(u)²;
9.         iff c(u) ≠ nil then
10.             link(u¹) ← fc(u)¹
11.        else
12.             link(u¹) ← u²;
13.        return(A)
```

```
14. endfor
15. return the linked list L(T) originating at r¹
16. end; {Construct-List}
```

What results is a linked list starting at r^1 and ending at r^2 with every edge of T traversed exactly once in each direction, implying that the total length of $L(T)$ is $O(n)$. Lin et al. [29] showed that $L(T)$ satisfies the following properties.

Lemma 4. For every node u of T, u^1 precedes u^2 in $L(T)$. Furthermore, if u and v belong to distinct subtrees T_i, T_j of T with $i < j$, then u^2 precedes v^1. $\square$

Lemma 5. If we remove the first (resp. second) occurrence of every node in $L(T)$, we obtain the postorder (resp. preorder) traversal of T. $\square$

Lemma 6. The linked list $L(T)$ uniquely determines the ordered tree T. $\square$

Lemma 7. $L(T)$ can be computed in $O(\log n)$ time using $O(n/\log n)$ processors in the EREW PRAM model of computation. $\square$

For later reference, the level of a node in an ordered tree T rooted at r is defined as follows:

$$level(u) = \begin{cases} 1 & \text{if } u = r \\ 1 + level(p(u)) & \text{otherwise} \end{cases}$$

Quite naturally, the level-order traversal [22, 25] of an ordered tree T involves enumerating the nodes of T in order of increasing level numbers and left to right within each level. Clearly, the level-order traversal of T is, in fact, a possible breadth-first traversal of T.

Lin et al. [29] showed that the encoding of T obtained above can be exploited to obtain the level-order traversal of T (equivalently, to produce a breadth-first traversal of T). To begin, we write $\tau(T) = b_1, b_2 \ldots b_{2n}$. With every bit b_i ($1 \le i \le 2n$) in $\tau(T)$ one can associate a weight x defined by the following scheme:

$$x_i = \begin{cases} 1 & \text{if } i = 1 \\ b_i + b_{i-1} - 1 & \text{otherwise} \end{cases}$$

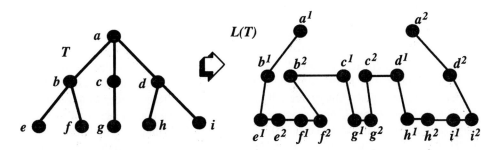

Figure 13.7 Illustration of procedure Construct-List

Now, compute the prefix sums of $x_1, x_2 \ldots x_{2n}$ and let the result be stored in $w_1, w_2 \ldots w_{2n}$. More precisely, for every $i (1 \leq i \leq 2n)$

$$w_i = \sum_{j=1}^{i} x_j$$

Lemma 8. Let u be an arbitrary node of T, and let the bits b_i and b_j correspond to u^1 and u^2 in $L(T)$. Then, $w_i = w_j$. Furthermore, w_i and w_j equal the level of node u in $T(G)$. $\square$

Lemma 9. Let a_p and a_q be matched bits in $\tau(T)$. Then $w_p = w_q$; furthermore, $q = \min\{t \mid t > p \; w_p = w_t\}$. $\square$

Note that by Lemmas 8 and 8, combined, if a_p and a_q are matched bits in $\tau(T)$, the corresponding nodes have the same level in T. This simple observation motivates the following way of obtaining a level-order traversal of the nodes of T.

Procedure Level-Order-Traversal(T)

```
{Input: an ordered tree T with n nodes;
Output: a linked list B(T) containing the nodes of T in a level-order
        traversal; the links in B(T) are denoted by →;}
1.    begin
2.    find the encoding τ(T) of T;
3.    flip every bit in τ(T) to obtain σ(T);
4.    find a maximum matching M of the bits in σ(T);
         {construct linked lists level by level}
5.    for all nodes u of T pardo
6.           set u → v whenever v = rs(u);
7.           let a_p and a_q be matched in σ(T) and let u and v be the
                corresponding nodes of T;
8.           set u → v
9.    endfor
10.   let u_1, u_2, ... ,u_d be the heads of the lists obtained in lines 4-8;
11.   let v_1, v_2, ... ,v_d be the last entries in each such list;
12.   for t = 1 to d- 1 pardo
13.          set v_t → u_{t+1};
14.   return the linked list B(T) originating at r
15.   end; {Level-Order-Traversal}
```

Theorem 3. Procedure Level-Order-Traversal correctly computes a level-order traversal of an n-node ordered tree in $O(\log n)$ time using $O(n/\log n)$ processors in the EREW-PRAM model. $\square$

Lin et al. [29] went on to show that the breadth-first traversal developed in Procedure Level-Order-Traversal can be used to obtain simple encodings of binary and ordered trees. A very simple and efficient encoding of an n-node binary tree into a bitstring of length $2n + 1$ was proposed by Knuth [25] and rediscovered later by several authors. The basic idea is to augment a given binary tree T by the addition of external nodes as illustrated in Fig. 13.8. In the new tree T' one associates with every internal node a label of 1 and with every leaf (i.e., external node) a label of 0. The encoding is obtained by simply concatenating the labels in level-by-level fashion, with the concatenation proceeding from left to right within each level. It is easy to see that with an n-node binary tree as input, the result-

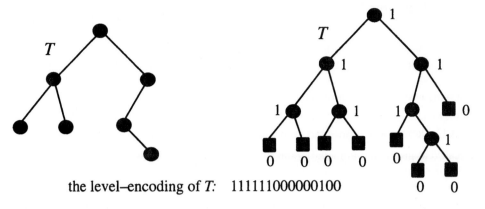

the level–encoding of T: 111111000000100

Figure 13.8 Illustration of the level-order encoding of an ordered tree

ing encoding takes $2n + 1$ bits and can be obtained by traversing T' in level-order manner [22]. Therefore, one can state the following result.

Theorem 4. A level-order encoding of an n-node binary tree can be obtained in $O(\log n)$ time using $O(n/\log n)$ processors in the EREW-PRAM model. □

A similar encoding has been recently proposed for ordered trees [22]. Specifically, given an n-node ordered tree T, one proceeds to augment T by the addition of a supernode adjacent to the root of T only, as illustrated in Fig. 13.9. In the newly created ordered tree T', every node with d children is labeled by a string $111\ldots10$ consisting of d 1s followed by a 0. Now the encoding of T is obtained by traversing T' in level-order fashion and by concatenating the labels of the nodes encountered. The reader is referred to Fig. 13.9 for an illustration of the encoding process. It is easy to confirm [22] that the encoding obtained is unique and contains $2n + 1$ bits whenever T has n nodes. By Theorem 4, the entire computation can be performed in $O(\log n)$ time. Thus, we have the following result.

Theorem 5. A level-order encoding of an n-node ordered tree can be obtained in $O(\log n)$ time using $O(n/\log n)$ processors in the EREW-PRAM. □

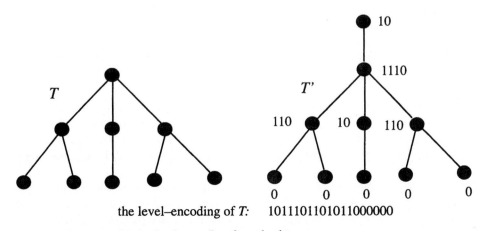

the level–encoding of T: 1011101101011000000

Figure 13.9 Illustration of the level-order encoding of an ordered tree

13.3 Algorithms for General Graphs

The purpose of this section is to acquaint the reader, at the elementary level, with a number of sample parallel algorithms that take as input an arbitrary graph and return artifacts characteristic of the input. Specifically, we discuss algorithms for classic problems on undirected graph including computing the connected components, computing a minimum spanning tree, and computing the biconnected components of a graph.

13.3.1 Computing connected components

Recall that given an undirected graph $G = (V, E)$, two vertices are in the same connected component of G whenever there is a path joining them. In many applications, one is interested in computing the connected components of a graph. Assuming that G has n vertices and m edges, there are well known sequential algorithms [2] that can solve this problem in $O(n + m)$ time. Developing efficient parallel algorithms for the connected component problem turned out to be quite challenging.

We will present the connected component algorithm of Awerbuch and Shiloah [6] that, although not cost optimal, is relatively simple and illustrates the beauty of parallel graph algorithms. We assume that an undirected graph G with n vertices and m edges is given as a list of vertices followed by a list of edges. For simplicity, we will assume that the vertices are integers from 1 through n. Every undirected edge $\{u, v\}$ is stored twice, once as (u, v) and once as (v, u). With every item (vertex or edge) we associate a processor. Thus, a total of $n + 2m$ processors are employed. We will let P_i stand for the processor associated with vertex i of the graph. If $\{i, j\}$ is an edge, then the processor associated with the edge $\{i, j\}$ will be denoted by P_{ij}. The processor associated with the edge $\{j, i\}$ is denoted by P_{ji}. The model of computation is the Arbitrary-CRCW where, in case of several processors attempting to write to the same memory location, one succeeds nondeterministically.

We assume that trees are represented by parent pointers; i.e. every node is pointing to its parent (the root points to itself). Specifically, for a node u of a rooted tree T, $D(u)$ denotes the parent of u. A rooted tree T is a star if for every node u in T we have $D(D(u)) = D(u)$. It is easy to see that if every node in T is endowed with its own processor, then the task of determining whether T is a star can be performed in $O(1)$ time. We shall, therefore, assume a function Star(i) that for every vertex i returns true if and only if i belongs to a star.

In outline, the connected component algorithm in Ref. [6] proceeds along the following lines. Initially, every vertex of the graph is in a singleton tree by itself. The main idea is to combine trees corresponding to adjacent vertices in the graph into larger trees until this operation can no longer be performed. At that moment, the algorithm terminates returning a set of rooted stars, each corresponding to a connected component in the graph. One interesting feature of the algorithm is the way in which trees are combined. Specifically, the algorithm relies on the following fundamental tree operations:

- *Shortcutting.* This is an operation performed in parallel by all the processors associated with a tree T that is not a star. It consists in updating, for every node u, the parent pointer by setting $D(u) \leftarrow D(D(u))$. The shortcut operation is very similar to pointer jumping except that it is performed in the context of a rooted tree. For an illustration, see Fig. 13.10.

- *Conditional star hooking.* This is an operation performed in parallel by processors assigned to nodes in a star. Specifically, if i is adjacent in G to j and node i is in a star, then the conditional star hooking performs the following assignment $D(D(i)) \leftarrow D(j)$ if $D(i) > D(j)$. If successful, this has for effect to "hook" the star containing node i onto the

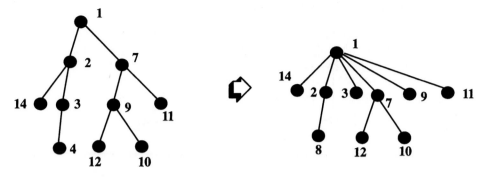

Figure 13.10 Illustration of the shortcut operation

rooted tree containing *j*. It is worth noting that in general several processors may attempt to perform conditional star hooking. Since the model is Arbitrary-CRCW, one processor will nondeterministically succeed in writing into $D(D(i))$, which may be the parent of a large number of nodes. As it turns out [6], the correctness of the algorithm is not affected by what processors succeed.

- *Unconditional star hooking*. This is an operation performed in parallel by processors assigned to nodes in a star. Specifically, if node *i* is adjacent in *G* to node *j* and node *i* is in a star, then the unconditional star hooking involves performing $D(D(i)) \leftarrow D(j)$ if $D(i) \neq D(j)$. We are now in a position to spell out the details of the connected component algorithm in Ref. [6].

Procedure Connected-Components(*G*)

```
{Input: an undirected graph G specified as a list of vertices and edges;
Output: a set of rooted stars each corresponding to a connected component in
     G;}
1.   begin
2.   for all i of G pardo D(i) ← i;
3.   repeat
4.         for edges ij in G pardo
5.                  if Star(i) and D(j) < D(i) then
6.                  D(D(i)) ← D(j);
7.            endfor;
8.         for edges ij in G pardo
9.              if Star(i) and D(j) ≠ D(i) then
10.                 D(D(i)) ← D(j);
11.           endfor;
12.        for all i of G pardo
13.                if not Star(i) then
14.                     D(D(i)) ← D(i)
15.  until forall i Star(i)
16.  end; {Connected-Components}
```

For a complete example, the reader is referred to Figs. 13.11 and 13.12. To summarize, we state the following result [6].

Theorem 6. The task of finding the connected components of an undirected graph with *n* vertices and *m* edges can be performed in $O(\log n)$ time using $O(n + m)$ processors in the Arbitrary CRCW. □

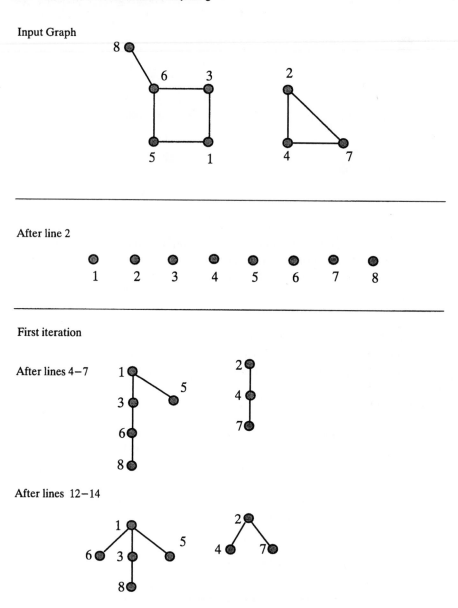

Figure 13.11 Illustration of procedure-connected components

A number of other parallel algorithms for connected components can be found in the literature. One of the earliest ones was the algorithm by Hirschberg et al. [21] that was running in $O(\log^2 n)$ time using n^2 processors in the CRCW model. Later, Chin et al. [10] produced an algorithm running in $O(\log^2 n)$ time using only $(n^2/\log^2 n)$ processors in the same model. Shiloah and Vishkin [36] obtained an algorithm very similar to that presented in this subsection and with the same complexity. Gazit [19] obtained the first cost-optimal randomized algorithm running in $O(\log n)$ time and using $O[(n + m)/\log n]$ processors. More recently, Cole and Vishkin [14] obtained a deterministic algorithm running in $O(\log n)$ time using $\{[(m + n)\,\alpha(m, n)]/\log n\}$, where $\alpha(m, n)$ is the inverse Ackerman function. For more information and a historical perspective the reader is referred to Ref. [33].

Second iteration

After lines 4−7

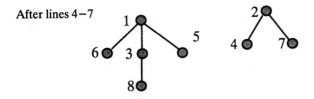

After lines 12−14

Figure 13.12 Illustration of procedure-connected components

13.3.2   Computing a minimum spanning tree

Consider a connected undirected graph $G = (V, E)$ along with a weight function $w : E \to R^+$ on the set of its edges. A spanning tree of G is a tree T such that every vertex in G is a node in T. A *minimum spanning tree* (MST) is a spanning tree of minimum total weight. The problem of computing the MST of an undirected graph is one of the classic problems of algorithmic graph theory and has received a great deal of well deserved attention in the literature [2, 16, 23].

Most of the sequential MST algorithms fall into one of the following categories: (1) algorithms that construct a MST by starting with a single-node tree and by adding new nodes to it, corresponding to minimum-weight edges joining a node already in the tree with a node outside; (2) algorithms that process edges in the order of their weight, retaining only those that do not create cycles with the already chosen edges; (3) algorithms that start with a forest consisting of all the vertices and join trees in the forest until only one tree remains. Prim's algorithm is the best known representative of the first category; Kruskal's algorithm is representative of the second one; and, finally, Sollin's algorithm best fits the third category.

In this subsection we will discuss the MST algorithm of Awerbuch and Shiloah [6] that closely resembles the connected component algorithm presented in the previous subsection. Since the algorithm in Ref. [6] uses Sollin's strategy for computing a MST of a graph, we now briefly review the basic idea of Sollin's algorithm. Given an weighted undirected graph $G = (V, E)$ as above, begin by creating a forest $F = (V, \phi)$, where every node is in a singleton tree by itself. Sollin's algorithm proceeds to make F into a tree in a number of stages. Every stage performs the following two steps:

Step 1   For every component of F, select an edge of minimum weight joining that component to a different component of F, with ties broken arbitrarily.

Step 2   Add the chosen edge to the minimum spanning tree.

The model of computation used in the MST algorithm of Ref. [6] is the Priority-CRCW where, in case several processors attempt to concurrently write into the same memory location, the processor with the highest priority succeeds. As in Ref. [6] we assume a weighted undirected graph $G = (V, E)$ with n vertices and m edges. As a simplifying assumption, the edge weights are considered distinct. With every vertex and with every edge of the graph, we associate one processor. The processors assigned to edges receive a priority equal to the weight of the corresponding edge. In this scheme, the lower the cost of the edge, the higher the priority of the processor. In particular, the processor storing the lowest-weight edge has the highest priority. The algorithm maintains a set of undirected edges termed *FOREST* which is always a subforest of the MST. In outline, the algorithm involves several iterations, each of them consisting of three steps as we are about to describe. Step 1 is performing unconditional star hooking as in the connected component algorithm of the previous subsection. All processors that correspond to edges emanating from a star attempt to "hook" the star onto a tree of the forest. Here, however, the processor that succeeds is the one with the highest priority, that is, the one corresponding to the edge of least weight. Since this edge must belong to the MST it will be added to *FOREST*.

As the reader will not fail to notice, Step 1 can create cycles in *FOREST*. The purpose of Step 2 is to identify and remove these cycles such that, when Step 3 begins, there are no cycles left. Finally, Step 3 is the shortcutting step where every processor associated with a node in the tree updates the corresponding parent pointer by pointer jumping. For an illustration, the reader is referred to Fig. 13.13. The details of the minimum spanning tree algorithm in [6] are spelled out as follows.

Procedure MST(*G*)

{**Input:** a weighted connected undirected graph G specified as a list of vertices and edges;

Output: a minimum spanning tree stored in the array FOREST;}

```
1.   begin
2.     for every edge (i, j) of G pardo FOREST((i, j)) ← 0;
3.     for all i of G pardo D(i) ← i;
4.   repeat
5.         for edges ij in G pardo
6.             if Star(i) and D(j) 7~ D(i) then
7.                 D(D(i)) ~ D(j);
8.                 WINNER(D(i)) ← (i, j)
9.             endif
10.            if WINNER(D(i)) = (i, j)) then
11.                FOREST((i, j)) ← (i, j)
12.    endfor;
13.    for all i of G pardo
14.        if i < D(i) and i = D(D(i)) then
15.            D(i) ← i
16.    endfor;
17.    for all i of G pardo
18.        if not Star(i) then
19.            D(D(i)) ← D(i)
20.    until forall i Star(i)
21.    end; {MST}
```

For a complete example, the see Figs. 13.14 and 13.15. The status of various variables associated with the algorithm is featured in Fig. 13.16, where the final result is also presented. To summarize, we state the following result [6].

Theorem 7. The task of finding a minimum spanning tree of a weighted undirected graph with n vertices and m edges can be performed in $O(\log n)$ time using $O(n + m)$ processors in the Priority-CRCW. □

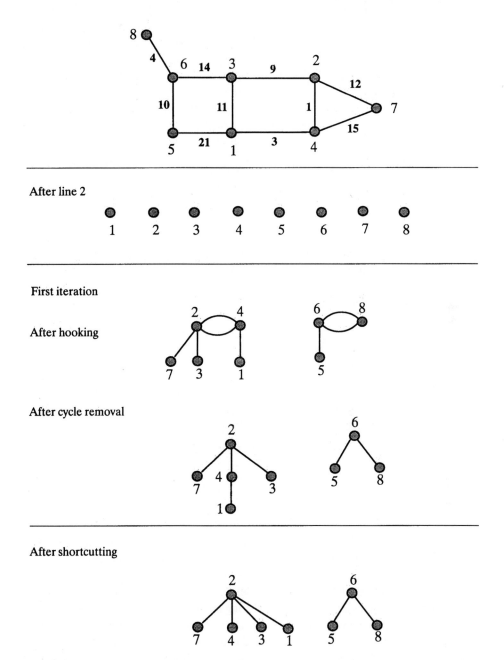

After line 2

First iteration

After hooking

After cycle removal

After shortcutting

Figure 13.13

A number of other parallel minimum spanning tree algorithms have been reported in the literature. For example the connected component algorithm of Hirschberg et al. [21] can be used to solve the MST problem in $O(\log^2 n)$ time using n^2 processors in the CRCW model. The same complexity was reported by Savage and JáJá [34]. Chin et al. [10] have proposed an MST algorithm running in $O(\log^2 n)$ time using only $(n^2/\log^2 n)$ processors in the same model. For more information and a historical perspective, see Ref. [33].

Second iteration

After hooking

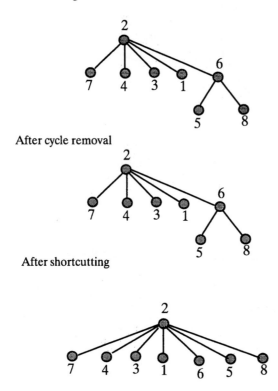

After cycle removal

After shortcutting

Figure 13.14

It is worth noting that, in many contexts, one is interested in computing an arbitrary spanning tree of a given undirected connected graph G. Clearly, the algorithm of Ref. [6] can be used for this purpose with all the weights assumed to be the same. However, as we are about to point out, this computation can be performed in the weaker Arbitrary-CRCW. Specifically, every time a hooking takes place, we would like to add the corresponding edge to the spanning tree. In this weaker model we do not know which processor has succeeded in performing the hooking. The following simple trick allows us to overcome this difficulty. When a processor $P_{i,j}$ succeeds in performing the hooking, it will write its "identity," i.e. (i, j), in a memory location associated with $D(i)$. Thus, the edge (i, j) can be safely added to the spanning tree. Consequently, we state the following well known result.

Theorem 8. The task of finding a spanning tree of an undirected graph with n vertices and m edges can be performed in $O(\log n)$ time using $O(n + m)$ processors in the Arbitrary-CRCW. $\square$

13.3.3 Computing biconnected components

A graph is said to be *biconnected* if the removal of any of its vertices does not disconnect the graph. A biconnected component of a graph is a maximal set of vertices that induces a

First iteration

After hooking

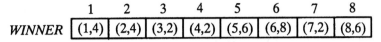

Edges (undirected) added to *FOREST*:

{1,4} {2,4} {3,2} {5,6} {6,8} {7,2}

Corresponding spanning forest

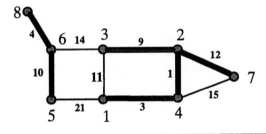

Second iteration

After hooking

	1	2	3	4	5	6	7	8
WINNER	(1,4)	(2,4)	(3,2)	(4,2)	(5,6)	(6,3)	(7,2)	(8,6)

Edges (undirected) added to *FOREST*:

{6,3}

Minimum spanning tree (in heavy edges)

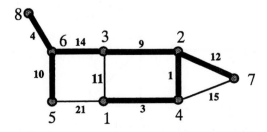

Figure 13.15 Status of various variables associated with the MST algorithm

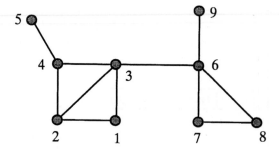

Figure 13.16 A graph G

biconnected graph. Figure 13.17 features all the biconnected components of the graph in Fig. 13.16. A vertex of a graph is termed a cut-vertex (also, an articulation point) if its removal disconnects the graph. It is not hard to see that the cut-vertices are precisely those vertices that belong to two or more of the biconnected components of the graph. As an illustration, vertices 3, 4, and 6 are cut-vertices of the graph in Fig. 13.16. An edge of a graph is termed a bridge if its removal disconnects the graph. The bridges of a graph are precisely the biconnected components with exactly two vertices. For example, the bridges of the graph in Fig. 13.16 are {4, 5}, {3, 6}, and {6, 9}.

The problems of computing the biconnected components, cut-vertices, and bridges of an undirected graph are classic problems of algorithmic graph theory and have received a great deal of attention in the literature [2, 16, 23]. Tarjan's well known sequential algorithm for computing the biconnected components of a graph relies heavily on *depth-first search* [2]. Unfortunately, depth first search seems to be inherently sequential, and so to obtain an efficient parallel algorithm, one has to looks elsewhere.

The purpose of this section is to show that the problem of computing the biconnected components of an undirected graph can be solved efficiently in parallel. For this purpose, we will present the details of the algorithm of Tarjan and Vishkin [38]. As it turns out, the biconnected component algorithm in Ref. [38] can be used to produce the cut-vertices and/ or bridges of the graph.

To make the presentation of the algorithm easier to understand, we take note of an equivalent definition of biconnectedness [38]. Specifically, a graph is biconnected if and only if every one of its edges belongs to a cycle in the graph. This new definition motivates associating with a given undirected graph G a new graph G' whose vertices are the edges

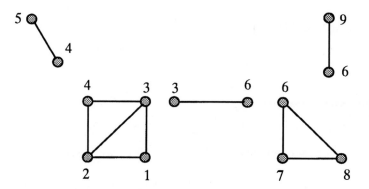

Figure 13.17 The biconnected components of the graph in Fig. 13.16

of G and such that two vertices are adjacent in G' whenever the corresponding edges of G belong to the same cycle. We will first present the sequential construction of G' and then show that the approach can be easily parallelized. Let T be an arbitrary rooted spanning tree of G and proceed to number the vertices of G in preorder as $1, 2, \ldots, n$. The edges of T will be described by ordered pairs (i, j) where $i = p(j)$; i.e., i is the parent of j.

The vertices of G' are (unordered) pairs of the type $\{i, j\}$ with vertices i and j adjacent in G. The construction of the edges of G' is governed by the following three rules:

Rule1: If (u, w) is an edge of T and $\{u, w\}$ is an edge in $G - T$, then $\{\{u, w\}, \{v, w\}\}$ is an edge in G'.

Rule2: If (u, v) and (x, w) are edges of T and $\{v, w\}$ is an edge in $G - T$ such that v and w are unrelated* in T, then $\{\{u, v\}, \{x, w\}\}$ is an edge in G'.

Rule3: If (u, v) and (v, w) are edges of T and some edge of G joins a descendent of w with a nondescendent of v, then $\{\{u, v\}, \{v, w\}\}$ is an edge in G'.

The intuition for these rules is that every edge of $G - T$ completes a cycle consisting of the edge and a path in T. The fundamental property of G' is that a set of edges of G belong to the same biconnected component if and only if the same edges (as vertices of G') belong to the same connected component of G'. Thus, the problem of computing the biconnected components of the graph G is reduced to the problem of computing the connected component of the derived graph G'. It is worth mentioning that the same idea was used in Ref. [40] in the authors' biconnected component algorithm. The reader is referred to Fig. 13.18 for the construction of the graph G' corresponding to the graph G in Fig. 13.16.

We are now in a position to present a sequential algorithm to compute the biconnected components of an undirected graph G.

Step 1 Find a (rooted) spanning tree T of G and denote the vertices of G by their pre-order number in T.

Step 2 Compute the number $nd(v)$ of descendents of every node v of T.

Step 3 For every node v of T, compute $low(v)$ and $high(v)$ defined, respectively, as the vertex with the lowest (resp. highest) preorder number that is a descendent of v or is adjacent to a descendent of v by an edge in $G - T$.

Comment: The computation of $low(v)$ and $high(v)$ can be done efficiently by traversing T in postorder and using the following recurrences:

•$low(v) = min(\{v\} \cup \{low(w) \mid (v, w) \in T\} \cup \{w \mid \{v, w\} \in G - T\})$;

•$high(v) = max(\{v\} \cup \{low(w) \mid (v, w) \in T\} \cup \{w \mid \{v, w\} \in G - T\})$.

Step 4 Construct the graph G'' (a subgraph of G') induced by the edges of T as follows:

•For every edge $\{w, v\}$ such that $v + nd(v) \le w$, add the edge $\{\{p(v), v\}, \{p(w), w\}\}$ to G''.

•For every edge $(v, w) \in T$ such that $v \ne 1$, add the edge $\{\{p(v), v\}, \{v, w\}\}$ to G''.

Comment: The edges added in this step correspond to those added by Rules 2 and 3, respectively.

---

*That is, neither is a descendent of the other.

Step 5   Find the connected components of G''.

Step 6   Complete the construction of G' by adding to G'' edges of the form $\{\{p(w), w\}, \{v, w\}\}$ whenever $\{v, w\}$ is an edge in $G - T$ and v precedes w in preorder.
Comment: The edges added in this step are those added by Rule 1.

It is not hard to see [38] that the above algorithm runs in $O(n + m)$ time whenever the input graph G has n vertices and m edges. We now point out how the various steps of this algorithm can be implemented efficiently in parallel.

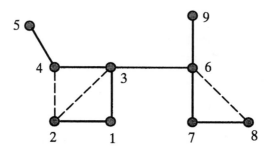

Edges added by Rule 1: $\{\{2,4\},\{3,4\}\}$

$\{\{1,3\},\{2,3\}\}$

$\{\{6,8\},\{7,8\}\}$

Edges added by Rule 2: $\{\{1,3\},\{1,2\}\}$

$\{\{1,2\},\{3,4\}\}$

Edges added by Rule 3: $\{\{6,7\},\{7,8\}\}$

$\{\{1,3\},\{3,4\}\}$

The graph G'

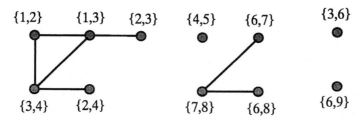

Figure 13.18 Constructing the graph G' corresponding to the graph G in Fig. 13.17

The model of parallel computation assumed is the Arbitrary-CRCW. The graph G is specified as a list of vertices and edges with every edge $\{i, j\}$ specified twice, once as (i, j) and once as (j, i). We assume that every input item (vertex or edge) has its own processor. To make the presentation easier to follow, a different representation for G is desirable. Specifically, we want G represented by its adjacency lists in the usual way. This latter representation can be obtained in $O(\log n)$ time using the processors available by a simple sort operation [38].

We now discuss how the various steps of the above algorithm can be implemented in parallel. Step 1 can be implemented in $O(\log n)$ time using the processors available. First, a rooted spanning tree T of G can be obtained an discussed at the end of the previous subsection. (The root of the tree T corresponds to the root of the resulting star.) Once T is available, we compute an Euler tour as discussed earlier in this chapter, number the nodes in preorder, and compute the number of descendents of every node. Thus, the tasks specific to Step 2 can be implemented in $O(\log n)$ time using the available processors (in fact, $O(n/\log n)$ processors are enough). Next, we argue that Step 3 can be implemented in $O(\log n)$ time using the processors available. We only outline the computation of $low(v)$ for every v. It should be clear that the function $high(v)$ can be computed similarly. Begin by determining for every node v the node $\tilde{v}$ defined by $min(\ \{v\} \cup \{u| \{u, v\} \} \in G - T)$. Once the adjacency list corresponding to every vertex v is available, the computation of $\tilde{v}$ reduces to traversing this list and computing the minimum preorder number for which the edge does not belong to T. Thus, $\tilde{v}$ can be computed in $O(\log n)$ using the processors associated with the adjacency list of v. Once $\tilde{v}$ is available, we compute $low(v) = min_{u \in T(v)}\{\tilde{u}\}$. As it turns out, this computation can be performed in $O(\log n)$ time using all together $O(n/\log n)$ processors by using tree contraction. Step 4 involves the construction of the graph G'' and can be done in $O(1)$ time using the processors available, since testing the appropriate condition for each possible edge of G'' takes $O(1)$ time. Step 5 involves computing the connected components of the graph G''. Using the algorithm of Ref. [6] or [36], this step runs in $O(\log n)$ time using the processors available. It is also easy to see that Step 6 can be performed in $O(1)$ time using $O(m)$ processors. To summarize our findings, we state the following result [38].

Theorem 9. The task of computing the biconnected components of an undirected graph with n vertices and m edges can be performed in $O(\log n)$ time using $n + 2m$ processors in the Arbitrary-CRCW model. $\square$

With simple modifications, the biconnected components algorithm just discussed can be used to detect all the cut-vertices of the graph as well as all the bridges. The reader is referred to the paper of Tarjan and Vishkin [38] for interesting details. We state the following result.

Theorem 10. The task of computing the cut-vertices and/or the bridges of an undirected graph with n vertices and m edges can be performed in $O(\log n)$ time using $n + 2m$ processors in the Arbitrary-CRCW model. $\square$

A number of other parallel algorithms for computing the biconnected components of an undirected and related problems have been reported in the literature. Savage and JáJá [34] as well as Tsin and Chin [40] have obtained $O(\log n)$ time algorithms for this problem of computing the biconnected components using $O(n^2)$ and $O(n^2/\log^2 n)$ processors, respectively. As pointed out by Tarjan and Vishkin [38], their algorithm can also be implemented to run in $O(\log^2 n)$ and $O(n^2/\log^2 n)$ processors in the CREW-PRAM model. A similar algorithm is discussed in Ref. [10]. For more information, see Refs. [23] and [33].

13.4   Algorithms for Particular Classes of Graphs

It is well known that many interesting algorithmic problems are NP-complete on general graphs. Yet, in many applications one does not have to contend with general graphs. Often, there is enough structure in the problem under investigation to restrict the class of graphs modeling the problem. The main goal of this section is to show that problems that are notoriously difficult in the context of general graphs become tractable for particular graph classes. Specifically, we demonstrate cost optimal parallel algorithms for some of these problems in the context of interval graphs and of cographs.

13.4.1   Algorithms for interval graphs

A number of applications in scheduling, VLSI, circuit design, and traffic control suggest associating a graph $G = (I, E)$ with a family $I = \{I_i = [a_i, b_i] \mid a_i \leq b_i, 1 \leq i \leq n\}$ of intervals on the real line. The vertices of G are precisely the intervals in I, with two vertices adjacent if and only if the corresponding intervals overlap. In this context, G is commonly referred to as the intersection graph of I or, simply, the interval graph corresponding to I. Refer to Fig. 13.19 for an example of a family of intervals and the corresponding interval graph. It is customary to say that the family I is the interval model of the graph G.

The purpose of this subsection is to present a number of efficient parallel algorithms for interval graphs. Specifically, we exhibit optimal parallel algorithms for such classic computational tasks as coloring, computing a maximum clique, a maximum independent set, a minimum covering by cliques, and a minimum dominating set. As it turns out [8, 20, 24, 32] that all these algorithms have a sequential lower bound of $\Omega(n \log n)$. For each of these problems, we will exhibit parallel algorithms running in $O(\log n)$ time and using $O(n)$ processors. In the light of the lower bound results mentioned, the algorithms are cost-optimal.

13.4.1.1 Coloring interval graphs. Our first algorithm involves coloring the vertices of an interval graph using the least number of colors. In VLSI, the same problem is known as

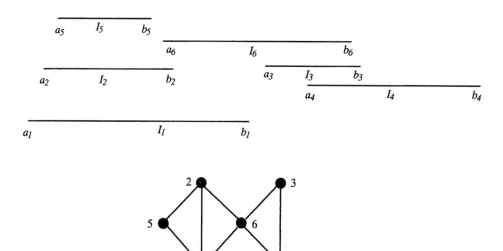

Figure 13.19 A family of intervals and the corresponding interval graph

channel assignment. Consider again a family $I = \{I_i = [a_i, b_i] \mid a_i \leq b_i, 1 \leq i \leq n\}$ of intervals on the real line. A color assignment for the family I is a partition of I into nonempty, disjoint subsets $C_1, C_2, \ldots, C_k$ such that within every set C_i the intervals are pairwise nonoverlapping. Intuitively, it is clear that if we assign the same color to all the intervals belonging to some C_i, then what results is a coloring of I such that no overlapping intervals share the same color. The problem of coloring an interval graph has been studied extensively in the literature. To make the presentation self-contained, we reproduce the details of the optimal sequential coloring algorithm in Ref. [20]. The idea of the algorithm is simple: the intervals are colored sequentially from left to right. As soon as an interval has ended, its color is released (pushed onto a stack) and can be reused for the first interval that starts after the current one ended. The details follow.

Sequential-Algorithm Optimal-Coloring(*I*)

```
{Input: a family I of intervals on the real line;
Output: an optimal coloring for the family I;}
Step 1.
     Let c(1),c(2),...,c(2n) be the left- and right-endpoints sorted in
     ascending order;*
Step 2.
     for i = 1to 2n do
          if c(i) = aₖ then begin
               avail ← pop();
          color(Iₖ) ← avail;
          end
     else {assume c(i) = bₖ′}
          push(color(Iₖ′));
Step 3.
     return(color);
```

We now present a parallel algorithm to color interval graphs that follows the idea of the sequential algorithm of Ref. [20]. There are, however, a number of differences worth noting. First, we need to introduce some terminology. Specifically, let $c(1), c(2), \ldots, c(2n)$ stand for the sorted sequence of the $2n$ endpoints in the family I of intervals. Assign to $c(1)$ the weight $w(1) = 0$, and for all i, ($2 \leq i \leq 2n$) assign to $c(i)$ the weight $w(i)$ defined as follows:

$$w(i) = \begin{cases} 1 & \text{if both } c(i-1) \text{ and } c(i) \text{ are left-endpoints} \\ -1 & \text{if both } c(i-1) \text{ and } c(i) \text{ are right-endpoints} \\ 0 & \text{otherwise} \end{cases} \qquad (13.1)$$

Next, compute the prefix sums of the sequence of weights assigned above and let $e(1)$, $e(2), \ldots, e(2n)$ be the result (i.e., $e(i) = w(1) + w(2) + \ldots + w(i)$). Call two intervals $Ii = [a_i, b_i]$ and $I_j = [a_j, b_j]$ related whenever the following conditions are satisfied: $b_i < a_j$, $e(a_j) = e(b_i)$, and for all endpoints u with $b_i < u < a_j$, $e(u)$ is distinct from $e(a_j)$ and $e(b_i)$. The following simple result follows directly from the definition of related intervals.

Lemma 10. Let $I_{i_1}, I_{i_2}, \ldots, I_{i_k}$ be a sequence of intervals such that every pair of consecutive intervals are related. Then no intervals in this sequence overlap. □

---

*If $c(i) = c(i+1) = \ldots$ we assume that left-endpoints precede right-endpoints.

In fact, it is easy to confirm that the sequential algorithm of Ref. [20] assigns the same color to a sequence $I_{i_1}, I_{i_2}, \ldots, I_{i_k}$ of intervals in the family I such that for all t ($1 \le t \le k$), I_{i_t} and $I_{i_{t+1}}$ are related. Sequentially, finding related intervals is done elegantly by using a stack. In parallel, however, emulating a stack is rather inefficient. Instead, our approach is based on the weighting scheme devised in Ref. [1].

The following result and its consequences are at the heart of our parallel algorithm for interval graph coloring.

Lemma 11. Consider the subsequence (i_1), $c(i_2), \ldots, c(i_r)$ obtained by scanning $c(1)$, $c(2), \ldots, c(2n)$ from left to right and retaining every it with $e(c(i_t)) = k$ for some fixed $k \le max\{e(i) \mid 1 \le i \le 2n\}$. Then, $c(i_t)$ is a left-endpoint or a right-endpoint, depending on whether t is odd or even. $\square$

Let $c_{i_1}, c_{i_2}, \ldots, c_{i_r}$ be as in the statement of Lemma 11. The reader should have no trouble confirming that all intervals I_k, $I_{k'}$ with $c(i_p) = b_k$ and $c(i_{p+1}) = a_{k'}$, are related. In addition, Lemma 11 motivates us to define a linear order $\prec$ on the set of ordered pairs $(e(i), c(i))$ ($1 \le i \le 2n$) such that

$$(e(i), c(i)) \prec (e(j), c(j)) \text{ whenever } e(i) < e(j) \text{ or } e(i) = e(j) \text{ and } c(i) < c(j)$$

For further reference we write $m = max\{e(i) \mid 1 \le i \le 2n\}$. It is not hard to see that the minimum number of colors in a coloring of the family I of intervals is $m + 1$. Now consider the sequence $\{(e(i), c(i)) \mid 1 \le i \le 2n\}$ sorted by $\prec$:

$$(e(i_1), c(i_1)), (e(i_2), c(i_2)), \ldots, (e(i_{2n}), c(i_{2n}))$$

The details of the parallel coloring algorithm for interval graphs are spelled out below.

Procedure Optimal-Coloring(*I*)

```
{Input: a family I of intervals on the real line;
Output: an optimal coloring for the family I;}
begin
1.    let c(1), c(2),...,c(2n) be the left and right-endpoints sorted in as-
      cending order;
2.    assign weights w(1),w(2),...,w(2n) as in Ref. [1];
3.    compute the prefix sums of the sequence w(1), w(2),...,w(2n) and let
```

$$\text{the result be } e(1), e(2), \ldots, e(2n) \; (\text{i.e., } e(i) = \sum_{j=1}^{i} w(i) ;$$

```
4.    let (e(i₁), c(i₁)), (e(i₂), c(i₂)),...,(e(i₂ₙ,), c(i₂ₙ)) be sorted as in
      Ref. [2];
5.    mark the interval having c(i₁) as its left-endpoint;
6.    set link(Iₜ) ← nil for bₜ = c(i₂ₙ);
7.    for all even j (2 ≤ j < 2n) pardo
8.    if e(iⱼ) ≠ e(iⱼ₊₁) then begin
9.          mark the interval having c(ij+1) as its left-endpoint;
10.         set link(Iₖ) ← nil where c(iⱼ) = bₖ
11.       end
12.   else
13.         set link(Iₖ) ← Iₖ' where C(iⱼ) = bₖ and c(iⱼ₊₁ = aₖ'
14.   endfor;
```

```
15.   let I_{k0}, I_{k1}, ..., I_{km} be the marked intervals obtained in lines 5 and 9;
16.   for i = 0 to m pardo
17.       transmit i to all the intervals on the linked list starting at I_k;
18.   end; {Optimal-Coloring}
```

The correctness being obvious we now briefly address the complexity. Lines 1 and 4 run in $O(\log n)$ time using $O(n)$ processors [11], line 3 runs in $O(\log n)$ time using n processors (in fact, $n/\log n$ processors suffice), and the loop in lines 7 through 14 runs in constant time. Finally, line 17 is a variant of list ranking and can be done in $O(\log n)$ time using $(n/\log n)$ processors. Given the $\Omega(n \log n)$ lower bound for the problem, it follows that the parallel algorithm is cost-optimal. The following result summarizes the previous discussion.

Theorem 11. Procedure Optimal-Coloring correctly colors the family I of n intervals with a minimum number of colors in $O(\log n)$ time using $O(n)$ processors in the EREW-PRAM model of computation. □

For the reader's benefit, a worked example is presented next. Consider the family of intervals featured in Fig. 13.20. The topmost part of Table 13.1 summarizes the various variables of procedure Optimal-Coloring at the end of line 3, i.e., after the prefix sums have been computed. The middle part of Table 13.1 shows the sequence (2) computed in line 4 of the procedure. Finally, the bottommost part of Table 13.1 features the values of the various link fields when the for loop in lines 7 through 14 is exited.

For the resulting color assignment, refer to Table 13.2.

13.4.1.2 Computing a maximum clique. Recall that a clique in a graph is a set of pairwise adjacent vertices. The problem of computing the largest size of a clique is one of the classic problems in algorithmic graph theory. The goal of this subsection is to present a cost-optimal algorithm for computing a maximum clique in an interval graph. Our presentation follows the algorithm of [17]. The algorithm proceeds in the following sequence of steps.

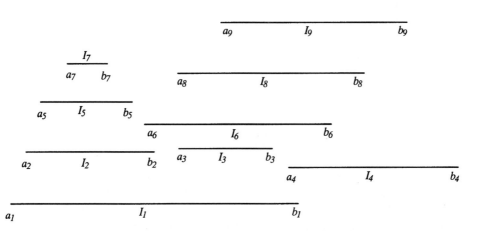

Figure 13.20 A family of intervals

TABLE 13.1 Illustration of Optimal-Coloring on the family of intervals in Figure 13.20

*	1	2	3	4	5	6	7	8	9	10	11	12	13	14	15	16	17	18
c	a_1	a_2	a_5	a_7	b_7	b_5	a_6	b_2	a_8	a_3	a_9	b_3	a_4	b_1	b_6	b_8	b_9	b_4
w	0	1	1	1	0	−1	0	0	0	1	1	0	0	0	−1	−1	−1	−1
e	0	1	2	3	3	2	2	2	2	2	3	4	4	4	3	2	1	0

*	1	2	3	4	5	6	7	8	9	10	11	12	13	14	15	16	17	18
e	0	0	1	1	2	2	2	2	2	2	3	3	3	3	4	4	4	4
c	a_1	b_4	a_2	b_9	a_5	b_5	a_6	b_2	a_8	b_8	a_7	b_7	a_3	b_6	a_9	b_3	a_4	b_1

I	I_{1*}	I_{2*}	I_{3*}	I_{4*}	I_{5*}	I_{6*}	I_{7*}	I_{8*}	I_{9*}
link	nil	I_8	I_4	nil	I_6	nil	I_3	nil	nil

TABLE 13.2 Colors Assigned by Optimal-Coloring

Color	Set of Intervals
0	$I_1 \rightarrow$ nil
1	$I_2 \rightarrow I_8 \rightarrow$ nil
2	$I_5 \rightarrow I_6 \rightarrow$ nil
3	$I_7 \rightarrow I_2 \rightarrow I_4 \rightarrow$ nil
4	$I_9 \rightarrow$ nil

Algorithm Max Clique(G)

Step 1. Sort all the 2n endpoints in increasing order as c_1, c_2, . . . ,c_{2n};
Step 2. Assign to each c_i a weight w_i defined by

$$
w_i = \begin{cases} 1 \text{ if } c_i = a_j \text{ for some } 1 \le j \le n; \\ -1 \text{ if } c_i = b_k \text{ for some } 1 \le k \le n; \end{cases}
$$

Step 3. Compute the prefix sums of the resulting weighted sequence and let
d_1, d_2, . . . ,d_{2n} be the result;
Step 4. Consider the sequence e_1, e_2, . . . ,e_{2n} obtained by replacing every d_j
corresponding to a right endpoint of an interval in I with −1 and compute
the leftmost maximum of the resulting sequence;
Step 5. Let the leftmost maximum returned in Step 4 occur at e_k and let c_k =
a_i; to determine all the intervals that belong to the maximum clique,
broadcast ai to all the processors containing intervals I_u satisfying
$a_u < a_i < b_u$. Every such processor marks itself;
Step 6. Finally, assigning every marked interval a label of 0 and every non-
marked interval a label of 1, we can use prefix sums to compact all the
marked intervals.

The correctness of this simple algorithm follows immediately from the correctness of
the algorithm of Ref. [17]. To argue for its running time, we note that the sorting operation
in Step 1 takes $O(\log n)$ time using $O(n)$ processors [11]. Steps 2 and 4 run in constant

time; Steps 3, 5, and 6 involve computing prefix operations and therefore run in $O(\log n)$ time using $O(n/\log n)$ processors. To summarize our findings we state the following result.

Theorem 12. The task of computing the maximum clique of an n-vertex interval graph specified by its interval model can be performed in $O(\log n)$ time using $O(n)$ processors in the EREW-PRAM model. □

The reader can find a worked example below. We refer again to the family of intervals depicted in Fig. 13.20. The status of the computation after the execution of Step 4 of the above algorithm is captured in Table 3. Referring to the last row of the table, we note that the leftmost maximum computed in Step 5 occurs at e_{11} and corresponds to a_9. (It is important to note that the value of the leftmost maximum computed in Step 5 provides the size of the largest clique in the corresponding interval graph; thus, in our case, the maximum clique contains 5 intervals.) To produce the intervals belonging to this maximum clique, certain intervals are marked in Step 5. In the case of the example, these are intervals I_1, I_3, I_6, I_8, and I_9.

TABLE 13.3 Computation of the Maximum Clique

*	1	2	3	4	5	6	7	8	9	10	11	12	13	14	15	16	17	18
c	a_1	a_2	a_5	a_7	b_7	b_5	a_6	b_2	a_8	a_3	a_9	b_3	a_4	b_1	b_6	b_8	b_9	b_4
w	1	1	1	1	−1	−1	1	−1	1	1	1	−1	1	−1	−1	−1	−1	−1
d	1	2	3	4	3	2	3	2	3	4	5	4	5	4	3	2	1	0
e	1	2	3	4	−1	−1	3	−1	3	4	5	−1	5	−1	−1	−1	−1	−1

13.4.1.3 Computing a maximum independent set.

Our goal is to present a cost optimal algorithm to compute a maximum size independent set in an interval G specified by its interval model $I = \{I_i = [a_i, b_i] \mid a_i \le b_i, 1 \le i \le n\}$. Equivalently, we are interested in finding a largest set of mutually non-overlapping intervals in the family I. In our arguments we shall find it convenient to rely on the parameter first(I) of the family I defined as follows:

$$\text{first}(I) = I_j \text{ with } b_j = \min\{b_i \mid 1 \le i \le n\}$$

In other words, first(I) is the interval of the family whose right-endpoint is farthest to the left. It is worth noting that the computation of first(I) reduces to computing prefix maxima and minima over the sequence of right endpoints of intervals in the family I. Consequently, we have the following result.

Lemma 12. Given a family $I = \{I_i = [a_i, b_i] \mid a_i \le b_i, 1 \le i \le n\}$ of intervals, first(I) can be computed in $O(\log n)$ time using $O(n/\log n)$ processors in the EREW-PRAM. □

In addition, for every interval Ii in the family I define the following parameter:

$$\text{next } (I_i) = \begin{cases} I_j & \text{if } b_j = \min \ \{b_k \mid (b_i < a_k)\} \\ \text{nil} & \text{otherwise} \end{cases}$$

In other words, next(I_i) is the interval that ends farthest to the left among all the intervals beginning after the end of I_i. We now show that for every i $(1 \leq i \leq n)$, next(I_i) can be computed in $O(\log n)$ time using $O(n)$ processors in the EREW-PRAM. The details of our algorithm for computing next(I_i) are spelled out as follows.

Algorithm Find Next(*I*)

```
Step 1. Sort the left-endpoints of intervals in I in ascending order as a'₁,
    a'₂,...,a'ₙ;
Step 2. Compute the prefix minimum of the sequence of right-endpoints b'ₙ,
    b'ₙ₋₁,...,b'₁ and let the result be b"₁, b"₂,...,b"ₙ
    (i.e., b"ₖ = minₖ≤ᵢ≤ₙ {b'ᵢ});
Step 3. For every i (1 ≤ i ≤ n) compute the rank r(i) of bᵢ with respect to
    a'₁, a'₂,...,a'ₙ;
Step 4. For every i (1 ≤ i ≤ n), in case 1 ≤ r(i) ≤ n - 1 set next(Iᵢ) = Iⱼ
    such that b"ᵣ₍ᵢ₎₊₁; set next(Iᵢ) = nil in case r(i) = n.
```

It is easy to see that the above steps implement the definition of next(I_i). To argue about the complexity, Step 1 can be performed in $O(\log n)$ time using the sorting algorithm in Ref. [11]. Step 2 involves a prefix computation, and Steps 3 and 4 take $O(1)$ time using n processors. Consequently, we have the following result.

Lemma 13. Given a family of intervals $I = \{I_i = [a_i, b_i] \mid 1 \leq i \leq n\}$, for all i $(1 \leq i \leq n)$, next(I_i) can be computed in $O(\log n)$ time using $O(n)$ processors in the EREW-PRAM model. □

We now return to the algorithm for computing a maximum independent set in an interval graph specified by a family of intervals. The idea of the algorithm is adapted from Ref. [8]. Specifically, it can be shown that there exists a maximum size independent set S in I such that first(I) $\in S$, and for every I_i that belongs to S, next(I_i) also belongs to S. This motivates us to proceed in as follows.

Algorithm Max_Independent_Set(*G*)

```
Step 1. Compute first(I);
Step 2. For every i (1 ≤ i ≤ n), compute next(Iᵢ);
{Comment: we shall find it convenient to interpret the family I as a forest
    F considering next(Iᵢ) to be a parent-pointer.}
Step 3. Mark all the elements on the path joining first(I) to the root of
    the tree containing first(I).
Step 4. Finally, assigning every marked interval a label of 0 and every non-
    marked interval a label of 1, we use sorting to compact all the marked
    intervals.
```

Let L be the linked list containing the marked elements at the end of Step 2. It is important to note that the definition of next(I_i) guarantees that all the entries in L are mutually non-overlapping. To argue for the complexity of this simple algorithm, note that Steps 1 and 2 run in $O(\log n)$ time using $O(n)$ processors. Step 3 involves marking the nodes on the path from a leaf to the root and can be done in $O(\log n)$ time using $O(n/\log n)$ processors as we discussed previously. Finally, Step 4 is an instance of prefix sums computation and runs in $O(\log n)$ time using $O(n/\log n)$ processors. To summarize, we state the following result.

Theorem 13. Given an arbitrary interval graph G specified by a family $I = \{I_i = [a_i, b_i] \mid a_i \le b_i, 1 \le i \le n\}$ of intervals, a maximum size independent set of G can computed in $O(\log n)$ time using $O(n)$ processors in the EREW-PRAM model. $\square$

As before, we will illustrate the working of this algorithm on the interval graph whose corresponding family of intervals is featured in Fig. 13.20. We begin by illustrating the construction of next(I_i). The details can be found in Table 13.4. Furthermore, Fig. 13.21 shows the tree resulting from the pointer assignment in Step 2 of Max_Independent_Set. Figure 13.21 also shows in heavy lines the path in the tree from first(I) = I_7 to the root. Thus, a maximum independent set in the corresponding interval graph consists of the intervals I_7, I_3, and I_4.

TABLE 13.4 Computation of next (I_i)

*	1	2	3	4	5	6	7	8	9
a'_i	a_1	a_2	a_5	a_7	a_6	a_8	a_3	a_9	a_4
b''_i	b_7	b_7	b_7	b_7	b_3	b_3	b_3	b_9	b_4
$r()$	9	5	8	9	4	9	4	9	9
next()	nil	I_3	I_4	nil	I_3	nil	I_3	nil	nil

13.4.1.4 Computing a minimum clique cover. Recall that a minimum clique cover in a graph G is a minimum cardinality partition of the vertices of G into non-empty sets such that each element of the partition is a clique. Since interval graphs are perfect [9], the cardinality of a minimum clique cover is equal to the cardinality of a maximum independent set. In addition, we can readily obtain a minimum clique cover once a maximum independent set is available (see Ref. [8] for details). Specifically, if $S = \{I_{k_1}, I_{k_2}, \ldots, I_{k_m}\}$ is a maximum independent set in I, then there exists a minimum clique cover $C_1, C_2, \ldots, C_m$ of size *exactly* m, such that

- $I_{k_i} \in C_i$ for all $1 \le i \le m$;
- for every interval $I_u \notin S$, I_u belongs to some clique C_3 if and only if $b_{k_j} = \max\{b_{k_i} \mid I_{k_t}$ and $a_u \le b_{k_t} \le b_u\}$.

The idea of the algorithm is motivated by the above discussion. Specifically, for every interval in I the algorithm will return the identity of the clique in a minimum cover by

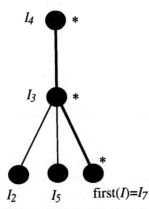

I_4 *

I_3 *

I_2 I_5 first(I)=I_7 *

Figure 13.21 The tree containing first(I) after Step 2

cliques that the interval belongs to. The algorithm proceeds in the following sequence of steps.

Algorithm Min_Clique_Cover(G)

```
Step 1. Compute a maximum independent set S = {I_k1, I_k2, ...,I_km} in I;
Step 2. Sort the sequence b1, b2, ...,bn of right endpoints in increasing
        order and call it d1, d2, ...,dn;
Step 3. Assign to every di a weight wi of 1 if Ii belongs to S and of 0 oth-
        erwise; compute the prefix sums of the resulting weighted sequence and let
        e1, e2, ...,en be the result;
Step 4. Finally, sort the intervals by the value of the corresponding ei com-
        puted in Step 3.
```

It is easy to see that for every interval I_i, the value of the prefix sum returned in Step 3 is the subscript of the clique in the minimum clique cover to which the interval belongs. Therefore, at the end of Step 4, a minimum clique cover is available in the first row of the mesh in left-to-right order. To argue for the running time, observe that by Theorem 13, Step 1 runs in $O(\log n)$ time using $O(n)$ processors, Steps 2 and 4 have the same complexity, and Step 3 involves prefix computations and runs in $O(\log n)$ time using $O(n/\log n)$ processors. Thus, we can state the following result.

Theorem 14. Given an arbitrary n-vertex interval graph specified by a family $I = \{I_i = [a_i, b_i] \mid a_i \le b_i, 1 \le i \le n\}$ of intervals, a minimum clique cover of the graph can be computed in $O(\log n)$ time using $O(n)$ processors in the EREW-PRAM model. □

For the reader's benefit, we shall illustrate the working of the minimum clique cover algorithm that we have just discussed on the interval graph whose corresponding family of intervals is featured in Fig. 13.20. We assume that a maximum independent set corresponding to this graph has been computed by the procedure presented in the previous subsection. Let this maximum independent set be $S = \{I_7, I_3, I_4\}$. A detailed account of all status of the computation can be found in Table 13.5. The last row in this table specifies the cliques in the resulting minimum clique cover. More specifically, the size of any minimum clique cover for the graph in Fig. 13.20 is 3. Algorithm Min_Clique_Cover returns the following cliques:

$$C_1 = \{I_7, I_5, I_2\};$$
$$C_2 = \{I_3, I_1, I_6, I_8, I_9\};$$
$$C_3 = \{I_4\}$$

TABLE 13.5 Computation of a minimum clique cover

*	1	2	3	4	5	6	7	8	9
d_i	b_7	b_5	b_2	b_3	b_1	b_6	b_8	b_9	b_4
w_i	1	0	0	1	0	0	0	0	1
e_i	1	1	1	2	2	2	2	2	3

13.4.1.5 Computing a minimum dominating set.

Recall that a subset D of vertices of a graph is said to be dominating if every vertex outside D is adjacent to some vertex in D. A dominating set is minimum if it has minimum cardinality among all dominating sets.

Consider an interval graph specified by a family $I = \{I_i = [a_i, b_i] \mid a_i \leq b_i, 1 \leq i \leq n\}$ of intervals on the real line. Following Ref. [8], define for every interval I_i:

$$\text{right}(I_i) = \begin{cases} I_j & \text{if } b_j \ \max\{b_k \mid (a_k \leq b_i < b_k)\} \\ \text{nil} & \text{otherwise} \end{cases}$$

That is, $\text{right}(I_i)$ is the interval ending farthest to the right among all intervals that overlap with I_i, if such an interval exists.

We now show that $\text{right}(I_i)$ can be computed in $O(\log n)$ time using $O(n)$ processors in the EREW-PRAM model. The details of our algorithm for computing $\text{right}(I_i)$ follow.

Algorithm Find_Right(*I*)

```
Step 1. Sort the left-endpoints of intervals in I in ascending order as a'₁,
    a'₂,...,a'ₙ;
Step 2. Compute the prefix maxima of the corresponding sequence b'₁,
    b'₂,...,b'ₙ of right endpoints and let the result be b"₁, b"₂,...,b"ₙ (i.e.,
    b"ₖ = max₁≤ᵢ≤ₖ {b'ᵢ});
Step 3. For every i (1 < i < n) compute the rank r(i) of bᵢ with respect to
    a'₁, a'₂,...,a'ₙ;
Step 4. For every i (1 < i < n), in case b"ᵢ > bᵢ set right(Iᵢ) = Iⱼ such that
    bⱼ = b"ᵣ₍ᵢ₎; otherwise set right(Iᵢ) = nil.
```

It is easy to see that the above steps implement the definition of $\text{right}(I_i)$ given above. We leave it to the reader to verify that all the steps can be performed in $O(\log n)$ time using $O(n)$ processors. Consequently, we have the following result.

Lemma 14. Given a family of intervals $I = \{I_i = [a_i, b_i] \mid 1 \leq i \leq n\}$, for all i $(1 \leq i \leq n)$, $\text{right}(I_i)$ can be computed in $O(\log n)$ time using $O(n)$ processors in the EREW-PRAM. $\square$

The idea behind our parallel algorithm to compute a minimum dominating set is borrowed from Ref. [8]. We implement the corresponding idea in the following sequence of steps.

Algorithm_Min_Dominating_Set(*G*)

```
Step 1. Compute first(I);
Step 2. For all i (1 ≤ i ≤ n), compute right(Iᵢ);
Step 3. For all i (1 ≤ i ≤ n), compute next(Iᵢ);
Step 4. For every interval Iⱼ compute
```

$$\text{link}(I_j) = \begin{cases} \text{right}(\text{next}(I_j)) & \text{if right}(\text{next}(I_j)) \neq \text{nil} \\ (\text{next}(I_j)) & \text{otherwise} \end{cases}$$

```
Comment: It is immediate that interpreting link(Iⱼ) (1 ≤ j ≤ n) as parent-
    pointers, I is organized as a collection F of trees.
Step 5. Let u denote right(first(I)) or first(I) depending on whether or not
    right(first(I)) ≠ nil; identify the nodes on the unique path l joining u
    and the root of the tree containing u.
Step 6. Now, assigning every marked interval a label of 0 and every non-
    marked interval a label of 1, we can use sorting to compact all the marked
    intervals.
```

The correctness of our algorithm follows immediately from the correctness of the sequential algorithm discussed in Ref. [8]. To address the complexity, note that Steps 1 and 2 take $O(\log n)$ time using at most $O(n)$ processors. Steps 3 and 4 run in constant time using $O(n)$ processors. Consequently, we have proven the following result.

Theorem 15. Given an arbitrary n-vertex interval graph specified by a family $I = \{I_i = [a_i, b_i] \mid a_i \leq b_i, 1 \leq i \leq n\}$ of intervals, a minimum dominating set can be computed in $O(\log n)$ time using $O(n)$ processors in the EREW-PRAM. $\square$

For the reader's benefit, we will illustrate the working of this algorithm on the interval graph whose corresponding family of intervals is featured in Fig. 13.20. We begin by illustrating the construction of right(I_j). The details can be found in Table 13.6. Furthermore, Table 13.7 gives the details of the pointer assignment in Step 4 of Min_Dominating_Set. Finally, Fig. 13.22 shows the tree resulting from the pointer assignment in Step 4 of Min_Dominating_Set. Figure 13.22 also shows in heavy lines the path in the tree from first(I) = I_7 to the root. Thus, a minimum dominating set in the corresponding interval graph consists of the intervals I_7 and I_9.

TABLE 13.6 Computation of right (I_i)

*	1	2	3	4	5	6	7	8	9
a'_i	a_1	a_2	a_5	a_7	a_6	a_8	a_3	a_9	a_4
b''_i	b_1	b_1	b_1	b_1	b_6	b_8	b_8	b_9	b_4
$r(i)$	9	5	8	9	4	9	4	9	9
right(I_i)	I_4	I_6	I_9	nil	I_1	I_4	I_1	I_4	I_4

TABLE 13.7 Pointer Assignment in Step 4

*	1	2	3	4	5	6	7	8	9
next(i)	nil	I_3	I_4	nil	I_3	nil	I_3	nil	nil
link(i)	nil	I_9	I_4	nil	I_4	nil	I_9	nil	nil

13.4.2 Algorithms for cographs

A well-known class of graphs arising in a wide spectrum of practical applications is the class of cographs, or complement-reducible graphs. The cographs are defined recursively as follows:

- A single-vertex graph is a cograph.
- If G is a cograph, then its complement G is also a cograph.
- If G and H are cographs, then their union is also a cograph.

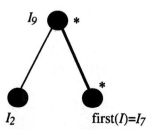

Figure 13.22 The tree containing first(I) after Step 2

As it turns out, the cographs admit a unique tree representation up to isomorphism. Specifically, one can associate with every cograph G a unique rooted tree $T(G)$ called the cotree of G, featuring the following properties:

(c1) Every internal node of $T(G)$ has at least two children.
(c2) The internal nodes of $T(G)$ are labeled by either 0 (0-nodes) or 1 (1-nodes) in such a way that labels alternate along every path starting at the root.
(c3) The leaves of $T(G)$ are precisely the vertices of G; vertices x and y are adjacent in G if and only if the lowest common ancestor of x and y in $T(G)$ is a 1-node.

Figure 13.23 features a cograph and the corresponding cotree.

13.4.2.1 A matching algorithm. Recall that a *matching* in a graph is a set of edges with the property that no two of them share a common endpoint. The size of a matching is the number of edges it contains. A matching is maximum if its size is as large as possible. The matching problem is to find a maximum matching in a given graph G. The literature on matching is extensive. Matching problems are related to flow problems, covering problems, and scheduling, to name just a few. Although in a sequential setting computing a maximum matching in a general graph can be solved in time polynomial in the size of the graph, parallel algorithms to compute a maximum matching in general graphs have been studied, but only with moderate success. In Ref. [27], a randomized parallel algorithm running in $O(\log^2 n)$ expected time determines a maximum matching for general n-vertex graphs. The fastest deterministic algorithm [23] for general graphs runs in $O(n \log n)$ time and uses $O(n^2)$ processors. Given the status of the maximum matching problem for general graphs, it makes sense to study parallel algorithms to compute a maximum matching in particular classes of graphs.

The purpose of this subsection is to show that the problem of finding a maximum matching in a cograph can be solved optimally in parallel by reducing it to tree contraction and to parenthesis matching. Specifically, quite recently Lin and Olariu [30] demonstrated that, for an n-vertex cograph represented by its cotree, the task of computing a maximum matching can be performed in $O(\log n)$ EREW-PRAM time using $O(n/\log n)$ processors and, therefore, it is cost-optimal.

To set the stage for the algorithm in Ref. [30], let $G = (V, E)$ be an arbitrary graph whose vertex set V partitions into disjoint sets A and B with $|A| \le |B|$, such that every vertex in A

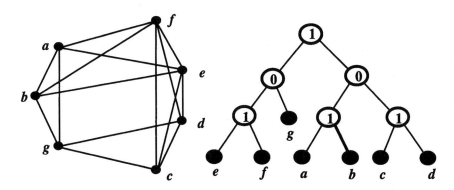

Figure 13.23 A cograph and the corresponding tree

is adjacent to all the vertices in B. Let $M(B)$ be a maximum matching in the subgraph of G induced by B, and let X stand for the set of all exposed vertices in B. Consider the set M of edges constructed by the following greedy procedure.

Procedure Greedy
```
begin
1.    if | A |≤| X | then
2.          find a bijection f from A to a subset of X;
3.          M ← M(B) ∪ {{a, f (a)} | a ∈ A}
4.    else
5.          let C be a subset of M(B) of size ⌈|A| − |X|/2⌉
6.          find a bijection f from A to a subset of X ∪ {u, v | the edge
            {u,v}belongs to C};
7.          M ← (M(B) − C) ∪ {{a, f(a)} | a ∈ A}
8.          return(M)
9.    end; {Greedy}
```

As it turns out, the following result is at the heart of our optimal parallel algorithm to compute a maximum matching in a cograph.

Theorem 16. The set M of edges returned by procedure Greedy is a maximum matching in G. □

We are now in a position to be more explicit about the matching algorithm in Ref. [30]. For this purpose, consider an n-vertex cograph G represented by its cotree $T(G)$. We binarize $T(G)$ as described in earlier and let $BT(G)$ stand for the resulting binary tree. For each node x of $BT(G)$, we let $BT(x)$ stand for the subtree of $BT(G)$ rooted at x; $L(x)$ stands for the set of all the leaves in $BT(x)$, and $G(x)$ represents the subgraph of G induced by $L(x)$. Throughout, we shall deal with a generic node x of $BT(G)$. Whenever this happens, it will be assumed that x has left and right children y and z, respectively. The algorithm in Ref. [30] proceeds in the four stages that we discuss next.

Stage 1. Perform an Euler-tour of $BT(G)$ and use the information returned to compute for every internal node 2 of $BT(G)$ the cardinality of $L(x)$. For every 1-node x of $BT(G)$, ensure that its left subtree contains at most as many leaves as the right subtree. Furthermore, every left child of a 1-node in $BT(G)$ is marked. For convenience, we continue to refer to the new tree as $BT(G)$. Call a node v of $BT(G)$ special if some ancestor of v has been marked above. Nodes of $BT(G)$, which are not special, are referred to as *regular*. Next, performing an Euler-tour, again mark all the special nodes in $BT(G)$, construct a linked list of all the leaves in $BT(G)$, number the leaves in preorder as $1, 2,\ldots,n$, and recompute the set $L(v)$ for every node v.

Stage 2. Invoke the tree contraction algorithm discussed previously to construct an optimal contraction sequence $T_1, T_2,\ldots,T \log n$ for $T(G)$. This sequence will be used explicitly in the next stage. In addition, we associate with every node x of $BT(G)$ a non-negative integer $m(x)$ as follows.

1. If x is a regular leaf, then $m(x) \leftarrow 0$.
2. If x is a regular 0-node, then $m(x) \leftarrow m(y) + m(z)$.

3. If x is a regular 1-node $\{y$ special$\}$, then

$$m(x) \leftarrow \begin{cases} m(z) + |L(y)| & \text{if } |L(y)| \leq |L(z)| - 2m(z) \\ \left\lfloor \left| \dfrac{L(x)}{2} \right| \right\rfloor & \text{otherwise} \end{cases}$$

4. If x is special, then $m(x) \leftarrow |L(x)|$.

The interpretation of $m(x)$ is given by the following result.

Lemma 15. For every regular node x of $BT(G)$, $m(x)$ represents the size of a maximum matching in $G(x)$. In particular, the value $m(root)$ computed at the root of $BT(G)$ gives the size of a maximum matching in G. $\square$

Stage 3. In the third stage, using the contraction sequence obtained in Stage 2, the tree $BT(G)$ is traversed from the root down to the leaves, and we distribute to various nodes in $BT(G)$ three types of objects:

- (α, β) units, which are pairs of α-tokens and β-tokens that are treated as a unit
- α-tokens
- β-tokens

During the execution of Stage 3 the (α, β)-units will be broken. When this happens, the tokens contained in the broken units are released. Every regular node x of $BT(G)$ will receive $U(x)$ (α, β)-units, $A(x)$ α-tokens, and $B(x)$ β-tokens. Eventually, every leaf of $BT(G)$ will receive at most one such token. To distribute these objects, we traverse the sequence $T_1, \ldots, T_{\log n}$ backward, allocating tokens to the nodes of $BT(G)$ in the following recursive way:

- $U(root) \leftarrow m(root)$ (α, β)-units
- $A(root) \leftarrow 0$ α-tokens
- $B(root) \leftarrow 0$ β-tokens

Every regular node x of $BT(G)$ transmits to its children:
if x is a 1-node **then**
    $U(y) \leftarrow 0$ (α, β)-units;
    $A(y) \leftarrow \min\{|L(y)|, U(x)\}$ α-tokens;
    $B(y) \leftarrow \min\{B(x), |L(y)| - A(y)\}$ β-tokens;
    $U(z) \leftarrow U(x) - A(y)$ (α, β)-units;
    $A(z) \leftarrow 0$ α-tokens;
    $B(z) \leftarrow B(x) - B(y) + A(y)$ β-tokens
else $\{$now x is a regular 0-node$\}$
    $U(y) \leftarrow \min\{m(y), U(x)\}$ (α, β)-units;
    $A(y) \leftarrow 0$ α-tokens;
    $B(y) \leftarrow \min\{B(x), |L(y)| - 2U(y)\}$ β-tokens;
    $U(z) \leftarrow U(x) - U(y)$ (α, β)-units;
    $A(z) \leftarrow 0$ α-tokens;
    $B(z) \leftarrow B(x) - B(y)$ β-tokens

A similar distribution scheme works for special nodes. Specifically, let x be a special node in $BT(G)$ with left and right children y and z, respectively. Now, x will distribute tokens to its children according to the following scheme.

- $B(y) \leftarrow \min\{B(x), |L(y)|\}$ β-tokens
- $A(y) \leftarrow \min\{A(x), |L(y)|B(y)\}$ α-tokens
- $A(z) \leftarrow A(x) - A(y)$ α-tokens
- $B(z) \leftarrow B(x) - B(y)$ β-tokens

Stage 4. The purpose of this stage is to match leaves that have received an α–token with leaves that have received a β–token in Stage 3. This is accomplished essentially by calling an optimal parenthesis matching algorithm. As proven in Ref. [30], what results is a maximum matching in G.

The correctness of this algorithm follows by a number of technical results in Ref. [30]. To argue for the complexity, note that Stage 1 can be trivially performed in $O(\log n)$ time using $O(n/\log n)$ processors. Next, Stage 3 can be performed in the same time and processor bounds since at every node the tokens can be distributed in $O(1)$ time. Furthermore, by using list ranking and an optimal parallel parenthesis matching algorithm, Stage 4 has the same complexity.

To make the presentation self-contained, we now argue that Stage 2 also requires $O(\log n)$ time and $O(n/\log n)$ processors. Specifically, we have to show that computing $m(root)$ can be done in the time of tree contraction. To show that this is the case, at each node u of $BT(G)$, we store an expression of the form $f_{u(x)}$ where x is an indeterminate representing the value of the unevaluated $m(u)$.

Referring to Fig. 13.24a, suppose that u has two children v and w, where w is a leaf (i.e., the value of $m(w)$ is a known constant c), and let $f_{u(x)}$ and $f_{v(x)}$ be the expressions stored by u and v, respectively. Let u contain an operation #. After prune(w) and bypass(u) have been performed, the reconstructed tree is as shown in Fig. 13.24b. Note that the expression stored at v is now $f_u(f_v(x)\#c)$.

As noted in Ref. [30], the crucial observation is that, for the purpose of computing $m(u)$ at every node u of $BT(G)$, the expression stored at u is of the form $f_u = \min\{a, x + b\}$, where a and b are constants. Initially, every node u of $BT(G)$ stores

$$f_u = \left\lfloor \frac{|L(u)|}{2} \right\rfloor$$

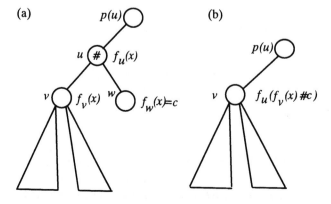

Figure 13.24 Process of updating a functional form

Note, in particular, that in this scheme every leaf of $BT(G)$ stores 0 as it should. To summarize this discussion, we state the following result [30]:

Theorem 17. Given an n-vertex cograph G specified by its cotree, a maximum matching in G can be computed in $O(\log n)$ time using $O(n/\log n)$ processors in the EREW-PRAM.

□

13.4.2.2 Computing a vertex cover for cographs. Recall that given a graph G, the vertex cover problem involves finding a smallest set C of vertices of G such that every edge in G in incident with at least one vertex in C. It is well known that the vertex cover problem is NP-complete for general graphs [18]. The goal of this subsection is to show that the vertex cover problem can be solved optimally in parallel for cographs. Specifically, with an n-vertex cograph G represented by its cotree as input, one can devise an algorithm that finds a minimum vertex cover in G in $O(\log n)$ time using $O(n/\log n)$ processors in the EREW-PRAM model.

The algorithm that we are about to discuss relies on a technical result that we present next. Let $G = (V, E)$ be a graph whose vertex set V partitions into disjoint sets X and Y, such that every vertex in X is adjacent to all the vertices in Y. Let $C(X), C(Y)$ be minimum vertex covers in the subgraphs of G induced by X and Y, respectively. As it turns out, a minimum vertex cover for G can be obtained from $C(X)$ and $C(Y)$ by the following greedy procedure.

Procedure Vertex_Cover(G)
```
begin
1.    if | X ∪ C(Y) |≤| X ∪ C(Y) | then
2.           C ← X ∪ C(Y);
3.    else
4.           C ← Y∪ C(X);
5.    return(C)
6.    end; {Vertex Cover}
```

The following result is at the heart of the optimal minimum vertex cover algorithm for cographs.

Lemma 16. The set C returned by Vertex_Cover is a minimum vertex cover for G. □

Lemma 16 suggests a simple algorithm to compute the vertex cover of a cograph. This algorithm proceeds in the following three stages.

Stage 1. Perform an Euler tour of $BT(G)$ and use the information returned to construct a preorder list of all the leaves in $BT(G)$. Also, number the leaves in preorder as $1, 2, \ldots, n$ and compute for every internal node x of $BT(G)$ the set $L(x)$ of descendents of x. (We refer the reader to the previous subsection where these operations have been described in detail.)

Stage 2. Invoke an optimal tree contraction algorithm to construct an optimal contraction sequence $T_1, T_2, \ldots, T_m$ for $T(G)$. In addition, we compute a certain algebraic expression associated with $BT(G)$; specifically, we associate with every node of $BT(G)$ a non-negative integer as follows. Let x be a node of BT with left and right children y and z, respectively.

1. If x is a leaf, then $c(x) \leftarrow 0$.
2. If x is a 0-node, then $c(x) \leftarrow c(y) + c(z)$.
3. If x is a 1-node, then $c(x) \leftarrow \max\{c(y)+ |L(z)|, c(z) + |L(y)|\}$.

The interpretation of $c(x)$ is given by the following result whose proof follows by a simple inductive argument.

Lemma 17. For every node x of $BT(G)$, $c(x)$ represents the size of a minimum vertex cover in $G(x)$. In particular, $c(root)$ computed at the root of $BT(G)$ gives the size of a minimum vertex cover in G. $\square$

To make our algorithm easier to follow and the computation less messy, we shall associate with every node x of $BT(G)$ a negative number $t(x)$ which is closely related to $c(x)$. To motivate our discussion, note that since $|L(x)| = |L(y)| + |L(z)|$, Statement 2, above, can be written as

$$c(x) - |L(x)| \leftarrow |(c(y) - |L(y)|, c(z) - |L(z)|)$$

Similarly, Statement 3 can be written as

$$c(x) - |L(x)| \leftarrow \min(c(y) - |L(y)|, c(z) - |L(z)|)$$

Now, setting for every node x of $BT(G)$,

$$t(x) \leftarrow c(x) - |L(x)|$$

Statements 1, 2, and 3 become, respectively,

4. If x is a leaf, then $t(x) \leftarrow -1$.
5. If x is a 0-node, then $t(x) \leftarrow t(y) + t(z)$.
6. If x is a 1-node, then $t(x) \leftarrow \min\{t(y), t(z)\}$.

By Lemma 16, for every node x of $BT(G)$, $-t(x)$ is the number of vertices in $G(x)$ which do not belong to a minimal vertex cover of $G(x)$. Actually, we use the tree contraction to compute the value $t(root)$. Once this value is known, $c(root)$ is obtained by a simple subtraction operation. We now argue that the value $t(root)$ can be computed efficiently in parallel. Thus, we only need show that Stage 2 requires $O(\log n)$ time and $O(n/\log n)$ processors. Specifically, we have to show that computing $t(root)$ can be done in the time of tree contraction. To show that this is the case, at each node u of $BT(G)$, we store an expression of the form $f_u(x)$, where x is an indeterminate representing the value of the unevaluated $t(u)$. Suppose that u has two children, v and w, where w is a leaf (i.e., the value of $t(w)$ is a known constant c), and let $f_u(x)$ and $f_v(x)$ be the expressions stored by u and v, respectively. Let u contain an operation #. It is not hard to confirm that after prune(w) and bypass(u) have been performed, the expression stored at v is now $f_u(f_u(x)\#c)$. The crucial observation is that, for the purpose of computing $t(u)$ at every node u of $BT(G)$, Statements 4 through 6 suggest that the expression stored at u is of the form $f_u(x) = \min\{a, x + b\}$, where a and b are constants. Initially, every node u of $BT(G)$ stores $f_u(x) = \min\{0, 1x + 0\}$ $= x$ (recall that x is negative), as it should. Note, in particular, that in this scheme every leaf of $BT(G)$ stores -1. Thus, we have the following result.

Theorem 18. Given an arbitrary n-vertex cograph G represented by its cotree, the size of a minimum vertex cover in G can be computed in $O(\log n)$ time using $O(n/\log n)$ processors in the EREW-PRAM model of computation. $\square$

Stage 3. In this stage, using the information gathered in Stage 2, we proceed to compute the actual vertex cover for G. The idea is very simple: we traverse the tree $BT(G)$ from the root down to the leaves, marking some nodes as we go. Specifically, a node x of $BT(G)$ will transmit to its children y and z the following information:

- If x is marked, then x marks both y and z.
- If x is unmarked, then
  - in case $|X \cup C(Y)| \le |Y \cup C(X)|$, x marks y.
  - in case $|X \cup C(Y)| > |Y \cup C(X)|$, x marks z.

Note that Lemma 16 guarantees that all the marked leaves must belong to a minimum vertex cover of G. In addition, it is easy to confirm that it is easy to confirm that the computation described previously can be performed in the time and processor bounds of the tree contraction, since at every node operation can be performed in $O(1)$ time. Now, using list ranking, we only need extract out of $L(root)$ the subset of all the marked leaves. Using an optimal list ranking algorithm this final operation takes $O(\log n)$ time using $O(n/\log n)$ processors. To summarize, we have proven the following result.

Theorem 19. Given an arbitrary n-vertex cograph G represented by its cotree, the vertex cover problem can be solved for G in $O(\log n)$ time using $O(n/\log n)$ processors in the EREW-PRAM model of computation. $\square$

It is important to note that given a graph $G = (V, E)$, once a minimum vertex cover C is available, one can obtain a maximum independent set directly. Indeed, it is not hard to see that $V - C$ is an independent set of maximum size. This simple observation along with Theorem 19, implies the following result.

Theorem 20. Given an arbitrary n-vertex cograph G represented by its cotree, the task of computing a maximum independent set in G can be solved in $O(\log n)$ time using $O(n/\log n)$ processors in the EREW-PRAM model of computation. $\square$

Furthermore, given a graph $G = (V, E)$, the complement $\overline{G}$ is the graph having the same vertices as G; a pair of vertices is adjacent in $\overline{G}$ if and only if it is not adjacent in G. It is an easy observation that a clique in G corresponds to an independent set in $\overline{G}$. Moreover, given a cograph G represented by its cotree, the complement $\overline{G}$ is obtained by swapping labels at the internal nodes: zeros become ones, and vice versa. Clearly, in the presence of an n-node cotree, this operation can be performed in $O(1)$ time with n processors or in $O(\log n)$ time using $O(n/\log n)$ processors. The reader is referred to Fig. 13.25, which shows the complement of the cograph in Fig. 13.23, along with the new cotree.

Thus, to compute the maximum clique in a cograph, one will convert to the complement by swapping labels as above and proceed to compute a maximum independent set on the complement. By Theorem 20, we have the following result.

Theorem 21. Given an arbitrary n-vertex cograph G represented by its cotree, the task of computing a maximum clique in G can be solved in $O(\log n)$ time using $O(n/\log n)$ processors in the EREW-PRAM model of computation. $\square$

13.5   Concluding Remarks

This chapter has attempted to present, at the elementary level, a number of sample algorithms from the extremely rich field of parallel graph algorithms. One of our main

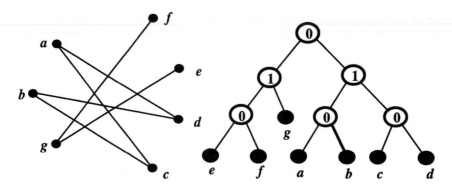

Figure 13.25 The complement of the cograph in Fig. 13.23

goals was to make the presentation as self-contained as possible. The material was specifically written with the nonspecialist in mind. Indeed, we wrote for the reader with a very basic knowledge about graph theory or algorithmics. We have presented preliminaries pertaining to both algorithmics and graph theory, including a brief description of the model of parallel computation adopted. We then went on to discuss tree algorithms, followed by algorithms for general graphs, followed in turn by algorithms for problems specialized to particular classes of graphs arising in practical applications. Each section contains pointers to sources of additional information, where each of the topics is presented in more detail.

Acknowledgment. This work is partly supported by the National Science Foundation under grant CCR-9407180, and by the office of Naval Research under grant N00014-95-1-0779.

13.6 References

1. Abrahamson, K, N. Dadoun, D. G. Kirkpatrick, and T. Przytycka. 1989. A simple parallel tree contraction algorithm. *Journal of Algorithms* 10, 287–302.
2. Aho, A. V., J. E. Hopcroft, and J. D. Ullman. 1974. *The Design and Analysis of Computer Algorithms.* Reading, Mass.: Addison Wesley .
3. Akl, S. G. 1989. *The Design and Analysis of Parallel Algorithms.* Englewood Cliffs, N.J.: Prentice Hall.
4. Anderson, R. J., and G. L. Miller. 1990. A simple randomized parallel algorithm for list ranking. *Information Processing Letters,* 33, 269–273.
5. Anderson, R.J., and G. L. Miller. 1991. Deterministic parallel list ranking. *Algorithmica,* 6 859–868.
6. Awerbuch, A., and Y. Shiloah. 1987. New connectivity and MSF algorithms for Shuffle-Exchange Network and PRAM. *IEEE Transactions on Computers,* C-36,1258–1263.
7. Berkman, O., D. Breslauer, Z. Galil, B. Schieber, and U. Vishkin. 1989. Highly parallelizable problems. *Proceedings of the Annual IEEE Symposium on Foundation of Computer Science,* 309–319.
8. Bertossi, A. A., and M. A. Bonuccelli. 1987. Some parallel algorithms on interval graphs. *Discrete Applied Mathematics,* 16, 101–111.
9. Bondy, J. A., and U. S. R. Murty. 1976. Graph Theory with Applications. Amsterdam: North-Holland.
10. Chin, F. Y., J. Lam, and I. Chen. 1982. Efficient parallel algorithms for some graph problems. *Communications of the ACM,* 25, 659–665.
11. Cole, R. 1988. Parallel merge sort. *SIAM Journal on Computing,* 17, 770–785.
12. Cole, R., and U. Vishkin. 1988. The accelerated centroid decomposition technique for optimal tree evaluation in logarithmic time. *Algorithmica,* 3, 329–346.

13. Cole, R., and U. Vishkin. 1988. Approximate parallel scheduling. Part 1: the basic technique with applications to optimal parallel list ranking in logarithmic time. *SIAM Journal on Computing,* 18, 128–142.

14. Cole, R., and U. Vishkin. 1991. Approximate parallel scheduling. Part 2: applications to optimal parallel graph algorithms in logarithmic time. *Information and Computation,* 91, 1–47.

15. Cole, R., and U. Vishkin. 1989. Faster optimal prefix sums and list ranking. *Information and Computation,* 81, 334–352.

16. Cormen, T. H., C. E. Leiserson, and R. L. Rivest. 1991. *Introduction to Algorithms.* New York: McGraw-Hill.

17. Dekel, E., and S. Sahni. 1983. Parallel scheduling algorithms. *Operations Research,* 31, 24–49.

18. Garey, M. R., and D. S. Johnson. 1979. *Computers and Intractability, a Guide to the Theory of NP-completeness.* San Francisco: W. H. Freeman.

19. Gazit, H. 1986. An optimal randomized parallel algorithm for finding connected components in a graph. *Proceedings of the 27th Annual Symposium on Foundations of Computer Science,* Toronto, Canada, 492–501.

20. Gupta, U. I., D. T. Lee, and J. Y. T. Leung. 1979. An optimal solution for the channel assignment problem. *IEEE Transactions on Computers,* C-28, 807–810.

21. Hirschberg, D. S., A. K. Chandra, and D. V. Sarwate. 1979. Computing connected components on parallel computers. *Communications of the ACM,* 22,461–464.

22. Jacobson, G. 1989. Space-efficient static trees and graphs. *Proceedings of the 30th Annual Symposium on Foundations of Computer Science,* Research Triangle Park, NC, 549–554.

23. JáJá, J. 1991. *An Introduction to Parallel Algorithms.* Reading, Mass.: Addison-Wesley.

24. Kim, S. K. 1988. Optimal parallel algorithms on sorted intervals. *Proceedings of the 27th Annual Allerton Conference on Communications, Control and Computing,* 766–774.

25. Knuth, D. E. 1973. *The Art of Computer Programming: Fundamental Algorithms,* Vol. 1, 2nd edition. Reading, Mass.: Addison-Wesley.

26. Miller, G. L., and J. Reif. 1985. Parallel tree contraction and its application. *Proceedings of the Annual IEEE Symposium on Foundation of Computer Science,* Portland, Oregon, 478–489.

27. Mulmuley, K, U. Vazirani, and V. Vazirani. 1987. Matching is as easy as matrix inversion. *Combinatorica,* 7, 105–113.

28. Lin, R., and S. Olariu. 1992. An optimal EREW algorithm to compute lowest common ancestors. *Parallel Computing,* 18, 511–516.

29. Lin, R., and S. Olariu. 1995. An optimal parallel breadth-first algorithm for ordered trees. *Parallel Algorithms and Applications,* Vol. 5, 187–197.

30. Lin, R., and S. Olariu. 1994. An optimal parallel matching algorithm for cographs. *Journal of Parallel and Distributed Computing,* 22, 26–37.

31. Olariu, S., C. M. Overstreet, and Z. Wen. 1995. An optimal parallel algorithm to reconstruct a binary tree from its traversals in doubly logarithmic CREW time. *Journal of Parallel and Distribute Computing,* Vol. 27, 100–105.

32. Olariu, S., J. L. Schwing, and J. Zhang. 1992. Optimal parallel algorithms for problems modeled by a family of intervals. *IEEE Transactions of Parallel and Distributed Systems,* 3, 364–373.

33. Quinn, M. J. 1994. *Parallel Computing: Theory and Practice.* New York: McGraw-Hill.

34. Savage, C. D., and J. JáJá. 1981. Fast efficient parallel algorithms for some graph problems. *SIAM Journal on Computing,* 10, 682 691.

35. Schieber, B., and U. Vishkin. 1988. On finding lowest common ancestors: simplification and parallelization. *SIAM Journal on Computing,* 17, 1253–1262.

36. Shiloah, Y., and U. Vishkin. 1982. An $O(\log n)$ parallel connectivity algorithm. *Journal of Algorithms,* 3, 57–67.

37. Springsteel, F., and I. Stojmenovic. 1989. Parallel general prefix computations with geometric, algebraic and other applications. *Proceedings of the International Conference on Fundamentals of Computation Theory,* Szeged, Hungary. Berlin: Springer-Verlag, 424–433.

38. Tarjan, R. E., and U. Vishkin. 1985. Finding biconnected components and computing tree functions in logarithmic parallel time. *SIAM Journal of Computing,* 14, 862–874.

39. Tsin, Y.H. 1986. Finding lowest common ancestors in parallel. *IEEE Transactions on Computers,* 35, 764–769.

40. Tsin, Y.H., and F. Y. Chin. 1984. Efficient parallel algorithms for a class of graph theoretic problems. *SIAM Journal on Computing,* 13, 580–599.

41. Wylie, J.C. 1979. The complexity of parallel computation. Doctoral thesis, Cornell University, 1979. Also available as Technical Report TR 79-387. Ithaca, N.Y.: Department of Computer Science, Cornell University.

14

Parallel Computational Geometry

Mikhail J. Atallah

The goal of computational geometry (CG for short) is the design of algorithms and data structures for the solution of geometric problems. Many of these geometric problems are relevant to important applications areas such as pattern recognition, computer graphics, statistics, operations research, computer-aided design, robotics, and so on. For example, consider the closest pair problem: given a set S of n points, find the two points of S with minimum distance between them. Finding efficient algorithms for this problem is interesting both from a practical point of view and from a purely intellectual one. In traditional (i.e., sequential) CG, the main question is: How fast can this problem be solved (hopefully better than the quadratic time taken by the obvious "brute force" algorithm, which looks at all pairs and chooses the closest among them)? The answer turns out to be $O(n \log n)$ time, and many sequential algorithms achieving this time bound are known. There is no hope of doing better than $O(n \log n)$ time, because there is a proof that $\Omega(n \log n)$ is a lower bound for the time to solve this problem. The questions "how fast can we do it" and "what is the best that can be hoped for" are the main ones for sequential CG. As we shall see in what follows, the situation is more complicated in a parallel framework, because we now have such additional parameters as the number of processors and how they operate and communicate with one another.

The rest of this introduction describes some typical problems in CG. While not exhaustive, the list will give the reader a flavor of the various types of problems with which CG has dealt.

1. *Convex hull.* Given a set S of n geometric objects, the *convex hull* of S is the smallest convex set that contains S. For example, if S is a set of points in the plane, then the convex hull of S is a convex polygon having a subset S' of S as vertices. The input is the set S, and the desired output is the convex hull of S.

2. *Maximal elements, dominance problems.* Given a set S of n d-dimensional points, a point $p \in S$ is said to *dominate* another point $q \in S$ if for all i, $1 \leq i \leq d$, the ith coordinate of p is larger than the ith coordinate of q. A point of S is *maximal* if it is not

dominated by any other point of S. The *maximal elements* problem is that of computing all the maximal points in a given set S.

The maximal elements problem is one of a class of problems derived from the above-mentioned notion of dominance of a point by another. Another such problem is the *two-set dominance counting problem,* where the input consists of two sets of points (say, a red set and a blue set), and the desired output is an array that contains, for each point, how many points of the other color it dominates.

3. *Visibility problems.* Given n opaque objects and a point p, the problem is to compute the region of space visible from p. For example, the opaque objects could be a collection of line segments in the plane, and the problem is then to compute a description of the region of the plane containing all the points q such that the line segment joining p to q does not intersect the interior of any of the n opaque segments. In one version of the problem, the n opaque segments are a polygonal chain (a contiguous portion of a polygon).

4. *Polygon triangulation.* Given an n-edge polygon P, the problem is to partition its interior into triangles by joining some pairs of its vertices by segments (the new segments do not intersect each other or the polygon's boundary).

5. *Voronoi diagram.* Given n points, the *Voronoi Diagram* problem is to compute a partition of the space into n regions, one region corresponding to each input point, where the region for point p consists of the set of points of the space whose distance to p is no larger than their distance to any other input point.

6. *All nearest neighbors.* This is a natural generalization of the already-mentioned closest pair problem. Here, the desired output is an array which contains, for every input point p, the input point nearest to p.

7. *Intersection detection and enumeration.* Given n geometric objects (e.g., line segments in the plane), detect whether any two intersect. The enumeration problem requires the listing of all intersecting pairs.

Other geometric problems will be defined as they arise in this text.

The rest of the chapter is organized as follows. The next section briefly discusses why parallelism is needed in CG and why it is not enough to perform a straightforward "parallelization" of the most successful sequential algorithms. Section 14.2 reviews basic subproblems that tend to arise in the solution of geometric problems on any parallel model, Section 14.3 discusses shared-memory techniques, Section 14.4 discusses techniques for mesh-connected arrays of processors, and Section 14.5 discusses the hypercube. Section 14.6 deals with other models such as coarse-grain and the hybrid RAM/ARRAY model, and Section 14.7 discusses the experimental work. Throughout, our main focus is on techniques; hence, no attempt is made to exhaustively cover all of the known parallel CG results.

14.1 Parallel CG: Why New Techniques Are Needed

Many of the application areas from which CG problems are drawn require real-time speeds, which makes parallelism a natural candidate for helping achieve the required performance. For many of these problems, we are already at the limits of what can be achieved through sequential computation. The traditional sequential methods can be inadequate for applications in which speed is important and the input can consist of a large number of geometric objects. Thus, it is important to study what kinds of speed-ups can be achieved through parallel computing. As an indication of the importance of this research direction, we note that four of the eleven problems used as benchmark problems to evaluate parallel architectures for the DARPA Architecture Workshop Benchmark Study of 1986 were CG problems.

Unfortunately, many of the powerful sequential techniques for CG do not translate well into a parallel setting. The difficulty is usually that these techniques use methods that either seem to be inherently sequential or would result in inefficient parallel implementations. We next illustrate this by using two sequential solution techniques and pointing out their inadequacy for the CREW-PRAM model (we refer the reader to the chapter on complexity for the definition of this parallel model). The central question for the closest pair problem on the CREW-PRAM is whether it can be solved in $O(\log n)$ parallel time using n processors, which would be asymptotically optimal since (1) the product of processors and time cannot be $o(n \log n)$, because such a time bound is known to be impossible to achieve sequentially for this problem [123], and (2) the time cannot be $o(\log n)$ because, in this model, it would take logarithmic time just to select the smallest of n entries (see Ref. [58] for a proof).

One $O(n \log n)$ time sequential solution to the closest pair problem works by two-way divide and conquer: the set S of points is divided into two roughly equal halves, S_1 and S_2 such that S_1 is geometrically to the left of S_2; that is, there is a vertical line V such that S_1 is to the left of V and S_2 is to the right of V. Then the closest pair is recursively computed for S_1, then for S_2; let the closest distance be δ_1 for S_1 and δ_2 for S_2. Finally, in the "combine" step that follows these two recursive calls, judicious use is made of the quantity $\delta = \min\{\delta_1, \delta_2\}$ to obtain, in linear time, the closest pair of points in $S_1 \times S_2$ whose corresponding distance is less than δ (if such a pair exists). The recurrence for the sequential time complexity $T_s(n)$ is therefore $T_s(n) = 2T_s(n/2) + c_1 n$ where c_1 is a constant. Trying to implement this algorithm in parallel, it is clear that the two recursive calls can be carried out in parallel, so that the recurrence for the parallel time becomes $T_p(n) = T_p(n/2) + c_2 f(n)$, where $f(n)$ is the time needed to perform the above-mentioned combining step in parallel. In this particular parallel model of computation, it is impossible for $f(n)$ to be better than logarithmic. This implies that the above two-way divide-and-conquer technique cannot give better than a discouraging $O(\log^2 n)$ time complexity (rather than the logarithmic time we are hoping to achieve). This impossibility to perform in constant time the "combining" step is by no means particular to the closest pair problem and is, in fact, a major reason why two-way divide and conquer has not been particularly successful for optimally solving geometric problems on the CREW-PRAM. We next review another sequential technique whose straightforward application typically fails to yield optimal algorithms: the *plane-sweep* method.

The plane-sweep paradigm has been extremely successful in sequential CG. We describe it in the case of a two-dimensional problem, but it also works for higher dimensions. In a nutshell, one imagines "sweeping" a line in the plane—say, a vertical line being swept from left to right. When that line encounters an object, such as a point or a segment endpoint, we call this an *event*. Such an event typically causes an update of a data structure associated with the sweep line (for example, the data structure could contain the segments currently intersected by the vertical sweep line). The list of events is not always determined ahead of time and can change dynamically as the left-to-right sweep proceeds, and so there is another data structure for the events (quite distinct from the above-mentioned data structure storing the status of the information associated with the sweep line's current position). Plane sweep has yielded elegant solutions to many problems, including the textbook example of intersection detection for line segments in the plane. In that case, there are $2n$ events (one for each segment endpoint), and the segments currently intersecting the vertical line are maintained in sorted order in a balanced tree data structure, etc. There is also an elegant line-sweep solution for the all nearest neighbors problem. From a parallel point of view, the above plane-sweep process may appear "inherently sequential," and this is indeed the case for some problems (one such problem will be mentioned later). For other problems, although it is not inherently sequential, parallelizing it poses major difficulties and results in suboptimal solutions. The cascading technique, to be sketched later, succeeds in the implementation of algorithms whose main idea is inspired from the plane-sweep paradigm, but the overall solution is so different from the sequential one that it can no longer be considered as a mere parallel implementation of sequential plane-sweep.

We could go on listing other techniques that are successful sequentially but disappointing in parallel; we refrain from doing so, because by now it should be clear that new paradigms are needed for CG—paradigms better suited for a parallel processing environment. Later on, we review the currently most successful techniques in parallel CG while simultaneously trying to give the flavor of the problematics of the area.

14.2   Basic Subproblems

This section reviews some basic subproblems that are ubiquitous tools in the design of parallel geometric algorithms, no matter what parallel model is used; these subproblems are not essentially geometric in nature, and they are also important for areas other than parallel CG. In many models, the complexity of these basic subproblems is well understood, but for some models (such as the hypercube), the complexity of some of these (e.g., sorting and list ranking) is still open, and in such situations no final statement about the complexity of the most common geometric problems can be made until these issues are resolved (especially given that many geometric problems are related to sorting). These basic operations are reviewed below. (The reader who has already reviewed these in other chapters of this handbook can skip the rest of this section.)

14.2.1   Sorting, merging

Sorting is probably the most frequently used subroutine in parallel geometric algorithms. Fortunately for PRAM models and for the mesh, we know how to sort optimally: $O(\log n)$ time and n processors on the EREW-PRAM [8, 52], and $O(\sqrt{n})$ time on a $\sqrt{n} \times \sqrt{n}$ mesh [111, 120, 102]. The parallel complexity of sorting on the hypercube is not known; the current best hypercube bound is $O(\log n \, \log \log n)$ with n processors [61]. On the mesh, the complexity of merging is the same as that of sorting, but on the hypercube and PRAM, it is easier than sorting [36, 94, 134, 140]; it is $O(\log n)$ time on an n-processor hypercube, and on the PRAM it is $O(\log \log n)$ time with $(n/\log n)$ processors [96].

14.2.2   Parallel prefix

Given an array A of n elements and an associative operation $+$, the parallel prefix problem is that of computing the array B of n elements such that

$$B(i) = \sum_{k=1}^{i} A(k)$$

Parallel prefix can be solved in $O(\log n)$ time and $(n/\log n)$ processors on an EREW-PRAM [105], $O(\log n/\log \log n)$ time and $(n \, \log \log n/\log n)$ processors on a CRCW-PRAM [56],

$O(\sqrt{n})$ time on the mesh (trivial), and in $O(\log n)$ time on an n-processor hypercube [108]. Computing the smallest element in the A array is a special case of parallel prefix; for the CRCW model, it can be done faster than general parallel prefix—in $O(\varepsilon^{-1})$ time with $n^{1+\varepsilon}$ processors for any positive constant ε or, alternatively, in $O(\log \log n)$ time with $(n/\log \log n)$ processors [134].

14.2.3 List ranking

List ranking is a more general version of the parallel prefix problem in that the elements are given as a linked list; i.e., we are given an array A each entry of which contains an element as well as a pointer to the entry of A containing the predecessor of that element in the linked list. The problem is to compute an array B such that $B(i)$ is the "sum" of the first i elements in the linked list. This problem is considerably harder than parallel prefix, and most tree computations as well as many graph computations reduce, via the *Euler tour technique* [138], to solving that problem. EREW-PRAM algorithms that run in $O(\log n)$ time and $(n/\log n)$ processors are known [55,11]. An $O(\sqrt{n})$ time mesh algorithm is also known [28]. Its complexity on the hypercube is still an open problem.

14.2.4 Tree contraction

Given a (not necessarily balanced) rooted tree, such that each internal node has two children, the problem is to reduce the tree to a 3-node tree by a sequence of *rake* operations. A rake operation can be applied to a leaf v by removing v and the parent of v and making the sibling of v the child of v's grandparent (note that the rake cannot be applied at a leaf whose parent is the root). This is done as follows. Number the leaves $1, 2, \ldots, n$ in left-to-right order, and apply the rake operation first to all the odd-numbered leaves that are left children, then to the other odd-numbered leaves. Repeat until done. The number of iterations is logarithmic because the number of leaves is divided by 2 at each iteration. Note that applying the rake to all the odd-numbered leaves at once would not work, as can be seen by considering the situation where v is an odd-numbered left child, w is an odd-numbered right child, and the parent of v is the grandparent of w (what goes wrong in that case is that v wants to remove its parent p and, simultaneously, w wants its sibling to become child of p). Tree contraction is an abstraction of many other problems, including that of evaluating an arithmetic expression tree [117]. Many elegant optimal EREW-PRAM algorithms for it are known [1, 55, 84, 99], running in $O(\log n)$ time with $(n/\log n)$ processors. It is easy to implement these in $O(\sqrt{n})$ time on a $\sqrt{n} \times \sqrt{n}$ mesh by using the techniques in Ref. [28].

14.2.5 Euler tour technique

"Wrapping a chain" around a tree defines an Euler tour of that tree. More formally, if T is an undirected tree, the Euler Tour of T is obtained by doing the following in parallel for every node v. Letting $w_0, \ldots, w_{k-1}$ be the nodes of T adjacent to v, in the order in which they appear in the adjacency list of v, we set the successor of each edge (w_i, v) equal to the edge $(v, w_{i+1 (mod\ k)})$. Most tree computations (as well as many graph computations) reduce via the Euler tour technique to the list ranking problem [138], as the following examples demonstrate.

1. Rooting an undirected tree at a designated node v. Create the Euler tour of the tree, "open" the tour at v (thus making the tour a path), then do a parallel prefix on the list

of arcs described by the successor function (with a weight of 1 for each arc). For each undirected arc $\{x, y\}$ of the tree, if the prefix sum of the directed edge (x, y) in the Euler tour is less than that of the directed edge (y, x), then set x to be the parent of y. This parallel computation of parents makes the tree rooted at v.

2. *Computing postorder numbers of a rooted tree, where the left-to-right ordering of the children of a vertex is the one implied by the Euler path.* Do a list ranking on the Euler path, where each directed edge (x, y) of the Euler path has a weight of unity if y is the parent of x, and a weight of zero if x is the parent of y.

3. *Computing the depth of each node.* Do a list ranking on the Euler path, where each directed edge (x, y) of the Euler path has a weight of -1 if y is the parent of x, and a weight of $+1$ if x is the parent of y.

4. *Number of descendants of each node.* This is the same computation as for postorder number, followed by the observation that the number of descendants of v equals the list rank of $(v, parent(v))$ minus that of $(parent(v), v)$.

These examples of reductions of so many problems to list ranking demonstrate the importance of the list ranking problem. See Refs. [96] and [138] for other examples, and for more details about the above.

14.2.6   Lowest common ancestors (LCA)

The problem is to preprocess a tree so that a lowest common ancestor (LCA) query of any two nodes x, y can be answered in constant time by one processor. The preprocessing is to be done in logarithmic time and $(n/\log n)$ processors. The problem is easy when the tree is a simple path (list ranking does the job) or when it is a complete binary tree (it is then solved by comparing the binary representations of the inorder numbers of x and y, specifically finding the leftmost position where they disagree). For a general tree, the problem is reduced to that of the range minima: create an Euler tour of the tree, where each v knows its first and last appearance on that tour and also its depth (distance from the root) in the tree. This reduces the problem of answering an LCA query to determining, in constant sequential time, the smallest entry in between two given indices i, j in an array. This last problem is called the *range-minima problem*. Reference [96] contains more details on the reduction of LCA to range minima, a solution to range minima, and references to the relevant literature for this problem.

The above list of basic subproblems is not exhaustive in that (1) many techniques that are basic for general combinatorial problems were omitted (we have focused only on those most relevant to geometric problems rather than to general combinatorial problems), and (2) among the techniques applicable to geometric problems, we have postponed coverage of the more specialized ones (they tend to be model-dependent).

14.3   CG on the PRAM

The PRAM has been extensively used as a theoretical model for designing parallel geometric algorithms. Although it captures important parameters of a parallel computation, the PRAM does not account for communication and synchronization costs. (More recent variations of the model do account for these, but we do not discuss them here since, essentially, no parallel geometric algorithms have yet been designed for them.) The PRAM enables the algorithm designer to focus on the structure of the problem itself without being distracted by architecture-specific issues. An advantage of the PRAM is that, if one can give strong evidence that a problem has no fast parallel solution on the PRAM, then there is no point in looking for a fast solution to it on more realistic parallel models (since these are weaker than the PRAM).

14.3.1   Inherently sequential geometric problems

We refer the reader to the chapter on complexity for the definition of the class NC, and of the notion of *P-complete* ("inherently sequential") problems. Most of the problems shown to be *P*-complete to date are not geometric (most are graph or algebra problems). This is no accident; geometric problems in the plane tend to have enough structure to enable membership in NC. Even the otherwise *P*-complete problem of linear programming [78, 79] is in NC when restricted to the plane. In the rest of this subsection, we mention one of the (very few) planar geometric problems that are known to be *P*-complete, and also a problem that is conjectured to be *P*-complete. Both problems belong to the class of *geometric layering* problems (also called *onion peeling* problems). We also mention one layering problem that is in NC.

In the *Visibility layers* problem, one is given a collection of n non-intersecting segments in the plane and asked to label each segment by its "depth" in terms of the following process (which starts with $i = 0$). Find the segments that are (partially) visible from $(0, +\infty)$, label each such segment as being at depth i, remove each such segment, increment i, and repeat until no segments are left. As mentioned previously, this is an example of a class of problems in computational geometry known as onion peeling problems [43, 106, 121], and it is P complete even if all the segments are horizontal.

The *P*-completeness of this problem was demonstrated in Refs. [15] and [91]. The proof consists of giving an NC reduction from a *P*-complete circuit value problem, and it involves the use of geometry to simulate a circuit by using the relative positions of objects in the plane.

One layering problem that remains open is the *convex layers* problem [43]. Given n points in the plane, mark the points on their convex hull as being layer zero, then remove layer zero and repeat the process, generating layers 1, 2, ... In view of the *P*-completeness of the above-mentioned visibility layers problem, it is a tempting conjecture that the convex layers problem is also *P*-complete. However, not all geometric layering problems are *P*-complete. For example, the *layers of maxima* problem (defined analogously to convex layers but with the words "maximal points" playing the role of "convex hull") is easily shown to be in NC by a straightforward reduction to the computation of longest paths on a directed acyclic graph derived from the geometric problem: every node corresponds to a point, and there is an edge from p to q iff p dominates q).

Once one has established that a geometric problem is in NC, the next step is to design a PRAM algorithm for it that runs as fast as possible while being efficient in the sense that it uses as few processors as possible. Ideally, the parallel time complexity should match the lower bound (assuming such a lower bound is known), and the *time $\times$ processors* product should match the best known sequential time bound for the problem. A parallel lower bound for a geometric problem is usually established by showing that it can be used to solve some other (perhaps nongeometric) problem having that lower bound. For example, it is well known [58] that computing the logical OR of n bits has an $\Omega(\log n)$ time lower bound on a CREW-PRAM. This can easily be used to show that detecting whether the boundaries of two convex polygons intersect also has an $\Omega(\log n)$ time lower bound on that same model, by encoding the n bits whose OR we wish to compute in two concentric regular n-gons such that the ith bit governs the relative positions of the ith vertices of the two

n-gons. Interestingly, if the word "boundaries" is removed from the previous sentence, then the lower bound argument falls apart, and it becomes possible to solve the problem in constant time on a CREW-PRAM, even using a sublinear number of processors [26].

Before reviewing the techniques that have resulted in many PRAM geometric algorithms that are fast and efficient in the above sense, a word of caution is in order. From a theoretical point of view, the class NC and the requirement that a "fast" parallel algorithm should run in polylogarithmic time are eminently reasonable. But from a more practical point of view, not having a polylogarithmic time algorithm does not entirely doom a problem to being "nonparallelizable." One can indeed argue that a problem of sequential complexity $\Theta(n)$ and that is solvable in $O(\sqrt{n})$ time by using $\sqrt{n}$ processors is "parallelizable" in a very real sense, even if no polylogarithmic time algorithm is known for it.

14.3.2 Parallel divide-and-conquer algorithms

The sequential divide-and-conquer algorithms that have efficient PRAM implementations are those for which the "conquer" step can be done extremely fast (e.g., in constant time). Take, for example, an $O(n \log n)$ time sequential algorithm that works by recursively solving two problems of size $n/2$ each and then combining the answers they return in linear time. For a PRAM implementation of such an algorithm to run in $O(\log n)$ time with n processors, the n processors must be capable of performing the "combine" stage in constant time. The time and processor complexities then obey the recurrences

$$T(n) \leq T\left(\frac{n}{2}\right) + c_1$$

$$P(n) \leq max\,\{n, 2P\left(\frac{n}{2}\right)\}$$

with $T(1) \leq c_2$ and $P(1) = 1$, where c_1 and c_2 are constants. These imply that $T(n) = O(\log n)$ and $P(n) = n$.

For some geometric problems this is indeed possible, but then the constant-time parallel combining step is typically much more complicated than (and indeed very different from) the combining step in the sequential divide-and-conquer solution (see, for example, the combining step in the parallel convex hull algorithms in Ref. [26]).

But for many problems (including, as we pointed out earlier, the closest pair problem), such an attempt at implementing a sequential algorithm fails because of the impossibility of performing the combining stage in constant time. For such problems, the approaches outlined next often work.

One alternative strategy is that of partitioning the problem into *more than two* subproblems, to be solved recursively in parallel. We illustrate why this often works by considering "rootish" divide and conquer, in which the problem is partitioned into $n^{1/k}$ subproblems to be solved recursively in parallel, for some constant integer k (usually, $k = 2$). For example, instead of dividing the problem into two subproblems of size $n/2$ each, we divide it into (say) $\sqrt{n}$ subproblems of size $\sqrt{n}$ each, which we recursively solve in parallel. That the conquer stage takes $O(\log n)$ time (assuming it does) causes no harm with this subdivision scheme, since the time and processor recurrences in that case would be

$$T(n) \leq T(\sqrt{n}) + c_1 \log n$$

$$P(n) \leq \max \{ n, \sqrt{n} P(\sqrt{n}) \}$$

with $T(1) \leq c_2$ and $P(1) = 1$, where c_1 and c_2 are constants. These imply that $T(n) = O(\log n)$ and $P(n) = n$.

The problems that can be solved using rootish divide and conquer include the convex hull [2, 25], the visibility of nonintersecting planar segments from a point [35], and the visibility of a polygonal chain from a point [19]. The scheme is useful in various ways and forms, and sometimes with recurrences different from the above-mentioned ones. For example, it was used in the form of a fourth-root divide and conquer to obtain (in a rather involved way) an optimal EREW algorithm for the visibility of a simple polygon from a point [19] (that is, $O(\log n)$ time with $(n/\log n)$ processors). See also Ref. [47] for a solution to the related problem of weak visibility.

There are instances where one has to use a hybrid of two-way divide and conquer and rootish divide and conquer to obtain the desired complexity bounds. For example, in Ref. [19], the recursive procedure takes two parameters (one of which is problem size) and uses either fourth-root divide and conquer or two-way divide and conquer, depending on the relative sizes of these two input parameters.

Another powerful technique for parallel divide and conquer is the cascading divide-and-conquer method. This sampling and iterative refinement method was introduced by Cole [52] for the sorting problem and was further developed in Ref. [20] for the solution of geometric problems. It has proved to be a fundamental technique, and one that enables optimal solutions when most other techniques fail. Its details are intricate even for sorting, but its main idea can be described easily. Since the technique works best for problems that are solved sequentially by divide and conquer, we use such a hypothetical problem to illustrate the discussion. Consider an $O(n \log n)$ time sequential algorithm that works by recursively solving two subproblems of size $n/2$ each, followed by an $O(n)$ time conquer stage. Let T be the tree of recursive calls for this algorithm, i.e., a node of this recursion tree at height h corresponds to a subproblem of size equal to the number of leaves in its subtree ($= 2^h$). A "natural" way of parallelizing such an algorithm would be to mimic it by using n processors to process T in a bottom-up fashion, one level at a time, completing level h before moving to level $h + 1$ of T (where by "level h" we mean the set of nodes of T whose height is h).

Such a parallelization will yield an $O(\log n)$ time algorithm only if processing each level can be done in constant time. It can be quite nontrivial to process one level in constant time, so this natural parallelization can be challenging. However, it is frequently the case that processing one level cannot be done in constant time, and it is precisely in these situations that the cascading idea can be useful. To be more specific when sketching this idea, we assume that the hypothetical problem being solved is about a set S of n points, with the points stored in the leaves of T.

In a nutshell, the general idea of cascading is as follows. The computation proceeds in a logarithmic number of stages, each of which takes constant time. Each stage involves activity by the n processors at more than one level, so the computation diffuses up the tree T rather than working on only one level at a time. For each node $v \in T$, let $h(v)$ be the height of v in T, $L(v)$ be the points stored in the leaves of the subtree of v in T, and let $I(L(v))$ be the information we seek to compute for node v (the precise definition of $I(\cdot)$ varies from problem to problem). The ultimate goal is for every $v \in T$ to compute the $I(L(v))$ array. Each $v \in T$ lies "dormant" and does nothing until the stage number exceeds a certain value (usually $h(v)$), at which time node v "wakes up" and starts computing, from stage to stage, $I(L')$ for a progressively larger subset L' of $L(v)$, a subset L' that (roughly) doubles in size from one stage to the next of the computation. $I(L')$ can be thought of as an approximation of the desired $I(L(v))$, an approximation that starts out being very rough (when L' consists of, say, a single point) but gets repeatedly refined from one stage to the next. When L' eventually becomes equal to $L(v)$, node v becomes inactive for all future stages (i.e., it is done with its computation, since it now has $I(L(v))$). There are many (often intricate) implementation details that vary from one problem to the next, and often the scheme substantially deviates from the above rough sketch, but our purpose was only to give the general idea of cascading.

The cascading technique has been used to solve many problems (not just geometric ones). Some of the geometric problems for which it has been used are:

- *Fractional cascading.* Given a directed graph G in which every node v contains a sorted list $C(v)$, construct a linear space data structure (that is, one whose size is at most a constant factor larger than the space taken by the input) that enables one processor to quickly locate any x in all the sorted lists stored along a given path $(v_1, v_2, \ldots, v_k)$ in G (by "quickly" we mean in $O(\log | C(v_1) | + k)$ time). This problem was introduced by Chazelle and Guibas [45], who gave an elegant sequential algorithm. An optimal $O(\log n)$ time and $(n/\log n)$ processor parallel algorithm for this problem was given in Ref. [20].
- *Trapezoidal decomposition.* Given a set S of n planar line segments, determine for each segment endpoint p the first segment encountered by starting at p and walking vertically upward (or downward). An $O(\log n)$ time and n processor CREW-PRAM algorithm is known [20]. This implies similar bounds for the polygon triangulation problem [142].
- *Topological sorting of n nonintersecting line segments.* This is the problem of ordering the segments so that, if a vertical line l intersects segments s_i and s_j and $i < j$, then the intersection between l and s_i is above the intersection between l and s_j. An $O(\log n)$ time, n-processor CREW-PRAM algorithm is easily obtained by implementing the main idea of the mesh algorithm of Ref. [29] (which reduces the problem to a trapezoidal decomposition computation followed by a tree computation).
- *Planar point location.* Given a subdivision of the plane into polygons, build a data structure that enables one processor to quickly locate, for any query point, the face containing it. Using n processors, cascading can be used to achieve $O(\log n)$ time for both construction and query [20, 57, 137]. The planar point location problem itself tends to arise rather frequently, even in geometric problems apparently unrelated to it.
- *Intersection detection, three-dimensional maximal elements, two-set dominance counting, visibility from a point, all nearest neighbors.* For all of these problems, cascading can be used to achieve $O(\log n)$ time with n processors [20, 53].

Alternative approaches have been proposed for some of the above problems; for example, see Refs. [35, 126, 131, 141] and also the elegant parallel hierarchical approach of Dadoun and Kirkpatrick, which is discussed later.

14.3.3 Matrix searching techniques

Many problems in CG can be formulated as searching problems in monotone matrices [4, 5]. Geometric problems amenable to such a formulation include the largest empty rectangle [6], various area minimization problems [5] (such as finding a minimum area circumscribing d-gon of a polygon), perimeter minimization [5] (finding a minimum perimeter triangle circumscribing a polygon), the layers of maxima problems [5], and rectilinear shortest paths in the presence of rectangular obstacles [16, 17]. Many more problems are likely to be formulated as such matrix searching problems in the future. We briefly review two of these matrix searching problems next.

14.3.3.1 Row minima. An important matrix searching formulation for geometric problems was introduced by Aggarwal et al. [4], where a linear time sequential solution was also given. The matrix searching problem, which we review next, has myriads of applications to geometric and combinatorial problems [5, 4].

Suppose we have an $m \times n$ matrix A, and we wish to compute the array θ_A such that, for every row index r ($1 \leq r \leq m$), $\theta_A(r)$ is the smallest column index c that minimizes $A(r, c)$ (that is, among all cs that minimize $A(r, c)$, $\theta_A(r)$ is the smallest). If θ_A satisfies the following *sorted* property:

$$\theta_A(r) \leq \theta_A(r+1)$$

and if for every submatrix A' of A, $\theta_{A'}$ also satisfies the sorted property, then matrix A is said to be *totally monotone* [5, 4].

Given a totally monotone matrix A, the problem of computing the θ_A array is known as that of "computing the row minima" of that matrix [5]. The best deterministic EREW PRAM algorithm for this problem runs in $O(\log n)$ time and n processors [30] (where $m = n$). Any improvement in this parallel complexity bound will also imply corresponding improvements on the parallel complexities of the many geometric applications of this problem. An optimal randomized algorithm was given in Ref. [125].

14.3.3.2 Tube minima. In what can be viewed as the three-dimensional version of the above row minima problem [5], one is given an $n_1 \times n_2 \times n_3$ matrix A, and one wishes to compute, for every $1 \leq i \leq n_1$ and $1 \leq j < n_3$, the $n_1 \times n_3$ matrix θ_A such that $\theta_A(i, j)$ is the smallest index k that minimizes $A(i, k, j)$ (that is, among all ks that minimize $A(i, k, j)$, $\theta_A(i, j)$ is the smallest). The matrix A is such that θ_A satisfies the following *sorted* property:

$$\theta_A(i,j) \leq \theta_A(i,j+1)$$

$$\theta_A(i,j) \leq \theta_A(i+1,j)$$

Furthermore, for any submatrix A' of A, $\theta_{A'}$ also satisfies the sorted property.

Given such a matrix A, the problem of computing the θ_A array is called, by Aggarwal and Park, [5] *computing the tube minima* of that matrix. Many geometric applications of this problem are mentioned in Ref. [5]. There are many nongeometric applications to this problem as well. These include *parallel string editing* [12], *constructing Huffmann codes in parallel* [31], and other tree-construction problems. (In Ref. [31], the problem was given the name *multiplying two concave matrices*.) The best CREW-PRAM algorithms for this problem run in $O(\log n)$ time and $(n^2/\log n)$ processors [5, 12], and the best CRCW-PRAM algorithm runs in $O(\log \log n)$ time and $(n^2/\log \log n)$ processors [14] (where $n = n_1 = n_2 = n_3$). Both the CREW and the CRCW bounds are asymptotically optimal.

14.3.4 Randomization

Reif and Sen [126, 127,128] have modified and applied to parallel geometric computation the randomization techniques that had proved their worth in sequential geometric computing (cf. the work of K. Clarkson, and also Haussler and Welzl, Mulmuley [119]) as well as in areas other than CG. Recall that a randomized algorithm is one that bases some of its decisions on the outcomes of coin flips. Thus, for a particular input, there are many possible executions of a randomized algorithm (which one actually happens depends on the outcomes of the coin flips). A good randomized algorithm must ensure that the number of "bad" possible executions (e.g., those that take too long to terminate) is a small fraction of all the possible executions. Algorithms that are not randomized are *deterministic* (although this adjective is usually omitted when the context does not leave room for confusion). Some deterministic algorithms have efficient expected time behavior for a randomly chosen set of input points, whereas randomized algorithms make no assumption about the input distribution.

Randomized algorithms have the disadvantage that they might fail, but if the probability of failure is made small enough then they can have advantages over deterministic ones. They are typically very simple (which makes them easy to program and to comprehend), and the multiplicative constant in their time complexity is usually small. For example, the algorithms given by Reif and Sen in [126] have a running time of $O(\log n)$ with n processors, *with high probability* (i.e., a probability that approaches 1 for very large n). The problems they deal with include planar point location and trapezoidal decomposition. The techniques they use there (and also in Ref. [128]) are somewhat reminiscent of the *flash-sort* algorithm of Reif and Valiant [129]. The *polling* technique of Reif and Sen [127] has yielded optimal randomized parallel bounds for problems that had long frustrated deterministic approaches. Also within the realm of randomized techniques, the *spherical separator* technique of Ref. [82] looks promising.

14.3.5   Other techniques, open problems

The paradigm of *geometric hierarchies* has proved extremely useful and general in CG, both sequential [100, 75, 76, 77] and parallel [63, 64]. Generally speaking, the method consists of building a sequence of descriptions of the geometric object under consideration, where an element w of the sequence is simpler and smaller (by a constant factor) than its predecessor v, and yet is "close" enough to v that information about w can be used to obtain in constant time information about v. This "information" could be, for example, the location of a query point in the subdivision, when the elements of the sequence are progressively simpler subdivisions of the plane. (In that case, pointers exist between faces of a subdivision and those of its predecessor—these pointers are part of the data structure representing the sequence of subdivisions). The technique turns out to be useful for other models than the PRAM as well.

A technique known as *Brent's theorem* [41] is frequently used to reduce the processor complexity without any increase in the time complexity. It states, subject to two qualifications, that any synchronous parallel algorithm taking time T that consists of a total of W operations can be simulated by P processors in time $O((W/P) + T)$. The two qualifications to the theorem before one can apply it to a PRAM are as follows:

1. At the beginning of the ith parallel step, we must be able to compute the amount of work W_i done by that step, in time $O(W_i/P)$ and with P processors.
2. We must know how to assign each processor to its task.

Both qualifications 1 and 2 are generally (but not always) easily satisfied in parallel geometric algorithms, so the hard part is usually achieving W operations in time T.

We now discuss a technique used for turning a CREW algorithm into an EREW one. To do so, obviously one needs to get rid of the read conflicts—the simultaneous reading from the same memory cell by many processors. Such read conflicts often occur in the conquer stage and can take the form of concurrent searching of a data structure by many processors (see, e.g., Ref. [19]). To avoid read conflicts during such concurrent searching, a scheme from Ref. [122] can be helpful. It states that if T is a 2-3 tree with m leaves, and $a_1, a_2, \ldots, a_k$, are data items that may or may not be stored in (the leaves of) T, and each processor P_j wants to search for a_j in $T, j = 1, 2, \ldots, k$, then in $O(\log m + \log k)$ time, the k processors can perform their respective searches without read conflicts. Many other types of searches can be accommodated by this technique, including searching for a particular item in the tree, searching for the tth item starting from item p, and searches where the tree is not a 2-3 tree (so long as each node of the tree has $O(1)$ children and the k searches are "sortable" according to their ranks in the sorted order of the leaves of the tree).

There are other PRAM techniques that we do not describe in detail. One such technique is the *array of trees* parallel data structure, originally designed in a nongeometric framework [27] but later used in Ref. [88] to establish geometric parallel bounds for such problems as hidden-line elimination, CSG evaluation, and computing the contour of a collection of rectangles. Another technique is the *stratified decomposition tree* used in Ref. [89] in the parallel solution of visibility and path problems in polygons. The novel approach of Ref. [87] leads to asymptotically faster and more efficient EREW PRAM parallel algorithms for a number of CG problems, including the development of the first optimal-work NC algorithm for the three-dimensional convex hull problem. The decomposition method in Ref. [42] will probably be useful for other geometric problems in higher dimensions. Another interesting area is the parallel solution of geometric problems in sublogarithmic time (see, e.g., Refs. [10, 34, 136]).

An optimal EREW-PRAM solution to linear programming in the plane remains elusive (for the CRCW model, see Ref. [74]). Much remains to be done in the development of output sensitive PRAM algorithms (where the complexity depends on the size of the output, as in Ref. [86]), and in the development of robust [112] parallel algorithms (whose correctness is not destroyed by roundoff error—many existing geometric algorithms misbehave if implemented with rounded arithmetic).

14.4 CG on the Mesh

In this section, for convenience, we limit the discussion to two-dimensional (that is, $\sqrt{n} \times \sqrt{n}$) meshes, but most of the results and techniques mentioned are known to easily generalize to higher-dimensional meshes as well. The geometric objects under consideration (e.g., points) are initially stored in the mesh, one object per processor. Therefore, we are implicitly assuming that the mesh has enough processors to store the problem description—the case where the problem size is too large to fit in the mesh is discussed in the next section.

Since it is known how to sort n items optimally (i.e., in $O(\sqrt{n})$ time) on a $\sqrt{n} \times \sqrt{n}$ mesh, sorting is not a bottleneck when trying to design $O(\sqrt{n})$ time solutions to geometric problems on the mesh. (Contrast this with the situation for the hypercube, a network in which the complexity of sorting is still unknown.) In fact, many of the classic problems of CG have been shown to be solvable on the mesh within the optimal $O(\sqrt{n})$ time bound (we mention some of these later). Most of these problems have an $O(n \log n)$ sequential time complexity, and since for the mesh the *time × processors* product is proportional to $n\sqrt{n}$, one might think that the word "optimal" is being abused here. However, this is not the case. Any nontrivial problem on a $\sqrt{n} \times \sqrt{n}$ mesh requires $\Omega(\sqrt{n})$ time (since it can take that long for two processors to communicate), and there is usually no hope of using $o(n)$ processors because of the already mentioned $O(1)$ storage limitation per processor (it takes $\Omega(n)$ space, and hence $\Omega(n)$ processors, just to store the input).

14.4.1 Mesh divide and conquer

Many geometric algorithms on the mesh use some form of divide and conquer. The problem gets partitioned into, for example, four pieces of size $n/4$ each, then each piece is moved into one of the four quadrants of the mesh, where it is solved recursively by the $(\sqrt{n}/2) \times \sqrt{n}/2)$ quadrant, after which the answers returned by the four recursive calls are combined to obtain the overall solution. The "combining" stage as well as the various bookkeeping steps usually involve sorting and take $O(\sqrt{n})$ time. Thus, the time recurrence of this scheme generally ends up being of the form

$$T(n) = T(n/4) + c_1\sqrt{n}$$

with $T(1) \leq c_2$ and where c_1, c_2 are constants, which implies that $T(n) = O(\sqrt{n})$. An example of this is the convex hull algorithm of Ref. [115]:

1. If n is small (say, $n \leq 4$), then solve the problem directly by brute force, otherwise proceed to Step 2 below.
2. Sort the n points whose convex hull we seek by x coordinates. Put those with the smallest $n/4$ x-coordinates in one of the four quadrants, those with the next $n/4$ smallest x-coordinates into another quadrant, etc. In fact the sorting itself can be done so that each quadrant automatically contains the appropriate $n/4$ points, i.e., no separate data movement is needed other than sorting. (See Ref. [115] for details.)
3. Recursively solve the problem for each of the four quadrants.
4. Combine the solutions returned by the four recursive calls, obtaining the hull of the whole point set. This involves finding the common tangents between pairs of disjoint convex polygons in $O(\sqrt{n})$ time.

The nontrivial part is usually the "combining" part (i.e., Step 4 in the above example). The data movement techniques of Ref. [120] often play a role in that stage, and sometimes the tree computation technique of Ref. [28] is needed (see Refs. [29] and [97]). There are other ways of solving the convex hull problem on the mesh (see, e.g., Ref. [95]).

14.4.2 Mesh prune and search

Some mesh algorithms use the parallel equivalent of what has been called, in sequential computation, the *prune-and-search* paradigm [106]. This paradigm consists of throwing away a subset of the input (after determining that it does not contribute to the answer) and then recursively searching the surviving portion of the input. The portion of the input thrown away is a fixed fraction of the input (i.e., a subset of size c_n of an input of size n, where c is a positive constant less than 1). Mesh implementations of this idea have the intriguing feature of advantageously keeping many of the processors idle during much of the computation. This is because, after doing the "pruning" (= decreasing problem size by a constant factor), the resulting (smaller) problem is compressed into a smaller submesh of the original mesh, where it is recursively solved while the processors not in this smaller submesh remain idle. This idling of many processors is a common occurrence in mesh algorithms. The disadvantage of having so many processors idle is more than compensated for by the decrease in the communication time among the active processors (because they form a smaller submesh).

Mesh algorithms might involve a sequence of many recursive calls (occurring after one another rather than in parallel) and still run in $O(\sqrt{n})$ time. So long as these successive recursive calls are on problems of sizes $c_1 n, c_2 n, \ldots, c_k n$ where k as well as the c_i variables are constants and $\sqrt{c_1} + \sqrt{c_2} + \ldots + \sqrt{c_k} < 1$, the time complexity is $O(\sqrt{n})$ (assuming that setting up each recursive call is done in $O(\sqrt{n})$ time, and that the other bookkeeping and combining of subsolutions also take $O(\sqrt{n})$ time). As an example, see the algorithm in Ref. [115] for computing the closest pair among a set of n planar points.

The parallel version of the prune-and-search technique has been far more useful for the mesh than for the PRAM because, in the mesh, we can afford to prune in $O(\sqrt{n})$ time and still end up with an optimal algorithm, whereas in the PRAM model, the technique typically yields superlogarithmic time bounds. (Reference [74] is one of the few instances in which it has been used for a PRAM geometric algorithm, and that was for the CRCW model.)

14.4.3 Multisearching

The following problem is often the bottleneck in the parallel solution of geometric problems on a network of processors. Given a search structure modeled as a graph G with n constant degree nodes, and given $O(n)$ search processes on that structure, the multisearch problem is that of performing, as quickly as possible, all of the search processes on that structure. The searches need not be processed in any particular order, and can simultaneously be processed in parallel by using, for example, one processor for each. However, the path that a search query will trace in G is *not* known ahead of time and must instead be determined *on-line;* only when a search query is at (say) node v of G can it determine which node of G it should visit next (it does so by comparing its own search key to the information stored at v—the nature of this information and of the comparison performed depend on the specific problem being solved).

The multisearch problem is a useful abstraction that can be used to solve many problems (more on this later). It is a challenging problem both for EREW-PRAMs and for networks of processors, since many searches might want to visit a single node of G, creating a "congestion" problem (with the added complication that one cannot even tally ahead of time how much congestion will occur at a node, since one does not know ahead of time the full search paths but only the nodes of G at which they start). When the parallel model used to solve the problem is a network of processors, the graph G is initially stored in the network in the natural way, with each processor containing one node of G and that node's adjacency list. It is important to keep in mind that the computational network's topology is not the same as that of the search structure G, so that a neighbor of node v in G need not be stored in a processor adjacent to the one containing v. Each processor also contains initially (at most) one of the search queries to be processed (in which case that search does not necessarily start at the node of G stored in that processor).

In the EREW-PRAM, the difficulty comes from the "exclusive read" restriction of the model; if k processes were to simultaneously access node v's information, the k processors assigned to these k search processes would be apparently unable to simultaneously access v's information. An elegant way around this problem was designed by Paul, Vishkin, and Wagener [122] for the case where G is a 2-3 tree (although they assume a linear ordering on the search keys, which usually does not hold in a geometric framework involving multidimensional search keys).

The multisearch problem is even more challenging for networks of processors because the data is distributed over a network and requires considerable time to be permuted to allow different processors access to different data items. Furthermore, each memory location can be accessed by only $O(1)$ query processes at a time, since a processor is unable to simultaneously store more than a constant number of search queries.

In Ref. [21], the multisearch problem is solved in $O[\sqrt{n} + r(\sqrt{n}/\log n)]$ time on a $\sqrt{n} \times \sqrt{n}$ mesh connected computer, where r is the length of the longest search path associated with a query. For most geometric data structures, the search path traversed when answering a query has length $r = O(\log n)$, and hence the time complexity is $O(\sqrt{n})$ time, which is asymptotically optimal. The classes of graphs for which this result holds contain most of the important cases of G that arise in practice, ranging from simple trees to the powerful Kirkpatrick hierarchical search DAG [100] that is so important in both sequential and parallel CG. Applications include interval trees and the related multiple interval intersection search, as well as hierarchical representations of polyhedra and its many applications, including lines-polyhedron intersection queries, multiple tangent plane determination, three-dimensional convex hull, and intersecting convex polyhedra.

14.4.4 Status of some problems on the mesh

Many geometric problems have $O(\sqrt{n})$ time mesh algorithms, including convex hull and all nearest neighbor problems for planar point sets [115], Voronoi diagram of a planar set of n points [97], minimum distance spanning tree for planar point sets, trapezoidal decomposition of a set of n (possibly intersecting) segments, polygon triangulation, visibility, topological sorting of nonintersecting line segments, computing the area of the union of iso-oriented rectangles, intersection detection between n planar line segments, computing the largest empty rectangle [65], three-dimensional convex hull, computing the intersection of two three-dimensional convex polyhedra [21], and many others. Other geometric problems considered in the literature include the computation of robot configuration space [69], visibility and separability [66], ECDF searching [73], multipoint and planar point location [97], and others.

The layering problems mentioned earlier remain open on the mesh, in the sense that no $O(\sqrt{n})$ time algorithm on a $\sqrt{n} \times \sqrt{n}$ mesh is known for them. (Recall that these include the planar convex layers problem and the layers of maxima problem.) Many open geometric problems on the mesh are in the hybrid model described later and will be mentioned in that context.

Another framework in which geometric problems have been considered on the mesh is that in which the input geometric figure is a binary image stored in the mesh in the natural way (the (i, j)th pixel is stored in the processor at row i and column j). The techniques needed in this image processing framework can be quite different from those we mentioned above and are not within the scope of this chapter (see, for example, Refs. [101] and [116]).

14.5 CG on the Hypercube

In view of the importance of the hypercube, surprisingly few geometric algorithms have been designed for this parallel model. (But see Refs. [40, 71, 72, 113, 114, 135, 109] for some of these).

Once the complexity of such basic operations as sorting and list ranking is settled for the hypercube model, algorithm design for geometric problems on that model and its relatives (called *hypercubic* networks) will probably receive increased attention. An important step in this direction is the sorting algorithm of Cypher and Plaxton [61].

One way around the *sorting bottleneck* for the hypercube would be to take the randomization approach, as did Reif and Sen [126]. (Sorting is then no longer a bottleneck, since there is an optimal randomized sorting algorithm for the hypercube [129].) In Ref. [93], randomized parallel algorithms on the butterfly network are given for trapezoidal decomposition, visibility, triangulation and 2-D convex hull.

Another way around the sorting bottleneck lies in exploiting special geometric properties of the input so as to optimally produce the desired sorted output. This is what was done very recently in Ref. [18], where hypercube techniques for solving optimally a class of polygon problems were given, resulting in optimal $O(\log n)$-time, n-processor hypercube algorithms for the problems of computing the portions of a polygonal chain C visible from a given source point, computing the convex hull of C, testing C for monotonicity, computing the kernel of C when C is closed (i.e., is a polygon), and computing the maximal elements in the set of vertices of C. The just-mentioned kernel and monotonicity problems are defined as follows. The kernel K of a polygon is the subset of the points in the polygon such that any point in K can "see" the whole polygon (K can be empty). A chain C is monotone with respect to a line L if and only if for every line L' that is orthogonal to L, $C \cap L'$ is either empty or a single point. A polygon P is monotone if there exists a line L such that the boundary of P can be partitioned into two chains each of which is monotone with respect to L. The problem of testing the monotonicity of P is that of finding a description for all the lines, with respect to each of which P is monotone, or reporting that no such lines exist (and hence P is not a monotone polygon). (See Ref. [49] for a PRAM solution.)

The multisearching problem was investigated in Ref. [107]. The case of locating n points in an n-region slab was also solved in Ref. [72] and, more recently, in Ref. [22]. These methods work for more general versions of this problem; the basic assumption is that any pair p, q in a processor can be compared in constant time if only one of p, q is a point, but not so if both x and y are points.

14.6   Other Parallel Models

We briefly describe some of the parallel CG work on other parallel models.

14.6.1   Coarse-grain models

Most commercially available parallel machines (e.g., the Intel Paragon, Intel iPSC/860, and CM-5) are *coarse-grained* in the sense that a processor is typically a state-of-the art unit with considerable processing power and local memory. This contrasts sharply with the $O(1)$ local memory traditionally assumed in *fine-grained* models and algorithms. Another feature of commercially available parallel machines is that basic communication primitives (e.g., routing, broadcasting, and sorting) are usually available as system calls or as highly optimized utilities. By using these primitives, an applications programmer can design solutions in an architecture-independent setting without having to be familiar with the specific communication patterns of the problem being solved. Nevertheless, the fact remains that communication primitives are typically much more expensive than local computations. These facts motivate the parallel model that has recently been used to develop new algorithms for computational geometry problems [67, 68]: the architecture-independent *coarse-grained, communication round model*. In this model, n inputs are evenly distributed among p processors (i.e., every processor stores n/p of the n input items). Each processor has a memory size of m with $m \geq n/p$. The processors communicate via an interconnection network in a *communication round* in which they specify the type of communication to occur. Typical operations performed in a communication round are broadcast or routing operations. Algorithms are designed by specifying the local computation done within each processor between two communication rounds, and by specifying the type of communication performed in a communication round. Algorithms designed within the coarse-grained, communication round model have the considerable advantage of being easier to program than algorithms that reimplement the communication operations. However, they still place the burden on the algorithm designer of designing schemes that use as few communication rounds as possible.

Various sorting algorithms on coarse-grain hypercubes have been presented [3, 59, 62, 98]. Plaxton provided efficient algorithms for selection and load balancing on coarse-grain hypercubes with one-port communication [59, 60]. Efficient coarse-grain algorithms for list ranking and several graph problems on hypercubes with all-port and one-port communication were given in Ref. [132]. In [133], efficient coarse-grain hypercube and shuffle exchange algorithms were devised for the multiple-prefix, data-dependent parallel prefix, image-component-labeling, and closest pair problems. Cohen et al. [51] designed and experimented parallel geometric algorithms for the convex hull and domination problems on coarse-grain hypercube architectures; their algorithms have efficient expected running time. The closest pair problem was studied in Ref. [133].

Overall, in the field of parallel CG, in spite of its importance in many application areas, there are not many results available on coarse-grain algorithms. But there is growing interest in the topic, and we will undoubtedly see many more geometric algorithms for coarse-grained models in the future.

14.6.2   A hybrid model: the RAM/ARRAY

The main justification for the hybrid model discussed in this section is that many existing parallel machines have a "front end" that is a conventional sequential computer, and that the number of processors in the parallel machine itself is typically the fixed number purchased rather than a function of the problem size, n. Suppose we have a parallel machine (such as a d-dimensional mesh-connected array of p processors) that can solve a problem of size p in time $O(p^{1/d})$ (this includes the time to input the data to the array as well as the actual computation time, a standard assumption in the literature of mesh-connected arrays, and certainly a reasonable one for the case $d = 1$). Suppose such a mesh-connected array of processors is attached to a conventional random access machine (RAM) that wishes to solve a problem of size $n > p$. We call such a machine a RAM/ARRAY(d). It is important to realize that the mesh alone cannot even store the description of the geometric problem because of the limitation that each processor has $O(1)$ storage registers and, hence, the sequential "front end" must play a role in the solution process. If the problem's sequential time complexity is, say, $\Theta(n \log n)$, then the mesh gives a factor of $s(p) = p^{1-1/d} \log p$ speedup *for a problem of size p*. However, if the RAM/ARRAY(d) is trying to solve a problem of size n, $n > p$, then it is not clear how it should use the mesh to achieve the factor of $s(p)$ speedup and obtain $O(n \log n/s(p))$ time performance.

The desired $O(n \log n/s(p))$ time bound can be achieved on a RAM/ARRAY(1) for the following geometric problems [32]: all nearest neighbors of a planar set of points, the measure and perimeter of a union of rectangles, the visibility of a set of non-intersecting line segments from a point, three-dimensional maxima, and dominance counting between two sets of points (and hence the related problem of counting intersections between recti-linear rectangles). Sorting was done in Refs. [7] and [23], which implies RAM/ARRAY(1) solutions for the geometric problems that can be solved in linear time after a pre-process-ing sorting step, like the planar convex hull and maximal elements problems.

We illustrate what is involved in a typical solution for the case $d = 1$ and for geometric problems whose sequential time complexity is $\Theta(n \log n)$, i.e., when the task is to design $O(n \log n/\log p)$ time algorithms on a RAM/ARRAY(1). In that case, the algorithm often follows the p-way divide-and-conquer paradigm. (An alternative method, based on a "lazy B-tree" approach, was used for sorting in Ref. [23], but we do not discuss it here.) That is, the algorithm divides the problem into p subproblems, then it recursively solves each of the p subproblems one after the other. After the p recursive calls return, it combines the subsolutions to form the final solution. The main difficulty is how to perform the combin-ing step in $O(n)$ time. If the combining step can be performed in $O(n)$ time, then the over-all time complexity $T(n)$ satisfies the recurrence $T(n) = p \cdot T(n/p) + O(n)$, which implies that $T(n) = O(n \log n/\log p)$. In the case of a RAM/ARRAY(d) where $d > 1$, instead of par-titioning the problem into p subproblems, the problem gets partitioned into $p^{1/(d+1)}$ sub-problems. In that case the $p^{1/(d+1)}$ subsolutions must be "combined" in $O(n/p^{1-1/d})$ time.

Remaining open problems on the RAM/ARRAY(1) include topological sorting of non-intersecting line segments, trapezoidal decomposition, Voronoi diagram of a planar point set, 3-dimensional convex hull, and computing the intersection of two 3-dimensional con-vex polyhedra. Negative results would also be interesting; which problems are inherently such that it is impossible to maintain the same speedup for $n > p$ as for $n = p$?

The techniques developed for RAM/ARRAY(d)s have also been used in Ref. [139] to achieve linear speedups on several hypercube-related computers that consist of p proces-sors each containing $O(n/p)$ local memory, provided that $n > p^{1+\varepsilon}$ for some constant $\varepsilon > 0$. The same speedup is known for sorting [3, 62]. Finally, there are close connections with the work on I/O complexity [7], and the techniques used to obtain the above results can also be used to obtain I/O complexity bounds for the same geometric problem considered [139]. There is increased interest in the related area of "external memory" algorithms for situations where the problem is too big to fit in internal memory (see Ref. [90]).

14.6.3   Reconfigurable models

Reconfigurable models are networks whose interconnections can advantageously be modi-fied during the algorithmic solution process. We do not discuss these here; for the inter-ested reader, Ref. [130] provides an illustration of a geometric problem solved on such a model.

14.7   Conclusions and Future Work

It should be clear from what preceded that previous work in parallel CG has been mostly theoretical so far. We expect to see more experimental work in the future. Some research-ers have already implemented geometric algorithms on various parallel architectures and reported interesting results. For example, Guy Blelloch [37, 38] has implemented parallel geometric algorithms on the Connection Machine (CM),[*] including convex hull (the $\sqrt{n}$-divide-and-conquer method we mentioned earlier). Blelloch argued that in the CM archi-

[*]Connection Machine and CM are registered trademarks of Thinking Machines Corporation.

tecture, scan operations (essentially, parallel prefix) are implemented so efficiently that one should solve problems on the CM architecture by using, whenever possible, calls to these built-in routines. He gave a detailed study of the implications of using scans as primitives for solving various problems. The experimental data obtained by Blelloch and by other researchers (e.g., [48]) seems to confirm the usefulness of Blelloch's approach.

Cohen, Miller, Sarraf, and Stout have implemented parallel geometric algorithms on hypercube architectures like the iPSC, including convex hulls and domination [51]. Miller and Miller have done so for convex hulls of digitized pictures [113].

The above-mentioned experimental work demonstrates, among other things, that algorithmic insights developed for abstract parallel models can be useful when programming "real" parallel machines, after suitable modification and fine-tuning. However, these experimental papers remain the exception rather than the rule, and work in parallel CG continues to be mostly theoretical. We believe this will change as researchers gain increased access to parallel machines, and as the size of the practical geometric problems to be solved keeps increasing.

14.8 References

1. Abrahamson, K., N. Dadoun, D. A. Kirpatrick, and T. Przytycka. 1989. A simple parallel tree contraction algorithm. *Journal of Algorithms,* Vol. 10, 287–302.
2. Aggarwal, A., B. Chazelle, L. Guibas, C. O'Dunlaing, and C. Yap. 1988. Parallel computational geometry. *Algorithmica,* Vol. 3, 293–328.
3. Aggarwal, A., and M.-D. Huang. 1988. Network complexity of sorting and graph problems and simulating CRCW PRAMS by interconnection networks. *Lecture Notes in Computer Science, 319: VLSI Algorithms and Architectures,* 3rd Aegean Workshop on Computing, AWOC 88, Springer Verlag, 339–350.
4. Aggarwal, A., M. M. Klawe, S. Moran, P. Shor and R. Wilber. 1987. Geometric applications of a matrix searching algorithm. *Algorithmica,* Vol. 2, 209–233.
5. Aggarwal, A., and J. Park. 1988. Parallel searching in multidimensional monotone arrays. *Proceedings of the 29th Annual IEEE Symposium on Foundations of Computer Science,* 497–512.
6. Aggarwal, A., and S. Suri. 1987. Fast algorithms for computing the largest empty rectangle. *Proceedings of the 3rd ACM Symposium on Computational Geometry,* 278–290.
7. Aggarwal, A., and J. S. Vitter. 1988. The input/output complexity of sorting and related problems. *Communications of the ACM,* Vol. 31, 1116–1127.
8. Ajtai, M., J. Komlos, and E. Szemeredi. 1983. Sorting in clog n parallel steps. *Combinatorica,* Vol. 3, 1–19.
9. Akl, S. G. 1982. A constant-time parallel algorithm for computing convex hulls. *BIT,* Vol. 22, 130–134.
10. Amato, N. M., and F. P. Preparata. 1993. NC^1 parallel 3D convex hull algorithm. *Proceedings of the 9th Annual Symposium on Computational Geometry,* 289–297.
11. Anderson, R., and G.L. Miller. 1991. Deterministic parallel list ranking. *Algorithmica,* Vol. 6, 859–868.
12. Apostolico, A., M. J. Atallah, L. Larmore, and H. S. McFaddin. 1990. Efficient parallel algorithms for string editing and related problems. *SIAM J. Comput.,* Vol. 19, 968–988.
13. Atallah., M. J. 1992. Parallel techniques for computational geometry. *Proceedings of the IEEE,* Vol. 80, 1425–1448.
14. Atallah, M. J. A. 1993. Faster parallel algorithm for a matrix searching problem. *Algorithmica,* Vol. 9, 156–167.
15. Atallah, M. J., P. Callahan and M. T. Goodrich. 1990. P-Complete geometric problems. *Proceedings of the 2d Annual ACM Symposium on Parallel Algorithms and Architectures,* Springer Verlag, 317–326.
16. Atallah, M. J. and D. Z. Chen. 1991. Parallel rectilinear shortest paths with rectangular obstacles. *Computational Geometry: Theory and Applications,* Vol. 1, 79–113.
17. Atallah, M. J. and D. Z. Chen. 1993. On parallel rectilinear obstacle-avoiding paths. *Proceedings of the 5th Canadian Conference on Computational Geometry,* Waterloo, Canada, 210–215.
18. Atallah, M. J. and D. Z. Chen. 1993. Optimal parallel hypercube algorithms for polygon problems. *Proceedings of the 5th IEEE Symposium on Parallel and Distributed Processing,* Dallas, Texas, 208–215.
19. Atallah, M. J., D. Z. Chen and H. Wagener. 1991. An optimal parallel algorithm for the visibility of a simple polygon from a point. *Journal of the ACM,* Vol. 38, 516–533.

20. Atallah, M. J., R. Cole, and M. T. Goodrich. 1989. Cascading divide-and-conquer: a technique for designing parallel algorithms. SIAM J. Comput., Vol. 18, 499–532.

21. Atallah, M. J., F. Dehne, R. Miller, A. Rau-Chaplin, J-J. Tsay. 1993. Multisearch techniques: parallel data structures on mesh-connected computers. *Journal of Parallel and Distributed Computing,* Vol. 19, 1–13.

22. Atallah, M. J. and A. Fabri. 1994. On the multisearching problem for hypercubes. *Proceedings of the 6th Parallel Architectures and Languages Europe Symposium,* Athens, Greece, Springer Verlag Lecture Notes in Computer Sci.: 817, 159–166.

23. Atallah, M. J., G. N. Frederickson, and S. R. Kosaraju. 1988. Sorting with efficient use of special purpose sorters. *Information Processing Letters,* Vol. 27, 13–15.

24. Atallah, M. J. and M. T. Goodrich. 1986. Efficient plane sweeping in parallel. *Proceedings of the 2nd Annual ACM Symposium on Computational Geometry,* Yorktown Heights, New York, 216–225.

25. Atallah, M. J. and M. T. Goodrich. 1986. Efficient parallel solutions to some geometric problems. *Journal of Parallel and Distributed Computing,* Vol. 3, 492–507.

26. Atallah, M. J. and M. T. Goodrich. 1988. Parallel algorithms for some functions of two convex polygons. *Algorithmica,* Vol. 3, 535–548.

27. Atallah, M. J., S.R. Kosaraju and M.T. Goodrich. 1988. On the parallel complexity of evaluating some sequences of set manipulation operations. *Lecture Notes in Computer Science, 319: VLSI Algorithms and Architectures,* 3rd Aegean Workshop on Computing, AWOC 88, Springer Verlag, 1–10.

28. Atallah, M. J. and S. E. Hambrusch. 1986. Solving tree problems on a mesh-connected processor array. *Info. and Control,* Vol. 69, 168–187.

29. Atallah, M. J., S. E. Hambrusch and L. E. TeWinkel. 1991. Parallel topological sorting of features in a binary image. *Algorithmica,* Vol. 6, 762–769.

30. Atallah, M. J. and S. R. Kosaraju. 1992. An efficient parallel algorithm for the row minima of a totally monotone matrix. *Journal of Algorithms,* Vol. 13, 394–413.

31. Atallah, M. J., S. R. Kosaraju, L. Larmore, G. L. Miller and S. Teng. 1989. Constructing trees in parallel. *Proceedings of the 1st Annual ACM Symposium on Parallel Algorithms and Architectures,* Santa Fe, New Mexico, 421–431.

32. Atallah, M. J., and J-J. Tsay. 1992. On the parallel-decomposibility of geometric problems. *Algorithmica,* Vol. 8, 209–231.

33. Beigel, R., and J. Gill. 1990. Sorting n objects with a k-sorter. *IEEE Transactions on Computers,* Vol. 39, 714–716.

34. Berkman, O., Breslauer, Z. Galil, B. Schieber, and U. Vishkin. 1989. Highly parallelizable problems. *Proceedings of the 21st Annual ACM Symposium on Theory of Computing,* 309–319.

35. Bertolazzi, P., C. Guerra and S. Salza. 1990. A parallel algorithm for the visibility problem from a point. *Journal of Parallel and Distributed Computing,* Vol. 9, 11–14.

36. Bilardi, G., and A. Nicolau. 1989. Adaptive bitonic sorting: an optimal parallel algorithm for shared memory machines. *SIAM J. Comput.,* Vol. 18, 216–228.

37. Blelloch, G. E. 1988. Scan primitives and parallel vector models. Ph.D. Thesis, Massachusetts Institute of Technology.

38. Blelloch, G. E. 1987. Scans as primitive parallel operations. *Proceedings of the of the Int. Conference on Parallel Processing,* 355–362.

39. Borodin, A., and J. E. Hopcroft. 1985. Routing, merging, and sorting on parallel models of computation. *Journal of Computer and System Sciences,* Vol. 30, 130–145.

40. Boxer, L. and R. Miller. 1989. Dynamic computational geometry on meshes and hypercubes. *Journal of Supercomputing,* Vol. 3, 161–191.

41. Brent, R. P. 1974. The parallel evaluation of general arithmetic expressions. *J. ACM,* Vol. 21, 201–206.

42. Callahan, P.B., and S. R. Kosaraju. 1992. Decomposition of multi-dimensional point-sets with applications to k-nearest-neighbors and n-body potential fields. *Proceedings of the 24th Ann. ACM Symposium on the Theory of Computing,* 546–556.

43. Chazelle, B. M. 1983. Optimal algorithms for computing depths and layers. *Proceedings of the of the 20th Allerton Conference on Communications, Control and Computing,* 427–436.

44. Chazelle, B. M. 1984. Computational geometry on a systolic chip. *IEEE Trans. on Computers,* Vol. C-33, 774–785.

45. Chazelle, B., and L. J. Guibas. 1986. Fractional cascading: I. A data structuring technique. *Algorithmica,* Vol. 1, 133–162.

46. Chen, D. Z. 1990. Efficient geometric algorithms in the EREW-PRAM. *Proceedings of the 28th Annual Allerton Conference on Communication, Control, and Computing,* Monticello, Illinois, 818–827.

47. Chen, D. Z. 1992. Optimal parallel algorithm for detecting weak visibility of a simple polygon. *Proceedings of the 8th Ann. Symposium On Computational Geometry,* 63–72.

48. Chen, L. T., and L. S. Davis. 1993. Parallel algorithm for the visibility of a simple polygon using scan operations. *CVGIP: Graphical Models and Image Processing,* Vol. 55, 192–202.

49. Chen, D. Z., and S. Guha. 1993. Testing a simple polygon for monotonicity optimally in parallel. *Information Processing Letters,* Vol. 47, 325–331.

50. Chow, A. 1980. Parallel algorithms for geometric problems. Ph.D. thesis. Computer Science Dept., Univ. of Illinois at Urbana-Champaign.

51. Cohen, E., R. Miller, E.M. Sarraf, and Q.F. Stout. 1992. Efficient convexity and domination algorithms for fine- and medium-grain hypercube computers. *Algorithmica,* Vol. 7, 51–75.

52. Cole, R. 1988. Parallel merge sort. *SIAM J. Comput.,* Vol. 17, 770–785.

53. Cole, R., and M. T. Goodrich. 1992. Optimal parallel algorithms for point-set and polygon problems. *Algorithmica,* Vol. 7, 3–23.

54. Cole, R., M. T. Goodrich, and C. O'Dunlaing. 1990. Merging free trees in parallel for efficient Voronoi diagram construction. *Proceedings of the 17th International Colloq. on Automata, Lang., and Programming,* Springer Verlag, 432–445.

55. Cole, R., and U. Vishkin. 1986. Approximate and exact parallel scheduling with applications to list, tree and graph problems. *Proceedings of the 27th Annual IEEE Symposium on Foundations of Comp. Sci.,* 487–491.

56. Cole, R., and U. Vishkin. 1989. Faster optimal parallel prefix sums and list ranking. *Info. and Control,* Vol. 81, 334–352.

57. Cole, R., and O. Zahicek. 1990. An optimal parallel algorithm for building a data structure for planar point location. *Journal of Parallel and Distributed Computing,* Vol. 8, 280–285.

58. Cook, S., and C. Dwork. 1982. Bounds on the time for parallel RAM's to compute simple functions. *Proceedings of the 14th ACM Annual Symposium on Theory of Computing,* 231–233.

59. Plaxton, C. G. 1989. Load balance, selection, and sorting on the hypercube. *Proceedings of the 1st Annual ACM Symposium on Parallel Algorithms and Architectures,* 64–73.

60. Plaxton, C. G. 1989. On the network complexity of selection. *Proceedings of the 30th Annual IEEE Symposium on Foundations of Computer Science,* 396–401.

61. Cypher, R., and C. G. Plaxton. 1990. Deterministic sorting in nearly logarithmic time on the hypercube and related computers. *Proceedings of the 22d Annual ACM Symposium on Theory of Computing,* 193–203.

62. Cypher, R., and J. L. C. Sanz. 1988. Optimal sorting on feasible parallel computers. *Proceedings of the International Conference on Parallel Processing,* 339–350.

63. Dadoun, N., and D. Kirkpatrick. 1987. Parallel processing for efficient subdivision search. *Proceedings of the 3rd ACM Symposium Computational Geom.,* 205–214.

64. Dadoun, N. 1990. Geometric hierarchies and parallel subdivision search. Ph.D. Thesis, University of British Columbia.

65. Dehne, F. 1990. Computing the largest empty rectangle on one and two dimensional processor arrays. *J. Parallel Dist. Comput.,* Vol. 9, 63–68.

66. Dehne, F. 1988. Solving visibility and separability problems on a mesh of processors. *Visual Computer,* Vol. 3, 356–370.

67. Dehne, F., A. Fabri, A., and A. Rau-Chaplin. 1993. Scalable parallel geometric algorithms for coarse grained multicomputers. *Proceedings of 9th ACM Symposium on Computational Geometry,* 298–307.

68. Devillers, O., and A. Fabri, A. 1993. Scalable algorithms for bichromatic line segment intersection problems on coarse grained multicomputers. *Proceedings of 1993 Workshop on Algorithms and Data Structures,* LNCS 709, Springer, Berlin, 277–288.

69. Dehne, F., A.-L. Hassenklover, and J.-R. Sack. 1989. Computing the configuration space for a robot on a mesh of processors. *Parallel Computing,* Vol. 12, 221–231.

70. Dehne, F., J.-R. Sack and I. Stojmenovic. 1988. A note on determining the 3-dimensional convex hull of a set of points on a mesh of processors. *Proceedings of the 1988 Scandinavian Workshop on Algorithms and Theory,* 154–162.

71. Dehne, F., A. Ferreira and A. Rau-Chaplin. 1992. Parallel fractional cascading on a hypercube multiprocessor. *Computational Geometry: Theory and Applications,* Vol. 2, 141–167.

72. Dehne, F., and A. Rau-Chaplin. 1990. Implementing data structures on a hypercube multiprocessor with applications in parallel geometry. *Journal of Parallel Distrib. Computing,* Vol. 8, 367–375.

73. Dehne, F., and I. Stojmenovic. 1988. An $O(\sqrt{n})$ time algorithm for the ecdf searching problem for arbitrary dimensions on a mesh of processors. *Information Processing Letters,* Vol. 28, 67–70.

74. Deng, X. 1990. An Optimal Parallel Algorithm for Linear Programming in the Plane. *Information Processing Letters,* Vol. 35, 213–217.

75. Dobkin, D. P., and D. G. Kirkpatrick. 1983. Fast detection of polyhedral intersections. *Theor. Comp. Sci.,* Vol. 27, 241–253.

76. Dobkin, D. P., and D. G. Kirkpatrick. 1985. A linear time algorithm for determining the separation of convex polyhedra, *Journal of Algorithms,* Vol. 6, 381–392.

77. Dobkin, D. P., and D. G. Kirkpatrick. 1990. Determining the separation of preprocessed polyhedra—a unified approach. *Proceedings of the of the Int. Colloq. on Automata, Lang., and Programming,* 154–165.

78. Dobkin, D., R. J. Lipton, and S. Reiss. 1979. Linear programming is log-space hard for P. *Information Processing Letters,* Vol. 9, 96–97.

79. Dobkin, D., and S. Reiss. 1980. The complexity of linear programming. *Theor. Comp. Sci.,* Vol. 11, 1–18.

80. ElGindy, H., and M. T. Goodrich. 1988. Parallel algorithms for shortest path problems in polygons. *The Visual Computer: International Journal of Computer Graphics,* Vol. 3, 371–378.

81. Frederickson, G. N., and D. B. Johnson. 1982. The complexity of selection and ranking in X + Y and matrices with sorted columns. *Journal of Computer and System Sciences,* Vol. 24, 197–208.

82. Frieze, A. M., G. L. Miller, and S-H. Teng. 1992. Separator based parallel divide-and-conquer in computational geometry. *Proceedings of the 4th Ann. ACM Symposium on Parallel Algorithms and Architectures,* 420–430.

83. Goldschlager, L. M. 1977. The monotone and planar circuit value problems are log space complete for P. *SIGACT News,* Vol. 9, 25–29.

84. Gibbons, A., and W. Rytter. 1986. An optimal parallel algorithm for dynamic expression evaluation and its applications. *Proceedings of the of Symposium on Found. of Software Technology and Theoretical Comp. Sci.,* Springer Verlag, 453–469.

85. Goodrich, M. T. 1987. Finding the convex hull of a sorted point set in parallel. *Information Processing Letters,* Vol. 26, 173–179.

86. Goodrich, M. T. 1991. Intersecting line segments in parallel with an output-sensitive number of processors. *SIAM J. Comput.,* Vol. 20, 737–755.

87. Goodrich, M. T. 1993. Geometric partitioning made easier, even in parallel. *Proceedings of the 9th Ann. Symposium on Computational Geometry,* 73–82.

88. Goodrich, M. T., M. Ghouse and J. Bright. 1990. Generalized sweep methods for parallel computational geometry. *Proceedings of the 2d Annual ACM Symposium on Parallel Algorithms and Architectures,* 280–289.

89. Goodrich, M. T., S. B. Shauck and S. Guha. 1990. Parallel method for visibility and shortest path problems in simple polygons. *Proceedings of the 6th Annual ACM Symposium on Computational Geometry,* 73–82.

90. Goodrich, M. T., J-J. Tsay, D. E. Vengroff, and J. S. Vitter. 1993. External-memory computational geometry. *Proceedings of the 34th Annual Symposium on Foundations of Computer Science,* 714–723.

91. Hershberger, J. 1990. Upper envelope onion peeling. *Proceedings of the 2d Scandinavian Workshop on Algorithm Theory,* Springer-Verlag, 368–379.

92. Hershberger, J. 1992. Optimal parallel algorithms for triangulated simple polygons. *Proceedings of the 8th Annual Symposium On Computational Geometry,* 33–42.

93. Hagerup, T. 1990. Efficient parallel computation of arrangements of hyperplanes in d dimensions. *Proceedings of the 2nd Annual ACM Symposium on Parallel Algorithms and Architectures,* 290–297.

94. Hagerup, T., and C. Rueb. 1989. Optimal merging and sorting on the EREW PRAM. *Information Processing Letters,* Vol. 33, 181–185.

95. Holey, J. A., and O. H. Ibarra. 1992. Iterative algorithms for the planar convex hull problem on mesh-connected arrays. *Parallel Comput.,* Vol. 18, 281–296.

96. JáJá, J. 1992. *An Introduction to Parallel Algorithms.* Reading, Mass.: Addison-Wesley.

97. Jeong, C. S., and D. T. Lee. 1990. Parallel geometric algorithms on a mesh connected computer. *Algorithmica,* Vol. 5, 155–177.

98. Johnsson, S. L. 1984. Combining parallel and sequential sorting on a Boolean *n*-cube. *Proceedings of the International Conference on Parallel Processing,* 444–448.

99. S. R. Kosaraju and A. Delcher. Optimal parallel evaluation of tree-structured computations by raking. *Lecture Notes in CS 319: AWOC 88,* Springer Verlag, 1988, 101–110.

100. Kirkpatrick, D. G. 1983. Optimal search in planar subdivisions. *SIAM Journal of Computing,* Vol. 12, 28–35.

101. Prasanna, V. K., M. M. Eshaghian. 1986. Parallel geometric algorithms for digitized pictures on mesh of trees. *Proceedings of the 1986 Int. Conference on Parallel Processing,* 270–273.

102. Kung, H. T., and C. D. Thompson. 1977. Sorting on a mesh-connected parallel computer. *Comm. ACM,* Vol. 20, 263–27.

103. Kruskal, C. P., L. Rudolph, and M. Snir. 1990. A complexity theory of efficient parallel algorithms. *Theoretical Comp. Sci.,* Vol. 71, 95–132.

104. Kruskal, C. P., L. Rudolph, and M. Snir. 1985. The power of parallel prefix. *Proceedings of the 1985 IEEE Int. Conference on Parallel Processing,* St. Charles, IL., 180–185.

105. Ladner, R. E., and M. J. Fischer. 1980. Parallel prefix computation. *J. ACM,* 831–838.

106. Lee, D. T., and F. P. Preparata. 1984. Computational geometry—a survey. *IEEE Trans. on Computers,* Vol. C-33, 872–1101.

107. Lee, D. T., and F. P. Preparata. 1989. Parallel batched planar point location on the CCC. *Information Processing Letters,* Vol. 33, 175–179.

108. Leighton, F. T. 1992. *An Introduction to Parallel Algorithms and Architectures: Arrays, Trees, Hypercubes.* San Mateo, Calif.: Morgan Kaufmann.

109. MacKenzie, P. D., and Q. F. Stout. 1990. Asymptotically efficient hypercube algorithms for computational geometry. *Proceedings of the 3rd Symposium on the Frontiers of Massively Parallel Computation,* 8–11.

110. MacKenzie, P. D., and Q. F. Stout. 1990. Practical hypercube algorithms for computational geometry. *Proceedings of the 3rd Symposium on the Frontiers of Massively Parallel Computation,* 75–78.

111. Marberg, J. M., and E. Gafni. 1986. Sorting in constant number of row and column phases on a mesh. *Proceedings of the 24th Annual Allerton Conference on Communication, Control, and Computing,* Monticello, Illinois, 603–612.

112. Li, Z., and V. Milenkovic. 1992. Constructing strongly convex hulls using exact or rounded arithmetic. *Algorithmica,* Vol. 8, 345–364.

113. Miller, R., and S. E. Miller. 1987. Using hypercube multiprocessors to determine geometric properties of digitized pictures. *Proceedings of the 1987 International Conference in Parallel Processing,* University Park, Pa., 638–640.

114. Miller, R., and Q.F. Stout. 1988. Efficient parallel convex hull algorithms. *IEEE Trans. Computers,* Vol. C-37, 1605–1618.

115. Miller, R., and Q.F. Stout. 1989. Mesh computer algorithms for computational geometry. *IEEE Trans. Computers,* Vol. C-38, 321–340.

116. Miller, R., and Q.F. Stout. 1985. Geometric algorithms for digitized pictures on a mesh connected computer. *IEEE Trans. PAMI,* Vol. 7, 216–228.

117. Miller, G. L., and J. H. Reif. 1985. Parallel tree contraction and its applications. *Proceedings of the 26th ACM Symposium on Foundations of Comp. Sci.,* 478–489.

118. Mueller, H. 1987. Sorting numbers using limited systolic coprocessors. *Information Processing Letters,* Vol. 24, 351–354.

119. Mulmuley, K. *Computational Geometry: An Introduction through Randomized Algorithms.* Englewood Cliffs, N.J.: Prentice Hall, 1994.

120. Nassimi, D., and S. Sahni. 1981. Data broadcasting in SIMD computers. *IEEE Trans. on Computers,* Vol, 30, 101–106.

121. O'Rourke, J. 1988. Computational geometry. *Ann. Rev. Comp. Sci.,* Vol. 3, 389–411.

122. Paul, W., U. Vishkin and H. Wagener. 1983. Parallel dictionaries on 2-3 trees. *Proceedings of the 10th Coll. on Autom., Lang., and Prog. (ICALP),* LNCS 154, Springer, Berlin, 597–609.

123. Preparata, F. P., and M. I. Shamos. 1985. *Computational Geometry: An Introduction.* New York: Springer Verlag.

124. Preparata, F. P., and R. Tamassia. 1988. Fully dynamic techniques for point location and transitive closure in planar structures. *Proceedings of the 29th ACM Symposium on Theory of Computing,* 558–567.

125. Raman, R., and U. Vishkin. 1994. optimal parallel algorithms for totally monotone matrix searching. *Proceedings of the 5th ACM-SIAM Symposium on Discrete Algorithms,* San Francisco, California, 613–621.

126. Reif, J. H., and S. Sen. 1992. Optimal randomized Parallel Algorithms for Computational Geometry. *Algorithmica,* Vol. 7, 91–117.

127. Reif, J. H., and S. Sen. 1989. Polling: a new random sampling technique for computational geometry. *Proceedings of the 21st Annual ACM Symposium on Theory of Computing,* 394–404.

128. Reif, J. H., and S. Sen. 1990. Randomized algorithms for binary search and load balancing on fixed connection networks with geometric applications (preliminary version). *Proceedings of the 2nd Annual ACM Symposium Parallel Algorithms and Architectures,* 327–337.

129. Reif, J. H., and L. Valiant. 1987. A logarithmic time sort for linear size networks. *J. ACM,* Vol. 34, 60–76.

130. Reisis, D. I. 1992. Efficient convex hull computation on the reconfigurable mesh. *Proceedings of the 6th Int. Parallel Proceedings of the Symposium,* 142–145.

131. Rueb, C. 1992. Line-segment intersection reporting in parallel. *Algorithmica,* Vol. 8, 119–144.

132. Ryu, K. W., and J. Jájá. 1990. Efficient algorithms for list ranking and for solving graph problems on the hypercube. *IEEE Trans. Parallel and Distributed Systems,* Vol. 1, 83–90.

133. Sanz, J. L. C., and R. Cypher. 1992. Data reduction and fast routing: A strategy for efficient algorithms for message-passing parallel computers. *Algorithmica,* 7, 77–89.

134. Shiloach, Y., and U. Vishkin. 1981. Finding the maximum, merging, and sorting in a parallel computation model. *J. Algorithms,* Vol. 2, 88–102.

135. Stojmenovic, I. 1988. Computational geometry on a hypercube. *Proceedings of the 1988 Int. Conference on Parallel Processing,* 100–103.

136. Stout, Q. F. 1988. Constant-time geometry on PRAMs. *Proceedings of the 1988 International Conference on Parallel Computing,* Vol. III, IEEE, 104–107.

137. Tamassia, R., and J. S. Vitter. 1991. Parallel transitive closure and point location in planar structures. *SIAM J. Comput.,* Vol 20, 708–725.

138. Tarjan, R. E., and U. Vishkin. 1985. Finding biconnected components and computing tree functions in logarithmic parallel time. *SIAM J. Comput.,* Vol. 14, 862–874.

139. Tsay, J.-J. 1992. Parallel algorithms for geometric problems on networks of processors. *Proceedings of the 5th IEEE Symposium on Parallel and Distributed Processing,* 200–207.

140. Valiant, L. 1975. Parallelism in comparison problems. *SIAM J. on Computing,* Vol. 4, 348–355.

141. Willard, D. E., and Y. C. Wee. 1988. Quasi-valid range querying and its implications for nearest neighbor problems. *Proceedings of the 4th Annual ACM Symposium on Computational Geometry,* 34

142. Yap, C. K.1988. Parallel triangulation of a polygon in two calls to the trapezoidal map. *Algorithmica,* Vol. 3, 279–288.

15

Data Structures
for Parallel Processing[*]

Sajal K. Das and Kwang-Bae Min

Parallel computing is one of the important areas of research in computer science. To make fruitful use of commercial multiprocessor architectures, one needs to develop efficient parallel algorithms that can be implemented on real machines with a minimum amount of overhead. Traditionally, data structures have played crucial roles in algorithm design, be it sequential or parallel. Therefore, without a rich collection of efficient parallel data structures, software advancement, and hence the widespread use of commercial parallel machines, would not be feasible.

In the context of sequential computing, there are numerous problems for which elegant solutions rely on efficient data structures (e.g., the use of a heap in computing minimum spanning trees or shortest paths). In such cases, a data structure helps achieve better scheduling and/or time complexity. In other cases, a data structure itself is the object of computation, (e.g., constructing and maintaining a dictionary machine). However, traditional data structures including stacks, queues, linked lists, and heaps that have served so well in sequential computing cannot be used easily for efficient parallel computing [25]. Therefore, there is an ever growing demand for efficient mechanisms to manipulate data structures in parallel.

In the context of parallel computing, so far the research emphasis has been mostly on the design of fast algorithms (such as polylogarithmic time) and on the identification of basic techniques such as balanced binary tree technique, pointer jumping, symmetry breaking, divide and conquer, partitioning, pipelining, accelerated cascading, tree decomposition, tree contraction, and so on [33]. Mostly, the focus has been on techniques to achieve logarithmic time scheduling. In this chapter, however, we will examine parallel algorithms from the viewpoint of data structures: (1) what kind of new data structures are required in parallel algorithms to achieve logarithmic time scheduling, and (2) how the data structures are constructed and maintained when they are themselves the objects of computation.

---

[*]This work is supported by Texas Advanced Research and Technology Program grants under Award Nos. TARP-003594003 and TATP-003594031. The authors can be reached by e-mail at {das,kbmin}@cs.unt.edu.

In our discussion, the term *parallel data structure* will be used to mean a single coherent data structure, different parts of which can be simultaneously accessed by different processors without conflicts. Thus, a parallel data structure may be an entirely new concept for solving a given problem (i.e., it does not necessarily have a sequential counterpart), or it may be a parallelized version of a data structure used in sequential computing. We will see that there are many problems, particularly in the domain of non-numeric computations, for which designing efficient parallel data structures is an important issue for enhancing performance of the proposed parallel algorithms. Of course, our objective is not to exhaust all possible parallel data structures proposed in the literature, but to provide an overview of several ideas that are supposed to be applicable to broader class of problems. In fact, we will mostly concentrate on various tree-related data structures.

Our model of computation is the widely accepted parallel random access machine (PRAM), which is an abstract model for synchronous, shared memory, parallel computations. Several variants of the PRAM family are distinguished by their memory access capabilities. In this chapter, we will be mostly considering the weakest (albeit most feasible) variant, called the exclusive-read, exclusive-write (EREW) PRAM model, which allows concurrent access to the global shared memory only if each processor is accessing a different memory location for reading or writing at any time. Another variant is the concurrent-read, exclusive-write (CREW) model, which allows concurrent reading from (but not concurrent writing into) a shared memory location by multiple processors.

This chapter is organized as follows. Sections 15.1 through 15.7 examine how the basic data structures are handled in parallel and what new data structures are introduced for parallel processing. The discussion is mainly on arrays, linked lists, trees, Euler tours, parentheses strings, stacks, and queues. Because some of the techniques, such as linked list ranking and Euler tour, are covered in other chapters of this book, we have kept their discussions as brief as possible, and only in the light of handling parallel data structures. Section 15.6 gives a detailed description of a parallel parentheses matching algorithm to demonstrate how to replace a traditional stack by another data structure, namely a binary matching tree, to achieve more parallelism. Section 15.8 shows in detail how to parallelize the heap (or priority queue) structure and perform maintenance operations (e.g., insert and delete) in parallel. Section 15.9 deals with a unified approach for the parallel construction of various kinds of search trees and also discusses maintenance operations on 2-3 trees. Concluding remarks are given in Section 15.10.

15.1   Arrays and Balanced Binary Trees

There are several problems in sequential processing, which can be easily solved by a linear scanning of the input array. These problems include broadcasting, finding minimum/maximum element, prefix-sums, segmented prefix-sums, etc. In parallel processing, it is well known that such problems can be solved optimally in logarithmic time (of the input size) on the EREW PRAM model by using the balanced binary tree technique [7, 16, 37]. Since these problems are probably the most widely used subalgorithms to solve many other problems in parallel computation, the balanced binary tree technique is thus fundamental. Given an array of $n = 2^k$ elements, this technique uses a balanced binary tree of n leaves and depth $k = \log n$. A parallel algorithm for these problems typically consists of upward and/or downward sweep of this tree, level by level, to ensure logarithmic time. From the viewpoint of data structures, the balanced binary tree used in the parallel algorithm design is an auxiliary data structure for scheduling purposes whereas, in sequential algorithms, no such additional data structure is needed for solving this class of problems.

Let us describe a prefix-sums algorithm to demonstrate explicitly the use of a balanced binary tree as a scheduling data structure. The notation BT[h, j] represents the jth node (from the left) at height h of the tree, where the leaf nodes are at height 0. During the upward sweep of the tree, the algorithm computes BT[h, j].SUB—the sum of the elements in the leaves of the subtree rooted at node BT[h, j]. Similarly, during the downward sweep, it computes BT[h, j].PRE—the sum of the elements in all the leaves that precede the leaf nodes of the subtree rooted at node BT[h, j]. Figure 15.1 shows this scheme.

Algorithm Prefix-sums;
Input: An array A[1 .. n] of n = 2k integers.
Output: An array S[1 .. n] such that S[j] is the jth prefix-sum of A for
    1 ≤ j ≤ n.
Declarations: BT[0 .. log n, 1 .. n] {balanced binary tree used as a scheduling data structure}
1.   {Copy the input elements to the leaves of the binary tree.}
      for 1 ≤ j ≤ n **do in parallel** BT[0, j].SUB ← A[j]
2.   {Upward sweep of the binary tree.}
      for 1 ≤ h ≤ log n **do**
            for 1 ≤ j ≤ n/2h **do in parallel**
                  BT[h,j].SUB ← BT[h − 1, 2j − 1].SUB + BT[h − 1, 2j].SUB
3.   {Assign identity element (with respect to + operation) to the root.}
      BT[log n, 1].PRE ← 0
4.   {Downward sweep of the binary tree.}
      for log n − 1 ≥ h ≥ 0 **do**
            for 1 ≤ j ≤ n/2h **do in parallel**
                  j = odd: BT[h, j].PRE ← BT[h + 1, ⌈j/2⌉].PRE
                  j = even: BT[h, j].PRE ← BT[h + 1, ⌈j/2⌉].PRE +
                        BT[h, j − 1].SUB
5.   {Compute the final results at the leaves.}
      for 1 ≤ j ≤ n **do in parallel** S[j] ← BT[0, j].PRE + BT[0, j].SUB

Other problems, including parentheses matching [44] and range minima/maxima [33], can be optimally solved in parallel by using the preceding technique as a way of achieving logarithmic time scheduling.

h=log n

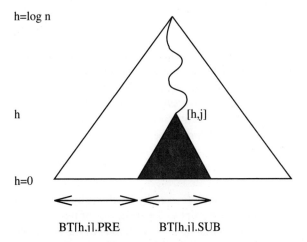

h

[h,j]

h=0

BT[h,j].PRE   BT[h,j].SUB

Figure 15.1 Balanced binary tree in prefix-sums computation

15.2   Linked Lists

A linked list in its natural form is an awkward data structure to be handled in parallel processing. This is because the pointers in a linked list give the logical ordering of only the next nodes rather than the logical ordering of the nodes in the entire list, and therefore sharing and partitioning among multiple processors without memory contention is difficult. To properly allocate processors to the nodes in a linked list while respecting their order of occurrences, we need to compute the order of appearance of each node. In other words, we need a parallel prefix/suffix algorithm for a linked list. With the help of an optimal linked list ranking algorithm [4, 16], a linked list can be easily converted to an array so that subsequent computations on the list boil down to array computations.

Given a linked list of length n, stored in an array such that the ith node, $1 \leq i \leq n$, has a pointer, NEXT(i), to its successor node in the list, the *linked list ranking* problem is to compute the *distance* (or *rank*) of each node from the end of the list. Formally, the rank of the ith node can be recursively defined as

$$RANK(i) = \begin{cases} 1 & \text{if } NEXT(i) = nil \\ RANK(NEXT(i)) + 1 & \text{otherwise} \end{cases}$$

In other words, RANK(i) provides the number of nodes succeeding the ith node (including itself) in the list. It can also be defined in terms of the number of nodes preceding each node in the list.

This problem can be solved in $O(n)$ sequential time by simply scanning or traversing the list. In sequential processing, we do not need the logical order number of each node, and the index of the successor is sufficient. In a multiprocessor environment, however, to process multiple nodes in parallel while respecting the logical structure of the list, we need to compute the mapping between the physical indices, i, of the nodes in the list and their logical indices, RANK(i). The parallel linked list ranking is an essential step for processing data structures such as trees and graphs that are representable by linked lists. It has been applied heavily to tree related problems, including Euler tours, computing the size of a subtree, various tree traversals, (accelerated) centroid decomposition, tree evaluation, and tree contraction. With respect to graphs, list ranking has been applied to various problems, including connected and biconnected components, forest decomposition, computing a spanning forest, ear decomposition, vertex connectivity, and bridge testing. For general use, list ranking has been applied to list packing, parentheses matching, radix sorting, etc. Discussions of many of these applications can be found in Refs. [19, 29, 33, 35, 44, 47].

Approximately 20 parallel algorithms have been proposed for the linked list ranking problem on various machine models. Variations among these algorithms deal primarily with the techniques used to schedule the processors to nodes in order to avoid conflicts and guarantee higher processor utilization. A comprehensive survey of the existing literature on the list ranking algorithms was done by Das and Halverson [19]. The algorithms proposed on the PRAM models are based on two major approaches sketched below.

15.2.1   List ranking by pointer jumping

The *pointer jumping* or *recursive doubling* technique, devised by Wyllie [50], repeatedly doubles the length of each link by having each node point to the successor of its successor until each node points to the end of the list. In each iteration, the RANK of the successor node is added to that of the current node.

Since at each iteration the distance between a node and its new successor is doubled. $\lceil \log n \rceil$ iterations are required for all the nodes to point to the end of the list (i.e., *nil*). Thus, the time complexity of the pointer jumping algorithm is $O(\log n)$ using n processors, for a total cost of $O(n \log n)$. This method, while quite simple and elegant, is not processor-efficient in that each processor continues to work (i.e., idling), even after the node to which it is assigned has had its rank computed. For example, on the first iteration, one processor is idle, on the second iteration two processors are idle, and on each successive iteration the number of idle processors doubles.

15.2.2   Reduce-rank-expand strategy

The list reduction strategy used by most cost-optimal parallel list ranking algorithms is a three-stage process referred to as *reduce-rank-expand* (RRE). The reduction stage starts by deleting nodes until the list is reduced to an acceptable size, generally $O(p)$, where $p = (n/\log n)$ is the number of available processors. The ranking stage computes the ranks of the nodes in the reduced list, commonly using pointer jumping. The expansion stage computes the ranks of the previously deleted nodes as they are reinserted into the list.

As discussed in Ref. [19], seven deterministic RRE list ranking algorithms have been developed for the EREW PRAM model, leading to successive improvements in terms of efficiency and/or simplicity [4, 14, 16]. The elegant algorithm devised by Anderson and Miller [4] efficiently incorporates several of the techniques proposed in earlier researches. The key idea in this optimal list ranking algorithm lies in the reduction phase, which is described below.

The input array, NEXT, is partitioned into p blocks, each assigned to a processor. Each processor attempts to process the nodes in its block independently as long as it does not have conflicts with other processors. When in conflict (i.e., the node currently being processed in one block is the predecessor or successor of a node currently being processed in another block), a symmetry-breaking algorithm such as coloring of a linear chain with $\log \log n$ colors, (for example, $(\log \log n)$-ruling set algorithm [14]), is invoked to cut a possibly very long chain of adjacent nodes into subchains of shorter length, say $O(\log \log n)$. Now, which processor will be in charge of processing the nodes in a subchain is determined by the number of remaining nodes in each block. A processor with fewer remaining nodes than both its predecessor and successor becomes responsible for deleting adjacent nodes. Thus, workload is fairly uniformly distributed among processors, and an overall logarithmic time scheduling is achieved. The complexity analysis of this cost-optimal algorithm is more difficult, but it can be shown that the algorithm runs in $O(\log n)$ time using $(n/\log n)$ processors on the EREW PRAM [4].

The reduction phase in this list ranking algorithm reveals an interesting point related to data structures. Each block can be viewed as a stack with its current stack pointer pointing to the top. When a stack is not in conflict with another stack, it proceeds very much like a stack in the sequential environment. Otherwise, the stacks in conflict with others cooperate together to decide which stack will be in charge of the top elements in other stacks. Since the linked list ranking problem can be solved sequentially using a single stack, the cooperative multiple stacks can be viewed as a parallel version of the sequential stack. The cooperation is achieved through symmetry breaking.

Figure 15.2 shows a situation where each stack needs to cooperate with logically adjacent stacks. For example, the collection of stacks S_2, S_5, S_3, and S_1 cooperate with one another. So do the stacks S_4, S_7, and S_6. A word of caution is appropriate. This view of cooperative multiple stacks can be applied only to the problems that can be sequentially solved by a simple linear scanning of a linked list—not to any problems solvable by using a stack.

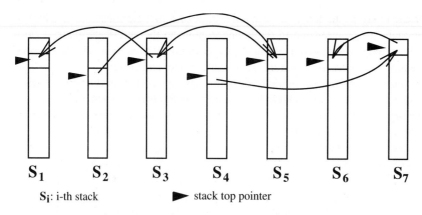

S_i: i-th stack   ► stack top pointer

Figure 15.2 Cooperative multiple stacks

15.3 Trees and Euler Tour

Trees are useful data structures to represent many real world applications. A rooted general (i.e., not necessarily binary) tree is one in which a vertex has an arbitrary number of children. Such a tree can be represented by one of the following relations:

1. parent-of
2. parent-of with explicit ordering of children
3. leftmost child and right sibling [2]

For a rooted tree represented by parent pointers, the most widely used technique is pointer jumping. But it has at least three limitations:

1. It causes read/write conflicts.
2. It solves a narrow class of problems, mostly restricted to prefix computations along the path from each vertex to the root.
3. The resulting algorithm is not work-optimal. For example, the level of each vertex can be computed using pointer jumping in $O(\log n)$ time on the CREW PRAM model, but incurring $O(n \log n)$ total number of operations in the worst case.

In a sense, these limitations are caused by representing the tree by parent pointers. The motivation behind the Euler tour technique [48] is to assign some linear order to the vertices of the tree while respecting its topology so that the problem of assigning processors to the vertices can be easily resolved. A natural linear ordering of a given tree structure is the depth-first traversal order.

Conceptually, the Euler tour is a linked list of edges or vertices, which preserves the depth first traversal of the tree. Since parallel linked list ranking can convert a linked list into a linear array, the input tree can now be converted to the most convenient data structure (i.e., an array) for manipulations in a multiprocessor environment.

Thus, any problem on trees that can be solved by depth-first traversal can also be solved by the Euler tour technique. Such problems include the depth-first—e.g., preorder, inorder, postorder—number of each vertex; the level of each vertex in a tree; the number of descendents of each vertex; the left-to-right numbering of leaf vertices; and so on. The underlying approach first constructs the Euler tour of the input tree and then applies linked list ranking. In fact, this process is equivalent to converting an input data structure representing a tree into a linear array and applying parallel prefix computation.

The Euler tour of a rooted tree can be constructed by replacing each edge by two arcs: a *forward arc* from a parent to a child vertex, and a *backward arc* connecting the arcs following the depth-first traversal order of vertices/edges in the tree as illustrated in Fig. 15.3. For a general tree of n vertices, this construction requires $O(\log n)$ time using $n/\log n$ processors on the EREW PRAM model [48]. For a detailed implementation, refer to Ref. [33]. The data structure used in Ref. [33] for the input tree is adjacency lists, each being a circular linked list of nodes representing the arcs emanating from a vertex in the tree. This implementation assumes that two list nodes, corresponding to the two directed arcs induced from each undirected edge of the input tree, are connected by cross pointers. It can be seen that either the explicit rank of each tree vertex among siblings in the input tree or the existence of cross pointers in the adjacency lists is required in order to construct the Euler tour of a tree.

15.4   General Trees and Binarized Trees

A (rooted) general tree is very awkward to handle in parallel processing, often causing read/write conflicts for nodes with more than two children. Also, a tree decomposition technique [15] is required to achieve a logarithmic time scheduling when the depth of the tree is greater than $O(\log n)$, for an n-node tree. On the other hand, a binary tree can avoid read/write conflicts and achieve a logarithmic time scheduling due to its simple regular structure. Hence, converting a general tree into an equivalent binary tree is a widely used technique, called *binarization*, for solving tree-related problems [1, 15, 36]. Given a general rooted tree T with n nodes, the conversion scheme works as follows: For a node u having d children (say, nodes $v_1,\ldots,v_i,\ldots,v_d$), first make $d + 1$ copies of node u (say, nodes $u^1,\ldots,u^{d+1}$). Then node u^{i+1} (resp. v_i) becomes the right (resp. left) child of node u^i. This process is applied to every node in T, as shown in Fig. 15.4. For example, node b having three children, $e, f,$ and g, is duplicated to four nodes $b_1, b_2, b_3,$ and b_4, where b_1 (resp. b_2, b_3) becomes the parent of b_2 (resp. b_3, b_4). Similarly, node e (resp. f, g) is now the child of

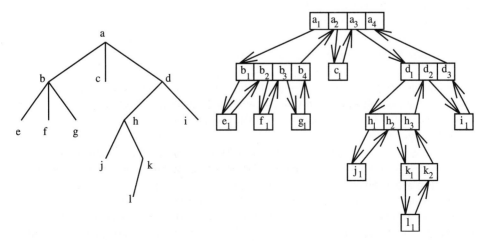

Figure 15.3 Euler tour construction

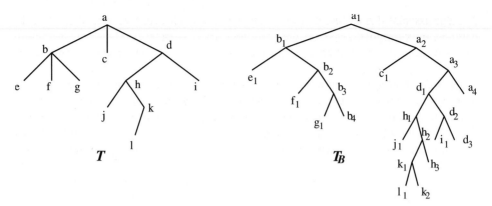

Figure 15.4 Binarization of a tree

b_1 (resp. b_2, b_3). Comparing Figs.15.3 and 15.4, it is clear that the binarized tree can be obtained from the Euler tour construction.

The resulting binary tree, T_B, is regular in the sense that any internal node has exactly two children. It has n leaves and $n-1$ internal nodes; i.e., the number of nodes is still $O(n)$, and also the left-to-right numbering of the leaf nodes in T_B is the same as the postorder numbering of the nodes in the general tree T. Note that the depth of T_B may be greater than $O(\log n)$. But this is not an obstacle in achieving a logarithmic time scheduling, since there is a simple *tree contraction* scheme [1, 41] on a binary tree for solving top-down or bottom-up tree computations in logarithmic time. In fact, for a wide variety of problems, binarization of a general rooted tree simplifies the scheduling process better than handling the general tree directly. In general, bottom-up or top-down algebraic tree computations can be optimally performed in logarithmic time by applying a tree contraction technique to the binarized tree [1, 31]. In comparison with the Euler tour technique, this approach can solve a wider class of problems. For example, the problem of finding the maximum or minimum value of each subtree in a rooted general tree can be easily and efficiently solved by the tree contraction technique applied to the binarized tree, while the Euler tour technique itself cannot handle it [1, 33]. We expect that the tree contraction technique and its applications will be presented elsewhere in this book, and hence we do not attempt to give any details here.

15.5 Euler Tour vs. Parentheses String

As mentioned in Section 15.3, there are three ways to represent a general tree. Here we recall only two data structures representing an input tree—one that specifies either the leftmost child and right sibling of each tree-node, and the other that specifies the parent and the rank of each node among siblings. In this subsection we discuss how a balanced string of parentheses can also be used to represent a rooted general tree or, for that matter, any problem having an implicit or underlying tree structure for its solution (such as an arithmetic expression evaluation) .

A tree can be encoded by a parenthesis string in three different ways [20]. A well-formed string of parentheses is *edge associated* if left (right) parentheses represent forward (backward) arcs, and it is *node associated* if left (right) parentheses represent the first (last) occurrences of the nodes in the Euler tour of the corresponding tree. A variation of the edge associated parentheses is called an *embedded string* representation, in which the parentheses correspond to the arcs and the node identifiers embedded between them correspond to the occurrences of nodes in the Euler tour. Figure 15.5 illustrates parentheses encodings of a general tree.

Given a well-formed string of *n* parentheses stored in an array, the parentheses matching problem is to determine the index of the mate for each parenthesis, which has an obvious (sequential) solution using a stack. However, a parallel solution to this problem is non-trivial, and various efficient algorithms have been proposed in the literature. A simple and elegant parallel parentheses matching algorithm has been suggested by Prasad, Das, and Chen [44]. It is based only on the prefix-sums computations on arrays and runs in $O(\log n)$ time using $O(n/\log n)$ processors on the EREW PRAM model. This algorithm will be discussed in Section 15.6 and, for the time being, we will assume that such an optimal algorithm exists. This section shows the equivalence between the Euler tour and parentheses string representations by solving several tree-related problems by parentheses matching, which are usually solved by the Euler tour technique.

For some problems, a parentheses string (stored in an array) is the natural representation of the input object. For example, an arithmetic expression has an underlying tree structure. In such cases, parentheses matching applied to the input string gives information equivalent to the Euler tour. Thus, parentheses representation combined with a parentheses-matching algorithm requires simple data structures (namely, arrays only), and hence for some applications is advantageous over the Euler tour, which uses more involved data structures such as cross pointers in linked lists or explicit ordering of sibling nodes (see Section 15.3).

Problems solvable by the Euler tour can also be solved by parentheses matching [20]; specifically, the depth-first traversal rank (such as preorder, inorder, postorder number) of each vertex in a tree, the level of each vertex in a tree (which is equivalent to the nesting level of an embedded node in the parentheses string), the number of descendents of each vertex (which is the difference of indices between matching parentheses), and the left-to-

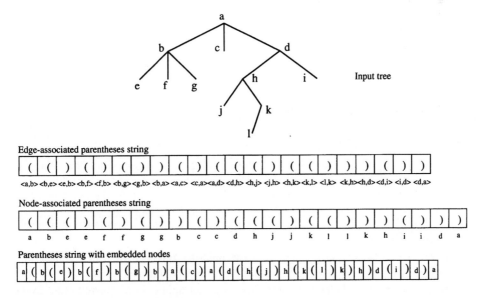

Figure 15.5 Parentheses string representations of a tree

right numbering of leaf vertices (which is equivalent to the numbering of terminal parentheses), and so on. In this context, it is worth mentioning that a few other problems have been efficiently solved in parallel by applying a parentheses matching algorithm. These problems include tree contraction and balancing binary trees [20], the breadth-first traversal of a general tree and sorting a restricted class of integers [12], approximate bin packing [3], and minimum coloring of an interval graph [17]. Many of these problems come from different domains of applications and do not have any obvious correlation with the parentheses string representation.

15.5.1   Converting parentheses to trees

As proposed by Das, Halverson, and Min [20], a well-formed string of parentheses can be converted into the corresponding (unique) tree form in which each node has a pointer to its parent. Given a string of "edge-associated parentheses," imagine dummy nodes embedded between parentheses, and also at the front and tail end of the string. Note that this is equivalent to "embedded string" encoding with dummy nodes. This way, the string represents the occurrences of nodes and edges as they would occur in the Euler tour of the corresponding tree. Also note that each dummy node following (preceding) each left (right) parenthesis corresponds to the first (last) occurrence of a node in the Euler tour. The dummy node at the front (tail) of the string corresponds to the first (last) occurrence of the root node. The consecutive occurrences of the same node are identified as follows: for each dummy node preceding (following) a left (right) parenthesis, the next occurrence of the same node is the dummy node following (pre ceding) the mate of the left (right) parenthesis. Thus the duplicate occurrences of the same nodes are linked together, forming several disjoint linked lists. Node identifiers are assigned in the order of the first (last) occurrences of the dummy nodes in the string, by prefix sums, resulting in preorder (postorder) numbering. The node identifiers assigned to the first or last occurrences are broadcast along the lists, by a multiple linked list ranking algorithm. Finally, parent-child relations are extracted from the node identifiers available on both sides of each parenthesis. The algorithm constructs the tree equivalent to the given parentheses string in $O(\log n)$ time using $(n/\log n)$ processors on the EREW model.

   Figure 15.6 shows how a tree can be constructed from a given edge associated string of parentheses. The dummy nodes enclosing a pair of matching parentheses are connected together, resulting multiple linked lists. Constructing a tree from a string of node-associated parentheses or a string with embedded node identifiers works in a similar way [20].

15.5.2   Tree reconstruction from traversals

Given preorder and postorder traversals of a general tree, the goal is to reconstruct the original tree. A linear-time sequential solution is stack based by recursively partitioning each of the two input traversal sequences into subsequences corresponding to the appropriate subtrees and identifying the roots of these subtrees. The parallel algorithm, proposed in Ref. [20], uses the parentheses matching paradigm. It transforms the reconstruction problem into easily parallelizable subproblems and is based on the following simple facts:

1. If two consecutive nodes u and v in the preorder traversal of a tree appear in a reversed order in the postorder, then v is the leftmost child of u.
2. If two consecutive nodes u and v in the postorder appear in a reversed order in the preorder, then u is the rightmost child of v.

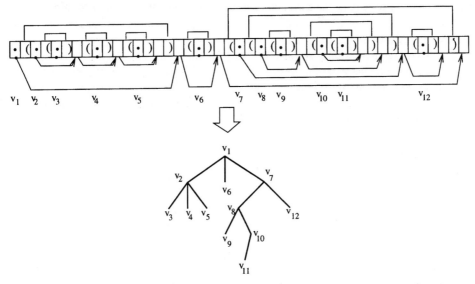

Figure 15.6 Converting a parentheses string to a tree

3. When the consecutive nodes with parent-to-leftmost child relations are linked together in the preorder, the tail nodes of the sublists are leaves in left-to-right ordering.
4. When the consecutive nodes with rightmost child-to-parent relations are linked together in the postorder, the head nodes of the sublists are leaf nodes appearing in left-to-right leaf ordering.
5. When the tail of the ith sublist in the postorder is linked to the head of the $(i + 1)$th sublist in the preorder, the latter is the right sibling of the former. (All sibling relations are identified by this.)

Based on these facts, parentheses can be properly assigned after forming the linked list indicated above. Assign '(' to parent-to-leftmost-child links, ')' to rightmost-child-to-parent links, and ') (' to the sibling links. Then, a linked list ranking constructs a string of parentheses with embedded node identifiers, which can be converted to the corresponding tree as discussed in Section 15.5.1. Clearly, the algorithm requires $O(\log n)$ time using $(n/\log n)$ processors on the EREW model. Figure 15.7 illustrates this idea, and the output tree is the same as that in Fig. 15.5.

A related problem is reconstructing a general tree from its inorder and preorder (or postorder) traversals. In an inorder traversal of a general tree, a node is visited after each of its subtrees. Since a node can appear more than once, we need to identify the first and last oc-

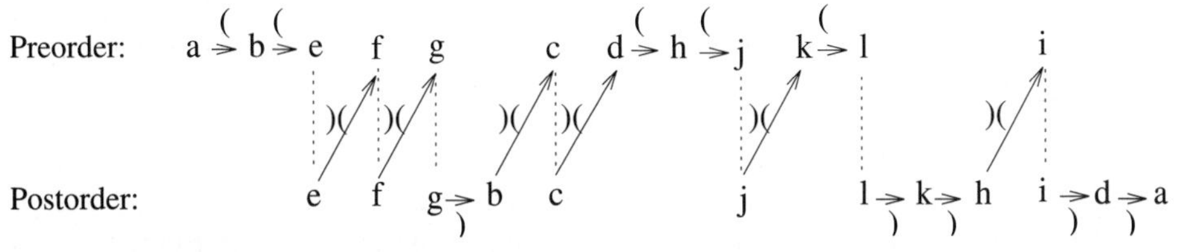

Preorder:

Postorder:

Figure 15.7 Reconstruction of a tree from traversals

currences of each node in the traversals. However, it is not clear how to identify the first and last occurrences of all the nodes in $O(\log n)$ time with $(n/\log n)$ processors on the EREW PRAM model.

The preceding discussions demonstrate a close relationship between the Euler tour technique and the parentheses string representation combined with the use of parallel parentheses matching (PPM) algorithm for solving tree-related problems, although the relationship may be less straightforward in some cases. As we apply the PPM technique [20] to problems for which the input is not in the form of an explicit tree but is represented by the corresponding string of balanced parentheses, the similarities and differences become clearer. It implies that, in such a case, the problem has an implicit underlying tree structure, and the parent-child relations can be extracted using parentheses matching.

15.6   Stacks

There exist problems that are easily solvable in sequential processing with the help of a stack. Such problems include parentheses matching, all nearest smaller values [8], depth-first traversals of a tree or depth-first search of a graph, top-down or bottom-up computations on a tree, etc. This section demonstrates how to get rid of an explicit stack; (which seems to be inherently sequential in nature) for processor scheduling and replace it with another data structure which lead to the desired performance for certain classes of problems. As a candidate example, let us consider the well known parentheses matching problem, described below.

15.6.1   Parentheses matching problem

Given a balanced string of n parentheses, we are to determine the mate of each parenthesis. It is an integral subproblem of parsing and evaluating expressions by computation-tree generation. Therefore, several cost-optimal parallel algorithms have been proposed for parentheses matching, using a variety of techniques [3, 6, 27, 39, 44]. All of these algorithms adopt other data structures (rather than stack) that offer more parallelism in terms of multiple access paths without conflicts.

Before presenting a cost-optimal parallel algorithm for this important problem, we briefly mention the techniques and the data structures used in some of the existing algorithms. The first cost-optimal algorithm due to Bar-On and Vishkin [6] requires $O(\log n)$ time, employing $O(n/\log n)$ processors on the CREW PRAM model. Each processor is initially assigned a $\log n$ size substring that is sequentially reduced after finding local mates. The second phase builds a binary tree. Each leaf is assigned a substring, and an internal node contains the nesting level of the leftmost left and the rightmost right parentheses in the substring spanned by that node. A parent node is easily computed from its children by concatenating their substrings and after finding possible mates. Then concurrent search is performed on this tree to find mates for the leftmost left and rightmost right parentheses of each substring. In the third phase, each processor uses the matching found in the second phase to sequentially scan and match other parentheses in the substring. The key data structure in this algorithm is an array representation of the binary tree.

The first EREW (optimal) algorithm for parentheses matching is from Anderson et al. [3], who also construct a matching tree of $\log n$ nodes such that each node is assigned a processor, and each leaf is assigned to a substring of $2 \log n$ parentheses. Although the implementation of the matching tree is simplified and concurrent reading is avoided, this algorithm uses extensive pipelining for communication of the unique indices given to the matching pairs. The EREW algorithm in Ref. [27] uses linked lists as the underlying data structure and employs a linked list ranking algorithm. The algorithm presented in Ref. [39] is unique in the sense that it does not rely on a binary matching tree but, as a major tool, it uses Cole's parallel mergesort [13], which has a large constant of proportionality. Recently Prasad, Das, and Chen [44] proposed four efficient algorithms for the parentheses matching problem on the EREW PRAM model. In comparison with other approaches, these algorithms use only arrays as their basic data structures and perform prefix-sums computations, and hence are easy to implement.

15.6.2 An optimal EREW PPM algorithm

The following describes a cost-optimal parallel parentheses matching (PPM) algorithm from Ref. [44], which also uses optimal space. The key idea is to avoid building the binary matching tree explicitly, because there are many empty (i.e., unused) spaces in the matching tree whose nodes represent the unmatched parentheses in their spanned substrings using an encoding scheme. Matching is performed at each pair of sibling nodes, and the remaining unmatched parentheses are copied into the parent node. In other words, a "virtual" copying/matching tree is built, and a positional address calculation is done via a parallel prefix-sums computation, thus achieving greater simplicity in less space on the EREW model. Let the input (balanced) string of n parentheses be stored in an array PAREN[$1 \ .. \ n$]. The output will be recorded in the array MATCH[$1 \ .. \ n$] such that MATCH[i] $= j$ if PAREN[i] and PAREN[j] are mates, for $1 \le i \ne j < n$. The three-phase algorithm uses p processors.

Phase I. Divide the input array into p blocks so that processor P_i sequentially finds the locally matched parentheses in the subarray

$$\text{PAREN}\left[(i-1)\frac{n}{p} + 1 \ .. \ i\frac{n}{p} \right]$$

The reduced substring in each block can be packed such that unmatched right parentheses are followed by unmatched left parentheses. In fact, processor P_i encodes its reduced substring into two *representative records*: *rrep* and *lrep*. Each representative record has three fields (*index*, *size*, and *u*) where *index* indicates the position of the unmatched rightmost right (resp. leftmost left) parenthesis in the substring of PAREN, and *size* represents the total number of right (left) parentheses in the reduced substring.

The third field *lrep.u* stores the total number of left parentheses represented by the record *lrep* and those represented by the records to the right of *lrep* in a node. Initially, *lrep* is the only record in a leaf node, and hence *lrep.u* = *lrep.size*. Similarly, *rrep.u* contains the total number of right parentheses represented by *rrep* and by those to its left in a node. Processor P_i is responsible for carrying its representative records along the matching tree. Depending on the current location of its representative records, P_i is associated with either a left or a right sibling. For a left sibling node (odd i), P_i keeps a copy of the total number of unmatched right (resp. left) parentheses in a local variable r_1 (resp. l_1.) Likewise, a right sibling processor P_i keeps the corresponding values in local variables r_2 and l_2.

Phase II. For simplicity, assume that $n = 2^m$ and $p = 2^q$, where $q = \lceil \log \lceil n/m \rceil \rceil$. Let us number the nodes of the matching tree starting from the leaf level down to the root level, and within each level, from left to right. Therefore, the leftmost leaf (at level 1) is node 1, and the rightmost leaf is node p. And the leftmost node at level 2 is node $p + 1$, while the rightmost is node $p + p/2$, and likewise, the root is node $2p - 1$. Thus an odd-numbered node (except the root) is a left sibling.

This phase uses two arrays, LMATCH[1 .. $n/2$] and RMATCH[1 .. $n/2$], to store and align the matching portions of the left and right siblings, respectively, in such a way that the left parenthesis at LMATCH[i] is the match of the right parenthesis at RMATCH[i], for $1 \leq i \leq n/2$. Thus, although each processor carries its two representative records level by level of the virtual matching tree, it does so only in its private memory while keeping track of the current level of the matching tree and the particular node at that level. Whenever the whole (or a part) of a representative record is found to be matched in the corresponding sibling, the representative (or a temporary) record is stored in an appropriate location of LMATCH if the current node is a left sibling. For a right sibling, this information is similarly recorded in RMATCH.

A processor associated with node i of the matching tree would need to know the exact number of parentheses matched at nodes 1 through i to store the parentheses matched at node i into the arrays LMATCH and RMATCH. For this purpose, a two-dimensional array INDEX[1 .. q][1 .. p] is used to precalculate and store the above information such that each processor P_j associated with the jth node (from left to right) at level 1, for $1 \leq j \leq (p/2^{l-1})$ and $(j - 1)2^{l-1} + 1 \leq i \leq j2^{l-1}$, has a private copy INDEX[l][i] as the index into arrays LMATCH and RMATCH. This avoids any potential concurrent reads.

To calculate and fill the array INDEX, the number of parentheses matched at each pair of left and right siblings, $(2j - 1)$th and $(2j)$th node at level l, for $1 \leq j \leq (p/2^l)$ and $1 \leq l \leq q$, is calculated and stored in INDEX[l][$2^l(2j - 1) + 1$]. This step does not require actual building of the matching tree; only the values r_1, l_1, r_2, and l_2 need to be calculated for each pair of siblings. Other entries of INDEX are initialized to zeros beforehand. Next, a parallel prefix sum algorithm is used on array INDEX treated as a linear array. The effect of this step is that now the entire subarray INDEX[l][$2^l(2j - 1) + 1$ .. $2^l(2j)$] contains the required index for the processors associated with the left and the right sibling. Figure 15.8 illustrates this process. In Fig. 15.8a, the number of parentheses matched at each node is depicted. The first (second) element of the tuple in this tree corresponds to the number of right (left) parentheses in each processor. Figure 15.8b shows the calculation of INDEX[1 .. *level*][1 .. p].

Next, the matching tree is constructed conceptually level by level. The matching left and right representative records of each pair of siblings are stored into the arrays LMATCH and RMATCH, respectively, by using the indices available in the array INDEX. Those records that remain unmatched are updated for the parent node. Figure 15.9 illustrates this step. The vertical dotted lines separate the private variables of each processor. A record *lrep* with fields *index*, *size*, and *u* is shown as ($^u_{sizeindex}$. Analogously, notation $size$)$^u_{index}$ is used for an *rrep*. The records matched are shown enclosed within '<' and '>'. In this step at every level, r and l values are exchanged among the processors associated with siblings, the unmatched records are updated for the parent node, and new rs and ls are computed.

Phase III. During this phase, processor P_i is assigned to the subarrays

$$\text{LMATCH}\left[(i - 1)\frac{n}{p} + 1 \ .. \ i\frac{n}{p} \right] \text{ and } \text{RMATCH}\left[(i + 1)\frac{n}{p} + 1 \ .. \ i\frac{n}{p} \right]$$

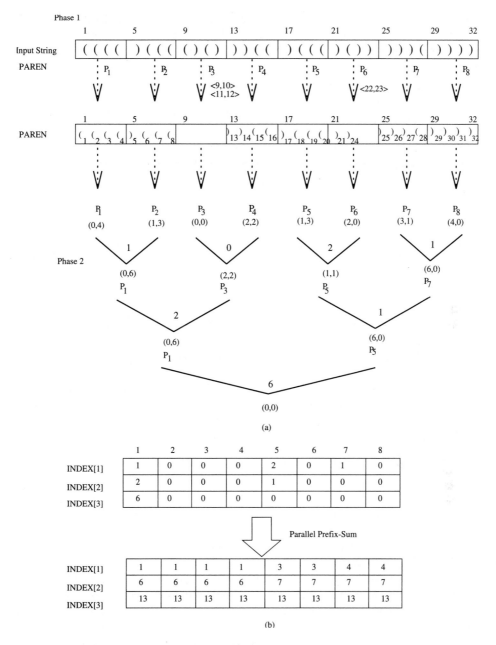

Figure 15.8 Illustration of Phases I and II of PPM algorithm

and it sequentially matches the parentheses. Each P_i (except for P_p) reads its rightmost left representative record *lrep* in its subarray of LMATCH. Let this record be at position j. If $lrep.size + j - 1 > i(n/p)$, then P_i writes a new temporary record at LMATCH[$i(n/p) + 1$]. The size of the record *lrep* is reduced so as to represent parentheses only through the sub-array boundary. Similarly, P_i fills in a temporary right representative record in RMATCH, if necessary. Now, each left and right subarray has its leftmost and rightmost representative records, respectively. Next, each P_i (including P_p) reads *lrep* = LMATCH[$(i - 1) n/p +$

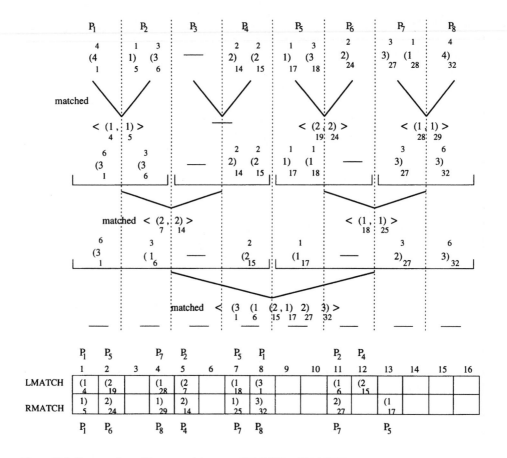

Figure 15.9 Conceptual matching tree and the arrays LMATCH and RMATCH

1] and $rrep = \text{RMATCH}[(i-1)\,n/p + 1]$ and matches the n/p left and right parentheses represented by these two subarrays. Figure 15.10 illustrates Phase III. In this figure, processors P_1 and P_4 fill in temporary records at respectively, the third the ninth locations of arrays LMATCH and RMATCH.

It is easy to see that this algorithm requires $O(n/p + \log p)$ time and $O(n + p \log p)$ space on the EREW PRAM model. Hence, for $P = O(n/\log n)$, it achieves an optimal speedup and the space complexity is $O(n)$.

	1	2	3	4	5	6	7	8	9	10	11	12	13	14	15	16
LMATCH	(1 4	(1 19	(1 20	(1 28	(2 7		(1 18	(1 1	(2 2		(1 6	(1 15	(1 16			
RMATCH	1) 5	1) 24	1) 23	1) 29	2) 14		1) 25	1) 32	2) 31		2) 27		1) 17			

P_1 P_2 P_3 P_4 P_5 P_6 P_7 P_8

<4,5>   <20,23>   <7,14>   <18,25>   <2,31>   <6,27>   <16,17>   —
<19,24>   <28,29>   <8,13>   <1,32>   <3,30>   <15,26>

Figure 15.10 Illustration of Phase III of PPM algorithm

Remarks. We have seen earlier (and will also see in the next section) that many problems related to trees and special graphs, (e.g., tree traversals, bottom-up computations, breadth-first traversal of a tree, interval graph coloring, and so on) that have optimal sequential solutions using stacks and/or queues can be efficiently solved in parallel by the application of a parentheses-matching algorithm. However, not all problems having optimal solutions based on stacks or queues have fast parallel solutions using an alternate data structure. A typical counterexample is the depth-first traversal of a general graph for which it is unlikely to have a polylogarithmic time parallel algorithm [46].

15.7 Queues

In sequential processing, queues play important roles as data structures. For example, breadth-first search of a graph can be efficiently implemented using a queue. The elegant parallel algorithm presented in this section for breadth-first tree traversal is an example of replacing a sequential data structure by another data structure suitable for parallel processing. We will also demonstrate how problems typically solvable by using stacks can be solved by parallelizing queue-based algorithms.

15.7.1 Breadth-first tree traversal

Chen and Das [12] designed a simple algorithm for the breadth-first traversal of a general tree with the help of the parentheses-matching algorithm. Since in a breadth-first traversal, nodes (or the edges) of the tree are visited in a level-by-level order, the algorithm works as follows. We first compute the level number of each arc corresponding to an Euler tour of the tree. In order to visit the arcs at any level from left to right, we delete the leftmost and rightmost arcs on each level, and then assign a left (resp. right) parenthesis to each remaining backward (resp. forward) arc. This assignment leads to a well formed string of parentheses. An application of the parentheses matching algorithm finds the mate of each parenthesis (i.e., arc) such that two arcs belonging to a level of the tree will form a matched pair if they are to be visited consecutively on that level. For jumping from one level to the next, let us define the match of the rightmost arc in level i to be the leftmost arc in level $i + 1$.

Let $A = (x, y)$ indicate an arc leaving node x and entering node y. A matched pair of arcs (A, B), where $A = (x, y)$ and $B = (w, z)$, is interpreted as $\text{NEXT}(x) = z$, which means that node z will be visited after node x in the breadth-first traversal. Since there is no level above the root, we need to specify that $\text{NEXT}(x) = y$ if arc $A = (x, y)$ is the leftmost arc in the first level. Clearly, the NEXT function converts the tree into a linked list of tree-nodes, and an application of the linked list ranking algorithm enumerates the nodes in the breadth-first traversal order. The algorithm runs in $O(\log n)$ time using $(n/\log n)$ processors on the EREW PRAM model, and it is formally described below.

Algorithm Breadth-First Traversal;

1. Compute the arc sequence corresponding to the Euler tour of an input tree.
2. Compute levels of these arcs.
3. Identify the leftmost and rightmost arcs at each level and delete them from the arc sequence by performing prefix-max and suffix-max operations on the levels of the arcs.
4. Assign a left (right) parenthesis to each backward (forward) arc in the sequence obtained in Step 3, and designate the parentheses sequence to be an *Euler sequence*.

5. Solve the parentheses matching problem corresponding to the Euler sequence to obtain the successor of each backward arc except the rightmost one at each level. The successor of the rightmost arc, a_i, at level $l(a_i)$, excluding the highest level, is the leftmost arc at level $l(a_i) + 1$.

6. Order the nodes by defining the NEXT function. Given a backward arc $\overline{X} = (x, y)$ and its successor $Y = (u, v)$, define $NEXT[x] = v$. Also define $NEXT[u] = v$ if $\langle u, v \rangle$ is the first arc in the arc sequence—this is for the root node, u.

7. Apply a linked list ranking algorithm to the NEXT function. The ranks of tree nodes yield the breadth-first traversal, where the rank of a node is the number of nodes preceding it (including itself) in the linked list, NEXT.

Figure 15.11 traces the breadth-first traversal algorithm. The Euler tour of the input tree is shown as a sequence of arcs (also labeled with capital letters). The levels of these arcs are computed as follows. If we assign a weight of +1 (resp. −1) to each forward (resp. backward) arc, the prefix-sum corresponding to a forward arc is its level, whereas the level of a backward is its prefix-sum plus one. The leftmost (or rightmost) arcs at each level can be easily identified by computing prefix-max (or suffix-max) on the levels and then deleted as required by the preceding algorithm.

After assigning appropriate parentheses to the remaining arcs (called Euler sequence) and applying a parentheses matching algorithm, the NEXT function (basically a linked list) is defined as in Step 6. An application of the linked list ranking yields the breadth-first traversal rank of the tree-nodes.

15.7.2   Sorting integers in a restricted class

Let us sort a sequence of integers such that any two consecutive elements in the sequence differ in value by at most unity. The sorting must be *stable*; i.e., two identical elements in

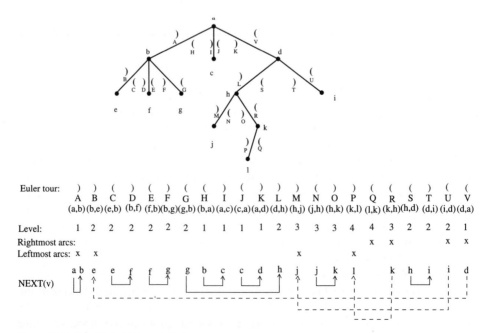

Figure 15.11 Breadth-first traversal of a tree

the input sequence must preserve their relative positions after sorting is complete. This kind of sequence is not artificial and can represent many common problems. For instance, the nesting levels of a string of balanced parentheses satisfy this property. Also, the level numbers of the nodes of a tree in its Euler tour follow this sequence.

Since this problem is a special case of bucket sorting, it can be trivially solved using a stack. However, there is no known parallel algorithm which (stable) sorts n integers in the range $[1 \ .. \ n]$, each representable in $O(\log n)$ bits, in $O(\log n)$ time using $O(n/\log n)$ processors on the EREW PRAM model. If we relax the *stability* property, the problem is simpler in nature.

A $O(\log n)$-time, work-optimal EREW algorithm for this restricted class of integers (RCI) has been proposed by Chen and Das [12] by applying the parallel parentheses matching technique. Again, the use of stack is conveniently avoided. Surprisingly, the breath-first traversal of a suitably constructed tree gives a simple solution. Thus, one can find a close relationship between stack- and queue-based problems (at least on trees) in the context of parallel computing. The underlying idea is to construct a tree such that the level numbers of nodes visited while traversing the tree according to its Euler tour, correspond to the given sequence of integers in the subclass. Starting with a node corresponding to the first element of the sequence, we assign a forward (parent-child) arc if two consecutive elements are in the increasing order, and a backward (child-parent) arc if they are in the decreasing order. For identical elements, we remain at the current node. Clearly, distinct elements are assigned distinct levels in such a tree, and the breadth-first traversal of this tree sorts the distinct elements in the desired order. However, identical elements belong to the same level and their stable sorting makes the parallel algorithm more nontrivial.

15.7.3   Minimum coloring of an interval graph

Given a collection of n intervals $\mathbf{I} = \{I_i = [a_i, b_i] \mid a_i \le b_i\}$, there exists an *interval graph* $G_I = (V, E)$ such that the node set is $V = \mathbf{I}\}$, and the edge set is $E = \{(I_i, I_j) \ I_i \cap I_j \ne 0\}$. The problem is to assign colors to the nodes of the interval graph using the fewest number of colors such that no two adjacent nodes have the same color.

A stack-based sequential algorithm for coloring an interval graph can be designed as follows. The endpoints a_i, b_i for $1 \le i \le n$ are sorted and processed in the ascending order. If a starting endpoint a_i of an interval is encountered, a color is popped off the stack and assigned to the corresponding node I_i. If the endpoint b_i is encountered, it indicates that all remaining nodes are not adjacent to this node and, therefore, the associated color is pushed back onto the stack for reuse.

The parallel algorithm from Das and Chen [17] again replaces the explicit stack as follows. This is probably another problem where the concept of sequential stacks and queues interplay with each other. After sorting the endpoints in ascending order, assign a left (resp. right) parentheses is assigned to each endpoint a_i (resp. b_i). Compute the nesting levels of these parentheses and sort them according to the levels, using the algorithm for sorting the restricted class of integers (RCI) discussed in Section 15.7.2. Then, build multiple linked lists, one for each level. Assign a distinct color to the first element of each list, and distribute it to the others in that list using a linked list ranking method.

The interval graph coloring problem is thus cost-optimally solved in $O(\log n)$ time using n processors on the EREW model. It can be shown that the algorithm performs correctly and the minimum number of colors required is equal to the deepest nesting level of a parenthesis.

Figure 15.12 illustrates the coloring algorithm for an input $I = \{I_1, I_2, \ldots, I_7\}$ of seven intervals. After sorting the endpoints and assigning them appropriate parentheses, the nesting levels of these parentheses are computed. Then the parentheses (and hence the endpoints) are sorted in the ascending order of their nesting levels. It is clear that the list of elements for each level of this sorted sequence must start with a_i and terminate with b_j, for some i and j. Moreover a_is and a_js alternate at each level.

Now, multiple linked lists are built as follows. For each b_i which is not the last element at $level(b_i)$, the interval I_i is linked to I_j if b_i is followed by a_j in the sorted sequence. At the final step, color an interval I_i with $level(a_i)$ if a_i is the first element at $level(a_i)$; otherwise color of I_i is assigned zero. Then apply a multiple linked list ranking algorithm such that the intervals belonging to the same list receive the same color.

15.8   Priority Queues (Heaps)

Priority queues and heaps are very useful data structures with numerous applications. For example, in a branch-and-bound (B&B) algorithm, a priority queue can be used to store the live nodes of a state-space tree. Since the insertion and deletion operations in a binary heap can be performed efficiently, a heap is often used for implementing a priority queue data structure. In the sequential version of a B&B algorithm, a *delete-think-insert* cycle is performed repeatedly. At the beginning of each cycle, a highest priority item is deleted from the top of the heap and some processing (or thinking) is done on that item which possibly generates new items. These items are then inserted back into the heap.

On the other hand, in a parallel B&B algorithm, p items (where p is the number of processors) of the highest priority are expected to be deleted at each cycle for the think phase to start. Also, the new items have to be inserted efficiently in parallel. Although it is appar-

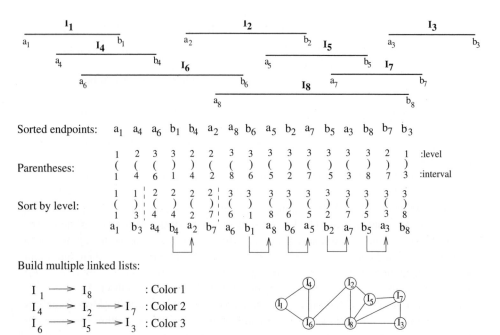

Figure 15.12 Minimum coloring of an interval graph

ent how to delete p items from the priority queue in constant time, it is not clear how to insert the generated items efficiently. Therefore, managing heaps in parallel is an important problem.

Various mechanisms have been proposed to manipulate heap operations on the shared memory parallel computers [11, 21, 26, 43, 45]. Some of these approaches [11, 45] deal with concurrent heaps (rather than truly parallel), which make use of only $O(\log n)$ processors effectively, thus offering a limited parallelism. Deo and Prasad [26] presented a new data structure, called *parallel heap,* which allows multiple inserts and deletes simultaneously for an EREW PRAM model. (The algorithm due to Pinotti and Pucci [43] is on the CREW model.) This algorithm requires $O(\log n)$ time for insertion and deletion of p items in an n-item binary heap and hence achieves a linear speedup using p processors. However, this approach suffers from a few limitations:

1. Linear speedup is not achieved if the think time is larger than $O(\log p)$.
2. At each delete-think-insert cycle, only r processors are active in the think phase or the insert-delete phase, where $r < p$ is the number of items per node in the heap.
3. The working memory space for managing the parallel heap is $O(n)$.

In this section, we describe a simplified yet efficient implementation of a parallel heap on the EREW PRAM model, overcoming the above limitations. This algorithm is proposed in Refs. [21, 22]. A parallel heap of n items will be represented as a complete binary tree of depth $d = \lceil \log (m + 1) \rceil - 1$ and $m = \lceil n/x \rceil$ nodes, each containing x items, for $1 \leq x \leq p$. The *parallel min-heap property* implies that a parent node does not contain larger values than its children.

The parallel heap algorithm will be described with respect to its application in the parallel branch-and-bound algorithms. Thus, we assume that a delete-think-insert cycle is performed repeatedly, a processed item generates at most two new items, and each processor requires the same amount of think time, which may be arbitrary. (For a more general case, refer to Ref. [21] in which a parallel heap is represented by a complete k-ary tree, and each processing item generates at most a constant number of new items.) In our approach, all processors are effectively utilized since p items of the highest priority are deleted from the parallel heap in $O(\log p)$ time and at most $2p$ new items are inserted in $O(\log n)$ time. With or without incorporating the think time, a linear speedup is guaranteed. Furthermore, the strict priority ordering of a sequential heap is retained.

15.8.1 Preliminaries

A heap of maximum size n can be implemented conveniently by a one-dimensional array such that the root (i.e., node 0) occupies the location 0. The left and right children of node i occupy locations $2i + 1$ and $2i + 2$, respectively, whereas its parent is at location $\lceil i/2 \rceil - 1$. A data field, TARGET, is associated with each level of a heap to indicate the index of the first available node (called *target* node) for inserting new items. (Note that the root is at level 0.) The value of TARGET is *nil* for a level that is full.

The *insertion path* of the target node is its path from the root [45]. Suppose the first non-nil target node is at level t, and FIRST$(t) = 2^t - 1$ is the index of the first (leftmost) node at that level. Then, the insertion path, *IP*, of the target node can be represented by the sequence of binary digits obtained by expressing TARGET—FIRST(t) in radix two. That is, $IP = (e_1 e_2 \ldots e_t)_2$, where $e_1 = 0$ or 1, respectively, implies that the left or right child of the root is in the insertion path. The interpretation of ei for $2 \leq i \leq t$, is similar.

Figure 15.13 shows an example of a parallel min-heap of 39 items with the insertion path from the root to the target node 13. Thus $x = 3$, TARGET = 13, FIRST(3) = 7, and the insertion path $IP = (110)_2$. Note that TARGET − FIRST(3) = $13 - 7 = 6 = (110)_2$.

Let x be the number of items in a heap-node, and let there be $p \leq n$ processors available. Each level of the parallel heap is associated with $\lceil x/\log x \rceil$ processors for maintenance operations, if $x \leq 2$, and one processor when $x = 1$. Since x (and hence the number of maintenance processors assigned per level) depends on the total number of available processors, the actual implementation of the heap maintenance algorithm is divided into three parts, although we apply variants of a single technique. Depending on the range of p, the following three cases are distinguished:

1. $\sqrt{n} < p \leq n$

2. $\lceil \log (n + 1) \rceil \leq p \leq \sqrt{n}$

3. $1 \leq p \leq \lceil \log (n + 1) \rceil$

15.8.2   Data structures associated with a parallel heap

The parallel heap is represented as an array PHEAP[0 .. m − 1] where m = $\lceil n/x \rceil$ is the total number of nodes and $1 \leq x \leq p$. Each heap-node contains a record of three fields: ITEM, #ITEMS, and EXPECTED. The array ITEM[0 .. $x - 1$] stores data items in the node, while #ITEMS and EXPECTED, respectively, indicate how many items are currently in it and eventually expected to be in it. An item that should be in a node but not yet available is called an *expected item*.

To maintain the parallel heap property at each level, a record called BLOCK is used. It consists of several fields: WNODE (index of a node to be processed), WTYPE (the type of reheapification performed at WNODE), INSERT_ITEM (an array of size LENGTH containing new items to be inserted into the TARGET node), and INSERTION_PATH (the re-

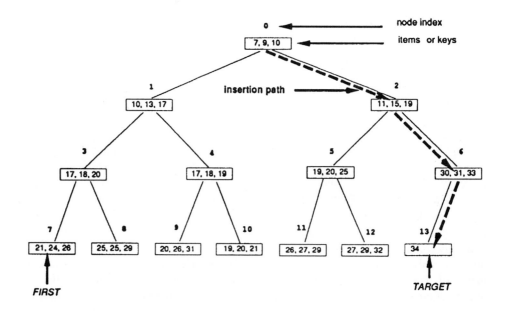

Figure 15.13 A parallel min-heap of 39 items

maining path from WNODE to TARGET). If WNODE = –1, then no operation is performed at that level during a pipeline cycle, defined later. If WTYPE = *delete*, then p smallest items are deleted (i.e., delete-reheapification). On the other hand, if WTYPE = *insert* then insert reheapification (i.e., insertion of at most $2p$ new items) takes place.

The remaining data structures include an integer variable LAST and three arrays, READY, NEW_ITEM, and WANTED_BY. The array READY contains p items (if the heap is not empty) of the highest priority for the next think phase to start. A processor P_i handles READY[i], for $0 \le i < p$, and after the current think phase is complete, P_i stores its two new generated items in NEW_ITEM[$2i$] and NEW_ITEM[$2i + 1$]. The field LAST contains the index of the first node i such that EXPECTED(i) $< p$. The array WANTED_BY is divided into $\lceil p/x \rceil + 1$ subarrays such that WANTED_BY[$x * \lceil p/x \rceil - k) .. x * (\lceil p/x \rceil - k+1) - 1$] is for the node LAST $- k$, for $0 \le k \le \lceil p/x \rceil$. This array, initialized to –1, indicating no data movement, is used by the processors at each level to check whether to move items from the bottom of the heap to the array NEW_ITEM.

Let $w \le p$ be the total number of items stored in the array NEW_ITEM after the think phase. Now the items belonging to this array and the root are merged into a single list, the smallest p items of which are sent to READY for the next think phase, while the next smallest x items are stored at the root. Although p processors participate in the think phase, some processors may generate less than two items and, hence, less than p items may be stored at NEW_ITEM. Under this situation, the required p–w items are brought from the heap. If these *wanted* items are already in their target nodes, then all of them are moved to NEW_ITEM. However, some of them may still be available somewhere along the insertion paths. If so, the corresponding elements in WANTED_BY point to the locations in NEW_ITEM where they should be moved to. In other words, an element j of WANTED_BY array is set to a non-negative integer g if WANTED_BY[j] is requested by NEW_ITEM[g].

At the next iteration of the insertion operation, the processors at each level l of the heap first check whether the items in INSERT_ITEM(l) are requested by the array WANTED_BY. This mechanism ensures that there is no need to wait for the arrival of the wanted items to the bottom of the heap. Instead, these items can be immediately fetched from the next iteration of the insertion process.

Example Let $p = 6$ and $x = 3$. Suppose NEW_ITEM contains only two items $< 15, 36 >$, lacking four more items. Figure 15.14 depicts that items $< 26 >$ and $< 24, 27 >$ are still along the insertion paths. Thus, in this figure, two levels are involved in the insertion operation. However, they follow different insertion paths because of two different target nodes, namely nodes 84 and 85. which are indexed by LAST – 1 and LAST, respectively.

The data fields #ITEMS and EXPECTED associated with LAST (i.e., node 85) indicate that one item has not arrived yet. Thus, we set the corresponding element of WANTED_BY to 5, indicating the index of NEW_ITEM. In the same manner, we set the elements of WANTED_BY. Since the first item pointed to by LAST – 1 is already there, it is immediately moved to NEW_ITEM[2]. So far the two arrays are

WANTED_BY = $< - 1, - 1, - 1, - 1, 3, 4, 5, - 1, - 1 >$, and
NEW_ITEM = $<15, 36, 15, -, -, -, -, -, -, >$

During the next iteration of the insertion process, the wanted items can be moved to NEW_ITEM directly in parallel, using at most p processors. Finally, WANTED_BY is reset for subsequent use, and thus NEW_ITEM = $< 15, 36, 15, 24, 27, 26, -, -, -, -, -, - >$.

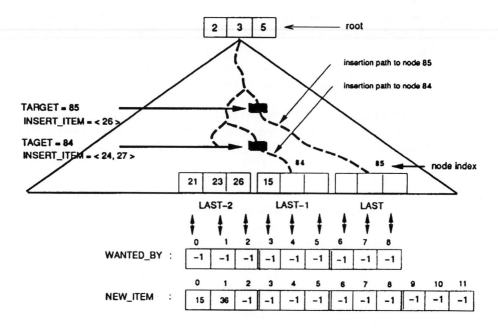

Figure 15.14 Moving wanted items

15.8.3 Generic Parallel Heap Operations

In order to maintain the parallel heap property, two generic operations—*delete-reheapification* and *insert-reheapification*—are performed. Since each iteration of the parallel reheapification is performed in a pipelined fashion, at most one node at each level may violate the parallel heap property. Such a configuration is called a *partial parallel heap* as long as the root contains x items with the highest priority from the entire heap. The parallel reheapification procedures described below are for partial parallel heaps. An iteration of the parallel delete-reheapification at level l is performed if WTYPE(l) = delete, implying that a node at level l is updated without satisfying the partial parallel heap property. The main objective of the reheapification procedure is to restore the parallel heap property at the current node.

Procedure Parallel_Delete_Reheapification(l);

```
current node ← WNODE(l).
if the current node is not a leaf then
Step 1: {Check if the current node satisfies the partial parallel heap
          property}
      WTYPE(l) ← null; WNODE(l) ← - 1 {reset}
      if the items at the current node are no larger than items at its
          children then return.
Step 2: {Readjust items in order to satisfy the partial parallel heap
          property at the current node}
      Let j be the index of a child containing the largest item and k the
          index of the other child.
      Merge the items at the current node with those at its children
          and move the smallest x items to the current node.
Step 3: {Fill in the children nodes}
```

```
            Move the next smallest 2 items to the node j. {Node j will satisfy
                the partial parallel heap property because the largest item at the
                new j is no larger than that at the old j.}
            Move the remaining items to the other child k.
Step 4: {Prepare for next iteration}
        WTYPE(l + 1) ← delete; WNODE(l + 1) ← k.
```

An iteration of the parallel insert-reheapification at level *l* is performed if WTYPE(*l*) = *insert*; i.e., whenever a set of new items are inserted at this level. Such an iteration is initiated at the root when a set of new items (i.e., array NEW_ITEM) is added to a parallel heap.

Procedure Parallel_Insert_Reheapification(l);

```
Step 1: {Move wanted items to NEW_ITEM, if necessary}
        do the following in parallel:
            for each item in INSERT_ITEM(l)
            if the corresponding WANTED_BY field is set to a
                non-negative integer j
            then move the item to NEW_ITEM[j].
Step 2: {Merge the items at INSERT_ITEM(l) and current node}
        current node ← WNODE(l); WTYPE(l) ← null; WNODE(l) ← - 1 {reset}
        if the buffer INSERT_ITEM(l) is empty then return.
            {This happens when all items at INSERT_ITEM(l) have moved to
                NEW_ITEM in Step 1}
        else merge the items at the current node and INSERT_ITEM(l) into a
            single list.
Step 3: {Move the merged items to the target nodes}
        if current node is a leaf then move the merged items to the current
            node and return.
        else
            Move the smallest x items to the current node and
            Move the remaining items to INSERT_ITEM(l + 1).
WTYPE(l + 1 ) ← insert
WNODE(l + 1) ← the index for the child lying on the insertion path
TARGET(l + 1) ← TARGET(l).
```

The processors at each level *l* of a partial parallel heap performs either an iteration of the insert-reheapification or delete-reheapification on WNODE(*l*) according to the value of WTYPE(*l*). One such iteration is called a pipeline cycle as described below.

Procedure Pipeline_Cycle(l);

```
If WTYPE(l) = delete then Parallel_Delete_Reheapification(l)
else if WTYPE(l) = insert then Parallel_Insert_Reheapification(l).
```

Note that our pipeline cycle is different from that in Ref. [26], where one cycle includes an iteration of the delete-reheapification and two iterations of the insert-reheapification.

15.8.4   Parallel heap implementation

The implementation uses the parallel prefix algorithm for packing the array NEW_ITEM, the optimal parallel mergesort from Cole [13] for sorting new generated items, and the adaptive bitonic merging algorithm [10] for maintaining the parallel heap property at each level of the heap. For input size $O(n)$, the bitonic merge algorithm requires $O(\log n)$ time using $(n/\log n)$ processors, whereas Cole's mergesort has $O(\log n)$ time complexity employing n processors on the EREW PRAM model. Thus, an iteration of the parallel delete- or insert-reheapification, and hence a pipeline cycle, requires $O(\log n)$ time. As mentioned earlier, the heap maintenance algorithm is divided into three distinct cases based on the number of processors available. Let us outline here the case for $\sqrt{n} < p \leq n$. (For implementation of other two cases, refer to Ref. [22].) In this case, $x = p$ and m $= \lceil n/p \rceil$. Each level in the heap is associated with $\lceil p/\log p \rceil$ processors.

Because the implementation keeps in mind the branch-and-bound algorithms, a delete-think-insert cycle is performed in two phases: *think phase* and *insert-delete* phase. As soon as the partial parallel heap property is satisfied at the root, all p processors are immediately switched to the think phase in which processor P_i, for $0 \leq i < p$, accesses the item READY[i]. After this phase is complete, P_i generates at most two new items and places them in the locations z_i and $z_i + 1$ of the array NEW_ITEM. Next. in the insert-delete phase, all processors are switched back to maintain the partial parallel heap property.

Procedure Insert_Delete_Phase;

```
Step 1: {Pack new generated items}
for l ← 0 to d do in parallel
     if l = 0 then Pack the w new items generated from the think phase {the
          root level}
     else Perform Pipeline Cycle(l).
Step 2: {Move the wanted items to NEW_ITEM, if necessary}
if (0 < w < p) then
Move last p − w items to NEW_ITEM if they are present at the bottom of
     the heap or set their corresponding WANTED_BY fields if they are still
     in their insertion paths. Update the related data fields accordingly.
     if any WANTED_BY field is set then
          for l ← 1 to d do in parallel Pipeline_Cycle(l).
     w ← p
Step 3: {Sort the items in NEW_ITEM}
if w = 0 then return else Perform Cole's parallel mergesort on NEW_ITEM.
Step 4: {Fill in the array READY}
for l ← 0 to d do in parallel
     if l ← 0 then Perform Pipeline_Cycle(l)
     else
        Perform bitonic merge on the items at the root and the first p items
             at NEW_ITEM if w ≥ p (or all items if w < p).
        Move smallest p items of merger to array READY and the remaining
             ones to the root.
        Update WNODE(0) and WTYPE(0) to perform one iteration of delete
             process.
        {This ensures the partial parallel heap property at the root}
for l ← 0 to d do in parallel Pipeline_Cycle(l).
Step 5: {Insert the remaining items in NEW_ITEM:}
if w > p then
     Let r be the number of remaining items in NEW_ITEM.
     v ← x − #ITEMS(LAST) − EXPECTED(LAST) {# items LAST node can
          accommodate}
     Update WNODE(0) and WTYPE(0) to perform one iteration of the insert
          process.
     TARGET(0) ← LAST
        Establish the insertion path according to LAST.
        if r < u then move remaining items in NEW_ITEM to INSERT_ITEM(0);
          r ← 0
```

```
else
    Move v of the remaining items at NEW_ITEM to INSERT_ITEM(0).
    LAST ← LAST + 1; EXPECTED(LAST) ← r − v; r ←r − v
for l ← 0 to d do in parallel Pipeline_Cycle(l).
if r > 0 then {second iteration of the insert process}
    Update WNODE(0) and WTYPE(0) and perform one iteration of
        insertion.
    TARGET(0) ← LAST
    Establish the insertion path according to LAST.
    Move the remaining r items in NEW_ITEM to INSERT_ITEM(0).
    for l ← 0 to d do in parallel Pipeline_Cycle(l).
```

Example   Figure 15.13 traces the basic flow of the preceding algorithm on a parallel min-heap. Let $p = 4$ and $x = 3$. We assume WTYPE(1) = *delete*, WNODE(1) = 1, WTYPE(2) = *insert*, WNODE(2) = 6, INSERT_ITEM = < 32, −, −, >, TARGET(2) = 13, and NEW_ITEM = < 11, 2, 17, 9, 6, 14, 12 >.

In Step 1 of the procedure Insert_Delete_Phase, several operations continue in parallel. For instance, NEW_ITEM is packed at node 0. Node 1 performs a *delete* operation and returns from Step 1 of the procedure Delete_Reheapification. An *insert* operation takes place at node 6 such that content of this node becomes < 30, 31, 32 >; INSERT_ITEM(3) ← < 33, −, −>; WNODE(3) ← 13; and WTYPE(3) ← insert.

In Step 3, the array NEW_ITEM is sorted as < 2, 6, 9, 11, 12, 14, 17, − >. Again, in Step 4, the following operations are performed in parallel. At node 0, the merged list is < 2, 6, 7, 9, 9, 8, 10, 11, 12, 14, 17, − >. The array is assigned as READY ← < 2, 6, 7, 9 >, the remaining items being < 12, 14, 17 >. Thus the content of node 0 becomes ← < 9, 10, 11 > along with WNODE(0) ← 0 and WTYPE(0) ← *delete*. At the same time an *insert* operation on node 13 yields its content < 33, 34, −>. Then a parallel delete operation takes place at node 0.

Step 5 yields LAST = 13; WNODE(0) ← 0; WNODE(0) ← *insert*; TARGET (0) ← 13; Insertion-Path = (node 0 → node 2 → node 6 → node 13). Since, in this case, the number r of remaining items is not less than the number v of items that LAST node can accommodate, INSERT_ITEM(0) ← < 12, −, − >; LAST ← 14; EXPECTED(14) ← 3 − 1; and r ← 3 − 1. Then parallel *insert* operation is performed at node 0. Since $r > 0$, the second iteration of the insertion takes place for the remaining items < 14, 17, − >. Therefore, we perform (in parallel) insert operations at nodes 0 and 2, and delete operation at node 4.

Figure 15.15 shows the current status of the heap after Step 5. Since the partial parallel heap property is satisfied at the root, all processors are switched to the think phase.

In can be shown that the parallel heap management algorithm is correct [22]. In particular, after each insert-delete phase in a partial parallel heap, the values of the items in the root are no larger than those in its descendents, and the items in the array READY are in the sorted order. Furthermore, the items in READY have no value larger than those in the root.

Recalling that one pipeline cycle requires $O(\log n)$ time for the considered range of p, and considering the time complexities of packing and mergesort, it is easy to show that the overall (parallel) time for an insert-delete phase to delete p smallest items and to insert at most $2p$ items is $O(\log n)$. Thus, an optimal speedup is achieved.

15.9   Search Trees/Dictionaries

A dictionary provides a very important data structure by keeping data items or keys in an ordered fashion. Typical operations on a dictionary include searching for a given key, insertion of new keys preserving dictionary order, and deletion of existing keys. For effi-

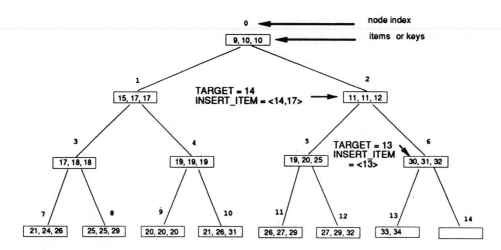

Figure 15.15 Trace of procedure Insert_Delete_Phase

cient manipulation of these operations (in logarithmic time), various kinds of search trees are used, which is the topic of this section. We will discuss how to construct and manipulate in parallel search trees corresponding to a sorted list of data items.

15.9.1   Search tree construction

Binary search trees, AVL trees, m-way search trees, 2-3 trees, 2-3-4 trees, B-trees, and (a, b)-trees are efficient data structures for representing sorted lists of items or keys. There exist parallel algorithms for constructing m-way search trees [24] and 2-3 trees [49]. This subsection focuses on a unified algorithm, devised by Das and Min [23], for constructing height and weight balanced search trees of various kinds on the PRAM model.

Basic definitions. An *m-way search tree* is a tree satisfying the following conditions:

1. Every node has at most m children.
2. An internal node with d children contains $d - 1$ keys.
3. A leaf node contains at most $m - 1$ keys.
4. The value of the kth key in a node is larger than the keys in the kth subtree and smaller than the keys in the $(k + 1)$th subtree, where $1 \le k \le m - 1$.

For $a \le 2$ and $b \le 2a - 1$, an (a, b)-tree is a search tree in which

1. All the leaves are on the same level.
2. The root node has at least two children, and every other internal node has at least a children.
3. Every node has at most b children.
4. An internal node with d children contains $d - 1$ keys.
5. The number of keys in a leaf node is at least $a - 1$.
6. The value of the kth key in a node is larger than the keys in the kth subtree and smaller than the keys in the $(k + 1)$th subtree.

Note that 2-3 trees, 2-3-4 trees, and B-trees are subclasses of (a, b)-trees, where $a = 2$ and $b = 3$; $a = 2$ and $b = 4$; and $a = \lceil b/2 \rceil$, respectively.

A tree is *height balanced* if the height of a subtree rooted at a node differs at most by one from that of another subtree rooted at a sibling node. A tree is *weight balanced* if the number of keys contained in a subtree rooted at a node differs at most by one from that in another subtree rooted at a sibling node. A tree of height h is full if every level has the maximum possible number of nodes. A tree of height h is *complete* if level i, $1 \le i \le h - 1$, has the maximum possible number of nodes, and the remaining nodes at level h occupy the contiguous positions from the left.

Bottom-level balanced search trees. Let us explain the bottom-level balancing scheme and derive properties of the m-way search trees and the (a, b)-trees constructed by this scheme. We also show that the characterization of the m-way search trees is incorporated into that of the (a, b)-trees, which makes the derivation of a unified algorithm possible.

The *bottom-level balancing scheme* is as follows: a full search tree of height h that can accommodate the given number (N) of keys is used as the underlying tree. The underlying tree is filled up with the maximum number of keys from levels 1 through $h - 1$. Let R be the number of remaining keys to be assigned to level h, and L be the number of nodes on level h of the underlying tree. Then the bottom level balancing scheme assigns

$$\left\lceil R \cdot \frac{j}{L} \right\rceil - \left\lceil R \cdot \frac{(j-1)}{L} \right\rceil$$

keys to the jth node on level h. This scheme constructs search trees that are height balanced as well as weight balanced.

m-way search trees. First, we describe how to construct the m-way search trees constructed by bottom-level balancing. For $N \in [m^{h-1}, m^h - 1]$, a full m-way search tree of height h is used as the underlying tree. The structure of this kind of m-way search trees is illustrated in Fig. 15.16. Note that $h = \lceil \log_m (N + 1) \rceil$ is the minimum height for any N in the above range. Also, an internal node contains $m - 1$ keys, while a leaf can have 0 to $m - 1$ keys.

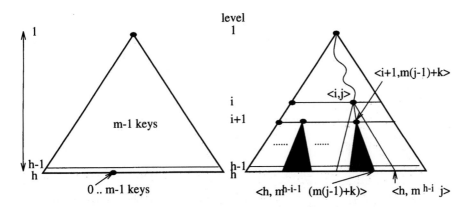

Figure 15.16 m-way search tree constructed by bottom-level balancing

Now, the goal is to find a mapping function between the indices of the keys in the m-way tree being constructed and the sorted ranks of the keys given as input. Since the distribution of the keys is predetermined in a bottom-level balancing scheme, we can easily derive the formulas for such a mapping. Thus, the problem of constructing an m-way search tree reduces to the problem of evaluating the formulas for such a mapping. For details, refer to Ref. [23].

(a, b)-trees. The problem of constructing a bottom-level balanced (a, b)-tree is a little subtler than m-way search trees. The difficulty comes from the fact that the number of children of the root varies from 2 to b. Thus it has to be properly adjusted to have all the leaf nodes on the same level. Here, we show that two simple types of (a, b)-trees are sufficient for covering any input size N. The structures of these two types are shown in Fig. 15.17.

In an (a, b)-*tree of Type I*, all the nodes on levels 1 through $h - 1$ are b-nodes, and the leaves are f-nodes where $a \le f \le b$. This type of trees covers the range $N \in [ab^{h-1} - 1, b^h - 1]$. For this range of N values, the (a, b)-trees of Type I have identical properties with the m-way search trees constructed by bottom-level balancing, where $m = b$. The ratio of the length of this range to that of the entire input size range $[b^{h-1}, b^h - 1]$ is approximately $(b - a)/(b - 1) > 1/2$. This implies that, for more than half of the times, the bottom-level balancing produces the same tree and also that the characterization of (a, b)-trees incorporates m-way search trees.

In an (a, b)-tree of Type II, the root is an r-node where $2 \le r \le b$, the nodes on levels 2 through $h - 2$ are b-nodes, the nodes on level $h - 1$ are a-nodes, and the leaves are f-nodes where $a \le f \le b$. This tree type with a fixed r value covers the range $[a^2b^{h-3}r - 1, ab^{h-2}r - 1]$. Since $2 \le r \le b$, the entire range is actually $[2a^2b^{h-3} - 1, ab^{h-1} - 1]$, which also covers the input size range $[b^{h-1}, ab^{h-1} - 1]$.

From the above properties, we know that for any $N \in [b^{h-1}, b^h - 1]$, there exists a simple bottom-level balanced (a, b)-tree of height h. Figure 15.18 illustrates the range of input size N covered by each type of (a, b)-trees. We have to calculate the r value for the root node, i.e., its number of children, in an (a, b)-tree of Type II, given input size $N \in [b^{h-1}, ab^{h-1} - 1]$. Since the subranges corresponding to adjacent r values overlap, we can arbitrarily select $ab^{h-2} - 1$ as a dividing boundary between subranges. Thus, the entire range for Type II (a, b)-trees is divided into subranges of $[ab^{h-2}(r - 1), ab^{h-2}r - 1]$ for $2 \le r \le b$, and the value r for the root node can be set to

$$\left\lceil \frac{N + 1}{ab^{h-2}} \right\rceil$$

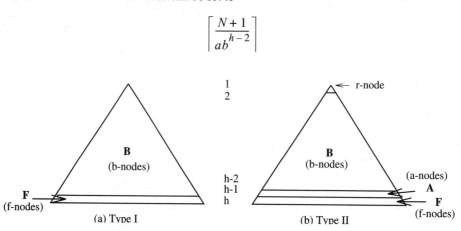

(a) Type I   (b) Type II

Figure 15.17 Bottom-level balanced (a, b)-trees

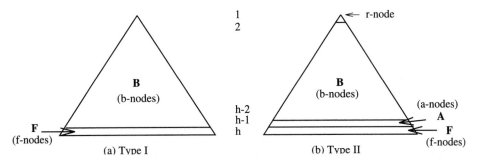

Figure 15.18 Bottom-level balanced (a, b)-trees

As in the m-way search trees, the mapping function between the indices of the keys in the tree and the sorted ranks of the input keys can be derived. Hence, the problem of constructing an (a, b)-tree again reduces to formula evaluations.

A unified algorithm. Since a bottom-level balanced m-way search tree has the same structure as the Type I (a, b)-tree, the algorithm for constructing an (a, b)-tree can be used to construct m-way search trees. This implies that AVL trees, height balanced m-way search trees, 2-3 trees, B-trees, and (a, b)-trees can all be constructed by the (a, b)-tree construction algorithm. The major work of the unified algorithm is evaluating the expressions in the formulas representing the mapping function between the sorted ranks of the keys and the indices of the keys in the tree being constructed. For detailed description of this approach, refer to Ref. [23], where it has been shown, that given N sorted keys, the formulas for the indices representing the positions of the keys in the tree can be evaluated in constant time using N processors on the CREW PRAM model, and $O(\log \log N)$ time using $O(N/\log \log N)$ processors on the EREW PRAM model.

15.9.2 Maintenance of 2-3 trees

For efficient retrieval and management of a dictionary, it is important that maintenance (e.g. insert, delete, merge, split) and search operations on the keys be performed as quickly as possible. Paul, Vishkin, and Wagener [42] developed efficient parallel algorithms for these operations on 2-3 trees in which all the data items are stored at the leaves. Their algorithms run in $O(\log n + \log k)$ time on the EREW PRAM model, where n is the total number of keys in the tree, and k is the number of keys involved in each operation. We will sketch these algorithms in this subsection.

Here, a 2-3 tree with n leaves contains n data items stored only at the leaf nodes from left to right in sorted order, one data item at each leaf node. An internal node v has either two or three children: the left child $l(v)$, the middle child $m(v)$ if it exists, and the right child $r(v)$. An internal node v also maintains the values of the largest data items stored in the subtrees rooted at its children, respectively denoted by $L(v)$, $M(v)$ and $R(v)$. Figure 15.19 shows the structure of a 2-3 tree as defined here. Given a query of k sorted data items, k processors will be used to process the query.

Search. Given a sorted sequence (or *chain*) of k keys to search for, the search algorithm proceeds from the root node to the bottom in several stages. The algorithm sends down a

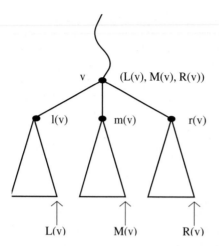

Figure 15.19 Example of a 2-3 tree

subsequence of keys until all the keys reach the proper leaf nodes. (See Fig. 15.20 for a general stage at an interior node.) Suppose a subsequence of keys is at node v at stage s. If the current sequence of keys falls entirely into one of the subtrees of the current node v, then it is sent directly to the appropriate child node n which is depicted in Fig. 15.20a. Otherwise, the current subsequence is halved into two subsequences. If the new half falls into a subtree, then it is sent to its appropriate subtree again, otherwise it remains at the current node to be processed in the next stage $s + 1$. Figure 15.20b illustrates how a chain $C = [a_f, a_{f+1}, \dots, a_l]$ is divided into two subchains $C_1 = [a_f, \dots, a_m]$ and $C_2 = [a_m + 1, \dots, a_l]$, where $m = \lceil (f + l)/2 \rceil$ is the middle element of the chain C.

One processor is allocated to each key in the query, and the processor corresponding to the first element in the current subsequence is responsible for all the keys in the current subsequence of keys. Initially, at stage 1, only one processor P_1 is active at the root node. As a sequence is halved into two subsequences, the processor corresponding to the first element in the second half of the subsequence is activated. To avoid actually sending all the elements in the current subsequence to its children, each active processor moves down the tree. (It suffices for each active processor to know only the first and the last element in

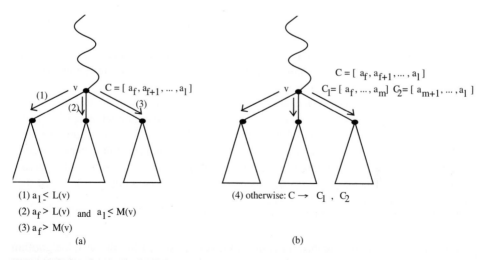

(1) $a_1 \le L(v)$

(2) $a_f > L(v)$ and $a_1 \le M(v)$

(3) $a_f > M(v)$

(a)

(4) otherwise: $C \rightarrow C_1$, C_2

(b)

Figure 15.20 Search operation in a 2-3 tree

the current subsequence.) It is proven in Ref. [42] that subsequences arriving at node v through different stages are always in consecutive order, and that no accumulation of sub-sequences take place at any node at any stage. Thus, no more than two chains may pass each edge of the 2-3 tree in any single stage. Thus, any possible read conflict can be avoided, and each stage takes constant time. Since the depth of the tree is $O(\log n)$, and the original chain can be halved $O(\log k)$ times at most, it takes $O(\log n + \log k)$ time for all the elements to arrive at the leaf level. When all the chains arrive at the leaves, the value of each leaf node where a chain has arrived is broadcast to all the elements in the chain. This takes $O(\log k)$ time. Hence, the overall time complexity for search operation is $O(\log n + \log k)$ on the EREW PRAM model using k processors.

Insertions. Let us insert k new (sorted) keys into a 2-3 tree with n leaves. The first phase of the insertion algorithm finds the proper places for the new keys on the leaf level, by the search algorithm, which results in a subchain of new keys between leaf nodes. Note that the number of elements in a subchain to be inserted between two old leaf nodes can be more than one.

In the second phase of the algorithm, the structure of the tree is adjusted from bottom up in a pipelined fashion as elements are inserted at the leaf level. This phase proceeds in stages. At each stage, the middle element of a chain is chosen for insertion at the leaf level, splitting the chain into halves. Since the insertion of a new element can result in more than three children for its parent node, the property of a 2-3 tree is violated. Therefore, another node (right sibling of the old parent) is created, and the children are split among two parents, followed by the updates of the values $L(v)$, $M(v)$ and $R(v)$. Since the new parent node can cause the same effect, the adjustment of the tree structure propagates upward in a pipelined fashion. Figure 15.21 shows two examples of inserting new nodes at height h of the tree. Figure 15.21a is a case where no further adjustment of the tree is needed and the height of the input tree remains unchanged. Figure 15.21b is a case in which new parent nodes need to be created, and the adjustment of the tree structure continues upward. In this case, the creation of three new parent nodes p', p'', and p''' implies a split of the node immediately above them. This process may continue until the root level, thereby increasing the height of the original tree by, at most, one.

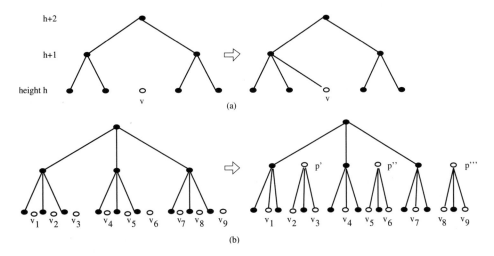

Figure 15.21 Insert operation in a 2-3 tree

The processor allocation scheme is similar to the search algorithm, i.e., one processor for each new key to be inserted. In the insertion phase, only the processors corresponding to the middle elements that are ready to be inserted become activated. When the processor reaches the root or a node from which the tree does not have to be readjusted, it becomes inactive. Since the depth of the tree is $O(\log n)$ and each chain can be split at most $O(\log k)$ times, the second phase takes $O(\log n + \log k)$ time. Hence, the overall time required for insertions is $O(\log n + \log k)$ on the EREW model.

Deletions. Deleting k sorted keys from the 2-3 tree also proceeds in two phases. In the first phase, the search algorithm is used to identify the leaves to be deleted. However, all those leaves are not deleted in a single step, because it may make the 2-3 tree unbalanced. Therefore, the second phase of the deletion algorithm initiates the deletion of some leaf nodes and adjusts the tree structure from bottom up in stages in a pipelined fashion. If the deletion of the leaves causes the parent node to have no children, then we mark all but one of those leaves for deletion and delay the deletion of the unmarked leaf node to the next stage. Then, the marked nodes are deleted, but this can cause the parent node to violate the property of a 2-3 tree, i.e., the parent node may have less than two nodes. In such a case, the only child of the parent node is merged with its cousin nodes and can be partitioned properly while maintaining the properties of 2-3 tree. The parent node that violated the 2-3 tree property now has no children and is marked to be deleted at a later stage. Thus, the deletion of nodes propagates upward in a pipelined fashion. The deletion of interior nodes follows the same process.

Figure 15.22 shows two examples of deleting nodes at height h. In Fig. 15.22a, nodes v_1 and v_2 are deleted, the remaining nodes are merged at the current stage, and node v will be deleted at the next stage. In Fig. 15.22b, nodes $v_1, \ldots, v_6$ are deleted, and the remaining nodes are merged at the current stage. Note that nodes v' and v'' wait to be deleted at the next stage.

The processor allocation scheme is similar to that of the insertion algorithm, i.e., a processor is assigned to each node to be deleted. In the deletion phase, only the processors corresponding to the nodes marked for deletion are activated. The deactivation of a proces-

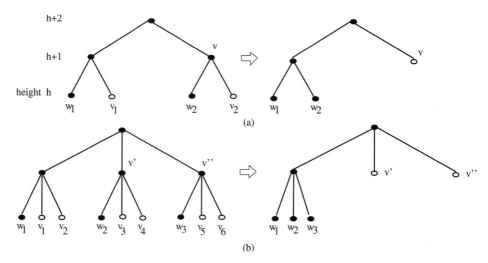

Figure 15.22 Delete operation in a 2-3 tree

sor is also similar to the insertion algorithm. Because the depth of the tree is $O(\log n)$, and at least half of the leaves among the remaining leaf nodes to be deleted are deleted at each stage, the second phase requires at most $O(\log k)$ steps. Hence the deletion of k keys also requires $O(\log n + \log k)$ time on the EREW model.

B-trees. Higham and Schenk [32] extended the search tree maintenance operations to B-trees in which data items are stored in all the nodes. Their algorithms interpret a B-tree as a binary tree, which is similar to the Euler tour constructed from a degree-bounded general tree. Thus, a B-tree of order b and height L is interpreted as a binary tree of height bL. The parallel search, insert, and delete algorithms presented in Ref. [32] are similar to the operations on 2-3 trees discussed in this section. Given a B-tree of order b containing m keys, search, insertion, and deletion of k keys require $O(\log n + b \log k)$, $O(b(\log n + \log k))$, and $O(b^2(10g\, n + \log k))$ time, respectively, on the EREW PRAM model using k processors. It is worth pointing out that in this chapter we have not considered existing literature on the concurrent (rather than truly parallel) maintenance of search structures [40], *concurrent* AVL trees [28], or B-trees [38], which are mostly used in concurrency control and locking protocols for large-scale database applications.

15.10   Conclusions

Viewing algorithms as a scheduling method helps us better understand an algorithm design strategy, which is nothing but a systematic way of finding a proper scheduling method for a given problem, leading to a solution. For example, consider the problem of a bottom-up computation on a tree, which is a combinatorial object, represented by parent pointers or adjacency lists. A sequential algorithm uses a stack/queue to schedule the tree computation, where the history of the stack/queue reflects the input data structure. On the other hand, in parallel processing, a binarized tree obtained from the original tree works as the data structure for scheduling purpose. For the linked list ranking problem, a sequential algorithm uses stack-based scheduling with monotonic stack history, but a parallel algorithm adopts a scheduling based on symmetry breaking on a linked list. Thus, on an intuitive level, it seems that the underlying topology of the data structure given by the problem dictates a proper scheduling method more in parallel processing than in sequential processing. Hence, while designing parallel algorithms, we may need to distinguish two different kinds of data structures—one for representing the input problem description and the other for scheduling purposes.

In this chapter, we have discussed various data structures from the viewpoint of their suitability in parallel computing environment. These include commonly used structures such as linked lists, trees, stacks, queues, heaps, search trees, and so on. For certain classes of problems, we have seen how to replace a sequential (or less parallelizable) data structure, for example stacks or queues, by an alternative method or simpler data structures that offer increased parallelism. In some other cases, it is shown how to efficiently manipulate parallel operations on the existing data structures, such as heaps or search trees, to facilitate implementation of efficient parallel algorithms based on them.

Due to space limitations, we have not been able to cover several other parallel data structures that have been proposed for developing efficient parallel algorithms. For example, Berkman et al. [9] proposed recursive star-trees for computing lowest common ancestors, restricted domain merging, parentheses matching, and other problems. Parallel construction of suffix trees is covered by Apostolico et al. [5] for pattern or string matching applications, and parallel construction of digital search trees used for sorting strings is presented by JáJá et al. [34]. Das and Ferragina [18] have recently applied the sparsification data structures, originally proposed by Eppstein et al. [30], to design improved parallel algorithms for dynamic updating of minimum spanning trees.

Acknowledgments. We would like to thank Pavel Tvrdik of the Czech Technical University, Prague, for his careful reading of the manuscript and suggestions that improved the quality of the presentation. Thanks are also due to Ranette Halverson of the Midwestern State University, Wichita Falls, Texas, for thoughtful discussions at an early stage of this chapter, and to Falguni Sarkar of the University of North Texas for his help in drawing some figures.

15.11 References

1. Abrahamson, K, N. Dadoun, D. A. Kirkpatrick, and T. Przytycka. 1989. A simple parallel tree contraction algorithm. *Journal of Algorithms,* 10(2):287–302.
2. Aho, A. V., J. E. Hopcroft, and J. D. Ullman. 1983. *Data Structures and Algorithms.* Reading, Mass.: Addison-Wesley.
3. Anderson, R., E. Mayr, and M. Warmuth. 1989. Parallel approximation algorithms for bin packing. *Information and Computing,* 82(3):262–277.
4. Anderson, R., and G. Miller. 1991. Deterministic parallel list ranking. *Algorithmica,* 859–868.
5. Apostolico, A., C. Iliopoulos, G. M. Landau, B. Scheiber, and U. Vishkin. 1988. Parallel construction of a suffix tree with applications. *Algorithmica,* 3(3):347–365.
6. Bar-On, I., and U. Vishkin. 1985. Optimal parallel generation of computation tree form. *ACM Transactions on Programming Languages and Systems,* 7(2):659–663.
7. Belloch, G. E. 1989. Scans as primitive parallel operations. *IEEE Transactions on Computers,* C-38:1526–1538.
8. Berkman, O., B. Schieber, and U. Vishkin. 1993. Optimal doubly logarithmic parallel algorithms based on finding all nearest smaller values. *Journal of Algorithms,* 14(3):344–370.
9. Berkman, O., and U. Vishkin. 1990. Recursive star-tree parallel data structure. *Tech. Rep. UMIACS-TR-90-40,* University of Maryland, College Park, MD.
10. Bilardi, G., and A. Nicolau. 1989. Adaptive bitonic sorting: An optimal algorithm for shared memory machines. *SIAM Journal of Computing,* 18(2):216–228.
11. Biswas, J., and J. C. Browne. 1987. An efficient algorithm for managing a parallel heap. *Proc. International Conference on Parallel Processing,* 124–131.
12. Chen, C., and S. K. Das. 1992. Breadth-first traversals of trees and integer sorting in parallel. *Information Processing Letters,* 41:39–49.
13. Cole, R. Parallel merge sort. 1988. *SIAM Journal on Computing,* 17:770–785
14. Cole, R., and U. Vishkin. 1986. Deterministic coin tossing with applications to optimal parallel list ranking. *Information and Control,* 70(1):32–53.
15. Cole, R., and U. Vishkin. 1988. The accelerated centroid decomposition technique for optimal parallel tree evaluation in logarithmic time. *Algorithmica,* 3(3):329–346.
16. Cole, R., and U. Vishkin. 1988. Approximate parallel scheduling. part i: The basic technique with applications to optimal list ranking in logarithmic time. *SIAM Journal on Computing,* 17(1):128–142.
17. Das, S. K., and C. Chen. 1992. Efficient algorithms on interval graphs. *Lecture Notes in Computer Science,* 605:131–143. New York: Springer-Verlag.
18. Das, S. K., and P. Ferragina. 1994. An $O(n)$ work EREW parallel algorithms for updating MST. *Proceedings of the 2nd European Symposium on Algorithms,* Utrecht, the Netherlands. Also in *Lecture Notes in Computer Science,* Vol. 855, 331–342. New York: Springer-Verlag.
19. Das, S. K., and R. H. Halverson. 1993. A comprehensive survey of parallel linked list ranking algorithms. *Tech. Rep. CRPDC-93-12,* University of North Texas, Denton, Texas.
20. Das, S. K., R. H. Halverson, and K. B. Min. 1993. Efficient parallel algorithms for tree-related problems using the parentheses matching strategy. *Tech. Rep. CRPDC-93-11,* University of North Texas, Denton, Texas, 1993. Also in *Proceedings of the 8th International Parallel Processing Symposium,* Cancun, Mexico. 362–367.

21. Das, S. K., and W.-B. Horng. 1991. Managing a parallel heap efficiently. In *Proceedings of Parallel Architectures and Languages Europe (PARLE), 1991*. Also in *Lecture Notes in Computer Science*, Vol. 505, 270–287.

22. Das, S. K., W.-B. Horng, and G. S. Moon. 1994. An efficient algorithms for managing a parallel heap. *Journal of Parallel Algorithms and Architectures,* 4:281–299.

23. Das, S. K., and K. B. Min. 1995. A unified approach to the parallel construction of search trees. *Journal of Parallel and Distributed Computing*, 27(1): 71–78.

24. Dekel, E., S. Peng, and S. S. Iyengar. 1986. Optimal parallel algorithms for constructing and maintaining a balanced *m*-way search tree. *International Journal of Parallel Programming,* 15:503–528.

25. Deo, N. 1990. Data structures for parallel computation on shared-memory machines. *NATO ASI Series Supercomputing,* F 62:341–355. Springer-Verlag.

26. Deo, N. and S. Prasad. 1992. Parallel heap: An optimal parallel priority queue. *Journal of Supercomputing,* 6:87–98.

27. Diks, K., and W. Rytter. 1991. On optimal parallel computations for sequence of brackets. *Theoretical Computer Science,* 87(2):251–262.

28. Ellis, C. 1980. Concurrent search and insertion in AVL trees. *IEEE Transactions on Computers.* C-29(9):811–817.

29. Eppstein, D., and Z. Galil. 1988. Parallel algorithmic techniques for combinatorial computation. *Annual Reviews of Computer Science,* 3:233–283.

30. Eppstein, D., Z. Galil, G. F. Italiano, and A. Nissenzweig. 1992. Sparcification—a technique for speeding up dynamic graph algorithms. *Proceedings of the IEEE Symposium on Foundations of Computer Science,* 60–69.

31. He, X., and Y. Yesha. 1988. Binary tree algebraic computation and parallel algorithms for simple graphs. *Journal of Algorithms,* 9(1):92–113.

32. Higham, L., and E. Schenk. 1992. Maintaining B-trees on an EREW PRAM. *Tech. Rep. 91/446/30,* University of Calgary, Calgary, Alberta, Canada.

33. JáJá, J. 1992. *An Introduction to Parallel Algorithms.* Reading, Mass.: Addison-Wesley.

34. JáJá, J., K. W. Ryu, and U. Vishkin. 1994. Sorting strings and constructing digital search trees in parallel. *Proceedings of the International Parallel Processing Symposium,* 349–356.

35. Karp, R., and V. Ramachandran. 1990. Parallel algorithms for shared-memory machines. In *Handbook of Theoretical Computer Science,* ed. J. van Leeuwen. New York: Elsevier, 869–941.

36. Knuth, D. 1968. *The Art of Computer Programming.* Vol. 1. Fundamental Algorithms, Vol. 1. Reading, Mass.: Addison-Wesley.

37. R. E. Ladner and M. J. Fischer. Parallel prefix computation. Journal of the Association for the Computing Machinery, 27:831–838, 1980.

38. Lehman, P. L., and S. B. Yao. 1981. Efficient locking for concurrent operations on b-trees. *ACM Transactions on Database Systems,* 6(4):650–670.

39. Levcopoulos, C., and O. Petersson. 1992. Matching parentheses in parallel. *Discrete Applied Mathematics,* 40(3):423–431.

40. Manber, U., and R. E. Ladner. 1984. Concurrency control in a dynamic search structure. *ACM Transactions on Database Systems,* 9(3):440–455.

41. Miller, G. L., and J. Reif. 1985. Parallel tree contraction and its applications. *Proceedings of the 22nd Annual IEEE Symposium on Foundations of Computer Science,* 478–489.

42. Paul, W., U. Vishkin, and H. Wagener. 1983. Parallel dictionaries on 2-3 trees. *Proceedings of ICALP, Lecture Notes in Computer Science,* volume 154, 597–609. Springer-Verlag.

43. Pinotti, M. C., and G. Pucci. 1991. Parallel priority queue. *Information Processing Letters,* 40:33–40.

44. Prasad, S., S. K. Das, and C. Chen. 1994. Efficient EREW PRAM algorithms for parentheses matching. *IEEE Trans. Parallel and Distributed Syst.,* 5(9):995–1008.

45. Rao, V. N., and V. Kumar. 1988. Concurrent access of priority queues. *IEEE Transactions on Computers,* 37:1657–1665.

46. Reif, J. 1985. Depth-first search is inherently sequential. *Information Processing Letters,* 20(5):229–234.

47. Reif, J., ed. 1993. *Synthesis of Parallel Algorithms.* San Mateo, Calif.: Morgan-Kaufmann.

48. Tarjan, R. E., and U. Vishkin. 1985. Finding biconnected components and computing tree functions in logarithmic parallel time. *SIAM J. Computing,* 14:641–644.

49. Wang, B.-F., and G.-H. Chen. 1991. Cost-optimal parallel algorithms for constructing 2-3 trees. *Journal of Parallel and Distributed Computing,* 11:257-261.

50. Wyllie, J. 1979. The complexity of parallel computation. *Tech. Rep. TR-79-387,* Cornell University.

16

Data Parallel Algorithms[*]

Howard Jay Siegel, Lee Wang, John John E. So, and Muthucumaru Maheswaran

As the size, hardware complexity, and programming diversity of parallel systems continue to evolve, the range of alternatives for implementing a task on these systems grows. Choosing a parallel algorithm and implementation becomes an important decision, and the choice has a significant impact on the execution time of the application.

Data parallelism is a model of parallel computing in which the same set of instructions is applied to all the elements in a data set [37, 56]. A sampling of data parallel algorithms is presented. The examples are certainly not exhaustive, but address many issues involved in designing data parallel algorithms. Case studies are used to illustrate some algorithm design techniques and to highlight some implementation decisions that influence the overall performance of a parallel algorithm. It is shown that the characteristics of a particular parallel machine to be used need to be considered in transforming a given task into a parallel algorithm that executes effectively.

16.1   Chapter Overview

This chapter focuses on algorithm design techniques for mapping data parallel algorithms onto large-scale (i.e., 2^6 to 2^{16} processors) distributed memory parallel machines. The *single instruction stream, multiple data stream (SIMD)* model of parallelism is used to demonstrate the techniques presented; however, the methods can be used with the *multiple instruction stream, multiple data stream (MIMD)*, the single program, multiple data stream (SPMD, which is a subclass of MIMD), and the *mixed-mode* (hybrid SIMD/ MIMD) models of parallelism as well. Related work is discussed in the chapter on the Fundamentals of Parallel Algorithms.

Data parallel programs typically exhibit a high degree of *uniformity* (operations to be performed are uniform across the data set) and are often well suited to the SIMD model because only a single instruction stream is necessary [8, 28]. Implementing these applications on SIMD machines is often more cost effective than solving them using MIMD machines [23]. Among the advantages of the SIMD mode of parallelism that can be exploited

[*]This research was supported by NRaD under subcontract number 20-950001-70 and by a Fulbright Scholarship.

are implicit synchronization that allows more efficient interprocessor communication, the ability to overlap scalar operations on the control unit with the operations on the processing elements, and the need for only one program [8, 47]. Mixed-mode processing and the trade-offs between the SIMD and MIMD modes of parallelism are discussed further in the Chapter 25, which focuses on heterogeneous computing.

In Section 16.2, the SIMD machine model used in this chapter and other machine models are discussed. The choices for data distribution among the processing elements are explored in Section 16.3. The effect of the data distribution on execution time is demonstrated using an image smoothing algorithm. In addition, the impact of the interconnection network topology on the number of data transfers required to perform the computations is also studied. It is shown that the optimum data distribution is dependent on the architecture of the machine in use and the application to be implemented in parallel. Section 16.4 examines overlapping the operations of the control unit and the processing elements and uses an image correlation algorithm to illustrate how this overlap can be optimized.

Section 16.5 discusses parallel reduction operations. A sampling of matrix and vector operations is covered in Section 16.6. Parallel implementations of matrix transpose, matrix-by-matrix multiplication, and matrix-by-vector multiplication are presented. In Section 16.7, the effect on execution time of increasing the number of processors used (scalability of the algorithm) and the impact of partitioning the system for subtask parallelism are explored. It is shown that increasing the number of processors used does not always yield a decrease in execution time or an increase in system efficiency. In Section 16.8, the computation of multiple quadratic forms is used as an illustrative example on how scalability can be achieved using a suite of algorithms. Some important future research areas in developing design techniques for data parallel algorithms are discussed in Section 16.9.

16.2 Machine Model

As stated previously, the data parallel algorithm studies will be based on the SIMD machine model [17], and most of the results will also be applicable to the SPMD model on MIMD machines and to both models on mixed-mode systems. In the SIMD model, only a single sequence of instructions is present, but each instruction is executed simultaneously in an arbitrary set of processors, with each processor operating on its own data. Typically, an SIMD machine consists of N processors, N memory modules, a *control unit (CU)*, and an interconnection network. Each processor is paired with a memory module to form a *PE (processing element)*. The CU broadcasts instructions in sequence to the PEs and all enabled PEs execute the same instruction at the same time. Thus, there is a single instruction stream. Each enabled PE executes the instructions on the data in its own associated memory module, resulting in multiple data streams. The interconnection network allows communication among the PEs [48]. Examples of SIMD systems that have been built include the Thinking Machines CM-2 [22], DAP [25], Illiac IV [4], MasPar MP-1 [10] and MP-2, MPP [7], and STARAN [5, 6].

Figure 16.1 shows the SIMD machine model used in this chapter. This model is referred to as a *PE-to-PE configuration* [48] or *physically distributed memory organization*. There are $N = 2^n$ PEs, and the PEs are numbered from 0 to $N - 1$. The PE's memory is used only to store data, not instructions.

The CU is made up of a processor, a memory, and an instruction broadcast queue. The CU fetches and decodes the program instructions from its own memory. Only one program exists, a portion of which is to be executed on the CU and the other portion of which is to

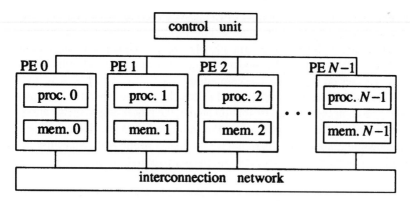

Figure 16.1 Distributed memory model of an SIMD machine

be executed on the PEs. Typically, the CU executes the control flow instructions (e.g., loop indexing) and broadcasts the data processing instructions to the PEs. The disabled PEs must remain idle while the instruction that was broadcast from the CU is executed in enabled PEs.

As an example, consider the execution of the *if-then-else* construct, where the if conditional involves data local to each PE. First, the enable status of the PEs prior to the execution of the if-then-else construct is saved. All disabled PEs will remain disabled during the execution of the if-then-else construct. An enabled PE is kept enabled for the "then" clause if the result of the if-conditional test is true for the PE. Otherwise, it is disabled for the "then" clause. Then, the instructions of the "then" clause are broadcast from the CU and executed by the enabled PEs. Next, the PEs that were enabled for the "then" clause are disabled, and the PEs enabled prior to the if-then-else construct, but disabled for the "then" clause, are enabled for the "else" clause. The CU broadcasts the instructions of the "else" clause, and the enabled PEs execute these instructions. At the end of the if-then-else construct, the previously saved enable status is restored; i.e., each PE returns to the enable state it was in before the execution of the if then-else construct.

The CU broadcasts a single stream of instructions from the instruction broadcast queue to the PEs in the computational engine. It will not issue the next instruction until all the enabled PEs have completed executing the current instruction, thereby implicitly synchronizing the PEs at instruction-level granularity. Once the CU processor determines the instructions that should be placed in the queue, it may proceed with its own execution while the queued instructions are broadcast to the PEs. This concurrent CU and PE execution is called *CU/PE overlap* [29, 51], which is discussed in Section 16.4.

Recall that most methods discussed in this chapter can also be adopted for SPMD operation on existing MIMD machines (e.g., Thinking Machines CM-5 [24], nCUBE 2 [21], Intel Paragon, and Cray T3D) and mixed-mode systems (e.g., OPSILA [15], PASM [51], and Triton/l [43]). In contrast to the SIMD machine model, the MIMD machine model [17] does not have a CU but consists of N independent PEs. The PEs are connected to one another via an interconnection network, as in the SIMD case. The PE memory stores the program in addition to the data. Each PE executes its own instruction stream, and all PEs operate asynchronously with respect to one another. Any synchronization among PEs needed has to be explicitly specified in the program. The SPMD model is a subclass of the MIMD model [14]. In the SPMD model, all PEs execute the same program asynchronously. Each of the PEs takes its own control path during the course of the execution. In a mixed-mode system [16], the PEs can switch between the SIMD and the MIMD modes of parallelism at instruction-level granularity.

There is a great variety of interconnection networks that can be used in parallel machines (e.g., see Refs. [40, 48, 54]). Considering the impact of network selection on each algorithm described is beyond the scope of this chapter. Therefore, unless otherwise specified, a network that efficiently supports the inter-PE communications needed is assumed for each algorithm. Furthermore, inter-PE communication will be measured in number of data words sent; other metrics, not considered here, include number of distinct network path establishments and size of the data items transferred. An example of the impact of the network selected is included in Section 16.3 to introduce the issues involved.

16.3   Impact of Data Distribution

16.3.1   Introduction

One issue that must be addressed in designing data parallel algorithms is how data should be mapped across the PEs. If the mapping of the data across the PEs is optimized, it is possible to gain significant performance improvements for a particular algorithm [11].

One class of problems that exhibit data parallelism are window-based tasks. Examples of these tasks include image correlation [3, 50], image smoothing [47, 49], and range image segmentation [19]. This section demonstrates the impact of data distribution across the PEs on the computation time and the communication time for a given algorithm. Image smoothing is used as a representative of window-based algorithms to illustrate how varying the mapping across the PEs can affect the execution time.

16.3.2   The serial image smoothing algorithm

Smoothing is a procedure applied to an image to reduce noise. An $M \times M$ image A is stored in memory as a two-dimensional array (matrix) where each element, called a picture element, or *pixel*, is an integer whose value represents the gray-level intensity of the corresponding point in the discretized image. For each non-edge pixel (i, j) in A, a 3×3 window centered at (i, j) is used to generate the corresponding pixel in the $M \times M$ smoothed image A'. A serial algorithm to accomplish this is:

```
for i = 1 to M - 2 do
   for j = 1 to M - 2 do
      A'(i, j) = [A(i - 1, j - 1) + A(i, j - 1) + A(i + 1, j - 1) + A(i - 1, j) +
                  A(i, j) + A(i + 1, j) + (i - 1, j + 1) + A (i, j + 1) +
                  A(i + 1, j + 1)]/9
```

In the case of an edge pixel, no calculation is performed, and the pixel itself is taken to be the smoothed value. Because there are $4M - 4$ edge pixels in the $M \times M$ image A, the serial time complexity is the time to execute $M^2 - (4M - 4) = O(M^2)$ smoothing operations. For $M = 4,096$, this is 16,760,836 smoothing operations (approximately M^2).

16.3.3   Parallel implementation using square subimages

In this chapter, it is assumed that when implementing an algorithm on a parallel machine, the goal is to minimize the execution time. Factors that, in general, help to do this include decreasing inter-PE communications and balancing the workload among the PEs so that as many PEs as possible are concurrently executing the algorithm.

Assume N PEs are logically arranged as a $\sqrt{N} \times \sqrt{N}$ grid. The $M \times M$ image A is mapped onto the PEs by superimposing the image onto the PE grid. Thus, each PE stores an $(M/\sqrt{N}) \times (M/\sqrt{N})$ square subimage, as shown in Fig. 16.2a. Therefore, each PE performs up to M^2/N smoothing operations. All PEs smooth their subimages simultaneously; i.e., at most N smoothing operations can be performed concurrently at each step, one in each PE.

To smooth the pixels at the edge of a subimage, pixels from spatially adjacent subimages must be transferred, as shown in Fig. 16.2b. PE i requires at most $M/\sqrt{N}$ pixels from each of the four PEs directly adjacent to it and one pixel from each of the four PEs diagonally adjacent to it. Thus, at most, $4M/\sqrt{N} + 4$ inter-PE data transfers are required per PE to smooth the entire image A. Just as the PEs can all smooth simultaneously, all N PEs can transfer data simultaneously. (Recall from Section 16.2 that an appropriate network is assumed to support this.) The time complexity of the parallel algorithm when operating on an $M \times M$ image A with N PEs is the sum of M^2/N smoothing operations and $4M/\sqrt{N} + 4$ inter-PE data transfers (where up to N smoothing operations or up to N data transfers can occur concurrently). Assuming that the time to perform one transfer is equal to τ times the time to perform one smoothing operation, the *speedup*, S, of the parallel version over that of the serial version of the algorithm is defined as:

$$ S = \frac{\text{serial time complexity}}{\text{parallel time complexity}} = \frac{(M-2)^2}{\frac{M^2}{N} + \tau\left(\frac{4M}{\sqrt{N}} + 4\right)} $$

For example, if $M = 4{,}096$ and $N = 256$, the number of inter-PE data transfers required is 1,028, and 65,536 smoothing operations are performed. If $\tau = 1$, the speedup is $4094^2/66{,}564 = 252$. (The value of τ is machine architecture dependent and could be less than, equal to, or greater than 1.)

In the above speedup calculation, the complexities are computed by only counting the smoothing operations and the inter-PE data transfers. It is assumed that the uniprocessor

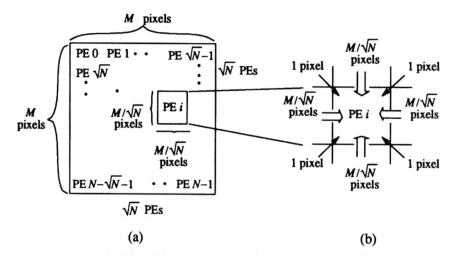

Figure 16.2 Data allocation (a) and pixel transfers (b) for image smoothing using square subimages

machine that executes the serial algorithm and each processor in the parallel machine are of equivalent computing power. Theoretically, the maximum possible speedup is N. This is achieved when the total workload is equally distributed among the PEs, no overhead (e.g., no inter-PE data transfers) is incurred, and all PEs are always active during the execution. If the time to perform an inter-PE data transfer becomes much less than the time to perform a smoothing operation, which is the case, for example, when using the Xnet in the MasPar MP-l [10], the speedup will be closer to N. The speedup will never equal N even if inter-PE communication is ignored, because the PEs containing the edge pixels of the $M \times M$ image A will be disabled for some smoothing operations and are therefore underutilized for some steps of the algorithm. This demonstrates that inter-PE data transfers and *masking* (i.e., disabling some PEs for some operations) result in a less than perfect speedup [47].

Consider the number of inter-PE data transfers required per PE. Instead of using square subimages, suppose the M^2/N pixels are mapped to each PE such that the subimage in a PE has r pixels per row and c pixels per column, where $rc = M^2/N$. To minimize the number of inter-PE data transfers per PE, which is $2r + 2c + 4$, replace r with $M^2/(cN)$ and minimize the expression with respect to c:

$$\frac{d}{dc}(2r + 2c + 4) = \frac{d}{dc}\left[\frac{2M^2}{cN} + 2c + 4\right] = 0$$

This yields $c = M/\sqrt{N}$, subject to the constraint that c is an integer. It follows that $r = M/\sqrt{N}$. Thus, $r = c$; i.e., each PE should contain a square subimage [50]. For example, if $M = 4{,}096$ and $N = 256$, each PE is assigned $M^2/N = (4{,}096)^2/256 = 65{,}536$ pixels. If the pixels are mapped to each PE as a 256×256 ($r = c = 256$) square subimage, the number of inter-PE data transfers needed is $4 \times 256 + 4 = 1{,}028$. If, instead, the 65,536 pixels are mapped to each PE as a $64 \times 1{,}024$ ($r = 64$, $c = 1{,}024$ or $r = 1{,}024$, $c = 64$) rectangular subimage, the number of inter-PE data transfers needed is $2 \times 1{,}024 + 2 \times 64 + 4 = 2{,}176$.

16.3.4 Parallel implementation using the horizontal (row) stripe method

In the preceding subsection, it was shown that the way data are distributed across the PEs dictates the number of inter-PE data transfers required. An alternative method of mapping data among the PEs is based on distributing consecutive rows of data, as opposed to square subimages, to the PEs. This approach results in a decreased number of calculations performed per PE compared to using square subimages.

Recall that in the image smoothing algorithm, the border (edge) pixels of image A do not require any smoothing operations. In general, for some window-based image processing tasks, these border pixels may play an important role in the selection of the data distribution method. The *horizontal stripe method* allows window-based algorithms to take advantage of the fact that no calculations need to be performed on the column border pixels by distributing the column border pixels evenly among all the PEs, thereby decreasing the total number of required calculations [19]. The square subimage method does not take advantage of this fact because the border pixels are distributed unevenly among the PEs; some PEs will not get any border pixels and consequently must perform the maximum number of smoothing operations (M^2/N). These PEs dictate the time required to perform the necessary computations to finish the task.

Using the same image smoothing algorithm discussed in the previous subsection, the $M \times M$ image A is divided into N rectangular $(M/N) \times M$ subimages, where $M \geq N$, as shown in Fig. 16.3. Each PE stores M/N rows by M columns of pixel data. The number of pixels assigned to each PE remains unchanged (still equal to M^2/N). However, PE i will now require a total of $2M$ pixels from its two neighboring PEs; M pixels will be required from PE $i-1$, and another M pixels will be required from PE $i+1$. Recall that it is always assumed that all N PEs can transfer data simultaneously. The number of inter-PE transfers increases, for $N > 4$, relative to the square subimage method discussed in previous subsection (from $4M/\sqrt{N} + 4$ to $2M$). However, for the stripe method, the border pixels located on the left side and the right side of the image are uniformly distributed among all the PEs. Because calculations on these column border pixels are not performed, once the inter-PE data transfers are complete, each PE will perform a total of at most $(M/N) \times (M-2)$ smoothing operations. Thus, each PE avoids smoothing at least $2M/N$ image edge pixels when using the horizontal stripe method compared with using square subimages.

Comparing the two schemes, the square distribution scheme requires fewer inter-PE transfers but more smoothing operations. Although the horizontal stripe method requires more transfers, it has the potential of decreasing the execution time due to the reduced number of smoothing operations resulting from not processing the column border pixels. For image smoothing, the square subimage method, as shown in Subsection 16.3.3, may be best, but for other window-based image processing algorithms (e.g., range image segmentation), it may be outperformed by the horizontal stripe method [19].

The results derived so far in this subsection used a 3×3 window. The derivations can be generalized for a $w \times w$ window, with more complex operations being performed on the set of data values within the window. For a task with a $w \times w$ window, the square subimage method performs M^2/N operations and the horizontal stripe method performs $(M^2/N) - 2\lfloor w/2 \rfloor (M/N)$ operations, where $\lfloor x \rfloor$ is the floor of x, i.e., the greatest integer less than or equal to x. In the case of the square subimage method, each PE must transfer $4\lfloor w/2 \rfloor \times (M/\sqrt{N}) + 4(\lfloor w/2 \rfloor)^2$ data elements. In the horizontal stripe data elements are transferred per PE. As the window size w increases, the number of operations required by the horizontal stripe method decreases. The time complexity of an operation increases as $O(w^2)$. The number of operations performed by the horizontal stripe method is $\Delta T_{op} = 2\lfloor w/2 \rfloor (M/N)$ less than that of the square subimage method. But the number of inter-PE transfers performed by the square subimage is $\Delta T_{transfer} = 2\lfloor w/2 \rfloor M - 4\lfloor w/2 \rfloor (M/\sqrt{N}) - 4(\lfloor w/2 \rfloor)^2$ less than that of the horizontal stripe method. If the computation time per operation is greater than or equal to $\lceil \Delta T_{transfer}/\Delta T_{op} \rceil \times$ the transfer time per data item, where $\lceil x \rceil$ is the ceiling of x, i.e.,

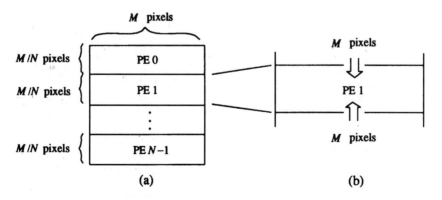

(a)   (b)

Figure 16.3 Data allocation (a) and pixel transfers (b) for image smoothing using the horizontal stripe method

the smallest integer greater than or equal to x, then the horizontal stripe method is better. For example, if $w = 21$, $M = 4,096$, and $N = 256$, then $\Delta T_{op} = 320$, and $\Delta T_{transfer} = 71,280$. If the computation time per operation equals $\lceil 71,280/320 \rceil = 223 \times$ transfer time, then the horizontal stripe method is better. For a task such as image correlation, where each operation involves more than $w^2 = 441$ multiplications and additions, this could be the case (it is machine architecture dependent).

Thus, for window-based tasks, distribution of the image across PEs should not always be done by using square subimages. The subimage window size and the task to be performed have to be considered when deciding which data distribution scheme to use. In general, the horizontal stripe method may result in a decreased execution time whenever calculations on the column border pixels are not needed and inter-PE transfers are relatively fast compared to calculations needed for each pixel position.

16.3.5 Impact of a specific interconnection network on execution time

Interconnection networks considered. The impact of a particular interconnection network on the time complexity of the parallel algorithm for a particular application is discussed next. The same data distribution across the PEs is used to compare the performance of an algorithm on two parallel machines with topologically distinct interconnection networks [46].

Consider two hypothetical parallel processing systems where the only difference between the two systems is the topological structure of their interconnection networks: one is a *mesh* with no wraparound connections from one edge to another, and the other is a *ring* (Fig. 16.4). All other system parameters are assumed identical, including the number of PEs, the computing and communication hardware, the operating system, the language, the communication protocols, and the link bandwidth. Both networks have N PEs labeled from O to $N - 1$. In the mesh topology, PE i is directly connected to PEs $i - 1$, $i + 1$, $i + \sqrt{N}$, and $i - \sqrt{N}$. In the ring topology, PE i only has direct connections to PEs $i - 1$ and $i + 1$. For both networks, all PEs can send data simultaneously in the same direction.

To illustrate the impact of different interconnection network topologies while holding all other factors identical, the same SIMD image smoothing algorithm is executed on both systems. The two distribution methods described in Subsections 16.3.3 and 16.3.4 are used to compare the two architectures.

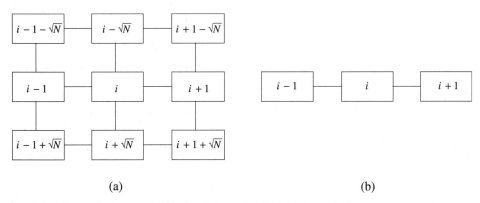

(a)   (b)

Figure 16.4 Connection patterns for PE i for (a) the mesh and (b) the ring topologies

Mesh versus ring using square subimages. First consider the system with a mesh interconnection network. Figure 16.2 depicts how pixels from an $M \times M$ image are mapped onto a $\sqrt{N} \times \sqrt{N}$ mesh of PEs as square subimages. Pixels from N subimages of size $(M/\sqrt{N}) \times (M/\sqrt{N})$ are mapped onto PEs so that adjacent subimages are mapped to adjacent PEs in the mesh. As discussed earlier, each PE performs at most M^2/N smoothing operations and $4(M/\sqrt{N}) + 4$ inter-PE data transfers. Figure 16.5a illustrates the required data transfers for PE 6 for $N = 16$.

Next, consider a system with a ring interconnection network. It is assumed that the image is distributed among the PEs as square subimages in the same way as described above for the mesh. For the general problem of smoothing $M \times M$ images using N PEs, it is shown below that a total of $2M\sqrt{N} + 2M + 4$ inter-PE data transfers are required for the ring. Figure 16.5b is an example showing the required data transfers for PE 6 for $N = 16$. This example will be used in the following general description.

The $M/\sqrt{N}$ pixels along the right vertical boundary of each subimage must be transferred from PE $i - 1$ to PE i (e.g., 5 to 6). This requires $M/\sqrt{N}$ inter-PE data transfers. Likewise, transferring the $M/\sqrt{N}$ pixels from the left vertical subimage boundary from PE $i + 1$ to PE i (e.g., 7 to 6) also requires $M/\sqrt{N}$ inter-PE data transfers. To transfer the $M/\sqrt{N}$ pixels along the upper horizontal subimage boundary from PE $i - \sqrt{N}$ to PE i (e.g., 2 to 6) requires $\sqrt{N} \times M/\sqrt{N} = M$ inter-PE data transfers because PE $i - \sqrt{N}$ and PE i are separated by links. Likewise, to transfer the $M/\sqrt{N}$ pixels along the lower horizontal subimage boundary from PE $i + \sqrt{N}$ to PE i (e.g., 10 to 6) requires M inter-PE data transfers. Next consider sending the corner pixels needed by PE i. Pixels from the diagonally adjacent PEs can be sent through intermediate PEs. The required pixels from $i - \sqrt{N} + 1$ (e.g., 3) and $i + \sqrt{N} + 1$ (e.g., 11) have already been transferred to PE $i + 1$ (e.g., 7), so only two transfers are needed to move these pixels to PE i. The case for the pixels from the upper and lower left diagonal PEs is similar. Thus, the total number of inter-PE data transfers for the ring is $2M/\sqrt{N} + 2M + 4$. For $N > 1$, this is greater than the number of required inter-PE data transfers for the mesh, given by $(4M/\sqrt{N}) + 4$. For example, if $M = 4,096$ and $N = 256$, the number of inter-PE transfers for the mesh is 1,028 and for the ring it is 8,708.

Mesh versus ring using the horizontal stripe method. Instead of using square subimages, consider the horizontal stripe method. A mapping of pixel values from the N rectangular $(M/N) \times M$ subimages onto the ring topology is shown in Fig. 16.6. Pixel values along the horizontal boundaries of the rectangular subimages are transferred to the neighboring PEs, which requires a total of $2M$ inter-PE data transfers for the ring topology; M pixels are transferred from PE $i - 1$ to PE i and M pixels are transferred from PE $i + 1$ to

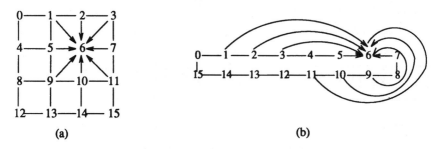

(a)   (b)

Figure 16.5 The required data transfers to PE 6 for (a) the mesh and (b) the ring topologies for $N = 16$

PE i. Thus, the total number of inter-PE data transfers is reduced from $2M/\sqrt{N} + 2M + 4$ using square subimages to $2M$ using the horizontal stripe method for the ring network.

For the mesh network with no wraparound connections between edges, if the horizontal stripe method is used, a total of $2M(\sqrt{N} + 1)$ inter-PE data transfers are required, as demonstrated in Fig. 16.7. For any PE i that is not on the right vertical edge of the N PE mesh, M transfers are needed to transfer pixels from PE i to PE $i + 1$. For any PE i that is not on the left vertical edge of the N PE mesh, another M transfers are required to transfer pixels from PE i to PE $i - 1$. These horizontal transfers are shown in Fig. 16.7a. In addition, M pixels have to be transferred from PE $i \times \sqrt{N} - 1$ to PE $i \times \sqrt{N}$, $1 \le i \le \sqrt{N} - 1$. This will require $\sqrt{N} \times M$ transfers, because the source PE is $\sqrt{N}$ links away from the destination PE. Similarly, PE $i \times \sqrt{N}$ will send M pixels to PE $i \times \sqrt{N} - 1$, $1 \le i \le \sqrt{N} - 1$. This will also require $\sqrt{N} \times M$ inter-PE data transfers. These transfers between PEs on the left and right vertical edges are depicted in Fig. 16.7b. Thus, the number of inter-PE transfers for the mesh network when using the horizontal stripe method is $M + M + M\sqrt{N} + M\sqrt{N} = 2M(\sqrt{N} + 1)$. This is much greater than that required for the ring network when using the same method for data distribution. For example, if for $M = 4{,}096$ and $N = 256$, the number

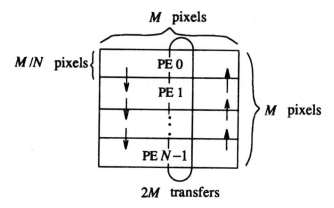

Figure 16.6 Mapping N subimages of size $(M/N) \times M$ each onto the PEs of the ring with the directions of transfers shown

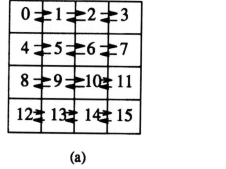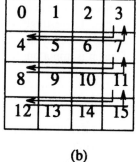

(a)   **(b)**

Figure 16.7 Mapping N subimages of size $(M/N) \times M$ each onto the PEs of the mesh with the directions of transfers shown: (a) 2M transfers, (b) $2(\sqrt{N} M)$ transfers

of inter-PE transfers for the ring is 8,192, and for the mesh it is 139,264. In this case, the ring outperforms the mesh.

Best case comparison. Based on the above analyses, in terms of the number of inter-PE data transfers, the square subimage data distribution is best for the mesh network, the horizontal stripe method is best for the ring network, and the best mesh approach is better than the best ring approach. However, recall from Subsection 16.3.4, the horizontal stripe method can also avoid processing the column border pixels, reducing the calculations performed by each PE. As was discussed in that subsection, consider a $w \times w$ window and a window-based image processing task more complex than smoothing. The comparison between the best mesh and best ring cases is the same as the comparison between the square subimage and horizontal stripe methods in that subsection. Thus, depending on the window size and the image processing task, the ring may be better than the mesh (as well as less expensive to implement).

Another aspect of the interaction of the algorithm, data distribution, and interconnection network is the number of distinct network settings (i.e., path changes). In certain types of network implementations, the number of network settings used can impact performance (even for a fixed total number of data items to be transferred). In many network implementations, due to software and/or hardware overhead, establishing the path for transmitting data is a significant portion of the transfer time, and is independent of the number of words to be transferred using that setting. Thus, it may be important to minimize the number of network settings used.

Consider image smoothing with 3×3 windows and the horizontal stripe method of Fig. 16.6. PE i stores rows $i \times M/N$ to $(i + 1) \times M/N - 1$ of the image. In general, it receives row $i \times M/N - 1$ from PE $i - 1$, and row $(i + 1) \times M/N$ from PE $i + 1$. PE i smooths the pixels in rows $i \times M/N$ to $(i + 1) \times M/N - 1$ of the image. Instead of using these two network settings, a single setting can be used by only sending rows $(i + 1) \times M/N$ and $(i + 1) \times M/N + 1$ from PE $i + 1$ to PE i. In this case, PE i smooths rows $i \times M/N + 1$ to $(i + 1) \times M/N$ of the image. With this approach, the number of data items transferred is still $2M$, and the maximum number of pixels smoothed in any PE is still M^2/N. This can be generalized for $w \times w$ windows. Also, this approach can be applied to square subimages, reducing the number of network settings from four to two.

16.3.6 Summary

As has been shown, the way in which data are distributed among the PEs may influence both the number of inter-PE transfers needed and the number of computations to be performed. Conversely, the topology of the inter-PE network of the machine may influence which data distribution is best. Interestingly, even if the problem domain is limited to window-based image processing tasks, no one data distribution is best for all window sizes and image operations. Thus, it is important to carefully analyze all of these factors to find the best data distribution.

16.4 CU/PE Overlap

In this section, the impact of CU/PE overlap on execution time in SIMD mode is examined. The CU CPU initiates parallel computations by sending blocks of SIMD code to the CU instruction broadcast queue. Once in the queue, each SIMD instruction is broadcast to all the PEs while the CU CPU can be performing its own computations. This property is called CU/PE overlap [3, 8, 29]. CU/PE overlap can improve the overall performance of a program because CU execution and PE execution can occur concurrently. This aspect of data parallel algorithms is the only one discussed in this chapter that is not directly applicable to MIMD machines.

Image correlation [3, 50] is used to examine the effects of CU/PE overlap. Image correlation involves determining the position at which an image template best matches a portion of an input image. Let x denote an $r \times c$ template array, let A be an $R \times C$ input image array, and let y be an $r \times c$ portion of A. Also, let a given image coordinate be (*row, col*) such that $0 \le row \le (R - r)$ and $0 \le col \le (C - c)$, and let $y (i, j) = A (row + i, col + j)$ for all coordinates (i, j) where $0 \le i < r$ and $0 \le j < c$. The *coefficient of determination*, D^2, is computed to determine the quality of fit of the template to the overlapped data of the input image. One part of this calculation is to compute $\Sigma\Sigma\, x\, (i, j)\, y(i, j)$ for each possible match position for $0 \le i < r$ and $0 \le j < c$.

CU/PE overlap is an analytical quantity that can be maximized. Consider the above computation. Two possible approaches to compute this two-dimensional summation are given in C-like notation in Figs. 16.8 and 16.9 (based on Ref. [3]). Assume that the instruction broadcast queue is empty when the CU starts execution of either code segment and that the queue is large enough such that it never overflows during execution. For simplicity, it is assumed that the match position is (0, 0), and the size of the template never exceeds that of the subimage. The image data are distributed among the PEs by segmenting

```
CU                                                                            PE

14      for (i = 0; i < r; i++) { /*4,8,6*/
17         send_short( &i, i ); /*send i (8 bits) from CU to PEs*/            8
14         for (j = 0; j < c; j++) { /*4,8,6*/
17            send_short( &j, j ); { /*send j (8 bits) from CU to PEs*/       8
6             simdbegin          /*load PE instruction block into queue */
                 xysum += temp[c × i + j] × subimage[C' × i + j];            90
              simdend

           }
        }
```

Figure 16.8 Code segment overworking the PEs

```
CU                                                                            PE

        tbase = temp[];            /*initialize template pointer*/
        ibase = subimage[];        /*initialize subimage pointer*/
14      for (i = 0; i < r; i++) {          /*4,8,6*/
32         tptr = tbase + c × i - 1;       /*increment template row pointer*/
32         iptr = ibase + C' × i - 1;      /*increment subimage row pointer*/
14         for (j = 0; j < c; j++)          /*4,8,6*/
10           tptr += 1;      /*increment template column pointer*/
10           iptr += 1;      /*increment subimage column pointer*/
17           send_int( &tptr, tptr );/*send template pointer (16 bits)
                                          from CU to PEs*/                     8
17           send_int( &iptr, iptr );/*send subimage pointer (16 bits)
                                          from CU to PEs*/                     8
6            simdbegin              /*load PE instruction block into queue*/
                xysum += (*tptr) × (*iptr);                                   34
             simdend
           }
        }
```

Figure 16.9 Code segment overworking the CU

the input image into rectangular subimages, each of which has the size $R' \times C'$. The template and subimage arrays are represented as one-dimensional arrays *temp*[] and *subimage*[], respectively. The scalar variable *xysum* accumulates the desired sum.

The same task is performed by both code segments, but the workload is distributed differently between the CU and the PEs. The first approach, shown in Fig. 16.8, overworks the PEs by performing all the array indexing calculations on the PEs, whereas the second approach, shown in Fig. 16.9, overworks the CU by letting the CU perform all the array indexing calculations. The numbers along the left and right sides of each figure provide approximate statement execution times in microseconds for the CU and for the PEs, respectively, as measured on the PASM prototype.

The time complexities can be determined by examining the code segments. For the "for" statements, the time to initialize the loop-control variable is 4, the time to test the end-of-loop condition is 8, and the time to increment the loop-control variable is 6. For the approach shown in Fig. 16.8, the time complexities of the CU and the PEs are

$$T_{CU}^{a} = r \times (14 + 17) + (4 + 8) + r \times (c \times (14 + 17 + 6) + (4 + 8)) = 37rc + 43r + 12$$

and

$$T_{PE}^{a} = r \times 8 + r \times c \times (90 + 8) = 98rc + 8r$$

respectively. For $r \geq 1$ and $c \geq 1$,

$$T_{PE}^{a} > T_{CU}^{a}$$

If $r = c = 7$, then

$$T_{CU}^{a} = 2{,}126, \text{ and } T_{PE}^{a} = 4{,}858$$

For the approach shown in Fig. 16.9,

$$T_{CU}^{b} = r \times (74 \times c + 90) + 12 = 74rc + 90r + 12$$

and

$$T_{PE}^{b} = 50cr$$

respectively. For $r \geq 1$ and $c \geq 1$, $T_{CU}^{b} > T_{PE}^{b}$.

If $r = c = 7$, $T_{CU}^{b} = 4{,}268$, and $T_{PE}^{b} = 2{,}450$.

Because the CU broadcasts the instructions to the PEs, the time before the PEs receive their first instructions and the time for the PEs to execute the final instructions broadcast to them have to be considered. Let T_{final} denote the time for the PEs to execute the last instructions broadcast to them minus the time to perform CU computation after the final SIMD block has been broadcast; i.e., T_{final} is the time some PE continues to execute after the CU is done. If the difference is less than zero, set $T_{final} = 0$. For the first approach, the CU increments and then checks the index value of i and j after it broadcasts the final SIMD block, and thus $T_{final}^a = (8 + 90) - (6 + 6 + 8 + 6 + 8) = 64$. Let $T_{startup}$ be the PE idle time during the first iteration. T_{final}^b and $T_{startup}^b$ can be calculated similarly. For the first approach, $T_{startup}^a = (4 + 8 + 17 + 4 + 8 + 17) - 8 = 50$. For $r = c = 7$, the total execution time of the first approach is

$$\max (T_{CU}^a + T_{final}^a, \ T_{startup}^a + T_{PE}^a) \ = \ \max (2{,}127 + 64, \ \ 50 + 4{,}858) \ = \ 4{,}908$$

and for the second approach, the total execution time is

$$\max (T_{CU}^b + T_{final}^b, \ T_{startup}^b + T_{PE}^b) \ = \ \max (4{,}268 + 8, \ \ 134 + 2{,}450) \ = \ 4{,}276$$

By balancing the workload between the CU and the PEs, CU/PE overlap can be maximized; i.e., $|T_{CU} + T_{final} - T_{PE} - T_{startup}|$ is minimized, and thus the execution time can be reduced. A third approach, shown in Fig. 16.10, attempts to minimize $|T_{CU} + T_{final} - T_{PE} - T_{startup}|$ by migrating two indexing operations from the CU to the PEs. The total execution time for the third approach is

$$\max (T_{CU}^c + T_{final}^c, \ T_{startup}^c + T_{PE}^c) \ = \ \max (1{,}860 + 26, \ \ 112 + 2{,}758) \ = \ 2{,}870$$

Thus, by balancing the operations performed on the CU and on the PEs, it is possible to achieve better performance.

In summary, this section has shown that, on SIMD machines that have CU/PE overlap capability, CU/PE overlap can be maximized to reduce the algorithm execution time, especially when loop-intensive computations are involved. Thus, to achieve the best performance, it is important to examine the code and to balance CU/PE overlap whenever possible.

CU   **PE**

```
      tbase = temp[];               /*initialize template pointer*/
      ibase = subimage[];           /*initialize subimage pointer*/
14    for (i = 0; i < r; i++) {      /*4,8,6*/
32       tptr = tbase + c × i - 1;   /*increment template row pointer*/
32       iptr = ibase + C' × i - 1;  /*increment subimage row pointer*/
17       send_int( &tptr, tptr );    /*send template pointer (16 bits)
                                         from CU to PEs*/            8
17       send_int( &iptr, iptr );    /*send subimage pointer (16 bits)
                                         from CU to PEs*/            8
14       for (j = 0; j < c; j++) {   /*4,8,6*/
6        simdbgegin                  /*load PE instruction block into queue*/
            tptr += 1;               /*increment template column pointer*/
            iptr += 1;               /*increment subimage column pointer*/
            xysum += (*tptr) × (*iptr);                             34
         simdend
         }
      }
```

Figure 16.10 Code segment optimizing CU/PE overlap

16.5   Parallel Reduction Operations

16.5.1   Introduction

Many problems require that all the elements of a data set be combined in some fashion, e.g., the sum or product of all elements of an array. These operations are known as *reduction operations;* i.e., a single value is computed as the result of an operation on the data set. Parallel reduction operations can only be applied to associative operations, e.g., sum, product, min, and max. It is important to study parallel algorithms for reduction operations, because data elements are generally distributed across the PEs on parallel machines. As examples, recursive doubling and parallel prefix are discussed in detail in this section.

16.5.2   A single reduction operation on a single set of data

Recursive doubling is used to demonstrate the concept of a single reduction operation on a single set of data in this subsection. The *recursive doubling* procedure, sometimes also called *tree summing,* is an algorithm that combines a set of operands distributed across PEs [52]. Consider the example of finding the sum of M numbers. Sequentially, this requires one load and $M - 1$ additions, or approximately M additions. However, if these M numbers are distributed across $N = M$ PEs, the parallel summing procedure requires $\log_2 N$ transfer-add steps, where a transfer add is composed of the transfer of a partial sum to a PE and the addition of that partial sum to the PE's local sum. Let t_{add} be the time required to execute an addition and $t_{transfer-add}$ be the time to execute a transfer-add. Then, the speedup of this algorithm is

$$\frac{M \times t_{add}}{\log_2 N \times t_{transfer-add}} = O\left(\frac{N}{\log_2 N}\right)$$

assuming t_{add} and $t_{transfer-add}$ are of the same order. This is demonstrated for $M = N = 8$ in Fig. 16.11. Assume that the goal is to calculate

$$\sum_{i=0}^{7} A(i)$$

and $A(i)$ is initially stored in PE i. First, each odd numbered PE i transfers its data item to PE $i - 1$ (at time t_0). Each PE receiving a data item adds it to the data item it is storing (at time t_1). PE 2 sends its partial sum $(A(2) + A(3))$ to PE 0, and PE 6 sends its partial sum $(A(6) + A(7))$ to PE 4 (at time t_2). Each of PEs 0 and 4 adds its received partial sum to its stored partial sum (at time t_3). Then, PE 4 sends its partial sum to PE 0 (at time t_4). PE 0 then computes the complete sum (at time t_5). All of the inter-PE transfers can be adjusted so that the complete sum could be computed by any one of the PEs.

The recursive doubling technique can also be applied to a set of operands whose size is greater than the number of PEs. Let M be the number of operands and let $N = 2^n$ be the number of PEs, addressed from 0 to $N - 1$, where PE P's address in binary is $p_{n-1} \ldots p_1 p_0$. Let each PE store $\lfloor M/N \rfloor$ of the operands, and let $M \bmod N$ PEs receive one additional operand from the remaining $M \bmod N$ operands. First, each PE sequentially sums its local data, requiring at most $\lceil M/N \rceil$ operations. Once the local sums are obtained, then $\log_2 N$ transfer-adds are performed. In general, in the jth inter-PE transfer, where j proceeds from 0 to $\log_2 N - 1$, PE $P = p_{n-1} \ldots p_{j+1} 1 0 \ldots 0$ sends its partial sum to PE $P' = p_{n-1} \ldots p_{j+1} 0 0 \ldots 0$. PE P' combines the received partial sum and its previously computed local partial

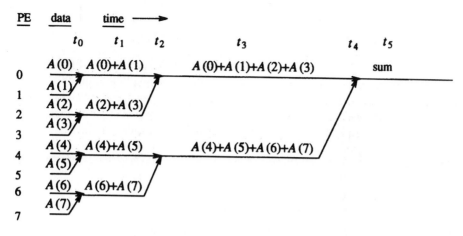

Figure 16.11 Recursive doubling using result-to-one-PE technique for $M = N = 8$

sum to form a new local partial sum. After $\log_2 N$ transfer-adds, PE 0 will contain the complete sum. The speedup is

$$\frac{M \times t_{add}}{\lceil M/N \rceil \times t_{add} + \log_2 N \times t_{transfer-add}}$$

The greater the difference between M and N, $M > N$ the closer the speedup is to N. If $M < N$, this technique can still be applied. In this situation, only M PEs would be used.

The above demonstrates the *result-to-one-PE* technique, where the global result will be available in a single PE. An alternative is the *result-to-every-PE* technique, in which each PE will have the global result. This is shown in Fig. 16.12 for $M = N = 8$. Let a *transfer-op* be composed of a transfer of an operand to a PE and an operation combining this operand and a local operand of this PE. As in the previous technique, once the local results have been obtained, $\log_2 N$ transfer-ops are made to find the global result. In transfer-op j, where

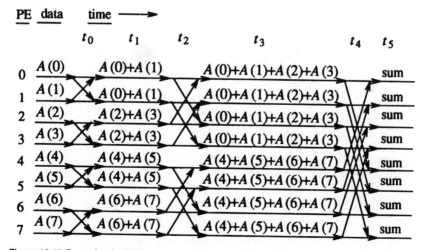

Figure 16.12 Recursive doubling using result-to-every-PE technique

j proceeds from 0 to $\log_2 N - 1$, PE $P = p_{n-1}\ldots p_1 p_0$ exchanges partial results with PE $P' = p_{n-1}\ldots \bar{p}_j\ldots p_1 p_0$. After $\log_2 N$ transfer-ops, each PE will have the global result in its local memory.

Consider the trade-offs between the result-to-one-PE and result-to-every-PE techniques. In SIMD mode, the result-to-one-PE approach has the overhead of disabling PEs in transfer-op steps other than the first step. Thus, even if the result is not needed in all PEs, the result-to-every-PE method may be faster because the masking overhead for disabling PEs is avoided. For SIMD or MIMD mode, when the result is only needed in one PE, using the result-to-every-PE method may cause unnecessary delays in the interconnection network. These trade-offs would have to be balanced for a particular SIMD architecture.

16.5.3 Multiple reduction operations on a single set of data

In the previous subsection, performing one reduction operation on a single data set was discussed. In this subsection, performing multiple reduction operations on a single set of data is reviewed. The min/max problem (i.e., finding the minimum and the maximum values from a set of data simultaneously) is analyzed as an illustrative example.

A straightforward method of solving the min/max problem in parallel is to use the recursive doubling procedure twice—first to find the maximum value and then to find the minimum value. A better technique is to divide the PEs into two groups such that one group finds the maximum value while the other group simultaneously finds the minimum value. Figure 16.13 demonstrates the process of finding the minimum and the maximum values of a data set of $M = 16$ elements on $N = 8$ PEs.

For simplicity, assume that the data set has M elements, where M is an integer multiple of N. Each PE is assigned M/N elements. In the first step, each PE finds the local minimum value and the local maximum value. Then in the next step, PEs are divided into two groups: the upper half and the lower half. A PE in the upper half has its high-order address bit $p_{n-1} = 0$, and a PE in the lower half has its high-order address bit $p_{n-1} = 1$.

In the third step, each PE $P = p_{n-1}\ldots p_1 p_0$ exchanges local values with PE $P' = \bar{p}_{n-1}\ldots p_1 p_0$. A PE in the lower half transfers its local minimum value, while a PE in the

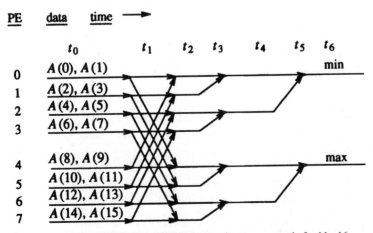

Figure 16.13 Finding the minimum and maximum values concurrently for $M = 16$ and $N = 8$ by adapting the result-to-one technique

upper half transfers its local maximum value. Each PE in the lower half computes a new local maximum value by choosing the larger value from its original local maximum value and the value it received. Each PE in the upper half computes a new local minimum value by choosing the smaller value from its original minimum value and the value it received. Then, the lower half of the PEs do transfer-max operations and the upper half of the PEs do transfer-min operations independently but concurrently (see Fig. 16.13). As a result of this step, the global maximum value is known by one or all PEs in the lower half (depending on which technique is selected—result-to-one-PE or result-to-all-PEs) and the global minimum value is known by one or all PEs in the upper half. If needed, the maximum and minimum values can be exchanged between the two groups (in a single inter-PE transfer) so that each group can have both the maximum and the minimum values. The total number of transfer-compares is $1 + \log_2(N/2)$, where the first term comes from the initial exchange of local values, and the second term comes from the concurrent recursive doubling processes.

In general, the technique can accommodate K reduction operations on a single set of data, where $K \geq 2$. The number of transfer-ops is given by $(K - 1) + \log_2(N/K)$, assuming that N and K are powers of two, and the number of operands is an integer multiple of N. Again, the first term and the second term come from the initial exchange of local values and from the concurrent recursive doubling processes, respectively.

16.5.4   A single reduction operation on multiple sets of data

Recursive doubling can also be applied to one reduction operation on multiple sets of data. An example is *global histogramming* [47, 49], also known as *vector summation*, which will be studied next. Let an $M \times M$ input image be mapped onto N PEs such that each PE holds M^2/N pixels, as in the image smoothing example discussed in Subsection 16.3.3. Global histogramming involves computing B bins, where each *bin* has two attributes associated with it: (1) the range of pixel values it represents, and (2) the number of pixels in the entire image that have values within that range. For this algorithm, it is assumed that N is an integer multiple of B. Each possible pixel value belongs to exactly one bin range. Each PE first computes a local B-bin histogram for the M^2/N pixels in its memory. Let $A(x, y)$ be the gray-level value of the subimage pixel in row x and column y, and let $bin(i)$, where $0 \leq i < B$, be initialized to 0. If each PE contains an $(M/\sqrt{N}) \times (M/\sqrt{N})$ subimage, then an algorithm to compute the local B-bin histograms is:

```
for x = 0 to (M/(√N)) - 1 do
    for y = 0 to (M/(√N)) - 1 do
        bin(A(x, y)) = bin(A(x, y)) + 1
```

(For the serial algorithm, set $N = 1$.) The PEs then combine their local histograms to obtain the global histogram. This is a process of a single combining operation on multiple bins. One straightforward approach to compute the global histogram is to combine one bin at a time using recursive doubling, requiring $B \times \log_2 N$ transfer-add steps.

Consider an overlapped recursive doubling procedure for combining local histograms, where all the bins are summed in an interleaved fashion. Figure 16.14 shows this process for $N = 16$ and $B = 4$. In the figure, $(0, \dots, 3)$ denotes the values of bins $0, \dots, 3$ accumulated in the PE. The N PEs are logically divided into N/B blocks of B PEs each. In the first $b = \log_2 B$ stages, each of the N/B blocks simultaneously combine their histograms. As a re-

PE

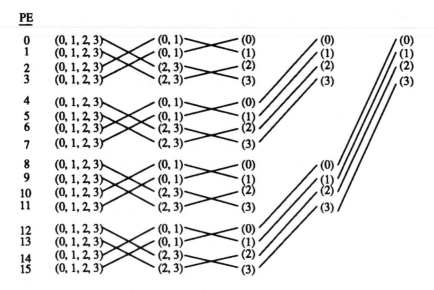

Figure 16.14 Global histogramming for $B = 4$ and $N = 16$

sult, each PE in a block holds a different bin computed by summing the values of the corresponding local bins of the PEs in the block. This is done by dividing each block of PEs in half such that the PEs with lower addresses form one group and the PEs with the higher addresses form the other group. Each group accumulates the sums for half of the bins and sends the bins it is not accumulating to the other group. For example, in stage 0, PE 0 accumulates bins 0 and 1 from PE 0 and PE 2. Simultaneously, PE 1 accumulates bins 0 and 1 from PE 1 and PE 3. The next phase involves dividing each group of $B/2$ PEs into two groups of $B/4$ PEs using the same rule for partitioning PEs as in the previous phase. These two new groups only exchange those bins for which they had accumulated sums in the previous phase. This subdividing process continues until each group has only one PE. Once the subdividing process terminates, each PE will contain only one bin, and this bin will have the accumulated value for the PEs in the block that originally included this PE. Continuing the example above, PE 0 accumulates bin 0 from PE 0 and PE 1, while PE 1 accumulates bin 1 from PE 0 and PE 1.

In the next $n - b$ stages, where $n = \log_2 N$, the partial histograms of all the blocks are combined by performing B simultaneous recursive-doubling operations. Each of these B recursive doubling operations involves the N/B PEs that store the bins of same index. For the example in Fig. 16.14, PEs 0, 4, 8, and 12 combine the bin 0 data, while PEs 1, 5, 9, and 13 combine the bin 1 data, etc. As a result, the histogram for the entire image is distributed over B PEs, where bin i is located in PE i, for $0 \le i < B$.

For the first b stages of the algorithm, $B/2^{j+1}$ transfer-adds are used at stage j, where $0 \le j < b$, for a total of $B - 1$. At each stage j, where $b \le j < n$, one transfer-add occurs. Thus, the final $n - b$ stages require $\log_2(N/B) = n - b$ transfer-adds. The total number of transfer-adds needed to merge the local histograms using the overlapped recursive-doubling scheme is then $B - 1 + \log_2(N/B)$. For realistic values of N and B (e.g., $N = 1,024$ and $B = 256$), this is approximately $\log_2 N$ times faster then the straightforward approach of performing B recursive doublings in sequence.

16.5.5   A generalized form of reduction operations

A generalization of the techniques discussed in Subsections 16.5.2 through 16.5.4 is to apply multiple reduction operations on multiple sets of data. A *distinct operation* is defined as an associative operation on one set of data. Both one associative operation on multiple sets of data and multiple associative operations on one set of data are considered multiple distinct operations. For example, if min and max operations are applied on one set of data, they are considered two distinct operations. In the global histogramming example, the combining operation is applied on four bins; thus, there are four distinct operations. It is obvious that the number of distinct operations is always greater than or equal to the number of associative operations. Assume that each data set is of the same size M. Let N be the number of PEs used and let K be the number of distinct operations. For simplicity, it is assumed that M, N, and K are all powers of two, where $M \geq N$ and $K \leq N$, and each PE has M/N elements of each data set.

The generalized reduction operation process can be divided into three phases: (1) computing local partial results, (2) grouping PEs for different distinct operations, and (3) combining partial results. In the first phase, each PE computes K local partial results, one for each distinct operation on its local data. There are no inter-PE transfers in this phase, and the number of concurrent local operations is $K \times (M/N)$. In the second phase, the PEs are logically divided into N/K groups. A single PE accumulates the partial results with respect to one distinct operation for the block during this phase. As shown in the example for global histogramming, $K - 1$ transfer-op steps are needed to accomplish this phase. The last phase is to perform K recursive doubling procedures concurrently, one for each distinct operation. This is the same as the last step shown in the min/max and the global histogramming examples and needs $\log_2(N/K)$ transfer-ops. Thus, the total number of transfer-ops is $K - 1 + \log_2(N/K)$.

When $K > N$ (i.e., when there are more distinct operations than the number of PEs), interleaved recursive doubling procedures can still be used to find the results. This approach is referred to as ρ-*recursive doubling* [55]. In this case, $\rho = K$, and the final results of K distinct operations can be computed in $\log_2 N$ steps, where $\lceil K/2^{j+1} \rceil$ transfer-op operations occur at step j, for $0 \leq j < \log_2 N$. In the ρ-recursive doubling procedure, K is not required to be a power of two. Figure 16.15 illustrates how $N = 4$ PEs sum the elements for $K = 6$ sets of data, assuming each set has N elements. This requires five transfer-adds, three of which are for the first step and two for the second step.

16.5.6   Parallel prefix

A parallel technique similar to recursive doubling is *parallel prefix*. For $A(i)$, $0 \leq i < N$, the parallel prefix method can compute

$$B(j) = \sum_{i=0}^{j} A(i), 0 \leq j < N$$

in $\log_2 N$ transfer-add steps [52].

It is assumed that PE i, $0 \leq i < N$, has $A(i)$ resident, and it will have $B(i)$ as the result. Figure 16.16 demonstrates this procedure for the case $N = 8$. Initially, every PE executes $B(i) = A(i)$. In transfer-add step j, where j proceeds from 0 to $\log_2 N - 1$, PE P sends its partial result $B(P)$ to PE $P' = P + 2^j$ only if $P' < N$. PE P' computes a new partial result $B(P')$ from its own partial result and the received value. After $\log_2 N$ transfer-op steps, PE i will have $B(i)$.

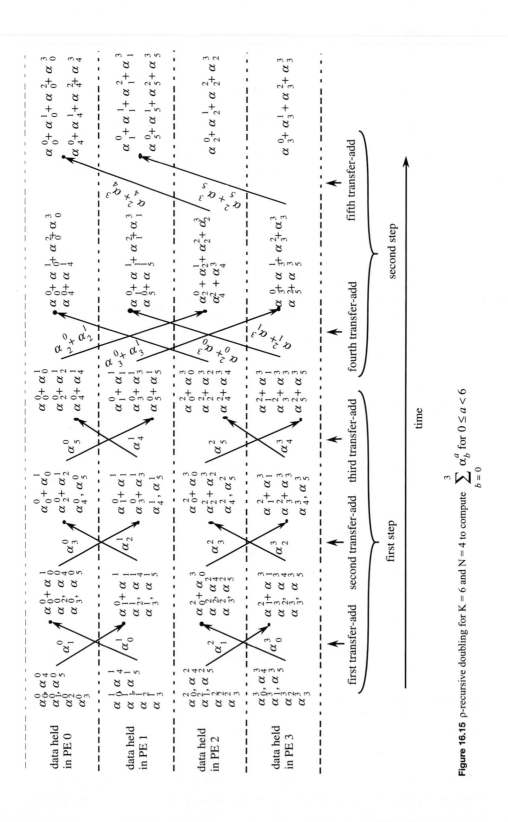

Figure 16.15 p-recursive doubling for K = 6 and N = 4 to compute $\displaystyle\sum_{b=0}^{3} \alpha_b^a$ for $0 \le a < 6$

PE	data	time $\longrightarrow$					

Figure 16.16 Parallel prefix for $N = M = 8$

16.5.7 Summary

Various parallel reduction operation techniques for associative operations were discussed in this section. The choice of technique will depend primarily on the application task to be performed. Reductions on machines with hypercube networks are mentioned in Chapter 19, on parallel and communication algorithms on hypercube multiprocessors.

Because computers use finite precision arithmetic, applying certain associative operations (such as floating point addition) on the same set of data but in different sequence could give different results. However, this problem exists for serial as well as parallel machines.

16.6 Matrix and Vector Operations

16.6.1 Introduction

This section examines parallel matrix operations by considering three examples: matrix transposition, matrix-by-matrix multiplication, and matrix-vector multiplication. In each case, only one of many possible approaches is presented.

16.6.2 Matrix transposition

Given an $M \times M$ matrix A, the transpose of A, A^T, is $A^T(i, j) = A(j, i)$, where $0 \leq i, j < M$. The parallel matrix transposition technique described here is part of a two-dimensional FFT algorithm presented in Ref. [27]. For simplicity, assume $M = N$, the number of the PEs used. The data are distributed across the PEs such that PE i contains row i of A, $0 \leq i < N$. The goal is to have row i of A^T (i.e., column i of A) stored in PE i, $0 \leq i < N$.

Let B be the base address of the location that holds the row of A in a PE and let B^T be the base address of the memory segment that is allocated to hold a row of A^T in a PE. The following code for PE i is executed by all PEs simultaneously to generate A^T.

```
(S1)  for j = 1 to M - 1 do
(S2)     fetch the value at location B + (i + j) mod M          /* A (i, i + j) */
(S3)     transfer from PE i to PE (i + j) mod M
(S4)     store the received value at location B^T + (i - j + M) mod M /* A^T(i + j, i) */
(S5)  copy the value at location B + i to B^T + i      /* move A (i, i) to A^T(i, i) */
```

In statement S2, each PE fetches the value to be sent from its memory location $B + (i + j)$ mod M. In S4, each PE stores the received value at location $B^T + (i - j + M)$ mod M. When

S1 through S4 are done, only the diagonal elements remain to be processed, which is done in S5. This is simply a matter of copying $A(i, i)$ to $A^T(i, i)$, which does not require any inter-PE transfers. Figure 16.17 shows the transposition process for $N = M = 4$ for $j = 1$. The total number of matrix elements that must be transferred is $M(M-1)$, because the M elements in the major diagonal are not transferred. At most, M elements can be moved in parallel in one transfer step. Thus, this algorithm uses $M-1$ transfers, which is the minimum possible given these data distributions. This technique can be directly extended for $M > N$, and each PE storing a square submatrix of A.

PE #

0	(0,0)	(0,1)	(0,2)	(0,3)
1	(1,0)	(1,1)	(1,2)	(1,3)
2	(2,0)	(2,1)	(2,2)	(2,3)
3	(3,0)	(3,1)	(3,2)	(3,3)

(a)

PE #

0	(0,0)	(1,0)	(2,0)	(3,0)
1	(0,1)	(1,1)	(2,1)	(3,1)
2	(0,2)	(1,2)	(2,2)	(3,2)
3	(0,3)	(1,3)	(2,3)	(3,3)

(b)

Figure 16.17 Matrix transpose for $N = M = 4$ for $j = 1$. The data transferred are underlined. (a) Matrix A and (b) Matrix A^T across PEs

16.6.3   Matrix-by-matrix multiplication

Given an $M \times M$ matrix A and an $M \times M$ matrix B, the product of A and B is an $M \times M$ matrix $C = A \times B$ whose elements are given by

$$C(q, r) = \sum_{h=0}^{M-1} A(q, h) \times B(h, r), \quad 0 \le q, r \le M-1$$

Using the straightforward approach, the serial execution time is $t_s = M^3 t_{op}$, where t_{op} is the time required for one addition and one multiplication.

Assume that N PEs are arranged in a logical $\sqrt{N} \times \sqrt{N}$ grid, and the PEs are labeled from PE(0, 0) to PE($\sqrt{N} - 1, \sqrt{N} - 1$). Consider two $M \times M$ matrices A and B partitioned into N submatrices A^{ij} and N submatrices B^{ij}, respectively, $0 \le i, j < \sqrt{N}$. Each of the submatrices is of size $M/\sqrt{N} \times M/\sqrt{N}$. Matrices A and B are superimposed onto the PE grid such that PE(i, j) has the submatrices A^{ij} and B^{ij}. PE(i, j) also allocates space for the submatrix C^{ij} of the result matrix C. To calculate C^{ij}, PE(i, j) needs the submatrices A^{ik} and B^{kj} for $0 \le k < \sqrt{N}$. That is, PE(i, j) needs data from PE(i, k) and PE(k, j), for $0 \le k < \sqrt{N}$. To acquire the required submatrices, all-to-all row and column broadcasts are performed by the PEs. Once the broadcasts are complete, PE(i, j) will have submatrices A^{ik} and submatrices B^{kj}, for $0 \le k < \sqrt{N}$. An algorithm that performs parallel matrix multiplication is shown below [33]. The following code for PE(i, j) is executed by all PEs simultaneously.

```
for k = 1 to √N - 1 do

    send Aⁱʲ to PE(i,(j + k) mod √N )   /* all-to-all row broadcast */
```

```
for k = 1 to √N - 1 do

    send Bⁱʲ to PE((i + k) mod √N , j)        /* all-to-all column broadcast */

initialize submatrix Cⁱʲ to all zeros

for k = 0 to √N - 1 do

    Cⁱʲ = Cⁱʲ + Aⁱᵏ × Bᵏʲ                    /* local multiplication of submatrices */
```

Consider the first "for" loop of the above algorithm. Each "send" takes $(M^2/N) \times t_{transfer}$ time units, where $t_{transfer}$ is the inter-PE transfer time for a matrix element. Hence, the time taken by the "for" loop is $(\sqrt{N} - 1)(M^2/N)t_{transfer}$. The total time for the first two "for" loops is $2(\sqrt{N} - 1)(M^2/N)t_{transfer}$. The final "for" loop performs the computations necessary to calculate the submatrix C^{ij} in time $\sqrt{N}(M/\sqrt{N})^3 t_{top} = (M^3/N)t_{top}$. The speedup achieved with N PEs is

$$ S = \frac{M^3 t_{op}}{(M^3/N)\, t_{op} + 2\,(\sqrt{N} - 1)\,(M^2/N)\, t_{transfer}} $$

More efficient (but more complex) algorithms for matrix-by-matrix multiplication can be found in the literature (e.g., Ref. [33]).

16.6.4   Matrix-by-vector multiplication

Matrix-by-vector multiplication is a special case of matrix-by-matrix multiplication. Given an $M \times M$ matrix A and an $M \times 1$ vector U, the product of A and U is an $M \times 1$ vector $V = A \times U$ whose elements are given by

$$ V(i) = \sum_{j=0}^{M-1} A(i,j) \times U(j), \quad 0 \le i < M $$

The straightforward serial implementation takes $O(M^2)$ time.

A parallel algorithm for matrix-by-vector multiplication is given below [33]. For simplicity, assume $M = N$, where N is the number of PEs used. The elements of matrix A are distributed among the PEs such that each PE i, $0 \le i < N$, is assigned row i of A. The vector U is stored in each PE. $V(i)$ is computed on PE i by multiplying the row i of A that is stored in PE i with the vector U; i.e., PE i computes the equation displayed above. This will require M multiplications and $M - 1$ additions per PE. Assuming that the time to perform the multiplication is of the same order as the time to perform an addition, the resulting parallel time complexity is $O(M)$. An $O(M)$ speedup is attained by this parallel algorithm. This technique can be adapted for $M > N$.

16.7   Mapping Algorithms onto Partitionable Machines

16.7.1   Introduction

Partitionable parallel machines are parallel processing systems that can be divided into independent or communicating subsystems, each having the same characteristics as the original machine [36, 49]. The ability to form multiple independent subsystems to execute multiple tasks in parallel provides such partitionable parallel machines with the potential of achieving better performance. Most MIMD machines can be partitioned, either under system or user control. A *multiple-SIMD* machine includes multiple CUs so that it can be partitioned into independent SIMD subsystems [41]. The CM-2 hardware design could support this [53]. PASM is a prototype partitionable system where each submachine can operate using mixed mode parallelism [51]. In this section, two aspects of mapping algorithms onto partitionable parallel machines are considered: the impact of increasing the number of PEs used, and the impact of subtask parallelism.

The time required to move data between the local PE memories and the system secondary memory (or external I/O devices) varies greatly among different configurations of parallel machines. Thus, for the analyses here, the simplifying assumption is made that there is not a significant difference in this memory transfer time due to varying the number of PEs used for a task. These analyses can be extended to include this memory transfer time if it is significant.

In this section, it is instructive to consider *overhead* and its impact on parallel efficiency. Overhead operations are those that are needed for parallel execution, but not for serial execution (e.g., inter-PE data transfers). The *parallel efficiency, E,* of a parallel implementation measures the amount of overhead that is incurred as follows:

$$E = \frac{\text{speedup}}{\text{number of PEs}} = \frac{\text{serial time}}{(\text{number of PEs}) \times \text{parallel time}}$$

That is, the efficiency is the speedup divided by the number of the PEs used to achieve this speedup. Obviously, efficiency could decrease if the increase in the speedup grows more slowly than the increase in the number of the PEs used; i.e., improved efficiency and increased speedup are not mutually implied. The efficiency measure will be used in the next two subsections.

16.7.2 Impact of increasing the number of PEs

Increasing the number of PEs may or may not decrease the execution time. Two examples, image smoothing and recursive doubling, are analyzed to demonstrate the impact of increasing the number of PEs on performance [31, 47].

Recall from Subsection 16.3.3 that smoothing an $M \times M$ image with N PEs using square subimages requires that each PE perform M^2/N smoothing operations and $4M/\sqrt{N} + 4$ inter-PE data transfers. The execution time decreases as N increases ($N \leq M^2$) because both the number of smoothing operations and the number of inter-PE transfers are inversely proportional to N.

Let t_{so} be the time to perform a smoothing operation and $t_{transfer}$ the time to perform an inter-PE data transfer. The parallel efficiency for the smoothing algorithm can be approximated as

$$\frac{(M-2)^2 t_{so}}{M^2 t_{so} + (4M\sqrt{N} + 4N) t_{transfer}}$$

As N increases, the efficiency decreases. Thus, increasing N reduces the total execution time but causes the efficiency to decrease. This is because as N increases, the PEs spend more time executing overhead operations (i.e., inter-PE transfers) relative to required computations (i.e., smoothing operations). For example, when $N = M^2$, the PEs may spend more time transferring data than smoothing. One must determine the metric that is important in a given situation: execution time or parallel efficiency.

The impact of increasing N has a different effect on the recursive doubling algorithm. Assume that M/N numbers are stored in each PE, where M is the number of operands and N is the number of PEs used. Both M and N are assumed to be powers of two, $M \geq N$, and $t_{transfer\text{-}add}$ and t_{add} are as defined in Subsection 16.5.2. The total execution time, t_{total}, is $(M/N)t_{add} + (\log_2 N)t_{transfer\text{-}add}$.

Assume that a transfer-add takes τ times as long as an addition, i.e., $t_{transfer\text{-}add} = \tau \times t_{add}$. In this case, as N increases the number of local additions decreases but the number of transfer-adds increases. Thus, as N increases, the execution time first decreases then increases. The minimum execution time is a function of τ. The efficiency of the parallel recursive doubling is $M/(M + \tau N \log_2 N)$. Thus, the parallel efficiency always decreases as N increases.

One way to find the number of PEs that will minimize the execution time is to determine the partial derivative of the execution time formula with respect to N. For the recursive doubling example, this yields $\partial T_{total}/\partial N = (-M/N^2 + \tau/(N\ln 2)) \times t_{add}$. The derivative is negative for $N < (M/\tau)\ln 2$ and is positive for $N > (M/\tau)\ln 2$. Thus, as N increases from 1 to $(M/\tau)\ln 2$, the execution time decreases, and as N increases beyond $(M/\tau)\ln 2$, the execution time increases. Let N_1 be the largest value such that $N_1 \leq (M/\tau)\ln 2$, and let N_2 be the smallest value such that $N_2 > (M/\tau)\ln 2$, where N_1 and N_2 are positive integers and are powers of two. Either N_1 or N_2, whichever gives a smaller T_{total}, is chosen as the number of the PEs to be used.

For example, if $\tau = 10$ and $M = 2^{14}$, then the execution time is $(2^{14}/N + 10\log_2 N) \times t_{add}$ time units. Figure 16.18 shows the execution times as N increases. Using the above procedure, 2^{10} PEs are chosen to achieve the minimum execution time. This result impacts the concept of maximizing machine utilization. Typically, one tries to make use of all the PEs available, with the goal of maximizing the number of concurrent operations to minimize the execution time. This study demonstrates that maximizing utilization (i.e., using the largest N possible) does not always mean minimizing execution time. It may be faster to partition the machine and use a subset of the PEs available.

Thus, the question of whether increasing N reduces overall execution time is algorithm dependent. Furthermore, increased parallel efficiency may not imply decreased execution time, and vice versa. Similarly, increased utilization of PEs may not imply decreased execution time, and vice versa.

16.7.3 Impact of subtask parallelism

The effect of partitioning a parallel task into smaller, concurrent subtasks can have an impact on performance. Consider the task of smoothing four images such that the total time to smooth all four is minimized. One way to do this is to smooth the four images one at a time, using all N PEs to smooth each image. Another way is to partition the task such that all four images are smoothed concurrently, each using $N/4$ PEs, as shown in Fig. 16.19.

N	2^7	2^8	2^9	2^{10}	2^{11}	2^{12}	2^{13}	2^{14}
time units	198	144	122	116	118	124	132	141

Figure 16.18 Execution time versus the number of PEs (N) for $M = 2^{14}$ and $\tau = 10$

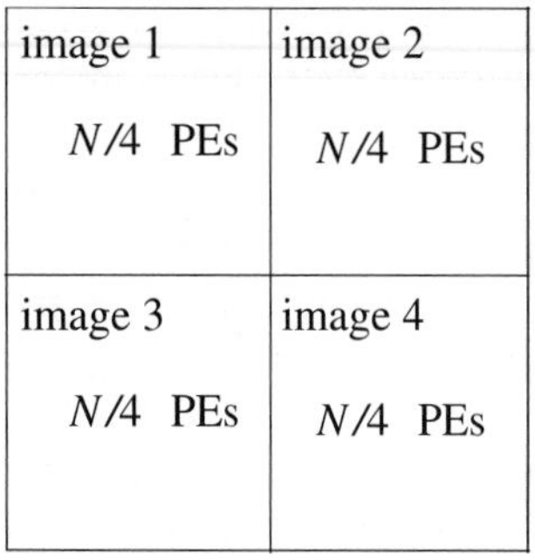

image 1	image 2
N/4 PEs	*N*/4 PEs
image 3	image 4
N/4 PEs	*N*/4 PEs

Figure 16.19 Smoothing four images concurrently on N PEs

The time required to smooth the four $M \times M$ images in sequence is

$$4((M^2/N)t_{so} + (4(M/\sqrt{N}) + 4)t_{transfer})$$

This is four times that to smooth a single image. For $M = 512$ and $N = 1{,}024$, this is

$$1{,}024 \times t_{so} + 272 \times t_{transfer}$$

The total time required for N PEs to smooth the four images concurrently, each on $N/4$ PEs, is $(M^2/(N/4))t_{so} + (4(M/\sqrt{N/4}) + 4) \, t_{transfer}$. (Because all four images are smoothed concurrently, it is the same as the time to smooth one image on $N/4$ PEs.) For $M = 512$ and $N = 1{,}024$, this is $1{,}024 \times t_{so} + 132 \times t_{transfer}$. Thus, partitioning the system and exploiting subtask parallelism decreases the execution time by reducing the number of inter-PE transfers.

The partitioning approach requires $4 \times (M/\sqrt{N/4}) + 4$ inter-PE transfers, whereas in the other approach, $4 \times (4(M/\sqrt{N}) + 4)$ inter-PE transfers are needed. This reduction in inter-PE transfers gives the partitioning approach a smaller execution time. For example, if $M = 512$ and $N = 1{,}024$, there are 132 versus 272 inter-PE transfers. Assuming that $t_{so} = t_{transfer} = 1$, the parallel efficiency of smoothing four 512×512 images in sequence, each using 1,024 PEs, is 78 percent, while the efficiency of smoothing all four images simultaneously on a system partitioned into four submachines of 256 PEs each is 88 percent. The efficiency of smoothing images concurrently is improved over that of smoothing images in sequence because the larger subimage size (32×32 versus 16×16) reduces the number of inter-PE data transfers. For $t_{so} = t_{transfer} = 1$, the percentage of time spent on performing inter-PE communication for smoothing images concurrently is $132/(32^2 + 132) = 11$ percent and for smoothing images sequentially is $68/(16^2 + 68) = 21$ percent. The same number of smoothing operations are performed in both schemes. Therefore, in this case, partitioning improves both execution time and efficiency.

16.7.4   Summary

In this section, two aspects of mapping algorithms onto partitionable machines were discussed. Partitioning can be used to select a subset of the PEs for the task when it is faster than using all of them. Partitioning can also be used to improve performance by executing multiple subtasks simultaneously [39].

16.8   Achieving Scalability Using a Set of Algorithms

A parallel algorithm is *scalable* if it is capable of delivering an increase in performance proportional to the increase of the number of processors utilized [56]. However, one algorithmic approach to a task may not always be able to give the best performance for various input-data parameters and system parameters. Scalability can be better achieved by selecting one algorithm or some algorithm combination from a suite of algorithms to perform the task effectively as these parameters vary.

The *set approach* is to have a set of algorithms from which the most appropriate algorithm or combination of algorithms is selected based on the ratio between the data size and the target machine size. Parallel algorithms for computing *multiple quadratic forms* (MQFs) are discussed in this section as a case study in the design of scalable algorithms [55]. Implementations of the MQF problem for various data-size/machine-size ratios were evaluated in great detail in Ref. [55]. The goal of this section is not to discuss the details of the MQF study, but rather to use the results to demonstrate the importance of the set approach to achieve scalability.

Let a steering vector (*s-vector*) be an $r \times 1$ vector of complex numbers and v be the total number of s-vectors. Define M to be an $r \times r$ matrix of complex numbers and $M(i, j)$ to be the element of M in row i and column j, where $0 \le i, j < r$. The qth s-vector is denoted by s_q, where $0 \le q < v$. Element m of the s-vector q is denoted by $s_q(m)$, where $0 \le m < r$. Let H denote the *Hermitian transpose,* i.e., the complex conjugate transpose of the s-vector. The MQF calculation can be defined as

$$w_q = s_q^H M s_q$$

where $0 \le q < v$.

One parallel implementation used to solve the MQF problem is the *uncoupled* method. In this approach, no inter-PE communication is needed. By distributing v s-vectors evenly among N PEs, $\lceil v/N \rceil$ quadratic forms will be calculated in each of $v \bmod N$ PEs, and $\lfloor v/N \rfloor$ on each remaining PE. Thus, the time to compute the MQFs is the same for $kN < v \le (k+1)N$, for $k \ge 1$.

Another approach used to solve the MQF problem is the *coupled* method, which does use inter-PE communications. Let $PE(i, j)$ denote the PE in row i and column j in an $a \times b$ logical PE grid, where $0 \le i < a$, $0 \le j < b$, and $a \times b = N$. The s-vectors are loaded into the PE memories such that an (r/a)-element subvector of the Hermitian of each s-vector, s_q^H, and an (r/b)-element subvector of at most $\lceil v/a \rceil$ s-vectors, s_q, are stored in each PE memory. Each PE also holds an $(r/a) \times (r/b)$ portion of M. After local computations are done, summations are performed, first within columns of PEs and later within rows of PEs, using the ρ-recursive doubling technique described in Subsection 16.5.5.

The results obtained from the nCUBE 2 MIMD machine for these two approaches are shown in Fig. 16.20, where $r = 16$, $N = 16$, 32, and 64, with b fixed at 16, and a varying such that $N = a \times b$. The execution times as a function of the number of steering vectors is shown for the uncoupled and coupled cases. As can be observed, the faster approach for a given N and r depends on v.

The uncoupled and coupled approaches can be combined by using the uncoupled approach to process $\lfloor v/N \rfloor \times N$ vectors and using the coupled approach to compute the remaining ones. For the nCUBE 2 implementation example of Fig. 16.20, if $v = 80$ and $N = 64$, then 64 vectors could be processed by the uncoupled method (0.0049 seconds) and 16 vectors by the coupled method (0.0032 seconds), for a total time of 0.0081 seconds. To process 80 vectors by the uncoupled method by itself takes 0.0097 seconds, and the coupled method by itself takes 0.01 seconds.

Another variation is to exploit subtask partitioning, as was presented in Subsection 16.7.3. For the nCUBE 2 implementation example of Fig. 16.20, processing 80 vectors us-

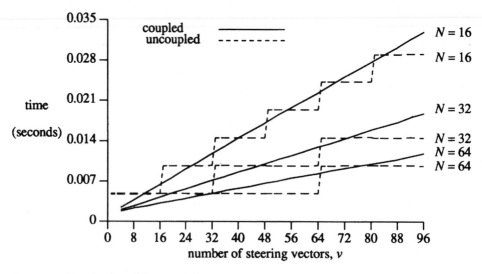

Figure 16.20 Execution time of the uncoupled and coupled data parallel methods on the nCUBE 2 for $r = 16$ and $N = 16, 32,$ and 64

ing 64 PEs with the coupled approach requires 0.01 seconds, but by partitioning the machine into four submachines of 16 PEs, each processing 20 vectors, requires only 0.0077 seconds. Whether partitioning is advantageous depends on the values of v, r, and N.

In summary, choosing an optimal algorithm for the MQF problem is dependent on the characteristics of the input-data and the machine. By having a set of algorithms that solve the problem efficiently for various input-data parameters and system parameters, an algorithm (or combined algorithm) selection methodology can be implemented.

16.9   Conclusions and Future Directions

In designing algorithms for large-scale parallel machines, many issues must be addressed to devise an effective implementation. The importance of these issues cannot be overemphasized, because of their direct impact on the performance of these algorithms. This chapter has surveyed some aspects of some of these issues. For more information, readers are encouraged to see the papers listed in the references and books on parallel algorithms, such as Refs. [2, 9, 12, 20, 26, 32–35, 38, 44]. Some potential future research problems in the design of data parallel algorithms are discussed below. However, the discussion is by no means exhaustive.

In Section 16.3, the impact of data distribution on the execution time was examined. It was shown that the best data distribution for one network may not be the best for another network. Future research is necessary to establish a methodology for determining an optimal data distribution for a given problem-machine combination. The factors that must be considered to design such a methodology for guiding data distribution decisions include (a) the problem parameters, (b) effectively utilizing the PEs, (c) the interconnection network topology, and (d) the inter-PE communication latency. The resulting methodology for data distribution should minimize the total overall execution time.

A machine-dependent issue that is not addressed in this chapter is the communication latency (i.e., the time needed to move a data item needed by one PE but located in another). Because this affects the overall communication cost, to reduce the execution time, it is necessary to develop design techniques that reduce or hide the communication latency. The communication latency is dependent on factors such as (a) the software and hardware overhead for formatting the data to be transferred and establishing the path in the network, (b) the distance from the source PE to destination PE in number of links, (c) the amount of communication and computation overlap possible, and (d) the message size. One way to hide the latency is to initiate the data transfer before the data item is actually needed by the destination PE. Incorporating this approach into one's data parallel programming style, or having a compiler do it automatically, is important.

None of the case studies used in this chapter involved real-time I/O. One of the challenging areas where data parallel algorithms could be effectively applied is real-time processing, e.g., real-time image processing, where real-time I/O operations are integrated into the algorithm. The overall goal is to minimize or to hide the I/O latency. The issues that must be considered include (a) quantifying the I/O characteristics of the application, (b) characterizing the target machine, and (c) scheduling the I/O operations.

Algorithm scalability is another key issue that must be addressed in designing an algorithm for massively parallel machines. Ideally, an algorithm should be designed so that it can be scaled to future-generation machines as they become available. Section 16.8 illustrated how to achieve scalability using a set of algorithms. One algorithm or combination of algorithms is selected based on the ratio between the data size and the target machine size to solve a given problem. Given the ratio, the algorithm(s) selected should achieve the best performance. Data parallel algorithm designers should consider this multiple algorithm approach to scalability. Research is needed to develop a methodology to automatically select an algorithm or combination of algorithms from a set of programs, given the information about the problem-size/machine-size ratio and other relevant information about the problem and the machine.

The examples of parallel matrix operations in Section 16.6 cover dense matrices. A related area that is receiving wide attention at present is operations on very large, very sparse matrices [33]. Sparse matrix representation schemes affect the efficiency of the solution methods. The best serial algorithms cannot be easily parallelized. Hence, it is necessary to develop new algorithms and new representation schemes for sparse matrices that would effectively make use of massively parallel machines. For example, much research is needed in parallel sparse matrix factorization. Current understanding of this area is based on empirical studies. Further research is necessary to theoretically analyze the parallelism and scalability of this operation.

Exploiting concurrent execution of multiple subtasks can be important in reducing the execution time as demonstrated in Section 16.7. Given a partitionable machine and an application, a systematic approach is necessary to determine whether employing subtask parallelism will decrease the execution time [39]. If subtask parallelism is to be employed, then it is necessary to determine the submachine sizes and the mapping of the subtasks onto the submachines that will result in the minimum execution time [13]. The issues that need to be considered in the analysis include (a) identifying the individual subtasks, (b) determining the characteristics (computational requirements and modes of parallelism) of each subtask, (c) mapping the subtasks onto the submachines, and (d) determining the optimal sizes for the submachines. In the case of heterogeneous computing, where there are different parallel machines connected by high-speed links, the analysis becomes even more complex (see Chapter 25).

An application that exhibits data parallelism can be implemented either on an SIMD machine or on an MIMD machine using SPMD mode. It is necessary to devise a systematic approach to aid the programmer in selecting the best type of machine architecture for the application under consideration. In making the choice, the trade-offs between SIMD and MIMD architectures discussed in Chapter 25 must be considered. Once an architecture is chosen, the algorithm for the given application should be optimized with respect to the target architecture.

A data parallel algorithm can also be mapped onto a hybrid SIMD/MIMD mixed-mode machine (a survey of mixed-mode machines is given in Chapter 25). When this is done, the programmer can specify which portion of the program should employ SIMD mode and which SPMD mode. In mapping an application onto a mixed-mode machine, two important issues that must be considered are: (a) where to switch modes within the program, and (b) how to identify the best execution mode (i.e., SIMD or SPMD) for each portion of the program. The use of both SIMD and SPMD modes to execute a single program becomes more complex when a suite of heterogeneous parallel machines is considered instead of a mixed-mode machine. In this case, mode switching implies switching between machines, which involves additional software and hardware overheads (see Chapter 25).

The previous paragraph discusses using both SIMD and SPMD modes within a single data parallel program. Mixed-mode machines and suites of heterogeneous parallel computers interconnected by high-speed links also offer the possibility of combining data parallelism and control parallelism (where each PE can follow a different program). The issues raised in the last paragraph can be extended to three choices: data parallelism using SIMD execution, data parallelism using SPMD execution, and control parallelism using MIMD execution. Designing such hybrid data/control parallel algorithms is another future direction for research.

An issue related to the data parallel algorithm design techniques is the development of an effective parallel programming environment [1, 42]. Such an environment is crucial in supporting the implementation of data parallel algorithms on massively parallel machines. The components of this environment include (a) user-friendly interfaces in both textual and graphical forms, (b) portable parallel programming languages, (c) compilers, (d) libraries that contains optimized machine-dependent data parallel algorithms, and (e) tools (e.g., parallel program debuggers, performance analyzers). An effective parallel programming environment will reduce program development time, increase programmer productivity, ensure program portability, and improve performance. Increased programmer productivity is achieved by making use of the scalable libraries and tools. The portability is ensured by supporting the same parallel language and library functions across multiple hardware platforms. The improved performance can be attained by fine tuning the program using the tools and libraries provided.

Researchers have been working for many years to develop fully automatic techniques for parallelizing sequential algorithms [1]. An alternative approach utilizes semi-automatic techniques for parallelization. The Data Parallel Meta Language (DPML) [18] is one such effort. The idea is to develop a meta language to specify the communication patterns, data distribution strategies, and coordination of subtasks. The meta language provides a model of the problem machine combination to the compiler so that the compiler can better parallelize the serial code. The input to the parallelizing compiler is the serial program augmented with the specification of the machine architecture and the problem characteristics in the meta language. This meta language approach is also promising for heterogeneous computing. Further research is necessary to identify the information needed in a meta language and to design a meta language for heterogeneous computing environments.

This chapter has provided an overview of data parallel algorithm design techniques. The case studies used to illustrate the techniques and to highlight the impact of implementation decisions were based on the message-passing distributed-memory SIMD machine model.

The analysis can be extended to other machine models as well, such as shared address space distributed-memory machines and various types of MIMD machines. However, additional analysis must be done to map the algorithms to those models efficiently, because of the different architectural properties of machines (e.g., multiple instruction streams). The remapping would not be necessary if the techniques discussed were based on an abstract machine model that subsumes all existing models. However, such a model does not presently exist. A future research problem is to develop one universal model or a small number of fundamental models that abstract the salient features of parallel machines and that will support machine-independent parallel programming. In addition, the model must provide realistic information on the relative costs of computation, communication, and synchronization [30]. Therein lies the difficulty, i.e., developing a general machine-independent model that is "'precise enough" about performance without being 'too explicit' about the implementation details" of any machine [45]. The algorithm design techniques based on such a standard parallel machine model would be applicable to all existing parallel machine models. Thus, the design techniques for parallel algorithms could be unified. The advantages of this unified approach would include algorithm portability, algorithm scalability, run-time migration of tasks across parallel machines, and reduction in algorithm development time.

In summary, there is a great deal of activity in the field of data parallel algorithms, as evidenced in the sampling of relevant references listed at the end of this chapter. The goal of this chapter was to provide an introduction to some of the issues germane to constructing effective data parallel programs. This section has discussed a variety of areas for future research directly or indirectly related to data parallel algorithms. In general, the future directions for research relating to data parallel algorithms should lead to the mapping of applications onto parallel machines in ways that most effectively exploit the computing power these machines provide. In many cases, this may enable the performance of application tasks that would be infeasible without the computing power of parallel machines. Finally, when contrasting the data parallel approach to the control parallel approach, many researchers feel that the data parallel paradigm must be employed, at least within subtasks, to take advantage of massively parallel machines with a thousand or more processors.

Acknowledgments

The authors thank K. H. Casey, R. Gupta, K. Liszka, J. M. Siegel, M. Tan, M. Theys, and A. Y. Zomaya for their comments.

16.10   References

1. Adre, V., A. Carle, E. Granston, S. Hiranandani, K. Kennedy, C. Koelbel, U. Kremer, J. Mellor-Crummey, S. Warren, and C. Tseng. 1994. Requirements for data-parallel programming environments. *IEEE Parallel and Distributed Technology Systems and Applications,* Vol.2, No. 3, 48–57.

2. Akl, S. G. 1989. *The Design and Analysis of Parallel Algorithms.* Englewood Cliffs, N.J.: Prentice Hall.

3. Armstrong, J. B., M. A. Nichols, H. J. Siegel, and L. H. Jamieson. 1991. Examining the effects of CU/PE overlap and synchronization overhead when using the complete sums approach to image correlation. *Proceedings of the Third IEEE Symposium on Parallel and Distributed Processing,* 224–232.

4. Barnes, G. H., R. Brown, M. Kato, D. J. Kuck, D. L. Slotnick, and R. A. Stokes. 1968. The Illiac IV Computer. *IEEE Trans. Computers,* Vol. C-17, No. 8, 746–757.

5. Batcher, K. E. 1974. STARAN parallel processor system hardware. *AFIPS 1974 National Computer Conf.,* 405–410.

6. Batcher, K. E. 1977. STARAN series E. *Proceedings of the 1977 Int'l Conf. on Parallel Processing,* 140–143.

7. Batcher, K. E. 1980. Design of a massively parallel processor. *IEEE Trans. Computers,* Vol. C-29, No. 9, 836–844.

8. Berg, T. B., and H. J. Siegel. 1991. Instruction execution trade-offs for SIMD vs. MIMD vs. mixed-mode parallelism. *Proceedings of the Fifth Int'l Parallel Processing Symposium,* 301–308.

9. Bertsekas, D. P., and J. N. Tsitsiklis. 1989. *Parallel and Distributed Computation.* Englewood Cliffs, N.J.: Prentice Hall.

10. Blank, T. 1990. The MasPar MP-1 architecture. *IEEE Compcon,* 20–24.

11. Blank, T., and J. R. Nickolls. 1992. A Grimm collection of MIMD fairy tales. *Proceedings of the Fourth Symposium on Frontiers of Massively Parallel Computation,* 448–457.

12. Chaudhuri, P. 1992. *Parallel Algorithms: Design and Analysis.* Englewood Cliffs, N.J.: Prentice Hall.

13. Chu, C. H., E. J. Delp, L. H. Jamieson, H. J. Siegel, F. J. Weil, and A. B. Whinston. 1989. A model for an intelligent operating system for executing image understanding tasks on a reconfigurable parallel architecture. *J. Parallel and Distributed Computing,* Vol. 21, No. 1, 598–622.

14. Darema, F., D. A. George, V. A. Norton, and G. F. Pfister. 1988. A single-program-multiple-data computational model for EPEXIFORTRAN. *Parallel Computing,* Vol 7, No. 1, 11–24.

15. Duclos, P., F. Boeri, M. Auguin, and G. Giraudon. 1988. Image processing on a SIMD/SPMD architecture: OPSILA. *Proceedings of the Ninth Int'l Conf. Pattern Recognition,* 430–433.

16. Fineberg, S. A., T. L. Casavant, and H. J. Siegel. 1991. Experimental analysis of a mixed-mode parallel architecture using bitonic sequence sorting. *J. Parallel and Distributed Computing,* Vol. 11, No. 3, 239–251.

17. Flynn, M. J. 1966. Very high-speed computing systems. *Proceedings of the IEEE,* Vol. 54, No. 12, 1901–1909.

18. Francis, R. S., I. D. Mathieson, P. G. Whiting, M. R. Dix, H. L. Davis, and L. D. Rotstayn. 1994. A data parallel scientific modelling language. *J. Parallel and Distributed Computing,* Vol. 21, No. 1, 46–60.

19. Giolmas, N., D. W. Watson, D. M. Chelberg, and H. J. Siegel. 1992. A parallel approach to hybrid range image segmentation. *Proceedings of the Sixth Int'l Parallel Processing Symposium,* 334–342.

20. Golub, G., and J. M. Ortega. 1993. *Scientific Computing: An Introduction with Parallel Computing.* Boston: Academic Press.

21. Hayes, J. P., and T. Mudge. Hypercube Supercomputers. *Proceedings of the IEEE,* Vol 77, No. 12, 1829–1841.

22. Hillis, W. D. 1985. *The Connection Machine.* Cambridge, Mass.: MIT Press.

23. Hillis, W. D., and G. L. Steele, Jr. 1986. Data parallel algorithms. *Communications of the ACM,* Vol. 29, No. 12, 1170–1183.

24. Hillis, W. D., and L. W. Tucker. 1993. The CM-5 Connection Machine: a scalable supercomputer. *Communications of the ACM,* Vol. 16, No. 11, 31–40

25. Hunt, D. J. 1989. AMT DAP—a processor array in a workstation environment. *Computer Systems Science and Engineering,* Vol. 4, No. 2, 107–114.

26. Jamieson, L. H., D. Gannon, and R. J. Douglass, eds. 1987. *The Characteristics of Parallel Algorithms.* Cambridge, Mass.: MIT Press.

27. Jamieson, L. H., P. H. Mueller, Jr., and H. J. Siegel. 1986. FFT algorithms for SIMD parallel processing systems. *J. Parallel and Distributed Computing,* Vol. 3, No. 1, 48–71.

28. Jamieson, L. H. 1987. Characterizing parallel algorithms. In *The Characteristics of Parallel Algorithms,* L. H. Jamieson, D. G. Gannon, and R. J. Douglass, eds. Cambridge, Mass.: MIT Press.

29. Kim, S. D., M. A. Nichols, and H. J. Siegel. 1991. Modeling overlapped operation between the control unit and processing elements in an SIMD machine. *J. Parallel and Distributed Computing,* Vol. 12, No. 4, 329–342.

30. Kowalik, J. S., and K. W. Neves. 1993. Software for Parallel Computing: Key Issues and Research Directions. In *Software for Parallel Computation,* J. S. Kowalik and L. Grandinetti, eds. Berlin: Springer-Verlag, 3–33.

31. Krishnamurti, R., and E. Ma. 1988. The processor partitioning problem in special-purpose partitionable systems. *Proceedings of the 1988 Int'l Conf. on Parallel Processing,* Vol. I, Aug. 434–443.

32. Kronsjo, L., and D. Shumsheruddin. 1992. *Advances in Parallel Algorithms.* New York: Halsted Press.

33. Kumar, V. A. Grama, A. Gupta, and G. Karypis. 1994. *Introduction to Parallel Computing: Design and Analysis.* Redwood City, Calif.: Benjamin/Cummings.

34. Kumar, V. K. P., ed. 1991. *Parallel Architectures and Algorithms for Image Understanding.* Boston: Academic Press.

35. Lakshmivarahan, S., and S. K. Dhall. 1990. *Analysis and Design of Parallel Algorithms: Arithmetic and Matrix Problems.* New York: McGraw-Hill.

36. Lipovski, G. J., and M. Malek. 1987. *Parallel Computing.* New York: John Wiley & Sons.

37. 1991. *Data-Parallel Programming Guide.* Sunnyvale, Calif.: MasPar Computer Corporation.

38. Modi, J. J. 1988. *Parallel Algorithms and Matrix Computation.* New York: Oxford University Press.

39. Nation, W. G., A. A. Maciejewski, and H. J. Siegel 1993. A methodology for exploiting concurrency among independent tasks in partitionable parallel processing systems. *J. Parallel and Distributed Computing,* Special Issue on Performance of Supercomputers, Vol. 19, No. 3, 271–278.

40. Nation, W. G., G. Saghib and H. J. Siegel. 1993. Properties of interconnection networks for large-scale parallel processing systems. *Proceedings of the Int'l Summer Institute on Parallel Architectures, Languages, and Algorithms,* 51–82.

41. Nutt, G. J. 1977. Multiprocessor implementation of a parallel processor. *Proceedings of the Fourth Annual Symposium on Computer Architecture,* 147–152.

42. Pancake, C. M. 1991. Software support for parallel computing: where are we headed? *Communications of the ACM,* Vol. 34, No. 11, 53–64.

43. Philippsen, M., T. Warschko, W. Tichy, and C. Herter. 1992. "Project Triton: towards improved programmability of parallel machines. *Proceedings of the 26th Hawaii Int'l Conf. System Sciences,* 192–201.

44. Pitas, I. 1993. *Parallel Algorithms For Digital Image Processing, Computer Vision, and Neural Networks.* New York: John Wiley & Sons.

45. Siegel, H. J., S. Abraham, W. L. Bain, K. E. Batcher, T. L. Casavant, D. DeGroot, J. B. Dennis, D. C. Douglas, T. Y. Feng, J. R. Goodman, A. Huang, H. F. Jordan, J. R. Jump, Y. N. Patt, A. J. Smith, J. E. Smith, L. Snyder, H. S. Stone, R. Tuck, and B. W. Wah. 1992. Report of the Purdue workshop on grand challenges in computer architecture for the support of high performance computing. *J. Parallel and Distributed Computing,* Vol 16, No. 3, 199–211.

46. Siegel, H. J., J. K. Antonio, and K. Liszka. 1992. Metrics for metrics: why is it difficult to compare interconnection networks or how would you compare an alligator to an armadillo? *Proceedings of the New Frontiers: A Workshop on Future Directions of Massively Parallel Processing,* 97–106.

47. Siegel, H. J., J. B. Armstrong, and D. W. Watson. Mapping Computer-Vision-Related Tasks Onto Reconfigurable Parallel Processing Systems," *Computer,* Vol. 25, No. 2, Feb. 1992, 54–63.

48. Siegel, H. J. 1990. *Interconnection Networks for Large-Scale Parallel Processing: Theory and Case Studies,* Second Edition. New York: McGraw-Hill.

49. Siegel, H. J., L. J. Siegel, F. C. Kemmerer, P. T. Mueller, Jr., H. E. Smalley, Jr., and S. D. Smith. 1981. PASM: a partitionable SIMD/MIMD system for image processing and pattern recognition. *IEEE Trans. Computers,* Vol. C-30, No. 12, 934–947.

50. Siegel, L. J., H. J. Siegel, and A. E. Feather. 1982. Parallel Processing Approaches to Image Correlation. *IEEE Trans. Computers,* Vol. C-31, 208–218.

51. Siegel, H. J., T. Schwederski, W. G. Nation, J. B. Armstrong, L. Wang, J. T. Kuehn, R. Gupta, M. D. Allemang, D. G. Meyer, and D. W. Watson. 1995. Chapter 3, The design and prototyping of the PASM reconfigurable parallel processing system. In *Parallel Computing: Paradigms and Applications,* A. Y. Zomaya, ed. London: International Thomson Computer Press, 78–114.

52. Stone, H. S. 1980. Parallel Computers. in *Introduction to Computer Architecture,* H. S. Stone, ed. Chicago: Science Research Associates.

53. Tucker, L. W., and G. G. Robertson. 1988. Architecture and applications of the connection machine. *Computer,* Vol. 21, No. 8, 26–38.

54. Varma, A. and C. S. Ragavendra, eds. 1994. *Interconnection Networks for Multiprocessors and Multicomputers: Theory and Practice.* Los Alamitos, Calif.: IEEE Computer Society Press.

55. Wang, M-C., W. G. Nation, J. B. Armstrong, H. J. Siegel, S. D. Kim, M. A. Nichols, and M. Gherrity. 1994. Multiple quadratic forms: a case study in the design of data-parallel algorithms. *J. Parallel and Distributed Computing,* Vol. 21, No. 1, 124–139.

56. Wilson, G. V. 1993. A glossary of parallel computing terminology. *IEEE Parallel and Distributed Technology, Vol 1,* No. 1, 52–67.

17

Systolic and VLSI Processor Arrays for Matrix Algorithms

D. J. Evans and M. Gusev

In this chapter the authors discuss the fundamentals of systolic and VLSI processor arrays, and their design and taxonomy. Then, some basic matrix algorithms and their systolic implementation are presented.

17.1 Processor Array Implementations

The ever-increasing demands for speed and performance in digital signal processing (DSP) clearly point to a large-scale computation requirement that may be satisfied only by revolutionary supercomputing technology. A very promising trend is the use of VLSI technology for building processor arrays on one chip [1]. In particular, this technology has inspired many innovative designs in processor architectures.

Modern DSP applications depend on high throughput and massive data used in computations. Sequential systems are inadequate to respond to the demands of real-time signal processing, and special-purpose processor arrays are the only appealing alternative to offer satisfactory processing speed and power [2]. There are two types of special-purpose computers. One type is characterized by inflexible and highly dedicated structures, whereas the other allows some flexibility, such as programmability and reconfigurability.

Facing the challenge of revolutionary VLSI technology, modern DSP should incorporate innovative but mature concepts and methods into its architecture/language design. The top-down design approach uses three different types of representations: functional, structural, and geometrical, as summarized by the Y-chart [2]. All representations possess different levels of design, such as the application, algorithm, basic operation, and Boolean logic level in the functional representation; the interconnection pattern, processor element, and bit-slice level in the structural representation; and finally WSI layout plan and cell description level in the geometrical representation. Only the mapping of the functional representation level to the structural representation level will be concerned here.

As long as VLSI technology is used, certain constraints should be analyzed and incorporated in the design methodology, such as the restrictive local interconnection (short communication paths), limited input/output interaction, data transfer time consumption, and synchronization. Along with the requirements of the restrictive VLSI technology, the process of designing incorporates a few other benefits, such as repetitive modular structure, easy expansion, reconfigurability, and flexibility. For the purpose of efficient use, the algorithms executed in these processor arrays should have a balanced distribution of the work and data and a regular data and information flow.

All these requirements serve as guidelines to the designer of VLSI architectures and algorithms and lead to new designs, solutions, and taxonomy of the parallel algorithms and architectures. The problem is to achieve a balance among many conflicting goals, such as the generality of the system versus ease of programming, flexibility versus efficiency, performance of the system versus its design and implementation costs, and the challenge of understanding the strengths and drawbacks of each approach used versus a selected suitable architecture for a given environment. After determining the parallel implementation of DSP problems and VLSI technology, the taxonomy of processor array architectures and algorithms will be given.

17.2   VLSI Processor Arrays

Current parallel computers can be divided into three structural classes: vector processors, multiprocessor systems, and processor arrays [3]. The first two classes belong to the general-purpose computer domain, and the third class is focused on special-purpose applications. The taxonomy [1] will be given according to the synchronicity timing scheme and data I/O used, as depicted in Fig. 17.1. A complete and objective comparison is very involved, as there exists a complex trade-off between numerous criteria, such as programmability, modularity, synchronization, and interconnection communication.

Systolic and wavefront arrays are determined by pipelining data concurrently with the processing. Wavefront arrays, on the other hand, use data-driven processing capability, while the systolic arrays use local instruction codes synchronized globally. In contrast to the pipelined data I/O, the single instruction multiple data (SIMD) and multiple instruction

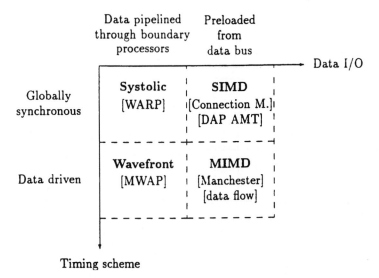

Figure 17.1 Classification of SIMD, MIMD, systolic, and wavefront arrays and typical examples

multiple data (MIMD) arrays use global data and control (instruction) access, allowing broadcasts from the memory and the control unit.[*] MIMD arrays are a kind of superset of all mentioned processing capabilities, such as asynchronous distributed control, data driven computations, preloaded data from global bus, and shared memory with a broadcast property.

A typical example of a SIMD array is depicted in Fig. 17.2, where the control unit (CU) controls the work of all processor units (PUs). As shown, the PU communicates with the memory via bidirectional links. New instructions are updated in the CU via the instruction stream (IS). The difference to the MIMD array is shown in Fig. 17.2, where the CU actually contains of different parts CU_1, CU_2,...,CU_n, each controlling one PU. The systolic and wavefront arrays are called VLSI processor arrays and are the topic of research here. The main difference between the SIMD and MIMD arrays is the use of local communications instead of global shared memory.

Several parameters that characterize the processor arrays are used in efficiency analysis, such as p, T_1, T_p, S_p, and E_p. These are :- the number of processors used in the processor array p; T_1 and T_p are the times used for the algorithm execution by one processor and the processor array, respectively; S_p is the speedup achieved by the processor array implementation; and E_p is the efficiency or processor utilization of the proposed implementation.

Although different processor array architectures are offered to execute the algorithm in parallel, there are certain bottlenecks that occur and cannot be eliminated [4]. The first one, called Amdahl's Law, is always used as an argument against parallel computations to express its limitations. Specifically, the algorithm can always be decomposed in two parts, a sequential part, denoted by S, and a parallel part, denoted by P. Assume that S and P are

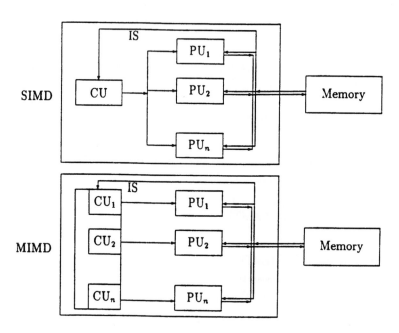

Figure 17.2 SIMD and MIMD processor array architectures

---

[*] These arrays are determined by the Flynn's taxonomy for computer architectures, according to the data and instruction streams.

relative values (i.e. $S + P = 1$) and denote by f their relative ratio S/P. Then, the time to execute the algorithm in parallel is equal to $T_p = fT_1 + (1 - f) T_1/p$, and the speedup achieved by the processor array implementation is therefore

$$S_p = \frac{1}{f + (1 - f)/p} \leq \frac{1}{f} \tag{17.1}$$

That is, the speedup is bounded, not linear. Moreover, it cannot be increased as the number of processors is increased, since it is limited by the algorithm, which determines the *decomposition* bottleneck problem.

The second bottleneck is the *Von Neumann* bottleneck, which is the result of communication limitations of the architecture model. At this stage of VLSI technology development, the processors perform their tasks much faster than the response speed of the memory or I/O rate. A typical system has disk memory with communication rate of up to 0.3 megawords per second (Mw/s), the processor communication rate begins at about 30 Mw/s and usually is more than 100 times faster. This is the reason why the VLSI processor arrays are preferred over the SIMD and MIMD designs that use global bus and shared memory. Specifically, the idea is to accept the data from memory and use it in all processors before an update is required in memory. This is actually the systolic principle, which will be discussed now in more detail.

17.2.1 Systolic arrays

The systolic array concept was invented by Kung and Leiserson [5]. A systolic system is a network of processors that rhythmically compute and pass data through the system. The analogy is made to the systole, where the data is pumped and passed regularly in the processors. The memory in a typical Von Neumann architecture is updated after processing a data element in each processor, producing a bottleneck. Once a data item is brought from memory to a systolic array, it can be used effectively in each cell as it passes while being "pumped" from cell to cell along the array, as depicted in Fig. 17.3. Even if the communication rate with the memory is slow compared to the processor rate, the data is processed by several processors and updated only once, so enhanced processing power is achieved.

The major factors of adopting systolic arrays for special-purpose processing architectures are simple and regular design, concurrency and communication, and balancing computation with I/O [6]. For example, Kung and Leiserson proposed a linear array for the matrix vector multiplication algorithm as shown in Fig. 17.4.

The original idea was based on the regular data flow of two data streams, x_i and y_i, for $i = 1,\ldots,n$ in opposite directions. Therefore, two consecutive elements must be separated

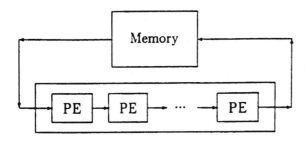

Figure 17.3 Basic configuration of systolic arrays

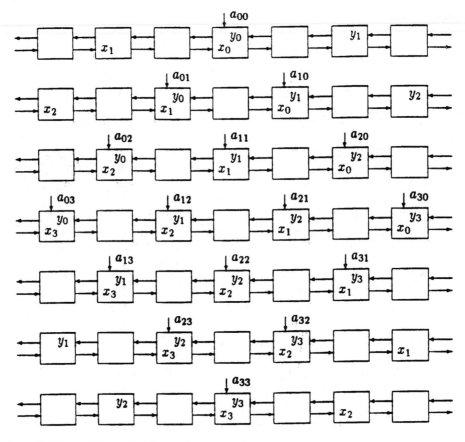

Figure 17.4 Time activities of the BLA matrix vector multiplication $(n = 4)$

by two units of time, since any of them (say, x_i) will meet all the elements of the other stream. Otherwise, if separated by a single unit of time then x_i will never meet all the elements, for example $y_{i+1}, y_{i+3}, \dots$. Then, the required computations are performed concurrently in the cells that compute an inner product step (IPS) operation, shown in Fig. 17.5. The value of y_{in} that enters the cell is updated by the product of the incoming values of x_{in} and a_{in}, and the result is passed as x_{out}. The value of x_{in} is also passed as x_{out} to be used by the remainder of the array. This implementation possesses a characteristic to work in alternate active and inactive time moments.

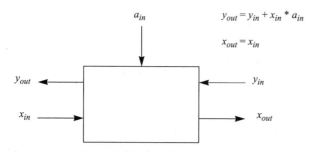

Figure 17.5 Functional description of an IPS cell

17.2.2 Wavefront arrays

The paths implemented in VLSI technology are thin, and the links between processing elements possess increased capacity. This is reflected in the time delay required for data to pass the link. Therefore, local communication is the strongest constraint in VLSI and, because it is ideal to build all the links with the same length, synchronicity becomes a great problem.

A solution to this problem is to use the asynchronous mode in systolic arrays, i.e., data-driven computations. In the wavefront arrays, the arrival of data from neighboring processors are interpreted as instructions to change the state and to activate new actions [7]. Such a data-driven phenomenon defines a wavefront propagation in the array, and the global clock used in systolic arrays is therefore avoided. In contrast to MIMD arrays, the wavefront arrays employ local communication and local instruction storage. The pipeline period may be decreased given the systolic arrays, since there may be some time-consuming operations that dictate the minimum processing rate. Wavefront arrays are preferred in applications where the cells are more complex than the IPS operations, when the synchronization of large array becomes impractical, or when a reliable computing technique such as fault-tolerant computing is essential.

17.2.3 Design criteria

Some basic principles and design criteria for VLSI processor arrays are as follows:

- *Modularity and simplicity.* The array consists of modular cells, each performing a simple operation on the incoming and resident data.
- *Locality and regularity.* The cells are interconnected using links that form a regular structure. The data flow is simple and regular.
- *Extensive pipelineability and parallelism.* This is exploited to achieve greater speedup.
- *Efficiency and speedup.* Each data item is used in multiple operations.

The sequence of data elements entering the array and flowing rhythmically is called a *data stream*. If the elements of a data stream are used only as input parameters in the computations performed in the processor array, then the stream is called an *input data stream*. An *output data stream* is a data stream whose elements are computed by the systolic array and need to be stored in the memory after the algorithm is executed. Actually, the output data stream represents the results computed by the algorithm. Notice that the elements in the data stream can be computed, but these values are used only as intermediate, partial results and are not stored in the memory after algorithm execution. This data stream is neither an input nor output data stream.

17.2.4 Typical VLSI processor arrays

Triangular, square, and hexagonal planar arrays with two or three nearest-neighbor interconnections in vertical, horizontal, and diagonal directions are the planar arrays usually used. Several types of linear arrays are determined according to the data flow, i.e., the number of data paths and the flow direction. Usually there are one, two, or three data paths (VLSI constraint) with the same or opposite direction. Some of them are labeled by ULA, BLA, and TLA as follows:

Definition 1. Unidirectional linear array (ULA). A ULA is a linear processor array with the data streams flowing in the same direction. The processors in a ULA can communicate by one, two, or three data paths.

Definition 2. Bidirectional linear array (BLA). A BLA is a linear processor array that uses two data streams flowing in opposite directions. A *regular* BLA is a BLA where one of the data streams is an output data stream.

Definition 3. Three-path communication linear array (TLA). A TLA is a linear processor array that uses three data streams flowing in different directions.

The ULA implementation of the matrix vector multiplication algorithm uses one data stream flowing to the right and one resident in the array, as depicted in Fig. 17.6. The processor used is a modification to the standard IPS cell, since only one data path is used, and the data elements from the other data paths are kept resident in the array. The timing diagrams of the matrix vector multiplication BLA are depicted in Fig. 17.4, and the data flow is shown in Fig. 17.7. The processors of the ULA implementation work more efficiently, i.e., in consecutive time moments, contrary to the active and inactive time moments in the BLA implementation.

An example of TLA is the ARMA filter implementation proposed by H.T. Kung [4]. It is depicted in Fig. 17.8. There are several versions of the TLA, varying according to the number of output data streams and resident memory values. (For example, the presented ARMA filter TLA uses one output data stream.) The processors perform two IPS operations and are usually called *double IPS cells*. Two versions of the double IPS cells are depicted in Fig. 17.9 according to whether it has one or two output data streams. These processor cells can be produced to use external input data instead of resident memory values, as depicted in Fig. 17.10. The TLAs are also called *double pipelines* [8], since they can be decomposed into two simple BLAs (linear pipelines). Similarly, it can be consid-

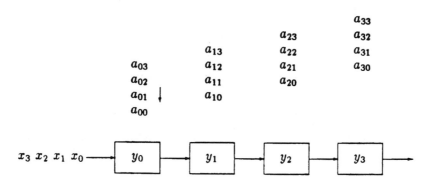

Figure 17.6 Data flow in the matrix vector multiplication ULA ($n = 4$)

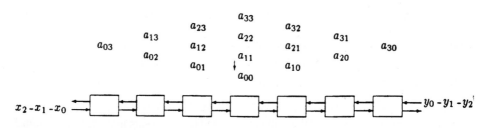

Figure 17.7 Data flow in the matrix vector multiplication BLA ($n = 4$)

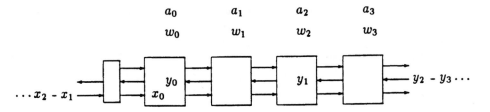

Figure 17.8 Data flow in the ARMA filter TLA ($n = 3$)

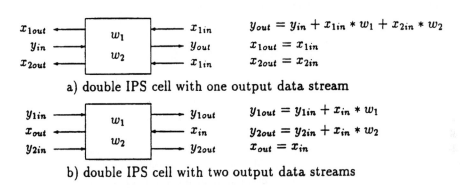

a) double IPS cell with one output data stream

b) double IPS cell with two output data streams

Figure 17.9 Two versions of the double IPS cell with two resident memory values

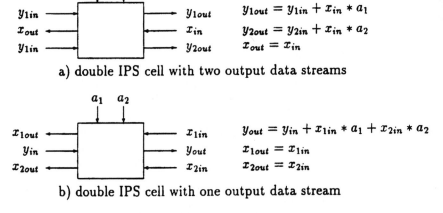

a) double IPS cell with two output data streams

b) double IPS cell with one output data stream

Figure 17.10 Two versions of the double IPS cell with external data input

ered that the double pipelines (TLAs) are obtained by a composition (grouping) of two BLAs with one equal data stream.

The presented ULA and BLA implementations of the matrix vector multiplication algorithm execute the algorithm in the same time, but the ULA uses half the processors of the

BLA implementation. However the ULA implementation uses resident data and additional load/store times are required for the total processing time. The BLA implementation also requires additional input/output time. Therefore the timings are a very important characteristic and are analyzed next.

17.2.5   Timings

Timings must be considered with a great care, since many authors neglect some very time-consuming actions. This may result in a false conclusion for a VLSI processor array implementation, and so the following definitions are particularly important:

Definition 4.   Timings for processor arrays.
- *Execution time* is the time required to execute all the computations in the processor array.
- *Input time* is the time required for the data elements to enter the processor where the first computation takes place.
- *Output time* is the time required for the data elements to exit the array from the processor where the last computation finished.
- *Load time* is the time required for the data elements to enter the array in the correct processor where they are resident.
- *Unload time* is the time required for the resulting data elements to exit the array if they were resident.
- *Total processing time* is the sum of each of the times for data loading, input, execution, output, and unloading.

Usually, the input and output time is equal to the processor size, since one data element must pass through all cells in this direction to enter (exit) the processor where the computation takes place (finished). The loading and unloading can be done by a special Boolean matrix flowing in the array using the input data stream. If it is done in a pipeline with the input string, then no additional time is required. Sometimes the pipeline data period is also required for a comparison between various solutions.

17.3   Systolic Array Algorithms

After identifying tasks and architectures to be implemented in VLSI, new algorithms with degrees of parallelism and regularity, low communication overheads, and adaptability to various I/O constraints imposed by the outside environment, must be developed. It is no longer important to minimize the number of multiplications, as designers once did; appropriate complexity measures should reflect trade-offs between chip area, time, and power. Many classical problems will thus require the design of new parallel algorithms or adaptation of existing algorithms, while the fast sequential approaches will be rejected as they possess no local or recursive structure.[*]

An array algorithm to be implemented on a VLSI processor array is a set of rules solving a problem in a finite number of steps by a multiple number of interconnected processors [2]. The most challenging problems to be targeted by an array algorithm are the compute bound problems [6], where the number of computations is much greater than the number of I/O communications, in contrast to the I/O bound algorithms.

---

[*] For example, DFT is preferred over FFT.

Array algorithms, as a subset of the class of parallel algorithms, are identified by the following criteria [9]:

- *Synchronicity.* Both the asynchronous and synchronous modes are used in wavefront and systolic arrays, respectively.
- *Concurrency control.* The computations should be pipelineable and parallel, which introduces the concept of dependency between the computations as one of the design criteria.
- *Granularity.* Each subtask performed by a separate cell should be as simple as possible but complex enough to be efficient. However, this is dependent on communication and the local memory in the cell.
- *Communication geometry.* Local interconnections require locally dependent subtasks to be performed in the neighboring cells.

These criteria determine a special class of algorithms that are recursive and locally dependent and are the subject of research in this chapter.

There are several definitions for VLSI systolic processor arrays as an abstract mathematical graph model [10], with hard and soft systolic classification of the algorithms according to the architecture used. This model can be traced back to the theory of cellular automata. A tool for the design of systolic algorithms has been proposed by Leiserson and Saxe [11] as a retiming theorem, and afterward the cut-set theorem was developed [12]. The approach used here is based on dependence analysis of the recursive algorithms.

17.3.1   Efficiency and speedup

Speedup and efficiency are measured according to the execution time and to the total processing time.

In all the algorithms, the time for processing a computation for passing the data to the neighboring cells is assumed to be equal to 1 unit time step. Therefore T_1 is equal to the space time product of the computations, i.e., the total number of operations. Otherwise, the standard definitions of speedup and efficiency will be used, i.e.,

$$S_p = \frac{T_1}{T_p}, E_p = \frac{S_p}{p} \tag{17.2}$$

17.3.2   Basic matrix algorithms

Many systolic designs have been proposed recently for different problems. The following is a list of the commonly used problems, with the reference where the design was proposed:

1. matrix vector multiplication [5]
2. matrix matrix multiplication [5]
3. solution of triangular systems [5]
4. LU decomposition [5]
5. QR decomposition [13]
6. preconditioned conjugate gradient [14]

Let $\Re$ denote the set of real and $\mathbf{Z}$ the set of integer numbers. A row vector is $x \in \Re^{1*n}$, and the vector $\vec{x} \in \Re^n$ is a vector column defined by

$$x = (x_1, ..., x_n), \vec{x} = \begin{bmatrix} x_1 \\ \vdots \\ x_n \end{bmatrix} \tag{17.3}$$

where $x_i \in \Re$.

The p-norm of the vector $\vec{x}$ is equal to

$$\left\| \vec{x} \right\| = \sqrt[p]{|x_1|p + ... + |x_n|p} \tag{17.4}$$

An $m*n$ real matrix $A \in \Re^{m*n}$ is defined by

$$A = [a_{ij}]_{m*n} = \begin{bmatrix} a_{11} & \cdots & a_{1n} \\ \vdots & & \vdots \\ a_{m1} & \cdots & a_{mn} \end{bmatrix} \tag{17.5}$$

The "x-0" notation in matrices will be used to represent the non-zero and zero elements in a given matrix respectively, and $\vec{v}$ $(j : m)$ is used to denote the essential part of $\vec{v}$, i.e., the vector $\vec{v} = [0...0v_j...v_m]^T$ where $v_k = 0$ for $k = 0,...,j-1$. The ijth minor of A is denoted by $a*_{ij}$, and the determinant by $det A$. An upper triangular matrix is the matrix whose elements below the diagonal are zero and a lower triangular with zero elements above the diagonal.

In the following definitions, $\alpha \in \Re$ will be used as a real scalar value; $c \in \Re$ as a real variable; $\vec{y}, \vec{z} \in \Re^n$, $\vec{v}, \vec{w} \in \Re^m$ vectors; $B \in \Re^{n*r}$, $C \in \Re^{m*r}$ real matrices; and $\vec{x}$ and A as defined above. The basic vector and matrix operations are summarized in Table 17.1.

The names saxpy and gaxpy are obtained as mnemonics to "scalar alpha $\vec{x}$ plus $\vec{y}$" and "generalized saxpy". Extensive descriptions and definitions can be found in Ref. [15]. The 2-norm of a given vector $\vec{x}$ is therefore

$$\left\| \vec{x} \right\|_2 = \sqrt{\vec{x}^T \vec{x}}$$

where $\vec{x}^T$ denotes the transposition of $\vec{x}$, i.e., a row vector.

Six different forms are obtained by reordering the loop indices while computing the matrix matrix multiplication, denoted as ijk, jik, ikj, jki, kij, and kji algorithm forms, each determined by one of the proposed basic vector operations [15].

17.4   Mathematical Methods in DSP

At this point, we will define and discuss, within the context of their relevance to DSP algorithms, some mathematical methods to solve a system of linear equations (SLE), decompose a given matrix, and minimize a given function or the error produced by over-determined data.

TABLE 17.1   Basic Matrix Vector Operations

Identification	Operation	Computation
saxpy	$\vec{z} = \alpha\vec{x} + \vec{y}$	$z_1 = \alpha x_i + y_i$
inner (dot) product	$c = \vec{x}^T\vec{y}$	$c = \sum_{i=1}^{n} x_i y_i$
outer product	$A = \vec{v}\vec{x}^T$	$a_{ij} = v_i x_j$
matrix vector multiply	$\vec{v} = A\vec{x}$	$z_i = \sum_{n=1}^{n} a_{ij} x_j$
gaxpy	$\vec{w} = A\vec{x} + \vec{v}$	$w_i = v_i + \sum_{j=1}^{n} a_{ij} x_j$
multiply	$C = AB$	$c_{ij} = \sum_{k=1}^{n} a_{ik} b_{kj}$

17.4.1   Solving systems of linear equations

An important problem that arises in DSP is solving a set of simultaneous linear equations. The problem is to find a vector $\vec{x} \in \mathfrak{R}^m$ for a given $m*n$ matrix A, $m \geq n$ and a vector $\vec{b} \in r^n$ such that,

$$A\vec{x} = \vec{b} \tag{17.6}$$

This problem can be solved by computing an inverse matrix A^{-1} if $m = n$ and $\vec{x} = A^{-1}\vec{y}$. This process is computationally extensive and usually the matrix A is transformed to triangular form and then the resulting system is solved by forward and back substitution. The approach using the matrix triangularization is best suited for systems where $m > n$.

A lower triangular system $L\vec{x} = \vec{b}$, determined by an $n*n$ lower triangular matrix L is solved by the forward-substitution algorithm,

$$x_i = \left(b_i - \sum_{j=1}^{i-1} l_{ij} x_j \right) / l_{ii} \tag{17.7}$$

An upper triangular system $U\vec{x} = \vec{b}$, where U is an $n*n$ upper triangular matrix, is solved by the back-substitution algorithm,

$$x_i = \left(b_i - \sum_{j=i+1}^{n} u_{ij} x_j \right) / u_{ii} \tag{17.8}$$

As shown, these algorithms use a matrix vector multiply (gaxpy) and a division operation.

17.4.2   Decomposition techniques

A given matrix can have different decompositions that use triangular matrices. The most usual methods are based on a linear transformation defined by a transformation matrix M such that a given matrix A is transformed into $B = MA$ (or, alternatively, $C = AM$). The zeroing process is defined by an appropriate computation of the transformation matrix M such that B (or C) has zero entries in selected places.

Let M_k be computed such that the elements of A below the diagonal in the kth column are transformed to zero, then the sequence of transformations for $k = 1,\ldots,n-1$ applied to A results in a triangular matrix $B = MA$ by the following matrix premultiplication approach

$$\text{for } k := 1 \text{ to } n-1 \text{ do}$$

$$B := M_k B$$

$$(17.9)$$

where B is initially A, and matrix M is obtained by the product $M = M_{n-1}\ldots M_2 M_1$. Matrix B obtained is upper triangular, and the process of transformation is shown in Fig. 17.11.

Alternatively, the matrix A can be triangularized by a postmultiplication instead of premultiplication used in the previous algorithm. Then, the matrix A is transformed into the matrix $C = AM$. In this case, the transformation matrix M affects only selected subrows instead of subcolumns defined by the algorithm (17.9) above. Let M_k be computed to eliminate the kth subrow consisting of the elements found on the right from the diagonal, then the sequence of transformations for $k = 1,\ldots,n-1$ applied to A results in a lower triangular matrix $C = AM$, by the following matrix postmultiplication

$$\text{for } k := 1 \text{ to } n-1 \text{ do}$$

$$C := CM_k$$

$$(17.10)$$

where C is initially A, and the matrix M is obtained by the product $M = M_1 M_2 \ldots M_{n-1}$. This process is illustrated in Fig. 17.12.

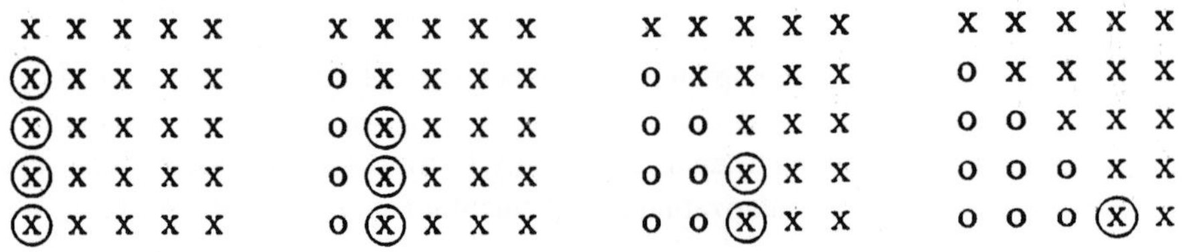

Figure 17.11 Triangularization of a non-zero matrix to an upper triangular form by matrix premultiplication $(n = 5)$

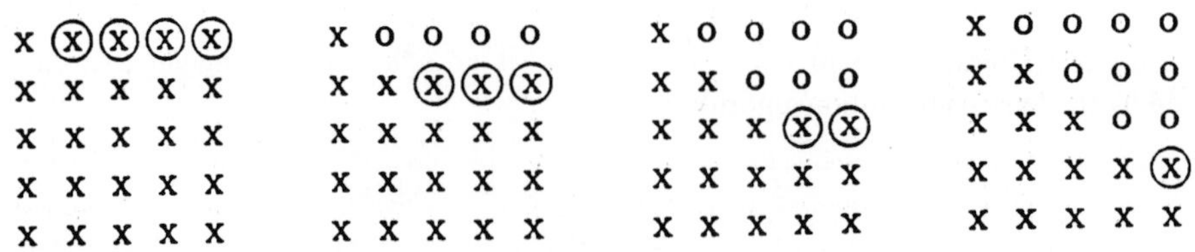

Figure 17.12 Triangularization of a non-zero matrix to a lower triangular form by matrix postmultiplication $(n = 5)$

Some decompositions use the transformation matrix M_k of the algorithm (17.9) formed by a sequence of transformation matrices M_{ki} for $i = k+1,\ldots,n$, each computed such that the element a_{ik} is zeroed. The following row transformation approach [15] is then more appropriate for the transformation of A:

$$\text{for } k := 1 \text{ to } n \text{ do}$$

$$\text{for } i := k+1 \text{ to } n \text{ do}$$

$$\begin{bmatrix} \text{row } (k) \\ \text{row } (i) \end{bmatrix} := M_{ki} \times \begin{bmatrix} \text{row } (k) \\ \text{row } (i) \end{bmatrix} \tag{17.11}$$

The process of transformation of the rows of A is depicted in Fig. 17.13.

A few different decompositions will now be described. The *LU and QR decompositions* result in triangular matrices and are obtained by one of the proposed triangularization algorithms (17.9) and (17.11). Also, the algorithms to compute a bidiagonal form and the *singular value decomposition* (SVD) are related.

17.4.3 LU decomposition

We now present a high-level algebraic description of the *Gaussian elimination transformation*. If

$$M_{ki} = \begin{bmatrix} 1 & 0 \\ -(a_{ik}/a_{kk}) & 1 \end{bmatrix}$$

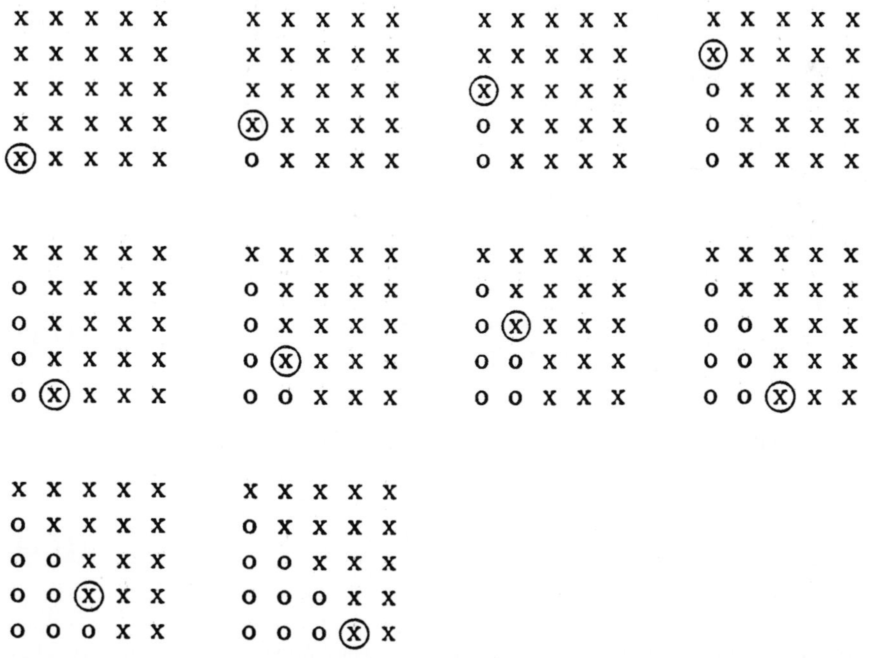

```
X X X X X      X X X X X      X X X X X      X X X X X
X X X X X      X X X X X      X X X X X     (X)X X X X
X X X X X      X X X X X     (X)X X X X      O X X X X
X X X X X     (X)X X X X      O X X X X      O X X X X
(X)X X X X     O X X X X      O X X X X      O X X X X


X X X X X      X X X X X      X X X X X      X X X X X
O X X X X      O X X X X      O X X X X      O X X X X
O X X X X      O X X X X      O(X)X X X      O O X X X
O X X X X      O(X)X X X      O O X X X      O O X X X
O(X)X X X      O O X X X      O O X X X      O O(X)X X


X X X X X      X X X X X
O X X X X      O X X X X
O O X X X      O O X X X
O O(X)X X      O O O X X
O O O X X      O O O(X)X
```

Figure 17.13 Triangularization of a non-zero matrix to an upper triangular form by row transformations $(n = 5)$

then the algorithm (17.11) leads to the Gaussian elimination algorithm and results in the LU decomposition of A. The matrix $L = M_1^{-1} \ldots M_{n-1}^{-1}$ is a unit lower triangular matrix, and $U = B$ is an upper triangular matrix obtained by this algorithm.

The SLE $A \vec{x} = \vec{b}$ reduces to LU $\vec{x} = \vec{b}$ and can be solved by the following two-step procedure:

1. $L \vec{y} = \vec{b}$ by the forward-substitution algorithm

2. $U \vec{x} = \vec{y}$ by the back-substitution algorithm

The algorithm to compute the LU decomposition can be represented in one of the forms *kji, jki, jik, ikj*, determined by a saxpy, gaxpy, outer, or inner product operation [15]. The computation of the LU decomposition of A is numerically unstable and is dependent on the distribution of the elements in the matrix A. For most DSP problems, this might cause difficulties, so the use of orthogonal transformations is recommended.

17.4.4 QR decomposition

The QR decomposition of an $m*n$ real matrix A, where $m \geq n$, is given by

$$A = Q \begin{pmatrix} R \\ 0 \end{pmatrix} \tag{17.12}$$

where Q is an $m*n$ orthogonal transformation matrix, and R is an $n*n$ upper triangular matrix.

If B and M are determined as a triangular decomposition of A, then the SLE $A \vec{x} = \vec{b}$ reduces to $B \vec{x} = MA \vec{x} = M \vec{b} = \vec{y}$, and the problem can be solved by a two-step procedure:

1. $\vec{y} = M \vec{b}$ matrix vector multiply (gaxpy)
2. $B \vec{x} = \vec{y}$ by the back-substitution algorithm

The QR decomposition is usually found by two methods; the first one uses a sequence of *Givens rotations,* and the second uses *Householder reflections.* A $2*2$ orthogonal matrix $Q^{(RT)}$ is rotation if it has the form

$$Q^{(RT)} = \begin{bmatrix} \cos(\theta) & \sin(\theta) \\ -\sin(\theta) & \cos(\theta) \end{bmatrix} \tag{17.13}$$

A $2*2$ orthogonal matrix $Q^{(RF)}$ is reflection if it has the form

$$Q^{(RF)} = \begin{bmatrix} \cos(\theta) & \sin(\theta) \\ \sin(\theta) & -\cos(\theta) \end{bmatrix} \tag{17.14}$$

Reflections and rotations are numerically stable methods to transform a given matrix. They are computationally attractive because they are easily constructed and efficiently used for the zeroing process by properly choosing the value of the angle θ.

QR Decompositions by Givens Rotations. If the transformation matrix is

$$M_{ki} = \begin{bmatrix} \cos{(\theta)} & \sin{(\theta)} \\ -\sin{(\theta)} & \cos{(\theta)} \end{bmatrix}$$

then the algorithm (17.11) leads to the QR decomposition of a given matrix A computed by Givens rotations, where $Q = M$ and $R = B$. The matrix M_{ki} possesses a property to affect only selected subrows of A if a premultiplication is performed, and it is then called a *row rotation*. A column rotation is obtained if a postmultiplication is applied, because only selected subcolumns are affected. Two operations are usually specified, i.e., a *Givens generation* (GG) for computation of the transformation matrix M_{ki} and the angle θ, and a *Givens rotation* (GR) for the pre- or post-multiplication operations.

QR decomposition by Householder reflections. An $n*n$ matrix P of the form

$$P = \frac{I - 2\vec{v}\vec{v}^T}{\vec{v}^T\vec{v}} \tag{17.15}$$

where I is an $n*n$ identity matrix and $\vec{v}$ is a non-zero n dimensional vector, is called *Householder reflection*. Usually, the Householder vector $\vec{v}$ is computed for a given vector $\vec{x}$, such that $P\vec{x}$ is zero in all but the first component. Let M_k be equal to the transformation matrix P such that $\vec{v} = A(k{:}m, k)$. Then algorithm (17.9) leads to the QR decomposition of a given matrix A, computed by Householder reflections, where $Q = M$ and $R = B$. Similar to the previous case, only selected subrows of A are affected if premultiplication is used, and selected subcolumns for postmultiplication, determining respective row and column reflections.

17.4.5   Bidiagonalization

Two bidiagonalization schemes are determined, and the first is obtained by Householder reflections. It is a combination of the two proposed algorithms (17.9) and (17.10) for matrix triangularization applied in a sequence. First the kth subcolumn is eliminated, and then the kth subrow, for $k = 1,\ldots,n-1$. The algorithm consists of

$$\text{for } k := 1 \text{ to } n - 1 \text{ do}$$

$$D := M_k^{(1)} D$$

$$D := D M_k^{(2)} \tag{17.16}$$

where D is initialized to A and $M_k^{(1)}$ is the premultiplication transformation matrix to eliminate the kth subcolumn below the diagonal, and $M_k^{(2)}$ is the postmultiplication transformation matrix to eliminate the subrow formed from the elements of the kth row to the right of the upper bidiagonal element.

The other scheme, called *R-bidiagonalization* [15], is obtained by applying a set of orthogonal rotations to reduce an upper triangular matrix to bidiagonal form. This algorithm applies a sequence of pair row and column rotations (premultiplying and postmultiplying the matrix), i.e., by the following algorithm:

for $k := 1$ to $n - 1$ do

  for $i := 1$ down to $k + 1$ do

$$R := RQ_{ki}^{(1)}$$

$$R := Q_{i\,(i-1)}^{(2)}\,R \tag{17.17}$$

where $Q^{(1)}$ is a postmultiplication rotation and $Q^{(2)}$ a premultiplication rotation. An example of this principle is shown for $n = 5$ in Fig. 17.14.

First, the last two columns are rotated to eliminate the element a_{1n}, then the last two rows are rotated to eliminate the unwanted element $a_{n(n-1)}$, depicted by +. This sequence of column and row rotations continues until the first row is zeroed in all but the first two elements, and then the whole process is repeated for the remaining rows until a bidiagonal form is achieved.

17.4.6   Conjugate gradient method and preconditioning

In circumstances when the methods based on matrix decomposition are not viable because the relevant matrix is too large or too dense, *conjugate gradient* (CG) methods are pre-

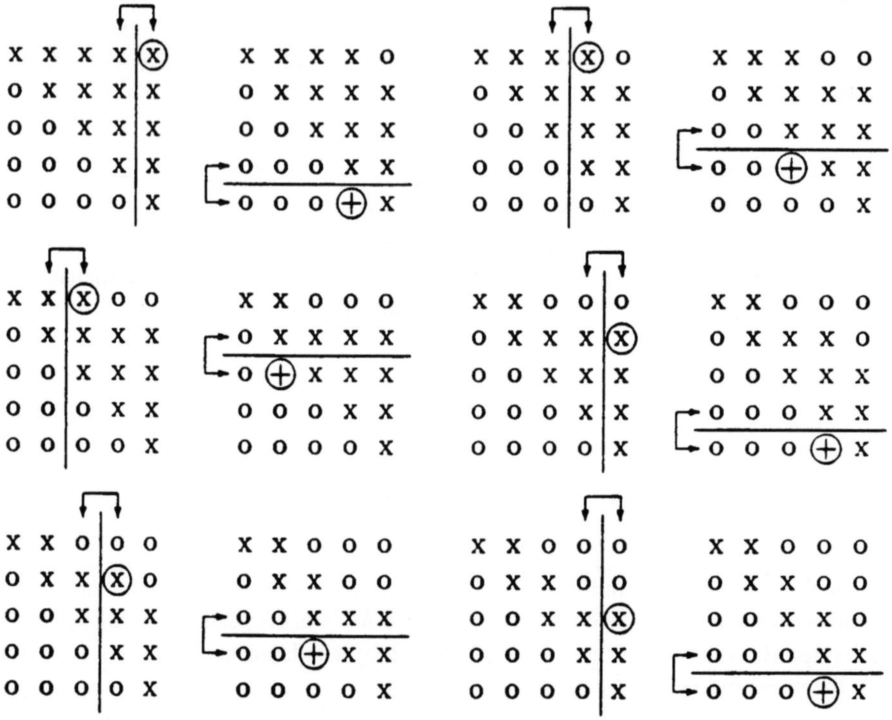

```
 X  X  X  X (X)      X  X  X  X  O      X  X  X (X) O      X  X  X  O  O
 O  X  X  X| X       O  X  X  X  X      O  X  X| X  X      O  X  X  X  X
 O  O  X  X| X       O  O  X  X  X      O  O  X| X  X      O  O  X  X  X
 O  O  O  X| X       O  O  O  X  X      O  O  O| X  X      O  O (+) X  X
 O  O  O  O| X       O  O  O (+) X      O  O  O| O  X      O  O  O  O  X

 X  X (X) O  O       X  X  O  O  O      X  X  O  O| O       X  X  O  O  O
 O  X| X  X  X       O  X  X  X  X      O  X  X  X|(X)      O  X  X  X  O
 O  O| X  X  X       O (+) X  X  X      O  O  X  X| X       O  O  X  X  X
 O  O| O  X  X       O  O  O  X  X      O  O  O  X| X       O  O  O  X  X
 O  O| O  O  X       O  O  O  O  X      O  O  O  O| X       O  O  O (+) X

 X  X  O| O  O       X  X  O  O  O      X  X  O  O| O       X  X  O  O  O
 O  X  X|(X) O       O  X  X  O  O      O  X  X  O| O       O  X  X  O  O
 O  O  X| X  X       O  O  X  X  X      O  O  X  X|(X)      O  O  X  X  O
 O  O  O| X  X       O  O (+) X  X      O  O  O  X| X       O  O  O  X  X
 O  O  O| O  X       O  O  O  O  X      O  O  O  O| X       O  O  O (+) X
```

Figure 17.14 Bidiagonalization of an upper tridiagonal matrix by Givens rotations ($n = 5$)

ferred. The *linear conjugate gradient* method can be viewed as a method for solving a set of positive definite symmetric linear equations. If applied to minimize the quadratic function

$$\vec{p}^T \vec{x} + \frac{1}{2} \vec{x}^T G \vec{x}$$

where G is symmetric and positive definite, it computes the solution of the system $G\vec{x} = \vec{p}$. The residual for the gradient vector $\vec{p} + G\vec{x}_j$ is denoted by $\vec{q}_j$, and the method performs the following iteration

$$\text{for } k := 0, 1, \ldots, \text{ until convergence}$$

$$\vec{s} = \vec{q}_k + \beta_{k-1}\vec{s}$$

$$\alpha_k = \frac{\|\vec{q}_k\|_2^2}{\vec{s}_k^T G \vec{s}_k}$$

$$\vec{x}_{k+1} = \vec{x}_k + \alpha_k \vec{s}_k$$

$$\vec{q}_{k+1} = \vec{q}_k - \alpha_k G \vec{s}_k$$

$$\beta_k = \frac{\|\vec{q}_{k+1}\|_2^2}{\|\vec{q}_k\|_2^2} \tag{17.18}$$

The linear conjugate gradient method will compute the exact solution within a fixed number of iterations, and if exact arithmetic is used, convergence will occur in n iterations where n is less than the number of eigenvalues of A. Therefore, preconditioning is recommended to decrease the number of iterations [16]. The *preconditioned conjugate gradient* (PCG) method for solving a linear system of equations $G\vec{x} = \vec{p}$ is expressed by solving $M^{-1} G\vec{x} = M^{-1}\vec{p}$, where M is a preconditioned matrix chosen for convenient parallel implementation and as close to G as possible. The properties requested for a preconditioner are:

- M is symmetric positive definite.
- M is sparse.
- M is easy to construct.
- $Mz = t$ is easy to solve on a parallel array.
- There is a "good" distribution of the eigenvalues of $M^{-1}G$.

The simplest preconditioner is the diagonal matrix $M = \text{diag}(G)$ which, although parallel, is not very efficient.

17.5 Implementation of Systolic Algorithms in DSP

Two examples will be presented to demonstrate the proposed algorithm transformation and implementations. Both are characteristic DSP problems, and their implementation requires complex solutions. The presented techniques improve the previously heuristically derived processor arrays.

The *QR decomposition algorithm* can be implemented in those problems where a stable method is required for a solution of a system of linear equations, such as the LS problem. The *preconditioned conjugate gradient method* can be implemented for the optimization problems, as a fast convergent and stable method.

17.5.1 QR decomposition algorithm

The QR decomposition algorithm can be expressed by the row transformation approach (17.11) using Givens rotations. It is rewritten in the following forms

$$\text{for } k := 1 \text{ to n}$$
$$\text{for } i := k+1 \text{ to } n$$
$$\text{row} (k) := F (\text{oldrow} (k), \text{oldrow} (i))$$
$$\text{row} (i) := G (\text{oldrow} (k), \text{oldrow} (i)) \tag{17.19}$$

$$\text{for } k := 1 \text{ to n}$$
$$\text{for } i := n \text{ to } k+1$$
$$\text{row} (k) := F (\text{oldrow} (i), \text{oldrow} (i-1))$$
$$\text{row} (i-1) := G (\text{oldrow} (i), \text{oldrow} (i-1)) \tag{17.20}$$

for the NN and AP Givens QR decompositions, respectively, where oldrow means the input value of row used in the proposed computations, and NN and AP denote nearest neighbor and all pair forms.

17.5.2 Algorithm transformation

The next task is to transform the algorithm into a localized form and a system of URE. Both algorithms are interrelated, and the procedure for both computational and data broadcast elimination will be used. Specifically, two different row variable identifications, $\text{row}_a(k, i)$ and $\text{row}_b(k, i)$ are used instead of $\text{row}(i)$ and $\text{row}(i-1)$ in the NN Givens QR algorithm (17.20), and instead of $\text{row}(k)$ and $\text{row}(i)$ in the AP Givens QR form (17.19). The AP Givens QR algorithm becomes

$$\text{for } k := 1 \text{ to n}$$
$$\text{row}_b (0, i) := \text{row} (i)$$
$$\text{for } k := 1 \text{ to n}$$

IP: $\quad \text{row}_a (k, k) := \text{row}_b (k-1, k)$

$$\text{for } i := k+1 \text{ to } n$$

CP: $\quad \text{row}_a (k, i) := F (\text{row}_a (k, i-1), \text{row}_b (k-1, i))$

$$\text{row}_b (k, i) := G (\text{row}_a (k, i-1), \text{row}_b (k-1, i)) \tag{17.21}$$

where $\mathrm{row}_a(i, n)$ is the output value in the algorithm. The NN Givens QR algorithm becomes

$$\text{for } i := 1 \text{ to n}$$
$$\mathrm{row}_a(0, i) := \mathrm{row}(i)$$
$$\text{for } k := 1 \text{ to n}$$

IP: $\qquad \mathrm{row}_b(k, n+1) := \mathrm{row}_a(k-1, n)$

$$\text{for } i := n \text{ down to } k+1$$

CP: $\qquad \mathrm{row}_a(k, i) := F(\mathrm{row}_b(k, i+1), \mathrm{row}_a(k-1, i-1))$

$$\mathrm{row}_b(k, i) := G(\mathrm{row}_b(k, i+1), \mathrm{row}_a(k-1, i-1)) \qquad (17.22)$$

where $\mathrm{row}_b(i, i+1)$ is the output value. Row operations are introduced by the loop index j, and the computational part (CP) is changed into a sequence of GG and GR operations where the row variables indexed by a and b are identified by the corresponding array variables.

GG: $\qquad$ Calculate $(c_{ki}, s_{ki}, a_{ik}, b_{ik})$

$$\text{for } j := k+1 \text{ to } n \text{ do}$$

GR: $\qquad a(k, i, j) := f(c_{ki}, s_{ki}, a(k-1, i-1, j), b(k, i+1, j))$

$$b(k, i, j) := g(c_{ki}, s_{ki}, a(k-1, i-1, j), b(k, i+1, j)) \qquad (17.23)$$

is the CP for the NN Givens QR algorithm and the following for the AP Givens QR algorithm

GG: $\qquad$ Calculate $(c_{ki}, s_{ki}, a_{kk}, b_{ik})$

$$\text{for } j := k+1 \text{ to } n \text{ do}$$

GR: $\qquad a(k, i, j) := f(c_{ki}, s_{ki}, a(k, i-1, j), b(k-1, i, j))$

$$b(k, i, j) := g(c_{ki}, s_{ki}, a(k, i-1, j), b(k-1, i, j)) \qquad (17.24)$$

A data broadcast occurs since c_{ki} and s_{ki} are transferred to all the computations in the inner loop, and also a_{kk} and b_{ik} are broadcast to index points where the GG operation is performed. All the results obtained by the broadcast elimination are summarized in Table 17.2.

The superscripts N and A are used to denote the difference between the NN Givens QR and AP Givens QR forms, respectively. The localized form is a system of URE and consists of three parts: the initialization (IP), the GG, and GR parts, each determined by one of the following data dependence vectors for the NN and AP Givens QR forms, respectively:

$$\vec{d_1}^N = \begin{bmatrix} 1 \\ 1 \\ 0 \end{bmatrix}, \vec{d_2}^N = \begin{bmatrix} 0 \\ -1 \\ 0 \end{bmatrix}, \vec{d_3}^N = \begin{bmatrix} 0 \\ 0 \\ 1 \end{bmatrix}$$

$$\vec{d_1}^A = \begin{bmatrix} 1 \\ 0 \\ 0 \end{bmatrix}, \vec{d_2}^A = \begin{bmatrix} 0 \\ 1 \\ 0 \end{bmatrix}, \vec{d_3}^A = \begin{bmatrix} 0 \\ 0 \\ 1 \end{bmatrix} \tag{17.25}$$

The initialization part is determined by the data dependence vector $\vec{d_1}$, the GG part by $\vec{d_1}$ and $\vec{d_2}$, and the GR part by $\vec{d_1}$, $\vec{d_2}$, and $\vec{d_3}$. Therefore, the system obtained is a system of URE identified by an unity data dependence matrix and pyramidal index set with structure similar to that of the LU decomposition.

17.5.3 Mapping onto VLSI processor arrays

The timing functions $\pi^N = [2 \ {-1} \ 1]$ and $\pi^A = [1 \ 1 \ 1]$ minimize the execution time to $3n - 2$ time steps. The following space allocation functions

$$S_1 = [1 \ 0 \ 0], S_2 = [0 \ 1 \ 0], S_3 = [0 \ 0 \ 1], S_4 = [-1 \ 1 \ 0], S_5 = [-1 \ 0 \ 1]$$

minimize the processor array dimensions and movements. The movement of the variable a is determined by the transformation of the data dependence vector $\vec{d_2}$ and the variable b by $\vec{d_1}$. The movement of c and s is determined by $\vec{d_3}$. Unless c and s remain resident in the array, it is better to transfer only one value θ instead of two data values. Therefore, the GG part becomes simpler, determining only the value of θ, and the GR part more complicated, because $\sin\theta$ and $\cos\theta$ should be calculated in all cells to perform the GR operation.

The following two transformation matrices are constructed:

$$T_r = \begin{bmatrix} 1 & 1 & 1 \\ 0 & 1 & 0 \\ 1 & 0 & 0 \end{bmatrix}, T_f = \begin{bmatrix} 1 & 1 & 1 \\ -1 & 1 & 0 \\ -1 & 0 & 1 \end{bmatrix} \tag{17.26}$$

according to the amount of the data movement. The transformation T_r produces one data stream to remain resident in the array, and T_f results in implementations where all data streams flow. The resident data stream can be chosen by performing column exchanges on T_r or T_f. Therefore, the solutions are classified by the amount of data movement.

TABLE 17.2 Transformations of the Variable Indices after Elimination of the Computational and Data Broadcast in the QR Decomposition Algorithm

Values	NN Givens QR		Both	AP Givens QR	
	b	a	c, s	b	a
new	(k, i, j)	(k, i, j)	(k, i, j)	(k, i, j)	(k, i, j)
used	$(k, i+1, j)$	$(k-1, i, j)$	$(k, i, j-1)$	$(k-1, i, j)$	$(k, i-1, j)$
initial	$(k, n+1, j)$	$(0, i, j)$	(k, i, k)	$(k-1, k, j)$	(k, k, j)
output	$(k, k+1, j)$	(n, i, j)	(k, i, n)	(n, i, j)	(k, n, j)

17.5.4   One resident data stream and two stream flows

A triangular processor array in which only the data streams b and a flow in the array and c and s are resident (as depicted in Fig. 17.15) is obtained by using the following space allocation function:

$$S^A_{ADM} = \begin{bmatrix} 1 & 0 & 0 \\ 0 & 1 & 0 \end{bmatrix}$$

The array is actually the *Ahmed-Delosme-Morf* (ADM) array, named after the authors of Ref. [17]. The data movements are mapped as follows ($T_{ADM} = [\pi^A \; S^A_{ADM}]^T$):

$b:$ $\quad T_{ADM} * [1\ 0\ 0]^T = [1\ 0\ 1]^T$ 1 time step/downward

$a:$ $\quad T_{ADM} * [0\ 1\ 0]^T = [1\ 1\ 0]^T$ 1 time step/to the right

$c,\ s:$ $\quad T_{ADM} * [0\ 0\ 1]^T = [1\ 0\ 0]^T$ 1 time step/resident

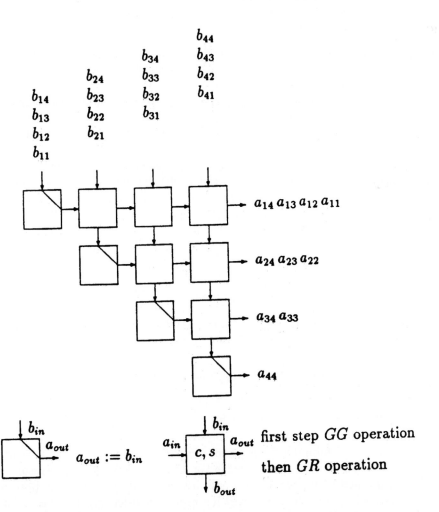

Figure 17.15 The Ahmed-Delosme-Morf triangular array for the QR decomposition algorithm ($n = 4$)

Two types of processors are used, as depicted in Fig. 17.15. The processors on the diagonal simply transfer the value accepted from the top to the right, and the remaining processors perform a GG operation when first activated and GR operations thereafter. The results exit the array from the rightmost column. Actually $n*(n-1)/2$ processors are used, since the n cells with marked upper right corner, found on the diagonal, perform only data transfer and can be considered as delay elements.

By using the space allocation function

$$S^N_{ADM} = \begin{bmatrix} 1 & 0 & 0 \\ 1 & -1 & 0 \end{bmatrix}$$

in the NN Givens QR algorithm, the data movements are mapped as follows ($T_{MADM} = [\pi^N \ S^N_{ADM}]^T$):

    a: $T_{MADM} * [1 \ 1 \ 0]^T = [1 \ 1 \ 0]^T$ 1 time step/downward

    b: $T_{MADM} * [0 \ -1 \ 0]^T = [1 \ 0 \ 1]^T$ 1 time step/to the right

    c, s: $T_{MADM} * [0 \ 0 \ 1]^T = [1 \ 0 \ 0]^T$ 1 time step/resident

The triangular processor array obtained uses data streams *b* and *a* flowing in the array, with *c* and *s* resident, as depicted in Fig. 17.16. The array obtained is actually the *modified Ahmed-Delosme-Morf* (MADM) array [18].

Two types of processors are used, as depicted in Fig. 17.16, and apart from that the processors perform the GG operation when first activated and the GR operation thereafter. The data transfer is from top to right and from the left downward. The results exit the array from the rightmost column.

The first step of solving the LS problem $A\vec{x} = \vec{x}$, is completed if the vector $\vec{x}$ follows the matrix data elements. Then the results A_T and $\vec{x}_T$ dynamically exit from the array. A special dedicated FIFO memory is required to store these elements and feedback for the back-substitution step since all the data is out of the array.

Another triangular processor array results in the Gentleman-Kung (GK) triangular array (named after the originators in Ref. [13]) by using the space allocation function

$$S^A_{GK} = \begin{bmatrix} 0 & 0 & 1 \\ 1 & 0 & 0 \end{bmatrix}$$

on the AP Givens QR algorithm as depicted in Fig. 17.17.

The data movements are mapped as follows ($T_{GK} = [\pi^A \ S^A_{GK}]^T$):

   b: $T_{GK} * [1 \ 0 \ 0]^T = [1 \ 0 \ 1]^T$ 1 time step/downward

   a: $T_{GK} * [0 \ 1 \ 0]^T = [1 \ 0 \ 0]^T$ 1 time step/resident

   c, s: $T_{GK} * [0 \ 0 \ 1]^T = [1 \ 1 \ 0]^T$ 1 time step/to the right

All the processors perform the IP operation on initialization, i.e., store the incoming value of *b* as the resident *a* value, as depicted in Fig. 17.17. After that, the *n* processors on the diagonal perform the GG operation and pass the *c* and *s* values to the right. Because the *c, s* data stream is not resident in the array, it is better to transfer only the value of θ instead of two data values. The remaining processors perform the GR operation after storing

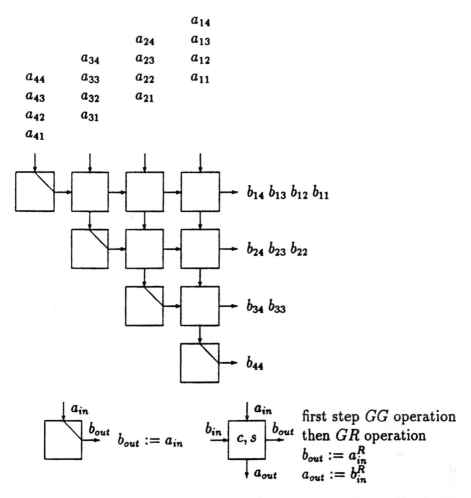

Figure 17.16 The modified Ahmed-Delosme-Morf triangular array for the QR decomposition algorithm $(n = 4)$

the IP data value. Because the results remain resident in the array, data unloading must be performed and this time must be added to the execution time to form the total processing time. A total of $n*(n + 1)/2$ processors are used.

The space allocation function

$$S^N_{MGK} = \begin{bmatrix} 0 & 0 & 1 \\ 1 & 0 & 0 \end{bmatrix}$$

applied to the NN Givens QR form gives a solution that is a modification of the Gentleman-Kung triangular array, denoted by (MGK).

The data movements are mapped as follows ($T_{MGK} = [\pi^A \ S^N_{MGK}]^T$):

a: $T_{MGK} * [1 \ 1 \ 0]^T = [1 \ 0 \ 1]^T$ 1 time step/downward

b: $T_{MGK} * [0 \ {-1} \ 0]^T = [1 \ 0 \ 0]^T$ 1 time step/resident

c, s: $T_{MGK} * [0 \ 0 \ 1]^T = [1 \ 1 \ 0]^T$ 1 time step/to the right

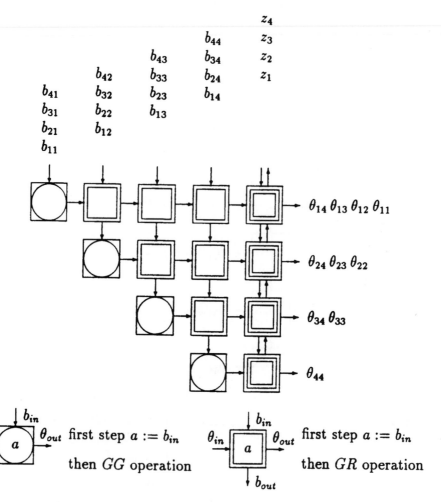

Figure 17.17 The Gentleman-Kung triangular array for the QR decomposition algorithm and backsubstitution step $(n = 4)$

The differences between the arrays obtained are that the data flow is in reversed order, and the processors perform a slightly different operation, as shown in Fig. 17.18. In the MGK array, there is no selection in the processors whether the IP is performed in the first step, or the remaining steps in which the processors perform a GG or GR operation. Since IP is a subset of GG and GR in this implementation, the processors are really specially dedicated—i.e., to perform the GG (or GR) operation, store the rotated value of the incoming variable a as b, and proceed to the rotated value of the last stored variable b as a. Also, c and s are transferred to the right, and it is better to transfer only the value of θ instead of two data values and force sin and cos functions to be computed in the GR cells.

Now, data unloading is much easier than in the original solution, since there is no resident value in the array; i.e., an appropriate set of numbers following after the incoming matrix will unload the array efficiently, without any special signals and a set of binary numbers coming into the array.[*]

---

[*] The incoming numbers are chosen to result in $c = 1$ and $s = 0$.

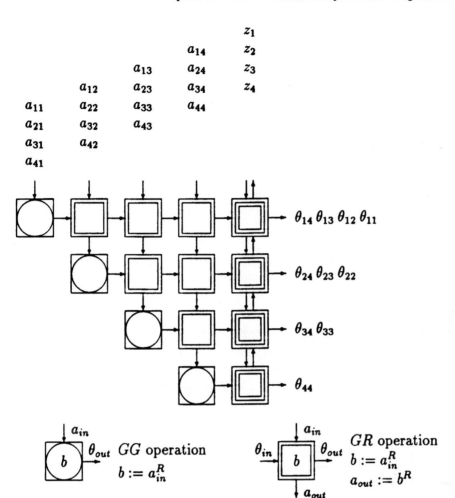

Figure 17.18 The modified Gentleman-Kung triangular array for the QR decomposition algorithm and backsubstitution step ($n = 4$)

If the vector $\vec{z}$ follows the input matrix in an additional column of processors, then it is also rotated by the same orthogonal elements. Data unloading in the MGK array is much easier than in the GK array because there is no resident value in the array; i.e., an appropriate set of numbers following after the incoming matrix will unload the array efficiently without any special signals and a set of binary numbers coming into the array. Now, the LS problem is solved by a simple back-substitution performed in the processors found on the additional column, depicted by three nested squares. While the dynamic unloading is performed in the array, the processors on the additional column act as standard IPS cells after the QR step, and a division operation is performed whenever the last data is received to compute the backsubstitution step.

17.5.5   All data stream flow

By choosing the transformation matrix T_j, all the data streams flow in the array. Six transformations are valid by column exchange, and all result in a square processor array. The solution denoted as a squared (SQ) implementation is similar to the LU decomposition algorithm implementation as depicted in Fig. 17.19. The data movements are mapped as follows:

b: $\quad T_f * [1\ 0\ 0]^T = [1\ -1\ -1]^T$ 1 time step/upward to the left

a: $\quad T_f * [0\ 1\ 0]^T = [1\ 1\ 0]^T$ 1 time step/to the right

c, s: $\quad T_f * [0\ 0\ 1]^T = [1\ 0\ 1]^T$ 1 time step/downward

It is better to transfer θ instead of two values c and s. The processors on the leftmost column are simply delay elements used to transfer the values of b coming diagonally from the bottom right corner to the right. Therefore, the array actually consists of $n*(n-1)$ processor cells. A total of $n-1$ processors on the top row perform the GG operation, while the remaining $(n-1)*(n-1)$ perform the GR operation as depicted in Fig. 17.19.

17.5.6 Folding transformations

The square implementation of the QR algorithm, where all data streams flow is similar to the LU decomposition algorithm implementation introduced by H.T. Kung [5] and dis-

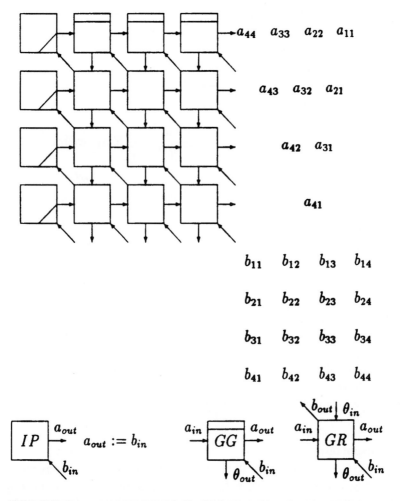

Figure 17.19 A square processor array for the QR decomposition algorithm ($n = 4$)

cussed by Moldovan [19]. The results obtained by the authors in Ref. [20] for folding the LU decomposition algorithm will be used to derive a new triangular array for the QR decomposition algorithm, as shown in Fig. 17.20. It will be denoted as a folded (FD) triangular array. The processors on the top row perform the IP operation and GG operation at alternate time moments. The leftmost cell on the top row performs only the IP operation, but it swaps the outputs at alternate time moments. The remaining processors perform the GR operation. These processors toggle between inputs and outputs at alternate time moments and then perform the proposed operation.

17.5.7 Grouping transformations

The number of processors can be reduced by 50 percent if the processor operations are grouped. In this case, a 2*2 GR operation is performed instead of a 2*1 operation. A 2*1 operation is an operation where two values a_{ij} and a_{kj} are computed. A 2*2 operation computes four variable values instead of two. The algorithm can be realized computing 2*2 operations for neighboring elements. (The elements a_{ij}, a_{ij+1}, a_{kj} and $a_{k,j+1}$ are computed at each processor.)

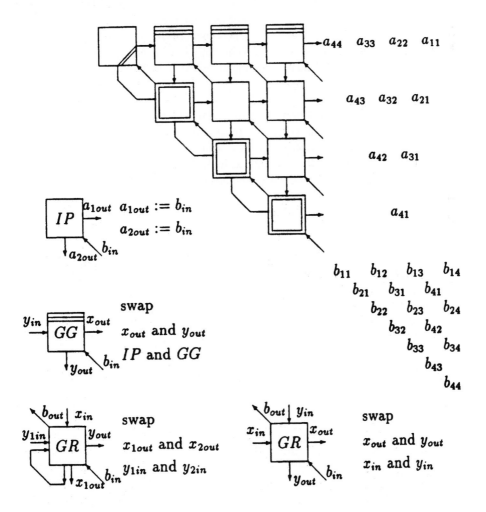

Figure 17.20 A new (folded) triangular processor array for the QR decomposition algorithm ($n = 4$)

The effect of grouping can be considered only as a technique that reduces the number of processors, since the solution does not possess an interlocking property. By applying the grouping technique, the resident memory is doubled and the communication geometry increased from two input/output links to three. Also the operations performed compute four values instead of two, and the processors on the diagonal actually compute three elements, since one is brought to zero.

Two types of grouping can be determined. The first one, called a *zig-zag grouping,* is illustrated in the Fig. 17.21. As shown, the processors denoted by 1 and 2 are grouped to processor *a,* 3 and 4 to *b,* 5 and 6 to *c,* 7 and 8 to *d,* and 9 and 10 to *e.* The links of processor *b* are now analyzed. Since processor 3 passes data to processor 8, and 4 to 9, then *b* communicates to *d* and *e.* Also, since processor 4 passes data to processor 5, then *b* communicates to *c.* Therefore, each of the grouped processors (for example *b*) passes data to (and receive from) three neighbors as shown in the figure.

Another grouping, called a *normal* grouping, is illustrated in Fig. 17.22. The processors notation for grouping is the same as in the previous discussion. Now, processor *b* has two links to processor *d,* since processors 3 and 4 pass data to processors 7 and 8. The final result involves three communication links between neighboring processors, but each processor passes data to two neighbors instead of three as in the previous case.

Both presented groupings result in planar arrays connected by three data links. Note that the constraint involved by using the VLSI technology is reduced communication. However, the implementations obtained by grouping the ADM and GK arrays possess the same amount of data transfer (three links) as the 2D hex solution for the matrix multiplication, or as the SQ or FD array.

If the processors from the GK array are grouped by the zig-zag grouping, then the resulting array is actually the Luk (LK) triangular array [21]. A segment of the array is shown in Fig. 17.22. The total number of processors used is $\lceil n/2 \rceil * \lceil (n+1)/2 \rceil$ instead of $(n(n+1))/2$. The data flow is the same as in the GK array.

A similar array can be obtained by using the normal grouping, as shown in Fig. 17.22. The processors used are the same as in the LK array, but the communication is much simpler. Although the same number of links are used, in this array each processor communicates to two neighbors instead of three as in the LK array.

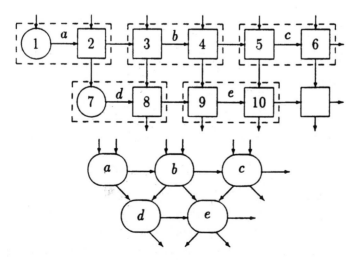

Figure 17.21 A zig-zag grouping for a triangular array

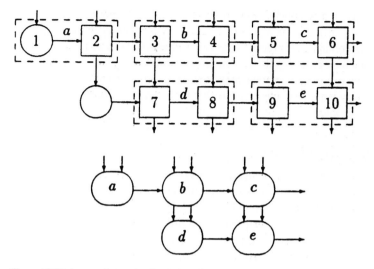

Figure 17.22 A normal grouping for a triangular array

Two more arrays can be obtained by applying the proposed grouping techniques on the ADM array. The figures illustrating these cases are the same as the previous ones, but the processors and the data flow as in the ADM array.

17.5.8 Comparisons of the implementations obtained

All the implementations have the same execution time. The ADM and SQ arrays require n delay elements instead of processors for the IP operation. In the GK array, the results are resident in the array, so an additional $n - 1$ time steps are required to form the total processing time. The number of processors used in triangular arrays is the same, since the ADM array uses n delay elements instead of processors. All these parameters are summarized in Table 17.3.

TABLE 17.3 Efficiency in Systolic Arrays for the QR Decomposition Algorithm

Systolic Array	Execution Time	Total Time	Processors Used	Delay Elements	Givens Operations	Comm. Links
ADM	$3n - 2$	$3n - 2$	$n*(n - 1)/2$	n	$2*1$	2
GK	$3n - 2$	$4n - 3$	$n*(n + 1)/2$	—	$2*1$	2
SQ	$3n - 2$	$3n - 2$	$n*(n - 1)$	n	$2*1$	2
FD	$3n - 2$	$3n - 2$	$n*(n + 1)/2$	—	$2*1$	3
LK	$3n - 2$	$4n - 3$	$n/2*(n/2 + 1)/2$	—	$2*2$	3

Only the SQ array uses specially dedicated processors; i.e., each processor performs only one operation. The ADM, GK, and FD triangular solutions require processors that change their operations during the time. In the ADM array, each processor performs the GG and, after that, the GR operations. In the GK array, each processor performs the IP and, after that, the GR (or GG) operations. In the folded solution, the processors on the top row switch between an IP and a GG operation at each time moment. The remaining processors perform the GR operation but swap between the inputs and outputs. The ADM array requires the least data transfer. The SQ and FD arrays use increased data flow, but both the Q and R matrices are obtained. Increased data flow (three links) is also obtained in the implementations obtained by grouping the GK and ADM arrays.

A discussion and comparison of various processor array implementations is given in Ref. [22]. The MGK array is preferred as an array that is most suitable for the parallel implementation of the LS problem.

17.6 Conjugate Gradient Method

It is obvious that both the preconditioned and direct CG algorithm are rather complex and consist of different tasks such as matrix vector multiplication, and vector constant multiplication and addition. A partial systolic solution exists for each of these tasks, but there is still the open question about how to synchronize the different stages of the algorithm, sketched in Fig. 17.23.

It is obvious that different computer architectures can be used at all levels of the computing system. However, there are only a few methods realized on supercomputer systems to solve the nonhomogeneous problem such as the PCG algorithm.

The major difficulty is the transformation of the original sequential algorithm to a form suitable for VLSI implementation. All the subalgorithms must be synchronized in a condensed pipelined form without bottlenecks or queues.

17.6.1 Algorithm transformation

All basic vector operations expressed by an iterative formula are transformed in the following recursive form.

```
      for k := 1 until convergence
          d_k := 0
          e_k := 0
          for j := 1 to n do
              t_kj := 0
              for i := 1 to n do
(a)               t_kj := t_kj + R_kj * s_(k-1)i
              d_k := d_k + s_(s-1)j * t_kj
              e_k := e_k + t_kj * t_kj
              a_k := c_k-1/d_k
(b)           b_k := a_k * e_k/d_k - 1
              c_k := c_k-1/b_k
          for j := 1 to n do
              w_kj := w_(k-1)j + a_k * s_(k-1)j
(c)           q_kj := q_(k-1)j - a_k * t_kj
              s_kj := q_kj + b_k * s_(k-1)j
```

Three index parameters are needed in loop (a) because of the matrix multiplication, and only two in loops (b) and (c).

17.6.2 Broadcast elimination

The computational broadcast of d_k, e_k, and t_{kj} in loop (a), and w_k, q_k, and s_k in loop (c), is eliminated by using the proposed algorithm transformation techniques. The results are summarized in Table 17.4.

matrix-vector
computation

inner product

vec.-const.
multiplication
and addition

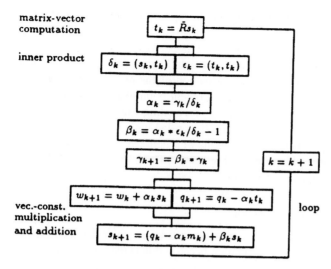

Figure 17.23 Different tasks and stages of the PCG method

The variable s is used in the matrix vector multiplication and also for the matrix multiplication, where it defines a data broadcast, so a new variable z is assigned to propagate this value. A new recurrence equation $z(k, j, i):= z(k, j - 1, i)$ is added to the system of equations, since the propagation vector obtained is [0 1 0]. Also, a_k and b_k define a data broadcast in the loop (c), and the used and initial values obtained by the elimination of data broadcast are summarized in Table 17.5.

Synchronization of data dependencies. The open problem is synchronizing the data dependence between the stages of the algorithm. For example, the values of z used in the computational part of the loop (a) are pipelined for matrix vector multiplication, but at the

TABLE 17.4 Transformations of the Variable Indexes after Elimination of the Computational Broadcast in the CG Algorithm

Values	Loop (a)		Loop (c)
	d_k, e_k	t_{kj}	w_k, q_k, s_k
new	(k, j)	(k, j, i)	(k, j)
used	$(k, j - 1)$	$(k, j, i - 1)$	$(k - 1, j)$
initial	$(k, 0)$	$(k, j, 0)$	$(0, j)$
output	(k, n)	(k, j, n)	(n, j)

TABLE 17.5 Transformations of the Variable Indexes after Elimination of the Data Broadcast in the CG Algorithm

Values	Loop (a)		Loop (c)
	$z_k (= s_k)$	$y_k (= s_k)$	a_k, b_k
new	(k, j, i)	(k, j, i)	(k, j)
used	$(k, j - 1, i)$	$(k, j, i - 1)$	$(k, j - 1)$
initial	$(k, 0, i)$	$(k, j, 0)$	$(k, 0)$
output	(k, n, i)	(k, j, n)	(k, n)

same time, the value of s (that was the initial value of z) is for the computation of d. The data dependencies $\vec{d_2}$ between variables $d(k, j, 0)$ and $z(k-1, 0, j)$ and $\vec{d_3}$ between $d(k, j, 0)$ and $t(k, j, n)$ must also be localized.

The data dependence $\vec{d_3} = [0\ 0\ -n]^T$ represents the spiral approach of the communication and is eliminated by choosing the opposite direction of the propagation vector and computing $d(k, j, n)$ instead of $d(k, j, 0)$. Another problem is the data dependence $\vec{d_2} = [1\ j\ -j]^T$, which can be resolved only by introducing propagating values. The variable y is chosen to pipeline the values of s in the inner loop (identified with i), and the results are summarized in Table 17.5. Thus, two representations of s denoted by y and z are introduced to pipeline its values for the loops indexed by i and j. The initial conditions for this new recurrence equation are also added to the system.

17.6.3 Mapping onto a VLSI processor array

The system of URE representing the loop (a) is expressed by an unity data dependence matrix. A 2D mesh array implementation is chosen because this is the most intense computational task of the algorithm, and a maximum speedup is required. By choosing the transformation

$$T = \begin{bmatrix} 1 & 1 & 1 \\ 0 & 1 & 0 \\ 0 & 0 & 1 \end{bmatrix} \tag{17.27}$$

the variables R are chosen to be resident in the array. This is used to avoid communication with memory as much as possible, since the matrix R is processed for all iterations. It is defined with the data dependence $\vec{d} = [1\ 0\ 0]^T$ and mapped on the same processor for every time moment, since the mapped data flow is $[0\ 0]^T$. Variables z flow in the direction $[1\ 0]^T$, that is, to the right, and variables t flow in the downward direction $[0\ 1]^T$ because they are mapped to these vector values.

The next step is the realization of the two dot products. As the data dependence is localized and the variables are pipelined, this can be done by using the transformation matrix

$$T_u = \begin{bmatrix} 1 & 1 \\ 0 & 1 \end{bmatrix}$$

The solution obtained is an ULA implementation and can be positioned below the 2D mesh array to obtain the required synchronization.

Variables e and d flow in the right direction, since $\vec{d} = [0\ 1]^T$ is mapped in the link associated by 1 (positive direction), and variables y and t are just input data in each processor. These are propagated downward by the previous stage and are received for correct computations.

The second stage of the algorithm is neither indexed nor parallelized, since only three operations are performed. In the algorithm, it is used only for initiating the values required for the third stage.

An ULA is also proposed for the system of URE expressing the loop (c), since the data dependence matrix is the same as in the loop (a). Now the ULA is positioned vertically to obtain the regular data flow of the variables s as the initial value for variables z and y in the loop (a).

It seems that the open problem for systolic organization is in synchronizing the algorithm stages or by using separate loops to obtain a regular data flow that preserves the validity of algorithm. All the procedures must not involve further new delays for variables, and the resulting implementation must preserve the data at the proper time moment in the correct processor. This is why the loop (c) is pipelined with the loop (a) implementation and the only data transfer, which is not local, is the transfer of variables a and b from the right bottom corner of the array to the left top corner of the array. This can be treated as the start of a new iteration (new value for index k) and actually represents a bottleneck that cannot be resolved.

The last problem is synchronizing the first stage with the third stage of the algorithm, and vice versa. As the variables from the loop (c) are pipelined to the loop (a), there is a problem of synchronizing the results for variables $t(k, j, n)$ coming down from the 2D mesh array and their use in the vertical array. For that purpose, the same pipeline route of s (to the right and downward) is used to pipeline t upward and to the left to the correct processor. A new variable identification v is used for the data propagation from bottom to top, and u from right to left. The diagonal elements are used as switches for the variables coming from the left to right to change their direction downward. They also switch the movement of variables coming from below upward to the left.

17.6.4   Systolic array implementation

The details of the processor array implementation are shown in Fig. 17.24. The description of processor operations is given in Fig. 17.25.

The processor elements depicted as simple squares perform an IPS operation and just pass data. The processors that are on the left border (simple squares) with signed top left corner pass the variable s left and to the right as variable z. The diagonal processors pass this value downward as the variable y.

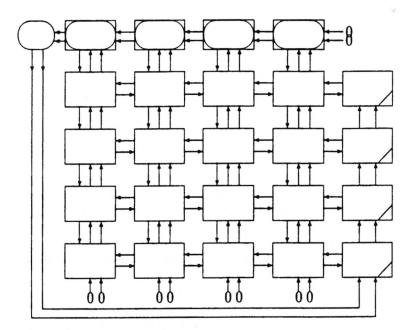

Figure 17.24 PCG systolic array ($n = 4$)

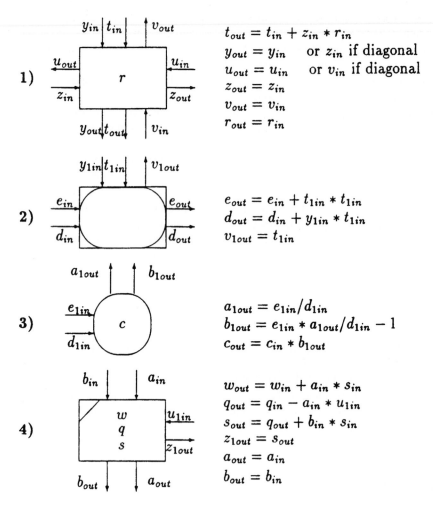

Figure 17.25 Processor operations for the PCG systolic array

The processors situated on the lower border of the array (rounded squares) perform two IPS operations, which means that they can be combined as two IPS cells (simple squares). They are used to calculate the scalar products for variables d and e.

The operations from the second stage are three multiplicative, additive, and division operations that can be done by one more complex processor (depicted as a rounded square) or three IPS cells (simple squares). Only the values for variables d and e come from the left; the new value for variable c is calculated and remains resident until the next iteration starts. The calculated values for a and b are transferred to the top left element.

The third stage is mapped on a linear array lying vertically to the left and shown as squares with a dashed left top triangle. These processors perform three IPS operations and can be realized as three IPS cells (simple squares). They use the variables a and b coming upward and the variable t coming from the left. This is why the values for the variable t are pipelined along the same route that s passes. The diagonal processors switch their movement from down to left. The calculated values for w, q, and s remain resident in the processor element, and the only transfer outward is the value of s.

Every iteration starts are the top left processor, and the computations are completed as a wavefront coming up to the bottom right processor as shown in Fig. 17.26. The total time is $2*n + 2$ time moments for one iteration. If complex processors are decomposed into simpler ones and the data is pipelined, then five time moments are added for every iteration. Since n iterations and $n - 1$ steps are needed for the calculation of w after the last iteration, the total time is $(2n + 2) * n + (n - 1) = (2n + 1)(n + 1)$ steps.

Parallel architectures like transputers using Occam as a hardware description language can be efficiently used for the simulation and logical testing of systolic arrays [23].

The PCG method implementation is precisely defined by all the necessary attributes, such as time activities, regular data flow, and synchronization. Occam is used in Ref. [14] for the simulation of the parallel execution in all processors. All operations and data transfer in the test program are done in parallel by Occam, using either the pseudo-parallel property of Occam or the true parallelism, in a transputer network. This really improves the simulation. The algorithm of the simulation consists of a two-step procedure iteratively repeated $2n + 2$ times for each step. The first step is the computational part and describes the processor activity in every time moment. In the second part, all the transfer activities are completed.

17.7   Summary

Dramatic growth in research and development of mapping various signal/image processing applications onto VLSI architectures has occurred during the last decade. In particular, VLSI technology has inspired many innovative designs in processor arrays, but economic

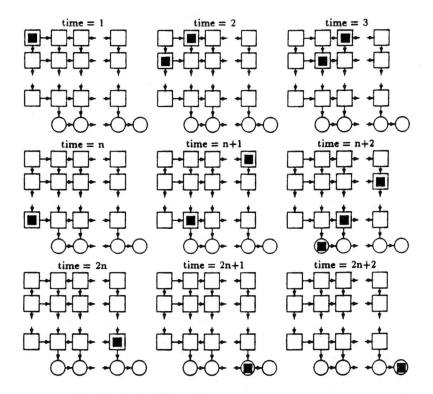

Figure 17.26 Snapshots for the PCG systolic array

factors such as low cost, modularity, and expansion capabilities have also entered the arena when assessing the final design. Consequently, many standard algorithms have become inconvenient for the new technology, and new concepts and designs must be developed.

In this chapter some of these issues are discussed for the important basic matrix algorithms occurring in digital signal processing (DSP): matrix vector multiplication, QR decomposition, back substitution, and the preconditioned conjugate gradient method. These result in many new innovative special-purpose VLSI processor array designs.

17.8   References

1. Mead, C., and L. Conway. 1980. *Introduction to VLSI Systems.* Reading, Mass.: Addison-Wesley.
2. Kung, S. Y. 1988. *VLSI Array Processors.* Englewood Cliffs, N.J.: Prentice Hall.
3. K. Hwang and F. Briggs. 1984. *Computer Architectures and Parallel Processing.* New York: McGraw Hill.
4. Kung, H. T. 1983. Notes on VLSI computation. In *Parallel Processing Systems,* D. J. Evans, ed. New York: Cambridge University Press, 339–356.
5. Kung, H. T., and C.E. Leiserson. 1978. Systolic arrays (for VLSI), *Tech. Rep. CS-79-103.* Pittsburgh, Pa.: Carnegie Mellon University.
6. Kung, H. T. 1982. Why Systolic Architectures? *IEEE Computer,* 15, 37–46.
7. Kung, S. Y. 1985. VLSI signal processing; from transversal filtering to concurrent array processing. In *VLSI and Modern Signal Processing,* S.Y. Kung, H.J. Whitehouse, and T. Kailath, eds. Englewood Cliffs, N.J.: Prentice Hall, 127-152.
8. Megson, G. M., and D.J. Evans. 1985. Soft-systolic pipelined matrix algorithms. *Proc. Int. Conf. on Parallel Computing 85.* Amsterdam: Elsevier Science Publishers B.V., North Holland, 171–180.
9. Kung, H. T. 1980. The structure of parallel algorithms. In *Advances in Computers,* M.C. Yovits, ed., Vol. 19, New York: Academic Press, 65–112.
10. Evans, D. J., ed. 1991. *Systolic Algorithms.* New York: Gordon and Breach.
11. Leiserson, C.E., and J.B. Saxe. 1983. Optimizing synchronous systems. *J. VLSI and Computer Systems,* 1, 41–67.
12. McEvoy, K., and P. M. Dew. 1988. The cut theorem—a tool for design of systolic algorithms. *International Journal of Computer Mathematics* 25, 203–233.
13. Gentleman, W.M., and H.T. Kung. 1981. Matrix triangularization by systolic arrays. *Proc. SPIE Real-Time Signal Processing IV,* Vol. 298, 19–26.
14. Gusev, M., and D.J. Evans. 1991. Systolic Array for Conjugate Gradient Method, *Tech. Rep. 639.* Loughborough University of Technology, PARC, Dept. of Computer Studies.
15. Golub, G. H., and C. F. V. Loan. 1983. *Matrix Computations.* Baltimore: John Hopkins University Press.
16. Evans, D. J., ed.1983. *Preconditioning Methods: Theory and Applications.* New York: Gordon and Breach.
17. Ahmed, H. M., J.-M. Delosme, and M. Morf. 1982. Highly concurrent structures for matrix arithmetic and signal processing. IEEE Computer, 15, 65–82.
18. Gusev, M., and D. J. Evans. 1992. Localization of 2*1 Output Algorithms. *Tech. Rep. 681.* Loughborough University of Technology, PARC, Dept. of Computer Studies.
19. Moldovan, D. I. 1982. On the analysis and synthesis of VLSI algorithms. *IEEE Trans.Comput.,* 31, 1121–1126.
20. Gusev, M. and D. Evans. 1991. Implementation of Folding Transformations on Planar Systolic or VLSI Processor Arrays, *Tech. Rep. 656.* Loughborough University of Technology, PARC, Dept. of Computer Studies.
21. Luk, F. T. 1986. A triangular processor array for computing singular values. *Linear Algebra and its Applications,* 77, 259–273.
22. Gusev, M., and D. J. Evans. 1992. Elimination of the Computational Broadcast and Application to the QR Decomposition Algorithm, *Tech. Rep. 676.* Loughborough University of Technology, PARC, Dept. of Computer Studies.
23. Gusev, M., and D.J. Evans. 1991. Simulation of the Matrix Multiplication Systolic Algorithm using Occam and Transputer Array, *Tech. Rep. 630.* Loughborough University of Technology, PARC, Dept. of Computer Studies.

18

Direct Interconnection Networks[*]

Ivan Stojmenović

18.1 Topological Properties of Interconnection Networks

An interconnection network consist of a set of processors, each with a local memory, and a set of bidirectional links that serve for the exchange of data between processors. A convenient representation of an interconnection network is by an undirected (in some cases directed) graph $G = (V, E)$ where each processor A_i is a vertex in V, and two vertices A_i and A_j are connected by an edge $(A_i, A_j) \in E$ if and only if there is a direct (bidirectional for undirected, and unidirectional for directed graphs) communication link between processors A_i and A_j. Such processors are called *neighbors*. The interconnection graph of a network is often referred to as its *topology*. We will use the terms *interconnection network* and *graph* interchangeably; similarly, we will use *node, vertex* and *processors* with the same meaning and, finally, terms *edge* and *link* are used herein as synonyms.

Usually, all processors in a network are identical, and each is assumed to have input and output ability. Sometimes a master processor is recognized as a leader and more powerful then others, serving as a communication port between user and parallel computer. Processors may execute the same or different programs, a network may be regular or irregular, and the number of neighbors of each processor may be a constant or a function of the size of the network. The time complexity of any algorithm has two components: *computation time,* which covers local computation by every processor, and *communication time,* which is the time needed for the exchange of data between processors. The present state of the art in computer technology is that a message exchange takes considerably more time that a computing step inside a processor. Thus, there is a higher demand for reducing communication than computation times. The situation may change with the development of faster communication technology.

---

[*] This research was supported by the Natural Sciences and Engineering Research Council of Canada.

Each processor is assumed to know its own coordinates within the network of n processors and to have a certain number of registers of size $O(\log n)$; in unit time, every processor performs some arithmetic or Boolean operation or communicates with one of its neighbors using a local link. Processors in a network may operate synchronously or asynchronously. Synchronous computation is also referred to as the single instruction multiple data (SIMD) mode. This means that, at each time unit, the same instruction is broadcast to all processors, which execute it and wait for the next instruction. We shall refer here to such computing mode as the *parallel computation. Asynchronous computation* corresponds to multiple instruction multiple data (MIMD) mode, where each processor has its own program and runs it on its own data; processors occasionally communicate each other to exchange data. This mode will be referred here as *distributed computation.*

In this chapter, we will outline some of the most widely used networks. For each of them, we discuss some of its basic topological properties. A topology is evaluated in terms of a number of parameters. The main ones are *degree* or *fan-out, diameter*, and *symmetry* (defined below). Further important measures of "goodness" of a network are bisection width (the minimum number of links that, when cut, separate the network into two parts with equal number of processors); wire length (or layout aspects); existence of optimal data communication techniques such as routing, broadcasting, gossiping, sorting, permutation, prefix scan, etc.; node disjoint paths (or fault tolerance); embeddability; and recursive decomposition (or scalability). In this survey, we are interested in symmetry, diameter, degree, recursive decomposition, and routing/broadcasting. Other chapters in this book deal with the remaining aspects of interconnection networks. Although all characteristics are important, it seems that the network cost, defined as the product of the degree and diameter (measured with respect to the number of nodes), is the most important parameter of a network.

The distance between two processors is the smallest number of communication links that a message has to traverse to be routed between the two processors. It corresponds to the number of edges in the shortest path between two processors (where each edge has unit weight). The diameter of a network is the farthest distance between any two processors. The fan-out (degree) of a network is the maximal number of neighbors of a processor.

A graph is vertex (edge) symmetric if the graph looks the same from each of its vertices (edges, respectively). The existence of vertex or edge symmetries in networks simplifies the design of algorithms. Asymmetry in networks causes premature saturation of certain links and large message queues at certain nodes under heavy message traffic, resulting in an increase in message delays.

In the coming subsections we will describe most popular interconnection network models. Here we will study only *direct networks,* which are networks that allow only direct communication between neighboring processors. Some networks, like meshes with multiple broadcasting, are enhanced by adding buses that allow quick broadcasting of a datum from one processor to a group of other processor. In *indirect networks,* nodes in underlying graph are distinguished as processors, or switches, or network controllers, etc. Switches are simple elements that can perform a comparison and forward the message in one or other direction. They are considerably less powerful than processors. Indirect networks, in general, have high bisection width, low degree, low diameter, and long wires, and are in practice considerably less popular than direct networks. Lately, one more kind of network has been introduced, called a *reconfigurable network*. These allow processors to dynamically change the links to their neighbors during the execution times, forming various buses for fast broadcasting of data. Indirect and reconfigurable networks will not be studied in this chapter.

Two extreme models of interconnection networks are *fully connected graphs* (with every node connected to every other node) and *independent* processors. (In the latter, no edge in the graph exists; in this model there exists a central processor that distributes the job to others and collects results from them. Therefore, in reality, it is a graph in which one node is connected to all other nodes). The implementation of fully connected graph and, in general, a graph with high degree has physical limits under current technology.

18.1.1   Linear array of processors

The simplest way to arrange n processors in a network is a one-dimensional array. If processors are numbered as $1, 2, \ldots, n$, the processor i is linked to its two neighbors $i-1$ and $i+1$ through a two-way communication link. End processors 1 and n each have one neighbor only. Figure 18.1 shows a linear array of n processors.

The diameter of a linear array of n processors is $n-1$. The total number of edges is $n-1$. The fan-out of a linear array is obviously 2. Linear array of processors is (arguably) the weakest model of parallel computation. Simpler models are necessarily disconnected, which obviously disallows communication between some processors.

If processors 1 and n are also connected, one obtains what is known as a *ring* network. It is a popular network in distributed computation.

18.1.2   Mesh-connected computer

Processors in a mesh-connected computer (MCC) occupy vertices of an $m \times m$ grid, where $n = m^2$. For easy reference, processor in row i and column j is denoted $p(i, j)$, $1 \le i, j \le m$. Processor $p(i, j)$ is connected via bidirectional links to its four neighbors $p(i+1, j)$, $p(i-1, j)$, $p(i, j+1)$ and $p(i, j-1)$, whichever of them exists; in other words, each node is connected with its four neighbors immediately above, below, to the left, and to right of it. Figure 18.2 shows an MCC with 16 processors.

An MCC has $n = m^2$ nodes, with fan-out 4 and diameter $2(m-1) = 2(\sqrt{n}-1) = O(\sqrt{n})$. The number of edges is $2(n - \sqrt{n})$.

In many applications, it is important to define an ordering among processors, i.e., an indexing from 1 to n (for example, to call some data *sorted*). There are two common indexing schemes. In row-major order, processor i is in row j and column k such that $i = (j-1)\sqrt{n} + k$. In snakelike row-major order, processor i is placed in row j and column k of the MCC such that $i = (j-1)\sqrt{n} + k$ when j is odd, and $i = (j-1)\sqrt{n} + \sqrt{n} - k + 1$ when j is even.

The definition of a two-dimensional MCC can be easily generalized to higher-dimensional arrays (or mesh-connected computers). A k-dimensional r-sided array has r^k nodes that correspond to vectors $(c_1, c_2, \ldots, c_k)$ where $0 \le c_j \le r-1$ for $1 \le j \le k$. Two nodes are linked by an edge if (a) they differ in precisely one coordinate and (b) the absolute value of the difference in that coordinate is 1. Also, an MCC may have rectangular or shape other than square. MCCs are asymmetric. The torus network [82] is obtained from a mesh-connected computer if the first and last processors in each row and column are connected. The torus network is vertex and edge symmetric.

Figure 18.1 Linear array of processors

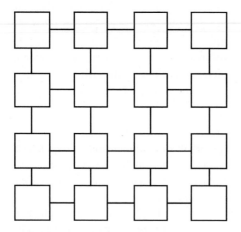

Figure 18.2 Mesh-connected computer

The MCC architecture (and its variants presented below) can be nicely embedded in the plane, making it useful for VLSI implementation. It is available in the commercial marketplace at present.

18.1.3 Trees

The processors in a tree network form a complete binary tree with d levels. The levels are numbered from 0 (the root) to $d-1$, and there are total of $n = 2^d - 1$ processors. A processor at level i is connected by a two-way line to its parent at level $i-1$ and to its two children at level $i-1$. The root processor has no parent, and the leaves, at level $d-1$, have no children. If the root is marked as processor 1 and other nodes are marked in a breadth-first order, the children of node i are $2i$ and $2i+1$, while the parent of node i is $i/2$ (integer division). Figure 18.3 shows a tree with 15 nodes. The diameter of a tree with n processors is $2\lfloor \log_2 n \rfloor = O(\log n)$, and the fan-out is 3.

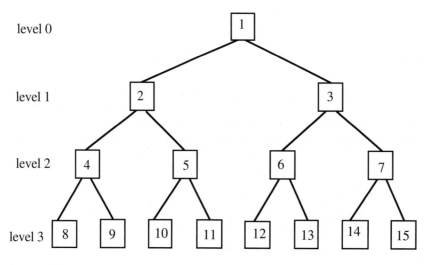

Figure 18.3 Tree structure

18.1.4   Mesh of trees

In a mesh of trees (MOT) network, m processors are placed in a square array with $R(m)$ rows and $R(m)$ columns. Each row and each column is organized in a tree network. There are several different variants of MOT networks. Four of them are based on whether these m processors form a mesh-connected computer network (i.e., each of them linked to its four neighbors), and whether additional processors are used to create tree networks. Figure 18.4 shows these variants for $m = 16$. Dashed lines mark links between neighboring processors (if they exist). Black circles are nodes in the tree network. These can be either separate processors, or it may be identical to one of its children. In the latter case, depending on how the identification is made, additional variants of MOTs can be designed. The total number of processors is m if no additional processors are used, and $n = 2m - 2\sqrt{m}$ otherwise. Other variants exist. The mesh of trees concept is introduced in Ref. [17].

18.1.5   Pyramid

If the nodes at each level of a tree are connected to form a linear array, a one-dimensional pyramid computer is obtained [24]. A two-dimensional pyramid consists of $(4^d-1)/3$ processors located on $d + 1$ levels. All processors at the same level are connected to form a mesh. There are 4^i processors at level i, $0 \le i \le d$, arranged in a $2^i \times 2^i$ MCC. A processor at level i, in addition to being connected to its four neighbors at the same level, also has connections to four children at level $i + 1$ and one parent at level $i - 1$ (if they exist). The only processor at level 0 is called the *apex*. An MCC at level d is called the *base*. Figure 18.5 shows a pyramid with 21 processors.

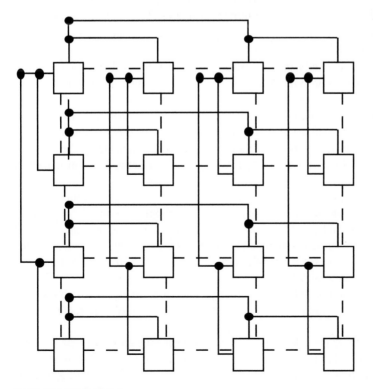

Figure 18.4 Mesh of trees

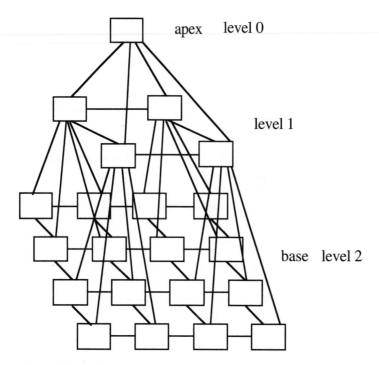

Figure 18.5 Pyramid

18.1.6 Star, pancake, arrangement, and rotator graphs

In this section we introduce some networks that are based on permutations and permutations of combinations.

Each processor of both star and pancake networks corresponds to a distinct permutation of k symbols; for simplicity, the symbols are $\{1, 2,\ldots,k\}$. Therefore, the number of processors is $n = k!$. In the star network, denoted by S_k, a processor u is connected to a processor v if and only if the corresponding permutation of u can be obtained from that of v by exchanging the first symbol with the ith symbol, for some i, $2 \le i \le k$. For example, processors 3421 and 2431 are neighbors because they are obtained by exchanging the first and third symbols from each other. Figure 18.6 shows S_4. The star network is introduced in Ref. [1].

In the pancake network, denoted by P_k, a processor v is connected to a processor u if and only if the label of u can be obtained from that of v by flipping the first i symbols, for some i, $2 \le i \le k$. For example, 3142 and 2413 are neighbors because they are obtained from each other by flipping all four symbols. Figure 18.7 shows P_4. The diameter of a pancake network with $n = k!$ processors is between $(5k + 5/3)$ and $(17k/16)$ (for k multiples of 16) [9], and the precise value is not known.

Star and pancakes are vertex and edge symmetric. The property allows for all the processors to be identical and minimizes the congestion when messages are routed in the graph.

Each vertex of the star or pancake has exactly $k - 1$ adjacent edges, and thus the total number of edges is $(k!(k - 1)/2)$. Since $k = O(\log k!)$, the fan-out of the star and pancake, being $k - 1$, is sublogarithmic in the number of processors. This is one of most appealing properties of these networks.

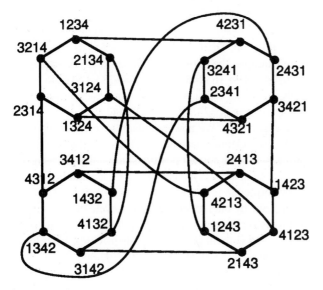

Figure 18.6 Star

Consider now the diameter of the star graph. Because stars are vertex and edge symmetric, the farthest distance from a node to any other node is the same as the farthest distance from the identity permutation to any other permutation. It is proven in Ref. [1] that the diameter of the star graph with $n = k!$ vertices is $\lfloor (3(k-1)/2) \rfloor$. The proof is as follows.

First, we present a greedy algorithm that will transform an arbitrary permutation to the identity one. Each step in the transformation will correspond to moving from a permutation to another one that is its neighbor in the star network. The whole process corresponds to the routing of a message from a given node to the node corresponding to the identity permutation.

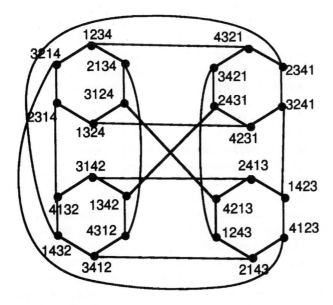

Figure 18.7 Pancake

In each step of the transform, if 1 is the first element, then exchange it with any position not occupied by the correct symbol (if any), else exchange the first element $x \neq 1$ in the permutation with the element at the xth position. The "else" case moves x to its position in the identity permutation. Consider, for example, the permutation 86324517. The sequence of transformations is: $86324517 \rightarrow 76324518 \rightarrow 16324578 \rightarrow 61324578 \rightarrow 51324678 \rightarrow 41325678 \rightarrow 21345678 \rightarrow 12345678$. Note that the first elements in these permutations follow the cyclic structure of initial permutation $86324517 = (871)(6542)$.

The correctness of the algorithm is easily seen. We now calculate the number of steps $d(p)$ needed to transfer permutation p to the identity permutation. According to the algorithm, elements that are invariant (already in their correct positions) are never exchanged. Let m be the total number of noninvariant (displaced) elements and c be the number of cycles of length ≥ 2. The "else" case is performed m times if $p_1 = 1$ and $m - 1$ times if $p_1 > 1$. Each "then" case is followed by placing, one by one, all elements of a cycle into their correct positions. Therefore, the "then" case is performed c times if $p_1 = 1$ and $c - 1$ times otherwise (in this case, the first cycle is resolved without preliminary "then" case). Therefore, $d(p) = c + m$ if $p_1 = 1$, and $d(p) = c + m - 2$ otherwise. When $p_1 = 1$, $m \leq k - 1$, and $c \leq (k-1)/2$, and $d(p) = m + c \leq 3(k-1)/2$. Otherwise, $m \leq k$, and $c \leq k/2$, and $d(p) = m + c - 2 \leq (3k - 4)/2$.

We now need to show that there are two permutation that are at distance $\lfloor 3(k-1)/2) \rfloor$. For k being odd, consider permutation $1325476 \ldots k(k-1)$, and for k being even, consider $214365 \ldots k(k-1)$. Every pair $t(t-1)$ for $t > 2$ requires at least three steps to be transformed to the correct position $(t-1)t$ (pair 21 needs only one): exchange the 1st and tth positions, exchange 1st and $(t-1)$th position, and exchange 1st and tth position. A simple, separate count for odd and even case verifies that the maximal distance is reached in both cases.

Let us look now at the structure of star and pancake networks. If the symbol in the last position is held fixed at value i, then there are $(k-1)!$ permutations and they, with all the other permissible interchanges, will constitute a $(k-1)$ star graph S_{k-1}. Thus, the vertices of S_k can be partitioned into k groups, each containing $(k-1)!$ vertices, based on the symbol in the last position. Each group is isomorphic to S_{k-1}. These groups will be interconnected by edges corresponding to interchanging the symbol in the first position with the symbol in the last position. Figure 18.6 shows S_4 as four interconnected copies of S_3. Analogous decomposition properties are valid for pancakes.

It should be noted that in S_k, the position at which we fix the symbols to get S_{k-1} values, could be at any i, $2 \leq i \leq k$ (and not just the last). Thus, in general, we can define

$$S^i_{k-1}(j)$$

to be a S_{k-1} such that all the vertices in it have the same symbol j at position i, $2 \leq i \leq k$, $1 \leq j \leq k$. Therefore $S_{k-1}(j) = s^k_{k-1}(j)$. It follows that there are $k - 1$ ways of decomposing S_k into k S_{k-1} values: $s^i_{k-1}(j)$, $2 \leq i \leq k$, $\leq j \leq k$ [15].

A new topology called the *arrangement graph* has been defined in Ref. [6] as a class of generalized star graphs. The arrangement graphs $A(k, m)$ are defined as graphs whose vertex set is the set $P(k, m)$ of all permutations of k out of m elements $\{1,2,\ldots,m\}$. Two vertices are adjacent if they differ in exactly one position. $A(k-1, k)$ is obviously the same as k-star (the equivalence is obtained when the first element in each permutation in the star nodes is dropped). The number of nodes in $A(k, m)$ is $m!/(m-k)!$, and each node has $k(m-k)$ neighbors. $A(k, m)$ is vertex and edge symmetric. Figure 18.8 shows $A(2, 4)$. The diameter of the arrangement graph $A(k, m)$ is $\lfloor 3k/2 \rfloor$ [6].

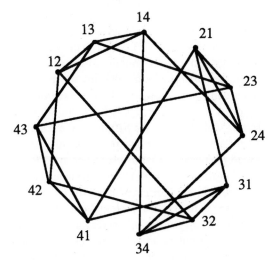

14
13
21
12
23
43
24
42
31
32

Figure 18.8 Arrangement graph

41
34

   The *k*-rotator graph is a directed graph with *k*! nodes that correspond to permutations of *k* symbols. The in-degree and out-degree of each node is $k - 1$. An edge exist from node $(a_1, a_2,\ldots,a_k)$ to the following nodes: $(a_2, a_3,\ldots,a_i, a_1, a_i + 1,\ldots,a_k)$ for $1 < i \le k$. Thus, there is an edge from one node to the other if the later corresponding permutation is obtained by inserting the first element of the permutation corresponding to the former node to any position, shifting to the left the elements which precede the position. For example, node 563214 is linked to nodes 653214, 635214, 632514, 632154, and 632145. Fig. 18.9 shows a three-rotator graph.

   Rotator graphs are introduced in Ref. [3] as a type of graph with reverse edges of cycle prefix digraphs [8], defined by Faber and Moore in 1988 (these two graphs are isomorphic). The generalized (k, m)-rotator graph $(m \le k)$ has $k!/(k - m)!)$ nodes that correspond to all permutations of *m* out of *k* elements $\{1,2,\ldots,k\}$. *A* node $(a_1,\ldots,a_m)$ is connected via outgoing links to nodes $(a_2, a_3,\ldots,a_i, a_1, a_i + 1,\ldots,a_m)$ for $1< i \le m$, and to nodes $(a_2, a_3, a_4,\ldots,a_m, a_j)$ where $m < j \le k$; i.e., the symbol a_j was not present in the original selection of *m* symbols. (The (k, k)-rotator graph is the same as the *k*-rotator graph. By definition, each node of the (k, m)-rotator graph is connected to $k - 1$ edges.

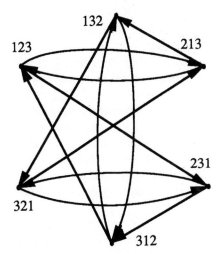

132
123
213

231

321

312

Figure 18.9 Rotator graph

The diameter of the k-rotator graph is $k - 1$, whereas the diameter of the (k, m)-rotator graph is m for $m < k$. The proof is as follows [3]. Since the graphs are symmetric, it suffices to consider the distance from any node $(a_1, \ldots, a_m)$ to the origin $123 \ldots m$. The problem corresponds to sorting the symbols in the permutations using the operations corresponding to existing edges, which here are rotations. The shortest path can be found using the insertion sort technique. Consider the complete permutation $(a_1, \ldots, a_m, a_{m+1}, \ldots, a_k)$ where the remaining elements are added in the increasing order ($a_{m+1} < \ldots < a_k$). Let i be chosen such that $a_i > a_{i+1} < a_{i+2} < \ldots < a_k$; i.e., the last $k - i$ elements in the permutation are sorted in increasing order (obviously, $i \le m$). The rotation moves a_1 to a new position while keeping the relative order of $a_2, a_3, \ldots, a_k$ unchanged. To place a_i after a_{i+1}, one has to perform i (left) rotations, placing the element at the first position, which allows it to be placed after a_{i+1} (more precisely, to its correct position). Therefore, the shortest path has the length $\ge i$. We now show that i such rotations actually suffices to sort the permutations. First, we place a_1 to its correct position in the sorted list $a_{i+1} < a_{i+2} < \ldots < a_k$, increasing the length of the monotone increasing sequence at the end of permutation by one. We repeat this operation i times, each time placing the first element to its correct position in the sorted sequence. At the end, the sequence will be sorted. Thus, the distance from node $(a_1, \ldots, a_m)$ to origin $123 \ldots m$ is i, where i is the greatest index of a symbol that is greater than its right neighbor. The greatest value of i is $k - 1$ for the k-rotator graph, and m for the (k, m)-rotator graph, $m < k$.

Consider an example of the routing in $(9, 6)$-rotator graph, from node 583492 to node 123456. Complete the source representation to 583492167 and route as follows: 583492167 → 834921567 → 349215678 → 492135678 → 921345678 → 213456789 → 123456789.

It is possible to determine the number of nodes at given distance from a given node. The last $i - 1$ symbols in the sorted order can be selected by choosing i symbols out of k symbols, putting any of $i - 1$ symbols (except the smallest) at the first position (in $i - 1$ ways), putting the rest of symbols in the sorted order, and putting nonchosen $k - i$ items in any order. Therefore, the number of nodes at distance $k - i + 1$ from an origin in a k-rotator graph, per Ref. [3], is

$$C(i, k)\,(i-1)\,(k-i)\,! \;=\; \frac{k!\,(i-1)}{i!}$$

18.1.7 Additional networks

We will briefly outline other existing networks in related literature, the sources for which may be found in the references at the end of this chapter.

Mesh-connected computers are based on regular squares, while hexagonal [39] and honeycomb [106] meshes and tori are based on regular triangles and hexagons, respectively.

An $O(n)$ processor network capable of sorting n numbers into nondecreasing order in $O(\log n)$ time is designed in Ref. [28,77]. It is referred to as the *AKS sorting network*. The original design has a huge constants; i.e., the constant c in the number of processors cn is very large. The number has subsequently been reduced, still remaining fairly large.

An incomplete star graph [96] is an induced subgraph of the star graph, containing the first m nodes of S_k with $(k - 1)! + 1 \le m < k!$. Find the total number of edges, the diameter, and a routing algorithm for the incomplete star graph.

A Stirling network [46] consist of n processors numbered $1, 2, \ldots, n$ such that two processors i and j are connected iff the number of permutations of i distinct elements that consist of exactly j cycles is odd (or j and i are connected; i.e., the underlying graph is undirected). The diameter of the network is $\lceil \log(n+1) \rceil - 1$, while the degree of vertices is $\leq \lceil F(n, 2) \rceil$.

An alternating group network [68] contains all even permutations as vertices with two vertices connected iff the corresponding permutations are obtained one from the other by applying cycle transformation $(12i)$ or $(1i2)$ for $3 \leq i \leq k$. The network is edge and vertex symmetric.

18.2 Hypercubic Networks

18.2.1 Hypercube

A hypercube computer consists of $n=2^d$ processors (or nodes, numbered 0 to $n - 1$), linked together in a d-dimensional binary cube network. Each node u $(0 \leq u < n)$ has associated local memory and is assigned its d-bit binary representation $u_{d-1}\ldots u_1 u_0$; that is

$$u = 2^{d-1} u_{d-1} + \ldots + 2u_1 + u_0$$

We define the relative address, $a \copyright b$, of two nodes a and b as the bitwise exclusive-OR of their binary representations, and define the Hamming distance $\rho(a, b)$ between two nodes a and b as the number of ones in $a \copyright b$. For example, $01101 \copyright 11001 = 10100$, $\rho(01101, 11001) = 2$. Two nodes a and b in a hypercube share a (communication) link if and only if $\rho(a, b) = 1$ (such nodes are called *neighbors*). The notation $\circledR(u, j)$ will be used to denote the node obtained from u by flipping jth bit of u. Flipping is a complement function, i.e., $\overline{0} = 1, \overline{1} = 0$. For example, $\circledR(01001, 3) = 00001$. The only possible neighbors of a node u are exactly $\circledR(u, 0),\ldots,\circledR(u, d - 1)$. A link has link number j if, for some u, it connects two nodes u and $\circledR(u, j)$.

Therefore, the vertices of a d-dimensional hypercube network correspond to all binary strings of length d. The d neighbors of processor i are those processors j such that the binary representation of the numbers i and j differs by exactly one bit. Figure 18.10 shows a hypercube with $n = 2^4 = 16$ processors.

Hypercubes have emerged as one of the most effective and popular architectures for parallel machines, and currently several parallel machines with hypercube architecture are commercially available. They have several highly desirable properties, including edge and vertex symmetric layout. The network allows very elegant and simple parallel solutions to a number of problems.

The diameter of a d-dimensional hypercube is d, achieved by any pair of nodes with complementary binary string representations. For example, 1001 and 0110 are at distance 4, and a path between them is given as follows: $1001 \rightarrow 0001 \rightarrow 0101 \rightarrow 0111 \rightarrow 0110$.

An important characteristic of a d-dimensional hypercube is that it is constructed recursively from two $(d - 1)$-dimensional ones. More precisely, consider two identical $(d - 1)$ cubes whose vertices are numbered from 0 to $2^{d-1} - 1$. By joining every vertex of the first $(d - 1)$-dimensional hypercube to the vertex of the second one having the same number, one obtains a d-dimensional hypercube. The nodes of the first cube receive a 0 at the beginning of their binary address, while the nodes of the second one receive a 1, which effectively increases their decimal address for 2^{d-1}. The subdivision of d-dimensional hypercube into two $(d - 1)$-dimensional ones can be made along any of d dimensions. For a given $i < d$, nodes having ith bit 1 belong to one $(d - 1)$-dimensional hypercube, and nodes having ith bit 0 belong to the other.

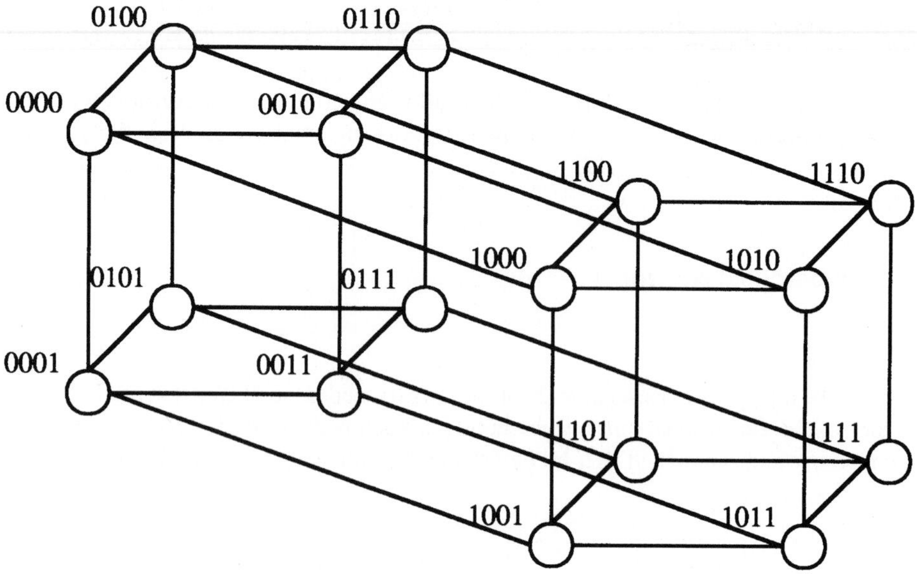

Figure 18.10 Hypercube

Although hypercubes are the most popular network, they still have some characteristics that can be improved. One of them is nonconstant fan-out, and the other is the fact that the number of processors must be a power of two. Modifications to the hypercube network therefore have been investigated to improve these characteristics. Some of these modifications will be described in subsequent subsections.

18.2.2 Cube-connected cycles

To obtain a cube-connected cycles (CCC) network (introduced in Ref. [19]), we "cut" each corner of a d-dimensional hypercube; i.e., we replace each of its 2^d corners with a cycle of d processors. Each processor has exactly three neighbors. Two of them are neighbors in its cycle of d processors corresponding to the same corner, and one neighbor corresponds to an edge of the original hypercube. Nodes can be conveniently denoted by adding one more number to their hypercube string notation, the number being the link number of corresponding edge of hypercube that connects the node to a (third) neighbor. We call the added number the *cycle index*. Thus, the neighbors of node $a_{d-1}\ a_{d-2}\ldots a_0 l$ of CCC are $a_{d-1}a_{d-2}\ldots a_0(l+1 \bmod d)$, $a_{d-1}a_{d-2}\ldots a_0(l-1 \bmod d)$ and ®*(a, l)l*= $a_{d-1}a_{d-2}\ldots a_{l+1}, \overline{a_l}, a_{l-1},\ldots a_0 l$. For example, node 010 of three-dimensional hypercube is replaced with three nodes of CCC (0100, 0101, 0102), and the neighbors of node 0102 are 0100, 0101, and 1102. Figure 18.11 shows a three-dimensional CCC with 24 processors.

The number of processors of a d-dimensional CCC is $2^d d$, and the Fan-out is 3. It is a nontrivial task to determine the diameter of a CCC. In Ref. [13], it is proven that the diameter of a d-dimensional CCC is $2d + \lfloor d/2)\rfloor - 2$ for $d > 3$, and 6 for $d = 6$; what follows is a brief outline of the proof. Since CCCs are symmetric, it suffices to consider the distances from node 00…00 ($d + 1$ zeros) to a node $a_{d-1}a_{d-2}\ldots a_0 l$ of CCC. For each 1 in the later node, say $a_i = 1$, a path must change two items in the node address: change the cycle index (except for $i = 0$ at the beginning) to i and change ith bit. Thus, any path to node 11…1i must be of length at least $2d - 1$ to change every bit in the address. The cycle index may be

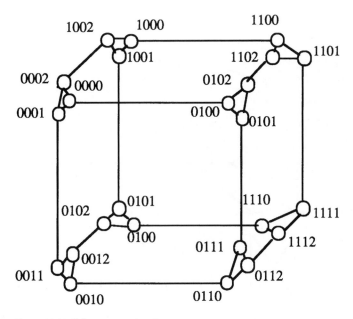

Figure 18.11 Cube-connected cycle

changed in cycle order only. Consider the following (possibly non-optimal) path: change the cycle index in one direction of the cycle, either $0, 1, 2,\ldots,d-1$ (if $d-l \leq l$) or $0, d-1, d-2,\ldots,2, 1$ (otherwise). After each change of cycle index, change the corresponding bit in the hypercube address if it is 1. It takes $\leq d$ changes in the hypercube bit address and $d-1$ changes in the cycle index. Next, change the cycle index to end up at l: either $d-1, d-2,\ldots,l$ or $1, 2,\ldots,l$, whichever is shorter), which takes $\min(l-1, d-1-l)$ steps, which is $\leq \lfloor d/2 \rfloor - 1$ for every l with equality for $l = \lfloor d/2 \rfloor$. The total number of steps is bounded by $2d + \lfloor d/2 \rfloor - 2$, and such diameter is achieved by the node $11,\ldots,1\lfloor d/2 \rfloor$. Let, for example, $d = 4$. The path between 00000 and 11112 is $00000 \Rightarrow 00010 \rightarrow 00011 \Rightarrow 00111 \rightarrow 00112 \Rightarrow 01112 \rightarrow 01113 \Rightarrow 11113 \rightarrow 11113$. Hypercube and cycle edges are denoted $\Rightarrow$ by and $\rightarrow$ respectively.

18.2.3   Butterfly

To obtain a butterfly, we replace each node of a hypercube with a linear array of $d+1$ processors. Processors are conveniently denoted as $al = a_{d-1}a_{d-2}\ldots a_0 l$ where $0 \leq l \leq d$. Node $a_{d-1}a_{d-2}\ldots a_0 l$ is connected to up to four nodes (whichever of them exists): $a(l-1) = a_{d-1}a_{d-2}\ldots a_0(l-1)$, $a(l+1) = a_{d-1}a_{d-2}\ldots a_0(l+1)$, $\circledR(a, l)(l+1) = a_{d-1}a_{d-2}\ldots a_l + 1, a_l)$, $\overline{a_l}, a_{l-1}\ldots a_0(l+1)$, and $\circledR(a, l-1)l$.

A butterfly consists therefore of $2^d(d+1)$ processors which are usually organized into $d+1$ rows and 2^d columns. The fan-out of the butterfly is 4. The network is not vertex or edge symmetric. Figure 18.12 gives a butterfly with $2^3(3+1) = 32$ processors. The diameter of the network is $2d$, which is the distance between the first and last processors in row 0. Every two processors can be connected via a processor in row d.

Note that a hypercube can be obtained by reducing each column to one processor and keeping all connections that existed before the reduction. The communication patterns of many hypercube algorithms can be conveniently represented by butterfly network, where each row corresponds to a stage in the hypercube algorithm. The butterfly network belongs

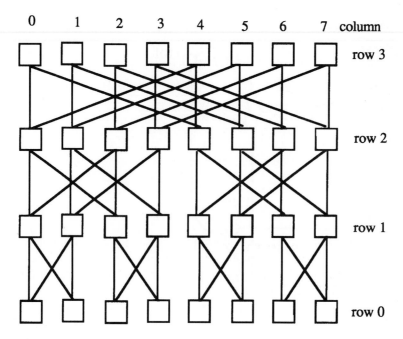

0   1   2   3   4   5   6   7   column

row 3

row 2

row 1

row 0

Figure 18.12 Butterfly network

to a class of multistage interconnection networks (each row of the butterfly corresponds to a different stage).

The butterfly network has a number of different names in the literature, including Banyan, baseline, bidelta, flip, omega, indirect binary cube network, and modified data manipulator.

18.2.4   de Bruijn network

The de Bruijn graph is defined by de Bruijn [4] as follows. The vertices of de Bruijn graph $D(m, k)$ are all sequences (called *variations*) $d_1 d_2 \ldots d_k$ such that $d_i \in \{0, 1, \ldots, m - 1\}$ for $1 \leq i \leq k$. Thus, the number of nodes is $n = m^k$. Two variations (nodes) are connected by an edge if and only if they are obtained from each other by a shift for one position to the left and appending arbitrary element to the position that becomes empty after shifting. More precisely, the neighbors of node $d_1 d_2 \ldots d_k$ are nodes $d_2 d_3 \ldots d_k p$ for each p, $0 \leq p \leq m - 1$. Therefore, each node has m neighbors. A binary de Bruijn graph $D(2, k)$ is shown in Fig. 18.13. For instance, node 110 is connected to the following nodes: 100 (left shift, append 0) and 101 (left shift, append 1). De Bruijn graph is a directed graph (with some loops), which facilitates the control of information flow and the design of some algorithms (it may be alternatively described as an undirected graph).

The diameter of $D(m, k)$ is $k = \log_m n$. This follows easily by observing that one can obtain any variation $u_1 u_2 \ldots u_k$ from any other $v_1 v_2 \ldots v_k$ by a sequence of k shifts as follows: $v_1 v_2 \ldots v_k \rightarrow v_2 v_3 \ldots v_k u_1 \rightarrow v_3 v_4 \ldots v_k u_1 u_2 \rightarrow \ldots \rightarrow v_k u_1 u_2 \ldots u_{k-1} \rightarrow u_1 u_2 \ldots u_k$. This transformation may also serve as a routing algorithm for sending messages from a source to a destination. On the other hand, the distance between nodes $00 \ldots 0$ and $11 \ldots 1$ is k, since in one step at most, one 0 can be turned to a 1; thus k bit changes are necessary to transform one node into the other.

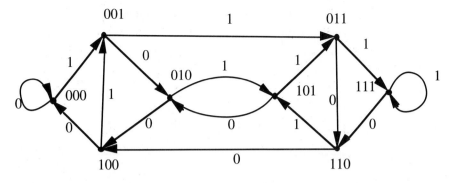

Figure 18.13 Binary de Bruijn graph

Binary de Bruijn graph has twice as many (directed) edges as the nodes. It also has a very interesting recursive structure. Nodes of $D(2, k + 1)$ correspond to edges of $D(2, k)$, and two nodes of $D(2, k + 1)$ are connected by a directed edge if the endpoint of the first coincides with the beginning of the second edge.

18.2.5 Incomplete and Gray code incomplete hypercubes

Because the number of nodes in a system must be a power of two, there are large gaps in the sizes of hypercubes that can be built. This restriction can be overcome by using an incomplete hypercube that can be constructed with any number of nodes. This type of hypercube has attracted research interest recently.

Katseff [11] proposed the incomplete hypercube model, i.e., a subgraph of a complete hypercube containing nodes numbered 0 to $n - 1$ (where n is not a power of two). Two nodes are neighbors in the incomplete hypercube if and only if they are neighbors in the corresponding complete hypercube. In other words, incomplete hypercube is a subgraph of hypercube induced by first n nodes.

Let $I = \{d_1, d_2, \ldots, d_k\}$, $d_1 > d_2 > \ldots > d_k$, be the set of indices j such that $n_j = 1$, i.e., $n_j = 1$ iff $j \in I$. An incomplete hypercube induced by an interval $[0, n-1]$ of nodes (Katseff model) can be decomposed into k subcubes (complete hypercubes) $H_1, \ldots, H_k$ such that the dimension of H_i is d_i, $1 \le i \le k$. A t-dimensional subcube contains 2^t nodes and is represented as a word over the alphabet $\{0, 1, *\}$ where exactly t symbols are *s or *do not care* elements. For example, a two-dimensional subcube $\{00010, 00110, 10010, 10110\}$ can be compactly written as *0*10. Clearly each node u of H_1 satisfies $u[t] = 0$ for $t \ge d_1$ and $u[t] = *$ for $d_1 > t$. Each node u of H_i for $1 < i \le k$ satisfies $u_j = n_j$ for $j > d_i$, $u_{d_i} = 0$ and $u_j = *$ for $d_i > j$. For example, for $n = 26 = 11010$, $I = \{4, 3, 1\}$, subcube decomposition is $0****$, $10***$, $1100*$. In other words, subcubes are intervals $[0, 15]$, $[16, 23]$, and $[24, 25]$. Subcubes of an incomplete hypercube are uniquely determined by n, and each processor may, by knowing n only, easily deduce information on the list of subcubes in the following way. Node u belongs to subcube H_j where d_j is the index of the most significant bit 1 in $u \odot n$, and $j = n_{d-1} + n_{d-2} + \ldots + n_{d_j})$. For instance, node $u = 19 = 10011$ belongs to subcube H_2 of incomplete hypercube with $n = 26 = 11010$ because $u \odot n = 01001$, $d_j = 3$, and $j = 1 + 1 = 2$.

Obviously, subcube H_1 is a complete d_1-dimensional hypercube and contains more than half of nodes of the incomplete hypercube. Each node u of subcube H_i for $i > 1$ is connected on link d_j to the node $v = ®(u, d_j)$ of H_j, for each $j < i$. For example, node

$25 = 11001$ of subcube H_3 of incomplete hypercube [0, 25] is connected to nodes 01001 of H_1 and node 10001 of H_2.

The Gray code order of nodes can be defined in the following way:

- For $d = 1$, the nodes are numbered $g(0) = 0$, and $g(1) = 1$, in this order,
- If $g(0), g(1),\ldots,g(2^d - 1)$ is the Gray code order of nodes of a d-dimensional hypercube, then $g(0) = 0g(0), g(1) = 0g(1),\ldots,g(2^d - 1) = 0g(2^d - 1), g(2^d) = 1g(2^d - 1), g(2^d + 1) = 1g(2^d - 2),\ldots,g(2^{d+1} - 2) = 1g(1), g(2^{d+1} - 1) = 1g(0)$ is a Gray code order of nodes of a $(d + 1)$-dimensional hypercube.

As an example, for $d = 3$ the order is $g(0) = 000, g(1) = 001, g(2) = 011, g(3) = 010$, $g(4) = 110, g(5) = 111, g(6) = 101, g(7) = 100$.

Any two integers a and b such that $0 \le a \le b < m = 2^d$ determine an interval of nodes $g(a), g(a + 1),\ldots,g(b)$. A subgraph of complete hypercube induced by these nodes is called a *Gray code incomplete hypercube* (introduced in Ref. [21]), shortly denoted GCIH, $[a, b]$-GCIH, or $[g(a), g(b)]$-GCIH. For example, a $[1, 10]$-GCIH, or $[0001, 1111]$-GCIH, contains nodes $g(1) = 0001, g(2) = 0011, g(3) = 0010, g(4) = 0110, g(5) = 0111$, $g(6) = 0101, g(7) = 0100, g(8) = 1100, g(9) = 1101, g(10) = 1111$. Applications of GCIH model in processor allocation and image processing tasks are described in Ref. [104].

Let us observe how two nodes u and v can be compared in Gray code order. Let i be the most significant bit, where u and v differ; i.e., $u_1 = v_1$ for $1 > i$, and $u_i \ne v_i$. Then, $u < v$ if and only if $u_{d-1} + u_{d-2} + \ldots + u_{i+1} + u_i$ is an even number. For instance, $11100 < 10100 < 10110$. The comparison method gives a way to find Gray code address t of a node u (satisfying $g(t) = u$), and the inverse operation; the corresponding two simple procedures are described in Ref. [21].

The important property of the Gray code order is that corresponding nodes of a hypercube are neighbors whenever they are neighbors in the Gray code order. (This property is not valid for the lexicographic order $0, 1, 2,\ldots, 2^d - 1$ of binary addresses.)

The nodes of an [a, b]-incomplete hypercube can be decomposed into $O(\log b)$ subcubes, each containing consecutive nodes (in Gray code order). For example, the [1, 10]-GCIH is divided into 5 subcubes: $[1, 1] = 0001, [2, 3] = 001*, [4, 7] = 01**, [8, 9] = 110*$, and $[10, 10] = 1111$ (see Fig. 18.14). Note that the dimensions of these subcubes are 0, 1, 2, 1, 0, which is a bimodal sequence (first increases and then decreases). This property is valid in general. Let h be the most significant bit where the address of two nodes of an [a, b]-incomplete hypercube differs. For example, $h = 3$ for [1, 10]-incomplete hypercube. Consider two parts, the one with nodes having hth bit 1 and 0, respectively. Each of these two parts, called the left and right GCIHs, is itself isomorphic to an incomplete hypercube and can be decomposed analogously. If, in addition, the largest subcubes in both parts have the same dimension, they together make a subcube of larger dimension.

18.2.6   Enhanced hypercubes

An enhanced hypercube is a hypercube with extra connections between nodes, using their fabricated but otherwise unused links. They achieve a considerable reduction in diameter but often lose some of desirable properties of hypercubes. We shall mention one enhanced hypercube model.

A folded hypercube of dimension k can be constructed from a standard hypercube by connecting each node to the unique node that is farthest from it. It is introduced independently in [7] and [23]. More precisely, a node $u = u_{k-1}u_{k-2}\ldots u_0$ is connected, in addition to hypercube connection, to node $\overline{u} = \overline{u}_{k-1}\ldots\overline{u}_0$. Figure 18.15 shows 3-folded hypercube,

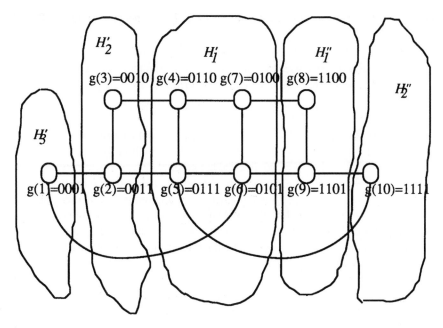

Figure 18.14 Gray code incomplete hypercube

i.e., a folded hypercube of dimension 3. Bold lines indicate new interconnections with respect to cube.

The fan-out of k-folded hypercube is $k + 1$. The network is obviously symmetric. It has the diameter $\lceil k/2 \rceil$ [7, 23], which can be shown as follows. Let $d(x, y)$ denote the distance between nodes x and y of a k-dimensional hypercube. Then $d(x, u) + d(x, \bar{u}) = k$, since for each bit in the binary address of x exactly one of u or $\bar{u}$ has a different bit at the same position. If $d(x, y) \leq \lceil k/2 \rceil$, then the shortest path between them in a k-folded hypercube is the same as the shortest path between them in k-dimensional hypercube. Otherwise, $d(x, \bar{u}) = k - d(x, y) \leq \lceil k/2 \rceil - 1$ and the path from x to $\bar{u}$ plus the edge $(u, \bar{u})$ has the length $\leq \lceil k/2 \rceil$.

The folded hypercube reduces network cost by almost half, as compared to the corresponding hypercube, and preserves most of hypercube's nice properties. One of drawbacks

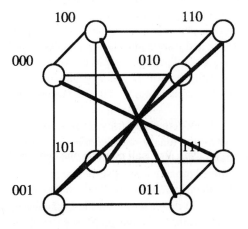

Figure 18.15 Folded hypercube

of folded hypercubes is that they do not have a direct decomposition into lower-dimensional folded hypercubes. In Ref. [7], it is shown how to partition k-folded hypercube into two $(k - 1)$-folded hypercubes, with opposite nodes connected by paths of length 2. Path length for these virtual links increases by 1 for each following dimension, which will limit their performance in practice for recursive types of algorithms.

18.2.7   Other hypercubic networks

A generalized hypercube [33] has $m_r * m_{r-1} * \ldots * m_1$ processors addressed $x_r x_{r-1} \ldots x_1$, $0 \le x_i \le m_i - 1$. Two processors are connected iff their addresses differ by exactly one position. Hypercubes are a special case of generalized hypercubes.

An (n, b, k)-cube [74], $n = b^k$, is a graph with n nodes labeled as base b integers with k digits (i.e., $x = x_k x_{k-1} \ldots x_1$, $0 \le x_i < b$) such that x and y are connected iff there exists j, $1 \le j \le k$ such that $x_j \ne y_j$ and $x_i = y_i$ for all $i \ne j$. Thus, two nodes are connected iff their labels differ in exactly one base b digit. It is a special case of generalized hypercubes. Also, hypercube is a special case of (n, b, k)-cube. The degree of each node is $(b - 1)\log_b n$, and the diameter of the network is $\log_b n$ [74].

An r-ary k-cube [20] has r^k nodes such that node $a_0 a_1 \ldots a_{k-1}$ $(0 \le a_i < r)$ is connected to nodes $a_0 \ldots a_i \pm 1 \pmod r \ldots a_{k-1}$, $0 \le i \le k - 1$. Toruses and hypercubes are special cases of r-ary k-cubes.

Fibonacci cubes, introduced in Ref. [66], are another type of incomplete hypercube. A Fibonacci cube of order k is a subgraph of hypercube graphs of dimension $k - 2$, induced by nodes that have no two consecutive ones in their binary representation. For example, nodes of Fibonacci cube of dimension 5 are 000, 001, 010, 100, and 101. The diameter of Fibonacci cube of order k is $k - 2$; for example, nodes $0101 \ldots 0(1)$ and $1010 \ldots 1(0)$ are at distance $k - 2$. We will now prove that the degree of each node is between $k - 2$ and $\lfloor (k - 2)/3 \rfloor$. Node $00 \ldots 0$ has degree $k - 2$, while node $001001 \ldots$ has degree $\lfloor (k - 2)/3 \rfloor$. We need to show that no node has less degree than $\lfloor (k - 2)/3 \rfloor$. It suffices to prove that, among any three consecutive bits in a node address, at least one of them corresponds to a neighbor. This is an easy to show since combinations 111, 011, and 110 are not permitted as part of a node address, in combination 000 the middle bit 0 can be turned to a 1, while for each of remaining combinations 101, 100, 010, 001 any of the ones can be turned to a zero, which will lead to a node address. Thus, at least one-third of bits link each node to a neighbor, which suffices to claim the lower bound on the degree.

The plus-minus interconnection network consists of n processors (n arbitrary integer); processor j is connected to processors p and q where $p = j + 2^i \pmod n$ and $q = j - 2^i \bmod n$ for each i, $0 \le i \le \log n$. In Ref. [105], the network is generalized as follows. The (r, k)-radix graph consists of $n = r^k$ vertices. Two vertices x and y are connected by an edge if and only if $|x - y| = r^i \bmod n$ for some integer i, $0 \le i \le k - 1$. The radix graphs are vertex symmetric and have logarithmic diameter and degree with the network cost (product of degree and diameter) which is better than for hypercubes for radix values $r < 20$. They can be easily decomposed into smaller radix graphs.

The perfect shuffle interconnection network [91,108] consists of $n = 2^d$ processors. A one-way line links processor i to processor j, where the binary representation of j is obtained by cyclically shifting that of i one position to the left. In addition to these shuffle links, two-way links connecting every even-numbered processor to its successor are sometimes added to the network, known then as the *shuffle-exchange network*. A binary de Bruijn graph $D(2, r)$ can be obtained from the $(r + 1)$-dimensional shuffle-exchange by contracting out all the exchange edges from the shuffle-exchange graph (i.e., merging nodes $u_1 \ldots u_r 0$ and $u_1 \ldots u_r 1$ into one node $u_1 \ldots u_r$).

An Omega network is a multistage interconnection network as defined in Ref. [73]. It consists of $d = \log n$ rows, each with n processors denoted as binary strings. The processors in row i are connected to those in row $i + 1$, for $i = 1, 2, \ldots, d - 1$, obtained by left shift and adding 0 or 1 to the last position to its corresponding binary string. In Ref. [115], it is shown that the butterfly, omega [73], baseline [115], and indirect binary n-cube [90] networks are topologically equivalent networks.

The hyper-de Bruijn network HD(m, k) [61] is the direct or cross product of an m-dimensional hypercube and binary de Bruijn graph $D(2, k)$.

The Banyan-hypercube network [116] is obtained from butterfly network when processors in each row are connected in a hypercube network (with nodes labeled lexicographically from 0 to $2^k - 1$ from left to right). The diameter of the network is k.

The composite hypercube model [35] is not restricted to the first n nodes of hypercube. The basic idea is that a composite hypercube is expressible as a collection of successively smaller and adjacent complete hypercubes. Composite hypercubes can be obtained by renumbering the nodes of an incomplete one.

A crossed cube architecture [52] is defined in the following way. There are 2^d nodes, addressed binary like hypercubes. Two nodes $(u_{d-1}, \ldots, u_0)$ and $(v_{d-1}, \ldots, v_0)$ are connected by an edge if and only if there exist j with $u_{d-1} \ldots u_j = v_{d-1} \ldots v_j$, $u_{j-1} \neq v_{j-1}$, $u_{j-2} = v_{j-2}$ if j is even, and for $0 \leq i < F(j - 1, 2)$, $(u_{2i+1}u_{2i}, v_{2i+1}v_{2i}) \in \{(00,00), (10,10), (01,11), (11,01)\}$. Each node has exactly d neighbors, and the diameter is $\lceil (d + 1)/2 \rceil$, i.e., about half the diameter of the hypercube.

Twisted k-cube [55] is obtained from k-dimensional hypercube when two independent edges in a cycle of length 4 are exchanged (i.e., a cycle $uvyx$ is replaced with cycle $uvxy$). This decreases the diameter of the hypercube by one. Also, $(k - 1)$-dimensional hypercube and complete binary tree with $2^k - 1$ nodes are subgraphs of the twisted cube.

Hyper Petersen graph [43] is cartesian (direct) product of a binary hypercube and the Petersen graph. A k-dimensional Hyper Petersen graph with $1.25 \cdot 2^k$ nodes has degree k and diameter $k - 1$.

A Hamming cube [41] consists of n nodes labeled $0, 1, \ldots, n - 1$ such that two nodes i and j, $i < j$, are connected by an edge iff their Hamming distance is either $\rho(i, j) = 1$ or $\rho(i, j) = \lceil \log(j + 1) \rceil$. The degrees of nodes are (when $n = 2^k$) between $k + 1$ and $2k - 1$. The diameter of the network is $\lceil k/2 \rceil$. A Hamming cube with $n = 2^k$ nodes can be decomposed into a Hamming cube with $n/2$ nodes and a hypercube with $n/2$ nodes.

A n node (n power of 2) hierarchical hypercube [83] consists of a father hypercube whose nodes are by themselves hypercubes rather than simple processors. The network has degree $O(\log \log n)$ and diameter $O(\log n)$.

The reading list at the end of this chapter contains references to additional hypercubic networks.

18.3 Routing and Broadcasting

In a *routing* problem a node *src* has to send a message to another node *dest*. The routing involves the selection of paths for the data and the specification of appropriate schedules for moving data along the paths. In the *broadcasting* problem, a node *src* wants to send a message to all other nodes. Routing and broadcasting are basic data communication techniques for any interconnection network model.

It is not difficult for any network to find a kind of routing or broadcasting algorithm. However, such operations are frequently used and should satisfy some desirable properties, first of which is speed. The time complexity of any algorithm on an interconnection network can be divided into the *computation time* inside each processor (which corresponds to the running time of the corresponding sequential program), and *communication time,* which is the time needed for the communication of data between processors. The current state of the art is such that the communication time is much more expensive than the computation time (i.e., has higher physical cost, or is considerably higher in comparison with the time complexity). Frequently, researchers consider only the communication time for the overall time complexity.

In the distributed or asynchronous computation mode of a network, messages can be sent at any time and from any sources to any destinations. It is possible that a processor wishes to send a message to one of its neighbors and at the same time that another neighbor seeks to delivery another message to it. Processors cannot send and receive data simultaneously, so the delivery of a message has to wait until the recipient is free to accept it. The *deadlock* is a situation when there is a cycle of processors, each attempting to send a message to its neighbor but unable to do so. This arises when a processor a_1 waits to send a message to its neighbor a_2 because node a_2 waits in turn for processor a_3 to free itself after delivering its message to processor $a_4, \ldots,$ and processor a_t waits to deliver a message to processor a_1 because a_1 is busy trying to delivery to a_2. There is a cycle of delivery and wait requests, and the network becomes deadlocked. A good routing or broadcasting procedure should never create a deadlock.

In practice, the information to be routed often consists of whole packet or record rather than a single datum. There are two types of routing strategies developed in literature: *wormhole* [98] and *store-and-forward* routing [22]. In store-and-forward, routing a packet is completely stored in a node before it is transmitted to the next node on the way to the destination. The wormhole routing operates by advancing the head of a packet directly from incoming to outgoing channels. Only a few flow control digits are buffered at each node. The message becomes spread out across the channels between the source and the destination. Therefore, the entire path from the source to its destination must be dedicated to the packet.

Any network causes store-and-forward deadlock, iff the directed graph that represents routing contains cycles. On the other hand, deadlock can be avoided if all the cycles are broken logically. In general, each vertex is separated into an appropriate number of distinct points using multiple set of buffers. Deadlock-free routing algorithms for the store-and-forward networks are based on the concept of a structured buffer pool, i.e., restricting the buffer allocation to define a partial order on buffer classes [14].

To break deadlock with wormhole routing, we must restrict the routing algorithm [5]. If the delivery requests are drawn as oriented edges, then the deadlock corresponds to a cycle in the this graph. A more formal study of the deadlock related graphs for the wormhole routing is given in Ref. [5] by introducing the link dependency graph. Let $L(l, r)$ denotes the link l going from node r. Links are vertices of the graph and two links $L(l', r',)$ and $L(l'', r'')$ define an directed edge in the link dependency graph if and only if there exist two nodes *src* and *dest* such that $L(l', r')$ and $L(l'', r'')$ are two consecutive links on the message path from *src* to *dest*. It easily follows [5] that a routing algorithm for an interconnection network is deadlock free iff there are no cycles in the link dependency graph. It should be noted here that the same edge joining two nodes a and b in the network corresponds to two different vertices in the link dependency graph. We may denote these vertices as ab and ba, corresponding to links from a to b, and from b to a, respectively.

We will now describe a deadlock-free routing procedure for an arbitrary graph [18]. First, we find a spanning tree with the property that for every node of the graph there exists a path within the spanning tree to a selected node (say, node 0) with the minimum possible number of edges (equal to the distance of the node from the selected node). Starting from a selected node as the source, follow the all-port broadcasting algorithm (or do a breadth-first search) and mark all edges that lead to a new node for the first time; if a node is reached simultaneously by several paths with the same distance from the selected node, choose only one connecting edge arbitrarily. All marked edges form the spanning tree.

Routing between any source and destination is now restricted to use only edges of the spanning tree. More precisely, the message is routed from node a to the selected node (node 0) and from the selected node to node b. The path can be shortened if the common edges in the two routings are discarded as follows. Processor a can calculate the lowest common ancestor $lca(a, b)$ of a and b in the spanning tree, and then the message is routed from a to $lca(a, b)$, followed by routing from $lca(a, b)$ to b.

We will show that the procedure is deadlock free. Let the root (node 0) be at level 0, its neighbors at level 1 in the spanning tree, etc. The level of each node is equal to its distance from the selected node. Let us analyze the link dependency graph for the routings along the spanning tree. Each edge of the spanning tree corresponds to two vertices of the link dependency graph. Suppose that the edge of the spanning tree joins a node at level i with a node at level $i + 1$. The corresponding vertices of the link dependency graph are labeled either $i(i + 1)$ or $(i + 1)i$, depending on the message direction. Let us call them *level increasing* and *level decreasing* vertices, respectively. All the vertices of the link dependency graph, when routing from a to $lca(a, b)$, are level decreasing, and similarly all the vertices on the path from $lca(a, b)$ to b are level increasing. The edges of the link dependency graph have one of the following form (the vertices are denoted by their labels only): $(i - 1)i \rightarrow i(i + 1)$, $(i + 1)i \rightarrow i(i - 1)$, $(i + 1)i \rightarrow i(i + 1)$. Suppose that there exist a cycle in the link dependency graph. The edges in the cycle are of the following form (again, vertices are replaced by their label): $a_1a_2 \rightarrow a_2a_3$, $a_2a_3 \rightarrow a_3a_4, \ldots, a_{t-2}a_{t-1} \rightarrow a_{t-1}a_t$, $a_{t-1}a_t \rightarrow a_ta_1$, $a_ta_1 \rightarrow a_1a_2$. Clearly $a_i = a_{i-1} \pm 1$ for each index i. Without loss of generality, let a_1 be the maximum number in the sequence $a_1, a_2, \ldots, a_t$. Then, $a_2 = a_2 - 1$, and $a_t = a_1 - 1$. Thus we have an edge of the form $(a_1 - 1)a_1 \rightarrow a_1(a_1 - 1)$, which is impossible. Note that such an edge would correspond to a path in routing a message that changes its direction, going first farther from root, and then turns and goes toward the root. The algorithm has no such routing path between any two nodes.

The longest path in the spanning tree is, in the worst case, twice as long as the diameter of the network. Hence, path between two nodes can be considerably longer than the minimum path between them. Another drawback of the procedure is that the nodes close to the root are congested. A good property is that the spanning tree may be changed at any time, for example, in case of node or edge failures. If a failure is detected, the spanning tree is redrawn using remaining nodes and edges, and new routing algorithm is established. The main advantage is that for some networks, like stars and cube-connected cycles, no alternative deadlock-free routing algorithm is known.

Finding the *shortest paths* for all messages is an important criterion of the efficiency of the routing algorithm, but such a criterion does not necessarily imply the completion of the message delivery in minimum time. Many paths may share the same edge or node, and minimizing the (edge or node) *congestion* is also important criterion. In the synchronous or parallel computation mode, the congestion criteria put a restriction on the each processor, during a number of simultaneous routing paths, that an interim processor shall not send or receive more than one message at a time (in certain advanced architectures a processor may have few channels for routing). To avoid congestion, messages often have to wait at particular nodes before proceeding, thus keeping their shortest paths but not necessarily minimum times.

A routing algorithm is said to be *oblivious* if the path traveled by each packet depends only on the origin and destination of the packet (and not on other packets or congestion). Routing algorithms presented in this chapter are oblivious. An inherently deadlock-free network is defined as a network for which there exists a deadlock-free, oblivious routing procedure. Mesh-connected computers and hypercubes are inherently deadlock-free networks. On the other hand, star, CCC, torus, pancake, and de Bruijn graphs are not known to be inherently deadlock free.

The broadcasting algorithms differ in whether they assume that a processor may send the message on all links simultaneously (all-port communication model) or is restricted to use one link at a time (one-port communication model). In Δ-port communication model, each node can simultaneously send a message on each of up to Δ links. Due to various physical limits, it is preferable to minimize Δ. the one-port communication model allows the use of broadcasting procedure in synchronous interconnection network model, in addition to the asynchronous one. We do not consider the length of the message, unit transmission cost, or latency, which is measured by some researchers [10].

The lower bound on the number of steps in one-port communication model is derived in Ref. [1]. In any step, each processor can send the information to at most one new processor. Thus, at every time step, the number of processors that have received the information being broadcasted can at most double. If a network has n vertices, any broadcasting algorithm in the one-port model requires $\Omega(\log n)$ steps.

An all-port broadcasting algorithm for any interconnection network can be described in the following way: all nodes that received the message for the first time send the message through all their links. Each node will clearly receive the message in the shortest possible time. The algorithm follows a well known breadth-first search paradigm. There are several possible problems with the algorithm. One is that no two neighboring nodes should attempt to communicate the message to one another. Such problem does not exist for bipartite graphs (hypercube, star, and so forth). The next problem is that the message can be communicated to a node from several neighbors simultaneously. Therefore, it is desirable that nodes make a list of links through which they send a message. This section will concentrate on the one-port model, which in most cases has equal or comparable broadcasting capabilities with the all-port communication model.

There exist several other kinds of routing or broadcasting problems that are not covered in this review. One of them is permutation routing, where processor i wishes to send its data to processor $\pi(i)$, where π is a permutation. The other is the multinode broadcast problem, where each node wants to send a package (the same package) to all the other nodes.

18.3.1 Hypercubes

The standard routing algorithm [20] for complete hypercubes is as follows:

```
for l ← d - 1 downto 0 do
    if src_l ≠ dest_l then send message on link l from node currently keeping it
```

Starting from the most significant bit $d - 1$, the algorithm checks the current bit and sends the message along the corresponding link if the bit in the source and destination differs. For example, a messages is routed from $src = 01100$ to $dest = 10110$ as follows: $01100 \rightarrow 11100 \rightarrow 10100 \rightarrow\rightarrow 10110 \rightarrow$. In this scheme, $\rightarrow$ corresponds to checking a bit in the address, and actual communication steps are recognized by giving the address of node currently receiving the message.

We give a simple proof that the algorithm does not exhibit deadlock when used in wormhole routing. The proof uses a link dependency graph [5]. Let an edge exist from $L(l', r')$ to $L(l'', r'')$. Then, in the standard routing algorithm, $l' > l''$ is satisfied without exception, and thus a cycle cannot be formed because, in any path through the link dependency graph, the corresponding link numbers strictly decrease.

This proof was used in Ref. [5] to design deadlock-free routing algorithms for some networks by introducing virtual channels (logical entities associated with physical links used to distinguish multiple data streams traversing the same physical channel; multiple virtual links are time-multiplexed across a physical channel). Consider, for example, a de Bruijn graph $D(m, k)$. The routing algorithm described in the previous section becomes deadlock free if every edge of the graph is subdivided into k channels, and the routing, which takes k steps, follows channels numbered $1, 2, \ldots, k$ in this order (at step i follow the channel i, $1 \leq i \leq k$). The channel numbers in the link dependency graph increase, and thus cycle cannot be formed.

The standard one-port communication model broadcasting algorithm [20] for hypercubes uses the same data paths as the routing one, and is therefore deadlock free. It can be described is follows:

```
for l ← d-1 downto 0 do
    send message on link l from node currently keeping it
```

Consider the number of nodes that have received the message after each step in the broadcasting algorithm. Initially, only one node (the source) "knows" the message. It sends the message on link $d - 1$, so that then two nodes, *src* and ®(*src, d – 1*) know the message. In the next step, two more nodes receive the message, which are the neighbors of the two on link $d - 2$. In each of next steps, the number of nodes that received the message doubles. This kind of doubling is one of main strategies in parallel computing. Another chapter [57] of this book describes additional hypercube data communication techniques.

18.3.2 Incomplete hypercubes

A routing algorithm that has optimal communication path $\rho(a, b)$ between any two nodes a and b and optimal computation time $O(\log n)$ is given in Ref. [21].

Note that if *src* ≥ *dest,* then the standard routing procedure sends the message first from src to ®(*src*, l), where l is the most significant bit 1 in *src*©dest, $src_l = 1$, $®(src, l)_l = 0$, and later the message is sent on links with link numbers less than l. Therefore, the addresses of all nodes on the message path in the standard routing procedure are less than *src*, so all intermediate nodes belong to incomplete hypercubes, and all links on the standard routing path exist. Thus, the message is "safely" delivered to *dest* using a standard routing procedure for complete hypercubes. All links exist also in case *src* < *dest* if they belong to the same subcube H_i. Otherwise, let *src* belong to H_i and *dest* to H_j, where $i < j$. Node ®(*dest*, d_i) belongs to H_i, and the message can be "safely" routed to it from *src* within subcube H_i. Then, the message can be sent to *dest* from ®(*dest*, d_i).

This procedure can be coded as follows.

Procedure routing_inc(src, dest);

```
Processor src determines i and j such that src belongs to Hi and dest to Hj.
    If i ≥ j then for l ← d_j downto 0 do if src_l ≠ dest_l then send message
        on link l from node currently keeping it;
```

```
else {for l ← dᵢ - 1 downto 0 do if src₁ ≠ dest₁ then send message
      on link l from node currently keeping it;
      Send the message from node ®(dest, dᵢ) to node dest.}
```

For example, let $n = 26 = 11010$, $src = 23 = 10111$ and $dest = 14 = 01110$. The message path is ($i = 2, d_i = 3, j = 1, d_j = 4$) 10111, 00111, 01111, 01110, i.e., the same as in the routing algorithm for complete hypercubes. Let $src = 14 = 01110$ and $dest = 23 = 10111$. Then, $i = 1, d_i = 4, j = 2, d_j = 3$, $®(dest, d_i) = 00111$, and the message path is 01110, 00110, 00111, 10111; i.e., the message is routed within complete hypercube H_1 to $®(dest, d_i) = 00111$ and then sent to $dest = 10111$.

The proof that the algorithm does not exhibit deadlock again uses a link dependency graph. Let an edge exist from $L(l', r')$ to $L(l'', r'')$. Then, in the standard routing algorithm, $l' > l''$ (in both **for** loops of the above algorithm); therefore the link number decreases in both **for** loops, and the only way a cycle can be formed is that the last line of procedure routing_inc produces an increase in link number. Let's say a message arrives from link $L(l, r)$ at node $®(dest, d_i)$ and is forwarded from node $®(dest, d_i)$ to node $dest$ on link d_i. However, in this case, the message terminates at $dest$, and there is no edge coming out of $L(d_i, ®(dest, d_i))$. Therefore $L(d_i, ®(dest, d_i))$ has outdegree 0 and cannot be part of any cycle. Thus, the routing procedure is deadlock free.

The broadcasting from a node src can be performed on an incomplete hypercube by the following procedure [21]. Let src belong to H_i. For $l = 1, 2, \ldots, i - 1$, src sends the message on link d_l to node $®(src, d_l)$. Nodes of all subcubes H_j for $j < i$ that received the message so far continue the broadcasting within the appropriate subcube, following the broadcasting algorithm for complete hypercubes, by sending the message to an appropriate neighbor. When the broadcasting within each subcube H_l for $l \leq i$ is complete, each node of H_i that has a neighbor in subcubes H_j for $j > i$ sends the message to that neighbor. This algorithm takes optimal $\lceil \log n \rceil$ communication steps because subcube H_j for $j < i$ receives the message with delay $j - 1$, while the dimension of H_j is $d_j \leq d - j$ ($d = \lceil \log n \rceil$). Therefore the total time for each node of H_j to get the message is $\leq j - 1 + d - j = d - 1 = \lceil \log n \rceil - 1$. The communication time for nodes of subcubes D_j for $j > i$ is $\leq d = \lceil \log n \rceil$ because it takes only one step for them to get the message after H_i has broadcast the message internally.

The algorithm does not exhibit deadlock because the message path between src and any other node is the same as for the routing algorithm (this can be easily verified).

We describe a routing procedure for GCIH from [21]. If both src and $dest$ belong to the left (right, respectively) GCIH, then the same procedure applied for routing in the incomplete hypercube can be used here (of course, d_i is replaced by either d'_i or d''_i). Otherwise, if $sym(src)$ exists, and then the message is first sent to this node and then routed to $dest$ as in an incomplete hypercube. If $sym(src)$ does not exist, then the message is first routed to $sym(dest)$ (which then exists as noted earlier) and then sent to $dest$ on link h. This can be coded as follows.

procedure routing_GCIH(src, dest, a, b);
```
if srcₕ = destₕ then routing_inc(src, dest) else {
  if sym(src) exists then {
    send message from src on link h to sym(src);
    route_inc(sym(src), dest)}
        else {
    route_inc(src, sym(dest));
    send message on link h from sym(dest) to dest}
```

The broadcasting algorithm may be given as follows [21]. If *sym(src)* exists, then the message is first sent to this node, and then two broadcasting procedures within the left and right GCIHs are performed in parallel using the same broadcasting procedure described earlier for incomplete hypercubes. Otherwise, the message is first broadcast within either left or right GCIH (the one that contains *src*) and then sent on link *h* from all nodes that received the message and have symmetric nodes.

For both algorithms, it can be easily verified that the message path between any two nodes *src* and *dest* is shortest possible $\rho(src, dest)$, communication time is optimal, and the computation time is asymptotically optimal $O(d)$.

18.3.3 Cube-connected cycles

In Ref. [13], an optimal routing algorithm for CCCs is presented. The algorithm routes a message from source to destination along the shortest possible path. The length of the path in the worst case is equal to the diameter $2d + \lfloor d/2 \rfloor - 2$ of the CCC. Although not very difficult to understand, it has tedious details. It shows first that the problem is equivalent to finding a shortest path between two points on a cycle consisting from zeros and ones (called *base cycle*) such that all ones are visited. These ones correspond to hypercube bits where source and destination differ. Next, Ref. [13] shows that there are eight types of possible shortest paths in the base cycle. These paths are the following: start from the source; go in a clockwise or counterclockwise direction; and make zero, one, or two turns (i.e., reversals in the direction).

We will describe here an optimal broadcasting algorithm for *d*-dimensional CCC on a one-port communication model. The algorithm is a modified version of the algorithm of Ref. [12]. Because of symmetry, we will simplify the exposition without loss of generality by assuming that the source is the node $00 \ldots 00$. There are two phases in the algorithm. In the first phase, the message is broadcast in CCC such that at least two (consecutive) processors in each cycle receive the message. It consists of *d* steps, each with two message passings: one on the *i*th link of underlying hypercube, and one on the cycle to increment *i*, $0 \le i \le d - 1$. In the second phase, these processors broadcasts the message, in opposite directions, to other processors in the same cycle. The procedure can be written as follows.

```
for i = 0 to d - 1 do {
    00...0xi-1...x0i → 00...01xi-1...x0i
    00...0xixi-1...x0i → 00...0xi...x0(i + 1 mod d) }
for i = 1 to ⌊d/2⌋ - 1 do in parallel {
    00...01xj-1...x0(j - i + 1 mod d) → 00...01xj-1...x0(j - i mod d)
    00...01xj-1...x0(j + i mod d) → 00...01xj-1...x0(j + i + 1 mod d)}
```

The first phase guaranties that nodes $x_{d-1} \ldots x_0 j$ and $x_{d-1} \ldots x_0 (j + 1 \bmod d)$ received the message. Here, *j* is the greatest index in sequence $x_{d-1}, \ldots, x_0$ such that $x_j = 1$. For $i = 0$ in the first loop, the node reduces to $00 \ldots 0i$ (similarly for *j*=0 in the second loop). In the second **for** loop, both message passings are done simultaneously, since they are performed by different nodes. All steps are performed simultaneously for all nodes of given form, e.g. for 2^i nodes of the form $00 \ldots 0x_{i-1} \ldots x_0 i$, and so on.

The correctness of the procedure is obvious. In the first phase, which takes 2*d* communication steps, clearly, two nodes on each cycle receive the message. In the second **for** loop, these two nodes broadcast the message along their cycles in the opposite directions, which requires $\lceil d/2 \rceil - 1$ communication steps. Thus the number of communication steps for the algorithm is $2d + \lceil d/2 \rceil - 1$. Since $\lceil d/2 \rceil - 1$ is $\lfloor d/2 \rfloor - 1$ for d even and $\lfloor d/2 \rfloor$ for d odd, the algorithm takes one more step than the diameter for d even and two more steps for d odd. It can be proved that the algorithm is optimal on one-port communication model; i.e., that one cannot find a broadcasting algorithm with fewer number of steps, and that the message does not arrive to a node from more than one node at any given step, which is also a desirable property of a broadcasting algorithm.

18.3.4 Stars, arrangement, and rotator graphs

A routing algorithm for stars has already been described in Section 18.1.6. We shall show here that this algorithm is not deadlock free. Let us fix all but the first three symbols in the permutation, i.e., consider routing in S_3 as part of routing in S_k. It is sufficient to consider the following paths of length 2: $123 \to 321 \to 231$, $321 \to 231 \to 132$, $231 \to 132 \to 312$, $132 \to 312 \to 213$, $312 \to 213 \to 123$, $213 \to 123 \to 321$. Here, all six nodes of S_3 send its message on link 3 (which is the correct position of the first symbol), and recipients are exactly these six nodes (cyclically shifted). They are supposed to forward the messages on link 2 to the same six nodes—thus each of them wants to send a message and at the same time is a recipient of another message, which is a deadlock situation. Let us derive the same conclusion using the link dependency graph. Denoting edge $a \to b$ simply as ab, the above six routing schedules correspond to the following edges in the link dependency graph: $123321 \to 321231$, $321231 \to 231132$, $231132 \to 132312$, $132312 \to 312213$, $312213 \to 213123$, $213123 \to 123321$. These six edges create a cycle.

We note here that the example used only one of two rules in the routing algorithm, namely to exchange the first symbol (if not in correct position) with the symbol in its correct position. It has a few implications. First, the algorithm cannot be made deadlock free by attempting to deterministically specify the second rule (i.e., to choose the symbol to be exchanged with the first one according to some test, if the first one is already in its correct position). Next, the example gave the only shortest paths (of length 2) between given pairs of nodes. Thus, any routing algorithm that preserves the shortest distances between any pair of vertices is not deadlock free.

Two asymptotically optimal one-port communication model broadcasting procedures for stars are known [16, 1]. From the general lower bound, it follows that any one-port broadcasting procedure on k-stars takes $\Omega(k \log k)$ steps. The algorithm of Ref. [1] uses a permutation switching network. The idea in Ref. [16] is as follows. Suppose that node $123 \ldots k \in X_{k-1}(k)$ wants to broadcast a message to all the nodes in X_k. In $O(\log k)$ time, it first sends this message to k nodes of the form $1*k, 2*k, \ldots, (k-1)*k$, where $a*b$ is an arbitrary permutation with the first symbol a and the last symbol b. This can be achieved by a standard doubling procedure, as follows. First, node $123 \ldots k$ sends the message to node $213 \ldots k$ using link 2. Next, node $123 \ldots k$ sends the message on link 3 while the node $213 \ldots k$ sends it on link 4. In the next step, four nodes that currently have the message will send it on links 5, 6, 7, and 8, respectively. This process continues, and in each step the number of processors that "learned" the message doubles. All these nodes are in $X_{k-1}(k)$. In one more step, all these $k-1$ nodes send this message, along dimension k, to nodes of the form $k*1 \in X_{k-1}(1)$, $k*2 \in X_{k-1}(2), \ldots, k*(k-1) \in X_{k-1}(k-1)$. Now every $X_{k-1}(i)$, $1 \le i \le k$, has at least one node containing the message, so the broadcasting can be done recursively in each X_{k-1}. Let $t(k)$ be the time needed to run this algorithm on X_k, then $t(k) = O(\log k) + t(k-1) = O(k \log k)$.

We will now describe the routing algorithm for the arrangement graphs from Ref. [6]. Due to symmetry, it suffices to route from arbitrary node p to the identity node $1, 2, \ldots, k$.

There are two kinds of cycles in p. Internal cycles are cycles with all values between 1 and k. External cycles are those which are interrupted by the appearance of an element greater than k (called also *foreign element*). Foreign elements are written at the end of cycle. For example, in $p = 6548371$ the cycle (1, 6, 7) is internal while (2, 5, 3, 4, 8) is external. Note that the first element in the external cycle is the index of an element of the permutation (and not the element itself).

The routing algorithm of Ref. [6] is an extension of the routing algorithm for the star network. Internal cycles are sorted by first exchanging one of its elements with a foreign element, then taking one by one elements from the cycle to their correct positions. An external cycle can be sorted by moving its elements, one by one, to their correct positions.

For example, routing from $p = 6548371$ to 1234567 is performed as follows:

- internal cycle (1, 6, 7): 6548371 → 6548379 → 1548379 → 1548369 → 1548367
- external cycle (2, 5, 3, 4, 8): 1548367 → 1248367 → 1248567 → 1238567 → 1234567

We will now consider the routing on the rotator graphs. The routing algorithm is a direct analog of the shortest path construction given in the proof that the diameter of the graphs is $k - 1$ for k-rotator graphs and m for (k, m)-rotator graphs ($m < k$) (see Section 18.1.6). For completeness, the algorithm for routing a message from node $(a_1, \ldots, a_m)$ to the origin $123 \ldots m$ is as follows [3]:

```
Complete the node representation to the full permutation
          (a₁, . . ., aₘ, aₘ₊₁, . . ., aₖ), aₘ₊₁ < . . . < aₖ
Let i be chosen such that aᵢ > aᵢ₊₁ < aᵢ₊₂ < . . . < aₖ
For s = 1 to i do {
    Rotate one place left the first j symbols of (a₁, a₂, . . ., aᵢ, bᵢ₊₁,
        bᵢ₊₂, . . ., bₘ, . . ., bₖ)
    where aᵢ > bᵢ₊₁ < bᵢ₊₂ . . . < k choosing j so that bⱼ < a₁ <bⱼ₊₁
        or j = i if a₁ < bᵢ₊₁.}
```

The algorithm is not deadlock free. For example, the routing paths 123 → 231 → 312, 231 → 312 → 123, and 312 → 123 → 213 create a deadlock.

The same constructive proof shows also that shortest path between any two vertices is unique. This means that, in the general all-port broadcasting algorithm, the message will arrive to a node for the first time from one node only. The node can close its links to prevent receiving the message again and forward the message through all its links to neighbors that wish to receive it.

The general lower bound for one-port broadcasting algorithm implies a lower bound of $\Omega(\log k!) = \Omega(k \log k)$ for broadcasting in k-rotator graphs. The bound is significantly bigger than the diameter of the network.

18.4   Conclusions

In this chapter, we described some of the interconnection networks. For each of them, we discussed some of its basic topological properties: degree diameter, recursive decomposition, and symmetry. The most important data communication routines, routing and broadcasting, are presented for some of these networks. The selection of the networks was made according to their popularity and satisfiability of some criteria. Although all characteristics are important, it seems that the network cost, defined as the product of the degree and diameter (measured with respect to the number of nodes), is the most important parameter of a network.

There are a number of open problems for each of the networks. To mention a few: What is the diameter of the pancake network? Is it possible to design optimal a one-port broadcasting algorithm for the rotator graphs? Can one design optimal sorting algorithm for the star graph? Designing new architectures remains an area of intensive investigation, given that there is no clear winner among existing ones. Next, cross products and the replacement of nodes by cycles will be used to obtain modifications and combinations of existing and newly created networks.

The following is an extensive but still incomplete reading list on interconnection networks.

18.5 References

1. Akers, S. B., Harel, D., and Krishnamurthy, B. 1987. The star graph: an attractive alternative to the n-cube. *Proc. Int. Conf. Parallel Processing,* 393–400.
2. Akers, S. B., and Krishnamurthy, B. 1989. A group theoretic model for symmetric interconnection networks. *IEEE Trans. Comp.,* C-38, 4, 555–566.
3. Corbett, P. F. 1992. Rotator graphs: An efficient topology for point-to-point multiprocessor networks. *IEEE Trans. Parallel and Distributed Systems,* 3, 5, 622–626.
4. de Bruijn N. G. 1946. A combinatorial problem. Koninklijke Netherlands: *Academie Van Wetenschappen, Proc. 49,* part 20, 758–764.
5. Dally, W. J., and Seitz, C. L. 1987. Deadlock-free message routing in multiprocessor interconnection networks. *IEEE Trans. Comp.* C-36, 5, 547–553.
6. Day, K., and Tripathi, A. 1992. Arrangement graphs: A class of generalized star graphs. *Information Processing Letters* 42, 235–241.
7. El-Amawy, A., and Latifi, S. 1991. Properties and performance of folded hypercubes. *IEEE Trans. Parallel and Distributed Systems,* 2, 1, 31–42.
8. Faber, V., Moore, J. W., and Chen W. Y. C.. 1993. Cycle prefix digraphs for symmetric interconnection networks. *Networks,* 23, 641–649.
9. Gates, W. H., and Papadimitriou C. H. 1979. Bounds for sorting by prefix reversal. *Discrete Mathematics,* 27, 47–57.
10. Graham, S. W., Seidel, S. R. 1993. The cost of broadcasting on star graphs and k-ary hypercubes. *IEEE Trans. Comp.,* 42, 6, 756–759.
11. Katseff, H. P. 1988. Incomplete hypercubes. *IEEE Trans. Comp.* C-37, No. 5, 604–608.
12. Liestman, A. L., and Peters J. G. 1988. Broadcast networks of bounded degree. *SIAM J. Discrete Math.,* 1, 531–540.
13. Meliksetian, D. S., and Chen, C. Y. R. 1993. Optimal routing algorithm and the diameter of the cube-connected cycles. *IEEE Trans. Parallel and Distributed Systems,* 4, 10, 1172–1178.
14. Merlin, P. M., and Schweitzer, P. J. 1980. Deadlock avoidance in store-and-forward networks. *IEEE Trans. Commun.,* COM-28, 345–354.
15. Menn, A., and Somani, A. K. 1990. An efficient sorting algorithm for the star graph interconnection network. *Proc. Int. Conf. Parallel Processing,* 1–8.
16. Mendia, V. E., and Sarkar, D. 1992. Optimal broadcasting on the star graph. *IEEE Trans. on Parallel and Distributed Systems,* 3, 4, 1992, 389–396.
17. Nath, D., Maheshwari, S. N., and Bhatt, P. C. P. 1983. Efficient VLSI networks for parallel processing based on orthogonal trees. *IEEE Trans. Comp.,* 32, 569–581.
18. Pritchard, D. J., and Nicole D. A. 1993. Cube connected Mobius ladders: An inherently deadlock free fixed degree network. *IEEE Trans. on Parallel and Distributed Systems,* 4, 1, 111–117.
19. Preparata, F. P., and Vuillemin, ?. 1981. The cube-connected cycles: A versatile network for parallel computation. *Commun. ACM,* 24, 5, 300–309.
20. Sullivan, H. and Bashkow, T. R. 1977. A large scale homogeneous, fully distributed parallel machine. *Proc. Fourth Symp. Comput. Architecture,* 105–117.
21. Stojmenović, I. 1993. Routing and broadcasting on incomplete and Gray code incomplete hypercubes. *Parallel Algorithms and Applications,* 1, 3, 167–177.
22. Tanenbaum, A. S. 1981. *Computer Networks.* Englewood Cliffs, N.J.: Prentice Hall.

23. Tzeng, N. F., and Wei, S. 1991. Enchanced hypercubes. *IEEE Trans. Comp.,* 40, 3, 284–294.
24. Uhr L. 1987. *Multicomputer Architectures for Artificial Intelligence.* New York: John Wiley & Sons.

18.6   Suggested Readings

25. Akl, S. G. 1985. *Parallel Sorting Algorithms.* Boston: Aademic Press.
26. Akl, S. G. 1989. *The Design and Analysis of Parallel Algorithms.* Englewood Cliffs, N.J.: Prentice Hall.
27. Akl, S. G., and Lyons K. A. 1993. *Parallel Computational Geometry.* Englewood Cliffs, N.J.: Prentice Hall.
28. Ajtai, M., Komlos J., and Szemeredi E. 1983. An $O(n \log n)$ sorting network. *Combinatorica,* 3, 1–19.
29. Arden, B. W., and Lee, H. 1982. Analysis of chordal ring networks. *IEEE Trans. Comp.,* 30, 4, 291–294.
30. Akl, S. G., and Qiu, K. 1993. A novel routing scheme on the star and pancake networks and its applications. *Parallel Computing,* 19, 95–101.
31. Akl, S. G., Qiu, Ke, and Stojmenovič, I. 1993. Fundamental algorithms for the star and pancake interconnection networks with applications to computational geometry. *Networks,* 23, 215–225.
32. Arden, B. W., and Lee, H. 1982. A regular network for multicomputer systems. *IEEE Trans. Comp.,* 31, 60–69.
33. Bhuyan, L. N., and Agrawal D.P. 1984. Generalized hypercube and hyperbus structures for a computer network. *IEEE Trans. Comp.,* 33, 4, 323–333.
34. Benes V. E. 1965. *Mathematical Theory of Connecting Networks and Telephone Traffic.* Boston: Academic Press.
35. Boals, A. J., Hashmi, J., Gupta, A. K., and Sherwani N. A. 1991. Incomplete hypercubes: Properties and recognition. *Proc. ICCI 1991,* Lecture Notes in Computer Science, 497.
36. Bhuyan L. N., Yang, Q., and Agrawal, D.P . 1989. Performance of multiprocessor interconnection networks. *IEEE Computer,* 22, 2, 25–37.
37. Chen, C., Agrawal, D. P., and Burke J. R. 1993. dBcube: a new class of hierarchical multiprocessor interconnection networks with area efficient layout. *IEEE Trans. Par. Distrib. Comp.,* 4, 12, 1332–1344.
38. Campbell, L., Carlsson, G., Dinneen, M. J., Faber, V., Fellows, M., Langston, M., Moore, J., Mullhaupt, A., Sexton, H. 1992. Small diameter symmetric networks from linear groups. *IEEE Trans. Comp.,* 41, 2, 218–220.
39. Chen, M. S., Shin, K. G., and Kandlur, D. D. 1990. Addressing, routing, and broadcasting in hexagonal mesh multiprocessors. *IEEE Trans. on Comp.,* 39, 1, 10–18.
40. Chen, H. L., and Tzeng, N. F. 1992. An effective approach to the enchancement of incomplete hypercube computers. *J. Par. Distr. Comput.,* 14, 163–174.
41. Das, S. K., and Mao, A. 1994. An interconnection network model and the Hamming cube networks. *Proc. 8th Int. Parallel Processing Symposium.*
42. Das, S. K. 1988. Wheel-augmented binary trees. *Intern. J. Comput. Math.,* 24, 199–211.
43. Das, S. K., and Banerjee, A. K. 1992. Hyper Petersen network, yet another hypercube like topology. *Proc. 4th Symp. Frontiers of Mass. Par. Comp.,* 270–277.
44. Das, S. K., and Deo, N. 1987. Rencontres graphs, a family of bipartite graphs. *Fibonacci Quart.,* 25, 250–262.
45. Das, S. K., Deo N., and Prasad S. 1991. Reverse binary digraphs and graphs. *J. Comb., Inf. & Syst. Sci.,* 16, 1, 107–128.
46. Das, S. K., Ghosh J., and Deo, N. 1992. Stirling networks, a versatile combinatorial topology for multiprocessor systems. *Discrete Applied Mathematics,* 37/38, 119–146.
47. Ghose K., and Desai, K. R. 1990. The design and evaluation of the hierarchical cube network. *Proc. Int. Conf. Parallel Proc.,* I, 355–362.
48. Despain, A. M., and Patterson, D. A. 1978. X-tree: A tree structured multi-processor computer architecture. *Proc. 5th Ann. Symp. Comput. Arch.,* 144–151.
49. Draper, R. N. 1991. An overview of supertoroidal networks. *Proc. ACM SPAA,* 95–102.
50. Day, K., and Tripathi, A. 1994. A comparative study of topological properties of hypercubes and star graphs, *IEEE Trans. Par. Distr. Sys.,* 5, 1, 1994, 31–38.
51. Duato J. 1992. Channel classes: a new concept for deadlock avoidance in wormhole networks. *Parallel processing Letters,* 2, 4, 347–354.
52. Efe, K. 1992. The crossed cube architecture for parallel computation. *IEEE Trans. Parallel and Distributed Systems,* 3, 5, 513–524.
53. El-Amawy, A., and Latifi S. 1990. Bridged hypercube networks. *J. of Parallel and Distributed Computing,* 10, 90–95.
54. Zheng, S. Q. 1994. Compressed tree machines. *IEEE Trans. Comp.,* 43, 2, 222–225.

55. Esfahanian A., Ni, L. M., and Sagan, B. E. 1991. The twisted N-cube with application to multiprocessing. *IEEE Trans. Comp.,* 40, 1, 1991, 88–93.

56. Feng, T. F. 1981. A survey on interconnection networks. *IEEE Comput. Mag.,* 12–27.

57. Ferreira, A. 1995. Parallel and communication algorithms on hypercube multiprocessors. Chapter 19 (this book).

58. Ferner, C. S., and Lee K. Y. 1992. Hyperbanyan networks: a new class of networks for distributed memory multiprocessors. *Proc. 4th Symp. Frontiers Mass. Par. Comp.,* 254–261.

59. Gessesse, G. A., and Chalasani, S. 1993. New degree four networks: Properties and performance. *Proc. Int. Parallel Processing Symposium,* 168–?.

60. Gowrisankaran, C. 1992. *Broadcasting on Recursively Decomposable Cayley Graphs.* Montreal: Depart. of Comp. Sci., Concordia Univ.

61. Ganesan, E., and Pradhan, D. K. 1993. The hyper-deBruijn networks: Scalable versatile architecture. *IEEE Trans. on Parallel and Distributed Systems,* 4, 9, 962–978.

62. Goodman, J. R., and Sequin, C. H. 1981. Hypertree: A multiprocessor interconnection topology. *IEEE Trans. Comp.,* 30, 12, 923–933.

63. Gaughan, P.T ., and Yalamanchili, S. 1993. Adaptive routing protocols for hypercube interconnection networks. *IEEE Computer,* 26, 2, 12–23.

64. Hwang, K., and Ghosh, J. 1987. Hypernet: A communication efficient architecture for constructing massively parallel computers. *IEEE Trans. Comp.,* 36, 12, 1450–1466.

65. Hilbers, P. A. J., Koopman, M. R. J., and Van der Sneapcheut, J. L. A. 1987. The twisted cube. *Proc. Parallel Archit. Alg. Europe,* 152–159.

66. Hsu, W. J. 1993. Fibonacci cubes—a new interconnection topology. *IEEE Trans. on Parallel and Distributed Systems,* 4, 1, 3–12.

67. Johnsson, S. L. 1987. Communication efficient basic linear algebra computations on hypercube architectures. *J. Parallel Distributed Computing,* 4, 2, 133–172.

68. Jwo, J. S., Lakshmivarahan, S. and Dhall, S. K. 1993. A new class of interconnection networks based on the alternating group. *Networks,* 23, 4, 315–326.

69. Yuan, S. M. 1991. Topological properties of supercube. *Information Processing Letters,* 37, 241–245.

70. Kranakis, E., Krizanc, D., and Ravi, S.S. 1993. *On multi-label linear interval routing schemes.* TR-220, SCS, Carleton University, Ottawa.

71. Kumar, V. P., and Raghavendra, C. S. 1987. Array processor with multiple broadcasting. *J. Parallel and Distributed Computing,* 2, 173–190.

72. Latifi, S., and Bagherzadeh, N. 1994. Incomplete star: An incrementally scalable network based on the star graph. *IEEE Trans. Par. Distr. Sys.,* 5, 1, 97–102.

73. Lawrie, D. H. 1976. Access and allignment of data in array processors. *IEEE Trans. Comp.,* C-24, 1145–1155.

74. Lakshmivarahan, S., and Dhall, S. K. 1990. *Analysis and Design of Parallel Algorithms.* New York: McGraw-Hill.

75. Lewis, T. G., and El-Rewini, H. 1992. *Introduction to Parallel Computing.* Englewood Cliffs, N.J.: Prentice Hall.

76. Leighton, F. T. 1992. *Introduction to Parallel Algorithms & Arcitectures: Array, Trees & Hypercubes.* San Mateo, Calif.: Morgan and Kaufmann.

77. Leighton, F. T. 1985. Tight bounds on the complexity of parallel sorting. *IEEE Trans. Comp.,* C-34, 4, 344–354.

78. Leiserson, C. E. 1985. Fat trees: Universal networks for hardware-efficient supercomputing. *IEEE Trans. Comp.,* 34, 10, 892–901.

79. Lee, K. Y., and Yoon, H. 1991. Indirect star-type networks for large multiprocessor systems. *IEEE Trans. Comp.,* 40, 11, 1277–1282.

80. Lakshmivarahan, Jwo S. K., and Dhall, S. K. 1993. Symmetry in interconnection networks based on Cayley graphs of permutation groups, a survey. *Parallel Computing,* 19, 4, 361–407.

81. Lee, K. Y., and Yoon, H. 1990. The B-network: A multistage interconnection network with backward links. *IEEE Trans. Comp.,* 39, 7, 966–969.

82. Martin, A. J. 1981. The torus, an exercise in constructing a processing surface. *Proc. 2nd Caltech Conf. VLSI,* 527–537.

83. Malluhi, Q. M., and Bayoumi, M. A. 1994. The hierarchical hypercube: A new interconnection topology for massively parallel systems. *IEEE Trans. Par. Distr. Sys.,* 5, 1, 17–30.

84. Misic, J. 1991. Multicomputer interconnection network based on a star graph. *Proc. IEEE HICSS,* Vol. 2, 373–381.

85. Nassimi, D., and Sahni, S. 1981. Data broadcasting in SIMD computers. *IEEE Trans. Comput.,* 30, 2, 101–106.

86. Ni, L. N., and McKinley, P. K. 1993. A survey of wormhole routing techniques in direct networks. *IEEE Computer,* 26, 2, 62–76.

87. Ohring, S. R., and Das, S. K. 1993. The folded Petersen network: a new communication efficient multiprocessor topology. *Proc. Int. Conf. Parall. Proc.,* I, 311–314.

88. Ohring, S. R., and Das, S. K. 1993. Folded Petersen Cube networks: New competitors for the hypercubes. *Proc. 5th Symp. Par. Distr. Proc.,* 582–589.

89. Padmanabhan, K. 1991. Design and analysis of even-sized binary shuffle-exchange networks for multiprocessors. *IEEE Trans. Parallel and Distributed Systems,* 2, 2, 385–397.

90. Pease, M. C. 1977. The indirect binary n-cube multiprocessor array. *IEEE Trans. Comput.,* C-26, 458–473.

91. Pease, M. C. 1968. An adaptation of the fast Fourier transform for parallel processing. *J. ACM* 15, 2, 252–264.

92. Parker, D. S., and Raghavendra, C. S. 1984. The gamma network. *IEEE Trans. Comput.,* 33, 367–373.

93. Qiu Ke. 1992. The star and pancake interconnection networks: Properties and algorithms. Ph.D. thesis, Queen's Univ., Kingston, Canada.

94. Qiu K., and Akl, S. G. 1994. Load balancing, selection and sorting on the star and pancake interconnection networks. *Parallel Algorithms and Applications,* 2, 2.

95. Reif, J.H., ed. 1993. *Synthesis of Parallel Algorithms.* San Mateo, Calif.: Morgan and Kaufmann.

96. Ravikumar, C. P., Kuchlous, A., and Manimaran, G. 1993. Incomplete star graph: An economical fault-tolerant interconnection network. *Int. Conf. on Parallel Processing,* I-83–90.

97. Scherson, I. D. 1991. Orhogonal graphs for the construction of a class of interconnection networks,. *IEEE Trans. Par. Distr. Sys.,* 2, 1, 3–19.

98. Seitz, C. L. et al. 1985. Wormhole chip project report.

99. Siegel, H. J. 1990. *Interconnection Networks for Large Scale Parallel Processing,* 2nd ed. New York: McGraw-Hill.

100. Singavi, N. K. 1993. The connection cubes: symmetric, low diameter interconnection networks with low node degree. *Proc. Int. Parallel Processing Symposium,* 260–265.

101. Sheu J. P., Liaw, W. H., and Chen T. S. 1993. A broadcasting algorithm in star graph interconnection networks. *Information Processing Letters,* 48, 237–241.

102. Saad, Y., and Schultz, M. H. 1988. Topological properties of hypercubes. *IEEE Trans. on Comput.,* 37, 7, 867–872.

103. Sen, A., Sengupta, A., and Bandyopadhyay, S. 1991. Generalized supercube: an incrementally expandable interconnection network. *J. Par. Distr. Comp.,* 13, 338–344.

104. Stojmenović, I. 1993. Job simulation techniques on incomplete and Gray code incomplete hypercubes. *Proc. IEEE Conf. on Computing and Information ICCI,* 225–229.

105. Stojmenović, I. 1993. *Radix networks: A new interconnection topology.* TR-93-24, Computer Science Department, University of Ottawa, Ottawa.

106. Stojmenović, I. 1995. Honeycomb networks, *Proc. Math. Found. Comp. Sci.*

107. Stohr E. 1991. Broadcasting in the Butterfly network. *Information Processing Letters,* 39, 1991, 41–43.

108. Stone H.S., Parallel processing with the perfect shuffle. *IEEE Trans. Comput.,* 1971, 153–161.

109. Sur, S., and Srimani, P. K. 1992. Incrementally extendible hypercube (IEH) graphs. *Proc. Phoenix Int. Conf. Comput. and Commun.,* 1–7.

110. Tien, J. Y., Ho, C. T., and Yang W. 1993. Broadcasting in incomplete hypercubes. *IEEE Trans. Comput.,* 42, 11, 1393–1403.

111. Tien, J. Y., and Yang, W. P. 1991. Hierarchical spanning trees and distributing on incomplete hypercubes. *Parallel Computing,* 17, 1343–1360.

112. Ullman, J. D. 1984. *Computational Aspects of VLSI.* Rockville, Md.: Computer Science Press.

113. Suaya, R., and Birtwistle, G., ed. 1990. *VLSI and Parallel Computation.* San Mateo, Calif.: Morgan and Kaufmann.

114. Wang, B. F., Chen, G. H., and Hsu C. C. 1991. Bitonic sort with an arbitrary number of keys. *Proc. 1991 Int. Conf. on Parallel Processing,* III 58–61.

115. Wu, C. L., and Feng, T. Y. 1980. On a class of multistage interconnection networks. *IEEE Trans. Comput.,* C-29, 694–702.

116. Youssef, A. S., and Narahari, B. 1990. The Banyan-hypercube networks. *IEEE Trans. Parallel and Distributed Systems,* 1, 2, 160–169.

19

Parallel and Communication Algorithms on Hypercube Multiprocessors

Afonso Ferreira

The most widely accepted model for the design of parallel algorithms is the *parallel random access machine,* known as *PRAM.* In this model, it is supposed that each processor (denoted PE, for processing element) has a local memory whose reduced capacity is compensated by the concurrent access to a large central memory, shared by all processors. In one cycle, a PE can gain access to any cell of the shared memory or execute a basic operation on its own local registers. Conflicts among processors arise when some or all of them try to access the same cell of the shared memory, and the different rules used to resolve such conflicts define the different models of PRAMs [23].

Unfortunately, implementing a shared memory in a parallel computer becomes unfeasible when the number of processors grows, due to the complexity of the hardware required to allow all processors to have access to every memory cell. Thus, in large-scale parallel systems, the memory is distributed among the processors. Interprocessor communication is then guaranteed by the exchange of messages through an interconnection network.

For the sake of simplicity, such distributed-memory parallel computers are modeled as graphs, where vertices represent processors and edges represent communication links. Two processors are said to be *neighbors* if they are connected by a direct communication link. A *path* of length λ is a sequence of neighbors $\{PE_{k_0}, PE_{k_1}, \ldots, PE_{k_\lambda}\}$, where k_i, $0 \leq i \leq \lambda$ is a valid processor label. The *distance* between any two processors is the length of the shortest path connecting them, and the *diameter* of a distributed-memory parallel computer is the maximum distance taken over all pairs of processors. Finally, the *degree* of a processor is the number of neighbors it has, and the *degree* of such a computer is the maximum of the degrees taken over all processors.

The time for communication between neighbors being proportional to the amount of information exchanged, the time for communication between any pair of processors is therefore proportional to their distance. Thus, it is of primary importance that the interconnection network have a small diameter. Graphs having the topology of *a hypercube* (to be defined in the next section) are ideally suited to modeling distributed-memory parallel computers, mainly because of the fact that the diameter of a hypercube is the logarithm of the number of vertices.

Several parallel computers based on such an architecture, and supporting different programming models (synchronous or asynchronous), were actually built and sold. These include the T-series by FPS, the Connection Machine CM-1 and CM-2 by Thinking Machines, the Intel iPSC, the NCube from NCube, and many others. As far as the choice of processor was concerned, it varied from company to company. Such computers could present, for instance, either 8 transputers on a 3-dimensional hypercube or 65,536 bit serial processors sold as a 16-dimensional hypercube.

With so many computers based on the hypercube architecture, models and algorithms were proposed to allow applications to be solved efficiently in parallel. For instance, it is interesting to note that restrictions on the access to specific memory positions yield a shared memory model for hypercube architectures. This can be seen as follows. Let $G(V, E)$ be a hypercube with N vertices. Then, we can imagine that the memory positions $m_1, m_2, \ldots, m_n$ are such that PE_i can access only memory positions m_j, where $j = i$, or the vertices v_j and v_i of V are neighbors in G [10].

In the remainder of this chapter, we will try to answer most of the questions that a nonspecialist reader may have. However, because of length considerations, some areas of "hypercube computing" will not be covered here. We will focus on the topological aspects of the hypercube, covered in Section 1, followed by some topics in communications and routing in Section 2. Sections 3 and 4 describe the main tools and some algorithms for the fine-grain hypercube (to be defined in Section 3). We close the chapter with some open problems and directions for further research.

19.1   Topological Aspects

For the sake of completeness, we define the notation to be used in this chapter. An interconnection network with the topology of a d-dimensional hypercube, denoted $H(d)$, is composed of $N = 2^d$ processors, labeled from 0 to $N - 1$, and $dN/2$ communication links. Let $(i)_2$ be the binary string representing i. Then, the neighbors of PE_i are all PE_j such that $(i)_2$ and $(j)_2$ differ in exactly one bit position – say k, $0 \leq k < d$. In this case, we say that PE_i and PE_j are neighbors along dimension k. For instance, PE_0 and PE_4 are neighbors along dimension 2 in any $d > 2$-dimensional hypercube. The processors are then the vertices of a hypercube of dimension d, each connected to d neighbors (see Fig. 19.1). It is not difficult to see that the maximum distance in a hypercube is given by those pairs of processors whose binary string differ in all d positions, implying that its diameter is $d = \log N$.[*] In the following, the terms vertices, nodes, and processors all refer to a computing element.

19.1.1   Graph properties

A *bipartition* of $H(d)$ is a partition of V into two sets, say A and B. Let $[A: B]$ denote the set of all edges of the hypercube with one endpoint in A and another in B. A *bisection* of $H(d)$ is defined as a bipartition where $|A| = |B| = N/2$. The *bisection width* of $H(d)$ is then the minimum cardinality of $[A: B]$ taken over all possible bisections of $H(d)$. Note that this is an important parameter for measuring the quality of an interconnection network, since any communication between the processors in A and B must traverse edges in $[A: B]$. Hence, the higher the bisection width of an interconnection network, the smaller the communica-

---

[*]All logarithms in this chapter are to the base 2, unless otherwise specified.

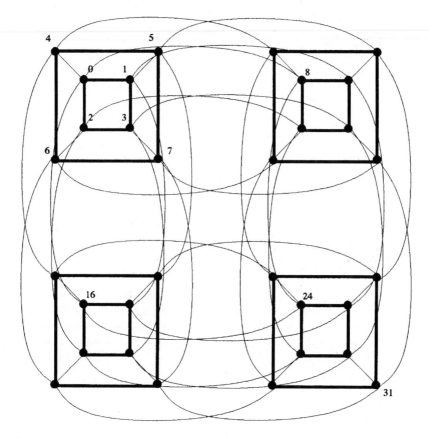

Figure 19.1 An $H(5)$ as four copies of $H(3)$ (in bold)

tion contention. Therefore, apart from having a small diameter, the hypercube is also a good choice for the interconnection network of distributed memory parallel computers because of its high bisection width, which equals $N/2$.

Furthermore, a hypercube $H(d)$ can be decomposed in d different ways into two copies of $H(d-1)$, with $N/2$ edges connecting them. To find one such decomposition, it suffices to fix any one bit position in the processors' addresses, say position k, $1 < k < d-1$. Then, the two copies of $H(d-1)$ are composed of the vertices $i_{d-1}\dots i_{k+1}0i_{k-1}\dots i_0$ and $i_{d-1}\dots i_{k+1}1i_{k-1}\dots i_0$, respectively.

Such a decomposition technique can be extended in such a way that a d-dimensional hypercube can be decomposed as 2^b copies of $H(d-b)$, $0 \le b \le d$. This is obtained by fixing an arbitrary number b of positions in the binary string of the processors' addresses, since the remaining non-fixed positions represent a $(d-b)$-dimensional hypercube. In Fig. 19.1, an $H(5)$ is depicted as four copies of $H(3)$. In that case, bit positions 3 and 4 are considered as fixed in each of the copies.

It is interesting to notice that one can use these decompositions to implement divide-and-conquer algorithms [24] in hypercubes. Moreover, they yield a recursive construction of the hypercube as follows. A 0-dimensional hypercube is composed of one node only. To build $H(1)$, we take two copies of $H(0)$ and join the two nodes by an edge. In general, we can build an $H(d)$ by taking two copies of $H(d-1)$ and connecting the isomorphic nodes by an edge (see Fig. 19.2).

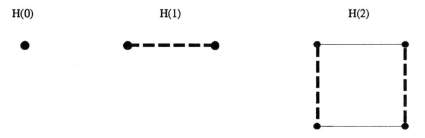

Figure 19.2 Recursive construction of $H(1)$ and $H(2)$

One further advantage of the hypercube, making it easy to use, is the fact that it is vertex and edge symmetric. This means that for every pair of edges (v_i, v_j) and (u_r, u_s) of $H(d)$, one can find an automorphism* f of $H(d)$ such that $((f(v_i), f(v_j)) = (u_r, u_s))$ (see Fig. 19.3) [25, 33]. Very informally, we could say that, as far as the neighbors are concerned, the hypercube looks the same from every node.

19.1.2 Embedding

Another attractive characteristic of this architecture is that it has many other architectures as its subgraphs. This implies that algorithms designed for several architectures can be implemented on hypercubes with no increase in the elapsed time. For instance, if $d = p + q$, then the two-dimensional torus with $2^p \times 2^q$ vertices is a subgraph of the hypercube.† Therefore, any algorithm running in such torus runs in the hypercube in exactly the same number of steps.

When a given topology $G(V, E)$ is not a subgraph of the hypercube, the goal is to find an embedding‡ that minimizes some parameters and that can be informally described as follows.

- *Dilation* is the maximum distance in the hypercube between any two neighbors in G.
- *Congestion* is the maximum number of different paths, representing edges of G, through a same edge of the hypercube.

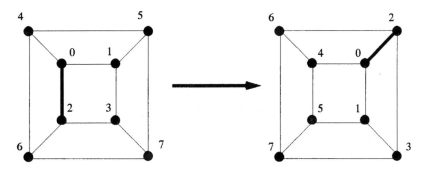

Figure 19.3 Example of symmetry in the hypercube

---

*An automorphism of $H(d)$ is a bijection from V onto V that induces a bijection from E onto E.
†A $2^p \times 2^q$ torus is a $2^p \times 2^q$ mesh with wraparound connections [32].
‡An embedding of a graph $G_1(V_1, E_1)$ into a graph $G_2(V_2, E_2)$, $|V_2| \geq |V_1|$ is an injective mapping $f : V_1 \rightarrow V_2$, along with a mapping g : $E_1 \rightarrow P$, where P is the set of all simple paths of G_2, such that $g((u, v))$ is a simple path that starts in $f(u)$ and ends in $f(v)$.

- *Expansion* is the ratio $N/|V|$. If hypercube is the smallest one such that $N \geq |V|$, it is called *optimal hypercube* with regard to G.

The general case of deciding whether a given graph is a subgraph of the hypercube is NP-complete [34]. Nevertheless, some results exist concerning specific topologies.

Cycles. Proving that a cycle of N vertices is a subgraph of $H(d)$ is done with the help of Gray codes. A *Gray code of order d*, G_d, is an ordered sequence of the 2^d binary strings of length d, such that two consecutive strings differ in exactly one position. It is clear that this definition implies that hypercube nodes represented by two consecutive strings in G_d are therefore neighbors. Moreover, the first and the last strings in the code also differ in only one position, yielding the cycle. The question is then how to build such a sequence.

Definition 1. A reflected Gray code of order d, G_d, is recursively defined as follows. Let $G_1 = \{0, 1\}$ and G_{d-1} be a Gray code of order $d - 1$, composed of the strings $\{\alpha_1, \alpha_2, \ldots, \alpha_{2^{d-1}}\}$. Then, $G_d = \{0\alpha_1, 0\alpha_2, \ldots, 0\alpha_{2^{d-1}}, 1\alpha_{2^{d-1}}, \ldots, 1\alpha_2, 1\alpha_1\}$.

As an important consequence, this constructively proves that the hypercube is *Hamiltonian** (see Fig. 19.4). As a matter of fact, all cycles of even length are its subgraphs.

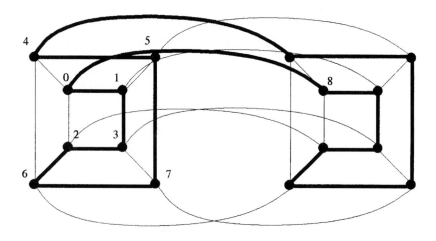

Corresponding reflected Gray code:

0000	0110	1100	1010
0001	0111	1101	1011
0011	0101	1111	1001
0010	0100	1110	1000

Figure 19.4 Example of symmetry in the hypercube

---

*A graph is said to be Hamiltonian if it has a Hamiltonian cycle; i.e., there is a path containing all vertices exactly once, such that the last vertex in the path is a neighbor of the first.

Binary trees. Another well studied embedding problem is the one of complete binary trees, noted $T(h)$, with $2^{h+1} - 1$ nodes. It is known how to embed them in $H(h + 2)$ with dilation 1, although it is impossible to find such an embedding in $H(h + 1)$, its optimal hypercube [25, 34]. On the other hand, a slightly different tree, namely the *double rooted complete binary tree DRT(h)*, with 2^{h+1} nodes, is a subgraph of $H(h + 1)$ (see Fig. 19.5).

Several more results can be found in the literature related to the embedding of different topologies in the hypercube, such as meshes, meshes of trees, pyramids, and others. For a thorough discussion on embedding in general, see Refs. [25, 34].

19.2 Communication Issues

In parallel applications, a large amount of processing time is spent routing data among the processors. Therefore, routing algorithms, techniques, and models are of primary importance for an efficient use of parallel computers. To design fast routing algorithms, several machine-related features, as well as the specific protocol to be used or even how the destination addresses are set, have to be taken into account.

Since the early days of parallel computing, packet routing has been implemented to support interprocessor communication. Suppose that a processor sends a message to a non-neighbor. Under the protocol known as *store-and-forward*, every processor in the path taken by the message first stores the message then forwards it to the next processor in the

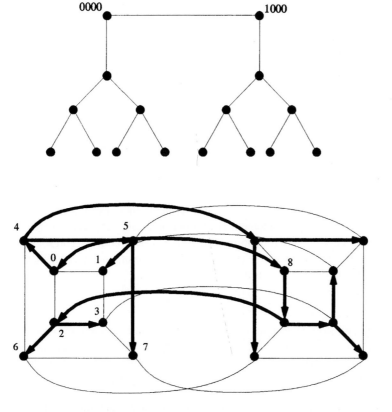

Figure 19.5 A *DRT*(3) and its embedding in *H*(4)

path. Practical experiments showed that the cost of a sending-to-neighbor operation could be modeled as a start-up time β (establishing the links, for instance) plus the time spent by the actual sending of the messages, which depends on the size of the message (L) and on the inverse of the link's bandwidth (τ). So, the time for such an operation could be expressed as $t = \beta + L\tau$, known as the *linear* model. However, in the case of short messages and large start-up times, it is important to understand the behavior of communication algorithms with regard to the number of steps taken only. Then, it could be assumed that a sending-to-neighbor operation takes constant time, regardless of the size of the message to be sent. This is the *constant* model [20].

Another important parameter for routing problems is the concurrency of communication links, i.e., the number of links that can be used concurrently by each processor [18]. Algorithms that, in a single step, allow communication with only one neighbor are said to respect the 1-*port* model. If, on the other hand, communication with all the d neighbors is allowed, then they are said to respect the d-*port* model.

When the routing pattern is known in advance, we call it *off-line* routing. This happens to be the case in many scientific applications. Also, for some applications for which this is not true, it could be interesting to precompute the communication pattern and store it in the processors (for instance, when the same communication pattern is to be repeated over and over during the execution of the algorithm). On the other hand, it could be crucial to be able to route *on-line* patterns also, i.e., patterns that are run-time dependent and thus known on the fly.

19.2.1   Off-line routing

Of particular interest in off-line routing are *global* communication procedures such as *one-to-all* or *all-to-all,* and *permutation* routing, where each processor has a piece of information to transmit to one processor, and all destinations form a permutation of the processors' addresses. Some examples are as follows.

- *Broadcasting.* One processor transmits a piece of information to all processors; also known as *one-to-all.*
- *Gossiping.* Every processor transmits a piece of information to all processors; also known as *all-to-all.*
- *Scattering.* One processor transmits a different piece of information to every other processor.
- *Gathering.* Every processor transmits a piece of information to the same processor.
- *BPC permutations.* Bit permute complement (BPC) permutations are those where the destination for each processor is a permutation of its own hit-string address, and some of the resulting bits may be complemented. Examples of BPC permutations include matrix transpose, shuffled row major, perfect shuffle, and others.

19.2.1.1 Constant model. In the constant model, the use of d-ports allows all permutations to be routed off-line within ($2d - 1$) steps. In the one-port, the communication patterns above can be routed optimally, in log N neighbor-to-neighbor communication steps. In the following, we show the interesting example of routing BPC permutations.

Definition 2. A BPC permutation on $H(d)$ is specified by a permutation vector $p = [p_{d-1}, p_{d-2}, \ldots, p_0]$, where $-d \leq p_i \leq d$, such that $[|p_{d-1}|, |p_{d-2}|, \ldots, |p_0|]$ is a permutation of

$[0, 1, \ldots, d-1]$. The destination processor PE_j of data stored in PE_i, $(i)_2 = i_{d-1}, i_{d-2}, \ldots, i_0$, is given by $j_{|p_k|} = i_k$, if $p_k \geq 0$, or $\overline{i_k}$ otherwise. We remark that -0 is considered to be less than 0.

The routing corrects the dimensions from right to left in the destination address, i.e., from j_0 to j_{d-1}. If $p_k = k$, then the kth bit of the address is already correct. If $p_k = \overline{k}$, then this register has its kth dimension corrected. If $|p_k| \neq k$, then a non-trivial cycle starts at bit k. The next position in the cycle is given by $c = |p_k|$, and the position following this one is given by $e = |p_k|$. Since this is a BPC permutation, $c \neq e$. Thus, if $e = k$, then c is the end of the cycle, otherwise e is the next position in the cycle. Repeating this reasoning, the complete cycle starting at position k can be identified.

Therefore, once all cycles have been identified, routing a BPC permutation is implemented so that cycles are routed one after another, with data changing dimension when necessary. When handling a cycle $c_0, c_1, \ldots, c_l$, routing is done, if necessary, with regard to dimension c_1 first, then c_2, eventually c_l, and finally c_0. Notice that we need queues of size two to store data that are to be routed along the next routing dimension [33].

19.2.1.2 Linear model. Under the linear model, the idea is to use pipelining to hide the diameter of the hypercube. For instance, the technique for broadcasting is based on the existence and construction of height h spanning trees[*] of the hypercube (see Fig. 19.6).

Assume that a message of length L is to be broadcast from PE_0. Before routing starts, the message is divided into packets of size s, whose optimal value can be analytically computed as $\sqrt{L\beta/(h-1)\tau}$. The basic idea is then to send the packets through the spanning tree in a pipelined way. Hence, the key point is to find small height spanning trees, because the time for broadcasting with this technique is then shown to be $(\sqrt{L\tau} + \sqrt{(h-1)\beta})^2$.

Note that this result holds for the d-port model, since processors send and receive messages onto two links concurrently. However, only two ports out of d are being used. Therefore, broadcasting can be improved if one considers edges as two arcs, one in each direction, and finds d arc-disjoint spanning trees in $H(d)$. By pipelining packets through different arc-disjoint spanning trees, the following can be shown.

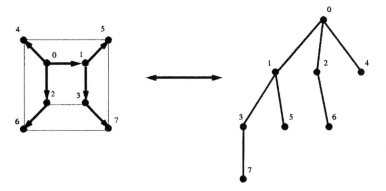

Figure 19.6 A spanning binomial tree of $H(3)$, rooted in node 0

---

[*]A *spanning tree* of a graph G is a tree that contains all the vertices of G and is a subgraph of G.

- Under the one-port model, broadcasting completes in $(\sqrt{L\tau} + \sqrt{d\beta})^2$ steps [21].
- Under the d-port model, the number of steps to broadcast completion [36] is

$$\left(\sqrt{\frac{L\tau}{d}} + \sqrt{(d-1)\,\beta} \right)^2$$

The spanning tree used is the *spanning binomial tree* (see Fig. 19.6), and the arc-disjoint spanning tree is the *rotated spanning binomial tree* (see Refs. [18, 19, 34] and references therein for more details). Recall that, because of the vertex symmetry of the hypercube discussed in Section 19.1.1, one needs to find only one spanning tree, rooted in any node, say i. If the broadcast is to be executed from another node, say j, it suffices to find an automorphism f that maps node i into node j. One spanning tree rooted in node j is then the image by f of the spanning tree rooted in node i.

19.2.2 On-line permutation routing

Packet routing algorithms for on-line permutation routing on the hypercube may be classified into *oblivious* or *adaptive.*

For *oblivious algorithms,* the routing path of a packet is uniquely determined from its source and final destination. Both the commonly used greedy algorithm, which uniformly corrects hypercube dimensions in either increasing or decreasing order, and also a variant that corrects dimensions in random order, belong to this class [7, 37]. Although these strategies are simple, their worst-case time complexity is $\Omega(\sqrt{N})$ [39]. An optimal oblivious algorithm, based on many-to-one and one-to-many routing in subhypercubes, achieves a worst-case time delay of $\Theta(\sqrt{N}/\log N)$ [22].

Even though optimal deterministic oblivious routing schemes on the N-node binary hypercube could be constructed for $N \leq 64$ [19], of major importance is the *randomized permutation routing* algorithm, first proposed by Valiant and Brebner [39] and later improved by Valiant [38]. For this algorithm, the probability that all packets have been routed correctly within $c \log N$ time is proved greater than $(1 - e^{-\delta \log N})$, where δ is a constant depending only on c.

The algorithm consists of a *randomization phase,* where packets are deterministically sent to randomly selected nodes through the network, followed by a *deterministic phase,* where packets follow (deterministically again) a shortest-path route to their final destination. Notice that, at the end of the first phase, packets are distributed randomly. The role of randomization is to reduce the gap between the average and the worst-case time complexities by avoiding worst-case congestions. In the context of generalized hypercubes,[*] this result has been slightly improved in [16].

Adaptive routing algorithms can use much more information about the messages and the system congestion; the destination of many packets, for instance. They are usually disguised as sorting algorithms. Therefore, the best known asymptotic result for deterministic routing is the same as for sorting (see Section 19.3.3).

19.2.3 Wormhole routing

The *wormhole* protocol was conceived to mask distances in the network. In fact, wormhole routing is used in most of the distributed-memory parallel computers built today. It is

---

[*]A b-ary k-dimensional generalized hypercube has $N = b^k$ nodes. A node x can be represented as $x = x_0 x_1 \ldots x_{k-1}$, where $x_i \in \{0,1,\ldots b-1\}$ for $0 \leq i \leq k-1$. Two nodes $x = x_0 x_1 \ldots x_{k-1}$ and $y = y_0 y_1 \ldots y_{k-1}$ are connected by an edge if and only if there is exactly one j, $0 \leq j \leq k-1$ such that $x_j \neq y_j$.

implemented through the use of special-purpose communication devices, named *routers,* incorporated with the computing elements. Routers perform only routing operations. Each message is composed of a *header,* coding the destination address; the body of the message, divided into a number of atomic units called *flits;* and a *tail,* indicating the end of the message.

The message is routed through the interconnection network so that intermediate computing elements in the message path do not have access to it. Only their associated routers deal with the message, treating it in a pipeline-like manner, flit after flit. Hence the name *wormhole* routing, since the delivering of messages reminds us of worms crawling in the network. The main feature of this protocol is that as soon as the tail of a message leaves a non-final router, the router is freed to perform other communication tasks.

For the one-port model, most information dissemination problems have log N as a lower bound on the number of steps required to complete. Thus, under such a model, wormhole routing is of little use in the hypercube. On the other hand, under the d-port model, the lower bound for this kind of problems is usually $\lfloor \log_{d+1} N \rfloor = \Omega (\log N / \log \log N)$. For optimal broadcasting, for instance, a tiling of the hypercube, such as the tiling of the two-dimensional torus proposed in Ref. [32], must be found.

19.3   Useful Algorithmic Tools

This section and the next focus on synchronous algorithms for *fine-grain* hypercubes. In this model, there are at least as many processors as the problem input size, and each processor has some local memory, of constant size, organized in $O(d)$-bit words. At each time step, a processor can simultaneously send a word of data to one of its neighbors, receive a word of data from one of its neighbors, and perform a local operation on word-sized operands.

Algorithms where, at each time step, only the edges associated with a single dimension are used, and consecutive dimensions are used on consecutive steps, are called *normal* algorithms (see Ref. [25], Section 3.1.4).

From the architectural point of view, one of the main drawbacks of the hypercube stems from its logarithmic degree, due to the large number of wires required to interconnect the processors in big systems. However, this problem can be solved through the use of bounded-degree hypercube-related networks, better known (along with hypercubes) as *hypercubic* networks. Any normal algorithm designed for the hypercube can be simulated in any of these networks with only a constant slowdown. The most studied members of this class, besides the hypercube itself, are the *butterfly,* the *shuffle-exchange,* and the *de Bruijn* graphs (see Ref. [25] for details on such interconnection networks). All the algorithms described henceforth are normal.

The procedures described next are well known and represent powerful tools for the design of hypercubic algorithms, although their most important feature may be that they can be combined to implement, in time proportional to sorting, *random access concurrent read* on hypercubes, as well as *random access write,* and even some forms of *concurrent* write. This implies that PRAM algorithms can be simulated in hypercubes with an overhead equal to the time complexity of the best sorting algorithm. We recall that the diameter of the hypercube yields a trivial $\Omega(\log N)$ time lower bound for solving any global sensitive problem.

For the sake of clarity, we will show the detailed implementation of some of these procedures. They all use the same strategy, correcting the bits in consecutive dimensions either from right to left or from left to right. More details can be found in Refs. [30, 33].

19.3.1 Semigroup operations

A semigroup computation applies an associative operation to all selected data items. The procedures that follow are computed in optimal time, i.e., they complete in $O(\log N)$ time.

Procedure Count.

Input: One flag per processor. **Output**: For all PE_i, the total number of processors PE_j whose flag was up.

Let i_b denote the bth digit, from right to left, in $(i)_2$. Let each processor have variables T( ) and ST( ), denoting total and subtotal, respectively, both of them initialized with the value 0. Counting is as follows.

```
Each PE_i :
  if flag is up then T(i):= 1;
  for b := 0 to d - 1 do
     exchange T(i) along dimension b and store it in ST(i);
     T(i) := ST(i) + T(i);
```

Procedure *Count* above can be extended to compute a special case of *parallel prefix*, known as *rank*, described in the following.

Procedure Rank.

Input: One flag per processor. **Output**: For each PE_i whose flag was up, the number of processors PE_j, $j < i$, such that PE_j's flag was up.

Let i_b denote the bth digit, from right to left, in $(i)_2$. Let each processor have variables R( ), T( ), ST( ), denoting rank, total, and subtotal, respectively, all of them initialized with the value 0. The *Rank* algorithm is as follows.

```
Each PE_i:
  if flag is up then T(i):= R(i):= 1;
  for b := 0 to d - 1 do
     exchange T(i) along dimension b and store it in ST(i);
     if i_b = 1 then R(i) := R(i) + ST(i); T(i) := ST(i) + T(i);
  if flag was down then R(i) := ∞.
```

It is not very difficult to modify the above procedures so that they can compute a *segmented parallel prefix,* defined as follows. Let PE_{b_b} and PE_{b_e} be the processors at the beginning and the end of a block respectively. (A block is a set of contiguous processors that have the same value stored in *block(i)*.) We leave as an exercise to the interested reader the description of a procedure *SegmentedPrefix* that takes two registers per PE_i, *block(i)* and *r(i)*, as input, and that completes $\oplus_{j=b_b}^{b_e} r(j)$, where $\oplus$ is an associative operation.

19.3.2 Data-movement operations

The procedures that follow can be seen as special cases of routing and were studied in Ref. [30]. Again, they are all time-optimal.

Procedure Concentrate.

Input: One register per processor, associated to a flag. **Output:** All registers whose flag was up are concentrated at the beginning of the hypercube, one per processor. The initial relative order remains unchanged.

Let i_b denote the bth digit, from right to left, in $(i)_2$. Let each processor have a variable G( ) to be concentrated. The record R( ) stores the value given by the rank procedure described previously.

Rank.

```
Each PE_i:
   for b := 0 to d - 1 do
      if R(i) ≠ ∞ and (R(i))_b ≠ i_b then
         exchange (G(i),R(i)) along dimension b.
```

Procedure Route.

Input: One register per processor, associated to a destination dest(i) and to a flag. The destinations should be monotone, i.e., for two processors PE_i and PE_j, it holds that i < j ⟺ dest(i) < dest(j). **Output**: All registers whose flag was up are routed to the processor given by dest( ).

Let i_b denote the bth digit, from right to left, in $(i)_2$. Let each processor have a variable G( ) to be concentrated. The record R( ) stores the value given by the rank procedure described previously.

Concentrate.

```
Each PE_i:
   if R(i) = ∞ then
      dest(i) := ∞
   for b:= d - 1 to 0 do
      if dest(i) ≠ ∞ and (dest(i))_b + i_b then
         exchange (G(i),dest(i)) along dimension b.
```

With the same input as for the *Route* procedure, we propose that the reader describe a procedure *RouteAndCopy* that, for all registers whose flag is up, routes a copy of the register of PE_i to processors $PE_{dest(i-1)+1}, \ldots, PE_{dest(i)}$.

19.3.3 Sorting

Below, we discuss the main deterministic sorting algorithms for fine-grain hypercubes. We describe the *bitonic* merge and sort, giving a recursive description of the latter, since this is the usual way algorithms are described for hypercubes. Computing the time complexity is then a matter of solving a recurrence relation. In the case of the bitonic sort, the relation to solve is $T(N) := T(N/2) + O(\log N)$, that equals $O(\log^2 N)$.

Procedure BitonicMerge.

Input: Two sorted lists, one in ascending order, the other in descending order. **Output**: The two lists merged into one sorted list. **Time**: $O(\log N)$.

Let i_b denote the bth digit, from right to left, in $(i)_2$. Let each processor have a variable M( ), where its corresponding element in the merged list is to be stored, and an auxiliary variable L′( ). The original lists are stored in the record L( ).

```
Each PE_i:
for b := d - 1 to 0 do
   exchange L(i) along dimension b and store it in L'(i).
```

```
if (i)_b = 0 then
   M(i) := min{L(i), L'(i)}
else M(i) := max{L(i), L'(i)}
```

Procedure BitonicSort.

Input: An unsorted list. **Output**: The sorted list. **Time**: $O(\log^2 N)$.
Do in parallel
```
   BitonicSort{first half of the list in ascending order}
   BitonicSort{second half of the list in descending order}
BitonicMerge{the two lists}
```

For more details on both algorithms, such as proof of correctness and time complexity, see Refs. [1] and [24].

In the case where there are more computing elements than data elements to be sorted, the *sparse enumeration* sort of Ref. [31] is able to sort N elements in a hypercube with $N^{1+1/k}$ processors, $1 \le k \le \log N$, in time $O(k \log N)$.

It is worth mentioning that the bitonic sort remained the best sorting algorithm for hypercubes for more than 20 years. Moreover, it is very simple to describe and implement, with a fairly small constant (approximately 1/2) hidden in its time complexity of $O(\log^2 N)$. Nevertheless, the asymptotic time complexity of the bitonic sort was finally improved by the *share* sort algorithm proposed in Ref. [11], that has another version described in Ref. [25]. To the best of our knowledge, the share sort is the asymptotically fastest existing sorting algorithm for sorting N data elements with N hypercube processors, in $O(\log N(\log \log N)^2)$ time.[*] We note that the share sort has a so-called *nonuniform* version that runs in $O(\log N \log \log N)$ time, where nonuniformity means that a routing table for each processor has to be computed beforehand, once and for all runs of the share sort on a given machine. A routing phase of the algorithm is then accelerated by table lookups. The problem is the exponential complexity of the algorithm building such tables. This is the reason why we preferred to list here the time complexity of the uniform version.

Finally, the sparse version of share sort was studied in Ref. [6]. It is therefore possible to sort N elements in a hypercube with $N^{1+1/k}$ processors, $2 \le k \le \log N$. in time $O(\log N \log^2 k)$.

19.3.4   Random access read

Suppose that each PE_i stores in a variable $A(i)$ the address from where data have to be read, and that processors that are not reading in this step have $A(\ ) = \infty$. Let $ID(i) = i$ be used to keep track of the reading processors. Suppose further that each processor shall work with tuples containing four variables, namely, $A(\ )$, $ID(\ )$, $D(\ )$, and $S(\ )$. The algorithm implementing random access read is as follows [30].

1. *Sort* tuples by $A(\ )$.
    All valid addresses are now concentrated at the beginning of the hypercube. Let $S(\ )$ take the value of the current address.
2. *Identify EndOfBlock.*
    Since we are implementing concurrent read, several processors may have the same value in $A(\ )$. Then we avoid routing conflicts by selecting end-of-block processors to be in charge of getting the data and distributing it to the others in its block. (Recall

---

[*]Unfortunately, the constant hidden in the *big Oh* notation is huge.

that a block is a set of contiguous processors that have the same value stored in *block(i)*.) This procedure is as follows.

Procedure IdentifyEndOfBlock.
```
Input: One register per PEᵢ, called block(i). Output: The last
  processor of each block is identified.
All PEᵢ,i > 0:
  Route{block(i) into aux(i - 1)};
Each PEᵢ:
if block(i) ≠ aux(i) then flag := up
```

3. *Route* tuples of end-of-block processors by $A(\ )$.
   The required data are then copied into $D(\ )$.
4. RouteAndCopy tuples by $S(\ )$.
   Every processor in each block receives a copy of the information required.
   The data now have to be sent to the original processors requiring them.
5. *Sort* tuples by $ID(\ )$.
   In order to make the destinations monotone.
6. *Route* tuples by $ID(\ )$.
   The reading step finishes.

It is easy to see that the time complexity of a random access read is dominated by the two calls to sorting algorithms in steps 1 and 6, since all the routing procedures are optimal because of the monotonicity of the addresses. Then, PRAM algorithms can be implemented in the hypercube with a time overhead proportional to the time complexity of sorting in the hypercube.

19.4   Solving Problems

In this section, we review some algorithms used for solving different problems in computational geometry because they are good representatives of the most difficult problems arising in the design of hypercube algorithms.

19.4.1   Classic problems

When it comes to discussing optimality of hypercube algorithms, one has to be somewhat careful. We can say that an algorithm is *time-optimal* if its number of steps is $O(\log N)$. However, this definition is not complete, since sparseness is not taken into account (note that the sparse enumeration sort, from Section 19.3, is time-optimal provided N^2 processors are available). Then, we define *dense* algorithms as those that solve problems of size N with N processors, as opposed to *sparse* algorithms, which lose more processors than the size of the input. For instance, in Section 19.3, we saw several time-optimal dense algorithms.

The classic problems we are going to address in this section have serial time complexity less or equal to that of sorting. For more algorithms see, for example, Refs. [2] and [28].

Efficient $O(\log^2 N)$ time-dense algorithms were introduced in Ref. [35] for several geometric problems like convex hull, ECDF search, and all points closest neighbors. Notice that, before the advent of the share sort [11], such algorithms matched the best known bound for sorting on a hypercube. However, since the serial time complexity of those problems is the same as for sorting, better hypercube algorithms were likely to exist.

19.4.1.1 Optimal dense algorithms for sorted data.

Indeed, in the literature we can find some algorithms for geometric problems that take $O(T_s + \log N)$ time, where T_s is the time complexity of the best sorting algorithm to date. Then, these are efficient dense algorithms, which are optimal in the case where the input is given in a sorted order. On the other hand, and in view of the time/processor trade-off obtained by the sparse share sort, these are also time-optimal sparse algorithms. This is the case of the solutions for the convex hull algorithm from Ref. [29], triangulation and visibility from a point in Ref. [26], and the algorithms for maximal points and smallest enclosing box in Ref. [9], that we describe in the following.

Convex hull. The algorithm for convex hull starts by sorting the points by x coordinates. The points are then divided into $N^{1/4}$ groups of size $N^{3/4}$ to be stored in consecutive sub-hypercubes. For each group, the solution is computed recursively. The partial solutions are then merged by routing, back and forth a number of times, $N^{1/4}$ slopes from each group to *all* other groups in order to compute the supporting lines between any two partial solutions. Thus, there are only $N^{3/4}$ messages being routed, and this call be accomplished within $O(\log N)$ steps if one uses the extra dimensions of the hypercube to avoid routing convicts. Once the points are sorted, the time complexity is given by $T(N) = T(N^{3/4}) + O(\log N) = O(\log N)$.

Triangulation. The solution for triangulation is based on the PRAM algorithm from Ref. [40]. First the points are sorted by x coordinates, and the partial solutions, of size $N^{3/4}$ each, are found recursively. The triangulation of the partial solutions uses the supporting lines between the convex hulls of each set. These can be computed with the method used for computing convex hull, as described above. This gives a time complexity as $T(N) = T(N^{3/4}) + O(T_s) = O(T_s)$.

Smallest enclosing box. Finally, the smallest enclosing box of a set of points is a box of minimum area of any of the smallest enclosing boxes that contains an edge collinear with a hull edge. Then, once the convex hull of the set has been computed, for each hull edge the support lines parallel and perpendicular to the edge are computed through search techniques where each point of the convex hull has an interval of support angles, and the edges are located in the interval containing their support angle. Such search operations can be accomplished in logarithmic time using the standard operations described in Section 19.3.

19.4.1.2 Time-optimal sparse algorithms.

A different approach for designing time-optimal sparse algorithms can be found in Ref. [5], where ECDF search, all points closest neighbors, and convex hull are solved in time $O(k \log N)$ with $N^{1+1/k}$ processors, $1 \leq k \leq \log N$. In the ECDF search, one computes for each point the number of points it dominates (i.e., the number of points with both x and y coordinates less than those of the point). The algorithm is based on the observation that, if subsets of points are sorted by x coordinates, then the number of points dominated by a given point can be computed as the sum of the points it dominates in each subset to its left, including its own subset. A divide-and-conquer strategy is used, where up to $2^{\log N/k}$ partial solutions are merged together at each step. Partial solutions are computed through the bitonic merge, and the extra dimensions are used to avoid collisions during the routing steps, ensuring that no more than $O(\log N)$ time is required in each one of the k steps.

19.4.1.3 Suboptimal dense algorithms

ECDF search. Other dense algorithms were proposed in Ref. [26]. They solve ECDF search, all points closest neighbors, and 3-dimensional maxima in time $O(T_s \log \log N)$. The idea is again to recursively solve problems whose size is a root function of the original one. For the ECDF search, the recursion is called twice, leading to a time complexity given by $T(N) = 2T(\sqrt{N}) + O(T_s) = O(T_s \log \log N)$. In the algorithm, data are first sorted by x coordinates, then by y coordinates. Then, $\sqrt{N}$ groups of $\sqrt{N}$ consecutive points are formed according to each direction. Let X_i and Y_j, $1 \le i, j \le \sqrt{N}$, denote such groups. This defines virtual cells $C_{i,j}$ composed of the points in the intersection of X_i and Y_j (see Fig. 19.7). The number of points dominated by a point in $C_{i,j}$ is then the sum of three values, namely, the number of points it dominates in its vertical region (X_i), the number of points it dominates in its horizontal region (Y_j), and the number of points completely to the left and completely below its cell. The first two values are computed recursively.

Selection. A sorting based $O(\log N \log^* N)$ time dense algorithm[*] for selecting the kth smallest element out of N unsorted items was given in Ref. [6]. The algorithm is a succession of phases where finer approximations for the kth smallest element are obtained, so

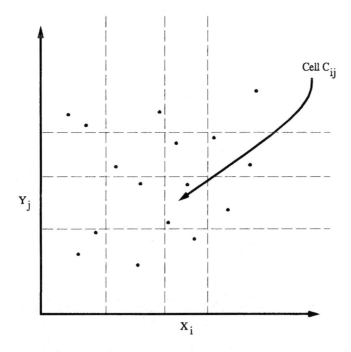

Figure 19.7 Groups and cells for solving ECDF search

---

[*]The function $\log^* N$ gives the number of iterations of the *logarithm* function to get the value 1 from N.

that the correct element is eventually computed. Each such approximate selection phase is identical to the others: sort sublists, then choose splitters as the new sublist. This is iterated until good bounds for the kth smallest element can be calculated. From one phase to the next, elements that are not between the bounds are definitely discarded, and the whole remaining list concentrated. As far as the time complexity is concerned, each phase is tuned so that it costs $O(\log N)$ time, for a total of h phases. We observe that the only costly operation used is sorting. At the beginning of the algorithm, sublists are sorted with the share sort, since the hypercube is full. However, in all remaining sort operations, it can be shown that the hypercube has enough free dimensions to allow the use of sparse sorts that can be used in different places during the algorithm. The best combination of the sorting algorithms yields $h = O(\log^* N)$.

Polygons. Finding shortest paths in rectilinear polygons was solved in time $O(T_t + \log N)$ with P_t processors in Ref. [17], where T_t and P_t are, respectively, the time and the number of processors required for solving trapezoidal decomposition (see Section 19.4.2). The algorithm is a combination of finding the visible pairs from special points (where the trapezoidal decomposition is used), and a technique for nested parenthesis elimination, along with the traversal of monotone rectilinear staircases.

Very recently, time-optimal dense algorithms were given for simple-polygon problems , such as visibility, convex hull, monotonicity, and others [3], and for tree contraction, a difficult problem arising in the solution of many graph algorithms [27].

19.4.2   Data structures

Aside from the diameter, one of the main differences between the PRAM and the hypercube is that the PRAM has a large shared memory similar to that of a sequential computer, while in the hypercube the memory is divided into blocks of constant size that are distributed among the processors. The concurrent access to the central memory has been extensively used in the design of efficient parallel search algorithms for computational geometry in the CREW PRAM,[*] where each processor is in charge of answering a different query. The shared memory paradigm allows the implementation of very sophisticated data structures, and once the data structure has been collectively built by all the processors, each one can independently gain access to it and execute a standard sequential search algorithm.

In hypercubes, concurrently answering several independent queries is not as straightforward. Direct approaches either use very simple data structures and concentrate on routing and collision avoidance issues, or simulate PRAM algorithms that use complex data structures, yielding more elegant, though less efficient, parallel algorithms for the hypercube.

One of the first tools for elegant and theoretically efficient use of data structures in hypercubes was the *multi-way search* (M-way search, for short) proposed in Ref. [15]. With the M-way search, several queries (whose maximum search path size is h) can traverse tree-like data structures simultaneously, with an overhead of only $O(\log N)$ with respect to CREW PRAM algorithms.

In the M-way search, the queries advance synchronously through the data structure, one level per phase. It is clear that if, during a phase, the queries were to be routed to the nodes they have to visit, then we would have severe congestion problems. Therefore, the basic idea is to create as many copies of each node of the data structure as necessary, and route

---

[*]In the concurrent-read, exclusive-write (CREW) model, all processors are allowed to read from the same memory cell in a given step.

them to the queries that requested them. If a certain order is kept among the queries, guaranteeing address monotonicity for the routing operations, then each phase can be implemented in $O(\log N)$ steps by the use of a constant number of calls to the procedures previously described in Section 19.3.

A powerful extension to this technique was proposed in Ref. [14]. *Hypercube cascading* is based on the *dynamic* M-way search in Ref. [12], which is in turn an extension of the M-way search that copes with dynamic structures. With hypercube cascading, most data structures represented by directed acyclic graphs can be traversed by concurrent queries that synchronously advance in their paths, step by step. Moreover, it keeps an important feature of the dynamic M-way search: it takes into account dynamic data structures, where at most $O(N)$ insertions and deletions can take place at each step, with structure update, with the same overhead as before. Finally, hypercube cascading allows more complex operations concerning the computation of the queries' paths.

Such a multi-iterative search paradigm consists of h phases such that, at the end of a phase x, $(1 \leq x \leq h)$, all queries are sorted with regard to the index of the xth node of the search path. Furthermore, each PE_i has a copy of the xth node of the search path of the ith query, as well as the information concerning the next node in the query's search path.

Phase 1 is implemented through a sorting. In case the directed acyclic graph representing the data structure is monotone (see definition below), then sorting can be avoided in the remainder of the algorithm and all the procedures used in a phase x, $(2 \leq x \leq h)$, take $O(\log N)$ time, implying that the time complexity of one such phase is also $O(\log N)$. Therefore, if we have a bounded degree monotone data structure of size s, then we can answer m iterative search queries whose search paths are no longer than h, in time $O((h + (\log \log N)^2) \log N)$ in hypercubes with $N = \max\{s, m\}$ processors [14].

The only condition imposed on the data structure so that this technique applies is that it should be *monotone,* as follows (see Fig. 19.8).

Definition 3. A directed acyclic graph $G(V, E)$ is *monotone* if there exists a numbering of the nodes $f: V \Rightarrow \{1,\ldots,n\}$ such that if (v, v') and (w, w') are two edges of G, then $f(v) < f(w) \Leftrightarrow f(v') < f(w')$.

In the following text, we briefly describe an application of hypercube cascading.

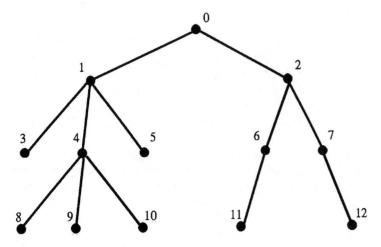

Figure 19.8 Example of a monotone numbering of a tree

19.4.2.1 Solving the multiple stabbing problem. This problem consists of determining all intersections of m lines with a simple polygonal path of size s (see Fig. 19.9). The solution shown in Ref. [14] presents a slowdown factor of only $O(\log N)$ with respect to the sequential version, where just one query line is answered.

With fractional cascading [8], the serial solution of the multiple stabbing problem is based on the following observation: a line l intersects a simple polygonal path P if and only if l intersects the convex hull, $CH(P)$, of P. Then, a decomposition tree of the convex hulls for a given P is built. Such a tree is a binary tree, T, with $CH(P)$ assigned to its root. The two children are the roots of the subtrees representing the first half and the second half of P respectively.

Hence, all intersections of P with a query line l can be found through the following traversal of T. At each node, if l intersects the convex hull assigned to the node, then the two subtrees are traversed recursively; otherwise, the traversal finishes at this point.

To solve the multiple stabbing problem on a hypercube, T has to be built and a strategy for its concurrent traversal defined. Hypercube cascading is used to implement both phases. Below, we give a brief description of the traversal of T. Notice that T is monotone and has bounded degree.

For traversing T, each query line could simply start at the root and be duplicated at each branching to simulate the sequential traversal. Unfortunately, this could lead to as many as (sm) queries by the end of the algorithm. The solution stems from a strategy where each line is represented by a single query that answers all intersections with P, traversing it in inorder.

Let $T'(l)$ be the subtree of T that shall be traversed in the sequential solution with fractional cascading for answering the query associated with l. In Ref. [8], it was shown that $|T'(l)| = O(k \log (s/k))$, where k is the number of intersections between l and P. Therefore, the length of the path (of an inorder traversal of T) with hypercube cascading shall also be $O(k \log (N/k))$, where $N = \max\{s, m\}$, yielding an overall time complexity of $O[(k \log (N/k) + (\log \log N)^2)\log N]$ to answer m queries simultaneously.

Figure 19.9 Example of a multiple stabbing problem

19.4.2.2 Summary. The problems that could be solved through the use of the paradigms cited above are as follows. (For more details on the algorithms, see Refs. [2] and [28].)

- trapezoidal decomposition and triangulation of polygons of size N in time $O(\log^2 N)$ with a hypercube of size $N \log N$ [15]
- parallel branch and bound on hypercube multiprocessors [12]
- multiple stabbing of a simple polygonal path, i.e., determining all k intersections of m lines with a simple polygonal path of length l, in time $O[(k \log (N/k) + (\log \log N)^2) \log N]$ with a hypercube of size N, where $N = \max\{l, m\}$ [14]
- multiple slanted range search: Let S be a set of l points in the plane. An *aligned trapezoid* is a trapezoid with one side on the x-axis and the two adjacent sides parallel to the y-axis. The *slanted range search problem* consists of reporting all k points contained in the aligned trapezoid. The multiple slanted range search problem consists of answering in parallel, m slanted range search queries on the set S. This can be done in time $O((k \log(N/k) + (\log \log N)^2) \log N)$ with a hypercube of size N, where $N = \max\{l, m\}$ [14].
- parallel processing of pointer-based quadtrees of height h in hypercube multiprocessors of size N [13]; namely,
  - Convert image or boundary code to quadtree in time $O(\log^2 N)$ or $O((h + (\log \log N)^2) \log N)$, respectively.
  - Determine neighbors of all leaf nodes and compute perimeter in time $O(h \log N)$.
  - Compute union and intersection of two quadtrees in time $O((h + (\log \log N)^2) \log N)$.

Recently, it was shown that the M-way search paradigm can be used to solve some query problems even faster. In the conference version of Ref. [4], an $O(\log N(\log \log N)^3)$ time algorithm is described to solve the multiple point location in a slab, and in the journal version of this paper, it is shown how to apply the same techniques to solve the trapezoidal decomposition in time $O(\log N(\log \log N)^3)$ with $N \log N$ processors in a hypercube.

19.5 Conclusions and Future Directions

19.5.1 Some open problems

Even though the design and analysis of hypercubic algorithms have been extensively studied, there are several problems still open that would be worth further scrutiny. The main one concerns the complexity of sorting and its implications on the complexity of many other problems in a number of other areas, such as convex hull in computational geometry [2] and on-line permutation adaptive routing [25], for instance. A related problem with open complexity is the problem of selecting the kth smallest element out of a nonsorted list. Its current upper bound [6] is better than the one of sorting [11], but no tight lower bounds exist for either problem for the fine-grain model. Finally, new techniques have recently emerged for optimally solving problems in this model [3, 27]. It would be very interesting to study their impact on the solution of graph and polygon related problems.

19.5.2 Summary

We started discussing some topics on topological aspects of the hypercube that naturally lead to the study of communication issues in parallel machines based on these topologies. Then, we described some of the main tools for algorithm design in the fine-grain hypercube, whose use was exemplified with the solution of problems in computational geometry and with the implementation of powerful and complex parallel data structures.

In fact, the existing research connected with hypercubes would constitute enough material for several books. Hence, in this chapter we covered only part of it, and we tried to give the reader some flavor of the main issues arising from the use of such an architecture in the design of parallel algorithms. On the other hand, extensive references to textbooks and to recent research have been provided, even when the specific subject was not discussed here because of space considerations.

Acknowledgments

I am indebted to Pascal Berthomé, Frank Dehne, Apostolos Gerasoulis, Pascal Koiran, Andrew Rau-Chaplin, Ivan Stojmenović, and A. Zomaya, whose invaluable help was most important for the completion of this work. I would like also to thank Olivier de Vet, Kevin O'Brien, and the anonymous referee for carefully reading previous versions of the manuscript, and the PRC PRS and Math-Info of the French CNRS, DRET, and DIMACS for partial financial support.

19.6   References

1. Akl, S. G. 1985. *Parallel sorting Algorithms.* Orlando, Fla.: Academic Press.
2. Akl, S. G., and K. Lyons. 1993. *Parallel Computational Geometry.* Englewood Cliffs, N.J.: Prentice Hall.
3. Atallah, M. J., and D. Z. Chen. Optimal parallel hypercube algorithms for polygon problems. *IEEE Transaction on Computers,* to appear.
4. Atallah, M. J., and A. Fabri. 1994. On the multisearching problem for hypercubes. *Proc. 6th Parallel Architectures and Languages Europe,* Lecture Notes in Computer Science. New York: Springer Verlag.
5. Berthomé, P., and A. Ferreira. 1992. Efficiently solving geometric problems on large hypercube multiprocessors. In *Parallel and Distributed Computing in Engineering Systems,* S. Tzafestas, P. Borne, and L. Grandinetti, eds. Amsterdam: IMACS-North Holland, 123–128.
6. Berthomé, P., A. Ferreira, S. Perennes, G. Plaxton, and B. Maggs. 1993. Sorting based selection algorithms on hypercubic networks. *Proceedings of the 7th IEEE International Parallel Processing Symposium.* New York: IEEE Press, 89–95.
7. Bhuyan, L. N., and D. En. Agrawal. 1984. Generalized hypercube and hyperbus structures for a computer network. *IEEE Transactions an Computers,* 4(C-33):323–334.
8. Chaselle, B. and L. J. Guibas. 1986. Fractional cascading: I. A data structuring technique. *Algorithmica,* 1(2):133–162.
9. Cohen, E, R. Miller, E.M. Sarraf, and Q. Stout. 1992. Efficient convexity and domination algorithms for fine- and medium-grain hypercube computers. *Algorithmica,* 7:51–75.
10. Cosnard, M., and A. Ferreira. 1991. On the real power of loosely coupled parallel architectures. *Parallel Processing Letters,* 1(2):103–111.
11. Cypher, R. E, and C. G. Plaxton. 1990. Deterministic sorting in nearly logarithmic time on the hypercube and related computers. *Proceedings of the 22nd Annual ACM Symposium on Theory of Computing,* 193–203.
12. Dehne, F., A. Ferreira, and A. Rau-Chaplin. 1990. Parallel branch and bound on fine grained hypercube multiprocessors. *Parallel Computing,* 15(1-3):201–209.
13. Dehne, F., A. Ferreira, and A. Rau-Chaplin. 1991. Efficient parallel construction and manipulation of pointer based quadtrees. *Proceedings of the International Conference on Parallel Processing,* 255–262.
14. Dehne, F., A. Ferreira, and A. Rau-Chaplin. 1992. Parallel fractional cascading on hypercube multiprocessors. *Computational Geometry—Theory and Applications,* 2:111–167.
15. Dehne, F., and A. Rau-Chaplin. 1990. Implementing data structures on a hypercube multiprocessor and applications in parallel computational geometry. *Journal of Parallel and Distributed Computing,* 8(4): 367–375.
16. Ferreira, A., and M. Grammatikakis. 1994. Improved probabilistic routing in generalized hypercubes. *Proc. 6th Parallel Architectures and Languages Europe,* Lecture Notes in Computer Science. Berlin: Springer Verlag.

17. Ferreira, A., and J. Peters. 1991. Finding the smallest path in a rectilinear polygon on a hypercube multiprocessor. *Proceedings of the Third Canadian Conference on Computational Geometry,* 162–165.
18. Fraigniaud, P., and E. Lazard. Methods and problems of communication in usual networks. *Discrete Applied Mathematics* (to appear in special issue on broadcasting).
19. Grammatikakis, M. D., D. F. Hsu, and F. K. Hwang. 1994. Adaptive and oblivious routing on the *d*-cube. *Proceedings of the 4th Annual International Symposium on Algorithms and Computation,* to appear in Lecture Notes in Computer Science. New York: Springer-Verlag.
20. Grammatikakis, M. D., D. F. Hsu, and M. Kraetzl. 1991. Multicomputer routing. *Technical Report 94-5.* School of Mathematics and Statistics, Curtin University of Technology, Perth, Australia.
21. Ho, C.-T., and S. L. Johnson. 1989. Optimum broadcasting and personalized communication in hypercubes. *IEEE Transactions on Computers,* 38(9):1249–1268.
22. Kaklamanis, C., D. Krizanc, and T. Tsantilas. 1990. Tight bounds for oblivious routing on the hypercube. *Proceedings of the 2nd Annual ACM Symposium of Parallel Algorithms and Architectures,* 31–36.
23. Karp, R. M, and V. Ramachandran. 1990. A survey of parallel algorithms for shared-memory machines. In *Handbook of Theoretical Computer Science, Volume A: Algorithms and Complexity,* J. van Leeuwen, ed., 869–911. New York: Elsevier/MIT Press.
24. Knuth, D. E. 1993. *The Art of Computer Programming,* Volume 3. Reading, Mass.: Addison-Wesley.
25. F. T. Leighton. 1991. *Introduction to Parallel Algorithms and Architectures: Arrays, Trees and Hypercubes.* San Mateo, Calif.: Morgan-Kauffman.
26. MacKenzie, P. D, and Q.F. Stout. 1990. Asymptotically efficient hypercube algorithms for computational geometry. *Proc. 3rd Symposium Frontiers of Massively Parallel Computation,* 8–11.
27. Mayr, E. W, and R. Werchner. 1993. Optimal tree contraction on the hypercube and related networks. In *DIMACS Workshop on Parallel Algorithms: from solving combinatorial problems to solving grand challenge problems,* J. Flanagan, Y. Matias, and V. Ramachandran, eds., 36.
28. R. Miller, A. Rau-Chaplin, and Q. Stout. *Parallel Algorithms for Regular Architectures.* Cambridge, Mass.: MIT Press (in press).
29. Miller, R., and Q. Stout. 1988. Efficient parallel convex hull algorithms. *IEEE Transactions on Computers,* 37(12):1605–1618.
30. Nassimi, D., and S. Sahni. 1981. Data broadcasting in SIMD computers. *IEEE Transactions on Computers,* C-30:101–107.
31. D. Nassimi and S. Sahni. 1982. Parallel permutation and sorting algorithms and a new generalized connection network. *J. of the ACM,* 29:642–667.
32. Peters, J. G., and M. Syska. 1993. Circuit-switched broadcasting in torus networks. *Tech. Rep. TR 93-4,* School of Computing Science, Simon Fraser Univ.
33. Ranka, S., and S. Sahni. 1990. *Hypercube Algorithms, with applications to image processing and pattern recognition. New York:* Springer-Verlag.
34. Rumeur, J. 1994. *Communications dans les Réseaux d'Interconnezions. Paris:* Masson.
35. Stojmenovic', I. 1988. Computational geometry on a hypercube. Pr*oceedings of the International Conference on Parallel Processing, Vol. III Algorithms and Applications,* 100–103.
36. Stout, Q. F., and B. Wagar. 1990. Intensive hypercube communication: prearranged communication in link-bound machines. *Journal of Parallel and Distributed Computing,* (10):167 181.
37. Szymanski, T. 1989. On the permutation capability of a circuit-switched hypercube. *Proceedings of the IEEE International Conference on Parallel Processing, Vol. I,* 103–110.
38. Valiant, L. G. 1985. A scheme for fast parallel communication. *SIAM J. of Computing,* 2(11):350–361.
39. Valiant, L. G., and G. J. Brebner. 1981. Universal schemes for parallel communication. *Proceedings of the 15th Annual ACM Symposium on Theory of Computing,* 263–277.
40. Wang, C. A., and Y. H. Tsin. 1987. An $O(\log n)$ time parallel algorithm for triangulating a set of points in the plane. *Information Processing Letters,* 25(1):55–60.

II

Architectures and Technologies

Architectures

20

RISC Architectures

Manolis Katevenis

This chapter is about processor architecture—it is not about parallel or distributed computer architecture. However, its contents are very relevant to the topic of this book. Three key hardware ingredients are needed to make high-performance parallel and distributed computer systems: (1) high-speed computation, (2) high-speed communication, and (3) high-speed peripheral (I/O) devices. This chapter talks about the first of these three hardware components, and specifically about how to make high-speed processors (at a reasonable cost, too), which are a necessary building block for high-performance parallel and distributed computers. Having reached this chapter, the reader will probably already realize that the overall performance of a parallel or distributed system will get worse if the system has f times more processors, but each processor has f times lower performance ($f >$ 1). This is so because every time we introduce more parallelism into a computation, more communication and synchronization overhead is also introduced. Thus, the way to high-performance parallel computers is *first* to choose the highest performance processors at the desired cost level, and then to interconnect as many of them as are needed to achieve the desired overall system performance. This chapter is about the "first step."

Reduced instruction set computer (RISC) architectures are the basis of modern high-performance processors. They have a simplified instruction set, with register-to-register arithmetic instructions; memory can only be accessed by *load* and *store* instructions, usually with a single, simple addressing mode; the instructions have a fixed size and few, simple formats. RISC processors use pipelining, and their instruction set allows tight control of their pipeline by the software. Optimizing compilers perform instruction scheduling so as to exploit the parallelism offered by the pipelined data path. RISC processors achieve high performance at low cost: the smaller and simpler circuits allow higher clock rates; the data path can use pipelining, and the compiler can exploit it. In addition, design time, design cost, and silicon area are reduced.

20.1   What is RISC?

RISC stands for *reduced instruction set computer,* a pun on the word "risk." Scientifically, RISC is a *style* or a *family* of processor architectures that share a certain number of characteristics, which are described and explained in this chapter. RISC is *not* one, single computer architecture. RISC is *not a strictly defined* set of architectures, either; there are many architectures "near the boundaries" of RISC that some people will say they are RISC and other people will say are not RISC.

The word RISC, or better *RISC concept,* has also come to be used in another way. It may refer, depending on the context, to the set of basic design principles that led to the development of these architectures. As technology changes, the particular design choices of RISC architectures may cease to be as good as they were when originally made (in the 1980s). However, the design principles that led to these choices should (hopefully) stay valid for a longer time and guide computer designers in the future on how to make good choices given a certain implementation technology. Also, these principles sometimes have a wider range of applications than just uniprocessor design—or than just hardware design.

The RISC ideas were developed mostly in the early 1980s and became popular in the second half of that decade. RISC architectures came, in part, as a reaction to the direction that computer architecture (except supercomputer architecture) had taken in the 1970s. Today, RISC is considered to be the basis for designing high-performance processors, and almost at any price level. In this introductory section, we will first discuss the basic design principles that led to RISC, and then we will list the most important architectural characteristics of RISC.

20.1.1   Design principles that led to RISC

Use quantitative engineering analysis rather than myths and marketing fads. It sounds quite obvious that the design of a technical device such as a processor should be based on engineering analysis rather than myths. However, it was not so with the state of the art in computer architecture in the 1970s. There was a widespread impression that if a task is executed *in hardware,* (i.e., as one computer instruction), then that task and the whole computer is faster than if this task is executed *in software* (i.e., by multiple instructions). Based on this groundless impression, the marketing strategy for several computers was based on how many tasks the computer could execute as single instructions, thus leading to *complex instruction set compilers* (*CISC*). RISC came as a reaction to this unfounded CISC trend.

Understand trade-offs across levels of abstraction. Computer processors are complex devices; hence, they can be mastered only by using engineering abstractions. Levels of these abstractions are, from bottom to top: integrated circuit (IC) polygon level, transistor level, logic gate level, register transfer level, hardware block diagram level, instruction set level, operating system (OS) and standard library level, programming language level, and application program level. Unfortunately, changing a design choice in one level often has repercussions at all other levels, and the repercussions are usually very difficult to evaluate accurately—even more so if the engineers doing this do not understand (or ignore!) some of these levels of abstraction. In the past, it was thought that, for the "purity" or the longevity of an architecture, this latter ought to be designed independently of the implementation hardware; it was even sometimes considered a bit pejorative for a computer architect to deal with such "low-level details" as anything below hardware block diagrams! Yet, parasitic capacitances and transistor currents are where the clock speed game is eventually won or lost. . . .

Make the common case fast. Overall performance is a weighted mean of performance on individual tasks, where the weight factors are the frequencies with which these individual tasks appear in the overall workload. Thus, speeding up the frequent operations pays back much more than doing so for the infrequent ones. Furthermore, if the acceleration of one operation has even slightly negative side effects on the speed of other operations, then such an acceleration can easily turn out to reduce overall performance if its target operation is one that rarely occurs.

Smaller is faster. In electronic circuits, the shorter a transmission line is, the faster electromagnetic waves will travel through it. When the dominant delay mechanism is not propagation delay through a transmission line, it is the charging or discharging of a parasitic capacitance by a given electric current. Reducing the parasitic capacitance will speed up the circuit. Shorter wires have less parasitic capacitance; also, wires with fewer transistors (fewer gates, simpler logic) connected on them have lower parasitic capacitance. As we see, in all cases, smaller circuits (i.e., simpler circuits and circuits with shorter wires) are faster. It is not surprising that, in processor design, fewer operations, simpler operations, and a more regular set of operations lead to smaller, and thus faster, circuits. The crucial point, of course, is where the golden mean lies; the rest of the chapter discusses this point.

Hardware is not necessarily faster than software. In some cases, a hardware implementation of a certain task is much faster than its software implementation. This is particularly the case when circuits that are specialized for this task are designed and added to the processor; for example, floating-point operations are much faster in hardware than when performed through repeated use of the integer arithmetic unit(s). In other cases, encoding a task as a single instruction is not (or cannot be) accompanied by any such circuit optimization. In many such cases, this single instruction takes as many clock cycles to execute as the multiple, simpler instructions that can execute the same task would; this will be further discussed in Section 20.3. In these cases, there is no "magic" in hardware that makes it any better than software. Providing a reduced instruction set in the processor and synthesizing the complex tasks in software is just as good as providing a complex instruction set—actually, it is better, because of the previous point: *smaller is faster* for those operations that remain in the reduced instruction set.

20.1.2   Architectural characteristics of RISC

As was said earlier, RISC is a style or a family of processor architectures that share some characteristics. Below, we list these characteristics, in order of importance, according to this author's view. Remember that RISC is not a strictly defined set of architectures, so what follows is not a "defining set" of characteristics.

Register orientation. Smaller is faster, as we said earlier, and smaller memories are faster than larger ones. This is the basis of memory hierarchies: register file, cache memory, main memory, disk, tape. The processor registers are the fastest level of this hierarchy, for three reasons: (1) short access time because of smaller size (shorter than level-1 (on-chip) cache memory, although the difference is often relatively small); (2) multiport access, i.e., multiple registers can be accessed in parallel (cache memories rarely have this capability, since it would greatly increase their cost and slightly lengthen their access time); and (3) low addressing overhead, because register specifiers consume only a few bits in the instruction, and addressing is simple and fast—no effective address calculation is involved, and no address translation exists. For these reasons, good exploitation of the registers is of capital importance in speeding up execution. Unfortunately, unlike the cache memory, the register file is not a transparent level of the memory hierarchy. (But this is also one reason why registers are faster than the cache!) Thus, it takes effort for the compiler to properly exploit the register file. The development of RISC was linked to the development of optimizing compilers that could allocate frequently used program variables in registers. Thus, in RISC architectures, the operands are more frequently in registers, or conversely, once the operands are more frequently in registers, RISC architectures make more sense than CISC. Also, RISCs tend to have more registers than CISCs.

Load/store architecture. Perhaps the most important concrete characteristic of RISC architectures is the fact that memory is only accessed via load (read) and store (write) instructions, while all other instructions operate exclusively on register-resident operands. To put it differently, transferring information between the memory hierarchy and the processor is separated from operating on this information. There are several reasons for this separation of work, and they will become apparent in the rest of this chapter; in brief, they are:

1. Operands are frequently in registers, so the demand for operations on memory operands is reduced.
2. Pipeline organization is easy.
3. Interruptibility and restartability is available for instructions in the pipeline.
4. Instruction size is fixed.
5. There is increased flexibility for instruction scheduling.

Fixed instruction size. The instruction fetch hardware of the processor is greatly simplified when instruction placement in memory follows a simple alignment pattern. Also, control transfer instructions (branch, jump) are sped up when their target instructions do not cross word boundaries. For these reasons, RISC architectures usually have a fixed instruction size—32 bits. In the early 1980s, RISC architectures were heavily criticized because this fixed instruction size requires less compact code. However, it became apparent with time that this drawback is less important than the aforementioned advantages. We will come back to these issues in Section 20.4.

Fixed position of source operands in instruction format. The latency of decoding and executing an instruction is reduced when the descriptors for the source operands (two source registers, usually) are in fixed positions within the instruction, independent of the instruction's opcode. When this property holds, as in RISC architectures, the source operands can be read *in parallel* with decoding of the opcode. Reducing the latency of instruction execution helps in reducing the branch execution time.

Single result per instruction. The pipeline is simplified if each instruction writes one result into one memory word or register and does not modify the user-visible machine state in any other complicated or intricate way. In particular, instruction interruptibility and restartability is made easier in this way. Most RISC architectures follow this; where condition codes exist, setting them is the most notable exception.

Regularity. As we saw, smaller is faster—in particular, simpler logic circuits with fewer gates are faster. The primary method to reduce the number of transistors connected to wires (other than the undesirable method of reducing the processor's capacity as, for example, the number of general-purpose registers) is to make the circuit more regular. Regularity means more similarity among the various operations so that they can all be handled by the same subsystems, and less special cases, thus reducing the number of circuits, buses, and multiplexers needed. RISC architectures strive for regularity in many respects: instructions perform a similar amount of work each, consisting of similar operations (read sources, perform one arithmetic operation, and so forth) on similar operands (two source registers, one destination register) in a fixed set of pipeline stages. There is usually one single memory addressing mode, and that resembles the rest of the operations.

Provide primitives—not solutions. The instruction set of an architecture is designed once and stays fixed, often for more than a decade. The software that runs on it can be changed at any time. Thus, hardware (the instruction set) should only provide general-purpose building blocks. Special features that match special programming language constructs should be left for the compiler or the libraries; if placed in the hardware, they are either too specialized and thus useless most of the time, or too general and often too slow (see Ref. [1], pp. 121, 124). Hardware instructions should each do one, simple thing well and fast, and should combine flexibly and fast with each other; combining them in intelligent ways should be left to the software.

Make performance visible and optimizable. The old, CISC view was that the hardware should provide an embellished, abstract interface (instruction set) that is appropriate for easy programming in assembly language by a human. The new, RISC view is that the hardware should provide a raw interface with visible performance characteristics, appropriate for use by an optimizing compiler. Accordingly, RISCs make pipeline characteristics visible at the instruction set level. Examples include delayed loads (Section 20.3.2) and delayed branches (Section 20.4.2).

The rest of this chapter is organized as follows. Section 20.2 explains pipelining, which is the basic speed-up technique for processor hardware. Section 20.3 discusses how a pipelined processor can be kept busy with useful work for as much of the time as possible and why RISC is better than CISC in this respect. Section 20.4 examines the issues related to instruction size and format, while Section 20.5 discusses other disadvantages of CISC: control and interrupt complexity, design time and cost, and clock frequency. A brief historical note and a general perspective follow in Section 20.6.

20.2 Pipelining and Bypassing

Pipelining is the principal method to achieve high performance in the operation of a processor. Up until the 1980s, substantial pipelining was used only in supercomputers and the high-end mainframes. Supercomputers often had a simple instruction set; for example, the machines designed by Seymour Cray have used register-oriented, load/store instruction sets. The lower-end computers did not use any appreciable amount of pipelining; these included some of the most CISC architectures, such as the DEC VAX [3]. Perhaps pipelining was considered too costly to be applied to low-end systems, or rather it was found too complicated and costly for the complex instruction sets that had evolved in the meanwhile. RISC emerged when it was realized that pipelining can perfectly well be used in low-cost processors (microprocessors) provided that the instruction set is simplified; furthermore, this approach led to significantly higher performance. For this reason, understanding pipelining is a prerequisite for understanding RISC. This section briefly explains pipelining; for a longer presentation, see Chapter 6 of Ref. [2], and for deeper treatments see Chapter 6 of Ref. [1] and Ref. [4].

20.2.1 The basic five-stage RISC pipeline

The basic pipeline employed by the integer part (i.e., non-floating-point) of RISC processors is the five-stage pipeline illustrated in Figs. 20.1 and 20.2. Some classic RISC processors follow this pipeline [5], while others use small variations of it [6, 7]. Figure 20.1

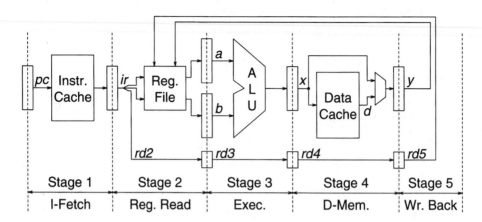

Figure 20.1 Simplified data path of a five-stage RISC pipeline

shows a simplified processor data path organized according to such a five-stage pipeline. Things missing from this figure are the control section, control transfer instruction paths, store data path, immediate constant paths, and bypass paths.

The vertical rectangles along the dashed lines in Fig. 20.1 are edge-triggered registers; they separate the stages of the pipeline. An instruction is fetched and executed by one pass of information flowing from left to right in this figure. It takes five clock cycles for this to complete; in each clock cycle, information advances from the input register(s) to the output register(s) of one pipeline stage. In the first stage, the program counter (pc) is used as address for reading the next instruction from the I-cache. In the second stage, the instruction register (ir) specifies the register numbers of the three operands of the instruction (as well as the opcode, which is not shown). Two of the operands are sources, so they are read immediately from the register file; the third register number, $rd2$, is the destination for the result y of the instruction, so it is held for later use. In the third stage, the two source operands, a and b, feed the arithmetic/logic unit (ALU), which computes their desired function. Two cases must be distinguished: if this is a data memory instruction (*load* or *store*), then the ALU computes $a + b$, which is the effective address for the memory access; otherwise, for the arithmetic or logic instructions, the ALU computes the desired final result. In the fourth stage, load or store instructions use the result of the ALU, x, as address for a data cache access; if this is a load instruction, the data read, d, constitute the result of the instruction. Non-memory instructions do nothing in this stage—they just pass x, the ready result, to the output; this may seem like a waste of time, but we will see that it is not, when we discuss bypassing. Finally, in the fifth stage, the final result y of the instruction, which is either x or d, is written back into the destination register, $rd5$, of the register file.

The data path of Fig. 20.1 is designed so that, in each clock cycle, the entire state of execution of an instruction is contained completely and exclusively within the circuits of *one, single* stage. This is the reason why the destination register number, rd, is copied (without modification) from $rd2$ to $rd3$ to $rd4$ to $rd5$, thus flowing from left to right in parallel with the data of the instruction. Given this property of the data path, at any time when an instruction is under processing inside it, four of the five stages of the circuit are free. Thus, if other *independent* instructions are available, they can also be processed, in parallel, within the other four stages. Obviously, the instructions that are executing in parallel will always be in different stages of completion each. This parallel processing in the style of an assembly line is called *pipelining*.

As a concrete example, consider five consecutive instructions, *I*1 through *I*5, from a program under execution. Figure 20.2 shows how their execution can proceed through the pipeline of Fig. 20.1, if the latter instructions are *independent* from the former. Assume that the data path was idle (the pipeline was empty); then, instruction *I*1 is fetched from (cache) memory. In the next cycle, say at time *t2*, *I*1 is in *ir*, and its two source registers are being read from the register file. Simultaneously, since the instruction cache is free from *I*1, and since *I*2 was assumed not to depend on *I*1 (i.e., in this case, *I*1 was not a branch or jump), *I*2 is fetched from the instruction cache.

In the next clock cycle, at time *t3*, the source operands of *I*1 are in *a* and *b*, and *I*1's operation is being performed in the ALU. Simultaneously, since the register file is free from *I*1, and since *I*2 was assumed not to depend on *I*1 (i.e., in this case, *I*2 does not read the result of *I*1), *I*2 can use the second stage; that is, its source registers are being read from the register file. At the same time, the instruction cache is free from *I*1 and *I*2, and since *I*3 was assumed not to depend on *I*1 or *I*2 (i.e. neither *I*1 nor *I*2 were branches), *I*3 is fetched from the instruction cache.

Similarly, in the next clock cycle, at time *t4*, *I*1 may be accessing the data cache if it is a load or store instruction, or else it may be doing nothing. Simultaneously, *I*2's operation is being performed in the ALU. At the same time, the source registers of *I*3 are being read from the register file, since we assumed that *I*3 does not read the results of *I*1 or *I*2. Also, at *t4*, *I*4 is fetched from the instruction cache, since none of its three previous instructions was a branch or a jump. In the fifth cycle (rightmost column in Fig. 20.2), the pipeline is full: the final result of *I*1 is being written back into the register file, while *I*2 is in the fourth stage, *I*3 is in execution, *I*4 is reading its source registers, and *I*5 is being fetched. Notice that three accesses to the register file need to be going on in parallel: one write (of an older instruction's result), and two reads (of a younger instruction's sources); thus, the register file needs to be *three-ported*.

From that cycle on, the pipeline may continue being full and operating in a similar manner, if every instruction is *independent of its four previous* instructions. Five consecutive instructions will always be in the pipeline, each at various stages of completion. Given our assumption of mutual independence, none of the instructions in the pipeline needs any of the results of a previous instruction that is also in the pipeline (at most, four such instructions exist), and thus execution can proceed unhindered at the rate of one instruction every clock cycle, rather than one instruction every four to five clock cycles if pipelining were

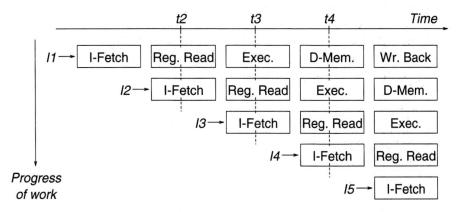

Figure 20.2 The basic five-stage RISC pipeline

not used. Of course, life is not as simple as the above ideal picture! Lots and lots of instruction dependences exist in real programs, so a real data path must be more complicated than Fig. 20.1, as explained next.

20.2.2 Dependences and bypassing

Consider the following simple example of dependence between two consecutive instructions in a program: $I1$: add $R1 + R2 \rightarrow R3$; $I2$: sub $R3 - R4 \rightarrow R5$; these compute the sum of two numbers minus a third number. Instruction $I2$ depends on $I1$, since it needs to read the result of $I1$ as a necessary source operand for its computation. Referring to Fig. 20.2, it would not be possible to execute $I2$ with the timing shown there, because that would produce the erroneous effect of $I2$ subtracting $R4$ from the *old* value of $R3$, which was read at $t2$, rather than from the new value, which will only be written into the register file in the fifth cycle. For correct execution, under the circumstances of Figs. 20.1 and 20.2, $I2$ would have to wait until the fifth cycle for its fetch stage (i.e., as $I5$ in Fig. 20.2), or at best until the fourth cycle (as $I4$) if we assume that the register file has the property of supplying the new value when the same register is simultaneously written and read. In the simple data path of Fig. 20.1, all dependences lead to underutilization of the pipeline and loss of performance.

However, there is often something better that can be done. Upon closer examination, one sees that $I2$ needs the new value of $R3$ (the result of the addition) when it starts performing its operation (subtraction), that is at the beginning of the fourth clock cycle (the cycle of $t4$ in Fig. 20.2). In the meanwhile, this desired value (the result of the addition) has *already been produced* by $I1$, during the third cycle, in the ALU. The only thing that has not yet been done is to write this result in its proper place, into $R3$ in the register file. If $I2$ uses the value that it reads at $t3$ from the register file, it will fail; however, if it *bypasses* stages 4 and 5 of $I1$ as well as its own stage 2, it can get the correct value directly from the source—from the output of the ALU.

Figure 20.3 shows the additions to the data path of Fig. 20.1 that are necessary for this bypassing to work. Only the second and third stages are shown, since they are the only ones that need modification. Two multiplexers are added after the two outputs of the register file, allowing one or both of its output values (the "old" values) to be discarded and replaced by the correct, new value that has just been computed by the ALU. The right-hand side of Fig. 20.3 shows the timing of this bypassing, by reference to Fig. 20.2. In the third

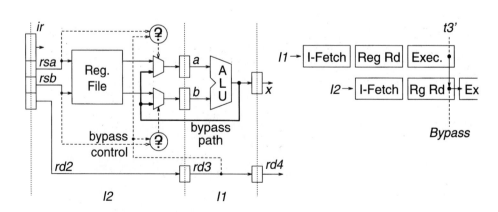

Figure 20.3 Bypassing from the third to the second stage

clock cycle, *I1* is in the third pipeline stage, computing the new value of *R3,* while *I2 is* in the second stage, reading the old value of *R3* from the top port of the register file (*rsa* = 3). What needs to be done is to switch the top multiplexer to its bypass input so that the old value is discarded and the new value is used instead. The control circuit necessary to make this decision is relatively simple, as shown in Fig. 20.3. The destination register of *I1, nd3,* is compared to the two source registers of *I2, rsa* and *rsb* (notice here the important distinction between *rd2* (*I2*'s destination) and *rd3* (*I1*'s destination)). Whenever equality is detected, this signals a dependence, and bypassing must be enabled. One additional detail to be checked is whether the instruction that is currently in the third (execution) stage has its write-destination-register flag asserted, since not all instructions write a result into a register. Time-wise, bypassing is decided during the second stage of the "receiving" instruction (*t3* in Fig. 20.2), and it actually occurs at the very end of that cycle (*t3'* in Fig. 20.3). Notice that this bypassing technique allows a series of consecutive dependent instructions to be executed at the maximum possible speed: the result of an operation in the ALU is fed back to it right away and is used as an input to the next operation, in the very next cycle. We will return to this point in Section 20.3.1. The clock cycle is determined by the delay of the ALU plus one multiplexer and one register delay.

Now consider a more complicated case. If *I1* is a *load* instruction, then its result is produced in the fourth pipeline stage, rather than in the third as for the *add* instruction above. This means that when *I2* enters its execution stage, at the beginning of the fourth clock cycle in Fig. 20.2, the result of *I1* is *not yet* available, so it cannot be bypassed! In other words, either *I2* must not use the result of *I1*, or *I2* will have to wait, with a resulting performance penalty. Figure 20.4 shows the former case, where *I2* does not read the register into which *I1* will write. Now, assume that the next instruction, *I3*, does need the result of *I1*, as in the example of Fig. 20.4: *I1*: load *M*[...] → *R3*; *I2*: ...; *I3*: sub *R3* − *R4* → *R5*. Can *I3* be executed without any pipeline delay? As we see, when *I3* enters its execution stage, in the fifth clock cycle, the desired new value of *R3* has been produced by *I1*, since it has been returned from the data cache to the processor in the fourth clock cycle; hence, *I3* can take and use it with bypassing. Figure 20.4 shows (in bold line) the new bypass path that must be added to the data path of Fig. 20.3 for this new bypassing to work. As shown

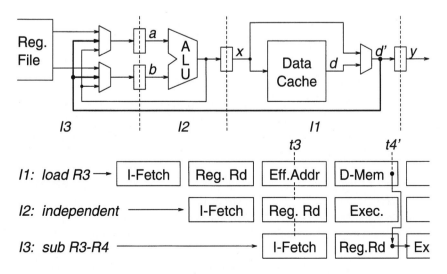

Figure 20.4 Bypassing from the fourth to the second stage

in the timing diagram, in the lower part of the figure, this new path is used at time *t4′*, right after the completion of the memory access of *I*1, and right before *I*3 enters the execution stage. As before, the bypass decision is made in the fourth cycle, when *I*3 is in the second stage; the control circuit that makes the decision compares the source register numbers of *I*3, *rsa* and *rsb* (Fig. 20.3), against the destination register number of *I*1, *rd4* (Fig. 20.1). Notice that bypassing works correctly whether *I*1 is a load or an arithmetic instruction: the value bypassed is *d′*, which will equal *d* if *I*1 is a load, or *x* if *I*1 is an arithmetic/logic instruction.

What happens if *I*1 is a *load* instruction, and *I*2 is not independent of it as assumed in Fig. 20.4? One of the original research RISC machines, the MIPS processor [8], would let *I*2 execute right away and use the *old* value of its source registers. It was left to the compiler to place a no-operation instruction after the load, if no other useful and independent instruction could be found, when the new value of the source registers was needed; this was called *delayed load*. The modern commercial RISC processors delay the execution of *I*2 for one cycle when they detect a dependence from its previous load instruction. After this one cycle delay, bypassing can be used to let *I*2 proceed, as in Fig. 20.4. All other instructions beyond *I*2 are also delayed by one cycle, so there is a performance penalty in this case (similar to the execution of a useless no-operation instruction in the research MIPS machine). In Section 20.3, we will discuss how the optimizing compiler should rearrange the object code of a RISC program in order to avoid these performance losses and how this relates to CISC architectures.

20.3   Dependences and Parallelism in CISC and in RISC

When there is no low-level parallelism in a program, the operation dependences are the limiting factor for performance. When low-level parallelism exists, the limiting factor for performance is the number and type of hardware resources that are available for performing the operations. In this section, we will see how CISC architectures try to approach these limits and how RISC architectures do it. The conclusion will be that the RISC approach is more flexible (it applies to more cases) and less costly. This section will also show why implementing a task "in hardware" (as a single instruction) is not necessarily faster than implementing it "in software" (using multiple, simpler instructions).

20.3.1   Dependence and resource limited performance

We will use a typical CISC instruction as a first, long example: a memory-to-memory, three-operand add; i.e., two words must be read from memory, added together, and the result written to a third location in memory. We will compare how this task is performed in a CISC machine and in a RISC machine. We assume a frequently used memory addressing mode (which is also the usual or the only RISC addressing mode): the address of each operand is the sum of the contents of a register plus a constant that is contained in the instruction. We assume *similar data path resources* for the two machines: a single-ported data cache, which is independent of the instruction cache, one general-purpose arithmetic/logic unit, and a register file. We count the number of cycles for execution, assuming that one cycle can contain a register file access, or an ALU operation, or a data cache access.

A first observation is that because our task requires three memory accesses, and because we have a single-ported data cache, it is impossible for the task to be completed in less than three cycles. This number—the number of operations divided by the number of functional units that can perform them—is one of the two fundamental performance limits that were mentioned above. The second observation is that the memory write access cannot be performed before the addition is completed, the addition cannot start before both memory read accesses have been completed, and the read accesses cannot start before their address calculations are done. Given that an address calculation takes two cycles (register read, add), and the two memory reads cannot be performed concurrently, our task cannot be completed in less than six cycles (two for the first address calculation, two for the memory reads, one for the data addition, and one for the memory write). This new lower limit also takes into consideration the second fundamental performance limit that was mentioned above: the data dependences, i.e., the fact that certain operations cannot be started before certain others have completed.

Now let us see how CISC and RISC architectures try to achieve high performance; that is, how they try to approach the above lower limit on the number of execution cycles. Figure 20.5 shows a possible implementation of our task as a single instruction in a CISC machine with the above hardware resources, and a typical implementation as a small piece of code in a RISC machine with the pipeline of Section 20.2. The idea behind complex instructions like this one in CISC architectures was to have a large enough task in the instruction that considerable parallelism would exist in it, and hence this parallelism could be exploited. Accordingly, Fig. 20.5 shows an implementation of this "add_memory" (*addm*) instruction with up to three operations going on in parallel. Not all CISC machines would do it in this way. There are several reasons [such as design complexity, structured microprogramming (e.g. microsubroutines), and so on] that have led several CISC machines to suboptimal implementations of instructions. However, the possibility shown in the figure is one of the claimed advantages of CISC, so it is of interest to examine how it relates to RISC.

The bottom part of Fig. 20.5 shows how this task would be implemented in RISC. Two *load* instructions are used to bring the desired addends from memory into temporary registers. The effective addresses of the *loads* are sums of a register and a constant, each; this is

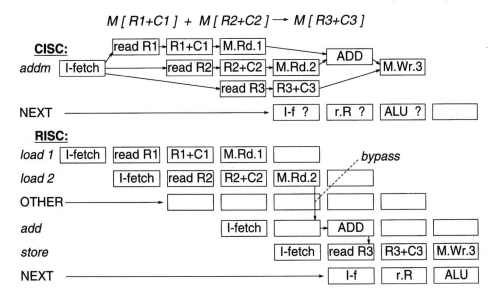

Figure 20.5 Memory-to-memory addition in CISC and RISC

slightly different from what was said (for simplicity) in Section 20.2 (sum of two registers), but the respective timing and pipeline organization remain the same. An *add* instruction follows after the two *loads,* adding the two above temporary registers and placing the result into a third one. This is a case like the one of Fig. 20.4, so we will either lose a cycle or insert another independent instruction between the second *load* and the *add.* Figure 20.5 shows the latter situation, labeling this instruction as *OTHER;* Section 20.3.2 discusses what this instruction can be. A *store* instruction follows after the *add,* writing the contents of the third temporary register into the desired memory address. As seen in Fig. 20.5, the load data are bypassed to the *add* instruction, and the result of the latter is bypassed to the *store* instruction. With these bypassings, the timing of the RISC implementation becomes very much similar to the (optimistic) CISC implementation shown in the same figure. The only difference is that the memory write access is performed one cycle later than in the CISC machine, due to the *store* instruction following after the *add* instruction in RISC.

To determine the actual performance, it is necessary to examine both cases in their overall execution context. This can be done by examining when the "next" instruction can start execution. Figure 20.5 shows such an instruction, labeled *NEXT.* For the RISC case, its timing is clear. For the CISC case, Fig. 20.5 shows an *aggressive* design (maximum possible pipelining for the given hardware resources), which many actual CISC machines do not follow. Many CISC machines use no pipelining, or use a modest amount of it, due to several design complexities that arise from the combination of complex instructions and pipelining (see Section 20.5.1). In the case of such an aggressive CISC design, the *NEXT* instruction is executed one cycle earlier in CISC than in RISC. However, RISC has executed one additional instruction in the meanwhile—the *OTHER* instruction. If this other instruction is a useful one, which it often is, then CISC will also have to execute that, and both machines will achieve exactly the same performance on this example.

No premature conclusions should be drawn based on this single example. In reality, many things can be different from the above simplified picture, as discussed further in Sections 20.4 and 20.5: the cycle time of the two machines may not be the same, the CISC machine may have less pipelining, the CISC implementation of the instruction may not be as good as in Fig. 20.5, there may be an additional "instruction decode" cycle in the CISC machine, the relative cycle counts may be different for other tasks, and so on. Still, several interesting facts can be discussed based on the above example, and we do so in the rest of this section.

First of all, as the above example illustrated, when performance is limited by dependences, there is no advantage to be gained by going to complex instructions. Even though the typical RISC (integer) instruction takes five cycles to complete (Section 20.2.1), *bypassing* (Section 20.2.2) allows dependent computations to be performed at the peak rate and with the minimum latency allowed by the ALU and data cache hardware delays. As seen in Fig. 20.3, an ALU computation using the result of a previous similar computation is performed in RISC with a delay relative to the previous computation as short as the ALU hardware delay plus one multiplexor and one register delay (the ALU delay being the dominant one among the three). Nothing better than that is possible—in CISC or in RISC. Notice that we do not discuss here issues of code size and instruction fetching; see Section 20.4 for that. Also, we assume well balanced delays among the ALU, the register file, and the cache memories; the slowest of them determines the clock cycle time, so we want all of them to be close to each other. This latter point affects RISC and CISC processors in a similar way (except that CISC may have a disadvantage if its (complex) control circuitry is slower than the above three data path components).

20.3.2 Instruction scheduling: RISC + compiler vs. CISC

To gain performance, we try to find useful operations to be performed in parallel with and while waiting for the results of other operations whose delay, due to dependences, cannot be reduced. In CISC, this is primarily done by defining compound instructions that contain a number of operations, some of which can be performed in parallel with the others. One example was seen in Fig. 20.5: while the first operand is being read from $M[R1 + C1]$ the address of the second operand is being computed, and so on. Consider another example: *auto-increment* (or autodecrement) addressing mode. In this example, while an operand is being read from memory, the register that was used for computing its address is incremented (or decremented) by the size of the operand. This is sometimes useful, e.g., when a program is sequentially stepping through an array of operands, and hence auto-incrementing prepares the register for the access to the next element in the array. (Statistics show that the number of cases where this applies is not very impressive: about 3 percent of the VAX memory operands in three benchmark programs used auto-increment/decrement mode (see Ref. [1] p. 170); yet, the point to be made here is independent of this low usage.)

In RISC, the execution of useful operations in parallel with and while waiting for the results of other operations is done first by exploiting the *pipelined* execution of instructions (as in Fig. 20.5), and second by the *compiler,* which properly *schedules* the sequence of instructions in a program. Thus, while in CISC the operations to be executed in parallel have to have some relation with each other so that it is realistic for them to be encoded as a single instruction, in RISC these operations can be *unrelated to* each other, since they are described by separate instructions each. This gives RISC a considerable advantage relative to CISC: RISC is *more flexible* than CISC in finding and exploiting parallelism. To put it in different words, CISC determines and defines the exploitable instruction-level parallelism at machine-design time—roughly, once per decade, and once globally—while RISC does so at compile time—once per program, and specifically for that program.

Instruction scheduling [9] is one of the basic techniques in modern compiling (see Ref. [1], p. 114). It appeared at about the same time as RISC architectures, and it enabled RISC to achieve better performance than CISC at lower cost. It applies to all operations whose execution takes more than one clock cycle. Typically, these are loads, branches, and floating-point operations. We will illustrate the technique using loads as an example. By carefully studying Fig. 20.4, the reader will see that, in a processor where a data cache access takes N clock cycles (usually $N = 1$), the N instructions after any *load* instruction should be independent of the data being loaded (and perform useful work) to avoid performance loss. These N (usually one) instructions are often called *load delay slots,* because they are instruction slots that the compiler should try to fill with operations independent of the loaded data for the load delay to be "hidden." Modern optimizing compilers do this by instruction rearrangement (code motion): some instructions are moved later or earlier than their previous position in a block of code. To preserve program correctness, this motion (usually) can be done only when the instructions being swapped are mutually independent and reside in the same basic block (a block of code always executed in its entirety, i.e., containing no branches or labels).

As a first example of instruction scheduling, and specifically of filling load delay slots with independent instructions, consider the above case of auto-increment addressing mode. Assume that we are in a loop that was originally compiled as: *load Rtmp ← M[Rp + offs]*; *compare Rtmp to Rkey; increment Rp; branch.* The *compare* and the *increment* instructions are independent of each other, and thus their positions can be interchanged. This interchange leads to better performance, since the incrementation is independent of the load—a fact that was not true for the comparison. In the new code sequence, the first two instructions, *load* and *increment,* are the equivalent of the CISC auto-increment addressing mode. Since the incrementation is done by a separate instruction, the address register can be changed by any amount—not just the size of the operand being loaded as is customary in CISC auto-increment addressing mode. For example, when stepping through an array of structures, this must be the size of the structure. As seen, RISC is more flexible than CISC here.

As a second example of filling load delay slots, consider a case that can be exploited in RISC, owing to its flexibility, but not in CISC. Assume that we wish to add three integers from memory, *A, B,* and *C* (with addressing mode, as in Fig. 20.5), and we wish the sum to be placed into register *Rd.* In CISC, this will be done using two addition instructions: *add Rd ← A + B; add Rd ← Rd + C.* The first of these instructions includes two reads from data memory; the processor stays idle during the second of them, waiting for the data to arrive from the cache (the timing is as in Fig. 20.5, but no *R3 + C3* needs to be computed now). The second instruction includes one read from data memory; the processor also stays idle during that period.

In RISC, the code sequence starts out as: *load Rd ← A; load Rt ← B; add Rd ← Rd + Rt; load Rt ← C; add Rd ← Rd + Rt.* The first *add is* in the delay slot of the second *load,* but it depends on that load, so it is desirable to move it farther down in the instruction block. At first sight, it appears not to be interchangeable with the third *load,* because that *load* alters *Rt*—one of the source registers of the *add.* However, *Rt* is just a temporary register, used first to hold *B* and then to hold *C.* There is no need to use the same temporary register for the two purposes; if we use *Rt*1 for *B* and *Rt*2 for *C,* the first *add* and the third *load* become mutually independent, hence interchangeable. This technique is called *anti-dependence elimination* in instruction scheduling terminology. Now, the RISC code sequence becomes: *load Rd ← A; load R1 ← B; load Rt2 ← C; add Rd ← Rd + Rt1; add Rd ← Rd + Rt2.* In this new code sequence, the third *load* fills the delay slot of the second *load,* and the first *add* does the same for the third *load;* the pipeline never stalls, and the processor never stays idle! The reader may draw, as an exercise, a diagram similar to Fig. 20.5 for this case. If we label as "1" the cycle when the first instruction of each code sequence is fetched, then CISC will fetch its second instruction in cycle 5, and it will fetch the *NEXT* instruction after this code block in cycle 9 (instructions cannot be overlapped so much that their ALU cycles coincide, since we assumed a single ALU, and no CISC overlaps instructions so much as to interchange their ALU cycles, because of the difficulties mentioned in Section 20.5.1). On the contrary, RISC will fetch each of the five instructions of its (new) code sequence in cycles 1, 2, 3, 4, and 5, respectively, and will fetch its *NEXT* instruction after the block in cycle 6: RISC is faster than CISC by 3 clock cycles (or 60 percent) in this example.

Instruction scheduling at compile time is called *static* because it is done once for all executions of a specific code segment. It is also possible to perform instruction scheduling *dynamically,* i.e., at run time (see Ref. [1] Section 6.7, and Ref. [10]); obviously, this cannot be done by the compiler—it must be done by the hardware (by the processor control unit). Dynamic scheduling is also called *out-of-order execution* of instructions; supercomputers of the past have supported this. Dynamic scheduling is obviously much more expensive than static scheduling, since the implementation of the former needs considerable hardware, while the latter is done purely in software and not at run time; among others, dynamic scheduling usually costs one extra pipeline stage, which increases the branch cost (see Section 20.4.3). Dynamic scheduling is useful only when static scheduling cannot fill all the delay slots because it is not known at compile time whether some instructions are mutually independent so as to be able to interchange their positions. In these cases, dynamic scheduling will interchange their execution at run time, only when they actually turn out to be independent, thus saving clock cycles in those cases. The most interesting of such cases of potential dependences that cannot be resolved at compile time are (1) instructions not in the same basic block (instruction scheduling across conditional branches), and (2) memory loads and stores whose effective addresses cannot be proven to differ from each other (compile-time "address disambiguation" fails). However, modern compiling has progressed considerably: the former case frequently can be treated by trace scheduling, loop unrolling, or software pipelining (see Ref. [1], Section 6.8), while sophisticated address disambiguation techniques [11] have reduced the frequency of occurrence of the latter case. It follows that dynamic scheduling, with its high cost, is rarely needed and thus rarely provided in modern processors. The general advice is: Do not postpone for run time what you can do at compile time!

20.3.3   Parallel hardware: superscalar RISC vs. CISC

In all of the preceding examples, we assumed that the CISC processor has one ALU, a three-port register file, and a single-port data cache, just like the typical RISC processor (Fig. 20.1). Was that a fair comparison? What if the CISC designer is willing to pay for more functional units in the data path so that the multiple operations in the complex instructions can all be performed in parallel and faster than the typical RISC processor of Section 20.2? In fact, the way we compared RISC and CISC up to now in this section *is* fair: the designer of any processor—CISC or RISC—can decide to include in its data path more functional units than what was assumed in Fig. 20.1. Usually, these are additional register file ports, additional ALUs, and more complicated instruction decoding and control logic. (Multiported data caches are usually too expensive to justify having them.) These additional functional units increase the cost of the processor (e.g., its silicon area), but also increase its performance, since more operations can now be performed in parallel. (Beyond a certain number of unit, performance gains become minuscule relative to the cost increase, so this is *not* a source of infinite performance improvement.)

CISC processors exploit their multiple functional units by performing in parallel the multiple operations in the currently executing complex instruction. RISC processors, on the other hand, exploit their multiple functional units by fetching, and subsequently executing, multiple *instructions* in each clock cycle of their pipeline. These are called *superscalar or* VLIW (very long instruction word) processors; a close relative of theirs are the *superpipelined* processors; see Refs. [7, 10, 12–16]. What was said for a single-ALU processor also applies to the case of multiple functional units: the RISC method of independently controlling each functional unit via a separate instruction in each clock cycle, and of scheduling the placement of these instructions during compilation, is more flexible and yields better results than the CISC method.

20.4   Instruction Alignment, Size, and Format

One of the differentiating characteristics of RISC relative to the earlier CISC architectures is the (almost always) fixed 32-bit instruction size, and the few, simplified instruction formats. In this section we explain the reasons behind these choices. Since they are related to the execution of control transfer instructions, we also treat the topic of delayed branches in Section 20.4.2.

20.4.1 Code size, I-cache Size, and fetch throughput

The size of usual RISC programs (binary code size) is considerably larger (usually 20 to 70 percent) than the size of the same programs in the popular CISC architectures of the 1970s and 1980s. Examples are measurements described in Ref. [1] (p. 79) of 35 to 70 percent, and Ref. [17] (p. 80) of 20 to 50 percent. The reason is twofold: (1) there are more instructions for a given task in RISC than in CISC (50 to 180 percent more instructions executed (dynamic measurements) by RISC relative to the VAX, according to Ref. [1], p. 123), and (2) more code space is lost in RISC due to "fragmentation," because RISC code is quantized in multiples of 32 bits, while CISC instruction size is often quantized in multiples of 8 bits. The former—more instructions in RISC—is a result of the smaller and simpler RISC instruction set, the advantages of which are explained throughout this chapter. The reason for the latter—the simple instruction format and fixed instruction size of RISC—is the saving of the alignment and decoding time, as detailed in Section 20.4.3. As for the disadvantages of the resulting increased code size in RISC, three of these exist or have been claimed at various times, and we proceed to discuss them here.

First, the increased code size of RISC means that more disk space is needed for the files containing the binary code of executable programs. A similar case holds for the main memory of the computer. The evolution of the RISC versus CISC controversy in the eight years that have passed since RISC appeared commercially shows that the customers prefer RISC for its higher performance and do not mind paying a little bit more for the increased disk and memory space. Factors in favor of RISC are: (1) the price per byte of disks and memory is steadily decreasing; (2) binary code occupies only a small fraction of disks and memory—data usually occupy the majority; and (3) dynamic linking of library routines has further reduced the above fraction.

Second, the increased code size of RISC means that a correspondingly larger instruction cache is needed to hold an equivalent working set, thus achieving a similar hit ratio. Given that it is desirable to have the (first level) instruction cache on the same chip as the processor, the chip area limitation poses a performance problem: for a given cache area (and, hence, capacity) RISC will suffer a lower hit ratio. On the other hand, the fixed instruction size, the simple instruction format, and the reduced instruction set of RISC result in large savings of processor chip area; no alignment is needed, instruction decoding is simplified (see Section 20.4.3), and no microcode memory is needed (see Section 20.5.2). Although it is difficult to precisely quantify these savings, in general they are more important than the increase needed in instruction cache size in order to maintain a similar hit ratio.

Third, it has been claimed that processor performance is limited to a large extent by the throughput of the path that feeds instructions to the processor, and hence, since RISC has a larger code size and needs a correspondingly larger instruction fetch throughput, the performance of RISC will be worse because of this effect. Today, when the instruction cache is on the same chip with the processor, it is certainly *not* true that the instruction fetch throughput is a bottleneck. But, even at the time when this claim was made—when there was no cache or the cache was not on the processor chip—this claim did not reflect reality. To see why, we must differentiate between *average* and *peak* instruction fetch throughput. CISC indeed needs less *average* fetch throughput than RISC, as we saw above. However, for an efficient (CISC) computer design, the *peak* throughput of the instruction fetch path must be significantly higher than the above average to avoid loss of performance on long but fast-executing instructions, such as to set a register to a 32-bit constant, where the constant is contained in the instruction. If the actual circuit paths can supply this higher instantaneous throughput when it is needed, there is no reason why they could not do so on a longer-term basis; the average throughput can be made equal to its peak instantaneous value at no extra cost. RISC does just that: its peak instantaneous requirements are for one instruction (32 bits) per cycle, and its average is the same in the lack of pipeline dependences. It is very hard to imagine a well performing CISC with a peak fetch throughput of less than 32 bits per cycle, because (1) most circuit path widths are quantized to powers of 2, and (2) there are several frequent instructions that need to execute in one cycle and that do not fit in 16 bits. The claim about reduced fetch throughput yielding higher performance looks more realistic if the processor contains an *instruction fetch buffer*—a FIFO queue that absorbs the fetch throughput fluctuations. However, such an instruction buffer must be invalidated after every successful branch instruction—roughly every 10 to 20 clock cycles—and thus its effect is greatly reduced. Finally, instructions whose size is not an integer multiple of the word width will present the problem of alignment, which is particularly manifest when the instruction at a branch target is split between multiple memory words, as discussed in Section 20.4.3.

20.4.2   Branches under pipelining: delayed branches

We divert briefly from our discussion of instruction size to see how branch instructions are executed in a pipelined processor, and especially to discuss delayed branches, which are a form of instruction scheduling (see Section 20.3.2). Figure 20.6 (top) shows the first two stages of the pipelined data path of Figs. 20.1 and 20.3, with the additional paths needed to fetch sequential instructions and to execute conditional branches. PC-relative addressing mode is assumed for the branches, as is customary in all modern processors: the branch target is the sum of the address of the branch instruction itself plus a (sign-extended) constant *offset,* which is contained in the instruction. Figure 20.6 shows a "fast" version of branch, which can be completely executed in the second stage of the pipeline. To achieve that, the branch condition can only be the result of testing a source register for sign and/or equality to zero; no register-to-register comparisons are provided, since they would require

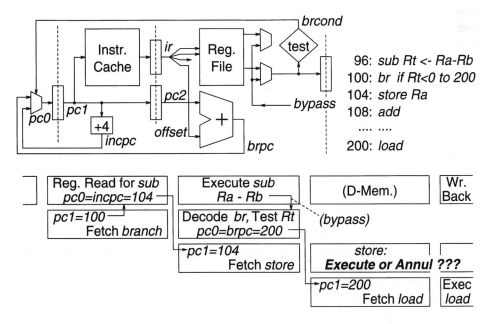

Figure 20.6 Conditional branch data path and timing

a full ALU carry-chain propagation delay. The register that is tested may contain the result of a previous full register-to-register comparison, in the form of a subtraction, as shown in the sample program fragment in Fig. 20.6. Using general-purpose registers rather than special-purpose *condition code* bits to store the result of comparisons increases the flexibility of instruction scheduling (see Section 20.3.2) and simplifies the hardware; see also Ref. [8] and Ref. [1], p. 106.

In Fig. 20.6, $pc1$ is the program counter (PC) of the instruction that is currently in the first stage of the pipeline: the address from which the instruction is being fetched; $pc2$ is the PC of the instruction currently in the second stage; $pc0$ is the address from which an instruction will be fetched in the next clock cycle. This latter instruction, from address $pc0$, can be determined in two different ways: either as the next sequential instruction after the one currently being fetched (*incpc*), or as the target of a branch instruction currently being decoded and executed in the second pipe stage (*brpc*). Due to pipelining, when a branch instruction is executed, it can only affect the instruction fetched *after its next* instruction. In the example program fragment shown in Fig. 20.6, the *store* instruction is fetched from address 104 while the *branch* is being decoded and executed—the *branch* can only affect whether the instruction fetched after the *store* will be the *add* from address 108 (branch failure) or the *load* from address 200 (branch success). As seen, branch instructions have a *delayed* effect that is quite similar to the delayed effect of load instructions seen in Section 20.2.2.

What should be done with the "extra" instruction that is fetched while the branch is being executed (*store* in Fig. 20.6) when the branch is successful? Should the processor always *annul* (squash, abort) it, like traditional processors (including CISC) did? But then, successful branch instructions will cost two clock cycles—one more than normal. Alternatively, the processor can always execute this "extra" instruction; this is called *delayed branch*, and the place occupied by the "extra" instruction is called *delay slot*. The advantage is obviously that no clock cycle is lost; the "problem" is that the instruction in the delay slot is always executed, regardless of the branch outcome, as if it were *before* the branch, yet the branch instruction cannot depend on it, since this instruction is actually executed *after* the branch. This strange behavior was considered undesirable in the old CISC machines, since the instruction set of those processors had to be "clean" to be appropriate for hand coding in assembly language. However, with the advent of optimizing compilers, the above behavior becomes merely another case of instruction scheduling, similar to the case of load instructions that was presented in Section 20.3.2. Hence, many RISC architectures do have delayed branches, and they similarly have delayed jumps, calls, and returns (i.e., unconditional control transfers).

There are many issues related to branch optimization, and much literature on them. For example, see Ref. [1], pp. 272–277 and 307–314, and Refs. [18] and [19]. Due to lack of space, we only mention some of the issues, and only briefly. In processors where branches take more cycles to be resolved, there are correspondingly more delay slots. Delay slots can be filled either (preferably) with instructions that are moved there *from before* the branch, provided there is independence that allows them to move, or else with instructions that are moved there from one of the two branch targets—preferably from the most likely *predicted target*, provided that these are *harmless* when the branch goes in the other direction, or that the hardware *annuls* them when that happens, or else with *noop* instructions (instructions that do nothing). In case the target address can be computed before the branch condition becomes resolved, the instructions fetched (and conditionally executed) in the meanwhile can be from that target, if the branch is predicted to usually succeed. The first (or only) delay slot after branches can be filled with useful instructions in many cases;

50 percent of the time, according to Ref. [1], p. 276, to 70 percent, according to Ref. [18]. Things are harder for the second delay slot (it could only be usefully filled 20 percent of the time in Ref. [18]), and become progressively harder beyond that. Some modern RISC processors [14] have abandoned delayed branches in favor of branch prediction because of the many delay slots and the corresponding inefficiency of filling them, especially in superscalar implementations, and because the number of delay slots may change with the pipeline, thus forcing an undesirable change in the instruction set between generations in a processor family.

20.4.3 Alignment and decoding vs. branch latency

We return now to the discussion of instruction size and format, with the goal of explaining the disadvantages of odd sizes and complex formats. When there are instructions whose size is not an integer multiple of the instruction cache word width, any instruction with size sz words, where $n - 1 < sz \leq n$ (n is integer), may be fragmented across $n + 1$ cache words (misaligned instructions). Figure 20.7a shows this, e.g. for instruction $t1$. This entails that $n + 1$ cache accesses may be required in order to fetch an instruction whose size does not exceed n words. In general, this means one extra clock cycle for fetching misaligned instructions. This is not precisely true, because a one-word instruction buffer (register) *Ibuf* can be used to save the remainder of the previously fetched word after the previous instruction is extracted from that word, as shown in Fig. 20.7b. However, this buffer is ineffective after successful branches: in Fig. 20.7a, when the *br* instruction is fetched and executed, *Ibuf* contains the remainder of word 72 (byte 75), which is useless for the target instruction $t1$ sought—two new accesses to words 88 and 92 will be required in order to fetch this 3-byte instruction.

When instruction sizes are not always integer multiples of words, an alignment network is needed to bring the bits of the instructions to their desirable places from whichever positions in the fetched words they may be in. Figure 20.7b shows the location of this alignment circuit, which is much like a barrel shifter, in the data path between the instruction cache and the register file. Its inputs are the remainder of the previously fetched word, from *Ibuf,* and the newly fetched word; its outputs are the aligned instruction and the new remainder.

When there are many and complicated instruction formats, another block of circuits is needed to at least partly decode the instruction and extract from it the fields of interest for further execution. This decoding must proceed at least to the point of determining the format of the particular instruction. Field extraction is another case of (basically) barrel

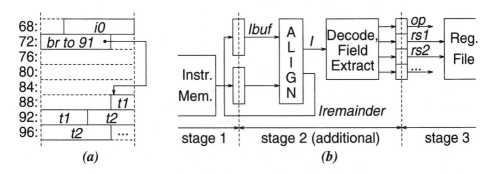

Figure 20.7 Instruction alignment and decoding: (a) fragmented branch target and (b) extra pipe stage

shifter(s). The decoding and field extraction circuit must follow the alignment network, because otherwise it would not know where the opcode to be decoded is located.

Alignment, decoding, and field extraction take time to be performed, and the instruction cannot start being executed before they have been completed at least to the point of knowing the first source operand(s). In complex instruction sets, this means an extra clock cycle or an *extra pipeline stage*. In non-pipelined processors, one extra clock cycle per instruction is a heavy performance penalty. In pipelined processors, the additional pipeline stage is not visible upon sequential instruction execution, because bypassing works as before—one cycle later for *all* instructions. However, the additional pipeline stage increases by one the number of *branch delay slots*. Given that successful branches are about 9 percent of the instructions executed by RISC processors, and that the last one of multiple delay slots usually cannot be filled, this additional pipeline stage would bring a performance penalty of about 6 to 8 percent. Avoiding this performance penalty and the additional penalty for misaligned branch targets, as well as the transistor (chip area) cost for the extra circuits, are the primary reasons why most RISC processors have instructions with a fixed 32-bit size and with simple formats as described below.

20.4.4 The style of RISC instruction formats

The first and foremost rule for a good RISC instruction format is: *keep the source register fields always at fixed positions*. As seen from Figs. 20.3 and 20.6, the critical timing path of the second pipeline stage is the reading from the register file. If the addresses for these accesses are not always at the same place in the instruction, (partial) opcode decoding and field extraction multiplexers will be needed before this read access can start, thus lengthening the clock cycle time or requiring a faster (hence more power-consuming) register file. Notice that some instructions have fewer than two source registers. However, since the register file has two read ports, RISC processors always read two registers and later throw away any unneeded values; thus, when only one source register exists, its address must be at one of the two standard source register locations.

Figure 20.8 shows the typical set of three instruction formats of most RISC processors; variations exist, of course, and a processor may have other formats as well, but Fig. 20.8 conveys the essence of the RISC formats. The format with the most stringent space requirements is the middle one, often called *immediate* (*I*); it contains the opcode, two register operands, and an immediate constant. We would like the immediate constant to be quite wide. In practice, it is as wide as the bits that are left over from the other three fields; Fig. 20.8 shows typical widths. In the top format, often called *register* (*R*), part of the immedi-

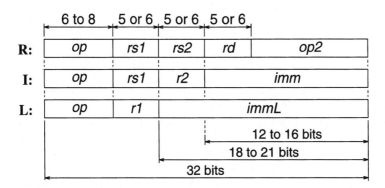

Figure 20.8 The typical set of RISC instruction formats

ate field is used as a third register operand, and the rest is available as extra opcode space. When only one register operand exists, the second register field can be joined with the immediate field to form a *long* (*L*) immediate format; format(s) with even longer immediate(s) may exist, when there is no register operand, and even possibly using a shorter opcode.

There is at least one instruction using the *I* format that has two register source operands the *store* instruction: one source register and the constant are used to form the address (see Section 20.3.1), while the other source register provides the data to be stored in memory. Since *rs*1 and *r*2 must be sources in this case, it follows according to our rule above that the two source register fields should always be at those same locations. Accordingly, the *R* format has its second source, *rs*2, at the place of *r*2 of the *I* format. Now, consider the *load* instruction: it needs the *I* format, due to its addressing mode; but, now, its second register operand, *r*2, is a destination—not a source. It follows that, unlike the source registers, the destination register of the instruction is *not* always at a fixed location: sometimes it is at *rd*, sometimes at *r*2, and it may even be at *rl*, in the *L* format. This is *not* a problem (as it would be in the case of the source registers), because the destination register number is needed quite later during execution, thus allowing plenty of time for opcode decoding and field selection.

20.5   Implementation Disadvantages of CISC

So far, we have seen that a well designed pipeline (Section 20.2), with instructions that execute small parcels of work each, allows better instruction scheduling and exploitation of parallelism during dependence delays than complex instructions (Section 20.3). We also saw that a fixed, one-word instruction size, with fixed source operand positions in the format, allows a shorter and faster pipeline, on less silicon area (Section 20.4). We now proceed to examine the other advantages of RISC, by describing the difficulties and costs of implementing CISC, and the lack of justification for paying these costs.

20.5.1   Pipeline and interrupt difficulties in CISC

All RISC instructions perform a similar amount of work each, in a fixed number of pipeline stages (see Ref. [1], Section 6.6, for the case of floating-point instructions). In contrast, CISC instructions perform variable amounts of work each. Thus, if one wants to design a CISC processor with substantial amounts of pipelining, its pipeline will have to have variable length. Besides the general hardware and design time cost of controlling such an irregular pipeline, there are a number of other problems to be solved, as outlined below. Although there are known solutions to these problems (see e.g., Ref. [1], section 6.5, and Ref. [10]), the reader will appreciate the amount of complexity that is saved by going to RISC architectures.

Figure 20.9 shows again the execution of a memory-to-memory add CISC instruction, as in Fig. 20.5. This time it is followed by two possible versions of execution of a simple instruction: clear (zero) register *R*3. Version 1 is very aggressive, but it helps illustrate a general problem that may easily arise in less aggressive cases: the old contents of *R*3 will be destroyed (cleared) by the latter instruction before the former instruction reads and uses them as it is supposed to do. Version 2 of the *clear R*3 execution is more conservative, and it illustrates a more subtle problem: What will happen if the third memory access of the addln instruction causes a page fault interrupt (the memory page that it tries to access is on disk)? By the time this page fault occurs (last clock cycle in Fig. 20.9), the subsequent instruction has already been executed, and it has destroyed *R*3. After the page fault has been

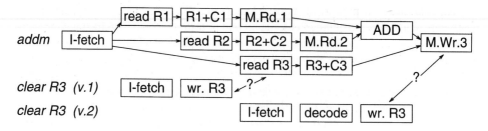

Figure 20.9 The typical set of RISC instruction formats

serviced (the operating system has brought the page in to memory), the *addm* instruction will have to be re-executed, but *R3* no longer contains the correct value! Another problem with page faults appears in instructions that write multiple results into registers and memory. Consider, e.g., a variation of the above *addm* instruction which uses autoincrement addressing mode for its first operand: *R1* is incremented, e.g., in parallel with the first memory access. Now, if the second or third memory access causes a page fault, *R1* will have an incorrect value when the instruction is re-executed after servicing the page fault.

20.5.2 Microinstruction memory versus instruction cache

Owing to the simple architecture, the control section of a RISC processor is quite small. In RISC II [17], it occupied 10 percent of the processor silicon area; the opcode decoder is even smaller (0.5 percent of the area in RISC II) and is usually implemented as a simple programmable logic array (PLA)—no microprogramming is used. By contrast, the large number and complexity of CISC instructions requires a large control section; e.g., about 40 percent of the Motorola 68000 processor silicon area, and 5 of the 9 processor chips of the VLSI VAX [20]. A large portion of the CISC control section is occupied by the *microcode ROM* (microinstruction memory); it is not realistic to implement complex instruction sets without microprogramming. The VAX had half a megabit of microcode memory.

In some sense, there is an analogy between RISC instructions and CISC microinstructions: the VAX microinstructions looked like RISC ALU instructions. Following this analogy, the microcode ROM could be compared to a RISC instruction cache. In this sense, an instruction cache is much preferable to microcode ROM: the former dynamically adjusts its contents at run time to the working set of the current program, while the contents of the latter are fixed once and for all. Furthermore, from the point of view of utilization of the precious resource that the processor silicon area represents, CISC makes a very poor choice of contents for their microcode ROM; the most complicated instructions and addressing modes account for most of the ROM bits and are executed the least frequently. For example, the microVAX-32 implementation [21] removed 85 percent of the original VAX microcode bits, implementing the corresponding instructions by trapping to software; the instructions that were thus removed from hardware accounted for about 2 percent of the execution.

20.5.3 Matching instructions to high-level languages

The main reason that led to more complex instructions in the CISC times was the desire to "match" the hardware to frequent programming constructs, to improve the performance of the latter and to ease the task of the assembly language programmer. Ironically, complex instructions had the opposite effect, especially with the advent of modern compilers. Today, optimizing compilers usually produce better RISC code than hand-written assembly, because they can perform a more thorough dependency analysis and hence better instruction scheduling. On the other hand, complex instructions are often either inappropriate for use, or they result in lower performance. This is due to the following phenomenon: If a complex instruction is made very general, so that the compiler can match it to many cases, then its performance always suffers from its generality, although in many cases this generality may not be exploited by the actual program. On the other hand, if a complex instruction is made specific to some cases that are considered to be frequent, then the compiler may have a hard time identifying these cases, or the belief that they are frequent may not materialize. Two educative examples are the VAX *call* and the IBM 360 *mvc* instructions: their most frequent case of application turned out to be (on the VAX/ IBM-360) slower than if implemented using multiple other, simpler (VAX/ IBM-360!) instructions. But it was virtually impossible to anticipate this at the time the hardware was being designed (see Ref. [1] pp. 21 and 124).

20.5.4 Clock frequency, design cost, and time to market

Up to now, our comparison of RISC versus CISC has been based mostly on performance, and we estimated performance mostly by counting clock cycles. But there is much more to it than that. First, in electronics, *smaller is faster.* Thus, because RISC leads to simpler, smaller circuits, its clock frequency is higher than that of a CISC in the same technology. Second, because a RISC processor occupies less area on the chip, there is more room left for (instruction and data) *cache memories,* resulting in better hit ratios. The net result is that RISC is faster than CISC not only because the former executes tasks in fewer clock cycles, but also because its clock cycles are shorter and because its on-chip caches are larger than those of the latter.

Next, performance is not the only important factor; *cost* is the other. One component of cost is processor chip area, since the production yield depends critically on it. RISC reduces the processor area, so it lowers this component of the cost. Another important contribution to cost comes from the engineering *design time.* Since RISC hardware is much simpler than CISC, design time is dramatically reduced. For example, the (student) design of RISC II took 3 person-years [17], as compared to 14 person-years for the Motorola 68000. When one is willing to put in more effort, design time can better be spent on circuit tuning to increase the clock frequency, rather than on (suboptimally) implementing complex instructions that rarely will be used.

Finally, in today's fiercely competitive computer market, *time to market* is a factor of utmost importance for success. When a new concept, idea, or opportunity appears, the company that can first present it as a product to the market enjoys significant advantages. RISC, with its much simpler design, allows a significantly shorter time to market. Additionally, contemporary integrated circuit (IC) technology is such that a reduced time to market also leads to higher performance, for the following reason. Among all products that appear at the market at a given time, those that had shorter design times started being designed more recently, so they were able to make use of a more recent IC technology, which is always faster than older IC technologies.

20.6 History, Perspective, and Conclusions

20.6.1 A brief RISC history

When RISC appeared in the early 1980s, the instructions and the features that it proposed were not new. Cray's supercomputers have always had load/store architectures and simple instruction sets, since the 1960s. Delayed branches had appeared in 1952, in the Ferranti-Manchester "Maniac I" computer. The contribution of the researchers who introduced RISC was the fact that they realized that mini- and microcomputer architecture should go back to those concepts, at a time when all other architects in that price range had gone in a much different direction. Between 1980, when RISC was first publicly advocated, and 1984, the computer architecture community was rather sceptical about it.

The first RISC research project was the IBM 801 machine, started in 1975 from a seminal idea of John Cocke [22]; not much of it was known outside IBM until the early 1980s. The first public RISC research was carried out at the University of California at Berkeley, in 1980-83, under Prof. David Patterson [17, 23]. This was the team that coined the acronym "RISC." Closely thereafter, in 1981-84, another team, at Stanford University, under Prof. John Hennessy, started the MIPS project, which was the second public RISC research project [8, 24]. It took another couple of years for some industrial companies to be convinced about RISC and to come up with products. The Stanford team was instrumental in the founding of the MIPS company, which introduced the homonymous processor in 1986 [5]. The Berkeley team influenced SUN Microsystems in their making of the SPARC architecture (delivered in 1987) [6]. Hewlett-Packard converted their minicomputer line to RISC [25]. In 1987, major semiconductor manufacturers began to supply RISC microprocessors: AMD with its 29000, then Motorola with its 88000 (1988), then Intel with the i860 (1989). IBM introduced the RT-PC in the late 1980s, and then the more successful RS/6000 in 1990 [13], which has now evolved into the PowerPC (jointly developed by IBM, Apple, and Motorola) [15] [7]. MIPS has evolved into the SGI TFP [16]. DEC started using MIPS microprocessors in 1989, and then developed its own RISC microprocessor, the Alpha [14] [7]. In the 1990s, the industry at large is convinced about the superiority of RISC over CISC, but it took about a decade for this idea to get from conception to widespread acceptance.

20.6.2 Recapitulation and perspective

> *"Everything should be made as simple as possible—*
> *but not any simpler."*
>
> *A. Einstein*

The heart of the data path of a processor performs arithmetic operations on data that may result from previous operations or contents of processor registers (Fig. 20.3) or data read from memory (Fig. 20.4). Pipelining and bypassing (Section 20.2) are used to make these loops operate at top speed. Another short pipeline (Fig. 20.6) feeds the above data path with instructions. These circuits have very high flexibility: they form a *general-purpose* computer. The patterns of consecutive operations performed on these circuits show great variations. Consequently, it is useless to introduce another layer of abstraction and a second interface between these circuits and the software. That is RISC: the raw data path operations are exposed to and controlled by the object code generated by the compiler and the optimizer. Thus, we achieve, at the same time, optimal utilization of the available hardware parallelism under the given data and program dependences, using instruction scheduling (Section 20.3), *and* simplicity, low cost, and high speed of circuit design, fabrication, and operation (Section 20.5).

Fast arithmetic operations is only one-third the story; fast access to operands and fast access to instructions are the other two-thirds. The general-purpose registers in the processor are the fastest level of the memory hierarchy. Hence, it is essential to use them as much as possible and to use them efficiently. For this reason, RISC architectures rely to a large extent on register-resident operands; memory is accessed only through separate instructions (*load* and *store*). Instruction fetch and decode latency is minimized when instructions have a fixed size, when they are aligned on word boundaries in memory, and when the source operand field positions are fixed in them; RISC architectures do all of these. Reduced instruction set computers were the result of deeply understanding both hardware and software, and of applying quantitative engineering analysis to the design of both, and to the design of their interactions and their interface—the instruction set.

After the "plain" RISC processors of the 1980s, superscalar RISCs (Section 20.3.3) were the next step in the evolution, in the early 1990s. Currently, in the mid-1990s, it looks like the hottest topic is *latency tolerance,* i.e., finding and performing useful work in parallel with waiting for long-running operations to complete. Interesting techniques in this direction include multiple-context processors [26], access/execute decoupling [27], and software pipelining [28].

20.6.3   Reading list

For those readers who know very little about computer organization and processor design but want to learn more, we suggest starting with Ref. [2]; for those who know basic computer organization but want to learn about pipelining, we propose Chapter 6 of that same book. For a more advanced treatment of computer architecture in general, readers are referred to Ref. [1]; Chapter 6 of that book treats pipelining in more depth than does Ref. [2]. RISC architectures, as seen in the mid-1980s, are discussed in Refs. [17, 29, 30]. Interesting articles on contemporary RISC processors are Refs. [7, 10, 13–16, 25].

20.7   References

1. Hennessy, J. L., and D. A. Patterson. 1990. *Computer Architecture: A Quantitative Approach.* San Mateo, Calif.: Morgan Kaufmann.
2. Patterson, D. A., and J. L. Hennessy. 1994. *Computer Organization and Design: The Hardware/Software Interface.* San Mateo, Calif.: Morgan Kaufmann.
3. *DEC VAX 11 Architecture Handbook.* 1979. Maynard, Mass.: Digital Equipment Corp.
4. Kogge, P. M. 1981. *The Architecture of Pipelined Computers.* New York: McGraw-Hill.
5. Kane, G. 1987. *MIPS R2000 RISC Architecture.* Englewood Cliffs, N. J.:Prentice Hall.
6. Garner, R. A. et al. 1988. Scalable processor architecture (SPARC). *Proceedings, IEEE COMPCON,* San Francisco, 278-283.
7. Smith, J., and S. Weiss. 1994. PowerPC 601 and Alpha 21064: a tale of two RISCs. *IEEE Computer,* Vol. 27, No. 6, 46–58.
8. Hennessy, J., N. Jouppi, F. Baskett, T. Gross, J. Gill. 1982. Hardware/software Trade-offs for increased performance. *Proc. Symp. on Architectural Support for Programming Languages and Operating Systems* (ASPLOS-I), ACM SIGARCH-10.2 SIGPLAN-17.4, 2–11.
9. Hennessy, J. L., and T. R. Gross. 1983. Postpass code optimization of pipeline constraints. *ACM Trans. on Programming Languages and Systems,* Vol. 5, No. 3, 422–448.
10. Popescu, V. , et al. 1991. The Metaflow architecture. *IEEE Micro,* Vol. 11, No. 3, 10.
11. Padua, D., and M. Wolfe. 1986. Advanced compiler optimizations for supercomputers. *Communications of the ACM,* Vol. 29, No. 12, 1184–1201.
12. Jouppi, N. P., and D. W. Wall. 1989. Available instruction-level parallelism for superscalar and superpipelined machines. *Proc. Third Conf. on Architectural Support for Programming Languages and Operating Systems (ASPLOS-III),* IEEE/ACM, Boston, 272–282.
13. 1990. The IBM RISC System/6000 processor (collection of papers). *IBM Journal of Research and Development,* Vol. 34, No. 1.
14. Sites, R. 1993. Alpha AXP Architecture. *Communications of the ACM,* Vol. 36, No. 2, 33–44.

15. Diefendorff, K, R. Oehler, and R. Hochsprung. 1994. Evolution of the PowerPC Architecture. *IEEE Micro,* Vol. 14, No. 2, 34–49.

16. Hsu, P. Yan-Tek. 1994. Designing the TFP Microprocessor. *IEEE Micro, Vol.* 14, No. 2, 23–33.

17. Katevenis, M. G. H. 1985. *Reduced Instruction Set Computer Architectures for VLSI* (ACM doctoral dissertation award 1984). Cambridge, Mass.: MIT Press.

18. McFarling, S., and J. Hennessy. 1986. Reducing the cost of branches. *Proceedings of the 13th Int. Symp. on Computer Architecture,* Tokyo, Japan, ACM SIGARCH Vol. 14, No. 2, . 396–403.

19. Lilja, D. 1988. Reducing the branch penalty in pipelined processors. *IEEE Computer,* Vol. 21, No. 7. 47–55.

20. Johnson, W. 1984. A VLSI superminicomputer CPU. *Proceedings of the IEEE Int. Solid-State Circuits Conf.,* San Francisco, 174–175.

21. Beck, J., et al. 1984. A 32b microprocessor with on-chip virtual memory management. *Proceedings of the IEEE Int. Solid State Circuits Conf.,* San Francisco, 178–179.

22. Radin, G. 1982. The 801 minicomputer. *Proceedings, Symp. on Architectural Support for Programming Languages and Operating Systems* (ASPLOS-I), ACM SIGARCH-10.2 SIGPLAN-17.4, 39–47.

23. Patterson, D. A., and C. H. Sequin. 1982. A VLSI RISC. *IEEE Computer Magazine,* Vol. 15, No. 9, 8–21.

24. Hennessy, J., N. Jouppi, S. Przybylski, C. Rowen, and T. Gross. 1983. Design of a high performance VLSI processor. *Proceedings, 3rd Caltech Conference on VLSI,* Pasadena, Calif., R.Bryant, ed. Comp. Sci. Press, 33–54.

25. Lee, R. B. 1989. Precision architecture. *IEEE Computer,* Vol. 22, No. 1, 78–91.

26. Agarwal, A., B-H. Lim, D. Kranz, and J. Kubiatowicz. 1990. APRIL: A Processor architecture for multiprocessing. *Proceedings of the 17th Int. Symp. on Computer Architecture,* Seattle, Wash. ACM SIGARCH Vol. 18, No. 2, 104–114.

27. Smith, J. 1984. Decoupled access/execute computer architectures. *ACM Trans. on Computer Systems*, Vol. 2, No.4, 289–308.

28. Tirumalai, P., M. Lee, and M. Schlansker. 1990. Parallelization of loops with exits on pipelined architectures. *Proceedings of the IEEE Supercomputing Conference,* 1990, pp. 200-212.

29. Patterson, D. A. 1985. Reduced instruction set computers. *Communications of the ACM, Vol.* 28, No. 1, 8–21.

30. Hennessy, J. L. 1984. VLSI processor architecture. *IEEE Transactions on Computers, Vol.* 33, No. 12, 1221–1246.

21

Superscalar and VLIW Processors

Thomas M. Conte

RISC processors use overlapped execution of instruction fetch, decode, execute and result write-back to achieve performance close to one instruction per cycle. The primary limit to achieving more than one instruction per cycle is inter-instruction dependencies due to the structure of the computation. Other limits are imposed by the hardware or the compiler, including the re-use of registers, limits on the number of functional units, and the instruction and data bandwidth into the processor.

A superscalar processor is one approach for executing more than one instruction per cycle. These processors can begin executing two or more instructions in parallel. This very fine level of parallelism is often referred to as *instruction-level parallelism.* Hardware is used to detect independent instructions and execute them in parallel, often using techniques developed for early supercomputers such as the Cray-1 and IBM 360/model 91. Contemporary superscalar processors have become a commercial success, with examples including the Intel Pentium [1], DEC Alpha 21064 [2], Motorola 88110 [3], IBM POWER1 (RS/6000) [4], and 604 processors [5]. This chapter discusses structures and design trade-offs for superscalar processors.

An alternative method for exploiting instruction-level parallelism is a direct extension of horizontal microcode compaction. These architectures employ an aggressive compiler to schedule multiple operations in a very long instruction word, or VLIW. The hardware issues one VLIW per cycle. The entire responsibility for finding and correctly scheduling parallel instructions is placed in the hands of the compiler. The primary advantage of VLIW processors over superscalar processors is the amount of parallelism they can detect. Hardware-based superscalar processors must detect independent instructions on-the-fly using a limited-sized hardware window. For VLIW processors, parallelism is detected before run time via the compiler. This effectively expands the scope of the window to encompass the entire program.

In spite of their advantages, VLIW processors have had only limited success at commercialization. The two examples are the Multiflow TRACE and the Cydrome Cydra-5. Neither of the companies producing these machines is in business today. Research into VLIW processors has continued, however. Recently, Hewlett-Packard Laboratories released *Play-Doh*, a new, experimental VLIW architecture for research purposes [6], and other indications suggest that VLIW processors are poised to make a reentrance into the commercial arena. For these reasons, VLIW processors are discussed in-depth in this chapter. This is followed by a comparison of superscalar versus VLIW processors. Throughout the chapter, the emphasis is on the underlying concepts and techniques that constitute superscalar and VLIW technology.

21.1   Superscalar Processors

A diagram of a generic superscalar processor is presented in Fig. 21.1. The instruction and data caches are referred to as the I-cache and D-cache, respectively. There are six units to a superscalar processor: (1) instruction fetch, (2) decode, (3) dispatch, (4) scheduling, (5) execution, and (6) state update. The *instruction fetch unit* acquires active instructions from the instruction memory hierarchy via the I-cache. This unit also maintains several PC values and handles branch instructions. Instructions leaving this stage are said to be successfully *fetched*. The decode unit determines the operation from the instruction encoding. Most commercial superscalars are encoded with fixed-width instructions. The most notable exceptions are the x86-series of processors produced by Intel and the Digital Equipment Corporation VAX architecture. For these processors, the fetch unit and the decode unit are often combined (see Ref. [1] for an example). Instructions that leave the decoder are said to be *decoded*.

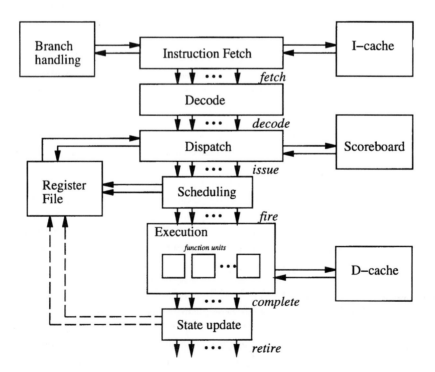

Figure 21.1 Architecture of a superscalar processor

The *dispatch* and *scheduling units* cooperate and act as the traffic controllers for a superscalar processor. Both units maintain queues of instructions. The scheduling unit's queue is often partitioned into entries for each of the execution unit's function units. Each entry is termed a *reservation station*. Reservation stations appear as virtual function units to the dispatcher (this is explained further in Section 21.1.2). Both queues are shown in Fig. 21.2. These queues are merged into one queue for some superscalar designs. In addition to its queue, the dispatch unit also maintains a list of free function units called the *scoreboard*. The scoreboard is used to keep track of the status of the scheduling queue. Once per cycle, the dispatcher removes instructions from its queue and reads their source operands, then places them in the scheduling unit's queue, depending on the status of the scoreboard. When an instruction makes this transition, it is said to be *issued*. Once per cycle, the scheduling unit checks each instruction in its queues for dependencies then begins executing instructions on their corresponding function units. Once an instruction has left the scheduling unit, it is said to have *fired*, using a term borrowed from data flow architectures.

The *execution unit* is composed of sets of function units. Example function units are integer ALU units, floating-point addition and multiplication, and data access operations. Each unit may be pipelined into several stages. When an instruction finishes execution, it is said to have *completed*. Once instructions have completed, their results are written back by the *state update unit* for use by instructions waiting in the scheduling unit's window. Some superscalar designs allow for out-of-order completion of instructions. This complicates exception and interrupt handling. Solutions to this problem may alter the state update unit's function so that it does not write back results into the register file, but rather uses additional hardware buffers or duplicated register files to maintain a consistent machine state (this is explained in Section 21.1.4). Instructions enter the consistent machine state as their last act before leaving the processor. This final step in an instruction's execution history is termed *retiring*.

The metric most frequently used to measure superscalar performance is IPC the number of instructions retiring per cycle (the *IPC*). The IPC is ultimately limited by the issue rate of the processor, since the overall rate of instructions leaving the processor cannot exceed the rate at which they enter. The rate at which instructions are issued is a critical parameter for superscalars. Typical issue rates range from one instruction per cycle for traditional architectures, to four or six instructions per cycle, for the PowerPC 604 and POWER2 architectures (respectively). It is interesting to note that IPC is the reciprocal of CPI (cycles per instruction), which in turn is the traditional metric for single-issue processors. Hence, IPC = 3 is equivalent to an average of 0.33 cycles per instruction.

The major aspects of the design of each of the six units are presented below, with a discussion of the trade-offs involved for each scheme.

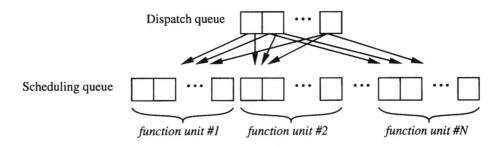

Figure 21.2 Dispatch and scheduling queues

21.1.1 Instruction fetch

The instruction fetch unit must predict for each cycle the next instruction to fetch from the I-cache, often tens of cycles before conditional branches have completed. With an average of four to six instructions between branch instructions, and issue rates in excess of four instructions per cycle, methods to predict and handle branch instructions are critical for superscalar processor design.

Branch handling schemes have been widely discussed in the literature. This is due in-part to the large number of problems involved in branch handling. For example, access-ing the I-cache is not instantaneous. Even when there is a cache hit, it typically takes two cycles to obtain the contents of the instruction address. The instruction fetch unit must therefore plan ahead. The next instruction accessed in a program is usually at the next se-quential location in memory from the current instruction. The instruction fetch unit can use this rule to help overcome the latency of the I-cache. Branch instructions are a nota-ble exception to the rule. Branches occur relatively frequently, typically every four to six instructions. An intelligent instruction fetch unit must predict (1) that an address contains a branch before it is decoded, (2) where in memory the branch transitions control to, and (3) in the case of conditional branches, whether the branch will be taken when the condi-tion is evaluated.

Branch handling involves the use of either hardware or software to predict branch be-havior. Superscalar processors with high issue rates place high demands on instruction fetch hardware. This in turn means that highly accurate branch prediction must be sup-ported by the hardware. In addition, branch prediction allows the use of *speculation*: the execution of instructions at the destination of a branch before the branch's true behavior is known. Speculation requires hardware support (specifically, support of the state update unit—see Section 21.1.4). Speculation can achieve much higher degrees of parallelism (i.e., higher IPC), which in turn results in faster program execution.

Software branch prediction. Software approaches rely on the compiler to predict the behavior of a branch. This is then encoded using a reserved bit in the instruction format, called the *likely-bit*. This bit is typically present in *all* formats, not just branch instructions. If the bit is set, the instruction is assumed to be a branch that will ultimately be taken. Unconditional branches (e.g., jump instructions) always have this bit set.

The compiler can use two general techniques for setting the likely bit. One technique performs analysis on the program source and decides whether each branch will be taken or not. For example, the branch used for common loop constructs, such as *for*/DO loops, will likely be taken when encountered because most loops execute for more than one iteration. This static approach to branch prediction suffers from relatively low accuracy (approxi-mately 60–80 percent for non-numeric programs).

An alternative to static branch prediction by the compiler is *profiled branch prediction*. In this approach, the program is run several times, perhaps using different program inputs. A profile of each branch's behavior is recorded during these tests. The profile information is then used by the compiler to recompile the program and predict the direction of each branch. Profiled branch prediction achieves prediction accuracy comparable to many hard-ware schemes [7].

Hardware branch prediction. Hardware schemes for branch prediction use previous exe-cutions of a branch instruction to predict its future execution. This information is typically held in the *branch history buffer,* which is a specialized cache indexed using the branch address. There are two classes of branch history buffers: o*ne-level* and *two-level buffers.* One-level buffer use the address of the branch instruction to index into a *branch target buffer* (BTB), which contains a small state machine for predicting the outcome of a branch. The nominal size for the one-level branch prediction buffer is between 256 to 1024 entries. When the branch completes execution, the actual outcome is used to update the state machine. Fig. 21.3 depicts this process. The most common state machine for one-level schemes is the two-bit counter predictor, described in Ref. [8]. The prediction information field of each entry in the BTB is two bits wide. These bits are used to implement the four-state *prediction counter,* shown in Fig. 21.4. Each time a branch is *taken,* the corresponding count in its BTB entry is incremented by one. Conversely, for *not-taken* branches, the count is decremented by one. The counter saturates at $0(00_2)$ and $3 (11_2)$. The prediction for a branch is made depending on the value of the counter. Counts 00_2 and 01_2 are predicted *not-taken,* while 10_2 and 11_2 are predicted *taken.* This mechanism can be realized very easily using the most significant bit (b_1) of the counter as the prediction bit.

The two-bit counter predictor has been widely used because of its high performance with modest hardware complexity (empirical accuracies are typically 85–90 percent). One advantage of this low complexity is a single-cycle or potentially subcycle prediction latency. The predictor is implemented in several contemporary processors, including the Intel Pentium and the PowerPC 604. On average, the two-bit counter is approximately as accurate as the software-based profiled branch predictor (this predictor is explained above).

Two-level schemes use two separate buffers. The first buffer is indexed similar to the BTB and stores the branch history as a binary string. The second is indexed using this

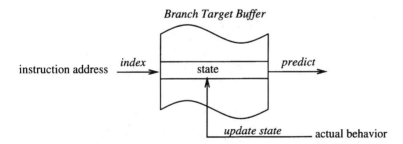

Figure 21.3 One-level branch prediction

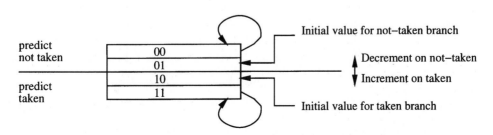

Figure 21.4 The two-bit counter predictor

branch history and stores the state of a predictor. This is depicted in Fig. 21.5. These schemes have been studied extensively by Yeh and Patt [9, 10] and we will use their nomenclature here. The first level buffer is termed the *history register table* (HRT). The HRT is *b* bits wide and stores a sequential, binary string of the branch's history, using 0 for not-taken and 1 for taken branches. A prediction is made by indexing into the HRT, then using the history string to index into a second table, the *pattern table* (PT). The PT stores the state of a small state machine used to predict the branch. This decouples the branch prediction from the address of the branch instruction. The effect of this decoupling is dramatic. Yeh's algorithm can achieve 96 percent branch prediction accuracy[*] for SPEC92 benchmarks [9, 10]. As of today, Yeh's algorithm has not been implemented in any commercially available microprocessor. However, the needs of high-issue superscalars will likely drive future implementations of this branch predictor.

21.1.2 Dispatch and scheduling

Ideally, a machine with an issue rate of N instructions per cycle should be able to achieve an IPC of N. That is, it should be able to retire N instructions each cycle. In practice, this is limited by dependencies between instructions and the available resources provided by the hardware. To see this, a further description of the dependencies between instructions is required. Table 21.1 presents the three classes of possible dependencies. *Flow dependencies* occur when a register is written by one instruction then read by a later instruction (e.g., $R1$ between instructions I1 and I2). An *anti-dependence* occurs when a register is written after it is read, preventing an out-of-order execution of the instructions (e.g., R2 between I3 and I4). An *output dependence* is similar to an anti-dependence and occurs when a two or more instructions write to the same register, also preventing an out-of-order execution (e.g., $R1$ between I5 and I6).

TABLE 21.1 Three Classes of Dependencies between Instructions

Flow dependence	Anti-dependence	Output dependence
I1: **R1** ← R2 + R3	I3: $R1$ ← R2 + R3	I5: **R1** ← R2 + R3
I2: R5 ← **R1** + R4	I4: **R2** ← R5 + R4	I6: **R1** ← R5 + R4

There are several schemes for exploiting instruction-level parallelism dynamically. The three most commonly implemented schemes are simple interlocking, the CDC 6600 scheme, and the Tomasulo algorithm. (For a survey of other schemes, see [11]). These

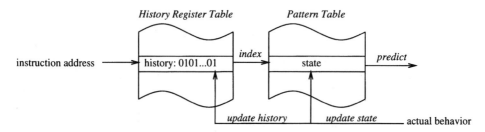

Figure 21.5 Two-level branch prediction

---

[*]When the "PAs" scheme with a 1024-entry HRT, (b = 12) is used.

schemes can be classified by their ability to schedule instructions out of order in spite of dependencies. This ability is often termed *resolving dependencies.* Table 21.2 presents the ability of these three schemes to resolve dependencies. Note that some schemes (notably Tomasulo and CDC 6600) can resolve output dependencies, whereas no scheme can schedule instructions around flow dependencies. For this reason, flow dependencies are sometimes called *pure-dependencies.* Simple interlocking and the Tomasulo algorithm schemes are explained below. Those interested in the CDC 6600 scheme are referred to Thornton [12].

TABLE 21.2 Three Common Scheduling Schemes

	Resolves dependencies of type:		
Scheme	Flow	Anti	Output
Simple	no	no	no
CDC 6600	no	no	yes
Tomasulo	no	yes	yes

In the discussions that follow, the code sequence presented in Fig. 21.6 is used as a running example. The execution unit is assumed to have one floating-point adder/subtracter and one floating-point multiplier. The multiplier is pipelined into two stages, meaning it can begin executing new instructions each cycle, but it takes two cycles to complete the multiplication. In addition, the issue rate is assumed to be two instructions per cycle.

The discussion also presents the interlocking algorithms in pseudocode format. Table 21.3 explains the nomenclature used for the algorithms.

Simple interlocking. Simple interlocking halts firing whenever any dependency occurs. To do this, it must detect dependencies in the dynamic instruction stream. This is done by adding a busy bit to each register entry in the register file. (The modified register file is sometimes referred to as a *register scoreboard,* but we reserve the term *scoreboard* to refer to function unit status information.) The interlocking algorithm semantics are shown in Fig. 21.7. If the function unit is unpipelined (i.e., it is used for multiple cycles), the algorithm must take this into account when an instruction completes. The dispatch and scheduling stages are essentially merged by this algorithm. The combined dispatch/scheduling queue is managed as a first-come, first-serve queue.

An example operation of the simple interlocking algorithm is shown in Fig. 21.8. The status of the busy bit in the register file is represented using the name of the instruction

I0 R0 ← R2 + R4

I1 R5 ← R6 * R7

I2 R1 ← R0 + R5

I3 R0 ← R2 * R4

I4 R0 ← R3 + R4

I5 R7 ← R2 - R0

Figure 21.6 Two-level branch prediction

TABLE 21.3 Nomenclature Used for Dispatch/Scheduling Algorithms

Name	Use
Inst	An instruction
Scoreboard	Function unit scoreboard
Common_Data_Bus	The common data bus (Tomasulo algorithm only)
busy/not busy	Boolean constants
ready/not ready	Boolean constants
Max_Source_reg	The maximum number of source registers the instruction encoding provides (e.g., 2)
Source_reg[i]	Source registers of a queue entry, instruction, or common data bus
Dest_Reg	Destination register of a queue entry, instruction, or common data bus
Function_Unit	Function unit of a queue entry, instruction, or common data bus
Busy	Busy bit; if = busy, then either a register or function unit is busy and cannot be accessed/ used
Sched_Queue	The scheduling queue
Dispatch_Queue	The dispatch queue
Queue_Tail	Tail of a queue
Queue_Entry	Entry in a queue
Ready	A register or function unit is ready (i.e., not busy)
Tag	The unique tag for the result of an instruction's execution (Tomasulo algorithm only)
Value	The result of an instruction's execution
Register_File [i]	The register file entry for register Ri

causing the busy status. The scoreboard is not explicitly represented in the example. In addition, the issue rate is assumed to be two instructions per cycle.

The limitations of simple interlocking occur for instruction sequences that contain dependent instructions followed by independent instructions. In such cases, the later group of instructions must wait to fire even though there are no dependencies or resource conflicts preventing them from firing. In this way, simple interlocking enforces an *in-order firing policy:* instructions must fire in sequential (program) order. Once instructions have fired, they can complete out-of-order, however.

The primary advantage of simple interlocking is its simplicity. The required control logic is uncomplicated. The additions to the register file and the scoreboard itself have minor hardware cost.

For each cycle of execution:

1. [**Scheduling unit**]

    (a) **for all** Function_Unit **do:**
        if (Function_Unit *is pipelined*) **then** Scoreboard[i].Busy = **not busy**

    (b) **for each** Queue_Entry **in** Sched_Queue **do:**
        if ((**for all** i: Queue_Entry.Source_Reg[i].Busy = **not busy**) **and**
            Queue_Entry.Dest_Reg = **not busy and**
            Scoreboard[Queue_Entry.Function_Unit] = **not busy**)
        then

            i. Scoreboard[Queue_Entry.Function_Unit].Busy ← **busy**
            ii. Register_File[Queue_Entry.Dest_Reg].Busy ← **busy**

        else exit loop *(halt issuing)*

2. [**Execution unit, at completion of instruction "Inst"**]

    (a) Register_File[Inst.Dest_Reg].Busy ← **not busy**
    (b) **if** (Inst.Function_Unit *is not pipelined*) **then** Scoreboard[Inst.Function_Unit].Busy ← **not busy**

Figure 21.7 The simple interlocking algorithm

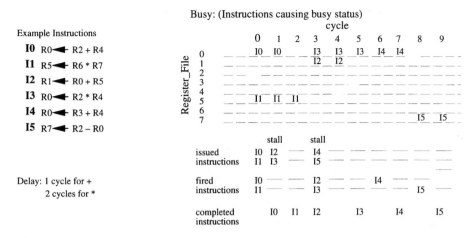

Figure 21.8 Example operation of the simple interlocking algorithm

Tomasulo algorithm. The example of Fig. 21.8 revealed several problems with simple interlocking. Program-defined register assignment interfered with parallel execution, and in-order firing did not exploit all potential parallelism. The solution is to decouple the program's values from the register names. This is termed *register renaming.* A hardware algorithm for register renaming was first described in 1967 by R. L. Tomasulo for the design of the IBM 360 model 91 [13]. This algorithm is therefore often referred to as the *Tomasulo algorithm.*

Register renaming in hardware is accomplished using several additional entries in the register file. (See Fig. 21.9.) The key idea is to associate a unique *tag* with the program value, then translate any subsequent reference to a register into a reference to the tag. This is accomplished by creating a unique tag for a register each time it is written. Other instructions that read a register must now have their register specifiers translated into tag references. This is accomplished by storing the tag with the register in the register file. The scheduling queue (Sched_Queue) is critical for the correct operation of the Tomasulo algorithm. Entries in this queue, termed *reservation stations,* have fields for the destination

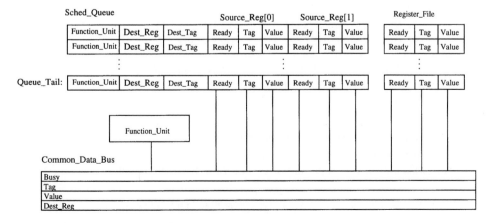

Figure 21.9 Hardware required for the Tomasulo algorithm

of the instruction that specifies the tag and register number of the instruction (the Tag and Dest_Reg fields in Fig. 21.9). Each source operand has a field containing the tag (the Tag field), a field to hold the actual value once it is available (the Value field), and a bit to indicate when the value is valid (the Ready field).

The scheduling queue, the function units, and the register file communicate to resolve dependencies. The mechanism for this is the Common_Data_Bus of Fig. 21.9, used to broadcast the value of a completed instruction, along with its tag, and its destination register number (i.e., Dest_Reg). Source entries in the scheduling window (e.g., Source_Reg[i] in Fig. 21.9) wait for the broadcast of their corresponding tag. Once this occurs, the source entries copy the Value lines from the Common_Data_Bus into the Value field in the scheduling queue and set their corresponding Ready bit. In addition, the register file also listens for broadcasts. When they occur, the Value field is copied from the Common_Data_Bus into the register and the Ready bit is also set. The Common_Data_Bus of the original Tomasulo algorithm is sometimes also referred to as the *result bus*.

The Tomasulo algorithm is divided between two figures. Steps 1 (dispatch unit), and 2 (scheduling unit) are shown in Fig. 21.10. Steps 3 and 4 (execution unit), and Step 5 (state update unit) are shown in Fig. 21.11.

For each cycle of execution:

1. **[Dispatch unit] for all** Inst **in** Dispatch_Queue **do:**
   if (Sched_Queue *is not full*) **then**

   (a) **Add** Inst *to tail of* Sched_Queue
   (b) **Delete** Inst *from* Dispatch_Queue
   (c) Queue_Tail.Function_Unit ← Inst.Function_Unit
   (d) Queue_Tail.Dest_Reg ← Inst.Dest_Reg
   (e) Register_File[Inst.Dest_Reg].Tag ← *unique tag id*
   (f) Queue_Tail.Tag ← Register_File[Inst.Dest_Reg].Tag
   (g) Register_File[Inst.Dest_Reg].Ready ← **not ready**
   (h) **for** i ← 0 **to** Max_Source_Reg − 1 **do:**
       if (Register_File[i].Ready = **ready**) **then**

       i. Queue_Tail.Value ← Register_File[i].Value
       ii. Queue_Tail.Source_Reg[i].Ready ← **ready**

       else

       i. Queue_Tail.Source_Reg[i].Tag ← Register_File[i].Tag
       ii. Queue_Tail.Source_Reg[i].Ready ← **not ready**

2. **[Scheduling unit] for each** Queue_Entry **in** Sched_Queue **do:**

   (a) **for** i ← 0 **to** Max_Source_Reg − 1 **do:**
       if (Common_Data_Bus.Tag = Queue_Entry.Source_Reg[i].Tag) **then**

       i. Queue_Entry.Source_Reg[i].Value = Common_Data_Bus.Value
       ii. Queue_Entry.Source_Reg[i].Ready = **ready**

   (b) **if** ((**for all** i: Queue_Entry.Source_Reg[i].Ready = **ready**)
       and Scoreboard[Queue_Entry.Function_Unit] = **not busy**) **then**

       i. Scoreboard[Queue_Entry.Function_Unit] ← **busy**
       ii. *Fire the instruction at* Queue_Entry *on* Function_Unit

Figure 21.10 The Tomasulo algorithm, part1

For each cycle of execution:

3. [**Execution unit, for function unit** Function_Unit]
   if Function_Unit *is pipelined* **then** Scoreboard[Inst.Function_Unit].Busy ← **not busy**

4. [**Execution unit, at completion of instruction** Inst]
   if (Common_Data_Bus.Busy = **not busy**) **then**

       (a) Common_Data_Bus.Busy ← **busy**

       (b) Common_Data_Bus.Tag ← Inst.Tag

       (c) Common_Data_Bus.Value ← Inst.Value

       (d) Common_Data_Bus.Dest_Reg ← Inst.Dest_Reg

       (e) **if** (Inst.Function_Unit *is not pipelined*) **then**
           Scoreboard[Inst.Function_Unit].Busy ← **not busy**

       (f) **Delete** Inst **from** Sched_Queue

   else *delay completion of* Inst

5. [**State update unit**]
   if (Common_Data_Bus.Busy = **busy and**
   Common_Data_Bus.Tag = Register_File[Common_Data_Bus.Dest_Reg].Tag) **then**

       (a) Register_File[Common_Data_Bus.Dest_Reg].Ready ← **ready**

       (b) Register_File[Common_Data_Bus.Dest_Reg].Value ← Common_Data_Bus.Value

Figure 21.11 The Tomasulo algorithm, part2

An example is shown in Fig. 21.12. The example of Fig. 21.6 has been reproduced here for convenience. The scheduling queue is assumed to be partitioned by function unit, and each entry is labeled as a reservation station. The notation used in the example is included at the bottom of the figure. Since every instruction has one destination register, the identity of the instruction is used for the tag name. Note that there is one adder (latency of one cycle) and multiplier (latency of two cycles). In addition, the issue rate is two instructions per cycle.

The Tomasulo algorithm is empirically much faster at executing code than simple interlocking. However, there are some drawbacks. For example, the register file is approximately 30 percent larger due to tags. In addition, the associative search of the scheduling queue is expensive to implement. Finally, the bottleneck for efficient operation is the common data bus (result bus). Several contemporary superscalars use multiple result buses to avoid this bottleneck.

21.1.3 Execution

The majority of the design of a superscalar execution unit is not significantly different from the design of traditional execution units and is a topic in *computer arithmetic,* which is beyond the scope of this chapter. (An interested reader is referred to Koren [14].)

One element of superscalar execution that is unique is the treatment of the D-cache as another function unit. To do so, the D-cache is often divided into two logical units: the *Store* unit and the *Load* unit. The store unit is a buffer that is used to write results to the cache. The primary advantage of this buffer is that it allows quick completion of store instructions, typically in one cycle. One complication occurs when a store is followed by a load. The hardware must either search the store buffer to see if the load instruction's data is present, or flush the contents of the store buffer to the data cache, then perform the load.

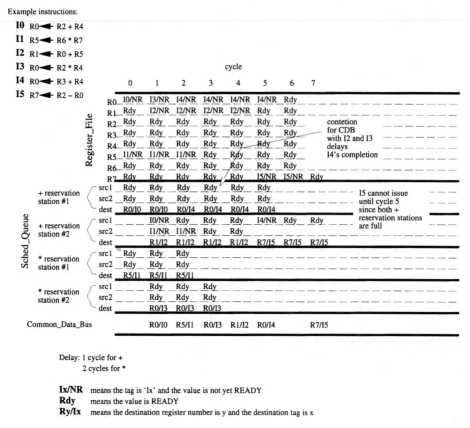

Delay: 1 cycle for +
2 cycles for *

Ix/NR means the tag is 'Ix' and the value is not yet READY
Rdy means the value is READY
Ry/Ix means the destination register number is y and the destination tag is x

Figure 21.12 Example operation of the Tomasulo algorithm

Load instructions must access the cache and cannot be buffered. When the load instruction's data is not present in the data cache, a miss occurs. Without modification to the cache, all subsequent load operations must stall until the cache fetches the missing data. This can degrade superscalar performance. To avoid this, some superscalar processors provide mechanisms to fetch data from main memory while executing other load instructions in parallel. This kind of cache is termed a *lockup-free* or *non-blocking cache,* and is described in Ref. [15]. Non-blocking caches are essential for high-performance superscalar design [16].

21.1.4 State update

Out-of-order parallel execution of instructions complicates interrupt and exception handling. When an interrupt occurs, the architectural state of the register file and memory may not correspond to any sequential state. No PC value exists to correctly restart program execution. Consider the example of Fig. 21.13. Assume that execution of the load instruction, I2, results in a page fault. After the page is brought into memory, where should the processor resume? I2 cannot be restarted since its sources (R3 and R4) where changed by instructions I3 and I4. This situation is termed an *imprecise fault.* To avoid this, some *consistent state* must be maintained somewhere in memory to allow the resumption of execution after a fault. This is the job of the state update unit. It performs instruction retire,

Example instructions: One possible execution:

Function units:

cycle	+	Load	*Notes:*
I1 R1◄ R4 − R5	0: I1	I2	I1 completes
I2 Load R2◄ M[R3+R4]	1: I3	I2	I3 completes
I3 R3◄ R1 + R5	2: I4	I2	I4 completes
I4 R4◄ R3 + R1	3:	I2	I2 page faults

Assumptions: Load delay is 5 cycles, + delay is 1 cycle

Figure 21.13 Example of an imprecise page fault

which is the act of inserting the results of an instruction in a consistent state of the machine.

The terms used to describe interrupts and faults are in no way universal. This chapter uses the following definitions:

- *Exception handler: a* routine invoked when an exception occurs
- *Fault:* an excepting instruction that should be retried after the exception handler executes (e.g., *page faults*)
- *Trap:* an excepting instruction that should not be retried after the exception handler executes; execution resumes after the trap (e.g., a *system call*) or not at all (e.g., a divide by zero)
- *Interrupt:* a fault or trap
- *Exception:* a condition that causes a *fault* or *trap*; also a generic term for a *fault* or *trap*

The architectural state is defined by three parameters: (1) its register contents, (2) the program counter value, and (3) its memory contents. An *instruction boundary* is any position in a program after the execution of instruction at location *i*, but before the execution of an instruction (*i* + 1). The *consistent state* for an instruction boundary is any architectural state such that (a) all instructions preceding the instruction boundary in the program have completed and modified the architectural state, and (b) all instructions following the instruction boundary in the program are unexecuted and have not modified the architectural state.

For example, the Tomasulo algorithm maintains an architectural state. Figure 21.13 demonstrates that this state is not always a consistent state. The state used by the dispatch and scheduling units is termed the *current state* or *messy state*. The register file that holds this state is termed the *messy register file.*

If an instruction boundary can be found for an excepting instruction and the consistent state restored for that instruction, the resulting interrupt is said to be *precise*. If such a boundary cannot be found, the interrupt is said to be imprecise. The original IBM 660 model 91 had imprecise interrupts. Precise interrupts are essential for exception handling. They are also important for debugging of traps (e.g., divide by 0). The hardware that implements precise interrupts often serves a dual purpose of recovering from mispredicted branches (see Section 21.1.1).

There are several schemes that have been developed for state update. Three general schemes have been implemented in commercial processors. They are the *reorder buffer,* the *future file,* and the *checkpoint-repair* schemes [17],[18].

A *reorder buffer* maintains a consistent state in memory and registers using a FIFO queue that holds completing instructions and retires them in program order. When instructions issue, they are allocated slots in the buffer. An excepting instruction indicates this condition by setting a bit in its buffer entry. The condition is not handled when the instruction completes, but when it reaches the head of the FIFO queue. In a simple reorder buffer, Tomasulo tags are broadcast on the common data bus at this time. However, this lengthens the time it takes for a result to reach instructions waiting in reservation stations, affecting the overall performance of the machine. To solve this problem, bypass logic is often added to check the reorder buffer for a register value at instruction issue time.

A *future file* is an extra register file used in conjunction with a reorder buffer, used to hold the architectural state of retired instructions. Because retired instructions update the future file instead of the messy register file, bypass logic can be removed. Instead, completing instructions broadcast their results on the common data bus and also write into the reorder buffer. Only instructions that reach the head of the reorder buffer have their results written back to the future file. After service of an exception, the contents of the future file are copied into the messy register file.

Checkpoint repair does not use a reorder buffer. Instead, periodic checkpoints are made of the instruction stream using multiple register files. Hwu and Patt have shown that at, at most, three register files are required [18]. One register file holds the messy state, a second holds a backup copy. A third register file is used to periodically build a new backup copy.

21.1.5 Summary

This section has discussed the design of superscalar processors. These processors execute instructions in parallel (out-of-program-order) using hardware techniques such as the simple interlocking scheme or the Tomasulo algorithm. Hardware or compiler assistance is used to handle branch instructions. The problem of imprecise interrupts can be solved using hardware buffers or backup copies of registers. The following section presents VLIW processors, which exploit instruction-level parallelism with minimal hardware support.

21.2 VLIW Processors

The architecture of a generic VLIW processor is shown in Fig. 21.14. One difference between Fig. 21.14 and 21.1 is the absence of the decode unit for the VLIW. Instruction decoding for a VLIW is extremely simple, and depicting a separate unit would overstate the importance of decoding. In this respect, VLIW follows the RISC philosophy. Also absent from the VLIW processor are the dispatch and scheduling units. In a strict VLIW processor design, hardware performs no interlocking whatsoever. Instead, it is the responsibility of the compiler to schedule code so that dependencies between instructions are never violated. This shift in emphasis from hardware to software is the main advantage of VLIW processors over superscalar processors. This section discusses the architecture of VLIW processors by focusing on the compiler techniques developed for their use.

21.2.1 The compiler's view of a program

A traditional compiler is composed of three phases: (1) parse the source language into an intermediate representation, (2) optimize the intermediate language, and (3) generate code

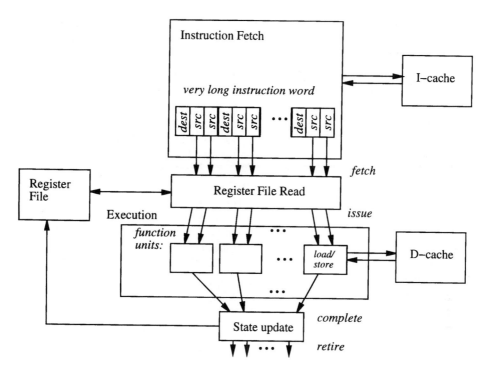

Figure 21.14 Architecture of a VLIW processor

of the target architecture from the intermediate language. The scheduling of resources is implicit in the code generation phase. For a VLIW architecture, scheduling becomes a primary function and is often merged with optimization.

A compiler views a program in a different way from the lines of source code or from the binary encoding of the instructions. The first difference is the intermediate code, which is typically an instruction set of a very simple architecture. The operations include rudimentary arithmetic and logical operations, floating-point operations, control flow, and memory load/store operations. Each intermediate-language operation has one destination register and several source registers. An exception to this is control transfers (e.g., branches), which specify a destination address but no destination register. The only access to memory is through load and store operations. There is also an unlimited supply of registers available. These registers are often termed *virtual registers* for this reason.

In addition to the intermediate language, a compiler also maintains two primary data structures that describe the program. They are the *control flow* and *data flow graphs*. As the names imply, these graphs describe the flow of control and data in the program. They do not completely characterize the program. That is to say that these two graphs do not contain enough information to describe the computation without the addition of the intermediate-language operations.

Figure 21.15a shows a short example list of intermediate-language operations. These operations can be partitioned into blocks of guaranteed-sequential operations. These groupings are known as *basic blocks*. To form basic blocks, the following procedure is used: the code is scanned, and a new basic block is started immediately after a branch or a code label. Operations are added to the block until another branch or code label is reached. Basic blocks are typically numbered sequentially, starting from the beginning of the

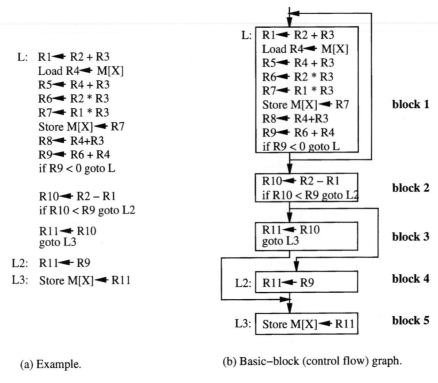

L: R1 ◂— R2 + R3
   Load R4 ◂— M[X]
   R5 ◂— R4 + R3
   R6 ◂— R2 * R3
   R7 ◂— R1 * R3
   Store M[X] ◂— R7
   R8 ◂— R4+R3
   R9 ◂— R6 + R4
   if R9 < 0 goto L

   R10 ◂— R2 – R1
   if R10 < R9 goto L2

   R11 ◂— R10
   goto L3
L2: R11 ◂— R9
L3: Store M[X] ◂— R11

(a) Example.

(b) Basic–block (control flow) graph.

Figure 21.15 Example (a) intermediate-language code and (b) intermediate-language code grouped into basic blocks

source file or function being compiled. If destinations of branches at the bottom of basic blocks are connected to their target blocks by arcs, a basic block graph or *control flow graph* is formed. This is depicted in Fig. 21.15b.

The definition of a data flow graph follows directly from the dependencies discussed in Table 21.1 of the previous section. Recall that the Tomasulo algorithm (Section 21.1.2) removed anti- and output dependencies using hardware renaming. The principal goal of renaming is to decouple the register names from their values so that register re-use in the program does not enforce a sequential execution order on the operations. For the compiler, the unlimited number of virtual registers accomplishes the same task, since each operation defines the value of a new virtual register name. Hence, of the three kinds of dependencies, only flow dependencies between operations remain in the intermediate-language representation. A data flow graph is a directed graph where the nodes are the operations, the source of each arc is a register-defining operation, and the destination is a use of the register value. A data flow graph for one of the basic blocks in the example is shown in Fig. 21.16.

21.2.2 Compiler-based scheduling

Scheduling of code to enhance parallelism in a VLIW is best illustrated with an example. This example uses a VLIW with two adders, a multiplier, a load unit, and a store unit. The operation latencies are assumed to be one cycle, except for the multiply operations (three cycles) and the load unit (two cycles). The VLIW instructions for the example block of

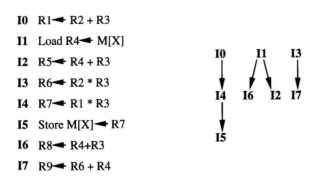

I0 R1 ◂— R2 + R3

I1 Load R4 ◂— M[X]

I2 R5 ◂— R4 + R3

I3 R6 ◂— R2 * R3

I4 R7 ◂— R1 * R3

I5 Store M[X] ◂— R7

I6 R8 ◂— R4+R3

I7 R9 ◂— R6 + R4

(a) Example operations.   (b) Data flow graph for example.

Figure 21.16 A basic block annotated to form a data flow graph

Fig. 21.16 are shown in Fig. 21.17. Empty spaces in the figure represent cycles in which no operations are executed.

There would be no difference in timing between execution of the intermediate-language operations of Fig. 21.17 on a superscalar employing the Tomasulo algorithm and execution of the VLIW instructions. In essence, the very long instruction words are entire scripts for the function units to follow in each cycle of execution. The dynamic responsibilities of the hardware have been reduced to obeying the dictates of the instruction format, without any hardware support to enforce or resolve dependencies. Nonetheless, the VLIW architecture can achieve the same or greater performance as a superscalar architecture. There are many reasons for this, including the cycle time advantage of simpler hardware and the compiler's ability to find more parallelism before execution than hardware can find during execution.

The above example illustrated the scheduling of operations from within one basic block. This is often termed *local scheduling,* since the scheduling is performed local to the block. Unfortunately, the size of basic blocks is typically only four to six operations long. This limits the amount of parallelism a VLIW can extract, in much the same way as branches limit the parallelism of superscalar processors (see Section 21.1.1). To extract more paral-

VLIW instruction encoding

+			+			*			Load	Store	
dest	*src*	*src*	*dest*	*src*	*src*	*dest*	*src*	*src*	*dest*	*address*	*address*	*dest*
R1	R2	R3				R6	R2	R3	R4	X		
						R7	R1	R3				
R5	R4	R3	R8	R4	R3							
R9	R6	R4										
											X	R7

Figure 21.17 The example basic block represented as VLIW instructions after scheduling

lelism, *global scheduling* is used. This technique moves operations between basic blocks to achieve a more-parallel schedule. Since instructions are moved above a branch, it is the VLIW analog of branch prediction for speculative execution in a superscalar (see Section 21.1.1).

Global scheduling can be broadly classified into *acyclic* and *cyclic* scheduling. Acyclic scheduling deals with sequential lists of blocks with control flow containing no loops. When loops occur, acyclic techniques can still be used by breaking one of the arcs and scheduling the loop body as a sequential code. However, better results are often obtained when cyclic scheduling techniques are employed. The following section reviews several techniques for acyclic and cyclic scheduling.

Acyclic scheduling. The first step in many acyclic scheduling techniques is to form larger groups of blocks out of basic blocks. There are many techniques for performing this grouping. The two most-researched techniques are *trace selection* and *superblock formation* [19,20].

Trace selection was first used by researchers at Yale University in the *Bulldog* compiler, then later extended for inclusion in the commercial compilers of Multiflow, one of the early VLIW processor vendors. Superblock formation is a an evolution of trace selection, used in the Illinois IMPACT project. The algorithms are similar. Superblock formation is described here, and then contrasted with trace selection.

Figure 21.18a shows a control flow graph composed of basic blocks. The numbers or weights beside the blocks and arcs are execution counts for each block or arc (respectively). Obtaining the weights can be performed using information from profiled runs of the program, via software estimates, or by use of branch prediction hardware [20–22]. A control flow graph annotated in this way is often referred to as a *weighted control flow graph.*

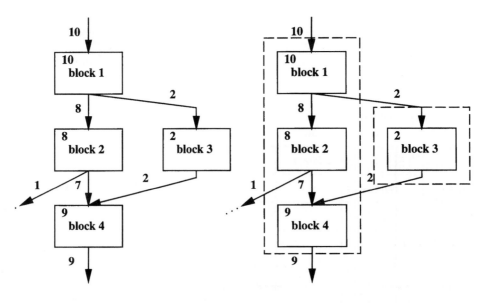

(a) Example control–flow graph.   (b) Example graph after trace selection.

Figure 21.18 Example (a) weighted control flow graph and (b) example after trace selection

Superblocks are formed first by grouping blocks together that tend to execute sequentially. Such groupings were termed *traces* by Fisher [19]. The result of trace selection is shown in Fig. 21.18b. The traces are represented as dashed rectangles in the figure. A superblock is a trace that has only one entrance at the top but any number of multiple exits. No side entrance into a superblock is allowed. Notice that the larger trace in Fig. 21.18b is not a superblock because *block 3* transfers into the middle of the trace. To solve this problem, tail duplication is performed. Specifically, *block 4* is duplicated. The overall result is shown in Fig. 21.19. Here, notice that *block 4* has been duplicated so that execution after *block 3* now flows to *block 4'*. (An excellent description of the complete superblock algorithm by its inventors is presented in Ref. [23].)

The interesting property of superblocks is that operations can be moved upward in a superblock across the boundaries of basic blocks. This is a direct consequence of the no-side-entrances rule for superblock formation. It allows code motion that can extend the scope of local scheduling. Since operations are moved above branches, they will be executed speculatively when the code is run. The compiler must insert patch-up code into basic blocks that are not in the superblock to undo the effects of this speculation. Some additional hardware modifications are also required to enable speculative execution of potentially excepting instructions. An extra bit is used in the VLIW encoding of each operation to indicate that the operation is being executed speculatively. If the speculative operation generates an exception, an imprecise interrupt problem exists in much the same way as it does for superscalar processors (see Section 21.1.4). This is solved via slight modifications to the register file to signal an exception when the result of an excepting operation is used. This modification to handle interrupts is referred to as *sentinel scheduling* [24].

Work has been done on scheduling algorithms that avoid all code duplication and allow for code motion in both directions. The most notable of these techniques is the hyperblock scheduling technique of the Illinois IMPACT project [25]. This technique relies on the use of *if-conversion* and *predicated execution*.

Predicates are one-bit registers that control whether the results of an operation are retired or discarded. Support for predicated execution requires the addition of a predicate register field to all operations in the VLIW instruction encoding. There are two ways to use these registers. In one technique, the predicate register is checked when an operation is

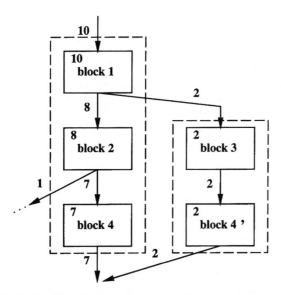

Figure 21.19 Example weighted control flow graph partitioned into superblocks

about to begin execution. If the register holds the value *false,* the operation is abandoned, otherwise it is executed normally. In the second technique, the specified predicate register is checked when the corresponding operation completes execution on its function unit. If the predicate holds the value *false,* the results of the operation are discarded instead of being written back. Some proposed VLIW architectures support the former technique, others support the latter.

Predicated execution is a mechanism for removing conditional, acyclic branches entirely from code sequences. To see this, consider the example of Fig. 21.20a. Here, a branch has been inserted in the middle of the code sequence of Fig. 21.16. Normally, this would severely limit code motion and reduce the amount of parallelism. Figure 21.20b shows a predicated version. The predicate specifier is represented by the keyword "if P2" in the intermediate language. Note the operation "P2 ← (R1 = 0)": this is a *predicate-define* operation. It tests the condition (e.g., R1 equals zero) and, if the condition is true, sets the predicate register P2 to *true,* else sets P2 to *false.* A scheduled version of the predicated code is shown in Fig. 21.21. Note how the branch has now been converted into a data dependence on the predicate register P2. This observation is the reason that conversion of code from branch-based control flow to predicated form is termed *if-conversion* [26].

Hyperblock formation uses if-conversion and predicated execution to remove short forward branches and create larger blocks of sequential code. Consider again the example of Fig. 21.19, where *block 4* had to be duplicated. If instead a predicate were used to merge *block 3* into the superblock, the resulting *hyperblock* would not require any code duplication. In addition, the scheduler can move operations in a hyperblock in any direction, as long as the dependencies are obeyed. As mentioned above, the control flow arc in the control flow graph is converted into a dependence arc in the data flow graph by if-conversion.

Before moving on to discuss cyclic scheduling, some final comments on acyclic scheduling research are in order. The technique for if-conversion mentioned above can be re-

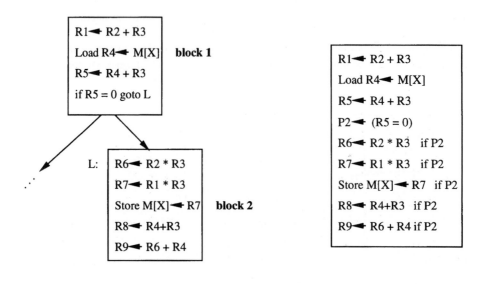

(a) Example.   (b) After predication.

Figure 21.20 Example (a) used to illustrate predicated execution and (b) after predication

VLIW instruction encoding

+				+				*				Load			Store			Predicate define			
dest	src	src	pred	dest	src	src	pred	dest	src	src	pred	dest	address	pred	address	dest	pred	dest	pred	dest	src	src	cond
R1	R2	R3						R6	R2	R3	P2	R4	X								
								R7	R1	R3	P2										
R5	R4	R3		R8	R4	R3	P2														
R9	R6	R4	P2															P2	R5	0	=
															X	R7	P2				

Figure 21.21 The example for predicated execution represented as VLIW instructions after scheduling

versed to convert predicated code back into code with branches [27]. This has an interesting consequence for local versus global scheduling. Since if-conversion makes an acyclic control-flow graph into a single, predicated basic block (i.e., a *hyperblock*), local scheduling techniques can be used after if-conversion as though the code were a single basic block. After scheduling, the block can be reverse-if-converted to execute on architectures that do not support predication. The combination of if-conversion and reverse-if-conversion form a transformation pair that converts global scheduling into local scheduling, much as the Fourier transform converts convolution into multiplication. Techniques to limit the number of branches in the code after reverse-if-conversion remain an active research topic.

Cyclic scheduling. Cyclic scheduling efficiently schedules loops to achieve high parallelism. An example loop is shown in Fig. 21.22a. The first step in most cyclic scheduling algorithms is to unroll the loop. An unrolled version of the loop is shown in Fig. 21.22b. Here, the body of the loop has been replicated four times.

If the loop is unrolled completely, as shown in Fig. 21.23, a pattern emerges. This pattern is evident in the boxed iterations between the heavy lines. This repeating pattern is termed the *kernel* of the loop. The loop can be rewritten using the kernel. This is shown in Fig. 21.24, where each line corresponds to a VLIW instruction (specifics of the VLIW encoding have been omitted for clarity). The instructions above the kernel are termed the *prolog,* and those after the kernel are termed the *epilog.* If each iteration of the loop is viewed as a single "macro-instruction," this kind of cyclic scheduling is equivalent to a pipelined execution of these macro instructions. In this case, the pipeline is composed of three stages: stage one performs a *load operation,* stage two performs the *multiplication,* and stage three performs the *store* of the results. The prologue loads the pipeline with the macro instructions, they are then performed in an overlapped fashion. When the end of the loop is near, the pipeline is drained by the epilogue. Because the pipeline does not exist in hardware, but rather is a construct of the compiler, this kind of cyclic scheduling is sometimes termed *software pipelining.* Because multiple iterations are executed at once, it is also referred to as *polycyclic scheduling.*

$$R1 \leftarrow 2.14$$
$$R3 \leftarrow X$$

L:
$$R2 \leftarrow Load\ M[R3]$$
$$R4 \leftarrow R2 * R1$$
$$Store\ M[R3] \leftarrow R4$$

$$R2 \leftarrow Load\ M[R3++]$$
$$R4 \leftarrow R2 * R1$$
$$Store\ M[R3] \leftarrow R4$$

$$R1 \leftarrow 2.14$$
$$R3 \leftarrow X$$

L:
$$R2 \leftarrow Load\ M[R3++]$$
$$R4 \leftarrow R2 * R1$$
$$Store\ M[R3] \leftarrow R4$$

$$R2 \leftarrow Load\ M[R3++]$$
$$R4 \leftarrow R2 * R1$$
$$Store\ M[R3] \leftarrow R4$$

$$if\ R3 < X+100\ goto\ L$$

$$R2 \leftarrow Load\ M[R3++]$$
$$R4 \leftarrow R2 * R1$$
$$Store\ M[R3] \leftarrow R4$$

(a) Example loop.

$$R2 \leftarrow Load\ M[R3++]$$
$$R4 \leftarrow R2 * R1$$
$$Store\ M[R3] \leftarrow R4$$

$$if\ R3 < X+100\ goto\ L$$

(b) Example loop unrolled four times.

Figure 21.22 An example loop used for cyclic scheduling (a) and the example unrolled four times (b)

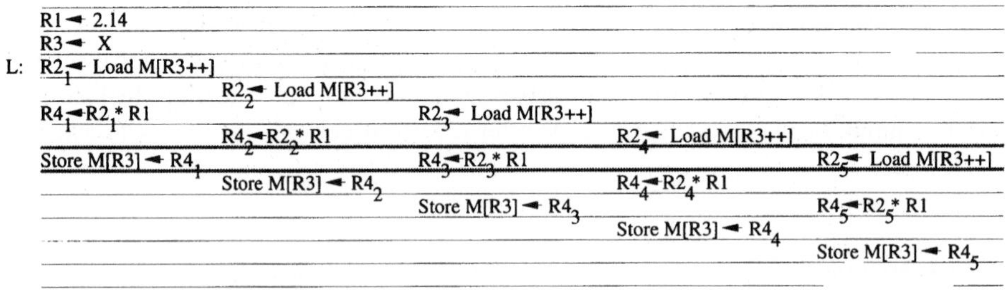

$$R1 \leftarrow 2.14$$
$$R3 \leftarrow X$$
L: $\ R2_1 \leftarrow Load\ M[R3++]$

$R2_2 \leftarrow Load\ M[R3++]$

$R4_1 \leftarrow R2_1 * R1$ $\qquad R2_3 \leftarrow Load\ M[R3++]$

$\qquad R4_2 \leftarrow R2_2 * R1$ $\qquad R2_4 \leftarrow Load\ M[R3++]$

$Store\ M[R3] \leftarrow R4_1$ $\qquad R4_3 \leftarrow R2_3 * R1$ $\qquad R2_5 \leftarrow Load\ M[R3++]$

$\qquad Store\ M[R3] \leftarrow R4_2$ $\qquad R4_4 \leftarrow R2_4 * R1$ $\qquad R4_5 \leftarrow R2_5 * R1$

$\qquad Store\ M[R3] \leftarrow R4_3$ $\qquad Store\ M[R3] \leftarrow R4_4$

$\qquad Store\ M[R3] \leftarrow R4_5$

Figure 21.23 An example loop unrolled completely with iterations overlapped in time (subscripts represent iteration numbers)

$$R1 \leftarrow 2.14$$
$$R3 \leftarrow X$$

$R2_1 \leftarrow Load\ M[R3++]$

$R2_2 \leftarrow Load\ M[R3++]$ $\quad R4_1 \leftarrow R2_1 * R1$

$R2_3 \leftarrow Load\ M[R3++]$ $\quad R4_2 \leftarrow R2_2 * R1$

L: $\ R2_{i+4} \leftarrow Load\ M[R3++]$ $\quad R4_{i+2} \leftarrow R2_{i+2} * R1$ $\quad Store\ M[R3] \leftarrow R4_i$ $\quad if\ R3 < X+97\ goto\ L$

$Store\ M[R3] \leftarrow R4_{97}$ $\quad R4_{99} \leftarrow R2_{99} * R1$

$Store\ M[R3] \leftarrow R4_{98}$ $\quad R4_{100} \leftarrow R2_{100} * R1$

$Store\ M[R3] \leftarrow R4_{99}$

$Store\ M[R3] \leftarrow R4_{100}$

Figure 21.24 The final cyclic schedule for the loop

There were several conditions that make polycyclic scheduling of the above example quite simple. These included the constant upper bound of 100 iterations, the lack of conditional code in the loop, and the well behaved register usage patterns and function unit requirements. When any of these conditions is not present, polycyclic scheduling can become extremely complicated. Each of these situations is illustrated in Fig. 21.25. In many cases, non-constant upper bound on the number of iterations and conditional code in the loop body can be handled using predicated execution via if-conversion (similar to hyperblock formation, see above).

Decoupling cross-iteration dependencies in a loop can be done either by using additional registers or by hardware support. Figure 21.26a demonstrates the use of additional registers for the loop of Fig. 21.25c. Here, the loop is unrolled to the degree required for polycyclic scheduling, then the cross-iteration dependence is removed by renaming R4 using R5, R6, and R7.

```
                                                        R1 ◂ 2.14
      R1 ◂ 2.14              R1 ◂ 2.14                   R3 ◂ X
      R3 ◂ X                 R3 ◂ X                      R4 ◂ 1.1

  L:  R2 ◂ Load M[R3++]  L:  R2 ◂ Load M[R3++]      L:  R2 ◂ Load M[R3++]
      R4 ◂ R2 * R1           R4 ◂ R2 * R1               R4 ◂ R4 * R1
      Store M[R3] ◂ R4                                  Store M[R3] ◂ R4
                             if R4 < 50 goto L2
      if R4 < 0 goto L                                  if R3 < X+100 goto L
                             Store M[R3] ◂ R4

                         L2: if R3 < X+100 goto L
```

(a) Non–constant upper bound.   (b) Conditional code in loop body.   (c) Cross–iteration register use (i.e., R4).

Figure 21.25 Loops that are difficult to cyclic-schedule because of (a) nonconstant upper bound on the number of iterations, (b) conditional code in loop body, and (c) cross-iteration dependence

```
      R1 ◂ 2.14
      R3 ◂ X
      R4 ◂ 1.1

  L:  R2 ◂ Load M[R3]
      R5 ◂ R4 * R1
      Store M[R3] ◂ R4

      R2 ◂ Load M[R3++]
      R6 ◂ R5 * R1                    R1 ◂ 2.14
      Store M[R3] ◂ R4                R3 ◂ X
                                      R4[1] ◂ 1.1
      R2 ◂ Load M[R3++]
      R7 ◂ R6 * R1             L:  R2 ◂ Load M[R3++]
      Store M[R3] ◂ R4             R4 ◂ R4[1] * R1
                                    Store M[R3] ◂ R4
      R2 ◂ Load M[R3++]
      R4 ◂ R7 * R1                remap R4
      Store M[R3] ◂ R4            if R3 < X+100 goto L

      if R3 < X+100 goto L
```

(a) Use of additional registers.   (b) Use of rotating registers.

Figure 21.26 Solution to the problem of cross-iteration register usage via (a) use of additional registers and (b) rotating registers

The primary hardware support for cross-iteration dependence removal was first proposed in the Cydrome Cydra-5 [26]. This technique is called *rotating registers*. A special register file is used that has multiple copies of each register. In essence, each register is replaced with a queue. The position in the queue is maintained using a base pointer. An operation, *remap*, is added to the architecture to advance the base pointer. Values of a register earlier in the queue can be accessed according to their distance from the current base pointer value. For example, if register R4 contains a value, then once *remap* R4 is executed, the value in R4 is undefined. The previous value of R4 can be accessed as R4[1]. Similarly, the initial value of R4 after *n remap* operations can be access as R4[n]. Because hardware is limited, remap operations can be defined only for limited scope, and each register is up of a finite, circular queue. This is the source of the rotating register name for this technique. If R4 in Fig. 21.25c is replaced with a rotating register, cyclic scheduling can again be used on the resulting code. The modified code is shown in Fig. 21.26b.

21.2.3 Compatibility between generations

VLIW processors exploit instruction-level parallelism using largely compiler-based techniques. The majority of these techniques require knowledge of the function unit latencies in order to schedule code. Without additional support, problems arise with the binary compatibility of code if these latencies change between generations of the same architecture.

Consider a pipelined multiplier that takes three cycles to execute operations (latency of three cycles). This was the assumption used to arrive at the schedules of Figs. 21.17 and 21.21. To see the problem with compatibility, consider what would happen if this code were executed on a VLIW with a longer-latency multiplier. The code would not execute correctly because operations that required the result of the multiplier would access its destination register earlier than they should (e.g., the last add operation would access R6, or the store operation would access R7 too early, in either Fig. 21.17 or 21.21). Rau has suggested a solution to this problem termed *split-issue* [28]. It is based on the observation that VLIW operations have two parts: the initiation part and the write-back part. Table 21.4 presents the example of 21.16a divided into its two parts. Here Vi represents a non-architected register: a register that exists in hardware but not accessible by the programmer.

TABLE 21.4 Example Operations Divided into Initiation and Write-Back Parts

	Initiation	Write-back
I0	V1 ← R2 + R3	V1 ← R1
I1	Load V4 ← M[X]	V4 ← R4
I2	V5 ← R4 + R3	V5 ← R5
I3	V6 ← R2 * R3	V6 ← R6
I4	V7 ← R1 * R3	V7 ← R7
I5	Store M[X] ← R7	(none)
I5	V8 ← R4 + R3	V8 ← R8
I7	V9 ← R6 + R4	V9 ← R9

There is a latency between the initiation of an operation and its completion. For example, in the original architecture, the multiplication performed by operation I3 would be initiated in cycle x. The write-back would execute in cycle $(x + 3)$. The split-issue technique uses simple interlocking to guarantee the write-back portion of each operation is correct. Simple interlocking is also used when the registers are read at operation issue. Any operation that encounters a busy register is stalled until the register becomes not busy. Rau argues that this extra hardware is very minimal and not used unless older code is run on newer hardware. It is a superscalar construct added to a VLIW architecture to support compatibility between generations.

A problem can also occur when function unit latencies get shorter from one generation to the next. This problem is illustrated in Fig. 21.27. For this example, the old machine has a latency of three cycles for its multiply operation, whereas the new machine has a latency of one cycle. When the code is executed on the new machine, the value of R3 is overwritten before the second addition can use the value of R3 generated by the first addition, and incorrect execution results. One solution to the problem is for the compiler to eliminate the anti-dependence by renaming R3. However, a limited number of actual hardware registers may prevent this solution. Another possibility is to constrain the scheduler so that it assumes a register value may be ready *earlier* than when the latency specifies, and adjust the schedule accordingly. This kind of VLIW architecture is termed a *less-than-or-equals* architecture, since the latency is assumed to be less than or equal to the published value. The original kind of VLIW is termed an *equals* architecture [6]. Less-than-or-equals architectures are identical to equals architectures in their hardware implementation. The only difference is the contract established between the programmer or compiler and the architecture designer. One drawback of a less-than-or-equals architecture is the slightly longer schedules that must be produced for it.

21.3 Superscalar vs. VLIW: Which Is Better?

Superscalar and VLIW designs both exploit instruction-level parallelism to achieve high performance from a single stream of execution. The techniques each architecture uses to do this are vastly different, however. Table 21.5 summarizes these differences.

Superscalar designs handle the challenge of anti- and output-dependencies using methods such as the Tomasulo algorithm. For VLIWs, this same problem is solved by use of a large number of registers, thereby approximating the virtual registers of the compiler's intermediate code. In addition, rotating registers can be used to break cross-iteration anti-

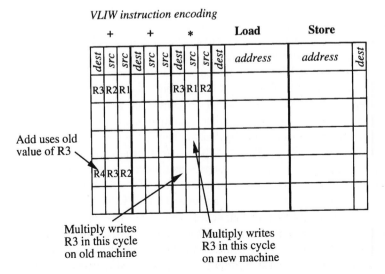

Figure 21.27 The problem with compatibility when latencies get shorter

TABLE 21.5   Differences between Superscalar and VLIW Solutions to Challenges in Exploiting Instruction-Level Parallelism

Challenge	Superscalar	VLIW
Anti-, output dependencies	Tomasulo algorithm	Large register file, rotating registers
Out-of-order execution	Hardware dispatch and scheduling units	Compiler-based local and global scheduling
Branch instructions	Hardware or profiled branch prediction	Profiled branch prediction Trace/superblock formation
Speculative execution	Branch prediction, state update hardware	If-conversion (predication), global code motion, sentinel scheduling
Precise interrupts	State update hardware	Sentinel scheduling

and output dependencies. Once registers are renamed, superscalar hardware uses hardware constructs such as the scheduling queue, the scoreboard, the common data bus, and multiple instruction issue to execute instructions out of program order. The VLIW approach uses compiler-based local and global scheduling techniques, the later including both acyclic (trace-, superblock, hyperblock-scheduling), and cyclic (software pipelining/polycyclic scheduling) techniques. As with dependencies, where superscalar uses a purely hardware-based solution, the VLIW solution is to rely on the compiler. Superscalars can use the compiler-assisted profiled branch prediction or hardware-based branch history buffers. VLIW processors use profile information to construct a weighted control flow graph, which is then scheduled. This process tends to place the most-likely target of a branch immediately after the branch instruction. Any situation where the successor to a basic block is not the most-likely target of its branch (e.g., at the end of the body of a loop) is handled with profiled branch prediction. The analog of superscalar state update hardware (e.g., the reorder buffer, future file, or checkpoint-repair) is the VLIW sentinel scheduling technique. Both aid in speculative execution. VLIWs also take advantage of predicated execution via if-conversion.

Which is better, then, a VLIW or a superscalar processor? Without question, compiler-based scheduling is far superior to hardware scheduling techniques alone. However, old executables often cannot be re-compiled to take advantage of new compiler scheduling techniques. In such situations, hardware scheduling has an advantage. In addition, the scheduling techniques used for VLIWs are not limited to VLIWs. A superscalar with simple interlocking can be viewed as a "forgiving VLIW," where it correctly executes unscheduled code but can achieve more substantial speedups for scheduled code. As mentioned at the end of the previous section, a similar technique (i.e., split issue) has been proposed for VLIW intergeneration compatibility. Such a machine is therefore part superscalar and part VLIW. What separates VLIW from superscalar is the programmer's view of the processor. The latencies of function units in a superscalar processor are not part of the instruction set architecture. For a VLIW, a programmer or compiler must know the latencies to correctly schedule code. This turns into an advantage for a VLIW, since known latencies result in accurate and highly-parallel code schedules.

In 1994, Hewlett-Packard (the number two workstation vendor after Sun Microsystems) and Intel (worldwide leader in microprocessor sales) announced they would join forces to create their next-generation processor design. Intel is widely believed to be transitioning away from its popular CISC x86 architecture. Many of the pioneers of VLIW, including founders from both Cydrome and Multiflow, now work for Hewlett-Packard. The announcement is seen by many to forecast the introduction of another commercial VLIW processor by the end of the decade. Other superscalar processor vendors have similarly stated that VLIW is the only way to achieve an issue rate greater than eight instructions per cycle. It may be too early to declare that VLIW is the clear winner over superscalar. The earlier success of RISC over CISC was largely due to RISC's superior ability to take advantage of compiler techniques. It seems that this trend toward complicated software with simplified hardware is likely to continue

Today, the majority of all processors used in systems ranging from personal-computer uniprocessors to massively-parallel multiprocessors exploit instruction-level parallelism. Whether superscalar will overcome VLIW or vice versa is not clear. It is clear is that instruction-level parallelism will continue to be the most-common and most-general form of parallel processing in use.

21.3.1   To probe further

An excellent advanced survey of research into instruction-level parallelism is presented by Fisher and Rau, two of the field's pioneers [29]. An equally excellent survey that focuses on superscalar architectures and their trade-offs is by Johnson [11].

The design of superscalar and VLIW processor architectures is an active research topic, and most of the ideas are first presented at annual conferences. A leading conference is the International Symposium on Microarchitecture, which is held every year and organized by the Association for Computing Machinery SIGMICRO special interest group and the IEEE TC-MICRO technical committee. Many of the ideas discussed in this chapter were first presented at this conference. Other conferences include the International Symposium on Computer Architecture and the International Conference on Architectural Support for Programming Languages and Operating Systems, both organized by the ACM SIGARCH special interest group and the IEEE Computer Architecture Technical Committee. In addition, readers interested in VLIW should also examine the journal *Software Practice & Experience,* and the Conference on Programming Language Design and Implementation (organized by the ACM SIGPLAN special interest group). The proceedings of the above conference are published by the IEEE and the ACM.

21.4   Bibliography

1. Alpert, D., and D. Avnon. 1993. Architecture of the Pentium microprocessor. *IEEE Micro,* vol. 13, 11–21.
2. McLellan, E. 1993. The Alpha AXP architecture and the 21064 processor. *IEEE Micro,* vol. 13, 36–47.
3. Diefendorff, K., and M. Allen. 1992. Organization of the Motorola 88110 superscalar RISC microprocessor. *IEEE Micro,* vol. 12, 40–63.
4. Markstein, P. W. 1990. Computation of elementary functions on the IBM RISC system/6000 processor. *IBM J. Research and Development,* vol. 34, 111–119.
5. Song, S. P., and M. Denman. 1994. The PowerPC 604 RISC microprocessor tech. rep. Austin, Tex.: Somerset Design Center.
6. Kathail, V., M. Schlansker, and B. R. Rau. 1994. HPL PlayDoh architecture specification: version 1.0. *Tech. Rep. HPL-93-80.* Palo Alto, Calif.: Hewlett-Packard Laboratories, Technical Publications Department.
7. Hwu., W. W., T. M. Conte, and P. P. Chang. 1991. Comparing software and hardware schemes for reducing the cost of branches. *Proc. 16th Ann. International Symposium Computer Architecture,* (Jerusalem, Israel), 224–233.
8. Smith, J. E. 1991. A study of branch prediction strategies. *Proc. 8th Ann. Int'l. Symp. Computer Architecture,* 135–148.
9. Yeh, T., and Y. N. Patt.1991. Two-level adaptive training branch prediction. *Proc. 24th Ann. International Symposium on Microarchitecture* (Albuquerque, NM), 51–61.

10. Yeh, T., and Y. N. Patt. 1993. A comparison of dynamic branch predictors that use two levels of branch history. *Proc. 20th Ann. International Symposium Computer Architecture* (Ann Arbor, Michigan), 257–266.

11. Johnson, M. *Superscalar Microprocessor Design.* Englewood Cliffs, N.J.: Prentice-Hall.

12. Thornton, J. E. 1970. *Design of a Computer—The Control Data 6600.* Glenview, Ill.: Scott, Foresman, and Co.

13. Tomasulo, R. M. 1967. An efficient algorithm for exploiting multiple arithmetic units. *IBM J. Research and Development,* Vol. 11, 34–53.

14. Koren, I. 1993. *Computer Arithmetic Algorithms.* Englewood Cliffs, N.J.: Prentice Hall.

15. Kroft, D. 1981. Lockup-free instruction fetch/prefetch cache organization. Proc. 8th Ann. Int'l. Symp. Computer Architecture, 81–87.

16. Conte, T. M. 1992. Tradeoffs in processor/memory interfaces for superscalar processors. *Proc. 25th Ann. Int'l. Symp. on Microarchitecture* (Portland, OR), 202–205.

17. Weiss, S., and J. E. Smith. 1984. Instruction issue logic for pipelined supercomputers. *IEEE Trans. Comput.,* Vol. C-33, 1013–1022.

18. Hwu, W. W., and Y. N. Patt. 1987. Checkpoint repair for high-performance out-of-order execution machines. *IEEE Trans. Comput.,* Vol. C-36, 1496–1514.

19. Fisher, J. A. 1981. Trace scheduling: A technique for global microcode compaction. *IEEE Trans. Comput.,* Vol. C-30, no. 7, 478–490.

20. Hwu, W. W., S. A. Mahlke, W. Y. Chen, P. P. Chang, N. J. Warter, R. A. Bringmann, R. G. Ouellette, R. E. Hank, T. Kiyohara, G. E. Haab, J. G. Holm, and D. M. Lavery. 1993. The superblock: An effective structure for VLIW and superscalar compilation. *Journal of Supercomputing,* Vol. 7, 229–248.

21. Hank, R. E., S. A. Mahlke, J. C. Gyllenhaal, R. Bringmann, and W. W. Hwu. 1993. Superblock formation using static program analysis. *Proc. 26th Ann. Int'l. Symp. on Microarchitecture* (Austin, TX), 247–255.

22. Conte, T. M., B. A. Patel, and J. S. Cox. 1994. Using branch handling hardware to support profile-driven optimization. *Proc. 27th Ann. International Symposium on Microarchitecture* (San Jose, CA).

23. Chang, P. P., S. A. Mahlke, and W. W. Hwu. 1991. Using profile information to assist classic code optimizations. *Software-Practice and Experience,* Vol. 21, 1301–1321.

24. Mahlke, S. A., W. Y. Chen, R. A. Bringmann, R. E. Hank, W. W. Hwu, B. R. Rau, and M. S. Schlansker. 1993. Sentinel scheduling: A model for compiler-controlled speculative execution. *ACM Trans. Comput. Sys.,* Vol. 11, 376–408.

25. Mahlke, S. A., D. C. Lin, W. Y. Chen, R. E. Hank, and R. A. Bringmann. 1992. Effective compiler support for predicated execution using the hyperblock. *Proc. 25th Ann. Int'l. Symp. on Microarchitecture* (Portland, OR), 45–54.

26. Rau, B. R., D. W. L. Yen, W. Yen, and R. A. Towle. 1989. The Cydra 5 departmental supercomputer. *Computer,* Vol. 22, 12–35.

27. Warter, N. J., S. A. Mahlke, W. W. Hwu, and B. R. Rau. 1993. Reverse if-conversion. *Proceedings of the ACM-SIGPLN '93 Conference on Programming Language Design and Implementation* (Albuquerque, N.M.), 290–299.

28. Rau, B. R. 1993. Dynamically scheduled VLIW processors. *Proc. 26th Ann, International Symposium on Microarchitecture* (Austin, TX), 80–90.

29. Rau, B. R., and J. A. Fisher. 1993. Instruction-level parallel processing: History, overview and perspective. *Journal of Supercomputing,* Vol. 7.

22

SIMD-Processing: Concepts and Systems

Michael Jurczyk and Thomas Schwederski

Single instruction stream, multiple data stream (SIMD) computers were the first systems to be implemented with a massive amount of processors (>104), and were among the first systems to provide computational power above the GFLOP range. In this chapter, one class of parallel computers, the *SIMD computers,* will be studied with respect to their architecture, hardware requirements, and field of application. SIMD computers consist of an array of processing elements that are controlled by a common control unit. Basic concepts including SIMD computer components and associative processing, existing SIMD systems, algorithms and applications well suitable for running on SIMD architectures, and the programming and available languages of SIMD machines will be examined and discussed.

22.1 Basic Concepts

The computer classifications introduced by Flynn [11] distinguishes computer architectures by the number of concurrent instruction and data streams. A machine in which many data items are operated upon simultaneously by the same data manipulations is thus classified as an SIMD machine. This rather general classification encompasses a wide variety of systems, all of which have some operational elements that synchronously execute identical computations on differing data items. Thus, many special-purpose VLSI array processors are SIMD machines, in which a number of arithmetic units (e.g., multiply/add elements) are controlled by a finite state machine to perform operations such as filtering. Such systolic arrays and wavefront processors are characterized by synchronous, high-speed operation as well as an optimized flow of data and results among neighboring elements. VLSI structures are discussed elsewhere in this book and have been studied in detail by Kung [23]. Many supercomputers also utilize the SIMD principle of operation. To efficiently process vectors and matrices, they employ vector processing units in which multiple arithmetic units (e.g., floating-point processors) operate on a large number of data items simultaneously. This SIMD-style operation is hidden from the user and will not be studied in this chapter.

Usually, VLSI array processors are specifically designed and optimized for the execution of a single algorithm. Topic of this chapter are programmable SIMD machines that can be used for a wide variety of applications. Such machines usually employ massive numbers of relatively simple processors, e.g., 64k processors in the Connection Machine CM-2 [42] and 16k processors in the MasPar MP-1 [5]. A trend towards more complex processors can be observed, e.g., 32-bit CPUs in the MasPar MP-2. This way, problems that require parallel manipulations of large data sets can be solved efficiently. Examples of such problems are algorithms based on vector and matrix operations [9, 12], and image processing algorithms where operations are performed on a large number of image pixels [34, 35]. For these well-suited SIMD algorithms, performances of several billion instructions per second have been reported, e.g., four billion operations per second for the inner loop of an image smoothing algorithm (see Section 22.5.2) running on a Connection Machine CM-2 with 64k processors [44]. Generally, SIMD machines have been used quite successfully in many scientific application areas.

An inherently limiting factor of SIMD machines is the sequentialization of conditional statements. Because all processors of the machine either execute the same instruction or are idle, conditionals must be serialized. Thus, significant loss of speedup potential is frequently observed. For this reason, many parallel systems deviate from the pure SIMD principle by permitting the execution of different instructions on the processors while SIMD-style synchronization is employed for communication. Examples of such machines are PASM (partitionable SIMD/MIMD [35]) and the Connection Machine CM-5 [29].

A generic SIMD machine model that concentrates on the interactions of the SIMD machine components is depicted in Fig. 22.1. The *processor array* operates on multiple data sets in parallel. This can be achieved, for example, by a content-addressable memory (CAM, see below and Section 22.3), or by multiple processors and memories connected by an interconnection network (see below). The single instruction stream that controls the processor array is generated by an *array control unit* (ACU). The ACU executes sequential program code, performs control flow operations, and broadcasts instructions, control sig-

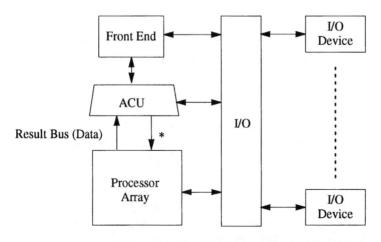

*) Broadcast Bus (Instruction, Data and Control)

Figure 22.1 Generic SIMD machine model

nals, and data to the processor array. The instructions are executed synchronously, in a lock-step-fashion, by the processors. The control signals are needed to synchronize instructions and communication, and control execution (enabling/disabling of processors, see Section 2.4.2). The results from control unit data operations can be distributed to the processor array via the *broadcast bus.* Many control operations depend on the result of data manipulation operations by the processor array, so that a *result bus* from the array to the ACU is included in many systems.

To provide a convenient user interface for program development, debugging, and graphical display, most SIMD systems utilize a conventional computer, e.g., a UNIX workstation, as a *front end.* For execution, programs are downloaded from the front end to the ACU, and data may be downloaded via a separate *I/O device,* e.g., a disk array, to the processor array. The ACU then begins program execution and broadcasts SIMD instructions for the processors via the broadcast bus.

Now consider the processor array. Assume that no CAM is used. Then, to operate on multiple data sets in an SIMD machine in parallel, multiple memory units are needed that store those data sets and multiple processors are necessary to manipulate the data sets. Also, data has to be exchanged between different data sets, so an interconnection network is necessary that either interconnects the processors or connects the processors with the memory units. The combined block of processors, memory units, and interconnection networks is termed a *processor array.*

To be flexible in instruction execution in the processor array, a mechanism has to be provided to select a subset of processors that execute a certain instruction. Compile-time and run-time processor masking schemes to enable/disable particular processors, along with masked and unmasked instructions, are used for this purpose. Unmasked instructions are executed unconditionally by all processors. Masked instructions, on the other hand, are executed by enabled processors only, while disabled processors remain idle during that instruction.

One important architectural aspect is the organization of the processor array. One such architecture is the *processing-element-to-processing-element organization,* which is depicted in Fig. 22.2a. In this configuration, N processing elements are connected via an interconnection network. Each *processing element* (PE) is a processor with local memory. The PEs execute the instructions that are distributed to the PEs by the ACU via the broadcast bus. Each PE operates on data stored in its own memory, and on data broadcast by the ACU. Data is exchanged among PEs via a unidirectional *interconnection network,* and the *I/0 bus* is used to transfer data from PEs to the I/O interface and vice versa. To transfer results from particular PEs to the ACU, the result bus is used. Because local memory can be employed, the hardware that is used in such a machine can be constructed efficiently. In many algorithms, communication is mostly local, e.g., among the nearest neighbors. For this reason, the nearest neighbor organizations have been very popular. Examples of SIMD machines with this architecture are the MasPar MP-I [5], Connection Machine CM-2 [42], GF11 [4], DAP [18], MPP [3], STARAN [2], PEPE [43], and Illiac IV [6] machines. The MPP, CM-2, GF11, and MP-1 machines are surveyed in Section 22.4 of this chapter.

A second SIMD architecture is the *processor-to-memory organization* as shown in Fig. 22.2b. In this configuration, a bidirectional interconnection network connects the N processors with M memory modules. The processors are controlled by the ACU via the broadcast bus. Data is exchanged between processors via the interconnection network and the memory modules. Again, data transfers between the memories and the I/O interface are handled via the I/O bus. To transfer data from particular memory units to the ACU, the re-

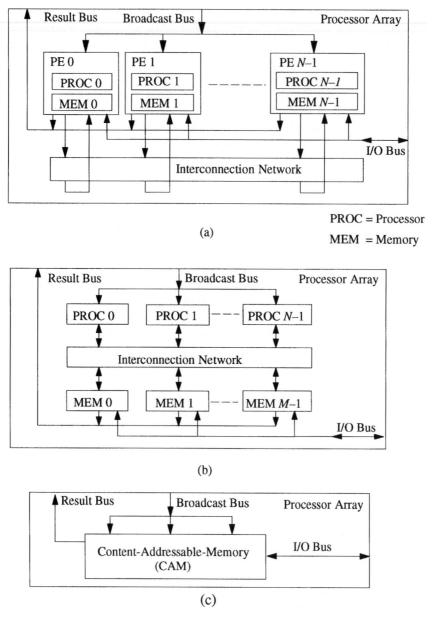

Figure 22.2 Processor array models: (a) processing-element-to-processing-element organization, (b) processor-to-memory organization, and (c) content-addressable-memory (CAM)

sult bus is used. Examples of this SIMD machine architecture are the Burroughs scientific processor BSP [22] and the Texas Reconfigurable Array Computer TRAC [33].

A third SIMD architecture is the *content-addressable memory* [8] as depicted in Fig. 22.2c. In contrast to a RAM, in which data can be accessed by providing data addresses serially, a CAM is content addressable; i.e., a data item is provided to the CAM, and those CAM cells that contain this value will set a flag (in parallel) to indicate whether the provided data item matches the value stored in its cell. Instead of employing separate processors in the processor array of an SIMD machine of this architecture, special compare and matching logic is present in each CAM bit cell. Thus, a CAM cell acts like a separate PE. SIMD machines consisting of a CAM are also called *associative processors.* Associative processing is discussed in Section 22.3.

Because, in an SIMD machine, a single ACU provides the instruction stream for all of the array processors, the system will frequently be underutilized whenever programs are run that require a few PEs only. To alleviate this problem, *multiple-SIMD* (MSIMD) machines were designed. They consist of multiple control units, each with its own program memory. As is shown in Fig. 22.3, the PEs are controlled by U control units that divide the machine into U independent *virtual SIMD machines* of various sizes. U is usually much smaller than N and determines the maximum number of SIMD programs that can operate simultaneously. The distribution of the PEs onto the ACUs can be either static or dynamic. Examples of MSIMD machines are MAP [26], PM4 [7], PASM [351], and TRAC [33]. Because a fully configured CM-2 contains four units with 16k PEs each that can be run individually or in a coordinated fashion (see Section 22.4.2), this machine also can be classified as an MSIMD system.

The MSIMD machine architecture has several advantages over normal SIMD machines [38], including

- *Efficiency.* If a program requires only a subset of the available PEs, the remaining PEs can be used for other programs.
- *Multiple users.* Up to U different users can execute different SIMD programs on the machine simultaneously.

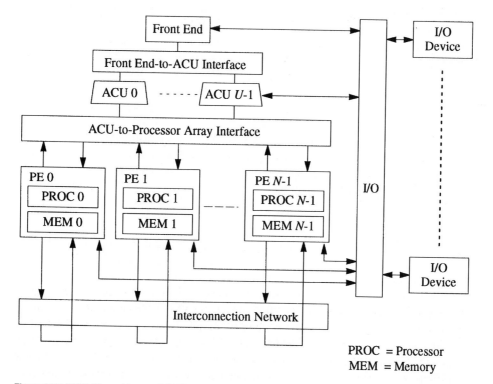

Figure 22.3 MSIMD machine model, PE-to-PE organization

- *Fault detection.* A program runs on two independent machine partitions, and errors are detected by result comparison.
- *Fault tolerance.* A faulty PE only affects one of the multiple SIMD machines, and other machines can still operate correctly.

A prerequisite of MSIMD operation is an independent operation of each of the multiple virtual SIMD machines. Thus, the interconnection network connecting the PEs must support this independence by providing methods for partitioning [37].

22.2   SIMD Machine Components

22.2.1   Introduction

In this section, the individual components of an SIMD machine will be studied. The PE-to-PE organization will be used, but most discussions are directly applicable to systems with processor-to-memory structure; associative processing will be surveyed in Section 22.3. The model shown in Fig. 22.1 is employed in the discussions. Functions and architectural alternatives, as well as actual implementations for each of the components, will be discussed.

22.2.2   Front end computer

The front end computer, e.g., a SUN-4 workstation in the CM-2 system, a DECstation 3000 in the MasPar system, or a DEC VAX-11/780 in the MPP system, connects the SIMD machine to the outside world via a network interface, e.g., an Ethernet interface as in the MasPar MP-1 machine. The front end provides the machine's operating system, which, in the majority of SIMD machines, is UNIX or an equivalent operating system such as ULTRIX. Frequently, the operating system supports time-sharing job execution so that multiple users can execute their application programs simultaneously on the SIMD machine.

On the front end, the machine users can develop and compile their application programs. During program execution, the program is first downloaded from the front end computer to the ACU. The ACU then executes the program and distributes instructions and data to the processor array. In some SIMD machines, like the Massively Parallel Computer MPP (Section 4.1), both the ACU and the front end computer can be used for program development, compilation, and debugging.

22.2.3   Array control unit

The ACU executes sequential program code, performs control flow operations such as loop counting, and broadcasts instructions and control signals to the PEs. Figure 22.4 shows one possible organization, which is similar to the PASM control unit [32].

When the execution of an application program is issued on the front end, the program is downloaded into the RAM of the ACU (*ACU-RAM*) via the I/O interface. PE program instructions and global mask data (determined at compile time) are downloaded to the *instruction and global mask RAM* (I&M-RAM), also via the I/O interface. Then, the ACU executes the program and fetches either a scalar instruction from the ACU-RAM or plural instructions from the I&M-RAM. Scalar instructions, i.e., instructions operating on scalar data and variables residing in the ACU, are executed by the CPU of the ACU, while parallel

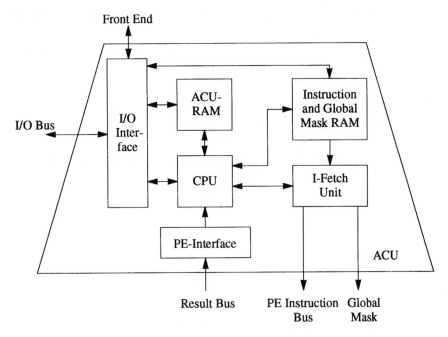

Figure 22.4 Array control unit (ACU) model

instructions that operate on parallel variables residing in each PE are transformed into simpler instructions (*nanoinstructions*) in the *instruction fetch unit* (*I-Fetch Unit*) and are sent together with the related mask data to the PE array via the PE instruction bus for execution. For example, a 32-bit addition instruction fetched by the ACU in the MPP machine is transformed into 32 1-bit addition nanoinstructions that are executed by each 1-bit PE serially.

In most algorithms, results and/or condition flags of data manipulation operations effect the overall program flow. To support such a mechanism in an SIMD machine, data or status information residing in the PEs must be combined into a single item that is sent to the ACU for program flow decisions. For example, in the construct

```
IF ALL (condition A) THEN DO B
```

statement B is executed (i.e., appropriate instructions are broadcast by the ACU) only if condition A is true for all PEs. To correctly enable/disable the PEs, the ACU therefore has to know the result of condition A on all PEs. For this reason, a unidirectional result bus, e.g., the GLOBAL line in the Connection Machine CM-2 or the SUM-OR tree in the MPP, is present in many SIMD machines. PEs place the results, e.g., the contents of a 1-bit register, onto the tree. These values are combined within the tree, e.g., by an OR operation on all data in the MPP SUM-OR tree, to a single result that is available for ACU operations.

22.2.4 Processing array

22.2.4.1 Processing element structure.
In most SIMD machines, simple reduced-instruction-set-computer (RISC) processors are used with a local memory of restricted

size. For example, the MasPar MP-1 PEs consist of 4-bit processors, each with up to 64 kB of memory [5]. In the MPP machine, l-bit processors, each with up to l kb of memory, are used [3], while a CM-2 PE consist of l-bit processors, each with up to 64 kb of local memory [42]. Because of the simplicity of the PEs, multiple PEs can be implemented on a single VLSI chip to reduce the number of inter-chip connections and the overall machine size. For example, 16 CM-2 processors (excluding the memory units) are implemented on one VLSI chip [42] while, in the MasPar MP-1, one chip contains 32 MasPar processors (excluding the memory units) [16]. A trend toward more complex processors can be observed, e.g., 32-bit processors, each with up to 256 kB of memory in the MasPar MP-2.

Essential PE components of most systems are an *arithmetic/logic unit* (ALU), *data registers,* a *network interface* (NI) that can include data-transfer registers, a unique *processor number,* an *enable flag,* F, and *local memory,* as shown in Fig. 22.5. The actual interconnections of these components are implementation dependent and therefore not shown in the figure. (Examples of interconnection structures can be found in the SIMD system case studies in Section 22.4.) Controlled by instructions that are broadcast by the ACU, a PE can fetch data from its local memory, store it in the data registers, manipulate memory and register contents by the use of the ALU, and store register data in its local memory. It can also operate on data that is broadcast by the ACU. Furthermore, it can transfer data to and receive data from other PEs via its network interface, which is connected to the NIs of other PEs via the interconnection network. In some SIMD machine architectures, as in the MasPar MP-1 [5], it is possible to transfer a data item from a source PE directly into a memory location of the destination PE, while in other machines, such as in the MPP [3], data items are transferred to a dedicated data-transfer register located in the NI. Transfer of data between a PE and I/O devices can be accomplished via the I/O bus of the SIMD ma-

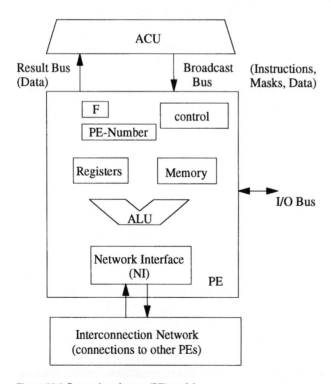

Figure 22.5 Processing element (PE) model

chine. In some SIMD machines, e.g., the MasPar MP-1, the PEs are connected to I/O devices via the inter-PE interconnection network and the machine's I/O channel. For the PE-to-ACU data transfer, individual PEs can place results, e.g., the contents of a 1-bit register, onto the result bus. These values are combined within the tree, e.g., by an OR operation on all data in the MPP SUM-OR tree, to a single result that is then available for ACU operations.

Each of the N PEs in a processor array is associated with a unique PE number, i.e., an integer ranging from 0 to $N-1$ that is also called the *PE address*. This number is needed so that a PE's location in the processor array can be determined at program run time. To permit enabling and disabling of PEs, an enable flag F is usually included which can be written by the ACU via control signals, by PE operations, or by both, as discussed in the next section.

One characteristic of SIMD machines is the synchronous operation of the PEs. Because all PEs receive and execute their instructions simultaneously, they always operate synchronously, i.e., in a lock-step fashion, independent of the program. This is most important for data transfer operations, which can be executed very efficiently. If an SIMD-system permits communication with its four nearest neighbors, for example, communication reduces to a register-to-register transfer operation among the NIs of neighboring PEs. The synchronous operation of a massively parallel SIMD system is also a serious implementation drawback due to the limited system clock speed that can be achieved.

22.2.4.2 Processing element enabling/disabling.

To execute selective program parts, as necessary for data conditional operations, for example, only a portion of the PEs should execute certain instructions, while the other PEs remain idle. This enabling/disabling of PEs can be controlled either globally by the ACU by using PE masking schemes that are determined at compile time (*global masking*), or locally by each PE determined at runtime (*data-conditional masking*) by using a special enable flag residing in each PE. While global masking can be performed exclusively under ACU control, data conditional masking poses additional requirements on the PE instruction set, as discussed below. Consider data conditional masking first. In this method, PEs determine their own enable/disable status. The instruction set contains *masked* and *unmasked* instructions. Masked instructions are executed depending on the status of the enable flag F shown in Fig. 22.5, while unmasked instructions ignore that flag. Consider the way the enabling/disabling is performed for an IF-THEN-ELSE construct. Assume that a variable x is local to (i.e., stored in) each PE. If, for example, the PE instruction:

```
IF (x > 0) DO <statement A> ELSE DO <statement B>
```

is executed in parallel, each PE evaluates the IF condition. Those PEs for which the IF condition is true, i.e., in which x is greater than 0, will set their enable flag, while the other PEs clear their flags. Then, the ACU will distribute <statement A> to all PEs. The instructions of that program block must be masked instructions, which will be executed by the enabled PEs only, i.e., the PEs for which the IF condition is true. Next, the ACU broadcasts the unmasked ELSE instruction, which forces all PEs to invert their enable flags. Then, the ACU broadcasts <statement B> to the PEs, which must again consist of masked instructions. This program part will be executed by the enabled PEs, i.e., the PEs for which the IF condition was false.

During the execution of a conventional IF-THEN-ELSE construct, i.e., an IF-THEN-ELSE construct the condition of which is evaluated by the ACU rather than by the PEs, either the THEN or the ELSE part will be executed, depending on the outcome of the boolean IF expression. If, instead, a plural boolean expression in the IF condition is used, i.e., the IF condition is evaluated by the PEs rather than by the ACU, both the THEN and the ELSE clauses may be executed, depending on the initial active PE set and the IF condition. Thus, in this case, the IF-THEN-ELSE construct is serialized.

IF-THEN-ELSE constructs can also be nested. Consider, for example, an IF construct (the inner IF) within an ELSE construct (the outer ELSE). In this case, the inner IF condition is evaluated only by those PEs that belong to the active set that perform the outer ELSE clause. Thus, each nested IF construct may reduce the number of active PEs within the overall nesting part. To implement this nesting scheme, an execution-time control stack must be present in each PE that stores the contents of the enable flag prior to each IF condition evaluation. The previous enable status must be taken into account for a nested ELSE, and the status must be restored after a nested construct has been completed.

If a global PE masking scheme is used, the ACU sends a PE enabling mask along with every instruction to the PEs. Each PE first decodes the mask and will then either execute the instruction or stay idle, depending on the mask. Several masking schemes exist which differ in the coding of the mask. In the general masking scheme that was implemented in the Illiac IV SIMD machine with 64 PEs with 64-bit words, an N-bit vector is used as the mask. Each vector bit represents the state of one PE. If the bit is a 1, the corresponding PE will be active; otherwise, the PE will be inactive. While this scheme permits enabling and disabling of arbitrary PE sets, it is prohibitively expensive for large N. In the PE address masking scheme [36], for example, an m-position mask is used ($m = \log_2 N$), in which each position represents a bit position of the binary PE address. A masking position contains either a 0, 1, or X. Thus, a mask requires $2m$ bits. If, for all i, $(0 \leq i < m)$, the ith position of the mask and the ith PE address bit are equal, or the ith mask position is X, the PE will be active. Thus, for example, the mask [000X1] represents PEs 1 and 3, while [XXXX0] represents all even PEs, out of a set of 32 PEs. With this masking scheme, only a subset of all possible PE combinations can be activated. But, since the subsets of active PEs used in actual algorithms are regular and not arbitrary, the PE address masking scheme can be used efficiently.

Global and local PE masking schemes can be combined. In this case, a PE is active if both the local F flag and the global masks provided with the IF and ELSE instructions enable the PE.

22.2.5 Networks for inter-PE communication

Fast and efficient inter-PE communication through the interconnection network is essential for high-performance SIMD machines. The network cost must be traded off against achievable communication speed and distance. Various network types with different topologies for SIMD machines have been proposed [10, 38]. An in-depth study of interconnection network architecture, performance, and implementation issues can be found elsewhere in this book.

In SIMD machines that contain thousands of processing elements, simple but efficient networks have to be employed to obtain a high performance/cost ratio. Each PE operates synchronously to a common clock in SIMD machines, so that they also communicate synchronously. The interconnection network therefore must support synchronous data transfers by accepting, transferring, and delivering data synchronously. One consequence of this synchronous communication is the special handling of simultaneous data transfers from different source PEs to a particular destination PE. If only a single data-transfer register for receiving data is implemented in the NI of a PE (see Section 22.2.4.1), multiple data items transferred simultaneously to a particular PE are overwritten; i.e., only a single data item is received by the destination PE after the data transfer, and the other data items are lost. To overcome this data loss, special mechanisms are implemented in some SIMD machines. In the Connection Machine CM-2, for example, hardware for combining messages destined to the same PE is implemented. This combining includes *Sum, Logical OR, Overwrite, Max,* and *Min* operations. In some SIMD machines, as in the MasPar MP- 1 [5], it is possible to transfer data items from source PEs to different memory locations in the destination PEs simultaneously so that data loss in such a transfer can be avoided.

Although data transfers via the interconnection network are issued by active PEs only, inactive PEs are also involved in those operations. If an active PE issues a read of a data item from a destination PE, the destination PE, which can be either active or inactive, has to send the data item to the issuing PE independent of its enable status. Also, if an active PE issues a write of a data item to a destination PE, the source PE sends the data to the destination PE, which again can be either active or inactive, and the destination PE has to store the received data item into the appropriate register or memory location.

Examples of networks used in SIMD machines are the *4-nearest-neighbor network* (*mesh*) as shown in Fig. 22.6a (in the Illiac IV, the MPP and the CM-2), *8-nearest-neighbor network* as shown in Fig. 22.6b (in the MasPar MP-1 and MP-2), *static cube networks* (in the CM-2), and *dynamic multistage interconnection networks* (in the MasPar MP-1, MP-2, and in the GF11). The nearest-neighbor networks are preferably employed in SIMD machines because of their simple implementation and scalability. These network topologies result in an efficient single-step communication between neighbor PEs, but communications of longer distances, e.g., from PE 0 to PE 15 in Fig. 22.6, need multiple steps. Thus, algorithms with global communication requirements profit from dynamic multistage interconnection networks.

22.2.6   Input/output

While a user program is stored in the memory of the front end computer, or sometimes in the ACU, input and output data of the PEs and the ACU can be stored on a mass storage

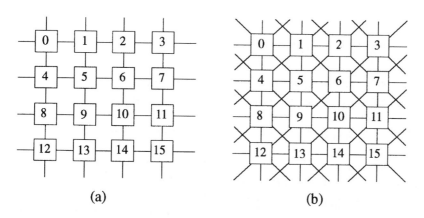

(a)   (b)

Figure 22.6 (a) *4-nearest neighbor mesh* and (b) *8-nearest neighbor mesh*, both for $N = 16$

device, such as a parallel disk array that may be connected to the PE array and/or the ACU via dedicated *I/O channels* as shown in Fig. 22.1. These I/O channels may also be used to connect external I/O devices, e.g., an image sensor array or a high-resolution graphic display. In some SIMD machines, special I/O processing units are employed to facilitate particular I/O applications. For example, in the Massively Parallel Processor (MPP) that was designed primarily for high-speed satellite imagery processing (see Section 22.4.1 for a detailed machine description), special staging memory units connect the I/O with the PE array to convert pixel streams originating from an image sensor to data streams that can be manipulated efficiently by its PE array.

22.3   Associative Processing

Conventional SIMD machines can be implemented by the use of multiple *random-access memory* (RAM) modules. An *associative processor* [24, 31, 19] can be implemented by the use of one *content-addressable memory* (CAM) unit [8]. The main distinction between RAM and CAM is that data in a RAM can be accessed by providing the addresses of those data items serially, while in a CAM a data item is provided and those CAM words that contain this value will set a flag (in parallel) to indicate a match. A mask can be provided to the CAM that determines the CAM portion that is to be searched.

One type of associative processor architecture is depicted in Fig. 22.7. It consists of a content-addressable memory array that stores n data words with a length of *m* bits. Every bit cell in the $m \times n$ array employs a flip-flop and some logic gates for read/write and comparison purposes. Furthermore, *comparand, mask,* and *match registers,* a *control unit*, and a *match resolver* are present in such a CAM.

To illustrate the operation of a CAM, consider a simplified data base of a company stored in a CAM as depicted in Fig. 22.8. Each data base entry (i.e., each CAM word) consists of the name, the age, and the salary of the employees of that company. Now assume that the names of those employees who are older than 35 are sought. For this CAM search, the comparand register is loaded with 35 in the age field, and the age field in the mask register is set to 1. The control unit issues a "greater than" search instruction to the CAM array. According to the search result, the match register is set for each CAM word in parallel. To extract the information, e.g., the names of the employees that are older than 35, the

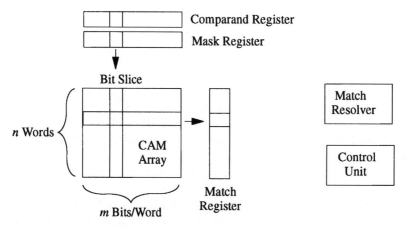

Figure 22.7 Associative processor architecture

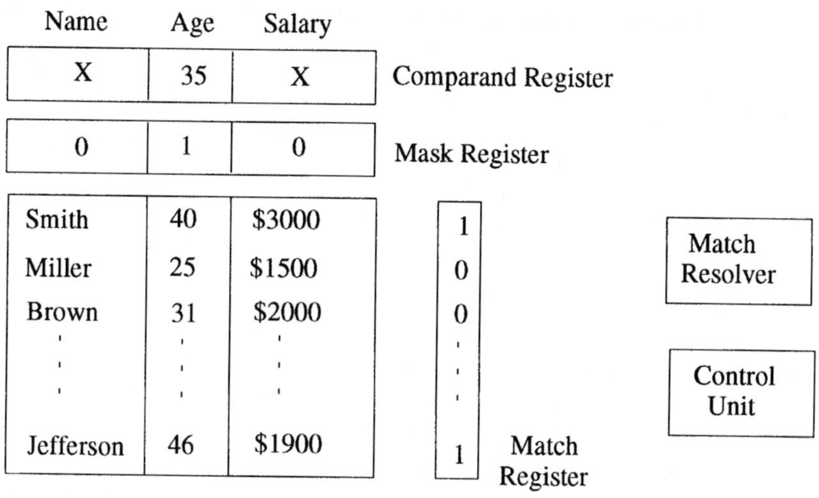

Name	Age	Salary	
X	35	X	Comparand Register

0	1	0	Mask Register

Smith	40	$3000	1
Miller	25	$1500	0
Brown	31	$2000	0
⋮	⋮	⋮	⋮
Jefferson	46	$1900	1

Match Resolver

Control Unit

Match Register

Figure 22.8 Example of a parallel search in a CAM

CAM words selected by the match register must be read out serially. This operation is performed by the match resolver unit. To search for values within a range, e.g., employees of age between 35 and 40, two consecutive search operations must be performed. After searching for employees older than 35, only those words that have found a match participate in the second search for ages less than 40.

With associative processors, not only can a variety of search operations be performed, but also logic and arithmetic operations on the data can be executed. Consider, for example, the addition of the value 1 to the contents of a particular 16-bit field A of all CAM words. Let A_i denote the ith bit of field A, and C denotes a 1-bit carry in each CAM word. Then, the addition of 1 to the A field can be implemented by the following algorithm (recall that the "select" and the "write" operations are performed in parallel).

```
select all CAM words
  write C = 1
for i = 0 to 1 = 15
  {
  select CAM words with C = 1, A_i = 0
    write C = 0, A_i = 1
  select CAM words with C = 1, A_i = 1
    write A_i = 0
  }.
```

For a CAM consisting of n data words, each of which is m bits long, the worst-case time to perform an addition is proportional to m but independent of n. In contrast, for most word parallel sequential machines, the time to perform that addition is proportional to n, but independent of m. Thus, if $n > m$, an associative processor will outperform a sequential machine of equal cycle time.

The above examples also demonstrate the SIMD characteristic of associative processors. A single instruction stream, e.g., a "greater than" search, is performed on multiple data sets (i.e., the contents of each CAM word) in parallel. Examples of associative processor arrays are the Goodyear Aerospace Corporation STARAN [2], the Parallel Element Processing Ensemble PEPE [43], and CAPRA [15].

22.4 Case Studies of SIMD Systems

Many large-scale SIMD parallel machines have been implemented, with systems such as the Illiac IV [6] being classical examples. In this section, architecture and implementation issues of four machines will be outlined. The machines range from the first machine to incorporate more than 16,000 PEs, the MPP, to more recent machines, namely, the GF11, the CM-2, and the MasPar machines MP-1 and MP-2.

22.4.1 The Massively Parallel Processor

The *Massively Parallel Processor* (MPP) was designed by Goodyear Aerospace Corporation primarily for high-speed satellite imagery processing [3]. The machine incorporates 16,384 bit-serial PEs and was delivered to NASA in 1982. It was the first SIMD machine built with such a massive number of processing elements.

The general structure of the machine is shown in Fig. 22.9 [3]. It consists of the *processor array unit* (ARU), the *array control unit* (ACU), the *program and data management unit* (PDMU), and *staging memory units*. The ACU controls the PE array and performs scalar arithmetic. The overall machine operation is controlled by the PDMU, which is a PDP-11 minicomputer running the RSX-11M real-time operating system. It is also used for program development and machine diagnostics. Data can be transferred to and from the PE array by the use of 128-bit wide I/O interfaces and through staging memory units

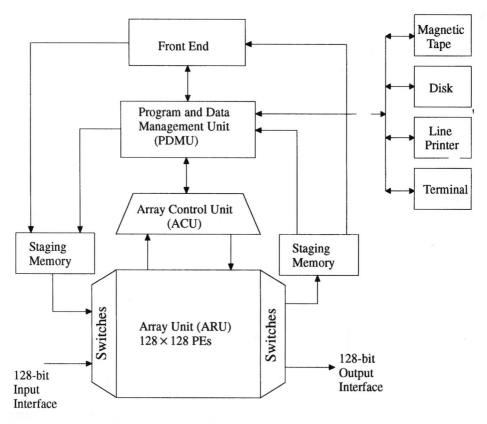

Figure 22.9 Architecture of the MPP

that connect the PEs with the external host computer (a DEC VAX-11/780), a terminal, a line printer, a magnetic tape, and a disk. The staging memory also reorders incoming and outgoing data streams to interface bit streams originating from image sensors, for example, to be stored in the ARU for processing.

The ARU consists of a matrix of 128×132 PEs. Each PE comprises a simple bit-serial processor and 1024 bits of RAM. A rectangle of 4×2 PEs, excluding the memory units, resides on one VLSI chip. A submatrix of 128×128 PEs performs array computations controlled by the ACU, while 4 columns of 128 PEs each are available to replace faulty columns of the PE array. In the case of a fault-free array, 4 columns of the 132 are disabled. Recall that a processor chip contains PEs of four different array columns. If a faulty processor chip within the array is detected, all PEs in the corresponding four columns are deactivated and bypassed, and the spare fault-free PE columns are activated instead, without loss of computational power. Because data might be corrupted due to the PE fault, a program running while a fault is detected has to be restarted from the beginning, or from the last fault-free checkpoint.

Because of the large number of PEs, a simple PE-to-PE interconnection network was implemented. Each PE is connected to its four nearest neighbors, and the interconnection patterns of the edge PEs can be chosen under software control. The top and bottom array edges can either be left open or can be connected within each column, while the left and right edges can be left open, or can be connected to form either an open or a closed spiral.

The bit-serial PEs forming the ARU have a simple architecture that is depicted in Fig. 22.10 [3]. Each PE consists of six 1-bit registers (A, B, C, P, G, and S), a shift-register of programmable length, a full-adder, RAM, and combinatorial logic. The inter-PE communication network can be accessed through register P. This register is connected to the P registers of its four neighboring PEs via bidirectional links. If, for example, the PEs execute a data movement to their left neighbors, each PE shifts the contents of its P register to its left neighbor, receives data from its right neighbor, and stores it in its P register. Boolean operations on the contents of the P register and the value currently on the data bus can be performed via logic incorporated in the P register. Arithmetic operations are performed by the

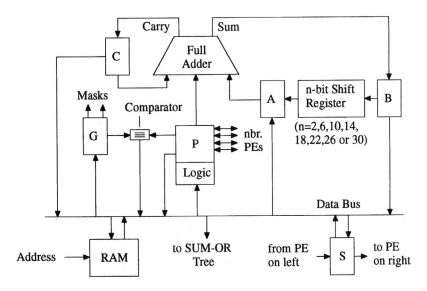

Figure 22.10 Architecture of an MPP processing element [3]

use of the full-adder; registers A, B, and C; and the shift register. With this combination, all basic arithmetic operations (e.g., addition, subtraction, multiplication, division, and floating point operations) can be performed.

MPP supports data conditional masking via the G register, which is used to select the active PE set. During the execution of a masked instruction, only PEs with a 1 in their G registers execute the instruction. Each PE is connected to the *SUM-OR tree,* i.e., a tree of inclusive-or gates. The tree output is connected to the ACU so that global minimum and maximum value searches, or other global operations can be performed.

The S register within each PE handles the I/O operations. Each S register is connected to the S register of the left and right neighbor PE so that a plane is formed that can shift data. During an I/O process, a 128×128 bit plane is shifted through either the I/O interfaces or the staging memory units into the S registers of the PE array in 128 cycles. Also, during this data shift, the old contents of the S registers are shifted out of the array. In an additional step, the contents of the register can be stored in the memory of each PE. This method of overlapping execution and data shifting provides an efficient way to move data into and out of the PE array with little effect on the PE operation.

22.4.2   The Connection Machine CM-2

The *Connection Machine CM-2* built by Thinking Machines Corporation and first delivered in 1987 is an improved version of the CM-1, an SIMD parallel machine the architecture of which Hillis proposed in his thesis at MIT [17]. A fully configured CM-2 machine comprises 65,536 PEs that are divided into four groups, as depicted in Fig. 22.11 [42]. Each group of 16,384 PEs is controlled by a *sequencer,* and the four sequencers are connected with up to four front end computers via a 4×4 *Nexus* switch. The front end computers can be Symbolics 3600 Lisp machines or DEC VAX 8000 series computers. The sequencer, which is an Advanced Micro Devices 2901/2910 bit-sliced machine, receives 32-bit instructions from a front end and converts them to nanoinstructions controlling its corresponding PE array. The PE array is connected to a parallel disk array *(Data Vault)* and to a graphic display via a high-speed I/O system.

Depending on the number of front end computers and the Nexus switch setting, different machine configurations are obtained. If, for example, one front end is present and is connected to all four sequencers, an SIMD machine is obtained with 64k PEs. If four front ends are available and each is connected to a different sequencer, a MSIMD machine is obtained with four independent SIMD machines consisting of 16k PEs each.

Sixteen processors, excluding the memory units, are implemented on one VLSI chip. Each PE consists of a simple bit-serial processor and 64 kb of RAM. A processor contains a single-bit ALU with three inputs and two outputs, flag and register bits, and memory and communication interfaces. The carry-out bits of all PEs can be combined with a logical OR on the GLOBAL line that connects the PEs with their related sequencers. Optionally, a floating-point accelerator consisting of one interface and one execution chip for each pair of processor chips can be employed. A fully configured machine therefore consists of 4k processor chips, 2k floating-point interface chips, 2k floating-point execution chips, and half a gigabyte of RAM.

The CM-2 machine supports two inter-PE communication network architectures, a *hypercube* and an *n-dimensional grid.* For a fully configured CM-2 machine, the hypercube is a 12-cube connecting 4k processor chips so that each 16-processor chip forms a vertex of a 12-cube. The nearest-neighbor grid is programmable in its dimension and uses a subset of the hypercube network connections. The programmer can employ a one-dimensional

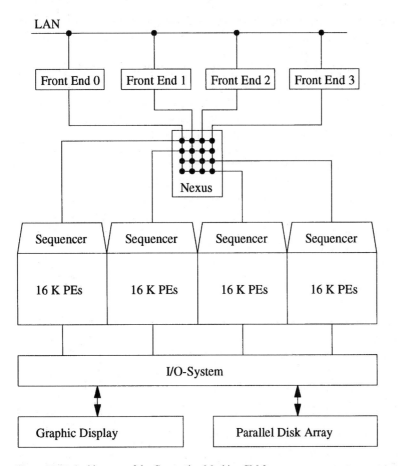

Figure 22.11 Architecture of the Connection Machine CM-2

to sixteen-dimensional grid, depending on the program task. Also, hardware for combining messages destined to the same PE is implemented. This combining includes *Sum, Logical OR, Overwrite, Max,* and *Min* operations.

22.4.3 The GF11 parallel computer

The GF11 SIMD machine, designed at the IBM T.J. Watson Research Center, contains 566 very powerful processing elements [4, 21]. The first machine version, with 256 processors, was completed in 1989, while the full machine became operational in 1990. The architecture of the GF11 machine is depicted in Fig. 22.12. The PE array consists of 566 32-bit processing elements; 512 PEs are for program use, while the remaining 54 PEs are hot spares. The PEs communicate via a three-stage nonblocking Benes network. This network is also used to connect an array of ten disks to the PE array.

The architecture of a processing element is shown in Fig. 22.13. It consists of two arithmetic units: a 20 MIPS 32-bit *integer unit* including an ALU and a barrel shifter, and a 20 MFLOPS 32-bit *floating-point unit* consisting of two ALUs and two multipliers that can be interconnected in a pipelined fashion. The *condition code registers* are controlled by the arithmetic units. The hierarchical memory of a GF11 PE consists of three units: a file of 256 32-bit registers, 64 kB of static RAM *(SRAM)* to store frequently used data, and 256

LAN

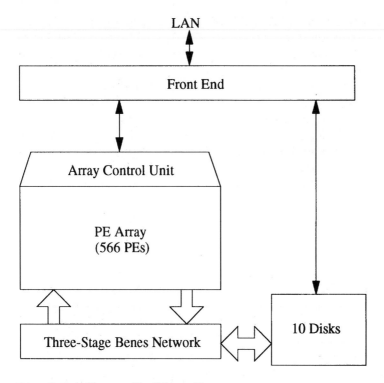

Figure 22.12 Architecture of the GF11 machine

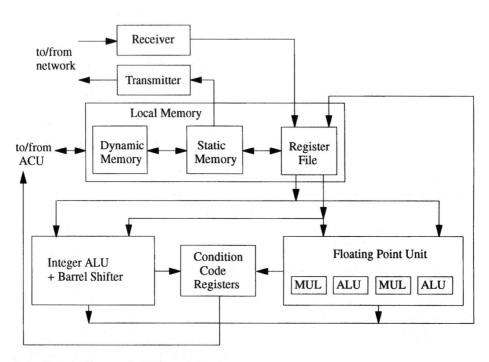

Figure 22.13 Architecture of a GF11 processing element

kB (expandable to 2 MB) of dynamic RAM (*DRAM*) to store long-term data. Special relocation hardware is used to offset memory addresses. This way, a PE can generate SRAM or DRAM addresses based on processor-specific data. Each PE is controlled by a 180-bit horizontal microcode that is distributed by the ACU.

The processors are interconnected via a three-stage *Benes network* constructed from 24×24 crossbar switches; each network stage consists of 24 switches. The network can be assigned to one of the 1,024 preloaded possibilities at each data transfer. This assignment, however, needs considerable time, so it is preferable to assign a network configuration at the beginning of a program run and leave it unchanged during program execution.

In contrast to other SIMD machines, the GF11 PEs are implemented by using off-the-shelf vendor components. For the Benes network, semicustom CMOS gate array chips were used. Due to the very powerful processing elements, a peak rate of 11.3 GFLOPS is obtained.

22.4.4   The MasPar MP-1 and MP-2

The MP-1 and MP-2 SIMD machines built by MasPar Computer Corporation contain up to 16,384 processing elements. The two machine types have the same basic architecture and differ mostly in the processing element complexity. While the PEs in the MP-1 are 4-bit wide, those of the MP-2 are 32-bit CPUs. Here, the MP-1 system will be surveyed first, followed be a short comparison of both machines. The architecture of the MP-1 machine is depicted in Fig. 22.14 [25]. It consists of the PE array, the array control unit (ACU), a global router, an I/O channel, and a front end computer.

The ACU performs operations on scalar data stored in its memory and controls interactions between the PE array, the front end, and the I/O channel. It consists of a 12-MIPS processor with 32 32-bit registers, 128 kB of data memory, and 1 MB of RAM as instruction memory. The front end is a DECstation 5000 running the ULTRIX operating system

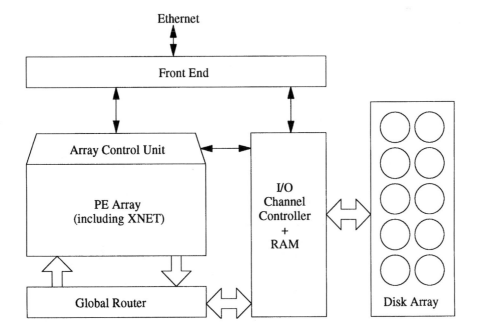

Figure 22.14 Architecture of the MasPar MP-1 machine

and is connected to an Ethernet LAN. The I/O channel provides high-speed data transmission between the PEs and external devices, such as a parallel disk array by the use of the global router.

Depending on the machine model, the PE array contains at least 1,024 PEs and can be expanded to up to 16,384 PEs. The array is divided into nonoverlapping PE clusters, each consisting of a square subarray of 4×4 PEs. Two PE clusters are implemented in one VLSI chip, and 64 PE clusters (i.e., 1024 PEs in 32 chips) fit on one PE board. Thus, a machine with 16k PEs consists of 16 PE boards.

Two independent inter-PE communication networks are implemented, the XNET and the *global router*. The XNET is an *8-nearest-neighbor network* as shown in Fig. 22.6b. The three-stage global router is a dynamic multistage interconnection network and connects the PE clusters. It can communicate with only one PE per cluster at a time, because at most one incoming and one outgoing connection to and from a specific cluster can be established simultaneously. While communication requests that do not share any source or destination PE clusters can be executed in parallel in a single step, requests within a cluster are serialized. The machine user has to be aware of this organization to obtain optimal inter-PE communication performance.

Each of the PEs used in a MasPar MP-1 machine consists of a 4-bit load/store 1.6-MIPS processor, and 16 or 64 kB of RAM (depending on the machine model). The PE architecture is depicted in Fig. 22.15. A 4-bit wide *Nibble Bus* and a single-bit *Bit Bus* connect the PE components. The ACU can broadcast data to these buses, and a global reduction of data on the 4-bit buses of the active PEs to the ACU can be performed. In addition to the RAM, 40 32-bit data registers are present in each PE. Special hardware is included in each PE for floating point arithmetic and logic operations.

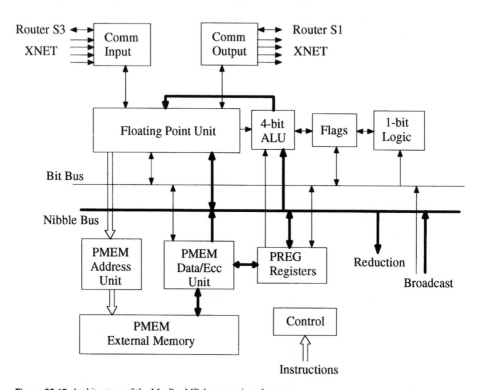

Figure 22.15 Architecture of the MasPar MP-1 processing element

Compared to the MasPar MP-1 PEs, the MasPar MP-2 PEs have an equivalent architecture but employ a 32-bit data path along with a 32-bit ALU, and 64 or 256 kB of RAM (depending on the machine model). These enhancements result in a processor speed of 4.25 MIPS. Performance figures of the MP-1 and MP-2 machines with 16,384 nodes each are listed below [25]:

Peak System Performance	MasPar MP-1	MasPar MP-2
32-bit integer MIPS	26,000	68,000
64-bit integer MIPS	13,000	34,000
32-bit MFLOPS	1,200	6,300
64-bit MFLOPS	550	2,400

22.5 Applications and Algorithms

22.5.1 Overview

The optimal choice of application classes running on a computer depends highly on the computer and application characteristics. The main characteristics of many SIMD machines are (1) the large number of processing elements (in the order of thousands of PEs), and (2) the synchronous execution of a single instruction stream on all PEs. Thus, to exploit the parallelism and power of SIMD machines, applications operating on large numbers of identical objects should be considered in which the interaction between objects and/or the object behavior are of importance. Each application object may be mapped onto a particular PE, and interactions among objects are implemented via inter-PE communication. Examples of such applications (as noted in Refs. [34, 45, 40, 42, 44]) include image processing (individual image pixels are mapped onto the PEs), climate and weather prediction (the atmosphere is divided into equal-sized blocks that are mapped onto the PEs), molecular dynamics (atoms and/or electrons are mapped onto the PEs), semiconductor design (charge or potential distributions are mapped onto the PEs), fluid-flow modeling (area of interest is divided into finite elements that are mapped onto the PEs), VLSI-design (transistors or gates are mapped onto the PEs for simulation or placement purposes), and information retrieval in large data bases (data units are mapped onto the PEs).

For algorithmic purposes, only parts of the generic SIMD machine model depicted in Fig. 22.1 are relevant, so a simpler SIMD programming model can be defined. An SIMD machine can then be modeled by a five-tuple (N, C, I, M, F) [38], where

1. N equals the number of PEs.
2. C equals the set of control-unit instructions used to control the program flow (for example, IF -THEN- ELSE constructs).
3. I equals the set of PE instructions executed within each PE for data manipulation, (e.g., addition or multiplication of two data items).
4. M equals the set of PE masking schemes that are used to enable/disable PEs.
5. F equals the set of interconnection functions provided by the interconnection network for inter-PE communication purposes.

The task of efficiently executing an algorithm on an SIMD machine requires mapping of the algorithm to the architecture as given by the model. The most important aspects are to allocate the data to the N PEs such that all PEs have to perform an equal amount of work, with as few PEs as possible being idle, and that the data exchanges among the PEs are well supported by the interconnection network. Many algorithms operate on data that is highly structured. Frequently, communication requirements are mostly local. In many image processing algorithms, for example, the image pixels are ordered in a matrix, and operations on a particular pixel most often need information of neighboring pixels only. These algorithms can be implemented efficiently on a massively parallel SIMD machine that employs a nearest-neighbor network for inter-PE communication. Frequently, a modification of data allocation permits more efficient communication [20].

Other algorithms, such as sorting, need the data to be distributed in a tree-like structure for optimum performance and would benefit from an SIMD machine that employs a tree-like network, or in whose network a tree structure can be embedded. While the distribution of data to the PEs poses no problems, the communication needs do, because, at best, a tree-like interconnection network should be available, which is not the case for most existing systems. The communication requirements are met rather easily if a dynamic multistage interconnection network as in the GF11 system is available. If, on the other hand, a direct network such as a mesh has to be used, methods of efficiently emulating the tree structure on that direct network have to be applied [14]. The choice of the network in an SIMD machine therefore effects the classes of algorithms that can be implemented optimally on that machine. Because in an SIMD machine, usually one or at most two different interconnection network structures are implemented for cost reasons, networks that can simulate other network structures efficiently are advantageous.

The emulation of a target network structure by a given network structure can be accomplished by executing one or more interconnection functions of set F on the given network combined with appropriate PE enabling/disabling by using set M to obtain the target network connection function. For example, to execute a shuffle-exchange interconnection function on a binary cube network, $\log_2 N$ cube interconnection functions must be performed, while the execution of a cube function on a mesh network needs $\sqrt{N/2} - 1$ mesh transfers [38].

To exemplify algorithms well suited for SIMD machines, three simple cases are studied the following section. The first one, recursive doubling, can be used in a broad variety of applications. The second one, a Poisson solver, can be used in semiconductor design, and the third one, image smoothing, is an important algorithm used in image processing.

22.5.2   Case studies of SIMD algorithms

22.5.2.1 Recursive doubling. In many applications, N values must be combined into a single value, e.g., the sum of the elements of a vector of size N. This combining can be any arithmetic or logic operation. A serial algorithm needs $N - 1$ combining steps to perform this task. On a parallel machine, this can be implemented more efficiently if the N values are stored in N PEs (N is assumed to be a power of 2).

As an example, consider the summation of a variable V residing in $N = 8$ PEs, as depicted in Fig. 22.16. In the first step, all odd-numbered PEs send their V values to an even-numbered PE, and the even-numbered PEs add the value they received to their own V variable. Thus, $N/2 = 4$ partial results are obtained. After repeating these two steps recursively twice as indicated in Fig. 22.16, variable V of PE 0 contains the result of the overall sum (see Fig. 22.16). Thus, only $\log_2 N = 3$ steps, each consisting of a transfer and an addition, are needed in the parallel algorithm, which is a significant improvement over the serial algorithm.

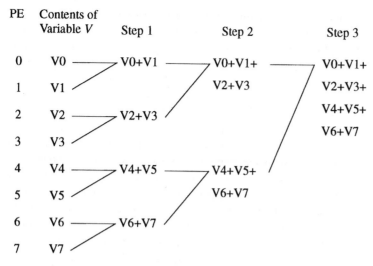

Figure 22.16 Recursive doubling with eight PEs

22.5.2.2 Poisson solver.

In material sciences, an important task is to calculate the distribution of the potential in a two-dimensional charged area, e.g., the electrical potential-distribution within a bipolar transistor. If the area of interest is divided into a two-dimensional mesh, and the charge of each mesh point (i, j) is known, the discrete Poisson equation shown below can be used to solve iteratively the distribution of the potential for each point:

$$V_{i,j}^{(t+1)} = \frac{V_{i+1,j}^{(t)} + V_{i-1,j}^{(t)} + V_{i,j+1}^{(t)} + V_{i,j-1}^{(t)}}{4} + C_{i,j} \qquad (22.1)$$

$C_{i,j}$ represents the charge located at point (i, j), and the superscript indicates the iteration index.

This Poisson solver approach is well suited to be implemented on an SIMD computer with mesh connections. Assume a two-dimensional charge distribution mesh of size $M \times M$ resulting in M^2 mesh points, and a system with M^2 PEs. Then, each mesh point can be mapped onto one PE of the machine. Furthermore, assume that the boundary conditions of the electrical potential V are stored in the edge PEs of the PE array. Then, the iteration loop for a Poisson solver would be similar to [39]:

```
IF (PE not edge-PE){
    tmp = V;
    V = tmp of east neighbor
    V += tmp of west neighbor;
    V += tmp of south neighbor;
    V += tmp of north neighbor;
    V = V/4 + C;}
```

The variables V, C, and tmp are variables stored in each PE. The variable tmp prevents the loss of the value $V^{(t)}$ during the calculation of $V^{(t+1)}$. Assuming an $M \times M$ charge mesh with fixed potential conditions at the edge of the mesh, the potential in $(M - 2)^2$ mesh points has to be iterated (not counting the edge points). Thus, a serial algorithm needs

$(M-2)^2$ averaging steps during one iteration. The parallel algorithm needs only one averaging step per iteration, but four inter-PE communication steps and one IF clause have to be introduced in addition. For large M, this overhead can be neglected so that the speedup of the parallel algorithm over the serial one is of the order of M^2, the number of processors, and linear speedup is achieved.

22.5.2.3 Image smoothing. For many image processing applications, the gray-level smoothing of an image is essential to reduce noise [34]. In the following example, gray level images are assumed that consist of $M \times M$ picture elements (pixels); the gray level of a pixel is represented by an unsigned 8-bit integer. The value 0 represents white, while black is represented by the value 255. The smoothing of an image is accomplished by averaging the gray level of a pixel with the gray levels of its eight neighbors. This averaging has to be performed on all pixels, excluding the edge pixels of an image. If the input image is denoted by I, the smoothed image is denoted by S, and the location of the pixel to be smoothed is indexed by (i, j), the following equation must be solved:

$$S(i,j) = [I(i,j) + I(i+1,j) + I(i-1,j) + I(i,j+1) + I(i,j-1)$$

$$+ I(i+1,j+1) + I(i+1,j-1) + I(i-1,j+1) + I(i-1,j-1)]/9$$

Because the $4M-4$ edge pixels of an image do not have eight neighbors, they must be treated specially. One alternative is to set their smoothed value to 0. Thus, on a serial machine, $(M-2)^2 = O(M^2)$ smoothing operations must be performed.

Now consider the implementation of image smoothing on an SIMD machine with N PEs arranged as a $\sqrt{N} \times \sqrt{N}$ grid (it is assumed that $N < M^2$). The $M \times M$ image is divided into N square subimages of size $(M/\sqrt{N}) \times (M/\sqrt{N})$. For example, on the MasPar MP-1 with 16k PEs, a 512×512 image would be divided into 16k subimages of size 4×4. Each subimage is stored in one PE, as depicted in Fig. 22.17 [34]. Thus, each PE has to perform M^2/N smoothing operations. Edge pixels from adjacent subimages must be transferred to a subimage to smooth the edge pixels of that image. To accomplish this, $4(M/\sqrt{N}) + 4$ data transfers with N pixels per transfer are required ($M/\sqrt{N}$ transfers for each subimage side

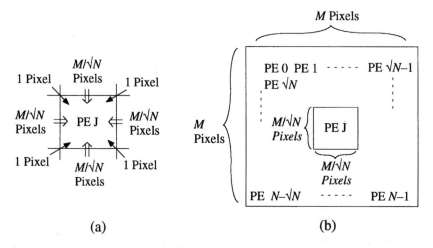

(a)   (b)

Figure 22.17 Smoothing an $M \times M$ image on N PEs [34]; (a) pixel transfers and (b) data allocation

and 4 transfers for the subimage corners, as shown in Fig. 22.17a). Assuming that a serial and an SIMD machine need the same time t_s to perform a smoothing operation, and the time to perform a data transfer operation on the SIMD machine equals t_t, the speedup of the parallel algorithm can be calculated to:

$$ S = \frac{(M-2)^2}{M^2/N + t_t/t_s\left(4\frac{m}{\sqrt{N}} + 4\right)} \tag{22.2} $$

which is in the order of N.

Consider the image smoothing algorithm implemented on, for example, a MasPar MP-1 machine with 16k PEs. For this machine, the factor t_t/t_s is approximately 0.2 if the 8-nearest-neighbor network XNET is used for data transfer. Thus, the smoothing of a 512×512 image on that machine ($M = 512$, $N = 16{,}384$) yields a speedup of $0.8 * N$.

The algorithms in this section have rather straight forward solutions on SIMD machines with nearest-neighbor connections. In the literature, many highly complex algorithms are described that were efficiently mapped to SIMD machines. Examples are event-driven net work simulations [1], object recognition in images [42], document retrieval [44] and more are added continuously.

22.6   Languages and Programming

22.6.1   Concepts and requirements

The most convenient approach to program a parallel machine is the use of a compiler that automatically parallelizes sequential programs. An early attempt was the language IVTRAN for the Illiac IV machine [30], where the program is coded in Fortran, and the machine compiler attempts to detect data parallelism inherent in the program's sequential DO loops. The resulting parallel operations are then distributed onto multiple PEs of the SIMD machine so that they can be executed in parallel. This approach is used extensively in vectorizing compilers for supercomputers. A serious disadvantage of this method is the compiler's limited ability to detect parallelism efficiently, which is dependent on the program structure. Even today, automatic parallelization of sequential programs to yield efficient parallel execution is not available. Therefore, the languages and programming issues surveyed in this section assume that the programmer is aware of the parallel nature of the algorithm and is responsible for mapping the algorithm to a parallel program.

Two approaches to the design of a high-level language for SIMD machines can be identified, which are characterized by the use of:

1. a machine-dependent extension of an existing language, or
2. a machine-independent extension of an existing language

In the first approach, which is widely used in today's commercial SIMD and MSIMD machines, existing sequential languages are extended by constructs that account for data parallelism. Examples of this language type are MPL (MasPar Programming Language, based on C) and MPF (MasPar Fortran) for the MasPar MP-1 machine [25], or Fortran, C*, and *LISP for the Connection Machine CM-2 [41]. These languages are machine dependent and reflect specific characteristics of the target machine to permit efficient programming. Examples of such specifics are data types, the PE array size, shape, and control, and the inter-PE communication capabilities.

The language Actus [27, 28], which is based on Pascal, belongs to the second programming language type. This language consists of constructs that are machine independent; i.e., they do not reflect any machine characteristics like actual number of PEs or PE interconnection functions implemented in the machine. In the following two sections, the MPL and Actus languages will be surveyed.

22.6.2   MasPar Programming Language (MPL)

MPL, which is based on ANSI C, was implemented for program development on the MasPar MP-1 system. Recall that the machine contains a PE array with up to 16k PEs and incorporates the 8-nearest-neighbor network XNET for fast local communication and the three-stage global router for efficient global inter-PE communication. To distinguish scalar and parallel data, the keyword *plural* was introduced. Consider, for example, the following program code:

```
{
int i;
plural int j, k;
i = 5;
j = 3;
k = j + 1;
}
```

Here, the variable i is scalar and resides in the ACU memory, while j and k are plural variables allocated in each PE. First, i is assigned the value 5 within the ACU. Then, the ACU broadcasts the value 3 to all PEs, and each PE assigns that value to its variable j. Finally, the ACU broadcasts the value of its variable i, which is 5, to all PEs and each PE adds that value to its variable j and stores the result in its variable k. Thus, after the program execution, the variable k of each PE equals to eight. Data conditional enabling/disabling of PEs can be accomplished by the use of extended IF-THEN-ELSE, WHILE, DO, and FOR constructs. If these constructs are controlled by scalar expressions, they are equivalent to the normal ANSI C statements. For example, if the IF part of an IF-THEN-ELSE construct is controlled by a scalar boolean expression, either the THEN or the ELSE part will be executed, depending on the result of the boolean expression. If a plural boolean expression is used, i.e., plural variables are used in the IF condition, both the THEN and the ELSE clauses may be executed, depending on the initial active PE set and the IF condition. Thus, the initial active PE set, i.e., the set of PEs that are active before the IF condition is evaluated, is divided into an active and an inactive subset by the IF condition. First, the active subset, i.e., the PEs for which the IF condition is true, will execute the THEN clause, and then the inactive subset, i.e., the PEs for which the IF condition is false, will execute the ELSE clause. After both clauses are executed, the initial active PE set is restored to continue program execution. If the reserved word ALL precedes a statement, this statement will be executed by both active and inactive PEs.

MPL provides predefined constants for PE addressing purposes. The size and shape of the PE array is reflected in the scalar constants NPROC (the total number of PEs in the array), NXPROC (the number of PEs in the x direction), and NYPROC (the number of PEs in the y direction). The position of each PE within the array is stored in the plural variables

IPROC (linear view of the PE array), IXPROC (PE's x coordinate within the array), and IYPROC (PE's *y* coordinate within the array). For example, if the program line

```
ALL IF (IPROC==1000) THEN <statement A>
```

is executed, only the PE with the linear address 1000 will execute <statement A>.

The XNET* and ROUTER constructs were introduced to enable the use of the XNET and the global router inter-PE communication networks. With the XNET* command, eight different communication directions can be used, which are north, south, west, east, northeast, northwest, southeast, and southwest. When executing the statement XNETN[2].j = i (*i* and *j* are plural variables), for example, the active PEs store the contents of their variable *i* into variable *j* of their second north neighbor. With the ROUTER command, a specific destination address has to be declared. For example, when executing the statement j = ROUTER[1000].i (*i* and *j* are plural variables), each active PE fetches the contents of variable *i* of PE 1000 and stores it into its variable *j*. Also, indirect addressing, i.e., using the contents of a singular or plural variable as the destination address, is supported. Thus, when executing the statement i = ROUTER[j].i (*i* and *j* are plural variables), each PE uses the value of its variable *j* as the destination PE address, fetches the contents of variable *i* of that destination PE, and stores it into its variable *i*. If two or more PEs try to send a data item to a particular destination PE during a single communication step, the individual data items received by the destination are overwritten and, after the communication, the contents of the destination variable will be the last data item received. In this case, data is lost during the inter-PE communication step. The programmer must be aware of this fact and has to ensure that this communication scenario will not occur unintentionally during program execution to guarantee predictable program behavior.

Because in MPL the allocation of parallel variables to the PEs has to be determined by the programmer during program development, this language is machine dependent. With the parallel extensions of MPL mentioned above, program development with efficient exploitation of the parallel nature of the MasPar MP-1 machine is possible.

22.6.3 Actus programming language

This language, proposed by Perrott [27], is based on PASCAL with machine-independent extensions for synchronous parallel programming. To express parallelism in a program, a special array notation is used. In general, singular arrays use a ".." in their size declaration, while plural arrays use a ":". Consider the following part of a declaration block:

```
singar : ARRAY [1 .. 20] OF INTEGER;
plurar : ARRAY [1 : 20] OF INTEGER; .
```

In this block, singar is defined to be a conventional singular array of size 20 in which the elements can be accessed one at a time. Plurar is defined as a plural array with 20 elements. These elements can be accessed and manipulated in parallel in one step. The size of a parallel array is called the extent of parallelism of this array. In Actus, the actual distribution of the elements of a parallel array onto the PEs of the target machine is performed implicitly by the language compiler and cannot be determined during program development. A parallel constant can be declared as

```
PARCONST identifier = start:[increment]end.
```

The values of start, increment, and end must be of integer type. This construct results in a parallel constant with a sequence of "start, start+increment, start+2*increment, ..., end."

To reference a subpart of a parallel array, an index can be defined. For example, the construct INDEX ind=2:5; declares ind to be an index that can be used to reference the elements 2 to 5 of a parallel array of size five and larger. An index cannot be redefined during run time but can be manipulated by the operators union (+), intersection (*), and difference (–). For example, if two indices $i1$ and $i2$ are defined to be:

```
INDEX   i1 = 2:5;
        i2 = 3:10;,
```

then the construct $il*i2$, i.e., the intersection of both indices, will select the elements 3 to 5 of parallel arrays of size five and larger. To move data within or between parallel arrays, the SHIFT and ROTATE data alignment constructs are available. If the following code is executed:

```
VAR arr: ARRAY[1:10] OF INTEGER;
INDEX ind = 2:4;
arr[ind] := arr[ind] + arr[ind SHIFT 2];,
```

the expressions $arr[2] = arr[2] + arr[4]$, $arr[3] = arr[3] + arr[5]$, and $arr[4] = arr[4] + arr[6]$ are evaluated in parallel by shifting the extent of parallelism. If a ROTATE construct is used, the extent of parallelism is circularly shifted with a modulo operation. Thus, if ROTATE is used in the above program part rather than the SHIFT construct, the expressions $arr[2] = arr[2] + arr[4]$, $arr[3] = arr[3] + arr[l]$, and $arr[4] = arr[4] + arr[2]$ are evaluated in parallel.

Each statement in Actus that uses parallel constants and/or variables must have a unique extent of parallelism. To simplify blocks of consecutive statements that operate on the same extent of parallelism, the "WITHIN specifier DO statement block" construct is available. This specifier defines the extent of parallelism the statement block has, and the declarations of parallelism in each statement within the statement block are each substituted by a "#". The specifier in the WITHIN statement can also consist of variables, for example, to dynamically change the extent of parallelism in program loops.

The IF, CASE, FOR, and WHILE statements can have singular and/or plural variables in their condition expression. Furthermore, two constructs, ALL and ANY, are introduced to extend the functionality of the conditional statements on parallel variables. Consider the two statements

```
IF ALL (A[1:N] > B[1:N]) THEN B[#] := 0;
IF ANY (A[1:N] > B[1:N]) THEN B[#] := 0;.
```

In the first statement, all the elements of B will be set to 0 if all elements of array A are greater than the corresponding elements of array B. In the second statement, on the other hand, all elements of B will be set to 0 if any element of array A is greater than its corresponding element of array B. The language Actus therefore provides means of expressing parallelism in programs in a general way without referring to any machine architecture the program might run on. The extent of parallelism of the program is defined in the data declarations, and data is manipulated in parallel (in part or in total) by the language statements. While this has the advantage of portability, it places a bigger burden on the compiler and may not permit an optimum mapping of algorithm to architecture.

22.7   Conclusions

22.7.1   Summary

In this chapter, one class of parallel computers, the SIMD computers, were discussed with respect to architecture, hardware requirements, and field of application. In Section 22.1, basic SIMD concepts, like machine and processor array models were introduced. Characteristics of the individual SIMD machine components were presented in Section 22.2, while the SIMD behavior of associative processing was studied in Section 22.3. Architecture and implementation of several commercial SIMD systems were presented in Section 22.4. Classes of applications and algorithms well suited for running on SIMD machines were discussed in Section 22.5, while language and programming issues were studied in Section 22.6.

22.7.2   Future directions

A broad variety of applications exist that are well suited for either pure SIMD or pure MIMD processing. However, many applications and algorithms have a more hybrid structure; i.e., rather than being pure SIMD or MIMD, they are partly SIMD and MIMD. Thus, machines that combine the advantages of SIMD and MIMD processing offer advantages, and for this reason, some companies have moved away from designing pure SIMD machines. For example, the Connection Machine CM-5, the successor of the Connection Machine CM-2, consists of very powerful independent SPARC PEs (MIMD characteristic) that can also execute a single program synchronously (SIMD characteristic). Other machines consisting of asynchronous processors can exploit the advantages of synchronous SIMD operation by utilizing barrier synchronization [13]. Another factor for future developments is the progress in VLSI technology. Restricted chip size and pin count have resulted in SIMD machines that consist of a high number of rather simple processing elements. Recent VLSI developments such as multichip modules and wafer scale integration enable the design of SIMD machines with more powerful processors. This is exemplified by the transition from 4-bit PEs in the MasPar MP-1 to 32-bit PEs in the MasPar MP-2.

22.8   References

1. Ayani, R., and B. Berkman. 1993. Parallel discrete event simulation on SIMD computers. *Journal of Parallel and Distributed Computing,* Vol. 18, No. 4, 501–508.
2. Batcher, K. E. 1977. The multidimensional access memory in STARAN. *IEEE Transactions on Computers,* Vol. C-26, No. 2, 174–177.
3. Batcher, K. E. 1980. Design of a massively parallel processor. *IEEE Transactions on Computers,* Vol. C-29, No. 9, 836–844.
4. Beetem, J., M. Denneau, and D. Weingarten. 1985. The GF11 supercomputer. *Proc. 12th Annual International Symposium on Computer Architecture,* 108–115.
5. Blank, T. 1990. The MasPar MP-1 architecture. *Proc. IEEE Compcon,* 20–24.
6. Bouknight, W. J., S. A. Denenberg, D. E. McIntyre, J. M. Randall, A. H. Sameh, and D. L. Slotnick. 1972. The Illiac IV system. *Proceedings of the IEEE,* Vol. 60, No. 4, 369–388.
7. Briggs, F. A., K.-S. Fu, K. Hwang, and J. H. Patel. 1979. PM4—a reconfigurable multimicroprocessor system for pattern recognition and image processing. *Proc. AFIPS 1979 National Computer Conference,* 255–265.
8. Chisvin, L., and R. J. Duckworth. 1989. Content-addressable and associative memory: alternatives to the ubiquitous RAM. *IEEE Computer,* 51–63.
9. Dekel, E., E. Nassimi, and S. Sahni. 1981. Parallel matrix and graph algorithms. *SIAM Journal of Computing,* Vol. 10, 657–675.
10. Feng, T. Y. 1981. A survey of interconnection networks. *Computer,* Vol. 14, No.12, 12–27.

11. Flynn, M. J. 1966. Very high-speed computing systems. *Proceedings of the IEEE,* Vol. 54, No. 12, 1901–1909.

12. Gentleman, W. M. 1978. Some complexity results for matrix computations on parallel processors. *Journal of the ACM,* Vol. 25, 112–115.

13. Ghose, K., and E.-D. Cheng. 1994. Efficient barrier synchronization techniques and their applications in large-scale shared memory multiprocessors. *Journal of Microcomputer Applications,* Vol. 17, No. 2, 197–206.

14. Gordon, D. 1987. Efficient embeddings of binary trees in VLSI arrays. *IEEE Transactions on Computers,* Vol. C-36, No. 9, 1009–1018.

15. Grosspietsch, K. E., and R. Reetz. 1992. The associative processor system CAPRA: Architecture and applications. *IEEE Micro,* Vol. 12, No. 6, 58–67.

16. Grondalski, R. 1987. A VLSI chip set for a massively parallel architecture. *Proc. International Solid States Conference,* 213–224.

17. Hillis, W.D. 1985. The Connection Machine. Cambridge, Mass.: MIT Press.

18. Hunt, D. J. 1981. The ICL DAP and its application to image processing. In *Languages and Architectures for Image Processing,* M. J. B. Duff and S. Levialdi, eds. London: Academic Press, 275–282.

19. Hwang, K., and F. A. Briggs. 1984. *Computer Architecture and Parallel Processing.* New York: McGraw-Hill.

20. Jurczyk, M., T. Schwederski, R. Born, H. J. Siegel, and S. Abraham. 1994. Strategies for the massively parallel simulation of interconnection networks. *Proc. 1994 International Conference on Parallel Processing.*

21. Kumar, M., Y. Baransky, and M. Denneau. 1993. The GF11 parallel computer. *Parallel Computing,* Vol. 19, 1393–1412.

22. Kuck, D. J., and R. A. Stokes. 1982. The Burroughs Scientific Processor (BSP). *IEEE Transactions on Computers,* Vol. C-31, 363–376.

23. Kung, S. Y. 1988. *VLSl Array Processors.* Englewood Cliffs, N.J.: Prentice Hall.

24. Lea, R. M., and I. P. Jalowiecki. 1991. Associative massively parallel computers. *Proceedings of the IEEE,* Vol. 79, No. 4, 469–479.

25. MasPar Computer Corporation, 1992. MasPar system overview, Part No. 9300-0100, Rev. A5, MasPar Computer Corporation.

26. Nutt, G. J. 1977. Microprocessor implementation of a parallel processor. *Proc. Fourth Annual Symposium on Computer Architecture,* 147–152.

27. Perrott, R. H. 1979. A language for array and vector processors. *ACM Transactions on Programming Languages and Systems,* Vol. 1, No. 2, 177–195.

28. Perrott, R. H. 1987. *Parallel Programming.* Reading, Mass.: Addison-Wesley.

29. Ponnusamy, R., R. Thakur, A. 1993. Choudhary, K. Velamakanni, Z. Bozkus, and G. Fox. Experimental performance evaluation of the CM-5. *Journal of Parallel and Distributed Computing,* Vol. 19, 192–202.

30. Presbreg, D. L., and N. W. Johnson. 1975. The Paralyzer: IVTRAN's parallelism analyzer and synthesizer. *Proc. ACM Conference on Programming Languages and Compilers for Parallel and Vector Machines,* 9–16.

31. Scherson, I. D., D. A. Kramer, and B. D. Alleyne. 1992. Bit-parallel arithmetic in a massively parallel associative processor. *IEEE Transactions on Computers,* Vol. 41, No. 10, 1201–1210.

32. Schwederski, T., W. G. Nation, H. J. Siegel, and D. G. Meyer. 1987. Design and implementation of the PASM prototype control hierarchy. *Proc. Second International Supercomputing Conference,* 418–427.

33. Sejnowski, M. C., E. T. Upchurch, R. N. Kapur, D. P. S. Charlu, and G. J. Lipovski. 1980. An overview of the Texas Reconfigurable Array Computer. *Proc. AFIPS 1980 National Computer Conference,* 631–641.

34. Siegel, H. J., J. B. Armstrong, and D. W. Watson. 1992. Mapping computer-vision-related tasks onto reconfigurable parallel-processing systems. *IEEE Computer,* Vol. 25, No. 2, , 54–63.

35. Siegel, H. J., L. J. Siegel, F. C. Kemmerer, P. T. Mueller, Jr., H. E. Smalley, Jr., and S. D. Smith. 1981. PASM: a partitionable SIMD/MIMD system for image processing and pattern recognition. *IEEE Transactions on Computers,* Vol. C-30, No. 12, 934–947.

36. Siegel, H. J. 1977. Analysis techniques for SIMD machine interconnection networks and the effects of processor address masks. *IEEE Transactions on Computers,* Vol. C-26, No. 2, 153–161.

37. Siegel, H. J. 1980. The theory underlying the partitioning of permutation networks. *IEEE Transactions on Computers,* Vol. C-29, No. 9, 791–801.

38. Siegel, H. J. 1990. *Interconnection Networks for Large-Scale Parallel Processing: Theory and Case Studies,* Second Edition. New York: McGraw-Hill.

39. Stone, H. S. 1987. *High Performance Computer Architecture.* Reading, Mass.: Addison-Wesley.

40. Strong, J. P. 1991. Computations on the Massively Parallel Processor at the Goddard Space Flight Center. *Proceedings of the IEEE,* Vol. 79, No. 4, 548–558.

41. Thinking Machines Corporation. 1990. Connection Machine Model CM-2 Technical Summary, Vers. 6.0. Thinking Machines Corporation.
42. Tucker, L. W., and G. G. Robertson. 1988. Architecture and applications of the Connection Machine. *Computer,* Vol. 21, 26–38.
43. Vick, C.R., and R.E. Merwin. 1973. An architecture description of a parallel processing element. *Proc. International Workshop on Computer Architecture.*
44. Waltz, D. L. 1987. Applications of the Connection Machine. *IEEE Computer,* 85–97.
45. Weems, C.C. 1991. Architectural requirements of image understanding with respect to parallel processing. *Proceedings of the lEEE,* Vol. 79, No. 4, 537–546.

23

MIMD Architectures: Shared and Distributed Memory Designs

Ralph Duncan

Since many of the parallel computer architectures introduced during the early 1990s have been multiple instruction stream, multiple data stream (MIMD) computers, understanding the different kinds of MIMD architectures is necessary for seeing how current machines fit into the spectrum of alternative parallel design approaches. However, this requires understanding many different parallel architecture designs. As a handbook contribution, this chapter's goals are to define MIMD architectures for the reader, suggest reasons for their diversity, and present the variety of MIMD architecture approaches as an orderly spectrum of machine design possibilities.

MIMD architectures have multiple processors that each execute an independent *stream* (*sequence*) of machine instructions. The processors execute these instructions by using any accessible data rather than being forced to operate upon a single, shared data stream. Hence, at any given time, a MIMD system can be using as many different instruction streams and data streams as there are processors (see Fig. 23.1).

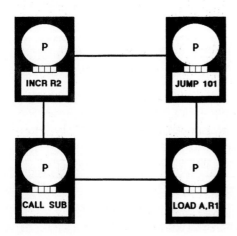

Figure 23.1 Using multiple instruction and data streams

Although software processes executing on MIMD architectures can be synchronized by passing data among processors through an interconnection network or by having processors examine data in a shared memory, the processors' autonomous execution makes MIMD architectures asynchronous machines. This distinguishes them from synchronous machines, such as:

- single instruction stream, multiple data stream (SIMD) array processors that broadcast instructions for processors to execute in lockstep
- systolic arrays that rhythmically move data through networks and control lockstep execution by broadcasting global clock pulses
- classic vector computers that *chain* the individual outputs of one functional unit's array operations into another unit as input as each result becomes available.

23.1 Proliferation of MIMD Designs

Increasing computer performance by speeding up how quickly a processor executes each single instruction continues to approach fundamental physical limits, such as the speed of light. However, even if one cannot increase individual processor speed a thousand-fold, one may be able to effectively employ a thousand processors. Thus, barriers to single processor speed-ups encourage designers to exploit parallelism to obtain large performance increases.

This interest in parallelism as a mechanism for general performance enhancement encourages looking for exploitable parallelism in many kinds of problems and extends the search for parallelizable applications past the obviously *vectorizable,* matrix-dominated applications that vector supercomputers successfully address.

Supporting effective parallel execution when the inherent application parallelism is not as structured as vector operations encourages machine characteristics that depart from the traditional synchronous machine approaches. Such designs may support parallelism at various levels or *granularities,* such as large tasks, medium-grained subroutines, or fine-grained loop iterations. Similarly, meeting the needs of large, event-driven applications such as airline reservation or military command and control systems requires asynchronous systems that exploit parallelism dynamically by creating software processes and assigning them to available processors.

The search for substantial performance increases has extended the scope of parallelizable domains to applications and algorithms with diverse characteristics. The effort to accommodate these new domains and to support more than one domain with a single machine has greatly encouraged the exploration of diverse MIMD architecture designs. To understand these varied approaches to MIMD architecture and how they relate to one another, it will be helpful to consider essential distinctions that differentiate MIMD designs.

A fundamental distinction among MIMD machines is made on the basis of whether an architecture provides a shared memory that all processors can access or distributes memory throughout the system in the form of local memories that are only accessed by an associated processor. In distributed memory machines, any data to be shared among processors must explicitly be sent from processor to processor in the form of data packets called *messages*. Section 23.2 further subdivides the shared memory MIMD category according to the mechanism used to provide shared memory access, while Section 23.3 categorizes distributed memory machines according to the topology of their message-passing interconnection networks (INs).

Figure 23.2 shows how MIMD architectures fit into the spectrum of parallel computing architectures that includes synchronous architectures and *MIMD-based paradigm* machines. The latter use MIMD principles of asynchronous operation but are geared to supporting special execution paradigms. For example, data-flow execution involves triggering an instruction's execution on the basis of operand availability. Although these execution paradigms often can be implemented on different MIMD architecture types, the most effective implementations are designed to support their special requirements. These execution paradigms and their hardware implementations are beyond the scope of our discussion. Instead, the sections below survey the major alternative designs for supporting general, asynchronous forms of parallel processing.

23.2   Shared Memory Architectures

Shared memory architectures allow each processor to access the contents of a global memory through an IN that provides a data pathway from each processor to each shared memory module. Processes executing in parallel can share data for a variety of purposes, including the using it to synchronize their activities. Such data sharing poses two fundamental problems for designers: synchronizing data access and maintaining memory cache consistency. We examine these problems below before discussing the principal shared memory design approaches.

23.2.1   Fundamentals: synchronization and cache consistency

Shared memory architectures must synchronize processes' memory accesses at a low level to prevent loss of data integrity, and at a high level to facilitate program event ordering [13]. These facilities are often supported by implementing synchronization primitives in

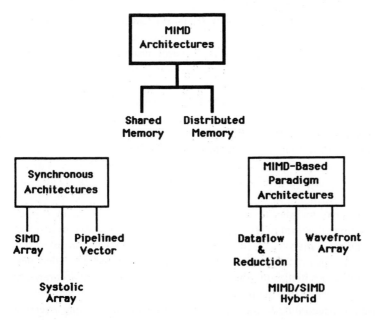

Figure 23.2 Major parallel architecture categories

hardware. For example, the popular *Test-and-Set* and *Reset* primitives shown below can ensure that only one process at a time can access a shared variable.

```
TEST-AND-SET (lock_variable)        RESET (lock_variable)
   temp := lock_variable;              lock_variable := 0;
   lock_variable := 1;                 END;
   RETURN (temp);
   END;
```

A process enjoys access only when its call to the Test-and-Set primitive is the sole call that returns a zero value. While one process has this access, other calls to the primitive return a value of one. A process calls the Reset primitive to relinquish control by setting the *lock variable* value back to zero. Blocked processes usually either *spin-lock* by repeatedly invoking the primitive or *suspend-lock* by waiting for an interrupt that signals that a Reset operation has been effected.

Memory controllers can implement such primitives in hardware to prevent data corruption through simultaneous access. The controller prevents one process from accessing data at a given memory location while another process is changing the contents of that location, thus leaving the data in an unpredictable state.

Access synchronization mechanisms, such as software routines, special machine instructions or control bits associated with memory words [13] can also be used to synchronize processes at a higher level. For example, consider a system in which routines on multiple processors either write sensor data to a buffer or read the buffer. Synchronization primitives can be used to ensure that each sensing event is processed by preventing a pair of WRITE operations without an intervening READ operation.

Maintaining memory *cache consistency* is equally important for modern shared memory computers. High performance memories are relatively expensive. However, since most programs exhibit localized memory reference patterns, placing small, expensive memories near each processor to act as *caches* is effective. During execution, a cache fills up with the memory items most recently referenced by its processor. Sometimes, copies of a single shared memory datum will exist in multiple caches. Ensuring that any processor's reference to such a datum returns a properly updated version of it is the cache *coherency* or *consistency* problem.

Hardware-based cache coherence can be provided by implementing write-invalidate or write-update *policies* with bus snooping, directory-based mechanisms, or special networks [48]. Figure 23.3 illustrates the two policies, beginning with Fig. 23.3a, which shows processors P1 and P2 both caching the value of shared variable, *x*. When a process running on P1 updates *x*'s value (Fig. 23.3b), the corresponding values in main memory, and P2's cache becomes obsolete. For a write-invalidate policy, P1 issues invalidate commands that cause the invalid copies in P2's cache and in memory to be marked as such (Fig. 23.3c). P1's cache, rather than main memory, can then service requests to access *x* . A write-update policy (Fig. 23.3d) would update P2's cache and, depending on the implementation, update memory.

In both coherence policies, altering a cached memory block results in commands for other caches with copies of the block to take action, i.e., to update their copy or mark it as invalid. Hardware-based *snooping* is an effective mechanism to implement this process for a small number of processors united by a common bus. Consistency commands are broadcast over the bus as all caches and the system memory monitor the bus traffic (snoop). Hence, caches and the memory recognize when they must act to maintain memory consistency. A cache also recognizes that it has the most recent version of a block and services an access in system memory's stead.

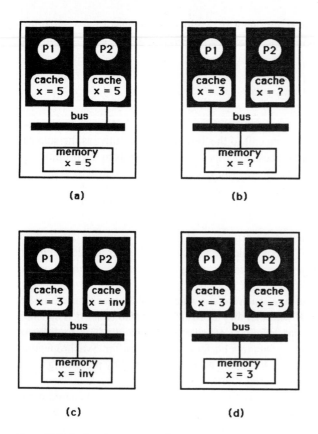

Figure 23.3 Cache coherency scenarios

Snooping schemes involve broadcasting each consistency command for all caches to monitor, rather than communicating only with caches that contain a copy of the altered memory block. This approach requires no overhead to track which caches have which memory blocks. However, for a large number of processors, such consistency command and value update broadcasts would saturate the bus and the caches' monitoring facilities. Thus, architectures with many processors need to restrict consistency communications to the affected caches only. This can be done with directories that track which caches have copies of which memory blocks. Three basic approaches to implementing such directories [7] are described below:

- *Full-map directories* store status information on each global memory block for each processor, so each processor can simultaneously cache a given block.
- *Limited directories* maintain a fixed amount of state information for each block regardless of the number of processors, thus constraining how many processors can simultaneously cache a given block.
- *Chained directories* maintain the full state information of full-map directories but decompose this information into multiple directories, which are distributed throughout the system and *chained* together with pointers.

Cache coherence approaches for large numbers of processors include special networks to implement cache coherence (see Section 23.2.6) and software-based approaches [8] that involve techniques such as predetermining data cachability or time-stamping data-structure updates [48]. Implementing an effective cache coherence mechanism is important for all the shared the shared memory architectures reviewed in the following sections.

23.2.2   Categories of shared memory architecture

The mechanism used to connect multiple processors to shared memory modules is typically the most important architectural characteristic that differentiates shared memory machines. In addition, the interconnection mechanism greatly influences an architecture's scalability and memory access latency. Thus, as Fig. 23.4 shows, most of our shared memory architecture categories are based on interconnection mechanisms such as buses or crossbar switches. A subcategory for newer machines that use only cache memories is also shown.

23.2.3   Bus interconnection architectures

In bus-based, shared memory systems, the bus is an electronic data pathway that connects processors to memory modules (Fig. 23.5). The bus typically "contains 50 to 100 wires and is physically realized by printed lines on a circuit board or by discrete (backplane) wiring" [29]. A processor places the address of a memory location onto the bus, and a memory module responds by sending a copy of the specified location's contents over the bus. For efficiency, instead of moving single datums, bus operations may move the contents of larger memory blocks containing the datum, such as a 16-byte *cache-line* or 128-byte *subpage*.

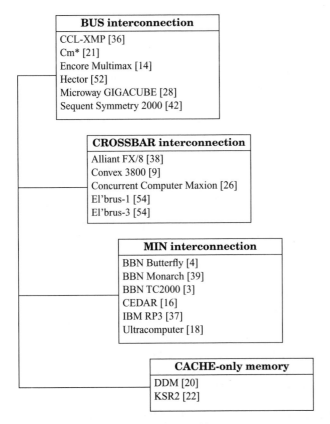

Figure 23.4 Examples of major shared memory architecture types

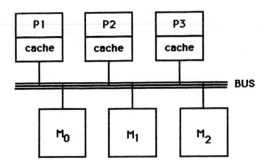

Figure 23.5 Example bus interconnection

Because buses offer a relatively simple and inexpensive way to give multiple processors access to a shared memory, they have been popular with commercial MIMD machine makers. A significant limitation of such systems with one or a few buses is their relative lack of 'scalability', which prevents them from maintaining adequate performance as the number of processors is *scaled up*. The limitation arises because a bus becomes saturated when its traffic overwhelms capacity and processors endure lengthy delays to effect data transfers. These problems are exacerbated by snooping techniques for cache coherency, since *consistency commands* (see Section 23.2.1) contribute substantially to the escalating bus traffic.

Researchers have proposed various multiple-bus solutions to these problems. The experimental Cm* architecture [21] used local buses to form processor clusters and a higher-level system bus to link special service processors associated with each cluster. More recently, the Stanford DASH architecture [24] has explored a bus-based, cache-coherent architecture that incorporates localized snooping and a directory-based approach to cache coherency. A Dash cluster uses a snooping bus to connect a few processors to a local shared memory. Cache directories (see Section 23.2.1) are partitioned among clusters to prevent access bottlenecks, while directory controllers restrict broadcasting memory requests to just the caches involved. The controllers (and their clusters) can be connected by a scalable IN, which need not utilize buses.

23.2.4   Crossbar interconnection architectures

Shared memory architectures can use a crossbar switch of n^2 crosspoints to connect n processors to n memories, as shown in Fig. 23.6. With a crossbar interconnection mechanism, two or more processors may contend for access to some specific memory location, but crossbars prevent contention for communication links by providing a dedicated pathway from each processor to each memory module.

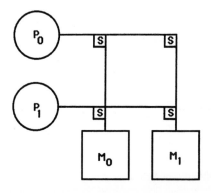

Figure 23.6 A 2×2 crossbar switch (Copyright © 1990, IEEE, reproduced by permission)

Crossbar interconnections offer high communications performance but are relatively expensive. Power, pinout, and size considerations usually limit crossbar architectures to using a small number of processors (i.e., 4 to 16). The 8-CPU Convex 3880 is a representative example of a commercial crossbar-based architecture.

Recent research suggests that crossbar INs might be extended to larger numbers of processors by using optical devices or hybrid IN approaches. For example, the Kyushu University RPP [30] accommodates 128 processors by arranging 256 8 × 8 LSI crossbar switches in a 16 × 16 matrix. Each processor has a memory request bus that provides one of the inputs to each of the 16 crossbars in one of the matrix rows and has a memory response bus that accepts one of the outputs from each of the crossbars in one of the matrix columns.

23.2.5 Multistage interconnection network architectures

A *multistage interconnection network* (MIN) connects multiple *stages* or banks of switches to form an IN pathway. MINs avoid requirements for switch complexity by having each of the switches in a stage handle only a small number of inputs and outputs. MIN designers can also maintain scalability by choosing switches and connectivity patterns that result in an IN pathway length that grows logarithmically as the number of processors and memories increases.

Figure 23.7 shows an omega MIN that uses three 4-switch stages of 2 × 2 switches to connect 8 processors to 8 memories. Thus, this MIN is designed to connect N processors to N memories by using $\log_2 N$ stages, each composed of $N/2$ modules that are 2 × 2 switches.

In our example MIN with 2 × 2 switches, a processor specifies a memory module to access (and the needed IN pathway to use) by providing a bit-value that consists of one control bit for each stage. The switch at stage i examines the ith bit of this value to determine whether to connect the input (requesting) port to the upper or lower output port. A control bit equal to zero indicates a connection to the upper output. Designers can use a variety of

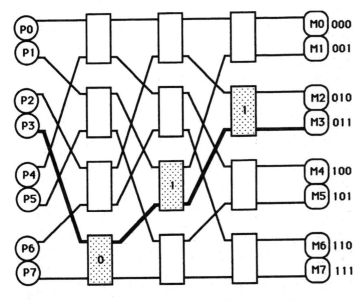

Figure 23.7 An omega MIN (Copyright © 1990, IEEE, reproduced by permission)

alternative components and connectivity patterns, such as those proposed for the omega, SW-banyan, multistage shuffle-exchange, and butterfly networks [5, 23, 45].

Bolt, Beranek and Newman (BBN) has constructed several commercial systems that use MINs to connect processors to memories, including the Butterfly, Monarch, and TC2000. The TC2000 [3], for example, uses 8×8 crossbar switches in its stages and connects them in a *butterfly* pattern. A 512-processor model can be built by using three stages of 64 modules each. The TC2000 automatically handles caching local and sharable read-only memory data.

23.2.6   Cache-only architectures

Two recent shared memory architectures use only cache memories, allowing a shared datum to be *attracted* to the cache of the processor that references it. Data movement is minimized by allowing multiple copies of a datum to exist for multiple-user read access.

The Swedish Institute of Computer Science's prototype Data Diffusion Machine (DDM) has the low-level structure shown in Fig. 23.8 [20]. A node controller snoops memory references on both its local bus and the bus above it. It responds to requests on the bus above it for shared items currently in its local attraction memory and it moves local bus requests for items not in its attraction memory onto the higher-level bus. The DDM team proposed using a hierarchy of buses to construct a tree-topology architecture in which leaves would consist of local attraction memory subsystems and internal tree nodes would consist of *directories* to serve the memory subsystems beneath them.

The KSR1 and KSR2 products [22] used local subsystems in which a slotted, ring-shaped pipeline carried data requests and shared data packets. The pipeline connected a *router directory cell* with *cell interconnect* units containing directories that described data in their caches. Router directory cells contained a directory of their subsystem's shared subpages and formed a hierarchy of rings through HIPPI-like communications links. The apparent demise of KSR products clouds the future of cache-only shared memory designs.

23.2.7   New designs with physically distributed shared memory

Although some of the interconnection mechanisms described in previous sections can be used to access memory modules that are physically distributed with a system's processors, they need not be used in this fashion. For example, a MIN could be used to access either memory units that are placed on circuit boards along with individual processors or memory modules that are grouped apart from the systems' processors.

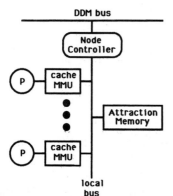

Figure 23.8 Prototype Swedish DDM node

The two architectures described below, however, make physically distributing shared memory modules among the system's processors a major aspect of the design. Both use special memory management units to obtain data from distant memories for their associated processors.

Cray Research's T3D [11] uses a 3D torus IN composed of bidirectional links and custom switches. Each 3D lattice switch is associated with a pair of *processing elements* (PEs), each consisting of a DEC Alpha processor, local DRAM memory, and a *memory control* unit. The memory control unit can obtain requested datums that are not currently in its processor's cache or local memory.

The Fujitsu VPP500 [15] promises a top configuration that connects 222 PEs (and 2 control processors) with a "conflict-free crossbar network." It is not clear how the large crossbar network is implemented (e.g., with many smaller crossbars). Each processing node contains a vector processor, RISC scalar processor, and *data transfer unit* (DTU). Like the T3D's memory control unit, the DTU interfaces with the IN and obtains memory datums for its processor.

23.3   Distributed Memory Architectures

In a distributed memory MIMD architecture (Fig. 23.9) the IN connects processors to processors (the actual links often connect auxiliary I/O processors). Because these systems lack shared memory, any data needed by more than one processor must be explicitly sent from processor to processor through the IN. When the IN does not directly connect two processors that share data, this message-sending process will require intermediary processors to receive and forward the data.

Many distributed memory architectures are geared to exploit medium-grained parallelism, because they often have more processors than task-level parallelism can effectively utilize, and because message-passing overhead would likely outweigh their execution speed-up for very fine-grained parallelism. In medium-grain parallelism, the *granularity* of parallel execution is a few executable statements or a small subroutine. Because medium-grain parallelism requires significant message-passing among processes, the processor-to-processor IN design is critical.

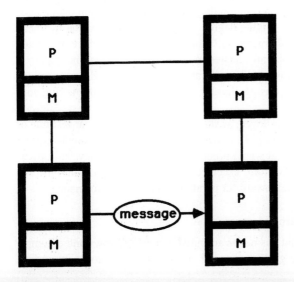

Figure 23.9 MIMD distributed memory architecture (Copyright © 1990, IEEE, reproduced by permission)

23.3.1 Fundamentals: interconnection network topology issues

A distributed memory machine's IN topology is especially important, due to its influence on architecture performance, feasibility, and scalability. IN topology drives performance and scalability because the connectivity pattern determines the length of message-passing pathways through intermediary nodes and links. This IN characteristic is captured by the *communication diameter* metric, which is the maximum number of links that a message must traverse between any source and any destination node while taking the shortest available path [5]. Communications diameter is the best-case pathway for the worst-case processor pairing. Since adding processors typically increases this measure, the most scalable topologies have communications diameters that grow slowly.

IN topology also determines how many IN links are required at each node, which strongly influences feasibility. As distributed memory architectures are *scaled up,* communications links are added to connect the additional processors to one another and to existing processors. Because current electronics technology cannot provide hundreds or thousands of pinout connections for a processor, each processor cannot be connected to every other one. Hence, the IN topology requires a feasible number of interconnections per processing node. This number of connections per node is referred to as the topology's *degree.* A scalable IN topology will exhibit minimal increases in node degree as the number of processors increases.

An ideal IN topology would have a small communications diameter and small node degree that grow slowly (if at all) as processors are added. In practice, it is difficult to attain all of these goals. Thus, the review that follows demonstrates that some topologies balance these metrics, while others clearly emphasize minimization of one of them.

Other IN attributes, such as the particular mechanisms used for message queueing and storing, can strongly affect performance. However, since these alternative message passing mechanics could be used with any topology, topology seems the most fundamental discriminator between IN designs and is used below to establish subcategories for distributed memory architectures, as shown in Fig. 23.10.

23.3.2 Ring topology architectures

A ring topology IN results in a fixed node degree of 2, since each processor is connected only to a pair of neighbors. For N nodes, a ring topology has a communication diameter of $N/2$. Although ring topologies provide a node degree that is fixed and low, the overhead imposed by the high communication diameter will usually outweigh this benefit for any applications that do not have predominantly nearest-neighbor communications patterns. Ring topologies' shortcomings can, however, be mitigated by adding chordal connections to decrease communication diameter or using chordal connections and multiple ring connections to improve fault tolerance.

Ring topologies can be effectively employed, as shown by the Swiss Federal Institute of Technology (SETH) MUSIC architecture [19]. MUSIC features pipelined, point-to-point connections that link 60 Motorola digital signal processors in a ring topology. This architecture has achieved a 3.6 GFlop maximum, executes physical system simulations, and helped its users win a 1993 Gordon Bell award for parallel computing performance.

23.3.3 Mesh and torus topology architectures

A 2D mesh is a planar structure that has nodes arranged in a grid of rows and columns, such that a node is connected by the grid's horizontal and vertical links to its immediate

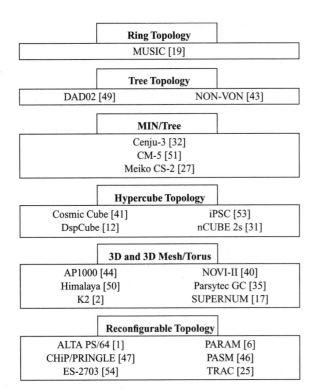

Ring Topology	
MUSIC [19]	

Tree Topology	
DAD02 [49]	NON-VON [43]

MIN/Tree	
Cenju-3 [32]	
CM-5 [51]	
Meiko CS-2 [27]	

Hypercube Topology	
Cosmic Cube [41]	iPSC [53]
DspCube [12]	nCUBE 2s [31]

3D and 3D Mesh/Torus	
AP1000 [44]	NOVI-II [40]
Himalaya [50]	Parsytec GC [35]
K2 [2]	SUPERNUM [17]

Reconfigurable Topology	
ALTA PS/64 [1]	PARAM [6]
CHiP/PRINGLE [47]	PASM [46]
ES-2703 [54]	TRAC [25]

Figure 23.10 Distributed memory architecture categories and examples

neighbors in the north, south, east and west directions. Although all nodes' nearest neighbors are logically equidistant, actual distances may vary when physical implementations involve off-board links.

A 2D mesh, or lattice, with n^2 nodes, has a communication diameter of $2(n-1)$. Figure 23.11 shows that edge nodes' lack of neighbors in some directions makes the worst-case node pairing involve nodes that lie across from one another along a diagonal running through the mesh's center. Adding wrap-around links for the edge nodes creates a torus and reduces the communication diameter to 2(INTEGER-part of $n/2$). Communication di-

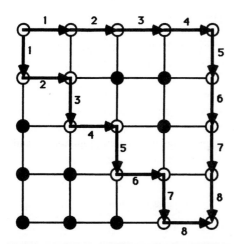

Figure 23.11 2D mesh topology and communication diameter

ameter reduction can also be obtained by adding links to nodes' diagonal neighbors or by connecting mesh rows and columns with buses.

Similarly, the communication diameter for a symmetrical 3D lattice with n^3 nodes and dimensions of length $n - 1$ is $3(n - 1)$, since the worst-case scenario involves moving the full length of each dimension. Again, adding wrap-around connections to create a torus reduces the communication diameter to 3(INTEGER-part of $n/2$), because worst-case movement along a dimension can use the wrap-around link to move no more than half the dimension's length.

The node degree of mesh and torus topology architectures is small and stays constant as processors are added, although their communication diameter is larger than that of some alternatives. For example, the communications diameter of a symmetrical 3D torus with 512 processors is 3(INT 8/2) = 12, while that of a 512-node hypercube is $\log_2 512 = 9$. However, mesh and torus architectures can show effective performance for applications with localized communications patterns.

As we saw in Fig. 23.10, several recent architectures have employed lattice topologies for message-passing INs. Including both 2D and 3D torus topologies, these architectures range from experimental university machines (K2), through corporate prototypes (AP1000, NOVI-II HiPIPE) to commercial products (Parsytec GC, Tandem Himalaya).

23.3.4 Hypercube topology architectures

A hypercube topology arranges processors in an n-dimensional cube. Usually, the number of processors is chosen to be a power of 2 so that there are $N = 2^n$ processors, and each node has $n = \log_2 N$ bidirectional links to adjacent nodes. Each node is associated with a numeric label that uniquely identifies it with an n-bit binary value ranging from 0 to $N - 1$. The nodes and labels are organized so that adjacent nodes' values will differ only by a single bit.

Messages travel through a hypercube topology in the following manner, as illustrated by Fig. 23.12. Each message includes the destination node's bit-label and a routing label initialized to the source node's bit-label. When a processor receives a message, it checks whether the destination label is identical to its own bit-label. If it is not, the processor selects an adjacent node that has a bit in common with the destination label that the current processor lacks, changes the value of that bit in the routing label, and sends the message to the adjacent processor.

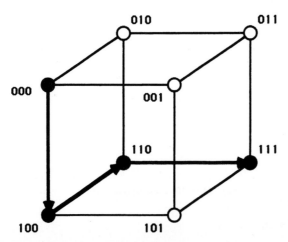

Figure 23.12 Hypercube message routing

Hypercube topologies offer an attractive degree of scalability because both node degree and communication diameter grow logarithmically as processors are added. Following typical conventions, both node degree and communication diameter will equal $\log_2 N$, for N processors.

The number of links a message traverses is equal to the number of bits that differ in the source and destination labels. In the worst case, they will differ by every one of the n bits in the label, so the communication diameter will be $n = \log_2 N$.

Since hypercube node degree is also equal to $\log_2 N$, the number of processors in a hypercube architecture can be doubled at the cost of increasing both the node degree (IN links per node) and the communication diameter by one. Thus, a hypercube architecture with 1024 processors has one more IN link per processor (10) than does a configuration with 512 processors.

The Cosmic Cube's development at the California Institute of Technology [41] ultimately spurred commercial hypercubes, such as the nCUBE 2s [31] and Intel Personal Supercomputer [53]. More recent experimental architectures using hypercube topology include the DspCube in Australia [12] and the PACESPARC in India [33].

23.3.5   Tree topology architectures

Classic tree topologies have a regular, hierarchical structure of processing node levels that creates a *tree* appearance by having each lower level contain more nodes than its predecessor above. Each node at level n has a single IN link to a parent node at higher level $n - 1$ and multiple links to descendent nodes at level $n + 1$. The node at the highest level is the *root* node, which has no parent. Nodes at the lowest level have no descendents. Most trees use a fixed *fan-out* scheme, where each internal node has the same number of descendents. Fig. 23.13 shows a classic binary tree.

Because each processing node has one parent and a fixed number of descendents, node degree is not a barrier to scalability. However, communication diameter and fault tolerance do pose problems. For example, the communication diameter of a binary tree with n levels and $2^n - 1$ processors is $2(n - 1)$, since some source and destination node pairings require messages to travel up the tree several levels, only to travel back down to the starting level or lower. Thus, a binary tree with 1023 nodes (10 levels) has a communication diameter of 18, compared to a diameter of 10 for a 1024-node hypercube. Binary tree communication shortcomings can also be characterized as poor *bisection bandwidth,* since moving data between tree halves must utilize a single node—the congested root.

If a single node or its link to the parent fails in a classic tree, communications between all that node's descendents and the rest of the tree cease. Proposed solutions to these problems include using buses or point-to-point links to unite tree nodes at the same level.

Several experimental architectures have been constructed that utilize classic tree or augmented tree topology INs, including DADO/DADO2 [49] and Non-Von [43].

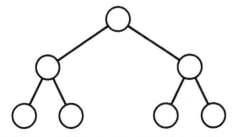

Figure 23.13 A binary tree

23.3.6   MIN/tree topology architectures

A few recent architectures feature INs with tree-shaped topologies that also have some MIN-like characteristics, such as redundant pathways and logarithmic growth attributes (see Section 23.2.5). These topologies for connecting processors can be formed from classic processor-to-memory MIN topologies by techniques such as collapsing multiple *stages* (that use simple switches and form strongly connected subgraphs) into a single stage that uses more complex switches.

The IN shown in Fig. 23.14 mirrors the structure of a two-stage network *layer* in Meiko's CS-2. Meiko insightfully relates MINs and trees, describing their IN as "a Benes network folded about its centre line, with each switch chip rolling up the functionality of eight of the unidirectional two way switches" [27]. Stage two of the Meiko layer shown in Fig. 23.14 can use its remaining four links per switch to connect with further stages (act as a stage in a $\log_4 N$ MIN) or double the number of processors by connecting to four new stage one switches.

Thinking Machines' CM-5 [51] uses a 4-ary tree IN (Fig. 23.15), that possesses MIN-like attributes and differs from classical tree topology in several ways. Only leaf nodes are processors; internal nodes contain only switches. The IN is a *fat tree*, where higher-level links have higher bandwidths than lower-level links. The number of switches per node increases with tree level. Switches in the three levels just above the leaves share each descendent with one other switch within the node, while those at higher levels share each descendent with three other switches within the node.

The CM-5's communication diameter is $2(\log_4 n)$. The descendent-sharing scheme ensures that no nodes are cut off by a single switch or link failure. Although switch degree does not exceed eight, the number of switches per node doubles each time the number of processors is quadrupled, so the root node needs $2(\log_4 n)$ switches for $\log_4 n$ processors.

The advantages of MIN topologies, such as low communications diameter and high scalability, have clearly influenced the design of Thinking Machine's CM-5 and Meiko's CS-2 products, while NEC's Cenju-3 uses an unalloyed MIN with butterfly-style connectivity for its processor-to-processor communication links [32].

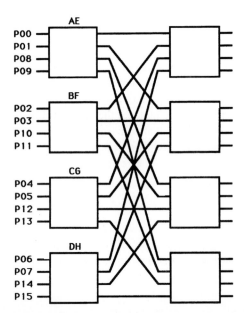

Figure 23.14 Meiko's modified Benes network topology

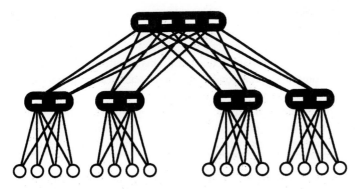

Figure 23.15 CM-5 4-ary, *fat* tree topology for 16 nodes

23.3.7   Reconfigurable topology architectures

Parallel applications exhibit a variety of interprocessor communications patterns. Having an IN topology closely match these patterns helps performance, since using dedicated physical links between processors to implement frequent data exchanges between application processes results in localized, reduced-latency communications. Hence, an important motivation for constructing reconfigurable topology architectures is to allow a single architecture to effectively support the communication patterns of many dissimilar algorithms or applications.

A reconfigurable architecture's IN typically uses programmable switches or routing processors to dynamically specify data pathways. By specifying that only a subset of available pathways are to be used, one can superimpose one kind of topology upon another. Fig. 23.16 demonstrates this principle by showing a binary tree mapped onto a 2D mesh.

Several reconfigurable architectures use a mesh or torus topology for the underlying IN, such as CHiP [47] and the ALTA PS/64 [1]. A few architectures use reconfigurable topologies to support specialized parallel execution, such as the Soviet ES-2703 for data-flow programming [54] or the PASM architecture for partitionable MIMD/SIMD hybrid execution [46].

23.4   Hybrid Shared/Distributed Memory Architectures

Although the vast majority of parallel architectures are either shared or distributed memory machines, hybrids do exist that use both shared and distributed memory. These machines are likely to be geared to special needs of a class of applications or a particular application.

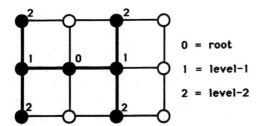

0 = **root**

1 = **level-1**

2 = **level-2**

Figure 23.16 Mapping a tree IN onto a reconfigurable mesh (Copyright © 1990, IEEE, reproduced by permission)

For example, the M^2 architecture under development at National Taiwan University [34] uses a Multibus II bus as a conduit for passing data in distributed memory fashion between processor clusters. Each cluster, however, is a small shared memory system with several CPUs and a local snooping bus to maintain cache coherency. This architecture aims to provide scalable design and use commodity components for large-scale transaction systems that are characterized by independent parallel processes.

The NEXUS architecture constructed at Australia's Swinburne University of Technology offers another example of hybrid memory design [10]. This architecture links 40 transputers through programmable crossbar switches for reconfigurable distributed memory processing and to a system bus for shared memory operation. This architecture is geared to support graphics applications such as animation.

23.5   Conclusion

23.5.1   Future directions

Despite occasional assertions that parallel architecture designs are converging on a single approach, the variety of recent designs provides abundant evidence to the contrary. Thus, this discussion assumes that parallel architecture diversity will persist. The difficulties that U.S. parallel machine vendors experienced in the early 1990s may foreshadow slower commercialization of newly proposed designs but research will continue to address open problem areas such as those suggested below.

Shared memory research in processor-to-memory interconnection mechanisms will probably be inseparably associated with exploring effective caching techniques. Areas of likely interest include hierarchical interconnection networks (e.g., multiple buses, crossbar trees), specialized cache management and routing processors, and optimizing architectures for physically distributed, but shared, memory.

Distributed memory architecture research will probably refine effective communication schemes for established topologies (e.g., mesh, torus, hypercube) and explore reconfigurable topology architectures. New topologies, some of them tree-oriented topologies derived from MIN research, may continue to emerge.

23.5.2   Summary

Our survey shows that the new approaches realized since 1990 for MIMD machines alone include cache-only architectures, large mesh INs and crossbars for physically distributed/shared memory machines, 2D and 3D torus products, hybrid MIN/tree IN topologies, and MINs for processor-to-processor connections. The recent implementation of so many new approaches shows that experimentation and innovation continue to characterize parallel architecture design.

Although new technologies naturally attract our interest, the survey shows the continuing commercial viability of more mature approaches, such as bus and crossbar-based shared memory machines or hypercube-based distributed memory architectures.

In sum, MIMD machines continue to present the parallel architecture student with a broad spectrum of design approaches and continue to show how new developments in IN mechanisms, caching, and topology can be translated into viable machine implementations.

23.5.3   Acknowledgments

The following people provided information on recent architectures: Sam Adams, Jim Baumgartner, Nick Bavaro, Gail Brady, Gordon Bynum, San-Cheng Chang, Wayne

Cosshall, Edward Forbes, Shing-Tsaan Huang, Ron Matlock, Krista Meriam, Bob Schmitz, and Larry Watts.

23.6   References

1. ALTA Technology. 1993. *PS/64 one GIGAFLOP compute engine* (brochure). Sandy, Utah: ALTA Technology.

2. Annaratone, M., M. Fillo, K. Nakabayashi, and M. Viredaz. 1990. The K2 parallel processor: architecture and hardware implementation. *Proc. Int'l Symp. Computer Architecture,* Seattle, Wash., 92–101.

3. BBN Advanced Computers Inc. 1989. *TC2000 Product Summary.* Cambridge, Mass.: BBN Advanced Computers Inc.

4. BBN Laboratories 1985. *Butterfly Parallel Processor Overview.* Cambridge, Mass.: BBN Laboratories.

5. Bhuyan, L. N. 1987. Interconnection networks for parallel and distributed processing. *Computer* 20(6) 9–12.

6. Centre for Development of Advanced Computing (C-DAC). 1991. *PARAM* (brochure). Pune, India: Centre for Development of Advanced Computing, Poona University.

7. Chaiken, D., C. Fields, K. Kurihara, and A. Agrawal. 1990. Directory-based cache coherence in large-scale multiprocessors. *Computer* 23 (6), 49–58.

8. Cheong, H., and A. Veidenbaum. 1990. Compiler-directed cache management in multiprocessors. *Computer* 23 (6), 39–47.

9. Convex Computer Corporation. 1991. *C3800 product overview* (doc. no. 080-002100-000) and related *C3800 product specification* (doc. no. 080-002101-000). Richardson, Tex.: Convex Computer Corp.

10. Cosshall, W. J. 1992. *A hybrid transputer-DSP parallel computer for computer graphics.* Tech. rep. no. SUT-CS-14/92. Victoria, Australia: Dept. Computer Science, Swinburne Univ. of Technology.

11. Cray Research, Inc. 1993. *Cray T3D: the right tool at the right time* (brochure MCPB-1330893). Eagan, Minn.: Cray Research, Inc.

12. Dick, C. 1992. DspCube: a hypercube multiprocessor for real-time signal processing. In *Directory of Parallel Processing Research Projects in Australia,* D. Abramson and P.G. Whiting, eds. Victoria, Australia: CITRI, 30.

13. Dubois, M., C. Scheurich, and F. A. Briggs. 1988. Synchronization, coherence, and event ordering in multiprocessors. *Computer* 21 (2), 9–21.

14. Encore Computer Corp. 1987. *Multimax technical summary,* Publ. No. 726-01759 Rev. D. Marlboro, Mass.: Encore Computer Corp.

15. Fujitsu Ltd. 1992. *VPP500 vector parallel processor* (brochure BP0009-1M, 1st. edition). Tokyo: Fujitsu Ltd.

16. Gajski, D. D., D. H. Lawrie, D. J. Kuck, and A. H. Sameh. 1987. CEDAR. In *Parallel Computing: Theory and Comparisons,* G. J. Lipovski and M. Malek, eds. New York: John Wiley & Sons, 284–291.

17. Giloi, W. K. 1989. Supercomputer development in Europe. *Proc. Int'l Conf. Supercomputing,* Crete, Greece, 391–397.

18. Gottlieb, A., R. Grishman, C. P. Kruskal, K. P. Mcauliffe, L. Rudolph, and M. Snir. 1983. The NYU Ultracomputer: designing an MIMD shared memory parallel computer. *IEEE Trans. Comput.* C-32(2), 175–189.

19. Gunzinger, A., U. A. Muller, W. Scott, B. Baumle, P. Kohler, H. R. vonder Muhll, F. , Muller-Plathe, W. F. van Gunsteren, and W. Guggenbuhl. 1992. Achieving supercomputer performance with a DSP array processor. *Proc. Supercomputing '92,* Minneapolis, Minnesota, 543–550.

20. Hagersten, E., A. Landin, and S. Haridi. 1992. DDM—a cache-only memory architecture. *Computer* 25 (9), 44–54.

21. Jones, A. K., and P. Schwarz. 1980. Experience using multiprocessor systems: a status report. *ACM Comput. Surveys* 12(2), 121–165.

22. Kendall Square Research. 1992. Kendall Square Research technical summary. Waltham, Mass.: KSR.

23. Kothari, S. C. 1987. Multistage interconnection networks for multiprocessor systems. In *Advances in Computers,* Vol. 26, M. C. Yovits, ed. New York: Academic Press, 155–199.

24. Lenoski, D., J. Laudon, K. Gharachorloo, W.-D. Weber, A. Gupta, J. Hennessey, M. Horowitz, and M. Lam. 1992. The Stanford Dash multiprocessor. *Computer* 25(3), 63–79.

25. Lipovski, G. J., and M. Malek. 1987. *Parallel Computing: Theory and Comparisons.* New York: John Wiley & Sons.

26. Mayer, J. H. 1995. The need for speed: real-time computers shatter bottlenecks. *Military & Aerospace Electronics* 6(1), 27–29.

27. Meiko. 1993. *Computing Surface 2: overview documentation set.* Waltham, Mass.: Meiko.

28. Microway. 1993. *GIGACUBE specifications* (brochure). Kingston, Mass.: Microway.

29. Mudge, T. N., J. P. Hayes, and D. C. Winsor. 1987. Multiple bus architectures. *Computer* 20(6), 42–48.

30. Murakami, K., S-i. Mori, A. Fukuda, T. Sueyoshi, and S. Tomita. 1989. The Kyushu University reconfigurable parallel processor design of memory and intercommunication architectures. *Proc. Int'l Conf. Supercomputing,* Crete, Greece, 351–360.

31. nCUBE Corporation. 1992. *nCUBE 2: Technical Overview.* Foster City, Calif.: nCube Corporation.

32. NEC. 1993. *Parallel Computer Cenju-3* (brochure). Boxborough, Mass.: HNSX Supercomputers.

33. Neelakantan, K., P. P. Ghosh, M. S. Ganagi, G. Athithan, M. V. Atre, and G. Venkataraman. 1990. Performance characteristics of a hypercube-type parallel computer. *Current Science* 59(20), 982–988.

34. Oyang, Y.-J., D. J. Sheu, C.-Y. Cheng, and C.-Z. Yang.1993. The M2 hierarchical multiprocessor. *Future Generation Computer Systems* 9(3), 235–240.

35. Parsytec Computer GmbH. 1991. *Parsytec GC: Beyond the Supercomputer.* Aachen, Germany: Parsytec Computer GmbH.

36. Peir, J.-K., L.-C. Chang, D.-Y. Chung, L.-P. Chen, S.-S. Huang, et al. 1993. CCL-XMP: a Pentium-based symmetric multiprocessor system. Hsinchu, Taiwan, R.O.C.: CCRL, Industrial Technology Research Institute.

37. Pfister, G. F., W. C. Brantley, D. A. George, S. L. Harvey, W. J. Kleinfelder, K. P. McAuliffe, E. A. Melton, V. A. Norton, and J. Weiss. 1987. An introduction to the IBM Research Parallel Processor Prototype (RP3). In *Experimental Parallel Computing Architectures,* J. J. Dongarra, ed. Amsterdam: Elsevier Science Publishers B. V. (North-Holland), 123–140.

38. Perron, R., and C. Mundie. 1986. The architecture of the Alliant FX/8 computer. In *Digest of Papers, COMPCON, Spring 1986*, A.G. Bell, ed. Silver Spring, Md.: IEEE Computer Society Press, 390–393.

39. Rettberg, R. D., W. R. Crowther, P. P. Carvey, and R. S. Tomlinson. 1990. The Monarch parallel processor hardware design. *Computer,* 23(4), 18–30.

40. Sawabe, T., T. Fuji, H. Nakada, N. Ohta, and S. Ono. 1992. A 15 GFLOPS parallel DSP system for super high definition image processing. *IEICE Trans. Fundamentals* E75-A(7), 786–793.

41. Seitz, C. L. 1985. The Cosmic Cube. *Comm. ACM* 28(1), 22–33.

42. Sequent Computer Systems. 1991. *Symmetry 2000/450 and 2000/ 750,* document PD-1070. Beaverton, Ore: Sequent Computer Systems.

43. Shaw, D. E. 1981. Non-von: a parallel machine architecture for knowledge based information processing. *Proc. 7th Int'l Joint Conf. on Artificial Intelligence,* 961–963.

44. Shimizu, T., T. Horie, and H. Ishihata. 1992. Low-Latency Message Communications Support for the AP1000. *Proc. Int'l. Symp. Computer Architecture,* Gold Coast, Australia, 288–297.

45. Siegel, H. J. 1990. *Interconnection Networks for Large-Scale Parallel Processing: Theory and Case Studies,* 2nd ed. New York: McGraw-Hill.

46. Siegel, H. J., T. Schwederski, J. T. Kuehn, and N. J. Davis. 1987. An overview of the PASM parallel processing system. In *Tutorial: Computer Architecture,* D. D. Gajski, V. M. Milutinovic, H. J. Siegel, and B. P. Furht, eds. Silver Spring, Md.: IEEE Computer Society Press, 387–407.

47. Snyder, L. 1982. Introduction to the Configurable Highly Parallel Computer. *Computer* 15(1), 47–56.

48. Stenstrom, P. 1990. A survey of cache coherence schemes for multiprocessors. *Computer*23(6), 12–24.

49. Stolfo, S. 1987. Initial performance of the DADO2 prototype. *Computer* 20(1), 75–83.

50. Tandem Computers Inc. 1993. *NonStop Himalaya Range,* order no. 300262. Cupertino, Calif.: Tandem Computers Inc.

51. Thinking Machines Corporation. 1992. *Connection Machine CM-5 technical summary* (chapter 19). Cambridge, Mass.: Thinking Machines Corporation.

52. Vranesic, Z., M. Stumm, D. Lewis, and R. White. 1991. Hector: a hierarchically structured shared-memory multiprocessor. *Computer* 24 (1), 72–79.

53. Wiley, P. 1987. A parallel architecture comes of age at last. *IEEE Spectrum* 24(6), 46–50.

54. Wolcott, P. and S. E. Goodman. 1988. High-speed computers of the Soviet Union. *Computer* 21 (9) 32–41.

24

Memory Models

Leonidas I. Kontothanassis and Michael L. Scott

The rapid evolution of RISC microprocessors has left memory speeds lagging behind: as of early 1994, typical main memory (DRAM) access times were on the order of 100 processor cycles, and this number is expected to increase for at least the next few years. Multiprocessors, like uniprocessors, use memory to store both instructions and data. The cost of the typical memory access can therefore have a major impact on program execution time.

To minimize the extent to which processors wait for memory, modern computer systems depend heavily on caches. Caches work extremely well for instructions: most programs spend most of their time in loops that fit entirely in the cache. As a result, instruction fetches usually have a negligible impact on a processor's average number of cycles per instruction (CPI). Caches also work well for data, but not as well. If we assume a simple RISC processor that can issue a single register-to-register instruction every cycle and that can hide the latency of a cache hit, then we can calculate CPI from the following formula:

*CPI = % non-memory instructions + % memory instructions * miss rate * memory access cost*

The percentages of memory and non-memory instructions vary with the type of application, the level of optimization, and the quality of the compiler. A good rule of thumb, however, is that somewhere between one-fifth and one-third of all dynamically executed program instructions have to access memory. For a memory access cost of 100 and a dynamic instruction mix consisting of one-fourth memory instructions, a 4 percent drop in the hit rate, from 95 to 91 percent, translates into a 50 percent increase in the CPI, from 2 to 3. Maximizing the hit rate (and minimizing the memory access cost) is clearly extremely important. As we shall see in Section 24.2, this goal is made substantially more difficult on shared-memory multiprocessors by the need to maintain a consistent view of data across all processors.

Performance issues aside, the memory model exported to the user can have a large effect on the ease with which parallel programs can be written. Most programmers are accustomed to a uniprocessor programming model in which memory accesses are all of equal cost, and in which every program datum can be accessed simply by referring to its name. Physical constraints on multiprocessors, however, dictate that data cannot be close to all processors at the same time. Multiprocessor hardware therefore presents a less convenient view of memory, either making some data accessible to a given processor only via a special interface (message passing), or making some data accesses substantially more expensive than others (distributed shared memory).

In this chapter we describe a variety of issues in the design and use of memory systems for parallel and distributed computing. We focus on tightly coupled multiprocessors, because these are the systems that display the greatest diversity of design alternatives and in which the choice among alternatives can have the greatest impact on performance and programmability.

We begin in Section 24.1 with technological issues: the types of memory chips currently available and their comparative advantages and costs. We then consider the major architectural alternatives in memory-system design (Section 24.2), and the impact of these alternatives on the programmer (Section 24.3). We argue that a shared-memory model is desirable from the programmer's point of view, regardless of the underlying hardware architecture. Assuming a shared-memory model, we turn in Section 24.4 to the issue of memory consistency models, which determine the extent to which processors must agree about the contents of memory at particular points in time. We discuss the implementation of various consistency models in Section 24.5. We conclude in Section 24.6 with a description of current trends in memory system design, and speculation as to what one may expect in future years.

24.1   Memory Hardware Technology

The ideal memory system, from a programmer's point of view, is one that is infinitely large and infinitely fast. Unfortunately, physics dictates that these two properties are mutually antagonistic: the larger a memory is, the more slowly it will tend to respond.

Memory systems are generally measured in terms of *size, latency*, and *bandwidth*. Size is simply the number of bytes of information that can be stored in the memory. Latency is usually characterized by two measures: *access time* and *cycle time*. Access time is the time it takes memory to produce a word of data requested by a processor, whereas cycle time is the minimum time between requests. Bandwidth measures the number of bytes that memory can supply per unit of time. Bandwidth is related to cycle time but is also dependent on other memory organization features discussed later in this section. These features include the width of the memory and the number of memory banks.

There are two basic types of memory chips available in the market today: *dynamic random access memory* (DRAM) and *static random-access memory* (SRAM). DRAMs require that the data contained in the chip be occasionally rewritten if they are not to be lost. DRAM chips must therefore be made unavailable to service requests every few milliseconds to refresh themselves. Furthermore, read accesses to DRAM chips must be followed by writes of the data read, because a read destroys the contents of the accessed location. As a result, DRAMs have a larger cycle time than access time. SRAMs, on the other hand, require neither refresh nor write-back, and thus have equal cycle and access times. Unfortunately, SRAM memories require more transistors per bit than DRAMs and thus have lower density. A general rule of thumb is that the product of size and speed is constant, given a particular technology level. Current SRAMs are approximately 16 times faster than DRAMs but have about 16 times less capacity.

SRAM and DRAM are not the only options when building a memory system. There are also hybrid options that take advantage of common memory-access patterns. Cached DRAMs [16] attempt to combine the best features of DRAM and SRAM (high density and fast cycle time) by integrating a small SRAM buffer and a larger DRAM memory on a single integrated circuit. Recently accessed data is cached in the SRAM buffer. If the memory access pattern displays a high degree of temporal locality (see Section 24.2), then many requests will be able to be serviced by the SRAM, for a faster average cycle time. The SRAM buffer in a cached DRAM is obviously slower than a processor-local cache (it is on the other side of the processor-memory interconnect), but it is automatically shared by all processors using the memory, and it does not introduce the need to maintain coherence (see Section 24.2.3).

Pagemode DRAMs [31] take advantage of the fact that addresses are presented in two steps to a DRAM chip. The first step specifies a row in the internal chip grid; the second step specifies the column. If successive accesses are to the same row, then the first addressing step can be omitted, improving the access time. Table 24.1 presents the basic performance characteristics of current technology RAM.

TABLE 24.1   Performance Characteristics of Different Memory

Memory Type	Size	Access Time	Cycle Time
DRAM	16 Mb	85 ns	140 ns
SRAM	1 Mb	8 ns	8 ns
Cached DRAM	16 Mb	8 ns (85 ns)	8 ns (140 ns)
Pagemode DRAM	16 Mb	30 ns (85 ns)	80 ns (140 ns)

Note: Numbers in parentheses denote worst-case response times.

The bandwidth of a memory system is related to the cycle time, but we can get high bandwidth out of slow memory chips by accessing more of them in parallel, or by *interleaving* successive memory references across different memory modules. Interleaving allows a pipelined processor to issue memory references to either DRAM or SRAM at a faster rate than the memory cycle time. Interleaving also allows a DRAM memory to rewrite the accessed data before being accessed again.

Most memory chips are one to eight bits wide, meaning that a given access returns from one to eight bits of data. Since most systems access data with at least word granularity (and many retrieve many-word cache lines), several chips are used in parallel to provide the desired granularity in a single memory cycle. A collection of memory chips that can satisfy one memory request is termed a *memory bank*.

There is a strong motivation to build a system from the densest available chips to minimize cost, physical bulk, power requirements, and heat dissipation. Unfortunately, the goals of high density and interleaving are at odds. If we use 16-Mb chips and assume that each chip can supply 4 bits of data in one access, then 8 chips in parallel are needed to supply a 32-bit word. This means that a single bank in our system would have 16 MB of memory, and our total memory size would have to increase in 16 MB increments. Using lower-density DRAM to increase interleaving sacrifices much of the cost advantage over SRAM.

In practice, most memory systems for microprocessors (including those employed in multiprocessors) are currently built from DRAM, with caches at the processor. Most memory systems for vector supercomputers (including modestly-parallel machines) are built from highly-interleaved SRAM without caches. The Cray Research C-90, for example, can have up 1024 banks of memory. Historically, supercomputer workloads have not displayed sufficient locality of reference to make good use of caches, and supercomputer customers have been willing to pay the extra cost for the SRAM memory.

24.2   Memory System Architecture

Beyond issues of chip technology, there are many architectural decisions that must be made in designing a multiprocessor memory system. In this section, we address three principal categories of decisions. We begin with the high-level organization of memory into modules, the physical location of those modules with respect to processors and to each other, the structure of the physical address space, and the role to be played by caches. We then turn to lower-level issues in cache design, most of which pertain equally well to uniprocessors. Finally, we consider the issue of coherence, which arises on shared-memory multiprocessors when copies of data to be modified may reside in more than one location.

24.2.1   High-level memory architecture

The most basic design decisions for multiprocessor memory systems hinge on the concept of locality. Programs that display a high degree of locality make heavier use of different portions of the address space during different intervals in time. Computer architects can take advantage of locality via caching, i.e., keeping copies of heavily used information near to the processors that use it, thereby presenting the illusion of a memory that is both large and fast.

There are two dimensions of locality of importance on a uniprocessor, and a third of importance on parallel machines:

1. *Temporal locality.* If a data item is referenced once, then it is likely to be referenced again in the near future.
2. *Spatial locality.* If a data item is referenced, then items with nearby addresses are likely to be referenced in the near future.
3. *Processor locality.* If a data item is referenced by a given processor, then that data item (and others at nearby addresses) are likely to be referenced by that same processor in the near future.

Hierarchical memory systems take advantage of locality by providing multiple levels of memory, usually organized so that each higher level contains a subset of the data in the level below it. Figure 24.1 presents the levels in a typical memory hierarchy, and Table 24.2 summarizes typical size and speed characteristics for each level. The performance of the memory hierarchy depends heavily on the success of the cache level(s) in satisfying re-

TABLE 24.2   Characteristics of Memory Hierarchy Levels

Level Name	Capacity	Response Time
Processors registers	64–256 B	0 cycles
First-level cache	4–64 kB	1–2 cycles
Second-level cache	256–4096 kB	10–20 cycles
Main memory	16–4096 MB	60–200 cycles
Swap & Disk	> 1 GB	> 200,000 cycles

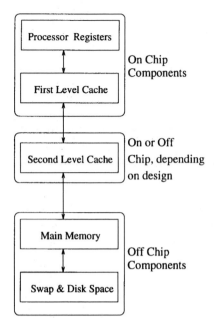

On Chip
Components

On or Off
Chip, depending
on design

Off Chip
Components

Figure 24.1 Levels in memory hierarchy

quests, so that the number of references that need to access slow main memory is minimized. Successful caches reduce both average memory latency and the bandwidth required of main memory and the interconnection network [36], allowing the use of cheaper parts.

Because it imposes lookup costs, a cache can actually reduce performance if the hit rate is very low. For programs without significant amounts of locality, it may make sense to build *flat* memory systems in which all references access main memory directly. Workloads on vector supercomputers, for example, have traditionally shown little locality and require more memory bandwidth than can be provided by a typical cache. As a result, supercomputers are often built with flat memory. For the sake of speed, this memory usually consists of highly-interleaved SRAM, and is a major—perhaps the dominant—component in the cost of these machines. Even so, supercomputer compilers must employ aggressive prefetching techniques, and supercomputer processors must be prepared to execute instructions out of order, to hide the latency of memory.

Current hardware and software trends suggest that caches are likely to become more effective for future supercomputer workloads. Hardware trends include the development of very large caches with multiple banks, which address the bandwidth problem. Software trends include the development of compilers that apply techniques such as blocking [7] to increase locality of reference.

Independent of the existence of caches, designers must address the question of where to locate main memory. They can choose to co-locate a memory module with each processor or group of processors, or to place all memory modules at one end of an interconnection network and all processors at the other. The first alternative is known as a *distributed memory architecture*; the second is known as a *dance-hall architecture.*[*] Figure 24.2 depicts the basic difference between the distributed and dance-hall designs. Distributed memory

---

[*]Dance-hall machines take their name from the image of boys and girls standing on opposite sides of a high school dance floor.

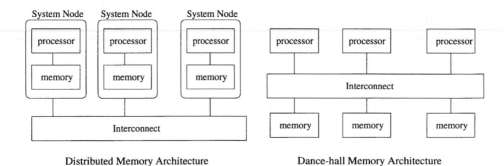

Distributed Memory Architecture   Dance-hall Memory Architecture

Figure 24.2 Simplified distributed and dance-hall memory architecture multiprocessors

architectures improve the scalability of systems when running applications with significant amounts of processor locality. This scalability comes at the expense of a slightly more complicated addressing scheme: node controllers must figure out where in the system to send each memory request that cannot be satisfied locally.

Dance-hall architectures dominate among small-scale multiprocessors, where all system components can share a single bus. The bus makes it easy for modules to monitor each other's activities, e.g., to maintain cache coherence. Large-scale dance-hall machines have also been designed. Examples include the Illinois Cedar machine [47], the BBN Monarch proposal [67], and the forthcoming Tera machine [4]. The dominant wisdom, however, holds that scalability issues dictate a hierarchical distributed memory organization for large-scale multiprocessors, and the trend toward such machines is likely to continue.

There are several important classes of distributed memory machines. The simplest are the so-called *multicomputers,* or *NORMA* (no remote access) machines. In these, each processor is able to access only its local memory. All communication between processors is by means of explicit *message passing.* Commercially-available NORMA machines include the Intel Paragon, the Thinking Machines CM-5, the NCube 2, and a variety of products based on the INMOS Transputer.

The remaining classes of distributed memory machines share with the dance-hall machines the provision of a single global physical address space. The machines in these classes are sometimes referred to as a group as *shared-memory multiprocessors.* Among them, the simplest are the so-called *NUMA* (non-uniform memory access) multiprocessors. Modern NUMA machines have caches, but the hardware does nothing to keep those caches consistent. Examples of NUMA machines include the BBN TC2000, the Hector machine at the University of Toronto [73], the Cray Research T3D, and the forthcoming Shrimp machine [12].

The most complex of the large-scale machines are those that maintain cache coherence in hardware beyond the confines of a single bus. Examples of such machines include the commercially-available Kendall Square KSR1 and 2, the Convex Exemplar (based on the IEEE Scalable Coherent Interface standard [43]), and the Dash [52], Flash [48], and Alewife [3] research projects. Alternative approaches to cache coherence are discussed in Section 24.4.

A cacheless dance-hall machine is sometimes called an *UMA* (uniform memory access) multiprocessor. Several companies produced such machines prior to the RISC revolution, when processor cycle times were closer to memory cycle times. The Tera machine will be an UMA, but will incorporate extremely aggressive techniques to hide memory latency.

The UMA name is sometimes applied to bus-based machines with caches, but this is an abuse of terminology: caches make memory highly non-uniform.

24.2.2   Basic issues in cache design

There are several metrics for judging the success of caches in a hierarchical memory system, the most common being *hit rate, miss rate,* and *mean cost per reference* (MCPR). Hit rate is defined as the percentage of a program's total memory references that are satisfied by the cache; miss rate indicates the percentage of references that could not be satisfied by the cache. The hit rate and miss rate sum to one.

Figure 24.3 illustrates a typical single-level cache organization and its cache lookup mechanism. When a tag check succeeds (the desired tag is found in the cache), we have a cache hit, and data is returned by the cache. When a tag check fails, we have a miss, and data must be fetched from main memory, possibly replacing some data already in the cache.

MCPR is defined to be the average cost of a memory access. It provides a more detailed performance measure for a cache, because it takes into account the cost of misses. If we assume a single level of cache, then the formula for MCPR is as follows:

$$MCPR = HitRate * HitTime + MissRate * MissTime$$

The actual MCPR observed by an application depends on many factors. Some of these are application dependent, while others are architecture dependent. The application dependent factors involve the locality properties of the application and will be discussed in more

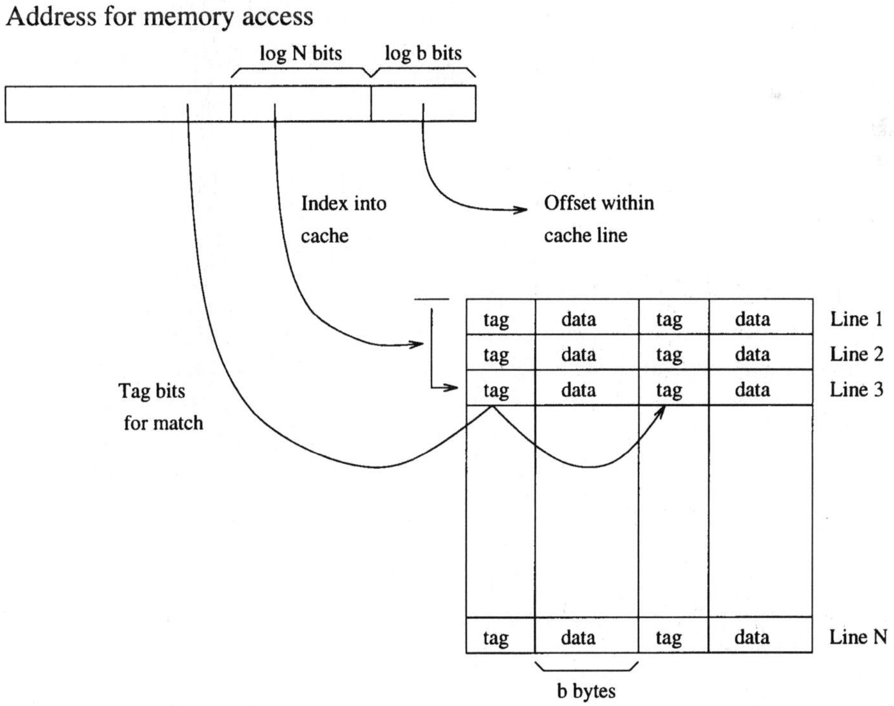

Figure 24.3 Lookup in a typical cache organization

detail in Section 24.3. The architecture dependent parameters are: *cache size, cache block size, associativity, tag and index organization, replacement policy, write policy,* and *choice of coherence mechanism*. These parameters can affect performance by changing the cost of cache hits,[*] changing the cost of cache misses, or changing the ratio of hits and misses.

With the exception of the coherence mechanism, which will be discussed in Section 24.2.3, the remaining parameters apply equally well to uniprocessor cache architectures and are discussed in most computer architecture books [41].

24.2.3   The cache coherence problem

Simply stated, the *coherence problem* is to ensure that no processor reads a stale copy of data in a system in which more than one copy may exist. The coherence problem arises on uniprocessors equipped with direct memory access (DMA) I/O devices, but it can generally be solved in this context via ad-hoc OS-level mechanisms. The problem also arises in multiprocessors with caches and is not as easily solved.

In small-scale multiprocessors, the coherence problem has been addressed by *snooping* on a shared bus. A processor writing a shared datum broadcasts its action on the bus. The remaining processors monitor the bus for all transactions on shared data and take whatever actions are needed to keep their caches consistent. The main problem with this approach is that the bus is a serial bottleneck, which limits system scalability. With current technology, buses can supply as much as 1.2 GB/s of data, while processors may consume data at a peak rate of as much as 200 MB/s. At these rates, fewer than ten processors suffice to saturate the bus. In practice, caches satisfy most of the processor requests but, even so, the number of processors that can successfully share a bus is limited to 20 or 30.

In the absence of a fast, system-wide broadcast mechanism, the cache coherence problem is addressed by maintaining some form of directory data structure [20]. A processor accessing a shared datum consults the directory, updates it if necessary, and sends individual messages to all processors that may be affected by its actions so that they can keep their caches consistent. Directory data structures have to maintain information for every cache line and every processor in the system. The obvious organization poses serious scalability problems, since the memory overhead for directory information increases with the square of the number of processors ($\Theta(P^2)$) [21]. Fortunately, studies indicate that most data are shared among a relatively small number of processors, even on very large machines [39], so a directory structure can be effective while maintaining only a small number of pointers for each sharable line [22]. In the exceptional case in which the small number of pointers is not sufficient, the system can emulate broadcast with point-to-point messages or use dynamically allocated (slower) memory to store the additional pointers.

Most directory-based machines store the information about a given cache line at the processor that holds the corresponding portion of main memory. Machines of this sort are said to have a *CC-NUMA* (cache-coherent NUMA) architecture. An alternative approach is to treat all of a processor's local memory as a secondary or tertiary cache, with a more dynamic directory structure and with no particular designated location for a given physical address. Machines of this sort are referred to as *COMA* (cache-only memory architecture) [40].

The Stanford Dash machine [52] uses the $\Theta(P^2)$ directory organization. The MIT Alewife machine has a limited number of pointers for each directory entry; it traps to software

---

[*]It is common, but not entirely accurate, to assume that cache hits cost a single cycle. In reality, most on-chip caches take 2 or 3 cycles to respond. The compiler attempts to hide cycles beyond the first by scheduling independent operations in the next 1 or 2 instructions, but it cannot always do so.

on overflow. The Convex Examplar is based on the IEEE SCI standard [43], which maintains a distributed directory whose total space overhead is linear in the size of the system's caches. The newer Stanford Flash machine [48] has a programmable cache controller; its directory structure is not fixed in hardware. The KSR-1 and 2 are COMA machines; their proprietary coherence protocol relies in part on hardware-supported broadcast over a hierarchical ring-based interconnection network. The Swedish Data Diffusion Machine [40] is also a COMA, with a hierarchical organization whose space overhead grows as $\Theta(P \log P)$.

Independent of the scalability issue is the question of the actual mechanism used to maintain coherence. The available alternatives include *write-invalidate* and *write-update* [29]. Write-invalidate makes sure that there is only a single copy of a datum before it can be written, by sending invalidation messages to processors that may currently have a copy of the cache line in which the datum resides. Subsequent accesses to this line by processors other than the writer will contact the writer for the latest value of the data. Write-update makes sure that all copies are consistent by sending new values as they are written to every processor with a copy of the line in its cache. The basic trade-off is that write-update leads to lower average latency for reads, but it generates more interprocessor communication traffic. In bus-based multiprocessors, the trade-off has traditionally been resolved in favor of write-invalidate [44, 63], because the interconnection network (the bus) is a scarce resource.

In large-scale multiprocessors with network-based interconnects, the latency for cache misses on read accesses is the most serious impediment to parallel program performance. Write update helps reduce this latency at the expense of higher network traffic. Depending on the amount of bandwidth available in the system and the sharing pattern exhibited by the program, write-update may yield better performance than write-invalidate. Hybrid protocols that choose between the two mechanisms dynamically have also been suggested [72] and have been shown to provide performance advantages over each of the individual mechanisms in isolation. The question of which is the best coherence mechanism for large-scale multiprocessors is still a topic of active research.

Techniques that tolerate (hide) latency can serve to tilt the balance toward write-invalidate protocols. Examples of such techniques include aggressive data prefetching [60] and the use of *micro-tasking* or *multi-threaded* processors [2, 5], which switch contexts on a cache miss. Extensive discussion of the issues encountered in designing coherence protocols can be found in several surveys [10, 38, 55, 71].

The choice of cache line size also has a significant impact on the efficiency of coherence mechanisms. Long cache lines amortize the cost of cache misses by transferring more data per cache miss in the hope of reducing the miss rate. In this sense, they constitute a form of hardware-initiated prefetching. At the same time, large cache lines can lead to *false sharing* [14, 27, 30] in which processors incur coherence overhead due to accesses to non-overlapping portions of a line. False sharing results in both higher miss rates and useless coherence traffic. Conventional wisdom has historically favored short cache lines (8–32 bytes) for multiprocessor systems, but the evolution of compiler techniques for partitioning data among cache lines, and the effort by programmers to produce programs with good locality, will likely allow future machines to use longer cache lines safely.

24.3 User-Level Memory Models

24.3.1 Shared memory vs. message passing

Memory models are an issue not only at the hardware level but also at the programmer level. The model of memory presented to the user can have a significant impact on the amount of effort required to produce a correct and efficient parallel program. Depending on system software, the programmer-level model may or may not resemble the hardware-level model.

We have seen that hardware memory architectures can be divided into *message-passing* and *shared-memory* classes. In a very analogous way, programmer-level memory models can be divided into message-passing and shared-memory classes as well. In a shared-memory model, processes can access variables irrespective of location (subject to the scoping rules of the programming language) simply by using the variables' names. In a message-passing model, variables are partitioned among the instances of some language-level abstraction of a processor. A process can access directly only those variables located on its own processor and must send messages to processes on other processors to access other variables.

There is a wide variety of programming models in both the message-passing and shared memory classes. In both cases, models may be implemented

- via library routines linked into programs written in a conventional sequential language,
- via features added to some existing sequential language by means of preprocessors or compiler extensions of various levels of complexity, or
- via special-purpose parallel or distributed programming languages.

Library packages have the advantage of portability and simplicity but are limited to a subroutine-call interface and cannot take advantage of compiler-based static analysis of program control and data flow. Languages and language extensions can use more elaborate syntax and can implement compile-time optimizations. Because message passing is based on *send* and *receive* operations, which can for the most part be expressed as subroutine calls, library-based implementations of message-passing models have tended to be more successful than library-based implementations of shared-memory models. The latter are generally forced to access all shared data indirectly through pointers obtained from library routines.

Andrews and Schneider [8] provide an excellent introduction to parallel programming models; a more recent book by Andrews [9] provides additional detail. Bal, Steiner, and Tanenbaum provide a survey of message-passing programming models [11]. A technical report by Cheng [23] surveys a large number of programming models and tools in use in the early 1990s. Probably the most widely used library-based shared-memory model is a set of macros for Fortran and C developed by researchers at Argonne National Laboratory [15]. Widespread experience with library-based message-passing models has led to efforts to standardize on a single interface [26].

It is widely believed that shared-memory programming models are easier to use than message-passing models. This belief is supported by the dominance of (small-scale) shared-memory multiprocessors in the market, and by the many efforts by compiler writers [32, 68, 56] and operating system developers [19, 62] to provide a shared-memory programming model on top of message-passing hardware. We focus on shared-memory models in the remainder of this chapter.

24.3.2 Implementation of shared-memory models

Any user-level parallel programming model must deal with several issues. These include

- specifying the computational tasks to be performed and the data structures to be used
- identifying data dependences among tasks

- allocating tasks to processes, and scheduling and synchronizing processes on processors, in a way that respects the dependences
- determining (which copies of) which data should be located at which processors at which points in time so that processes have the data they need when they need it
- arranging for communication among processors to effect the location decisions

Programming models and their implementations differ greatly in the extent to which the user, compiler, run-time system, operating system, and hardware are responsible for each of these issues. The programmer's view of memory is intimately connected to the final two issues, and more peripherally to all of the others.

At one extreme, some programming models make the user responsible for all aspects of parallelization and data placement. Message-passing models fall in this camp, as do some models developed for non-cache-coherent machines. Split-C [25], for example, provides a global namespace for C programs on multicomputers, with a wealth of mechanisms for data placement, remote load/store, prefetch, bulk data transfer, and so on. Other simple models [e.g., Sun's LWP (Light Weight Processes) and OSF's pthreads] require the user to manage processes explicitly, but rely on hardware cache coherence for data placement.

At the next level of implementation complexity, several models employ optimizing compilers to reduce the burden of process management. Early work on parallel Fortran dialects [65] focused on the problem of restructuring loops to minimize data dependences and maximize the amount of work that could be performed in parallel. These dialects were primarily intended for vector supercomputers with an UMA memory model at both the user and hardware levels. Gelernter and Carriero's work on *Linda* has focused on minimizing communication costs for an explicitly-parallel language in which processes communicate via loads and stores in an associative global *tuple space* [18]. More recently, several groups have addressed the goal of increasing processor locality on cache-coherent machines by adopting scheduling techniques that attempt to place processes close to the expected current location of the data they access most often [57].

For large-scale scientific computing, where multicomputers dominate the market, much attention recently has been devoted to the development of efficient shared-memory programming models. Most notably, many research groups have cooperated in the development of *High Performance Fortran* (HPF) [56]. HPF is intended not as an ideal language but as a common starting point for future developments. To first approximation, it combines the syntax of Fortran-90 (including operations on whole arrays and slices) with the data distribution and alignment concepts developed in Fortran-D [42]. Users specify which elements of which arrays are to be located on which processors and which loops are to be executed in parallel. The compiler then bases its parallelization on the *owner computes* rule, which specifies that the computations on the right-hand side of an assignment statement are performed on the processor at which the variable on the left-hand side of the assignment is located. Similar approaches are being taken in the development of pC++, a parallel C++ dialect [32], and Jade, a parallel dialect of C [68].

Many alternatives and extensions to the owner-computes approach are currently under development. Li and Pingali, for example, have addressed the goal of restructuring loops and assigning computations to processors in order to maximize data locality and minimize communication on machines with a single physical address space [54]. Saltz et al. have developed code generation techniques that defer dependence analysis to the execution of loop prologs at run time to maximize parallelism and balance load in programs whose dependence patterns are determined by input data [69].

Ideally, programmers would presumably prefer a programming model in which *both* parallelization *and* locality were managed by the compiler, with no need for the user to specify process creation, synchronization, or data distribution. As of 1994, no system approached this ideal. It is possible that full automation of process and locality management will require that users program in a non-imperative style [17].

24.3.3 Performance of shared-memory models

While bus-based shared memory machines dominate the market for general-purpose multiprocessors, message-passing machines have tended to dominate the market for large-scale high-performance machines. There are two main reasons for this contrast. First, large-scale shared-memory multiprocessors are much harder to build than either their small-scale bus-based counterparts or their large-scale message-based competitors. Second, widespread opinion (supported by some good research—see e.g. [61]) holds that the shared-memory programming paradigm is inefficient, even on machines that provide hardware support for it. We believe, however, that the observed performance problems of shared memory models are actually artifacts of the way in which those models have been used and are not intrinsic to shared memory itself. Specifically:

- Shared-memory models have evolved in large part out of concurrent (*quasi-parallel*) computing on uniprocessors, in which it is natural to associate a separate process with each logically distinct piece of work. To the extent, then, that shared-memory programs are written with a large number of processes, they will display high overhead for process creation and management. This overhead can be minimized by adopting a programming style with fewer processes or by using more lightweight thread management techniques [6].
- The shared-memory programming style requires use of explicit synchronization. Naive implementations of synchronization based on busy-waiting can perform extremely badly on large machines. Recent advances in both hardware and software synchronization techniques have eliminated the need to spin on non-local locations, making synchronization a much less significant factor in shared-memory program performance [1, 58].
- Communication is a dominant source of overhead in parallel programs. Communication in shared-memory models is implicit and happens as a side-effect of ordinary references to variables. It is consequently easy to write shared-memory programs that have very bad locality and that generate large amounts of communication as a result. This communication was much less costly in early shared-memory machines than it is today, leading many shared-memory programmers to assume incorrectly that locality was not a concern. This should be viewed as an indictment not of shared memory but of the naive use of shared memory. Recent studies have shown that, when programmed in a locality conscious way, shared memory can provide performance equal to or better than that of message passing [51, 57].

Reducing communication is probably the most important task for a programmer/compiler that aims to get good performance out of a shared memory parallel program. Improving locality is a multidimensional problem that is a topic of active research. Some of its most important aspects (but by no means the only ones) are:

Mismatch between the program data structures and the coherence units. If data structures are not aligned with the coherence units, accesses to unrelated data structures will cause unnecessary communication. This effect is known as *false sharing* [14, 27, 30] and can be addressed using sophisticated compilers, advanced coherence protocols, or careful data allocation and alignment by the programmer.

- *Relative location of computational tasks and the data they access.* Random placement of tasks (as achieved by centralized work queues) results in very high communication rates. Programmers/compilers should attempt to place all tasks that access the same data on the same processor to take advantage of processor locality.
- *Poor spatial locality.* In some cases, programs access array data with non-unit strides. In that case, a potentially large amount of data fetched in the cache remains untouched. Blocking techniques and algorithmic restructuring can often help alleviate this problem.
- *Poor temporal locality.* Data is invalidated or evicted from a processor's cache before it is touched again. As in the previous case, restructuring the program can help improve its access pattern and, as a consequence, its temporal locality.
- *Conflicts between data structures.* Data structures that map into the same cache lines will cause this problem in caches with limited associativity. If these data structures are used in the same computational phase, performance can suffer severe degradation. Skewing or relocating data structures can help to solve this problem.
- *Communication intrinsic in the algorithm.* Such communication is necessary for correctness and cannot be removed. However, it may be tolerated if data is pre-fetched before it is needed, or if the processor is able to switch to other computation while a miss is being serviced.ors. S

One might argue that if considerable effort is going to be needed to tune programs to run well under a shared-memory programming model, then the conceptual simplicity argument in favor of shared memory is severely weakened. This is not the case, however, since tuning effort is required only for the computationally intensive parts of a shared-memory program, whereas message passing would require similar effort throughout the program text, including initialization, debugging, error recovery, and statistics gathering and printing. Furthermore, most people find it easier to write a correct program and then spend effort in refining and tuning it, as opposed to writing a correct and efficient program from the outset. The principal argument for shared memory is that it provides *referential transparency:* everything can be referenced with a common syntax, and a simple and naive programming style can be used in non-performance-critical sections of code.

24.4   Memory Consistency Models

Shared-memory coherence protocols, whether implemented in hardware or in software, must resolve conflicting accesses—accesses to the same coherence block that are made by different processors at approximately the same time. The precise semantics of this conflict resolution determine the *memory consistency model* for the coherence protocol. Put another way, the memory consistency model determines the values that may be returned by read operations in a given set of parallel reads and writes. The goal of most parallel system architects has been to exhibit behavior as close as possible to that of sequential machines. Therefore, the model of choice has traditionally been *sequential consistency* [50] (see definition later in the section). Unfortunately, sequential consistency imposes a strict ordering on memory access operations and precludes many potentially valuable performance optimizations in coherence protocols. Relaxing the constraints of sequential consistency offers the opportunity to achieve significant performance gains in parallel programs. The memory models discussed in this section achieve these gains at the expense of a slightly more complicated programming model.

Several attributes of memory references can be considered when defining a consistency model. These attributes include the locations of the data and the accessing processor, the direction of access (read, write, or both), the value transmitted in the access, the causality of the access (what other operations caused control to reach this point and produced the value(s) involved), and the *category* of the access [59]. Non-uniform or *hybrid* consistency models distinguish among memory accesses in different categories; *uniform* models do not.

Of the uniform memory models the two most important are *sequential consistency* [50] and *processor consistency* [37]. Others include atomic consistency, causal consistency, pipelined RAM, and cache consistency. A concise definition of all memory models mentioned in this section can be found in a technical report by Mosberger [59].

- *Sequential consistency.* A memory system is sequentially consistent if the result of any execution is the same as if the operations of all the processors were executed in some sequential order, and the operations of each individual processor appear in this sequence in the order specified by its program. In simpler words, this definition requires all processors to agree on the order of memory events. If, for example, a memory location is written by two different processors, then all processors must agree as to which of the writes happened first.
- *Processor consistency.* A memory system is processor consistent if all the reads issued by a given processor appear to all other processors to occur before any of the given processor's subsequent reads and writes, and all the writes issued by a given processor appear to all other processors to occur before any of the given processor's subsequent writes. Processors can disagree on the order of reads and writes by different processors, and a write can be delayed past some of its processor's subsequent reads.[*] (*Subsequent* here refers to the ordering specified by the processor's program.)

Figure 24.4 shows a code segment that has a potential outcome under processor consistency that is impossible under sequential consistency. Under sequential consistency, regardless of the interleaving of instructions, variable A, variable B, or both will have the value 10, so at least one 10 will be printed. Under processor consistency, however, processors do not have to agree on the order of writes by different processors. As a result, both processors may fail to see the other's update in our example, so that neither prints a 10. While processor consistency seems unintuitive, it is easier to implement than sequential consistency, and it allows several important performance optimizations. Most programs written with the sequential consistency model in mind execute correctly under processor consistency.

By constraining the order in which memory accesses can be observed by other processors, sequential and processor consistency force a processor to stall until its previous accesses have completed.[†] This prevents or severely limits the pipelining of memory

Processor 0	Processor 1
X = 10	Y = 10
A = Y	B = X
Print A	Print B

Figure 24.4 Code segment that may yield different results under sequential and processor consistency

---

[*]This definition is from the Stanford Dash project [33]. Processor consistency was originally defined informally by Goodman and has been formalized differently by other researchers.

requests, sacrificing a potentially significant performance improvement. More relaxed consistency models solve this problem by taking advantage of the fact that user programs do not depend on the coherence protocol alone to impose an ordering on the operations of their programs. Even with the strictest of consistency models (i.e., sequential consistency) there are still too many valid interleavings of a parallel execution. Higher-level synchronization primitives are therefore used to impose a desired ordering. Most relaxed consistency models distinguish between regular and synchronization accesses, and sometimes between different categories of synchronization accesses. They then impose different ordering constraints on the accesses in different categories. Figure 24.5 shows a possible categorization of memory accesses [59]. Other categorizations can be found in other papers [33, 35].

Relaxed consistency models can be viewed either from an architectural point of view (What does the coherence protocol do?) or from the programmer's point of view (What does the program have to do to ensure the appearance of sequential consistency?). The two most important relaxed consistency models, from the architectural point of view, are *weak consistency* [28] and *release consistency* [33]. The most important models from the programmer's point of view are DRF0 (data-race-free 0), DRF1, and PLpc (properly-labeled processor consistent) [35]. Other relaxed models include TSO (total store ordering), PSO (partial store ordering), and entry consistency.

- *Weak consistency.* A system is weakly consistent if (a) accesses to synchronization variables are sequentially consistent, (b) accesses to synchronization variables are issued (made visible to other processors) only after all of the local processor's previous data accesses have completed, and (c) accesses to data are issued only after all of the local processor's previous synchronization accesses have completed.
- *Release consistency.* A system is release consistent if (a) accesses to data are issued only after all of the local processor's previous acquire accesses have completed, (b) release accesses are issued only after all previous data accesses have completed, and (c) special (synchronization) accesses are processor consistent.

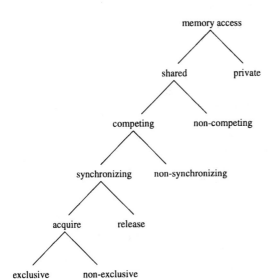

Figure 24.5 Categories of memory references

---

†A read access is considered completed when no future write can change the value returned by the read; a write access is considered completed when all future reads will return the value of this or a future write.

Both weak and release consistency violate the constraints of sequential consistency as shown in Fig. 24.4. The difference between the relaxed models and processor consistency stems from the ordering of writes. Under processor consistency writes by a single processor must be observed by all processors in the order they were issued. Under the weaker models, writes that occur between a matched pair of synchronization operations can be observed in any order as long as they all complete before the second operation of the pair. This relaxation allows a given processor's writes to be pipelined and their ordering to be changed. Release consistency improves on weak consistency by allowing acquire accesses to occur immediately (regardless of what data accesses may be outstanding at the moment) and by allowing data accesses to proceed even if there are release accesses outstanding.

The programmer-centric consistency models take the form of a contract between the programmer and the coherence system. The contract defines a notion of *conflicting accesses* and specifies rules that determine a partial order on the operations in a program. A program obeys the contract if, for every pair of conflicting accesses, one access comes before the other in the partial order. A coherence system obeys the contract if it provides the appearance of sequential consistency to programs that obey the contract.

DRF0 says that two accesses conflict if they access the same location, and if at least one of them is a write. It defines an order on a pair of accesses if they are issued by the same processor, or if they are synchronization references to the same location (or if there is a transitive chain of such orderings between them). DRF1 extends DRF0 by distinguishing between acquire and release operations. Synchronization reference A precedes synchronization reference B iff A is a release operation, B is an acquire operation, and A and B are *paired* (e.g., B returns a value written by A). PLpc extends DRF1 by further distinguishing *loop* and *non-loop* synchronization accesses. Informally, the final read in a busy-wait loop is ordered with respect to the write that changed the value, but earlier reads in the loop are not ordered with respect to that write, or with respect to previous loop-terminating writes made by the spinning processor.

24.5   Implementation and Performance of Memory Consistency Models

The implementation of memory consistency models presents architects with several design decisions:

- Should coherence be implemented in hardware, software, or some combination of the two?
- What should be the size of coherence blocks?
- Should copies of data be updated or invalidated on a write?
- Should the updates/invalidates be performed in an eager or lazy fashion?

While all combinations of answers to the above questions are possible, the choice between hardware and software has a strong influence on the remaining questions. For this reason, we will discuss hardware and software implementation of consistency models separately.

24.5.1   Hardware implementations

The coherence unit of choice under a hardware implementation is usually a cache line. (Cache line sizes currently vary between about 32 and 256 bytes.) Existing hardware systems use an *eager* coherence protocol, performing updates or invalidations as soon as inconsistencies arise. A *lazy* protocol would delay updates or invalidations until an inconsistency might be detected by the user program. It would require memory and logic that generally have been considered too expensive to implement in hardware. The choice between updating and invalidating has generally been made in favor of invalidation, due to its lesser amount of communication (see Section 24.2), but recent studies have shown that hybrid protocols using an update mechanism on some program data may provide performance benefits [72].

Hardware coherence protocols have been implemented with a wide range of consistency models. The differences between the models are mainly seen in the design and use of write buffers.[*] Specifically, the consistency model dictates answers to the following questions:

- Can reads bypass writes?
- Can writes be pipelined?
- Can writes that belong to the same cache line be merged?

Sequential consistency does not allow any of these optimizations, since the concept of sequential execution requires that each operation complete before the next one can be started.

All of the relaxed consistency models described in Section 24.4 allow reads to bypass writes. That is, they allow a read instruction to complete (e.g. from the cache) before one or more previous write instruction(s) have completed. The non-uniform models relax the constraints even further by allowing writes between synchronization accesses to be pipelined and merged. The models that distinguish between different categories of synchronization accesses allow some of those accesses to be pipelined with neighboring data accesses.

Pipelining of writes means more than simply requesting another cache line before a previous one has arrived from memory:[†] it also means temporally overlapping the coherence protocol messages required to obtain writable copies of the lines. In an invalidate-based protocol, a processor performing a write must obtain an exclusive copy of the relevant line. If the protocol is sequentially or processor consistent, then acknowledgments must be received from all former holders of copies of the line before invalidations are sent out for lines written by any subsequent instructions; otherwise, different processors might see the invalidations, and hence the writes, in different orders. Non-uniform relaxed models allow a processor to retire a write from its write buffer once all the invalidation messages have been sent, and to issue a subsequent write before acknowledgments of the earlier invalidations have been received. Table 24.3 summarizes the differences in behavior of the four most common consistency models encountered in the literature. A more detailed version of the table, along with a complete description of the behavior of hardware implementations of different memory consistency models, can be found in a paper by Gharachorloo et al. [34].

Allowing reads to bypass writes is crucial for achieving good performance, and all relaxed memory models (processor consistency, weak consistency, release consistency, etc.)

---

[*]A write buffer allows a processor to continue executing immediately after a write miss. It holds the written data until the appropriate line can be retrieved. A several-entry-deep write buffer can hide the performance impact of a burst of write instructions.

[†]Note that cache lines generally must be fetched from memory on a write miss because only one part of the line is being written at the moment, and the entire line will be written back (under a write-back policy) when it is again evicted.

provide significant performance benefits over sequential consistency. The pipelining of writes and the ability to overlap some of the synchronization operations with data accesses is of secondary importance, and the benefits depend on specific program patterns that are not very common in practice. The ability to merge writes provides performance improvements mainly for update protocols. It allows one to collect writes to the same line and then send them in one operation. Such an approach reduces interconnect traffic and, therefore, the potential latency of other operations.

TABLE 24.3 Implementation Issues of Consistency Models (Reproduced with permission from Ref. [34])

Action	SC	PC	WC	RC
1. Read	Processor stalls for pending writes to perform (or, in very aggressive implementations, to gain ownership). Processor issues read and stalls for read to perform.	Processor issues read and stalls for read to perform. *Note:* Reads are allowed to bypass pending writes.	Processor issues read and stalls for read to perform. *Note:* Reads are allowed to bypass pending writes. For interaction with pending releases, see point 4.	Processor issues read and stalls for read to perform. *Note:* Reads are allowed to bypass pending writes and releases.
2. Write	Processor sends write to buffer (stalls if write buffer is full). *Note:* Write buffer retires a write only after the write is performed, or in very aggressive implementations, when ownership is gained.		Processor sends write to write buffer (stalls if write buffer is full). *Notes:* Write buffer does not require ownership to be gained before retiring a write. For interactions with acquires/releases, see points 3 and 4.	
3. Acquire	Treated as read	Treated as read	Processor stalls for pending writes and releases to perform. Processor issues acquire and stalls for acquire to perform.	Processor issues acquire and stalls for acquire to perform. *Note:* Processor does not need to stall for pending writes and releases.
4. Release	Treated as write	Treated as write	Processor sends release to write buffer (stalls if write buffer is full). *Notes:* Write buffer cannot retire the release until all previous writes are performed. Write buffer stalls for release to perform. Processor stalls at next read after release for release to perform.	Processor sends release to write buffer (stalls if write buffer is full). *Note:* Write buffer cannot retire the release until all previous writes and releases are performed.

The importance of letting reads bypass writes stems from the fact that current processors stall on a read miss. Any latency added to reads is completely lost to the processors. Bypassing becomes less important if processors are able to perform non-blocking loads (i.e., prefetch operations) or to switch to a different context. In this case, pipelining and merging of writes become significantly more important, and the hybrid consistency models provide significant performance advantages over both sequential and processor consistency.

24.5.2   Software implementations

Coherence can be implemented in software in two very different ways: via compiler code generation, as in HPF (see Section 24.3), or via run-time or OS-level observation of program behavior. We focus here on the *behavior-driven* approach, which is analogous to the functioning of hardware cache coherence.

Most behavior-driven software coherence schemes are the intellectual descendants of Li and Hudak's Ivy [53]. Ivy was designed to run on a network of workstations, and it uses conventional virtual memory protection bits to implement a simple single-writer sequentially consistent coherence scheme for pages. *Uncached* pages are marked invalid in the (per processor) TLB and page table. On a page fault, the OS interrupt handler (or user-level signal handler) uses message passing to obtain a copy of the page and place it in local memory, possibly invalidating other copies, and modifying the page table to make the new copy accessible.

Most systems since Ivy have abandoned sequential consistency and exclusive writers, but continue to implement consistency at the level of pages, using VM hardware. These systems are generally referred to as *DSM* (distributed shared memory) or *SVM* (shared virtual memory) emulations. They have been implemented both on networks of workstations and on more tightly coupled NORMA machines. Recent work suggests that it may be feasible to implement finer-grain software coherence by exploiting special hardware features (e.g. error-correcting codes) or by automatically modifying programs to perform in-line coherence checks prior to accessing memory [70, 74].

While an eager hardware protocol can run in parallel with the execution of the application, a software protocol must compete with the application for processor cycles. An eager software protocol therefore incurs as much overhead per transaction as a lazy protocol and may introduce a larger number of transactions [45]. As a result, most recent DSM systems use lazy protocols. The choice of update versus invalidate is less clear-cut in software than it is in hardware protocols; the large (page sized) coherence blocks imply that reacquiring an invalidated block can be very expensive, whereas updating it a small piece at a time can be comparatively cheap.

Page sized coherence blocks also tend to result in a large amount of false sharing. To minimize the performance impact, recent software systems (e.g. Munin [19]) permit a page to be written concurrently by multiple processors as long as none of them executes a synchronization operation. The assumption is that the processors would use explicit synchronization to protect any truly shared data, so any non-synchronized accesses to the same page must constitute false sharing. At synchronization points, the protocol must determine the set of changes made by each processor and must combine these changes to obtain a new consistent copy of the page. This process usually involves a word-by-word comparison of the modified copies of the page with respect to an unmodified, shadow copy.

The existence of a shared physical address space on NUMA machines permits several optimizations in software coherence systems [64]. Simple directory operations can be performed via (uncached) remote reference rather than by sending a message. Pages can also be mapped remotely so that cache fills bring lines into the local node on demand, eliminating the need for a copy in local memory. If the caches use write-through or write-back of only dirty words, then remote mapping of a single main-memory copy eliminates the need to merge inconsistent copies. It is also possible to map pages remotely and uncached so that each individual reference traverses the interconnection network. This option was heavily used in so-called *NUMA memory management* for older CISC-based NUMA machines without caches [13, 24, 49], and it may still be desirable in unusual cases for pages with very poor processor locality.

The main performance benefits of relaxed consistency models over sequential consistency, when implemented in software, stem from the reduction in coherence transactions due to the removal of the single-writer restriction. Processors perform protocol actions only at synchronization points and thus spend more time doing useful work and less time processing protocol operations. Reduced coherence traffic, in turn, can help alleviate memory and network congestion, thereby speeding up other memory operations. The choice between software and hardware implementation of the different consistency models is a topic of active research. Recent work suggests that software coherence with lazy relaxed consistency can be competitive with the fastest hardware alternatives (eager relaxed consistency). Hardware architects also seem to be concluding that the flexibility available in software can be a significant performance advantage; some designs are beginning to incorporate programmable protocol engines [48, 66].

24.6   Conclusions and Trends

The way we think of memory has an impact at several layers of system design, starting with hardware and spanning architecture, choice of programming model, and choice and implementation of memory consistency model. Three important concepts can help us build cheap, efficient, and convenient memory systems:

1. *Referential transparency.* Shared memory is closer to the sequential world to which most programmers are accustomed. Message passing can make an already difficult task (parallel programming) even harder. Message passing provides performance advantages in some situations, however, and some newer machines and systems support both classes of programming model [12, 46, 48].
2. *The principle of locality.* The importance of locality is evident throughout the memory hierarchy, with page mode DRAMs at the hardware level, hierarchical memory systems at the memory architecture level, and NUMA programming models at the user level. Taking advantage of locality can help us build cheap memory systems, and having locality principles in mind when programming can help us get good performance out of the systems we build.
3. *Sequentially consistent ordering of events.* When imposed on arbitrary parallel programs, sequential consistency precludes a large number of hardware/software optimizations. Weaker models allow programs to run faster but make programming harder. A promising way to deal with the complexity is to adopt programmer-centric consistency models, which guarantee sequential consistency when certain synchronization rules are obeyed.

Effective memory system design and usage is probably the most active research topic in multiprocessors today. Several trends can be discerned. Deeper memory hierarchies are likely, with two-level on-chip caches and the possibility of a third level off chip. Locality will become even more important as faster superscalar processors increase the cost of communication relative to computation. Current compiler research focuses on techniques that can improve the temporal and spatial locality of programs by restructuring the code that accesses data. The advent of programmable cache controllers may allow us to customize protocols and models to the needs of the program at hand, rather than imposing a system-wide decision. Finally, as the relative cost of memory systems and processors tilts farther toward expensive memory, it may make sense to change our processor-centric view of the world. Processor utilization may become a secondary performance metric, and memory (cache) utilization may start to dominate performance studies. If program execution time depends primarily on memory response time, then peak performance will be achieved when the memory system is fully utilized, regardless of the processor utilization.

24.7 References

1. Aboulenein, N. M., J. R. Goodman, S. Gjessing, and P. J. Woest. 1994. Hardware support for synchronization in the scalable coherent interface (SCI). *Proceedings of the Eighth International Parallel Processing Symposium,* Cancun, Mexico.

2. A. Agarwal. 1992. Performance tradeoffs in multithreaded processors. *IEEE Transactions on Parallel and Distributed Systems,* 3(5):525–539.

3. Agarwal, A., et al. 1992. The MIT Alewife Machine: a large-scale distributed memory multiprocessor. In *Scalable Shared Memory Multiprocessors,* M. Dubois and S. S. Thakkar, eds. Kluwer Academic Publishers.

4. Alverson, R., D. Callahan, D. Cummings, B. Koblenz, A. Porterfield, and B. Smith. 1990. The Tera computer system. *Proceedings 1990 ACM International Conference on Supercomputing,* 1–6, Amsterdam, The Netherlands, and *ACM SIGARCH Computer Architecture News,* 18:3.

5. Alverson, G. A., R. Alverson, D. Callahan, B. Koblenz, A. Porterfield, and B. Smith. 1992. Exploiting heterogeneous parallelism on a multithreaded multiprocessor. *Proceedings 1992 ACM International Conference on Supercomputing,* Washington, D.C.

6. Anderson, T. E., E. D. Lazowska, and H. M. Levy. 1989. The performance implications of thread management alternatives for shared-memory multiprocessors. *IEEE Transactions on Computers,* 38(12):1631–1644.

7. Anderson, J. M., and M. S. Lam. 1993. Global optimizations for parallelism and locality on scalable parallel machines. *Proceedings of the SIGPLAN '93 Conference on Programming Language Design and Implementation,* Albuquerque, N.M.

8. Andrews, G. R, and F. B. Schneider. 1983. Concepts and notations for concurrent programming. *ACM Computing Surveys,* 15(1):3–44.

9. Andrews, G. R. 1991. *Concurrent Programming: Principles and Practice.* Redwood City, Calif.: Benjamin/Cummings.

10. Archibald, J., and J. Baer. 1986. Cache coherence protocols: evaluation using a multiprocessor simulation model. *ACM Transactions on Computer Systems,* 4(4):273–298.

11. Bal, H. E., J. G. Steiner, and A. S. Tanenbaum. 1989. Programming languages for distributed computing systems. *ACM Computing Surveys,* 21(3):261–322.

12. Blumrich, M., K. Li, R. Alpert, C. Dubnicki, E. Felten, and J. Sandberg. 1994. Virtual memory mapped network interface for the SHRIMP multicomputer. *Proceedings of the Twenty-First International Symposium on Computer Architecture,* Chicago, Ill.

13. Bolosky, W. J., M. L. Scott, R. P. Fitzgerald, R. J. Fowler, and A. L. Cox. 1991. NUMA policies and their relation to memory architecture. *Proceedings of the Fourth International Conference on Architectural Support for Programming Languages and Operating Systems,* 212–221, Santa Clara, Calif.

14. Bolosky, W. J., and M. L. Scott. 1993. False sharing and its effect on shared memory performance. *Proceedings of the Fourth USENIX Symposium on Experiences with Distributed and Multiprocessor Systems,* 57–71,. Also available as MSR-TR-93-1, Microsoft Research Laboratory.

15. Bomans, L., D. Roose, and R. Hempel. 1990. The Argonne/GMD macros in FORTRAN for portable parallel programming and their implementation on the Intel iPSC/2. *Parallel Computing,* 15:119–132.

16. Bondurand, D. 1992. Enhanced dynamic RAM. *IEEE Spectrum,* 29(10):49.

17. Cann, D. Retire Fortran? a debate rekindled. *Communications of the ACM,* 35(8):81–89. Originally presented at Supercomputing '91.

18. Carriero, N., and D. Gelernter. 1989. Linda in context. *Communications of the ACM,* 32(4):444–458. Relevant correspondence appears in Vol. 32, No. 10.

19. Carter, J. B., J. K. Bennett, and W. Zwaenepoel. 1991. Implementation and performance of Munin. *Proceedings of the Thirteenth ACM Symposium on Operating Systems Principles,* Pacific Grove, Calif., 152–164.

20. Censier, L. M., and P. Feautrier. 1978. A new solution to coherence problems in multicache systems. *IEEE Transactions on Computers,* C-27(12):1112–1118.

21. Chaiken, D., C. Fields, K. Kurihara, and A. Agarwal. 1990. Directory-based cache coherence in large-scale multiprocessors. *Computer,* 23(6):49–58.

22. Chaiken, D., J. Kubiatowicz, and A. Agarwal. 1991. LimitLESS directories: a scalable cache coherence scheme. *Proceedings of the Fourth International Conference on Architectural Support for Programming languages and Operating Systems,* Santa Clara, Calif., 224–234.

23. Cheng, D. Y. 1993. A Survey of Parallel Programming Languages and Tools. RND-93-005. Moffet Field, Calif., NASA Ames Research Center.

24. Cox, A. L., and R. J. Fowler. 1989. The implementation of a coherent memory abstraction on a NUMA Multiprocessor: experiences with PLATINUM. *Proceedings of the Twelfth ACM Symposium on Operating Systems Principles,* Litchfield Park, Ariz., 32–44.

25. Culler, D., A. Dusseau, S. Goldstein, A. Krishnamurthy, S. Lumetta, T. von Eicken, and K. Yelick. 1993. Parallel programming in Split-C. *Proceedings Supercomputing '93,* Portland, Ore.

26. Dongarra, J. J., R. Hempel, A. J. G. Hey, and D. W. Walker. 1992. A proposal for a user level, message-passing interface in a distributed memory environment. *ORNL/TM 12231.* Oak Ridge, Tenn.: Oak Ridge National Laboratory.

27. Dubois, M., J. Skeppstedt, L. Ricciulli, K. Ramamurthy, and P. Stenstrom. 1993. The detection and elimination of useless misses in multiprocessors. *Proceedings of the Twentieth International Symposium on Computer Architecture,* San Diego, Calif., 88–97.

28. Dubois, M., C. Scheurich, and F. A. Briggs. 1986. Memory access buffering in multiprocessors. *Proceedings of the Thirteenth International Symposium on Computer Architecture,* 434–442.

29. Eggers, S. J., and R. H. Katz. 1989. Evaluation of the performance of four snooping cache coherency protocols. *Proceedings of the Sixteenth International Symposium on Computer Architecture,* 2–15.

30. Eggers, S. J., and R. H. Katz. 1989. The effect of sharing on the cache and bus performance of parallel programs. Proceedings *of the Third International Conference on Architectural Support for Programming Languages and Operating Systems,* Boston, Mass., 257–270.

31. Farmwald, M., and D. Mooring. 1992. A fast path to one memory. *IEEE Spectrum,* 29(10):50–51.

32. Gannon, D., and J. K. Lee. 1991. Object oriented parallelism: pC++ ideas and experiments. *Proceedings of the Japan Society for Parallel Processing,* 13–23.

33. Gharachorloo, K., D. Lenoski, J. Laudon, P. Gibbons, A. Gupta, and J. L. Hennessy. 1990. Memory consistency and event ordering in scalable shared-memory multiprocessors. *Proceedings of the Seventeenth International Symposium on Computer Architecture,* Seattle, Wash., 15–26.

34. Gharachorloo, K., A. Gupta, and J. Hennessy. 1991. Performance evaluation of memory consistency models for shared memory multiprocessors. *Proceedings of the Fourth International Conference on Architectural Support for Programming Languages and Operating Systems,* Santa Clara, Calif., 245–257.

35. Gharachorloo, K., S. V. Adve, A. Gupta, J. L. Hennessy, and M. D. Hill. 1992. Programming for different memory consistency models. *Journal of Parallel and Distributed Computing,* 15:399–407.

36. Goodman, J. R. 1983. Using cache memory to reduce processor/memory traffic. *Proceedings of the Tenth International Symposium on Computer Architecture,* 124–131.

37. Goodman, J. R. 1991. Cache consistency and sequential consistency. *Computer Sciences Technical Report #1006,* University of Wisconsin, Madison.

38. Gupta, A., J. Hennessy, K. Gharachorloo, T. Mowry, and W. Weber. 1991. Comparative evaluation of latency reducing and tolerating techniques. *Proceedings of the Eighteenth International Symposium on Computer Architecture,* Toronto, Canada, 254–263.

39. Gupta A., and W. Weber. 1992. Cache invalidation patterns in shared-memory multiprocessors. *IEEE Transactions on Computers,* 41(7):794–810.

40. Hagersten, E., A. Landin, and S. Haridi. 1992. DDM—a cache-only memory architecture. *Computer,* 25(9):44–54.

41. Hennessy, J. L., and D. A. Patterson. 1990. *Computer Architecture: A Quantitative Approach.* San Mateo, Calif.: Morgan Kaufmann.

42. Hiranandani, S., K. Kennedy, and C. Tseng. 1992. Compiling Fortran D for MIMD distributed-memory machines. *Communications of the ACM,* 35(8):66–80. Originally presented at Supercomputing '91.

43. James, D. V., A. T. Laundrie, S. Gjessing, and G. S. Sohi. 1990. Scalable coherent interface. *Computer,* 23(6):74–77.

44. Katz, R., S. Eggers, D. A. Wood, C. Perkins, and R. G. Sheldon. 1985. Implementing a cache consistency protocol. *Proceedings of the Twelfth International Symposium on Computer Architecture,* Boston, Mass., 276–283.

45. Keleher, P., A. Cox, and W. Zwaenepoel. 1992. Lazy consistency for software distributed shared memory. *Proceedings of the Nineteenth International Symposium on Computer Architecture,* Gold Coast, Australia, 13–21.

46. Koch, P. T., and R. Fowler. 1994. Integrating message-passing with lazy release consistent distributed shared memory. *Proceedings of the First Symposium on Operating Systems Design and Implementation,* Monterey, Calif.

47. Kuck, D. J., E. S. Davidson, D. H. Lawrie, and A. H. Sameh. 1986. Parallel supercomputing today and the Cedar approach. *Science,* 231:967–974.

48. Kuskin, J., D. Ofelt, M. Heinrich, J. Heinlein, R. Simoni, K. Gharachorloo, J. Chapin, D. Nakahira, J. Baxter, M. Horowitz, A. Gupta, M. Rosenblum, and J. Hennessy. 1994. The FLASH multiprocessor. *Proceedings of the Twenty-First International Symposium on Computer Architecture,* Chicago, Ill.

49. LaRowe, Jr., R. P., and C. S. Ellis. 1991. Experimental comparison of memory management policies for NUMA multiprocessors. *ACM Transactions on Computer Systems,* 9(4):319–363.

50. Lamport, L. 1979. How to make a multiprocessor computer that correctly executes multiprocess programs. *IEEE Transactions on Computers,* C-28(9):241–248.

51. LeBlanc, T. J., and E. P. Markatos. 1992. Shared memory vs. message passing in shared memory multiprocessors. *Proceedings of the Fourth IEEE Symposium on Parallel and Distributed Processing,* Dallas, Tex., 254–263.

52. Lenoski, D., J. Laudon, K. Gharachorloo, W. Weber, A. Gupta, J. Hennessy, M. Horowitz, and M. S. Lam. 1992. The Stanford Dash multiprocessor. *Computer,* 25(3):63 79.

53. Li, K., and P. Hudak. 1989. Memory coherence in shared virtual memory systems. *ACM Transactions on Computer Systems,* 7(4):321–359. Originally presented at the Fifth ACM Symposium on Principles of Distributed Computing, August 1986.

54. Li, W., and K. Pingali. 1993. Access normalization: loop restructuring for NUMA compilers. *ACM Transactions on Computer Systems,* 11(4):353–375. Earlier version presented at the Fifth International Conference on Architectural Support for Programming Languages and Operating Systems, October 1992.

55. Lilja, D. J. 1993. Cache coherence in large-scale shared-memory multiprocessors: issues and comparisons. *ACM Computing Surveys,* 25(3):303–338.

56. Loveman, D. B. 1993. High performance Fortran. *IEEE Parallel and Distributed Technology,* 1(1):25–42.

57. Markatos, E. P., and T. J. LeBlanc. 1994. Using processor affinity in loop scheduling on shared-memory multiprocessors. *IEEE Transactions on Parallel and Distributed Systems,* 5(4):379–400. Earlier version presented at Supercomputing '92.

58. Mellor-Crummey, J. M., and M. L. Scott. 1991. Algorithms for scalable synchronization on shared-memory multiprocessors. *ACM Transactions on Computer Systems,* 9(1):21–65.

59. Mosberger, D. 1993. Memory Consistency Models. *ACM SIGOPS Operating Systems Review,* 27(1):18–26. Relevant correspondence appears in Vol. 27, No. 3; revised version available Technical Report 92/11, Department of Computer Science, University of Arizona.

60. Mowry, T., and A. Gupta. 1991. Tolerating latency through software-controlled prefetching in shared-memory multiprocessors. *Journal of Parallel and Distributed Computing,* 12(2):87–106.

61. Ngo, T., and L. Snyder. 1992. On the influence of programming models on shared memory computer performance. *Proceedings of the 1992 Scalable High Performance Computing Conference,* Williamsburg, Va.

62. Nitzberg, B., and V. Lo. Distributed shared memory: a survey of issues and algorithms. *Computer,* 24(8): 52–60.

63. Papamarcos, M., and J. Patel. 1984. A low overhead coherence solution for multiprocessors with private cache memories. *Proceedings of the Eleventh International Symposium on Computer Architecture,* 348–354.

64. Petersen, K., and K. Li. 1993. Cache coherence for shared memory multiprocessors based on virtual memory support. *Proceedings of the Seventh International Parallel Processing Symposium,* Newport Beach, Calif.

65. Polychronopoulos, C. D., et al. 1989. Parafrase-2: A multilingual compiler for optimizing, partitioning, and scheduling ordinary programs. *Proceedings of the 1989 International Conference on Parallel Processing,* St. Charles, Ill.

66. Reinhardt, S. K., J. R. Larus, and D. A. Wood. 1994. Tempest and Typhoon: user-level shared-memory. *Proceedings of the Twenty-First International Symposium on Computer Architecture,* Chicago, Ill., 325–336.

67. Rettberg, R. D., W. R. Crowther, P. P. Carvey, and R. S. Tomlinson. 1990. The Monarch parallel processor hardware design. *Computer,* 23(4):18–30.

68. Rinard, M. C., D. J. Scales, and M. S. Lam. 1993. Jade: a high-level machine-independent language for parallel programming. *Computer,* 26(6):28–38.

69. Saltz, J. H., R. Mirchandaney, and K. Crowley. 1991. Run-time parallelization and scheduling of loops. IEEE *Transactions on Computers,* 40(5):603–612.

70. Schoinas, I., B. Falsafi, A. R. Lebeck, S. K. Reinhardt, J. R. Larus, and D. A. Wood. 1994. Fine-grain access control for distributed shared memory. *Proceedings of the Sixth International Conference on Architectural Support for Programming Languages and Operating Systems,* San Jose, Calif.

71. Stenstrom, P. 1990. A survey of cache coherence schemes for multiprocessors. *Computer,* 23(6) :12–24.

72. Veenstra, J. E., and R. J. Fowler. 1992. A performance evaluation of optimal hybrid cache coherency protocols. *Proceedings of the Fifth International Conference on Architectural Support for Programming Languages and Operating Systems,* Boston, Mass., 149–160.

73. Vranesic, Z. G., M. Stumm, D. M. Lewis, and R. White. 1991. Hector: a hierarchically structured shared-memory multiprocessor. *Computer,* 24(1):72–79.

74. Zekauskas, M. J., W. A. Sawdon, and B. N. Bershad. 1994. Software write detection for distributed shared memory. *Proceedings of the First Symposium on Operating Systems Design and Implementation,* Monterey, Calif.

Technologies

25

Heterogeneous Computing

Howard Jay Siegel,[*] John K. Antonio,[*]
Richard C. Metzger,[†] Min Tan,[*] and Yan Alexander Li[*]

A single application task often requires a variety of different types of computation (e.g., operations on arrays versus operations on scalars). Existing supercomputers generally achieve only a fraction of their peak performance on certain portions of such application programs. This is because different subtasks of an application can have very different computational requirements that result in different needs for machine capabilities. In general, it is currently impossible for a single machine architecture to satisfy all the computational requirements of various subtasks in certain applications equally well [36]. Thus, a more appropriate approach for high-performance computing is to construct a heterogeneous computing environment.

25.1 Introduction

A *heterogeneous computing* (*HC*) system provides a variety of architectural capabilities, orchestrated to perform an application whose subtasks have diverse execution requirements. A *mixed-mode* HC system is a single parallel processing machine that is capable of operating in either the synchronous SIMD or asynchronous MIMD mode of parallelism, and can dynamically switch between modes at instruction-level granularity with generally negligible overhead [30]. A *mixed-machine* HC system is a heterogeneous suite of independent machines of different types interconnected by a high-speed network [81]. Unlike mixed-mode machines, switching execution among machines in a mixed-machine system requires measurable overhead because data may need to be transferred among machines. Thus, the mixed-machine systems considered in this chapter are assumed to have high-speed connections among machines that make decomposition at the subtask level feasible.

[*]Supported by Rome Laboratory under contract number F30602-94-C-0022 and by NRaD under subcontract number 20-950001-70. Some of the research discussed used equipment supported by the National Science Foundation under grant number CDA-9015696.
[†]Supported by AFOSR under RL JON 2304F2TK.

Another difference is that, in mixed-machine systems, the set of subtasks may be executed as an ordered sequence and/or concurrently on multiple machines. Mixed-machine HC has also been referred to as *metacomputing* [49, 51].

To fully exploit HC systems, a task must be decomposed into subtasks, where each subtask is computationally homogeneous, and different subtasks may have different machine architectural requirements. The subtasks are then assigned to and executed with the machines (or modes) that will result in a minimal overall execution time for the task. Currently, users typically must specify this decomposition and assignment. One long-term pursuit in the field of heterogeneous computing is to do this automatically.

Figure 25.1 shows a hypothetical example of an application program whose various subtasks are best suited for execution on different machine architectures, i.e., vector, SIMD, MIMD, data-flow, and special purpose [33]. Executing the whole program on a vector supercomputer only gives twice the performance achieved by a baseline serial machine. The vector portion of the program can be executed significantly faster. However, the non-vector portions of the program may only have a slight improvement in execution time due to the mismatch between each subtask's unique computational requirement and the machine architecture being used. Alternatively, the use of five different machines, each matched with the computational requirements of the subtasks for which it is used, can result in an execution 20 times as fast as the baseline serial machine.

A programming language used in an HC environment must be portable. To allow full flexibility of execution targets, the language must be compilable into efficient code for any machine in the mixed-machine suite or any mode available in a mixed-mode machine. Thus, ideally, this portable programming language must be machine/mode-independent and supply the compiler with the information it needs to produce efficient code for different target architectures and/or modes of parallelism. In this chapter, the existence of such a language is assumed. More about this topic is in Ref. [84].

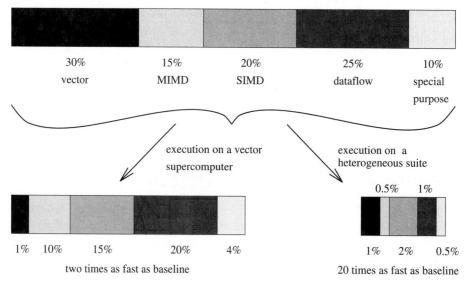

Figure 25.1 A hypothetical example of the advantage of using heterogeneous computing [33] in which the execution time for the heterogeneous suite includes inter-machine communications. Percentages are based on 100% being the total execution time on the baseline serial system but are not drawn to scale.

This chapter is a brief introduction to HC. In Section 25.2, mixed-mode systems are discussed. After Section 25.2, "HC system" will imply "mixed-machine system," as it is more commonly used in that way. Examples of existing mixed-machine systems are presented in Section 25.3. Section 25.4 describes some existing software tools for HC systems. A conceptual model for HC is introduced in Section 25.5. Existing literature that presents explicit frameworks for performing task profiling and analytical benchmarking, a stage in the conceptual model, is overviewed in Section 25.6. In Section 25.7, matching and scheduling techniques for selecting machines for each subtask based on certain cost metrics are overviewed. Finally, open problems in the field of HC are explored in Section 25.8.

25.2 Mixed-Mode Systems

25.2.1 Trade-offs among SIMD, MIMD, and mixed-mode

Two types of parallel processing systems are the *SIMD* (single instruction stream, multiple data stream) machine and the *MIMD* (multiple instruction stream, multiple data stream) machine [31]. Figure 25.2a shows a distributed memory SIMD architecture in which each processor is paired with a memory module to form *N processing elements* (*PEs*). In SIMD mode, there is a single program and the control unit broadcasts instructions of this program in sequence to the *N* PEs. All enabled PEs execute the same instruction (broadcast by the control unit) at the same time, but each PE operates on data from its own local memory and registers. The interconnection network provides inter-PE communication.

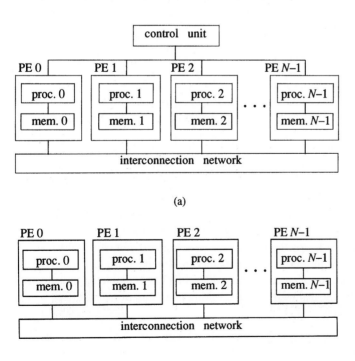

(a)

(b)

Figure 25.2 (a) Distributed memory SIMD machine model and (b) distributed memory MIMD machine model

In an MIMD machine, each PE stores its own instructions and data. Distributed memory MIMD systems are typically structured like SIMD systems without the control unit (see Fig. 25.2b). Each PE executes its own program asynchronously with respect to the other PEs. The use of SIMD and MIMD machines is discussed further in Chapter 16.

There are many trade-offs between SIMD and MIMD machines. The advantages of SIMD mode include:

1. The single instruction stream and implicit synchronization of the PEs make SIMD programs easier to create, understand, and debug.
2. In SIMD mode, the PEs are implicitly synchronized at the instruction level. Explicit synchronization primitives may be required in MIMD mode and generally incur overhead.
3. In SIMD mode, if the PEs communicate through messages, during a given transfer, all enabled PEs send a message to distinct PEs, thereby implicitly synchronizing the "send" and "receive" commands and implicitly identifying the message. MIMD architectures require the overhead of message identification protocols and a scheme to signal when a message has been sent and received.
4. Control flow instructions and scalar operations that are common to all PEs (e.g., computing common local subimage data point addresses) can be executed on the *control unit* (*CU*) while the processors are executing other instructions (this is implementation dependent); this is referred to as *CU/PE overlap* [4, 50].
5. Only a single copy of the instructions needs to be stored in the system memory, thus possibly reducing memory cost and size, allowing for more data storage, and/or reducing communication between primary and secondary memory.
6. Cost is reduced by the need for only a single instruction decoder in the CU (versus one in each PE for MIMD mode).

The advantages of MIMD mode include:

1. MIMD allows different operations to be performed on different PEs simultaneously (i.e., multiple threads of control). Thus, MIMD can support both *functional* (*control*) *parallelism*, where separate and relatively independent processes or functions are assigned to and executed on different sets of processors simultaneously, and *data parallelism*, where the same set of instructions is applied to all the elements in a data set [58, 85]. SIMD is limited to data parallelism.
2. When executing conditional statements (e.g., "if-then-else") based on data local to PEs in MIMD mode, each PE can independently follow either decision path. In SIMD mode, all of the instructions for the "then" block must be broadcast, followed by all of the "else" block, with the appropriate PEs enabled for each block.
3. Consider a sequence of instructions each of whose execution time is data dependent (e.g., a sequence of "while" loops whose bounds are dependent on data local to PEs). In SIMD mode, a PE must wait until all the other PEs have completed an instruction before continuing to the next instruction, resulting in a "sum of maxs" effect:

$$T_{\text{SIMD}} = \sum_{\text{instrs}} \max_{\text{PEs}} \ (\text{instr. time})$$

MIMD mode allows each PE to execute the sequence of instructions independently, resulting in a "max of sums" effect (see Fig. 25.3):

$$T_{\text{MIMD}} = \max_{\text{PEs}} \sum_{\text{instrs}} (\text{instr. time}) \leq T_{\text{SIMD}}$$

4. MIMD machines do not need the SIMD instruction broadcasting hardware.

The trade-offs above are summarized from Refs. [13, 45, 69]. Because both modes have advantages, mixed-mode systems have been proposed. Various algorithm case studies have shown that the use of mixed-mode can outperform the use of a single-mode (e.g., Refs. [39, 67, 79]).

As a simple example, consider the bitonic sorting [7] of sequences on the mixed-mode PASM prototype [30], where L/N numbers are stored in each of N PEs and the numbers within a PE are sorted. The goal is to have each PE contain a sorted list of L/N elements, where each element in PE i is less than or equal to all of the elements in PE k, for $i < k$. The regular bitonic sorting algorithm for $L = N$ is modified for $L > N$ (see Fig. 25.4). An ordered merge is done between the local PE sequence X and the transferred sequence Y using local data conditional statements in merge(X, Y). The lesser half of the merged sequence is assigned the pointer X and the greater half is assigned the pointer Y. These pointers may be swapped by swap(X, Y), based on a precomputed data-independent mask.

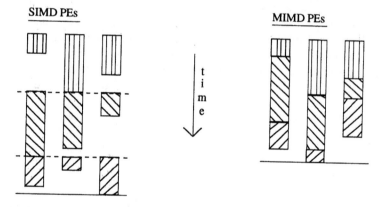

Figure 25.3 *Sum of maxs* vs. *max of sums* effects

---

for $k = 1$ to $\log_2 N$ do

    for $i = 1$ to k do

        { for $q = 1$ to L/N do

            { load $X[q]$ into network

            send to PE whose number differs in bit $(k - i)$

            $Y[q] \leftarrow$ network output }

        merge(X, Y)

        swap(X, Y) }

Figure 25.4 Bitonic sequence-sorting algorithm [30]

The ordered merge involves many comparisons, which can be more efficiently computed in MIMD mode. The innermost loop of the algorithm requires many network transfers, which are better performed in SIMD mode. In a mixed-mode implementation, the ordered merge and swap routines can be executed in MIMD mode, while the other operations are performed in SIMD mode. This approach has an advantage over pure SIMD or pure MIMD mode implementations because all comparisons are done in MIMD mode and all network transfers are done in SIMD mode. Additionally, there is opportunity in SIMD mode for CU/PE overlap. It is shown in Ref. [30] that there is a noticeable improvement in execution time for the mixed-mode implementation as a result of properties inherent to the modes of parallelism.

Most of the advantages of SIMD and MIMD modes can be realized with a mixed-mode architecture. Disadvantages of mixed-mode parallelism include higher hardware cost (because mixed-mode machines must have the hardware needed for both modes), more complicated use (because the mode switching ability adds another dimension of complexity for the programmer), and, when switching from MIMD to SIMD mode, some PEs may remain idle while they wait for the other PEs to reach the switch point (which they may not need to do if only MIMD mode was used) [12]. Very brief descriptions of five existing mixed-mode systems follow, emphasizing the mode-switching mechanism.

25.2.2   Partitionable SIMD/MIMD

PASM is a *PA*rtitionable-*SI*MD/*MI*MD system concept being developed as a design for a large-scale distributed memory dynamically reconfigurable parallel machine [70, 71]. PASM can be dynamically partitioned to form independent mixed-mode submachines of various sizes. PASM uses a fault-tolerant implementation of the flexible multistage cube network [68] (the Extra Stage Cube [1]), for inter-PE communication. Thus, PASM is dynamically reconfigurable along three dimensions: partitionability, mode of parallelism, and connections among PEs. A small-scale proof-of-concept prototype (30 processors, 16 PEs in the computational engine) has been built at Purdue University, in the USA. The prototype is a constantly evolving tool for studying the design and use of reconfigurable parallel machines.

Consider a single submachine. In SIMD mode, a PE fetches SIMD instructions by issuing a read to a reserved segment of the logical address space (that does not correspond to physical PE memory). Each memory access made by a PE's processor is monitored by the *instruction broadcast unit* (IBU). The IBU sends an SIMD instruction request to the control unit, and when all enabled PEs in a submachine have requested a new instruction, it is broadcast from a queue in the control unit. In MIMD, a PE fetches instructions from its local memory. A PE can switch from SIMD mode to an MIMD program located at some address A in its local memory by receiving a "branch to A" instruction in SIMD mode. Similarly, a PE can change from MIMD mode to SIMD mode by executing a branch to the logical SIMD instruction space. Such flexibility in mode switching allows mixed-mode programs to be written that change modes at instruction-level granularity with generally nominal overhead.

25.2.3   Texas Reconfigurable Array Computer

The *Texas Reconfigurable Array Computer* (TRAC) is a dynamically partitionable mixed-mode shared-memory parallel machine which was developed at the University of Texas at Austin, in the U.S.A. [57]. The TRAC prototype consisted of four microprocessors connected to nine memory modules by an SW-Banyan network with fan-out of three, spread of two, and two levels (see Fig. 25.5). *Data trees* connect data memories with their corresponding processors. An *instruction tree* connects a specific program memory with processors to form a virtual SIMD machine. In MIMD mode, each processor can independently fetch its own instructions from a memory module associated with it. Mode switching between SIMD and MIMD is implemented by changing the source of the instructions for the processors.

25.2.4   OPSILA

OPSILA is a limited mixed-mode 16-PE prototype built at the University of Nice, in France [27]. It operates in SIMD and *SPMD* (single program - multiple data stream) mode, a special form of MIMD mode where all the PEs execute the same program in an asynchronous fashion, each on its own data [24]. A synchronous Omega network [53] (a member of the multistage cube family [68]) is used for inter-PE communication.

The central control unit consists of the *scalar processor* (*SP*) and the *instruction processor* (*IP*). In SIMD mode, the program is stored entirely in the *scalar memory* (*SM*) managed by the SP. The IP broadcasts SIMD instructions to the PEs. In SPMD mode, the same program is duplicated in each PE memory. SPMD mode is initialized by the IP, which provides each PE with the starting SPMD code address. The synchronization mechanism for initializing the SPMD mode and for returning to SIMD mode is a fork-join operation executed over the set of PEs. The transition from SPMD to SIMD mode is made in one machine cycle after the last PE executes the join. Inter-PE data transfers can only occur in SIMD mode.

25.2.5   Triton

Triton is a mixed-mode machine being developed at the University of Karlsruhe, in Germany [44, 64]. The Triton/1 prototype will consist of 260 nodes (four are for fault tolerance), though the Triton concept is scalable up to 4096 nodes. Each node consists of a processor/memory pair, a memory management unit, a numeric coprocessor, a SCSI interface, and a network processor. Triton uses a generalized De Bruijn network for inter-PE communication.

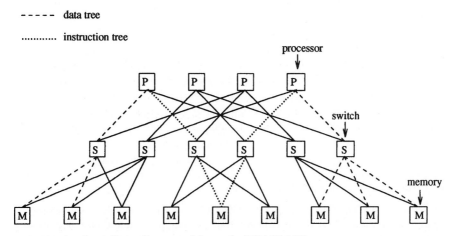

Figure 25.5 A task tree (instruction tree and data tree) of TRAC 1.1 [3]

In SIMD mode, a single front-end processor produces the instruction stream for all PEs. If a PE is not selected to execute an instruction, a local signal for the instruction stream is turned off and the corresponding PE is disabled. To switch to MIMD mode, the program must be downloaded to the local memory of the PEs. This is done via load instructions in SIMD mode. The switch from SIMD to MIMD mode is accomplished by setting the program counter to the MIMD program location and deactivating the SIMD request bit for each PE. To switch from MIMD to SIMD mode, the SIMD request bit for each PE is activated. The result of a global-wired-or operation of all PEs' SIMD request bits instructs the front-end processor to activate SIMD mode. Then each PE switches to SIMD mode and the next instruction is from the instruction stream broadcast by the front-end processor.

25.2.6   EXECUBE chip

The *EXECUBE chip* consists of eight 16-bit CPU mixed-mode PEs, each with a 64 kB memory module [52]. A hypercube interconnection network is used for inter-PE communication. This is all contained on a single chip developed by IBM Federal Systems Division, in the USA. A system with 64 EXECUBE chips (512 CPUs) has been constructed.

In SIMD mode, instructions are sent into each PE's instruction register by a separate controller via the SIMD broadcast bus. In MIMD mode, each PE obtains its own instructions from its local memory. Because the only way of accessing the memory system of each PE is through its CPU, MIMD instructions are sent and stored into participating PEs' local memory in SIMD mode via the SIMD broadcast bus. Arbitrary collections of PEs can be in either mode simultaneously. Mode switching instructions are machine operation codes that activate special hardware functions. By executing an instruction to "switch to MIMD mode," participating PEs begin execution at a specified address in local memory. After executing a switching instruction, the participating PEs stop fetching instructions from the SIMD broadcast bus and start to execute the instructions stored in local memory. A "switch to SIMD mode" instruction causes PEs to fetch instructions from the SIMD broadcast bus. A collective signal from the PEs is sent to the controller that sends SIMD instructions to each PE's instruction register. If any PE in the PE group that is changing to SIMD mode is still in MIMD execution, then the controller will wait until the collective signal from the PEs is set, at which point SIMD execution is started.

25.2.7   Conclusions

Mixed-mode machines are one extreme form of HC. Decomposing a task for mixed-mode execution is easier than for mixed-machine execution because three major problems in the use of mixed-machine HC are not present: moving data among machines, concurrent use of multiple machines, and determining machine loads. The study of mixed-mode machines provides valuable information about the trade-offs between SIMD and MIMD parallelism, explores the advantages and disadvantages of mixed-mode computation as a mode of parallelism, and establishes a relatively simpler environment for developing algorithm mapping techniques that may possibly be adapted to the mixed-machine arena. For example, a block-based mode selection methodology developed for mixed-mode machines, presented in Ref. [82], was then extended for use as a heuristic for the mixed-machine case [81] (see Section 25.7.3). Sections 25.3 through 25.7 focus on mixed-machine HC.

A variation on mixed-mode is a *mixed-component* HC system, where each separate component of a single machine represents one mode of parallelism, and two or more distinct modes are present in the machine. Examples are the SIMD/MIMD Image Understanding Architecture [83] and the SIMD/vector Cray-3/SSS Super Scalable System [23]. Mixed-component systems are outside the scope of this chapter.

25.3 Examples of Existing Mixed-Machine HC Systems

25.3.1 Simulation of mixing in turbulent convection at the Minnesota Supercomputer Center

In Ref. [51], the usefulness of an HC system developed at the Minnesota Supercomputer Center is demonstrated through a particular application involving the simulation of mixing in turbulent convection in three dimensions. The particular HC system developed consists of Thinking Machines' CM-200 and CM-5, a CRAY 2, and a Silicon Graphics VGX workstation, all interconnected over a high-speed *HiPPI* (high-performance parallel interface) network.

The required calculations for the simulation were divided into three phases: (1) calculation of velocity and temperature fields, (2) calculation of particle traces, and (3) calculation of particle distribution statistics and refinement of the temperature field. The velocity and temperature fields associated with the phase 1 calculations are governed by two second order partial differential equations. Three-dimensional cubic splines (over a grid of size 128 $\times$ 128 $\times$ 64) were used to approximate the velocity and temperature fields in these equations, resulting in a linear system of equations for the unknown spline coefficients. A conjugate gradient method was applied to solve this system of equations. These computations were done on the CM-5. At each time step, the grid of 128 $\times$ 128 $\times$ 64 spline coefficients were transferred to the CRAY 2, where the calculation of the particle traces were done.

The particle traces (phase 2) were calculated by solving a set of ordinary differential equations based on the velocity field solution from phase 1. This computation was attempted on the CM-200 by employing an Eulerian approach. Although this approach worked well for a two-dimensional instance of the problem, the same approach could not be used for the three-dimensional simulations because a prohibitive amount of memory was required. Instead, the three-dimensional simulations were implemented using a vectorized Lagrangian approach on the CRAY 2, which required substantially less memory than the parallel Eulerian scheme. The coordinates of the particles and the spline coefficients of the temperature field were then sent from the CRAY 2 to the CM-200.

The CM-200 was used to calculate statistics of the particle distribution and to assemble a three-dimensional temperature field from the associated spline coefficients (phase 3). A $256 \times 256 \times 128$ point temperature field file was produced from the $128 \times 128 \times 64$ grid of splines, representing a volume of eight million voxels (a *voxel* is a three-dimensional element). This file of voxels and the coordinates of the particles (one million particles were used) were then sent to an SGI VGX workstation, where they were visualized using an interactive volume renderer.

The application was successful in demonstrating the benefits of HC. However, in Ref. [51], it is noted that there is still much work to be done to improve the environment for developing HC applications.

25.3.2 Interactive rendering of multiple Earth science data sets on the CASA testbed

In 1990, the National Science Foundation (NSF), in conjunction with the Defense Advanced Research Projects Agency (DARPA), established a program to conduct research in the area of networking at gigabit per second speeds [72]. In this and the next subsection, two applications that utilize the HC resources available on two of the testbeds are overviewed.

The CASA testbed interconnects several remote sites including the California Institute of Technology, San Diego Supercomputer Center, Jet Propulsion Laboratory (JPL), and Los Alamos National Laboratory. One of the applications developed on the CASA testbed involves interactive three-dimensional rendering of multiple Earth science data sets. Geology can be regarded as a "three-dimensional science," in the sense that both surface and subsurface data from the Earth are collected and studied. In the past, these two types of data were generally collected and analyzed separately. By making effective use of the computing and networking resources of the CASA testbed, researchers can construct a more complete image of the Earth's surface and subsurface, together, by combining multiple sets of data from various sources. The required processing and communication for merging these data sets should be fast enough to enable interactive manipulation of the associated image. According to Ref. [14], researchers can rotate, slice, zoom, and "fly over" a full-color view of the Earth's surface and subsurface while sitting at a workstation.

The software for the application is divided into three categories: (1) a collection of functionally distinct two-dimensional image processing modules that generate and/or manipulate color images and elevation data, (2) a rendering process that combines data and creates an electronic rendered image, and (3) the network and control software that coordinate the various processes. The two-dimensional modules are implemented using Network Express, which is a portable, message passing, programming environment developed by the ParaSoft Corporation. Initially, raw data sets are transferred to one of the two-dimensional functional modules for processing. The two-dimensional modules manipulate image and/or elevation data via a number of different algorithms. Most of the two-dimensional modules were developed for the CRAY Y-MP/232 at JPL and the CRAY Y-MP8/864 at the San Diego Supercomputer Center. Two of the two-dimensional modules were implemented on the CM-5 and CM-200 located at Los Alamos. Output from the two-dimensional modules are sent over the network to the three-dimensional rendering process, which was implemented on the Intel Touchstone Delta located at the California Institute of Technology.

In the current implementation of the CASA testbed, there are high-speed HiPPI connections only among machines located at a common geographical site. The current connections among the distributed sites, which utilize lower speed networks, will be upgraded by using HiPPI-SONET gateways to interconnect each site's local HiPPI network to a wide area high-speed SONET network. Future work includes experimenting with this application over this new high-speed HiPPI/SONET network.

25.3.3   Using VISTAnet to compute radiation treatment planning for cancer patients

VISTAnet is also in the group of gigabit testbeds mentioned in the last subsection. The VISTAnet testbed sites include the Center for Communications and Signal Processing at North Carolina State University, BellSouth, GTE, and three organizations within the University of North Carolina at Chapel Hill (the Graphics and Image Laboratory in the Department of Computer Science, the Microelectronics Systems Laboratory in the Department of Computer Science, and the Department of Radiation Oncology) [73]. The machines connected to the testbed include a CRAY Y-MP, a Pixel-Planes 5, a MasPar MP-1, and Silicon Graphics workstations.

A major application focus for this testbed has been the computation of radiation treatment planning for cancer patients [66]. Radiation is effective in treating the cancer only if it is delivered to the tumorous cells in a high dose while sparing the nontumorous cells. The physician must determine the number of treatment beams to be used, the beam angles and shapes, the time the beam is to be activated, and which custom filters to use to alter the beam. This process is know as *radiation treatment planning* and in the past was carried out in only two spatial dimensions; however, some types of cancer require that the planning take place in three dimensions to achieve maximum effectiveness. This requires advanced modeling of human anatomy (rendered from tomography scans) as well as three-dimensional modeling of the radiation beam (i.e., the treatment plan). In the application, the treatment plan model is superimposed onto the anatomical model. One of the objectives is to provide a visualization of these models that can be rotated, zoomed, and/or modified interactively.

The CRAY Y-MP was demonstrated to be ideal for radiation dose calculation and interpolation throughout the entire model. The Pixel-Planes 5 machine (which contains a quarter-million custom one-bit processors) is designed for rendering images and is used for shading and merging large amounts of image data. The physician interacts with the system via a medical workstation hosted on a Silicon Graphics 340 VGX. From this workstation, the physician can modify the treatment plan based on the current dosage patterns and can adjust the view by rotating the image. When an image viewpoint is adjusted, the new viewpoint information is sent to the Pixel-Planes 5, which renders the otherwise unchanged data according to the new viewing angle and presents the new image to the physician at the workstation. If the treatment plan is modified, the new treatment plan information is sent to the CRAY Y-MP, which computes the new three-dimensional dose distribution and sends the information to the Pixel-Planes 5 for rendering.

In the future, a MasPar MP-1 will be integrated into the application and will receive the three-dimensional dose distribution generated by the CRAY Y-MP. With this information, the MP-1 will be used to compute a statistical analysis of the treatment plan in relation to the anatomical data. This computed information will provide the physician with a quantitative measure of merit for each treatment plan.

25.4 Examples of Software Tools for Mixed-Machine HC Systems

25.4.1 Overview

A variety of software tools and environments have been implemented to assist programmers in developing applications to execute on a mixed-machine HC system. A common feature among most of the existing tools is that they create a layer of abstraction between programmers and the suite of machines. Some also provide explicit constructs needed to express synchronization and communication among tasks within the application. The following subsections discuss examples of software tools that exist or are being developed for HC systems. The functionalities of most of the tools described in this section tend to evolve and change rapidly; the descriptions here are based on the references given. A survey of distributed queueing and clustering systems, some of which can be applied to HC, is given in Ref. [46].

25.4.2 Linda

Linda was originally implemented for *homogeneous* computing environments such as shared memory parallel computers (e.g., the Sequent Symmetry), distributed memory computers (e.g., the Intel iPSC/2), and local area networks (e.g., a network of workstations). As suggested in Ref. [19], it is an attractive choice for HC systems as well. In Linda, processes communicate via persistent objects called *tuples*, and not through transient events such as message passing or procedure calls. A process can generate a tuple and place it in a globally shared collection of tuples, the *tuple space*. Tuples can be also removed, read, and evaluated from the tuple space. *Process tuples* incorporate executable code and *data tuples* are passive, ordered collections of data items [16]. Although the current version of Linda does not support concurrent interaction among machines in an HC system, the issues that must be resolved to do this are outlined and discussed in Ref. [19].

25.4.3   p4

p4 is a set of parallel programming tools designed to support portability across a wide range of architectures [16–18]. p4 includes high-level operations that allow certain procedure calls to be replaced with the equivalent p4 calls that are implemented by utilizing system-specific procedures. The long-term goal of this project is to allow a single program to be written for an entire class of systems (e.g., message passing) without requiring the explicit utilization of constructs of the specific system (e.g., Intel Paragon versus nCUBE 2) in the source code. The p4 function library is linked with the source code to provide functions for message passing, shared memory monitoring, process management, debugging, and language interfacing.

Three classes of architectures are supported by p4. The first consists of shared memory multiprocessors (e.g., the Alliant FX/8). p4 provides monitor data types for encapsulating shared data and controlling access. The second consists of distributed memory systems that implement communication through message passing, specifically distributed memory multiprocessors and groups of workstations that communicate over a network [15]. The third consists of "communicating clusters," which can include multiprocessor machines that communicate via shared-memory and/or through the exchange of messages. Therefore, p4 can support communication within and among both shared-memory and message-passing machines.

The process of executing a p4 program begins with the user compiling the code for the desired set of machines. The configuration of the system is specified by a *procgroup file*, which defines the number, names, and target machines of the programs to be executed. This allows the user to experiment with different configurations and machines. p4 also has a utility called ALOG, which creates a log of time-stamped events captured during program execution, and can be used with C or FORTRAN. This event log can then be used as an input file for a graphical tool called Upshot [43]. With Upshot, the log file can be examined in detail to detect computational and/or communication bottlenecks.

The developers of p4 stress that it is not an "abstract tool" and that various components of p4 evolved through the development of real applications. As an example, p4 was used in developing a piezoelectric crystal simulation program to coordinate the computations and communications among an Intel Touchstone Delta, the graphical output on a Stardent Titan, and a Solbourne workstation (which was used as an I/O server). Current and future research directions for p4 include the implementation of Linda with p4 to provide a single high-level programming model.

25.4.4   Mentat

Mentat consists of execution time support facilities and language abstractions that provide a clear separation between the user's application and the target machine [41]. This separation is achieved by using an object-oriented language to specify parallelism within the application and compiler technology to handle many of the tedious and time consuming bookkeeping tasks. Mentat combines a medium-grain dataflow computation model with the object-oriented programming paradigm to produce a system that facilitates hierarchies of parallelism [40]. Programs are characterized as directed graphs. The vertices represent computational elements (e.g., class member functions) and the edges model data dependencies between these elements. The idea behind Mentat is to allow the programmer to express the problem in a C++ based language, called *MPL* (Mentat programming language), which provides many popular features of the C++ language. Mentat uses the dataflow model to exploit the inherent medium-grain parallelism of the program; in addition, the programmer can specify those C++ classes which are themselves of sufficient computational complexity to warrant parallel execution [40].

The use of object-oriented programming languages, such as MPL, masks much of the underlying complexity from the user and is the basis for "separating" the user from the various machines in the HC system. The basic unit of computation in MPL is the Mentat class instance, which consists of objects (e.g., local and member variables), their procedures, and a thread of control [41]. In MPL, the standard object-oriented notions of data encapsulation and method encapsulation have been extended to include "parallelism encapsulation" [40]. MPL supports two types of parallelism encapsulation that are hidden from the user: intraobject (within a member function) and interobject (among member-function invocations). For interobject, it is the responsibility of the MPL compiler to ensure that data dependencies between invocations are satisfied and that communication and synchronization are handled correctly [40]. The MPL compiler maps MPL programs onto the dataflow model by translating the MPL programs into C++ programs with embedded calls to the Mentat execution time system. These C++ programs are then compiled by the host C++ compiler resulting in executable object code.

A distinguishing feature of MPL is its implementation of a construct called *rtf* (return-to-future) [40], which is analogous to the "return" function commonly found in imperative languages such as C. The rtf construct allows Mentat member functions to return values to successor nodes in the macro-dataflow graph. These returned values are forwarded to all member functions (of the successor nodes) that are dependent on the result. The rtf function differs from a standard return in three ways. First, a member function may "rtf a value" from a Mentat-object member function that has not completed execution. Second, the execution of rtf indicates only that the associated values are ready (additional computation may be carried out after the rtf call). Finally, depending on the program's data dependency structure, rtf may not return data to its caller. In particular, if the caller does not use the resulting values locally, then the caller does not receive a copy of the values.

The *RTS* (run time system) of Mentat, which initially supported execution on homogeneous parallel machines, has been extended to support HC systems. The RTS uses a *virtual macro-dataflow machine* that provides support routines to perform execution time data dependence detection, program graph construction, program graph execution, scheduling, communication, and synchronization [40, 41]. The virtual macro-dataflow machine contains a set of machine-independent components and libraries and a set of machine-dependent components. The virtual macro-dataflow machine can be ported to any supported machine in the HC system by changing only the machine-dependent components, which allows the user to port the application source code to any supported machine without changes.

The RTS has been implemented for several platforms including a network of Sun workstations, the Silicon Graphics Iris, and the Intel iPSC/2. Matrix multiplication and Gaussian elimination programs have been coded in MPL and executed on a network of eight Sun workstations and a 32 node iPSC/2. While MPL improved the ease of use of the HC system, it was indicated that the performance may not be as good as hand-coded versions that use send and receive protocols. Thus, there is a trade-off between ease of use and some performance degradation. Future work includes the implementation of several optimizations for the MPL compiler.

25.4.5   PVM, Xab, and HeNCE

PVM (parallel virtual machine). The *parallel virtual machine* (*PVM*) is a software system that enables a mixed-machine HC system to be used as a coherent, flexible, and concurrent computational resource [11, 74, 75]. The PVM package includes system level daemons, called *pvmds*, which reside on each computer in the HC system, and a library of PVM interface routines.

The pvmds provide services to both local processes and remote processes on other platforms in the HC system. Together, the entire collection of pvmds form what is called a "virtual machine" by enabling the HC system to be viewed as a single "meta-computer." Two of the major services provided by the pvmds are communication and synchronization. Processes communicate via the use of messages, which are exchanged asynchronously so that a sending process may continue execution without waiting for an acknowledgment from the receiving process. The other major service provided is the synchronization among processes, which can be accomplished by using barriers or by using event rendezvous. The synchronizations may be among multiple processes that are executing on a local machine or may be among processes on different machines.

The second part of the PVM package is a library of interface routines. Applications to be executed on one or more computing platforms in the HC system are able to access these platforms via library calls embedded in imperative procedural languages such as C or FORTRAN. The library routines interact with the pvmd (resident on each machine) to provide services such as communication, synchronization, and process management. The pvmd may provide the requested service alone or in cooperation with other pvmds in the HC system.

From the user's point of view, the PVM system can be conceptualized as a three-level hierarchy. At the uppermost layer, which is the interface to the programmer, is the concept of an *instance* (or process), which is the basic unit of computational abstraction in PVM. Applications developed with PVM generally consist of several instances (possibly executing concurrently) that cooperate across machine boundaries. The middle layer is defined as the virtual machine layer. The virtual machine layer consists of the pvmds that reside on the machines of the HC system. The lowest layer is the actual set of machines in the HC system.

The computational resources in the HC system may be accessed using three different modes: (1) the transparent mode, in which instances are automatically located at the most appropriate sites based upon a user-specified cost matrix, (2) the architecture-dependent mode, in which the user can indicate specific architecture types on which particular instances are to execute, and (3) the low-level mode, in which particular machines may be specified by the user. The PVM tools described next can be used in any of these access modes.

Xab (X-window analysis and debugging). *Xab* (X-window analysis and debugging) is a tool developed for the execution time monitoring of PVM programs [8, 11]. The Xab tool gives the user direct feedback on what PVM functions the program is executing and how the program is performing in an HC environment. Xab consists of three parts: the Xab library, which contains instrumented PVM routines that are linked to the user's code, a special monitoring process called *admon*, which receives trace messages from the library routines, and a front-end process, which graphically displays trace events.

Xab monitors a user's program by instrumenting calls to the PVM library. Each instrumented call not only performs its intended PVM function, but also sends an Xab event message to the admon process. (The Xab event message is itself a PVM message.) An Xab event message generally includes an event type, a time stamp, and event-specific information.

The admon process receives Xab event messages to be sent either to a file or to the Xab display process, which displays each event captured in an X-window. The user can single-step through these events or allow Xab to replay the events continuously in real-time.

HeNCE (heterogeneous network computing environment). *HeNCE* (heterogeneous network computing environment) aids users of PVM in decomposing their application into subtasks and deciding how to allocate these subtasks onto the available machines in the HC system [9–11, 75]. In HeNCE, the programmer explicitly specifies the parallelism for an application by drawing a directed graph, where nodes represent subtasks written in either FORTRAN or C, and arcs represent dependencies and flow control. There are also four types of control constructs: conditional, looping, fan-out, and pipelining.

The user must specify a cost matrix, which represents the cost of executing each subtask on each machine in the HC system. The meaning of the cost parameters are defined by the user (e.g., estimated execution times or utilization costs in terms of dollars). At execution time, HeNCE uses the cost matrix to estimate the most cost effective machine on which to execute each subtask.

Given the graph and the matrix, HeNCE configures a subset of the machines defined in the cost matrix as a "virtual machine" using PVM constructs. Then HeNCE begins execution of the program. Each node in the graph is realized by a distinct process on some machine. The nodes communicate with each other by sending parameter values needed for execution of a given node, which are specified by the user for each node (subtask). A node obtains parameter values needed to begin execution from predecessors nodes. If the immediate predecessors do not have all the required parameters for a node, earlier predecessors are checked until all required parameters are found. Then the node (subtask) is executed and passes the appropriate parameters onto descendant nodes.

HeNCE can trace the execution of the application for display in real-time or replay later. The trace tool displays active machines in the network as icons whose colors change depending on whether they are computing or communicating. The tool also displays the user's directed graph and dynamically illustrates paths of execution. This can enable the programmer to detect bottlenecks in the application by displaying the states of the subtasks during execution. The trace animation can also be used for performance tuning, i.e., the programmer can reallocate subtasks across the machines in the HC system and tune the application's behavior to match the environment for subsequent executions of the application.

25.5 A Conceptual Model for Automatic Mixed-Machine HC

A conceptual model for automatic task decomposition, matching of subtasks to machines, and scheduling of subtasks in a mixed-machine HC environment is shown in Fig. 25.6. It

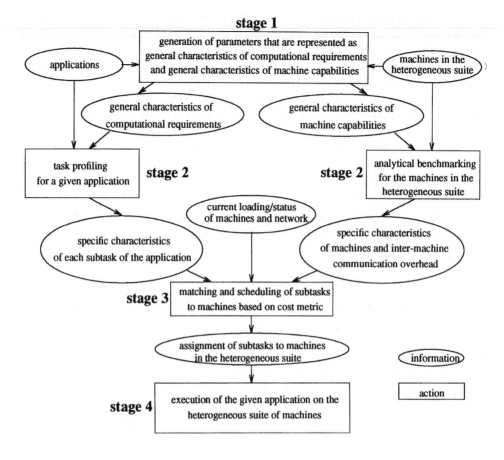

Figure 25.6 Conceptual model of the automatic assignment of subtasks to machines in an HC environment

builds on the one presented in Ref. [36] and is referred to as a "conceptual" model because no complete automatic implementation currently exists. Automatic implementation of HC is a long-term goal that is important for encouraging and facilitating the use of HC, and for improving HC system performance.

In stage 1, using information about the expected types of application tasks and about the machines in the HC suite, a set of parameters is generated that is relevant to both the computational requirements of the applications and the machine capabilities of the HC system. These parameters define the multidimensional decision space to be used for describing and matching subtasks and machines. For each parameter, computational requirements and machine architecture features are derived. For example, considering the parameter "floating point operations," the computational requirements of the application tasks to be quantified are the number and types of the floating point operations needed to perform the calculation. The architecture features of the machines in the HC suite to be quantified are the speeds for these different types of floating point operations. Irrelevant parameters are excluded. For example, if the given applications have no floating point operations, then it is not necessary to evaluate the machine capabilities for executing floating point operations; if there is no vector machine available in the suite, vectorizable code may be excluded from the set of the computational requirements considered.

The total number of parameters enumerated determines the complexity of this automation problem. The chosen parameters evolve dynamically when new types of applications and/or new types of machines are added.

In stage 2, *task profiling* decomposes the application task into subtasks, each of which is computationally homogeneous. The computational requirements for each subtask are then quantified. The term often used for this step in the literature is code profiling. The reason for using task profiling in this chapter instead is that, to identify the types of computational requirements present in a specific task, both the code and data upon which the specified HC system will operate must be profiled. *Analytical benchmarking* is used in stage 2 to quantify how effectively each of the available machines in the suite performs on each of the types of computations being considered. Existing literature that presents explicit methodologies for performing task profiling and analytical benchmarking in the context of HC is reviewed in Section 25.6 of this chapter.

One of the functions of stage 3 is to use the information from stage 2 to derive, for a given application, the estimated execution time of each subtask on each machine in the HC suite [55] and the inter-machine communication overhead associated with each assignment of subtasks to machines. In stage 3, these static results and the dynamic information about the current loading and "status" of the machines and inter-machine network are used to generate an assignment of the subtasks to machines in the HC system based on certain cost metrics. The "status" could include whether the machines/network are fully or partially functioning due to faults, and when other tasks using the machines/network are expected to complete. The most common cost metric for HC is to minimize the overall execution time (including the inter-machine communication time) of a given application task on a particular HC system. Another problem is to find the most appropriate suite of heterogeneous machines for a given collection of applications, such that the cost of the corresponding HC system is minimized for a given set of execution time constraints [32]. Section 25.7 of this chapter presents a variety of techniques in the literature for matching subtasks and machines.

Stage 4 is the execution of the given applications on the HC suite. Because the loading/status of the machines/network may change, sometimes it is necessary to reselect machines for certain subtasks by reactivating stage 3. Techniques for the migration of a subtask from one type of machine to another in the middle of execution present a difficult problem; one approach is described in Ref. [5].

Automatic HC is a relatively new field. The task profiling, analytical benchmarking, and matching and scheduling techniques discussed in Sections 25.6 and 25.7 are representative frameworks that require further research before they are practical tools.

25.6   Task Profiling and Analytical Benchmarking

25.6.1   Overview

To execute a task on a mixed-machine HC system, the task must be decomposed into a collection of subtasks, where each subtask is a homogeneous code *block*, such that the computations within a given code block have similar processing requirements (e.g., see Refs. [21, 32, 35, 49, 75, 80]). These homogeneous code blocks are then assigned to different types of machines to minimize the overall execution time. In some cases, it is better not to use the best-matched machine because of the overhead involved in any inter-machine data transfer that may be needed. Thus, it is important to know the degree to which a code block matches each machine.

This section presents example methodologies for the task profiling and analytical benchmarking that comprise stage 2 of the conceptual model in Section 25.5. These methodologies make many simplifying operating assumptions and are frameworks, rather than fully implemented systems. Further research is needed before they are practical tools that can provide the quantitative results needed for matching and scheduling (stage 3).

25.6.2   Definitions of task profiling and analytical benchmarking

Task profiling is a method used to quantify the types of computations that are present in the application program [32]. Task profiling divides the source program into homogeneous code blocks based on the types of computations required. The definition of the set of code-types is based on the features of the machine architectures available and the computational requirements of the applications being considered for execution on the HC system. This is done in stage 1 of the conceptual model in Section 25.5.

Analytical benchmarking is a procedure that provides a measure of how well each of the available machines in the heterogeneous suite performs on each of the given code-types [32]. Together, the task profiling and analytical benchmarking steps provide the information needed for the matching and scheduling step, which is described in Section 25.7. The performance of a particular kind of machine on a specific code-type is a multivariable function. The parameters (i.e., variables) for this performance function can include the requirements (e.g., data precision) of the application, the size of the data set to be processed, the algorithm to be applied, the programmer's and compiler's efforts to optimize the program, and the operating system and architecture of the machine that will execute the specific code-type [38].

There are a variety of mathematical formulations, collectively called *selection theory*, that have been proposed to choose the appropriate machine for each code block. Many (e.g., Refs. [21, 48, 80]) define analytical benchmarking as a method of measuring the optimal speedup a particular kind of machine can achieve compared to a baseline system when the best-matched code-type for that machine is executed. The ratio between the actual speedup and the optimal speedup defines how well a code block is matched with each machine type, and the actual speedup, in general, is less than the optimal speedup.

25.6.3   Methodologies for performing task profiling and analytical benchmarking

A comparison between traditional benchmarking and analytical benchmarking. One of the most widely used traditional benchmarking techniques is to execute a set of well-studied programs on a machine (e.g., see Refs. [22] and [26]), using the total execution time as the final measure to compare that specific machine's performance with that of others. But in the context of HC, only code blocks, rather than a whole program, are executed on a specific type of computer. Traditional benchmarking techniques do not reflect the performance of a particular kind of machine on a specific code-type. The problem with these traditional benchmarking techniques is that they are not *analytical*.

The techniques for analytical benchmarking should not only be able to show the overall execution time of a specific kind of machine on a certain type of code, but should also be able to predict future capabilities when new types of machines and/or new types of applications are added [33]. As introduced in Ref. [33], the goal of analytical benchmarking is to construct a class of relatively basic benchmarking programs for each type of computer available in the HC suite. A set of benchmarking programs can be used to derive the performance metrics of the system for a range of conditions, such as the size of the input data file and the type of calculations required. This is in contrast to the usual benchmarking program, whose result is just the execution time.

Parallel assessment window system. *Parallel assessment window system* (*PAWS*) is an experimental platform capable of performing machine and application evaluations for task profiling and analytical benchmarking. It consists of four tools: the application characterization tool, the architecture characterization tool, the performance assessment tool, and the interactive graphical display tool [63].

The *application characterization tool* transforms a given program written in Ada into IF1, an acyclic graphical language that illustrates the program's data dependencies. In IF1, basic operations, such as addition and multiplication, are represented by simple nodes, and complex constructs, such as conditional branches and loops, are represented by compound nodes. By grouping sets of nodes and edges into functions and procedures, the application characterization tool can describe the execution behavior of a given program at various levels.

The *architecture characterization tool* partitions the architecture of a specific type of machine into four categories: computation, data movement and communication, I/O, and control. Each category can be repeatedly partitioned into subsystems until the subsystems in the lowest level can be described by raw timing information. This hierarchical organization of architectural parameters for a specific machine provides a detailed model for determining the operational behavior of each subsystem. The raw timing information of each leaf node of the tree can be user specified or obtained by low-level benchmarking.

The *performance assessment tool* obtains information from the architecture characterization tool and generates timing information for operations on a given machine upon request. Timings for primitive operations are stored within the architecture characterization tool; the performance assessment tool uses these to determine timings for more complicated operations (e.g., complex floating point multiplication).

Two sets of performance parameters for an application, parallelism profiles and execution profiles, are generated by the performance assessment tool. *Parallelism profiles* represent the applications' theoretical upper bounds of performance (e.g., the maximal number of operations that can be parallelized). *Execution profiles* represent the estimated performance of the applications after they have been partitioned and mapped onto one particular machine. Both parallelism and execution profiles are produced by traversing the applications' task-flow graph and then computing and recording each node's performance and statistically based execution time estimates. The objectives of parallelism and execution profiles are very similar to those of task profiling and analytical benchmarking. The *interactive graphical display tool* is the menu-driven user interface for accessing the other tools.

Distributed heterogeneous supercomputing management system. In Ref. [38], a framework called the *distributed heterogeneous supercomputing management system* (*DHSMS*) introduces a systematic methodology for performing both task profiling and analytical benchmarking. The basic approach in DHSMS is to generate a *USC* (universal set of codes) for task profiling. The USC can also be viewed as a standardized set of benchmarking programs used in analytical benchmarking. Because the method of generating a USC is architecture-driven, these programs can provide information about a machine's hardware features.

The construction of a USC is based on an architecture-dependent hierarchical structure that is a detailed architectural characterization of machines available in an HC system, and is similar to the hardware organization generated by the architectural characterization tool in PAWS. At the highest level of this hierarchical structure, the modes of parallelism for classifying machine architectures are selected. At the second level, finer architectural characteristics, such as the organization of the memory system, can be chosen. This hierarchical structure is organized in such a way that the architectural characteristics at any level are choices for a given category, e.g., type of interconnection network used.

To generate a USC, DHSMS assigns a code-type to each path from the root of the hierarchical structure to a leaf node. Every such path defines a set of architectural features corresponding to the nodes traversed by that path. Mathematically, a USC is defined as a set of code-types $C = \{C_i\}$, where $1 \leq i \leq K$, and K is the total number of paths from the root to a leaf node. In this proposed framework, conceptually each C_i represents the type of code ideally suited for the architectural features indicated by the ith path. Thus, K is also the number of code-types in C. Let $v_0(j)$ be the size of the parallelism (e.g., maximum possible number of concurrent threads of execution) in the given code block S_j, and let $v_i(j)$ $(1 \leq i \leq K)$ be a real number between 0 and 1 that indicates how well the code block S_j is matched with the code-type C_i. Then, a *task profiling vector* V_j for a given code block S_j is defined as $V_j = [v_0(j), v_1(j), v_2(j), \ldots, v_K(j)]$. The objective of task profiling in DHSMS is to estimate V_j for each S_j.

The element $v_0(j)$ that quantifies the size of parallelism for code block S_j is very important. Benchmarking results for supercomputers show that the size of parallelism can affect the choice of machines used to achieve the best performance on certain programs [22, 26]. As an example, consider the study in Ref. [33] where the performances of a SIMD machine and a vector machine on SAXPY code (i.e., matrix-vector calculation of the form $S = AX + Y$) are evaluated and compared. Even for a code block that is perfectly matched with the vectorizable code-type, the SIMD machine outperforms the vector machine on vectors with length (the size of the parallelism) longer than the optimal length for the vector machine. Hence, the suggestion in Section 25.5 that the term "task profiling" be used instead of code profiling is very appropriate, because both code and data must be considered.

The task profiling process must be repeated for each application. A fine-grained task profiling, with all levels of architectural features incorporated into the hierarchical structure of machine characteristics mentioned above, will certainly generate a more accurate task profiling vector V_j, but the overhead associated with it increases significantly. Alternatively, a coarse-grained task profiling, which chooses only a few levels of architectural features in the corresponding hierarchical structure, can result in relatively low overhead, but the information obtained from task profiling may not be accurate enough for the subsequent procedures of matching and scheduling. Thus, there is a trade-off between the accuracy of the task profiling and the complexity of the overhead incurred, and this choice can be user-specified [86].

Let $b_q(n)$ be the speedup that machine q can achieve compared to a baseline system by executing optimally matched benchmarking programs with the size of parallelism equal to n. Then, analytical benchmarking can be formally defined as a vector $B(n) = [b_q(n)]$, $q = 1, 2, \ldots, M$, where M is the number of machines available in the HC suite.

$B(n)$ does not evaluate the inter-machine communication overheads of the application program and is categorized as *computation benchmarking* in DHSMS. *I/O benchmarking* estimates the I/O overhead of a given architecture as a performance metric that is a function of the amount of data being transmitted through the I/O subsystem. *Network-interface profiles* estimate the overhead of the network due to the protocols for communication and media access. Let $d_q(a_m)$ be the destination-independent expected I/O and network-interface overhead of machine q, when there are a_m units of data transmitted through the mth edge of the data dependence graph of the original program. Then, I/O benchmarking and network-interface profiles are defined by the communication overhead vector $D(a_m) = [d_1(a_m), d_2(a_m),\ldots,d_M(a_m)]$. If the exact value of a_m is not known at compile time, some stochastic performance measures are required.

DHSMS begins with a *task-flow graph (TFG)*, which provides the execution time of each code block S_j on a baseline system and the amount of data transferred between code blocks due to data-dependencies. A task profiling vector V_j is assigned to each code block S_j in the TFG, forming an intermediate *code-flow graph (CFG)*. The length of V_j and the complexity of task profiling each depend on the number of levels of the hierarchical structure selected by the user. In the final CFG, each code block S_j in the intermediate CFG is associated with an estimated computation time vector $E_j = [e_1, e_2, \ldots, e_M]$, where e_q is the estimated computation time of code block S_j on machine q and is a function of V_j and $B(n)$.

Let $d^*_{p,q}(a_m)$ be the expected I/O and network-interface overhead when there are a_m units of data transmitted between machine p and machine q (this is a function of $d_p(a_m)$ and $d_q(a_m)$). Then, in the final CFG, each communication link m between two code blocks in the original TFG is associated with a communication overhead matrix $D^*(a_m) = [d^*_{p,q}(a_m)], 1 \le p, q \le M$. (In Ref. [38], an asterisk is used to distinguish the communication overhead *matrix D* from the communication overhead *vector D*.) The data format conversion overhead also can be added to $d^*_{p,q}(a_m)$. The $M \times M$ matrix $D^*(a_m)$ is assumed to be symmetric along the diagonal. The final CFG can be used in matching and scheduling.

Extensions to the DHSMS approach are presented in Ref. [86]. Two techniques, *augmented task profiling* and *augmented analytical benchmarking*, are proposed to characterize the applications (as well as the machines available in the corresponding HC system) during the construction of the set of code-types. The new augmented approach is a two level framework that combines both fine-grained and coarse-grained characterization techniques. This framework of task profiling and analytical benchmarking is based on generating a representative set of templates *(RST)* that can characterize the execution behavior of the programs at variant levels of details.

Parametric task profiling and parametric analytical benchmarking. In the above two methodologies for performing task profiling and analytical benchmarking, a task profiling vector is defined as a function that maps each combination of the subtasks in the application program and the elements in the set of code-types to a real number in the range [0, 1]. This real number quantifies the degree of the match between the specific subtask and the code-type. Analytical benchmarking is defined as a method of measuring the optimal speedup a certain kind of machine can achieve compared to a baseline system when the best-matched code-type for that machine is executed. By combining the results from the above two characterization steps as discussed in DHSMS, the estimation of the execution times of the subtasks on the available machines in the HC system can be obtained. Most of the selection theories of HC use similar mathematical formulation for task profiling and analytical benchmarking (e.g., see Refs. [21, 59, 80]). Section 25.6.4 presents that mathematical formulation in detail.

The goal of Ref. [87] is to predict the execution time of a task on a single machine. The parametric task profiling and parametric analytical benchmarking proposed in Ref. [87] adopt different mathematical formulations for these two characterization steps in PAWS and DHSMS, but is still compatible with the conceptual model of Section 25.5. First, a set of parameters of cardinality P is defined such that each parameter represents a distinct category of low-level operations performed in a task. This step corresponds to stage 1 of the conceptual model. Let v_i be the operation count for parameter i for a given task. Then, in parametric task profiling, the task profiling of stage 2 is defined as a parametric task profiling vector $V_t = [v_1, v_2, ..., v_P]$ for an application task t. The handling of data-dependent loop parameters and conditionals is not included in this formulation.

Let $b_{mi} (1 \le i \le P)$ be the execution time of machine m, when that specific kind of machine is used to execute one occurrence of parameter i. Then, in parametric analytical benchmarking, a parametric computation benchmarking vector

$$B^m = [b_{m1}, b_{m2}, ..., b_{mP}]$$

is defined.

Let the estimated computational time for a given task on machine m be

$$e_m^{comp} = \sum_{i=1}^{P} v_i b_{mi}$$

Then, a *computation estimation vector* for a given application task t is

$$E_t^{comp} = [e_1^{comp}, e_2^{comp}, ..., e_M^{comp}]$$

where M is the number of machines available in the HC system. Thus, E_t^{comp} can be used to select a machine for task t.

In Ref. [25], a prototype software system called Automatic Heterogeneous Supercomputing (*AHS*) is introduced. AHS uses a method similar to the V_t and B^m vectors in Ref. [87] to predict execution time. It differs from Ref. [87] in several ways. Data-dependent loop parameters and conditional branch probabilities are approximated by constant values. AHS can use information about the current load on a machine to appropriately weight the expected execution time to account for the load. AHS can estimate the execution time of a specific application program on a group of networked sequential UNIX machines (it is not limited to a single machine). The inter-machine data transfers are handled by asynchronous communication through a UDP (user datagram protocol) socket. AHS can generate the code for inter-machine communication automatically. A proof-of-concept functioning AHS prototype for the MasPar MP-1 and some Unix-based workstations has demonstrated the usefulness of this approach.

25.6.4   A general mathematical formulation for task profiling and analytical benchmarking

One possible general mathematical formulation for task profiling and analytical benchmarking can now be presented. Let CS be a code space spanned by C, where $C = \{C_i\} (1 \le i \le K)$ is a set of code-types generated as dimensions for task profiling and analytical benchmarking. CS is a K-dimensional space, where K is the number of code-types in C. The contents of C depend on the characteristics of the applications as well as the machine architectures in a given HC system. For example, in DHSMS [38], a USC is generated to be C, where C is a set of code-types for characterizing the architectures of machines in the corresponding HC system. In Refs. [87] and [25], the code-types are individual machine instructions.

Let $S = \{S_j\}$ be a set of computationally homogeneous code blocks (subtasks) generated by decomposing a given application program. After task profiling, for each code block S_j, a K-dimensional vector $\Omega(j) = [\Omega_1(j), \Omega_2(j), ..., \Omega_k(j)]$ is generated, where $\Omega_1(j)$ is a real number in the interval $[0, 1]$ that quantifies the degree of match between S_j and the ith dimension of the code space CS.

Let $R = \{m_i\}$ be a set of machines in the HC system. A computation cost-coefficient vector $T = \{t_i\}$ can also be defined, where t_i is the maximal speedup a machine i can achieve compared to a baseline system when it executes the best-matched code-type. The purpose of analytical benchmarking is to estimate t_i as a function of a set of parameters, such as types of operations and length of data vectors.

The amount of communication overhead depends on many factors, such as the bandwidth of the memory channels of the source and destination machines, the topology and bandwidth of the interconnection network, and the complexity of the data format conversion. Let $\delta_{r,s}^*(a)$ represent the expected communication overhead incurred when there are a units of data transmitted from machine r to machine s [48]. Then, a communication cost-coefficient matrix $B^*(a) = [\delta_{r,s}^*(a)]$ is also part of analytical benchmarking. It is possible for $B^*(a)$ to be affected during execution time by network usage of other tasks.

The above formulation is based on the ideas presented in several papers [21, 33, 48, 59, 80]. Methods for automatically determining C, S, Ω, T, and B^* are still largely open problems.

25.6.5 Summary

Example methodologies for the task profiling and analytical benchmarking stage were presented. The information generated by this stage is used in the matching and scheduling stage, discussed in the next section.

25.7 Matching and Scheduling for Mixed-Machine HC Systems

25.7.1 Overview

For mixed-machine HC systems, *matching* involves deciding on which machine(s) each code block should be executed and *scheduling* involves deciding when to execute a code block on the machine to which it was mapped [77]. Matching and scheduling constitute stage 3 of the conceptual model in Section 25.5.

Mapping and scheduling research for general parallel and distributed computing systems, which are closely related to matching and scheduling problems for HC systems, has focused on how to effectively execute multiple subtasks across a network of sequential processors (e.g., see Refs. [6, 20, 60]). In such an environment, load balancing can be an effective way to improve response time and throughput. Although some of these existing mapping and scheduling concepts and techniques can be (and have been) applied to matching and scheduling for HC systems, there is a fundamental distinction between mapping and scheduling subtasks for a network of sequential processors (e.g., a network of workstations) and matching and scheduling subtasks for an HC system consisting of various types of parallel computers (e.g., MIMD, SIMD, and vector). In the latter case, a subtask may execute most effectively on a particular type of parallel architecture and matching subtasks to machines of the appropriate type is a more important factor than merely balancing the load among all machines in the suite.

This section describes some basic characteristics of matching and scheduling for HC systems and overviews some existing techniques and formulations for matching and scheduling. Although some of the proposed techniques make simplifying assumptions that may be difficult to justify in practice, the body of work reviewed represents solid research that is being conducted as important first steps in a relatively new field.

25.7.2   Characterizing matching and scheduling for HC systems

In HC systems, the total execution time of a task depends on the matching and scheduling techniques used as well as the local mapping and local scheduling employed on each machine in the HC system. *Local mapping* involves the assignment of a code block and its associated data to the processors/memories of a given parallel architecture. Formulating and solving local mapping problems for specific types of parallel architectures is a subject of extensive research within the parallel processing community (e.g., see Ref. [62]). The choice of the local mapping will impact the execution time of a block, which influences matching/scheduling decisions [21, 59]. *Local scheduling* is typically performed by the individual operating system of each machine to decide when to execute multiple jobs that are assigned to run on that machine. Matching/scheduling techniques for HC systems often assume that load information, such as start time and percentage of cycles available, can be obtained from local schedulers [6].

In a broad sense, matching and scheduling problems can be viewed as resource management problems consisting of three main components: consumers, resources, and policy [20]. In the context of HC systems, the *consumers* are represented by the code blocks, which are identified by task profiling. The *resources* include the suite of computers, the network(s) that interconnect these computers, and the I/O devices. The *policy* is the set of rules used by the matcher/scheduler to determine how to allocate resources to consumers based on knowledge of the availability of the resources and the suitability of the available resources for each consumer.

Matching/scheduling policies are generally designed to optimize an objective function subject to a set of constraints. Minimizing the overall execution time under a cost constraint or minimizing cost under a performance constraint are two commonly used formulations for HC systems [21, 32, 80]. Cost can be defined in different ways, including as a weighted sum of execution times for the machines in an existing HC system, or as the total system price (in terms of dollars) for prospective purchases to build an HC suite. Execution time can be estimated through the task profiling and analytical benchmarking techniques discussed in Section 25.6, or from user supplied information or empirical measurements based on typical input data sets [55]. The I/O time and network delay among machines can also be incorporated in the formulation (e.g., Refs. [38, 81]). Once the objective function and constraints are defined, the associated matching/scheduling problem can be solved. In many cases, matching and scheduling problems are NP-complete, thus heuristics and approximation algorithms are often used in practice to obtain solutions (e.g., Ref. [78]).

Matching/scheduling techniques (i.e., policies) can be classified as either static or dynamic. *Static* refers to the case where the decisions of where/when to execute the various code blocks of the given task are made at compile time, and information about the code blocks (e.g., code types and execution time estimates) are available. Either no information on the load of the machines in the HC system is used, or statistically-based models and/or assumptions for these loads may be incorporated. *Dynamic* matching/scheduling decisions are made at execution time, utilizing static and execution time information, such as measured load. Dynamic techniques can either be non-preemptive assignments or can allow dynamic reassignments. They can be adaptive or non-adaptive, depending on whether feedback about the effectiveness of the matching/scheduling policy is used to modify the policy itself.

In the next subsection, some of the existing matching and scheduling techniques and formulations are reviewed. This is not an exhaustive review; however, it demonstrates the range of issues involved.

25.7.3   Examples of techniques and formulations for matching and scheduling for HC systems

Block-based SIMD/SPMD mode selection technique and its extension. In Ref. [82], a *block-based mode selection* (*BBMS*) technique is proposed that uses static source code analysis of data-parallel program behavior to assign each code block to SIMD mode or SPMD mode in a single mixed-mode machine (e.g., PASM [71]). BBMS is used as a basis for a heuristic for machine selection for SIMD/SPMD mixed-machine systems (i.e., a network of SIMD and MIMD machines) in Ref. [81].

In the mixed-mode BBMS framework developed in Ref. [82], the application program is assumed to be written in a mode-independent language (e.g., Ref. [61]). In a mode-independent language, *operations* represent the most explicit level at which program representation is identical for each mode of parallelism. Task profiling is done by dividing the program into code blocks. *Code blocks* are identified by their leading statements, called *leaders*. The first statement in a program is a leader and any statement that is a target of a branch at the machine code level is a leader, any statement following a conditional branch at the machine code level is a leader [2]. In addition, any statement requiring a synchronization or an inter-PE data transfer and the statement that follows it are leaders.

After the code blocks are defined, the program is transformed into a *flow analysis tree*, whose structure represents the scope levels within the program. The root of the tree represents the scope of the whole program. The non-leaf nodes represent control and data-conditional constructs. Code blocks are represented by leaf nodes of the tree. An example program segment and its associated flow analysis tree are shown in Fig. 25.7.

It is assumed that leaf blocks (i.e., code blocks) are executed either completely in SIMD or completely in SPMD mode, and mode changes are allowed only at inter-block boundaries. Also, the leaf blocks are executed in an ordered sequence (from left to right) as they appear in the flow analysis tree. Thus, the schedule for executing the code blocks is static and is defined by the program itself. If a block is to be executed more than once, such as in a loop, then the mode of parallelism for that block is the same for all loop iterations. Each iteration of a loop body must begin and end execution in the same mode of parallelism (but can change modes within the body). All blocks that are descendants of a data-conditional construct are implemented in the same mode of parallelism (this is to avoid complex execution time bookkeeping overhead, and is independent of BBMS).

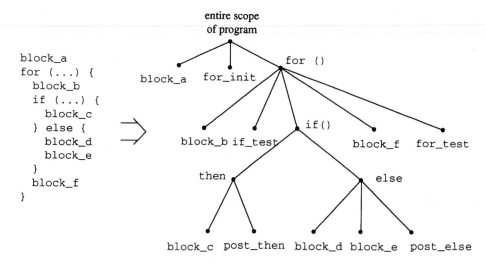

```
block_a
for (...) {
   block_b
   if (...) {
      block_c
   } else {
      block_d
      block_e
   }
   block_f
}
```

Figure 25.7 Example program segment and its associated flow-analysis tree [82]

Execution time estimates are assumed to be known (e.g., based on the results of analytical benchmarking) for the leaf blocks in both SIMD and SPMD modes, and are denoted by T_l^{SIMD} and T_l^{SPMD} for the lth leaf block, respectively. It is also assumed that the number of iterations for each looping construct and the probability that a PE executes the "then" clause of each data conditional construct are known or estimated at compile time (e.g., through compiler directives). In general, the information associated with sibling nodes at each level of the tree is combined to determine the minimum execution times for starting in each mode and ending in each mode.

Figure 25.8 shows how mode selection for a sequence of sibling code blocks is transformed into a multistage optimization graph. The parameters C^{SIMD} and C^{SPMD} repre-

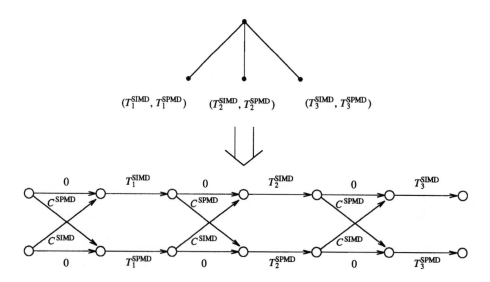

Figure 25.8 Transformation from flow-analysis tree to multistage optimization graph [82]

sent the times for switching to SIMD and SPMD modes, respectively. From the multistage optimization graph, four shortest (in terms of time) paths, corresponding to the four minimum execution times mentioned earlier, are determined. The algorithm for the multistage optimization problem reduces a sequence of three stages to two stages by determining the shortest four paths associated with all possible starting and ending mode choices (starting at the first stage and ending at the third stage). This is repeated until only the initial and final stages remain.

If the parent node is a looping construct, then the known (or approximated) number of iterations is utilized to estimate the total time for the loop. If the parent node is a data conditional construct, then the known (or approximated) probability of executing the "then" clause is used to estimate the total time for the data conditional. The time of the shortest of the four paths at the root is the optimal mixed-mode execution time. The mode assignments corresponding to this path are then made.

Using BBMS as a heuristic for machine selection in a mixed-machine system with two machines is considered in Ref. [81]. The time to switch execution from one machine to the other depends on the time to transfer the required data between machines. Thus, in contrast to the assumed constant time associated with switching modes in a mixed-mode machine, the time of switching execution from one machine to another is dependent on which machine(s) contain the data sets that are required to execute the next block, which depends on the machine choices made for executing the previous blocks, and on the size of the data set to be transferred. A given machine may contain a data set because it was initially loaded there, it was received from another machine, or it was generated by that machine.

Consider a program segment consisting of a sequence of blocks $S_0, S_1, S_2, \ldots$ (subtasks). For each machine, there is an associated execution time that is assumed to be known for each block. The data structures used by each block are assumed to be known and are stored in a *data use* (*DU*) table, denoted by DU_i for block S_i. Each DU_i entry is labeled as read, create, or modify.

Each data structure is assigned a *cost* attribute, which corresponds to the time required to transfer the data structure between the two machines (to simplify the presentation, this cost is assumed to be machine independent). A *location* attribute is used to track the availability of each data structure for each machine. A *data location* (*DL*) table stores these as the cost and location attributes for each data structure. If the data structure is on one machine only, then the cost to transfer the data structure to the other machine is tabulated; if the data structure is located on both machines, then a cost of zero is used. DL_i is used to denote the state of the data location table just before executing block S_i.

Figure 25.9 shows example DU and DL tables for a program segment, where blocks S_0 and S_1 are assigned machine Y, and block S_2 is assigned to machine X (this assignment is arbitrary). T_i^X is the time required to execute block S_i on machine X.

Given the information specified above, the goal is to find an assignment of blocks to machines that results in the minimum overall execution time. In Ref. [81], this problem is transformed into a multistage optimization problem similar to the one used in Ref. [82]. Each time the graph is reduced, a separate DL table is updated for each of the four aggregate paths generated in the reduction step (see Fig. 25.10). Because the time to switch between machines depends on past machine selections, the proposed approach may not always produce optimal assignments. For example, the algorithm may make a machine assignment for a given block that will either require a later block to read a large data structure from the other machine or use a machine that is not well suited for that block. However, simulation studies of program behaviors indicate that the proposed approach, which has a polynomial time complexity, typically produces assignments with overall exe-

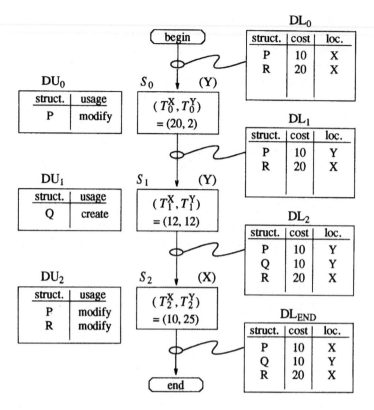

Figure 25.9 Simplified model of parallel program behavior with an arbitrary choice of machine for each code block [81]

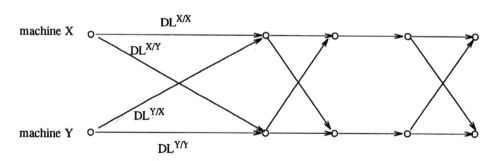

Figure 25.10 Heuristic building on the multistage technique [81]

cution times that are within 1 percent of the optimal assignments, which are determined using an exhaustive search that has an exponential time complexity. This research is currently being extended to more than two machines.

Optimal selection theory and its extensions. A mathematical programming formulation for selecting an optimal heterogeneous configuration of machines for a given set of problems under a fixed cost constraint, known as *Optimal Selection Theory (OST)* [32, 33], is overviewed in this subsection. An extension of OST, called *Augmented Optimal Selection Theory (AOST)* [80], is presented (in considerable detail) to illustrate the various components of the mathematical model. Two other extensions of OST, *Heterogeneous Optimal Selection Theory (HOST)* [21] and *Generalized Optimal Selection Theory (GOST)* [59] are also reviewed.

In the OST framework [32, 33], the application is assumed to consist of a set of non-overlapping *code segments* that are totally ordered in time. Thus, the total execution time of the application is equal to the sum of the execution times of all its code segments. These code segments are identified by task profiling such that each segment is homogeneous in computational requirements. A code segment is defined to be *decomposable* if it can be partitioned into different code blocks that can be executed on different machines of the same type concurrently. A nondecomposable code segment is a code block. The OST formulation assumes, for simplicity, linear speedup when a decomposable code segment is executed on multiple copies of a best matched machine type, and that there are always a sufficient number of machines of each type available. Information about the code blocks and machines is assumed known, as was the case for the methodologies described in Section 25.6. It is noted in Ref. [33] that integer programming techniques can be used with the OST formulation to solve the problem of minimizing the execution time of the application under a fixed dollar cost constraint to purchase the machines that will compose the HC suite, or minimizing the purchase cost under a fixed execution time constraint.

AOST [80] augments OST by incorporating the performance of code segments for all available machine choices (not just the best matched machine type) and by considering non-uniform decompositions of code segments. The issue of considering all available choices of machines is important in practice because the best matched machine may be unavailable. Different machine types are considered, and each machine type may include different models (e.g., the SIMD machine type may include multiple copies of Thinking Machine's CM-2 and/or MasPar's MP-1). Unlike the OST formulation, the number of available machines for each type is limited. For ease of presentation and without loss of generality, the case of having only one model (perhaps multiple copies) for every machine type is considered here.

The optimal speedup $\theta[\tau]$ with respect to a baseline sequential system (e.g., a Sun SPARCstation 5), is assumed to be estimated by analytical benchmarking based on the best matched code type for each machine type τ. For an M-machine suite, for each code segment j, an M-tuple is assumed to be known from task profiling:

$$\omega[j] = (\pi[1,j], \pi[2,j], \ldots \pi[M,j])$$

where $0 \le \pi[\tau,j] \le 1$ is an indicator of how well code segment j can be matched with machine type τ. Let S be the set of $|S|$ non-overlapping code segments of the application task. Let μ be the number of different machine types to be considered.

The maximum number of independent code blocks into which code segment j can be decomposed for concurrent execution on machines of type τ is defined as $v[\tau,j]$ and is assumed to be known. Let $\beta[\tau]$ = number of machines of type τ available (or possible to purchase). Thus, the number of code blocks into which code segment j can be decomposed is $\gamma[\tau,j] = \min(v[\tau,j], \beta[t])$. Assume that on the baseline system $p[j]$ = fraction of time spent executing code segment j relative to the overall execution time of S, and $p[j,i]$ = fraction of time spent executing code block i relative to the execution time of code segment j. Thus,

$$\sum_{j=1}^{|S|} p[j] = 1 \quad \text{and} \quad \sum_{i=1}^{\gamma[\tau,j]} p[j,i] = 1, \quad \text{for all } \tau, j$$

The *available parallelism* of a code segment is defined to be the minimum number of processors that results in the optimal execution time with respect to its assumed machine model. Let $\Lambda[\tau,j]$ denote the utilization factor when executing a code segment (or block) j on a machine of type τ. $\Lambda[\tau,j] = 1$ if the available parallelism of code segment j with respect to machine type τ is greater than or equal to the number of processors within machine type τ; otherwise $\Lambda[\tau,j] = $ (available parallelism)/(total number of processors) < 1. Thus, the expected actual speedup of code segment j on machine τ is $q[\tau] \times \pi[\tau,j] \times \Lambda[\tau,j]$. The execution time of a decomposable code segment is the longest execution time among all its code blocks executing on the selected machines. The relative execution time for code segment j on machine type τ, i.e., the time to execute code segment j on a machine of type τ divided by the time to execute the entire task on the baseline machine, is given by

$$\lambda[\tau,j] = \max_{1 \le i \le \gamma[\tau,j]} \left\{ \frac{p[j] \times p[j,i]}{\theta[\tau] \times \pi[\tau,j] \times \Lambda[\tau,i]} \right\}$$

Code segment j is assumed to be executed on machines of type $\tau[j]$, $1 \le \tau[j] \le \mu$, for each $1 \le j \le |S|$. Thus, for a given matching of code segments to machine types (i.e., $\tau[j]$ values), the relative execution time of S is given by:

$$ET\{\tau[1], \tau[2], \ldots, \tau[|S|]\} = \sum_{j=1}^{|S|} \lambda[\tau[j], j]$$

Given the overall cost constraint, H, and the cost of a machine of type τ, $h[\tau]$, AOST is formulated as:

$$\min_{\substack{1 \le [\tau,j] \le \mu \\ 1 \le j \le |S|}} ET\{\tau[1], \tau[2], \ldots \tau[|S|]\}$$

subject to

$$\sum_{\tau=1}^{\mu} \left(\max_{1 \le j \le |S|} \gamma[\tau,j] \right) \times h[\tau] \le H$$

HOST [21] extends AOST by incorporating the effects of various local mapping techniques and allowing concurrent execution of mutually independent code segments on different types of machines. The "Hierarchical Cluster-M" model [28] is discussed in Ref. [21] as a way to simplify the matching process by exploiting the hierarchically clustered structure of both the system architecture and the application's communication graph.

In the formulation of HOST, it is assumed that a particular application task is divided into *subtasks*. Subtasks are executed serially. Each subtask may consist of a collection of code segments (as defined earlier) that can be executed concurrently. A code segment consists of homogeneous parallel instructions. Each code segment is further decomposed into several code blocks that can be executed concurrently on machines of the same type. The execution time of a subtask is equal to the longest execution time among all code segments in that subtask. Similarly, the execution time of a code segment is equal to the longest execution time among all code blocks in that segment. The underlying mathematical formulation of HOST is similar to (and a generalization of) that of AOST.

GOST [59] generalizes OST and its extensions to include tasks modeled by general dependency graphs. In GOST, it is assumed that there are ω different machine types and an unlimited number of machines in each type. Different machine models are treated as different types.

In GOST, the most basic code element·is a *process,* which corresponds to a block or a nondecomposable code segment (as defined by AOST). It is assumed that an application task consists of several processes modeled by a dependency graph, which could be generated by task profiling. Each node η_i of the graph represents a process and has a number of weights corresponding to the execution times of that process on each machine type for each mapping available on that machine. An edge of the graph represents dependencies between two processes that require communication. Each edge (η_i, η_j) has a number of weights (communication times)—one for each reasonable communication path between each possible pair of host machines for processes η_i and η_j. The weights for nodes and edges are assumed to be derivable from analytical benchmarking. The objective is to determine the optimal matching/scheduling in which each process node in the dependency graph is assigned one machine type and a start time, and the completion time of the whole application is minimized using polynomial time heuristics.

Other formulations and solution techniques. SmartNet is a real-time look-ahead/look-back, near optimal scheduler/planner for HC systems that is being designed and developed at NRaD (a Naval Laboratory) [34]. SmartNet's goal is to minimize the total time to execute a task set, taking into account the times to compute each task i on each machine j, as well as the latency time for any needed data transfers involved in computing task i on machine j. It is assumed that each machine may have previously scheduled tasks either executing or awaiting execution. The objective is not to minimize the compute time for any specific task, but rather to maximize the overall throughput. SmartNet is currently operational and performs the functions discussed above. Various extensions to SmartNet are under development.

In Ref. [78], the HC system is represented by an architecture graph, in which the nodes represent the machines and the edges represent the interconnections among the machines. The application task, which is also modeled with a graph, uses nodes to represent the interacting code blocks and edges to represent data communication dependencies among the code blocks. It is assumed that the bandwidth of each link and the interface overhead between each pair of machines are known. It is also assumed that the computation time of each code block on each machine and the amount of communication required between each pair of code blocks are known. Mapping interacting code blocks of the given application task to machines in the HC system is done by assigning each code block to a machine (i.e., node in the architecture graph). The objective is to minimize the completion time of the whole program. An initial mapping is assumed at the beginning of the search. The basic actions of the proposed graph-based search are called moves. An example of a move is swapping the current locations of two code blocks. Three types of heuristics are used for attempting to find the optimal mapping. Simulations on randomly generated models are conducted to compare the solution quality and execution times among the three approaches.

Another graph-based method for representing problems for automatically matching code blocks to machines in an HC environment is presented in Ref. [54]. In this work, a "generalized virtual fully-connected architecture graph" is proposed as the machine abstraction and a "Meta Graph" is proposed as the abstraction for the task. In the architecture graph, each node represents a machine in the HC system and contains various machine characteristics. Each edge represents the virtual communication link between every pair of machines, and includes information such as connectivity (i.e., direct versus indirect), connection bandwidth, physical distance, and node-pair heterogeneity (i.e., data-reformatting requirements). In the Meta Graph, the nodes represent code blocks, and edges represent control and data flows between code blocks. Classical list scheduling [65] is augmented to utilize the node-pair heterogeneity representation and is used in simulations on randomly generated problems to match code blocks to machines. Based on several hundred simulations, an average improvement of approximately 70% is obtained from this implementation over the regular weighted graph implementation (i.e., without the node-pair heterogeneity information).

A crossover strategy for assigning tasks on a simple HC system consisting of two machines is proposed in Ref. [56]. It is assumed that the two machines work in a client/server mode. The proposed strategy is used by the client to decide when the speedup of executing a subtask on the server can compensate for the communication/interface overhead involved. When deemed to be beneficial, a remote procedure call is used to execute this subtask on the server. Two experiments were conducted on an actual HC system consisting of a Sun workstation, which functioned as the client, and a Thinking Machines CM-200, which operated as the server. The first experiment was an implementation of the "maximum subvector problem," which involves finding the maximum sum of elements of any contiguous subvector of a given real input vector. The second experiment was based on an implementation of the shallow weather prediction benchmark [76]. The proposed crossover strategy was shown to make the correct choice for executing these applications (i.e., executing entirely on the client or using both the client and the server). In the first application, using both the client and the server was shown to be the proper choice provided that the vector size was larger than a critical value. For the second application, the choice was to always use (only) the client because of high communication requirements.

25.7.4 Summary

Some existing matching and scheduling techniques for stage 3 of the conceptual model were overviewed. More research is needed to integrate all of the stages of the conceptual model into a practical system.

25.8 Conclusions and Future Directions

Although the underlying goal of HC is straightforward—to support computationally intensive applications with diverse computing requirements—there are a great many open problems that need to be solved before HC can be made available to the average applications programmer in a transparent way. Many (possibly even most) need to be addressed just to facilitate near-optimal practical use of mixed-machine HC systems in a "visible" (i.e., user specified) way. Below is a brief informal discussion of some of these open problems for mixed-machine HC systems; the first few also apply to mixed-mode HC systems. While this list is far from exhaustive, it will convey the types of issues that need to be addressed. Others may be found in Refs. [49, 75].

Implementation of an automatic HC programming environment, such as that envisioned in Section 25.5, will require a great deal of research for devising practical and theoretically sound methodologies for each component of each stage. A general open question that is particularly applicable to stages 1 and 2 of the conceptual model is: "What information should (must) the user provide and what information should (can) be determined automatically?" For example, should the user specify the subtasks within an application or can this be done automatically? Future HC systems will probably not completely automate all of the steps in the conceptual model. A key to the future success of HC hinges on striking a proper balance between the amount of information expected from the user (i.e., effort) and the level of performance delivered by the system.

To program an HC system, it would be best to have one or more machine-independent programming languages that allow the user to augment the code with compiler directives. The programming language and user specified directives should be designed to facilitate (1) the compilation of the program into efficient code for any of the machines in the suite, (2) the decomposition of tasks into homogeneous subtasks, (3) the determination of computational requirements of each subtask, and (4) the use of machine-dependent subroutine libraries.

Along with programming languages, there is a need for debugging and performance tuning tools that can be used across an HC suite of machines. This involves research in the areas of distributed programming environments and visualization tools.

Operating system support for HC is needed. This includes techniques applicable at both the local machine level and at the system-wide network level.

Ideally, information about the current loading and status of the machines in the HC suite and the network that is linking these machines should be incorporated into the matching and scheduling decisions. Many questions arise here: what information to include in the status (e.g., faulty or not, pending tasks), how to measure current loading, how to effectively incorporate current loading and status information into matching and scheduling decisions, how to communicate and structure the loading and status information in the other machines, how often to update this information, and how to estimate task/transfer completion time.

There is much ongoing research in the area of inter-machine data transport. This research includes the hardware support required, the software protocols required, designing the network topology, computing the minimum time path between two machines, and devising rerouting schemes in case of faults or heavy loads. Related to this is the data reformatting problem, involving issues such as data type storage formats and sizes, byte ordering within data types, and machines' network-interface buffer sizes.

Another area of research pertains to methods for dynamic task migration between different parallel machines at execution time. Current research in this area involves how to move an executing task between different machines and determining how and when to use dynamic task migration for load rebalancing or fault tolerance.

Lastly, there are policy issues that require system support. These include what to do with priority tasks, what to do with priority users, what to do with interactive tasks, and security.

Thus, although the uses of existing HC systems demonstrate the significant benefit of HC, the amount of effort currently required to implement an application on an HC system can be substantial. Future research on the above open problems will improve this situation and allow HC to realize its inherent potential.

Acknowledgments

We thank K. H. Casey, H. G. Dietz, R. F. Freund, A. Ghafoor, A. S. Grimshaw, R. Gupta, E. L. Lusk, R. W. Quong, J. Rosenman, J. M. Siegel, V. S. Sunderam, and J. Yang for their valuable comments. We especially thank J. M. Siegel for her careful reading of the manuscript.

25.9   References

1. Adams III, G. B., and H. J. Siegel. 1982. The extra stage cube: a fault tolerant interconnection network for supersystems. *IEEE Transactions on Computers,* Vol. C-31, No. 5, 443–454.

2. Aho, A., R. Sethi, and J. D. Ullman. 1986. *Compilers: Principles, Techniques, and Tools.* Reading, Mass.: Addison-Wesley.

3. Almasi, G. S., and A. Gottlieb. 1989. *Highly Parallel Computing.* Redwood City, Calif.: Benjamin/Cummings.

4. Armstrong, J. B., M. A. Nichols, H. J. Siegel, and L. H. Jamieson. 1991. Examining the effects of CU/PE overlap and synchronization overhead when using the complete sums approach to image correlation. *Proceedings of the Third IEEE Symposium on Parallel and Distributed Processing,* 224–232.

5. Armstrong, J. B., H. J. Siegel, W. E. Cohen, M. Tan, H. G. Dietz, and J. A. B. Fortes. 1994. Dynamic task migration from SPMD to SIMD virtual machines. *Proceedings of the 1994 International Conference on Parallel Processing,* Vol. II, 160–169.

6. Atallah, M. J., C. L. Black, D. C. Marinescu, H. J. Siegel, and T. L Casavant. 1992. Models and algorithms for coscheduling compute-intensive tasks on a network of workstations. *J. Parallel and Distributed Computing,* Vol. 16, No. 4, 319–327.

7. Batcher, K. E. 1968. Sorting networks and their applications. *Proceedings of the AFIPS 1968 Spring Joint Computer Conference,* 307–314.

8. Beguelin, A. L. 1993. Xab: a tool for monitoring PVM programs. *Proceedings of the Workshop on Heterogeneous Processing,* 92–97.

9. Beguelin, A., J. Dongarra, G. A. Geist, R. Manchek, K. Moore, R. Wade, J. Plank, and V. Sunderam. 1992. *HeNCE: A User's Guide, Version 1.2.* Oak Ridge, Tenn.: Oak Ridge National Laboratory, Engineering Physics and Mathematics Division, Dec. 1992, 28 pp.

10. Beguelin, A., J. Dongarra, G. A. Geist, R. Manchek, and V. Sunderam. 1991. *A User's Guide to PVM: Parallel Virtual Machine.* Technical Report ORNLITM-11826. Oak Ridge, Tenn.: Oak Ridge National Laboratory, Engineering Physics and Mathematics Division, 13 pp.

11. Beguelin, A., J. Dongarra, A. Geist, R. Manchek, and V. Sunderam. 1993. Visualization and debugging in a heterogeneous environment. *IEEE Computer,* Vol. 26, No. 6, 88–95.

12. Berg, T. B., S.-D. Kim, and H. J. Siegel. 1991. Limitations imposed on mixed-mode performance of optimized phases due to temporal juxtaposition. *J. Parallel and Distributed Computing,* Vol. 13, No. 2, 154–169.

13. Berg, T. B., and H. J. Siegel. 1991. Instruction execution trade-offs for SIMD vs. MIMD vs. mixed-mode parallelism. *Proceedings of the Fifth International Parallel Processing Symposium,* 301–308.

14. Bergman, L., H-W. Braun, B. Chinoy, A. Kolawa, A. Kuppermann, P. Lyster, C. R. Mechoso, P. Messina, J. Morrison, D. Stanfill, W. St. John, and S. Tenbrink. 1993. *CASA Gigabit Testbed 1993 Annual Report: A Testbed for Distributed Supercomputing.* Technical Report CCSF-33. Pasadena, Calif.: Caltech Concurrent Supercomputing Facilities, California Institute of Technology, 68 pp.

15. Butler, R., W. Gropp, and E. Lusk. 1993. Developing applications for a heterogeneous computing environment. *Proceedings of the Workshop on Heterogeneous Processing,* 77–83.

16. Butler, R. M., A. L. Leveton, and E. L. Lusk. 1993. p4-Linda: a portable implementation of Linda. *Proceedings of the Second International Symposium on High Performance Distributed Computing,* 50–58.

17. Butler, R. M., and E. L. Lusk. User's guide to the p4 parallel programming system. *Technical Report ANL-92117.* Argonne, Ill.: Argonne National Laboratory, Mathematics and Computer Science Division, 34 pp.

18. Butler, R. M., and E. L. Lusk. 1994. Monitors, messages, and clusters: The p4 parallel programming system. *Parallel Computing,* Vol. 20, 547–564.

19. Carriero, N., D. Gelernter, and T. G. Mattson. 1992. Linda in heterogeneous computing environments. *Proceedings of the Workshop on Heterogeneous Processing,* 43–46.

20. Casavant, T. L., and J. G. Kuhl. 1988. A taxonomy of scheduling in general-purpose distributed computing systems. *IEEE Trans. Software Engineering,* Vol. 14, No. 2, 141–154.

21. Chen, S., M. M. Eshaghian, A. Khokhar, and M. E. Shaaban. 1993. A selection theory and methodology for heterogeneous supercomputing. *Proceedings of the Workshop on Heterogeneous Processing,* 3, 15–22.

22. Conte, T. M., and W. W. Hwu. 1991. Benchmark characterization. *IEEE Computer,* Vol. 24, No.1, 48–56.

23. Cray Corp. 1994. The Cray-3/Super Scalable System, preliminary specifications. Cray Computer Corporation, 10 pp.

24. Darema, F., D. A. George, V. A. Norton, and G. F. Pfister. 1988. A single-program-multiple-data computational model for EPEX/FORTRAN. *Parallel Computing,* Vol. 7, 11–24.

25. Dietz, H. G., W. E. Cohen, and B. K. Grant. 1993. Would you run it here... or there? (AHS: automatic heterogeneous supercomputing). *Proceedings of 1993 International Conference on Parallel Processing,* Vol. II, 217–221.

26. Dongarra, J., J. L. Martin, and J. Worlton. 1987. Computer benchmarking: paths and pitfalls. *IEEE Spectrum,* 38–43.

27. Duclos, P., F. Boeri, M. Auguin, and G. Giraudon. 1988. Image processing on a SIMD/SPMD architecture: OPSILA. *Proceedings of the 9th International Conference on Pattern Recognition,* 14–17.

28. Eshaghian, M. M., and R. F. Freund. 1992. Cluster-M paradigms for high-order heterogeneous procedural specification computing. *Proceedings of the Workshop on Heterogeneous Processing,* 47–49.

29. Parasoft Corporation. 1990. *Express User's Guide Version 3.0.* Pasadena, Calif.: Parasoft Corporation.

30. Fineberg, S. A., T. L. Casavant, and H. J. Siegel. 1991. Experimental analysis of a mixed-mode parallel architecture using bitonic sequence sorting. *J. Parallel and Distributed Computing,* Vol. 11, No. 3, 239–251.

31. Flynn, M. J. 1966. Very high-speed computing systems. *Proceedings of the IEEE,* Vol. 54, No. 12, 1901–1909.

32. Freund, R. F. 1989. Optimal selection theory for superconcurrency. *Proceedings of Supercomputing '89,* 699–703.

33. Freund, R. F. 1991. SuperC or distributed heterogeneous HPC. *Computing Systems in Engineering,* Vol. 2, No. 4, 349–355.

34. Freund, R. F. 1994. The challenges of heterogeneous computing. *Proceedings of the Parallel Systems Fair at the 8th International Parallel Processing Symposium,* 84–91.

35. Freund, R. F., and D. S. Conwell. 1990. Superconcurrency: a form of distributed heterogeneous supercomputing. *Supercomputing Review,* Vol. 3, No. 10, 47–50.

36. Freund, R. F., and H. J. Siegel. 1993. Heterogeneous processing. *IEEE Computer,* Vol. 26, No. 6, 13–17.

37. Geist, G. A., M. T. Heath, B. W. Peyton, and P. H. Worley. 1990. *User's Guide to PICL: A Portable Instrumented Communication Library.* Technical Report ORNLITM-11616. Oak Ridge, Tenn.: Oak Ridge National Laboratory, Engineering Physics and Mathematics Division, 22 pp.

38. Ghafoor, A., and J. Yang. 1993. Distributed heterogeneous supercomputing management system. *IEEE Computer,* Vol. 26, No. 6, 78–86.

39. Giolmas, N., D. W. Watson, D. M. Chelberg, and H. J. Siegel. 1992. A parallel approach to hybrid range image segmentation. *Proceedings of the 6th International Parallel Processing Symposium,* 334–342.

40. Grimshaw, A. S. 1993. Easy-to-use object-oriented parallel processing with Mentat. *IEEE Computer,* Vol. 26, No. 5, 39–51.

41. Grimshaw, A. S., J. B. Weissman, E. A. West, and E. Loyot. 1994. Meta Systems: an approach combining parallel processing and heterogeneous distributed computing systems. *J. Parallel and Distributed Computing,* Vol. 21, No. 3, 257–270.

42. Haddad, E. 1993. Load distribution optimization in heterogeneous multiple processor systems. *Proceedings of the Workshop on Heterogeneous Processing,* 42–47.

43. Herrarte, V., and E. Lusk. 1991. *Studying Parallel Program Behavior with Upshot.* Technical Report ANL-91115. Argonne, Ill.: Argonne National Laboratory, Mathematics and Computer Science Division.

44. Herter, C. G., T. M. Warschko, W. F. Tichy, and M. Philippsen. 1993. Triton/1: a massively parallel mixed-mode computer designed to support high level languages. *Proceedings of the Workshop on Heterogeneous Processing,* 65–70.

45. Jamieson, L. H. 1987. Characterizing parallel algorithms. In *The Characteristics of Parallel Algorithms,* L. H. Jamieson, D. B. Gannon, and R. J. Douglass, eds. Cambridge, Mass.: MIT Press, 65–100.

46. Kaplan, J. A., and M. L. Nelson. 1993. A comparison of queueing, cluster and distributed computing systems. *NASA Technical Memorandum 109025.* Langley Research Center, Hampton, Va.: National Aeronautics and Space Administration, 47 pp.

47. Karpovich, J. K., M. Judd, W. T. Strayer, and A. S. Grimshaw. 1993. A parallel object-oriented framework for stencil algorithms. *Proceedings of the Second International Symposium on High Performance Distributed Computing,* 34–41.

48. Khokhar, A., V. K. Prasanna, M. Shaaban, and C. L. Wang. 1992. Heterogeneous supercomputing: problems and issues. *Proceedings of Workshop on Heterogeneous Processing,* 3–12.

49. Khokhar, A., V. K. Prasanna, M. Shaaban, and C. L. Wang. 1993. Heterogeneous computing: challenges and opportunities. *IEEE Computer,* Vol. 26, No. 6, 18–27.

50. Kim, S.-D., M. A. Nichols, and H. J. Siegel. 1991. Modeling overlapped operation between the control unit and processing elements in an SIMD machine. *J. Parallel and Distributed Computing,* Vol. 12, No. 4, 329–342.

51. Klietz, A. E., A. V. Malevsky, and K. Chin-Purcell. 1993. A case study in metacomputing: distributed simulations of mixing in turbulent convection. *Proceedings of the Workshop on Heterogeneous Processing,* 101–106.

52. Kogge, P. M. 1994. EXECUBE—a new architecture for scalable MPPs. *Proceedings of 1994 International Conference on Parallel Processing,* Vol. I, 77–84.

53. Lowrie, D. H 1975. Access and alignment of data in an array processor. IEEE Transactions on Computers, Vol. C-24 (December), 1145–1155.

54. Leangsuksun, C., and J. Potter. 1993. Problem representations for an automatic mapping algorithm on heterogeneous processing environments. *Proceedings of the Workshop on Heterogeneous Processing,* 48–53.

55. Li, Y. A., J. K. Antonio, H. J. Siegel, M. Tan, and D. W. Watson. 1995. Estimating the distribution of execution times for SIMD/SPMD mixed-mode programs. *Proceedings of Heterogeneous Computing Workshop,* 35–46.

56. Lilja, D. J. 1993. Experiments with a task partitioning model for heterogeneous computing. *Proceedings of the Workshop on Heterogeneous Processing,* 29–35.

57. Lipovski, G. J., and M. Malek. 1987. *Parallel Computing: Theory and Comparisons.* New York: John Wiley & Sons.

58. MasPar Computer Corp. 1993. *Data-Parallel Programming Guide.* Sunnyvale, Calif.: MasPar Computer Corporation.

59. Narahari, B., A. Youssef, and H. A. Choi. 1994. Matching and scheduling in a generalized optimal selection theory. *Proceedings of the Heterogeneous Computing Workshop,* 3–8.

60. Ni, L. M., and K. Hwang. 1981. Optimal load balancing strategies for a multiple processor system. *Proceeding of the 1981 International Conference on Parallel Processing,* 352–357.

61. Nichols, M. A., H. J. Siegel, and H. G. Dietz. 1993. Data management and control-flow aspects of an SIMD/SPMD parallel language/compiler. *IEEE Trans. on Parallel and Distributed Systems,* Vol. 4, No. 2, 222–234.

62. Norman, M. G., and P. Thanisch. 1993. Models of machines and computation for mapping in multicomputers. *ACM Computing Surveys,* Vol. 25, No. 3, 263–302.

63. Pease, D., A. Ghafoor, I. Ahmad, D. L. Andrews, K. Foudil-Bey, T. E. Karpinski, M. A. Mikki, and M. Zerrouki. 1991. PAWS: a performance evaluation tool for parallel computing systems. *IEEE Computer,* Vol. 24, No. 1, 18–29.

64. Philippsen, M., T. Warschko, W. F. Tichy, and C. Herter. 1993. Project Triton: towards improved programmability of parallel machines. *Proceedings of the 26th Hawaii International Conference on System Sciences,* 192–201.

65. Polychronopoulos, C. D. 1988. *Parallel Programming and Compilers.* Norwell, Mass.: Kluwer Academic Publishers.

66. Rosenman, J., and T. Cullip. 1992. High-performance computing in radiation cancer treatment. *CRC Critical Reviews in Biomedical Engineering,* Vol. 20, Nos. 5–6, 391–402.

67. Saghi, G., H. J. Siegel, and J. L. Gray. 1993. Predicting performance and selecting modes of parallelism: a case study using cyclic reduction on three parallel machines. *J. Parallel and Distributed Computing,* Vol. 19, No. 3, 219–233.

68. Siegel., H. J. 1990. *Interconnection Networks for Large-Scale Parallel Processing: Theory and Case Studies,* ed. 2. New York: McGraw-Hill.

69. Siegel, H. J., J. B. Armstrong, and D. W. Watson. 1992. Mapping computer-vision-related tasks onto reconfigurable parallel processing systems. *IEEE Computer,* Vol. 25, No. 2, 54–63.

70. Siegel, H. J., T. Schwederski, J. T. Kuehn, and N. J. Davis IV. An Overview of the PASM Parallel Processing System. In *Computer Architecture,* D. D. Gajski, V. M. Milutinovic, H. J. Siegel, and B. P. Furht, eds., IEEE Computer Society Press, Washington, DC, 1987, 387–407.

71. Siegel, H. J., T. Schwederski, W. G. Nation, J. B. Armstrong, L. Wang, J. T. Kuehn, R. Gupta, M. D. Allemang, D. G. Meyer, and D. W. Watson. 1995. Chapter 25, The design and prototyping of the PASM reconfig-

urable parallel processing system. To appear in *Parallel Computing: Paradigms and Applications,* 2nd ed., A. Y. Zomaya, ed. London: International Thomson Computer Press, 78–114.

72. 1990. Special report: gigabit network testbeds. *IEEE Computer,* Vol. 23, No. 9, 77–80.

73. Stevenson, D., and K. Antell, eds. 1993. *VISTAnet Annual Report,* 138 pp.

74. Sunderam, V. S. 1990. PVM: a framework for parallel distributed computing. *Concurrency: Practice and Experience,* Vol. 2, No. 4, 315–339.

75. Sunderam, V. S. 1992. Design issues in heterogeneous network computing. *Proceedings of the Workshop on Heterogeneous Processing,* revised edition, 101–112.

76. Swartzrauber, P. N. 1984. *The Shallow Benchmark Weather Prediction Program.* National Center for Atmospheric Research.

77. Tan, M., J. K. Antonio, H. J. Siegel, and Y. A. Li. Scheduling and data relocation for sequentially executed subtasks in a heterogeneous computing system. *Proceedings of the Heterogeneous Computing Workshop,* 109–120.

78. Tao, L., B. Narahari, and Y. C. Zhao. 1993. Heuristics for mapping parallel computations to heterogeneous parallel architectures. *Proceedings of the Workshop on Heterogeneous Processing,* 36–41.

79. Ulrey, R. R., A. A. Maciejewski, and H. J. Siegel. 1994. Parallel algorithms for singular value decomposition. *Proceedings of the 8th International Parallel Processing Symposium,* 524–533.

80. Wang, M., S.-D. Kim, M. A. Nichols, R. F. Freund, H. J. Siegel, and W. G. Nation. 1992. Augmenting the optimal selection theory for superconcurrency. *Proceedings of the Workshop on Heterogeneous Processing,* 13–22.

81. Watson, D. W., J. K. Antonio, H. J. Siegel, and M. J. Atallah. 1994. Static program decomposition among machines in an SIMD/SPMD heterogeneous environment with nonconstant mode switching costs. *Proceedings of the Heterogeneous Computing Workshop,* 58–65.

82. Watson, D. W., H. J. Siegel, J. K. Antonio, M. A. Nichols, and M. J. Atallah. 1994. A block based mode selection model for SIMD/SPMD parallel environments. *J. Parallel and Distributed Computing,* Vol. 21, No. 3, 271–288.

83. Weems, C. C., E. M. Riseman, and A. R. Hanson. 1992. Image understanding architecture: exploiting potential parallelism in machine vision. *IEEE Computer,* Vol. 25, No. 2, 65–68.

84. Weems, C. C., G. E. Weaver, and S. G. Dropsho. 1994. Linguistic support for heterogeneous parallel processing: a survey and an approach. *Proceedings of the Heterogeneous Computing Workshop,* 81–88.

85. Wilson, G. V. 1993. A glossary of parallel terminology. *IEEE Parallel and Distributed Technology,* Vol. 1, No. 1, 52–67.

86. Yang, J., I. Ahmad, and A. Ghafoor. 1993. Estimation of execution times on heterogeneous supercomputer architecture. *Proceedings of the 1993 International Conference on Parallel Processing,* Vol. I, 219–225.

87. Yang, J., A. Khokhar, S. Sheikh, and A. Ghafoor. 1994. Estimating execution time for parallel tasks in heterogeneous processing (HP) environment. *Proceedings of the Heterogeneous Computing Workshop,* 23–28.

26

Cluster Computing

Louis H. Turcotte

Traditionally, high-performance computing (HPC) has been synonymous with the deployment of multimillion dollar vector and/or parallel computers. Recent advances in both microprocessor performance and network bandwidth are radically altering HPC environments. Over the last four years, HPC environments created using workstations interconnected via high-speed networks have enjoyed considerable success as complementary components of traditional HPC systems. The term *cluster computing* has been used most frequently to describe this new form of computing. Other terms such as *workstation cluster, ensemble computing, hypercomputing, network-based concurrent computing, workstation farm, workstation array,* and *ultracomputing* have also been used to describe the same concept. Cluster computing is becoming widely successful for several reasons:

- Anyone with two or more workstations connected via a network can create a cluster. Little or no additional cost is involved.
- A cluster provides a readily available environment for research into parallel computing. Previously, an expensive proprietary parallel computer (*proprietary* meaning systems such as CM5, Paragon, T3D, and so on) was required.
- Unused computing cycles can be scavenged, which provides additional computing capacity at no additional cost.
- Clusters leverage the enormous cost performance benefit offered by commodity microprocessor systems .
- Robust/stable first generation software systems for clustering are commonly available [27]. The majority of these are freely available via the Internet.

This chapter begins with an analysis of technological factors that have led to the deployment of workstation clusters. This analysis presents the advances in hardware and the marketplace trends that have motivated clustering. Next, an overview of issues directly related to clustering is presented. This section is followed by a discussion of the three most common categories of cluster use: enterprise clusters, dedicated clusters, and clusters used for parallel computing. Finally, this chapter is concluded with a discussion of open issues related to clustering.

26.1   Technological Evolution

The evolution of workstation clustering is worthy of review and analysis. The trend occurred independent of the normal "vendor offering" process which has pervaded the computing industry since its inception. As a matter of fact, vendors only begin to recognize and offer cluster solutions after the trend was well underway. Workstation clustering is symbolic of the expansion in user independence resulting from the increase in computer literacy—a direct result of the personal computer. Two factors have primarily been responsible for the attention which workstation clustering has received:

- rapid advances in microprocessor technology
- the change in marketplace dynamics

Baskett and Hennessey [2] and Hardenbergh [11] have provided comprehensive perspectives of the microprocessor evolution and noted that one of the main factors required to increase computational performance is directly related to the number of transistors in a microprocessor die. Figure 26.1 summarizes the phenomenal rate of increase for transistors per die for both microprocessor and memory chips. Both types of chips double in transistor per die count every two years, with the Pentium chip having 3.1 million transistors. Figure 26.2 indicates that, along with the number of transistors per die, microprocessor manufacturers have also been enlarging the die size. Over the last 20 years, die sizes have averaged doubling every 5 years. Another important characteristic of microprocessors is the minimum feature size. Feature size has been halving every seven years, which means that the same size die produced ten years ago has ten times the transistors today.

Other factors, related directly to improvements from microelectronics, have radically affected the computer marketplace. In 1975, the Intel 8080 generation of microprocessors had an 8-bit data bus width. Practically all the RISC microprocessors that will be shipped

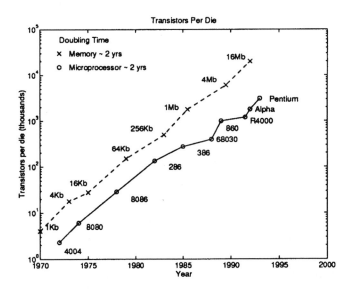

Figure 26.1 Transistors per die

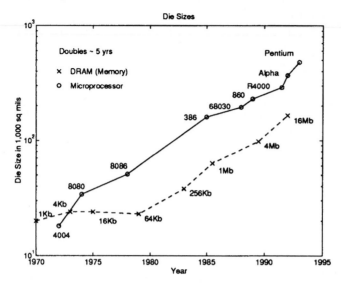

Figure 26.2 Die sizes

in 1994 will have 64-bit data buses, a doubling factor of every five years. Additionally, backplane bus speeds have been doubling every ten years, which means that systems can move larger quantities of data efficiently between internal subsystems. Memory chip capacity has also advanced at rates comparable to microprocessors and data buses. Figure 26.3 illustrates the advances in memory chip capacities over the last 20 years and the prospectus for memory chips through the year 2000. Memory chip capacity has been doubling at the amazing rate of every 1.5 years with 16-Mb chips commonly available today. Finally, Fig. 26.4 summarizes a final important characteristic of computers, the internal clock rate, or heartbeat, of the system. Some very important conclusions can be realized by studying the curves for supercomputers and microprocessors. The microprocessor

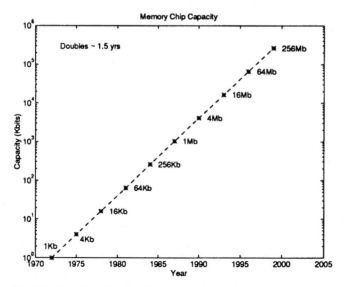

Figure 26.3 Memory chip capacity

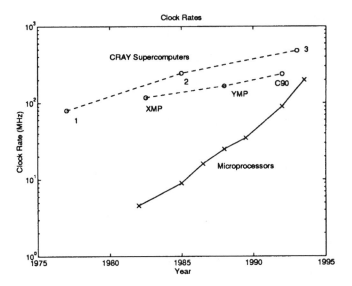

Figure 26.4 Clock rates

curve (a collection of data points from various vendor chips) is approaching the clock rates of the most expensive, liquid-cooled CRAY systems commercially available.

While raw manufacturing specifications of today's microprocessors are important and certainly offer a clue to the advances in computing capability, one of the most important characteristics of any computer used for scientific investigation is the ability to perform floating-point calculations.

Figure 26.5 illustrates graphically results from the standard LINPACK 1000 benchmark* report published regularly by Dongarra [9]. This figure indicates that a microprocessor-based workstation costing around $50,000 can achieve approximately one-sixth the performance of a multimillion dollar CRAY supercomputer.

The microprocessor trends presented above have radically altered the computing industry. Amazingly, these trends are forecast to continue at least until the turn of the century. Intel has frequently indicated that it expects the following characteristics for a microprocessor in the year 2000:

- 50 million transistors per chip
- 250 MHz clock rate
- 750 MIPS and 500 MFLOPS peak per processor
- 4 processors per chip
- multi-megabyte cache

This chip, with a peak performance of 2 GFLOPS, will be delivered on 1 square inch of silicon. Often, industries are slow to respond to changes in technologies. It is always important to consider the economic realities of the marketplace when studying the effect of new solutions and to remember that technologies thrive, ultimately, because of their ability to make their companies profitable. Figure 26.6, published by the U.S. Congressional Budget Office [8], provides some interesting clues related to future high-performance comput-

---

*The LINPACK 1000 benchmark involves solving a set of 1000 equations.

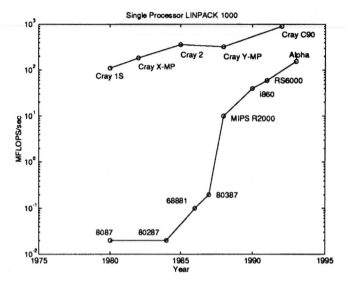

Figure 26.5 Single-processor LINPACK 1000

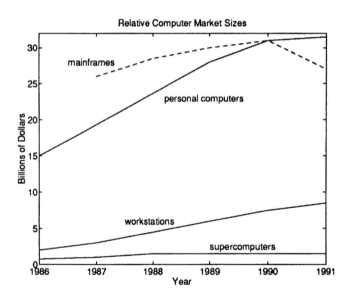

Figure 26.6 Relative computer market sizes

ing technologies. Analysis of this figure for the mainframe and personal computer curves clearly reveals what happened in the business computing marketplace. In 1990, the large mainframe marketplace was drastically reduced by the recognition by users that highly capable, networked, distributed personal computers could serve a large percentage of their computing needs at a fraction of the cost of a mainframe.[*] A similar phenomenon is rap-

---

[*]Most large mainframe providers were late reacting to this trend.

idly occurring in the scientific marketplace. The figure indicates that the market for super-computers has remained flat for several years but the workstation market is growing rapidly. There are two primary reasons for the stagnation in the supercomputer market-place: (1) the decline in spending by federal agencies on supercomputers, mainly due to reduced federal budgets, and (2) the realization that microprocessor-based solutions are simply more cost effective for the majority of workloads.

The combination of advances in microprocessor technology and marketplace dynamics has led to experimentation with new computing environments. Cluster computing represents one of the new environments based on commodity microprocessor components. The emergence of high-speed (e.g., in the gigabit/second range) networks means that significant computational resources can be configured using, in many cases, powerful workstations that already exist but are poorly utilized. The remainder of this chapter will discuss the unique characteristics of cluster computing.

26.2   Overview of Clustering

Workstation clusters offer many benefits when compared to other computing solutions. The most favorable characteristic is that, in principal and implementation, clustering is relatively simple. This *lack of complexity* has also led to many exaggerated claims, including the statement that workstation clusters are practical replacements for general-purpose supercomputers. Clusters, however, should be seen as complementary components of high-performance computing environments. In essence, the rapid increase in microprocessor performance and network bandwidth has made clustering a practical, cost effective computing solution which is readily available to the masses.[*]

At the hardware level, a cluster is simply a collection of independent systems, typically workstations, connected via a commodity network.[†] Workstations in a cluster communicate via one of the two common transport protocols: connection oriented or connection-less. The connection model, based on *transmission control protocol* (TCP), processes streams of bytes with the feature that reliability of message delivery is assured. The connection-less model, based on user datagram protocol (UDP), sends packets of data that are *attempted* to be delivered (e.g. reliability of message delivery is *not* assured.). Two software methods are commonly used to communicate information: message passing and distributed shared memory. Message passing relies on explicitly transmitting messages containing the information between systems. Distributed shared memory (DSM) [2, 17, 20, 25], which is usually implemented using message passing, provides a programming model where data is accessed without concern for physical location. The actual movement of data between cluster nodes is hidden from the application programmer and handled by the DSM implementation.

Because clusters are relatively simple to configure, it is important to categorize which jobs are most conducive to this environment. Studies by several supercomputing centers have shown that many applications presently executing on supercomputers do not absolutely require supercomputing resources. Therefore, it is important to understand the two contrasting categories of demand for scientific applications:

---

[*]This chapter is directed at scientific and not commercial uses of clusters. Commercial clusters are also becoming popular but require greater attention to issues such as fault tolerance, data replication, and security.
[†]More and more dedicated clusters are using expensive high-speed networks.

- *Capability demand.* This is aimed at coping with megaproblems that require all the computational capabilities of any available system, including memory and CPU. "Grand challenge" applications fall into this bracket, which has historically been used as the principal justification for adoption of massively parallel processing (MPP).
- *Capacity demand.* This aims at applications requiring substantial, but far from ultimate, performance and making moderate demands on memory. This class of job is large in total volume and is needed on a regular day-to-day basis. These jobs are excellent candidates for workstation clusters.

Clusters, which should be considered complementary to supercomputers, provide the most cost-effective solution for *capacity demand* problems for many organizations.

26.2.1   Benefits of clustering

Workstation clusters offer many benefits over traditional computing environments. Some of the motivations for workstation clusters are summarized below.

- The demand to provide a large number of cost-effective CPU cycles to serve scientific applications with moderate computational requirements continues to challenge organizations. As numerical methods have grown in popularity, more and more scientists have grown to depend on readily accessible, inexpensive computing cycles to perform their studies.
- Government budgets are under greater pressure to deliver more services at less cost. Industry has likewise been subject to greater competitive pressures as a result of a more global economy. Therefore, users are paying serious attention to any computing solution, whether based on personal computers or workstation clusters, that is more cost effective.
- In the previous section, the technological characteristics of microprocessors were presented. The fact that RISC microprocessors are commodity components (e.g., several hundred thousand workstations are shipped each year that use these microprocessors) means that they offer excellent price/performance benefits when compared with traditional vector and parallel computing solutions.
- Workstations are designed to be excellent at serving the interactive job requirements of users. They are especially useful for performing data interpretation (e.g. scientific visualization) of large, complex data sets. However, most workstations are used for interactive work during normal business hours. This means that an abundant number of *batch CPU cycles* are available in nonprime hours. A properly configured cluster can provide both prime-time interactive cycles and nonprime-time batch cycles, yielding a more cost-effective computing environment .
- Parallel computing has been accepted as the most likely approach toward achieving necessary advances in computational capability. Prior to clusters, only dedicated, proprietary, expensive parallel computers could be used for fundamental work necessary to implement applications that exploit parallelism. Clusters offer a platform to develop and explore parallel computing without the investment in specialized hardware. This increases the access to parallel computing, particularly in academic settings, and should subsequently lead to more rapid advances.
- Historically, computing hardware to service scientific workloads has been both expensive and risky. Clusters rely on commodity, consumer-based technology (e.g. workstations and networks), which minimizes economic risk and provides a very simple, low-cost, incremental growth path.

- Fast computers address only one requirement of numerical modeling. Problems continue to grow in complexity and detail. For many problems, the availability of a large memory is as important as, if not greater than, than processor speed. Workstations can provide large, cost-effective memory previously unavailable in most computing environments.
- Clusters offer the best features of sequential and parallel processing. Sequential processing is dependent on the processing speed of the individual processor, and RISC workstations have the most powerful, inexpensive processors available at any time. Parallel processing is inherent in clustering, and additional processing elements can be added incrementally.
- Most dedicated parallel computers have immature software environments. Operating systems, compilers, libraries, and software tools (debuggers, etc.) have yet to develop to a point of generally acceptable stability. Users of workstation clusters for parallel computing have the benefit of operating within mature, stable software environments. This results in a more productive development environment.
- Many individuals now believe that future HPC environments will achieve the maximum performance capabilities only by exploiting the benefits of heterogeneous computing environments (e.g., using a combination of vector, parallel, and other computing approaches). Clusters provide a cost-effective environment to study topics related to heterogeneous computing (e.g. different floating-point data representations between systems, improved networking protocols, etc.).
- Clusters provide an environment that has graceful degradability. When *a single* system in a cluster goes down, the entire cluster is not lost during the maintenance period. Additionally, because clusters are created using commodity components, maintenance costs may be much less than for an equivalent investment in a vector or parallel computer.

26.2.2 Clustering infrastructure

Clusters provide unique infrastructure challenges. They may be geographically dispersed, include workstations owned by multiple individuals and groups, include hardware from multiple vendors, and have various software differences (different operating system releases, different commercial software availability, etc.). A cluster should be configured to address the following requirements.

- Most clusters are presently configured to use an Ethernet network (10 Mb/s). However, to achieve improved functionality, especially for parallel applications, clusters should be configured with some type of higher-speed network (e.g., 100–800 Mb/s per host).
- One of the principal impediments of implementing parallel applications that have good performance characteristics is communication latency. Workstation clusters exacerbate this problem. Interconnects should have a minimal latency of 100–500 μs between hosts.
- Network topology has a considerable influence on the scaling characteristics of clusters. I/O traffic (via NFS or DFS file systems), utilization as a parallel computer, and response to interactive and batch workloads affect the performance characteristics of a cluster. Clusters should be configured to achieve good scalability (e.g., 10–100 hosts).
- A generous amount of globally available (via NFS or DFS) disk storage is required.
- A reliable job scheduler/controller is required to achieve high efficiency. This system should automatically allocate interactive cycles, allocate serial batch jobs, and allocate parallel batch jobs in a manner that achieves good *load balancing* between the cluster components. Additionally, this system should be capable of reacting to node/network

TABLE 26.1   Cluster Interconnects

Connection	Max. bandwidth (Mb/s)	Measured bandwidth (MB/s)	Latency	Approx. cost per node ($)
Ethernet	10	1	1–3 ms	<1,000
Token ring	16	1.5	1–3 ms	<1,600
Fiber distributed data interface (FDDI)	100	3.5–8	1–3 ms	<5,000
Serial optical channel converter (SOCC) (IBM)	220•	3–14*	1–3 ms	>10,000
Fiber channel standard (FCS)	250–1000*	TBD	1–3 ms	>10,000
HiPPI	800	55	1–3 ms	>30,000
ATM (OC-3)	155*	–	–	>3,000
ATM (OC-12)	622*	–	–	>5,000
VME Bit3	20	6*	10–100 μs	>5,000
Allnode (IBM)	8*	3.5*	15–100 μs	>6,700
GIGAswitch (DEC)	100	–	–	>5,000

*per node

failures by either reassigning the active job or rescheduling the entire job to another system.[*]

26.2.3   Networking

One of the obvious limitations of clusters is created by the relatively slow network interconnection typically employed. The interface will depend on the bandwidth requirements, latency requirements, distance limitations, and budget constraints. Ethernet is the most commonly implemented network for clusters and transmits information at 10 Mb/s.[†] Many dedicated clusters are interconnected by more expensive technologies to overcome the limitations induced by the speed of Ethernet. Table 26.1 provides a summary of several interconnection solutions for clusters. The need to maximize the network performance (high bandwidth and low latency) for workstation clusters, particularly for parallel applications, has yielded unique solutions. IBM offers a VME Micro Channel card (manufactured by Bit3 Computer Corporation) as an interconnection for IBM workstation clusters. Both IBM and DEC offer high-performance network multiport switches that provide excellent performance for limited size cluster configurations (albeit at a high price). Finally, the CONVEX Meta Series cluster (based on HP workstations) has a proprietary "Shared Memory Interconnect," which enhances internode data access. All of these proprietary solutions provide performance at a cost—dependence on a vendor-specific solution.

Figure 26.7 summarizes the three most typical network topologies commonly employed with workstation clusters. Ethernet, or bus, connected clusters are most frequently configured. Switch based interconnects typically are configured in a star arrangement and are used exclusively with dedicated clusters. Recently, hierarchical designs have been employed in which multiple types of interconnects are utilized [22]. A simple implementation of multiple level interconnects might consist of the following:

- All nodes connected via Ethernet. Ethernet used for low volume traffic.
- All nodes also connected via FDDI to an NFS mounted disk farm that serves file I/O. This improves file I/O and eliminates bandwidth competition with low-volume traffic.

---

[*]True fault tolerance is a more challenging issue for clusters. However, several vendors of clusters directed at business markets have developed *high availability systems.* Historically, scientific users have not been willing to accept the overhead and performance penalties associated with true fault tolerance.

[†]First-generation Ethernet cards operating at 100 Mb/s are now commercially available.

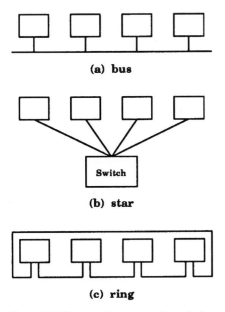

Figure 26.7 Common cluster network topologies

- All nodes also connected via HiPPI. The high-speed HiPPI channels are used when large volumes of data are transmitted between nodes. Messages passed over the Ethernet trigger the bulk data transfers over HiPPI.

So far, general guidelines for interconnect strategies for clusters have not emerged, and extensive research and experimentation is still required to determine the most appropriate interconnect strategy and topology.

26.3   Distinct Uses of Clusters

26.3.1   Enterprise clusters

One of the prime attractions of workstation clusters is the ability to utilize idle resources distributed across an enterprise network (sometimes called *cycle scavenging*). This type of clustering has also been called *ad hoc, carpet,* and *opportunistic* clustering. Several studies at national laboratories have indicated that most workstations lie idle more than 75 percent of the time. These "free" computing resources are a primary motivation for enterprise clustering. Enterprise clusters have the following characteristics .

- Workstations composing the cluster are typically geographically dispersed. Additionally, the ownership of workstations in the cluster may be held by multiple parties. Therefore, this type of clustering relies on independent workstation owners contributing their unused computing cycles to a shared pool. Most participants in the cluster will have access to greater computational resources than they had previously.
- An enterprise cluster by its nature will normally consist of heterogeneous (multivendor) systems. Heterogeneity is prevalent in both hardware and software (For example, even systems from the same vendor may be at different software releases, etc.) Success of an enterprise cluster is dependent on assembling a critical mass of similar systems.

- Since ownership is distributed, there is limited control of individual system configuration and participation .
- Enterprise clusters are almost exclusively connected via Ethernet.

The key requirement to configure an enterprise cluster is management software, which includes the implementation of an *availability policy.* Because an enterprise cluster is configured with workstations that are individually owned, owners must have a motivation to participate in the cluster. This frequently means that individual owners expect to receive more resources than they contribute. However, individual owners do not want to have their system saturated while they are trying to use the system. An availability policy allows the individual owner to specify how the system will participate in the resource pool. Some of the characteristics of management software for an enterprise cluster are:

- a user-specified time period that the system must be inactive before receiving enterprise-queued jobs
- the interruption/suspension of an enterprise job if the workstation owner issues a keystroke
- the use of a local hardware/software configuration specification table to determine whether the enterprise job meets the defined parameters (e.g. appropriate type of hardware, sufficient local memory, necessary operating system release level, availability of commercial software or compilers, etc.)
- the ability for the individual workstation owner to dynamically remove or include the system in the pool

Two types of management software exist for enterprise clusters:

- dynamic resource sharing
- automatic resource sharing

Systems that support dynamic resource sharing have the capability to breakpoint a job (possibly when an owner becomes active) and actually move the job to another system in the pool [5, 14, 15, 28]. Dynamic resource sharing systems typically require the user to link programs with a run-time library that supports this mechanism.[*] Automatic resource sharing systems will dispatch jobs within the pool, but once the job is initiated on a particular machine, it is simply suspended when the owner becomes active. Normally, automatic resource sharing systems do not require any modifications to the user's software.

26.3.2   Dedicated clusters

The price/performance characteristics offered by workstations are causing a significant number of organizations to consider alternatives to the traditional monolithic mainframe solution for interactive and batch jobs. *Dedicated* workstation clusters have been installed as substitutes or replacements for traditional computing solutions and represent a *corporate,* or centralized, solution that is a cost effective method of delivering both interactive and batch CPU cycles. Dedicated workstation clusters typically have the following characteristics:

- Individual workstations are installed in 19-inch racks in a central computer room.
- They are homogeneous configurations (within the cluster unit).

---

[*]To support job migration, it is necessary to relink an application with a run-time library that traps all I/O and maintains the necessary system tables to support this relocation.

- They are managed by a single group that administers the cluster like a central mainframe.
- They are frequently interconnected by networks with better performance than Ethernet (e.g., FDDI, SOCC, FSC, HiPPI, etc.).
- They utilize management software that partitions the cluster to accommodate interactive, serial batch, and parallel batch jobs. This software also provides the capability to automatically change the number of systems serving interactive and batch jobs—a characteristic that satisfies the need to provide more interactive systems during traditional working hours.

Dedicated clusters are most often configured to increase overall batch computing capabilities. They achieve increased throughput by relying on parallelization at the single job level (i.e., individual jobs are submitted to a single workstation in the cluster). Dedicated clusters typically have a *control* workstation that manages the job queue, provides extended disk space for I/O, and isolates users from the cluster. Several excellent software products have been developed to provide the job queueing capabilities for clusters [12, 23]. However, only a few of these systems support resource partitioning and scheduling necessary to support parallel applications.

Figure 26.8 shows a generic 32-node cluster configuration. The cluster consists of four strings of eight nodes. Each string has one node (the lowest in the stack) that contains suitable disk facilities to support the entire eight-node string's normal NFS requirements. A central system, also a workstation, is used as the master controller for the entire 32-node cluster and is connected to the rest of the cluster by a high-speed network. One of the advantages of this type of cluster configuration is that the control system can be used to dynamically modify the resource allocation mix. For example, one eight-node string could be allocated for jobs that have been parallelized, another eight-node string could be allocated for interactive jobs (code development, graphics, etc.), and the remaining strings could be

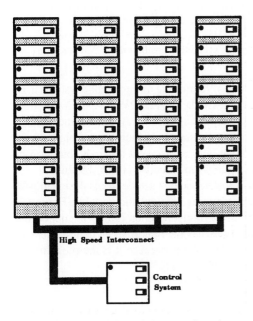

Figure 26.8 Example dedicated cluster configuration

allocated to serve batch jobs. Obviously, this job mix could be changed to reflect resource requirements.

Clusters also provide many administrative benefits. A cluster can be incrementally expanded as resource requirements grow. Additionally, a single-node hardware failure affects only a small portion of the cluster and can be isolated and replaced with minimum effects on the operational activities of the cluster. And, as microprocessors continue to improve in performance capabilities, more capable nodes can be added incrementally .

26.3.3   Parallel computing

Workstation clusters provide a unique flexibility to prototype and experiment with parallel programming. Previously, parallel programming required investing in computers designed and marketed for this specific purpose. Now, any researcher with access to multiple workstations can develop and experiment with parallel programs. Both enterprise and dedicated clusters can be configured as parallel computing environments. Beguelin et al. [3], Porter et al. [22], and Nakanishi et al. [19] have even demonstrated the capability to solve *Grand Challenge* problems using a workstation cluster. Two approaches are used to implement parallelization over a workstation cluster (sometimes called *network orchestration* environments):

- Extend existing sequential languages such as C and FORTRAN to handle necessary communications and synchronization (e.g., message passing, distributed shared memory, etc.).
- Define new programming languages based on object-oriented, functional, or logical paradigms.

Mattson [16] has more succinctly described these two approaches:

> Programmers usually code in terms of high level constructs with communication, synchronization, and sometimes even computation rolled into a single routine. These three functions are collectively referred to as *coordination* operations—a term borrowed from the coordination languages research community. Hence, a more pragmatic term for message passing libraries such as p4, PVM, and TCGMSG is *coordination library*. The use of the term coordination library also serves to more accurately position these systems with respect to coordination languages such as Linda.

This section will briefly discuss characteristics of four of the more popular systems used to create parallel applications in clusters: Express, Linda, p4, and PVM. A short introduction for a potential new standard, MPI (*message-passing interface*), is also included, A more comprehensive review of software for cluster systems, including parallel programming systems, is contained in a report by Turcotte [27], and several in-depth comparisons of parallel environments for clusters have been published recently [10, 13, 16, 26].

Express [21] was developed by ParaSoft Corporation and is a set of tools designed with the goal of making parallel and distributed programming efficient and portable. Express is based on the message-passing model and includes a robust function library to support developing cluster-based parallel applications. Express has the richest set of complementary components of any of the cluster message-passing systems because it is the most mature of any of the commercial products, and includes the following:

- a hardware independent communication system
- a high-level library of communication facilities to implement most of the commonly used parallel processing strategies

- heterogeneous systems capability
- global data operations (no explicit message passing required)
- an automated decomposition system that distributes programs among the nodes and supports run-time configuration
- a parallel I/O and graphics system tailored to parallel applications
- an excellent set of performance analysis tools
- an interactive, source-level debugger for distributed programs
- an automatic parallelizer that takes conventional, sequential C programs and converts them to parallel programs

Linda [7] was originally developed by Yale University. There have been several commercial implementations of Linda. However, Scientific Computing Associates, Inc. [24] has been the most successful vendor of Linda products. Linda is a parallel programming language created by extending a sequential language, C or FORTRAN, with five additional statements, making it easy to learn. The Linda model is based on a concept called *associative, virtual shared memory.* The benefit of this concept is that all of the additional information necessary to achieve parallelization is expressed in terms of objects. These objects include both data and information on how to process the data. Linda refers to these objects as *tuples.* A collection of tuples is called *tuple space.* Since tuples exist in virtual shared memory, several independent systems can access tuples concurrently. Simplistically, parallelization is achieved by placing tuples (which contain data and the associated operations) in shared tuple space and allowing individual processors to extract tuples, perform the work, and return the results to tuple space. The benefit of Linda is that the user is not required to program using an explicit message-passing model.

Portable Programs for Parallel Processors (p4) was developed by Argonne National Laboratory [6] and is a library of macros and functions for programming a variety of parallel systems (including networked workstations). It supports both C and FORTRAN and includes monitors for shared memory models, message passing for the distributed memory model, and support for combining the two models. p4 has the following features:

- a portable programming system, dependent only on a small number of computational models (shared memory, distributed memory, cluster)
- an efficient implementation on each architecture supported (The p4 distribution has architecture-specific implementations that optimize performance.)
- shared-memory programming at the monitor level (not locks)
- message passing at an efficient yet portable level
- support for the cluster model of parallel programming
- a foundation for high-level programming systems and libraries

Parallel Virtual Machine (PVM) [4] was developed by Oak Ridge National Laboratory, University of Tennessee at Knoxville, and Emory University and is the most popular system for developing parallel applications on clusters, both homogeneous and heterogeneous.[*] PVM, based on the message passing model, is a software library that allows the utilization of a heterogeneous network of parallel and serial computers as a single computational resource. In PVM, an application consists of multiple *components.* Each of these components implements a particular functional process. Four categories of components exist in PVM: process management, interprocess communication, synchronization, and

---

[*]PVM is considered a de facto standard for message passing and is available on most integrated parallel systems such as the Paragon, nCUBE, CRAY T3D, KSR, etc.

service (status checking, buffer manipulation, etc.). The PVM model is based on asynchronous processes that are typically executed as individual programs (e.g., heavyweight Unix Processes) on each system in the cluster. Communication between processes occurs via explicit message passing.

Message Passing Interface (MPI) [18] has recently emerged as the future standard for message passing in both distributed and parallel computing environments. This standard is the result of contributions from more than 40 organizations (hardware/software vendors, federal laboratories, and universities) and offers, for the first time, an accepted standard for message passing. MPI is based on point-to-point communication between pairs of processes and collective communications within groups of processes. Additionally, MPI includes the specification of a much richer set of features than previous message-passing models such as p4 and PVM. These allow the programmer to manipulate entire process groups, define topological structure for process groups, and explicit facilities to aid in the development and use of parallel libraries.

26.3.4 Performance observations

One justification for using clusters is based on the tremendous performance improvements of microprocessors during the last decade. Table 26.2 summarizes the performance of several computer systems. This data was extracted from the industry accepted LINPACK 1000 performance table[*] produced periodically by Dongarra [9]. This table also includes the "peak" performance for each system, or the theoretical maximum performance that the system can achieve.

TABLE 26.2 Performance Statistics

Computer	LINPACK 1000 (MFLOPS/s)	Peak (MFLOPS/s)
Cray C90 (1 proc., 4.2 ns)	902	952
Cray Y-MP (1 proc., 6 ns)	324	333
Cray X-MP (1 proc., 8.5 ns)	218	235
DEC Alpha 10000-610 (200 MHz)	155	200
HP 9000/735 (99 MHz)	120	198
IBM 6000-580 (62.5 MHz)	104	125
IBM 6000-580 cluster (2 proc.)	144	250
IBM 6000-580 cluster (4 proc.)	206	500
Kendall Square (32 proc.)	513	1,280
Kendall Square (4 proc.)	47	160
Kendall Square (1 proc.)	31	40
Intel Delta (512 proc.)	446	20,480
nCUBE 2 (512 proc.)	204	1,205

Table 26.2 indicates that workstations offered by HP, DEC, and IBM for approximately $30,000 to $50,000 can achieve performance results within the same order of magnitude as a $30 million CRAY supercomputer for certain classes of problems.[†] Even more interesting is the comparison between workstations and parallel systems offered by Kendall Square, Intel, and nCUBE. The fact that a 512-node nCUBE offers less than twice the per-

---

[*]The LINPACK 1000 benchmark involves solving a set of 1000 equations.
[†]The use of select benchmarks for comparisons does not adequately reflect the capabilities of various systems. Such important issues as memory bandwidth, I/O bandwidth, etc., are not reflected by the LINPACK benchmark.

formance of single-processor workstations should not be overlooked. Additionally, effi-ciency[*] also provides insight into the competitiveness of clusters with integrated parallel systems. The four-processor IBM cluster achieves a 41 percent efficiency, the four-processor Kendall Square a 29 percent efficiency (the 512-processor Kendall Square achieves only a 40 percent efficiency), and the 512-processor nCUBE a 17 percent efficiency.

A second performance factor which has been promoted as a major weakness of clusters is latency. Figure 26.9 charts the memory latency for several systems reported by researchers at Oak Ridge National Laboratory. It is interesting to note that expensive integrated parallel systems such as the CM5 from Thinking Machines are only twice as fast as clusters with special interconnects.[†]

The NAS Parallel Benchmarks [1] are one of the few sets of benchmarks that have been developed to study the performance characteristics of parallel computers. This set of benchmarks, which is actually a suite of several applications, is unique in that the applications are described algorithmically. Programmers may utilize the most appropriate programming model for their target architecture. The suite includes applications that span from being "embarrassingly parallel" to being difficult to parallelize. Sukup [26] has recently reported results from implementing four of the NAS benchmarks on clusters. Sukup used a cluster of nine IBM RS6000-320H workstations (25 MHz) connected via a token ring network (16 Mb/s). Table 26.3 presents results for these four benchmarks for different software systems for clusters. Table 26.4 compares results from two of these benchmarks for clusters with results from vector and parallel computers.

TABLE 26.3   Some NAS Benchmark Results for Clusters

Application	PVM 2.x	PVM 3.x	p4	Express	Linda
Embarrassingly parallel	584	586	589	589	586
Multigrid (128^3)	22.5	34.8	24	24.2	44.7
Conjugate gradient	442	685	623	277	322
Integer sort	175	208	277	175	179

Note: All timings are average time for execution in seconds.

TABLE 26.4   NAS Benchmark Comparison Using PVM 2.x

System	No. of processors	Embarrassingly parallel benchmark	Conjugate gradient benchmark
Cray Y-MP	1	126.2	11.92
	8	15.9	2.38
TMC CM5	16	42.4	–
	32	21.5	–
KSR-1	32	69.8	21.7
RS6000 320H cluster	9	584	442

26.4   Open Issues

Clustering has emerged as a practical, cost-effective complement to high-performance computing environments. However, present clustering solutions are quite immature, and

---

[*]The ratio of measured over peak performance.
[†]The measurements were: KSR/32 (7), Intel Paragon (10), CM5 (73), RS6000 V-7 (140), Convex HP Meta (150), RS6000 Bit3 (240), and RS6000 Ethernet (3500).

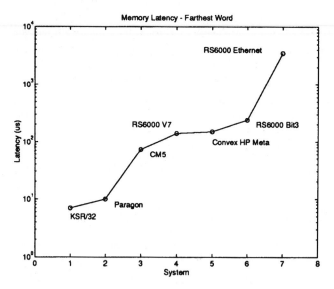

Figure 26.9 Memory Latency

significant effort is still required to address many of the open issues. Many of these issues are not unique to clusters but also plague the general acceptance of parallel and distributed computing. Some of the issues that need to be addressed are:

- availability and implementation of standard low-level primitives for communications, synchronization, and scheduling across architectures
- heterogeneous remote procedure calls that hide architecture, protocol, and system differences
- availability of real-time performance monitors
- improved cluster application development tools
- support for traditional high-level languages for heterogeneous computing
- applications which are capable of exploiting workstation clusters
- new system administration tools to address system management issues for clusters; particularly improved operating system and management software for managing multivendor cluster environments (e.g., user management, distributed program loading, load balancing, network management, dynamic reconfiguration, etc.)
- development of improved standards for developing parallel applications that protect software investments.
- the ability to ensure security and user authentication in clusters
- the need to provide accurate resource utilization accounting
- improved system transparency to the user (e.g., architecture/processor, network/communication, task location, network fault, system fault, etc.)
- the need for common, transparent programming tools (e.g., debuggers, performance analyzers, monitors, etc.)

As many of these issues are resolved, the concept of clustering will slowly disappear. Computing environments will slowly migrate toward the concept that "the network is the computer." No longer will users know, or even care, where their computer jobs are executed. The transition from clusters to truly transparent network computing will have occurred.

26.5   References

1. Bailey, D., Barton, J., Lasinski, T., and Simon, H. 1991. *The NAS Parallel Benchmarks.* Report RNR-91002 Revision 2. Moffett Field, Calif.: Numerical Aerodynamic Simulation (NAS) Systems Division, NASA Ames Research Center.

2. Baskett, F., and Hennessy, J. L. 1993. Microprocessors: from desktops to supercomputers. *Science,* Vol. 261, 864–871.

3. Beguelin, A., Dongarra, J., Geist, A., Manchek, R., and Sunderam, V. S. 1991. Solving computational grand challenges using a network of heterogeneous supercomputers. *Proceedings of Fifth SIAM Conference on Parallel Processing,* Philadelphia, PA.

4. Beguelin, A., Dongarra, J., Geist, A., Jiang, W., Manchek, R., and Sunderam, V. 1992. *PVM 3 User's Guide and Reference Manual.* ORNL/TM-12187. Oak Ridge, Tenn.: Oak Ridge National Laboratory.

5. Bricker, A., Litzkow, M., and Livny, M. 1994. *Condor Technical Summary.* Madison, Wis.: Computer Sciences Department, University of Wisconsin.

6. Butler, R. and Lusk, E. 1992. *User's Guide to the p4 Programming System.* ANL-92/17. Argonne National Laboratory.

7. Carriero, N. J., Gelernter, D., Mattson, T. G., and Sherman, A. H. 1994. The Linda alternative to message-passing systems. *Parallel Computing,* Vol. 20, 633–655.

8. CBO Office. 1993. *Promoting High-Performance Computing and Communication.* U.S. Congressional Budget Office Study. Washington, D.C.: Government Printing Office.

9. Dongarra, J. J. 1994. *Performance of Various Computers Using Standard Linear Equations Software,* CS-89-85. Computer Science Department, University of Tennessee.

10. Douglas, C. C., Mattson, T. G., and Schultz, M. H. 1993. *Parallel Programming Systems for Workstation Clusters,* technical report YALEU/DCS/TR-975. Yale University, Computer Science Department.

11. Hardenbergh, H. H. 1994. CPU Performance: where are we headed? *Dr. Dobb's Journal,* (January) 30–38.

12. Kaplan, J. A., and Nelson, M. L. 1993. *A Comparison of Queueing, Cluster and Distributed Computing Systems,* NASA-TM-109025. NASA Langley Research Center.

13. Larrabee, A.R. 1994. The p4 parallel programming system, the Linda environment, and some experiences with parallel computation. *Scientific Programming,* Vol. 2, 23–26.

14. Litzkow, M. J. 1987. Remote Unix—turning idle workstations into cycle servers. *Proceedings of the Summer 1987 Usenix Conference,* Phoenix, Ariz.

15. Litzkow, M. J., Livny, M., and Mutka, M. W. 1988. Condor—a hunter of idle workstations. *Proceedings of the 8th International Conference on Distributed Computing Systems,* San Jose, CA, 104–111.

16. Mattson, T. G. Programming environments for parallel and distributed computing: a comparison of p4, PVM, Linda, and TCGMSG. To appear, *International Journal of Supercomputing Applications.*

17. Mohindra, A., and Ramachandran, U. 1991. *A Survey of Distributed Shared Memory in Loosely-coupled Systems,* GITCC-91/01. Atlanta, Ga.: College of Computing, Georgia Institute of Technology.

18. 1994. *Message Passing Interface Forum Document for a Standard Message-Passing Interface.* Computer Science Department, University of Tennessee.

19. Nakanishi, H. and Sunderam, V. S. 1992. Superconcurrent simulation of polymer chains on heterogeneous networks. *Proceedings of IEEE Supercomputing Symposium, 1992.*

20. Nitzberg, B. and Lo, V. 1991. Distributed shared memory: a survey of issues and algorithms. *IEEE Computer,* Vol. 24, No. 8, 52–60.

21. Parasoft Corp. 1992. *Express User's Guide.* Pasadena, Calif.: Parasoft Corp.

22. Porter, D. H., Woodward, P. R., Anderson, S., MacDonald, J., Chin-Purcell, K., Hessel, R., Perro, D., Zacharov, I., Ryan, J., Wildra, L., and Galles, M. 1993. Attacking a grand challenge in computational fluid dynamics on a cluster of Silicon Graphics Challenge Machines. *Proc. Cluster Workshop 1993.* Tallahassee, Fla.: Supercomputing Research Institute, Florida State University.

23. Revor, L. S. 1992. *DQS Users Guide.* Internal report, Computing and Telecommunications Division, Argonne National Laboratory.

24. Scientific Computing Associates Inc. 1992. Linda Users's Guide and Reference Manual. New Haven, Conn.: Scientific Computing Associates Inc.

25. Stumm, M. and Zhou, S. 1990. Algorithms *Implementing Distributed Shared Mem*ory. *IEEE Computer,* Vol. 23, No. 5, 54–64.

26. Sukup, F. 1994. *Efficiency Evaluation of Some Parallelization Tools on a Workstation Cluster Using the NAS Parallel Benchmarks,* ACPC/TR 942. Computing Center, Vienna University of Technology.

27. Turcotte, L. H. 1993. *A Survey of Software Environments for Exploiting Networked Computing Resources,* MSSU-EIRS-ERC-932. Engineering Research Center, Mississippi State University.

28. Songnian, Z. 1992. LSF: load sharing in large-scale heterogeneous distributed systems. *Proceedings of the Workshop on Cluster Computing.* Tallahassee, Fla.: Supercomputing Computations Research Institute, Florida State University.

27

Massively Parallel Processing with Optical Interconnections

Eugen Schenfeld

This chapter reviews optical networks that have been suggested for use in parallel processing. It also examines some of the interesting electronic-based architectures that have special features that are important if an optical network is to be used.

Parallel processing with free-space optical interconnections may be feasible in the near future, so it is appropriate to review massively parallel processing (MPP) architectures and some optical interconnections proposed for MPP systems. By *massively* we refer to parallel systems with hundreds to thousands of processing elements (PEs). This chapter discusses the pertinent issues:

- some thoughts about general-purpose parallel processing
- an adaptable topology for better fitting parallel programs
- several aspects of optical interconnections—high-bandwidth, power-budget, and manufacturing ease
- adapting the underlying interconnection structure to the parallel program's natural communication structure.

This chapter does not address some other issues that might be important for the realization of an MPP using an optical interconnection working system. These issues can be the scope of future investigation, but they are more of an "engineering nature" and their solution will not affect the main viewpoints expressed herein. Some of these issue are:

- investigating the alignment problem
- cost of components and the system versus the overall processing power obtained
- reliability (MTBF)
- what happens to the scattered light (background optical noise)

It appears that the most promising technology for MPP (versus telecommunications or local area networks) is free space optics, as it offers connectivity as well as bandwidth (which is exactly what is needed for MPP systems with 1,000 or 10,000 PEs or more).

Many issues can be further investigated, such as the search for better materials with better optical qualities to implement the global switches, but these are oriented more toward laboratory work and practical experiments. A more fundamental study concerning the particular structure that the communication and computation processors should possess is also an interesting subject. In a more theoretical direction, a model to balance the whole system is a very important topic. The communication balancing, particular with regard to distributed algorithms, can be investigated, as well as other regular graphs to be embedded with a minimal cluster size. From a practical view point, reliability and fault tolerance are very important and might impose upper bounds on the size and complexity of the PEs in any MPP system.

Introduction

For the last several decades, scientists have been looking for new ways to increase the computational power available on electronic supercomputers. Although much has been done to find new and improved algorithms to solve many problems faster, the desire for more computation power seems to be endless. New applications and larger data volumes demand better computers, making this search a necessity. Technology can help accelerate computations through the use of newer microchip fabrication processes, making logic circuits with a denser integration and a faster operation. Scientists and engineers are looking into ultra-high integration densities, with millions of transistors on one chip. The size and speed of the transistor and the reduction in its power consummation keeps improving all the time. New semiconducting materials, such as GaAs, appear to be a promising way to improve performance.

The power of supercomputers seems to double every two to four years. Since the time of the first electronic computer, we have witnessed a computation revolution and evolution that influences our daily life. The computation power once found only in supercomputers is now available on our desk as a powerful, yet affordable, personal computer (PC). Single-chip CPUs become more powerful each year. Since it was first introduced in 1971, the "one chip" CPU has threatened the low end of the so-called *minicomputers*. The distinction between micro, mini, and mainframe is not as clear cut as it was ten years ago, in terms of computation performance. It looks like the definition of these terms should carry a time stamp indicating the year they were valid.

Judging from the technological progress made in improving the computational power of electronic computers, one might expect that this could go on forever and that electronic computers could get better and better as long as technology continues to develop. At least, it seems, this process could go on for a substantial period of time. It seems that by advancing the technology alone, it will be possible to achieve faster and faster computers while keeping the basic computer architecture the same. So the question is: Why should one trouble himself with finding new architectures and processing concepts if more processing power can be achieved through technology advancement alone? Why should we look into, so-called *optical computing* unless we believe that computing with light can be substantially different from electronic computing? Is electricity the ideal means to carry data and perform computations, or might there be other, better means (light, chemicals as in living tissues, magnetic fields, etc.)? Should one look into an optical computer or stay with the safe, proven electronic technology? And if we look into optical computing, should we mimic the existing electronic architectures, or think about a completely new architecture?

The above questions, and many others, can be asked when suggestions for new computer architectures are made. The answers are not always clear cut. Some might argue that there is only a narrow application niche for a parallel processor, and even this niche might exist only for certain applications. Optical computers present even a more difficult situation. At least for their current status, it seems that optics cannot compete economically with electronic machines as a general-purpose computer. Special-purpose optical systems, making a direct use of optical analog processing, are commonly available. Some of these systems replace digital electronic machines, offering better price-performance ratios.

Parallel processing is an architectural concept to provide higher performance by using many, independent PEs working together on one problem. One of the major challenges facing parallel processing is how to provide large-scale information interchange. To build a network capable of interchanging information between many PEs, one must solve conceptual problems while staying within the limits of the physical laws of nature. It currently appears that information transfer based on the medium of light is more appropriate for very large (more than a 1,000 PEs) interconnection networks. Optical interconnects offer potential major improvements over electronic networks. Light is a different means of communication, with better qualities than electrons. The performance of parallel processing architectures is strongly limited by the performance of their underlying interconnection networks. Light beams offer some important advantages for parallel processing communication: larger bandwidth, minimal propagation delays, lack of interference, and higher connectivity using free-space communication (no wires).

The terms *parallel* and *massively parallel* need some clarification. Parallel processing can be done also with small number of processors, but we do not want to address the problems of such smaller numbers in this discussion. We use the term *massively parallel* to indicate a system with hundreds of CPUs interconnected by a network. It is likely that the most promising technology for MPP (versus telecommunication or local area networks) is free space optics, as it offers connectivity as well as bandwidth. Other technologies exist as well, but because their development does not appear imminent, they are not reviewed them in this chapter. One example is wavelength division multiplexing (WDM), which uses different wavelengths of light beams to separate different communication channels. It will require either variable (tunable) wavelength light sources (laser diodes, for example) or light filters. Neither of these technologies is mature enough, fast enough, or cheap enough to be practical for MPPs (although they may be used first in telecommunication application to take advantage of the existing bandwidth capacities of optical fibers in existing networks).

Guided wave interconnection is another technology that integrates light-conducting guides integrated on chips. Again, this technology seems to be presently out of reach for application to MPPs. Since each these technologies could be the subject of a chapter by itself, we will focus here on free-space optics, which seems most promising for MPP applications.

27.1   Parallel Processing Motivations

27.1.1   Goals and definitions

The purpose of a definition is to state and describe the meaning of a word, a concept, or an idea. In thinking of a proper definition, we need to look at the purpose it could serve and the context in which it might be used.

Defining *parallel processing* is not an easy task. There are many definitions of this term. One reason is that such a definition serves many purposes. Parallel processing has many facets, ranging from system architecture to research in theoretical computer science. For some, the emphasis should be on the way the programmer views the computer. Others may regard the price/performance ratio as more important. Some classify parallel processing from the viewpoint of the program and data, as in Flynn's classification [13]. Other definitions just point to the partitioning of a job between several processors that cooperate and coordinate their activities to carry out the task. Another definition looks at the ratio of computation versus communication overhead. In a *fine-grain parallelism*, there is a relatively a large amount of communication (overhead or data transfer) for the computation done. In a *coarse-grain parallelism,* there is very little communication activity relative to the computation performed.

Models of parallel processing are used for qualifying and quantifying the special properties of parallel processing. In deriving these models, we look to determine the main characteristics of the phenomenon we try to model. In the case of the simple, single processor, the Von-Neumann model seems to well describe the most important considerations of processing done by this architecture. The Von-Neumann model is mainly an architectural view of the uniprocessor. Yet, the Turing machine is a simplified version that is used as a model to study algorithm complexity, in spite of the fact that no real computer is build like this. Flynn's taxonomy is one of the most widely used classifications for parallel computational models, enhancing the simple, Von-Neumann view [13, 25, 531 15]. An assumption in this chapter is that the main characteristic of parallel processing is switching, i.e., the need to change the communicating parties (source to destination PEs) in certain ways, patterns, rates, and so forth. Therefore, we will adopt the definition describing many parallel processor systems, suggested by Almasi and Gottlieb [2], stating that:

Definition 1. A *parallel processor* is large collection of processing elements that can communicate and cooperate to solve large problems quickly.

This definition reflects the general idea of parallel processing: the use of several PEs that work together on one problem so that a result is obtained faster than could be obtained from one processor. Of course, this definition raises a lot of questions, but it is hard to find one universally accepted definition. One common aspect is present in this and other definitions. This aspect is the notion of several PEs working on various parts of a problem. This definition is not complete, as there are other elements involved in the architecture of a multiprocessor computer. Take, for example, the issue of communication. This has many aspects:

- the model of programming reflecting the view of the computer architecture as seen by the user (shared memory or message passing)
- the amount of knowledge each processor has about the other PEs (tightly or loosely coupled systems)
- the physical distance between the PEs and the cost of communication (as in distributed processing)
- the amount of speed-up; i.e., if a job takes a certain time t_1 on a single PE, will it take less time on N PEs according to a linear ratio: $t_N = t_1/N$?
- other issues relevant to concurrent processing data by several PEs communicating among themselves

The most important aspect of parallel processing is communication. Communication is there for many reasons: sharing data, exchanging partial results, coordinating and synchronizing the various tasks and PE, and so on. The interconnection network is a crucial ingredient of the parallel computer system. In fact, communication is present even in the uniprocessor architecture, although with less impact and importance.

27.1.2   Technology and economics

Technology influences many areas of our life as well as computer science. Over the years, changes in technology are reflected in the computer devices and architectures available to the mankind. The first "computing device" could be the fingers on the hand, used for counting. These fingers are still in use by young children (and some adults) as an aid for taking the first steps of performing simple arithmetic calculations. The "technology" used in this case is very simple. The "computing devices" are only a means of remembering and counting, while the "processing" itself is done by the human brain. The brain performs a very simple algorithm when the notion of addition and subscription is very simple: counting forward or backward using physical elements (fingers). The brain, in this case, is the processor. The fingers (or the making of marks) is the memory. Therefore, the technology is based on living tissues.

A more advanced device is the abacus, which uses rows of small beads. Here, the technology is more sophisticated, using objects (beads) on strings. The algorithm is more complex as well. Of course, the results are better, as quite complex calculations can be made quickly.

Computing machines have come a long way since the days of the abacus. The mechanical era produced several computing machines. Built from the state-of-the-art technology of those days (metal wheels, gears, cogs and other mechanical parts), some were just counters (i.e., they did the addition of two numbers by counting as many cycles as the sum of them), but others used novel ideas (e.g., Newton's mechanical calculator) and algorithms suited to the available technology. Programming was first just a development of Jecard's weaving machine, controlled by perforated cards. Electricity gave a new possibility to computing devices. Using relays, it was possible to replace the mechanical gears with a more compact design. The vacuum tube eliminated the need to use moving parts and sped up calculations. The transistor further enhanced the scope of problems that were solvable by the computer, as a result of working faster, handling larger problems, and offering a much higher reliability level. The integrated circuit changed some of the basic assumptions that influenced the design of computer architectures as well as algorithms and programs. Memory became less expensive and rare. Some functions, previously implemented by a program, were made a part of the architecture using special-purpose circuits. In all of the above development history of computing devices, technology was the driving force for new inventions and designs.

The Von-Neumann model emerges from the technology used in the days when wires (or input/output pins) were expensive compared to memory devices. Although the Von-Neumann model is not intended to impose any limitations, but it is worth noting that the use of memory for storage of both data and the program (known as the "Von-Neumann bottleneck) is a hindrance to calculation time reduction.

It is not possible to consider parallel processing without mentioning economic concerns. One motivation for doing parallel processing is the quest to achieve higher performance at a lower cost. Technology plays an important role in determining this cost. The role of the computer architect is to determine the right blend of components using the appropriate technologies so that the whole system is optimized. This optimization is not always just based on cost minimization. Sometimes we are willing to pay a higher price to achieve an important goal. If there is a need for specific high-performance processing, we might pay the cost if there is no alternative. But no one will consider parallel processing if *Grosch's law,* stating that the best price/performance is obtained with a uniprocessor architecture, is to be found valid. It is only technological breakthroughs, as were made in the VLSI technology that can lead to the repeal of Grosch's law [11].

We assume that no computer could emerge without economic justification. If the computer were a very expensive device (and the meaning of "very expensive" is relative to the computer's benefits versus other ways of deriving the same benefits), it would not have a right to exist. Only because the human mind is so limited in processing large amounts of data, and because there are problems that need such processing, was there a driving force to invent such a device.

We do not consider *performance* simply in terms of achieving more millions of instructions per second (MIPS). In a modern computer architecture, there are many functions and tasks to perform. No one technology can implement them all. For example, implementing a larger memory means using a technology other than the one used for achieving faster access times (i.e., main memory vs. cache memory). The task of the computer architect is to wisely choose the right technology for the specific job. The goals can be to make a cost-effective computer (i.e., one that offers moderate performance but is affordable and applicable to a wide variety of applications—the personal computer being a good example) or to make the fastest, most high-performance supercomputer in the world. In either case, the considerations and the trade-offs are different, and hence the sort of technology used will be different. Today, no one will use gallium arsenide (GaAs) technology for PCs, simply because it is too expensive. Some functions can be implemented (at a certain time in history) using a mix of technologies, but the cost will be different. Suppose that we would like to build the equivalent of a Cray 1 computer using mechanical devices. Even if mechanical engineering were to advance considerably and be capable (in theory, at least) to produce the required component devices, the price would be astronomical, and the low reliability might prevent the actual use of such a machine.

As we discuss the economics aspects of parallel processing, there is a very important note to be made. The price of goods are set according to market forces. A new technology is very expensive in the beginning. Only after it is widely used in many applications does its price drop. This is mentioned only to observe that a computer architect should have a far-sighted vision and not hesitate when it comes to using new technologies, if he believes that they have inherent properties needed for a certain component of the architecture.

In this chapter we will not further discuss the economic problem. However we note that it has been estimated that, as parallel processing includes "data processing" and "data flow" (i.e. communication), a reasonable price for the communication system of a multiple instruction, multiple data (MIMD) parallel computer, as per Flynn's taxonomy [13]), should be in the range of half the total system price [10]. As economics itself is a very complex issue, it is not the scope of this chapter to treat this problem as related to the proposed architecture. No doubt that before a practical machine is to be constructed, the economic feasibility issue should be carefully studied.

27.2 General-Purpose Parallel Computers

The term *general-purpose* has many aspects. It is a term describing the wide functionality of something in a variety of applications. These applications can be completely different in nature, yet they might be carried out using a common denominator task or device. The term *general purpose* also has an economical meaning. It means that such a common functionality can be realized using mass-production techniques, yielding a low-cost product or service used by the masses. Yet another inference can be that a general-purpose solution might not be as fully optimized as a special-purpose or "made to order" solution. There is a price to pay (in terms of performance) for the gain in the price (in terms of economics). We would like to briefly examine general-purpose parallel processing systems approaches and to argue that a massively parallel processor should be made for general-purpose applications so that many can use it economically.

A good example of the way general-purpose products entered our life and became widely used is in the history of PCs. The story goes back to the days of Hoff's invention of the microprocessor at Intel in 1971 [37]. The way in which a general-purpose item is recognized can involve a process that consumes quite a bit of time and money. The story is that Intel, a newborn company of the late 1960s, was asked by a Japanese company to design an electronic controller chip for the automatic operation of washing machines. Intel, then a company mainly producing memory chips, realized that there is a good market potential in producing a general-purpose central processing unit (CPU) on one chip. Such a device could be used in many applications that did not need massive computational power (the technology of those days could not deliver many MIPS from a single chip). In December 1971, the Intel 4004 CPU was introduced, which opened the era of what can be called "computation for the masses." This observation of Intel had a crucial influence on the company's success. By spotting general-purpose building blocks such as the CPU, a push was given to using them in applications that were dominated by older technologies (mechanical assemblies). This was described by Gordon Moore, one of Intel's founders, who saw microprocessor development as "finding a broadly applicable, complex integrated logic circuit that can be produced in volume and that utilizes the technology advantages" [35]. (This is exactly what we are looking for today: a general-purpose, technologically advanced vehicle, with the "right" system architecture: a parallel computer.) The new CPU device was seen by Nichols, also of Intel, as having several advantages over the use of transistor-to-transistor logic (TTL) circuits: lower cost, decreased development time, design flexibility, increased capability and reliability, and self diagnostic features [36]. Not just the CPUs but other function units of a computer (memory, input and output, CRT controllers, communications etc.) became general-purpose units, used in many diverse applications. In time, newer technologies enhanced the performance of these units, making them suitable for even higher-performance and more demanding, applications.

It is difficult to identify the first personal computer. But the companies that had the most notable influence, and that still have not gone into bankruptcy, are certainly Apple and IBM. Some point to the IBM 5100 as the first portable computer, based on an IBM-designed 8-bit microprocessor. Roberson called it a "personal computer," meaning one that is operated by and for the person wishing to solve a problem [43]. However, it is not easily viewed as a PC, given its high cost ($11,000) and limited functionality. It was treated by IBM with the traditional marketing tactics used in the mainframe business, with the monthly software leases and support, and was offered as late as May 1981. Home personal computers began to take off commercially only when Steve Wozniak and Steve Jobs introduced the Apple II in April 1977 [14].

The PC owes a lot to diverse factors, with technology being one of them. In addition, there are many common, daily applications of our life that (fortunately) can be worked out using what present technology can offer. The reputation of IBM in the mainframe market, as a company that stands behind its customers, as well the size and total value of a PC, marked an important turning point back in 1981, when IBM introduced the first PC. Many companies were quick to follow in an attempt to develop products that were either similar (PC clones) or add-ons for the PC (application software and hardware). As more and more company joined in, the prices fell, more people could afford to buy a PC or a clone. This process had an affect not only on the PC market but also on all the components used to assemble PCs (memory, VLSI technology, disks, monitors, etc.). This, in turn, led to an improvement of the state-of-the-art technologies from which supercomputers are built. Today, high-end PCs are a threat to the manufacturers of so-called *minicomputers*. The competition is fierce, forcing a continued straggle for getting higher-performance computers while improving the cost/performance ratio.

One question that might be raised is why the PC had such a significant influence spreading outside its direct domain. The answer might be that this happened because the PC computer system is relatively low priced and doesn't require the latest technology. There are companies competing with IBM in producing compatible products (software and hardware) in the mainframe market. But there are not many such companies, as they themselves have to be giant corporations with the financial depth necessary to sustain such an activity. From the other point of view, there is a larger potential market that has a high enough volume to make investing in PC-related activity very attractive. IBM itself was successful because it came up with the right answer for a problem that, at that time, was looking for a solution. IBM made use of the contemporary technology to make a computing machine capable of processing data at a cost that compared favorably with any other available solution. At the same time, the applications were important enough and had a large demand, making it economical worthwhile to come with a solution. These last words relate not only to the PC story but also to the first computer IBM made, and it can be seen as a common pattern for many other cases (the car mass-produced by Ford, the telephone, the light bulb, etc.). One common aspect of all these cases is that they all were general-purpose, mass produced devices or systems.

The PC story was related as an example to remember while thinking of parallel processing activity. Although there are differences between a PC and a parallel computer, there are some similarities common to many new products and technologies. A general-purpose parallel computer has a better chance to be developed into a cost-effective product. As in the PC story, technology can then advance in a "bootstrapping" process, yielding even better and cheaper computers. Better technologies can result in improved architectures and languages. The major problem facing parallel processing today is the availability of proven, cost effective uniprocessor solutions. At least for now, Grosch's law seems to be valid. When thinking of a parallel computer, we have to think of either the most powerful computer as of the time it will be constructed, or the best cost/performance computing solution. The technology of uniprocessor systems keeps advancing, so if one were to begin a project to build a parallel computer today and finish it in about three years, he should be convinced that, by that time, the design will not be obsolete by what then will be state-of-the-art unicomputer systems.

27.2.1 Review of parallel systems

Today, there are hundreds of parallel systems offered as commercial products or assembled for research purposes. We would like to focus on previous architectures that have been attempted to fit the interconnection network to the particular needs of running each algorithm. Such attempts were made using electronic VLSI technology, so they suffer from severe performance limits. Nevertheless, these examples serve our purpose of demonstrating the importance of interconnection flexibility. Since this chapter presents optical interconnections for MPP system, it is useful to examine the weak points of electronic communication first. Then we will look into early attempts of optical technology for communication purposes. In this subsection several of these parallel architectures are presented, with some comments regarding their fitness to make a general-purpose parallel computer. Of course, we cannot review all of the different types of machines built but rather will focus on the parallel systems that seem to exhibit special features where optical interconnection can be used. The initial principles set by these systems contribute to our understanding and to future attempts to use optics as the interconnection technology.

27.2.2 The CHiP computer

The *configurable highly parallel computer* (CHiP) [50] computer used electronic VLSI circuit technology for communication. The approach to the problem and the basic ideas are general. In Ref. [50], Snyder analyzes the problem of fitting "algorithmically specialized processing" to a flat, VLSI reconfigurable architecture. The conclusions of this work might be summarized in the following observations:

- There are some algorithms that can work better if we provide such a reconfigurable interconnection network. The CHiP computer is limited in providing only a "flat" architecture. It cannot accommodate full interconnection, thus only *some* algorithms (such as presenting mesh, tree or pipelining communication pattern structures) might benefit from it.
- For multi-problem solving, CHiP can be adapted to the problem size. Two small problems might run in parallel, with different interconnection structures.
- In terms of fault tolerance, by the reconfiguration process, faulty processors can be mapped out so as to not be allocated for any task. The interconnection network will skip these faulty processors so the computer can be operational, even with some faulty PEs.

27.2.3 The PASM

Another example of a reconfigurable type computer is the *partitionable SIMD/MIMD (PASM) system for image processing and pattern recognition* [47]. This large-scale, multi-microprocessor system was designed at Purdue University for image processing and pattern recognition. It can be dynamically reconfigured to operate as one or more independent SIMD and/or MIMD machines. The numbers of PEs in PASM can be something like 1024 PEs and 16 controllers of the dynamic network configuration process. Again, this project was based on electronic technology. It is different from the CHiP in that its PE unit is more powerful, and the approach is of medium- to large-grain parallel processing (not fine-grain as with CHiP). The main conclusion of the PASM concept, in addition to similar conclusions found in the CHiP project, is that a dynamic reconfigurable system should be a valuable tool for both image processing/pattern recognition and *parallel processing* research.

27.2.4 The NYU Ultracomputer and the IBM RP3

The NYU Ultracomputer [16] is a general-purpose MIMD computer designed at Courant Institute of New York University. In its original design, this was supposed to be a 4096-PE machine with the 1990s technology in mind. The interconnection network used was based on a *multilevel interconnection network (MIN) Omega* that connected the PEs and the memory modules. The network employed a message-passing policy and was the first to incorporate the fetch-and-op concurrent operand.

The interconnection network consist of 24,576 2×2 bidirectional switches with buffers and combining logic. The conclusion of this architecture can be summarized by the following observations:

Complex MIN. The interconnection network, although incorporating some novel ideas (combining requests and fetch-and-op logic), is too complex and cannot sustain the performance of the PEs. In other words, the system is not balanced. It is doubtful how well these ideas suit the real needs of practical parallel programs. As described in the RP3 architecture below, analysis of "hot-spots" raises some doubts on the viability of increasing the functionality of each basic switch in the MIN on account of decreasing the basic throughput. In the RP3, this problem was recognized, and the solution was to build two networks. One had a simple switch (no combining or synchronization primitives built in) using fast bipolar technology. The second used CMOS technology, incorporated smart switches, but with a much slower throughput. Still, as we will see, this did not help. Under what conditions we should build in combining, and which types of combining to use, are currently open questions. The cost-effectiveness of different types of combining will depend on the task domain and the operating system [48].

Model and architecture. There is no doubt that a shared memory system is easier to work with from the viewpoint of the programmers. But at question is what is best and balanced as a physical architecture. In our opinion, shared memory, where each memory access may pass the entire length of a MIN network, could be quite wasteful, even with the pipelining of requests through the network. The reasons could be either because parallel programs do not make a use of this pipelining (i.e., they need the previous request to be served before issuing the next one) or because if each memory request has to go over $\log(N)$ stages of switches (where N is the total number of PEs), the overhead (latency and switching time) is too high. We believe that there should be a separation between the physical architecture (which should use circuit switching and a one-stage network—crossbar) and the way a programmer views the architecture (i.e., the parallel language).

The IBM Research Parallel Prototype (RP3) [40] is a follow-up of the NYU Ultracomputer described above. It was designed to have 512 then state-of-the-art PEs and 512 memory modules each with 2 to 4 MB capacity. Two multistage networks are used to connect the PEs to the memory modules in a shared-memory type of architecture. These networks are based on an expanded Omega network topology that provides dual source-sink paths. One network is built of four stages of 4×4 switching elements with extensive buffering, but without the combining and fetch-and-op functions (as in the NYU Ultracomputer). The second network is based on a more complex basic switch, with combining and synchronization primitives (fetch-and-op) functions. It is built of six stages of 2×2 bidirectional switches. The RP3 was designed as a general-purpose MIMD mode of parallel computer, to support a variety of applications. The conclusions and observations on this architecture are summarized as follows.

Network latency and functionality. As learned from the NYU Ultracomputer, the RP3 design incorporates two parallel networks. One has "muscles," i.e., low latency time by using fast bipolar technology (which at the same time prevents high functionality integration). The second has "brains," i.e., is capable of doing smart operations on the information interchange flow (fetch-and-op and combining). A study done by Pfister and Norton [41] gives an important insight on the hot-spot issue, although the results of it are rather negative for the fetch-and-op mechanism. As the functionality build in the network grows, the performance decreases. It is debatable whether the use of "smartness" in interconnection networks rather than "muscles" is a good choice. Under what conditions to build in combining and which types of combining to use are currently open questions. The cost-effectiveness of different types of combining will depend on the task domain and the operating system [48].

Technology. The use of a bipolar technology for the faster network, and the slower CMOS technology for the combining network, raises the question of what technology to choose so that the system is balanced. It is clear that, even with the bipolar network, the RP3 performances was not so great. This may be due to the network architecture, i.e., the use of a multistage structure. It seems that to make a faster network (compared to the PE's throughput) a different approach should be taken. If the PE is built of the state-of-the-art technology, it is hard to see how we can use the same technology for the interconnection network. Optics may not be good for building PEs, but it seems ready (and presently is in use) for communication.

Contributions. There is no doubt that the RP3 project made a major contribution to our understanding of parallel processing. The development of usable software and reliable hardware make it a key project. The hot-spot analyses increases the understanding of where we should concentrate our resources. Under what conditions to build in combining and which types of combining to use are currently open questions. The cost-effectiveness of different types of combining will depend on the task domain and the operating system [48].

27.2.5   The Denelcor HEP

The Denelcor HEP [24, 49] is a shared-memory, pipelined multiprocessor of MIMD parallelism, having up to 256 concurrent (potentially parallel) processes. The HEP consists of up to 16 PEs called *process execution modules* (PEMs). Each PEM has special hardware consisting of multiple task queues, each of which can be the source of the next executable instruction. This mechanism resembles a data-flow-like approach where the ready-to-execute instruction (op-code and operands) is taken out of each queue when all its relevant data has been fetched from memory. The fast task-switching mechanism makes the memory and pipeline latency time invisible, making the maximal use of the executable hardware circuits. The main features and key observation of the HEP architecture can be summarized as follows:

Pipelining. A key feature of the HEP architecture is the use of pipelining for both memory and arithmetic logic unit (ALU) operations. This is done to hide the time taken to access memory or execute an instruction. The HEP uses instruction-level multiprogramming and multitasking to ensure that executed instructions come from different processes, thus having a high probability of being independent .

System balance. As only 4 PEMs, connected by a network of 3 input and 3 output switches, were built, it is hard to say if the system is balanced, i.e., if the network can serve the PEs. The network bandwidth was 10 million words per link, while each PEM could execute one instruction every 100 ns, for a 10 MIPS throughput. Because of the extensive instruction-level multitasking and multiprogramming, this figures seems quite balanced.

Technology. At the time, each PEM was a state-of-the-art engineering achievement by itself. Of course, a 10 MIPS computer doesn't seem so impressive today, where there are 100 MIPS CPUs on one chip. The extensive use of registers no doubt played an important part on the overall performance. It might be interesting to consider a one-chip architecture CPU today, when many of the HEP PEM functions can be integrated on a single chip.

27.2.6   The Goodyear MPP and the TMC CM-1

The Goodyear's Massively Parallel Processor (MPP) [4] has several similar features in common, as well as some differences from, the Thinking Machine's CM (Connection Machine) series that it predates. MPP is a 128×128 grid of bit-serial processors. It operates in an SIMD structure, primary processing images obtained by NASA satellites. The PEs can be daisy chained together, as in the CM, to form larger words. Each CMOS chip has eight bit-serial PEs operating at a 10 MHz clock cycle. Each PE has one mask bit (the CM-1 has 16) to represent the local state. Each PE has 1024 bits of RAM with 45 ns access time.

The CM-1 was originally intended to do "symbol processing" in applications such as AI and expert systems [21]. CM-1 has no hardware floating point processing (it was added in the next-generation model, the CM-2 [54]). The thought was to provide an efficient parallel pattern matching search and not to target algorithms that make heavy use of floating-point operations. The CM-1 has 4 times the number of PEs as the MPP. Similar to the MPP, they are arranged in a 256×256 grid. However, there is a second interconnection network, forming a hypercube topology between clumps of 16 single-bit PEs. The MPP and CM machines represent a trend to do very fine-grain parallel processing. Yet, one interesting observation is that the MPP achieves 6553 IMPS of 8-bit integer additions, compared to 4,600 IMPS on the CM-1. At least from this example, one might conclude that finer-grain parallel processing does not always result in better performance. The extra overhead associated with the CM-1, and the extra hardware complexity, result in an overall slowdown. Of course, it is very hard to really compare these machines, as they were built using different technologies, at different times. The CM-1 even has an extra hypercube network that can offer better communication. Yet, the problem of balancing an architecture is very important. In the extreme, if a powerful uniprocessor computer can achieve higher or comparable performance in comparison to a parallel machine, the later solution cannot be considered practical.

27.2.7   The IBM GF11

The IBM GF11 is another SIMD-type parallel computer with 566 PEs [5, 6]. The PEs are interconnected by a Benes network. This network has a major innovation idea, as it was designed to reconfigure among 1024 preselected permutations in one processor cycle. It takes a much longer time to set the network to an arbitrary permutation that is not included in the pre-loaded set of 1024.

It was observed that this number of pre-settings is quite enough for many useful topologies. A hypercube of dimension d requires $2d$ switch settings. A QCD (quantum chromodynamics) computation in which the GF11 is configured as an $8 \times 8 \times 8$ array uses only 6 of the available 1024 settings.

It is easier to switch between 1024 possible permutations in an SIMD computer, as all the control is centralized. Yet the importance of the GF11 is to show how the network is a critical part of a parallel system, as well as how several real and useful applications can be solved within the constrains of the 1024 preset permutations.

27.2.8 The BBN Butterfly

The BBN Butterfly parallel processor [9] is a shared-memory architecture. It has up to 256 PEs connected via a multistage packet-switching Omega network. It uses a hardware implementation of the "fetch-and-add" primitive (see the NYU Ultracomputer) at the memory, but without message-combining hardware. Originally, it was designed for satellite message-processing applications had little communication between the PEs. In this application, the network traffic was infrequently sent large massages (compressed phone conversations) .

The BBN approach was to keep the hardware simpler (and thus, maybe, get more throughput) and use software and programs that place fewer demands on the network. There are those who claim that the BBN machine has no network tree-saturation while having a nonuniform "hot spot" memory access patterns. Others claim that the tests showing the above result were done without placing high enough traffic load on the network, and that this can change when using different tests.

27.2.9 The MasPar MP-1

The MasPar MP-1 parallel processor is an SIMD type machine with up to 16,384 PEs. The PEs are RISC-like CPUs with direct connection to local memory. Each CMOS VLSI chip has 32 PEs that can do operations on up to 64-bit operands (including a floating-point unit), but internally it has a 4-bit-wide construction. The machine is arranged as a rectangular two-dimensional lattice. Sixteen PEs form clusters of 4×4 grid. Each cluster has its own memory and connection to the global communication network.

The interprocessor communication can use one of two available paths. Interestingly (for our purpose), it depends on the communication regularity an application exhibit. For regular data transfers, the X-Net communication mesh is more efficient. The X-Net mesh is a one-bit-wide network linking each PE with its eight nearest neighbors. Random communications between arbitrary PEs are possible via a three-stage global router emulating a crossbar switch. Each PE cluster has a connection to the router stage of the switch, and another to the target stage. The PEs of each cluster share these two connections. The router and target units are connected by an intermediate stage.

The interesting characteristic of the MasPar machines is separation between random and regular communications. Still, they are able to closely match various regular communication patterns on their X-Net.

27.2.10 The Cray X-MP and Y-MP

The Cray Y-MP evolved from the X-MP model and is quit similar to it. It has up to 8 PEs (4 in the X-MP), each with an instruction set compatible with the Cray-1. Both have shared-memory, tightly coupled architectures, using a circuit-switching crossbar network to connect between the PEs and the memory banks. Special-purpose, hardware supported, registers allow the PEs fast intercommunication.

These multiprocessor machines represent the opposite trend in parallel processing. The MasPar and the Connection Machine favor fine-grain parallelism, both of which (and others) have an SIMD operating mode. In contrast, the Cray machine (and others as well) favor tight coupling between a small number of supercomputers (in this case, compatible to the Cray-1). In both cases, the bottom line is cost-performance! From the viewpoint of the user, one important aspect is how much how much processing power costs. Between these two extremes, there are many intermediate solutions. We favor a solution that tries to couple thousands of the fastest available one-chip CPUs using an optical network.

27.3   How Much Interconnection?

The interprocessor communication structures required to support parallel processing have been recognized as a key issue in the design of parallel machines [52]. This section reviews the problem and algorithmic sides. We examine several algorithms that present some recognizable interconnection patterns (mesh, tree, etc.). The significance of this recognizable interconnection pattern is to make a good choice of the interconnection network topology of the parallel computer. As we can suspect, the diversity of interconnection patterns implies that, for a general-purpose parallel computer (i.e., one that should run different applications), we need an interconnection network that can efficiently support all these communication patterns. For optics to play a major role in MPP communication, it seems that some special feature of the applications must be known. As optics seems not yet to offer good processing features (including routing, buffering, and so on directly applied to optical signals), non-processing or non-logic operation of such a network may be needed. We review some examples of applications that exhibit some communication patterns. In such cases, optical networks can be used as a point-to-point, reconfigurable structures. Such arrangements, coupled with electronic processing, may lead to a more efficient role for optics in future MPP systems. In this section, we review a small number of such applications.

27.3.1   Parallel applications: the topology choice

The question of what is the best network topology has no one simple answer. It seems to be directly concerned with the type of the parallel algorithm one tries to solve. We can recognize several classes of parallel behavior. These classes might be seen by analyzing the communication patterns observed, at run time, between the PEs. Examples of such common structures must be classified by parallel programming paradigms such as those listed below.

27.3.1.1   Compute-aggregate-broadcast. Compute-aggregate-broadcast (CAB) algorithms are composed of three basic phases:

- *Compute phase.* Some computation is done on the data already present in the PE. This could be a very simple or very complex processing, depending on the application.
- *Aggregate phase.* This is the combination of local data into one or a few global values.
- *Broadcast phase.* The global information is sent back to each PE.

Examples of CAB behavior can be found in algorithms solving problems such as:

- the Jacobi iterative method for solving Laplace's equation on a square (for example, the electric field problem).

- the parallel conjugate gradient algorithm investigated by Adams [1], used to solve any linear system of: $Ax = y$ where A is a symmetric, positive definite
- topological sort where directed acyclic graph G is given, and an ordering on the nodes is needed such that if node i is before node j, there is no path from node j to i

27.3.1.2 Systolic and pipeline computations. Systolic computation in machines and algorithms is perhaps one of the best examples of clear communication patterns. This type of behavior is important for possible structured implementation in silicon VLSI circuits, and as a method to speed up "sequential" processors.

Pipelining is a known method to divide a complex work into some simple tasks done serially. Computers use this to speed up their instruction execution times by dividing a complex instruction into several simple operations. Each of these simple operations is done by a separate, dedicated hardware unit. After one unit is done with processing a part of one instruction, it continues to perform a part of the next instruction. In this way, the pipe might hold parts from several instructions, each one in a different level of execution. This is a form of parallel processing, adding hardware (separate execution units) to gain speed.

The systolic type of computation was defined in several papers (e.g., Refs. [26, 28, 27]). It can be characterized by several observations:

- *Flow of data.* The processing elements do computations on data that "flows" through them. Each processor makes some computation based on the data stream passing through it and (maybe) some previous results it keeps in local memory.
- *Flow of results.* The results (and partial results) flow out (in direction different from that of the input data) of the processing array. Some partial results might be combined in the way with incoming data. Not all *systolic algorithms* must meet this type of observation.
- *Locality of communication.* Each PE communicates with its neighbors in a mesh-like interconnection network (not exactly a mesh; there are diagonal connections, too).
- *Regular communication structure.* The direction of flow is maintained throughout all of the computation process (left to right, right to left, diagonals, etc.).
- *Simple operations.* Each PE usually does simple operations on the data flow.

There are many examples of systolic and pipeline algorithms, such as:

- The band matrix multiplication [26]. This algorithm multiplies two $n \times n$ matrices, A and B, producing the result C.
- The dynamic programming algorithm [18].
- Solving graph problems in parallel environments [22]. This algorithm uses funneled pipelines to solve several graph theoretic problems.

27.3.1.3 Divide-and-conquer. This is a well known technique for solving problems in both sequential and parallel types of execution vehicles. The given problem is divided recursively into smaller and smaller problems of the same kind (only with less data). Then these small problems are solved, and their results are recursively combined, giving results for larger problems each time. This splitting into smaller problems and combining the results needs processing power, also in parallel. Most of the divide-and-conquer algorithms need a binary n-cube interconnection network (hypercube). Some, however, need mesh interconnection for their operation.

There are many examples of this behavior for information pattern flow:

- In Batcher's bitonic merge sort [3], the problem is to sort n data items using n PEs. The algorithm divides the n items several times. Each time a partial sort takes place, done by the n PEs. Then the bitonic sequence is transformed into a monotonic sequence. This type of algorithm presents a binary n-cube structure of communication. The binary n-cube is the minimum connection graph needed by this type of algorithms.
- The fast Fourier transform (FFT) is another example where this binary n-cube type of interconnection is needed [29] .
- Several image processing algorithms use divide-and-conquer techniques but require only a mesh type of interconnection [51].

The above examples demonstrate that at least some parallel algorithms present well structured patterns of communication between the PEs. These patterns might impose certain interconnection topologies. Using any other topology (that cannot implement the required one in a direct way) might result in very poor speed-up. A general-purpose parallel computer might need to implement many interconnections for different applications.

27.4   Considerations in Choosing the Interconnection Topology

The previous section outlined the need for a massively interconnected parallel architecture in view of communication patterns seen in the behavior of some classes of parallel algorithms. There are other issues important for having a general communication topology. The following is a list of factors influencing the need for such a network.

1. *Multi-user architecture.* Building a multi-user, general-purpose parallel computer implies that this machine will have to run different algorithms at the same time. Fixed, non-reconfigurable type of architectures, might impose severe performance penalties.
2. *One problem, many solutions.* For one problem, there might be several possible solutions. Better algorithms might need a more "sophisticated" network. One cannot naturally expected that, with a bus topology, we will get the performance that is possible with the full network. There might be problems that could not benefit from this extra power, but this is not the general rule.
3. *Full interconnection algorithms.* There are several types of algorithms that need the full interconnection for their operation. An example of this need might be found in the 2-dimensional FFT algorithm presented in Ref. [46].
4. *Fine-/medium-grain parallel processing.* This is a straightforward concept. It is not the same problem to build an interconnection network for a 100 PEs machine as it is for a 100,000-PE machine. Even what seem to be simple topologies like the bus and shuffle networks get very complex and impossible with many PEs, using electrical networking. Probably the main and most noticeable feature in which optics seem to offer is in the communication domain. Even with MIN-type solutions, it will be very difficult to build so many intermediate elements (on the order of $N \times \log(N)$; there are about 500,000 routing elements for $N = 100, 000$ PEs). The PEs can be very powerful—not the type built for the CHiP. So, in conclusion, if we assume that the optical side of the problem will come out successfully, a new magnitude of processing power could be achieved.

5. *Operating system (OS) considerations.* The operation system's main job is to manage the resources of the computer and try to keep the computer as busy as possible (no idle time). We have to provide the OS with the flexibility it needs for managing the valuable computation potentials hidden inside. Multi-tasking, multi-user operations might be possible in parallel, with different interconnection setups.

6. *The mapping problem.* Simply stated, the mapping problem is the problem of fitting an algorithm or program (with the communication structures it needs, and the number of PEs it was designed for) to a particular given architecture. This is not an easy problem, and there is no known complete solution to solve it (but there are some practical solutions that are close to the theoretically best solution possible). This problem also deals with the way parallel programs are generated. It seems that it will be very difficult having to know the particular architecture on which we'll run our application. We would like to present the programmer some "virtual" picture of the machine and let automatic processing worry about the translations .

27.5   Optical Communication: Free-Space Interconnection

This section presents a short review of the research on optical free-space light beam routing for parallel processing interconnection networks. The optical networks reviewed in this section are examples of some basic techniques that were adopted in an effort to reach a free-space interconnection network. There are many other proposals for building interconnection networks using free-space optical communication. We choose to review here some representative designs. Most of the other designs fall into one or several classes of the designs presented here. We will try to generalize all the aspects.

27.5.1   Motivations for optical interconnection

Sufficient communication bandwidth is one of the major bottlenecks, if not *the* major bottleneck, in the construction of massive parallel computers. The electrical medium appears to limit the performance possible in an electronic-based interconnection network. Optics seem to offer better features, among which the important ones are wide bandwidth, lack of crosstalk and interference, minimal space propagation delays, large communication channel densities, and an easy way of producing directional modulated beams (lasers).

It is interesting to see where, in the parallel computer system domain, it is possible to use optics instead of electronic communications. One way is to make a comparison between optical and electrical interconnects based on power and speed considerations. In their paper, Feldman et al. [12] present the conditions for which optical interconnects can transmit information at a higher data rate and consume less power than the equivalent electrical interconnections. Their conclusion is that optical free-space interconnection is a better candidate for intra-chip communication in large area VLSI or wafer-scale integrated circuits, especially when high data rates and/or large fan-outs are required.

In another paper by Miller [34], a similar comparison between optics and electrical communication is presented based on detectors, sources, and modulator operating parameters. This comparison suggests that all except the shortest intra-chip communications should be optical.

We suggest a different approach looking at three parameters: size, communication bandwidth, and interconnection needs. A similar study recently published by Guha et al. [17], examines the communication needs of the state-of-the-art components (microprocessors and memories) forming a parallel computer. We present a comparison from the basic gate to the complete PE module based on three factors.

Table 1 presents a coarse view of the relations in the size, data throughput and functionality. The throughput of the basic gate reflects the state-of-the-art technology used for VLSI fabrication. The cited figure of 2–4 Gb/s reflects present technology. However, this rate is possible in a medium scale integration using bipolar circuits. The basic gate is connected in a certain fixed way to other gates. Due to using several levels of logic (making a state machine for example) and the packaging of the basic chip, the data rate drops to 100 MB/s at the chip level. Again, this reflects present state-of-the-art technology. The chip usually is connected in a fixed way to certain other chips, making a PE. The PE data rate is similar to the chip, as it is made of several chips connected together. The PE uses a dynamic interconnection, depending on the parallel architecture (MIMD versus SIMD) and, as previously presented, can benefit from the optical technology.

TABLE 27.1   Size, Throughput, and Functionality

Circuit	Size	Throughput	Connectivity
Gate	10×10 microns^2	2–4 Gb/s	Static
Chip	1×1 cm^2	100 MB/s	Static
PE	10×20 cm^2	100 MB/s	Dynamic

The topology of an optical interconnection network may be reconfigurable. This reconfiguration need not be done in every machine cycle. Observing the behavior of several parallel algorithms, we can notice that the communication patterns related to them remains stable for quite a long time—up to the entire running session. This parallel programming paradigm was observed in non-shared-memory computers [23]. Optics seems to be a technology appropriate for such parallel architectures. It is easier to build large parallel computers that are message-based and nonsynchronous rather then the classical synchronous PRAM CRCW model. These stable communication patterns constitute a very important observation. Its meaning is that we need not construct interconnection network capable of reconfiguration in every memory cycle. Such a fast network might be difficult to implement, particularly for a large number of PEs.

27.5.2   Optical interconnections

Designing a large parallel interconnection network is one of the important problems looking for good solutions [20]. There are many proposals to build optical interconnection networks capable of connecting a large number of processing elements (PEs) [31–33, 42, 45]. The present suggestions employ one or few of the following principles of operation:

1. *Centralization.* This has two distinct aspects. One is the use of a central arbitration scheme. For large interconnection networks (10,000 or more) this could be a major bottleneck. The other is the use of a central device to make the routing [such as one common cathode ray tube (CRT), computer generated hologram (CGH), or one photorefractive crystal]. This might limit the total number of PEs that can be accommodated.
2. *Massive computations.* To perform the routing, it might be necessary to make heavy calculations (for example, generating a CGH according to the particular interconnection pattern).

3. *Bulk optics.* The CRTs or very large lenses and mirrors that are used become critical components.

4. *Slow switching rates.* These occur due to the lack of fast spatial light modulators (SLMs). In view of the above, a diligent effort has been made to find good interconnection networks using optics. We review some of the previous work done in the following subsections. The structure of this review includes two major parts:

   (a) *Description.* We briefly describe the way the design works, its components, and the setup of the network. For a detailed description, the citation of the refereed work should be read.

   (b) *Comments.* We bring our comments for every design we review. As far as we know, many current designs suffer from one or several drawbacks which we think prevent their use in a general-purpose, parallel computing interconnection network. We will briefly state what these drawbacks are, in our opinion.

27.5.3   The Holoswitch

The Holoswitch described in Ref. [8] is based on an array of optical switches (liquid crystal polarizing beam splitters) that direct a set of optical beams toward any of a selection of holograms. Each hologram, when selected, deflects the input beams toward an output array with any desired prerecorded permutation. Every switch acts on the N input beams in parallel. These N beams are directed as one to the prerecorded routing hologram. N input channels might give $N!$ possible permutations. But only a small fraction of these might be useful for parallel processing. These permutations are prerecorded in a hologram. Every hologram has a number of holographic spots, one for each channel, that individually deflect the beams in the desired direction.

The new characteristic of the Holoswitch design is the use of a switch array to address different holograms to change the interconnection pattern. The Holoswitch can be used for large interconnection networks. Considering the effects that limit the number of channels, the number of channels the Holoswitch can accommodate can be up to 1000. The prerecorded routing holograms can be made with an automatic recording setup, thus making the mass-production process simple.

In our opinion, the Holoswitch design can be used only in special-purpose parallel computing. The following constrains prevent this design from being a good choice in parallel optical interconnection networks:

1. *Central operation.* The reconfiguration of the network can be done only to predefined communication patterns and by a central unit. This means five separate things:

   (a) The parallel computer can run those algorithms that use a prerecorded pattern, and that this pattern should be used by all parts of the algorithm at a given time.

   (b) The pattern is changed by a central controller. A PE cannot initiate a separate change in the global interconnection pattern. The PEs must either wait until all of them finish and ask the controller to change the pattern (complex coordination for large number of PEs), or the pattern remains unchanged throughout all the session (and is set as part of the program loading process established during compile time).

   (c) The parallel machine cannot run multiple, different parallel programs using part of the PEs for each. If these programs need different communication patterns at one time, they cannot be served using the Holoswitch design. This is a most limiting feature since it suggest that the machine powers cannot be shared among several users at a time (only batch-mode operation, resulting in a waste of computation power).

(d) Since all the routing is done by one central component, there is a question of how the information from the PEs reaches this component. Using optical fibers would introduce a wiring complexity that behave in a 3-D space as $O(P^{3/2})$ where P is the number of PEs. Free-space routing could be another possibility, but it was not shown in the paper.

(e) Lack of locality in the optical paths of the data means that the length of the paths will be always the same (resulting in high latency times in the network). Some topologies (like the mesh) could benefit from a more local orientation in the network configuration.

2. *Scale-up.* The design is limited in the maximum number of channels it can handle. This number is quite small in principle (1000) and can be less in practice. The problem of centralized operation limits further the number of channels. It will be hard to route the information channels to this one device, considering that the total space occupied by the entire system could span a large distance relative to the HoloSwitch size.

3. *Parallel input.* The input to the HoloSwitch should be a perfectly parallel light beam array. This seems to be a difficult task, considering that the beams might come from relatively remote sites.

27.5.4 Diffractive-reflective optical interconnects (DROI)

The diffractive-reflective optical interconnects (DROI) described in Ref. [7] present a different approach. These interconnects consist of a sandwich of a holographic plane and a reflective plane. This setup presents various possibilities like beam relaying, connection switching, and broadcasting. The DROI can work in two possible configurations:

▪ *Focused beam.* In the focused beam method, a lens is used to image the light emitting source onto a surface of the mirror. After reflection, the beam passes a second lens (hologram) that is used to focused it back onto the detector.

▪ *Collimated beam.* In this method, the light beam coming out of the source is collimated and deflected by a holographic lens. The collimated deflected beam is then reflected by a mirror onto a second holographic lens. This lens focuses the beam onto a detector.

The above configurations can be used to form dynamically reconfigurable interconnection networks. To build such networks we need to apply the following extensions:

▪ *Beam relaying.* To allow the beam to travel over a greater distance than is possible with the original two holographic lens schemes, a multi-hopping method can be used. In this method, the beam bounces several times between the hologram and mirror planes until it reaches the desired destination. The holographic plane has reflective spots at selective places, allowing the light path to bounce up and down in a zigzag way. Thus the interconnection distance is increased without going to extreme deflection angles or large system thicknesses (because of the distance between the mirror and the holographic plates)

▪ *Nearest-neighbor broadcasting.* By using the hologram for splitting the source beam, it is possible to achieve any desired neighborhood configuration. Thus "flat" communication patterns (such that their topologies form planar graphs) can be built. Grid, tree, and some other systolic patterns are examples of such neighborhood communications.

- *Connection switch.* Building reconfigurable interconnection networks requires the use of a dynamic element for routing changes. This can be done by using an active element that can change its state from transmission to reflection according to a control signal. We can replace the mirror with such a switch, thus making the path a beam travels to change dynamically.

In our opinion, the limitations of the DROI design are as follows:

- *Fixed topologies.* The reconfiguration done by using the connection switch is limited to topologies that can be derived from the basic routing paths by interrupting the beam at various locations along the way. This could be suitable for a splitting bus interconnection network.
- *Routing scheme.* It is difficult to use this scheme in a parallel computing system. This is because the way a PE works with information is by addressing, at run-time, a location in memory or by sending a message to one (or several) PEs. If we examine the way this design achieves reconfiguration, we can observe that any connection switch should decide whether to pass (not break the path) or block ("swallow" the information for its PE). This implies that the path should pass through all the potential PEs (making a bus-like structure) or that the network reconfiguration should be done by a central controller (see the observations regarding the HoloSwitch design above).
- *Making use of 3-D space.* The basic design use only 2-D space. To decrease latency times it is desirable to use 3-D space efficiently.

27.5.5   Variable wavelength recording for dynamic gratings

In Ref. [38], a dynamic hologram is incorporated to deflect a source beam in the desired direction. Based on a Bragg diffraction, the beam is steered by a dynamic grating in which spacing and orientation are changeable in real time. This is done by acting on the optical frequency of the control beams used to record the dynamic gratings. A deflection angle of up to $30°$ can be achieved with a 50 nm change in the control wavelengths. By combining a 1-D scanning element (mechanical scanning mirror or 1-D acousto-optic devices such as described by Harris and VanderLugt [19]) with the varying wavelength recording, a 2-light beam steering can be made.

An index grating is optically recorded in a dynamic photosensitive material by interference of two coherent plane waves at wavelength λ, emitted by a dye laser. Varying the wavelengths of these beams modifies the grating period, thus producing a change in the diffraction direction of a light beam at wavelength λ_0 (reading beam). The control beam is split into two recording beams in the path of which fixed dispersive elements are placed. The combining of these two beams forms the interference patterns in the photosensitive crystal. By using several multi-wavelength control beams and simultaneously recording different grating inside the photosensitive crystal, it is possible to split the reading beam so that multiple directions can be addressed. Some practical and conceptual limitations of this design prevent it from being the solution for a free-space optical interconnection network device. These includes the following issues:

- *Bulk devices.* The use of dye lasers with the λ monitoring circuitry prevents the use of this device as a building block for interconnection networks. These devices could occupy a large volume, increasing the total space the system needs (resulting in other problems such as large latency, energy consumption, etc.).

- *Complex control.* There is a need for a relatively complex computation to convert the addressing information to the wavelength difference needed for the dye laser beam frequency modulation.
- *Limited multicasting.* When multicasting is needed, there is a need for an extra control beam for every added channel. The reading beam's energy will be split by the hologram recorded in the photorefractive crystal. This means a separate dye laser for every such added channel.
- *Extra components.* To get the 2-D scanning, it is necessary to use some external device to deflect the light at the second dimension in space (the frequency modulation can be used for one dimension scanning only). Using the device in a 1-D reduces the possible addressable output destinations (from a possible of N^2 to only N; spreading over a certain deflection angle)

27.5.6   Computer-generated routing holograms

The use of a computer-generated hologram (CGH) to form a free-space dynamically reconfigurable interconnection scheme is presented in Ref. [31]. The setup uses a double-pass reflective configuration. A light source collimated by a lens is reflected by a liquid crystal light valve (LCLV), with phase modulation on a certain polarization, toward a mirror. The LCLV is controlled by a CRT in real time. The CRT displays different connection patterns by using an electronic computing element. A two-dimensional array interconnection is formed by computing the appropriated routing hologram to connect N sources with M destinations.

The light reflected and diffracted from the LCLV is imaged (focused) on a mirror. The reflected light (from the LCLV) focused onto the mirror consists of the original beam and the diffracted beam acquired from the reflective grating. These beams, on their way back, undergo another reflection from the LCLV, thus generating a second diffraction pattern from the same grating. A quarter wave plate is introduced in front of the mirror to prevent this.

Computer-generated hologram schemes suffer from some common problems. In the following, we describe some of these problems and others related to this particular scheme:

1. *CGH problems.* CGH used for free-space reconfigurable interconnection networks have the following drawbacks (some of which are related to the above scheme):
   (a) *Complex computation.* Computing holograms for an arbitrary routing pattern requires great computational power. To get any reasonably useful hologram, a resolution of for about 1000 to 2000 line pairs per millimeter is needed. This means that the process of generating a new network configuration could be very slow or could consume a large amount of memory (if it was precomputed for some fix communication patterns).
   (b) *Recording the CGH.* A device capable of recording the CGH is needed. This presents two problems—the device that can write the CGH information with the required density, and the material to be used for holding the written CGH. The CRT solution partially fulfills this requirement, since it only displays a resolution reaching the lower end of what is needed for a good hologram quality (good efficiency and minimum crosstalk between channels). This hologram is limited in the number of interconnection points it can handle.
   (c) *Efficiency and crosstalk.* The algorithms used for CGH make some approximations, resulting in poor efficiency (less than 10 percent; usually 3 to 5 percent).

The efficiency is especially low for a one to many (multicasting) operation. Another problem is crosstalk due to these approximations (some of the beam energy intended for one target is scattered to other targets).

2. *Serialization.* The CRT is a sequential device. Each source must pass to a central controller to ask for a new destination. This cause to all the drawbacks of a central controller for the routing, referred to at the previous designs.

3. *Bulk optics.* The CRT and the lenses and mirrors that are used in the setup are bulky and hard to work

27.5.7   Incoherent optical interconnection

Communication doesn't have to be made using coherent light. Such an incoherent optical setup is presented in Ref. [39]. The device is based on a lenslet array and an optical mask assembly. The lenslet array creates multiple images of the input. The mask is used to control the interconnection pattern by blocking parts of theses images.

The light of each source is spread and imagine by the lenslet array on the mask. For an interconnection network of N sources and M targets, there is a need for M sub-masks, each with N addressable pixels. Turning on or off the appropriate pixel will establish a connection between the source and destination. Although this design can use incoherent light, two severe drawbacks might be observed:

- *Broadcast operation.* Any source must broadcast the information over an array of M by N pixels. In high bandwidth operation, there is a problem of getting low beam energy after the mask. This could limit the bandwidth of such a scheme. On the other hand, there is a limit to the minimum size a pixel can be made. Small pixels might present crosstalk, diffraction (from a small aperture) and other problems
- *Bulk optics.* The design calls for bulk optics. The system of mirrors used to deflect the beam might present problems when the size of the overall system increases. Practical parallel processing systems might expand over a large space, thus the use of large mirrors seems impractical.

27.5.8   Adaptive optical networks

A different type of interconnection network used for neural networks applications is examined in Ref. [42]. The optical setup includes the following items:

- *Input neural plane.* On this plane, there is an array of light sources emitting signal beams. Each source represent a neuron cell.
- *Training neural plane.* This plane is used in the process of forming the interconnection holograms into a photorefractive crystal. The combined interference patterns of the training and the input planes are recorded inside this crystal to form a volume interconnecting hologram.
- *Photorefractive crystal.* In this crystal, a volume hologram is embedded. The use of such a crystal can provide the real-time creation of reconfigurable interconnection patterns.
- *Fourier lens.* There are two of them in this setup. The first transforms the spatial position of each neuron into a spatial frequency associated with light emitted by or incident on that neuron. After the beam passes the routing hologram embedded in the photorefractive crystal, a second Fourier lens translates the spatial frequencies back to spatial positions. Thus the beams hit the target at the required locations.

- *Output neural plane*. This plane is the target of the routing hologram. Each neuron on this plane gets the beam energy sum of the input neural plane.

An interconnection between the ith neuron in the input plane and the jth neuron in the output plane is formed by interfering light emitted by the input neuron with light emitted by the jth neuron in the training plane. The image of the jth training neurons lies at, in the writing phase (creating the routing hologram), the position of the jth neuron in the output plane. The interference of the training signal and the input creates a grating in the recording medium (the photorefractive crystal). In the reading phase (using the routing hologram), the jth training neuron is turned of. The light emitted by the ith input neuron interferes with the previously recorded routing hologram. The beam is refracted to the jth neuron in the output plane. We might suggest the following observations regarding some weak aspects of this scheme:

- *Control*. The formation of the hologram in the photorefractive crystal requires communication between the input and the training plane. For large neuron array, this could lead to a high density between these planes.
- *Serialization*. For each pair of neurons, the formation of the routing hologram connecting them must be established separately. If two pairs, (i_1, j_1) and (i_2, j_2), transmit the beams simultaneously, the routing hologram will establish a *combined mixed* connection between them such that the light emitted by the input neuron i_1 will be split between output neurons j_1 and j_2. The same will happen with input neuron i_2. The output neurons j_1 and $j2$ will get a sum of the combined transmitted beams from i_1 and i_2 neurons. This limitation also adds to the control complexity mentioned above, since it is needed to synchronize the operation of the neurons at the hologram writing phase.
- *Hologram erasing*. Using the same wavelengths for reading and writing the routing hologram limits the time it is possible to use this hologram. After some time, it will be necessary to "refresh" *all* the routing holograms written in the crystal.

27.5.9 3-D multistage optical interconnection network

The dynamic reconfigurable interconnection network described in Ref. [44] interconnects two $N \times N$ arrays on 2-D planes using 3-D space. Using the volume (non-planar) nature of optics, this design attempts to build interconnection networks that will not be possible with standard electronic parts and technology.

This design proposes to use a multistage interconnection network (MIN) to reduce the complexity needed to build a non-blocking, crossbar-like network. MINs were and are considered to be an economical and efficient way to build networks with full access (any input can communicate with any output) capabilities. Instead of the $O(N^2)$ switches needed by the crossbar-like networks, these networks need only $O(N \times \log(N))$ switches.

The design presented in the above citation implements a version of an Omega network with non-blocking operation. The components used by the design are:

- *Basic switch*. This switch is the basic building board of the MIN. It is a 4×4 switching module providing a one-to-one crossbar interconnection between 4 input and 4 output lines. The control to the switching operation, provided by a 5-bit control code is provided to select one of $4! = 24$ possible permutations. It is suggested to implement each switch using optoelectronics (for the near term) as a 4-element detector array providing input to electronic switching which drives 4 output sources (lasers or LEDs). For the long term, there is a thought of using passive optical components under electronic control to switch and route the optical data paths (thus providing more possible bandwidth than the optoelectronic approach).

- *Boards.* Each level of the MIN can be thought of as a plane or a huge board. Actually, one plane could be made of several smaller boards. Each of these board or planes has $O(N)$ switches. There are $O(\log(N))$ such boards—one for each level in the MIN. The communication between the boards is done by *static* or *passive* optics. MINs have fixed topologies between the levels, and the switching is done by routing the data through various segments between the levels, switching them according to the required connections.

- *Interconnection optics.* Between the boards, there are passive interconnection optics. One way to make the needed interconnection between the levels is first to make a planar shuffle prism that provides the appropriate shuffle deflection in one direction and has no spatial variation in the orthogonal direction. A sandwich of two of these devices, oriented orthogonally, is placed at the output of the 4 x 4 switching modules on the board. A CGH may be used also for the interconnection optics, but the low efficiency of a thin hologram is a drawback. The CGH can be copied onto a volume phase material such that higher efficiency could be achieved.

We claim that building MIN in general, and this particular system, could not be a solution for large interconnection networks. The following is a list of the reasons why we assert that MIN are not a practical solution. We bring also the drawbacks of the above particular scheme.

- *Number of active elements.* Using $O(N \times \log(N))$ active elements in a large network is too expensive. For example, an interconnection network of 10^6 input and output lines will need about 20×10^6 *switches*. We claim that this is to high. In this paper, we show a way to implement a crossbar like network using only N active elements. The complexity of each element is higher than it is for the MIN switch, but multiplying it by $\log(N)$ stages, as well as other considerations, make our design much more attractive.

- *Control.* Passing the switching information to every switch is a common problem found in every MIN. Two possible solutions could be adopted:
    (a) *Separate control.* The switching control information is provided by some external controller, not by the message itself. This means that a central controller must be used with all the problems that such a solution raises (see previous designs).
    (b) *In-message control.* The message itself carries the needed routing information. This simplifies the control scheme but introduces an extra complexity in the basic switch. The switch will have now to detect (in real time!) each incoming message and decide quickly where to route it. This could severely limit the bandwidth such a switch could handle. Using passive elements controlled by electronic circuits (as suggested in the scheme above) cannot provide this option at all.

- *Non-blocking.* For non-blocking operation, we need to increase the number of basic switches. As stated in the cited paper above, it requires $\log_2(N)^2$ stages to do this. In the above example, this means a total of 400×10^6 switches.

- *System volume.* The total volume an optical MIN will need might be very large. Taking into account some limited viewing angle (meaning that there is a limit to the maximum deflection angle the interconnection optics can provide) and the fact that the interconnection channels between the stages must (sometimes) cross all the space from one side of the board to the opposite side, the ratio these facts impose on the geometrically minimum dimensions lead to huge overall system dimensions (with all the drawbacks such as long latency times, etc.). It might be that beam diffraction in space might be too high when passing the distance between the levels. This will create large spots at the detectors, increasing further the overall system volume.

- *Bad component distribution.* The MIN, using all of the volume, just interconnects two arrays of PEs. We would like to get a more even distribution of the PEs in space. We will bring forth a design that meets this goal.

27.5.10   Parallel N^4 weighted optical interconnection

A weighted optical interconnection network for neural systems is described in Ref. [8]. This network can form arbitrarily variable-weight interconnections between neurons in a system. The components of the system are:

- *Input array.* This neuron array is formed by a matrix of $N \times N$ neurons. Each neuron has its own reflective spatial light modulator (SLM) in the for of an $N \times N$ pixel matrix. The SLM can either reflect or transmit (absorbed—the useful light beam is just the reflected one) any one of N beams reaching its pixel plane. Each pixel in the SLM represents the appropriate output neuron. If a connection is to be established to any particular neuron, its corresponding pixel is turn to the reflective state.
- *Output array.* The output array consists also of an $N \times N$ matrix of neurons. Each neuron will receive the beam energy sum of up to N beams coming from the input array.
- *Beam splitter.* The beam splitter reflects the light beams coming from the input array to the output array. It passes beams with one polarity and reflect beams with the orthogonal one. The unmodulated, weighted beams pass through it the first time, hitting the SLMs on the input neuron array. Then, the reflected beams return back (following the original path in reverse) and hit the beam splitter again. This time (with an orthogonal polarity) they are reflected on to the output neuron array.
- *Mask hologram.* The mask hologram is an $N \times N$ array of subholograms, each with an $N \times N$ array of facets (there diameters are between 1 and 2 mm). Each of these facets is used to apply a weight (in the form of modulating the reconstruction beam used to illuminating the entire mask array).

This network can be used, in principle, to connect PEs a parallel computer. By modulating the beams coming from the mask hologram, it is possible to pass more information (rather than just summing beam weights) to the output array. The SLM of the input array should work at appropriately high rates to allow a reasonable bandwidths. This is one of the main drawbacks of this system. It is hard to visualize a good SLM, working at high rates (10 GHz) appearing in the near future. Another problem is the critical component needed to connect the input and output arrays (the beam splitter). Since the physical dimensions of each array could be quite large (each PE must have its own reflective SLM, having some minimal dimensions), it might require use of a large beam splitter.

27.5.11   Two-wavelength photorefractive dynamic interconnect

A novel switch geometry using routing holograms formed in photorefractive crystals is presented in Ref. [33]. This design features different wavelength beams for reading and writing to prevent hologram destruction by the reading beam. The setup of this system is formed using the following components:

- *Routing element.* The routing element used by this scheme is a hologram held inside a photorefractive crystal. The hologram (working in the transmissive mode) diffracts N input signals to M outputs using N by M gratings.

- *Writing beams.* Using a conical geometry, it is possible to make arbitrary $N \times M$ gratings using only $N + M$ writing beams. This is done by simulating the formation of the hologram in the crystal using N writing beams for the input sources simulation and M writing beams for simulating the output destinations. The wavelength of a writing beam, λ_w, is shorter than the wavelength of a reading beam λ_r. The writing beams wavelength is in the sensitivity domain of the photorefractive crystal to excite the creation of photocarriers. These photocarriers will be transported and trapped in the crystal structure, causing the formation of internal electric fields. This formation corresponds to the interference pattern of the writing beams and changes the index of refraction in the crystal in a pattern corresponding to the interference of the beams. This process forms a phase hologram used for routing.

- *Reading beams.* The reading beams are diffracted by the phase hologram created in the crystal. Since the wavelength of the reading beam is chosen so that it is longer that the writing beam wavelength so that no photocarriers are produced, the index gratings are not affected by the reading process. The reading beams can diffract off the gratings created by the writing beams but, because the difference between the wavelengths, a special geometry must be used.

This switch uses a conical geometry to compensate the difference between the wavelengths of the reading and writing beams. Since separated beams, having different wavelengths, are used to prevent the distraction of the routing hologram, it is clear that these beams cannot come from the same geometrical place in space. The conical geometry ensures that any grating in which a writing beam participates during formation will have the correct Bragg condition for the associated input or output signal. The writing and reading beam sources are placed in two circles proportional to the relative difference between their wavelengths and the distance to the hologram plane formed in the crystal.

This design, although solving some problems (destructive reading, compact size, etc.), still has some basic problems, several of which were mentioned regarding the other designs. The following is a list of these problems:

- *Central operation.* The recording of the routing holograms are done on the *same* crystal. Therefore, it is not possible to record two separate routing holograms in parallel (because the interference patterns of their writing beams will mix, forming the wrong interconnection). A *central* arbiter must decide who gets the access to record a new interconnection pattern in the photorefractive crystal. As explained for the other schemes, this central operation is a real bottleneck.

- *Scale-up.* Because all the interconnection patterns are mixed in the same crystal, it is hard to record too many of them in parallel. The authors of the paper cite a number of 100 as the capacity of their design to interconnect between sources and destinations. There isn't any scheme on how it is possible to use multiples of such switches and combine them into one interconnection network.

- *System dimensions.* Because only one small crystal (it is not easy to grow big crystals of 1 to 3 cm^3) is used for routing, and the sources and destinations span relatively large distances, the crystal should be located quite remotely over the source plane (otherwise it would be necessary to go to extreme angles of diffraction). It is then unclear how efficient such a design could be in terms of the power needed by a writing beam and reading beam to pass information at a certain rate, the overall system dimensions, etc.

27.5.12   A Dynamic holographic switch for parallel processing

A dynamic holographic optical interconnect scheme involving SLMs and volume holograms is presented in Ref. [30]. The basic idea is to use a multiple-exposure technique to store several routing holograms in a volume recording medium. The basic idea has the following components:

- *Routing element.* The routing elements used by this scheme are multiple-exposure holograms stored inside a photorefractive crystal. The holograms (working in the transmissive mode) associate the address of a destination PE by encoding an SLM with a distinct reference beam. All the routing is done within one crystal.
- *Writing beams.* The writing procedure is done using different exposure times to compensate for different efficiencies. Thus, the hologram that was created last will have the same efficiency as the one that was created first. The holograms are written serially in the crystal (i.e., they have to be pre-written, and if a change is needed, it is necessary to write a new set of holograms). The holograms are created using different angular reference beams so that a reading beam could be deflected to the corresponding angles (representing different target PEs).
- *Reading beams.* The address of a destination PE is encoded into an orthogonal code. For N PEs, M bits are needed. The actual number of M depends on the acceptable crosstalk. For crosstalk less than one-half, M is about $\sqrt{2N}$ bits (or pixels) long. An SLM of $N \times M$ pixels is used. Each PE controls a row of M pixels. The PE encodes the destination PE address using its M pixels. The reading beam of this PE passes through M pixels and reconstructs the appropriate reference beam in the angle corresponding to the target PE for communication.

This design call for storing a number of hologram that is of the order $O(N^2)$ in one photorefractive crystal (N PEs, each one storing $N-1$ holograms, corresponding to $N-1$ possible destinations). The idea uses a smaller SLM than was used by other central-type, optical crossbar switches ($O(N \times \sqrt{2N})$). In our opinion, although solving some problems, the idea still has the following drawbacks:

- *Central operation.* The recording of the routing holograms is done using one crystal. This is a general problem (found in other schemes as well) that prevents us from building a really large system.
- *Scale-up.* Since all the interconnection holograms are multiplexed inside one crystal, it is difficult to use this scheme for large number of PEs. The scheme calls for multiplexing of $O(N^2)$ holograms! The efficiency of these holograms is (at least) inversely proportional to N, due to fan-in losses (as reported in the paper).
- *Slow operation.* Using pre-stored holograms inside the photorefractive crystal does eliminate the long writing time needed to create them dynamically. But the use of an SLM still is the switching-time bottleneck. Even the fast SLMs, will take few microseconds to switch (not counting the control time).
- *Large SLM.* Even with the reduction from $O(N^2)$ to only $O(N^{3/2})$, it is still difficult to use this scheme for massively parallel processing systems with more than 10,000 PEs! The poor contrast ratio of big SLMs will force the value of M to be even larger than was mentioned above.
- *System dimensions.* Because only one small crystal (it is not easy to grow big crystals of 1 to 3 cm^3) is used for routing, and the sources and destinations span over relatively

large distances, the crystal should be located quite remotely over the source plane (otherwise it would be necessary to go to extreme angles of diffraction). It is then unclear how efficient such a design could be in terms of the power needed by a writing beam and reading beam to pass information at a certain rate, the overall system dimensions, etc.

27.6 Conclusions and Future Work

Parallel processing with free-space optical interconnections may be feasible in the near future. We reviewed some MPP architectures and some optical interconnections proposed for MPP systems. We have discussed the following items:

- some thoughts about general-purpose parallel processing
- an adaptable topology for better fitting parallel programs
- several aspects of the optical interconnection—important considerations such as high bandwidth, power budget, and manufacturing ease
- adaptation of the underlying interconnection structure with parallel programming's natural communication structure

We have not addressed some other issues that might be important for the realization of an MPP using an optical interconnection in a working system. These issues can be in the scope of future investigation. However, we think that these issues are more of an "engineering nature" and could be solved without affecting our main review. Some of these issue are:

- investigating the alignment problem
- cost of components and the system versus the overall processing power obtained
- reliability (MTBF)
- what happens to the scattered light (optical background noise)

There are many issues that can be further investigated. Technological topics such as finding better materials with better optical qualities to implement the global switches is one issue, but it fits more in the realm of laboratory work and practical experiments. Investigating the implication of OPAM on the software (languages, compilers, OS, etc.) is another possible direction. A more fundamental study concerning the particular structure required of the communication and computation processors would also be interesting. As a more theoretical direction, a model to balance the whole system would be fascinating. The communication balancing for a particular distributed algorithm can be investigated, as well as other regular graphs to be embedded with a minimal cluster size. From a practical viewpoint, reliability and fault tolerance are very important and might impose upper bounds on the size and complexity of the PEs in any MPP system.

27.7 References

1. Adams, L. M. 1982. Iterative algorithms for large sparse linear systems on parallel computers. Ph.D. thesis. Charlottesville, Va.: University of Virginia.
2. Almasi, G. S., and A. Gottlieb. 1989. *Highly Parallel Computing.* Benjamin/Cummings.
3. Batcher, K. E. 1968. Sorting networks and their applications. *Proc. of the AFIPS Spring Joint Computer Conf.,* Vol. 32, 307–314.
4. Batcher, K. E. 1980. Design of a massively parallel processor. *IEEE Transactions on Computers,* Vol. 29, No. 9, 836–840.

5. Beetem, J., M. Denneau, and D. Weingarten. 1985. GF11—a supercomputer for scientific applications. *Proceedings of the 12th International Symposium on Computer Architecture.* Boston: IEEE Computer Society.

6. Beetem, J. M. Denneau, and D. Weingarten. 1987. The GF11 parallel computer. In J. J. Dongarra, *Experimental Parallel Computer Architectures.* Amsterdam: North-Holland.

7. Brenner, K. H., and F. Sauer. 1988. Diffractive-reflective optical interconnects. *Applied Optics,* 27, 4251.

8. Caulfield, H. J. 1987. Parallel n4 weighted optical interconnections. *Applied Optics,* 26, 4039.

9. Crowther, W., J. Goodhue, R. Gurwitz, R. Rettberg, and R. Thomas. 1985. The Butterfly parallel processor. *IEEE Computer Architecture Technical Committee Newsletter,* 18–45.

10. Dekker, L., and E. E. E. Frietman. 1988. Optical link and processor clustering in the delft parallel processor. *Proceedings, Second European Simulation Multiconference,* Nice, France. ACM, 25–37.

11. Ein-Dor, P. 1985. Grosch's law revisited. *Communications of the ACM,* 28(2), 142–151.

12. Feldman, M. R., S. C. Esener, C. C. Guest, and S. H. Lee. 1988. Comparison between optical and electrical interconnects based on power and speed considerations. *Applied Optics,* 27(9), 1742–1751.

13. Flynn, M. J. 1972. Some computer organizations and their effectiveness. *IEEE Transaction on Computers,* C(21), 948–960.

14. Freiberger, P., and M. Swaine. *Fire in the Valley: The Making of the Personal Computer.* Berkeley, Calif.: University of California Press, 1984.

15. Gajski, D. D., and J-K. Peir. 1985. Comparison of five multiprocessors systems. *Parallel Computing,* 2, 265–282.

16. Gottlieb, A., R. Grishman, C. Kruskal, P. McAuliffe, L. Rudolph, and M. Snir. 1982. The NYU Ultracomputer—designing an MIMD shared memory parallel computer. *Proc. of the 9th Annual Int. Conf. on Comp. Arch.,* 27–42.

17. Guha, A., J. Bristow, C. Sullivan, and A. Husain. 1990. Optical interconnections for massively parallel architectures. *Applied Optics,* 29(8), 1077–1093.

18. Guibas, L. J., H. T. Kung, and C. D. Thompson. 1979. Direct VLSI implementation of combinatorial algorithms. *Proc. of the Conf. on Very large Scale Integration: Architecture, Design, Fabrication,* CalTech, 255–264.

19. Harris, D. O., and A. VanderLugt. 1988. Acousto-optic photonic switch. *Opt. Lett.,* 14(21), 1177–1179.

20. Hartmann, A., and S. Redfield. Ox—optical crossbar switch designs for parallel processing. *SPIE Proc. on Optical Computing Conf.,* 963, 218–222.

21. Hillis, W. D. 1985. *The Connection Machine.* Cambridge, Mass.: MIT Press.

22. Hochschild, P. H., E. W. Mayr, and A. R. Siegel. 1983. Techniques for solving graph problems in parallel environments. *Proc. of the 1983 Int. Conf. on Parallel Processing,* 351–359.

23. Jamieson, L. H., D. B. Gannon, and R. J. Douglass, eds. 1987. *The Characteristics of Parallel Algorithms.* Cambridge, Mass.: MIT Press.

24. Jordan, H. F., and B. Smith. 1985. *Parallel MIMD Computation: HEP Supercomputer and Its Applications.* Scientific Computation Series. Cambridge, Mass.: MIT Press.

25. Kuck, D. J. . 1978. *The Structure of Computers and Computations.* New York: John Wiley & Sons.

26. Kung, H. T., and C. E. Leiserson. 1979. Systolic arrays (for VLSI) in sparse matrix. *Proceedings Society for Industrial and Applied Mathematics,* 256–283.

27. Lee, H. F., and R. Jayakumar. 1986. Systolic structures: a notion and characterization. *Journal of Parallel and Distributed Computing,* 3, 373–397.

28. Leiserson, C. E. 1983. *Area-Efficient VLSI Computation.* Cambridge, Mass.: MIT Press.

29. J. D. Lipson, Elements of Algebra Computing. Addison-Wesley, 1981.

30. Maniloff, E. S., K. M. Johnson, and J. Reif. 1989. Holographic routing networks for parallel processing machines. *Proc. SPIE—Int. Soc. Opt. Eng.,* Vol. 1136, 283–289.

31. Maromand, E., N. Konforti. 1986. Programming optical interconnects. *Proc. Int. Optical Computing Conf. SPIE* Vol. 700, 209–213.

32. McManus, J. B., R S. Putman, and H. J. Caulfield. 1988. Switched holograms for reconfigurable optical interconnections: demonstration of a prototype device. *Applied Optics,* 27(20), 4244–4250.

33. McRuer, R., J. Wilde, L. Hesselink, and J. Goodman. 1988. Two wavelength photorefractive dynamic optical interconnect. *Proc. Soc. Photo-Opt. Instrum. Eng.*

34. Miller, D. A. B. 1989. Optics for low-energy communication inside digital processors. *Optics Letters,* 14(2), 146–148.

35. Moore, G.E. 1976. Microprocessors and integrated electronic technology. *Proceedings of the IEEE,* 64(6), 837 841.

36. Nichols, A. J. 1976. An overview of microprocessor applications. *Proceedings of the IEEE,* 64(6), 951–953.

37. Noyce, R., and M. Hoff. 1981. A history of microprocessor development at Intel. *IEEE Micro,* 1(1), 8–21.

38. Pauliat, G., and G. Roosen. 1986. Large scale interconnection using dynamic gratings. *Proc. of the 1986 Inter. Optical Computing Conf.,* SPIE Vol. 700, 202.

39. Perelmutte, L., and I. Glaser. 1986. Digital incoherent optical interconnections. *Proc. 1986 Inter. Optical Computing Conf.,* SPIE Vol. 700, 214.

40. Pfister, G. F., W. C. Brantley, D. A. George, S. L. Harvey, W. J. Kleinfelder, K. P. McAuliffe, and E. A. Melton. 1985. The IBM research parallel processor prototype (RP3): introduction and architecture. *Proc. of 1985 Int. Conf. on Parallel Processing,* 764–771.

41. Pfister, G. F., and V. A. Norton. 1985. Hot spot contention and the combining in multistage interconnection networks. *IEEE Transaction on Computers,* C-34, 943–948.

42. Psaltis, D., D. Brady, and K. Wagner. 1988. Adaptive optical networks using photorefractive crystals. *Applied Optics,* 27(9), 1752–1759.

43. Robertson, D. A. 1976. A microprocessor-based portable computer: the IBM 5100. *Proceedings of the IEEE,* 64(6), 994–999.

44. Sawchuk, A. A. 1987. 3-d optical interconnection networks. *Proc. of the 14th. Congress of the Int. Commission for Optics.*

45. Sawchuk, A. A., B. J. Jenkins, C. S. Raghavendra, and A. Varma. 1987. Optical crossbar networks. *Computer,* 20(6), 50–60.

46. Siegel, H. J. 1981. Languages and architectures for image processing. In *Image Processing on a Partitionable SIMD Machine.* London: Academic Press, 294–300.

47. Siegel, H. J. 1981. PASM: a partitionable SIMD/MIMD system for image processing and pattern recognition. *IEEE Trans. on Computers,* C-30, 934–947.

48. Siegel, H. J., W. G. Nation, C. P. Kruskal, and L. M. N. Jr. 1989. Using the multistage cube network topology in parallel supercomputers. *Proc. of the IEEE,* 77(12), 1932–1953.

49. Smith, B. J. 1981. Architecture and applications of the HEP multiprocessor computer system. *SPIE Real-Time Signal Processing IV,* Vol. 298, T. F. Tao, ed. Bellingham, Wash.: Society of Photo-Optical Instrumentation Engineers, 241–248.

50. Snyder, L. 1982. Introduction to the configurable, highly parallel computer. *Computer,* 15, 47–56.

51. Stout, Q. F. 1985. Properties of divide and conquer algorithms for image processing. *Proc. of the 1985 IEEE Computer Society Workshop on Computer Architecture for Pattern Analysis and Image Database Management,* 203–209.

52. Suterland, I. E., and C. A. Mead. 1977. Microelectronics and computer science. *Scientific American,* 237(3), 210–229.

53. Treleaven, P. C. 1985. Control-driven data-driven, and demand-driven computer architecture (abstract). *Parallel Computing,* 2, 1985.

54. Tucker, L. W., and G. G. Robertson. 1988. Architecture and applications of the connection machine. *Computer,* 21, 26–38.

28

ATM-Based Parallel and Distributed Computing

Salim Hariri and Bei Lu

Advances in processing and networking technology will make it feasible to develop parallel and distributed computing applications that run on a variety of networked computing resources. Using high-speed networks, these resources can work together to speed up both computations and I/Os, and can incorporate a large amount of memory. The asynchronous transfer mode (ATM) network has the potential to achieve parallel and distributed computing across a network of computers interconnected by LANs, MANs, and WANs.

In this chapter, we review issues related to ATM technology such as ATM basic concepts, protocol layers, congestion control, switches, interfaces, and application programming interface. We also discuss how to use ATM networks to build parallel and distributed computing systems. We conclude this chapter by identifying the research activities required to achieve high-performance parallel and distributed computing over ATM networks.

28.1 Introduction

A wide range of applications require high-performance distributed computing systems. These applications include scientific computing, *Grand Challenge* applications, remote visualization, high-resolution medical imaging, three-dimensional volumetric animation, full motion video conferencing, video-on-demand, and computer imaging. The critical performance criteria for most of these applications are low delay and high bandwidth.

One approach to support high-performance distributed computing is based on using advances in devices and fabrication technologies and merging several architectural designs into a new high-performance computer. However, this approach is not practical, considering the costs involved in designing and building new hardware and developing new software to support this design.

Another approach is to interconnect existing computers using a high-speed, low latency communication network. Such a network-based computing system can be viewed as an extension of bus-based computer system, in which application programs run transparently on a collection of computers that range from supercomputers and/or massively parallel computers down to high-performance desktop/laptop computers.

The current advances in fiber technology have been able to stretch transmission rates from kilobit/second range in 1970s to over a gigabit/second in the 1990s. Furthermore it, has been established that clusters of high-performance workstations interconnected by high-speed networks (e.g. ATM/SONET, HUB-based LAN) have an aggregate computing power similar to that offered by supercomputers and massively parallel computers [1]; a cluster of 1024 workstations based on the DEC Alpha processor has a peak performance of 150 GFLOPS, while the same size configuration of the CM-5 from Thinking Machines Inc. has a peak rating of 128 GFLOPS [2].

Consequently, the future of high-performance computing systems lies in the integration of existing architectures and technologies into a powerful, scalable, general-purpose computing environment that is efficient, cost-effective, and capitalizes on current advances in computing and communication technologies. Such a network-based distributed computing environment will be capable of delivering the required performance levels for general classes of applications. In such an environment, the high-speed communication network plays an important role in the success and the widespread acceptance of such computing environments. It is important to note that existing standard LAN technology (e.g., Ethernet, token ring) is not suitable for *parallel and distributed computing* (also interchangeably referred to as *high-performance distributed computing* (HPDC) in this text) because of the low bandwidth and high latency. The current communication systems can be improved by applying high-speed technology to one or a combination of the areas of (1) networks, (2) protocols, and (3) switches and network interfaces [3].

High-speed networks include *local input/output networks* (LINs), *local area networks* (LANs), *metropolitan area networks* (MANs) and *wide area networks* (WANs). Higher-speed communication can be achieved in these networks by using high-speed technologies such as the *high-performance parallel interface* (HiPPI), fiber channels in LINs, *fiber distributed data interface* (FDDI) and *asynchronous transfer mode* (ATM) in LANs; *distributed queue dual bus* (DQDB), ATM and *switched multi-megabit data service* (SMDS) in MANs; and *frame relay* and ATM in WANs. The high-speed protocols proposed in the literature can be classified into three categories: (1) improved structures (such as XTP), (2) hardware implementations (e.g., protocol engine and HOP), and (3) new protocols (e.g., VMTP, XTP, and NETBLT). High-speed switches include ATM switches and HUBs. Examples of high-speed network adapters are CAB, NAB, and HIP [3].

Among these, the ATM technology is rapidly being accepted as viable for building high-speed communication systems for local area networks as well as for wide area networks. We do believe that ATM-based communication systems will play an important role in the development of high-performance parallel and distributed systems.

In this chapter, we discuss the main design issues related to ATM protocols, ATM switch architectures, and ATM host-network interface designs. Furthermore, we briefly describe how to build an ATM-based high-performance distributed system.

28.2 Broadband Integrated Service Data Network (B-ISDN)

The CCITT (International Telegraph and Telephone Consultative Committee) standard defines an *integrated service data network* (ISDN) as the one that provides end-to-end digital connectivity to support a wide range of services, including voice and non-voice services, to which users have access by a limited set of standard multipurpose user-network interfaces [5]. ISDN can be viewed as a digital pipeline into which multiple sources (e.g. voice, data, and facsimile) are multiplexed. There are several communication channels that can be multiplexed over this pipeline, and they are as follows:

1. *B channel.* This operates at a 64 kbps rate and is used to provide circuit switched, packet switched, and semipermanent circuit interconnections. It is used to carry digital data, digitized voice, and mixtures of lower-rate digital traffic.
2. *D channel.* This operates at 16 or 64 kbps, and it is used for signaling purposes in conjunction with circuit-switched calls on associated B channels, and as a pipeline to carry packet-switched or slow-speed telemetry information.
3. *H channels.* These are three hybrid channels with speeds identified as follows: H0 channel operates at 384 kbps, H11 channel operates at 1.536 Mbps, and H12 channel operates at 1.92 Mbps. These channels are used for providing higher bit rates for applications such as fast facsimile, high-speed data, high quality audio, and video.

Two characteristics of these channels have been standardized: basic access rate and primary access rate. The basic access consists of 2B + D channels, providing 192 kbps (including 48 kbps overhead). The primary access mode is intended for higher data rate communication requirements, which typically fall under the category of nB + D channels. In this mode, the user can use all or part of the B channels and the D channel. This primary access rate service is provided using time-division multiplexed signals over four-wire copper circuits or other media. Each B channel can be switched independently; some B channels may be permanently connected, depending on the service application. The H channels can also be considered to fall into this category.

With the explosive growth of network applications and services, it has been recognized that ISDN cannot deliver required bandwidth for these emerging applications. Consequently, the majority of the delegates within CCITT COM XVIII agreed in 1985 that there is a need for a *broadband ISDN* (B-ISDN) that allows total integration of broadband services [6]. Since then, the original ISDN is called *narrow-band ISDN* (N-ISDN).

The B-ISDN standards defined by the CCITT integrate services and networks, and network access are based on a single optical fiber for each customer. It targets both business applications and residential subscribers. Hence, it is required to be highly flexible to support high-speed data transmission, video conferencing, and high-definition television distribution.

The selected transfer mode for B-ISDN has changed several times since its inception. So far, two types of transfer modes have been used for digital data transmission: synchronous transfer mode (STM) and asynchronous transfer mode (ATM).

STM is suitable for traffic that has severe real-time requirements (e.g., voice and video traffic). This mode is based on circuit switching service in which the network bandwidth is divided into periodic *slots*. Each slot is assigned to a call according to the peak rate of the call. However, this protocol is rigid and does not support bursty traffic. The size of data packets transmitted on a computer network varies dynamically, depending on the current activity of the system. Furthermore, some traffic on a data communication network is time insensitive. Therefore, the STM is not selected for B-ISDN and the ATM, a packet switching technique, is selected.

ATM technology divides voice, data, image, and video into short packets and transmits these packets by interleaving them across an ATM link. The packet transmission time is equal to the slot length. In ATM, the slots are allocated on demand while, for STM, periodic slots are allocated for every call. In ATM, therefore, no bandwidth is consumed unless information is actually transmitted.

For ATM, an important parameter is whether the packet or cell size should be fixed or variable. The main factors that need to be taken into consideration when we compare fixed packet size versus variable packet sizes are the transmission bandwidth efficiency, the switching performance (i.e., the switching speed, and the switch's complexity), and the delay.

Variable packet length is preferred to achieve high transmission efficiency. This is because, with fixed packet length, a long message has to be divided into several data packets, and each data packet is transmitted with overhead. Consequently, the total transmission efficiency is low. However, with variable packet length, a long message can be transmitted with only one overhead unit.

Since the speed of switching depends on the functions to be performed, with fixed packet length, the header processing is simplified, and therefore the processing time is reduced. Consequently, from switching point of view, fixed packet length is preferable.

From a delay perspective, the packets with fixed small size result in minimal functionalities at intermediate switches and take less time in queue memory management. As a result, fixed-size packets reduce delays in the overall network.

For a broadband network with large bandwidth, the transmission efficiency is not as critical as high-speed throughput and low latency. The gain in the transmission efficiency brought by the variable packet length strategy is traded off for the gain in the speed and the complexity of switching and the low latency brought by the fixed packet length strategy. In 1988, the CCITT decided to use fixed-size cells in ATM.

Another important parameter that the CCITT had to determine, once it decided to adopt fixed size cells, was the length of cells. Two options were debated in the choice of the cell length: 32 bytes and 64 bytes. The choice was mainly influenced by the overall network delay and the transmission efficiency. The overall end-to-end delay has to be limited in voice connections to avoid echo cancellers. For a short cell length like 32 bytes, voice connections can be supported without using echo cancellers. However, for 64 byte cells, echo cancellers need to be installed. From this point of view, Europe was more in favor of 32 bytes so echo cancellers could be eliminated. But longer cell length increases transmission efficiency, which was an important concern to the U.S. and Japan. Finally, a compromise of 48 bytes was reached in the CCITT SGX VIII meeting of June 1989 in Geneva [4].

In summary, ATM network traffic is transmitted in fixed cells with 48 bytes as payload and another 5 bytes as header for routing through the network. The network bandwidth is allocated on demand; that is, asynchronously. The cells of different types of traffic (voice, video, imaging, data, etc.) are interleaved on a single digital transmission pipe. This allows statistical multiplexing of different types of traffic if burst rate exceeds available bandwidth for a certain traffic type. An ATM network is highly flexible and can support high-speed data transmission as well as real-time voice and video applications.

28.3   ATM Protocols

28.3.1   Virtual connections

Fundamentally, ATM is a connection-oriented technology, different from other connectionless LAN technologies. Before data transmission takes place in an ATM network, a connection needs to be established between the two ends using a signaling protocol. Cells then can be routed to their destinations with minimal information required in their headers. The source and destination addresses, which are the necessary fields of a data packet in a connectionless network, are not required in an ATM network. The logical connections in ATM are called *virtual connections*.

Two layers of virtual connections are defined by CCITT: virtual channel (VC) connections (VCC) and virtual path (VP) connections (VPC). One transmission path contains several VPs, as shown in Fig. 28.1, and some of them can be permanent or semi-permanent. Furthermore, each VP contains bundles of VCs (also shown in the figure).

By defining VPC and VCC, a virtual connection is identified by two fields in the header of an ATM cell: *virtual path identifier* (VPI) and *virtual channel identifier* (VCI). The VPI/VCIs have only local significance per link in the virtual connection. They are not addresses and are used just for multiplexing and switching packets from different traffic sources. Hence, ATM does not have the overhead associated with LANs and other packet-switched networks where packets are forwarded based on the headers and addresses that vary in location and size, depending on the protocol used. Instead, an ATM switch only needs to perform a mapping between the VPI/VCI of a cell on the input link and an appropriate VPI/VCI value on the output link.

Virtual channel connection. A virtual channel connection is a logical end-to-end connection. It is analogous to a virtual circuit in X.25 connection. It is the concatenation of virtual channel links which exist between two switching points. For example, as depicted in Fig. 28.2 [7], a virtual channel connection between user A and user B consists of three concatenated virtual channel links with VCI values 1, 3, and 5 respectively. A virtual channel has traffic usage parameters associated with it, such as cell loss rate, peak rate, bandwidth, quality of service, and so on.

Virtual path connection. A virtual path connection is meant to contain bundles of virtual channel connections that are switched together as one unit. The use of virtual paths can simplify network architecture and increase network performance and reliability, since the network deals with fewer aggregated entities.

The VPI/VCI fields in an ATM cell can be used to support two types of switching: VP switching and VP/VC switching. In a VP switch, the VPI field is used to route the cells in

Figure 28.1 Relationship between transmission path, VPs, and VCs

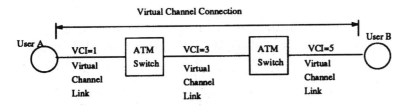

Figure 28.2 Virtual channel connection

the ATM switch while the VCI values are not changed. For example, in Fig. 28.3a, there are two virtual path connections, one with VPI = 5 and another with VPI = 10. The virtual path with the VPI value of 5 contains three virtual channel connections that have the VCI values of 1, 2, and 3 respectively, while the one with the VPI value of 10 contains two virtual channel connections that have the VCI values of 5 and 6. The VP switch changed the values of the VPIs to VPI = 20 and VPI = 8 and kept the VCIs unchanged.

But a VP/VC switch examines both VPI and VCI and can change or terminate both VC and VP. Figure 28.3b shows an example where a virtual connection's identifiers VPI = 5/VCI = 2 are changed to VPI = 8/VCI = 8 while another virtual connection's identifiers VPI = 5/VCI = 4 are changed to VPI = 6/VCI = 10. Figure 28.3c shows a network with VP and VC/VP switches.

28.3.2   ATM reference model

The ATM protocol reference model is shown in Fig. 28.4. It consists of the higher layers, the ATM adaptation layer (AAL), the ATM layer, and the physical layer. The ATM reference/stack model differs from the open system interconnection (OSI) model in its use of planes. The portion of the architecture used for user-to-user or end-to-end data transfer is called the *user plane* (U-plane). The *control plane* (C-plane) performs call connection control. The *management plane* (M-plane) performs functions related to resources and parameters residing in its protocol entities. ATM is connection oriented, and it uses out-of-

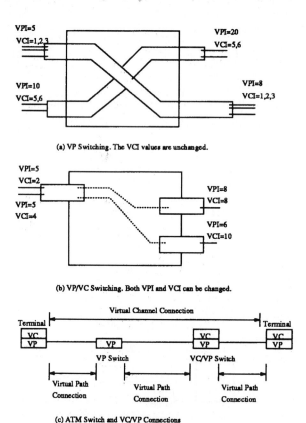

(a) VP Switching. The VCI values are unchanged.

(b) VP/VC Switching. Both VPI and VCI can be changed.

(c) ATM Switch and VC/VP Connections

Figure 28.3 VP and VC switching in an ATM switch

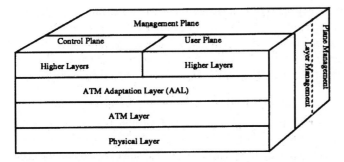

Figure 28.4 B-ISDN reference model

band signaling. This is in contrast with the in-band signaling mode of the OSI protocols (X.25) where control packets are intermixed with data packets. Therefore, during virtual channel connection setup, only the control plane is active. In the OSI model, the two planes are merged and are indistinguishable.

28.3.2.1 Physical layer

Functions of physical layer. The *physical layer* provides the transport of ATM cells between two ATM entities. Based on its functionalities, the physical layer is segmented into two sublayers, namely *physical medium dependent* (PMD) sublayer and the *transmission convergence* (TC) sublayer. This sublayering separates transmission from physical interfacing, and allows ATM interfaces to be built on variety of physical interfaces. The PMD sublayer is device dependent. Its typical functions include bit timing and physical-medium-like connectors. TC sublayer generates and recovers transmission frames. The sending TC sublayer performs the mapping of ATM cells to the transmission system. The receiving TC sublayer receives a bit stream from PMD, extracts the cells, and passes them on to the ATM layer. It generates and checks HEC (header error control) field in the ATM header, and it also performs cell rate decoupling through deletion and insertion of idle cells.

SONET transmission format. The *synchronous optical network* (SONET), also known internationally as *synchronous digital hierarchy* (SDH), is the physical layer transmission standard for B-ISDN originally proposed by Bellcore for specifying standards for optical fiber-based transmission line equipment. The corresponding hierarchy for electrical signaling is *synchronous transport signal* (STS). The lowest SONET frame rate, called STS-1, defines 8 kHz frames of 9 rows and 90 bytes. First 3 bytes are used for *operation, administration and management* (OAM) purposes, and the remaining 87 bytes are used for data. This gives a data rate of 51.84 Mbps. The next highest frame rate standard is STS-3, with 9 bytes for OAM and 261 bytes for data, providing a 155.52 Mbps transmission system. There are other, higher-speed SONET standards available: STS-12, 622.08 Mbps; STS-24, 1.244 Gbps; STS-48, 2.488 Gbps; and so on (STS-N, $N \times 51.84$ Mbps). The STS-3 (155.52 Mbps) and STS-12 (622.08 Mbps) have been designated as the customer access rates in future B-ISDN networks.

28.3.2.2 ATM layer. The ATM layer performs multiplexing and demultiplexing of cells from different connections (identified by different VPIs/VCIs) onto a single cell stream. It extracts cell headers from received cells and adds cell headers to the cells being transmitted. Translation of VCI/VPI may be required at ATM switches.

Figure 28.5a shows the ATM cell format. Cell header formats for *user-network interface* (UNI) and *network-network interface* (NNI) are shown in Figs. 28.5b and 28.5c, respectively. The function of the various fields in the ATM cell headers are as follows:

- *Generic flow control (GFC).* This is a 4-bit field used only across UNI to control traffic flow across the UNI and alleviate short-term overload conditions, particularly when multiple terminals are supported across a single UNI.
- *Virtual path identifier (VPI).* This is an 8-bit field across UNI and 12-bits across NNI. For idle cells or cells with no information the VPIs are set to zero, this is also the default value for VPI. The use of non-zero values of VPI across NNI is well understood (for trunking purposes), however the procedures for accomplishing this are under study.
- *Virtual circuit identifier (VCI).* The 16-bit VCI is used to identify the virtual circuit in a UNI or an NNI. The default value for VCI is also zero. Typically, VPI/VCI values are assigned symmetrically; that is, the same values are reserved for both directions across a link.
- *Payload type identifier (PTI).* This is a 3-bit field for identifying the payload type as well as for identifying the control procedures. When bit 4 in the octet is set to 0, it means it is a user cell. For user cells, if bit 3 is set to 0, it means that the cell did not experience any congestion in the relay between two nodes. Bit 2 for the user cell is used to indicate the type of user cell. When bit 4 is set to 1, it implies the cell is used for management functions as error indications across the UNI.
- *Cell loss priority (CLP).* This field is used to provide guidance to the network in the event of congestion. The CLP bit is set to 1 if a cell can be discarded during congestion. The CLP bit can be set by the user or by the network. An example for the network setting is when the user exceeds the committed bandwidth, and the link is under-utilized.
- *Header error check (HEC).* This is an 8-bit *cyclic redundancy code* (CRC) computed over all fields in the ATM cell header. It is capable of detecting all single bit errors and certain multiple bit errors. It can also be used to correct single bit errors, but this is not mandatory.

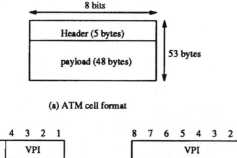

(a) ATM cell format

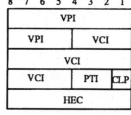

(b) ATM cell header format acroos UNI   **(c) ATM cell header format across NNI**

Figure 28.5 ATM cell and header formats

28.3.2.3 ATM adaptation layer (AAL). The AAL layer provides the proper interface between the ATM layer and the higher layers. If an application can use the ATM layer directly, there is no need to have the AAL layer in between. But applications usually have specific requirements (e.g., real-time, constant bit rate, or variable bit rate) that cannot be satisfied using pure ATM services. Accordingly, the AAL layer provides four classes of services to meet the requirements of different applications. The AAL layer has five types of protocols to support the four classes of traffic patterns, and they are as follows:

1. *Type 1.* Supports Class A applications, which require constant bit rate (CBR) services with time relation between source and destination. Error recovery is not supported. Examples include real-time voice messages, video traffic, and some current data video systems.
2. *Type 2.* Supports Class B applications, which require variable bit rate (VBR) services with time relation between source and destination. Error recovery is also not supported. Examples are teleconferencing and encoded image transmission.
3. *Type 3.* Supports Class C applications, which are *connection-oriented* (CO) data transmission applications. Time relation between source and destination is not required. It is intended to provide services to the applications that use a network service like X.25.
4. *Type 4.* Supports Class D applications, which are *connectionless* (CL) data transmission applications. Time relation between source and destination is not required. The current datagram networking applications such as TCP/IP or TP4/CLNP belong to Class D. Since the protocol formats of AAL type 3 and type 4 are similar, they have been merged to AAL type 3/4 [8, 9].
5. *Type 5.* This type was developed to reduce the overhead related to AAL type 3/4. It supports connection-oriented services more efficiently. It is more often referred to as *simple and efficient AAL,* and it is used for Class C applications.

The AAL layer is further divided into 2 sublayers: *the convergence sublayer* (CS) and the *segmentation-and-reassembly sublayer* (SAR). The CS is service dependent and provides the functions needed to support specific applications using AAL. The SAR sublayer is responsible for packing information received from CS into cells for transmission and unpacking the information at the other end. The services provided by the ATM and AAL Layers are shown in Fig. 28.6.

In the following text, we describe the *protocol data units* (PDUs) for different AAL protocols.

AAL type 1. The SAR sublayer of AAL type 1 accepts 47 octet data blocks from its CS sublayer and adds 1 octet header to form the SAR-PDU. The header contains a 1-bit CS indication (CSI) field, 3-bit sequence number (SN) field, and 4-bit sequence number protection (SNP) field, as shown in Fig. 28.7a. The CSI is used to indicate the existence of the CS layer. The SAR sublayer receives it from the CS sublayer and conveys it to the peer CS sublayer. The SN field is used for detecting lost or misordered cells. The SNP field is used for error detection and correction against the SN field. If the SN field is corrupted and cannot be corrected by the SAR sublayer, the CS is informed.

AAL type 2. The SAR-PDU of AAL type 2 is shown in Fig. 28.7b, where an information-type field (IT) and a length-indicator (LI) field are used to allow the segmentation and reassembly of information from higher levels. The IT field is used to indicate the beginning, continuation, or end of a message (BOM, COM, and EOM). The LI field indicates

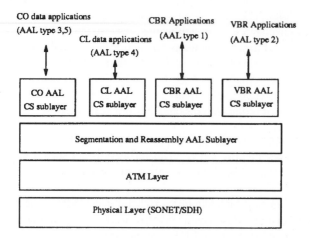

Figure 28.6 Services- and protocols-based ATM and AAL layers

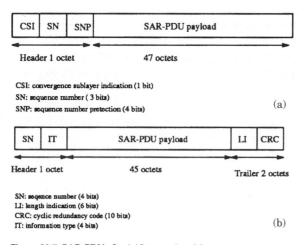

Figure 28.7 SAR-PDUs for AAL types 1 and 2

the number of useful octets in a partially filled cell. The cyclic redundancy code (CRC) field is for correcting bit errors in the SAR-PDU.

AAL type 3/4

- *SAR-PDU for AAL type 3/4.* The AAL type 3/4 share the same SAR-PDU format as shown in Fig. 28.8a. The 10-bit MID field allows us to multiplex 210 AAL user-to-user connections on a single user-to-user ATM connection for connection-oriented data communication. For connectionless data communications, in AAL type 4, e.g., the *switched megabit data service* (SMDS), the MID field allows us to interleave SAR-PDUs of up to 210 CS-PDUs on the same semi-permanent ATM layer virtual connection. The CS-PDUs for the same message have the same MID field. The *segment type* (ST) field is used in AAL type 3/4 to indicate the beginning of message, continuations of message, end of message, or single-segmented message.

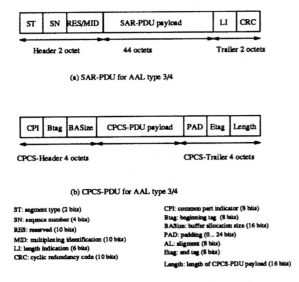

(a) SAR-PDU for AAL type 3/4

(b) CPCS-PDU for AAL type 3/4

ST: segment type (2 bits)
SN: seqence number (4 bits)
RES: reserved (10 bits)
MID: multiplexing identification (10 bits)
LI: length indication (6 bits)
CRC: cyclic redundancy code (10 bits)

CPI: common part indicator (8 bits)
Btag: beginning tag (8 bits)
BASize: buffer allocation size (16 bits)
PAD: padding (0...24 bits)
AL: aligment (8 bits)
Etag: end tag (8 bits)
Length: length of CPCS-PDU payload (16 bits)

Figure 28.8 SAR-PDU and CPCS-PDU for AAL type 3/4

- *Convergence sublayer functions in AAL 3/4.* The CS layer is further divided into two sublayers: the common part CS (CPCS) and service-specific CS (SSCS). The AAL 3 and AAL 4 share the SAR and CPCS data format, as shown in Fig. 28.8b.The beginning tag (Btag)/end tag (Etag) field is used to associate the CPCS-PDU header and trailer at the receiver. The same value is inserted for Btag and Etag for a given CSPD-PDU. The buffer allocation size indication (BASize) field informs the receiving peer entity of the maximum buffer required to receive this CPCS-PDU. The alignment (AL) field achieves 32-bit alignment of the CPCS-PDU trailer.

For AAL type 3/4, two modes of service are defined. One is *message-mode service,* which is used for framed data. Any of the OSI-related protocols and applications would fit into this category. Another mode is *streaming-mode service,* which is used for low-speed continuous data with low delay requirement. The data are presented to AAL in fixed size blocks, which may be as small as 1 octet. One block is transferred per cell. Figures 28.9 and 28.10 show how these two modes of services operate.

AAL type 5. Although AAL type 3/4 is designed for connection-oriented and connection-less data transmission, some fields are not necessary for connection-oriented service. For instance the MID field is not needed because connection-oriented services never require MID routing, the VCI/VPI routing can achieve cell-by-cell multiplexing. Even the CRC code for every SAR-PDU may be unnecessary if the bit error for every CS-PDU is checked. Further, AAL type 3/4 has high overhead of 4 bytes per SAR-PDU of 48 bytes. AAL type 5 is designed to reduce protocol processing overhead, reduce transmission overhead, and to ensure adaptability to existing transport protocols.

The SAR sublayer accepts variable length CPCS-PDUs which are integral multiples of 48 bytes from CPCS and generates SAR-PDUs containing 48 octets of SAR data. The last bit of the payload type field in ATM header is used to indicate whether the cell is the end of the packet. The receiving side can simply queue cells until it receives a cell with the end-of-packet bit set. Figure 28.11 shows the formats of ATM-PDU and CPCS-PDU for AAL

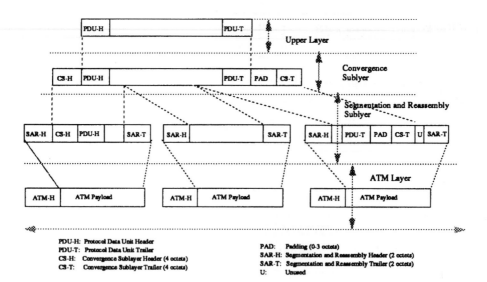

PDU-H: Protocol Data Unit Header
PDU-T: Protocol Data Unit Trailer
CS-H: Convergence Sublayer Header (4 octets)
CS-T: Convergence Sublayer Trailer (4 octets)

PAD: Padding (0-3 octets)
SAR-H: Segmentation and Reassembly Header (2 octets)
SAR-T: Segmentation and Reassembly Trailer (2 octets)
U: Unused

Figure 28.9 Example of ATM-adaptation layer (message-mode service)

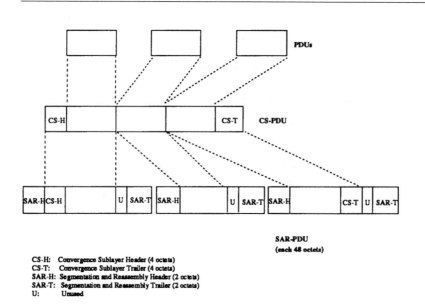

CS-H: Convergence Sublayer Header (4 octets)
CS-T: Convergence Sublayer Trailer (4 octets)
SAR-H: Segmentation and Reassembly Header (2 octets)
SAR-T: Segmentation and Reassembly Trailer (2 octets)
U: Unused

Figure 28.10 Example of ATM-adaptation layer (stream-mode service)

type 5 [12, 4]. The SSCS (service specific convergence sublayer) for AAL type 5 has yet to be further studied [11].

The trailer has four fields: *user-to-user indication* (UU) field (1 octet), *common part indicator* (CPI) (1 octet), *length field* (2 octets) and *CRC* (4 octets). The UU and CPI fields are currently unused and are set to 0. The length field indicates the number of bytes of data in the packet (not including the padding). The CRC is a 32-bit CRC over the entire convergence layer packet, including the padding and the trailer. It has been shown that the 32-bit CRC field can provide robust detection to guard against cell error and misordering [13].

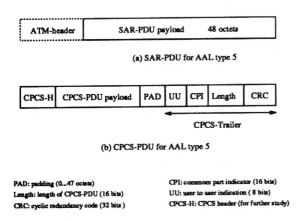

(a) SAR-PDU for AAL type 5

(b) CPCS-PDU for AAL type 5

PAD: padding (0...47 octets)   CPI: common part indicator (16 bits)
Length: length of CPCS-PDU (16 bits)   UU: user to user indication (8 bits)
CRC: cyclic redundancy code (32 bits)   CPCS-H: CPCS header (for further study)

Figure 28.11 ATM-PDU and CPCS-PDU for AAL type 5

28.3.3   traffic management and congestion control

An important character in ATM traffic is its *burstiness,* meaning that some traffic sources may generate cells at a near-peak rate for a very short period of time, and immediately afterward it may become inactive, generating no cells. Such a bursty traffic source will not require continuous allocation of bandwidth at its peak rate. Since an ATM network supports a large number of such bursty traffic sources, statistical multiplexing can be used to gain bandwidth efficiency, allowing more traffic sources to share the bandwidth. But if a large number of traffic sources become active simultaneously, severe network congestion can result.

The propagation delay is independent of optical link rate, and it becomes significant at B-ISDN rates. The window flow control is not feasible for services that have real time constrains (voice, video). Thus, some of the congestion schemes developed for existing networks may no longer be applicable in ATM high-speed networks.

In an ATM network, congestion control is performed by monitoring the connection usage. It is called *source policing.* Every virtual connection (VPC or VCC) is associated with a traffic contract which defines some traffic characteristics such as peak bit rate, mean bit rate, and duration of burst time. The network monitors all connections for possible contract violation. It is also a *preventive control* strategy. Preventive control does not wait until congestion actually occurs. It tries to prevent the network from reaching an unacceptable level of congestion by controlling traffic flow at entry points to the network.

28.3.3.1 Preventive control. Preventive control for ATM can be performed in two ways: *admission control* and *bandwidth enforcement* [14]. Admission control determines whether to accept or reject a new connection at the time of a call setup. This decision is based on traffic characteristics of the new connection and the current network load. Bandwidth enforcement monitors individual connections to ensure that the actual traffic flow conforms with that reported at call establishment. The *leaky bucket* method [14] can be used to implement enforcement mechanisms for ATM networks. One possible implementation of a leaky bucket method is to control the traffic flow by means of tokens, and that can be briefly explained as follows.

An arriving cell first enters a queue. If the queue is full, cells are simply discarded. To enter the network, a cell must first obtain a token from the token-pool. If there is no token, a cell must wait in the queue until a new token is generated. Tokens are generated at a fixed rate corresponding to the average rate of the connection. If the number of tokens in the token-pool exceeds some predefined threshold value, the process of token generation stops. This threshold value corresponds to the burstiness of the transmission.

In the leaky bucket method, violating cells are either discarded or stored in a buffer, even when the traffic on the network is light, and thus network resources are wasted. The marking method presented in Ref. [15] improves total network throughput. In this scheme, violating cells, rather than being discarded, are permitted to enter the network with violation tags in their cell headers. These cells are discarded only when they arrive at a congested node, otherwise they will be transferred through the network.

28.3.3.2 Cell loss priority. Preventive control is implemented via a single indicator in the ATM cell header, termed the *cell loss priority* (CLP) indicator. It is set to 1 to indicate that a cell has low priority and may be selectively discarded if the connection experiences congestion. This CLP indicator serves a dual purpose: setting of the CLP indicator of a cell to 1 by the transmitting terminal to indicate that the cell carries nonessential information (and thus the cell is selectively discardable under congestion conditions); and setting of the CLP indicator of a cell to 1 by the network during the cell's access to the network to indicate that the cell is not in compliance with the traffic limits agreed to in the service contract.

The CLP indicator can also be used to provide different *qualities of services* (QOS). For instance, the connections used for signaling can have higher priority in case of congestion.

28.4   ATM Switches

The function of a switch in a network is to route and switch packets or messages being transmitted on the network. The design of ATM packet switches that are capable of switching relatively small packets at rates of 100,000 to 1,000,000 packets per second per line is a challenging task. There have been a number of ATM switch architectures proposed in the literature, and an important subset of the switches will be discussed in this section.

An ATM switch consists of a set of N input and N output ports, a switching transfer medium, and a *management and control processor* (MCP). A generic model for an ATM switch is shown in Fig. 28.12. The functions of each components are as follows [19].

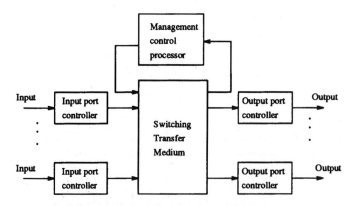

Figure 28.12 Generic structure of an ATM switch

- *Input/output controllers*. Each I/O port is managed by an intelligent controller. Input port controllers typically provide buffering, cell duplication for multicasting, cell processing, VCI/VPI translation, multiplexing traffic from several low-speed devices, and path connection requests and reservations through the switching transfer medium. Similarly, the output controllers can provide buffering and VCI/VPI translation, demutiplexing, and an $N{:}R$ selection (selecting R packets from a maximum of N for buffering).
- *Switching transfer medium*. this consists of the hardware and software that commonly deal with the issues such as establishing a path between an input and an output port within the switch, service discipline for input ports, contention resolution schemes for cells contending for the link(s), or other internal resources of the switch such as internal blocking.
- *Management and control processor*. The MCP's function is to communicate with port controllers and facilitate switch operation, administration, and management. For example, MCP can be the residence of a simple network-management protocol (SNMP) or OSI management agent.
- *Buffering*. Buffering strategies are needed to prevent congestion and deadlock issues. When buffers are placed at the input ports, cell-scheduling strategies are needed at these ports. First-in-first-out (FIFO) input queues are subject to head-of-line (HOL) blocking, meaning that even if the output links are available, packets behind the blocked packet (the head of the queue) cannot be transferred. HOL blocking can be removed by a windowing mechanism at the input port (where buffers in a window size of w are examined for the selection of the candidate cells). Buffering can also be provided at the output ports. If more than one cell is destined to the same output port, the output port can only serve one cell and the rest of the cells must be stored.
- *Cell processing*. The input port controller may maintain information such as cell priority and loading factors. This information may be appended to the cell for optimal routing within the fabric. This adds an extra header to the cell that must be removed by the output controllers.

There are several ways to classify ATM switch architectures, some based on the internal structures and some based on the applications. We follow the classification proposed by Newman [16] as shown in Fig. 28.13, which is based on switch structures. Newman classifies ATM switches into time-division and space-division types. Time-division switches can be classified into shared memory and shared medium. The space-division switches can be divided into two basic classes: single- and multiple-path networks. A single-path network has a unique path through the interconnection network between any given input and output pair. Examples are matrix, Banyan, Batcher-Banyan, and Delta networks. A multiple-path network has a number of different paths available between any input and output. Examples include Augmented Banyan, Parallel Planes, Clos Networks, and Recirculation Networks. In what follows, we discuss some representative ATM switch architectures.

28.4.1   Time-division architecture

In a time-division switch, all cells flow across a single communication highway shared by all input and output ports. The communication highway could be either a shared memory or a shared medium.

28.4.1.1 Shared-memory architectures. A shared-memory ATM switch architecture consists of a set of N inputs, N outputs, and a dual-ported memory as shown in Fig. 28.14.

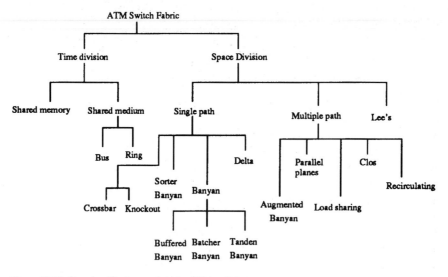

Figure 28.13 One classification method for ATM switches

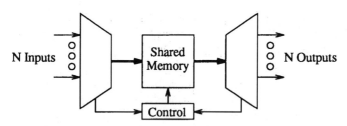

Transfer rate of each input and output = V bits/second

Figure 28.14 Shared-memory switch architecture

Packets received at the inputs are time multiplexed into a single stream of data and written to the dual-ported memory. Concurrently, a stream of packets is formed by reading the dual-ported memory. The packets forming this data stream are demultiplexed and written to the set of N outputs.

Shared-memory architectures are inherently free of internal blocking but not free from output blocking. It is possible during any time slot that two incoming packets may be destined for the same output. Consequently, to reduce packet loss, each output port has an output buffer. The required size of each of the output buffers depends on the desired packet loss probability for the switch. The potential bottleneck in the implementation of this architecture is the bandwidth of the dual-ported memory and its control circuitry. The bandwidth of the dual-ported memory must be $2NV$ bps, where N is the number of inputs and V is the data rate at each input. The number of ports and the port speeds of a switching module using a shared-memory architecture is thus bounded by the bandwidth of the shared memory and the control circuitry used in its implementation.

28.4.1.2 Shared-medium architectures. A shared-medium ATM switch architecture consists of a set of N inputs, N outputs, and a common high-speed medium such as a parallel bus as shown in Fig. 28.15. Incoming packets to the switch are time multiplexed onto this common high-speed medium. Each switch output has an address filter and a FIFO to store outgoing packets. As the time-multiplexed packets appear on the shared medium, each address filter discards packets that are not destined for that output port and propagate those that are destined for that output port. The packets that are passed by the address filter are stored in the output FIFO and then transmitted on the network.

The shared-medium ATM switch architecture is similar to the shared-memory switch architecture in that incoming packets are time multiplexed onto a single packet stream and then demultiplexed into separate streams. The difference in the architectures lies in partitioning the memory for the output channels. In the shared-memory architecture, each parallel channel utilizes the dual-ported memory for packet storage whereas, in the shared-medium architecture, each output port has its own memory. The memory partitioning of the shared-medium architecture implies the use of a FIFO memory as opposed to a dual-ported memory in its implementation.

For a shared-medium switch, the number of ports and the port speeds of a switching module is bounded by the bandwidth of the shared medium and FIFO memories used in its implementation. The aggregate speed of the bus and FIFO memories must be at least NV bps or packets may be lost. Shared-medium architectures, like the shared-memory architectures, are inherently free of internal blocking which is characteristic of many Banyan-based and space-division-based switches. However, shared-memory architectures are not free from output blocking. There must be buffers for each output port. Consequently, output buffering is needed for each output port, and its size depends on the desired packet loss probability of the switch.

28.4.2 Space-division architectures

In a space-division switch, concurrent paths are established from the switch inputs to the switch outputs, each path with a data rate of V bits/second. An abstract model for a space-division switch is shown in Fig. 28.16. Space-division architectures avoid the memory bottleneck of the shared-memory and shared-medium architectures since no memory component in the switching fabric has to run at a rate higher than $2V$. Another distinct feature of the space-division architecture is that the control of the switch need not be centralized but may be distributed throughout the switching fabric.

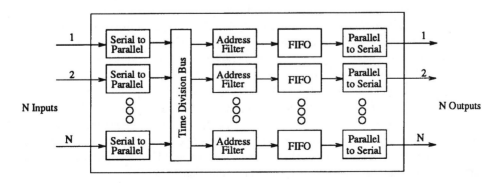

Transfer rate of each input and output = V bits/second

Figure 28.15 Shared-medium switch architecture

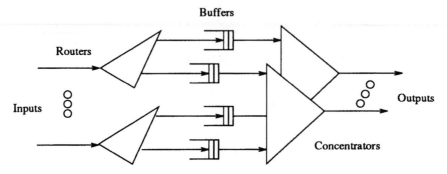

Transfer rate of each input and output = V bits/second

Figure 28.16 Shared-medium switch architecture

Depending on the particular internal switching fabric used and the resources available to establish paths from the inputs to the outputs, it may not be possible for all required paths to be set simultaneously. This characteristic, commonly referred to as *internal blocking* potentially limits the throughput of the switch and thus becomes the central performance limitation of space-division switch architectures.

An important issue in space-division architectures is buffering. There are different buffering strategies in designing an ATM switch, namely internally buffered and externally buffered. Internally buffered architectures include buffered banyan and buffered switch modules, and externally buffered architectures are input buffered, output buffered, input and output buffered, and recirculation buffered.

The input buffered strategy suffers from an HOL blocking problem and has a throughput of about 58 percent [17]. The performance can be improved by a technique known as *input queue bypass,* which means access to other cells in the queue. But this technique complicates hardware implementation of the input buffer.

An output buffer is to accept cells from every input simultaneously. The size of the buffer depends on the required cell loss probability. Output buffer could be a single buffer shared by all input and output ports, or each output has its own buffer shared by all input ports that wish to access the output port. For matrix interconnection network, a separate buffer can be placed in each crosspoint.

Input and output buffered switches combine the two approaches of input and output buffering. The cells that can't be handled at the output buffer due to transient traffic pattern are retained in input buffers. This strategy is favored by many large switch designs.

Recirculation buffering is used to solve output port contention by recirculating those cells that cannot be output during the current slot back to the input ports via a set of recirculation buffers. It offers the performance of an output buffered switch but suffers from out-of-sequence errors unless steps are taken to prevent it. The Sunshine [20] switch and Tandem Banyan [21] switch combine recirculation buffering with output buffering. In what follows, we discuss some representative space-division ATM switches.

28.4.2.1 Knockout switch architecture. A knockout switch ATM architecture, a special case of crossbar architecture, as illustrated in Fig. 28.17 [18], consists of a set of N inputs and N outputs. Each switch input has its own packet routing circuit associated with it, and this routes each input packet to its destined output interface. Each output interface contains N address filters, one for each input line, which drop packets not destined for that

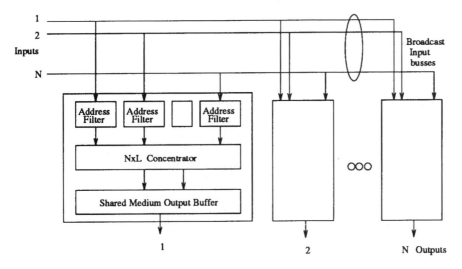

Figure 28.17 Knockout switch basic structure

output port. The outputs of each of the address filters for a particular switch output are connected to an $N \times L$ concentrator which selects up to L packets of those passing through the N address filters in a given slot. If more than L packets are present at the input of the concentrator in a given slot, only L will be selected, and the remaining packets will be dropped. The use of the $N \times L$ concentrator simplifies the output interface circuitry by reducing the number of output buffer FIFO memories and the control circuit complexity without exceeding the required packet loss probability of the switch. It has been shown that a packet loss rate of 10^{-6} is achieved with L as small as 8, regardless of the switch load and size [18].

The knockout switch architecture is practically free of internal blocking. The potential for packet loss in the $N \times L$ concentrator implies the presence of internal blocking since the presence of $L + 1$ packets at the input of the $N \times L$ concentrator will result in packet loss. This loss could conceivably be diminished with the use of buffers at the input of the $N \times L$ concentrator. The knockout switch architecture is not free from output blocking and therefore must provide buffering for each output. The bandwidth requirements of the knockout switch is LV as compared to the NV bandwidth requirement of the shared-medium switch. This would yield a potential performance gain of N/L for the knockout switch over the shared-medium switch, where memory bandwidth is concerned.

Although the knockout switch is a space-division switch like the crossbar switch, each of the inputs has a unique path to the output buffers, much like the shared-medium switch, and thus it circumvents the internal blocking problem. The lack of internal blocking sets the knockout switch apart from other space-division switch architectures such as the crossbar switch architecture.

28.4.2.2 Banyan-based fabrics. Multistage interconnection networks, generally referred to as Banyan networks, represent an important approach to build large ATM network switches. A multistage interconnection network for N inputs and N outputs, where N is a power of 2, consists of $\log_2 N$ stages each comprising $N/2$ binary switching elements, and interconnection lines between the stages placed in such a way as to allow a path from each input to each output as shown in Fig. 28.18.

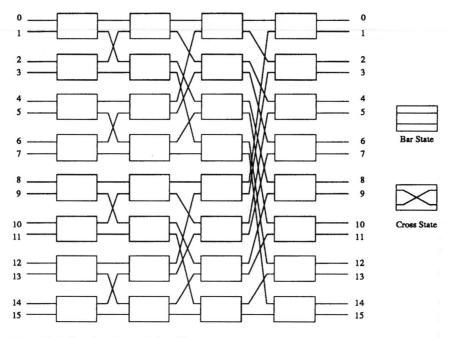

Figure 28.18 Shared-medium switch architecture

The Banyan network has a complexity of order $N\log N$, comparing with the crossbar structure of order N^2. But the Banyan network is an internal blocking network and its performance degrades rapidly as the size of the network increases. Also, the degree of blocking is related to the specific combination of destination requests present in the incident traffic. Thus, some instantaneous patterns of incident traffic will give much poorer performance than others. The performance can be improved by employing switching elements larger than 2×2 to build multistage interconnection network (referred to as Delta networks) [22]. The internal blocking can be eliminated by sorting the incoming cells into an ascending order based upon their output port requests. This type of networks is referred to as Batcher-Banyan network (see Fig. 28.19).

In a Batcher-Banyan network, the incoming packets are sorted according to their requested output addresses by a Batcher sorter, which is based on a bitonic sorting algorithm and has a multistage structure similar to an interconnection network [23]. The sorted list is

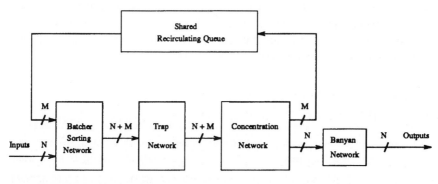

Figure 28.19 Batcher-banyan switch architecture

passed to the *trap network*. The trap network detects the cells with same destination addresses and selects one cell for each destination to pass on to the concentration network. The concentration network routes the selected cells to the input of the Banyan network. The packets that are not selected by the trap network are recirculated and are fed back into the fabric at later slots. A certain number of input ports, M, are reserved for this purpose, thus reducing the number of input/output lines the switching fabric can serve. Since the number of recirculated packets may exceed M, buffering of these packets may still be required. M and the buffer size are selected so as to not exceed a given loss rate [21].

28.5   Host-to-Network Interfaces

With the development of high-speed optical fibers, network speeds are shifting toward gigabits per second. Given this bandwidth, the slowest part of computer communication is no longer the physical transmission. The host-to-network interface is considered the bottleneck to deliver high bandwidth and low latency to user applications. There are intensive efforts to eliminate host-to network interface bottleneck [25–28].

Various studies have been undertaken to analyze the cost of running protocol stacks on the host. They have indicated that data transfer over host's system bus is the most costly operation, especially when sending or receiving large packets of data. The path taken by data as it passes through a conventional protocol stack is as follows [25]. The application writes data into its buffer, the *application buffer,* then invokes a system call to send the data. The socket layer copies the application data into a buffer in the kernel space called the *socket buffer.* Then the transport layer reads the data in the socket buffer to compute a checksum and, finally the data is copied out to the network interface using DMA. Consequently, the memory bus system is accessed five times for each word of data sent.

On the receive path, data is copied first from the network interface into kernel memory using DMA. The transport layer checksum is verified and, when the application is ready, the socket layer copies the data from the socket buffer in kernel memory to the application buffer in the user memory. Finally, the application reads the data. Thus, the memory bus system is also accessed five times for each word of received data. Therefore, most of the new interface designs aim at reducing the number of data copy operations incurred during transmission and receiving.

Network interfaces can be classified into two classes: *programmed I/O* and *direct memory access* (DMA). In programmed I/O interface, the I/O operations are performed by the host processor, while in the DMA technique, the data is copied without interrupting the processor. The network interfaces can also be classified based on the intelligence of the interface. Some interfaces are only slaves with respect to the host processor, as in programmed I/O. In intelligent interfaces, the processor(s) on the interface commands the DMA controller to perform data transfer. In what follows, we discuss important representative host-network interface architectures.

28.5.1   Programmed I/O host-to-network interface

An experimental FDDI interface called Medusa is based on programmed I/O and is implemented at Hewlett-Packard. Its goal is to realize "one copy interface." The design is illustrated in Fig. 28.20 [26]. The interface board is plugged into a 32-bit memory bus, the graphic bus, and maps its memory into the host processor's I/O memory space. At the center of the interface is a triple-ported video memory with two serial ports and one parallel port. The parallel port is used to support fast access to the graphic bus, and the two serial

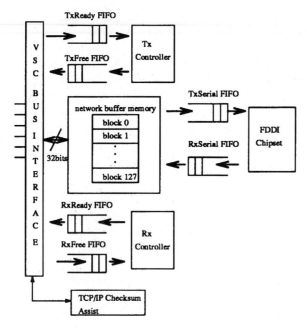

Figure 28.20 The Medusa FDDI interface

ports provide transmitting and receiving paths to the transmitting and receiving FIFOs, which are interfaced to network-specific hardware.

In the Medusa FDDI interface, all network, transport, and socket layer processing is performed by the host, and it is therefore classified as a *slave interface.* In data transmission, a user process presents data to the socket layer, and the socket layer copies the data to the buffers on the interface. Some space is left at the front of the network buffer for the protocol headers. On the receiving side, the socket layer copies data from the interface buffer to the user memory space after checksum is verified.

28.5.2 ATM interface with DMA

An example of ATM interface with DMA is the one Designed for AURORA Testbed [27]. In this design, the DEC 5000 workstation is chosen as the host workstation because it was a high performance workstation and had a high-speed I/O bus, TURBOchannel, which had a peak bandwidth of 800 Mbps. Direct Memory Access (DMA) is used to move data between host and network interface.

The main functions of the host interface are depicted in Fig. 28.21. Microprocessors are used to control the interface, because this allows for an arbitrary variety of information to be exchanged between the host and the interface. Furthermore, this enables flexibility in the choice of the algorithm used to decide which piece of data to send at a given instant (the segmentation algorithm).

In the case of transmission, the host decides that a certain area of its memory is ready to be transmitted and notifies the processor in the transmission side of the interface. When the transmission is completed, the interface processor notifies the host of the fact. It is desirable to multiplex several areas of data concurrently. When receiving data, the processor in the host interface must decide where the payload of each received cell is to be stored in the host's memory. This decision will be made based on the VCI of each cell (and the se-

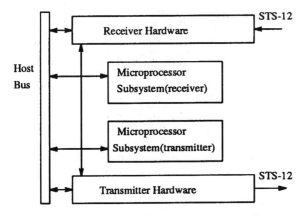

Figure 28.21 Block diagram of host interface for AURORA testbed

quence number if used). The host, therefore, must provide the interface processor with a list of buffers into which cells may be placed. When the buffer is filled, the processor must notify the host.

The author proposed to use certain dedicated *data path hardware* to move data blocks between host and interface. The microprocessors are not directly involved in moving data, as this would consume too many instructions to leave time for the decision-making process.

The communication between the host and the microprocessors on the interface is through shared memory between the host and the microprocessors. Two versions of host bus interface are proposed. One is a DMA version, and another is a dual-ported memory version.

The experimental results showed that, when no other devices were using bandwidth on the TURBOchannel, the maximum sustainable throughout was 367 Mbps in the transmission direction and 462 Mbps in the receive direction. The available bandwidth for payload in a 622 Mbps ATM/SONET stream is approximately 500 Mbps. Thus, the measured figures correspond to 74 percent link utilization in the transmit direction and 93 percent in the receive direction. Efforts are being made to improve the performance in the transmit direction.

28.5.3   ATM interface with direct memory/cache access

The ATM interface board (AIB) is a host-to-ATM-network interface developed at Syracuse University [29]. It implements the functions of the ATM layer and AAL layer. It communicates with the upper layer protocols through shared memory. For high-speed data transfer, the AIB moves data between host's cache/memory and network. Figure 28.22 shows the block diagram of the AIB and its connection with host. The AIB consists of two units, a message transmitting unit (MTU) and message receiving unit (MRU).

In this interface design, both the MTU and the MRU have a direct memory/cache access controller to move data between network and host. The DM/CA controller can be considered to have two parts according to its functionalities. The first part is associated with the main memory and performs the DMA mechanism. The second part is associated with the cache and implements the DCA (direct cache access) mechanism. On the transmitting side, the DCA controller communicates with the memory management units (MMUs) in

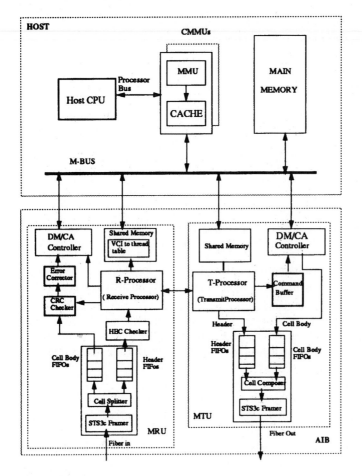

Figure 28.22 AIB structure

the host and requests data from cache. When the DCA controller gets a cache hit, it will move data from the cache to the network FIFOs. If a cache miss occurs, the DMA controller will transfer data from the main memory to the network FIFOs. On the receiving side, the DCA controller writes data into cache if the addressed line is already in cache; otherwise, the DMA mechanism is initialized to store data into main memory. The DM/CA is not allowed to cause cache update to avoid cache miss.

To process arriving messages efficiently, the MRU is designed to be a message-driven receiver. In other words, the operation of the MRU is not controlled by the host but by the received messages. This concept is referred to as *active message dispatching* [33]. Active message dispatching means that messages are dispatched to the corresponding message handling processes by extracting the information carried in the cell header. The VPI/VCI field is mapped to the thread associated with a requested service. The mapping between a VPI/VCI to a thread is performed by a VPI/VCI-to-thread table. This table is established during the signaling period and stored in the AIB. Based on the thread, the data portion of a message is transferred from the network FIFOs into a certain area of host's cache or memory. The host will be notified when data transfer is completed.

Furthermore, the DM/CA controller is designed to have an internal pipeline to allow overlap between read and write operations. Assuming that cache access rate is similar to

FIFO access rate, the best performance can be achieved when the DM/CA gets a cache hit, as reads from the host and writes to the FIFO on interface are overlapped.

28.6   Parallel and Distributed Computing Environment Over ATM

The main objective of an HPDC environment is to provide parallel and distributed applications with the message passing primitives required to achieve efficient interprocess communications over the emerging high-speed ATM networks. An application programming interface (API), a run-time system, is needed to provide an interface between a parallel distributed programming tool and the ATM AAL protocol. In this section, we take the Fore's System API as an example to show how to develop parallel and distributed applications over ATM networks.

The API of Fore systems provides a portable API to the ATM data link layer [30]. It basically performs the functions that the socket layer performs in a conventional communication software system. Figure 28.23 shows the architecture of the host system software. The boxes on the left side represent unmodified applications and operating system modules, and the ones on the right represent the new modules that Fore supplies with its ATM host interfaces. The left side of the diagram shows that Fore's ATM LAN can be utilized immediately via any existing network application. The right side, however, shows that a new interface is required for ATM applications that need to exploit capabilities not found in other LANs and protocols, such as guaranteed bandwidth reservation, selection of a specific AAL type, multicasting, and other ATM-specific features. The following describes ATM API.

The ATM API is implemented as an ATM communications library. The library routines for Sun OS and IRIX platforms are STREAMS-based and those for ULTRIX platforms are socket based. Although the internals differ, the same application programming interface is provided on all platforms and is portable across all platforms. The ATM API provides a connection oriented client-server model of distributed computing.

When a client program attempts to establish an ATM connection, it specifies the ATM address, the ASAP (application service access point) of the destination, the desired AAL

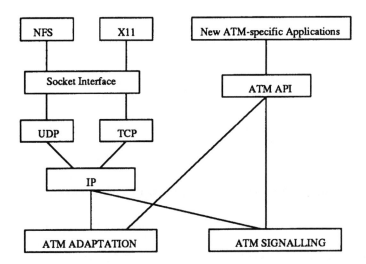

Figure 28.23 Fore's system application programming interface

protocols (AAL type 3/4, 5, or null) and acceptable bandwidth requirement through an ATM-connection routine. And a server program receives requests for incoming ATM connections then chooses to accept or reject the connection. End-to-end flow control and retransmission are left to the application layer. < *atm_send* > and < *atm_receive* > routines are provided with buffers which are used for data transfer. The data given to the < *atm_send* > is segmented by the appropriate AAL protocol. Each cell is prefixed with the outgoing VC (virtual connection) identifier for its connection as it is transmitted. On the receiving side, the data is reassembled and delivered via the connection descriptor corresponding to the incoming VC identifier.

To support TCP/IP over ATM, IP packets are encapsulated in ATM PDUs. The AAL is used to segment, reassemble, and frame the IP packet. Internet addresses are mapped to ATM-layer addresses for connection establishment. And the IP layer, with its connectionless service model, interfaces to the ATM data link layer, with its connection-oriented model.

28.7   Conclusions and Future Directions

The asynchronous transfer mode (ATM) technology will play an important role in the widespread deployment of high-speed networks. The ATM is the first technology that has been widely accepted as a viable technology for LANs, MANs, and WANs. This will allow applications to run on computers inter connected by different combinations of these networks. An important feature of ATM is its ability to divide different types of traffic into fixed small cells and multiplex them over a single fiber optic transmission line. In this environment, users will be charged based on actual usage of bandwidth and not on the number of leased lines.

In this chapter, we reviewed the basic concepts of ATM, AAL protocols, and ATM network flow control strategies and discussed the design issues associated with network-based parallel and distributed computing systems. High-speed networks in general, and ATM networks in particular, will have profound impact on the proliferation of parallel and distributed computing over networks. However, to utilize the existing computing power available on an ATM network, more research is needed to reduce delay and to improve throughput. The research activities can be grouped into the following four areas: (1) high-speed switches, (2) high-speed protocols, (3) host-to-network interfaces, and (4) application programming interfaces. In what follows, we briefly outline the research issues associated with each area.

1. *Designing high-speed ATM switches.* ATM switches should synchronize to the incoming cell streams and identify the virtual connections, and derive routing and control information. The research activities here should investigate novel techniques to design high-performance, large-scale ATM switches that can operate at OC-3, OC-12, and even OC-48 rates. The main features for such switching systems are as follows:

    (a) High-speed processing rate. These switch systems need to receive, decode, and route ATM cells at rates of 2.7 cell/μs for 155 Mbps (OC-3) networks and 0.68 cell/μs for 622 Mb/s (OC-12) networks.

    (b) Scalability. The switch systems need to handle large number of I/O ports, say 1024 × 1024. The existing switches can handle far fewer input/output ports (e.g., 16 or 64 I/O ports).

    (c) Efficient multiplexing and group communications.

    (d) Capability of maintaining QOS associated with parallel and distributed applications.

    (e) Efficient policing techniques.

    (f) Non-blocking and re-configuration switch systems.

2. Designing high-performance host-to-network interfaces that can utilize the increasing high data transmission rates. The research activities include:

    (a) reducing data movement by using DMA or programmed I/O schemes

    (b) using hardware checksumming elements to save one trip to main memory

    (c) remapping pages associated with the application's virtual address to the operating system's virtual addresses to avoid data copying overhead

    (d) reducing OS overhead by minimizing interrupts and context switches

3. Designing high-speed transport protocols that can perform connection management, data transfer/reception, flow control, and error recovery at data rates of hundred megabits to gigabits per second. The research activities include:

    (a) implementing transport protocol in hardware, and investigating the use of parallel processing methodology to implement communication protocols

    (b) developing novel flow control methods which are optimal for ATM based parallel and distributed computing

    (c) exploring efficient connection management strategies

    (d) developing new buffer management schemes to meet the requirements of parallel and distributed applications running on high-speed networks

4. Designing application programming interfaces (API):

    (a) allowing parallel and distributed applications to bypass the conventional protocol stack and directly access streamlined communication protocols running on ATM networks

    (b) implementing an ATM-based message-passing library to provide all services required in parallel computing, such as point-to-point communications, group communications, process synchronization, system control and management, error handling, and reduction applications (i.e., summing, maximizing, and minimizing)

28.8   References

1. Kung, H. T. 1992. Gigabit local area networks: a system perspective. *IEEE Communications* (April).
2. Bell, G. 1992. Ultracomputer: a teraflop before its time. *Communications of the ACM*, 27–47.
3. Hariri, S. 1995. *High Performance Distributed Systems: Concepts and Design.* Norwood, Mass.: Artech House.
4. Prycher, M. D. 1993. *Asynchronous Transfer Mode, Solution for Broadband ISDN.* Ellis Horwood.
5. CCITT COM XVIII-228-E, March 1984.
6. CCITT COM XVIII-R1-E, February 1985.
7. Boudec, J. L. 1992. The asynchronous transfer mode: a tutorial. *Computer Networks and ISDN Systems*, 24, 279–309.
8. ITU-T 1992 Recommendation I.362. *B-ISDN ATM Adaptation Layer (AAL) Functional Description.* Study Group XVIII, Geneva, June 1992.
9. ITU-T 1992 Recommendation I.363. *B-ISDN ATM Adaptation Layer (AAL) Specification.* Study Group XVIII, Geneva, June 1992.
10. Akyildiz, L. F. 1993. Broadband-ISDN (ATM) networks. *Proc. 2nd International Symposium on High Performance Distributed Computing.*
11. Suzuki, T. 1994. ATM adaptation layer protocol. *IEEE Communications* (April).
12. Partridge, C. *Gigabit Networking.* Reading, Mass.: Addison Wesley, 1994.
13. Wang, Z., and J. Crowcroft. 1992. SEAL detects cell misordering. *IEEE Network*, 8–9.

14. Bae, J. J., et. al. 1991. Survey of traffic control schemes and protocols in ATM networks. *Proceedings IEEE*, 170–188.

15. Gallassi, G., G. Rigolio, and L. Fratta. 1989. ATM: bandwidth assignment and bandwidth enforcement policies. *Proceedings of IEEE GLOBECOM 1989*, 49.6.1–49.6.6.

16. Newman, P. 1992. ATM technology for corporate networks. *IEEE Communications* (April) 90–101.

17. Hluchy, M. G., and M. J. Karol. 1988. Queueing in high-performance packet switching. *IEEE J. Selected Areas in Communications*, Vol. 6, No. 9, 1587–1597.

18. Yeh, Y. S., M. G. Hluchyi, and A. S. Acampora. 1987. The knockout switch: a simple modular architecture for high-performance packet switching. *IEEE J. Selected Areas in Communications*, 1274–1283.

19. Rooholamini, R., V. Cherkassky, and M. Garver. 1994. Finding the right ATM Switch for the market. *IEEE Computer* (April) 16–28.

20. Giacopelli, J. N. et al. 1991. Sunshine: a high-performance self-routing broadband packet switch architecture. *IEEE J. Selected Areas in Communications*, Vol. 9, No. 8, 1289–1298.

21. Tobagi, F. A., T. Kwok, and F. M. Chiussi. 1991. Architecture, performance, and implementation of the tandem banyan fast packet switch. *IEEE J. Selected Areas in Communications*, Vol. 9, No. 8, 1173–93.

22. Dias, D. M., and M. Kumar. 1984. Packet switching in $N\log N$ multistage networks. *Proc. IEEE Globecom*, 114–120.

23. Batcher, K. E. 1968. Sorting networks and their applications. *AFIPS Proceedings of the 1968 Spring Joint Computer Conference*, Vol. 32, 307–314.

24. Zegura, E. W. 1993. Architecture for ATM switching systems. *IEEE Communications* (February), 28–37.

25. Banks, D., and M. Prudence. 1993. A high-performance network architecture for a PA-RISC workstation. *IEEE J. Selected Areas in Communications*, Vol. 11, No. 2, 191–202.

26. Cooper, E. et al. 1991. Host interface design for ATM LANs. *Proceedings of the 16th Conference on Local Computer Networks*, IEEE Computer Society, 247–258

27. Davie, B. S. 1993. The architecture and implementation of a high-speed host interface. *IEEE J. Selected Areas in Communications*, Vol. 11, No. 2, 228–239.

28. Traw, C., and J. Smith. 1993. Hardware/software organization of a high-performance ATM host interface. *IEEE J. Selected Areas in Communications*, Vol. 11, No. 2, 240–253.

29. Lu, B., and S. Hariri. 1993. An ATM based hypercube distributed system. *Proceedings of the Sixth Symposium on Parallel and Distributed Computing*, International Society of Computer Applications.

30. Biagioni, E., E. Cooper, and R. Sansom. 1993. Designing a practical ATM LAN. *IEEE Network* (March), 32–39.

31. Parasoft Corporation. 1988. *Express Reference Manual*.

32. Sunderam, V. S. 1991. *PVM: A framework for parallel distributed computing*. Technical report. Oak Ridge, Tenn.: Oak Ridge National Laboratory.

33. Eicken, T., D. E. Culler, S. C. Goldstein, and K. E. Schauser. 1992. Active messages: a mechanism for integrated communication and computation. *Computer Architecture News*, Vol. 20, No. 2, 256–266.

Tools and Applications

Development Tools

29

Parallel Languages

R. H. Perrott

One of the features of parallelism that influences the acceptance of parallel computing technology among potential users is the programming language. This chapter considers the programming languages that have been proposed or used for the programming of parallel computers; the emphasis is on languages which have been used or proposed for scientific applications since this is the area which has been in the vanguard of parallel computing. Example languages are used to illustrate the concepts.

29.1   Introduction

Sequential computing has benefited from the fact that there has been a single model of computation, widely known as the von Neumann model, on which architects and software and algorithm designers have based their work. Parallel computing has not been so fortunate in that there has not been a single model of computation; as a result a variety of different architectures and programming paradigms have been proposed [33].

The main issue affecting the architectural model for parallel computers is how to organize multiple processors to execute in parallel. This has given rise to a wide range of models, chief of which is the multiple instruction, multiple data (MIMD) model. These MIMD systems employ multiple processors that execute independent instruction streams accessing data autonomously. The design of such systems requires careful consideration of the number of processors and their interconnection network topology.

The shared-memory model, where the processors are connected to a number of memory modules to form a common global shared memory, was the basis of the earliest MIMD machines. One of the major problems with this model was that severe memory contention can occur when processors try to access data residing in the same memory module. This memory contention problem can be solved by using the distributed-memory model where the memory is distributed among the processors. If data needs to be accessed from the local memory of another processor, then it is transferred using the interconnection network. Such a configuration is scalable to higher orders of parallelism than that of the shared-memory model. Initial programming experience indicates that shared-memory systems are easier to use and program compared to distributed-memory systems which require more intricate programming.

The construction of MIMD systems is now commercially viable, and it is feasible to build systems involving large numbers of components. However, consistent with previous trends, effective software methods for programming such systems lag the hardware developments, resulting in their inefficient use in many instances.

Software for parallel systems has followed the paradigms of sequential software with the choice of programming languages for such systems being based on either the imperative or declarative approach; the declarative group can be further divided into logic and functional languages while the imperative group consists of procedural and object oriented languages as follows:

- Imperative languages
  - Procedural
  - Object oriented
- Declarative languages
  - Logic
  - Functional

The third component in the effective use of parallel systems that requires careful consideration is the choice of algorithm. Studies have shown that transferring an efficient sequential algorithm to a parallel machine can result in an inefficient parallel algorithm. The design and construction of a new parallel algorithm for a particular application area can produce major performance improvements. The following sections consider the various approaches which have been proposed or implemented for languages to programming parallel systems and uses examples to emphasize the concepts.

29.2   Language Categories

The programming languages used for programming parallel systems, at present, can be broadly categorized into the *imperative* and *declarative* groups [32], both of which support a variety of design approaches.

29.2.1   Imperative languages

An imperative language is based on the control-driven approach where users specify solutions in terms of a sequence of operations that change the state of the machine. The central operation is that of *assignment,* which enables any memory location in the machine to be changed, hence such languages are described as being *state oriented* languages. These languages may be further subdivided into two groups; namely, *procedural* and *object-oriented* languages.

Procedural languages provide the procedure/function unit as the basic means for program development. This is usually augmented with the module/package facility for modularization and data abstraction, as well as communication and synchronization features for specifying parallel activities. Examples are Ada [42], Concurrent C [18], Occam 2 [10] and Joyce [8].

Object-oriented languages, on the other hand, provide the object as the basis for program development, where an object is represented as a composition of data describing its state as well as a set of operations that characterize the actions it performs. Objects have restricted visibility and interact by passing messages. An object can be accessed only through a well defined interface, and the data contained within an object are visible only within the object itself; this facilitates data hiding and independence. Examples are ABCL/1 [44], Concurrent Smalltalk [43], Distributed Smalltalk [7], POOL-T [3], and PO (Parallel Objects) [13].

29.2.2   Declarative languages

A declarative language is based on the principle of representing a program in terms of *what* is to be calculated rather than *how* the calculation is to be performed. A major difference between the declarative and imperative approaches is that, in the declarative approach, there is no concept of state and program control structures. Such languages can be subdivided into two groups; namely, *functional* languages and *logic* languages.

Functional languages use expressions and functions as a basis for developing programs where the application of a function to its arguments is the only control structure; there are no features that deal with assignment, conditional, or looping structures as in imperative languages. An important property of functional programs is that of *referential transparency,* which means that the value of a function is determined solely by the values of its arguments; this, in turn, means that function evaluation does not produce side effects. If functions have no side effects, then it makes no difference in which order they are executed, and it is therefore possible to evaluate them in parallel. As a consequence, functional languages do not require explicit constructs to specify processes, communication, and synchronization; that is, the parallelism is implicit. To allow users the possibility of influencing the behavior of the program, various researchers have designed extensions to functional languages; such languages are referred to as *parafunctional* programming languages. The extensions amount to a meta language that facilitates the expression of the order of evaluations, task priorities, and mapping of processes to processors. Examples are Blaze [29], Concurrent LISP [40], Haskell [23], Multilisp [20], Miranda [41] and ParAlfl [22].

Logic languages use relations as a basis for developing programs. These languages are goal oriented, where a goal is solved by using a depth-first, left-to-right search of relations and facts to solve a goal. When a dead end is reached, the implementation backtracks; that is, it retraces its steps and then goes down another path. Parallelism is introduced by evaluating various subgoal components using the following techniques:

1. *AND-parallelism* comes from trying to solve many parts of the problem simultaneously, that is, pursuing solutions to a number of subgoals at the same time. The main goal is solved if the first subgoal is solved AND, the second AND, the third, etc. For example, given

    ```
    A <- B,C,D
    ```

    the subgoals are processed in parallel until they all succeed or any one of them fails. If at any time a subgoal cannot be solved, backtracking must occur. Thus, in AND-parallelism, all subgoals are working on the same problem, and each must succeed if the goal is to succeed. Because the subgoals are working on the same problem, they share a common set of data and variables. A change to one variable in a subgoal may affect the computation of another subgoal.

2. *OR-parallelism* involves the evaluation of a problem in several ways at the same time. The problem is solved if the first clause is solved OR, the second OR, the third, etc. For example, given

```
A  <-  B,C,D
A  <-  E,F
```

the two clauses for A can be worked on in parallel until one of them succeeds or they both fail. Even though only one clause needs to succeed, more than one answer may be produced for a given query. Databases are examples of systems that can benefit from OR-parallelism. All queries can be worked on in parallel as long as no cooperation or communication between the activities is required.

Examples are Concurrent Prolog [38], PARLOG [11] and Strand [15].

An argument in favor of the declarative approach, compared to the imperative approach, is that the notation is at a higher level of abstraction so that a user does not, for example, have to be concerned with architectural details. It is claimed that, in comparison to imperative languages, declarative languages have simpler semantics, are easier to write and understand as well as manipulate, and facilitate program correctness proofs. However, a major criticism is that declarative languages have not been efficiently implemented; the advent of parallel architectures represents an opportunity to explore their efficient implementation. The challenge at present is to produce implementations (on parallel machines) that are comparable in efficiency to imperative language implementations.

29.3 Programming Languages

The remaining sections of this chapter concentrate on the approaches that have been used for imperative languages.

29.3.1 Design approaches

Four main approaches have been suggested as a basis for designing programming languages that would promote the wider use of parallel systems [31, 32]; these are:

1. *Invent a completely new language.* This approach ignores all existing languages and applications in the formulation of the new language. The rationale is that it will facilitate the coherent treatment of all aspects of parallel programming, allowing the user to develop and to express a parallel solution explicitly. The drawback, however, is that existing software will have to be reprogrammed in the new language, which is a labor-intensive and error-prone procedure. Typical of this approach has been the introduction and adoption of the Ada language [42].

2. *Use a coordination language.* This approach integrates a sequential language with a coordination language; the latter provides operations that create parallel activities and supports communication. The sequential language allows the programmer to develop single computation activities and the coordination language binds these activities together to produce a parallel program. At present, this approach has been used with Fortran, Pascal, and C, and the coordination language Linda [1], and implemented on distributed-memory systems such as the Inmos Transputer [39] and Intel iPSC [19].

3. *Enhance an existing sequential language compiler to detect parallelism.* In this approach, the responsibility is placed on the compiler to detect which parts of a sequentially constructed program can be executed in parallel. The principal advantage of this approach is that existing sequential programs can be moved to the target parallel machine exploiting its parallel facilities relatively inexpensively and quickly. This approach has been applied to Fortran where the DO loops of a program are examined to determine if it is possible to spread the iterations of the loop across different processors [2].

4. *Introduce features to an existing language that deal explicitly with parallelism.* This approach should enable existing software to be adapted and transferred to parallel machines, where appropriate, by existing programmers. Many of the extensions have been developed by different groups using the same language base, leading to the definition of nonstandard variants; this makes the production of a standard for such languages difficult. This approach is reflected in several Fortran proposals recently proposed for distributed-memory systems programming, for example, Intel Fortran [25], Vienna Fortran [45], and Fortran D [16]. High Performance Fortran [24] is an attempt to stop this proliferation of Fortran language extensions.

29.3.2   Invent a completely new language

This section considers language constructs that have been introduced to enable the programmer to specify parallelism directly. The main features that have been used in recent languages are processes (or tasks) and nondeterminism as based on a concept known as *guarded commands* (considered shortly). Languages incorporating theses constructs are generally referred to as languages based on the *message-passing paradigm*; Ada is considered to be a representative example of this group of languages.

In this approach, a parallel program is represented as a number of computation units, each unit being normally referred to as a *process* or *task*. These processes execute in parallel and at unpredictable speeds, giving rise to unpredictable interaction between the processes; interactions that are usually described as *nondeterministic*. In the course of program execution, there will be situations when processes need to be coordinated, and that requires facilitating process *communication* (the transfer of data values from one process to another) and *synchronization* (the transmission of information concerning the state of a process). Process coordination can be achieved by passing messages between the processes.

Since the order in which interactions can take place is not deterministic (that is, it is unpredictable), the notion of *guarded commands* was proposed by Dijkstra [14] to introduce non-determinism into process selection. A guarded command basically consists of several guards, each followed by an associated process; a guard can constitute either a Boolean expression, an input command, or a combination of a Boolean expression followed by an input command. The selection of a process from several alternatives can be made dependent on one or more of the guards being satisfied, that is, becoming ready. The process associated with the first guard to become ready is chosen for execution. If multiple guards become ready simultaneously, then either a random selection is made or the selection can be prioritized according to assigned process priorities. In the situation when none of the guards are ready, then the execution is usually delayed and the guards repeatedly re-evaluated until one of them is ready; this procedure can be terminated after a certain period of time to avoid deadlock conditions.

In a parallel system, when communication is possible, two basic operations are necessary to effect the operation, that is, *send* and *receive*. These are used symmetrically between communicating processes to pass messages. One of the first programming notations to use the message passing concept was *communicating sequential processes* (CSP) [21], which employed synchronous message transfers between pairs of named processes. The essential idea of CSP is that synchronization is set up using input/output commands where an output command in a sender process specifies the destination process, and all input commands in a receiver process specify the source process. A pair of input and output commands match if two processes name each other and the message types of the command parameter lists are compatible. If this is the case, then the command is executed simultaneously, and a message transfer between the processes takes place. Processes involved in a message transfer are delayed until the matching input or output command is issued by each process. The coming together of the two processes to transfer a message is referred to as a *rendezvous*.

In Ada, the rendezvous is based on three concepts, namely, the *entry declaration, the entry call,* and the *accept statement.* The entry declaration and accept statement are part of the called process, while the entry call is part of the calling process. An entry call has a syntax similar to a procedure call statement, which names the entry and the process containing the entry and supplies any actual parameters. An accept statement for the entry may contain a list of statements to be executed when the entry is called; the accept statement has the following structure:

```
accept service (in value_parameters; out result_parameters)
   body
```

If a message has not arrived, the execution of the accept statement delays. When a message arrives, the body of the accept statement is then executed. After execution of the accept statement is completed, a reply is sent back to the calling process via the output result parameters, and the called process then continues its execution. These actions of the processes are referred to as a *rendezvous* because both the calling and called processes meet for the duration of the execution of the accept statement and then go their separate ways. Therefore, the rendezvous technique enables a called process to choose when it wants to serve calls.

A feature of the rendezvous approach is that the interaction is fully synchronous; that is, the first process that is ready to interact waits for the other to become ready. A consequence of this is that process synchronization is also accomplished with the message passing. To illustrate these features, the problem or sharing a single resource among a number of competing processes is expressed in Ada. Each task in Ada is expressed in two parts;

1. a specification part, which is the interface presented to the other tasks; it may contain entry specifications, that is, a list of the services provided by the task
2. the task body, which contains a sequence of statements to be executed when any of its services are requested

In Ada, the resource must be represented as a task, and the calls accepted by this task are those listed in the specification part; in this case, calls to ACQUIRE or RELEASE. The statements of the task body following the reserved word **accept** are executed whenever another task calls the appropriate entry procedure and that call is accepted by the called task. Thus, whichever task reaches its communication statement first must wait for the other task; that is, either the calling task calls an entry procedure such as RESOURCE.ACQUIRE or the called task reaches the appropriate **accept** statement, in this case, **accept** ACQUIRE **do**.

The body of the task RESOURCE has been set up as an infinite loop that contains a **select** statement with two limbs. Each time the **select** statement is encountered, if both limbs are ready for selection, then either one of the limbs is selected arbitrarily; this is how nondeterministic behavior is expressed in Ada. The first limb has a condition attached in the form of a **when** clause; this limb can not be selected unless the condition is true. In this case, the resource is available.

```
task RESOURCE is -- specification
   entry ACQUIRE;
   entry REMOVE;
end RESOURCE;
task body RESOURCE is -- body
   FREE: BOOLEAN := TRUE;
begin
   loop
      select
      when FREE =>
         accept ACQUIRE do
            FREE := FALSE;
         end;
   or
         accept RELEASE do
            FREE := TRUE;
         end ;
      end select ;
   end loop;
end RESOURCE;

task type PRODUCER is
   -- specification
end;

task type PRODUCER is
begin
   -- statements
   RESOURCE.ACQUIRE;
   -- use resource
   RESOURCE.RELEASE;
   -- statements
end;

BEES: array(1..N) of PRODUCER;
   -- N tasks are declared with the lifestyle of PRODUCER
```

The technique of message passing had a profound effect on language design for MIMD parallel processing and has resulted in the development of significant parallel programming languages for distributed-memory systems.

29.3.3   Use a coordination language

Coordination languages are based on using distributed data structures for synchronization and communication where a distributed data structure contains separate groups of data objects that can be manipulated simultaneously by several processes. Process communication involves accessing the data structure contents (read operations), depositing new components into the data structure (write operations), or extracting existing components for updating and then returning them to the structure (read-write operations). Parallelism is introduced by allowing such actions to be performed by multiple processes on different components of the data structure. If a process tries either a read or read-write operation on a nonexistent data structure component, the process is suspended until another process adds the required component to the data structure; this introduces process synchronization. In contrast with message passing, communication involving distributed data structures is anonymous.

The use of distributed data structures for parallel programming was first introduced in the coordination language Linda [12], which uses the concept of a *tuple space* to implement the distributed data structures. Linda consists of a few simple operations that support the tuple space model of parallel programming. These operations are independent of any base computing language and are designed to support as well as simplify the construction of parallel programs. When the simple operators that Linda provides are included in a host language X, then X is turned into a Linda-based parallel programming language, i.e., a parallel programming dialect of the host language. Examples of Linda-based implementation include C-Linda, C++-Linda, Modula-2-Linda, and Fortran-Linda; these implementations also consist of a run-time kernel that implements interprocess communication and process management.

The elements of tuple space are referred to as *tuples,* which are simply ordered sequences of typed values, e.g., ("program", 9, true) is a three-tuple that consists of a string, an integer, and a Boolean. Tuples do not have addresses and are accessed using associative look-up methods by searching on any combination of tuple field values. Three basic operations are defined on tuple space, namely, **out** adds a tuple to tuple space, **read** or **rd** reads a tuple contained in tuple space, and **in** reads a tuple and also deletes it from tuple space. The operation **rd(s)** is identical to **in(s)** except that the matched tuple remains in tuple space; this is used in situations where the target tuple **s** does not require modification, and also when it is required to be read by other process(es). Both the **in** and **rd** operations try to find a target tuple, and if several matching tuples exist, then one is chosen nondeterministically. If there are no matching tuples, then the operation and the process involved is delayed until another process adds the tuple using the **out** operation. These operations form the basis for implementing process communication and synchronization.

For example, in the client-server paradigm, many client processes communicate with a server process, which renders a service to the client processes involved. A client can request that a service be performed by sending a message to the server process, which performs the service and, if necessary, returns a result to the client concerned; the server repeatedly receives a request for service from a client. The following is a C-Linda solution of the multiple client-single server paradigm that illustrates the use of the **in** and **out** operators.

```
client ()
{
...
   while (1) {
      for (i = 0; i <= maxclients, ++i) {
         out ("request", i, demand);
         in ("response", i, ?demand);
      }
   }
...
}
```

```
server ()
}
   while (1) {
   in ("request", clientno, ?req);
   ...
   out ("response", clientno, value);
   }
}
```

The **eval(t)** operator is used to introduce parallelism into programs; it is identical to **out(t)** except that the values corresponding to tuple **t** are evaluated "after" rather than before it is entered into tuple space, i.e., it creates a live tuple. For example, the operation

```
eval("compute", 0.0, manipulate (9.0))
```

results in the generation of a three-element live tuple consisting of a process that evaluates "compute", 0.0 and the function call manipulate (9.0).

When the computation concerned with the evaluation of tuple **t** is complete, then **t** becomes an ordinary, passive data tuple which may then be read. Therefore, to create processes, live tuples are generated which eventually turn into data tuples directly. The operation

```
   rd ("compute", 0.0, ?result)
```

can thus be used to read the value generated by the **eval** operation, that is, after the live tuple has been transformed into a passive data tuple in tuple space; execution before this point will block the statement until the tuple has been added to the tuple space by the completion of the **eval** operation.

The use of **eval** results in process creation as well as communication. The effect of **eval** is to implicitly fork a new process to perform the tuple evaluation and its addition to tuple space, i.e., the communication. This feature can be used to create any number of processes. For example, the creation of 50 parallel processes to evaluate the tuples for square root of numbers in the range 1 to 50 can be accomplished as follows:

```
for (i = 0; i <= 50; ++i) {
   eval("square root", i, sqrt(i))
}
```

Other languages were introduced after Linda that support the distributed data structure concept; these include Orca [5], SDL [35], and Tuple Space Smalltalk [28].

29.3.4 Enhance an existing sequential language compiler to detect parallelism

In this case, a conventional sequential programming language is used to specify the problem solution, and the code is subsequently compiled for parallel execution. The methods used require the use of sophisticated compilation techniques to detect which parts of the sequentially written program can be executed in parallel.

The automatic parallelization by a compiler of a given sequential program for parallel processing poses many challenges. The difficulties inherent are concerned primarily with problem decomposition into parallel processes as well as the identification and management of process communication and synchronization.

Compiler-based parallelization of sequential programs was first introduced success-fully in Fortran compilers for the early supercomputers to vectorize DO loops of a program [31]. The technique of data dependency analysis involving looping statements has been extended to the parallelization of sequential programs for both shared and distributed-memory systems. The compiler detects those loops that have no direct data dependency between the iterations; in such a case, the iterations are assigned to the processors for execution either individually or in groups, depending on the parallelization strategy used.

The data dependency analysis technique, when applied to complete programs, poses difficulties with regard to dependency analysis across procedure calls of whole programs. At present, some experimental systems exist, such as the IBM PTRAN system [2], PARAFRASE 2 [34], and SIGMA 11 [17], that perform inter-procedural data dependency analysis and the restructuring of Fortran programs for execution on parallel architectures.

Essentially, it requires the examination of the statements of a program to determine if and what kind of dependencies exist among a group or statements. For example, a straight-forward situation is illustrated in the following group of statements, which involve only scalar variables and contain a number of dependencies that can be easily identified. For example, in the following statements

```
S1:   A=1.0
S2:   B=A+3.14
S3:   A = 1/3 * (C-D)
      . . . . . . . . . . . . .
S4:   A = (B* 3.8) / 2.7
```

A true dependence exists between statements $S1$ and $S2$ since the value of A computed in $S1$ is used in $S2$; therefore $S1$ must be executed before $S2$; i.e., the two statements have to be executed in the order in which they appear. But an antidependence exists between $S2$ and $S3$, since the value of A used in $S2$ must be fetched before the variable A is assigned in $S3$. An output dependence occurs between $S3$ and $S4$, since the value assigned to A in $S3$ must occur before the assignment in $S4$. It may be possible to remove the effects of the last two dependencies by, for example, introducing temporary storage, or reordering statements, and so on. The dependencies dictated the order in which statements must be executed and whether it is possible to reorder statements to improve the possibilities of introducing parallelism.

In the presence of array references, computing the dependence relationship can be considerably more difficult. For example, consider the loop

```
      DO 3 I = 1,N
S1:      X(a* I + b) = ....
S2:      .............. = X(c*I+d)
3     CONTINUE
```

To determine if there is a dependence between statements $S1$ and $S2$, it is necessary to determine whether the equation $a*i1+ b = c*i2 + d$ has a solution in $i1$ and $i2$ both within the loop limits 1 to N. There will be a true dependency between $S1$ and $S2$ if there is a solution satisfying the constraint $i1 <= i2$; there will be an anti-dependency on $S1$ if there is a solution satisfying $i2 < i1$. Hence, a considerable effort is required in solving all dependency equations that can occur in a complete program involving many, possibly deeply nested, loops.

In addition, the type of hardware that is available for program execution can effect the detection required. For example, if the hardware supports vector operations (that is, the complete execution of a vector of operands), then only detection of such operations in a sequentially written program is necessary.

For such an environment, loops of the following kind would need to be examined carefully because, if the programmer had written in sequential Fortran the loop

```
    DO 6 I = 2,4
      B(I) = A(I-1)
      A(I) = C(I)
6   CONTINUE
```

the intended sequence of execution steps is

```
B(2) = A(1), A(2) = C(2), B(3) = A(2),
A(3) = C(3), B(4) = A(3), A(4) = C(4),
```

With vectorization, this would become

```
B(2) = A(1), B(3) = A(2), B(4) = A(3)
A(2) = C(2), A(3) = C(3), A(4) = C(4)
```

which was not what was intended as illustrated by the underlined statements.

On a parallel machine, that is, one that has many processors capable of independent execution, again loops must be carefully considered. In the following loop, each iteration requires data from a preceding loop.

```
    DO 2 I = 1,N
      X = A(I) + DF
      A(I+1) = X+C(I)
      STATS(I)
2   CONTINUE
```

as shown when the loop is unrolled

```
iteration 1          iteration 2          iteration 3          etc.

X = A(1+DF
A(1+1) = X+C(1)      X = A(2)+DF
STATS(1)            A(2+1)= X+C(2)        X = A(3)+DF          etc.
                    STATS(2)             A(3+1) = X+C(3)
                                         STATS(3)
```

the value of $A(2)$ is computed in iteration 1 and used in iteration 2, etc., which is a true dependency. If the loop is to be executed on a parallel machine, then synchronization must be introduced to ensure the correct updating of the array A. The user, in some systems, can indicate such situations and thus help the detection process.

When a parallel machine with vector facilities is the intended target, then a decision as to whether to vectorize on I (or J) or to parallelize on J (or I) in a nested loop is required. For example, given the loop

```
     DO 1 I = 1, N
       DO 2 J = 1, N
S1:      A(I, J + 1) = B(I, J) + C(I)
```

```
S2:        D(I, J) = A(I, J)*2
2          CONTINUE
1          CONTINUE
```

Vectorizing on index *I* and parallelizing on index *J* gives

```
      DOACROSS J = 1 ,N
S1:   A(1:N, J+1) = B(1:N, J) + C(1:N, J)
         SIGNAL(J)
         IF J > 1 WAIT(J-1)
S2:        D(1:N, J) = A(1:N, J)*2
      END DO
```

where the WAIT statement blocks execution until the SIGNAL statement with the same number is executed, and the DOACROSS imposes a partial order across iterations

However, if it is decided to vectorize across index *J* and parallelize across index *I*, then the following situation emerges:

```
      DOALL I = 1,N
S1:    A(I, 2:N+1) = B(I, 1:N) + C(I, 1:N)
S2:    D(I, 1:N) = A(I, 1:N)*2
      END DOALL
```

where the DOALL imposes no order across the iterations. Essentially, the characteristics of the underlying hardware dictate which scenario to adopt.

There are in existence systems that assist in the transformation of sequential programs for execution on distributed-memory systems exploiting program data parallelism features; these include SUPERB [46], PANDORE [4], OXYGEN [36], MIMDizer [37] and Kali [26].

29.3.5 Introduce features to an existing language to deal explicitly with parallelism

The language category receiving most attention at present is that of extensions to existing sequential languages. This is a consequence of the fact that the main users of parallel machines have been scientists and engineers whose main working language is Fortran. They are therefore persuaded to use these machines if language extensions to Fortran are provided. Two main paradigms have been proposed, namely, virtual shared memory (VSM), and single program, multiple data (SPMD).

29.3.5.1 Shared virtual memory. A virtual shared-memory method involves process communication and synchronIzation characteristics based on shared variables; it is equivalent to a single address space that is shared by a number of processes executing on different processors of a distributed-memory system. Most of the languages using this method have been based on the proposals of the Parallel Computing Forum (PCF) which has attempted to establish a standard for parallel extensions to Fortran targeted at shared-memory systems; the extensions cover the following constructs:

- PARALLEL DO. This construct describes the parallelism between iterations of a DO loop that may be initiated in any order; this, however, requires that the loop iterations have no data dependencies.

```
PARALLEL DO [<qualifiers>,] <identifier> = <start>, <finish> [,<step>]
  <declarations>
  <executable statements>
END PARALLEL DO
```

For example,

```
REAL A(N), B(N), C(N), D(N)
PARALLEL DO I = 1 N
  A(I) = B(I) + C(I)
  D(I) = A(I) - X
END PARALLEL DO
```

The qualifiers specify when ordering is to be imposed on the execution of loop statements and the number of loop iterations that should correspond to a process.

- PARALLEL SECTIONS. This construct defines a number of sections of code that may be executed in parallel. For example,

```
PARALLEL SECTIONS [<qualifiers>]
  <section specifications>
END PARALLEL SECTIONS
```

The specification of each section takes the following form:

```
SECTION [<section name>] [WAIT(<section names>)]
  <declarations>
  <executable statements>
```

The construct defines a number of sections of code that may be executed in parallel with the WAIT qualifier being used to introduce partial ordering among the sections of the code; that is, it indicates that a section will be initiated only when certain other named sections have completed their execution. Communication between sections is effected through shared variables. For example,

```
REAL Z(4)
PARALLEL SECTIONS(ORDERED)
SECTION A
  P1 = AFUNC(Z(1))
SECTION B
  P2 = 2*AFUNC(Z(2))
SECTION C, WAIT(A)
  P3 = P1 + AFUNC(Z(3))
SECTION D, WAIT(A, B)
  P4 = P1 * P2 + AFUNC(Z(4))
END PARALLEL SECTIONS
```

Partial ordering is introduced in the above code such that section C must wait for the completion of section A, and section D must wait for section A as well as B. Both sections A and B are executed in parallel followed by sections C and D; possible parallelism may also exist between sections B and C, since there is no waiting requirement for section C. Communication between the sections is effected through the shared variables $P1$ and $P2$.

Other means of synchronization can be specified as follows: *lock/unlock* operations applied to GATE variables to provide mutual exclusion of shared data; *critical sections* where only one process can execute the associated portion of code achieving mutual exclusion; and *events,* which are used to signify that some event has occurred, such as communication needed by more than one process. Languages that support the virtual shared-memory method at present includes Fortran-S [9], KSR Fortran [27] and TC2000 Fortran [6].

29.3.5.2 Single program, multiple data parallelism.

The main idea behind data parallelism is that parallel processing is achieved using a single thread of program control operating over a large set of data elements. Essentially, there are two data parallel models, namely, the single instruction, multiple data (SIMD) model, where the same instruction is applied to different sets of data in lock-step synchronous fashion, and the single program, multiple data (SPMD) model, where the same operations are applied in asynchronous fashion to different sets of data. SIMD usually involves array processor systems, and SPMD involves distributed-memory systems. The SPMD model is the one that is currently receiving most attention and is considered further in the following sections.

The performance of an SPMD program depends to a large extent on the mapping of data to the processors in a system as this ultimately determines the amount of communication necessary for the accessing of non-local data. Thus, problem decomposition for SPMD data parallel computation is concerned primarily with specifying data decomposition that subsequently can be executed in parallel; for imperative languages, the basic means of specifying data decomposition would be the array structure. There are two main issues related to data decomposition involving arrays, namely, their *alignment* and *distribution.*

Data alignment is concerned with the alignment of arrays with respect to each other and is directly related to the structure of the underlying computation. It provides a mapping between the data components to be manipulated together which enables the compiler to assign array elements to the same processor. This, in turn, reduces the data transfers necessary between processors for accessing non-local data. Alignment can be based on complete or part array dimensions and involve contiguous or non-contiguous elements; it is essentially independent of the underlying machine architecture.

The direct alignment approach is where the elements concerned are explicitly aligned with respect to each other; this is considered to be a natural means of specifying array alignments for parallel manipulation since it encourages users to program directly in terms of the arrays. Indirect alignments require the use of *template* arrays that provide an abstract indexing space where arrays to be aligned are first declared to be aligned to a template. The consequence of the indirect approach is that once a number of arrays are aligned with respect to the same template, then they can be later realigned onto a processor arrangement using a single statement. It also facilitates program modularity and increased portability since the compiler for different architectures need only adapt its code generation to map templates onto the actual processor configuration. The declaration of a template does not result in any storage being allocated for it since a template represents an abstract entity.

Data distribution is concerned with the assignment of arrays onto the finite resources of the target parallel machine; it must take into account the number of processors available, the individual processor memory sizes, and the interconnection topology employed so that the distribution will minimize data transfers between the processors as well as accomplish load balancing.

Two basic forms of array data distributions are needed; that is, *regular* and *irregular* distributions. Regular distributions are classified into two main types; namely, *block* and *cyclic.* Block distribution involves the allocation of blocks of array elements amongst the processors; this is based on the array size and the number of processors available. Cyclic distribution, on the other hand, involves the round-robin allocation of array elements among the available processors. A variation of these two distributions is a combined form of *block-cyclic* distribution where the array elements are divided into blocks and then allocated cyclically between the processors. Irregular distributions occur when the distribution specification for an array cannot be specified at compile time because it is dependent on the input data or on the values of another array. In such cases, the array is indirectly distributed at run time using a mapping array whose values are used to map the array elements to processors.

High Performance Fortran. Recently, language researchers have proposed imperative languages with abstractions for data parallelism based on sequential constructs. The approach generally adopted is to develop extensions to Fortran that could exploit the full capability of a wide variety of parallel architectures; the languages proposed include Vienna Fortran [45], Fortran D [16] and High Performance Fortran [24], which are based on the SPMD data parallel computation model. Fortran D is a joint development between Rice University (Dallas, Texas) and Syracuse University (Syracuse, New York), and Vienna Fortran involves the University of Vienna (Austria). High Performance Fortran is based on proposals from a forum consisting of a broad coalition of industrial and academic groups and is based on Fortran 90. It is widely considered as a fusion between Fortran D and Vienna Fortran and has become an unofficial standard [24].

In the following sections, salient features related to data parallel computation in High Performance Fortran are examined.

Processor declarations. The processor configuration of the distributed system needs to be conveyed to the compilation system in order that the compiler may generate appropriate SPMD code. This information is required for the compiler to carry out data decomposition, as specified by the program, across the target machine and effect message passing for non-local data accesses.

High Performance Fortran allows the specification of processors in program units through a processor declaration using directive statements; a directive is introduced by the characters !HPF$ in the first columns of a statement as follows:

```
        PARAMETER (P1=10, P2=100)
!HPF$   PROCESSORS DM(P1, P2)
```

The directive introduces the identifier DM as a two-dimensional 1000 processor structure whose scope is local to the program unit in which it is declared. Other processor declarations may be introduced that allow a declared set of processors to be viewed as having a different shape; this is introduced through the VIEW directive as follows:

```
!HPF$   PROCESSORS DM(P1, P2), DMC(P1*P2)
!HPF$   VIEW OF DM :: DMC
```

The VIEW directive designates the processor arrays DM and DMC to be equivalent; that is, they refer to the same set of processors.

The processor array declaration does not imply any actual underlying hardware interconnection topology; its scope is concerned primarily with algorithm design and is limited to the program unit in which it is declared.

Data alignments. High Performance Fortran supports indirect data alignments involving template declarations; these introduce the name, size and dimensionality of the alignment template.

Templates are declared using directives appearing only in the declaration part of a program unit. For example,

```
!HPF$    TEMPLATE VECTOR_TEMPLATE(N)
!HPF$    TEMPLATE, DIMENSION(N, N) :: MATRIX_TEMPLATE
```

Templates that are declared in different program units will be distinct entities, and a template may not be passed as a parameter to a subprogram. The template to which a formal parameter is aligned is distinct from the template to which an actual parameter is aligned; because of its local scope, a template becomes undefined when exit is made from a subprogram.

The major forms of array alignments are explained by using a set of examples illustrating the alignment specifications.

Identical alignment. Identical alignment occurs when there is an exact match of array indices to be aligned. For example,

```
!HPF$    TEMPLATE VECTOR_TEMPLATE(N)
         INTEGER VECTOR_A(N), VECTOR_B(N)
!HPF$    ALIGN WITH VECTOR_TEMPLATE :: VECTOR_A, VECTOR_B
```

The statements effect the following identical alignments

```
VECTOR_A(1) to VECTOR_B(1); VECTOR_A(2) to VECTOR_B(2); ...
     VECTOR_A(N) to VECTOR_B(N)
```

Non-identical alignment. Non-identical alignment occurs when there is an inexact match of array indices to be aligned. For example,

```
!HPF$    TEMPLATE VECTOR_TEMPLATE(N)
         REAL VECTOR_A(N), VECTOR_B(N)
!HPF$    ALIGN VECTOR_A(I) WITH VECTOR_TEMPLATE(I)
!HPF$    ALIGN VECTOR_B(I) WITH VECTOR_TEMPLATE(I+1)
```

Non-identical alignments occur between the elements of vectors VECTOR_A and VECTOR_B due to the addition in the decomposition VECTOR_TEMPLATE of an offset; the alignments effected are

```
VECTOR_A(2) to VECTOR_B(1); VECTOR_A(3) to VECTOR_B(2);
   VECTOR_A(4) to VECTOR_B(3); VECTOR_A(5) to VECTOR_B(4); ...
     VECTOR_A(N) to VECTOR_B(N-1)
```

Other forms of integer subscript expressions may be used, resulting in irregular array alignments.

Permutation alignment. Permutation in the alignment of arrays occurs when the array indices for alignment are arbitrarily changed. These are expressed within the alignment specifications as shown in the following declarations, which effect array transpositions.

```
!HPF$    TEMPLATE MATRIX_TEMPLATE(N, N)
         INTEGER MATRIX_A(N, N), MATRIX_B(N, N)
!HPF$    ALIGN MATRIX_A(I, J) WITH MATRIX_TEMPLATE(I, J)
!HPF$    ALIGN MATRIX_B(J, I) WITH MATRIX_TEMPLATE(I, ,J)
```

Collapsed alignment. Complete array dimensions can be mapped onto a single position involving arrays with non-identical dimension sizes; this corresponds to the collapse of the larger array dimension. For example,

```
!HPF$    TEMPLATE VECTOR_TEMPLATE(N)
         INTEGER MATRIX_A(N, N), VECTOR_A(N)
!HPF$    ALIGN MATRIX_A(I, J) WITH VECTOR_TEMPLATE(I)
!HPF$    ALIGN VECTOR_A(I) WITH VECTOR_TEMPLATE(I)
```

Alignments between matrix MATRIX_A and vector VECTOR_A are effected such that the row elements of MATRIX_A are collapsed and aligned with a single element of VECTOR_A whose index value corresponds to the row number.

Embedding alignment. Arrays may be aligned with larger dimensional arrays whereby the smaller dimensional array is mapped onto the specific dimensions of the larger array; this corresponds to the embedding of the smaller array in the larger array. For example,

```
!HPF$    TEMPLATE MATRIX_TEMPLATE(N, N)
         INTEGER VECTOR_A(N), MATRIX_B(N, N)
!HPF$    ALIGN MATRIX_B(:,J) WITH MATRIX_TEMPLATE(I, J)
!HPF$    ALIGN VECTOR_A(J) WITH MATRIX_TEMPLATE(I, J)
```

Alignments occur between vector VECTOR_A and matrix MATRIX_B such that the elements of VECTOR_A are aligned to only the row elements of MATRIX_B corresponding to row 1. This would correspond to the embedding of vector VECTOR_A into row I of matrix MATRIX_B resulting in the following alignments:

```
VECTOR_A(1) to MATRIX_B(I, 1); VECTOR_A(2) to MATRIX_B(I, 2)
   VECTOR_A(3) to MATRIX_B(I, 3); ...VECTOR_A(N) to MATRIX_B(I, N)
```

Replication alignment. Array elements may be duplicated and aligned with a set of elements from another array; this represents alignments by replication. For example

```
!HPF$    TEMPLATE MATRIX_TEMPLATE(N, N)
         DIMENSION VECTOR_A(N), MATRIX_B(N, N)
!HPF$    ALIGN VECTOR_A(I) WITH MATRIX_TEMPLATE(I, 1 :N) !HPF$
!HPF$    ALIGN MATRIX_B(I, J) WITH MATRIX_TEMPLATE(I, 1:N)
```

The elements of vector VECTOR_A are replicated and aligned with the column elements of matrix MATRIX_B.

The alignment specifications outlined above are categorized as being of type *static* since they are included as program declaration statements. However, alignments can also be specified such that they take on values during execution time; such alignments are referred to as *dynamic* alignments.

Data distribution. Data distribution involves the assignment of array elements to the available processors. The type of distribution specified by a program provides opportunities to minimize communications between processors as well as achieving load balancing.

Distributed arrays can be classified into two categories, namely, *static* and *dynamic*. Statically distributed arrays have distributions that are fixed within the scope of the array declaration. Dynamically distributed arrays have distributions that may be specified and modified during program execution.

Static data distributions. Static data distributions may be classified as regular and irregular distributions based on the distribution pattern involved. The following is an outline of the various forms of static data distributions supported.

1. Regular distributions
   (a) Block distribution. The mapping of evenly sized groups of array elements to processors or templates is referred to as *block* data distributions.

   ```
   !HPF$    PROCESSORS P(4)
   !HPF$    TEMPLATE T(12)
   !HPF$    DISTRIBUTE T(BLOCK)
   ```

   The T elements are divided into evenly sized blocks for assignment to the processor array which is indicated by the identifier BLOCK; the following distributions are effected

Processor	1	2	3	4
	T(1)	T(4)	T(7)	T(10)
	T(2)	T(5)	T(8)	T(11)
	T(3)	T(6)	T(9)	T(12)

   (b) Cyclic distribution. The mapping of array elements to processors in round-robin fashion is referred to as *cyclic* data distributions.

   ```
   !HPF$ DISTRIBUTE T(CYCLIC)
   ```

   The T elements are assigned cyclically to the processor array which is indicated by the identifier CYCLIC; the following distributions are effected

Processor	D	2	3	4
	T(1)	T(2)	T(3)	T(4)
	T(5)	T(6)	T(7)	T(8)
	T(9)	T(10)	T(11)	T(12)

   (c) Block-cyclic distribution. The mapping of array elements to processors that have been divided into evenly sized blocks and then allocated in round-robin fashion is referred to as a *block-cyclic* data distribution.

   ```
   !HPF$ DISTRIBUTE TT(BLOCK, CYCLIC)
   ```

   TT has a distribution pattern in which the rows are divided into evenly sized groups; the divided elements of TT are then mapped onto the processor array in row-wise fashion.

Processor	1	2	3	4
	TT(1,1)	TT(3,1)	TT(1,2)	TT(3,2)
	TT(2,1)	TT(4,1)	TT(2,2)	TT(4,2)
	TT(1,3)	TT(3,3)	TT(1,4)	TT(3,4)
	TT(2,3)	TT(4,3)	TT(2,4)	TT(4,4)

2. Irregular distributions

Irregular distributions are not explicitly supported in High Performance Fortran.

Dynamic data distributions. In High Performance Fortran, dynamic data distributions require the explicit declaration of templates as being dynamic, using program directives as follows:

```
!HPF$ DYNAMIC TT
```

The directive specifics that the template TT can be subject to redistribution during execution time using the REDISTRIBUTEI directive. The redistribution of the TT template such that both dimensions have block distribution features is specified as follows:

```
!HPF$ REDISTRIBUTE TT(BLOCK, BLOCK)
```

- *Parallel constructs.* The specification of parallel loops indicates to the compiler explicitly that the iterations of a loop are independent and therefore different instantiations of the loop body can be executed in parallel without the need for communication.

   A necessary requirement for parallel loops is that there be no inter-iteration dependencies; this makes it possible to execute the iterations in any order. The basic assumption for parallel loops is that when a statement in an iteration accesses a memory location it can only use values defined before the loop or within the current iteration. The explicit specification of parallel loops is useful in situations where the compiler cannot detect the data independence of loop iterations, for example, through the application of indirect array reference using index arrays.

   A parallel loop is specified as a combination of the INDEPENDENT directive and a DO loop as follows:

```
!HPF$ INDEPENDENT
   DO I = 1, N
      A(I) = ...
      ...
      ... = A(I)
   END DO
```

The actual assignment of loop iterations to processors is implementation dependent and users cannot influence the assignment procedure.

29.4   Summary

The topic of parallel computing or high-performance computing is receiving considerable attention in the U.S., Europe, and Japan. It is being hailed as a technology that can make major contributions to the improvement of the quality of life in many sectors of human activity. It is now possible to construct parallel hardware that is both efficient and reliable, and it is confidentially predicted that a teraflop machine will become a reality within a short period of time. However, as with previous developments in computing, it is the software that is proving the most difficult to master, and the widespread use of parallel computing will not occur if this problem is not solved. The situation is further complicated in that there is no widely accepted single model of parallel computation on which software and algorithm designers can base their products. This results in a range of different types of languages being proposed or used to efficiently exploit the underlying parallel hardware.

The imperative language approach is the one that has received the most attention; in particular, Fortran-based languages reflect the initial use of parallel computing in the engineering and scientific fields. In the arena of imperative languages, several different approaches have been developed, chief of which are languages with explicit parallel features and parallelizing compilers that take a sequentially constructed program and detect any inherent parallelism. These language approaches have been developed for machines based on the shared-memory model and on the distributed-memory model. The distributed-memory model is currently receiving considerable attention because it appears to be scalable to higher orders of parallelism in a way that the other model is not. However, the techniques to divide up a program to utilize the underlying hardware are not well understood or efficiently implemented.

To stop the proliferation of extensions to Fortran, an informal group of language designers and implementers, users, and parallel computer vendors produced the High Performance Fortran "standard." High Performance Fortran is based on the SPMD data distribution model, which requires the user to introduce array distribution annotations based on a familiarity with how the data is manipulated; eventually it is hoped that this can be automated.

Another approach is the explicit method of programming using message passing; however, the effort required for implementing the message-passing model can become large due to the inherent complexities of the technique. In addition, the porting of such software between parallel machines is a difficult task due to the close relationship between the language features and the underlying architecture. The SPMD model overcomes such drawbacks by requiring the specification of data distribution annotations. However, a programmer is still required to have some understanding of how the compiler manipulates the program to determine which distribution annotations will enable the compiler to produce an efficient parallel program.

In the long term, if High Performance Fortran and similar languages are to become the dominant method of programming parallel computers, then programming techniques, efficient parallelizing compilers, and appropriate program development tools will need to be provided, these are still areas of ongoing research.

Acknowledgments

The author wishes to express his thanks to Adib Zarea-Aliabadi for help in the development of this chapter. Thanks are also due to the European Commission through their funding of the Esprit Project APPARC, in which much of this work was carried out.

29.5   References

1. Ahuja, S., N. Carriero and D. Gelernter. 1986. Linda and friends. *IEEE Computer,* Vol. 19, No. 8, 26–34.

2. Allen, F., M. P. Charles, R. Cytron and J. Ferrante. 1988. An overview of the PTRAN Analysis System for multiprocessing. *Journal of Parallel and Distributed Computing,* Vol. 5, 617–640.

3. America, P., and T. Pool. 1987. A parallel object-oriented language. In *Object-Oriented Concurrent Programming,* A. Yonezawa and M. Tokoro, eds. Cambridge, Mass.: MIT Press, 100–220.

4. Andre, F., J. Pazat, and H. Thomas. 1990. PANDORE: a system to manage data distribution. *Proceedings ACM Int. Conference on Supercomputing,* Amsterdam.

5. Bal, H. E., M. F. Kaashoek, and A. S. Tanenbaum. 1992. Orca: a language for parallel programming of distributed systems. *IEEE Trans. Software Engineering,* Vol. 18, No. 3, 190–205.

6. BBN. 1989. *TC2000 Technical Product Summary.* Cambridge, Mass.: BBN Advanced Computers Inc.

7. Bennett, J. K. 1987. The design and implementation of distributed Smalltalk. *Proceedings of OOPSLA, ACM SIGPLAN Notices,* Vol. 22, No. 12, 318–330.

8. Hansen, P. B. 1987. Joyce—a programming language for distributed systems. *Software, Practice, Experience,* Vol. 17, No. 1, 29–50.

9. Bodin, F., and T. Priol. 1992. Overview of the KOAN programming environment for the iPSC/2 and performance evaluation of the BECAUSE test program 2.5.1. *Proceedings BECAUSE European Workshop,* Sophia-Antipolis.

10. Burns, A. 1988. *Programming in Occam 2.* Wokingham, England: Addison-Wesley.

11. Clark, K. L., and S. Gregory. 1986. PARLOG: parallel programming in logic. *ACM Trans. Programming Language Systems,* Vol. 8, No. 1, 1–49.

12. Carriero, N., D. Gelernler, and J. Leichter. 1986. Distributed data structures in Linda. *Proceedings 13th ACM Symposium Principles Programming Language,* 236–242.

13. Corradi, A., and L. Leonardi. 1988. The specification of concurrency: an object-base approach. *IEEE Conference on Computers and Communications,* Phoenix, Ariz.

14. Dijkstra, E. W. 1975. Guarded commands, nondeterminacy, and formal derivation of programs. *Communications ACM,* Vol. 18, No. 8, 453–457.

15. Foster, I., and S. Taylor. 1990. *Strand: New Concepts in Parallel Programming.* Englewood Cliffs, N.J.: Prentice Hall.

16. Fox, G., S. Hiranandani, K. Kennedy, C. Koelbel, U. Kremer, C. Tsang, and M. Wu. 1991. Fortran D language specification. Report TR 900079. Dept. of Computer Science, Rice University.

17. Gannon, D., J. Q. Lee, B. Shei, et al. 1992. SIGMA Il: A tool kit for building parallelising compilers and performance analysis systems. *Proceedings of IFIP Working Conference on Parallel Programming Environments for Parallel Computing,* Edinburgh.

18. Gehani, N., and W. D. Roome. 1989. *The Concurrent C Programming Language.* N.J.: Silicon Press.

19. Gelernter, D., and N. Carriero. 1986. Linda on Hypercube multicomputers. *Proceedings of the SIAM Conference,* Texas, 45–55.

20. Halstead Jr., R. H. 1985. Multilisp: A language for concurrent symbolic computation. *ACM Trans. Programming Language Systems,* Vol. 7, No. 4, 501–538.

21. Hoare, C. A. R. 1978. Communicating sequential processes. *Communications ACM,* Vol. 21, No. 8, 666–677.

22. Hudak, P., and L. Smith. 1986. Parafunctional programming: a paradigm for programming multiprocessors. *Proceedings of the 13th ACM Symposium on Principles of Programming Languages,* 243–254.

23. Hudak, P., and P. Wadler, eds. 1988. *Report on the Functional Programming Language Haskell.* Technical Report DCS/RR656, Yale University.

24. HPFF. 1993. High performance Fortran language specification. *Scientific Programming,* Vol. 2, No. l, 1–165.

25. Intel. 1992. *iPSC/840 Fortran System Calls.* Reference manual, order number. 312234–001.

26. Koelbel, C., and P. Mehrotra. 1991. Compiling global name-space parallel loops for distributed execution. *IEEE Trans. on Parallel and Distributed Systems,* Vol. 2, No. 4, 440–451.

27. KSR. 1991. Kendall Square Product Description. Waltham, Mass.: Kendall Square Research Corporation.

28. Matsuoka, S., and S. Kawai. 1988. Using tuple space communication in distributed object oriented languages. *Proceedings OOPSLA, ACM SIGPLAN Notices,* Vol. 23, No. 11, 276–284.

29. Mehrotra, P. and J. Van Rosendale. 1985. The Blaze language: a parallel language for scientific programming. *Parallel Computing,* Vol. 5, No. 3, 339–361.

30. Perrott, R. H., and A. Zarea-Aliabadi. 1986. Supercomputer languages. *ACM Computing Surveys,* Vol. 18, No. 1, 5–22.

31. Perrott, R. H., and A. Zarea-Aliabadi. 1993. Languages for programming distributed memory systems. *IEEE Computing and Control,* Vol. 4, No. 6, 261–268.

32. Perrott, R. H. 1992. Parallel language developments in Europe: an overview. *Concurrency: Practice and Experience,* Vol. 4, No. 8, 589–617.

33. Perrott, R. H. 1987. *Parallel Programming.* Wokingham, England: Addison-Wesley.

34. Polychronopoulos, C., M. Girkar, M. Haghigghat, et al. 1990. The structure of Parafrase 2: an advanced parallelising compiler for C and Fortran. In *Languages and Compilers for Parallel Computing.* Cambridge, Mass.: MIT Press.

35. Roman, G. C., K. C. Cunningham, and M. E. Ehlers. 1988. A shared dataspace language supporting large-scale concurrency. *Proceedings 8th Int. Conference on Distributed Computing Systems.* New York: IEEE, 265–272.

36. Ruhl, R., and M. Annaratone. 1990. Parallelization of Fortran code on distributed memory parallel processors. *Proceedings ACM Int. Conference on Supercomputing,* Amsterdam.

37. Sawdayi, R., G. Wagenbreth, and J. Williamson. 1991. MlMDizer: Functional and data decomposition; creating parallel programs from scratch, transforming existing Fortran programs to parallel. In *Computers and Runtime Software for Scalable Multiprocessors,* J. Saltz and P. Mehrotra, eds. Amsterdam: Elsevier.

38. Shapiro, E. 1987. *Concurrent Prolog: Collected Papers.* Cambridge, Mass.: MIT Press.

39. Shekhar, K. H, and Y. N. Srikant. 1992. Linda sub-system on transputers. *Computer Languages,* Vol. 18, No. 2, 125–136.

40. Sugimoto, S., K. Agusa, K. Tabata, and Y. Ohno. 1983. A multi-processor system for concurrent Lisp. *Proceedings Int. Conference on Parallel Processing,* Belaire, Michigan., 135–143.

41. Turner, D. A. 1985. Miranda: a non-strict functional language with polymorphic types. In *Functional Languages and Computer Architecture,* Lecture Notes in Computer Science. New York: Springer-Verlag.

42. U.S. Department of Defense. 1981. *Programming Language Ada: Reference Manual,* Lecture Notes in Computer Science, 106. New York: Springer-Verlag.

43. Yokote, Y. and M. Tokoro. 1986. The design and implementation of concurrent Smalltalk. *Proceedings OOPSLA, ACM SIGPLAN Notices,* Vol. 21, No. 11, 331–340.

44. Yonezawa, A., J. P. Briot, and E. Shibayama. 1986. Object-oriented concurrent programming in ABCL/1. *Proceedings OOPSLA, ACM SIGPLAN Notices,* Vol. 21, No. 1, 258–268.

45. Zima, H., P. Brezany, B. Chapman, and A. Schwald. 1992. Programming in Vienna Fortran. *Scientific Programming,* Vol. 1, No. 1, 31–50.

46. Zima, H., H. Bast and M. Gerndt. 1988. SUPERB: A tool for semi-automatic MIMD/SIMD parallelization. *Parallel Computing,* Vol. 6, 1–18.

30

Tools for Portable High-Performance Parallel Computing[*]

Doreen Y. Cheng

High-performance application developers often use machine-specific, non-portable programming interfaces, even though many tools have been developed to support application portability. This chapter summarizes the results of a user survey and a tool developer survey and points out the mismatch between user priorities and current tool capabilities. From the user survey, the author has derived three criteria and a list of priorities for evaluating tool support for portable high-performance computing. Guided by the criteria, the author has examined the status of and the issues involved in the design and development of the tools that support portable parallel programming. Unless the tools satisfy these criteria, they are unlikely to be accepted by the real-world application developers. On the other hand, achieving scalable high performance on massively parallel systems is difficult. Tools are not a panacea; users will probably have to take an active role in parallel programming for the foreseeable future.

30.1 Introduction

Today, parallel machines and programming models are evolving rapidly, and there are no clear winners. Therefore, applications frequently must be ported to take advantage of improved hardware and software. To reduce the porting effort, many tools have been developed to support portable parallel programming [54]. However, many users in the high-performance community use machine-specific programming interfaces. This chapter examines the reasons behind this phenomenon and the issues involved in the design and implementation of the tools that support portable parallel program development.

In the sequential world, high-level languages, such as C and Fortran, and their compilers and run-time libraries, provide support for application portability. The meaning of *portability* seems to have been implicitly understood until real-world parallel applications began to be developed. Portability often comes with the price of reduced performance and tool-

---

[*]Work was done when the author was with Computer Sciences Corporation at NASA Ames Research Center.

introduced bugs that affect program correctness. This results in difficult trade-offs for the developers of parallel programs. The implicit understanding of portability is no longer sufficient, and we need criteria for evaluating the tools that support portable, parallel application development.

The criteria for measuring portability can vary from one application domain to another. For example, in high-performance computing, accuracy and performance often take precedence over source code preservation, whereas developers for distributed client-server applications may desire to minimize source code modification even at the expense of performance.

With the intent of understanding user priorities in high-performance computing, the author has surveyed supercomputer users of the Numerical Aerodynamic Simulation (NAS) facility at NASA Ames Research Center and the users of other national laboratories and supercomputing centers. From the results, the author has derived three criteria for evaluating tools that support portable parallel programming: (1) machine-independent semantics, (2) reasonable performance, and (3) machine-independent syntax. Section 30.2 describes these criteria and discusses the difficulties in achieving them.

Guided by the criteria, the author has surveyed the developers of the tools that support portable parallel program development. Sections 30.3 through 30.5 summarize the results and examine the tools that use one of the three common approaches: portable libraries, language-centered tools, and parallelizing compilers and preprocessors. While standardization and careful implementation can help tools achieve machine-independent semantics, difficult research issues must be addressed for the tools to deliver reasonable performance—especially for the tools that support implicit parallelism. But as stated previously, achieving scalable high performance on massively parallel systems is difficult, and users will probably have to take an active role in parallel programming for the foreseeable future.

Because of the limitations in current tools, real-world application developers frequently use low-level tools such as libraries, and many of them have reported good performance. Perhaps high-performance implementations of MPI will provide a reasonable interface to portable high-performance parallel programming. While developers and users of language-centered tools have reported encouraging performance results, more work is needed for these tools to deliver scalable high performance for real-world applications on massively parallel machines. Probably the most attractive alternative in terms of application development is to use parallelizing compilers. However, the performance delivered by current parallelizing compilers is limited by the fact that they only parallelize loops. The author believes that algorithm-level manipulations, currently beyond the state of the art, will be required for these tools to achieve scalable, high, parallel performance.

The tool status reported in this chapter is up-to-date as of June 1994. While the design and implementation of the tools that support portable parallel program development are rapidly improving, the criteria and the user priorities will remain the same for a longer time. The author hopes that the criteria, the user priorities, and the issues examined in achieving machine-independent semantics and reasonable performance will help users evaluate and select the tools, and will help developers make trade-offs in the design and development of such tools.

The following terms will be used to report the performance in this chapter.

- overhead: $(T_t - T_b) / T_b$
- speedup: T_s / T_p
- efficiency: $T_s / (N * T_p)$

where T_t is the execution time of a program written using a tool t, T_b is a base execution time for comparison, T_s is sequential execution time, T_p is parallel execution time, and N is the number of processors used.

30.2   Criteria for Evaluating Portability Support

30.2.1   The criteria

Three criteria for evaluating tool support for portability have been derived from the survey of high-performance application developers:

1. *Machine-independent semantics.* The tool ensures that application programs have the same logical and numerical behavior on all platforms.
2. *Reasonable performance.* The tool allows the user to develop programs with an acceptable level of performance.
3. *Machine-independent syntax.* The tool enables application programs to be compiled on different platforms with an acceptably low level of source code modification.

Of these criteria, the easiest to achieve is machine-independent syntax. All tools considered in this chapter satisfy this criterion, and this issue will not be discussed further.

This chapter focuses only on the needs of scientific computations. In many scientific applications, the order of arithmetic operations must be preserved to maintain numerical accuracy. The survey results show that user requirements range from matching four significant digits to limiting the variation to the order of the machine precision. If the mutual exclusion or message-passing semantics differ from platform to platform, an application may be deadlock-free on one machine yet deadlock on another. The users surveyed unanimously consider such a situation unacceptable. In exchange for portability, the users are willing to accept performance degradation of from 2 to 50 percent compared with a version of their program developed using vendor-specific tools. While some users are willing to modify up to 5 percent of their source code, others want to limit their modifications to no more than a few hundred source lines, regardless of the size of their program.

From the user survey results, the author has defined four user priorities for portability support in a decreasing order:

1. machine-independent semantics to guarantee consistent logical behavior across platforms
2. control over numerical variation caused by nondeterministic concurrent execution to ensure consistent numerical results
3. reasonable performance
4. small amount of source code modifications required

30.2.2   Achieving machine-independent semantics

Naive code translation may cause a program to behave inconsistently on different machines and may result in errors for numerically sensitive programs. This section describes the main issues that must be addressed for tools to avoid these problems.

30.2.2.1 Machine-independent message-passing semantics. Message-passing is the most common programming model used in today's high-performance computing. The dramatic differences between the semantics of the various vendor-supported message-passing facilities can make porting applications very difficult. For example, tagged message passing, used in PVM [40, 75] and NX [66], is difficult to implement in a CSP-like [10] channel-based message-passing system.

Most of today's portable tools support tagged message passing. The discussion in this chapter, therefore, will focus on the different semantics used in this type of message passing and how the differences affect the behavior of programs. The message-passing semantics defined by MPI [32] listed below are used in this chapter.

A send can be blocking or non-blocking, each of which can use four modes: standard, ready, synchronous, and buffered:

- A *blocking send* can start whether or not a matching receive has been posted, and it completes when the send buffer[*] can be reused. These semantics give flexibility for optimization. For example, an implementation can choose to use buffered mode to avoid deadlock, or write the message directly into receiver's buffer whenever possible. However, it gives the user little control over how the message should be sent.
- A *non-blocking send* can start whether or not a matching receive has been posted. It initiates the send operation and returns before the message is copied out from the sender's buffer. A separate call must be made to check its completion. These semantics allow the actual message transmission to occur concurrently with computation. However, its use may make an application more difficult to understand.
- A *standard send* can start whether or not a matching receive has been posted. If the outgoing message is buffered, the send operation completes when the buffer can be reused. If buffer space is unavailable, the send operation completes after a matching receive has been posted and the data has been moved to the receiver.
- A *ready send* can start only if the matching receive is already posted; otherwise the operation is erroneous. It completes when the send buffer can be reused. This mode allows some systems to remove a hand-shake operation that is otherwise required.
- A *synchronous send* can start whether or not a matching receive has been posted. It completes after the send buffer can be reused and the receiver has started the matching receive operation. This semantics eliminate the need of buffering of messages, but it is the users' responsibility to avoid deadlock.
- A *buffered send* guarantees that a message will be buffered in the system if there is no currently matching receive. A buffered send can start whether or not a matching receive has been posted, and its completion does not depend on the occurrence of a matching receive. This mode can avoid deadlock that is possible in certain implementations of standard send, but it may have an adverse effects on performance [44].

A receive can be either blocking or non-blocking:

- A *blocking receive* can be initiated at any tim, and completes after the message is delivered to the receiver's buffer.
- A *non-blocking receive* initiates a receive operation and returns before the message is stored into the user's receive buffer. A separate call is needed to test the completion of the operation.[†] This mode allows the receiver to perform computation concurrently with the message delivery to take advantage of independent communication hardware.

---

[*]The buffer here refers to a user-allocated memory space for storing message contents.

[†]Some systems, such as Charm [82], provide *probe receive*, a non-blocking receive that checks the availability of an incoming message, gets the message length, and receives it if the message is available. If the message is not available, the operation returns a NULL.

Message-passing hardware vendors usually support a combination of the above semantics. Tables 30.1 and 30.2 show examples of the semantics of send and receive operations on the Intel Paragon (NX) and the TMC CM5 (CMMD). A "*yes*" means that the operation supports the semantics, an empty entry means not, and a number leads to the corresponding explanation following the table.

TABLE 30.1   Semantics of Point-to-Point Operations on the Intel Paragon

type	csend	isend/msgwait	crecv	irecv/msgwait
blocking	yes		yes [2]	
non-blocking		yes		yes[3]
standard	yes	yes		
ready				
synchronous				
buffered	1	1		yes [4]

1. If matching receive has been posted, no buffering occurs and the data is delivered directly to the user's receiving buffer. Otherwise, the message is stored in the system buffer of the receiver.
2. If the message is in the receiver's system buffer, it is copied to the user space. Otherwise, the receiving process blocks.
3. Under OSF/1, if the message is in the receiver's system buffer, the msgwait() copies the message into the user space.
4. If a crecv() is called between calls to irecv() and msgwait(), the crecv copies the arrived message out while waiting for its own message to arrive. This means a call to gsync(), which is implemented using crecv() and csend() may cause additional buffering [44].

TABLE 30.2   Semantics of Point-to-Point Operations on the TMC CM5

type	send_block	send_noblock	send_async/msg_wait	receive_block	receive_async/msg_wait
blocking	yes	yes		yes	
non-blocking			yes		yes
standard		yes	yes		
ready					
synchronous	yes				
buffered		1	2	3	3

1. Same as csend except that the message is buffered at the sender side if no matching receive has been posted. The message will be sent concurrently with computation.
2. The message is not sent until the matching receive is posted. No buffering is performed.
3. No buffering in the receiver's side; messages are always delivered to a user-specified area.

The above tables show that the semantics of send and receive are quite different on these machines, and there is no one-to-one correspondence at this level of the interface. The following code fragment in which two processes exchange data shows an example of the impact of these differences on the logical behavior of a program.

```
proc1 {                                      proc2 {
   send(proc2, msg1);                           send(proc1, msg2);
   receive(msg2);                               receive(msg1);
}                                            }
```

The code will not deadlock if the buffered mode is used in send, assuming that there is sufficient space in the buffer. However, it will deadlock if the send is synchronous. If a code translator naively translates the send to a vendor-supported counterpart without enforcing consistent semantics, a deterministic program can run deadlock-free on one machine and deadlock on another. An example of such a naive mapping is to map the send to *csend*, the blocking send on a Paragon and *CMMD_send_block* on a CM5, and to map the receive to *crecv* and *CMMD_receive_block*.

30.2.2.2 Machine-independent numerical semantics. Numerical analysts often take considerable care in choosing the order of operations in order to minimize numerical errors in a sequential program. If non-determinism introduced by parallel execution cannot be controlled, the error caused by the order of execution may eventually exceed the user's tolerance. For example, if n processes $P_1, P_2, \ldots, P_n$ all send data to process P. Process P issues n receives with no order specified and it performs a mathematically commutative and associative operation on the data received. In the worst case, the errors caused by finite precision of computer representations can be magnified, accumulated, and eventually produce wrong results. To avoid such errors, tools must maintain machine-independent numerical semantics. One approach is to allow users to avoid such errors by choosing the execution that preserves the specified order of arithmetic operations.

30.2.3 Achieving reasonable performance

Maintaining good performance while porting a program from one architecture to another is equally or more difficult. Factors that affect performance include application characteristics, system capabilities, and the mapping between them. Examples of application characteristics include granularity (the amount of work done between synchronizations [94]), the frequency of synchronization, and the volume and frequency of data communication. The system capabilities include processing power, synchronization cost, interconnect latency and bandwidth, memory capacity, latency and bandwidth, and the software overhead incurred in concurrency.

   To achieve high performance, a user must make trade-offs and intelligent decisions. The user must partition the application so that the useful computation is large enough compared with the overhead introduced by parallel execution. The user must partition and place the data to maximize locality. In addition, the tasks must be arranged to minimize communication and synchronization costs which, in turn, may require minimizing communication distance and overlapping communication with computation.

   To complicate the matter further, the choice of message-passing semantics can also have a significant impact on performance. Hardware features may support some semantics better than others. For example, the direct remote memory access to (from) user address space on a Meiko CS-2 [9] makes the ready send, the synchronous send, and no-buffering send efficient operations. Special hardware may also be available for synchronization and collective operations as in the case of a CM5, and a CS-2. If hardware supports concurrent message processing (for instance, the communication processor on each Paragon node), using non-blocking operations can give significant performance advantages for applications whose computation can be overlapped with communication. In addition, the particular system balance may also make some semantics better choices than others. When interprocessor communication bandwidth approaches local memory bandwidth, the buffering of a message can dramatically reduce the performance. For example, on the Intel Paragon running OSF/1 R1.1 (without the communication coprocessor), the bandwidth of unbuffered NX messages is about 35 MBps. When the messages are buffered, the bandwidth is reduced to less than 5 MBps.* The bandwidth is further reduced to about 2 MBps when PVM 3.2 is used, since this version of PVM buffers messages at least twice [44]†. In general, for a tool to achieve high performance, it must make the same trade-offs and decisions that users make. Naive code generation generally produces poor code; sophisticated code analysis, evaluation, and transformation techniques must be employed for acceptable performance. However, many of these techniques are still active research topics [2, 22]. In many cases, it is desirable for tools to allow the users to guide them in making their decisions. For example, a user may choose synchronous mode send operations to eliminate buffering on the machines where buffering messages is an expensive operation.

Sometimes, different algorithms and/or implementation strategies may be required for efficient use of different architectures. For simplicity, the discussions in this chapter assume that the same algorithms and implementation strategies are used when a program is ported from one machine to another.

The rest of this chapter will describe three common approaches that support portable parallel program development: portable libraries, language-centered tools, and parallelizing compilers and preprocessors.

30.3 Portable Message-Passing Libraries

Distributed-memory parallel systems usually provide libraries that support the message-passing programming model. These machine-specific libraries usually have a unique combination of syntax and semantics, making application programs non-portable. To reduce programming effort, portable libraries have been developed to hide vendor specifics in a unified message-passing interface [74]. It is hoped that modifications to application source code can be significantly reduced once the library is ported. This section discusses the portability support provided by these libraries.

30.3.1 Example tools

Portable libraries provide routines that create and terminate processes, send and receive messages, and perform collective operations. Portable libraries include APPL [41], Chameleon [20], Express [63], MPI [32], p4 [4], PARMACS [61], PICL [16], PVM [75], and TCGMSG [21].

The functions supported by these libraries are quite different, as shown in the following two tables. The first row of the table lists the names of the tools in the alphabetic order. The second row indicates how a tool supports process creation. An "*I*" means implementation

---

*When the communication coprocessor is used under OSF/1 R1.2, the bandwidth of unbuffered NX messages is increased to 90 MBps, and the bandwidth of buffered messages is 70 MBps.

†PVM 3.3 has achieved 19.7 MBps for 80 kB messages on the Paragon [38].

dependent, a "*D*" stands for *dynamic* and means that processes can be spawned at any time during an execution, and an "*S*" stands for *static* and means that all processes must be created together, usually at the beginning of program execution. In the case of Express, all processes running on a parallel machine must be created at the beginning of the program, but processes on different machines can be spawned dynamically. The other rows show whether the tool provide a particular function. A "*yes*" means that the function is supported, and a blank entry means not. The row "*process group*" shows whether the tool supports dynamic creation and management of process groups. The row "*safe message passing*" indicates whether the tool supports safe message passing (the meaning of which will be described later in this section). The row "*global operation*" shows whether the tool supports operations participated by a group of processes, where the "+" sign means that the operations can be performed based on process groups, and the row "*virtual topology*" indicates whether the tool provides facilities for the user to specify process topology (e.g., mesh or torus) and the mapping between processes and processors.

TABLE 30.3   Functions Provided by Portable Message-Passing Libraries

function	appl	chameleon	express	mpi	p4
process creation	S	S	S, D	I	S
process group		yes	yes	yes	
safe message passing				yes	
global operation	yes	yes	yes	yes+	yes
virtual topology		yes	yes	yes	
parallel I/O		yes	yes		

TABLE 30.4   Functions Provided by Portable Message-Passing Libraries

function	parmacs	picl	pvm	tcgmsg
process creation	S	S	D	S
process group	yes		yes	
safe message passing				
global operation	yes	yes	yes	yes
virtual topology	yes			
parallel I/O			yes	

While most of these libraries provide trace facilities, some provide additional tools to aid debugging and performance optimization. For example, the trace generated by a PICL program and a PARMACS program can be visualized using versions of ParaGraph [64], the performance of a program written using Chameleon or p4 can be tuned using Upshot [72], and a PVM program can be composed graphically using HeNCE [65] and monitored using XPVM [75]. Express provides tools such as ASPAR for automatic parallelization, NDBtool for debugging, and Vtool, Ctool, Etool, and Xtool for performance tuning.

The first-generation libraries have made it easier for users to develop parallel programs. User experience and developer experience with them have formed the foundation for the standardization of message-passing interfaces. The experience has also revealed three main weaknesses in these libraries: lack of safe message-passing, lack of flexible group operations, and lack of support for parallel I/O, where safe message-passing means that message exchange in one part of a program is guaranteed not to interfere with the message exchange in other parts of the program unless such interference is intentional. The next paragraph explains the need for safe message passing and flexible group operations.

Applications often need to call subprograms, such as numerical solvers, and perform global operations, such as global summation. These subprograms may be developed as libraries by different organizations. In these cases, messages used in the subprograms must not unintentionally interfere with the messages used in the other parts of the program. The pseudo code segment listed below is extracted from a common pattern used in many application programs running on the NAS parallel systems [83]. It demonstrates how the interference of message passing in different parts of the program may result in errors. Consider eight processes connected in a ring, all executing the this code:

```
process i:
    while (not done) {
        compute;                                            (0)
        exchange boundary data with right_neighbor;         (1)
        exchange boundary data with left_neighbor           (2)
        perform a global operation;                         (3)
    }
```

To achieve high performance, it is a common practice for the user to use a receive that matches any incoming messages in steps (1), (2) and (3), i.e.,

```
        receive (fromAnyProcess, withAnyTag);
```

Assuming that, for some reason, process 4 has been lagging, and it is in step (0). Process 5 is waiting for process 4 to send its boundary data at step (2). Now process 1 has finished steps (1) and (2) and broadcasts a message in step (3). One possibility is that the data broadcast in step (3) has the same shape as the boundary data. In this case, Process 5 would successfully receive the message from process 1 instead of process 4, and it uses it as the boundary data for further computation. Obviously, this will lead to incorrect results. Furthermore, this kind of intermittent, time-dependent error can be very difficult to reproduce and locate. However, none of the first-generation libraries can guarantee that this kind of error will be prevented. Applications also frequently require organizing processes into groups and performing global operations over only the members of a particular group. The first-generation libraries do not support such needs, either.

To address these problems, the library developers and parallel computer vendors have formed a forum and drafted the first version of the Message Passing Interface standard (MPI) [32]. Having recognized the difficulties in standardizing process control and parallel I/O, MPI focuses on the standardization of process communication, synchronization, and group operations. MPI uses a tool-managed tag, or *context*, to divide a communication domain into non-interfering subdomains. A message sent in one context can be received only by processes in the same context, thus guaranteeing non-interfering message passing. It supports the management of hierarchical, possibly overlapping process groups and global operations participated by the processes of a particular group. In addition, it supports specifications of the virtual topology of processes and the mapping of processes to processors. Currently, four versions of MPI implementation are available: one developed at Argonne National Laboratory and the Mississippi State University, one developed at the Ohio Supercomputer Center, one at IBM, and one at Edinburgh Parallel Computing Centre. Although they are currently in the early testing stage, the wide support from vendors as well as research institutes has made MPI a good candidate tool for portable, high-performance computing.

30.3.2   Performance

Most portable libraries support a large number of parallel machines, such as the Intel Paragon, the Delta, the iPSC/860, the TMC CM5, the KSR-1, and the nCube, as well as networks of workstations. It is easier for the libraries to achieve good performance than language-centered tools and parallelizing compilers, owing to their simplicity and their close relationship to low-level primitives. As a result, many real-world applications developed using these libraries have shown from reasonable to very good performance:

1. The most impressive example is the 1992 Gordon Bell Prize winner, a solver for large sparse linear systems written in Chameleon [25].
2. A conjugate gradient program developed using PARMACS [6] has less than 3 percent overhead on a 32-node iPSC/860 and a 128 node nCube-2. A red-black relaxation program using the same tool and running on the same machines has an overhead of less than 15 percent.
3. A 2-D Euler code developed using APPL [41] has less than 4 percent overhead on the 16-node iPSC/860.
4. A communication-intensive kernel of Parallel Spectral Transform Shallow Water Model code developed using PICL is just as fast as a version using NX [38].

The libraries designed for MPPs usually add from a few percent to 24 percent overhead to native point-to-point communication primitives. For example, Chameleon adds virtually no run-time overhead since its macro-based communication interface is inlined at compile time. On the Paragon (OSF/1) using blocking send and receive, PICL adds less than 4 percent overhead to passing messages of up to 1000 bytes between neighboring processors and about 3 percent overhead to exchange of such messages. PICL adds 8 to 15 percent overhead if non-blocking send and receive are used [16]. In a 16-processor ring test on the iPSC/860 and the Delta, APPL adds less than 24 percent overhead for 8-byte messages, and less than 4 percent for messages longer than 2 kB. On the iPSC/860 and the nCube 2, PARMACS adds 15 to 20 percent overhead for short messages and less than 5 percent for messages longer than 256 bytes [6]. On the iPSC/860, Express adds 16 to 24 percent overhead for messages shorter than 100 bytes and less than 7 percent for longer messages. It adds 1 to 8 percent overhead for exchanging a message of less than 100 bytes and 15 to 22 percent for longer messages [1].

While good point-to-point communication performance is clearly necessary, being able to scale the performance to a large number of processors and to complex communication patterns is also critical. Section 30.2.3 has shown that message-passing semantics used by a library can significantly affect the performance of an application. The following results indicate that the impact of semantics on performance may increase with increasing number of processors and with more complex message-passing patterns. An experiment [11] comparing several tools shows that while TCGMSG is 50 percent faster than tools like p4 and PVM 2.4.1 in two-node communication, it is only 20 percent faster in a four-node ring communication. Its performance reduces to about the same as the slowest package for a communication-bound molecular dynamics code. The main reason indicated in Ref. [11] is that TCGMSG supports only synchronous communication and, therefore, it cannot take advantage of overlapping communication and computation.

Another important factor that impacts scalability is the semantics and implementation of global operations, such as synchronization, broadcast, and reduction. Compared with native support on the iPSC/860, Express adds 50 to 58 percent overhead for messages passed through a 4-node ring and 100 percent for an all-to-all broadcast [1]; the overhead increases when messages become larger and more processors are involved. This experience has revealed the importance of taking advantage of the interconnect topology on certain architectures. The experience of Charm users has demonstrated the advantage of using non-blocking semantics in global operations. Supporting non-blocking global reduction operations in the Charm run-time system has resulted in constant execution time for such operations on up to 64 processors, whereas the required time increases almost linearly in a native version [52]. Unfortunately, most of the libraries that support global operations do not perform such optimizations.

The early versions of packages that are primarily designed for networks of heterogeneous workstations, such as PVM, have significantly eased parallel programming in a heterogeneous environment. The complexity of providing a uniform system image has understandably taken precedence over performance in these early versions. As a result, they have demonstrated good performance only for applications with little or no communication. Unsatisfactory performance has been observed for applications that require communication [47, 49]. When using these tools for parallel processing on a network of workstations, the communication cost is high compared with the achievable network capacity. For example, in the initial port of the five kernels of the NAS Parallel Benchmarks onto a network of workstations, 61 to 91 percent of the execution time was spent in communication [49]. On MPPs, PVM3.2's high overhead makes it unsuitable for programs with any communication. Fineberg compared the performance of PVM3.2 and NX on the Paragon using point-to-point message passing, broadcast, and synchronization [14]. Messages of 1 byte to 1 MB were used in the point-to-point measurements. The results have shown that PVM3.2 is 4 to 20 times slower than NX, and the overhead added by PVM3.2 increases linearly with the size of a message. As a result, the highest effective bandwidth using PVM3.2 is less than 1/15 of the NX counterpart. The time required by PVM3.2 to broadcast a 1 kB message on 2 to 208 processors ranges from 15 to 36 times the time required by NX, and the time required for synchronization is 12 to 300 times longer.

The importance of performance has become evident when increasing number of real-world application developers start to use PVM-like tools. The developers have therefore devoted significant amount of effort to improving performance. For example, an optimized version of PVM3.2 has been developed that has achieved a 1.2 to 5 times performance increase over the standard PVM3.2 for the five kernels of the NAS parallel benchmarks [49]. This version does not perform packing and unpacking for every message, assuming that most messages have a single data type and are sent from a single data area. It does not perform data representation transformation, assuming that many machines use a standard data representation. It also bypasses the network transport layer by implementing a subset of transport functions within the PVM. These optimizations have been absorbed in the newly released version PVM3.3. The initial performance results measured on two SS1+ connected through Ethernet shows that for messages of more than 8 kB, PVM3.3 has achieved more than 84 percent of the sustained Ethernet bandwidth. When passing messages of size ranging from 2 bytes to 131 kB between two neighboring nodes of the Paragon, PVM3.3 has achieved latency and bandwidth comparable to NX [38].

30.3.3 Machine-independent semantics of portable libraries

Portable libraries are normally implemented in terms of machine-specific primitives. Enforcing exact semantic equivalence is difficult, owing to vastly diversified semantics supported by vendor-defined message-passing interfaces. The semantics of many libraries was determined by early vendor-defined interfaces such as NX. Library developers have found it difficult to enforce chosen semantics on machines that support very different semantics, especially when trade-offs between high performance and semantic consistency must be made. When this happens, one approach is not to support the particular library calls on these machines or to report an error. For example, the PICL developer chose to treat the use of non-blocking send on an nCube as an error to avoid performance penalties. Another approach is to emulate the chosen semantics if they are not directly supported. For example, Chameleon emulates non-blocking buffered sends on the IBM SP-1, and PARMACS uses mailboxes to allow matching sender to be used for selecting incoming messages. Emulation can give a program a higher probability of consistent logical behavior. For example, supporting buffering semantics has helped Chameleon eliminate the most common source of deadlock. However, emulation often reduces performance. For example, allowing receivers to select a messages by sender accounts for a significant portion of the 20 percent overhead [6] added by PARMACS in point-to-point communication [38].

For numerically sensitive applications, it is often desirable to maintain the user-specified order of arithmetic operations, but this has been largely neglected by most libraries. One source of numerical inconsistency is the possible non-determinism in global operations. While some libraries do not support these operations, those that do support them merely provide global operations as convenience functions. In other words, these libraries do not optimize the performance of global operations, nor do they support numerical consistency. Chameleon is the only exception; it provides a version that preserves the order of operations without exploiting associativity.

A major source of inconsistent logical behavior is the unsafe message passing in first-generation libraries. As shown in Section 30.3.1, messages passed in different parts of a program may interfere with each other, owing to the varying order of events during concurrent execution. This interference, in turn, may produce erroneous results, which may vary from time to time and from platform to platform. The birth of MPI is a major step toward supporting portable parallel computing. However, a good implementation of MPI must enforce the defined semantics. For any third party to be able to enforce the semantics, vendors must make their semantic definition available if they support interfaces other than MPI. (Today, some vendors do not provide adequate documentation for message-passing semantics.) In addition, a good MPI implementation should also give the user choices to avoid possible numerical errors in collective operations.

30.3.4   Summary—portable libraries

The following two tables summarize the status of portable libraries presented in this section. The first row lists the names of the tools in the alphabetic order. The second, third, and fourth rows show the platforms of each architecture on which a tool currently runs. The meaning of the letter code is listed after the tables. These entries are based on the list of the machines given by the developers. The fifth row indicates whether a tool supports networks of workstations. A "*yes*" means the tool does support computing on networks of workstations, and a blank means it does not. The sixth row shows the type of applications that have been developed using the tool. A "*real*" means real-world applications have been developed using the tool (see references listed in the *Performance* sections). The seventh row indicates whether the performance achieved using the tool has been compared with the performance achieved by versions of the same application using native primitives. A "*native*" means the comparison has been performed, whereas a blank means it has not. The last row shows whether there are additional tools to assist program development using the library, A "*deb*" means a debugger is available, a "*perf*" means performance tools are available, "*par*" indicates that parallelization tools are available, and "*ge*" means that a graphical editor is available. We can see from the tables that these libraries have been ported to many platforms and have been used in developing real-world applications. Furthermore, their performance has been compared with the native version programs.

TABLE 30.5   Summary—Portable Libraries

	appl	chameleon	express	p4
shared memory	A,SG	A,B,K,R,S,X,	HP,IB,SG, SU,T3,X,Y	A,B,K,R,S,X,
distributed memory	D,G,N,SP	5,D,G,N,O,P,SP	5,D,G,I,N,SP	5,D,G,N,O,P,SP
SIMD				
network of workstations	yes	yes	yes	yes
application	real	real	real	real
performance comparison	native	native	native	native
other tools		perf	par, deb, perf	perf

TABLE 30.6   Summary—Portable Libraries

	parmacs	picl	pvm	tcgmsg
shared memory	C,HP,IB,K,S,G,SU,T3,Y	T	A,E,HP,IB,K,S,SG,SU,V	A,K,R,V
distributed memory	5,G,I,N,O,P, SP,V	D,I,G,N,P	5,G,I,O,P	D,G
SIMD				
network of workstations	yes	yes	yes	yes
application	real	real	real	real
performance comparison	native	native	native	native
other tools		perf	ge, deb, perf	

The meaning of the letter code in the tables is listed in the alphabetic order, and SM means shared-memory:

2:	TMC CM2		5:	TMC CM5
A:	Alliant		B:	BBN Butterfly
C:	CRAY C90		D:	Intel Delta
E:	Encore		G:	Intel iPSC/860
H:	Dash		I:	Intel iPSC/2
IB:	IBM SM		K:	KSR-1
M:	MasPar		N:	nCube
O:	Meiko CS-2		P:	Intel Paragon
R:	Ardent		S:	Sequent
SG:	SGI SM		SP:	IBM SP-1
SU:	SUN SM		T:	Cogent
T3:	CRAY T3D		V:	Convex
X:	CRAY X-MP		Y:	CRAY Y-M

The table below indicates the availability of the portable libraries on the platforms. The platforms are listed in the alphabetic order. An "x" indicates that the tools is available on the system, and a blank means it is not.

TABLE 30.7   Portable Libraries on Platforms

platform	appl	chameleon	express	p4	parmacs	picl	pvm	tcgmsg
alliant	x	x		x			x	x
Ardent		x		x				x
butterfly		x		x				
c90					x			
cm2								
cm5		x	x	x	x		x	
cogent						x		
convex					x		x	x
cs-2		x		x	x		x	
dash								
delta	x	x	x	x		x		x
encore							x	
hp sm			x		x		x	
ibm sm			x		x		x	
ipsc/2			x		x	x	x	
ipsc/860	x	x	x	x	x	x	x	x
ksr		x		x	x		x	x
maspar								
ncube	x	x	x	x	x	x		
paragon		x		x	x	x	x	
sequent		x		x			x	
sgi sm	x		x		x		x	
sp-1	x	x	x	x	x			
sun sm			x		x		x	
t3d			x		x			
x-mp		x	x	x				
y-mp			x		x			
net wks	x	x	x	x	x	x	x	x

In Table 30.7, *net wks* means networks of workstations.

30.4   Language-Centered Tools

Many researchers believe that providing libraries is not a viable long-term solution for machine-independent parallel programming because library-based approaches require the users operate at too low a level and therefore are error prone and tedious. By using modified languages, on the other hand, compilers can automatically manage concurrency. One advantage of the language-based approach is that languages can define the semantics to preclude some hazards caused by parallel execution. Another advantage is that languages can be designed to capture user-specified information necessary for better performance. This section presents the encouraging results of these tools and valuable experiences of their developers.

30.4.1   New Languages

Ideally, programs should be written expressing only application algorithms, not implementation issues dictated by hardware and system software. The parallelism inherent in an application should be available to the tools so that the potential parallelism can be utilized to take advantage of system resources.

Functional languages [53], logical programming languages [45], and data flow languages [59, 60] have more of the above desirable characteristics than conventional programming languages. Of these languages, Sisal [31] is the best known in high-performance computing community. Sisal is a functional language for parallel numerical computation. It provides constructs to express scientific algorithms in a form close to their mathematical formulation. When using Sisal, the user does not specify program parallelism explicitly; the compiler automatically determines the parallelism in the program.

These languages usually impose limitations such as requiring that program elements have no side effects. While these restrictions make compiler analysis and transformations much easier, they often make it difficult to express commonly needed functions such as I/O, and they require the users to change their programming habits. In addition, these restrictions cause programs to use significantly more memory than conventional languages. Since memory is still a precious resource for many applications, techniques must be developed for the compilers to optimize memory usage in addition to optimizing performance [5, 18]. As a result, developing compilers for these languages can be a long and costly process.

After much effort, the developers of Sisal have reported examples with equal or better performance compared with the Fortran counterparts on the CRAY Y-MP and the SGI shared-memory machines using up to 8 processors [31].[*] However, the performance on message-passing machines remains to be seen.

A group of users at NASA Ames Research Center has indicated that for them to adopt a new language, the performance should be at least 5 to 10 times the performance of the current languages, or the development time should be 1/2 of what is required by the current practice [38]. Owing to the high cost and high risk involved, vendors will support a new language only if many users request it. On the other hand, many users will only try a new language if there are mature vendor-supported compilers. The above situation has hindered the application and the development of these languages and has made evolutionary approaches more practical.

30.4.2   Tools for Extending Sequential Languages

One evolutionary approach is to extend existing sequential languages such as Fortran, C, and C++. The first set of extensions were added by computer vendors for expressing task creation, termination, communication, and synchronization. While these early activities have provided valuable experience in developing compilers, the proliferation of these

---

[*]Sisal compilers are currently being developed for IBM and Tera computers, and Sisal applications are being developed for pharmaceutical and oil companies [38].

extensions has made writing portable programs difficult and kept many users away from using them. In addition to many research language extensions, parallel language standards, such as Fortran90 [12] and High-Performance Fortran (HPF) [24], have been proposed. Since Fortran90 contains extensions that are not specifically for parallel programming, this chapter will consider only the subset of Fortran90 that has gained increasing acceptance from parallel machine vendors (see Table 30.8) and users, and *Fortran90-* is used to denote this subset.

Many tools that support extended languages have been developed for many different architectures [76, 84, 85]. The tools reported in this section are the ones that have been used in developing real-world high-performance applications, and the application performance has been compared with hand-coded versions of corresponding programs.

30.4.2.1   Data parallel languages.

Many researchers and users have found that proper management of data locality is critical to achieving high application performance. For this reason, language extensions have been developed to express data parallelism. Users can use these extensions to specify how data is to be used in a segment of code, how arrays should be aligned with respect to each other, and how they should be distributed to processor memories. This section presents the functions and performance results of the tools that support data parallelism.

Example Tools. Fortran90- supports implicit parallelism by allowing basic Fortran77 operations and functions to operate on array-valued operands. It provides intrinsic procedures to manipulate and construct arrays, and to perform gather and scatter operations. In addition, it supports extended computational capabilities involving arrays (for example, dot product and matrix computations). The compilers partition the arrays, distribute them onto multiple processor memories, and generate concurrent code for computations using these arrays. Recognizing the difficulties for the compilers to achieve high performance, vendors often provide directives for users to specify data partitioning and distribution.

HPF is a result of a standardization effort participated by many vendors and research institutes. It adds extensions to Fortran90. It provides directives for specifying array alignment and distribution to increase data locality, and for specifying user assertions on code properties to aid code transformations. In addition, it provides directives for specifying processor arrangement to assist task distribution. Research activities (such as FortranD [13] and Fortran90D [3] and their compilation systems) are being conducted to develop compiler and tool technologies. Commercial products such as Forge 90 [15] are also available.

Jade [86] provides constructs for specifying how a program written in a standard sequential, imperative programming language accesses variables (for example: read, write, or both). Its compilation system preserves serial semantics and a single address space while distributing data to multiple processors and managing concurrency, communication and synchronization.

pC++ [30] adds data-parallel extensions to C++. It supports the notion of a collection class for describing data structures such as vectors, arrays, matrices, grids, trees, dags, and other aggregates. These aggregate data structures can be distributed over processor memories in a parallel system. Parallel operations on collections can be generated either by the concurrent application of a class method over the entire aggregate or by the application of a parallel operator associated with the structure of the collection itself. pC++ allows specification of data alignment and distribution as in HPF, and provides a collection library that supports Fortran90-style arrays, array operations, and parallel matrix computations. In addition, it provides profiling and analysis tools for performance optimization.

Performance. Commercial compilers for data parallel languages have been available on the single instruction, multiple data (SIMD) machines for several years [77] and have delivered good performance for some applications [43]. Many potentially useful features from them have become part of Fortran90 and HPF. However, achieving scalable high performance on multiple instruction, multiple data (MIMD) distributed-memory machines requires further research effort in sophisticated compiler analysis and transformation techniques. The three examples below show the preliminary performance results.

Computation kernels have been compiled using FortranD and Fortran90D on the iPSC/860. The difference in performance compared with hand-optimized code has been reported to range from a few percent to a factor of two [23, 3]. Their experience indicates that more sophisticated techniques are required for optimizing programs with complicated communication patterns. Performance results of HPF programs compiled using Forge 90 are presented in Section 30.5.2.

Simulation programs, programs of the SPLASH parallel benchmark suite [89], and programs of the Harwell-Boeing sparse matrix test set [92] have been implemented using Jade. The speedup of a Jade program reported below was obtained on 32 processors and was compared with the version implemented using a conventional sequential language. The Jade program that evaluates forces and potentials in a system of liquid water molecules achieved a speedup of 26.3 on the iPSC/860, and a speedup of 27.5 on the Stanford DASH machine. The program that computes a velocity model of the geology between two oil wells achieved a speedup of 28.9 on the iPSC/860, and a speedup of 27.4 on the DASH. The program that simulates the interaction of electron beams with solids [93] achieved a speedup of 27.9 on the iPSC/860, and a speedup of 31.2 on the DASH. The Jade version of the Ocean program achieved a speedup of 9.34 on the DASH machine, and did not perform well on the iPSC/860. The Jade program that factors a sparse positive definite BCSSTK15 [91] achieved a speedup of 5.0 on the DASH machine, but no speed up on the iPSC/860. The developer attributed the poor performance of the last two programs on the iPSC/860 to small grain size compared with the overhead of communication and task distribution.

Two kernels of the NAS parallel benchmarks [67] (EP and CG) have been implemented using pC++ on shared-memory machines (the KSR-1, the BBN/TC-2000 and the Sequent) and distributed-memory machines (the CM5 and the Paragon) [30]. EP achieved good performance on both types of machines. On the CM5, the performance is within 10 percent of the best published and manually optimized Fortran results. The performance of CG on the 256-node CM5 matched the performance of a Y-MP/1 untuned Fortran code. On the Paragon, however, CG performed poorly. The developers attributed the poor performance to intensive communication and high latency in an early version of the Paragon system software. To improve performance, they indicated the necessity of increasing data locality on shared-memory machines and overlapping communication with computation on distributed-memory machines.

30.4.2.2 Parallel object-oriented languages.

Proponents of object-oriented programming claim that object-oriented languages provide more support for information hiding and software reuse than conventional languages. To explore whether object orientation can help reduce the difficulty of programming parallel machines, sequential languages have been extended to support parallel objects. This section describes tools that support these languages.

Example Tools. In addition to pC++ described above, there are several packages that support portable parallel programming in C++. Mentat [19] adds *Mentat classes* to C++ to allow users to specify objects whose instances can be executed in parallel. Parallelism is encapsulated by parallel implementation of a member function and concurrent execution of member functions. Mentat uses the macro data flow model [55] as its run-time computation model. Its compiler automatically detects data and control dependencies between Mentat class instances, and manages invocation, communication, and synchronization of objects. A tool, *MentatView,* is provided to visualize the queue size, processor load, and object-to-processor mapping.

Object-oriented Fortran [35] extends Fortran77 to support declaration, creation, and management of objects. (It does not support strong typing.) Parallelism is implicitly encapsulated in concurrent invocations of object member functions on different object instances. Its run-time kernel[*] automatically manages concurrent execution of object instances and the communication and synchronization between them.

Charm [82] provides extensions to C for concurrent objects. In addition to supporting message passing among concurrent objects, it supports data parallelism by providing constructs for declaring specifically shared variables. Charm supports a message-driven scheduling strategy for latency tolerance and employs user-specified message priorities and automatic load balancing strategies for performance optimization. A C++ based version of Charm called Charm++ [26], which provides inheritance and polymorphism, and a Fortran based version of Charm called fortCharm, are also available. A graphical editor, *Dagger* [81], is provided for constructing Charm programs, and a tool, *Projections* [79], is available for performance analysis.

Performance. Tools that support object-oriented parallel programming are relatively straightforward to create. As a result, there are quite a few of them, including several that have generated code for multiple architectures. Some of them have demonstrated reasonable to good performance as shown in the first two examples below.

Applications in biochemistry, genetic algorithms, image processing, finite element analysis, and computer aided design have been developed using Mentat and run on the Intel Paragon, the iPSC/860, the iPSC/2, and a network of workstations [62, 73]. For a DNA sequence comparison program on the iPSC/2, performance comparable to the hand-coded version has been obtained.

CFD applications have been developed in Object-Oriented Fortran on the Intel Delta, the iPSC/860, and the CRAY Y-MP. These applications have demonstrated near linear speedup over the sequential versions [87]. Performance results of the NAS Parallel Benchmarks indicates an average of about 50 percent overhead compared with native versions on a 32-node iPSC/860. The developers attributed the overhead mostly to the extra copy operations required [88].

A molecular dynamics program, *EGO*, and VLSI CAD applications have been developed using Charm. The CAD applications[†] achieved speedups of from 3.7 to 7.2 on 8 Encore processors, from 4.6 to 7.9 on 8 Sequent processors, and from 2.5 to 7.4 on 8 iPSC/

---

[*]A C++ interface to the same rum-time kernel is also available [95].

[†]The time reported excluded I/O.

860 processors [78]. Small kernels have been developed using Charm++ on the CM5, the nCube-2, and the Sequent Symmetry [26]. A speedup of 19 on 64 CM5 processors was reported for a 64×64 Jacobi solver, a speedup of 25 was achieved for the Primes program, and a speedup of 20 was achieved for a Travel Salesman Problem program.

30.4.2.3 Functional, dataflow, and task parallel languages. Often, invocations of subprograms can be executed concurrently as long as the dependencies between these subprograms are satisfied. This type of parallelism is often referred to as *task* or *functional* parallelism and is used to encapsulate large grain of computation in concurrent subprograms. Hybrid models are often used to support this type of parallelism; sequential languages (usually C and Fortran) are used to express the computation in a component, and a new language is used to specify the interaction between the components. The components are usually forbidden to have side effects.

Example tools. PCN [39, 96] uses a logic-based language for constructing parallel programs from simpler components. The system preserves the properties of the components. For example, deterministic compositions of deterministic components are guaranteed to be deterministic. It provides single-assignment *definitional* variables for communication and synchronization and provides constructs for specifying the mapping of computation to physical processors. In addition to compilers, a source-level debugger, *PDB*, a profile collection and visualization tool, *Gauge*, and a performance analysis tool and visualization tool, *Upshot*, are available.

CODE2 [34] uses a language based on the concept of large-grain data flow for composing parallel programs from subprogram components. Graphical tools are provided for specifying the composition. CODE 2 users can use the graphical constructs *node* to represent a graph or a call to a sequential component, and *arc* to indicate dataflow between the connecting nodes. In addition, it provides tools for specifying attributes of the nodes and arcs, for example, specifying the condition (firing rule) under which the subprogram represented by a node can start execution.

GLU [17] uses a high-level language derived from the dataflow language *Lucid* [57] and the visual language *Operator Nets* [58] to integrate computational components written in an imperative language such as C. GLU programs implicitly express data, pipeline, and tree parallelism in addition to functional parallelism. GLU users can specify a program as either a structured set of equations or an equivalent visual dependency graph that operate on multidimensional data streams. GLU compilers decide on an appropriate partitioning, mapping, and load balancing strategy and generate target-specific communication and synchronization code. The execution model is demand-driven, meaning an operator (computational component) is executable if its required values are available and its output has been demanded.

TOPSYS [56] provides a graphical user interface to support the message-based programming model. Its Multiprocessor Multitasking Kernel (*MMK*) uses objects such as *tasks* for concurrency, *mailboxes* for communication, and *semaphores* for synchronization. Each task can be written in C, C++, or Fortran. Predefined operators can be used to manipulate these objects, and its compiler translates the explicit parallelism in a TOPSYS program into native primitives on each platform. A debugger, *DECTOP*, a performance analyzer, *PATOP*, and a thread-interaction animation tool are provided to assist program development.

PYRROS [100] is a software tool for automatic task scheduling and code generation on message-passing, distributed-memory architectures. It uses a task graph language to specify macro-dataflow computation. Each task can be written in C or Fortran. The system provides an automatic scheduling mechanism to map tasks onto the given processors by eliminating unnecessary communication, overlapping computation with communication and balancing processor load. Tools are provided to predict the performance of parallelization and display task execution schedule.

FortranM [97] adds extensions to Fortran77 for dynamic creation and destruction of *processes* and communication *channels*. Processes can encapsulate state and communicate by sending and receiving messages on channels; references to channels, called *ports*, can be passed as arguments or transferred between processes in messages, providing a restricted global address space. The compiler can guarantee deterministic execution, even in programs with dynamic process and channel structures. In addition to supporting the modular construction of large parallel programs, it supports the development of libraries implementing other programming paradigms. Current libraries allow the integration of message-passing programs, distributed data structures, and HPF programs into a task-parallel frame-work [101]. An interface to the Pablo performance tools [102] provides performance visualization capabilities.

Performance. Although optimizing performance is one of the design goals of these languages and tools, some of them have been applied only to computational kernels and do not provide a comparison to the implementation of the same applications using native facilities.

Applications in fluid dynamics, atmospheric modeling, oil reservoir modeling, protein structure prediction, and genetic algorithms have been developed using PCN and run on the iPSC/860, the Delta, the Sequent Symmetry, and a variety of workstations [39]. A full-scale icosahedral grid solver achieved 2.5 GFLOPS on 492 Delta processors, with efficiency of 75 percent relative to the pure Fortran code running on a single i860 processor.

A block triangular solver and an N-body particle simulation program developed using Code 2 and run on the Sequent Symmetry has achieved performance within a few percentage points of the performance of hand-written versions [34].

A structural stress analysis package and a cubic image convolution program have been developed using GLU for the CM5, multiprocessor workstations, and networks of workstations [106]. The parallel cubic image convolution program ran as fast as a version using PVM on workstation networks. On a network of 16 Sun Sparcs, a 512×512 Laplacian relaxation program has achieved a speedup of about 11, and a 1200×1200 LU decomposition [17] program has achieved a speedup of 5, compared with the single-processor performance.

Programs for discrete event simulations, database management, multigrid methods, and VLSI layout and placement have been developed using Topsys [48] on a network of workstations and the iPSC/860. It was found that the performance on iPSC/860 was not satisfactory for programs that frequently exchange short messages. One reason is that the setup time for an MMK communication is about nine times of what is required in NX [38].

Computation kernels such as LU factorization, matrix multiplication, Laplace equation solver, and an n-body particle simulation program have been developed using PYRROS for the nCUBE-2 and the iPSC/860 [99]. The performance for dense matrix computation and Laplace equation solver is comparable to that of an optimized hand-written parallel program. The n-body program achieves near-linear speedups for 1 million particles on 128 nCUBE-2 processors. The scheduler of the system has also been used in for solving sparse triangular systems on CM-5 [98].

Applications in airshed modeling, computational biology, and computational chemistry have been developed using FortranM. On the IBM SP1, FortranM communication performance is within 10 to 15 percent of vendor libraries. The performance is competitive with p4 and is roughly three times faster than PVM2.4 for larger messages, with both running over TCP/IP, and about 50 percent faster for smaller messages. An untuned Hartree-Fock computational chemistry code achieved performance on the IBM SP1 within ten percent of a non-portable hand-coded implementation [38].

30.4.2.4 Virtual Shared-Memory Extensions for Conventional Languages. Shared-memory is considered to be an easier programming model than message passing. Several tools have extended sequential languages to provide a virtual shared memory on distributed-memory systems. This section presents two of them.

Example Tools. Linda [27] provides the *tuple space* as a virtual shared repository for tasks and data. It provides operations such as *rd* (read), *in*, and *out* to deposit tuples into and retrieve them from the tuple space, and provides the operation *eval* for creating processes that evaluate tuples. On a system with a distributed memory, the Linda compiler partitions the tuple space and distributes it to processor memories. If a tuple being accessed resides in a remote processor memory, the compiler will generate code to pass messages. For Linda programs to achieve good performance on distributed-memory systems, a compiler must maximize data locality by optimizing tuple space partitioning and placement in addition to reducing expensive searches in the tuple space.

Split-C [104, 105] provides parallel extensions to C. *Global pointers* are introduced for accessing remote memory locations, while traditional pointers are used for accessing local memory. The *split-phase assignment* is introduced to allow overlapping of communication with computation. A *get* (*put*) operation initiates the remote access and the computation performs a *sync* to ensure the completion of the remote access. The *signaling-store assignment* operator signals the processor which owns the updated location that the write operation has occurred. Split-C uses an *active message*[*] layer [103] on distributed memory machines and provides a graphical debugger for program development.

Performance. Applications have been developed using Linda on the Intel iPSC/860, the iPSC/2, the CM5, the nCube, the Sequent Symmetry, and multiprocessor workstations, as well as networks of workstations. A 2D FFT program achieved 95 percent of the performance of a native version on the iPSC/2 and the iPSC/860. The performance data shows that Linda imposes a long latency on distributed-memory systems [28]. When moving data around a ring of ten processors on the IBM/SP-1, Linda adds 400 percent overhead compared with EUI, the IBM native message-passing interface, for a one-byte message. The overhead reduces to 15 percent for a 1 MB message. In a multi-token ring test, the one-byte message overhead is reduced to 99 percent, and Linda performs better than EUI by 10 to 13 percent for messages longer than 260 kB. The results of the Shallow Water benchmark using up to 64 processors show that Linda adds less than 3 percent overhead on the iPSC/2 and less than 12 percent on the iPSC/860. However, the overhead increases with the number of processors used on both machines. The performance difference of running the ping-pong tests and Shallow Water benchmark on a network of workstations is less than 5 percent compared with using PVM.

---

[*]An active message differs from a traditional message in that an active message contains the address of a handler which will deposit the message in the receiver's address space or provide a quick reply.

Programs for cancer cell simulation, human genome matching, as well as numeric and non-numeric kernels, have been developed using Split-C [38]. A matrix multiplication program (256 256) has achieved 413 MFLOPS on 1024 CM5 processors [105].

30.4.3 Machine-independent semantics for language-centered tools

Many languages define semantics to preclude some potential problems of parallel processing. For example, by preserving sequential semantics in Jade, and guaranteeing deterministic execution in Fortran90, HPF, and FortranM [21], these languages avoid hazards such as deadlocks. Sometimes, additional restrictions are imposed for this purpose. For example, HPF requires the pure functions to have no side effects, and Fortran M requires communication to occur only along the channels explicitly declared and created by users. (Fortran M provides additional constructs for expressing user-intended non-determinism for potential performance advantages).

Many language-centered tools translate a program into a base language, such as C or Fortran, plus machine-specific primitives. On a distributed-memory machine, these tools must generate code to pass messages between processes. One approach used is to directly invoke vendor-specific send and receive primitives whenever a message-passing is required. This approach has to enforce consistent message-passing semantics at every calling site. Most tools hide vendor-specific semantics in a portable interface. Read and write operations, remote procedure calls, and a portable message passing layer are a few examples of such portable interfaces. Careful implementation is still necessary to ensure the semantics specified for the interface. Providing machine-independent semantics can be difficult when the chosen semantics are quite different from the native semantics supported by a machine. For example, enforcing the semantics of the message-passing layer of Mentat onto the CM5 has been difficult [38]. Section discusses in more detail the enforcing of consistent semantics defined by a portable message-passing layer.

30.4.4 Summary—language-centered tools

Tables 30.8 through 30.10 summarize the status of language-centered programming tools presented in this chapter. The first row lists the languages in the alphabetic order. The meaning of the other entries are the same as in Section 30.3.4. The new entry in the sixth row, "*kernel*," means only kernels and small applications have been developed using the tool. The parenthesis indicates that the entry applies only to the indicated architecture. The "-" sign means a subset of the language has been implemented.

Tables 30.11 and 30.12 indicate the availability of the language-centered tools on the platforms. The platforms are listed in the alphabetic order. An "x" indicates that the tools is available on the system, and a blank means not. The "*net wks*" means networks of workstations.

30.5 Parallelizing Compilers and Preprocessors

Providing tools to parallelize sequential programs is another approach to supporting portability. These tools analyze the control and data flow of a sequential program, analyze dependencies between loop iterations, and transform loops to eliminate the dependencies. If possible, they parallelize loops by inserting compiler directives and necessary code for correctness and concurrency.

TABLE 30.8   Summary—Language-Centered Tools

	charm	code2	fortran90-	fortran M	glu
shared memory	A,E,S	S	C,K,Y	SG,SU	SU
distributed memory	5,G,I,N,P,SP	G	5,G	P,SP	5
SIMD			2,M		
network of workstations	yes	yes	yes	yes	yes
applications	real	kernel	real (SIMD), kernel (MIMD)	real	real
performance comparison		native	native	native	
other tools	ge, perf	ge	deb, perf (SIMD)	deb, perf	

TABLE 30.9   Summary—Language-Centered Tools

	jade	hpf-	linda	mentat	o.o. fortran	pc++
shared memory	SG	T3	E,S	SG,SU	SG,SU,Y	B,K,S
distributed memory	H,G	G	5,G,I,N,SP	5,G,I,P	D,G	5,O,P,SP
SIMD						
network of workstations		yes	yes	yes	yes	yes
applications	kernel	kernel	real	real	real	real
performance comparison	native	native	native	native	native	native
other tools			deb, perf	deb, perf	ge	perf

TABLE 30.10   Summary—Language-Centered Tools

	pcn	pyrros	sisal	splitC	topsys
shared memory	S		A,B,C,E,IB,S,SG,V,X,Y	SU	
distributed memory	D,G,I,SP	G,N		5,P,SP	G
SIMD					
network of workstations	yes		yes	yes	
applications	real	kernel	real	kernel	real
performance comparison		native	native		native
other tools	deb, perf	perf	deb	deb	deb, perf

30.5.1   Example tools

Parallelization tools can be classified into two types: *batch tools* (also known as *automatic parallelizing compilers*) and *interactive tools*. When using batch tools, a user manually inserts directives into their source files, compiles the code using the tools, and inspects the output after the entire file has been processed. Hardware vendors often provide batch parallelization tools (e.g. *fpp* [68] and *cmax* [70]). However, each of these tools only generates code for the platforms offered by the vendor. Batch tools provided by third-party software vendors, such as KAP [69], support multiple platforms and thus support the development of portable parallel programs.

Interactive parallelization tools provide assistance to and accept guidance from a user while parallelizing a loop. They typically provide profiling and browsing facilities in addi-

TABLE 30.11   Language-Centered Tools on Platforms

platform	charm	code2	fortran 90	fortranM	glu	hpf	jade	linda	mentat
alliant	x								
Ardent									
butterfly									
c90			x						
cm2			x						
cm5	x				x			x	x
cogent									
convex									
cs-2									
dash							x		
delta									
encore	x							x	
hp sm									
ibm sm									
ipsc/2	x							x	x
ipsc/860	x	x				x	x	x	x
ksr			x						
maspar			x						
ncube	x							x	
paragon	x		x	x					x
sequent	x	x						x	
sgi sm				x			x		x
sp-1	x		x	x				x	
sun sm				x	x				x
t3d						x			
x-mp									
y-mp			x						
net wks	x	x	x	x	x	x		x	x

tion to analysis tools. For example, when obstacles to parallelization are detected, Forge 90 [15] immediately informs the user by displaying the reasons along with the source code. The user can then use the database browsing tools to examine whether the causes of the obstacle should be of real concern. Tools such as ParaScope [37] also guide users to choose code transformations and estimate whether the transformation would improve performance.

30.5.2   Performance

Performance of the code generated by automatic parallelization tools *fpp* and *KAP* has been evaluated on a dedicated CRAY Y-MP using 1 to 8 processors [7]. The execution time was compared with the time required by the best performing vector code running on a single Y-MP processor. Twenty-five benchmarks, representing common algorithms used at

TABLE 30.12   Language-Centered Tools on Platforms

platform	o.o. fortran	pc++	pcn	pyrros	sisal	splitC	topsys
alliant					x		
Ardent							
butterfly		x			x		
c90					x		
cm2							
cm5		x				x	
cogent							
convex					x		
cs-2		x					
dash							
delta			x				
encore					x		
hp sm							
ibm sm					x		
ipsc/2			x				
ipsc/860			x	x			x
ksr		x					
maspar							
ncube				x			
paragon		x				x	
sequent		x	x		x		
sgi sm	x				x		
sp-1		x	x			x	
sun sm	x					x	
t3d							
x-mp					x		
y-mp	x				x		
net. wks	x	x	x		x	x	

NAS, were employed in the evaluation. Using only default options, it was found that only six codes which had been heavily vectorized achieved a speedup of more than 2 on 4 processors. The maximum speedup was below 3 on 4 processors and below 4.5 on 8 processors. Six of them even slowed down by from 2 to 35 percent. An independent study indicated that after 1 to 3 person-months effort of inserting directives to guide *fpp*, speedup on 8 processors for two highly vectorized programs was increased to 5 [38]. It was found that interactive tools add no significant performance advantage over automatic tools. On the contrary, when interactive tools do not optimize performance for a particular platform, the overhead added for concurrency management can significantly reduce the performance [7].

Performance of 16 benchmarks parallelized using the interactive tool Forge 90 and run on the Paragon using up to 32 processors has been reported by the vendor [50]. The speed-ups of eight programs increased when more processors were used. Using 32 processors, one of these eight programs achieved 93 percent efficiency, three achieved 50 to 58 percent efficiency, and the other four achieved 26 to 33 percent efficiency. The rest of the programs gained little or no performance improvement after from four to eight processors. The execution time of one program actually increased with the number of processors. The vendor attributed the poor performance of some programs to intensive communication and small granularity. After changing the array allocation strategy from allocating only the portion of the array that is distributed to the processor to allocating the entire array on all processors, the performance of five programs has been improved by a factor of 1.1 to 5.6. An independent evaluation on the iPSC/860 at NAS also found the execution time to increase with the number of processors [38].

The performance deliverable by a parallelization tool is determined by three factors: the amount of parallelism that exists in a program, the amount of parallelism that can be discovered by the tool, and the characteristics of the parallel system on which the application is to run. The main reason for the low scalability of the performance on the Y-MP is that most of the parallelism in the program discovered by the tools is consumed by the vector processors, leaving very little for execution on multiple processors.

The scalability of the performance deliverable by parallelization tools is limited by the fact that these tools only parallelize loops. The loops in dusty decks typically are not structured for efficient parallel processing. An analysis [8] of the loops in five typical NAS applications has shown that only 64 to 90 percent of loops are parallelizable; up to 62 percent of them consume less than 1 percent of the execution time. Small to medium loop sizes severely limit the achievable performance when only loops are parallelized. Worse yet, all the benchmarks contain serial loops[*] whose bounds are proportional to the problem size, which makes it impossible to improve performance by simply increasing the problem size.

30.5.3   Machine-independent semantics for parallelizing preprocessors

Parallelization tools provided by hardware vendors only support their own machines. Most research and third-party tools only support a small number of shared-memory machines. For distributed-memory machines, the best implementation approach for parallelizing preprocessors is to isolate the message-passing semantics in a portable interface. The tools must then enforce the defined semantics the same way as portable libraries described in Section 30.3.3.

30.5.4   Summary—parallelizing preprocessors

Table 30.13 summarizes the status of parallelization tools presented in this section. The meanings of the entries are the same as in Section 30.3.4. The new entry *"se"* means that a structure and text editor is provided, a blank means not supported, and *"NA"* means *not applicable*, since these tools are native tools.

Table 30.14 indicates the availability of parallelization tools on the platforms. The entries mean the same as other similar tables.

---

[*]These are the loops that contain serial computation such as reduction.

TABLE 30.13   Summary of Parallelizing Preprocessors

	cmax	fpp	kap	forge 90	parascope
shared memory		C,X,Y	K,Y	T3,Y	S,Y
distributed memory	5			5,D,G,P	5,G,P
SIMD				2	
network of workstations				yes	
applications	real	real	real	real	real
performance comparison	NA	NA	native	native	
other tools		deb, perf		perf	se, deb

TABLE 30.14   Parallelizing Preprocessors on Platforms

platform	cmax	fpp	kap	forge 90	parascope
alliant					
Ardent					
butterfly					
c90		x			
cm2				x	
cm5	x			x	x
cogent					
convex					
cs-2					
dash					
delta				x	
encore					
hp sm					
ibm sm					
ipsc/2					
ipsc/860				x	x
ksr			x		
maspar					
ncube					
paragon				x	x
sequent					x
sgi sm					
sp-1					
sun sm					
t3d				x	
x-mp		x			
y-mp		x	x	x	x
net wks				x	

30.6   Conclusion

While a large number of tools have been developed to support portable parallel program development, many application developers today use machine-specific, non-portable programming interfaces. This is because of the mismatch between user priorities and current tool capabilities. Two surveys were conducted to understand the issues involved: a user survey and a tool developer survey.

From the results of the user survey, the author has derived three criteria and a list of priorities for evaluating tools that support portable parallel programming. The three criteria are machine-independent semantics, reasonable application performance, and machine-independent syntax. The user priorities in a decreasing order are guaranteeing consistent logical behavior across platforms, controlling numerical variation caused by non-deterministic concurrent execution, delivering reasonable performance, and limited source code modification.

Guided by the criteria, the author has evaluated the current tool support for portable parallel programming, and examined the issues involved in design and development of the tools for portable parallel programming. While standardization and careful implementation of tools can begin to address user priorities 1 and 2, there are difficult research issues that must be solved for the tools to deliver high performance, especially for the tools that support implicit parallelism. Unless the tools satisfy these criteria, they are unlikely to be accepted by real-world, high-performance application developers. On the other hand, achieving scalable high performance on massively parallel systems is difficult. Tools are not a panacea; users will probably have to take an active role in parallel programming for the foreseeable future.

Because of the limitations in current tools, real-world application developers frequently use low-level tools such as libraries, and many of them have reported good performance. Perhaps high-performance implementations of MPI will provide a reasonable interface to portable high-performance parallel programming. While developers and users of language-centered tools have reported encouraging performance results, more work is needed for these tools to deliver scalable high performance for real-world applications on massively parallel machines. Probably the most attractive alternative, in terms of application development, is to use parallelizing compilers. However, the performance delivered by current parallelizing compilers is limited by the fact that they parallelize only loops. The author believes that algorithm level manipulations, currently beyond the state of the art, will be required for these tools to achieve scalable, high parallel performance.

While the design and implementation of the tools that support portable parallel program development are rapidly evolving, the criteria and the user priorities will remain the same for longer time. The author hopes that the criteria, the user priorities, and the issues examined in achieving machine-independent semantics and reasonable performance would help users evaluate and select the tools, and would help developers make trade-offs in the design and development of such tools.

Acknowledgment

This work would not be possible without the help of application and tool developers who responded to my survey and reviewed this chapter. I would like to express my appreciation to all of them. I would like to thank James Cownie of Meiko for helpful discussions. I would like to express special thanks to Dr. Jeffrey Deutsch of Deutsch Research for his insightful comments and suggestions about the organization and presentation of this chapter.

30.7   References

1. Ahmad, I, M. Y. Wu, J. Yung, A. Ghafoor. 1994. *A Performance Assessment of Express on the iPSC/2 and iPSC/860 Hypercube Computers*. Technical report, 94-08, Dept. of Computer Science, State University of New York at Buffalo, and Syracuse University, Syracuse, N.Y.

2. Amarasinghe, S., and M. Lam. 1993. Communication optimization and code generation for distributed-memory machines. *Proceedings of the SIGPLAN'93 Conference on Parallel language Design and Implementation*, SIGPLAN Notices 28(7).

3. Bozkus, Z., A. Choudhary, G. Fox, T. Haupt, and S. Ranka. 1993. Fortran 90D/HPF compiler for distributed memory MIMD computers: design, implementation, and performance results. *Proceedings of Supercomputing'93*, Portland, Oregon.

4. Butler, R., and E. Lusk. *Users' Guide to the P4 Programming System*. Tech. report. Argonne National Laboratory ANL-92/17, 1992.

5. Cann, D. Retire Fortran? a debate rekindled. 1992. *Communications of the ACM*, Vol 35, Number 8.

6. Calkin, R., R. Hempel, H. Hoppe, and P. Wypior. 1994. Portable programming with the PARMACS message-passing library. Special issue on message--passing interfaces, *Parallel Computing*, Vol. 20.

7. Cheng, D., and D. Pase. 1991. An evaluation of automatic and interactive parallel programming tools. *Proceedings of Supercomputing'91*, Albuquerque, N.M.

8. Cheng, D. 1992. Evaluation of Forge: an interactive parallelization tool. *Proceedings of Hawaii International Conference on System Sciences*, Vol. II.

9. Barton, E., J. Cownie, and M. McLaren. 1994. Message Passing on Meiko CS-2. *Parallel Computing*, Vol. 20, No. 4.

10. Hoare, C. A. R. 1985. *Communicating Sequential Processes*. Englewood Cliffs, N.J.: Prentice Hall.

11. Douglas, C., T. Mattson, and M. Schultz. 1994. Experimental results for data sharing on high performance multiprocessors. *Proceedings of the NSF LSU Mardigras teraFLOPS Conference*, Baton Rouge, La.

12. 1991. *Information technology—programming languages—Fortran*. International Standard, Ref. no. ISO/IEC 1539: 1991(E).

13. Fox, G., S. Hiranandani, K. Kennedy, C. Koelbel, U. Kremer, C. Tseng, and M. Wu. 1991. *Fortran D Language Specification*. Tech. Report, COMP TR90-141 (Rice Univ.) and SCCS-42c (Syracuse Univ.), Dept. of Comp. Sci.

14. Fineberg, S. A. 1994. *The Map Library—A Flexible Group Mechanism for Message Passing Systems*. NAS RND tech report, 1993, and its revision, 1994.

15. Applied Parallel Research. 1992. *Forge 90 Version 8.0 Baseline System Users' Guide*.

16. Geist, G., M. Heath, B. Peyton, and P. Worley. 1990. *PICL, A Portable Instrumented Communication Library, C Reference Manual*. Tech report ORNL/TM-11130.

17. Jagannathan, R. 1994. Coarse-grain dataflow programming of conventional parallel computers. In *Advanced Topics in Dataflow Computing and Multithreading*. New York: IEEE Computer Society Press.

18. Gopinath, K., and J. Hennessy. 1989. Copy elimination on functional languages. *Proceedings of ACM Symposium on Prin. of Prog. Lang.*

19. Grimshaw, A. 1993. Easy to use object-orient parallel programming with Mentat. *IEEE Computer* (May), 39–51.

20. Gropp, W., and B. Smith. 1993. *Chameleon Parallel Programming Tools Users Manual*. Technical report ANL-93/23, Argonne National Laboratory.

21. Harrison, R. 1991. *Moving Beyond Message Passing: Experiments with A Distributed-Data Model*. Tech. report, Argonne National Laboratory.

22. Hiranandani, S., K. Kennedy, and C. Tseng. 1992. Compiling Fortran D for MIMD distributed-memory machines. *Communications of ACM*, 35(8):66-80.

23. Hiranandani, S., K. Kennedy, and C. Tseng. 1993. Preliminary experiences with the Fortran D compiler. *Proceedings of Supercomputing'93*, Portland, Oregon.

24. Koelbel, C., D. Loveman, R. Schreiber, G. Steele, and M. Zosel. 1993. *The High Performance Fortran Handbook*. Scientific and Engineering Computation Series. Cambridge Mass.: MIT Press.

25. Jones, M., and P. Plassmann. 1992. Solution of large, sparse systems of linear equations in massively parallel applications. *Proceedings of Supercomputing'92*, Minneapolis, Minn., 551–560.

26. Kale, L., and S. Krishnan. 1993. Charm++: A Portable Concurrent Object-Oriented System Based On C++. *Proceedings of the Conference on Object Oriented Programming Systems, Languages and Applications*.

27. Carriero, N., and D. Gelernter. 1989. Linda in Context. *CACM*, 32(4):444-458.

28. Carriero, N., D. Gelernter, T. Mattson, and A. Sherman. 1994. The Linda alternative to message-passing systems. *Parallel Computing*, Vol. 20, 633–655.

29. Ashcroft, E. A., A. Faustini, R. Jagannathan, and W. W. Wadge. 1994. *Multidimensional Declarative Programming*. Oxford University Press.

30. Malony, A., B. Mohr, P. Beckman, D. Gannon, S. Yang, and F. Bodin. 1994. Performance analysis of pC++: A portable data-parallel programming system for scalable parallel computers. *Proceedings of IEEE International Parallel Processing Symposium*, Cancun, Mexico.

31. McGraw, J. 1993. Parallel functional programming in Sisal: fictions, facts, and future. *Proceedings of Advanced Workshop: Programming Tools for Parallel Machines,* Otranto, Italy.

32. Message Passing Interface Forum. 1994. *MPI: A Message-Passing Interface Standard.* Computer Science Dept. Technical Report CS-94-230, University of Tennessee, and *International Journal of Supercomputer Applications,* Volume 8, Number 3/4.

33. MPI Specification, Chapter 1, Version of Jan. 1994.

34. Newton, P., and J. Browne. 1992. The CODE 2.0 Parallel programming language. *Proceedings of ACM International Conf. on Supercomputing.*

35. Reese, D., and E. Luck. 1991. Object-oriented Fortran for portable, parallel programs. *Proceedings of The Third IEEE Symposium on Parallel and Distributed Processing*, 608–615.

36. Gannon, D., S. Yang P. Bode, V. Menkov, and S. Srinivas. 1994. Object-oriented methods for parallel execution of astrophysics simulations. *Proceedings of the NSF LSU Mardigras teraFLOPS Conference,* Baton Rouge, La.

37. Kennedy, K., K. McKinley, and C. Tseng. 1991. Interactive parallel programming using the ParaScope editor. *TOPDS*, Vol. 2 No. 3, 329–341.

38. Bergeron, R., M. Merriam, T. Pulliam, A. Grimshaw, P. Luksch, D. Reese, I. Stockdale, P. Worley, J. Feo, I. Foster, D. Culler, and V. Sunderam (private communication).

39. Foster, I., and S. Taylor (to appear). A compiler approach to scalable concurrent program design. *ACM Trans. Parallel Programming Languages and Systems.*

40. Sunderam, V. 1990. PVM: a framework for parallel distributed computing. *Concurrency Practice and Experience*, 2(4):315–339.

41. Quealy, A., G. Cole, R. Blech. 1993. Portable programming on parallel networked computers using the application portable parallel library (APPL). *NASA TM 106238.*

42. Rinard, M., and M. Lam. 1992. Semantic Foundations of Jade. *Record of the 19th Annual ACM Symposium on Principles of Programming languages,* 105–118.

43. Sabot, G., L. Tennies, A. Vasilevsky, and R. Shapiro. 1991. Compiler parallelization of an elliptic grid generator for 1990 Gordon Bell prize. *Proceedings of Supoercomputing'91*, Albuquerque, N.M., 338–346.

44. Saphir, W. 1994. *The Effect of Message Buffering on Communication Performance on Parallel Computers.* NAS RNS Tech. Report.

45. Shapiro, E. 1986. Concurrent Prolog: a progress report. *IEEE Computer.*

46. Shadid, J. N., and R. S. Tuminaro. 1991. Iterative methods for nonsymmetric systems on MIMD machines. *Proceedings of the Fifth SIAM Conference on Parallel Processing for Scientific Computing*, Houston, Tex., 123–129.

47. Smith, M. H., and J. Pallis. 1993. MEDUSA—An overset grid flow solver for network based parallel computer systems. *Proc. AIAA-93-3312CP*, Orlando, Fla.

48. Bemmerl, T., and A. Bode. 1991. An integrated tool environment for programming distributed memory multiprocessors. *Distributed Memory Computing, Lecture Notes in Computer Science*, ed. A. Bode, Vol. 487. New York: Springer-Verlag, 130–142.

49. White, S., A. Alund, and V. Sunderam. 1994. *Performance of the NAS Parallel Benchmarks on PVM Based Networks.* NAS Technical. Report RNR-94-008.

50. 1993. Applied parallel research. *Marketing Material* (March).

51. Foster, I., R. Olson, and S. Tuecke. 1993. *Programming in Fortran M*. Tech. Report, ANL-93/26, Argonne National Laboratory.

52. Kale, L., and A. Gursoy. 1994. Performance benefits of message driven executions. *Proc. Intel User Group Conference.*

53. Hudak, P. 1989. Concept, evolution, and application of functional programming languages. *ACM Computing Surveys,* 21(3):359–411.

54. Cheng, D. Y. 1993. *A Survey of Parallel Programming Languages and Tools.* Technical Report RND-93-005, NAS Division, NASA Ames Research Center.

55. Grimshaw, A. 1993. *The Mentat Computation Model—Data-Driven Support for Dynamic Object-Oriented Parallel Processing.* Computer Science Technical Report, CS-93-30, University of Virginia.

56. Bemmerl, T., C. Kasperbauer, M. Mairandres, and B. Ries. 1991. *Programming Tools for Distributed Multiprocessor Computing Environments.* Tech. report, Technische universitat Munchen, Germany.

57. Wadge, W., and E. Ashcroft. *Lucid, the Dataflow Programming Language.* New York: Academic Press, 1985.

58. Ashcroft, E., and R. Jagannathan.1986. Operator nets. In *Fifth Generation Computer Architectures,* J. Woods, ed. Amsterdam: North-Holland, 177–202.

59. Ackermann, W. B. 1982. Data flow languages. *Computer* (February), 15–24.

60. Sharp, J. A., ed. 1993. *Data Flow Computing: Theory and Practice.* Ablex Publishing Co.

61. Hempel, R. 1991. The ANL/GMD macros (PARMACS) in Fortran for portable parallel programming using the message passing programming model. *User's Guide and Reference Manual,* version 5.1.

62. Weissman, J., A. Grimshaw, and R. Ferraro. 1994. Parallel object-oriented computation applied to a finite element problem. *Scientific Computing,* Vol. 2 No. 4, 133–144.

63. Parasoft Corporation. 1992. Express C: User's Guide 3.0, Reference 3.0, Release Notes 3.2.

64. Heath, M., and J. Etheridge. 1991. Visualizing the performance of parallel programs. *IEEE Software,* Vol. 8, No. 5, 29–39.

65. Dongarra, J., G. Geist, R. Manchek, K. Moore, R. Wade, and V. Sunderam. 1992. HeNCE: graphical development tool for network-based concurrent computers. *Proceedings of the Scalable High Performance Computing Conference.* IEEE Computer Society Press, 129–136.

66. Intel Corporation. 1992. *Paragon OSF/1 Fortran System Call Reference Manual.*

67. Bailey, D., J. Barton, T. Lasinski, and H. Simon, eds. 1993. *The NAS Parallel Benchmarks.* NASA Technical Memorandum 103863. Moffett Field, Calif.: Ames Research Center.

68. Cray Research, Inc. 1990. *CF77 Compiling System Volume 4: Parallel Processing Guide.*

69. Kuck & Associates, Inc. *KAP/CRAY User's Guide.* Champaign, Ill.: Kuck & Assoc.

70. Thinking Machines Co. cmax *man page.*

71. Beguelin, A. 1993. *Xab: A Tool for Monitoring PVM Programs.* Technical report CMU-CS-93-105, Carnegie Mellon University.

72. Herrarte, V., and E. Lusk. 1991. *Studying Parallel Programming Behavior with Upshot.* Technical report, ANL-91/15, Argonne National Laboratory.

73. Grimshaw, A., E. West, and W. Pearson. 1993. No Pain and Gain!—Experiences with Mentat on Biological Application. *Concurrency: Practice & Experience,* Vol. 5, No. 4, 309–328.

74. Hempel, R., A. Hey, O. McBryan, and D. Walker. 1994. Special Issue: Message Passing Interfaces. *Parallel Computing* (April).

75. Geist, A. G. A., A. L. Beguelin, A. J. J. Dongarra, A. W. Jiang, A. R. J. Manchek, and A. V. S. Sunderam. 1994. *PVM: Parallel Virtual Machine—A Users Guide and Tutorial for Network Parallel Computing.* Cambridge, Mass.: MIT Press.

76. Brandes, T. 1992. *Effective Data Parallel Program without Explicit Message Passing for Distributed Memory Multiprocessors.* GMD Technical Report, TR92-4.

77. Thinking Machine Co. 1991. CM Fortran Release Notes.

78. Ramkumar, B., and P. Banerjee. 1994. ProperCAD: a portable object-oriented parallel environment for VLSI CAD. *IEEE Trans. on CAD* (June).

79. Kale, L. V., and A. B. Sinha. 1993. Projections: A Scalable Performance Tool. *Proc. Parallel Systems Fair, International Parallel Processing Symposium,* Los Angeles, Calif.

80. Sinha, A. B., and L. V. Kale. 1993. A Load Balancing Strategy for Prioritized Execution of Tasks. *Proc. International Parallel Processing Symposium,* Newport Beach, Calif.

81. Gursoy, A., and L. V. Kale. 1994. Dagger: combining the benefits of synchronous and asynchronous communication styles. *Proc. International Parallel Processing Symposium,* Cancun, Mexico.

82. Fenton, W., B. Ramkumar, V. Saletore, A. B. Sinha, and L. V. Kale. 1991. Supporting machine independent programming on diverse parallel architectures. *ICPP* (August).

83. Saphir, W. 1993. Collective communication in PVM3. *Distributed Computing in Aerospace Applications.* NASA Ames Research Center.

84. Hatcher, P., and M. Quinn. 1991. Data-Parallel Programming on MIMD Computers. Cambridge, Mass.: MIT Press.

85. Averbuch, A., E. Gabbe, and A. Yehudai. 1993. Portable, parallelizing Pascal compiler. *IEEE Software,* Vol. 10, No. 2, 71–81.

86. Rinard, M. 1994. *The Design, Implementation and Evaluation of Jade, a Portable, Implicitly Parallel Programming Language.* Ph.D. Thesis, Stanford University.

87. Luke, E. 1993. *The Development of a Scalable Parallel 3-D CFD Algorithm for Turbomachinery.* Master's Thesis, Mississippi State University.

88. Korlakunta, A. 1994. *Object-Oriented Implementation for the NAS Parallel Benchmarks.* Master's Thesis, Mississippi State University, 1994.

89. J. Singh, W. Weber, and A. Gupta. 1991. *SPLASH: Stanford Parallel Applications for Shared Memory.* Technical report, Stanford University, CSL-TR-91-469.

90. Browning, R., T. Li, B. Chui, J. Ye, R. Pease, Z. Czyzewski, and D. Joy. Empirical forms for the electron/atom elastic scattering cross sections from 0.1–30 keV. *Journal of Applied Physics* (to appear).

91. Rothberg, E. 1993. *Exploiting the Memory Hierarchy in Sequential and Parallel Sparse Cholesky Factorization.* Ph.D. Thesis, Stanford University.

92. Duff, I., R. Grimes, and J. Lewis. 1989. Sparse Matrix Problem. *ACMTMS*, Vol. 15, No. 1, 1–14.

93. Browning, R., T. Li, B. Chui, J. Ye, R. Pease, Z. Czyzewski, and D. Joy. Empirical forms for the electron/atom elastic scattering cross sections from 0.1–30 keV. *Journal of Applied Physics* (to appear).

94. Quinn, M. J. 1987. *Designing Efficient Algorithms for Parallel Computers.* New York: McGraw-Hill, 61.

95. Doss, N. 1992. *An Introduction to CPC - C++ With Parallel Classes.* Technical. Report. MSSU-EIRS-ERC-93-8, Engineering Research Center for Computational Field Simulation.

96. Foster, I., R. Olson, and S. Tuecke. 1993. Productive parallel programming: the PCN approach. *Scientific Programming*, 1(1).

97. Foster, I., and K. M. Chandy. 1994. Fortran M: a language for modular parallel programming. *JPDC*.

98. Chong, F. T., S. D. Sharma, E. Brewer, and J. Saltz. 1994. *Multiprocessor Runtime Support for Fine Grain, irregular DAGs.* Technical Report, Computer Science Dept., University of Maryland, CS-TR-3266.

99. Gerasoulis, A., J. Jiao and T. Yang. 1994. *Scheduling of Structured and Unstructured Computation.* Technical Report, Rutgers University.

100. Yang, T., and A. Gerasoulis. 1992. PYRROS: static task scheduling and code generation for message-passing multiprocessors. *Proc. of 6th ACM Inter. Conf. on Supercomputing*, Washington D.C., 428–437.

101. Foster, I., B. Avalani, A. Choudhary, and M. Xu. 1994. A compilation system that integrates High Performance Fortran and FortranM. *Proc. 1994 Scalable High Performance Computing Conf.*, Knoxville, Tenn.

102. Reed, D., R. Aydt, T. Madhyastha, R. Noe, K. Shield, and B. Schwartz. *The Pablo Performance Analysis Environment.* Technical Report, University of Illinois.

103. von Eicken, T., D. Culler, S. Goldstein, and K. Schauser. 1992. Active messages: a mechanism for integrated communication and computation. *Proc. of the 19th Int'l Symposium on Computer Architecture*, Gold Coast, Australia.

104. Luna, S. 1994. *Implementing an Efficient Portable Global Memory Layer on Distributed Memory Multiprocessors.* Technical report UCB/CSD/-94-810, University of California, Computers Science Division.

105. Culler, D., A. Dusseau, S. Goldstein, A. Krishnamurthy, S. Lumetta, T. von Eicken, and K. Yelick. 1993. Parallel programming in Split-C. *Proceedings of Supercomputing 93,* Portland, Oregon.

106. Moreno, E. 1993. GLU-BRIDGE. *A Structural Engineering Application in the GLU Programming Language.* Master's Thesis, Computer Science Department, Arizona State University, Tempe, Arizona.

31

Visualization of Parallel and Distributed Systems

Michael T. Heath

Parallel and distributed systems typically exhibit substantially more complicated behavior than conventional serial systems. This greater complexity is usually manifested at many levels, including hardware, system software, algorithms, and application programs. Consequently, parallel and distributed systems, and their constituent components, tend to be much more difficult to design and implement. In particular, debugging parallel systems, both for correctness and for optimal performance, presents formidable challenges to both designer and user.

Achieving optimal performance requires that all facets of a system—architecture, operating system, programming language, compiler, application program, etc.—work efficiently together. The techniques discussed in this chapter can be useful at any of these levels, from allowing system designers to assess the parallel efficiency of code generated by a parallelizing compiler, to helping end-users understand the behavior and performance of application programs.

Debugging and performance tuning in conventional serial systems are often based on execution trace or profile information, supplied either automatically by the system or through diagnostic printing inserted by the user. The volume of data produced by execution tracing can be overwhelming, particularly with parallel systems. Moreover, when trace data are produced by multiple processors, they can be extremely difficult to interpret. One proven method for dealing with large volumes of complex data is through visualization, graphically depicting the data for easier human comprehension. In effect, a picture can be worth a thousand words (or a million numbers). This chapter reviews techniques for visualizing the behavior of parallel and distributed systems. The main focus is on graphical techniques for evaluating and improving performance, but many of the same ideas are applicable to parallel debugging as well.

It is beyond the scope of this chapter to provide a complete history or bibliography of work on visualization of parallel and distributed systems. One way for the interested reader to trace this history would be through the numerous conference proceedings [10,

17, 34, 35] and special journal issues [21, 22] devoted in whole or in part to this topic, and the references cited in turn by the papers therein.

The earliest performance visualization tools of which the author is aware were Seecube [6] and HyperView [26]. Other such tools of more recent vintage include AIMS [43], Gauge [23], Pablo [31], ParaGraph [19], Seeplex [7], SIEVE [33], TOPSYS [3, 4], Traceview [25], and Upshot [20], as well as tools aimed at specific experimental machines, such as RP3 [24] and Monsoon [30], and vendor-specific tools from many of the manufacturers of parallel machines. This chapter is intended to he a discussion of general techniques and issues in visualizing parallel and distributed systems rather than specific tools. Most of the illustrative examples used were generated by ParaGraph, simply because of its greater availability and familiarity to the author, who was its principal developer.

The publication medium for this volume limits the illustrations to monochrome still pictures, whereas most performance visualization tools gain much of their effectiveness from color and motion. Despite the restrictive medium, some feeling should still be conveyed for the kind of useful information that performance visualization can provide to aid in improving the performance of parallel systems.

31.1   Performance Monitoring

Any type of visualization requires data describing the phenomenon to be depicted. In the case of performance visualization, it is the computation itself that is to be displayed graphically rather than the results of the computation. Such data are usually obtained by an execution tracing facility, which may be implemented at various levels and take various forms. Some systems have performance monitoring facilities built into the hardware, while others must rely on software instrumentation. The latter may he in the form of macros inserted at compile time or may be embedded in system libraries (e.g., for interprocessor communication) and invoked at run time. Notable examples of such monitoring and tracing systems include AIMS [43], Alog [20], IPS-2 [28], Pablo [31], and PICL [12].

In addition to these more or less portable monitoring facilities, some vendors of parallel systems supply system-specific performance monitoring facilities as part of their system software. There are also some higher level, application-oriented programming interfaces that offer tracing capability, such as Chameleon [l4]. The emerging MPI message-passing standard [13, 27] includes a profiling interface that could be used for instrumentation.

Execution trace data are usually stored in a file for postmortem analysis but may also be processed on the fly in a real-time system. Typically, various monitoring systems use varied trace formats that are usually not directly interchangeable but often can be converted for compatibility with various tools that may use the data. There appears to be some potential for convergence on a standard trace file format, such as Ref. [41], but there may never be universal agreement on this issue. An alternative approach is to use a self-describing format, such as SDDF [1], in which the structure and semantics of the data in a trace file are determined by a header that can be interpreted by a simple parser. This approach permits much greater flexibility in targeting a given trace file for use in various performance tools and for filtering out data that may not be needed in a given context.

The type of data collected depends on the type of architecture and the level of tracing. Very low-level tracing on a shared-memory architecture, for example, might trace individual memory references, cache misses, etc., while higher-level systems may trace subroutine calls, etc. Message-passing systems typically trace sends, receives, and other similar communication functions. Some systems also support high-level tracing of application-specific functions and data structures.

There are numerous additional important issues in performance monitoring, including the intrusiveness of performance probes, synchronization of clocks across multiple processors, and limiting the potentially large volume of data that may be generated, but a detailed discussion of these issues is beyond the scope of this chapter.

31.2   Performance Visualization

Assuming that valid execution trace data are available, how can they be depicted graphically for easier human comprehension? Visual displays of performance data can be classified in a number of ways. They can be processor (hardware) oriented or process (software) oriented. They can depict low-level events (load/store, send/receive, cache miss, context switch, etc.) or show higher-level activities that are more directly meaningful for the application at hand. They can show individual events on individual processors as a function of time, or only summaries aggregated over space and/or time. They can provide a static image or an animation of the parallel system's dynamic behavior.

Additional distinctions result from architectural considerations. For example, memory-oriented displays may be most appropriate for a shared memory system, while communication-oriented displays may he most appropriate for a distributed-memory system. In a shared-memory symmetric multiprocessor, or a system in which parallelism is entirely automated or implicit, the identities of individual processors may be of little interest, whereas, for explicitly parallel programs in a distributed system, processor identities may be of vital importance. Cache effects may be of critical interest in some systems but of only secondary importance in others.

Ultimately, the whole point of visualization is to provide insight into the factors affecting performance (and ideally some indication of how to improve it), not just to produce pretty pictures or flashy graphics. Thus, the visual displays must be designed to convey a maximum amount of relevant information that the user can readily relate to the application at hand. Unfortunately, the events or qualities that are easiest to collect and depict are often not easy to relate directly to the user's application. For example, to show that a processor is idle at some moment does not necessarily reveal why the processor is idle, or even the location in the user's code at which the undesirable behavior takes place.

Bridging the gap between low-level system information (which is relatively easy to monitor automatically) and application-level constructs that are meaningful to the user is a major challenge, and it is exacerbated by the desire for portability and broad applicability. Almost by definition, any high-level abstraction that relates readily to a particular application is likely to be application dependent and require custom-written data collection and display techniques. Conversely, generic visual displays based on universally available, low-level execution trace data are almost inevitably only indirectly related to the application level.

In view of this conundrum, most performance visualization tools have either featured only generic displays, leaving it up to the user to interpret their significance in a given application, or else the tool is explicitly application dependent and of correspondingly restricted applicability. Some performance visualization tools are extensible, however, providing hooks for users to add their own application-dependent displays that can then be viewed alongside the generic views already provided. The usefulness of this approach depends on the convenience with which new displays can be constructed. Various approaches to easing the latter problem include the use of "metatools," or tools for building tools, tool kits, specification languages, and general scientific visualization packages that already provide for relatively easy construction of displays by the user. (See, e.g., Refs. [5, 16, 29, 40])

Another key issue in performance visualization is real-time versus postmortem display. Real time display has the undeniable appeal of immediacy, and the potential for interaction with and "steering" of the application, but is fraught with practical difficulties such as how to extract the necessary data unintrusively and to transport them to a graphical workstation in real time. With postmortem display, on the other hand, the user cannot interact with the program during execution but gains the considerable advantages of multiple replays to study intricate behavior in detail. There are also trade-offs in resources required. Real-time display requires higher network bandwidth, while postmortem display requires greater storage volume. Most performance tools for tightly-coupled parallel systems have focused on postmortem display. Real-time display has become increasingly important, however, for visual monitoring of more loosely coupled, heterogeneous networks and clusters (e.g, Ref. [2]).

31.2.1 Processor displays

Attention is now turned to specific examples of visual displays of performance information, starting with processor-oriented displays. Suppose that a given computation requires W units of work on a serial machine. Using P processors, ideal parallel performance would require

1. *Load balance.* Each processor does the same amount of work (in this case W/P).
2. *Concurrency.* The work is done simultaneously on all processors (i.e., the work is completely overlapped, not staggered) .
3. *Overhead.* No additional work is done beyond that required for the serial solution.

Under these ideal conditions, the parallel computation would require the same execution time as W/P units of serial work (barring any "superlinear speedup" effects, such as increased cache capacity when using more processors), giving a speedup of P and an efficiency of 100 percent.

Each of these three criteria is difficult (often impossible) to achieve in practice.

1. *Load balance.* It may not be possible to spread the computational work perfectly evenly among the processors using the given level of computational granularity (e.g., P may not divide W).
2. *Concurrency.* It may not be possible to overlap all of the work completely, as most parallel computations contain at least some serial portion, and typically ramp up to and down from fully parallel mode, possibly several times.
3. *Overhead.* There is almost always some system overhead in a parallel computation because of communication, synchronization, contention, etc., and the parallel algorithm may also require additional work, either redundant or speculative, that is unnecessary in the serial algorithm (e.g., it may be cheaper to compute a shared quantity redundantly on each processor than to compute it on one processor and then communicate it to the others).

In performance tuning, one seeks to minimize load imbalance, maximize concurrency, and minimize overhead. Processor-oriented displays graphically depict these quantities in terms of low-level state information about individual processors, such as whether a processor is busy or idle. System overhead, such as that which is due to interprocessor communication, is relatively easy to monitor, but application-dependent overhead, such as redundant or speculative computation, is not easily monitored without some assistance from the user. For this reason, generic processor-oriented displays often depict only system overhead and do not attempt to capture possible application-level overhead inherent in the algorithm.

Figure 31.1 illustrates perhaps the most basic type of processor-oriented display. It shows the percentage of execution time that each processor spent in each of three states (busy, overhead, and idle) during a run. This type of display provides feedback on the overall efficiency of the program and on the load balance across processors. More insight into how well the computation is overlapped is provided by Fig. 31.2, which shows the concur-

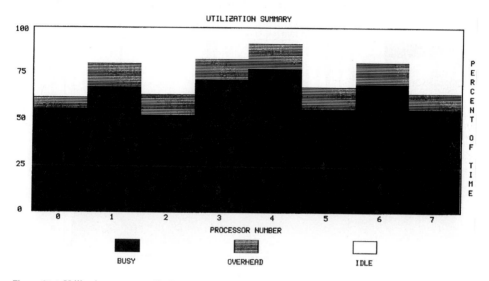

Figure 31.1 Utilization summary display

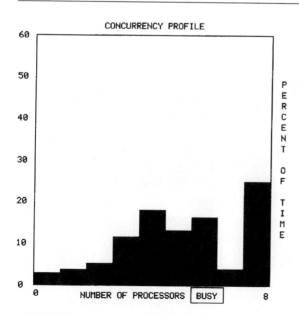

Figure 31.2 Concurrency profile display

rency profile of the application (i.e., the percentage of execution time that each possible number of processors was, in fact, busy). These displays are integrated over time, and hence can reveal the presence of poor load balance or concurrency, or excessive overhead, but do not pinpoint the specific time of their occurrence.

Detailed performance tuning requires information on the dynamic, time-dependent behavior of the program. Typical examples of time-dependent processor-oriented displays are shown in Figs. 31.3 and 31.4. Figure 31.3 shows the aggregate number of processors in each of three states (busy, overhead, idle) as a function of time, which is plotted along one axis. Figure 31.4, often called a Gantt chart, breaks this same information down on an in-

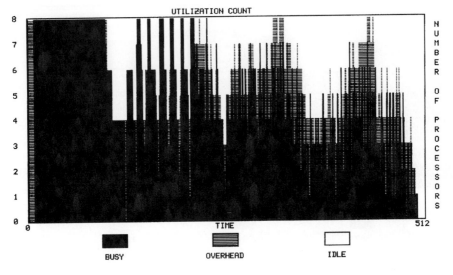

Figure 31.3 Utilization count display

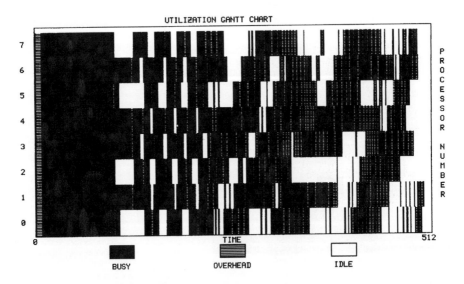

Figure 31.4 Utilization Gantt chart display

dividual processor basis for detailed study. Such displays provide feedback to the user on the effectiveness with which the processors are being used—how much idle time and overhead are present, how well processor activity is overlapped, and at what point in the execution any problems occurred.

Another dynamic depiction of load balance is given by a Kiviat diagram such as the one shown in Fig. 31.5. Here, the use of each processor is plotted as a point on the corresponding spoke of a wheel, with distance from the hub corresponding to the percentage of use. Connecting the resulting points by straight lines gives a polygon whose size and shape portray the current load balance, which changes dynamically with time.

These displays, and many others of a similar nature, are effective in providing feedback on the concurrency, load balance, and overhead in a parallel computation. It is up to the user, however, to relate this low-level, processor-oriented information back to the application program. Thus, these displays can alert the user to performance problems, but they do not necessarily make the solution obvious.

31.2.2 Communication displays

In distributed systems, the major source of overhead is usually interprocessor communication. It is of critical importance for performance tuning, therefore, to provide feedback to the user on the communication activity in the parallel program. The frequency, volume, and pattern of communication largely determine the overall efficiency of distributed parallel programs. For example, a distributed parallel application could have perfect load balance and concurrency, yet be inefficient because of high communication overhead resulting, for example, from poor data locality. Thus, the user needs to know who is communicating with whom, how often, and in what quantity.

Interprocessor communication can be depicted either logically (i.e., without regard to the underlying interconnection network) or physically (i.e., in terms of the actual paths traveled in a specific network). Logical display of communication is independent of any

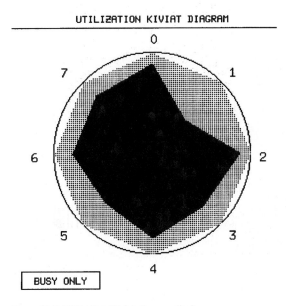

Figure 31.5 Utilization Kiviat diagram display

particular network topology and therefore can be portable across parallel architectures, but it does not account for possible network contention.

A typical display of logical communication is shown in Fig. 31.6, where processors are represented by numbered circles, and logical communication between processors is represented by lines between the corresponding circles. The communication lines are drawn and erased as messages are sent and received, providing an animation of program behavior over time. The arrangement of the processors and lines bears no necessary relation to the actual network topology on which the program runs, but the user still obtains useful information on the pattern of communication implied by the algorithm. Another way of depicting the communication pattern uses a square array, with the sending and receiving processors along the two dimensions, as illustrated in Fig. 31.7. Yet another alternative animation is in terms of message queues, as shown in Fig. 31.8, which helps the user determine whether the communication load is well balanced and whether messages are being consumed as rapidly as they are generated.

Another way of depicting time in displaying logical communication is with a spacetime diagram, as shown in Fig. 31.9. Here, time is shown along one screen dimension. Processor activity is depicted by horizontal lines and interprocessor communication by diagonal lines. Such a display gives a more precise indication of time-dependent program behavior, while still providing feedback on the overall pattern of communication.

To display actual physical network communication traffic requires more detailed information than just the source and destination processors for each message. Such detailed information could conceivably be provided by trace data, but a more practical alternative is to infer the message paths based on user-supplied information about the network topology and routing scheme in use. An example is shown in Fig. 31.10, where message traffic is displayed in terms of the paths that would be taken in a given network topology, in this case a binary tree. This particular display offers the user a choice of several popular network topologies (tree, mesh, butterfly, crossbar, etc.) and routing schemes, so the user can visually simulate the program's communication behavior on other topologies in addition to the one actually used, all based on the same underlying trace data.

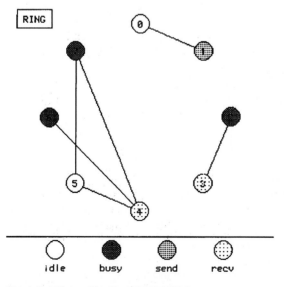

Figure 31.6 Communication animation display

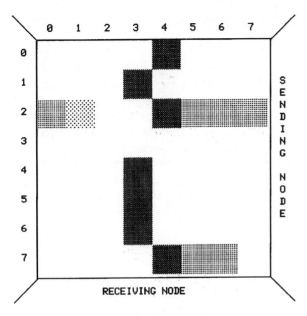

Figure 31.7 Communication matrix display

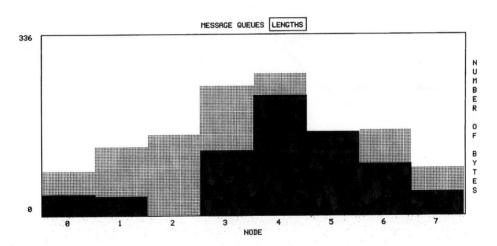

Figure 31.8 Message queues display

31.2.3 Task displays

The displays considered thus far show what the processors are doing (e.g., busy or idle, sending or receiving), but not in terms that the user can necessarily relate directly to the program. The user is more likely to think in terms of the subroutines, data structures, etc., that reflect the nature of the application problem being solved. Such information can be gathered by a variety of means, including profiling procedure entry/exit, incorporating compiler-supplied information (e.g., line numbers in the program text) into trace data, or by explicitly delimiting user-defined program tasks with appropriate instrumentation. An

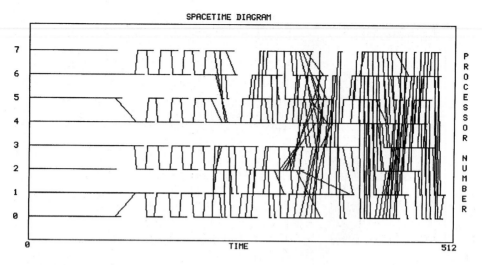

Figure 31.9 Spacetime display

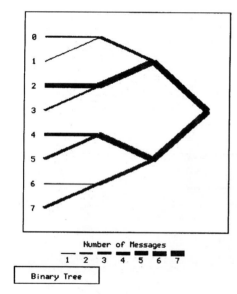

Figure 31.10 Network display

example of how such information can be displayed is shown in Fig. 31.11, which is analogous to the Gantt chart shown earlier but, instead of showing processor busy/idle status, it indicates by a color code (or shading) which task each processor is executing at any given time. This display reflects the SPMD programming model, in which all processors collaborate on each task. There are many other effective ways of displaying the interaction of processes or threads in parallel computations. (See, e.g., Ref. [24].)

31.2.4 Memory displays

Memory-oriented displays obviously provide critical information for performance tuning on shared-memory parallel systems, since contention among processors for memory

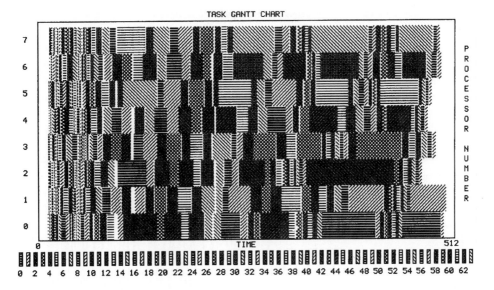

Figure 31.11 Task Gantt chart display

access can be a major bottleneck. Such displays may seem somewhat less important for distributed systems, but taking a higher-level perspective, such as that of application-level algorithms or high-level programming languages (e.g., High Performance Fortran), memory access patterns are among the most crucial factors affecting performance for any type of parallel system. In particular, achieving good performance depends on effective exploitation of the memory hierarchy, which may include remote memory references in distributed systems, as well as nonuniform memory access times in shared systems.

Although generic memory-oriented displays are feasible (e.g., Ref. [36]), such displays are most effective when they relate directly to data structures that are meaningful to the user in terms of the application program. A good example is SHMAP [9], which animates array-oriented computations in terms of memory references, so that the effects of various memory partitions and cache policies can be assessed. Other examples are the performance visualization systems under development for data parallel languages, such as HPF and pC++, which depict performance data in terms of user-level data structures, such as arrays, meshes, and particles (e.g., Refs. [15, 37]).

31.2.5   Other displays

In addition to standard displays that depict basic system statistics—processor use, interprocessor communication, load balance, memory accesses, etc.—by means of strip charts, histograms, etc., there is also ample room for the invention of displays that attempt to synthesize such detailed information in terms of higher-level concepts that provide an overall impression of the performance characteristics of a parallel program. One such concept is the *critical path*, which is the longest serial thread (i.e., chain of dependences) running through a parallel computation. Obviously, any improvement in performance must shorten the critical path, so it is a plausible place to look for bottlenecks. Moreover, the stability of the critical path across successive executions of the program may also provide useful information about possible systematic performance problems. One way of depicting the critical path is shown in Fig. 31.12, which is an adaptation of the space-time diagram illustrated earlier, but with the critical path highlighted by a heavier line.

Another key concept in most programs is repetitive behavior, such as that governed by iterative loops. Of course, such repetitive behavior is usually evident in any time-dependent display, such as a spacetime diagram. One way of displaying program behavior that may reveal more subtle cycles is a type of *phase portrait* obtained by plotting processor utilization and message traffic as a point in a two-dimensional plane. Over time, this point traces out a trajectory that conveys information about the (often inverse) relationship between computation and communication. Repetitive computations show up as orbits or cycles in the phase portrait. (See Fig. 31.13.) Another way to detect cyclic behavior is

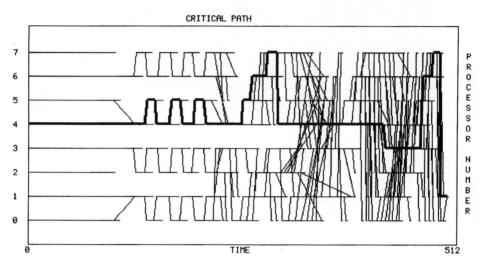

Figure 31.12 Critical path display

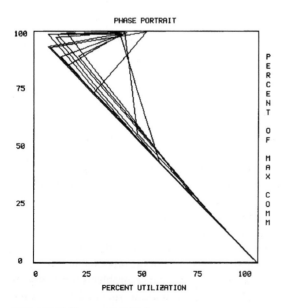

Figure 31.13 Phase portrait display

through standard signal processing techniques, such as autocorrelation, which can then be plotted graphically [39].

31.2.6   Application-specific displays

The generic displays considered thus far are equally applicable to any algorithm or application program. This lack of specificity means that such tools are widely applicable, which is an advantage, but it also means that they do not necessarily display performance information in the most meaningful manner for any specific application. For example, a generic communication display may show the message passing required for data exchanges between processors in a parallel sorting program, but it may have no way to show (or even to know) which particular data items are being exchanged. The reader can undoubtedly imagine many other instances in which performance data would be most meaningfully presented in terms of application-specific objects, such as graphs (in the combinatorial sense), trees, lists, particles, meshes, etc. Similarly, particular algorithmic paradigms, such as divide-and-conquer, domain decomposition, etc., may lend themselves to more revealing displays than a generic system would permit

For these reasons, developers of performance visualization tools are somewhat torn between the wide applicability of generic views and the potentially greater effectiveness of application-specific displays. Since there is no way to anticipate every possible application anyway, an appealing alternative is to make the generic system extensible so that application-specific displays can he added as needed and viewed along with any generic displays already provided. Such extensibility may be achieved by means of a toolkit or a special language or graphical interface for building new application-specific displays.

31.2.7   General scientific visualization

The displays discussed thus far are all specifically designed to depict performance data from parallel systems. Obviously, such custom designed displays are capable of depicting performance data in minute detail, but their construction can be extremely labor intensive, often requiring graphical programming that is tedious and error-prone. An alternative is to employ standard scientific visualization tools, such as AVS or Data Explorer, or the graphical facilities of general scientific problem-solving environments, such as Matlab or Mathematica. Such an approach has been explored, for example in Ref. [16, 29, 40].

Although the use of general-purpose graphical tools for performance visualization may sacrifice some potential customization and detail, it gains several advantages, including

- easy implementation, compared to developing a custom performance display, since little or no programming may be required
- easy scalability, since scientific visualization packages and the views they provide are usually designed to deal with very large quantities of data
- easy access to three-dimensional views, which are already built into most scientific visualization tools
- easy integration of views of performance data with views of computational results produced by the program

On the other hand, although general visualization tools can be effective in conveying an overall impression of program performance, their lack of detail can make it difficult to relate this information to a specific cause in the program.

31.2.8 Three-dimensional views

One of the advantages of using general-purpose scientific visualization tools for performance visualization is the ease with which three-dimensional views can be constructed, and these views tend to scale naturally to very large data sets. In addition, such views can easily be rotated, zoomed, etc., to enable closer inspection from a variety of perspectives. These capabilities of general-purpose scientific visualization packages have been put to good use by a number of researchers. A good example is the "do-loop surface" [29]. In this case, for a given loop of interest in a parallel program, the processor number and iteration number are plotted along two horizontal axes, while execution time per iteration is plotted on the vertical axis. (See Fig. 31.14.) The resulting surface helps pinpoint performance bottlenecks such as load imbalance, communication delays, or ineffective overlapping of communication and computation.

Another example of an effective three-dimensional display is the Kiviat "tube" [16]. The usual Kiviat diagram plots processor number and use in two dimensions (in polar coordinates) at a given time. By plotting time in the third dimension, the resulting three-dimensional object depicts the load balance over time, and it can be rotated and otherwise manipulated to be viewed from various perspectives. (See Fig. 31.15.)

31.3 Example

An example is now given to illustrate the use of visual feedback in performance tuning of parallel programs. The application is a spectral transform method for solving the shallow water equations on a sphere [42], which is a nonlinear system of time-dependent partial differential equations typical of many global climate models. In one phase of the parallel computation, the work load should be perfectly balanced across the processors, yet the performance was observed to be poor. Upon looking at the computation visually in Fig. 31.16, it is seen that some tasks take much longer on some processors than on others, even though the same amount of work is involved, causing the first four processors to fall behind and disrupting the otherwise regular behavior of the algorithm. Closer examina-

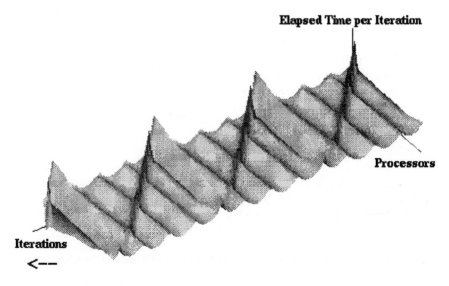

Figure 31.14 Do-loop surface display

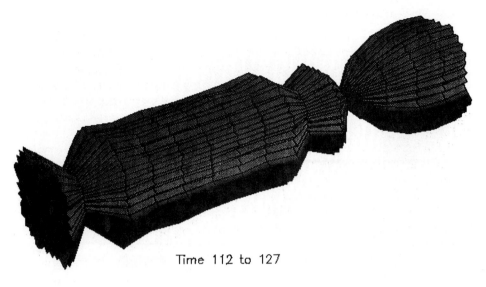

Time 112 to 127

Figure 31.15 Kiviat tube display

tion revealed that this anomaly was due to denormalized floating-point numbers in some ranges of the problem domain, and these are handled in software at a much slower speed by this system (an iPSC/860). Setting denormals to zero produced the improved behavior shown in Fig. 31.17.

Another source of performance degradation was communication overhead. Initially the implementation used the default three-trip communication protocol for this system (request to send, wait for acknowledgment, then send the message), which resulted in significant idle time, as evidenced by the white streaks in Fig. 31.18. Preallocating message buffers so that a one-trip communication protocol could be used resulted in the improved performance shown in Fig. 31.19.

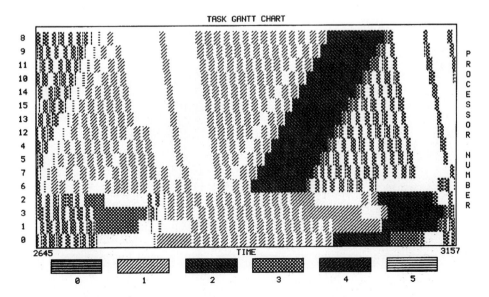

Figure 31.16 Task Gantt chart for initial implementation

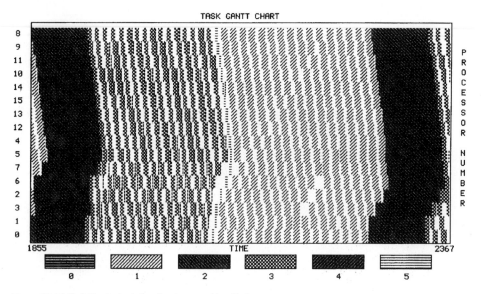

Figure 31.17 Task Gantt chart showing improved load balance

In both of these instances, the initial poor performance was first observed simply from unexpectedly high overall run times and disappointing parallel speedup. The visualization tool then helped pinpoint the portion of the compilation that was limiting performance, but did not explicitly identify the specific cause (e.g., denormalized numbers or the communication protocol). This is typical of performance tuning, with the visualization tool assisting the user in identifying performance bottlenecks, but with the user's insight playing a necessary role in actually correcting the problems. For some additional examples of visual performance tuning, see Refs. [18, 32, 38].

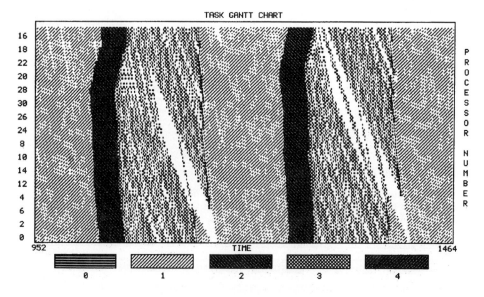

Figure 31.18 Task Gantt chart showing effect of three-trip message protocol

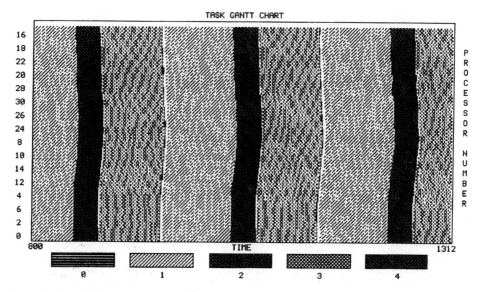

Figure 31.19 Task Gantt chart showing effect of one-trip message protocol

31.4   Future Directions

Although visualization has been shown to be a generally effective tool in improving the performance of parallel programs, many challenges remain, some of which have already been mentioned.

Perhaps the greatest challenge in performance visualization is *scalability*—the ability to deal with vary large data sets, arising from large numbers of processors or long runs, or both. Many of the most effective graphical displays for small numbers of processors do not scale to larger numbers of processors because there simply are not enough pixels on the screen to support the level of detail required. One possible alternative is to analyze and present the data by statistical methods, using means, maxima, minima, medians, etc., instead of the underlying raw data. Another alternative is to display cross sections or other subsets of the data or, more generally, to use hierarchical displays that provide a high-level view that can he zoomed in to reveal finer detail in a specific area of interest. Another novel approach to scalability is through "categories and context" [8].

Visual resolution is not the only scalability problem, however, as the volume of trace data required to specify the finest level of detail may be unwieldy or impractical to store or process. This suggests that whatever approach is taken—statistical, hierarchical, etc.— should be reflected in the data collection process as well as in the visualization. Indeed, the processing of performance data may be so computationally intensive as to require a parallel machine of similar power to the one under study, which, in turn, suggests that perhaps a client-server model would he most appropriate for performance visualization, with the parallel machine responding to queries initiated by the user at a graphical workstation, either in real-time or postmortem. The proliferation of distributed database techniques may provide the necessary support for this approach.

Detecting a performance problem is only the first step toward eliminating it. As noted earlier, another major challenge in performance visualization is to relate the performance information displayed to the source code or, better still, to concepts that the user can easily understand in the context of a given application, and that provide a basis to enable the user to correct any problems. Unfortunately, it is difficult to provide visualizations that are both widely applicable and easy to relate to specific applications. Perhaps the best one can do is to make generic displays as helpful and intuitively meaningful as possible, while also making it easy for the user to customize existing displays or to create new ones that are tailored to specific needs.

This trade-off between portability and ease of use on one hand, and application-specific customization on the other, is not unique to performance visualization but permeates software design in many areas. One popular way of dealing with it is to provide a tool kit or interface that permits the user to mix and match the components needed for a given situation. This approach may imply a fairly steep learning curve, however, which might be overcome by a hierarchical system whose high-level defaults are widely applicable, but whose deeper layers can be exploited, as the user gains experience, to provide greater customization for given applications. One can expect to see further experimentation with new ideas in user interfaces for performance evaluation systems, such as spreadsheets or simple specification languages.

Another challenge in performance visualization is dealing with the wide diversity of parallel architectures and programming paradigms that are currently prevalent. There may eventually be a convergence on a "standard" architecture, but for the near term one must continue to deal with a variety of memory organizations (e.g., shared versus distributed), control structures (e.g., SIMD versus MIMD), and network topologies. Some standards are emerging in parallel software (e.g., HPF, MPI, PVM), that will permit some degree of focus in performance visualization efforts, but diversity in parallel programming paradigms is likely to persist for some time. It is doubtful that any single performance visualization tool can (or should) embrace the full range of hardware and software options, but it would still be helpful to users if some commonality could be achieved in the way performance information is presented and the way such tools are used.

Many technical challenges remain for more effective performance visualization. The difficulty in providing accurate and consistent clock readings across processors has already been mentioned. The intrusiveness of execution tracing will continue to be an issue unless it is provided in hardware by vendors. The effectiveness of visual animation could be enhanced by better control of animation speed, so that it could he made more nearly proportional to the original execution speed, and also to provide better support for zooming the time scale up or down for more or less detail, or for backing up to replay interesting activity.

Another new direction that is already underway is the exploration of additional sensory modalities, such as sound or virtual reality, to enhance the visual presentation of performance data. Auralization of performance data can provide helpful feedback in its own right, but is perhaps most effective as a "soundtrack" accompanying the visual depiction of the data [11]. Virtual reality techniques suggest promising new paradigms for dealing with large and complex data sets, such as "flying through" the data to provide a natural transition between the big picture and the finer details, and many of these will undoubtedly prove useful for performance data.

Most current performance visualization tools are aimed primarily at depicting a single run of a parallel program. The ability to make comparisons across multiple runs, with varying problem sizes and numbers of processors, is essential in studying the speedup, efficiency, and scalability of parallel systems. The increasing emphasis on scalable algorithms and architectures will require appropriate support from performance visualization tools for making such comparisons.

Many of the issues cited here—dealing with very large data sets, giving the user greater control over how performance data are displayed, etc.—can be addressed, at least in part, by using standard scientific visualization techniques and packages, with their inherent ability to process a large volume of data and display it in a scalable way. Thus, one can expect to see greater borrowing of techniques from scientific visualization as well as direct use of such visualization packages for displaying performance data. Standard mechanisms for importing and exporting data, such as those in AVS, could enhance the interoperability of performance visualization tools with standard graphics environments.

One hopes that effective parallel programming will eventually become much more automated, with architectures, operating systems, programming languages, and compilers that will enable even casual users to achieve a reasonable fraction of the peak performance of which the underlying hardware is capable. Until that dream is realized, performance visualization techniques can play a vital role in helping users tune their codes manually, and they can also provide helpful feedback to system builders on the effectiveness of the hardware and software that they design.

38.5 References

1. Aydt, R. 1993. *SDDF: the Pablo self-describing data format.* Technical report. Urbana-Champaign, Ill.: Dept. of Computer Science, University of Illinois.

2. Beguelin, A., J. Dongarra, A. Geist, and V. Sunderam. 1993. Visualization and debugging in a heterogeneous environment. *IEEE Computer,* 26(6):88–95.

3. Bemmerl, T., and P. Braun. 1993. Visualization of message passing parallel programs with the TOPSYS parallel programming environment. *Journal of Parallel and Distributed Computing,* 18(2):118–128.

4. Bode, A., and P. Braun. 1993. Monitoring and visualization in TOPSYS. In *Performance Measurement and Visualization of Parallel Systems,* G. Haring and G. Kotsis, eds. Amsterdam: Elsevier Science Publishers, 97–118.

5. Casavant, T. L., and J. A. Kohl. 1993. The IMPROV meta-tool design methodology for visualization of parallel programs. *Proceedings of the International Workshop on Modeling, Analysis, and Simulation of Computer and Telecommunication Systems (MASCOTS).*

6. Couch, A. L. 1988. *Graphical representations of program performance on hypercube message-passing multiprocessors.* Technical Report 88-4. Medford, Mass.: Dept. Of Computer Science, Tufts University.

7. Couch, A. L. 1990. Monitoring parallel executions in real time. *Proceedings of the Fifth Distributed Memory Computing Conference,* volume II, D. W. Walker and Q. F. Stout, eds. Los Alamitos, Calif.: IEEE Computer Society Press, 1187–1196.

8. Couch, A. L. 1993. Categories and context in scalable execution visualization. *Journal of Parallel and Distributed Computing,* 18(2):195–204.

9. Dongarra, J. J., O. Brewer, J. A. Kohl, and S. Fineberg. 1990. A tool to aid in the design, implementation, and understanding of matrix algorithms for parallel processors. *Journal of Parallel and Distributed Computing,* 9:185–202.

10. Dongarra, J. J., and B. Tourancheau, eds. 1993. Environments and Tools for Parallel Scientific Computing. Amsterdam: Elsevier Science Publishers.

11. Francioni, J. M., and J. A. Jackson. 1993. Breaking the silence: auralization of parallel program behavior. *Journal of Parallel and Distributed Computing,* 18(2):181–194.

12. Geist, G. A., M. T. Heath, B. W. Peyton, and P. H. Worley. 1990. *PICL: a portable instrumented communication library, C reference manual.* Technical Report ORNL/TM-11130, Oak Ridge, Tenn.: Oak Ridge National Laboratory.

13. Gropp, W., E. Lusk, and A. Skjellum. 1994. *Using MPI: Portable Parallel Programming with the Message Passing Interface.* Cambridge, Mass.: MIT Press.

14. Gropp, W., and B. Smith. 1993. *Users manual for the Chameleon parallel programming tools.* Technical Report ANL-93/23. Argonne, Ill.: Argonne National Laboratory.

15. Hackstadt, S., and A. Malony. 1993. *Data distribution visualization for performance evaluation.* Technical Report CIS-TR-93-21, Dept. of Computer and Information Science, University of Oregon, Eugene, Oregon.

16. Hackstadt, S., and A. Malony. Next-generation parallel performance visualization. Technical Report CIS-TR93-23, Dept. of Computer and Information Science, University of Oregon, Eugene, Oregon.

17. Haring, G., and G. Kotsis, eds. 1993. *Performance Measurement and Visualization of Parallel Systems.* Amsterdam: Elsevier Science Publishers.

18. Heath, M. T. 1993. Recent developments and case studies in performance visualization using ParaGraph. In G. Haring and G. Kotsis, eds. 1993. *Performance Measurement and Visualization of Parallel Systems.*, Amsterdam: Elsevier Science Publishers, 175–200.

19. Heath, M. T., and J. A. Etheridge. 1991. Visualizing the performance of parallel programs. *IEEE Software,* 8(5):29–39.

20. Herrarte, V., and E. Lusk. *Studying parallel program behavior with Upshot.* Technical Report ANL-91/15, Argonne, Ill.: Argonne National Laboratory.

21. 1991. *IEEE Software* (September). K. Nichols and P. W. Oman, guest editors.

22. 1993. *Journal of Parallel and Distributed Computing* (June). T. L. Casavant, guest editor.

23. Kesselman, C. 1991. *Tools and techniques for performance measurement and performance improvement in parallel programs.* Technical Report UCLA-CS-TR-91-03. Los Angeles: Univ. of California at Los Angeles.

24. Kimelman, D. N., and T. A. Ngo. 1991. The RP3 program visualization environment. *IBM Journal of Research and Development,* 35(5/6):635–651.

25. Malony, A. D., D. H. Hammerslag, and D. J. Jablonowski. 1991. Traceview: a trace visualization tool. *IEEE Software,* 8(5):19–28.

26. Malony, A. D., and D. A. Reed. 1988. Visualizing parallel computer system performance. Technical Report UIUCDCS-R-88-1465. Urbana-Champaign, Ill.: Dept. of Computer Science, University of Illinois.

27. Message Passing Interface Forum. 1994. *MPI: a message-passing interface standard.* Technical Report CS-94-230. Knoxville, Tenn.: Dept. of Computer Science, University of Tennessee.

28. Miller, B. P., M. Clark, J. Hollingsworth, S. Kierstead, S. Lim, and T. Torzewski. 1990. IPS-2: the second generation of a parallel program measurement system. *IEEE Transactions on Parallel and Distributed Systems,* 1:206–217.

29. Naim, O., and A. J. G. Hey. 1994. Integration of performance analysis of parallel programs and scientific data visualization (unpublished).

30. Natarajan, V., D. Chiou, and B. S. Ang. 1993. Performance visualization on Monsoon. *Journal of Parallel and Distributed Computing,* 18(2):169–180.

31. D. A. Reed et al. 1993. Scalable performance analysis: the Pablo performance analysis environment. *Proceedings of the Scalable Parallel Libraries Conference.* Los Alamitos, Calif.: IEEE Computer Society Press, 104–113.

32. Rover, D. T., and M. B. Carter. 1991. Performance visualization of SLALOM. In *Proceedings of the Sixth Distributed Memory Computing Conference,* Q. F. Stout and M. Wolfe, eds. Los Alamitos, Calif.: IEEE Computer Society Press, 543–550.

33. Sarukkai, S. R., and D. Gannon. 1993. SIEVE: a performance debugging environment for parallel programs. *Journal of Parallel and Distributed Computing,* 18(2):147–168.

34. Simmons, M., R. Koskela, and I. Bucher, eds. 1989. *Instrumentation for Future Parallel Computing Systems.* New York: ACM Press.

35. Simmons, M., R. Koskela, and I. Bucher, eds. 1990. *Parallel Computer Systems: Performance Instrumentation and Visualization.* New York: ACM Press.

36. Speight, F. J., and J. K. Bennett. 1994. ParaView: performance debugging of shared-memory parallel programs (unpublished).

37. Srinivas, S., and D. Gannon. 1994. *Visualizing distributed data structures.* Technical Report TR-406. Bloomington, Ind.: Dept. of Computer Science, Indiana University.

38. Tomas, G., and C. W. Ueberhuber. 1993. *Visualization of scientific parallel programs.* Technical report, Vienna, Austria: Institute for Applied and Numerical Mathematics, Technical University of Vienna.

39. Waheed, A., and D. T. Rover. 1993. *Estimation of repetitive patterns and loops in parallel programs.* Technical Report TRMSU-ESSCSL-014 93. East Lansing, Mich.: Dept. of Electrical Engineering, Michigan State University.

40. Waheed, A., and D. T. Rover. 1993. Performance visualization of parallel programs. In *Proceedings of Visualization '95,* G. M. Nielson, ed. San Jose, Calif.

41. Worley, P. H. 1992. *A new PICL trace file format.* Technical Report ORNL/TM-12125. Oak Ridge, Tenn.: Oak Ridge National Laboratory.

42. Worley, P. H., and J. B. Drake. 1992. Parallelizing the spectral transform method. *Concurrency: Practice and Experience,* 4(4):269–291.

43. Yan, J., P. Hontalas, S. Listgarten, et al. 1993. *The automated instrumentation and monitoring system AIMS reference manual.* Technical Report 108795. Moffett Field, Calif.: NASA Ames Research Center.

32

Constructing Numerical Software Libraries for High-Performance Computer Environments

Jack J. Dongarra and David W. Walker

This chapter discusses the design of linear algebra libraries for high-performance computers. Particular emphasis is placed on the development of scalable algorithms for MIMD distributed memory concurrent computers. A brief description of the EISPACK, LINPACK, and LAPACK libraries is given, followed by an outline of ScaLAPACK, which is a distributed memory version of LAPACK currently under development. The importance of block-partitioned algorithms in reducing the frequency of data movement between various levels of hierarchical memory is stressed. The use of such algorithms helps reduce the message start-up costs on distributed memory concurrent computers. Other key ideas in our approach are:

1. the use of distributed versions of the Level 3 Basic Linear Algebra Subprograms (BLAS) as computational building blocks
2. the use of Basic Linear Algebra Communication Subprograms (BLACS) as communication building blocks

Together, the distributed BLAS and the BLACS can be used to construct higher-level algorithms and to hide many details of the parallelism from the application developer. The block-cyclic data distribution is described and is adopted as a good way of distributing block partitioned matrices. Block-partitioned versions of the Cholesky and LU factorizations are presented, and optimization issues associated with the implementation of the LU factorization algorithm on distributed memory concurrent computers are discussed, together with its performance on the Intel Delta system. Finally, approaches to the design of library interfaces are reviewed.

32.1 Introduction

The increasing availability of advanced-architecture computers is having a very significant effect on all spheres of scientific computation, including algorithm research and software development in numerical linear algebra. Linear algebra—in particular, the solution of linear systems of equations—lies at the heart of most calculations in scientific computing. This chapter discusses some of the recent developments in linear algebra designed to exploit these advanced-architecture computers. Particular attention will be paid to dense factorization routines such as the Cholesky and LU factorizations. These will be used as examples to highlight the most important factors that must be considered in designing linear algebra software for advanced-architecture computers. We use these factorization routines for illustrative purposes not only because they are relatively simple, but also because of their importance in several scientific and engineering applications that make use of boundary element methods. These applications include electromagnetic scattering and computational fluid dynamics problems, as discussed in more detail in Section 32.4.1.

Much of the work in developing linear algebra software for advanced-architecture computers is motivated by the need to solve large problems on the fastest computers available. In this chapter, we focus on four basic issues:

1. the motivation for the work
2. the development of standards for use in linear algebra and the building blocks for a library
3. aspects of algorithm design and parallel implementation
4. future directions for research

For the past 15 years or so, there has been a great deal of activity in the area of algorithms and software for solving linear algebra problems. The linear algebra community has long recognized the need for help in developing algorithms into software libraries, and several years ago, as a community effort, put together a *de facto* standard for identifying basic operations required in linear algebra algorithms and software. The hope was that the routines making up this standard, known collectively as the Basic Linear Algebra Subprograms (BLAS), would be efficiently implemented on advanced-architecture computers by many manufacturers, making it possible to reap the portability benefits of having them efficiently implemented on a wide range of machines. This goal has been largely realized.

The key insight of our approach to designing linear algebra algorithms for advanced architecture computers is that the frequency with which data are moved between different levels of the memory hierarchy must be minimized in order to attain high performance. Thus, our main algorithmic approach for exploiting both vectorization and parallelism in our implementations is the use of block-partitioned algorithms, particularly in conjunction with highly-tuned kernels for performing matrix-vector and matrix-matrix operations (the Level 2 and 3 BLAS). In general, the use of block-partitioned algorithms requires data to be moved as blocks, rather than as vectors or scalars, so that although the total amount of data moved is unchanged, the latency (or startup cost) associated with the moment is greatly reduced because fewer messages are needed to move the data.

A second key idea is that the performance of an algorithm can be tuned by a user by varying the parameters that specify the data layout. On shared memory machines, this is controlled by the block size, while on distributed memory machines it is controlled by the block size and the configuration of the logical process mesh, as described in more detail in Section 32.5.

Section 32.1 gives an overview of some of the major software projects aimed at solving dense linear algebra problems. It then describes the types of machines that benefit most from the use of block-partitioned algorithms, and discusses what is meant by high-quality, reusable software for advanced-architecture computers. Section 32.2 discusses the role of the BLAS in portability and performance on high-performance computers. The design of these building blocks, and their use in block-partitioned algorithms, are covered in Section 32.3. Section 32.4 focuses on the design of a block-partitioned algorithm for LU factorization, and Sections 32.5, 32.6, and 32.7 use this example to illustrate the most important factors in implementing dense linear algebra routines on MIMD distributed memory concurrent computers. Section 32.5 deals with the issue of mapping the data onto the hierarchical memory of a concurrent computer. The layout of an application's data is crucial in determining the performance and scalability of the parallel code. In Sections 32.6 and 32.7, details of the parallel implementation and optimization issues are discussed. Section 32.8 presents some future directions for investigation.

32.1.1   Dense linear algebra libraries

Over the past 25 years, the first author has been directly involved in the development of several important packages of dense linear algebra software: EISPACK, LINPACK, LAPACK, and the BLAS. In addition, both authors are currently involved in the development of ScaLAPACK, a scalable version of LAPACK for distributed memory concurrent computers. In this section, we give a brief review of these packages—their history, their advantages, and their limitations on high-performance computers.

32.1.1.1 EISPACK. EISPACK is a collection of Fortran subroutines that compute the eigenvalues and eigenvectors of nine classes of matrices: complex general, complex Hermitian, real general, real symmetric, real symmetric banded, real symmetric tridiagonal, special real tridiagonal, generalized real, and generalized real symmetric matrices. In addition, two routines are included that use singular value decomposition to solve certain least-squares problems.

EISPACK is primarily based on a collection of Algol procedures developed in the 1960s and collected by J. H. Wilkinson and C. Reinsch in a volume entitled in the *Handbook for Automatic Computation* [57] series. This volume was not designed to cover every possible method of solution; rather, algorithms were chosen on the basis of their generality, elegance, accuracy, speed, or economy of storage.

Since the release of EISPACK in 1972, over ten thousand copies of the collection have been distributed worldwide.

32.1.1.2 LINPACK. LINPACK is a collection of Fortran subroutines that analyze and solve linear equations and linear least-squares problems. The package solves linear systems whose matrices are general, banded, symmetric indefinite, symmetric positive definite, triangular, and tridiagonal square. In addition, the package computes the QR and singular value decompositions of rectangular matrices and applies them to least-squares problems.

LINPACK is organized around four matrix factorizations: LU factorization, pivoted Cholesky factorization, QR factorization, and singular value decomposition. The term *LU factorization* is used here in a very general sense to mean the factorization of a square matrix into a lower triangular part and an upper triangular part, perhaps with pivoting. These factorizations will be treated at greater length later, when the actual LINPACK subroutines are discussed. But first a digression on organization and factors influencing LINPACK's efficiency is necessary.

LINPACK uses column-oriented algorithms to increase efficiency by preserving locality of reference. This means that if a program references an item in a particular block, the next reference is likely to be in the same block. By column orientation we mean that the LIN-PACK codes always reference arrays down columns, not across rows. This works because Fortran stores arrays in column major order. Thus, as one proceeds down a column of an array, the memory references proceed sequentially in memory. On the other hand, as one proceeds across a row, the memory references jump across memory, the length of the jump being proportional to the length of a column. The effects of column orientation are quite dramatic: on systems with virtual or cache memories, the LINPACK codes will signifi-cantly outperform codes that are not column oriented. We note, however, that textbook ex-amples of matrix algorithms are seldom column oriented.

Another important factor influencing the efficiency of LINPACK is the use of the Level 1 BLAS; there are three effects.

First, the overhead entailed in calling the BLAS reduces the efficiency of the code. This reduction is negligible for large matrices, but it can be quite significant for small matrices. The matrix size at which it becomes unimportant varies from system to system; for square matrices it is typically between $n = 25$ and $n = 100$. If this seems like an unacceptably large overhead, remember that on many modern systems the solution of a system of order 25 or less is itself a negligible calculation. Nonetheless, it cannot be denied that a person whose programs depend critically on solving small matrix problems in inner loops will be better off with BLAS-less versions of the LINPACK codes. Fortunately, the BLAS can be re-moved from the smaller, more frequently used program in a short editing session.

Second, the BLAS improve the efficiency of programs when they are run on nonop-tomizing compilers. This is because doubly subscripted array references in the inner loop of the algorithm are replaced by singly subscripted array references in the appropriate BLAS. The effect can be seen for matrices of quite small order, and for large orders the savings are quite significant.

Finally, improved efficiency can be achieved by coding a set of BLAS [17] to take ad-vantage of the special features of the computers on which LINPACK is being run. For most computers, this simply means producing machine-language versions. However, the code can also take advantage of more exotic architectural features, such as vector opera-tions. Further details about the BLAS are presented in Section 32.2.

32.1.1.3 LAPACK. LAPACK [14] provides routines for solving systems of simultaneous linear equations, least-squares solutions of linear systems of equations, eigenvalue prob-lems, and singular value problems. The associated matrix factorizations (LU, Cholesky, QR, SVD, Schur, generalized Schur) are also provided, as are related computations such as reordering the Schur factorizations and estimating condition numbers. Dense and banded matrices are handled, but not general sparse matrices. In all areas, similar functionality is provided for real and complex matrices, in both single and double precision.

The original goal of the LAPACK project was to make the widely used EISPACK and LINPACK libraries run efficiently on shared-memory vector and parallel processors. On these machines, LINPACK and EISPACK are inefficient because their memory access pat-terns disregard the multilayered memory hierarchies of the machines, thereby spending too much time moving data instead of doing useful floating-point operations. LAPACK addresses this problem by reorganizing the algorithms to use block matrix operations, such as matrix multiplication, in the innermost loops [3, 14]. These block operations can be op-timized for each architecture to account for the memory hierarchy [2], and so provide a transportable way to achieve high efficiency on diverse modern machines. Here we use the term *transportable* instead of *portable* because, for fastest possible performance,

LAPACK requires that highly optimized block matrix operations be already implemented on each machine. In other words, the correctness of the code is portable, but high performance is not—if we limit ourselves to a single Fortran source code.

LAPACK can be regarded as a successor to LINPACK and EISPACK. It has virtually all the capabilities of these two packages and much more. LAPACK improves on LINPACK and EISPACK in four main respects: speed, accuracy, robustness and functionality. While LINPACK and EISPACK are based on the vector operation kernels of the Level 1 BLAS, LAPACK was designed at the outset to exploit the Level 3 BLAS—a set of specifications for Fortran subprograms that do various types of matrix multiplication and the solution of triangular systems with multiple right-hand sides. Because of the course granularity of the Level 3 BLAS operations, their use tends to promote high efficiency on many high-performance computers, particularly if specially coded implementations are provided by the manufacturer.

32.1.1.4 ScaLAPACK. The ScaLAPACK software library, scheduled for completion by the end of 1994, will extend the LAPACK library to run scalably on MIMD distributed memory concurrent computers [10, 11]. For such machines the memory hierarchy includes the off-processor memory of other processors, in addition to the hierarchy of registers, cache, and local memory on each processor. Like LAPACK, the ScaLAPACK routines are based on block-partitioned algorithms in order to minimize the frequency of data movement among various levels of the memory hierarchy. The fundamental building blocks of the ScaLAPACK library are distributed memory versions of the Level 2 and Level 3 BLAS, and a set of Basic Linear Algebra Communication Subprograms (BLACS) [16, 26] for communication tasks that arise frequently in parallel linear algebra computations. In the ScaLAPACK routines, all interprocessor communication occurs within the distributed BLAS and the BLACS, so the source code of the top software layer of ScaLAPACK looks very similar to that of LAPACK.

We envisage a number of user interfaces to ScaLAPACK. Initially, the interface will be similar to that of LAPACK, with some additional arguments passed to each routine to specify the data layout. Once this is in place, we intend to modify the interface so the arguments to each ScaLAPACK routine are the same as in LAPACK. This will require information about the data distribution of each matrix and vector to be hidden from the user. This may be done by means of a ScaLAPACK initialization routine. This interface will be fully compatible with LAPACK. Provided "dummy" versions of the ScaLAPACK initialization routine and the BLACS are added to LAPACK, there will be no distinction between LAPACK and ScaLAPACK at the application level, though each will link to various versions of the BLAS and BLACS. Following this, we will experiment with object-based interfaces for LAPACK and ScaLAPACK, with the goal of developing interfaces compatible with Fortran 90 [10] and C++ [24].

32.1.2 Target architecture

The EISPACK and LINPACK software libraries were designed for supercomputers used in the 1970s and early 1980s, such as the CDC-7600, Cyber 205, and Cray-1. These machines featured multiple functional units pipelined for good performance [43]. The CDC-7600 was basically a high-performance scalar computer, while the Cyber 205 and Cray-1 were early vector computers.

The development of LAPACK in the late 1980s was intended to make the EISPACK and LINPACK libraries run efficiently on shared memory vector supercomputers. The ScaLA-

PACK software library will extend the use of LAPACK to distributed memory concurrent supercomputers. The development of ScaLAPACK began in 1991 and was scheduled to be completed by the end of 1994.

The underlying concept of both the LAPACK and ScaLAPACK libraries is the use of block partitioned algorithms to minimize data movement between various levels in hierarchical memory. Thus, the ideas discussed in this chapter for developing a library for dense linear algebra computations are applicable to any computer with a hierarchical memory that (1) imposes a sufficiently large start-up cost on the movement of data among various levels in the hierarchy, and for which (2) the cost of a context switch is too great to make fine grain size multithreading worthwhile. Our target machines are, therefore, medium and large grain size advanced-architecture computers. These include "traditional" shared memory vector supercomputers, such as the Cray Y-MP and C90, and MIMD distributed memory concurrent supercomputers, such as the Intel Paragon and Thinking Machines' CM-5, and the more recently announced IBM SP1 and Cray T3D concurrent systems. Since these machines have only very recently become available, most of the ongoing development of the ScaLAPACK library is being done on a 128-node Intel iPSC/860 hypercube and on the 520-node Intel Delta system.

The Intel Paragon supercomputer can have up to 2000 nodes, each consisting of an i860 processor and a communications processor. The nodes each have at least 16 MB of memory, and are connected by a high-speed network with the topology of a two-dimensional mesh. The CM-5 from Thinking Machines Corporation [53] supports both SIMD and MIMD programming models, and may have up to 16k processors, though the largest CM-5 currently installed has 1024 processors. Each CM-5 node is a Sparc processor and up to 4 associated vector processors. Point-to-point communication between nodes is supported by a data network with the topology of a "fat tree" [46]. Global communication operations, such as synchronization and reduction, are supported by a separate control network. The IBM SP1 system is based on the same RISC chip used in the IBM RS/6000 workstations and uses a multistage switch to connect processors. The Cray T3D uses the Alpha chip from Digital Equipment Corporation, and connects the processors in a three-dimensional torus.

Future advances in compiler and hardware technologies in the mid to late 1990s are expected to make multithreading a viable approach for masking communication costs. Since the blocks in a block-partitioned algorithm can be regarded as separate threads, our approach will still be applicable on machines that exploit medium and coarse grain size multithreading.

32.1.3   High-quality, reusable, mathematical software

In developing a library of high-quality subroutines for dense linear algebra computations the design goals fall into three broad classes:

- performance
- ease-of-use
- range-of-use

32.1.3.1 Performance. Two important performance metrics are concurrent efficiency and scalability. We seek good performance characteristics in our algorithms by eliminating, as much as possible, overhead that is due to load imbalance, data movement, and algorithm restructuring. The way the data are distributed (or decomposed) over the memory hierarchy of a computer is of fundamental importance to these factors. Concurrent efficiency, ε, defined as the concurrent speedup per processor [32], where the concurrent speedup is the execution time, T_{seq}, for the best sequential algorithm running on one processor of the concurrent computer, divided by the execution time, T, of the parallel algorithm running on N_p processors. When direct methods are used, as in LU factorization, the concurrent efficiency depends on the problem size and the number of processors, so on a given parallel computer and for a fixed number of processors, the running time should not vary greatly for problems of the same size. Thus, we may write

$$\varepsilon\,(N, N_p)\; =\; \frac{1}{N_p}\times\frac{T_{seq}\,(N)}{N_p\,T\,(N, N_p)} \tag{32.1}$$

where N represents the problem size. In dense linear algebra computations, the execution time is usually dominated by the floating-point operation count, so the concurrent efficiency is related to performance, G, measured in floating-point operations per second by

$$G\,(\dot{N}, N_p)\; =\; \frac{N_p}{t_{calc}}\times\varepsilon\,(N, N_p) \tag{32.2}$$

where t_{calc} is the time for one floating-point operation. For iterative routines, such as eigensolvers, the number of iterations, and hence the execution time, depends not only on the problem size, but also on other characteristics of the input data, such as condition number. A parallel algorithm is said to be scalable [37] if the concurrent efficiency depends on the problem size and number of processors only through their ratio. This ratio is simply the problem size per processor, often referred to as the *granularity*. Thus, for a scalable algorithm, the concurrent efficiency is constant as the number of processors increases while keeping the granularity fixed. Alternatively, Eq. 32.2 shows that this is equivalent to saying that, for a scalable algorithm, the performance depends linearly on the number of processors for fixed granularity.

32.1.3.2 Ease-of-use. Ease-of-use is concerned with factors such as portability and the user interface to the library. Portability, in its most inclusive sense, means that the code is written in a standard language, such as Fortran, and that the source code can be compiled on an arbitrary machine to produce a program that will run correctly. We call this the "mail-order software" model of portability, since it reflects the model used by software servers such as *netlib* [20]. This notion of portability is quite demanding. It requires that all relevant properties of the computer's arithmetic and architecture be discovered at run time within the confines of a Fortran code. For example, if it is important to know the overflow threshold for scaling purposes, it must be determined at run time *without overflowing*, since overflow is generally fatal. Such demands have resulted in quite large and sophisticated programs [28, 44] that must be modified frequently to deal with new architectures and software releases. This "mail-order" notion of software portability also means that codes generally must be written for the worst possible machine expected to be used, thereby often degrading performance on all others. Ease-of-use is also enhanced if implementation details are largely hidden from the user, for example, through the use of an object-based interface to the library [24].

32.1.3.3 Range-of-use. Range-of-use may be gauged by how numerically stable the algorithms are over a range of input problems, and the range of data structures the library will support. For example, LINPACK and EISPACK deal with dense matrices stored in a rectangular array, packed matrices where only the upper or lower half of a symmetric matrix is stored, and banded matrices where only the nonzero bands are stored. In addition, some special formats such as Householder vectors are used internally to represent orthogonal matrices. There are also sparse matrices, which may be stored in many different ways; but in this chapter we focus on dense and banded matrices, the mathematical types addressed by LINPACK, EISPACK, and LAPACK.

32.2 The BLAS as the Key to Portability

At least three factors affect the performance of portable Fortran code.

1. *Vectorization.* Designing vectorizable algorithms in linear algebra is usually straightforward. Indeed, for many computations there are several variants, all vectorizable, but with varied characteristics in performance. (See, for example, [15].) Linear algebra algorithms can approach the peak performance of many machines—principally because peak performance depends on some form of chaining of vector addition and multiplication operations, and this is just what the algorithms require. However, when the algorithms are realized in straightforward Fortran 77 code, the performance may fall well short of the expected level, usually because vectorizing Fortran compilers fail to minimize the number of memory references—that is, the number of vector load and store operations.

2. *Data movement.* What often limits the actual performance of a vector or scalar floating-point unit is the rate of transfer of data between various levels of memory in the machine. Examples include the transfer of vector operands in and out of vector registers, the transfer of scalar operands in and out of a high-speed scalar processor, the movements of data between main memory and a high-speed cache or local memory, paging between actual memory and disk storage in a virtual memory system, and interprocessor communication on a distributed memory concurrent computer.

3. *Parallelism.* The nested loop structure of most linear algebra algorithms offers considerable scope for loop-based parallelism. This is the principal type of parallelism that LAPACK and ScaLAPACK presently aim to exploit. On shared memory concurrent computers, this type of parallelism can sometimes be generated automatically by a compiler, but often requires the insertion of compiler directives. On distributed memory concurrent computers, data must be moved between processors. This is usually done by explicit calls to message passing routines, although parallel language extensions such as Coherent Parallel C [31] and Split-C [13] do the message passing implicitly.

The question arises, "How can we achieve sufficient control over these three factors to obtain the levels of performance that machines can offer?" The answer is through use of the BLAS.

There are now three levels of BLAS:

Level 1 BLAS [45]: for vector operations, such as $y \leftarrow \alpha x + y$
Level 2 BLAS [18]: for matrix-vector operations, such as $y \leftarrow \alpha A x + \beta y$
Level 3 BLAS [17]: for matrix-matrix ←operations, such as $C \leftarrow \alpha A B + \beta C$.

Here, A, B and C are matrices, x and y are vectors, and α and β are scalars.

The Level 1 BLAS are used in LAPACK, but for convenience rather than for performance. They perform an insignificant fraction of the computation, and they cannot achieve high efficiency on most modern supercomputers.

The Level 2 BLAS can achieve near-peak performance on many vector processors, such as a single processor of a CRAY X-MP or Y-MP, or Convex C-2 machine. However, on other vector processors such as a CRAY-2 or an IBM 3090 VF, the performance of the Level 2 BLAS is limited by the rate of data movement among various levels of memory.

The Level 3 BLAS overcome this limitation. This third level of BLAS performs $O(n^3)$ floating point operations on $O(n^2)$ data, whereas the Level 2 BLAS perform only $O(n^2)$ operations on $O(n^2)$ data. The Level 3 BLAS also allow us to exploit parallelism in a way that is transparent to the software that calls them. While the Level 2 BLAS offer some scope for exploiting parallelism, greater scope is provided by the Level 3 BLAS, as Table 32.1 illustrates.

TABLE 32.1 Speed (Megaflops) of Level 2 and Level 3 BLAS Operations on a Cray Y-MP. (All matrices are of order 500; U is upper triangular.)

Number of processors:	1	2	4	8
Level 2: $y \leftarrow \alpha Ax + \beta y$	311	611	1197	2285
Level 3: $C \leftarrow \alpha AB + \beta C$	312	623	1247	2425
Level 2: $x \leftarrow Ux$	293	544	898	1613
Level 3: $B \leftarrow UB$	310	374	479	584
Level 2: $x \leftarrow U^{-1}x$	272	374	479	584
Level 3: $B \leftarrow U^{-1}B$	309	618	1235	2398

32.3 Block Algorithms and Their Derivation

It is comparatively straightforward to recode many of the algorithms in LINPACK and EISPACK so that they call Level 2 BLAS. Indeed, in the simplest cases the same floating-point operations are done, possibly even in the same order. It is just a matter of reorganizing the software. To illustrate this point, we consider the Cholesky factorization algorithm used in the LINPACK routine SPOFA, which factorizes a symmetric positive definite matrix as $A = UTU$. We consider Cholesky factorization because the algorithm is simple, and no pivoting is required. In Section 32.4 we shall consider the slightly more complicated example of LU factorization.

Suppose that after $j-1$ steps the block A_{oo} in the upper lefthand corner of A has been factored as $A_{oo} = U^T_{oo} U_{oo}$. The next row and column of the factorization can then be computed by writing $A = U^T U$ as

$$\begin{bmatrix} A_{oo} & b_j & A_{02} \\ \cdot & a_{jj} & c_j^T \\ \cdot & \cdot & A_{22} \end{bmatrix} = \begin{bmatrix} U^T_{oo} & 0 & 0 \\ v_j^T & u_{jj} & 0 \\ U^T_{02} & w_j^T & U^T_{22} \end{bmatrix} \begin{bmatrix} U_{oo} & v_j & U_{02} \\ 0 & u_{jj} & w_j^T \\ 0 & 0 & U_{22} \end{bmatrix}$$

where b_j, c_j, v_j, and w_j are column vectors of length $j-1$, and a_{jj} and u_{jj} are scalars. Equating coefficients of the jth column, we obtain

$$b_j = U_{oo}^T v_j$$

$$a_{jj} = v_j^T v_j + u_{jj}^2$$

Since U_{oo} has already been computed, we can compute v_j and u_{jj} from the equations

$$U_{oo}^T v_j = b_j$$

$$u_{jj}^2 = a_{jj} - v_j^T v_j$$

The body of the code of the LINPACK routine SPOFA that implements the above method is shown in Fig. 32.1. The same computation recoded in "LAPACK-style" to use the Level 2 BLAS routine STRSV (which solves a triangular system of equations) is shown in Fig. 32.2. The call to STRSV has replaced the loop over K that made several calls to the Level 1 BLAS routine SDOT. (For reasons given below, this is not the actual code used in LAPACK—hence the term "LAPACK-style.")

This change by itself is sufficient to result in large gains in performance on a number of machines—for example, from 72 to 251 megaflops for a matrix of order 500 on one processor of a CRAY Y-MP. Since this is 81 percent of the peak speed of matrix-matrix multiplication on this processor, we cannot hope to do very much better by using Level 3 BLAS.

```
do j = 0, n-1
   info = j + 1
   s = 0.0e0
   jm1 = j
   if (jm1 .ge.  1) then
     do k = 0, jm1 - 1
        t = a(k,j) - sdot(k,a(0,k),1,a(0,j),1)
        t = t/a(k,k)
        a(k,j) = t
        s = s + t*t
     end do
   end if
   s = a(j,j) - s
   if (s .le.  0.0e0) go to 40
   a(j,j) = sqrt(s)
end do
```

Figure 32.1 Body of the LINPACK routing SPOFA for Cholesky factorization

```
do j = 0, n - 1
  call strsv( 'upper', 'transpose', 'non-unit', j, a, lda,
a(0,j), 1 )
  s = a(j,j) - sdot( j, a(0,j), 1, a(0,j), 1 )
  if ( s .le.  zero ) go to 20
  a(j,j) = sqrt( s )
end do
```

Figure 32.2 Body of the "LINPACK-style" routing SPOFA for Cholesky factorization

We can, however, restructure the algorithm at a deeper level to exploit the faster speed of the Level 3 BLAS. This restructuring involves recasting the algorithm as a block algorithm—that is, an algorithm that operates on blocks or submatrices of the original matrix.

32.3.1   Deriving a block algorithm

To derive a block form of Cholesky factorization, we partition the matrices as shown in Fig. 32.3, in which the diagonal blocks of A and U are square, but of differing sizes. We assume that the first block has already been factored as $A_{00} = U_{00}^T U_{00}$, and that we now want to determine the second block column of U consisting of the blocks U_{01} and U_{11}. Equating submatrices in the second block of columns, we obtain

$$A_{01} = U_{00}^T U_{01}$$

$$A_{11} = U_{01}^T U_{01} = U_{11}^T U_{11}$$

Hence, since U_{00} has already been computed, we can compute U_{01} as the solution to the equation

$$U_{00}^T U_{01} = A_{01}$$

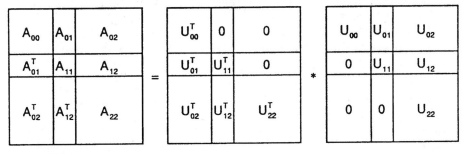

Figure 32.3 Partitioning of A, UT, and U into blocks. It is assumed that the first block has already been factored as $A_{00} = U_{00}^T U_{00}$, and we want to determine the block column consisting of U_{01} and U_{11}. Note that the diagonal blocks of A and U are square matrices.

by a call to the Level 3 BLAS routine STRSM; and then we can compute U_{11} from

$$U_{11}^T U_{11} = A_{11} - U_{01}^T U_{01}$$

This involves first updating the symmetric submatrix A_{11} by a call to the Level 3 BLAS routine SSYRK, and then computing its Cholesky factorization. Since Fortran does not allow recursion, a separate routine must be called (using Level 2 BLAS rather than Level 3), named SPOTF2 in Fig. 32.4. In this way, successive blocks of columns of U are computed. The LAPACK-style code for the block algorithm is shown in Fig. 32.4. This code runs at 49 megaflops on an IBM 3090, more than double the speed of the LINPACK code. On a CRAY Y-MP, the use of Level 3 BLAS squeezes a little more performance out of one processor, but makes a large improvement when using all eight processors.

But that is not the end of the story, and the code given above is not the code actually used in the LAPACK routine SPOTRF. We mentioned earlier that for many linear algebra computations there are several algorithmic variants, often referred to as i-, j-, and k-variants, according to a convention introduced in [15] and used in [36]. The same is true of the corresponding block algorithms.

It turns out that the j-variant chosen for LINPACK, and used in the above examples, is not the fastest on many machines, because it performs most of the work in solving triangular systems of equations, which can be significantly slower than matrix-matrix multiplication. The variant actually used in LAPACK is the i-variant, which relies on matrix-matrix multiplication for most of the work. Table 32.2 summarizes the results.

TABLE 32.2   Speed (Megaflops) of Cholesky Factorization $A = U^T U$ for $n = 500$

	IBM 3090 VF, 1 processor	Cray Y-MP, 1 processor	Cray Y-MP, 8 processors
j-variant: LINPACK	23	72	72
j-variant: using level 2 BLAS	24	251	378
j-variant: using level 3BLAS	49	287	1225
i-variant: using level 3BLAS	50	290	1414

```
do j = 0, n-1, nb
  jb = min( nb, n-j )
  call strsm( 'left', 'upper', 'transpose', 'non-unit', j, jb,
one,
              a, lda, a(0,j), lda )
  call ssyrk( 'upper', 'transpose', jb, j, -one, a(0,j), lda,
one,
              a(j,j), lda )
  call spotf2( 'upper', jb, a(j,j), lda, info )
  if( info .ne.  0 ) go to 20
end do
```

Figure 32.4 The body of the "LAPACK-style" routine SPOFA for block Cholesky factorization. In this code segment, nb denotes the width of the blocks.

32.3.2 Examples of block algorithms in LAPACK

Having discussed in detail the derivation of one particular block algorithm, we now describe examples of the performance achieved with two well known block algorithms: LU and Cholesky factorizations. No extra floating-point operations nor extra working storage is required for either of these simple block algorithms. (See Gallivan et al. [33] and Dongarra et al. [19] for surveys of algorithms for dense linear algebra on high-performance computers.)

Table 32.3 illustrates the speed of the LAPACK routine for LU factorization of a real matrix, SGETRF in single precision on CRAY machines, and DGETRF in double precision on all other machines. Thus, 64-bit floating-point arithmetic is used on all machines tested. A block size of 1 means that the unblocked algorithm is used, since it is faster than—or at least as fast as—a block algorithm.

TABLE 32.3 Speed (Megaflops) of SGETRF/DGETRF for Square Matrices of Order n

Machine	No. of processors	Block size	Values of n				
			100	200	300	400	500
IBM RISC/6000-530	1	32	19	25	29	31	33
Alliant FX/8	8	16	9	26	32	46	57
IBM 3090J VF	1	64	23	41	52	58	63
Convex C-240	4	64	31	60	82	100	112
Cray Y-MP	1	1	132	219	254	272	283
Cray-2	1	64	110	211	292	318	358
Siemens/Fujitsu VP 400-EX	1	64	46	132	222	309	397
NEC SX2	1	1	118	274	412	504	577
Cray Y-MP	8	64	195	556	920	1188	1408

LAPACK is designed to give high efficiency on vector processors, high-performance "superscalar" workstations, and shared memory multiprocessors. LAPACK in its present form is less likely to give good performance on other types of parallel architectures (for example, massively parallel SIMD machines or MIMD distributed memory machines), but the ScaLAPACK project, described in Section 32.1.1.4, is intended to adapt LAPACK to these new architectures. LAPACK can also be used satisfactorily on all types of scalar machines (PCs, workstations, mainframes). Table 32.4 gives similar results for Cholesky factorization, extending the results given in Table 32.2.

TABLE 32.4 Speed (Megaflops) of SPOTRF/DPOTRF for Matrices of Order n. Here, UPLO = "U," so the factorization is of the form $A = U^T U$.

Machine	No. of processors	Block size	Values of n				
			100	200	300	400	500
IBM RISC/6000-530	1	32	21	29	34	36	38
Alliant FX/8	8	16	10	27	40	49	52
IBM 3090J VF	1	48	26	43	56	62	67
Convex C-240	4	64	32	63	82	96	103
Cray Y-MP	1	1	126	219	257	275	285
Cray-2	1	64	109	213	294	318	362
Siemens/Fujitsu VP 400-EX	1	64	53	145	237	312	369
NEC SX2	1	1	155	387	589	719	819
Cray Y-MP	8	32	146	479	845	1164	1393

LAPACK, like LINPACK, provides LU and Cholesky factorizations of band matrices. The LINPACK, algorithms can easily be restructured to use Level 2 BLAS, though restructuring has little effect on performance for matrices of very narrow bandwidth. It is also possible to use Level 3 BLAS, at the price of doing some extra work with zero elements outside the band [22]. This process becomes worthwhile for large matrices and semi-bandwidth greater than 100 or so.

32.4   LU Factorization

In this section, we first discuss the uses of dense LU factorization in several fields. We next develop a block-partitioned version of the k, or right-looking, variant of the LU factorization algorithm. In subsequent sections, the parallelization of this algorithm is described in detail in order to highlight the issues and considerations that must be taken into account in developing an efficient, scalable, and transportable dense linear algebra library for MIMD distributed memory concurrent computers.

32.4.1   Uses of LU factorization in science and engineering

A major source of large dense linear systems is that of problems involving the solution of boundary integral equations. These are integral equations defined on the boundary of a region of interest. All examples of practical interest compute some intermediate quantity on a two-dimensional boundary and then use this information to compute the final desired quantity in three-dimensional space. The price one pays for replacing three dimensions with two is that what started as a sparse problem in $O(n^3)$ variables is replaced by a dense problem in $O(n^2)$.

Dense systems of linear equations are found in numerous applications, including:

- airplane wing design
- radar cross-section studies
- flow around ships and other off-shore constructions
- diffusion of solid bodies in a liquid
- noise reduction
- diffusion of light through small particles

The electromagnetics community is a major user of dense linear systems solvers. Of particular interest to this community is the solution of the so-called radar cross-section problem. In this problem, a signal of fixed frequency bounces off an object. The goal is to determine the intensity of the reflected signal in all possible directions. The underlying differential equation may vary, depending on the specific problem. In the design of stealth aircraft, the principal equation is the Helmholtz equation. To solve this equation, researchers use the *method of moments* [38, 56]. In the case of fluid flow, the problem often involves solving the Laplace or Poisson equation. Here, the boundary integral solution is known as the *panel method* [40, 41], so named from the quadrilaterals that discretize and approximate a structure such as an airplane. Generally, these methods are called *boundary element methods*.

Use of these methods produces a dense linear system of size $O(N)$ by $O(N)$, where N is the number of boundary points (or panels) being used. It is not unusual to see size $3N$ by $3N$, because of three physical quantities of interest at every boundary element.

A typical approach to solving such systems is to use LU factorization. Each entry of the matrix is computed as an interaction of two boundary elements. Often, many integrals must be computed. In many instances, the time required to compute the matrix is considerably larger than the time for solution.

Only the builders of stealth technology who are interested in radar cross sections are considering using direct Gaussian elimination methods for solving dense linear systems. These systems are always symmetric and complex, but not Hermitian.

For further information on various methods for solving large dense linear algebra problems that arise in computational fluid dynamics, see the report by Edelman [30].

32.4.2   Derivation of a block algorithm for LU factorization

Suppose the $M \times N$ matrix A is partitioned as shown in Fig. 32.5, and we seek a factorization $A = LU$, where the partitioning of L and U is also shown in Fig. 32.5. Then we may write,

$$L_{00}U_{00} = A_{00} \tag{38.3}$$

$$L_{10}U_{00} = A_{10} \tag{38.4}$$

$$L_{00}U_{01} = A_{01} \tag{38.5}$$

$$L_{10}U_{01} + L_{11}U_{11} = A_{11} \tag{38.6}$$

where A_{00} is $r \times r$, A_{01} is $r \times (N-r)$, A_{10} is $(M-r) \times r$, and A_{11} is $(M-r) \times (N-r)$. L_{00} and L_{11} are lower triangular matrices with ones on the main diagonal, and U_{00} and U_{11} are upper triangular matrices.

Equations (32.3) and (32.4), taken together, perform an LU factorization on the first $M \times r$ panel of A (i.e., A_{00} and A_{10}). Once this is completed, the matrices, L_{00}, L_{10}, and U_{00}, are known, and the lower triangular system in Eq. (32.5) can be solved to give U_{01}. Finally, we rearrange Eq. (32.6) as

$$A'_{11} = A_{11} - L_{10}U_{01} = L_{11}U_{11} \tag{32.7}$$

From this equation, we see that the problem of finding L_{11} and U_{11} reduces to finding the LU factorization of the $(M-r) \times (N-r)$ matrix A'_{11}. This can be done by applying the steps outlined above to A_{11} instead of to A. Repeating these steps K times, where

$$K = \min\left(\lceil M/r \rceil, \lceil N/r \rceil\right) \tag{32.8}$$

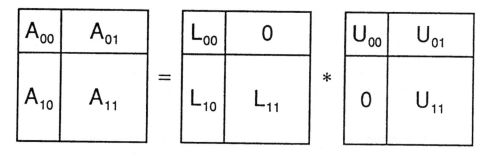

Figure 32.5 Block LU factorization of the partitioned matrix A. A_{00} is $r \times r$, A_{01} is $r \times (N-r)$, A_{10} is $(M-r) \times r$, and A_{11} is $(M-r) \times (N-r)$. L_{00} and L_{11} are lower triangular matrices with ones on the main diagonal, and U_{00} and U_{11} are upper triangular matrices.

we obtain the LU factorization of the original $M \times N$ matrix A. For an in-place algorithm, A is overwritten by L and U—the ones on the diagonal of L do not need to be stored explicitly. Similarly, when A is updated by Eq. (32.7), this may also be done in place.

After k of these K steps, the first kr columns of L and the first kr rows of U have been evaluated, and matrix A has been updated to the form shown in Fig. 32.6, in which panel B is $(M - kr) \times r$ and C is $r \times (N - (k - 1) r)$. Step $k + 1$ then proceeds as follows:

1. Factor B to form the next panel of L, performing partial pivoting over rows if necessary. (See Fig. 32.14.) This evaluates the matrices L_0, L_1, and U_0 in Fig. 32.6.
2. Solve the triangular system $L_0 U_1 = C$ to get the next row of blocks of U.
3. Do a rank-r update on the trailing submatrix E, replacing it with $E' = E - L_1 U_1$.

The LAPACK implementation of this form of LU factorization uses the Level 3 BLAS routines xTRSM and xGEMM to perform the triangular solve and rank-r update. We can regard the algorithm as acting on matrices that have been partitioned into blocks of $r \times r$ elements, as shown in Fig. 32.7.

32.5   Data Distribution

The fundamental data object in the LU factorization algorithm presented in Section 32.4.2 is a block-partitioned matrix. In this section, we described the block-cyclic method for distributing such a matrix over a two-dimensional mesh of processes, or template. In general, each process has an independent thread of control, and with each process is associated some local memory directly accessible only by that process. The assignment of these processes to physical processors is a machine-dependent optimization issue, and will be considered later in Section 32.7.

An important property of the class of data distribution we shall use is that independent decompositions are applied over rows and columns. We shall, therefore, begin by considering the distribution of a vector of M data objects over P processes. This can be described by a mapping of the global index, of a data object to an index pair, (p,i), where p specifies the process to which the data object is assigned, and i specifies the location in the local memory of p at which it is stored. We shall assume $0 \le m < M$ and $0 \le p < P$.

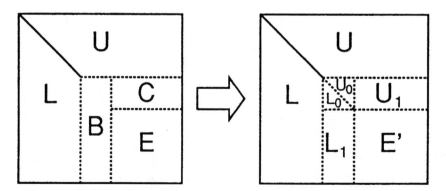

Figure 32.6 Stage $k + 1$ of the block LU factorization algorithm showing how the panels B and C, and the trailing submatrix E are updated. The trapezoidal submatrices L and U have already been factored in previous steps. L has kr columns, and U has kr rows. In the step shown another r columns of L and r rows of U are evaluated.

$A_{0,0}$	$A_{0,1}$	$A_{0,2}$	$A_{0,3}$	$A_{0,4}$	$A_{0,5}$
$A_{1,0}$	$A_{1,1}$	$A_{1,2}$	$A_{1,3}$	$A_{1,4}$	$A_{1,5}$
$A_{2,0}$	$A_{2,1}$	$A_{2,2}$	$A_{2,3}$	$A_{2,4}$	$A_{2,5}$
$A_{3,0}$	$A_{3,1}$	$A_{3,2}$	$A_{3,3}$	$A_{3,4}$	$A_{3,5}$
$A_{4,0}$	$A_{4,1}$	$A_{4,2}$	$A_{4,3}$	$A_{4,4}$	$A_{4,5}$
$A_{5,0}$	$A_{5,1}$	$A_{5,2}$	$A_{5,3}$	$A_{5,4}$	$A_{5,5}$

Figure 32.7 Block-partitioned matrix A. Each block $A_{i,j}$ consists of $r \times r$ matrix elements.

Two common decompositions are the *block* and the *cyclic* decompositions [32, 55]. The block decomposition, which is often used when the computational load is distributed homogeneously over a regular data structure such as a Cartesian grid, assigns contiguous entries in the global vector to the processes in blocks.

$$m \rightarrow (\lfloor m/L \rfloor, m \bmod L), \tag{32.9}$$

where $L = \lceil M/P \rceil$. The cyclic decomposition (also known as the wrapped or scattered decomposition) is commonly used to improve load balance when the computational load is distributed inhomogeneously over a regular data structure. The cyclic decomposition assigns consecutive entries in the global vector to successive different processes,

$$m \rightarrow (m \bmod P, \lfloor m/P \rfloor) \tag{32.10}$$

Examples of the block and cyclic decompositions are shown in Fig. 32.8.

The block cyclic decomposition is a generalization of the block and cyclic decompositions in which blocks of consecutive data objects are distributed cyclically over the processes. In the block cyclic decomposition the mapping of the global index, m, can be expressed as $m \rightarrow (p, b, i)$, where p is the process number, b is the block number in process p, and i is the index within block b to which m is mapped. Thus, if the number of data objects in a block is r, the block cyclic decomposition may be written,

$$m \rightarrow \left(\left\lfloor \frac{m \bmod T}{r} \right\rfloor, \left\lfloor \frac{m}{T} \right\rfloor, m \bmod r \right) \tag{32.11}$$

m	0	1	2	3	4	5	6	7	8	9
p	0	0	0	0	1	1	1	1	2	2
i	0	1	2	3	0	1	2	3	0	1

(a) Block

m	0	1	2	3	4	5	6	7	8	9
p	0	1	2	0	1	2	0	1	2	0
i	0	0	0	1	1	1	2	2	2	3

(b) Cyclic

Figure 32.8 Examples of (a) block and (b) cyclic decompositions of $M = 10$ data objects over $P = 3$ processes.

where $T = rP$. It should be noted that this reverts to the cyclic decomposition when $r = 1$, with local index $i = 0$ for all blocks. A block decomposition is recovered when $r = L$, in which case there is a single block in each process with block number $b = 0$. The inverse mapping of the triplet (p, b, i) to a global index is given by,

$$(p,b,i) \rightarrow Br+i = pr+bT+i \tag{32.12}$$

where $B = p + bP$ is the global block number. The block cyclic decomposition is one of the data distributions supported by High Performance Fortran (HPF) [42], and has been previously used, in one form or another, by several researchers. (See [1, 4, 5, 9, 23, 27, 50, 52, 54] for examples of its use.) The block cyclic decomposition is illustrated with an example in Fig. 32.9.

The form of the block cyclic decomposition given by Eq. (32.11) ensures that the block with global index 0 is placed in process 0, the next block is placed in process 1, and so on. However it is sometimes necessary to offset the processes relative to the global block index so that, in general, the first block is placed in process p_0, the next in process $p_0 + 1$, and so on. We, therefore, generalize the block cyclic decomposition by replacing m on the righthand side of Eq. (32.11) by $m' = m + rp_0$ to give,

$$m \rightarrow \left(\left\lfloor \frac{m' \bmod T}{r} \right\rfloor , \left\lfloor \frac{m'}{T} \right\rfloor , m' \bmod r \right) \tag{32.13}$$

$$= \left(\left(\left\lfloor \frac{m \bmod T}{r} \right\rfloor + p_0 \right) \bmod P , \left\lfloor \frac{m + rp_0}{T} \right\rfloor , m \bmod r \right)$$

Equation (32.12) may also be generalized to,

$$(p, b, i) \rightarrow Br + i = (p - p_0)r + bT + i \tag{32.14}$$

where now the global block number is given by $B = (p - p_0) + bP$. It should be noted that in processes with $p < p_0$, block 0 is not within the range of the block cyclic mapping and it is, therefore, an error to reference it in any way.

m	0	1	2	3	4	5	6	7	8	9	10	11	12	13	14	15	16	17	18	19	20	21	22
p	0	0	1	1	2	2	0	0	1	1	2	2	0	0	1	1	2	2	0	0	1	1	2
b	0	0	0	0	0	0	1	1	1	1	1	1	2	2	2	2	2	2	3	3	3	3	3
i	0	1	0	1	0	1	0	1	0	1	0	1	0	1	0	1	0	1	0	1	0	1	0

(a) $m \mapsto (p, b, i)$

p	0	0	0	0	0	0	0	0	1	1	1	1	1	1	1	1	2	2	2	2	2	2	2
b	0	0	1	1	2	2	3	3	0	0	1	1	2	2	3	3	0	0	1	1	2	2	3
i	0	1	0	1	0	1	0	1	0	1	0	1	0	1	0	1	0	1	0	1	0	1	0
m	0	1	6	7	12	13	18	19	2	3	8	9	14	15	20	21	4	5	10	11	16	17	22

(b) $(p, b, i) \mapsto m$

Figure 32.9 An example of the block cyclic decomposition of $M = 23$ data objects over $P = 3$ processes for a block size of $r = 2$; (top) mapping from global index, m, to the triplet (p, b, i) and (bottom) the inverse mapping.

1. In decomposing an $M \times N$ matrix we apply independent block cyclic decompositions in the row and column directions. Thus, suppose the matrix rows are distributed with block size r and offset p_0 over P processes by the block cyclic mapping $\mu_{r, p_0, P}$, and the matrix columns are distributed with block size s and offset q_0 over Q processes by the block cyclic mapping $v_{s, q_0 Q}$. Then the matrix element indexed globally by (m, n) is mapped as follows,

$$
\begin{aligned}
m &\overset{\mu}{\to} (p, b, i) \\
n &\overset{v}{\to} (q, d, j)
\end{aligned}
\tag{32.15}
$$

The decomposition of the matrix can be regarded as the tensor product of the row and column decompositions, and we can write,

$$
(m, n) \to ((p, q), (b, d), (i, j)) \tag{32.16}
$$

The block cyclic matrix decomposition given by Eqs. (32.15) and (32.16) distributes blocks of size $r \times s$ to a mesh of $P \times Q$ processes. We shall refer to this mesh as the process template, and refer to processes by their position in the template. Equation (32.16) says that global index (m, n) is mapped to process (p, q), where it is stored in the block at location (b, d) in a two-dimensional array of blocks. Within this block it is stored at location (i, j). The decomposition is completely specified by the parameters r, s, p_0, q_0, P, and Q. In Fig. 32.10, an example is given of the block cyclic decomposition of a 36×80 matrix for block size 4×5, a process template 3×4, and a template offset $(p_0, q_0) = (0,0)$. Figure 32.11 shows the same example but for a template offset of $(1,2)$.

The block cyclic decomposition can reproduce most of the data distributions commonly used in linear algebra computations on parallel computers. For example, if $Q = 1$ and $r = \lceil M/P \rceil$ the block row decomposition is obtained. Similarly $P = 1$ and $s = \lceil N/Q \rceil$ gives a block column decomposition. These compositions, together with row and column cyclic decompositions, are shown in Fig. 32.12. Other commonly used block cyclic matrix decompositions are shown in Fig. 32.13.

32.6 Parallel Implementation

In this section, we describe the parallel implementation of LU factorization, with partial pivoting over rows, for a block-partitioned matrix. The matrix A to be factored is assumed to have a block cyclic decomposition, and at the end of the computation it is overwritten by the lower and upper triangular factors, L and U. This implicitly determines the decomposition of L and U. Quite a high-level description is given here since the details of the parallel implementation involve optimization issues that will be addressed in Section 32.7.

The sequential LU factorization algorithm described in Section 32.4.2 uses square blocks. Although in the parallel algorithm we could choose to decompose the matrix using nonsquare blocks, this would result in a more complicated code, and additional sources of concurrent overhead. For LU factorization we, therefore, restrict the decomposition to use only square blocks, so that the blocks used to decompose the matrix are the same as those used to partition the computation. If the block size is $r \times r$, then an $M \times N$ matrix consists of $M_b \times N_b$ blocks, where $M_b = \lceil M/r \rceil$ and $N_b = \lceil N/r \rceil$.

As discussed in Section 32.4.2, LU factorization proceeds in a series of sequential steps indexed by $k = 0$, min $(M_b, N_b) - 1$, in each of which the following three tasks are performed:

p,q							D									
	0	1	2	3	4	5	6	7	8	9	10	11	12	13	14	15
0	0,0	0,1	0,2	0,3	0,0	0,1	0,2	0,3	0,0	0,1	0,2	0,3	0,0	0,1	0,2	0,3
1	1,0	1,1	1,2	1,3	1,0	1,1	1,2	1,3	1,0	1,1	1,2	1,3	1,0	1,1	1,2	1,3
2	2,0	2,1	2,2	2,3	2,0	2,1	2,2	2,3	2,0	2,1	2,2	2,3	2,0	2,1	2,2	2,3
3	0,0	0,1	0,2	0,3	0,0	0,1	0,2	0,3	0,0	0,1	0,2	0,3	0,0	0,1	0,2	0,3
4	1,0	1,1	1,2	1,3	1,0	1,1	1,2	1,3	1,0	1,1	1,2	1,3	1,0	1,1	1,2	1,3
B 5	2,0	2,1	2,2	2,3	2,0	2,1	2,2	2,3	2,0	2,1	2,2	2,3	2,0	2,1	2,2	2,3
6	0,0	0,1	0,2	0,3	0,0	0,1	0,2	0,3	0,0	0,1	0,2	0,3	0,0	0,1	0,2	0,3
7	1,0	1,1	1,2	1,3	1,0	1,1	1,2	1,3	1,0	1,1	1,2	1,3	1,0	1,1	1,2	1,3
8	2,0	2,1	2,2	2,3	2,0	2,1	2,2	2,3	2,0	2,1	2,2	2,3	2,0	2,1	2,2	2,3
9	0,0	0,1	0,2	0,3	0,0	0,1	0,2	0,3	0,0	0,1	0,2	0,3	0,0	0,1	0,2	0,3
10	1,0	1,1	1,2	1,3	1,0	1,1	1,2	1,3	1,0	1,1	1,2	1,3	1,0	1,1	1,2	1,3
11	2,0	2,1	2,2	2,3	2,0	2,1	2,2	2,3	2,0	2,1	2,2	2,3	2,0	2,1	2,2	2,3

(a) Assignment of global block indices (*B, D*), to processes (*p, q*)

B,D	0				1				q 2				3			
0	0,0	0,4	0,8	0,12	0,1	0,5	0,9	0,13	0,2	0,6	0,10	0,14	0,3	0,7	0,11	0,15
	3,0	3,4	3,8	3,12	3,1	3,5	3,9	3,13	3,2	3,6	3,10	3,14	3,3	3,7	3,11	3,15
	6,0	6,4	6,8	6,12	6,1	6,5	6,9	6,13	6,2	6,6	6,10	6,14	6,3	6,7	6,11	6,15
	9,0	9,4	9,8	9,12	9,1	9,5	9,9	9,13	9,2	9,6	9,10	9,14	9,3	9,7	9,11	9,15
p 1	1,0	1,4	1,8	1,12	1,1	1,5	1,9	1,13	1,2	1,6	1,10	1,14	1,3	1,7	1,11	1,15
	4,0	4,4	4,8	4,12	4,1	4,5	4,9	4,13	4,2	4,6	4,10	4,14	4,3	4,7	4,11	4,15
	7,0	7,4	7,8	7,12	7,1	7,5	7,9	7,13	7,2	7,6	7,10	7,14	7,3	7,7	7,11	7,15
	10,0	10,4	10,8	10,12	10,1	10,5	10,9	10,13	10,2	10,6	10,10	10,14	10,3	10,7	10,11	10,15
2	2,0	2,4	2,8	2,12	2,1	2,5	2,9	2,13	2,2	2,6	2,10	2,14	2,3	2,7	2,11	2,15
	5,0	5,4	5,8	5,12	5,1	5,5	5,9	5,13	5,2	5,6	5,10	5,14	5,3	5,7	5,11	5,15
	8,0	8,4	8,8	8,12	8,1	8,5	8,9	8,13	8,2	8,6	8,10	8,14	8,3	8,7	8,11	8,15
	11,0	11,4	11,8	11,12	11,1	11,5	11,9	11,13	11,2	11,6	11,10	11,14	11,3	11,7	11,11	11,15

(b) Global blocks (*B, D*) in each process (*p, q*)

Figure 32.10 Block cyclic decomposition of a 36×80 matrix with a block size of 4×5, onto a 3×4 process template. Each small rectangle represents one matrix block—individual matrix elements are not shown. In (a), shading is used to emphasize the process template that is periodically stamped over the matrix, and each block is labeled with the process to which it is assigned. In (b), each shaded region shows the blocks in one process and is labeled with the corresponding global block indices. In both figures, the black rectangles indicate the blocks assigned to process (0,0).

(a) Assignment of global block indices (B, D), to processes (p, q)

D

p,q \	0	1	2	3	4	5	6	7	8	9	10	11	12	13	14	15
0	1,2	1,3	1,0	1,1	1,2	1,3	1,0	1,1	1,2	1,3	1,0	1,1	1,2	1,3	1,0	1,1
1	2,2	2,3	2,0	2,1	2,2	2,3	2,0	2,1	2,2	2,3	2,0	2,1	2,2	2,3	2,0	2,1
2	0,2	0,3	0,0	0,1	0,2	0,3	0,0	0,1	0,2	0,3	0,0	0,1	0,2	0,3	0,0	0,1
3	1,2	1,3	1,0	1,1	1,2	1,3	1,0	1,1	1,2	1,3	1,0	1,1	1,2	1,3	1,0	1,1
4	2,2	2,3	2,0	2,1	2,2	2,3	2,0	2,1	2,2	2,3	2,0	2,1	2,2	2,3	2,0	2,1
5	0,2	0,3	0,0	0,1	0,2	0,3	0,0	0,1	0,2	0,3	0,0	0,1	0,2	0,3	0,0	0,1
6	1,2	1,3	1,0	1,1	1,2	1,3	1,0	1,1	1,2	1,3	1,0	1,1	1,2	1,3	1,0	1,1
7	2,2	2,3	2,0	2,1	2,2	2,3	2,0	2,1	2,2	2,3	2,0	2,1	2,2	2,3	2,0	2,1
8	0,2	0,3	0,0	0,1	0,2	0,3	0,0	0,1	0,2	0,3	0,0	0,1	0,2	0,3	0,0	0,1
9	1,2	1,3	1,0	1,1	1,2	1,3	1,0	1,1	1,2	1,3	1,0	1,1	1,2	1,3	1,0	1,1
10	2,2	2,3	2,0	2,1	2,2	2,3	2,0	2,1	2,2	2,3	2,0	2,1	2,2	2,3	2,0	2,1
11	0,2	0,3	0,0	0,1	0,2	0,3	0,0	0,1	0,2	0,3	0,0	0,1	0,2	0,3	0,0	0,1

(B label at left, rows 0–11)

q

B,D	0					1				2					3				
	—	—	—	—	—	—	—	—	—	—	—	—	—	—	—	—	—	—	—
0	2,2	2,6	2,10	2,14	—	2,3	2,7	2,11	2,15	2,0	2,4	2,8	2,12	—	2,1	2,5	2,9	2,13	—
	5,2	5,6	5,10	5,14	—	5,3	5,7	5,11	5,15	5,0	5,4	5,8	5,12	—	5,1	5,5	5,9	5,13	—
	8,2	8,6	8,10	8,14	—	8,3	8,7	8,11	8,15	8,0	8,4	8,8	8,12	—	8,1	8,5	8,9	8,13	—
	11,2	11,6	11,10	11,14	—	11,3	11,7	11,11	11,15	11,0	11,4	11,8	11,12	—	11,1	11,5	11,9	11,13	—
p 1	0,2	0,6	0,10	0,14	—	0,3	0,7	0,11	0,15	0,0	0,4	0,8	0,12	—	0,1	0,5	0,9	0,13	—
	3,2	3,6	3,10	3,14	—	3,3	3,7	3,11	3,15	3,0	3,4	3,8	3,12	—	3,1	3,5	3,9	3,13	—
	6,2	6,6	6,10	6,14	—	6,3	6,7	6,11	6,15	6,0	6,4	6,8	6,12	—	6,1	6,5	6,9	6,13	—
	9,2	9,6	9,10	9,14	—	9,3	9,7	9,11	9,15	9,0	9,4	9,8	9,12	—	9,1	9,5	9,9	9,13	—
	—	—	—	—	—	—	—	—	—	—	—	—	—	—	—	—	—	—	—
2	1,2	1,6	1,10	1,14	—	1,3	1,7	1,11	1,15	1,0	1,4	1,8	1,12	—	1,1	1,5	1,9	1,13	—
	4,2	4,6	4,10	4,14	—	4,3	4,7	4,11	4,15	4,0	4,4	4,8	4,12	—	4,1	4,5	4,9	4,13	—
	7,2	7,6	7,10	7,14	—	7,3	7,7	7,11	7,15	7,0	7,4	7,8	7,12	—	7,1	7,5	7,9	7,13	—
	10,2	10,6	10,10	10,14	—	10,3	10,7	10,11	10,15	10,0	10,4	10,8	10,12	—	10,1	10,5	10,9	10,13	—
	—	—	—	—	—	—	—	—	—	—	—	—	—	—	—	—	—	—	—

(b) Global blocks (B, D) in each process (p, q)

Figure 32.11 The same matrix decomposition as shown in Fig. 32.10 but for a template offset of $(p_0, q_0) = (1, 2)$. Dashed entries in (b) indicate that the block does not contain any data. In both figures, the black rectangles indicate the blocks assigned to process (0, 0).

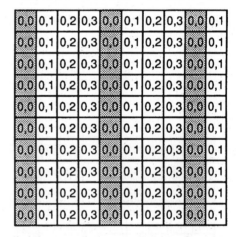

(a) $r = 3$, $s = 10$, $P = 4$, $Q = 1$

0,0	0,0	0,0	0,0	0,0	0,0	0,0	0,0	0,0	0,0
0,0	0,0	0,0	0,0	0,0	0,0	0,0	0,0	0,0	0,0
0,0	0,0	0,0	0,0	0,0	0,0	0,0	0,0	0,0	0,0
1,0	1,0	1,0	1,0	1,0	1,0	1,0	1,0	1,0	1,0
1,0	1,0	1,0	1,0	1,0	1,0	1,0	1,0	1,0	1,0
1,0	1,0	1,0	1,0	1,0	1,0	1,0	1,0	1,0	1,0
2,0	2,0	2,0	2,0	2,0	2,0	2,0	2,0	2,0	2,0
2,0	2,0	2,0	2,0	2,0	2,0	2,0	2,0	2,0	2,0
2,0	2,0	2,0	2,0	2,0	2,0	2,0	2,0	2,0	2,0
3,0	3,0	3,0	3,0	3,0	3,0	3,0	3,0	3,0	3,0

(b) $r = 1$, $s = 10$, $P = 4$, $Q = 1$

0,0	0,0	0,0	0,0	0,0	0,0	0,0	0,0	0,0	0,0
1,0	1,0	1,0	1,0	1,0	1,0	1,0	1,0	1,0	1,0
2,0	2,0	2,0	2,0	2,0	2,0	2,0	2,0	2,0	2,0
3,0	3,0	3,0	3,0	3,0	3,0	3,0	3,0	3,0	3,0
0,0	0,0	0,0	0,0	0,0	0,0	0,0	0,0	0,0	0,0
1,0	1,0	1,0	1,0	1,0	1,0	1,0	1,0	1,0	1,0
2,0	2,0	2,0	2,0	2,0	2,0	2,0	2,0	2,0	2,0
3,0	3,0	3,0	3,0	3,0	3,0	3,0	3,0	3,0	3,0
0,0	0,0	0,0	0,0	0,0	0,0	0,0	0,0	0,0	0,0
1,0	1,0	1,0	1,0	1,0	1,0	1,0	1,0	1,0	1,0

(c) $r = 10$, $s = 3$, $P = 1$, $Q = 4$

0,0	0,0	0,0	0,1	0,1	0,1	0,2	0,2	0,2	0,3
0,0	0,0	0,0	0,1	0,1	0,1	0,2	0,2	0,2	0,3
0,0	0,0	0,0	0,1	0,1	0,1	0,2	0,2	0,2	0,3
0,0	0,0	0,0	0,1	0,1	0,1	0,2	0,2	0,2	0,3
0,0	0,0	0,0	0,1	0,1	0,1	0,2	0,2	0,2	0,3
0,0	0,0	0,0	0,1	0,1	0,1	0,2	0,2	0,2	0,3
0,0	0,0	0,0	0,1	0,1	0,1	0,2	0,2	0,2	0,3
0,0	0,0	0,0	0,1	0,1	0,1	0,2	0,2	0,2	0,3
0,0	0,0	0,0	0,1	0,1	0,1	0,2	0,2	0,2	0,3
0,0	0,0	0,0	0,1	0,1	0,1	0,2	0,2	0,2	0,3

(d) $r = 10$, $s = 1$, $P = 1$, $Q = 4$

0,0	0,1	0,2	0,3	0,0	0,1	0,2	0,3	0,0	0,1
0,0	0,1	0,2	0,3	0,0	0,1	0,2	0,3	0,0	0,1
0,0	0,1	0,2	0,3	0,0	0,1	0,2	0,3	0,0	0,1
0,0	0,1	0,2	0,3	0,0	0,1	0,2	0,3	0,0	0,1
0,0	0,1	0,2	0,3	0,0	0,1	0,2	0,3	0,0	0,1
0,0	0,1	0,2	0,3	0,0	0,1	0,2	0,3	0,0	0,1
0,0	0,1	0,2	0,3	0,0	0,1	0,2	0,3	0,0	0,1
0,0	0,1	0,2	0,3	0,0	0,1	0,2	0,3	0,0	0,1
0,0	0,1	0,2	0,3	0,0	0,1	0,2	0,3	0,0	0,1
0,0	0,1	0,2	0,3	0,0	0,1	0,2	0,3	0,0	0,1

Figure 32.12 These four figures show different ways of decomposing a 10×10 matrix. Each cell represents a matrix element and is labeled by the position (p, q) in the template of the process to which it is assigned. To emphasize the pattern of decomposition, the matrix entries assigned to the process in the first row and column of the templates are shaded, and each separate shaded region represents a matrix block. (a) and (b) show block and cyclic row-oriented decompositions, respectively, for four nodes. In (c) and (d), the corresponding column-oriented decompositions are shown. Below each figure we give the values of r, s, P, and Q corresponding to the decomposition. In all cases, $p_0 = q_0 = 0$.

1. Factor the kth column of blocks, performing pivoting if necessary. This evaluates the matrices L_0, L_1, and U_0 in Fig. 32.6.
2. Evaluate the kth block row of U by solving the lower triangular system $L_0 U_1 = C$.
3. Do a rank-r update on the trailing submatrix E, replacing it with $E' = E - L_1 U_1$.

We now consider the parallel implementation of each of these tasks. The computation in the factorization step involves a single column of blocks, and these lie in a single column of the process template. In the kth factorization step, each of the r columns in block column k is processed in turn. Consider the ith column in block column k. The pivot is se-

0,0	0,0	0,0	0,1	0,1	0,1	0,2	0,2	0,2	0,3
0,0	0,0	0,0	0,1	0,1	0,1	0,2	0,2	0,2	0,3
0,0	0,0	0,0	0,1	0,1	0,1	0,2	0,2	0,2	0,3
1,0	1,0	1,0	1,1	1,1	1,1	1,2	1,2	1,2	1,3
1,0	1,0	1,0	1,1	1,1	1,1	1,2	1,2	1,2	1,3
1,0	1,0	1,0	1,1	1,1	1,1	1,2	1,2	1,2	1,3
2,0	2,0	2,0	2,1	2,1	2,1	2,2	2,2	2,2	2,3
2,0	2,0	2,0	2,1	2,1	2,1	2,2	2,2	2,2	2,3
2,0	2,0	2,0	2,1	2,1	2,1	2,2	2,2	2,2	2,3
3,0	3,0	3,0	3,1	3,1	3,1	3,2	3,2	3,2	3,3

(a) $r = 3,\ s = 3,\ P = 4,\ Q = 4$

0,0	0,1	0,2	0,3	0,0	0,1	0,2	0,3	0,0	0,1
0,0	0,1	0,2	0,3	0,0	0,1	0,2	0,3	0,0	0,1
0,0	0,1	0,2	0,3	0,0	0,1	0,2	0,3	0,0	0,1
1,0	1,1	1,2	1,3	1,0	1,1	1,2	1,3	1,0	1,1
1,0	1,1	1,2	1,3	1,0	1,1	1,2	1,3	1,0	1,1
1,0	1,1	1,2	1,3	1,0	1,1	1,2	1,3	1,0	1,1
2,0	2,1	2,2	2,3	2,0	2,1	2,2	2,3	2,0	2,1
2,0	2,1	2,2	2,3	2,0	2,1	2,2	2,3	2,0	2,1
2,0	2,1	2,2	2,3	2,0	2,1	2,2	2,3	2,0	2,1
3,0	3,1	3,2	3,3	3,0	3,1	3,2	3,3	3,0	3,1

(b) $r = 3,\ s = 1,\ P = 4,\ Q = 4$

0,0	0,0	0,0	0,1	0,1	0,1	0,2	0,2	0,2	0,3
1,0	1,0	1,0	1,1	1,1	1,1	1,2	1,2	1,2	1,3
2,0	2,0	2,0	2,1	2,1	2,1	2,2	2,2	2,2	2,3
3,0	3,0	3,0	3,1	3,1	3,1	3,2	3,2	3,2	3,3
0,0	0,0	0,0	0,1	0,1	0,1	0,2	0,2	0,2	0,3
1,0	1,0	1,0	1,1	1,1	1,1	1,2	1,2	1,2	1,3
2,0	2,0	2,0	2,1	2,1	2,1	2,2	2,2	2,2	2,3
3,0	3,0	3,0	3,1	3,1	3,1	3,2	3,2	3,2	3,3
0,0	0,0	0,0	0,1	0,1	0,1	0,2	0,2	0,2	0,3
1,0	1,0	1,0	1,1	1,1	1,1	1,2	1,2	1,2	1,3

(c) $r = 1,\ s = 3,\ P = 4,\ Q = 4$

0,0	0,1	0,2	0,3	0,0	0,1	0,2	0,3	0,0	0,1
1,0	1,1	1,2	1,3	1,0	1,1	1,2	1,3	1,0	1,1
2,0	2,1	2,2	2,3	2,0	2,1	2,2	2,3	2,0	2,1
3,0	3,1	3,2	3,3	3,0	3,1	3,2	3,3	3,0	3,1
0,0	0,1	0,2	0,3	0,0	0,1	0,2	0,3	0,0	0,1
1,0	1,1	1,2	1,3	1,0	1,1	1,2	1,3	1,0	1,1
2,0	2,1	2,2	2,3	2,0	2,1	2,2	2,3	2,0	2,1
3,0	3,1	3,2	3,3	3,0	3,1	3,2	3,3	3,0	3,1
0,0	0,1	0,2	0,3	0,0	0,1	0,2	0,3	0,0	0,1
1,0	1,1	1,2	1,3	1,0	1,1	1,2	1,3	1,0	1,1

(d) $r = 1,\ s = 1,\ P = 4,\ Q = 4$

Figure 32.13 These four figures show different ways of decomposing a 10×10 matrix over 16 processes arranged as a 4×4 template. Below each figure we give the values of $r,\ s,\ P,$ and Q corresponding to the decomposition. In all cases, $p_0 = q_0 = 0$.

lected by finding the element with largest absolute value in this column between row $kr + i$ and the last row, inclusive. The elements involved in the pivot search at this stage are shown shaded in Fig. 32.14. Having selected the pivot, the value of the pivot and its row are broadcast to all other processors. Next, pivoting is performed by exchanging the entire row $kr + i$ with the row containing the pivot. We exchange entire rows, rather than just the part to the right of the columns already factored, in order to simplify the application of the pivots to the righthand side in any subsequent solve phase. Finally, each value in the column below the pivot is divided by the pivot. If a cyclic column decomposition is used, as that shown in Fig. 32.12d, only one processor is involved in the factorization of the block column, and no communication is necessary between the processes. However, in general, P processes are involved, and communication is necessary in selecting the pivot, and in exchanging the pivot rows.

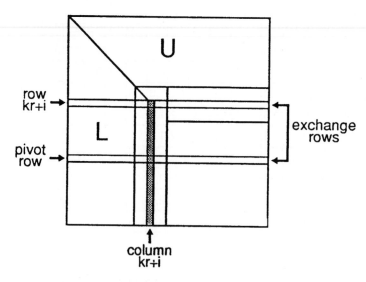

Figure 32.14 This figure shows pivoting for step i of the kth stage of LU factorization. The element with the largest absolute value in the gray shaded part of column $kr + i$ is found, and the row containing it is exchanged with row $kr + i$. If the rows exchanged lie in different processes, communication may be necessary.

The solution of the lower triangular system $L_0 U_1 = C$ to evaluate the kth block row of U involves a single row of blocks, and these lie in a single row of the process template. If a cyclic row decomposition is used, as that shown in Fig. 32.12a, only one processor is involved in the triangular solve, and no communication is necessary between the processes. However, in general Q processes are involved, and communication is necessary to broadcast the lower triangular matrix, L_0, to all processes in the row. Once this has been done, each process in the row independently performs a lower triangular solve for the blocks of C that it holds.

The communication necessary to update the trailing submatrix at step k takes place in two steps. First, each process holding part of L_1 broadcasts these blocks to the other processes in the same row of the template. This may be done in conjunction with the broadcast of L_0, mentioned in the preceding paragraph, so that all of the factored panel is broadcast together. Next, each process holding part of U_1 broadcasts these blocks to the other processes in the same column of the template. Each process can then complete the update of the blocks that it holds with no further communication.

A pseudocode outline of the parallel LU factorization algorithm is given in Fig. 32.15. There are two points worth noting in Fig. 32.15. First, the triangular solve and update phases operate on matrix blocks and may, therefore, be done with parallel versions of the Level 3 BLAS (specifically, xTRSM and xGEMM, respectively). The factorization of the column of blocks, however, involves a loop over matrix columns. Hence it is not a block-oriented computation, and cannot be performed using the Level 3 BLAS. The second point to note is that most of the parallelism in the code comes from updating the trailing submatrix since this is the only phase in which all the processes are busy.

Figure 32.15 also shows quite clearly where communication is required; namely, in finding the pivot, exchanging pivot rows, and performing various types of broadcast. The exact way in which these communications are done and interleaved with computation generally has an important effect on performance, as will be discussed in more detail in Section 32.7.

```
pcol= q₀
prow= p₀
do k= 0, min (Mᵦ, Nᵦ) − 1

    ┌─────────────────────────────────────────────────────────────┐
    │ do i= 0, r − 1                                               │
    │     if (q =pcol) find pivot value and location              │
    │     broadcast pivot value and location to all processes     │
    │     exchange pivot rows                                      │
    │     if (q =pcol) divide column r below diagonal by pivot    │
    │ end do                                                       │
    └─────────────────────────────────────────────────────────────┘

    ┌─────────────────────────────────────────────────────────────┐
    │ if (p =prow) then                                            │
    │     broadcast L₀ to all process in same template row        │
    │     solve L₀U₁ = C                                           │
    │ end if                                                       │
    └─────────────────────────────────────────────────────────────┘

    ┌─────────────────────────────────────────────────────────────┐
    │ broadcast L₁ to all processes in same template row          │
    │ broadcast U₁ to all processes in same template column       │
    │ update E ← E − L₁U₁                                          │
    └─────────────────────────────────────────────────────────────┘

    pcol= (pcol + 1) mod Q
    prow= (prow + 1) mod P
end do
```

Figure 32.15 Pseudocode for the basic parallel block-partitioned LU factorization algorithm. This code is executed by each process. The first box inside the k loop factors the kth column of blocks. The second box solves a lower triangular system to evaluate the kth row of blocks of U, and the third box updates the trailing submatrix. The template offset is given by (p_0, q_0), and (p, q) is the position of a process in the template.

Figure 32.15 refers to broadcasting data to all processes in the same row or column of the template. This is a common operation in parallel linear algebra algorithms, so the idea will be described here in a little more detail. Consider, for example, the task of broadcasting the lower triangular block, L_0, to all processes in the same row of the template, as required before solving $L_0U_1 = C$. If L_0 is in process (p, q), then it will be broadcast to all processes in row p of the process template. As a second example, consider the broadcast of L_1 to all processes in the same template row, as required before updating the trailing submatrix. This type of "rowcast" is shown schematically in Fig. 32.16a. If L_1 is in column q of the template, then each process (p, q) broadcasts its blocks of L_1 to the other processes in row p of the template. Loosely speaking, we can say that L_0 and L_1 are broadcast along the rows of the template. This type of data movement is the same as that performed by the Fortran 90 routine SPREAD [7]. The broadcast of U_1 to all processes in the same template column is very similar. This type of communication is sometimes referred to as a "colcast," and is shown in Fig. 32.16a.

32.7 Optimization, Tuning, and Trade-Offs

In this section, we examine techniques for optimizing the basic LU factorization code presented in Section 32.4.2. Among the issues to be considered are the assignment of processes to physical processors, the arrangement of the data in the local memory of each

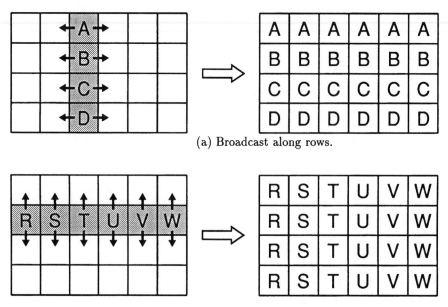

(a) Broadcast along rows.

(b) Broadcast along columns.

Figure 32.16 Schematic representation of broadcast along rows and columns of a 4×6 process template. In (a), each shaded process broadcasts to the processes in the same row of the process template. In (b), each shaded process broadcasts to the processes in the same column of the process template.

process, the trade-off between load imbalance and communication latency, the potential for overlapping communication and calculation, and the type of algorithm used to broadcast data. Many of these issues are interdependent, and in addition the portability and ease of code maintenance and use must be considered. For further details of the optimization of parallel LU factorization algorithms for specific concurrent machines, together with timing results, the reader is referred to the work of Chu and George [12], Geist and Heath [34], Geist and Romine [35], Van de Velde [55], Brent [8], Hendrickson and Womble [39], Lichtenstein and Johnsson [47], and Dongarra and coworkers [10, 25].

32.7.1 Mapping logical memory to physical memory

In Section 32.5, a logical (or virtual) matrix decomposition was described in which the global index (m,n) is mapped to a position, (p,q), in a logical process template, a position, (b,d), in a logical array of blocks local to the process, and a position, (i,j), in a logical array of matrix elements local to the block. Thus, the block cyclic decomposition is hierarchical, and attempts to represent the hierarchical memory of advanced-architecture computers. Although the parallel LU factorization algorithm can be specified solely in terms of this logical hierarchical memory, its performance depends on how the logical memory is mapped to physical memory.

32.7.1.1 Assignment of processes to processors. Consider, first, the assignment of processes, (p, q), to physical processors. In general, more than one process may be assigned to a processor, so the problem may be overdecomposed. To avoid load imbalance the same number of processes should be assigned to each processor as nearly as possible. If this condition is satisfied, the assignment of processes to processors can still affect performance by influencing the communication overhead. On recent distributed memory machines, such as the Intel Delta and CM-5, the time to send a single message between two processors is largely independent of their physical location [29, 48, 49], and hence the assignment of processes to processors does not have much direct effect on performance. However, when a collective communication task, such as a broadcast, is being done, contention for physical resources can degrade performance. Thus, the way in which processes are assigned to processors can affect performance if some assignments result in differing amounts of contention. Logarithmic contention-free broadcast algorithms have been developed for processors connected as a two dimensional mesh [6, 51], so on such machines process (p, q) is usually mapped to the processor at position (p, q) in the mesh of processors. Such an assignment also ensures that the multiple one dimensional broadcasts of L_1 and U_1 along the rows and columns of the template, respectively, do not give rise to contention.

32.7.1.2 Layout of local process memory. The layout of matrix blocks in the local memory of a process, and the arrangement of matrix elements within each block, can also affect performance. Here, trade-offs among several factors need to be taken into account. When communicating matrix blocks, for example in the broadcast of L_1 and U_1, we would like the data in each block to be contiguous in physical memory so there is no need to pack them into a communication buffer before sending them. On the other hand, when updating the trailing submatrix, E, each process multiplies a column of blocks by a row of blocks, to do a rank-r update on the part of E that it contains. If this were done as a series of separate block-block matrix multiplications, as shown in Fig. 32.17a, the performance would be poor except for sufficiently large block sizes, r, since the vector and/or pipeline units on most processors would not be fully used, as may be seen in Fig. 32.18 for the i860 processor. Instead, we arrange the loops of the computation as shown in Fig. 32.17b. Now, if the data are laid out in physical memory first by running over the i index and then over the d index the inner two loops can be merged, so that the length of the inner loop is now rd_{max}. This generally results in much better vector/pipeline performance. The b and j loops in Fig. 32.17b can also be merged, giving the algorithm shown in Fig. 32.17c. This is just the outer product form of the multiplication of an $rd_{max} \times r$ by an $r \times rb_{max}$ matrix, and would usually be done by a call to the Level 3 BLAS routine xGEMM of which an assembly coded sequential version is available on most machines. Note that in Fig. 32.17c the order of the inner two loops is appropriate for a Fortran implementation—for the C language this order should be reversed, and the data should be stored in each process by rows instead of by columns.

We have found in our work on the Intel iPSC/860 hypercube and the Delta system that it is better to optimize for the sequential matrix multiplication with an (i, d, j, b) ordering of memory in each process, rather than adopting an (i, j, d, b) ordering to avoid buffer copies when communicating blocks. However, there is another reason for doing this. On most distributed memory computers the message startup cost is sufficiently large that it is preferable wherever possible to send data as one large message rather than as several smaller messages. Thus, when communicating L_1 and U_1 the blocks to be broadcast would be amalgamated into a single message, which requires a buffer copy. The emerging Message Passing Interface (MPI) standard [21] provides support for noncontiguous messages, so in the future the need to avoid buffer copies will not be of such concern to the application developer.

```
do b = 0, b_max − 1
  do d = 0, d_max − 1
    do i = 0, r − 1
      do j = 0, r − 1
        do k = 0, r − 1
          E(b, d; i, j) = E(b, d; i, j) − L₁(b, d; i, k) U₁(b, d; k, j)
end all do loops
```

$$E(b, d; i, j) = E(b, d; i, j) - L_1(b, d; i, k)\, U_1(b, d; k, j)$$

(a) Block-block multiplication

```
do k = 0, r − 1
  do b = 0, b_max − 1
    do j = 0, r − 1
      do d = 0, d_max − 1
        do i = 0, r − 1
          E(b, d; i, j) = E(b, d; i, j) − L₁(b, d; i, k) U₁(b, d; k, j)
end all do loops
```

$$E(b, d; i, j) = E(b, d; i, j) - L_1(b, d; i, k)\, U_1(b, d; k, j)$$

(b) Intermediate form of algorithm

```
do k = 0, r − 1
  do x = 0, rb_max − 1
    do y = 0, rd_max − 1
      E(x, y) = E(x, y) − L₁(x, k) U₁(k, y)
end all do loops
```

$$E(x, y) = E(x, y) - L_1(x, k)\, U_1(k, y)$$

(c) Outer product form of algorithm

Figure 32.17 Pseudocode for different versions of the rank-r update, $E \leftarrow E - L_1 U_1$ for one process. The number of row and column blocks per process is given by b_{max} and d_{max}, respectively; r is the block size. Blocks are indexed by (b, d), and the elements within a block by (i, j). In version (a), the $r \times r$ blocks are multiplied one at a time, giving an inner loop of length r. (b) shows the loops rearranged before merging the i and d loops, and the j and b loops. This leads to the outer product form of the algorithm shown in (c) in which the inner loop is now of length rd_{max}.

32.7.2 Trade-offs between load balance and communication latency

We have discussed the mapping of the logical hierarchical memory to physical memory. In addition, we have pointed out the importance of maintaining long inner loops to get good sequential performance for each process, and the desirability of sending a few large messages rather than many smaller ones. We next consider load balance issues. Assuming that equal numbers of processes have been assigned to each processor, load imbalance arises in two phases of the parallel LU factorization algorithm; namely, in factoring each column block, which involves only P processes, and in solving the lower triangular system to evaluate each row block of U, which involves only Q processes. If the time for data movement is negligible, the aspect ratio of the template that minimizes load imbalance in step k of the algorithm is,

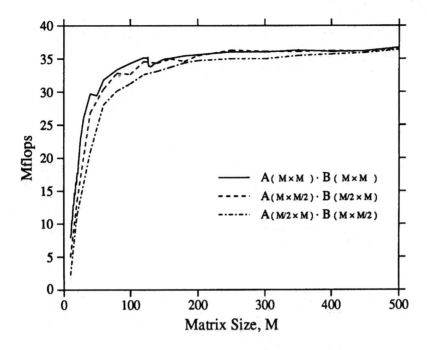

Figure 32.18 Performance of the assembly-coded Level 3 BLAS matrix multiplication routine DGEMM on one i860 processor of the Intel Delta system. Results for square and rectangular matrices are shown. Note that the peak performance of about 35 MFLOPS is attained only for matrices whose smallest dimension exceeds 100. Thus, performance is improved if a few large matrices are multiplied by each process rather than many small ones.

$$\frac{P}{Q} = \frac{\text{Sequential time to factor column block}}{\text{Sequential time for triangular solve}}$$

$$= \frac{M_b - k - 1/3 + O(1/r^2)}{N_b - k - 1 + O(1/r^2)} \tag{38.17}$$

where $Mb \times Nb$, is the matrix size in blocks, and r the block size. Thus, the optimal aspect ratio of the template should be the same as the aspect ratio of the matrix, i.e., $Mb \times Nb$ in blocks, or M/N in elements. If the effect of communication time is included then we must take into account the relative times taken to locate and broadcast the pivot information, and the time to broadcast the lower triangular matrix, L_0, along a row of the template. For both tasks the communication time increases with the number of processes involved, and since the communication time associated with the pivoting is greater than that associated with the triangular solve, we would expect the optimum aspect ratio of the template to be less than M/N. In fact, for our runs on the Intel Delta system we found an aspect ratio, P/Q, of between 1/4 and 1/8 to be optimal for most problems with square matrices, and that performance depends rather weakly on the aspect ratio, particularly for large grain sizes. Some typical results are illustrated by Fig. 32.19 for 256 processors, which shows a variation of less than 20 percent in performance as P/Q varies between 1/16 and 1 for the largest problem.

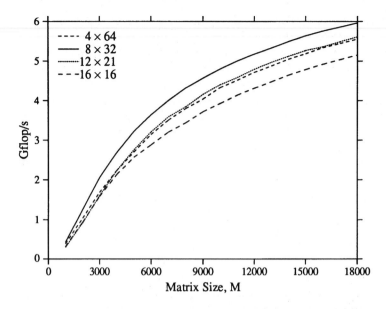

Figure 32.19 Performance of LU factorization on the Intel Delta as a function of square matrix size for different processor templates containing approximately 256 processors. The best performance is for an aspect ratio of 1/4, though the dependence on aspect ratio is rather weak.

The block size, r, also affects load balance. Here the trade-off is between the load imbalance that arises as rows and columns of the matrix are eliminated as the algorithm progresses, and communication start-up costs. The block cyclic decomposition seeks to maintain good load balance by cyclically assigning blocks to processes, and the load balance is best if the blocks are small. On the other hand, cumulative communication start-up costs are less if the block size is large since, in this case, fewer messages must be sent (although the total volume of data sent is independent of the block size). Thus, there is a block size that optimally balances the load imbalance and communication start-up costs.

32.7.3 Optimality and pipelining trade-offs

The communication algorithms used also influence performance. In the LU factorization algorithm, all the communication can be done by moving data along rows and/or columns of the process template. This type of communication can be done by passing from one process to the next along the row or column. We shall call this a "ring" algorithm, although the ring may, or may not, be closed. An alternative is to use a spanning tree algorithm, of which there are several varieties. The complexity of the ring algorithm is linear in the number of processes involved, whereas that of spanning tree algorithms is logarithmic. (For example, see Ref. [6].) Thus, considered in isolation, the spanning tree algorithms are preferable to a ring algorithm. However, in a spanning tree algorithm, a process may take part in several of the logarithmic steps, and in some implementations these algorithms act as a barrier. In a ring algorithm, each process needs to communicate only once, and can then continue to compute, in effect overlapping the communication with computation. An algorithm that interleaves communication and calculation in this way is often referred to as a pipelined algorithm. In a pipelined LU factorization algorithm with no pivoting, communication and calculation would flow in waves across the matrix. Pivoting tends to inhibit this advantage of pipelining.

In the pseudocode in Fig. 32.15, we do not specify how the pivot information should be broadcast. In an optimized implementation, we need to finish with the pivot phase, and the triangular solve phase, as soon as possible in order to begin the update phase that is richest in parallelism. Thus, it is not a good idea to broadcast the pivot information from a single source process using a spanning tree algorithm, since this may occupy some of the processes involved in the panel factorization for too long. It is important to get the pivot information to the other processes in this template column as soon as possible, so the pivot information is first sent to these processes that subsequently broadcast it along the template rows to the other processes not involved in the panel factorization. In addition, the exchange of the parts of the pivot rows lying within the panel is done separately from that of the parts outside the pivot panel. Another factor to consider here is when the pivot information should be broadcast along the template columns. In Fig. 32.15, the information is broadcast, and rows exchanged, immediately after the pivot is found. An alternative is to store up the sequence of r pivots for a panel and to broadcast them along the template rows when panel factorization is complete. This defers the exchange of pivot rows for the parts outside the panel until the panel factorization has been done, as shown in the pseudocode fragment in Fig. 32.20. An advantage of this second approach is that only one message is used to send the pivot information for the panel along the template rows, instead of r messages.

In our implementation of LU factorization on the Intel Delta system, we used a spanning tree algorithm to locate the pivot and to broadcast it within the column of the process template performing the panel factorization. This ensures that pivoting, which involves only P processes, is completed as quickly as possible. A ring broadcast is used to pipeline the pivot information and the factored panel along the template rows. Finally, after the triangular solve phase has completed, a spanning tree broadcast is used to send the newly-formed block row of U along the template columns. Results for square matrices from runs on the Intel Delta system are shown in Fig. 32.21. For each curve the results for the best process template configuration are shown. Recalling that for a scalable algorithm the performance should depend linearly on the number of processors for fixed granularity [see

```
if (q =pcol) then
    do i= 0, r − 1
        find pivot value and location
        exchange pivot rows lying within panel
        divide column r below diagonal by pivot
    end do
end if
broadcast pivot information for r pivots along template rows
exchange pivot rows lying outside the panel for each of r pivots
```

Figure 32.20 Pseudocode fragment for a partial pivoting over rows. This may be regarded as replacing the first box inside the k loop in Figure 32.15. In the above code, pivot information is first disseminated within the template column doing the factorization. The pivoting of the parts of the rows lying outside the panel is deferred until the panel factorization has been completed.

Eq. (32.2)], it is apparent that scalability may be assessed by the extent to which isogranularity curves differ from linearity. An isogranularity curve is a plot of performance against number of processors for a fixed granularity. The results in Fig. 32.21 can be used to generate the isogranularity curves shown in Fig. 32.22 that show that on the Delta system the LU factorization routine starts to lose scalability when the granularity falls below about 0.2×10^6. This corresponds to a matrix size of about $M = 10000$ on 512 processors, or about 13 percent of the memory available to applications on the Delta, indicating that LU factorization scales rather well on the Intel Delta system.

38.8   Conclusions and Future Research Directions

Portability of programs has always been an important consideration. Portability was easy to achieve when there was a single architectural paradigm (the serial von Neumann machine) and a single programming language for scientific programming (Fortran) embodying that common model of computation. Architectural and linguistic diversity have made portability much more difficult, but no less important, to attain. Users simply do not wish to invest significant amounts of time to create large-scale application codes for each new machine. Our answer is to develop portable software libraries that hide machine-specific details.

38.8.1   Portability, scalability, and standards

To be truly portable, parallel software libraries must be standardized. In a parallel computing environment in which the higher-level routines and/or abstractions are built upon lower-level computation and message-passing routines, the benefits of standardization are particularly apparent. Furthermore, the definition of computational and message-passing standards provides vendors with a clearly defined base set of routines that they can implement efficiently.

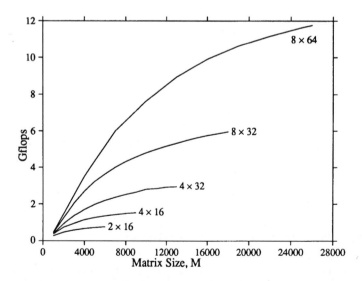

Figure 32.21 Performance of LU factorization on the Intel Delta as a function of square matrix size for different numbers of processors. For each curve, results are shown for the process template configuration that gave the best performance for that number of processors.

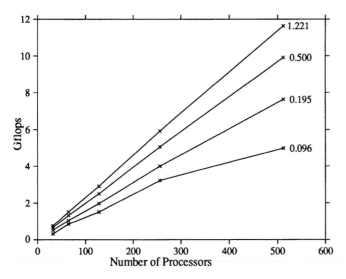

Figure 32.22 Isogranularity curves in the (Np, G) plane for the LU factorization of square matrices on the Intel Delta system. The curves are labeled by the granularity in units of 10^6 matrix elements per processor. The linearity of the plots for granularities exceeding about 0.2×10^6 indicates that the LU factorization algorithm scales well on the Delta.

From the user's point of view, portability means that, as new machines are developed, they are simply added to the network, supplying cycles where they are most appropriate.

From the mathematical software developer's point of view, portability may require significant effort. Economy in development and maintenance of mathematical software demands that such development effort be leveraged over as many varied computer systems as possible. Given the great diversity of parallel architectures, this type of portability is attainable to only a limited degree, but machine dependences can at least be isolated.

LAPACK is an example of a mathematical software package whose highest-level components are portable, while machine dependences are hidden in lower-level modules. Such a hierarchical approach is probably the closest one can come to software portability across diverse parallel architectures. And the BLAS that are used so heavily in LAPACK provide a portable, efficient, and flexible standard for applications programmers.

Like portability, *scalability* demands that a program be reasonably effective over a wide range of number of processors. The scalability of parallel algorithms, and software libraries based on them, over a wide range of architectural designs and numbers of processors will likely require that the fundamental granularity of computation be adjustable to suit the particular circumstances in which the software may happen to execute. Our approach to this problem is block algorithms with adjustable block size. In many cases, however, polyalgorithms[*] may be required to deal with the full range of architectures and processor multiplicity likely to be available in the future.

Scalable parallel architectures of the future are likely to be based on a distributed memory architectural paradigm. In the longer term, progress in hardware development, operating systems, languages, compilers, and communications may make it possible for users to view such distributed architectures (without significant loss of efficiency) as having a

---

[*]In a polyalgorithm the actual algorithm used depends on the computing environment and the input data. The optimal algorithm in a particular instance is automatically selected at runtime.

shared memory with a global address space. For the near term, however, the distributed nature of the underlying hardware will continue to be visible at the programming level; therefore, efficient procedures for explicit communication will continue to be necessary. Given this fact, standards for basic message passing (send/receive), as well as higher-level communication constructs (such as global summation and broadcast), become essential to the development of scalable libraries that have any degree of portability. In addition to standardizing general communication primitives, it may also be advantageous to establish standards for problem-specific constructs in commonly occurring areas such as linear algebra.

The BLACS (Basic Linear Algebra Communication Subprograms) [16, 26] is a package that provides the same ease of use and portability for MIMD message-passing linear algebra communication that the BLAS [17, 18, 45] provide for linear algebra computation. Therefore, we recommend that future software for dense linear algebra on MIMD platforms consist of calls to the BLAS for computation and calls to the BLACS for communication. Since both packages will have been optimized for a particular platform, good performance should be achieved with relatively little effort. Also, since both packages will be available on a wide variety of machines, code modifications required to change platforms should be minimal.

38.8.2   Alternative approaches

Traditionally, large, general-purpose mathematical software libraries have required users to write their own programs that call library routines to solve specific subproblems that arise during a computation. Adapted to a shared-memory parallel environment, this conventional interface still offers some potential for hiding underlying complexity. For example, the LAPACK project incorporates parallelism in the Level 3 BLAS, where it is not directly visible to the user.

But when going from shared-memory systems to the more readily scalable distributed memory systems, the complexity of the distributed data structures required is more difficult to hide from the user. Not only must the problem decomposition and data layout be specified, but different phases of the user's problem may require transformations between various distributed data structures.

These deficiencies in the conventional user interface have prompted extensive discussion of alternative approaches for scalable parallel software libraries of the future. Possibilities include:

1. Traditional functional library (i.e., minimum possible change to the *status quo* in going from serial to parallel environment). This will allow one to protect the programming investment that has been made.
2. Reactive servers on the network. A user would be able to send a computational problem to a server that was specialized in dealing with the problem. This fits well with the concepts of a networked, heterogeneous computing environment with various specialized hardware resources (or even the heterogeneous partitioning of a single homogeneous parallel machine).
3. General interactive environments like Matlab or Mathematica, perhaps with "expert" drivers (i.e., knowledge-based systems). With the growing popularity of the many integrated packages based on this idea, this approach would provide an interactive, graphical interface for specifying and solving scientific problems. Both the algorithms and data structures are hidden from the user, because the package itself is responsible for storing and retrieving the problem data in an efficient, distributed

manner. In a heterogeneous networked environment, such interfaces could provide seamless access to computational engines that would be invoked selectively for various parts of the user's computation according to which machine is most appropriate for a particular subproblem.

4. Domain-specific problem solving environments, such as those for structural analysis. Environments like Matlab and Mathematica have proven to be especially attractive for rapid prototyping of new algorithms and systems that may subsequently be implemented in a more customized manner for higher performance.

5. Reusable templates (i.e., users adapt "source code" to their particular applications). A template is a description of a general algorithm rather than the executable object code or the source code more commonly found in a conventional software library. Nevertheless, although templates are general descriptions of key data structures, they offer whatever degree of customization the user may desire.

Novel user interfaces that hide the complexity of scalable parallelism will require new concepts and mechanisms for representing scientific computational problems and for specifying how those problems relate to each other. Very high-level languages and systems, perhaps graphically based, not only would facilitate the use of mathematical software from the user's point of view, but also would help to automate the determination of effective partitioning, mapping, granularity, data structures, etc. However, new concepts in problem specification and representation may also require new mathematical research on the analytic, algebraic, and topological properties of problems (e.g., existence and uniqueness).

We have already begun work on developing such templates for sparse matrix computations. Future work will focus on extending the use of templates to dense matrix computations.

We hope the insight we gained from our work will influence future developers of hardware, compilers, and systems software so that they provide tools to facilitate development of high-quality portable numerical software.

The EISPACK, LINPACK, and LAPACK linear algebra libraries are in the public domain, and are available from netlib. For example, for more information on how to obtain LAPACK, send the following one-line e-mail message to netlib@ornl.gov:

send index from lapack

Information for EISPACK and LINPACK can be similarly obtained. We expect to make a preliminary version of the ScaLAPACK library available from netlib in 1993.

Acknowledgments

This research was performed in part using the Intel Touchstone Delta System operated by the California Institute of Technology on behalf of the Concurrent Supercomputing Consortium. Access to this facility was provided through the Center for Research on Parallel Computing.

38.9   References

1. Anderson, E., A. Benzoni, J. J. Dongarra, S. Moulton, S. Ostrouchov, B. Tourancheau, and R. van de Geijn. 1991. LAPACK for distributed memory architectures: Progress report. In *Parallel Processing for Scientific Computing, Fifth SIAM Conference.*

2. Anderson, E., and J. Dongarra. 1989. *Results from the initial release of LAPACK.* Technical Report LAPACK working note 16. Knoxville, Tenn.: Computer Science Department, University of Tennessee.

3. Anderson, E., and J. Dongarra. 1990. *Evaluating block algorithm variants in LAPACK.* Technical Report LAPACK working note 19, Knoxville, Tenn.: Computer Science Department, University of Tennessee.

4. Ashcraft, C. C. 1990. *The distributed solution of linear systems using the torus wrap data mapping.* Engineering Computing and Analysis Technical Report ECA-TR-147, Boeing Computer Services.

5. Ashcraft, C. C. 1991. *A taxonomy of distributed dense LU factorization methods.* Engineering Computing and Analysis Technical Report ECA-TR161, Boeing Computer Services.

6. Barnett, M., D. G. Payne, and R. van de Geijn. 1993. *Broadcasting on meshes with worm-hole routing.* Technical report. Austin, Tex.: Department of Computer Science, University of Texas at Austin.

7. Brainerd, W. S., C. H. Goldbergs, and J. C. Adams. 1990. *Programmers guide to Fortran 90.* New York: McGraw-Hill.

8. Brent, R. P. 1992. The LINPACK benchmark for the Fujitsu AP 1000. *Proceedings of the Fourth Symposium on the Frontiers of Massively Parallel Computation.* Los Alamitos, Calif.: IEEE Computer Society Press, 128–135.

9. Brent, R. P. 1991. The LINPACK benchmark on the AP 1000: Preliminary report. *Proceedings of the 2nd CAP Workshop.*

10. Choi, J., J. J. Dongarra, R. Pozo, and D. W. Walker. 1992. Scalapack: A scalable linear algebra library for distributed memory concurrent computers. *Proceedings of the Fourth Symposium on the Frontiers of Massively Parallel Computation.* Los Alamitos, Calif.: IEEE Computer Society Press, 120–127.

11. Choi, J., J. J. Dongarra, and D. W. Walker. 1993. The design of scalable software libraries for distributed memory concurrent computers. In *Environments and Tools for Parallel Scientific Computing,* J. J. Dongarra and B. Tourancheau, eds. Amsterdam: Elsevier Science Publishers.

12. Chu, E., and A. George. 1987. Gaussian elimination with partial pivoting and load balancing on a multiprocessor. *Parallel Computing,* 5:65–74.

13. Culler, D. E., A. Dusseau, S. C. Goldstein, A. Krishnamurthy, S. Lumetta, T. von Eicken, and K. Yelick. 1993. *Introduction to Split-C: Version 0.9.* Technical report. Berkeley, Calif.: Computer Science Division, EECS, University of California.

14. Demmel, J. 1989. LAPACK: A portable linear algebra library for supercomputers. *Proceedings of the 1989 IEEE Control Systems Society Workshop on Computer-Aided Control System Design.*

15. Dongarra, J. J. 1984. Increasing the performance of mathematical software through high-level modularity. *Proc. Sixth Int. Symp. Comp. Methods in Eng. & Applied Sciences, Versailles, France.* Amsterdam: North-Holland, 239–248.

16. Dongarra, J. J. 1991. *LAPACK Working Note 34: Workshop on the BLACS.* Computer Science Dept. Technical Report CS-91-134. Knoxville, Tenn.: University of Tennessee at Knoxville, Computer Science Dept.

17. Dongarra, J. J., J. Du Croz, S. Hammarling, and I. Duff. 1990. A set of level 3 basic linear algebra subprograms. *ACM Transactions on Mathematical Software,* 16(1):1–17.

18. Dongarra, J. J., J. Du Croz, S. Hammarling, and R. Hanson. 1988. An extended set of Fortran basic linear algebra subroutines. *ACM Transactions on Mathematical Software,* 14(1):1–17.

19. Dongarra, J. J., I. S. Duff, D. C. Sorensen, and H. A. Van der Vorst. 1991. *Solving Linear Systems on Vector and Shared Memory Computers.* Philadelphia, Pa.: SIAM Publications.

20. Dongarra, J. J., and E. Grosse. 1987. Distribution of mathematical software via electronic mail. *Communications of the ACM,* 30(5):403–407.

21. Dongarra, J. J., R. Hempel, A. J. G. Hey, and D. W. Walker. 1993. *A proposal for a user-level message passing interface in a distributed memory environment.* Technical Report TM 12231. Oak Ridge, Tenn.: Oak Ridge National Laboratory.

22. Dongarra, J. J., P. Mayes, and G. R. di Brozolo. 1991. The IBM RISC System/6000 and linear algebra operations. *Supercomputer,* 44(VIII-4):15–30.

23. Dongarra, J. J., and S. Ostrouchov. 1990. *LAPACK block factorization algorithms on the Intel iPSC/860.* Technical Report CS-90-115. Knoxville, Tenn.: University of Tennessee at Knoxville, Computer Science Department.

24. Dongarra, J. J., R. Pozo, and D. W. Walker. 1993. An object oriented design for high performance linear algebra on distributed memory architectures. *Proceedings of the Object Oriented Numerics Conference.*

25. Dongarra, J. J., R. van de Geijn, and D. W. Walker. 1992. A look at scalable dense linear algebra libraries. *Proceedings of the Scalable High-Performance Computing Conference. New York: IEEE,* 372–379.

26. Dongarra, J. J., and R. A. van de Geijn. 1991. *Two-dimensional basic linear algebra communication subprograms.* Technical Report LAPACK working note 37. Knoxville, Tenn.: University of Tennessee at Knoxville, Computer Science Department.

27. Dongarra, J. J., and R. A. van de Geijn. 1992. Reduction to condensed form for the eigenvalue problem on distributed memory architectures. *Parallel Computing*, 18:973–982.

28. Du Croz, J., and M. Pont. 1987. The development of a floating-point validation package. *Proceedings of the 8th Symposium on Computer Arithmetic*, M. J. Irwin and R. Stefanelli, eds. Los Alamitos, Calif.: IEEE Computer Society Press.

29. Dunigan, T. H. 1992. *Communication performance of the Intel Touchstone Delta mesh*. Technical Report T M-11983, Oak Ridge, Tenn.: Oak Ridge National Laboratory.

30. Edelman, A. 1993. Large dense numerical linear algebra in 1993: the parallel computing influence. *International Journal Supercomputer Applications*, 7(2):113–128.

31. Felten, E. W ., and S. W. Otto. 1988. Coherent parallel C. *Proceedings of the Third Conference on Hypercube Concurrent Computers and Applications,* G. C. Fox, ed. ACM Press, 440–450.

32. Fox, G. C., M. A. Johnson, G. A. Lyzenga, S. W. Otto, J. K. Salmon, and D. W. Walker. 1988. *Solving Problems on Concurrent Processors,* Volume 1. Englewood Cliffs, N.J.: Prentice Hall.

33. Gallivan, K., R. Plemmons, and A. Sameh. 1990. Parallel algorithms for dense linear algebra computations. *SIAM Review,* 32(1):54–135.

34. Geist, A., and M. Heath. 1986. Matrix factorization on a hypercube multiprocessor. In *Hypercube Multiprocessors*, M. Heath, ed. Philadelphia, Pa:. Society for Industrial and Applied Mathematics, 161–180.

35. Geist, A., and C. Romine. 1988. LU factorization algorithms on distributed-memory multiprocessor architectures. *SIAM J. Sci. Statist. Comput.*, 9(4):639–649.

36. Golub, G. H., and C. F. Van Loan. 1989. *Matrix Computations*, 2nd ed. Baltimore: The Johns Hopkins Press.

37. Gupta, A., and V. Kumar. 1990. On the scalability of FFT on parallel computers. *Proceedings of the Frontiers 90 Conference on Massively Parallel Computation.* Los Alamitos, Calif.: IEEE Computer Society Press. Also available as technical report TR 90-20 from the Computer Science Department, University of Minnesota, Minneapolis, Minn.

38. Harrington, R. 1990. Origin and development of the method of moments for field computation. *IEEE Antennas and Propagation Magazine* (June).

39. Hendrickson, B., and D. Womble. 1992. *The torus-wrap mapping for dense matrix computations on massively parallel computers.* Technical Report SAND92-0792. Sandia National Laboratories.

40. Hess, J. L. 1990. Panel methods in computational fluid dynamics. *Annual Reviews of Fluid Mechanics,* 22:255–274.

41. Hess, J. L., and M. O. Smith. 1967. Calculation of potential flows about arbitrary bodies. In *Progress in Aeronautical Sciences*, D. Kuchemann, ed.,Volume 8. New York: Pergamon Press.

42. High Performance Fortran Forum. 1993. *High Performance Fortran Language Specification,* Version 1.0.

43. Hockney, R. W., and C. R. Jesshope. 1981. *Parallel Computers*. Bristol, U.K.: Adam Hilger Ltd.

44. Kahan, W. *Paranoia*. Available from netlib [20].

45. Lawson, C., R. Hanson, D. Kincaid, and F. Krogh. 1979. Basic linear algebra subprograms for Fortran usage. *ACM Trans. Math. Softw.*, 5:308–323.

46. Leiserson, C. 1985. Fat trees: universal networks for hardware-efficient supercomputing. *IEEE Transactions on Computers*, C-34(10):892–901.

47. Lichtenstein, W., and S. L. Johnsson. 1992. *Block-cyclic dense linear algebra*. Technical Report TR-04-92. Harvard University, Center for Research in Computing Technology.

48. Lin, M., D. Du, A. E. Klietz, and S. Saroff. 1992. *Performance evaluation of the CM-5 interconnection network.* Technical report, Department of Computer Science, University of Minnesota.

49. Ponnusamy, R., A. Choudhary, and G. Fox. 1992. Communication overhead on CM-5: An experimental performance evaluation. *Proceedings of the Fourth Symposium on the Frontiers of Massively Parallel Computation*. Los Alamitos, Calif.: IEEE Computer Society Press, 108–115.

50. Saad, Y., and M. H. Schultz. 1985. *Parallel direct methods for solving banded linear systems*. Technical Report YALEU/DCS/RR387. Department of Computer Science, Yale University.

51. Seidel, S. R. 1993. *Broadcasting on linear arrays and meshes*. Technical Report TM-12356. Oak Ridge, Tenn.: Oak Ridge National Laboratory.

52. Skjellum, A., and A. Leung. 1990. LU factorization of sparse, unsymmetric, Jacobian matrices on multicomputers. *Proceedings of the Fifth Distributed Memory Concurrent Computing Conference,* D. W. Walker and Q. F. Stout, eds. New York: IEEE Press, 328–337.

53. Thinking Machines Corporation. 1991. *CM-5 Technical Summary.* Cambridge, Mass.: Thinking Machines Corp.

54. van de Geijn, R. A. 1991. *Massively parallel LINPACK benchmark on the Intel Touchstone Delta and iPSC/860 systems.* Computer Science report TR91-28, Univ. of Texas.

55. Van de Velde, E. F. 1990. Data redistribution and concurrency. *Parallel Computing*, 16 (December).

56. Wang, J. J. H. 1991. *Generalized Moment Methods in Electromagnetics.* New York: John Wiley & Sons.
57. Wilkinson, J., and C. Reinsch. 1971. *Handbook for Automatic Computation: Volume II—Linear Algebra.* New York: Springer-Verlag.

33

Testing of Distributed Programs

K. C. Tai [*] *and Richard H. Carver* [†]

The life cycle of a program includes a testing and debugging phase. Testing is the process of executing the program with selected tests in order to detect faults.[‡] Debugging is the process of locating and correcting the faults detected during testing. Together, testing and debugging account for at least half of the cost of developing a large software system [1].

The conventional approach to testing a program is to execute the program with each selected test input once and then compare the test results with the expected results. If a test input detects the existence of a fault, the program is usually executed again with the same input in order to collect debugging information. After the fault has been located and corrected, the program is executed again with each of the previously tested inputs to verify that the fault has been corrected and that in doing so no new faults have been introduced. (Such testing, called *regression testing*, is also needed after the program has been modified during the maintenance phase.) This cyclical process of testing, followed by debugging, followed by more testing, is commonly applied to computer programs.

Let P be a distributed program. Multiple executions of P with the same input may produce different results.[§] This *nondeterministic execution behavior* creates the following problems during the testing and debugging cycle of P:

- *Problem 1*. When testing P with input X, a single execution is insufficient to determine the correctness of P with X. Even if P with input X has been executed successfully many times, it is possible that a future execution of P with X will produce an incorrect result.
- *Problem 2*. When debugging an erroneous execution of P with input X, there is no guarantee that this execution will be repeated by executing P with X.

---

[*] This work was supported in part by NSF grants CCR-8907807 and CCR-9320992.
[†] This work was supported in part by NSF grant CCR-9309043.
[‡] A fault is IEEE standard terminology for what is popularly called a "bug."
[§] In some parallel programs, any nondeterminism is considered to be the result of a programming fault. In this chapter, it is assumed that nondeterminism is intentional and is a desirable property of distributed programs.

- *Problem 3*. After P has been modified to correct the fault detected during an erroneous execution of P with input X, one or more successful executions of P with X during regression testing do not imply that the detected fault has been corrected.

This chapter addresses issues related to the testing of distributed programs. Information about debugging distributed programs can be found in Ref. [2].

A distributed program consists of two or more processes that communicate and synchronize with each other via message passing. An execution of a message passing construct generates one or more *synchronization events* (or *SYN-events*). An actual or expected execution of a distributed program can be characterized by its sequence of synchronization events, referred to as a *synchronization sequence* (or *SYN-sequence*). The definitions of SYN-events and SYN-sequences of a distributed program are based on the distributed programming constructs used in the program. How to define SYN-events and SYN-sequences is the focus of Section 33.1. In later sections, general issues in testing distributed programs are discussed in terms of SYN-events and SYN-sequences. By doing so, this discussion can be applied to any distributed program.

33.1   SYN-Sequences of Distributed Programs

Section 33.1.1 gives a brief overview of distributed programming constructs and languages. Section 33.1.2 describes general formats for information about SYN-events and SYN-sequences. Section 33.1.3 defines totally-ordered SYN-sequences for two distributed programming constructs. Section 33.1.4 discusses how to define partially-ordered SYN-sequences.

33.1.1   Distributed programming constructs and languages

Detailed discussion about distributed programming constructs can be found in Ref. [3]. Since the send and receive constructs and the Ada language will be used later, a brief description of each is given below. Send and receive are primitive constructs that are supported by many distributed languages and operating systems. The Ada language contains a rich set of high-level distributed programming constructs.

Send and receive. A message is sent by executing "*send* message *to* designation_designator." The destination_designator controls where the message goes. A message is received by executing "*receive* message *from* source_designator." The source_designator controls where the message comes from.

Messages are delivered through *communication channels* defined by the destination (source) designators in send (receive) commands. There exist various types of naming schemes for communication channels. In *direct naming*, destination and source designators are process names. In *mailbox naming*, destination and source designators are mailbox names. Messages sent to a given mailbox can be received by any process that executes a *receive* naming that mailbox. A special case of mailbox naming is *port naming*, in which messages sent to a given mailbox can be received by only one specific process. A *blocking* send (receive) delays its invoker until the sent message (until a message) is received. The use of nonblocking send and blocking receive is called *asynchronous message passing*, and the use of blocking send and receive is called *synchronous message passing*.

Distributed programming constructs in Ada

In the *Ada* language, processes are called *tasks*. Tasks communicate through ports, which are called *entries*. Messages are sent using blocking entry call statements, which have a form similar to that used for procedure call statements in sequential languages. Messages are received using blocking *accept* statements. An execution of an *accept* statement is like that of a procedure: it receives the message sent to it, executes a sequence of statements, and returns a reply message to the calling task. The calling task is delayed at its entry call statement until this reply is returned. This type of synchronous message passing is called a *rendezvous*.

Ada also provides a *select* statement that allows a task to make a choice among accept statements that are eligible for execution. Each *accept* statement may be preceded by a Boolean expression that specifies the conditions under which the accept statement can be selected. An accept statement is said to be eligible for execution if it has at least one waiting call and either it has no preceding Boolean expression or this expression is true. If two or more *accept* statements are eligible for execution, then one of them is selected.

Figure 33.1 shows an Ada program, called Bounded_Buffer, for solving the two-slot bounded buffer problem. Tasks Producer and Consumer call task Buffer_Control to de-

```
task body Producer is              task body Consumer is
     H : character;                     H : character;
begin                              begin
     Get(H);                            Withdraw(H);
     Deposit(H);                        Put(H);
     Get(H);                            Withdraw(H);
     Deposit(H);                        Put(H);
     Get(H);                            Withdraw(H);
     Deposit(H);                        Put(H);
end Producer;                      end Consumer;

task body Buffer_Control is
     Size:        constant integer := 2;      --Buffer holds at most 2 items
     Buffer:      array(1..Size) of character;
     Count:       integer range 0..Size := 0; --current no. of items in buffer
                  --Count = 0 implies that Buffer is empty
                  --Count = Size implies that Buffer is full
     In_Index, Out_Index : integer range 1..Size := 1;
     --In_Index is the location for the next deposit
     --Out_Index is the location of the first item, if any

begin
     loop
          select
               when Count <= Size =>
                    accept Deposit(C : character) do
                         Buffer(In_Index) := C;
                    end;
                    In_Index := (In_Index mod Size) + 1;
                    Count := Count + 1;
               or when Count > 0 =>
                    accept Withdraw(C : out character) do
                         C := Buffer(Out_Index);
                    end;
                    Out_Index := (Out_Index mod Size) + 1;
                    Count := Count - 1;
               or terminate;
          end select;
     end loop;
end Buffer_Control;
```

Figure 33.1 An incorrect Ada program for the two-slot bounded buffer problem

posit and withdraw characters instead of interacting directly with each other. Buffer_Control has a buffering capacity of two characters. Task Producer reads in three characters and calls entry Deposit three times to deposit these characters. Task Consumer calls entry Withdraw three times to withdraw the characters deposited by task Producer and prints out these characters. Task Buffer_Control contains a *select* statement having three alternatives: an *accept* alternative to accept a call to entry Deposit when the buffer is not full, an *accept* alternative to accept a call to entry Withdraw when the buffer is not empty, and a *terminate* alternative to terminate task Buffer_Control when tasks Producer and Consumer have terminated. When the buffer is neither full nor empty and when calls from tasks Producer and Consumer are waiting, one of the calls is chosen arbitrarily. Program Bounded_Buffer contains a fault that is described later.

33.1.2 SYN-events and SYN-sequences of distributed programs

The definition of a SYN-event or SYN-sequence can be *language-based* or *implementation-based*, or a combination of both. A language-based definition is based on the distributed programming constructs available in a given distributed language. An implementation-based definition is based on the implementation of these constructs, including the interfaces with the run-time and operating systems. This section discusses language-based definitions of SYN-events and SYN-sequences. A discussion about language-based and implementation-based approaches to building testing tools for distributed programs will be given in Section 33.6. In the following discussion, it is assumed that a distributed program has no real-time constraints and that all non-synchronization constructs in a distributed program are deterministic.

An execution of a distributed programming construct (send, receive, accept, etc.) gives rise to one or more *SYN-events*. The communication channels accessed by SYN-events are referred to as *SYN-objects*. Thus, a distributed program consists of distributed processes performing SYN-events on SYN-objects. The relative order in which SYN-events of different processes are executed is nondeterministic. Consider a program with mailbox naming. If two or more processes intend to send messages to the same mailbox, the order in which these messages will arrive at the mailbox is unpredictable. Also, if two or more processes intend to receive a message from the same mailbox, the order in which these processes will withdraw messages from this mailbox is unpredictable. A complete characterization of an execution of a distributed program requires a definition of the SYN-events involved in this program.

For a distributed programming language or construct, its *complete SYN-event set* is the set of all types of SYN-events, and its *complete SYN-event format* is the format for information about each type of SYN-event. The complete SYN-event set and format are not necessarily unique, but they must provide sufficient information to describe the possible synchronizations between distributed processes. In general, the complete SYN-event format is (event type, process name, object name, other necessary information), indicating that a process executes a SYN-event of a specific type on a specific SYN-object. (The ordering of components in a SYN-event format is immaterial.)

The complete SYN-sequence of an execution of a distributed program P is an ordered sequence of all SYN-events of this execution, where the information about each SYN-event appears in the complete SYN-event format. Thus, the complete SYN-sequence of an execution of P provides sufficient information to resolve any sources of nondeterminism during execution. *Consequently, the result of an execution of P can be determined by P and the input and complete SYN-sequence of this execution.* In the following discussion, SYN-sequences are assumed to be complete unless otherwise specified, and only SYN-sequences of finite length are considered.

A SYN-sequence can be a *total* or *partial* ordering of SYN-events. A totally-ordered SYN-sequence is a sequence of SYN-events with the general format shown above. When a distributed program is tested or debugged according to totally-ordered SYN-sequences, the serialization of SYN-events could have a significant impact on performance. This problem can be alleviated if partially-ordered SYN-sequences are used instead. Note that the totally-ordered and partially-ordered SYN-sequences of an execution of a distributed program should have the same "happened before" relation. Section 33.1.3 shows complete SYN-event sets and formats and totally-ordered complete SYN-sequences for the send and receive constructs and the Ada language. Section 33.1.4 discusses how to define partially-ordered SYN-sequences.

33.1.3 Totally-Ordered SYN-sequences

SRM-sequences for Send and Receive statements with mailbox naming. Assume that send and receive statements are blocking and that messages are deposited into and withdrawn from mailboxes in FIFO order. Each mailbox is considered to be a SYN-object. The SYN-event set for send and receive statements with mailbox naming includes the following types of SYN-events:

(a) the *arrival* of a *send* from a process at a *mailbox*
(b) the *arrival* of a *receive* from a process at a *mailbox*
(c) the *start* of execution of a *send* that results in an exception (and thus no arrival at its destination)
(d) the *start* of execution of a *send* that results in a deadlock situation
(e) the *start* of execution of a *receive* that results in an exception
(f) the *start* of execution of a *receive* that results in a deadlock situation

(The start of execution of a send or receive that eventually arrives at a mailbox is not considered to be a SYN-event because the order in which send and receive statements start is not necessarily the order in which these statements arrive at mailboxes.) A sequence of events of the above types for a set of processes is called a *send-receive-mailbox sequence* (or *SRM-sequence*), with format

$$((V1,C1,M1), (V2,C2,M2), \ldots)$$

where (Vi, Ci, Mi) denotes the ith, $i > 0$, SYN-event in the sequence, with Vi being the type of this SYN-event, Ci the sender or receiver of this event, and Mi the mailbox if this event is of type (a) or (b), respectively.

R-sequences for Entry Call, Accept, and Select Statements in Ada Each entry in an Ada program is considered to be a SYN-object. The SYN-event set for the distributed programming constructs in Ada contains seventeen types of SYN-events [4]. Below is a partial list of these types:

(a1) the *start* of execution of an *accept* statement that is not at the beginning of an alternative of a selective wait and eventually accepts an entry call (a2) the *end* of execution of an *accept* statement that is not at the beginning of an alternative of a selective wait

(b1) the *selection* of an *accept* statement that is at the beginning of an alternative of a selective wait statement

(b2) the *end* of execution of an *accept* statement that is at the beginning of an alternative of a selective wait

(c1) the *selection* of a *delay* alternative of a selective wait

(c2) the *end* of execution of a *delay* alternative of a selective wait

(d1) the *selection* of the *else* part of a selective wait

(d2) the *end* of execution of the *else* part of a selective wait

(e) the *selection* of a *terminate* alternative of a selective wait

(The eight other types of SYN-events cover unsuccessful executions of accept, selective wait, and entry call statements that are due to exceptions or deadlocks, and they also cover other Ada constructs.)

The format of a SYN-event in Ada is (V,C,U,N,D), where V denotes the type of this event, C the calling task, U the accepting task, N the entry name, and D the other necessary information (if any) about this event. The calling task refers to the caller of an entry call; it does not exist for events of type (c1)-(e). The accepting task refers to the task that executes an accept or select statement. Note that the start and end of execution of a successful entry call are not considered as SYN-events for Ada since these events are implied by the execution of the corresponding accept statement. Also, arrivals of entry calls are not considered as SYN-events since the arrival order of entry calls for an entry can be determined by the order of those SYN-events that correspond to executions of accept statements for this entry. A sequence of SYN-events in Ada is referred to as a *rendezvous sequence* (or *R-sequence*).

33.1.4 Partially-ordered SYN-sequences

Assume that a distributed program P consists of distributed process P1, P2,...,Pm, m > 0, and SYN-objects O1, O2,...,and On, n > 0. The two basic approaches to defining a partially-ordered SYN-sequence of P are:

- *Process-based.* The general format of a process-based, partially-ordered SYN-sequence of P is (R1, R2,...,Rm), where R_i, $0 < i < = m$, denotes a totally-ordered sequence of SYN-events. Each event in R_i has P_i as the executing process and has the general format: (event type, object name, object order number, other necessary information). The *object order number* is used to indicate the relative order of this event among all of the events executed on the SYN-object. Process-based partially-ordered R-sequences for Ada were defined in Ref. [4]. These R-sequences contain SYN-events of the format (V,C,N,D) and do not contain object order numbers since an entry belongs to only one task and thus the order number is implicit. Process-based partially ordered SYN-sequences were defined for monitor-based programs [5] and for programs using read and write operations to access shared variables [6].
- *Object-based.* The general format of an object-based, partially-ordered SYN-sequence of P is (B1, B2,...,Bn), where B_j, $0 < j < = n$, denotes a totally-ordered sequence of SYN-events. Each event in B_j has object O_j as the SYN-object and has the general format: (event type, process name, *process order number*, other necessary information). The process order number is used to indicate the relative order of this event among all of the events executed by the process.

Totally-ordered SYN-sequences can be converted into object- and process-based, partially-ordered SYN-sequences, and the latter two can be converted into each other. Unless otherwise specified, SYN-sequences of a distributed program are assumed to be either totally- or partially-ordered.

33.2 Definitions of Correctness and Faults for Distributed Programs

Because of their nondeterministic execution behavior, distributed programs have definitions of correctness and faults that are different from those for sequential programs. These definitions are needed to address the issues involved in testing and debugging distributed programs. Let P be a distributed program. A SYN-sequence is said to be *feasible* for P with input X if this SYN-sequence can possibly be exercised during an execution of P with input X. Because of the unpredictable rates of progress of distributed processes and the use of nondeterministic constructs, the implementation (or code) of P may allow the existence of two or more distinct, feasible SYN-sequences for P with the same input. Let

Feasible(P, X) = the set of feasible SYN-sequences of P with input X

A SYN-sequence is said to be *valid* for P with input X if, according to the specification of P, this SYN-sequence is expected to be the SYN-sequence of an execution of P with input X. Since nondeterministic execution behavior is expected, the specification of P may allow the existence of two or more distinct, valid SYN-sequences of P with the same input. Let

Valid(P, X) = the set of valid SYN-sequences of P with input X

Although the Feasible and Valid sets are, in general, impossible to determine, they can be used for defining the correctness of distributed programs, classifying types of faults in distributed programs, and comparing various validation techniques for distributed programs. Note that P could be a stand-alone program (i.e., it reads data from and writes data to files) or a program that interfaces with other processes. In the latter case, an input of P may contain SYN-events. The above discussion of feasible and valid SYN-sequences of P with input X is independent of the existence of SYN-events in X.

P is said to be *correct for input X* (with respect to the specification of P) if

(a) Feasible(P, X) = Valid(P, X), and
(b) Every execution of P with input X produces the correct (or expected) result. (The result of an execution includes the output and termination condition of the execution. Possible types of abnormal termination include divide-by-zero errors, deadlock, expiration of allocated CPU-time, etc. It is assumed that if a process in P has an abnormal termination, other processes in P that can continue to execute will do so.)

P is said to be *correct* (with respect to the specification of P) if and only if P is correct for every possible input.

Under certain circumstances, condition (a) can be modified. Several possible modifications of condition (a) are given below:

(a1) Feasible(P, X) is a proper subset of Valid(P, X). This condition is used when the specification of P uses nondeterminism to model implementation decisions that are to be made later by the implementor. Making implementation decisions amounts to a *reduction* of the nondeterminism in the specification.

(a2)   Valid(P, X) is a proper subset of Feasible(P,X). This condition is used when the specification of P is incomplete and thus is *extended* by the implementation.

(a3)   Valid(P, X) = Spec_Feasible(P, X), where Spec_Feasible(P, X) is Feasible(P, X) modified by deleting "nonspecification" events from each feasible SYN-sequence of P with input X. Nonspecification events refer to events that are not mentioned in the specification of P. This condition is used when the SYN-event set for P's specification is a proper subset of that for P's implementation and when only the external behavior (i.e., the behavior observed by a user) of P is specified.

Based on the earlier definition of correctness, *P is incorrect for input X* if and only if one or more of the following conditions hold:

(a)   Feasible(P,X) is not equal to Valid(P,X). Thus, either
   (a.1)   there exists at least one SYN-sequence that is feasible but invalid for P with input X, or
   (a.2)   there exists at least one SYN-sequence that is valid but infeasible for P with input X. (Or both (a.1) and (a.2) are true.)
(b)   There exists an execution of P with input X that exercises a valid (and feasible) SYN-sequence, but produces an incorrect result.

The existence of condition (a) is referred to as a *synchronization fault or inter-process fault*. (The term "synchronization fault" has several synonyms such as "timing error" and "time-dependent error.") The existence of condition (b) is referred to as an *intra-process fault*. (For a sequential program, intra-process faults exist, but inter-process faults do not.)

After an execution of P with input X completes, it is necessary to determine whether any synchronization or intra-process faults exist. Assume that this execution exercises a (feasible) SYN-sequence, say S, and produces a result, say Y. Then one of the following conditions holds:

(a) S is valid and Y is correct.
(b) S is valid and Y is incorrect.
(c) S is invalid and Y is incorrect.
(d) S is invalid and Y is correct.

Condition (b) implies the existence of an intra-process fault. Each of conditions (c) and (d) implies the existence of a synchronization fault. Note that condition (d) implies the production of a correct result Y from an incorrect SYN-sequence S. If the programmer checks only the correctness of Y, the invalidity of S would go undetected and might cause condition (c) to occur in the future for the same or a different input. Thus, during the testing of P, it is important to collect the SYN-sequences of executions of P and then to determine the validity of each collected SYN-sequence with respect to the corresponding input. (As shown later, the collected SYN-sequences are also needed for testing and debugging.)

To illustrate the types of faults in a distributed program, consider again the Ada program called Bounded_Buffer shown in Fig. 33.1. Note that all valid inputs of Bounded_Buffer have the same set of feasible (valid) SYN-sequences. Bounded_Buffer contains a fault. The following statement in task Buffer_Control

"when Count <= Size => accept Deposit(C : in character) do...;"

is incorrect because of the wrong relational operator "<=" (versus "<"). This fault is a synchronization fault since it allows the deposit of an item when the buffer is full. Consider an execution of Bounded_Buffer with input ('A','B','C') that exercises the R-sequence in Fig. 33.2. This R-sequence starts with three consecutive rendezvous involving entry Deposit (D), followed by three consecutive rendezvous involving entry Withdraw (W). The output of this execution is ('C','B','C'), not the expected output ('A','B','C'). This is an example of fault condition (c) above, since this R-sequence is invalid and the output ('C','B','C') is incorrect. If an execution of Bounded_Buffer with input ('C','B','C') exercises the R-sequence in Fig. 33.2, then the output of this execution is ('C','B','C') as expected. This is an example of fault condition (d) above, since this R-sequence is invalid but the output ('C','B','C') is correct. Note that if an execution of Bounded_Buffer with any valid input does not exercise the R-sequence in Fig. 33.2, this execution does not produce an invalid SYN-sequence or an incorrect result.

Finally, assume that the incorrect statement in task Buffer_Control is modified to

"when Count+1 < Size => accept Deposit(C : in character) do ...;"

Now task Buffer_Control allows at most only one character in the buffer. In this case, the set of feasible R-sequences of Bounded_Buffer is a proper subset of the set of valid R-sequences of Bounded_Buffer. Thus, Bounded_Buffer still contains a synchronization fault. However, this fault cannot be detected by any execution of Bounded_Buffer.

33.3   Approaches to Testing Distributed Programs

Typically, a distributed program P, like a sequential program, is subjected to two types of testing:

- *Black-box testing*: access to P's implementation is not allowed. Thus, only the specification of P can be used for test generation, and only the result (including the output and termination condition) of each execution of P can be collected.
- *White-box testing*: access to P's implementation is allowed and only the implementation of P can be used for test generation. Any desired information about each execution of P can be collected.

	V	C	U	N
1.	b1	producer	buffer_control	deposit
2.	b2	producer	buffer_control	deposit
3.	b1	producer	buffer_control	deposit
4.	b2	producer	buffer_control	deposit
5.	b1	producer	buffer_control	deposit
6.	b2	producer	buffer_control	deposit
7.	b1	consumer	buffer_control	withdraw
8.	b2	consumer	buffer_control	withdraw
9.	b1	consumer	buffer_control	withdraw
10.	b2	consumer	buffer_control	withdraw
11.	b1	consumer	buffer_control	withdraw
12.	b2	consumer	buffer_control	withdraw

Figure 33.2 An R-sequence of the bounded buffer program

White-box testing is usually not appropriate for system or acceptance testing because of the complexity and size of the code or the inability to access the code. Below is a third type of testing:

- *Extended black-box testing*: During an execution of P, only the result and SYN-sequence of this execution can be collected. Thus, only the specification and collected SYN-sequences of P can be used for test generation. Also, an input and a SYN-sequence can be used to control the execution of P.

The remainder of this section describes several approaches to testing distributed programs. For each testing approach, its relationship with the above types of testing is discussed.

33.3.1 Nondeterministic testing

Nondeterministic testing of a distributed program P involves the following steps:

1. Select a set of inputs of P.
2. For each selected input X, execute P with X many times and examine the result of each execution.

Multiple, nondeterministic executions of P with input X may exercise different feasible SYN-sequences and thus may detect more faults than a single execution of P with input X. Nondeterministic testing was referred to as multiple execution testing in Ref. [4, 7].

Since the purpose of nondeterministic testing is to exercise as many distinct SYN-sequences of P as possible, the following variations of nondeterministic testing have been proposed to increase the likelihood of exercising different SYN-sequences:

(a) Control the scheduling of processes in the ready queue of the operating system. One commonly used scheduling method is round-robin scheduling. Changing the value of the time quantum for round-robin scheduling may produce different SYN-sequences.
(b) Insert delay statements into P with the delay amount randomly chosen [8, 9]. Executing a delay statement forces a context switch and indirectly affects process scheduling. This method requires that delay statements are supported by the underlying distributed language or operating system.
(c) Insert simulated delay statements into P with the delay amount randomly chosen. Simulated delay statements are based on the use of a virtual-time clock [5]. One implementation strategy is to create an additional process that has lower priority than any other process in P. This lowest-priority process runs only when all of the other processes in P are blocked, at which time it awakens the blocked processes with the smallest wake-up time. A slightly different strategy is to let this lowest-priority process perform "tick" operations such that the ith, $i > 0$, tick operation awakens the processes that requested to be delayed until time i.

Nondeterministic testing of P with input X has two major problems. First, some feasible SYN-sequences of P with input X may never be executed. Second, some feasible SYN-sequence of P with input X may be executed many times. For example, nondeterministic testing was applied to the final concurrent Ada program in Ref. [9] for solving the gas station problem. This program was executed many times without showing a deadlock. (An execution resulting in a deadlock indicates the existence of a feasible, but invalid SYN-sequence.) But a deadlock in this program was found later. Another example is the experiment reported in Ref. [10]. This experiment used a semaphore-based program called

Prod_Cons that implemented an incorrect solution to the bounded buffer problem. Prod_Cons contained two producer processes and one consumer process. A buffer of size three was used, but the overflow or underflow of the buffer was not checked. Each producer deposited two items into the buffer, and the consumer withdrew four items from the buffer. A semaphore variable was used to ensure mutual exclusion while accessing the buffer, but there was no control over the order in which items were deposited into and withdrawn from the buffer. The experiment observed the sequence in which the two producers and one consumer entered their critical sections, referred to as a CS-sequence. Prod_Cons had a total of 420 different possible CS-sequences. Using nondeterministic testing without any delay statements, 1,000 executions of Prod_Cons collected about 10 distinct CS-sequences. Using nondeterministic testing with delay statements inserted to force context switches, 1,000 executions of Prod_Cons collected about 15 distinct CS-sequences. Thus, less than 4 percent of the CS-sequences of Prod_Cons were collected by nondeterministic testing.

Nondeterministic testing of P is allowed in black-box, white-box, or extended black-box testing. The latter two types of nondeterministic testing permit the collection of SYN-sequences of P that can be analyzed to detect the existence of feasible, but invalid SYN-sequences of P. The collection and analysis of SYN-sequences provide useful information, but they involve the following problems:

- *SYN-sequence definition*. This problem was discussed in Section 33.1.
- *SYN-sequence collection*. The collection of SYN-sequences and other information is commonly referred to as monitoring or tracing [9, 11]. The interference caused by monitoring creates the probe effect. It is possible that a fault detected by an interference-free execution of P would not be detected by an execution of P with interference. However, by varying the degree of interference (as discussed earlier in this section), repeated executions of P generally exercise a larger number of different SYN-sequences and detect more faults than interference-free executions of P. Approaches to building tools for SYN-sequence collection will be discussed in Section 33.6.
- *SYN-sequence validity*. This problem is to determine whether a SYN-sequence is valid for P with a given input. If the specification of P is informal, then the validity problem has to be solved manually. If the specification of P is formal, and the mapping between the SYN-events in the specification and the SYN-events in P is formally defined, then the validity problem can be solved automatically.

33.3.2   Deterministic testing

Deterministic testing of a distributed program P involves the following steps:

1. Select a set of tests, each of the form (X, S), where X and S are an input and a SYN-sequence of P respectively.
2. For each selected test (X, S),
   - force a deterministic execution of P with input X according to S. This forced execution determines whether S is feasible for P with input X.
   - compare the actual and expected results (including the output, feasibility of S, and termination condition) of the forced execution. If the actual and expected results are different, a fault is detected.

Note that for deterministic testing, a test for P is not just an input of P; it consists of an input and a SYN-sequence of P, and is referred to as an *IN-SYN test*. Since deterministic testing requires the control of SYN-events during a program's execution, it must be white-box or extended black-box testing. If deterministic testing of P uses a SYN-sequence that is known to be feasible for P with input X, then the SYN-sequence is definitely repeated and such testing is referred to as *replay* [6] or *reproducible testing* [5]. Deterministic testing was referred to as deterministic execution testing in Ref. [4,7]; it is similar to the concept of "forcing a path" mentioned in Ref. [12] (see Section 33.4.1 for more discussion).

Deterministic testing provides several advantages over nondeterministic testing:

(a) Nondeterministic testing may leave certain portions or paths of P uncovered. Deterministic testing allows carefully selected SYN-sequences to be used to test specific portions or paths of P.

(b) Nondeterministic testing exercises feasible SYN-sequences only; thus, it can detect the existence of invalid, feasible SYN-sequences of P, but not the existence of valid, infeasible SYN-sequences of P. Deterministic testing can detect both types of faults.

(c) After P has been modified for correction or enhancement, deterministic regression testing with the inputs and SYN-sequences of previous executions of P provides more confidence about the correctness of P than nondeterministic testing of P with only the inputs of previous executions.

Deterministic testing involves the following problems, in addition to the problems of SYN-sequence definition, collection, and validity that were discussed earlier:

- *SYN-Sequence feasibility:* This problem is to determine whether a SYN-sequence S is feasible for P with input X. The feasibility of S is determined by attempting to force a deterministic execution of P with input X according to S. *S is feasible for P with input X if and only if the forced execution exercises exactly S before P terminates.* If the forced execution of P with (X,S) terminates, then the feasibility of S for P with input X can be determined. However, the problem of determining whether a distributed program terminates on a given input and SYN-sequence is undecidable, as this problem can be reduced to the program halting problem, which is undecidable. A practical technique for coping with this problem and approaches to building tools for determining SYN-sequence feasibility will be discussed in Section 33.6.

- *IN-SYN test selection:* The selection of IN-SYN tests for P can be done in various ways:

  (a) Select inputs and then select a set of SYN-sequences for each input.
  (b) Select SYN-sequences and then select a set of inputs for each SYN-sequence.
  (c) Select inputs and SYN-sequences separately and then combine them.
  (d) Select pairs of inputs and SYN-sequences together.

The inputs and SYN-sequences of P can be selected according to the specification and implementation of P. However, if only extended black-box testing is allowed, the available information about the implementation P is the collected SYN-sequences of P. The test generation problem for distributed programs will be discussed later in Section 33.4.

33.3.3 Combinations of deterministic and nondeterministic testing

Although deterministic testing has advantages over nondeterministic testing, it requires additional effort for selecting SYN-sequences and determining their feasibility. Such effort can be reduced by combining deterministic and nondeterministic testing. Below are several possible combined strategies for testing a distributed program P:

(a) Apply nondeterministic testing first until the test coverage of P has reached a certain level. Then apply deterministic testing to achieve a higher test coverage.

(b) Apply deterministic testing during module and integration testing of P and nondeterministic testing during system and acceptance testing of P.

(c) *Prefix-based testing:* For P with input X, first apply deterministic testing according to a given SYN-sequence. Then, if deterministic testing succeeds, apply nondeterministic testing immediately after the completion of the given SYN-sequence. The purpose of prefix-based testing is to start nondeterministic testing at a state other than the initial state of P. If prefix-based testing of P with input X uses a prefix of a SYN-sequence that is known to be feasible for P with input X, such prefix-based testing is referred to as *prefix-based replay.*

(d) *Event-subset testing:* Event-subset testing of P is to apply deterministic testing to a subset of the SYN-events of P. As mentioned in Section 33.1, a SYN-event is associated with a particular process, object, and event type. *Process-subset testing* of P is to apply deterministic testing to a subset of the processes in P, and nondeterministic testing to the remaining processes. Similarly, object-subset and event-type-subset testing are applied to a subset of the objects and event-types, respectively, in P. These three types of subset testing can be combined.

In Ref. [13] an experiment was described in which both nondeterministic and deterministic testing were applied. For a correct version of Bounded_Buffer in Fig. 33.1, 81 mutants were generated. (Each mutant was created by making a single syntactic change to a program statement in Bounded_Buffer, inducing a typical programming fault.) First, nondeterministic testing with the use of random delays was applied to Bounded_Buffer to collect 4 (feasible) R-sequences. (Bounded_Buffer has only 4 feasible R-sequences.) Next, the collected R-sequences were used to perform deterministic testing on the 81 mutants of Bounded_Buffer. 58 of the mutants were distinguished from Bounded_Buffer by the collected R-sequences, because each of these 58 mutants either produced an incorrect result or had at least one of the collected R-sequences as an infeasible R-sequence. The remaining 23 mutants, however, produced correct results for the collected R-sequences. Finally, two of the collected R-sequences were modified to create 2 new R-sequences that were infeasible for Bounded_Buffer. These new R-sequences distinguished the remaining 23 mutants from Bounded_Buffer during deterministic testing, as each of the remaining mutants had at least one of the new R-sequences as a feasible R-sequence. This experiment indicates that effective testing of distributed programs requires a combination of nondeterministic and deterministic testing.

33.3.4   Reachability testing

Assume that every execution of a distributed program P with input X terminates. The number of feasible SYN-sequences of P with input X is finite. As shown in Section 33.3.1, nondeterministic testing of P with input X is not guaranteed to derive all feasible SYN-sequences. One approach to deriving all feasible SYN-sequences of P with input X is to perform *reachability analysis*, which is to derive all reachable states of P with input X (see Section 33.4.1). An alternative approach, called *reachability testing* [10], combines nondeterministic and deterministic testing and can derive all feasible SYN-sequences of P with input X by testing.

Assume S is a feasible SYN-sequence of P with input X. Reachability testing of P with input X and SYN-sequence S involves the following steps:

1. Use S to derive a set of prefixes of other feasible SYN-sequences of P with input X. Such prefixes, called race-variants of S, are derived by changing the outcomes of race conditions in S.
2. Perform prefix-based replay of P with input X and race-variants of S. By doing so, new feasible SYN-sequences of P with input X may be derived.
3. For each new feasible SYN-sequence of P with input X, repeat steps 1 and 2.

The above procedure derives all feasible SYN-sequences of P with input X, reaches all possible states of P with input X, and produces all possible results of P with input X; thus, it determines the correctness of P with input X. Ref. [10] showed a refinement of the above procedure for a concurrent program using read and write operations and implemented the refinement to generate all 420 distinct CS-sequences for program Prod_Cons mentioned in Section 33.3.1.

For a distributed program with a large or infinite number of SYN-sequences, the above procedure can be modified to derive a reasonable number of SYN-sequences. Reachability testing is allowed in white-box or extended black-box testing. Reference [14] showed how to identify race conditions in SYN-sequences and derive race-variants for various types of message passing constructs.

33.3.5 Conformance testing

Conformance testing of a distributed program P refers to black-box testing with test sequences generated from P's specification. Conformance testing is commonly used to test communication protocols that have been specified using the *finite state machine* (FSM) model. Conformance testing of a protocol implementation verifies that the control structure of the protocol implementation conforms to the FSM-based protocol specification. Such testing focuses on the states and transitions in the FSM-based specification (see Section 33.4.2).

A conformance test sequence for P is a sequence of SYN-events between P and its environment, not including SYN-events internal to P. Therefore, conformance testing can be viewed as black-box, nondeterministic testing with an input being a specification-based SYN-sequence. If P allows nondeterministic selections (e.g., selecting one message from among two or more message queues), the expected behavior of P with a specification-based SYN-sequence may be nondeterministic. For an FSM-based protocol specification with multiple message queues, Ref. [15] defined *synchronizable test sequences* such that the behavior of the FSM with a synchronizable test sequence is deterministic. To deal with nondeterminism during conformance testing, the following approaches have been suggested: prohibit the use of non-synchronizable test sequences, avoid nondeterminism in protocol specifications, or apply multiple executions during nondeterministic testing.

33.4 Test Generation for Distributed Programs

As mentioned earlier, a test for nondeterministic testing may be a SYN-sequence, and a test for deterministic testing consists of an input and a SYN-sequence. This section discusses test generation for distributed programs, with emphasis on the selection of SYN-sequences. The selection of inputs is a problem for testing sequential as well as distributed programs, and details about this subject can be found in Ref. [1]. Sections 33.4.1 and 33.4.2 discuss program-based and specification-based SYN-sequence selection, respectively. Note that "specification" and "program" are relative terms. A program may be used as the specification of a lower-level implementation, and if a specification is executable, it can be viewed as a high-level or "abstract" program.

33.4.1 Program-based SYN-sequence selection

Program-based test selection for a distributed program P often focuses on the selection of a set of paths of P. Thus, the relationship between the paths and SYN-sequences of P needs to be addressed. A *totally-ordered path* of P is a sequence of SYN-events and non-synchronization statements (or blocks) of P. A *partially-ordered path* of P contains one path for each process in P, where a path of a process is a sequence of SYN-events and non-synchronization statements (or blocks) involving this process. In the following discussion, unless otherwise specified, paths of P are totally-ordered or partially-ordered, and SYN-sequences of P are process-based totally-ordered or partially-ordered. A path (or SYN-sequence) of P is said to be *feasible for P with input X* if this path can be executed by some execution of P with input X. A path (or SYN-sequence) of P is said to be *feasible for P* if this path (or SYN-sequence) can be executed by some execution of P with some input. The *domain* of a path (or SYN-sequence) S of P is {X | S is feasible for P with input X }. The domain of an infeasible path or SYN-sequence of P is empty.

The following relationships exist between the paths and SYN-sequences of P:

- If a path is feasible for P with input X, the SYN-sequence of this path is feasible for P with input X. However, the converse is not necessarily true.
- A partially- (or totally-) ordered, feasible path of P with input X corresponds to a unique partially- (or totally-) ordered SYN-sequence of P with input X, and vice versa.
- If different totally-ordered, feasible paths of P with input X have the same partially-ordered SYN-sequence, then these paths produce the same result and thus are equivalent.
- The domains of different partially- (or totally-) ordered, feasible paths of P are not necessarily mutually disjoint. The reason is that an input of P may have different partially- (or totally-) ordered, feasible SYN-sequences.
- If different partially- (or totally-) ordered, feasible paths of P have the same partially- (or totally-) ordered SYN-sequence, then their input domains are mutually disjoint.

Based on a reachability or concurrency graph. The *reachability graph* of program P, RG(P), contains all possible states of P, where a state of P contains the location of the next statement to be executed by each process in P. (Other information such as the values of variables and the contents of message queues are usually omitted to reduce the number of states in RG(P).) When the variables and predicates in P are ignored during the construction of RG(P), some states in RG(P) are not reachable and also some paths in RG(P) are infeasible. The *concurrency graph* of P, CG(P), is a simplified reachability graph in which only synchronization related statements are considered. Each state in CG(P) is called a concurrency state and each path in CG(P) a concurrency history [16]. (A concurrency history defines a totally-ordered SYN-sequence.) Both RG(P) and CG(P) suffer from the state explosion problem, since the size of RG(P) or CG(P) grows exponentially with the number of processes in P. Reachability and concurrency graphs can be used to verify certain properties such as freedom from deadlock [16,17] and to select paths for testing or symbolic execution.

In Ref. [12] the following five structural criteria for the selection of paths in CG(P) were defined:

- *all-concurrency-paths*. This criterion is to cover all paths in CG(P); it cannot be satisfied if CG(P) contains infinite loops.
- *all-proper-concurrency-histories*. This criterion is to cover all proper concurrency histories in CG(P), which are paths in CG(P) that do not contain cycles.
- *all-edges-between-concurrency-states*. This criterion is to cover all edges in CG(P).
- *all-cc-states*. This criterion is to cover all states in CG(P).
- *all-possible-rendezvous criterion*. This criterion is to cover all states in CG(P) that involve rendezvous.

Similar criteria can be defined for the selection of paths in RG(P). Reachability (or concurrency) graphs contain totally-ordered sequences. Since totally-ordered sequences that have the same partial order are equivalent, several methods can be applied to reduce the number of totally-ordered sequences in a reachability graph [18]. A different approach to SYN-sequence selection is to incrementally derive and reduce reachability graphs [19, 20].

Based on paths of individual processes. In Ref. [21] the following strategy was proposed for selecting paths and SYN-sequences of P:

(a) Select paths of individual processes of P, where a path of a process is a sequence of statements in this process, not SYN-events involving this and other processes. For example, a path of an Ada task may contain accept statements, but not rendezvous events. (An accept statement does not specify the calling task, while a rendezvous event does.)

(b) Combine the selected paths of individual processes of P to produce C-paths of P, where a C-path is the same as a partially-ordered path except that the former does not include SYN-events (see the example below).

(c) For each selected C-path of P, generate its possible SYN-sequences. Below is a C-path of an Ada program containing tasks T1, T2 and T3:

```
for task T1: (begin, T3.E1, end)
for task T2: (begin, T3.E1, end)
for task T3: (begin, accept E1, accept E1, end)
```

The above C-path has two different SYN-sequences: one containing a rendezvous between the entry call in T1 (T2) and the first (second) *accept* in T3 and the other containing a rendezvous between the entry call in T1 (T2) and the second (first) accept in T3. If a C-path of P contains SYN-statements, but does not have SYN-sequences, it is infeasible for P.

Although this approach avoids the generation of the concurrency graph of P, the selection of paths of individual processes of P and the combination of such paths are still difficult problems.

Generation of IN-SYN tests for selected SYN-sequences. After a SYN-sequence S of P is selected, one or more inputs for S need to be generated. The following two approaches are possible:

- Identify a set of paths of P that have S as their SYN-sequence. For each selected path H of P, determine whether H is feasible and, if so, find a set of inputs of H. However, the path feasibility problem, like the SYN-sequence feasibility problem, is undecidable. Assume that H is feasible and X is an input for H. "Forcing path H by using input X" can be accomplished by deterministic testing of P with (X, S).

- Select a set of inputs of P. For each selected input X, perform deterministic testing of P with (X, S) to determine whether S is feasible for P with input X. If S is infeasible for P with input X, the approach described in Ref. [22] can be applied, which attempts to change the value of X in order to exercise S.

33.4.2   Specification-based SYN-sequence selection

As mentioned earlier in Section 33.3, specification-based SYN-sequences of program P can be used for nondeterministic or deterministic testing of P. In order to do so, SYN-events in the specification of P must be mapped into SYN-events in P; however, this mapping may not be simple or direct. Also, P may contain "internal SYN-events" that do not exist in P's specification. If nondeterministic testing is used, another issue is that the behavior of P for a specification-based SYN-sequence may be nondeterministic. A number of models and languages have been developed for specifying concurrency. However, only a few of them have been considered for test generation.

Based on finite state machine specifications. The FSM model is commonly used for software specification, especially for the specification of communications protocols. Assume that the specification of program P is a set of FSMs communicating with each other via message passing. The program-based methods for SYN-sequence selection that were discussed in Section 33.4.1, can be applied to the FSM-based specification of P to select specification-based SYN-sequences. These methods can also be applied to extended FSMs that are FSMs with variables and predicates.

Conformance testing with an FSM-based specification (see Section 33.3.5) attempts to verify that an implementation contains the same states and transitions as the specification. The general approach to conformance testing based on an FSM M is the following: for each transition T in M, a SYN-sequence is constructed to first reach the head state of T, then execute T, and then execute an additional input sequence to verify that the tail state of T is correct. The basic methods for selecting SYN-sequences for transitions in M are the following:

- *T-method.* Traverse each transition in M at least once (i.e., no verification of the tail state of a transition is required.)
- *D-method.* Verify the tail state of a transition in M by applying a *distinguishing sequence* that is an input sequence that produces different output sequences for different states of M.
- *U-method.* Verify the tail state of a transition in M by applying an input sequence that produces a unique output sequence for the tail state of the transition. The combination of the input and output sequences is referred to as a unique-input-output-sequence.
- *W-method.* Construct sequences of input and output messages that guarantee the detection of the following types of faults in an FSM: incorrect output, incorrect tail state, and extra or missing states.

After constructing SYN-sequences for the transitions in M, these SYN-sequences are combined to form a test sequence for M. Details about conformance test generation can be found in Ref. [17].

Based on abstract programs. Since the FSM model has limited expressive power, a number of languages have been developed for describing the specification and design of a distributed system. In particular, the languages LOTOS, Estelle, and SDL have been used to specify communication protocols [23]. Specifications written in executable specification languages are commonly referred to as *abstract programs*. Program-based SYN-sequence selection methods (see Section 33.4.1) can be applied to abstract programs to generate specification-based test sequences. The problem of test sequence generation for programs written in LOTOS, Estelle, or SDL has been studied in recent years.

Based on sequencing constraints. *Sequencing constraints* for program P specify restrictions on the valid SYN-sequences of P by defining patterns of events that are valid or invalid for P. (If the specification of P does not contain sequencing constraints, such constraints can be derived from the specification or high-level design of P.) The SYN-sequences collected during nondeterministic testing of P can be examined to determine whether they satisfy or violate the constraints for P. Also, sequencing constraints for P can guide the selection of SYN-sequences for deterministic testing of P.

A constraint notation called TSL *(Task Specification Language)*, was described in Refs. [24, 25]. One basic form of a TSL statement is

when activating_event *then* specified_event *before* terminating_event

A SYN-sequence is checked against the above TSL statement as follows. After the activating_event matches,

- If the specified_event matches before the terminating_event, the TSL statement is satisfied.
- If the terminating_event matches before the specified_event, the TSL statement is violated.

TSL also allows the definitions of *properties*. A property is a function of the events that occur during an execution; it has an initial value that changes when specific events happen.

For the 2-slot bounded buffer problem described in Section 33.1.1, the following property can be defined:

```
property Buffer_Count return Natural := 0 is
begin
     when Deposit then set Buffer_Count := Buffer_Count + 1;
     when Withdraw then set Buffer_Count := Buffer_Count - 1;
end Buffer_Count;
```

The following two TSL statements are valid for the 2-slot bounded buffer:

```
when Deposit where Buffer_Count = 1 then Withdraw before Deposit;
when Withdraw where Buffer_Count = 1 then Deposit before Withdraw;
```

CSPE *(Constraints on Succeeding and Preceding Events)* is another notation for specifying sequencing constraints [26]. The following shows three basic types of successor constraints for a program unit U. Each constraint is preceded by a constraint operator "op" that is explained later.

- *Entrance Constraint.* op[# ;→ E1]U : a constraint on event E1 being the *first* event in a SYN-sequence exercised during an execution of U.
- *Successor Constraint.* op[E1 ;→ E2]U: a constraint on event E1 being *succeeded immediately* by event E2 in a SYN-sequence exercised during an execution of U.
- *Exit Constraint.* op[E1 ;→ $]U: a constraint on event E1 being the last event in a SYN-sequence exercised during an execution of U.

The constraint operator "op" is one of the following: "a" for "always," "p" for "possibly," and "~" for "never." The meaning of these operators is explained below:

- a[E1 ;→ E2]U: During an execution of U, immediately after an occurrence of event E1, an occurrence of event E2 is always valid (but E2 need not necessarily occur).
- ~[E1 ;→ E2]U: During an execution of U, immediately after an occurrence of event E1, an occurrence of event E2 is never valid.
- p[E1 ;→ E2]U: During an execution of U, immediately after an occurrence of event E1, an occurrence of event E2 is possibly valid (i.e., neither always valid nor never valid).

For the 2-slot bounded buffer problem, the following CSPE constraints are valid:

a[# ;→ deposit]   a[deposit ;→ withdraw]   a[withdraw ;→ deposit]
~[# ;→ withdraw]   p[deposit ;→ deposit]   p[withdraw ;→ withdraw]

Reference [26] defined constraint coverage and violation for CSPE and showed how to accomplish coverage and detect violations of CSPE constraints by nondeterministic and deterministic testing. Reference [27] showed how to automatically derive CSPE constraints from a formal specification and select SYN-sequences from the specification.

33.5   Analysis and Replay of Program Executions

After the completion of one or more executions of a distributed program P, the collected execution information, including the output, SYN-sequences, termination conditions, etc., should be analyzed to detect the existence of faults in P. The collected SYN-sequences of P can be subjected to different types of analysis, including

- detecting deadlocks. Such detection is referred to as *dynamic deadlock detection* (in contrast to *static deadlock detection*, which analyzes a program without executing it). Let S be a collected SYN-sequence of P. S results in a *global deadlock*, also called an *infinite wait* [16] or *global blocking* [9, 11], if at the end of S, every process in P is either blocked at a synchronization statement or terminated, and at least one process in P is blocked. In particular, S results in a *circular deadlock* if, at the end of S, there exists a circular list of two or more processes such that each process is waiting to synchronize with the next process on the list.
- detecting the satisfaction and violation of sequencing constraints [24, 25, 27]
- detecting the satisfaction and violation of predicates involving program states
- determining causal relationships between events

A number of papers on the last two types of analysis are included in Ref. [28].

If additional debugging information for P is needed, previous executions of P can be repeated to collect more information. The *replay* of a previous execution of P can be accomplished by deterministic testing of P with the input and complete SYN-sequence of the execution. However, the replay of an execution does not require the complete SYN-sequence of this execution. The complete SYN-sequence of an execution of P can be reduced into a shorter SYN-sequence, called the *simple SYN-sequence* of the execution, such that the result of the execution can be determined from P and the input and simple SYN-sequence of the execution. This implies that the complete partially-ordered (not totally-ordered) SYN-sequence of an execution of P can be determined by P and the input and simple SYN-sequence of this execution. Often the information needed for replay can be reduced even further since the SYN-events of an execution of P that are not involved in race conditions are not needed for the replay of the execution [29]. The use of simple SYN-sequences instead of complete SYN-sequences results in simpler solutions for replay, with less execution overhead.

To illustrate the difference between complete and simple SYN-sequences, consider the types of SYN-events for the Ada language mentioned in Section 33.1.3. Types (a2), (b2), (c2), and (d2) involve the end of execution of an *accept* statement or a *select* alternative. To replay a previous execution of an Ada program, events of types (a2), (b2), (c2), and (d2) are not needed since such events would occur as expected if they occurred in the previous execution. Furthermore, the complete format (V,C,U,N,D) for the remaining types can be simplified by removing type (V) and entry name (N). By deleting and simplifying SYN-events, an R-sequence is reduced to a *simple R-sequence* (or *SR-sequence*). Below is the SR-sequence of the R-sequence in Fig. 33.2 (component "D" is not needed for this SR-sequence):

	C	U
1.	producer	buffer_control
2.	producer	buffer_control
3.	producer	buffer_control
4.	consumer	buffer_control
5.	consumer	buffer_control
6.	consumer	buffer_control

The result of a forced execution of P with an input and a simple SYN-sequence of P is deterministic. Thus, simple SYN-sequences of P can be used for deterministic testing of P. The selection of simple SYN-sequences of P is easier than that of complete SYN-sequences since simple SYN-sequences have fewer SYN-events and a simpler SYN-event format than complete SYN-sequences. However, the role of complete SYN-sequences in testing distributed programs cannot be completely replaced by simple SYN-sequences.

Assume that S is a complete SYN-sequence of P and S′ is the simple SYN-sequence of S. If S is feasible for P with input X, then S′ is also feasible for P with input X, but the converse is not true. To illustrate this, consider an Ada program Q consisting of the two tasks shown in Fig. 33.3a. Assume that statements S1 and S2 in task T2 do not contain synchronization statements. Let program Q′ be Q modified by changing task T2 as shown in Fig. 33.3b. Each of programs Q and Q′ has only one R-sequence and thus only one SR-sequence. Although Q and Q′ have distinct R-sequences, they have the same SR-sequence. (Note that the end of execution of an accept statement appears in an R-sequence, but not in

```
    task T1;              task T2 is              task T2 is
                              entry E1(...);          entry E1(...);
                              entry E2(...);          entry E2(...);
    task body T1 is       end;                    end;
    begin                 task body T2 is         task body T2 is
        T2.E1(...);       begin                   begin
    end T1;                   accept E1(...) do       accept E1(...) do
                                  S1;                     S1;
    task T3;                  end E1;                     accept E2(...) do
    task body T3 is          accept E2(...) do               S2;
    begin                         S2;                    end E2;
        T2.E2(...);           end E2;                 end E1;
    end T3;               end T2;                 end T2;
```

(a) program Q   (b) modified task T2

Figure 33.3 Ada program with two tasks

an SR-sequence.) Now let program Q″ be Q modified by reversing the order of the two accept statements for E1 and E2. Again, Q and Q″ have distinct R-sequences, but they have the same SR-sequence. (Entry name information is not included in an SR-sequence.) If R-sequences are used for deterministic testing, then Q, Q′, and Q″ can be distinguished, but not so if SR-sequences are used. Since complete SYN-sequences provide more synchronization-related information than simple SYN-sequences, the former allow more checking than the latter during deterministic testing.

One major problem in distributed computing is the recovery of a distributed system from failures. Replay can be combined with a scheme for recovering from failures in a distributed system. As an example, Ref. [30] proposed the following steps to handle recovery from a failed execution of a distributed program P:

- *Recording mode*: During a normal execution of P, each process in P logs its SYN-events, maintains ordering information among the local states, and occasionally checkpoints its local state.
- *Analysis mode*: After an execution of P, the saved information is used to restore consistent global states of the previous execution, as requested by the user. A *consistent global state* of P is a state in which the component local states of P are mutually concurrent.
- *Replay mode:* After a consistent global state of a previous execution of P has been restored, use the saved information to replay the remaining portion of the previous execution.

33.6   Building Testing Tools for Distributed Programs

For a distributed program, tools can be used to collect SYN-sequences, determine the feasibility of SYN-sequences, and replay feasible SYN-sequences. The SYN-sequence feasibility and replay problems require forced executions of SYN-sequences and are referred to collectively as the SYN-sequence execution problem. To solve the SYN-sequence collection and execution problems for a distributed programming language L, two basic approaches are discussed below.

An implementation-based approach is to modify one or more of the three components in the implementation of L: the compiler, the run-time system, and the operating system. These modifications enable the collection and forced execution of SYN-sequences during an execution of a program written in L. For example, many implementation-based debuggers allow the programmer to directly control execution by performing "scheduler-control" operations such as setting breakpoints, selecting the next running process, rearranging processes in various queues, and so on [2]. Also, in Refs. [12] and [22], "forcing a path" is accomplished by controlling the run-time scheduler. Such human control is tedious and error-prone. In addition, since an implementation-based solution to SYN-sequence collection and execution is dependent on a particular compiler, run-time system or operating system, applying such a solution from one implementation to another may be difficult if not impossible, forcing developers to "reinvent the wheel" for each implementation.

A *language-based approach* to solving the SYN-sequence collection and execution problems for L has two steps. The first step is to define the formats of complete and simple SYN-sequences for L in terms of the synchronization constructs available in L (see Section 33.1). The second step is to develop program transformation tools for L in order to support SYN-sequence collection and execution. Below are brief descriptions of two tools for transforming a distributed program P written in L. (These two tools can be combined into one.)

SYN-sequence collection tool for L. This tool transforms P into a slightly different program P′, which is also written in L. P′ is equivalent to P except that during an execution of P′, SYN-events involving statements in P are collected. (SYN-events involving statements in P′, but not in P, are not collected.) The basic transformation strategy is to first identify the statements in P that execute SYN-events. Then, immediately before each such statement, insert an additional statement to record the event. The use of source transformation for collecting traces of a program is common (e.g., Ref. [11]).

SYN-sequence execution tool for L. This tool transforms P into a slightly different program P″, which is also written in L. An execution of P″ with (X, S) as input, where X is an input of P and S a SYN-sequence of P, is an attempt to force the execution of P with input X according to S. *This forced execution determines the feasibility of S and, if S is feasible, produces the same result as P with input X and SYN-sequence S would.* Thus, P″ is used to solve the SYN-sequence feasibility and replay problems for P. The basic transformation strategy is to first identify the statements in P that execute SYN-events. Then, immediately before (after) each such statement in P, insert an additional statement to request (release) a permit, referred to as a SYN-*permit*, for the event. The requests and releases of SYN-permits are handled at run-time according to a given SYN-sequence. As mentioned in Section 33.3.2, the problem of determining whether a distributed program terminates, and hence whether a SYN-sequence is feasible, is in general undecidable. A practical way to deal with this problem is to set a maximum allowed interval between two consecutive SYN-events, referred to as a SYN-event interval. If an execution of P″ with (X, S) as input results in a violation of the SYN-event interval, then P″ is forced to terminate abnormally and S is assumed to be infeasible.

A language-based approach to SYN-sequence collection and execution has several advantages. This approach is independent of any particular implementation of a distributed language and does not create a portability problem. Also, language-based solutions to the SYN-sequence collection and execution problems for a distributed language can serve as a high-level design for implementation-based solutions. A toolset called *TDCAda* (Testing and Debugging Concurrent Ada) was implemented using the language-based approach [4]. This toolset supports nondeterministic and deterministic testing of concurrent Ada programs; it also supports prefix-based and process-subset testing that are combinations of nondeterministic and deterministic testing. Ref. [7] showed how to implement language-based tools for supporting nondeterministic and deterministic testing of concurrent programs using the semaphore and monitor constructs.

33.7   Conclusions and Future Work

Although protocol conformance testing has been an active research area for more than one decade, how to test other types of distributed programs has received less attention. This chapter has examined general issues involved in testing distributed programs and surveyed existing approaches to solving important problems. Many problems either have not been solved yet or need more studies. Below are some problems for future research:

- derivation of tests from formal or semi-formal specifications of concurrency,
- combination of reachability graph reduction techniques and incremental selection of SYN-sequences
- definition and evaluation of test coverage criteria for detecting synchronization faults in a distributed program
- generation of new SYN-sequences from existing SYN-sequences of a distributed program.

Distributed software systems are becoming more common and generally require very high reliability. Cost-effective testing techniques are critical for reducing the development cost of and improving the quality of distributed software. Research on testing distributed software can be expected to continue for a long time.

33.8   References

1. Beizer, B. 1990. *Software Testing Techniques,* 2nd edition. New York: Van Nostrand Reinhold.
2. McDowell, E., and D. P. Helmbold. 1989. Debugging concurrent programs. *ACM Computing Surveys,* Vol. 21, No. 4, 593–622.
3. Andrews, G. 1991. *Concurrent Programming: Principles and Practice.* Redwood, Calif.: Benjamin Cummings.
4. Tai, C., R. H. Carver, and E. Obaid. 1991. Debugging concurrent Ada programs by deterministic execution. *IEEE Trans. on Software Engineering,* Vol. 17, No. 1, 45–63.
5. Hansen, P. B. 1978. Reproducible testing of monitors. *Software-Practice and Experience,* Vol. 8, 721–729.
6. LeBlanc, J., and J. M. Mellor-Crummey. 1987. Debugging parallel programs with instant replay. *IEEE Trans. Computers*, Vol. C-36, No. 4, 471–482.
7. Carver, R., and K. C. Tai. 1991. Replay and testing for concurrent programs. *IEEE Software,* Vol. 8. No. 2, 66–74.
8. Gait, J. 1986. A probe effect in concurrent programs. *Software-Practice and Experience,* Vol. 16, No. 3, 225–233.
9. Helmbold, D., and D. Luckham. 1985. Debugging Ada tasking programs. *IEEE Software,* Vol. 2, No. 2, 47–57.
10. Hwang, G. H., K. C. Tai, and T. L. Huang. 1995. Reachability testing: an approach to testing concurrent software. *Int. Journal on Software Engineering and Knowledge Engineering*, Vol. 5, No. 4.
11. German, S. M. 1984. Monitoring for deadlock and blocking in Ada tasking. *IEEE Trans. Software Engineering.*, Vol. SE-10, No. 6, 764–777.
12. Taylor, R. N., D. L. Levine, and C. D. Kelly. 1992. Structural testing of concurrent programs. *IEEE Trans. on Software Engineering*, Vol. 18, No. 3, 206–215.
13. Carver, R. H. 1993. Mutation-based testing of concurrent Ada programs. *Proc. of Int. Test Conference,* 845–853.
14. Tai, K. C. and Y. C. Young. 1995. Reachability testing of message-passing programs. *Technical Report 95-12.* Raleigh, N.C.: North Carolina State University, Dept. of Computer Science.

15. Tai, K. C. 1995. Port-synchronizable test sequences for communication protocols. *Proc. IFIP 8th Int. Workshop on Protocol Test Systems,* 379–394.

16. Taylor, R. N. 1983. A general-purpose algorithm for analyzing concurrent programs. *Communications of the ACM,* Vol.26, No.5, 362–376.

17. Holzman, G. J. 1991. *Design and Validation of Computer Protocols.* Englewood Cliffs, N.J.: Prentice Hall.

18. Duri, S., U. Buy, R. Devarapalli, and S. M. Shatz. 1994. Application and experimental evaluation of state space reduction methods for deadlock analysis in Ada. *ACM Trans. Software Engineering and Methodology,* Vol. 3, No. 4, 340–380.

19. Koppol, P. V., and K. C. Tai. 1995. Conformance testing of protocols specified as labeled transition systems. *Proc. IFIP 8th Int. Workshop on Protocol Test Systems,* 143–158.

20. Koppol., P. V., and K.C. Tai. 1996. An incremental approach to structural testing of concurrent programs. *Proc. ACM Int. Symp. Software Testing and Analysis.*

21. Yang, R. D., and C. G. Chung. 1992. Path analysis testing of concurrent programs. *Information and Software Technology,* Vol. 34. No. 1, 43–56.

22. Korel, B., H. Wedde, and R. Ferguson. 1992. Dynamic methods of test generation for distributed software. *Information and Software Technology,* Vol. 34, No. 8, 523–531.

23. Turner, K. J., ed. 1993. *Using Formal Description Techniques: An Introduction to Estelle, LOTOS and SDL.* New York: John Wiley & Sons.

24. Helmbold, D., and D. Luckham. 1985. TSL: task sequencing language. *Proc. of Ada Int. Conference,* 255–274.

25. Rosenblum, D. 1991. Specifying concurrent systems with TSL. *IEEE Software,* Vol. 8. No. 3, 52–61.

26. Tai., K. C., and R. H. Carver. 1995. A specification-based methodology for testing concurrent programs. *Proc. European Software Engineering Conference,* Lecture Notes in Computer Science, Vol. 989, W. Schafer and P. Botella, eds. New York: Springer-Verlag, 154–172.

27. Carver, R. H., and K. C. Tai. 1995. Test sequence generation from formal specification of distributed programs. *Proc. IEEE Int. Conf. Distributed Computing Systems,* 360–367.

28. Yang, Z., and T. A. Marsland, eds. 1994. *Global States and Time in Distributed Systems.* Los Alamitos, Calif.: IEEE Computer Society Press.

29. Netzer, R. H. B., and B. P. Miller. 1994. Optimal tracing and replay for debugging message-passing parallel programs. *Journal of Supercomputing,* Vol. 8, No. 4, 371–388.

30. Goldberg, A., A. Gopal, A. Lowry, and R. Strom. 1991. Restoring consistent global states of distributed computations. *Proc. of ACM/ONR Workshop on Parallel and Distributed Debugging,* 140–150.

Applications

34

Scientific Computation

Timothy G. Mattson

The goal of *scientific computing* is to answer scientific questions. To the scientific programmer, the computer hardware and software are tools, not ends in themselves. Rather, it is the science embodied in a computation—not the computer science—that really matters.

Science has proven to be such a great motivator for users of high performance computers, that the literature of parallel computing is dominated by scientific applications. The result is more material about parallel scientific computing than can be summarized in a single chapter. The best that can be done is to provide a framework for understanding parallel scientific computing and a sampling of important methods. To this end, we will focus on two of the most important issues:

- parallel algorithms
- programming environments

The number of algorithms used in parallel scientific computing is overwhelming. We will approach this complex problem at two levels. First, we will consider abstract algorithm classes for scientific computing. These classes are useful when reasoning about algorithms at a high level. Second, we will look at the implementation of algorithms by considering algorithmic motifs (i.e., software constructs that frequently appear together in programs).

An algorithm, of course, is useless until it is expressed in a program. While programming environments have been discussed elsewhere, we will discuss them briefly to show the importance of using a portable programming environment. Portability is vital for any type of computing, but with its emphasis on application software, portability is especially important to scientific computing. It will be argued in this chapter that with a portable programming environment and some planning by the programmer, it is possible to write a single program that can run on most parallel systems, from workstation clusters to scalable parallel computers.

Following the programming environment and algorithms sections, we will describe a number of case studies. Selecting cases to study was difficult. After all, parallel computing has played an important role in practically every field of science, including weather/climate modeling, computational chemistry, fluid dynamics, computational biology, and structural mechanics. Rather than brief summaries representing the full range of parallel scientific applications, this discussion will provide in-depth descriptions of three problems. These problems all come from the field of molecular modeling thereby showing how various parallel algorithms are applied to similar problems. Of course, no prior knowledge of molecular modeling is assumed, so any programmer—not just computational chemists—will benefit fully from these case studies.

Finally, a complex discipline such as parallel scientific programming can not be fully addressed in a single chapter. To master this field, additional reading is required. Therefore, several trends to watch and a few key references will be pointed out.

34.1   Programming Models for Parallel Computing

Programmers view a computer in terms of a high-level abstraction called a *programming model*. For sequential computers, the von Neumann model serves as a universal programming model. Parallel computing, however, lacks a single universal programming model, forcing each programmer to choose a model.

Choosing a programming model is complicated and involves subjective decisions (e.g., expressiveness and ease of use) as well as objective matters (e.g., performance and portability). In this discussion, we will focus on a single objective issue—the ability of a programming model to support the writing of portable software.

Some programmers view the phrase "portable software for parallel and distributed computing" as an oxymoron. They claim that a standard high-level model of a parallel computer doesn't exist, so it is premature to talk about portability. This isn't necessarily true. To see this point, we will have to digress for a moment and take a look at the architecture of parallel computers.

There are countless ways to combine CPUs into a single system. The most common approach is to organize these architectures in terms of instruction streams and data streams [3]. Three cases have become everyday terms to the parallel programmer, SISD, MIMD, and SIMD.

Lets begin with the SISD case. SISD computers have a single instruction stream and a single data stream and are the basis of traditional von Neumann computers. All operations on a SISD architecture are logically sequential.

The SIMD model views the parallel computer in terms of a single instruction stream that is applied to multiple data streams. It is legitimate to view vector processors or superscaler CPUs as SIMD computers, but this term is generally reserved for large multiprocessor systems with many simple processors.

Finally, there is the most general model. This is the MIMD, or multiple instruction, multiple data model. Most typically, a MIMD computer has multiple processing elements (called *nodes*) each of which is a complete computer in its own right.

Of these three models, only the SIMD model will execute on all parallel computers. Therefore, it would be tempting to base a portable programming model on this lowest common dominator. In fact, this is the strategy taken by the High Performance Fortran Forum and their proposed HPF language standard [4]. SIMD computers, however, have not done well in the market. While these systems are easy to program, it turns out that optimizing SIMD programs to yield acceptable performance is very difficult. Because of this, only one company currently manufactures general-purpose supercomputers based on the SIMD architecture, making MIMD systems the overwhelming majority of parallel systems—especially when workstation networks are viewed as single MIMD computers. Hence, it is reasonable to ignore SIMD computers and base one's programming activities on the most general computer architecture model, the MIMD model.

One last factor must be incorporated into our common parallel computer model. A MIMD computer has processors and memory. The memory can be shared among the processors, or it can distributed with the processors. In some ways, it is worth while to consider two distinct programming models—shared memory MIMD and distributed memory MIMD. However, in both cases, the same issues of data locality and concurrency arise, suggesting that any MIMD computer can be viewed in terms of a common model. One such model is the *coordination model*. Within this model, a parallel computation is viewed as a collection of distinct processes that interact at discrete points through a coordination operation. The term *coordination* refers to the fundamental operations to control a parallel computer (i.e., information exchange, synchronization, and process management). The coordination operations may vary in speed and structure, but the overall model is essentially the same.

Describing parallel and distributed computers in terms of a coordination model is not universally accepted in the same way that the von Neumann model is. The point is, such a model can be stated and used to program parallel computers within a universal programming model. Computer systems differ, but the difference is the granularity (ratio of computation to communication), not the fundamental programming model.

A programming model is only half a solution. To be useful, the model must be implemented as a programming environment. Numerous programming environments exist that support various incarnations of the coordination model and run well on workstation clusters as well as parallel computers [1, 2]. One can use elegant high-level languages designed specifically to support parallel and distributed computing such as PCN [5], or one could use a sequential language combined with a coordination library (sometimes incorrectly called a *message-passing library*) such as PVM [6] or TCGMSG [7]. There are even intermediate systems that use extensions to sequential languages to provide the benefits of both approaches (Fortran-M [8] or Linda [9]). The key is to study your options, choose a portable environment, and then develop portable code. Architecture dependent details can be important, but only in terms of final optimization of the code.

38.2 Algorithms for Parallel Scientific Computing

A computer program is an expression of one or more algorithms. We will take two views of scientific algorithms. First, we will consider classes of algorithms common to scientific computing. This classification scheme is useful when reasoning about algorithms at a high level. Second, we will look at the implementation of algorithms by considering *algorithmic motifs*. An algorithmic motif is a collection of software constructs that frequently appear together in a program.

Before we can discuss algorithms, however, two general concepts must be described. These are *SPMD* (single programs, multiple data) and *granularity*.

Most programs written for parallel computers fall into the SPMD category. For these programs, the same program is loaded onto each node of the parallel computer. SPMD is much less restrictive than SIMD. Unlike a SIMD program, the instructions executed may vary widely from node to node because of conditional statements within the code.

The advantages of SPMD are clear. Since there is only one program, the programmer needs only to maintain a single source code. Furthermore, when the various concurrent tasks must be tightly integrated, it is easier to code in an SPMD style because the code for the various tasks fall within a single case structure.

The second general issue to keep in mind when thinking about any algorithm is granularity. This is a subtle matter as the term granularity is used in a few various ways. Precisely, granularity is the ratio of computation time to communication time. An algorithm that requires data exchange after a small number of computations is called *fine grained*. At the other extreme are *coarse-grained* algorithms in which computation continues for a long time before communication is required.

Granularity is important to consider when mapping algorithms onto varied systems. If the hardware is inherently coarse grained (such as a workstation cluster connected by ethernet), fine-grained algorithms will not run well. The opposite mapping works, though, since a coarse-grained algorithm performs well on a fine-grained architecture.

Granularity is also used to describe the number of concurrent tasks used within a program. If an algorithm can only use a small number of tasks (on the order of tens), the program is considered to be coarse grained even if it requires a great amount of communication relative to computation. Usually, a program with a small degree of potential concurrency is also coarse in terms of the communication/computation ratio. This is not always the case; hence, the confusion over the term *granularity*.

It is essential to remember that it is the granularity—not the amount of communication—that governs the effectiveness of an algorithm. This fact is easy to forget. For example, many parallel programmers mistakenly assume that a collection of workstations connected by Ethernet cannot be used for algorithms that require significant communication. This isn't true! If, for an increasing problem size, computation grows faster than communication, the granularity of the algorithm can increase to a point at which the cluster will be effective. Hence, it isn't the amount of communication but the ratio of computation to communication (granularity) that matters.

38.2.1 Algorithm classes

It is easy to be overwhelmed by the breadth of algorithms found in scientific programs. Most if not all of these algorithms, however, can be classified [10] in terms of the regularity of the underlying data structures (space) and the synchronization required as these data elements are updated (time). Based on this classification scheme, four general classes of parallel algorithms exist:

1. Synchronous—tightly coupled manipulation of identical data elements, regular in space and time
2. Loosely synchronous—tightly coupled as with the Synchronous case, but the data elements are not identical, irregular in space, regular in time
3. Asynchronous—unpredictable or nonexistent coupling between tasks, irregular in time and usually (though not always) irregular in space
4. Embarrassingly parallel—independent execution of uncoupled tasks (a subset of the Asynchronous class)

A single program may utilize more than one algorithm class at various times. These classes, however, capture the key elements of a parallel algorithm. As an added bonus, this set of algorithm classes is complete since it includes all combinations of spatial regularity/irregularity with temporal regularity/irregularity.

38.2.1.1 Synchronous. Synchronous algorithms are those in which regular data elements are updated at regular time intervals. Since the synchronization occurs at regular intervals and the data elements have the same form, it is reasonable to express these algorithms in terms of a single instruction stream. Therefore, synchronous algorithms are natural for SIMD computers and are sometimes called SIMD algorithms.

Any programming environment can express synchronous algorithms. An appealing option, however, is to express the parallelism strictly in terms of the decomposition of the data. In essence, the data drives the parallelism, hence the name *data parallelism*. Note that data parallelism is more general than SIMD parallelism, since data parallelism doesn't insist on a single instruction stream.

The classic example of a synchronous algorithm is explicit finite difference methods for solving partial differential equations [11]. In these algorithms, the problem's domain is divided into a number of similar regions that are mapped to the nodes of the parallel computer. Within each node, the finite difference stencil is applied to the interior region. At the boundaries of the regions, boundary data is communicated and finally updated. This process continues—usually inside of a time stepping loop—to represent the evolution of the system.

These methods tend to run well on all parallel systems. The computation scales as the area (or volume in the 3-D case) of the domain, while communication scales as the size of the edges (faces in the 3-D case). This means that the granularity can be smoothly increased to match hardware requirements.

38.2.1.2 Loosely synchronous. A loosely synchronous algorithm is one that synchronously updates data elements that differ from one node to another (i.e., regular in time, irregular in space). These problems share a great deal with the synchronous class of problems in that both involve some type of regular iterative or time stepping process. Since the data elements vary, however, loosely synchronous algorithms face a unique set of problems.

Data elements that vary from node to node imply work loads that can vary from node to node. To make matters worse, many loosely synchronous problems have data elements that not only vary across the nodes, but also vary in the course of the computation. This means that a key part of a loosely synchronous algorithm is some mechanism to balance the computational load among the nodes of the parallel computer.

Good examples of loosely synchronous algorithms are problems based on adaptive grids [12]. These calculations are similar to the synchronous, explicit finite difference problems in that the problem is viewed in terms of a spatial decomposition of the problem's domain with communication of boundary data taking place at predictable and synchronous points. The grid decomposition (and therefore the data elements) varies in the course of the computation, however, so over time the data elements become different on each node.

38.2.1.3 Asynchronous. Asynchronous algorithms lack regular data updates, so the system proceeds with nonuniform and sometimes random synchronization. This class of problem—other than the embarrassingly parallel subset described next—is the most rare. This is not due to a lack of problems that would benefit from asynchronous algorithms; it is just that the programs are so difficult to construct that they aren't commonly written.

While synchronous algorithms and loosely synchronous algorithms are usually made parallel by focusing on the decomposition of the data, asynchronous algorithms are usually made parallel based on decomposition of control. This is referred to as *functional* or *control parallelism.*

A good source for asynchronous problems arises in the simulation and monitoring of complex systems. For example, the Process-Trellis program [13] is a general parallel system for monitoring a collection of asynchronous processes. It has been used in a number of projects, the most extensive of which was to model a cardiac intensive care unit.

38.2.1.4 Embarrassingly parallel. Embarrassingly parallel algorithms are those asynchronous problems for which the tasks are completely independent. In this case, the parallelism is trivial, hence the term *embarrassingly parallel.*

Embarrassingly parallel programs usually utilize an SPMD model combined with some mechanism for load balancing (i.e., evenly distributing the iterations about the nodes of the computer). Load balancing schemes can be static or dynamic (i.e., modify the load as the computation proceeds). In either case, the programs are among the simplest parallel programs to construct.

Problems of this class are very common in parallel computing because they are so easy to code but also because many simulations really do map into this model. Any time a program consists of a loop with compute-intensive and independent iterations, an embarrassingly parallel solution is possible.

For example, some differential equations can be transformed into a form in which one or more degrees of freedom completely decouple. This trick was used in Ref. [14] to make parallel a seismic migration algorithm. Using a Fourier transform, they represented the seismic data in terms of frequency components that could be migrated independently in an embarrassingly parallel program. An even greater source of embarrassingly parallel programs is Monte Carlo simulations in which a process is computed many times for randomly selected initial conditions. These problems arise in almost every branch of scientific computing.

38.2.2   Algorithmic motifs

Regardless of the algorithm class, the scientific programmer is still faced with a number of options. In the literature of scientific programming, just about every algorithmic trick is used at some point. Most parallel scientific programs, however, utilize a rather small set of recurring algorithm constructs. We call a collection of program constructs that commonly appear together an *algorithmic motif.* The three most common algorithmic motifs in parallel scientific programming are:

- loop splitting
- domain decomposition
- master/slave

In the following subsections, each of these motifs will be described.

38.2.2.1 Loop splitting. In a *loop splitting* program, the parallelism is expressed by assigning loop iterations to various nodes. The best way to see what is meant by loop splitting is to consider an example. Consider the following program loop:

```
do i = 0, NUM_ITERS
    Do_some_work()
end do
```

If the loop iterations are independent, the program can be made parallel as follows. First, the same program is loaded onto each node. This is an instance of a single program, multiple data or SPMD strategy. The central data structures manipulated by the program are replicated on each of the nodes. The parallelism is then expressed in terms of assigning loop iteration to the various nodes as:

```
do i = ID, NUM_ITERS, NUM_NODES
     Do_some_work()
end do
GLOBAL_COMBINE()
```

Note that throughout this chapter, we will assume a programming environment in which each of NUM_NODES processors has a unique node ID ranging in value from 0 to NUM_NODES – 1. The operation at the end of the loop, GLOBAL_COMBINE(), puts the results from the distributed loop iterations back together into a form that the rest of the program can use. The most typical form for GLOBAL_COMBINE() is a global summation. This takes an array on each node and sums corresponding elements across the nodes. At the conclusion of the GLOBAL_COMBINE() operation, the identical, summed vector is present on each node of the parallel system. This operation is vital to many algorithms (not just loop splitting algorithms) and is therefore included as a standard element in many parallel programming environments.

When reuse of data out of a cache is important, it can be more efficient to express the loop splitting in terms of contiguous blocks of loop indices. This uses an array indexed by node ID that gives the first and last loop index for each node. This leads to loops with the form:

```
do i = first(ID), last(ID)
     Do_some_work()
end do
GLOBAL_COMBINE()
```

There are many other variations of loop splitting. In every case, however, the distribution of loop iterations is the driving force in the parallel algorithm. This is almost always followed by some kind of global combination to build identical copies of replicated data structures. The weakness of loop splitting is that the memory available on each node must satisfy the demands of the full data structure. This restriction can be relaxed, but only with substantial modifications to the code to appropriately organize the data into blocks.

38.2.2.2 Domain decomposition.
Domain decomposition (or geometric decomposition) methods are very common in parallel computing. The idea is to represent the parallelism by breaking some domain into subdomains that can be explicitly farmed out to various processors. Communication takes place at the subdomain boundaries. This algorithm most naturally maps onto synchronous and loosely synchronous problems within the SPMD programming model.

The key parameter governing the effectiveness of these methods is the granularity. Granularity describes how much work is associated with each task relative to communication. In the case of domain decomposition, the time required to update the interior of a region must be long compared to the time to communication the boundaries.

For example, domain decomposition algorithms for matrix multiplication map quite well onto parallel computers. This holds because for each block of size N^2 there are N^3

computations to carry out. Matrix vector computations, on the other hand, are far more difficult to effectively carry out on parallel computers since the computation and communication is of order N^2.

Note that the distinction between loop splitting and domain decomposition can be blurry. The key differentiating factor is whether the distribution of data or the splitting of loops drives the design of the parallel algorithm. Also, loop-splitting algorithms usually replicate key data structures on the nodes of the parallel computer, whereas domain decomposition algorithms just work on local data blocks. This is only a general guideline, however, and it is possible for two programmers to assign the same code to one case or to the other.

38.2.2.3 Master-worker. The master-worker algorithmic motif is most commonly used to provide dynamic load balancing within embarrassingly parallel algorithms. There are exceptions to this generalization for systems, such as Linda [9], which support anonymous communication (i.e., communication between nodes that do not know each other's identity) and therefore support the master worker algorithm for loosely coupled problems as well.

In its simplest form, the master worker program consists of two types of processes—a master and a worker. The master is responsible for managing the overall computation and does the following:

- sets up the computation
- creates and manages a collection of tasks
- consumes results

If the operation of consuming results is trivial or easily delayed to the end of the computation, it is quite simple to modify the master to turn into a worker after setting up the collection of tasks. In another variation, the generation of tasks can be spread among the workers as well.

Returning to the simplest case, the second process—the worker— sets up the computation and then enters an infinite loop within which it:

- grabs a task
- carries out the indicated computation
- returns the result to the master

This continues until a termination condition is encountered. This condition either is detected by the master or by any worker's generation of a *poison pill* (i.e., a task that indicates to the consuming worker that it should shut down).

There are several advantages associated with these methods. First, they are very easy to code. Notice that the worker requires only a small amount of code to manage the task queue. The bulk of the work within the worker can be handled by a function call within the worker's infinite loop; the function called is essentially the original sequential program.

Ease of programming is an important advantage. Even without this advantage, there is a compelling reason to use this algorithmic motif when it is possible to do so: a master-worker program can be constructed such that it automatically balances the load among the nodes of the parallel computer.

For example, consider a parallel system in which the nodes have various processing speeds. Just to make matters even more complicated, also let the computational requirements of each task vary significantly and unpredictably. In this case, a static distribution of tasks is guaranteed to be suboptimal. However, master-worker algorithms deal quite easily with this situation. The workers grab tasks and compute them at their own paces. A faster node will naturally grab more tasks and therefore balance the load. Furthermore, nodes that happen to grab more complex tasks will take more time, and so they will access the task pool less frequently. Once again, the number of tasks is naturally reduced for these more heavily loaded nodes.

The result is that the master-worker algorithm is not only simple to write, but it solves the load balancing problem as well. To be effective, however, two conditions must be met. First, there must be more tasks than the number of nodes—preferably many more. The reason for this is easy to see. The parallelism is expressed strictly in terms of the number of tasks. Hence, once the tasks are all assigned, no further parallelism is available to the system.

Second, to be effective, it is best if the longest tasks are handled first. If the long tasks are not handled until late in the computation, a single process can be executing a long task while no other tasks remain for the other nodes. By handling the long tasks first, the odds are greatest that work will be available for the other nodes during computation on the long tasks.

While the minimal requirements for a master worker algorithm are provided by most any programming environment, to really excel, the programming environment needs two capabilities:

- anonymous communication
- shared counter

Anonymous communication refers to the ability for two nodes to communication without knowing each other's identity. This isn't required for master worker algorithms, but it can make a big difference in terms of efficiency and ease of coding. If a specific process must serve as an intermediary for all interactions, then that process becomes a bottleneck that will limit the overall performance. If the nodes can interact arbitrarily, the bottlenecks can be avoided, and efficiency is enhanced.

The globally maintained counter helps the programmer to conveniently express management of the task queue (or bag of tasks) at the core of this algorithmic motif. The global counter is a simple example of a distributed data structure (i.e., a coherent data structure available to all nodes of the parallel system). This simple structure and its use in master-slave algorithms is so important that some programming environments [7, 15] include it as an integral part of the package. This is also available in any system that supports general distributed data structures such as Linda [9, 16].

38.3 Case Studies: Molecular Modeling

To make this discussion more concrete, it is useful to consider some case studies. Rather than a quick summary of many different examples, we will consider three related problems in some detail. These examples are all taken from the field of molecular modeling.

Molecular modeling [17] computations represent a molecular system in terms of a high-level, simple model. These models are usually based on a classical force field [18], but they can also utilize other simplified representations of the molecular system. These methods are of great scientific interest since they are the only way to deal with the large systems of interest to molecular biologists and materials scientists.

Molecular modeling programs frequently map well onto parallel computers and provide classic examples of each of the above-mentioned algorithmic motifs. For the sake of this discussion, two different problems will be described: *molecular dynamics* and *distance geometry*.

38.3.1 Molecular dynamics simulation

Molecular dynamics programs compute classical trajectories for atoms moving on a simple potential energy surface. The potential energy is composed of two types of interactions: bonded and nonbonded. As the name implies, the bonded interactions represent the energy associated with bond rotation, stretching, and bending. These terms are usually represented by Hooke's law type potentials with one or two anharmonic terms and are easy to compute (complexity of order $O(N)$ in which N is the number of atoms). The nonbonded interactions consist of the van der Waals and electrostatic terms and, in principle, represent the interaction of each atom with every other atom (complexity of order $O(N^2)$).

Given the potential energy surface, the computation is basically the same in every molecular dynamics code.

```
input initial geometry
loop over each time step
    compute forces on each atom
    update atomic coordinates
end loop
```

The effort required to compute the forces on each atom varies from atom to atom. The time steps, however, define a regular synchronization schedule, hence this is an example of a loosely synchronous problem. To make matters worse, the differential equations solved to update the atomic positions are stiff and require very small time steps. Therefore, this is not just a loosely synchronous problem but a problem for which synchronization occurs frequently.

Eighty percent or more of a molecular dynamics calculation is taken up by the nonbonded term, so molecular dynamics codes can be made parallel by focusing primarily on the nonbonded calculation. This interaction energy is computed as a double sum over all pairs of atoms. To make this code parallel, all that must be done is to somehow decompose the outermost loop. This is called a *parallel direct sum method.*

The direct sum computation has a computational complexity of $O(N^2)$ and is almost never done in practice. Most production molecular dynamics programs utilize a cutoff radius beyond which the interaction of atom pairs is neglected. The underlying physics permits this because the electrostatic potentials drop off fairly rapidly. From a parallel computing point of view, however, the cutoff greatly complicates life. Basically, with the cutoff, the computational complexity scales as $O(N)$, which is of the same order of complexity as the communication. Hence, these cutoff oriented molecular dynamics programs can be very difficult to port to parallel systems—at least in a manner that supports good scalable performance.

38.3.1.1 Molecular dynamics: loop splitting. The most common method for making MD programs parallel is loop splitting. This algorithmic motif has been used to make parallel a number of MD programs for the full range of parallel and distributed computers: from workstation clusters to tightly coupled MIMD computers [19, 20]. In this section, we describe an effort to make parallel GROMOS [20] for the iPSC/860 parallel computer. GROMOS is a popular MD program consisting of over 74,000 lines of Fortran.

The parallel version of GROMOS (known as UHGROMOS) is a clear instance of a loop splitting code. UHGROMOS uses an SPMD structure with replication of the force and coordinate arrays. They made the two most time-consuming steps in the program parallel: the nonbonded force and nonbond pair list update (i.e., computing for each atom which atoms are within a given cutoff radius). These two steps are of $O(N^2)$ complexity. The rest of the computation was redundantly computed on each node. This simplified the program but also was a sound strategy, since communicating these terms would require a significant fraction of the time required to compute them.

As far as the computational kernel of GROMOS is concerned, the first step in the program is to compute a nonbonded pair list. This list contains for atom, i, the indices for all atoms that interact with atom i. The packing into the list occurs contiguously for each i starting at i = 1 up to the number of atoms. Once the list is computed, we can, for each atom, loop over the appropriate portion of the nonbond list. Consider the code fragment:

```
Do i=1,Natoms
     do J = firstJ(i), lastJ(i)
         Call Update_Force (F, i, j)
     end do
end do
```

The arrays firstJ(i) and lastJ(i) delineate the block of values in the nonbonded lists that are relevant for index, i. The core of the calculation is not shown and is hidden in a "call Update_Force()" statement. All we need to know is that no loop carried dependencies exist within Update_Force(), so the loop iterations are completely independent. Therefore, to make this parallel, we only need to split up the outermost loop:

```
Do i=FirstI(ID), LastI(ID)
     do J = firstJ(i), lastJ(i)
         Call Update_Force (F, i, j)
     end do
end do
call global_sum (j)
```

In this pseudo code, ID is a unique integer for each node ranging from 0 to the number of nodes minus 1. The arrays FirstI and LastI indicate which iterations belong to which node. As the computation proceeds, a subset of the force array is filled in on each node. The fill-in for the force array is scattered because of symmetry relations between atoms. For example, the force on atom, i, from atom, j, is the opposite of the force on atom, j, by atom, i. Rather than compute these twice, the (j, i) element is filled in when the (i, j) element is computed. The last step in this parallel construct is to combine the force array elements into a global, summed force array with a call to a global_sum() routine.

Hence, with changes to the loop structure and the addition of a simple subroutine call, they were able to make parallel the most computationally significant portion of GROMOS. The next bottleneck is the computation of the nonbond list. This computation is less important to make parallel than the nonbond forces themselves since the atoms do not move far at each time step and the list needs updating only once out of every 10 to 50 time steps. In fact, a number of parallel MD codes don't make the nonbond list update parallel at all—especially if the target systems are workstation clusters for which only a small number of nodes can be used.

The UHGROMOS group, however, did make parallel the nonbond list computation. This computation was also made parallel with a loop splitting algorithm. In this case, the GROMOS data structure for the nonbond pair list created a loop-ordered dependence (i.e., you don't know where to place the pair indices for atom i + 1 until you know how many pairs would be required for atom i). To remove this loop-carried dependence, they used a temporary array to store the pair list data independently for each i and then, at the end of the loops, globally distributed the temporary arrays and packed them into a contiguous pair list on each node.

With these simple changes, UHGROMOS was able to execute effectively for modest numbers of nodes. For example, in Ref. [20], results are given for 500 time steps of molecular dynamics simulations for the 6968 atom enzyme superoxide Dismutase. The portions of the code made parallel displayed nearly linear speedup out to 32 nodes of an iPSC/860. The overall results are shown in Figure 34.1 along with the results that would follow from a perfectly parallel code (i.e., linear speedup).

The performance levels off around 32 nodes were due to the portions of the code that were not made parallel (Amdahl's law). Of course, it is possible to make these parallel, as well, to achieve much better performance. For example, in Ref. [19], the loop-splitting strategy is used for every facet of the CHARMM molecular dynamics program. They were able to achieve speedups of 80 on 128 nodes of an iPSC/860.

The reason loop splitting so dominates parallel molecular dynamics is the simplicity of this algorithmic motif. Since the parallel programming requires only a small amount of coding to assign loop iterations in an SPMD program, it is relatively easy to make parallel MD codes this way. The disadvantage is that these loop-splitting methods usually replicate data structures on the parallel machine and therefore do not efficiently utilize system memory. In addition, the global_sum operation requires global communication. These two factors restrict the number of nodes that can be utilized with these algorithms.

38.3.1.2 Molecular dynamics: domain decomposition. To utilize large numbers of nodes, the loop-splitting methods will not be effective. The large problem sizes required to utilize large numbers of nodes make the memory required to replicate force and coordinate arrays unacceptable. Furthermore, even with sophisticated global communication algorithms, the need to globally communicate force data eventually becomes a bottleneck.

What is needed is an algorithm that efficiently utilizes memory and only depends on local communication. The algorithm of choice in this case is based on a domain decomposition [21]. The idea is that, instead of distributing the atoms about the parallel computer (i.e., the loop iterations), one distributes regions of space.

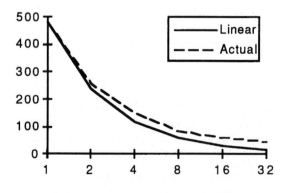

Figure 34.1 UHGROMOS run times in minutes on an iPSC/860 (dotted line is the run time assuming perfect linear speedup, solid line is for the measured run times)

Consider the domain decomposition example in Ref. [21]. Based on the atomic density in the simulated system, regions of space are divided among the nodes of the parallel computer. The program at each time step carries out the following operations on each node:

- Move atoms as needed.
- Update interaction lists moving atom data as needed.
- Compute forces on all atoms in your region.
- Update atomic coordinates for atoms in your region.
- Exchange coordinates for atoms that have moved across domain boundaries.

The only need for communication is when moving atoms or when computing non-bonded force contribution across domain boundaries. These interactions require only local communications, hence the better performance for large numbers of nodes with these methods. In addition, memory is required only for the maximum number of atoms that can fit in a spatial domain—not the entire force, coordinate, and nonbond list arrays.

The test problem considered in Ref. [21] was a system of atoms evenly distributed in a box interacting through a Lennard-Jones potential. The potential is attractive at long distances and repulsive at short distances:

$$\Phi(r) = 4 \; \varepsilon \; [\; (\sigma/r)^{12} - (\sigma/r)^{6}]$$

in which r is distance between the two atoms. Periodic boundary conditions and a cutoff of 2.5σ were utilized. In Table 34.1, we report results in seconds per time step for this spatial decomposition algorithm.

TABLE 34.1   Total Run Time per Time Step in Seconds for Domain Decomposition Based Molecular Dynamics

Number of atoms	nCube 2		Intel Delta	
	P = 512	P = 1024	P = 256	P = 512
500	0.130	0.0119	0.0070	0.0059
6912	0.0374	0.0250	0.0159	0.011
50,000	0.160	0.0967	0.0664	0.0380
100,000	0.298	0.165	0.119	0.0678

For larger problems, the results display optimal O(N/P) scaling (in which P is the number of nodes). Also notice that the number of nodes is much larger than reported for loop-splitting based methods. As mentioned earlier, this is a general trend—algorithms based ondata replication do not scale as well as domain decomposition methods for large numbers of nodes.

The performance advantages of the domain decomposition MD method must be counterbalanced with the disadvantages of these methods. These algorithms are very sensitive to load imbalances which are quite common since real molecular systems are usually not evenly distributed in 3D space (unlike the test case from [21]). Therefore, the programmer must expend significant effort to balance the computational load dynamically in the course of the computation.

Domain decomposition methods are usually not structured around simple replicated data approaches. They typically force the programmer to manage distributed data structures corresponding to the domain decomposition. Between the load balancing and the need to distribute the data structures, the demands placed on the programmer are much greater with domain decomposition codes than loop splitting codes. Hence, it is no surprise that the overwhelming majority of parallel MD codes utilize loop splitting rather than domain decomposition.

38.3.2   Distance geometry

Distance geometry programs [22] take a molecule represented in terms of a collection of interatomic distances and convert it into a representation in terms of Cartesian coordinates [23]. The interatomic distances are represented as distance-bounds (i.e., a range of possible separations) which come from user-provided constraints or are derived from input structural data.

A typical distance geometry calculation proceeds as follows. After the upper and lower distance bounds are generated, the triangle inequality (the length of one side of a triangle must be less than or equal to the sum of the other two sides) is used to refine or smooth the lower and upper bounds. These smoothed distance bounds are placed in a matrix with lower bounds below and upper bounds above the main diagonal. Once the distance matrix has been produced, the program generates a number of structures as follows:

- Randomly select a specific distance within each distance range.
- Convert the set of distances into Cartesian coordinates using a method called embedding [22].
- Optimize the structure by conjugate gradient minimization.
- Repeat up to the desired number of structures.

The function optimized in Step 3 minimizes the degree to which the original distance bounds are violated. The final result is a set of molecular configurations that satisfy (or come close to satisfying) the original distance constraints.

Distance geometry methods are used to elucidate a structure from incomplete data. For the purpose of this discussion, however, this code is important as an instance of a major class of algorithms called *Monte Carlo* or *stochastic* optimization methods. The basic problem is a nonlinear, global optimization problem in which the presence of local minima are handled through the use of randomly selected initial conditions. This type of problem occurs in practically every branch of scientific computing.

38.3.3   DGEOM: master-worker

One of the most commonly used programs for distance geometry is DGEOM [23]. DGEOM uses the algorithm described in the last section which is apparent from the DGEOM pseudocode in Fig. 34.2. Notice that the program structure is similar to that of most stochastic optimization codes; the core of this program is a loop, the body of which:

1. is compute intensive
2. contains no loop carried dependencies

Given that hundreds of structures are generated in a typical DGEOM application, this program is embarrassingly parallel. The parallel version of DGEOM (called pDGEOM [24]) was made parallel using the Linda programming environment [9]. Therefore, before we discuss the pseudocode for pDGEOM, we will need to describe the basics of Linda.

Linda uses virtual shared memory (i.e., the memory does not have to be shared at the hardware level) to handle all interaction between processes. These memory operations are

```
program DGEOM
PROCESS_USER_INPUT
GENERATE_DISTANCE_BOUNDS_MATRIX(dist_matrix)
do istrct = 1, number_of_structures
     EMBED_AND_OPTIMIZE (istrct, dist_matrix, results)
     EVALUATE_AND_OUTPUT_RESULTS(istrct, results)
end do
end DGEOM
```

Figure 34.2 Sequential DGEOM program, pseudocode

added to a sequential language to create a hybrid parallel programming language. For example, in pDGEOM, the combination of C and Linda was used (C-Linda).

Linda consists of four basic operations. The first operation is called eval(). This operation creates one or more processes to generate data in the shared memory. To place data directly into shared memory, the out() command is used. If some process wishes to fetch this data and remove it from shared memory (so only one process can grab the item), the in() instruction is used. Finally, the rd() operations grabs the data but leaves a copy behind for other processes. With these four basic operations, the reader knows enough Linda to read the pseudocode descriptions of pDGEOM.

pDGEOM was made parallel rather quickly using a master-worker algorithmic motif. The structure of this program is described in the following pseudocode. Note that within the pseudocode, terms written using all capital letters refer to code taken directly from the original DGEOM program.

First consider the master program in Fig.34.3. The master program sets up the overall calculation by processing user input and generating the smoothed distance bounds matrix. It then initiates the worker processes and puts the calculation's basic data (i.e., the

```
program pDGEOM
PROCESS_USER_INPUT
GENERATE_DISTANCE_BOUNDS_MATRIX(dist_matrix)
do i = 1, number_of_workers           ! create workers
     eval(worker())
end do

out(structure_data)                    ! Put basic data
out(dist_matrix)                       ! into shared memory

do istrct = 1, number_of_structures    ! Create tasks
     out(istrct)
end do

do istrct = 1, number_of_structures    ! Fetch results
     in(result)
     EVALUATE_AND_OUTPUT_RESULTS(istrct, results)
end do
end PDGEOM
```

Figure 34.3 pDGEOM parallel program, pseudocode for the master routine

smoothed distance bounds matrix and the structure data) into the shared memory. The master then creates the collection of tasks for the workers to consume and waits to pickup and analyze the final results.

The worker code in Fig. 34.4 is even simpler. It consists of an infinite loop within which the worker grabs a task (an istrct value), carries out the indicated operations, and puts the result out to the shared memory.

Notice that in both of these pseudocode routines, other than some constructs to manage the master-worker algorithm, the bulk of the code is taken directly from the original DGEOM program.

The pDGEOM program was ported to several different MIMD computers: from workstation clusters to shared memory multiprocessor computers. The same source code was used in each case with any nonportability originating in the Fortran code or the C-Fortran interface.

In Table 34.2, pDGEOM execution times are given for the standard DGEOM benchmark in which 10 Cyclosporin A structures are generated. Results are given for the following computer systems [24]:

- Cray XMP running DGEOM
- Ethernet-connected network of IBM RISC System 6000 model 560 workstations (RS/6000 560)
- Shared memory, 32 node IBM Power Visualization Server (PVS)

pDGEOM displayed excellent parallel performance for each system in the one- to four-node range. Beyond four nodes, however, the run time levels off. The leveling effect occurs because the amount of parallelism is fixed by the number of generated structures (10), leading to what is commonly referred to a "bin-packing" problem.

Bin packing is one way to think of parallel programs expressed in terms of a "bag of tasks." Each processor is a bin to hold and consume tasks. The tasks are then shuffled among the bins. For example, the most even distribution of ten tasks among four nodes would place two tasks in two nodes and three tasks in the other two nodes. The total program time is dependent on the bin with the most tasks. In Ref. [24], the pDGEOM program is analyzed in terms of this bin-packing model and matches the model exactly.

It is important to note the role of automatic load balancing in this program. Each task varies significantly in the amount of time required for the computation. In addition, the effort required can not be predicted ahead of time. Therefore, it is essential to use algorithms (such as the Master Worker algorithm) that support dynamic load balancing.

```
subroutine worker()
infinite loop
    in (istrct)
    rd (structure_data)
    rd (distance_matrix)
    EMBED_AND_OPTIMIZE (istrct, results)
    out (results)
end infinite loop
end worker
```

Figure 34.4 Sequential DGEOM program, pseudocode

TABLE 34.2   pDGEOM Run Times in Seconds for Several Systems

Machine	Number of workers	Wall time (seconds)
Cray X-MP	serial	155.1
IBM RS/6000 560	1	137.4
	2	73.0
	3	52.0
	4	42.0
	5	32.0
IBM PVS	2	374.4
	4	147.6
	6	81.63

Linda is not the only system that supports master worker programming with dynamic load balancing. TCGMSG [7] and p4 [15] also support this model by providing a globally accessible counter. The fact that they include such a construct is more evidence of the importance of this class of problems.

38.4   Trends

Scientific programmers are an innovative lot. Whenever new and promising technologies become available, they are rapidly evaluated and if appropriate, adopted. This makes predicting trends risky since the state of the art can change so rapidly . Reservations aside, however, a few noteworthy trends have emerged concerning programming environments for parallel scientific computing.

38.4.1   Data parallel programming: HPF

Data parallel programming models are receiving a great deal of attention. This attention is motivated by the belief that programming environments for parallel computers need to resemble programming environments for traditional supercomputers: (i.e., vector computers). For vector computers, the compiler generally vectorizes the program. The programmer may restructure the code or add directives to help the compiler, but the actual vectorization is done by the compiler.

It appears that this is not possible with general parallelism. However, what if a restricted model of parallelism were adopted? Maybe if a simple enough model were selected, the compiler could do the parallelization with the programmer only adding some directives or restructuring some sequential code. The hope is that the data parallel model will indeed let this happen.

A number of vendors and end-users have joined together to make this technology a reality by producing the High Performance Fortran (HPF) standard [4]. This is a strict data parallel model and therefore only supports synchronous and embarrassingly parallel algorithms. For this reason, there is skepticism about HPF's long-term impact because it lacks generality. In recognition of this criticism, a number of research groups are working to extend HPF to support irregular data structures. This would add loosely synchronous algorithms to the set of algorithms expressible within HPF.

It is too early to determine whether data parallel methods will succeed. With so many research groups and vendors supporting this approach, however, it is bound to be an important force in the parallel computing community.

38.4.2 Object-oriented programming

There is a tired joke that has been floating around the scientific programming community for a number of years. It can be paraphrased as, "I don't know what language I will be using in the next century, but it will be called Fortran."

This clearly will be the case for those with large bodies of old code to maintain, but for new code development, a new answer is emerging. In this new form of the statement, we have, "I don't know what language I will be using in the next century, but it will be called C++."

So what has happened to make C++ a contender for scientific programming language of the future? Basically, object-oriented programming really is as powerful as the computer scientists claim. While a discourse on object-oriented programming is well beyond the scope of this chapter, I would like to focus on two reasons why object-oriented programming (as expressed in C++) will take off in scientific programming:

- overloading function names and operators
- encapsulation

Overloading lets multiple functions share a calling sequence. The actual function used within a program is selected based on the number and types of parameters to the function. Overloading can even be applied to standard operators such as "+". This is provided to some degree in Fortran90, but its cleanest and most extensive expression is in C++. The best way to see the value of operator overloading is with an example.

Consider the use of alternative arithmetic systems in scientific computing. One example is the use of an interval rather than a floating-point number to represent real numbers. For example, consider the statement:

```
force = Const/(r*r);
```

With operator overloading, all the programmer would need is to change the types on "Const", "force", and "r", and the system would automatically call the user's interval arithmetic functions. Without operator overloading, however, this interval expression would have to be written out by explicitly calling the appropriate interval functions:

```
temp = interval_mult(r, r)
force = interval_divide(Const, temp)
```

This rewrite is error prone and time consuming.

Another advantage of object oriented programming is its support of detailed control over data and function visibility (i.e., encapsulation). For example, a C++ programmer can define public and private parts of an object and therefore carefully control access to the an object's internal state. This makes it much easier to share modules between programs since, given the interface, the programmer can use an object class without being concerned with how the class was implemented.

Finally, data can be very abstract and contain knowledge about itself embedded in its definition. When combined with operator overloading, this lets one write code that looks like the underlying physics. For example, a C++ statement to compute the force on a system in terms of mass and acceleration could be:

```
F = m * a;
```

which is the common expression of Newton's second law. This simple code would cover all cases: scalar forces acting on the center of gravity, individual forces on an ensemble of particles, or more complex tensor expressions for a rigid body. All of these can be covered with this one expression rather than a nesting of case statements.

These are just a few of the many features of C++ that make it one of the fastest-growing programming languages in the world. The techniques available to C++ programmers let them write safer, more complex code. This has been shown to be a real benefit, not just a theory [25], so C++ is most likely here to stay and will make a major impact on scientific programming.

38.4.3 Virtual shared memory

Shared memory helps the programmer express complex parallel algorithms. It lets the programmer think in terms of distributed data structures, so parallel programming becomes a natural extension of sequential programming. Stated another way, programmers are used to thinking:

$$\text{Algorithms} + \text{data Structures} = \text{program}$$

With shared memory, this same approach carries over to parallel computing as:

$$\text{Algorithms} + \text{distributed data Structures} = \text{parallel programs}$$

Virtual shared memory (i.e., providing the illusion of a shared memory even when the physical memory is distributed), has been around for awhile. However, the difficulty of providing a general shared address space without unduly sacrificing efficiency has made adoption of this technology slow.

The response to this is to add a specialized virtual shared memory. The Linda coordination language [9] is the most famous example of a restricted, virtual shared memory. Another example is the shared counter within TCGMSG [7]. This selective inclusion of shared memory is becoming more common and will start appearing in more programming environments.

38.5 Further Reading

This chapter has tackled the almost impossible task of summarizing the key issues in scientific computing on parallel computers. Really, to appreciate parallel scientific computing, there are a number of additional sources to consult.

Of the many journals dedicated to parallel and distributed computing, two stand out for parallel scientific computing. These are the *Journal of Scientific Programming* (published by John Wiley & Sons, Inc.) and *Concurrency: Practice and Experience* (also published by Wiley). While these journals are important, the pace of development in parallel scientific computing is so fast that, to really keep up, one needs to utilize services over the Internet such as the comp.parallel news groups and netlib (to get started with netlib, send e-mail to netlib@ornl.gov with a blank message body and a subject line reading *send index*).

A relatively old reference (at least by the standards of parallel computing) is the classic book by Geoffrey Fox et al. [26] summarizing the results of the California Institution of Technology concurrent computation project. This two-volume set provides a wide range of important case studies as well as a useful collection of practical algorithms.

Any time one gets involved with parallel computing, the issue of describing performance comes up. This topic hasn't been discussed in this chapter, but only because of space limitations. Required reading for everyone involved with parallel computing is the paper by Bailey [27] on fooling the masses with parallel benchmarks.

Coming up to speed in any new field require mastering the jargon of that discipline. An invaluable guide to the jargon of parallel computing is the glossary by G. Wilson [28].

38.6   Conclusion

Writing parallel programs is a complex undertaking. Parallel programmers must define the concurrency within an application as well as when and where the program's data is needed. This chapter set out to help scientific programmers manage this complexity and start writing parallel scientific programs. To do this, we concentrated on two of the most important issues for parallel computing: portable parallel programming environments and parallel algorithms.

The portability argument was simple and direct. Only with portable software can a programmer avoid a software maintenance nightmare, since parallel hardware is both diverse and rapidly evolving. From a pragmatic point of view, portability is possible at a small cost in decreased performance. Therefore, parallel programmers have little or no excuse for writing nonportable parallel programs.

While portable software development is important, this topic has been discussed elsewhere in this book. Hence, the bulk of this chapter focused on parallel algorithms for scientific computing. We began by defining classes of parallel algorithms in terms of the regularity of both the central data structures and the timing with which these data structures are updated:

- synchronous
- loosely synchronous
- asynchronous
- embarrassingly parallel

These classes view a parallel program at a high level and are an effective place to start when thinking about porting a scientific program to a parallel computer.

An algorithm is of little value, however, until it is represented in terms of program code. Hence, we spent a great deal of time discussing the common algorithmic tricks used by parallel scientific programmers. While *every trick in the book* has been used in parallel scientific computing, most parallel scientific programs use a small number of software constructs. We call these *algorithmic motifs*. The most common motifs are:

- domain decomposition
- master-worker
- loop splitting

At this point, two paths were available to us. One path would survey all of the *hot* applications of parallel computing to scientific problems. Under this approach, no single application could be described in enough detail to provide practical guidance to the scientific programmer. Since the goal of this chapter is to help these programmers get started with parallel computing, we took a second path. In this approach, three problems were described in enough detail so the programming itself could be appreciated. Furthermore, these case studies came from the same general field—molecular modeling—which served to demonstrate different algorithms applied to related problems.

Reading a single chapter does not make one a parallel programming expert. Therefore, we closed this chapter with two discussions designed to help programmers take the next step on the road to parallel computing proficiency. First, we discussed a number of emerging trends in parallel scientific computing. Second, we provided several key references to pursue for further information about this rapidly changing field.

It is hoped that the reader will leave this chapter with an appreciation for the issues involved in programming parallel computers. It is a difficult specialization and demands a great deal from its practitioners. At the same time, the field is mature enough that standard algorithmic tricks have emerged. By learning what these tricks are and when they can be used, the difficult task of writing parallel programs becomes much easier—so much so that there is no reason why any competent programmer cannot make scientific programs parallel and take advantage of parallel and distributed computers.

38.7 References

1. Turcotte, L. 1993. *A survey of software environments for exploiting networked computing resources.* Tech Report # MSM-EIRS-ERC-93-2, Mississippi State University.
2. Cheng, D. Y. 1993. *A survey of parallel programming languages and tools.* NASA Ames Research Center Technical Report RND-93-005.
3. Flynn, M. J. 1972. Some computer organizations and their effectiveness. *IEEE Trans. Computers, Vol. C-21,* No. 9.
4. The HPF Forum. 1993. High performance fortran, journal of development. Special issue of *Scientific Programming*, Vol. 2, No. 1, 2, 1–170.
5. Foster, I., R. Olson, and S. Tuecke. 1992. Productive parallel programming: the PCN approach. *Scientific Programming*, Vol. 1, 51.
6. Sunderam, V. 1990. PVM: a framework for parallel distributed computing. *Concurrency: Practice and Experience,* Vol. 2, 315–339.
7. Harrison, R. J. 1991. Portable tools and applications for parallel computers. *Int. J. Quantum Chem.,* Vol. 40, 847–863.
8. Foster, I., R. Olson, and S. Tuecke. 1993. *Programming in Fortran M.* Technical Report ANL-93/26. Argonne National laboratory.
9. Carriero, N., and D. Gelernter. 1991. *How to Write Parallel Programs: A First Course.* Cambridge, Mass.: MIT Press.
10. Fox, G. C. 1989. Parallel computing comes of age: supercomputer level parallel computations at Caltech. *Concurrency: Practice and Experience,* Vol. 1, No. 1, 63.
11. Deshpande, A., and M.H. Schultz. 1992. Efficient parallel programming with Linda. *Proceedings of Supercomputing '92*, Minneapolis, Minn., 238.
12. Simon, H. D. 1991. Partitioning of Unstructured Problems for Parallel Processing. *Computing Systems in Engineering,* Vol 2, 135.
13. Factor, M., D. Gelernter, C. Kolb, P. Miller, and D. Sittig. 1991. Real time data fusion in the intensive care unit. *Computer,* 45 (November).
14. Almasi, G. S., T. McLuckie, J. Bell, A. Gordon, and D. Hale. 1992. Parallel distributed seismic migration. *Concurrency: Practice and Experience,* Vol 5, 105.
15. Butler, R., and E. Lusk. 1992. *User's guide to the p4 programming system.* Technical Report ANL-92/17, Argonne National laboratory.
16. Schoinas, G. 1992. *Issues on the implementation of programming systems for distributed applications.* University of Create, Technical Report.
17. Cohen, N. C., J. M. Blaney, C. Humblet, P. Gund, and D. C. Barry. 1990. Molecular modeling software and methods for medicinal chemistry. *Journal of Medicinal Chemistry,* Vol. 33, 883.
18. Buckert, R., and N.L. Allinger. 1982. *Molecular Mechanics.* Washington, D.C.: American Chemical Society.
19. Brooks, B.R., and M. Hodoscek. 1992. Parallelization of CHARMM for MIMD machines. *Chemical Design and Automation News,* Vol 7, 16.
20. Clark, T. W. , R. V. Hanxleden, K. Kennedy, C. Koelbel, and L. R. Scott. 1992. Evaluating parallel languages for molecular dynamics computations. *Proceedings of the Scalable High Performance Computing Conference,* (SHPCC-92), 98

21. Plimpton, S. 1993. *Fast Parallel Algorithms for Short-Range Molecular Dynamics.* Sandia Technical Report, SAND91-1144. Sandia National Laboratories.

22. Crippen, G. M., and T. F. Havel. 1988. *Distance Geometry and Molecular Conformation.* New York: John Wiley & Sons.

23. Blaney, J. M., and G. M. Crippen. DGEOM. Quantum Chemistry Program exchange.

24. Mattson, T. G., and R. Judsen. 1993. *pDGEOM: a parallel distance geometry program.* Yale University Research Report.

25. Vermeulen, A., and M. Chapman, eds. 1994. The First Annual Object-Oriented Numerics Conference (OON-SKI '93). Special issue of *Scientific Programming,* Vol. 2, No. 3, 109–246.

26. Fox, G. C., M. A. Johnson, G. A. Lyzenga, S. W. Otto, J. K. Salmon, and D. W. Walker. 1988. *Solving Problems on Concurrent Processors: Volumes 1 and 2.* Englewood Cliffs, N.J.: Prentice Hall.

27. Bailey, D. 1991. Twelve ways to fool the masses when giving performance results on parallel computers. *Supercomputing Review* (August), 54. Also published in *Supercomputer* (September), 4.

28. Wilson, G. V. 1993. A glossary of parallel computing terminology. *Parallel and Distributed Technology,* Vol. 1, 52.

35

Parallel and Distributed Simulation of Discrete Event Systems

Alois Ferscha

Modeling and analysis of the time behavior of dynamic systems is of wide interest in various fields of science and engineering. Common to *realistic* models of time dynamic systems is their complexity, very often prohibiting numerical or analytical evaluation. Consequently, for those cases, simulation remains the only tractable evaluation methodology. Simulation experiments, however, can take exceedingly long to execute. For statistical reasons it might, for example, be necessary to perform a whole series of simulation runs to establish the required confidence in the performance parameters obtained by the simulation (i.e., make confidence intervals sufficiently small). Another natural reason why simulation should be as fast as possible comes from the objective of exploring large parameter spaces.

One of the possibilities for resolving these shortcomings is the use of statistical knowledge to prune the number of required simulation runs (e.g., variance reduction techniques or importance sampling methods). More naturally, however, faster simulations can be obtained by using more computational resources—particularly multiple processors operating in parallel. It seems obvious, at least for simulation models reflecting real life systems constituted by components operating in parallel, that this inherent model parallelism could be exploited to make the use of a parallel computer potentially more effective. Moreover, for the execution of independent replications of the same simulation model with varied parametrizations , the parallelization appears to be trivial. In this chapter, we systematically describe ways of accelerating simulations using multiprocessor systems with focus on the synchronization of *logical simulation processes* executing in parallel on various processing nodes in a parallel or distributed environment.

35.1 Simulation Principles

Basically, every simulation model is a description of a physical system (or at least some of its components) in terms of a set of *states* and *events*. Performing a simulation thus means mimicking the occurrence of events as they evolve in time and recognizing their effects as represented by states. Future event occurrences induced by states have to be planned (scheduled). In a *continuous* simulation, state changes occur continuously in time, while in a *discrete* simulation the occurrence of an event is instantaneous and fixed to a selected point in time. Because of the convertibility of continuous simulation models into discrete models by just considering the start instant as well as the end instant of the event occurrence, we shall subsequently only consider the discrete case.

35.1.1 Time-driven vs. event-driven simulation

Two kinds of discrete simulation have emerged that can be distinguished with respect to the way simulation time is progressed. In *time-driven* discrete simulation, simulated time is advanced in *time steps* (or ticks) of constant size Δ, (i.e., the observation of the simulated dynamic system is made discrete by unitary time intervals). The choice of Δ interchanges simulation accuracy and elapsed simulation time; ticks short enough to guarantee the required precision generally imply longer simulation time. Intuitively, for event structures irregularly dispersed over time, the time-driven concept generates inefficient simulation algorithms.

Event-driven discrete simulation makes discrete the observation of the simulated system at event occurrence instants. We shall refer to this kind of simulation as *discrete event simulation* (DES) subsequently. A DES, when executed sequentially, repeatedly processes the occurrence of events in simulated time (often called *virtual time*, or VT) by maintaining a time ordered *event list* (EVL) holding time-stamped event scheduled to occur in the future, a (global) clock indicating the current time, and *state variables* $S = (s_1, s_2, \ldots s_n)$ defining the current state of the system. A *simulation engine* (SE) drives the simulation by continuously taking the first event out of the event list (i.e., the one with the lowest time-stamp), simulating the effect of the event by changing the state variables and/or scheduling new events in EVL—possibly also removing obsolete events. This is performed until some predefined end time is reached or there are no further events to occur.

The time-driven DES increments VT by one time unit each step and collects the state vector S as observed at that time. Due to time resolution and non-time-consuming state changes of the system, not all the relevant information can be collected with this simulation strategy.

35.1.2 Levels of parallelism/distribution

Application level. The most obvious acceleration of simulation experiments that aim to explore large search spaces is to assign independent replications of the same simulation model with possibly different input parameters to the available processors. Since no coordination is required between processors during their execution, high efficiency can be expected. The sequential simulation code can be reused, avoiding costly program parallelization, and problem scalability is unlimited. Distributing whole simulation experiments, however, might not be possible because of memory space limitations in the individual processing nodes.

Subroutine level. Simulation studies in which experiments must be sequenced because of iteration dependencies among the replications (i.e., input parameters of replication i are determined by the output values of replication $i - 1$) naturally preclude application-level distribution. The distribution of subroutines constituting a simulation experiment, such as random number generation, event processing, state update, and statistics collection might be effective for acceleration in this case. Because of a rather small number of simulation engine subtasks, the number of processors that can be employed, and thus the degree of attainable speed-up, is limited with a subroutine-level distribution.

35.1.2.1 Component level. Neither of the two distribution levels above makes use of the parallelism available in the physical system being modelled. For that, the simulation model has to be decomposed into *model components* or submodels, such that the decomposition directly reflects the inherent model parallelism or at least preserves the chance to gain from it during the simulation run. A natural simulation problem decomposition could be the result of an object oriented system design, where object class instances corresponding to (real) system components represent computational tasks to be assigned to parallel processors for execution. A queueing network workflow model of a business organization for example, that directly reflects organizational units like offices or agents as single queues, defines in a natural way the decomposition and assignment of the simulation experiment to a multiprocessor. The processing of documents by an agent then could be simulated by a processor, while the document propagation to another agent in the physical system could be simulated by sending a message from one processor to the other.

35.1.2.2 Event-level, centralized EVL. Model parallelism exploitation at the next lower level aims at a distribution of single events among processors for their concurrent execution. In a scheme where EVL is a centralized data structure maintained by a master processor, acceleration can be achieved by distributing (heavy weighted) concurrent events to a pool of slave processors dedicated to execute them. The master processor in this case takes care that consistency in the event structure is preserved (i.e., prohibits the execution of events potentially yielding causality violations because of overlapping effects of events being concurrently processed). This, however, requires knowledge about the event structure that must be extracted from the simulation model. The distribution at the event level with a centralized EVL is particularly appropriate for shared memory multiprocessors where EVL can be implemented as a shared data structure accessed by all processors. The events processed in parallel are typically the ones located at the same time moment (or small epoch) of the space-time plane.

35.1.2.3 Event-level, decentralized EVL. The most permissive way of conducting simulation in parallel is at the level where events from arbitrary points of the space-time are assigned to various processors, either in a regular or an unstructured way. Indeed, a higher degree of parallelism can be expected to be exploitable in strategies that allow the concurrent simulation of events with different timestamps. Schemes following this idea require protocols for local synchronization, which may in turn cause increased communication costs depending on the event dispersion over space and time in the underlying simulation model. Such synchronization protocols have been the objective of *parallel and distributed simulation* research, which has received significant attention since the proliferation of massively parallel and distributed computing platforms.

35.1.3   Parallel vs. distributed simulation

In a SIMD operated environment, a set of processors perform identical operations on various data in lock-step. Each processor has its own local memory for private data and programs and executes an instruction stream controlled by a central unit, which forces synchronism among the independent computations. Whenever the synchronism imposed by the SIMD operational principle is exploited to conduct simulation with P processors (under central control), we will talk about *parallel simulation*. Alternatively, if a collection of simulation *processes* is assigned to processors that operate *asynchronously* in parallel, usually employing message passing as a means for communication and synchronization, we talk about *distributed simulation*.

35.1.4   Logical process simulation

Common to all simulation strategies with distribution at the event level is their aim to divide a global simulation task into a set of communicating *logical processes* (LPs), trying to exploit the parallelism inherent among the respective model components with the concurrent execution of these processes. We can thus view a *logical process simulation* (LP simulation) as the cooperation of an arrangement of interacting LPs, each of them simulating a subspace of the space-time which we will call an event structure *region*. Generally a region is represented by the set of all events in a sub-epoch of the simulation time, or the set of all events in a certain subspace of the simulation space. The basic architecture of an LP simulation, as shown in Fig. 35.1, can be viewed as follows:

- A *set of LPs* is devised to execute event occurrences synchronously or asynchronously in parallel.
- A *communication system* (CS) provides the possibility to LPs to exchange local data, but also to synchronize local activities.
- Every LP_i has assigned a *region* R_i as part of the simulation model, upon which a simulation engine SE_i operating in event-driven mode executes *local* (and generates *remote*) event occurrences, thus progressing a *local clock* (local virtual time, LVT).
- Each LP_i (SE_i) has access only to a statically partitioned *subset of the state variables S_i* $\subset S$, disjoint to state variables assigned to other LPs.
- Two kinds of events are processed in each LP_i: internal events which have causal impact only to $S_i \subset S$, and *external events*, which also affect $S_j \subset S$ ($i \neq j$) the local states of other LPs.

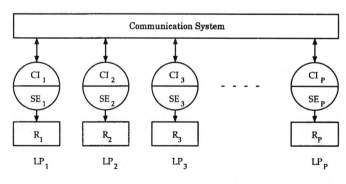

CI ... Communication Interface   SE ... Simulation Engine
R ... Region, Simulation Sub-Model   LP ... Logical Process

Figure 35.1 Architecture of a logical process simulation

- A communication interface CI_i attached to the SE takes care for the propagation of effects causal to events to be simulated by remote LPs, and the proper inclusion of causal effects to the local simulation as produced by remote LPs. The main mechanism for this is the sending, receiving and processing of event messages piggybacked with copies of the senders LVT at the sending instant.

Basically, two classes of CIs have been studied for LP simulation, either taking a *conservative* or an *optimistic* position with respect to the advancement of event executions. Both are based on the sending of messages carrying causality information that has been created by one LP and affects one or more other LPs. On the other hand, the CI is also responsible for preventing global event causality violations. In the first case, the conservative protocol, the CI triggers the SE in a way which prevents from causality errors ever occurring (by blocking the SE if there is the chance to process an *unsafe* event, i.e., one for which causal dependencies are still pending). In the optimistic protocol, the CI triggers the SE to redo the simulation of an event should it detect that premature processing of local events is inconsistent with causality conditions produced by other LPs. In both cases, messages are invoked and collected by the CIs of LPs, the propagation of which consumes real time dependent on the technology the communication system is based on. The practical impact of the CI protocols developed in theory therefore is highly related to the effective technology used in target multiprocessor architectures. (We will avoid presenting the achievements of research in the light of readily available technology, which is permanently subject to change.)

For the representation and advancement of simulated time (VT) in an LP simulation, we can devise two possibilities: a *synchronous LP simulation* implements VT as a global clock, which is either represented explicitly as a centralized data structure or implicitly implemented by a time stepped execution procedure—the key characteristic being that each LP (at any point in real time) faces the same VT. This restriction is relaxed in an *asynchronous LP simulation*, where every LP maintains a *local* VT (LVT) with generally different clock values at a given point in real time.

35.1.4.1 Synchronous LP simulation. In a *time-stepped* LP simulation, all the LPs' local clocks are kept at the same value at every point in real time, i.e., every local clock evolves on a sequence of discrete values $(0, \Delta, 2\Delta, 3\Delta, \ldots)$. In other words, simulation proceeds according to a global clock, since all local clocks appear to be just a copy of the global clock value. Every LP must process all events in the time interval $[i\Delta, (i + 1)\Delta)$ (time step i) before any of the LPs are allowed to begin processing events with occurrence time $(i + 1)\Delta$ and later. This strategy considerably simplifies the implementation of correct simulations by avoiding problems of deadlock and possibly overwhelming message traffic and/or memory requirements as will be seen with synchronization protocols for asynchronous simulation. Moreover, it can efficiently use the barrier synchronization mechanisms available in almost every parallel processing environment. The imbalance of work across the LPs in certain time steps on the other hand naturally leads to idle times and thus represents a source of inefficiency.

Both centralized and decentralized approaches of implementing global clocks have been followed. In a centralized implementation, one dedicated processor controls the global clock. To overcome stepping the time at instances where no events are occurring, algorithms have been developed to determine for every LP at what point in time the next interaction with another LP shall occur. Once the minimum timestamp of possible next external events is determined, the global clock can be advanced by $\Delta(S)$, i.e., an amount that depends on the particular state S. For a distributed implementation of a global clock, a structured (hierarchical) LP organization can be used to determine the minimum next event time. If for example the LPs are organized in a tree, then a parallel minimum reduction operation can bring this timestamp to the root LP, which would then broadcast it down the tree to the other LPs. Another possibility is to apply a distributed snapshot algorithm [8] to avoid the bottleneck of a centralized global clock coordinator. Combinations of synchronous LP simulation with event-driven global clock progression have also been studied. Although the global clock is advanced to the minimum next event time as in the event-driven scheme, LPs are only allowed to simulate within a Δ-tick of time, called a *bounded lag* by Lubachevsky [31] or a *Moving Time Window* by Sokol et al. [40].

35.1.4.2 Asynchronous LP simulation. Asynchronous LP simulation relies on the presence of events occurring at various simulated times that do not affect one another. Concurrent processing of those events thus effectively accelerates sequential simulation execution time.

The critical problem, however, that asynchronous LP simulation poses is the chance of causality errors. Indeed, an asynchronous LP simulation ensures correctness if the (total) event ordering as produced by a sequential DES is consistent with the (partial) event ordering as generated by the distributed execution. Jefferson [27] recognized this problem to be the inverse of Lamport's logical clock problem [28], i.e., providing clock values for events occurring in a distributed system such that all events appear ordered in logical time.

It is intuitively convincing and has been shown in Ref. [34] that no causality error can ever occur in an asynchronous LP simulation if and only if every LP adheres to processing events in nondecreasing timestamp order only (*local causality constraint (lcc)* as formulated in Ref. [19]). Although sufficient, it is not always necessary to obey the *lcc*, because two events occurring within one and the same LP may be concurrent (independent of each other) and thus could be processed in any order. The two main categories of mechanisms for asynchronous LP simulation already mentioned adhere to the *lcc* in various ways: conservative methods strictly avoid *lcc* violations, even if there is some non-zero probability that an event ordering mismatch will *not* occur; whereas optimistic methods hazardously use the chance of processing events even if there *is* non-zero probability for an event ordering mismatch. The variety of mechanisms around these schemes will be the main body of this review.

In a comparison of synchronous and asynchronous LP simulation schemes it has been shown [16] that the potential performance improvement of an asynchronous LP simulation strategy over the time-stepped variant is at most $O(logP)$, P being the number of LPs executing concurrently on independent processors. The analysis assumes each time step to take an exponentially distributed amount of execution time

$$T_{step,i} \sim \exp(\lambda) \text{ in every LP}_i \ (\text{E}\,[T_{step,i}] = \frac{1}{\lambda}\ )$$

As a consequence, the expected simulation time $E[T^{sync}]$ for a k time step-synchronous simulation is

$$E\,[T^{synch}] = k\,\text{E}\,[max_{i\,=\,1..P}\,(T_{step,i}\,)] = k\frac{1}{\lambda}\sum_{i=1}^{P}\frac{1}{i} \leq \frac{k}{\lambda}\log\,(P)$$

Relaxing now the synchronization constraint (as an asynchronous simulation would), the expected simulation time would be

$$E[T^{async}] = E\left([max_{i=1..P} \langle k \ T_{step,i} \rangle] > \frac{k}{\lambda} \right)$$

We have

$$lim_{k \to \infty, P \to \infty} \frac{E[T^{sync}]}{E[T^{async}]} \approx \log(P)$$

saying that with increasing simulation size k, an asynchronous simulation could complete (at most) $log(P)$ times as fast as the synchronous simulation, and the maximum attainable speedup of any time stepped simulation is $P/log(P)$. These results, however, are a direct consequence of the exponential step execution time assumption, i.e., comparing the expectation of the k-fold sum over the max of exponential random variates (synchronous) with the expectation of the max over P k-stage Erlang random variates. For a step execution time uniformly distributed over $[l, u]$ we have

$$lim_{k \to \infty, P \to \infty} \frac{E[T^{sync}]}{E[T^{async}]} \approx 2$$

or intuitively with

$$T^{sync} \le ku \text{ and } E[T^{async}] \ge k\frac{\langle l + u \rangle}{2}$$

the ratio of synchronous to asynchronous finishing times is

$$\frac{2}{(ku) \ (k \ (l + u))} \le 2$$

(i.e., constant). Therefore for a local event processing time distribution with finite support the improvement of an asynchronous strategy reduces to an amount independent of P.

Certainly the model assumptions are far from what would be observed in real implementations on certain platforms, but the results might help to rank the two approaches at least from a statistical viewpoint.

35.2   "Classical" LP Simulation Protocols

35.2.1   Conservative logical processes

LP simulations following a conservative strategy date back to original works by Chandy and Misra [9] and Bryant [6] and are often referred to as the Chandy-Misra-Bryant (CMB) protocols. As described by Ref. [34], in CMB, causality of events across LPs is preserved by sending time-stamped (external) event messages of type $\langle ee @ t \rangle$, where ee denotes the event and t is a copy of LVT of the sending LP at (@) the instant when the message was created and sent. Variable $t = ts(ee)$ is also called the *timestamp* of the event. A logical process following the conservative protocol (subsequently denoted by LP^{cons}) is allowed to process *safe* events only, i.e., events up to an LVT for which the LP has been guaranteed not to receive (external event) messages with LVT $< t$ (timestamp "in the past"). Moreover, all events (internal and external) must be processed in chronological order. This guarantees that the message stream produced by an LP^{cons} is, in turn, in chronological order. It also guarantees that a communication system (Fig. 35.1) preserving the order of messages sent from LP_i^{cons} to LP_j^{cons} (FIFO) is sufficient to guarantee that no out of chronological order message can ever arrive in any LP_i^{cons} (necessary for correctness). A conservative LP simulation can thus be seen as a set of all LPs $LP^{cons} = \cup_k LP_k^{cons}$ together with a set of directed, reliable, FIFO communication channels $CH = \cup_{k,i\ (k\neq i)} ch_{k,i} = (LP_k, LP_i)$ that constitute the *Graph of Logical Processes* $GLP^{cons} = (LP, CH)$. (It is important to note, that GLP^{cons} has a *static* topology, which compared to optimistic protocols, prohibits dynamic scheduling (or rescheduling) of LPs in a set of physical processors. Note at the same time, that this view of a conservative simulation is based on a logical process model. To be conservative, a parallel simulation does not necessarily need to employ this model, nor is the message transmission order assumption required [31].)

The communication interface CI^{cons} of an LP^{cons} on the input side maintains one input buffer IB[i] and a channel (or link) clock CC[i] for every channel $ch_{i,k} \in CH$ pointing to LP_k^{cons} (Fig. 35.2). IB[i] intermediately stores arriving messages in FIFO order, whereas CC[i] holds a copy of the timestamp of the message at the head of IB[i]; initially CC[i] is set to zero. LVTH = $\min_i$ CC[i] is the local virtual time horizon up until which LVT is allowed to progress by simulating internal or external events, since no external event can arrive with a timestamp smaller than LVTH. CI now triggers the SE to conduct event processing similar to a (sequential) event-driven SE based on (internal) events in the EVL, but also to process (external) events from the corresponding IBs respecting chronological order and only up until LVT meets LVTH. During this, SE might have produced future events for remote LPs. For each of those, a message is constructed by adding a copy of LVT to the event, and deposited into FIFO output buffers OB[i] to be picked up there and delivered by the communi-

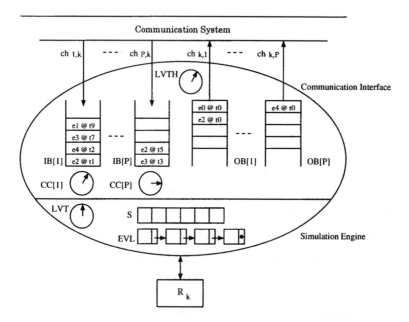

Figure 35.2 Architecture of a conservative logical process

cation system. CI maintains individual output buffers OB[i] for every outgoing channel $ch_{k,l}$ ∈ CH to subsequent LPs LP$_l$. The basic algorithm is sketched in Fig. 35.3.

Given now that within the horizon LVTH neither internal nor external events are available to process, then LP$_k^{cons}$ blocks processing, and idles to receive new messages potentially widening the time horizon. Two key problems appear with this policy of "blocking-until-safe-to-process," namely *deadlock* and *memory overflow* as explained with Fig. 35.4. Each LP is waiting for a message to arrive, however, awaiting it from an LP that is blocked itself (deadlock). Moreover, the cyclic waiting of the LPs involved in deadlock leaves events unprocessed in their respective input buffers, the amount of which can grow unpredictably, thus causing memory overflow. This is possible even in the absence of deadlock. Several methods have been proposed to overcome the vulnerability of the CMB protocol to deadlock, falling into the two principle categories, *deadlock avoidance* and *deadlock detection/recovery*.

35.2.1.1 Deadlock avoidance. Deadlock as in Fig. 35.4 can be prevented by modifying the communication protocol based on the sending of *null messages* [34] of the form (0@t), where 0 denotes a null event (event without effect). A null message is *not* related to the simulated model and only serves for synchronization purposes. Essentially it is sent on every output channel as a promise not send any other message with smaller timestamp in the future. It is launched whenever an LP processed an event that did not generate an event message for some corresponding target LP. The receiver LP can use this implicit information to extend its LVTH and by that become unblocked. In our example (Fig. 35.4), after the LP in the middle would have broadcasted (0@19) to the neighboring LPs, both of them would have chance to progress their LVT up until time 19, and in turn issue new event

```
program LP^cons(R_k)
S1        LVT = 0; EVL = {}; S = initialstate();
S2        for all CC[i] do (CC[i] = 0) od;
S3        for all ie_i caused by S do chronological_insert(⟨ie_i@occurrence_time(ie_i)⟩, EVL) od;
S4        while LVT ≤ endtime do
S4.1          for all IB[i] do await not_empty(IB[i]) od;
S4.2          for all CC[i] do CC[i] = ts(first(IB[i])) od;
S4.3          LVTH = min_i(CC[i]);
S4.4          min_channel_index = i | CC[i] == min_channel_clock;
S4.5          if ts(first(EVL)) ≤ LVTH
                  then /* select first internal event*/
                       e = remove_first(EVL) ;
                  else /* select first external event*/
                       e = remove_first(IB[min_channel_index]);
              end if;
              /* now process the selected event */
S4.6          LVT = ts(e);
S4.7          if not nullmessage(e) then
S4.7.1            S = modified_by_occurrence_of(e);
S4.7.2            for all ie_i caused by S do chronological_insert(⟨ie_i@occurrence_time(ie_i)⟩, EVL) od;
S4.7.3            for all ie_i preempted by S do remove(ie_i, EVL) od;
S4.7.4            for all ee_i caused by S do deposit(⟨ee_i@LVT⟩, corresponding(OB[j])) od;
              end if;
S4.11         for all empty(OB[i]) do deposit(⟨0@LVT + lookahead(ch_{k,i})⟩, OB[i]) od;
S4.12         for all OB[i] do send_out_contents(OB[i]) od;
          od while;
```

Figure 35.3 Conservative LP simulation algorithm sketch

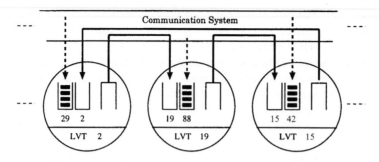

Figure 35.4 Deadlock and memory overflow

messages expanding the LVTHs of other LPs etc. The null-message-based protocol can be guaranteed to be deadlock-free as long as there are no closed cycles of channels, for which a message traversing this cycle cannot increment its timestamp. This implies, that simulation models whose event structure cannot be decomposed into regions such that for every directed channel cycle there is at least one LP to put a non-zero time increment on traversing messages cannot be simulated using CMB with null messages.

Although the protocol extension is straightforward to implement, it can put a dramatic burden of null-message overhead on the performance of the LP simulation. Optimizations of the protocol to reduce the frequency and amount of null messages, e.g., sending them only on demand (upon request), delayed until some time-out, or only when an LP becomes blocked have been proposed [34]. An approach where additional information (essentially the routing path as observed during traversal) is attached to the null message, the *carrier null-message protocol* [7] will be investigated in more detail later.

One problem that still remains with conservative LPs is the determination of when it is safe to process an event. The degree to which LPs can *look ahead* and predict future events plays a critical role in the safety verification and as a consequence for the performance of conservative LP simulations. In the example in Fig. 35.4, if the LP with LVT 19 could know that processing the next event will certainly increment LVT to 22, then null messages (0@22) (so-called *look-ahead* of 3) could have been broadcasted as further improvement on the LVTH of the receivers.

Look-ahead must come directly from the underlying simulation model, and it enhances the prediction of future events, which is necessary to determine when it is safe to process an event. The ability to exploit look-ahead from FCFS queueing network simulations was originally demonstrated by Nicol [35], the basic idea being that the simulation of a job arriving at a FCFS queue will certainly increment LVT by the service time, which can already be determined (e.g., by random variate presampling) upon arrival, since the number of queued jobs is known and preemption is not possible.

35.2.1.2 Deadlock detection/recovery. An alternative to the Chandy-Misra-Bryant protocol avoiding null messages has also been proposed by Chandy and Misra [10], allowing a deadlock to occur but providing a mechanism to detect it and recover from it. Their algorithm runs in two phases: (1) *parallel phase*, in which the simulation runs until it deadlocks, and (2) *phase interface*, which initiates a computation allowing some LP to advance LVT. They prove that, in every parallel phase, at least one event will be processed generating at least one event message, which will also be propagated before the next deadlock. A central *controller* is assumed in their algorithm, thus violating a distributed computing principle. To avoid a single resource (controller) becoming a communication performance bottleneck during deadlock detection, any general distributed termination detection algorithm or distributed deadlock detection algorithm could be used instead.

In an algorithm described by Misra [34], a special message called *marker* circulates through GLP to detect and correct deadlock. A cyclic path for traversing all $ch_{i,j} \in CH$ is precomputed and LPs are initially colored *white*. An LP that received the marker takes the color *white* and is supposed to route it along the cycle in *finite* time. When an LP has either received or sent an event message since passing the marker, it turns to *red*. The marker identifies deadlock if the last N LPs visited were all *white*. Deadlock is properly detected as long as for any $ch_{i,j} \in CH$ all messages sent over $ch_{i,j}$ arrive at LP_j in the time order as sent by LP_i. If the marker also carries the next event times of visited *white* LPs, it knows upon detection of deadlock the smallest next event time as well as the LP in which this event is supposed to occur. To recover from deadlock, this LP is invoked to process its first event. Obviously, message lengths in this algorithm grow proportionally to the number of nodes in GLP.

Bain and Scott [3] propose an algorithm for demand-driven deadlock-free synchronization in conservative LP simulation that avoids message lengths growth with the size of GLP. If an LP wants to process an event with timestamp t but is prohibited from doing so because $CC[j] < t$ for some j, then it sends *time requests* containing the sender's process ID and the requested time t to all predecessor LPs with this property. (The predecessors, however, may have already advanced their LVT in the mean time.) Predecessors are supposed to inform the sender LP when they can guarantee that they will not emit an event message at a time lower than the requested time t. Three types of reply types are used to avoid repeated polling in the presence of cycles: a *yes* indicates that the predecessor has reached the requested time, a *no* indicates that it has not (in which case another request must be made), and an *ryes* (reflected yes) indicates that it has conditionally reached t. Ryes replies, together with a *request queue* maintained in every LP, essentially have the purpose of detecting cycles and minimizing the number of subsequent requests sent to predecessors. If the process ID and time of a request received match any request already in the request queue, a cycle is detected and *ryes* is replied. Otherwise, if the LP's LVT equals or exceeds the requested time, a *yes* is replied, whereas if the LP's LVT is less the requested time, the request is enqueued in the request queue, and request copies are recursively sent to the receiver's predecessors with $CC[i]$s $< t$, etc. The request is complete when all channels have responded, and the request reached the head of the request queue. At this time, the request is removed from the request queue and a reply is sent to the requesting LP. The reply to the successor from which the request was received is no (*ryes*), if any request to a predecessor was answered with no (*ryes*), otherwise *yes* is sent. If *no* was received in an LP initiating a request, the LP has to restart the time request with lower channel clocks.

The *time-of-next-event algorithm* as proposed by Groselj and Tropper [24] assumes more than one LP mapped onto a single physical processor, and computes the greatest lower bound of the timestamps of the event messages expected to arrive next at *all empty* links on the LPs located at that processor. It thus helps to unblock LPs within one processor, but does not prevent deadlocks across processors. The lower bound algorithm is an instance of the single source shortest path problem.

35.2.1.3 Conservative time windows.
Conservative LP simulations as presented above are distributed in nature since LPs can operate in a totally asynchronous way. One way to

make these algorithms more synchronous to gain from the availability of fast synchronization hardware in multiprocessors is to introduce a *window*, W_i, in simulated time for each LP_i such that events within this *time window* are *safe* (events in W_i are independent of events in W_j, $i \neq j$) and can be processed concurrently across all LP_i [31, 36]. A conservative time window (CTW) *parallel* LP simulation synchronously operates in two phases. In phase (1) (*window identification*), for every LP_i a chronological set of events W_i is identified such that for every event, $e \in W_i$, e is causally independent of any $e' \in W_j, j \neq i$. Phase (1) is accomplished by a barrier synchronization over all LPs. In phase (2) (*event processing*), every LP_i processes events $e \in W_i$ sequentially in chronological order. Again, phase (2) is accomplished by a barrier synchronization. Since the algorithm iteratively lock-steps over the two consecutive phases, the hope to gain speedup over a purely sequential DES heavily depends on the efficiency of the synchronization operation on the target architecture, but also on the event structure in the simulation model. Different windows will generally have different cardinality of the covered event set, maybe some windows will remain empty after the identification phase for one cycle. In this case, the corresponding LPs would idle for that cycle.

A considerable overhead can be imposed on the algorithm by the identification of when it is safe to process an event within LP_i (window identification phase). Lubachevsky [31] proposes to reduce the complexity of this operation by restricting the *lag* on the LP simulation, i.e., the difference in occurrence time of events being processed concurrently is bounded from above by a known finite constant (*bounded lag* protocol). By this restriction, and assuming a "reasonable" amount of dispersion of events in space and time, the execution of the algorithm on N processors in parallel will have one event processed in $O(logN)$ time, on average. An idealized message passing architecture with a tree-structured synchronization network supporting an efficient realization of the bounded lag restriction is assumed for the analysis.

35.2.1.4 The carrier null-message protocol. Another approach to reduce the overwhelming number of null messages occurring with the CMB protocol is to add more information to the null messages. The *carrier null-message protocol* (CNM) [7] uses null messages to advance CC[i]s and to acquire/propagate knowledge global to the participating LPs, with the goal of improving the ability of look-ahead to reduce the message traffic.

Indeed, good look-ahead can reduce the number of null messages as is motivated by the example in Fig. 35.5, where a *source* process produces objects in constant time intervals $\omega = 50$. The *join, pass,* and *split* processes manipulate objects, consuming two time units per object. Eventually, objects are released from *split* into *sink*. For the example, we have look-ahead per channel of $la(ch_{i,j}) = 2 \; \forall_{i,j} \in$ {source, join, pass, split, sink}, $(i \neq j)$, except $la(ch_{source,join}) = 50$. After the first object release into LP_{join}, all LPs except LP_{source} are blocked and therefore start propagating local look-ahead via null messages. After the propagation of (overall) four null messages all LPs beyond LP_{source} have progressed LVTs and CCs to two. It will take another 96 null messages until LP_{join} can make its first object manipulation, and after that another 100 for the second object, etc. If LP_{join} could have learned that it was just waiting for itself, it could have immediately simulated the external event (with VT 50). Besides the importance of the availability of global information within the LPs, the impact of look-ahead onto LP simulation performance is now also easily seen—the smaller the look-ahead in the successor LPs, the higher the communication overhead caused by null messages, and the higher the performance degradation.

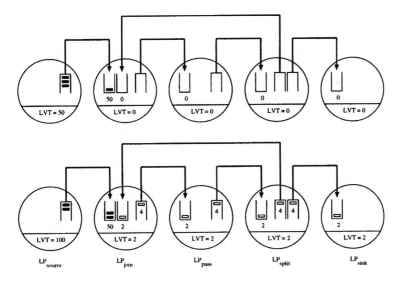

Figure 35.5 Motivation for look-ahead propagation using CNM

To realize such a waiting dependency across LPs, generally the CNM protocol employs additional null messages of type $(c0, t, \Re, la.inf)$, where $c0$ is an identification as a *carrier null message, t* is the timestamp, $\Re$ is information about the traveling route of the message and *la.inf* is look-ahead information. Once LP$_{join}$ had received a carrier null message with its ID as source and sink in $\Re$, it can be sure (but only in the particular example) not to receive an event message via that path unless LP$_{join}$ itself had sent an event message along that path. So it can—without further waiting—after having received the first carrier null message, process the event message from LP$_{source}$ and thus increment the CCs and LVTs of all successors on the route in $\Re$ considerably.

Should there be any other "source-like" LP entering event messages into the waiting dependency loop, the arguments above are no longer valid. For this case, it is in fact not sufficient only to carry out the route information with the null message, but also the earliest time of possible event messages that would break the cyclic waiting dependency. Exactly this information is carried by *la.inf*, the last component in the carrier null message.

35.2.2 Optimistic logical processes

Optimistic LP simulation strategies, in contrast to conservative ones, do not strictly adhere to the local causality constraint *lcc* (see Section 35.1.4.2) but allow the occurrence of causality errors and provide a mechanism to recover from *lcc* violations. To avoid *blocking* and *safe-to-process* determination, which are serious performance pitfalls in the conservative approach, an optimistic LP progresses simulation (and by that advances LVT) as far into the simulated future as possible, without any warranty that the set of generated (internal and external) events is consistent with *lcc*, and regardless of the possibility of the arrival of an external event with a timestamp in the local past.

35.2.2.1 Time Warp. Pioneering work in optimistic LP simulation was done by Jefferson and Sowizral [27] in the definition of the *Time Warp* (TW) mechanism which, like the

Chandy-Misra-Bryant protocol, uses the sending of messages for synchronization. Time Warp employs a rollback (in time) mechanism to take care of proper synchronization with respect to *lcc*. If an external event arrives with timestamp in the local past, i.e., out of chronological order (*straggler message*), then the Time Warp scheme *rolls back* to the most recently saved state in the simulation history consistent with the timestamp of the arriving external event, and restarts simulation from that state on as a matter of *lcc* violation correction. Rollback, however, requires a record of the LP's history with respect to the simulation of internal *and* external events. Hence, an LP^{opt} has to keep sufficient internal state information, say a *state stack* SS, which allows for restoring a past state. Furthermore, it has to administrate an *input queue* IQ and an *output queue* OQ for storing messages received and sent. For reasons to be seen, this logging of the LP's communication history must be done in chronological order. Since the arrival of event messages in increasing timestamp order cannot be guaranteed, two different kinds of messages are necessary to implement the synchronization protocol. First, there are the usual external event messages $(m^+ = \langle ee@t,+\rangle)$ (where again *ee* is the external event and t *is* a copy of the senders LVT at the sending instant), which will subsequently call *positive* messages. Opposed to that are messages of type $(m^- = \langle ee@t,-\rangle)$ called *negative* or *antimessages*, which are transmitted among LPs as a request to annihilate a previous positive message containing *ee*, but for which it meanwhile turned out as computed based on a causally erroneous state.

The basic architecture of an optimistic LP employing the Time Warp rollback mechanism is outlined in Fig. 35.6. External events are brought to some LP_k by the communication system in much the same way as in the conservative protocol. Messages, however, are

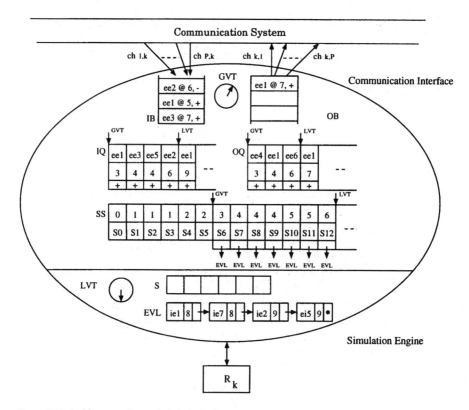

Figure 35.6 Architecture of an optimistic logical process

not required to arrive in the sending order (FIFO) in the optimistic protocol, which weakens the hardware requirements for executing Time Warp. Moreover, the separation of arrival streams is also not necessary, and so there is only a single IB and a single OB (assuming that the routing path can be deduced from the message itself). The communication related history of LP_k is kept in IQ and OQ, whereas the state related history is maintained in the SS data structure. All those together represent CI_k; SE_k is an event-driven simulation engine equivalent to the one in LP^{cons}.

The triggering of CI to SE is sketched with the basic algorithm for LP^{opt} in Fig. 35.7. The LP mainly loops (*S3*) over four parts: (1) an input-synchronization to other LPs (*S3.1*), (2) local event processing (*S3.2–S3.8*), (3) the propagation of external effects (*S3.9*), and (4) the (global) confirmation of locally simulated events (*S3.10–S3.11*). Parts (2) and (3) are almost the same, as was seen with LP^{cons}. The input synchronization (*rollback and annihilation*) and confirmation (GVT) parts, however, are the key mechanisms in optimistic LP simulation.

```
program LP^opt(R_k)
S1        GVT = 0; LVT = 0; EVL = {}; S = initialstate();
S2        for all ie_i caused by S do chronological_insert(⟨ie_i@occurrence_time(ie_i)⟩, EVL) od;
S3        while GVT ≤ endtime do
S3.1          for all m ∈ IB do
S3.1.1            if ts(m) ≤ LVT /* m potentially affects local past */
                    then
                            if (positive(m) and dual(m) ∉ IQ) or (negative(m) and dual(m) ∈ IQ)
                              then /* rollback */
                                    restore_earliest_state_before(ts(m));
                                    generate_and_sendout(antimessages);
                                    LVT = earliest_state_timestamp_before(m);
                            endif;
                  endif;
                  /* irrespective of how m is related to LVT */
S3.1.2            if dual(m) ∈ IQ
                    then remove(dual(m), IQ);  /* annihilate */
                    else  chronological_insert(external_event(m)@ts(m), sign(m)), IQ);
                  endif;
               od;
S3.2          if ts(first(EVL)) ≤ ts(first_nonnegative(IQ))
                    then e = remove_first(EVL);              /* select first internal event*/
                    else  e = first_nonnegative(IQ);         /* select first external event*/
               endif;
               /* now process the selected event */
S3.3          LVT = ts(e);
S3.4          S = modified_by_occurrence_of(e);
S3.5          for all ie_i caused by S do chronological_insert(⟨ie_i@occurrence_time(ie_i)⟩, EVL) od;
S3.6          for all ie_i preempted by S do remove(ie_i, EVL) od;
S3.7          log_new_state(⟨LVT, S, copy_of(EVL)⟩, SS);
S3.8          for all ee_i caused by S do
                    deposit(⟨ee_i@LVT, +⟩, OB);
                    chronological_insert(⟨ie_i@LVT, +⟩, OQ);
               od;
S3.9          send_out_contents(OB);
S3.10         GVT = advance_GVT();
S3.11         fossil_collection(GVT);
          od while;
```

Figure 35.7 Optimistic LP simulation algorithm sketch

35.2.2.2 Rollback and annihilation mechanisms. The input synchronization of LP^{opt} (rollback mechanism) relates arriving messages to the current value of the LP's LVT and reacts accordingly. (See Fig. 35.8.) A message affecting the LP's "localfuture" is moved from the IB to the IQ respecting the timestamp order, and the encoded external event will be processed as soon as LVT advances to that time. (*ee3@7,+*) in the IB in Fig. 35.6 is an example of such an unproblematic message (LVT = 6). A message timestamped in the "local past" however is an indicator of a causality violation because of tentative event processing. The rollback mechanism (*S3.1.1*) in this case restores the most recent *lcc*-consistent state, by reconstructing S and EVL in the simulation engine from copies attached to the SS in the communication interface. Also LVT is warped back to the timestamp of the straggler message. This so far has compensated the *local* effects of the *lcc* violation; the *external* effects are annihilated by sending an antimessage for all previously sent output messages. In the example (*ee6@6,–*) and (*eel@7,–*) are generated and sent out, while at the same time (*ee6@6,+*) and (*eel@7, +*) are removed from OQ. Finally, if a negative message is received (e.g., (*ee2@6,–*)) it is used to annihilate the dual positive message ((*ee2@6,+*)) in the local IQ. Two cases for the negative messages must be distinguished. In case (1) if the dual positive message is present in the receiver IQ, then this entry is deleted as an annihilation. This can be done easily if the positive message has not yet been processed, but requires rollback if it was. In case (2), if a dual positive message is not present (this case can only arise if the communication system does not deliver messages in a FIFO fashion), then the negative message is inserted in IQ (irrespective of its relation to LVT) to be annihilated later by the (delayed) positive message still in traffic.

As is now apparent, the rollback mechanism requires a periodic saving of the states of SE (LVT, S, and EVL) to able to restore a past state (*S3.7*), and to log output messages in OQ to be able to undo propagated external events (*S3.8*). Since antimessages can also cause rollback, there is the chance of *rollback chains*, and even recursive rollback if the cascade unrolls sufficiently deep on a directed cycle of GLP. The protocol, however, (although consuming considerable memory and communication resources) guarantees that *any* rollback chain eventually terminates, whatever its length or recursive depth.

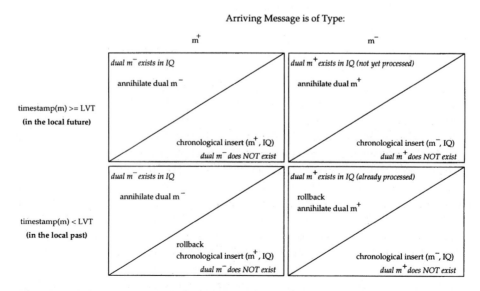

Figure 35.8 The Time Warp message based on synchronization mechanism

Related to the possibility of rollbacks at any time of the simulation is the problem of termination detection. An alternative to the termination criterion in statement *S3* in the algorithm in Fig. 35.7 is to introduce the timestamp ∞. An LP that completes the local simulation sets LVT= ∞, and every incoming message will induce a rollback. Once GVT has reached the time ∞ (i.e., LVT= ∞ for every LP), termination is detected.

Lazy cancellation. In the original Time Warp protocol as described above, an LP receiving a *straggler message* initiates sending antimessages immediately when executing the rollback procedure. This behavior is called *aggressive cancellation*. As a performance improvement over *aggressive cancellation*, the *lazy cancellation* policy does not send an antimessage (m^-) for m^+ immediately upon receipt of a straggler. Instead, it delays its propagation until the resimulation after rollback has progressed to LVT = $ts(m^+)$ producing $m^{+'} \neq m^+$. If the resimulation produced $m^{+'} = m^+$, no antimessage has to be sent at all [21]. Lazy cancellation thus avoids unnecessary cancelling of correct messages but has the liability of additional memory and bookkeeping overhead (potential antimessages must be maintained in a rollback queue) and delaying the annihilation of actually wrong simulations.

Lazy cancellation can also be based on the use of look-ahead available in the simulation model. If a straggler $m^+ <$ LVT is received, than obviously antimessages do not have to be sent for messages m with timestamp, $ts(m^+) \leq ts(m) < ts(m^+) + la$. Moreover, if $ts(m^+) + la \geq$ LVT, even rollback does not need to be invoked. As opposed to look-ahead computation in the CMB protocol, lazy cancellation can exploit implicit look-ahead, i.e., does not require its explicit computation.

It has been shown [26] that Time Warp with lazy cancellation can produce so-called *supercritical speedup*, i.e., surpass the simulations critical path by the chance of having wrong computations produce correct results. By immediately discarding rolled back computations, this chance is lost for the aggressive cancellation policy. A performance comparison of the two, however, is related to the simulation model. Analysis showed that lazy cancellation can arbitrarily outperform aggressive cancellation and vice versa, i.e., one can construct extreme cases for lazy and aggressive cancellation such that if one protocol executes in α time using N processors, the other uses αN time. Nevertheless, empirical evidence is reported *slightly* in favor of lazy cancellation for certain simulation applications.

Lazy re-evaluation. Much as *lazy cancellation* delays the annihilation of external effects upon receiving a straggler at LVT, *lazy reevaluation* delays discarding entries on the state stack SS. Should the recomputation after rollback to time $t <$ LVT reach a state that exactly matches one logged in SS *and* the IQ is the same as the one at that state, then the simulation can immediately *jump forward* to LVT, the time before rollback occurred. Thus, lazy reevaluation prevents the unnecessary recomputation of correct states and is therefore promising in simulation models where events do not modify states ("read-only" events). A serious liability of this optimization is again additional memory and bookkeeping overhead, but also (and mainly) the considerable complication of the Time Warp code [19]. To verify equivalence of IQs, the protocol must draw and log copies of the IQ in every state, saving step (*S3.7*). In a weaker *lazy reevaluation* strategy, one could allow jumping forward only if no message has arrived since rollback.

Lazy rollback. The difference of virtual time in between the straggler m^*, $ts(m^*)$, and its actual effect at time $ts(m^*) + la(ee) \geq$ LVT can again be overjumped, saving the computation time for the resimulation of events in between $[ts(m^*), ts(m^*) + la(ee))$. $la(ee)$ is the look-ahead imposed by the external event carried by m^*.

Breaking/preventing rollback chains. Besides the postponing of erroneous messages and state annihilation until it turns out that they are not reproduced in the repeated simulation, other techniques have been studied to break cascades of rollbacks as early as possible. Prakash and Subramanian, in a method comparable to the carrier null-message approach, attach a limited amount of state information to messages to prevent recursive rollbacks in cyclic GLPs. This information allows LPs to filter out messages based on preempted (obsolete) states to be eventually annihilated by chasing antimessages currently in transit. Related to the (conservative) *bounded lag* algorithm, Lubachevsky, Shwartz, and Weiss have developed a *filtered rollback* protocol [32] that allows optimistically crossing the lag bound, but only up to a time window upper edge. Causality violations can only affect the time period in between the window edge and the lag bound, thus limiting (the relative) length of rollback chains. The SRADS protocol by Reynolds [14], although allowing optimistic simulation progression, prohibits the propagation of uncommitted events to other LPs. Therefore, rollback can be *local* only to some LP and cascades of rollback can never occur. Madisetti, Walrand, and Messerschmitt with their protocol called Wolf-calls freeze the spatial spreading of uncommitted events in so-called *spheres of influence* $W(LP_i, \tau)$, defined as the set of LPs that can be influenced by a message from LP_i at time $ts(m) + \tau$ respecting computation and communication times. The Wolf algorithm ensures that the effects of an uncommitted event generated by LP_i are limited to a sphere of a computable (or selectable) radius around LP_i, and the number of broadcasts necessary for a complete annihilation within the sphere is bounded by a computable (or chooseable) number of steps B (B being provably smaller than for the standard Time Warp protocol).

35.2.2.3 Optimistic time windows.
A similar idea of "limiting the optimism" to overcome rollback overhead potentials is to advance computations by "windows" moving over simulated time. In the original work of Sokol, Briscoe, and Wieland [40], the *moving time window* (MTW) protocol, neither internal nor external events e with $ts(e) > t + \Delta$ are allowed to be simulated in the time window $[t, t + \Delta)$, but are postponed for the next time window $[t + \Delta, t + 2\Delta)$. Two events e and e' timestamped $ts(e)$ and $ts(e')$, respectively, therefore can be simulated in parallel only if $|ts(e) - ts(e')| < \Delta$. Naturally, the protocol is in favor of simulation models with a low variation of event occurrence distances relative to the window size. Compared to a time-stepped simulation, MTW does not await the completion of *all* events e with $t \leq ts(e) < t + \Delta$ which would cause idle processors at the end of each time window, but invokes an attempt to move the window as soon as the number of events to be executed falls below a certain threshold. To keep moving the time window, LPs are polled for the timestamp of their earliest next event $t_i(e)$ (polling takes place simultaneously with event processing) and the window is advanced to $\min_i t_i(e)$, $\min_i t_i(e) + \Delta$. [The next section will show the equivalence of the window lower edge determination to global visual time (GVT) computation.] Obviously, the advantage of MTW and related protocols is the potential effective implementation as a *parallel* LP simulation, either on a SIMD architecture or in a MIMD environment where the reduction operation $\min_i t_i(e)$ can be computed utilizing synchronization hardware. Points of criticism are the assumption of approximately uniform distribution of event occurrence times in space and the ignorance with respect to potentially "good" optimism beyond the upper window edge. Furthermore, a natural difficulty is the determination of the Δ admitting enough events to make the simulation efficient.

The latter is addressed with the *adaptive Time Warp concurrency control algorithm* (ATW) proposed by Ball and Hoyt [4], allowing the window size $\Delta(t)$ be adapted at any point t in simulation time. ATW aims to temporarily suspend event processing if it has observed a certain amount of *lcc* violations in the past. In this case, the LP would conclude that it progresses LVT too fast compared to the predecessor LPs and would therefore stop LVT advancement for a time period called the *blocking window* (BW). BW is determined based on the minimum of a function describing wasted computation in terms of time spent in a (conservatively) blocked mode, or a fault recovery mode as induced by the Time Warp rollback mechanism.

35.2.2.4 The limited memory dilemma. All arguments on the execution of the Time Warp protocol so far assumed the availability of a *sufficient* amount of free memory to record internal and external effect history for pending rollbacks, and all arguments were related to the time complexity. Indeed, Time Warp with certain memory management strategies to be described in the sequel can be proven to work correctly when executed with $O(M^{seq})$ memory, where M^{seq} is the number of memory locations utilized by the corresponding sequential DES. Opposed to that, the CMB protocol may require $O(kM^{seq})$ space, but may also use less storage than sequential simulation, depending on the simulation model (it can even be proven that simulation models exist such that the space complexity of CMB is $O((M^{seq})^k)$). Time Warp *always* consumes more memory than sequential simulation [30], and a memory limitation imposes a performance decrease on Time Warp. Providing just the minimum of memory necessary may cause the protocol to execute fairly slowly, such that the memory/performance trade-off becomes an issue.

Memory management in Time Warp follows two goals: (1) to make the protocol *operable* on real multiprocessors with bounded memory, and (2) to make the execution of Time Warp *performance efficient* by providing "sufficient" memory. An *infrequent* or *incremental* saving of history information in some cases can prevent aggressive memory consumption—possibly more effectively than one of the techniques presented for *limiting the optimism* in Time Warp. Once, despite the application of those techniques, available memory is exhausted, *fossil collection* could be applied as a technique to recover memory used for history recording that will definitely not be used anymore due to an assured lower bound on the timestamp of any possible future rollback (GVT). Finally, if even fossil collection fails to recover enough memory to proceed with the protocol, additional memory could be freed by returning messages from the IQ (*message sendback, cancelback*) or invoking an *artificial rollback* reducing space used for storing the OQ.

35.2.2.5 Incremental and interleaved state saving. Minimizing the storage space required for simulation models with complex sets of state variables $S_i \subset S$, S_i being the subset stored and maintained by LP_i that do not extensively change values over LVT progression, can be accomplished effectively by just saving the variables $s_j \in S_i$ affected by a state change. This is mainly an implementation optimization upon step *S3.4* in the algorithm in Fig. 35.7. This *incremental* state saving can also improve the execution complexity in step *S3.7*, since generally less data has to be copied into the logrecord. Obviously, the same strategy could be followed for the EVL, or the IQ in a lazy reevaluation protocol. Alternatively, imposing a condition upon step *S3.7*:

S3.7 **if** (step_count modulo π) = 0 **then** log_new_state((LVT, S, copy_of(EVL)), SS);

could be used to interleave the continuity of saved states and thus on the average reduce the storage requirement to $1/\pi$ of the noninterleaved case.

Both optimizations, however, increase the execution complexity of rollback. In incremental state saving protocols, desired states have to be reconstructed from increments following back a path further into the simulated past than required by rollback itself. The same is true for interleaved state saving, where the most recent *saved* state older than the straggler must be searched for, a re-execution up until the timestamp of the straggler (*coast forward*) must be started which is a clear waste of CPU cycles since it just reproduces states that have already been computed but were not saved and, finally, the straggler integration and usual re-execution are necessary (Fig. 35.9). The trade-off between state saving costs and the coast forward overhead has been studied (as reported in Ref. [37] in reference to Lin and Lazowska) based on expected event processing time ($\varepsilon = E[exec(e)]$) and state saving costs (σ), giving an optimal *interleaving factor* π^* as

$$\lfloor \sqrt{(\alpha - 1)\beta} \rfloor < \pi^* < \lceil \sqrt{(2\alpha + 1)\beta} \rceil$$

where α is the average number of rollbacks with $\pi = 1$ and $\beta = \sigma/\varepsilon$. The result expresses that an overestimation of π^* is more severe to performance than an underestimation by the same (absolute) amount. In a study of the optimal checkpointing interval explicitly considering state saving and restoration costs while assuming π does neither affect the number of rollbacks nor the number of rolled back events in [29], an algorithm is developed that, integrated into the protocol, "on-the fly," within a few iterations, automatically adjusts π to π^*. It has been shown that some π, though increasing the rollback overhead, can reduce overall execution time.

35.2.2.6 Fossil collection. Opposed to techniques that reclaim memory temporarily used for storing events and messages related to the *future* of some LP, fossil collection aims to return space used by history records that will no longer be used by the rollback synchronization mechanism. To determine from which state in the history (and back) computations can be considered fully committed, the determination of the value of *global virtual time* (GVT) is necessary.

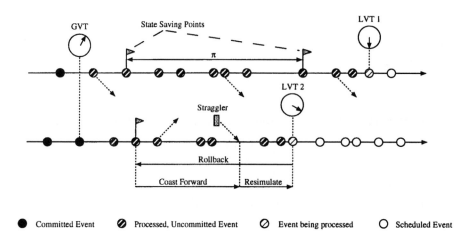

Figure 35.9 Interleaved state saving

Consider the tuple $\Sigma_i(T) = (LVT_i(T), IQ_i(T), SS_i(T), OQ_i(T))$ to be a *local snapshot* of LP_i at *real time T*, i.e., LVT, IQ_i is the input queue as seen by an external observer at *real time T*, etc., and $\Sigma(T) = \cup^N_{i=1} \Sigma_i(T) \cup CS(T)$ be the *global snapshot* of GLP. Further let $LVT_i(T)$ be the local virtual time in LP_i, i.e., the timestamp of the event being processed at the observation instant *T*, and $UM_{i,j}(T)$ the set of external events imposed by LP_i upon LP_j encoded as messages m in the snapshot Σ. This means *m* is either in transit on channel $ch_{i,j}$ in CS or stored in some IQ_j, but not yet processed at time *T*. Then the GVT at real time *T* is defined to be:

$$(GVT(T) = \min\,(\min_i LVT_i(T))), \quad \min_{i,j;m \,\in\, UM_{i,j}(T)} ts\,(m)$$

It should be clear even by intuition, that at *any* (real time) *T*, GVT(*T*) represents the maximum lower bound to which any rollback could ever backdate LVT_i ($\forall i$). An obvious consequence is that any processed event *e* with $ts(e) < $ GVT(*T*) can never (at no instant *T*) be rolled back and therefore can be considered as (*irrevocably*) *committed* [30] (see Fig. 35.9). Further consequences (for all LP_i) are that:

1. Messages $m \in IQ_i$ with $ts(m) \leq$ GVT(T), as well as messages $m \in OQ_i$ with $ts(m) \leq$ GVT(T) are obsolete and can be discarded (from IQ, OQ) after real time *T*.
2. State variables $s \in S_i$ stored in SS_i as with $ts(s) \geq$ GVT(*T*) are obsolete and can be discarded after real time *T*.

Making use of these possibilities, i.e., getting rid of now unneeded external event history according to (1) and of internal history according to (2) to reclaim memory space is the idea behind *fossil collection*. It is called as a procedure in the abstracted Time Warp algorithm (Fig. 35.7) in step *S3.11*. The idea of reclaiming memory for history earlier than GVT is also expressed in Fig. 35.6, which shows IQ and OQ sections for entries with timestamp later than GVT only, and copies of EVL in SS if not older than GVT (also the rest of SS beyond GVT could be purged as irrelevant for Time Warp, but we assume here that the state trace is required for a post-simulation analysis).

Generally, a combination of fossil collection with any of the incremental/interleaved state saving schemes is recommended. Related to interleaving, however, rollback might be induced to events beyond the momentary committed GVT, with an average overhead directly proportional π. Not only that the interleaving of state recording is prohibiting fossil collection for states timestamped in the gap between GVT and the most recent saved state chronologically before GVT, it is also contraproductive to GVT computation, which is comparably more expensive than state saving, as will be seen shortly.

35.2.2.7 Freeing memory by returning messages. Previous strategies (interleaved, incremental state saving as well as fossil collection) are merely able to reduce the chance of memory exhaustion; they cannot actually prevent such situations from occurring. In cases where memory is already completely allocated, only additional techniques, mostly based on returning messages to senders or artificially initiating rollback, can help to escape from deadlocks due to waiting for free memory:

Message sendback. The first approach to recover from memory overflow in Time Warp was proposed by the *message sendback* mechanism by Jefferson [27]. Here, whenever the system runs out of memory on the occasion of an arriving message, part or all of the space used for saving the *input history* is used to recover free memory by returning unprocessed input messages (not necessarily including the one just received) back to the sender and relocating the freed (local) memory. By intuition, input messages with the highest send timestamps should be returned first, since they are more likely to carry incorrect information than "older" input messages, and since the annihilation of their effects can be expected not to disperse as much in virtual time, thus restricting annihilation influence spheres. Related to the original definition of the Time Warp protocol which distinguishes the *send time* (ST) and *receive time* (RT) $(ST(m) \leq RT(m))$ of messages, only messages with $ST(m) > LVT$ (local *future messages*) are considered for returning.

An indirect effect of the sendback could also be storage release in remote LPs due to annihilation of messages triggered by the original sender's rollback procedure.

Gafni's protocol. In a message traffic study of *aggressive* and *lazy cancellation*, Gafni [21] notes that *past* $(RT(m) < GVT)$ *and present messages* $(ST(m) < GVT < RT(m))$ and events accumulate in IQ, OQ, SS for the two annihilation mechanisms at the same rate, pointing out also the interweaving of messages and events in memory consumption. Past messages and events can be fossil collected as soon as a new value of GVT is available. The amount of "present" messages and events present in LP_i reflects the difference of LVT_i to the global GVT directly expressing the asynchrony or "imbalance" of LVT progression. This fact gives an intuitive explanation of the *message sendback's* attempt to *balance* LVT progression across LPs, i.e., intentionally roll back those LPs that have progressed LVT ahead of others. Gafni, considering this asynchrony to be exactly the source from which Time Warp can gain real execution speedup, states that LVT progression balancing does not solve the storage overflow problem. His algorithm reclaims memory by relocating space used for saving the *input* or *state* or *output* history in the following way. Whether or not the overflow condition is raised by an arriving input message, the request to log a new state or the creation of a new output message, the element (message or event) with the largest timestamp is selected irrespective of its type.

- If it is an *output message*, a corresponding antimessage is sent, the element is removed from OQ and the state before sending the original message is restored. The antimessage arriving at the receiver will find its annihilation partner in the receiver's IQ upon arrival (at least in FIFO CSs), so memory is also reclaimed in the receiver LP.
- If it is an *input message*, it is removed from IQ and returned to the original sender to be annihilated with its dual in the OQ, perhaps invoking rollback there. Again, the receiver LP also relocates memory.
- If it is a *state* in SS, it is discarded (and will be recomputed in case of local rollback).

The desirable property of both *message sendback* and Gafni's protocol is that LPs that ran out of memory can be relieved without shifting the overflow condition to another LP. So, given a certain minimum but limited amount of memory, both protocols make Time Warp "operable."

Cancelback. An LP simulation memory management scheme is considered to be storage optimal iff it consumes $O(M^{seq})$ constant bounded memory [30]. The worst-case space complexity of Gafni's protocol is $O(NM^{seq}) = O(N^2)$ (irrespective of whether memory is shared or distributed), the reason for this being that it can only cancel elements within the individual LPs. *Cancelback* is the first optimal memory management protocol [25], and it was developed targeting Time Warp implementations on shared memory systems. As opposed to Gafni's protocol, in cancelback elements can be *canceled* in *any* LP_i (not necessarily in the one that observed memory overflow), whereas the element selection scheme is the same. Cancelback thus allows to selectively reclaim those memory spaces that are used for the *very* most recent (globally seen) input-, state- or output-history records, whichever LP maintains this data. An obvious implementation of cancelback is therefore for shared-memory environments and making use of system-level interrupts. A Markov model of cancelback [1], predicting speedup as the amount of available memory beyond M^{seq} is varied, revealed that even with small fractions of additional memory, the protocol performs about as well as with unlimited memory. The model assumes totally symmetric workload and a constant number of messages, but it is verified with empirical observations.

Artificial rollback. Although cancelback theoretically solves the memory management dilemma of Time Warp (since it produces correct simulations in real, limited memory environments with the same order of storage requirement as the sequential DES), it has been criticized for its implementation not being straightforward, especially in distributed memory environments. Lin [30] describes a memory management scheme that is, in turn memory optimal (there exists a shared memory implementation of Time Warp with space complexity $O(M^{seq})$,[*]) but it has a simpler implementation. Lin's protocol is called *artificial rollback* for provoking the rollback procedure not only for the purpose of *lcc*-violation restoration, but also for its side effect of reclaiming memory. (Since rollback as such does not affect operational correctness of Time Warp, it can also be invoked *artificially*, i.e., even in the absence of a straggler.) Equivalent to cancelback in effect (cancelling an element generated by LP_j from IQ_i is equivalent to a rollback in LP_j, whereas cancelling an element from OQ_i or SS_i is equivalent to a rollback in LP_i), artificial rollback has a simpler implementation, since the rollback procedure already available can be used together with an artificial rollback trigger. Determining, however, in which LP_i to invoke artificial rollback, to what LVT to rollback and at what instant of real time T to trigger it is not trivial (except the triggering, which can be related to the overflow condition and the failure of fossil collection). In the implementation proposed by Lin and Preiss [30], the two other issues are coupled to a processor scheduling policy in order to guarantee a certain amount of free memory (called *salvage parameter* in [37]), while following the "cancel-furthest-ahead" principle.

Adaptive memory management. The *adaptive memory management* (AMM) scheme proposed by Das and Fujimoto [12] attempts a combination of controlling optimism in Time Warp and an *automatic* adjustment of the amount of memory in order to optimize fossil collection, cancelback, and rollback overheads. Analytical performance models of Time Warp with cancelback [1] for homogeneous (artificial) workloads have shown that at a certain amount of available free memory fossil collection is sufficient to allocate enough memory. With a decreasing amount of available memory, absolute execution performance decreases due to more frequent cancelbacks until it be comes frozen at some point. Strong empirical evidence has been given as a support to this analytical observations. The motivation now for an *adaptive* mechanism to control memory is twofold:

---

[*]For implementations in distributed memory environments, Time Warp with artificial rollback cannot guarantee a space complexity of $O(M^{seq})$. Cancelback and artificial rollback in achieving the sequential DES storage complexity bound rely on the availability of a global, shared pool of (free) memory.

1. Absolute performance is supposed to have negative increments after reducing memory even further. Indeed, one would like to run Time Warp in the area of the "knee-point" of absolute performance. A successive adaptation to that performance optimal point is desired.
2. The location of the knee might vary during the course of simulation because of the underlying simulation model. A run-time adaptation to follow movements of the knee point is desired.

The AMM protocol for automatic adjustment of available storage uses a memory flow model that divides the available (limited) memory space M into three "pools," $M = M^c + M^{uc} + M^f$. M^c is the set of all memory locations used to store committed elements ($t(e) \leq$ GVT), M^{uc} is its analogy for uncommitted events (in IQ, OQ, or SS with $t(e) >$ GVT) and M^f holds temporarily unused (free) memory. The behavior of Time Warp can now be described in terms of flows of (fixed sized) memory buffers (able to record one message or event for simplicity) from one pool into the other (see Fig. 35.10). Free memory must be allocated for every message created/sent, and every state logged or any future event scheduled, causing buffer moves from M^f to M^{uc}. Fossil collection, on the other hand, returns buffers from M^c as invoked upon exhaustion of M^f, whereas M^c is being supplied by the progression of GVT. Buffers move from M^{uc} to M^f with each message annihilation, either incurred by rollback or by cancelback. A *cancelback cycle* is defined by two consecutive invocations of cancelback. A cycle starts where cancelback was called because of failure of fossil collection to reclaim memory; at this point there are no buffers in M^c. Progression of LVT will move buffers to M^{uc}, rollback of LVT will occasionally return free memory, and progression of GVT will deposit into M^c to be depleted again by fossil collection, but tendentially the free pool will be drained, thus necessitating a new cancelback.

Time Warp can now be controlled by two (mutually dependent) parameters: (1) α, the number of processed but uncommitted buffers left behind after cancelback, as a parameter to control optimism; and (2) β, the number of buffers freed by cancelback, as a parameter to control the cycle length. Obviously, α has to be chosen small enough to avoid rollback thrashing and overly aggressive memory consumption, but not too small in order to prevent rollbacks of states that are most likely to be confirmed (events on the critical path). β should be chosen in such a way as to minimize the overhead caused by unnecessary frequent cancelback (and fossil collection) calls. The AMM protocol now attempts (by moni-

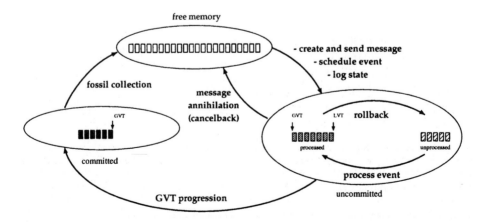

Figure 35.10 "Flow of buffers" in the AMM protocol

toring the Time Warp execution behavior during one cycle) to simultaneously minimize the values of α and β, but respecting the constraints above. It assumes cancelback (and fossil collection) overhead to be directly proportional to the cancelback invocation frequency. Let

$$\rho M^{uc} = \frac{e_{committed} N}{T_{process}} - \rho FC$$

be the rate of growth of M^{uc}, where $e_{committed}$ is the fraction of processed events also committed during the last cycle, $T_{process}$ is the average (real) time to process an event, and ρFC is the rate of depletion of M^{uc} because of fossil collection. (Estimates for the right-hand side are generated from monitoring the simulation execution.) β is then approximated by

$$\beta = (T_{cycle} - T_{CB,FC}) \rho M^{uc}$$

where $T_{CB,FC}$ is the overhead incurred by cancelback and fossil collection in real time units, and T_{cycle} is the current invocation interval. Indeed, α is a parameter to control the upper tolerable bound for the progression of LVT. To set α appropriately, AMM records by a marking mechanism whether an event was rolled back by cancelback. A global counting mechanism (across all LPs) lets AMM determine the number $\#(e_{cp})$ of events that should not have been rolled back by cancelback, since they were located on the critical path, and by that causing a definitive performance degradation.[*] Starting now with a high parameter value for α (which will give an observation $\#(e_{cp}) \cong 0$, α is continuously reduced as long as $\#(e_{cp})$ remains negligible. Rollback thrashing is explicitly tackled by a third mechanism that monitors $e_{committed}$ and reduces α and β to their halves when the decrease of $e_{committed}$ hits a predefined threshold.

Experiments with the AMM protocol have shown that both the claimed needs can be achieved: limiting optimism in Time Warp *indirectly* by controlling the rate of drain of free memory can be accomplished effectively by a dynamically adaptive mechanism. AMM adapts this rate towards the performance knee-point automatically and adjusts it to follow dynamical movements of that point because of workloads varying (linearly) over time.

35.2.2.8 Algorithms for GVT computation. So far, *global virtual time* has been assumed to be available at any instant of real time T in any LP_i, e.g,. for fossil collection (*S3.11*) or in the simulation stopping criterion (*S3*). The definition of GVT(T) has been given in Section 35.2.2.6. An essential property of GVT(T) not mentioned yet is that it is nondecreasing over (real time) T and therefore can guarantee that Time Warp eventually progresses

---

[*] The *critical path* of a DES is computed in terms of the (*real*) processing time on a certain target architecture respecting *lcc*. Traditionally, *critical path analysis* has been used to study the performance of distributed DES as reference to an "ideal," fastest possible asynchronous distributed execution of the simulation model. Indeed, it has been shown that the length of the critical path is a lower bound on the execution time of *any* conservative protocol, but *some* optimistic protocols do exist (Time Warp with lazy cancellation, Time Warp with lazy rollback, Time Warp with phase decomposition, and the Chandy-Sherman Space-Time Method [26]), which can surpass the critical path. The resulting possibility of so-called *supercritical speedup*, and as a consequence its nonsuitability as an *absolute* lower bound reference, however, has made critical path analysis less attractive.

the simulation by committing intermediate simulation work. Efficient algorithms to compute GVT therefore are another foundational issue to make Time Warp "operable."

The computation of $GVT(T)$ (S3.10) generally is hard, such that in practice only estimates $\widehat{GVT}(T) \geq GVT(T)$ are attempted. Estimates $\widehat{GVT}(T)$, however, (as a necessity to be practically useful) are guaranteed to not overestimate the *actual* $GVT(T)$ and to improve past estimates eventually.

GVT computations employing a central GVT manager. Basically $\widehat{GVT}(T)$ can be computed by a central GVT manager broadcasting a request to all LPs for their current LVT and while collecting those values perform a min-reduction. Clearly, the two main problems are that (1) messages in transit potentially rolling back a reported LVT are not taken into consideration, and (2) all reported $LVT_i(T_i)$ values were drawn at different real times T_i. Problem 1 can be tackled by message acknowledging and FIFO message passing in the CS, problem 2 is generally approached by computing GVT using real time intervals $[T_i^>, T_i^<]$ for every LP_i such that $T_i^> \leq T_i = T^* \leq T_i^<$ for all LP_i. T^*, thus is an instant of real time that happens to lie within every LP's interval.

Samadi's algorithm follows the idea of GVT triggering via a central GVT manager sending out a GVT-*start* message to announce a new GVT computation epoch. After all LPs have prompted the request, the manager computes and broadcasts the new GVT value and completes the GVT epoch. The "message-in-transit" problem is solved by acknowledging *every* message and reporting the minimum over all timestamps of *unacknowledged* messages in one LP's *OQ*, together with the timestamp of *first(EVL)* (as the LP's *local* GVT estimate, $LGVT_i(T_i)$) to the GVT master. An improvement on Samadi's algorithm by *Lin and Lazowska* does not acknowledge every single message. Instead, to every message a sequence number is piggybacked such that LP_i can identify missing messages as gaps in the arriving sequence numbers. Upon receipt of a control message, the protocol sends out to (all) LP_j the smallest sequence number still demanded from LP_j as an implicit acknowledgment of all the previous messages with a smaller sequence number. LP_j receiving smallest sequence numbers from other LPs can determine the messages still in transit and compute a lower bound on their timestamps.

To reduce communication complexity, *Bellenot's algorithm* [5] embeds GLP in a *message routing graph* MRG, which is mainly a composition of two binary trees with arcs interconnecting their leaves. The MRG for a GLP with $N = 10$ LPs, e.g., would be a three level binary tree mirrored along its four node leaf base (a MRG construction procedure for arbitrary N is given in Ref. [5]). The algorithm efficiently utilizes the static MRG topology and operates in three steps:

1. *MRG forward phase.* LP_0 (GVT manager) sends a GVT-start to the (one or) two successor LPs on the MRG. Once an LP_i has received GVT-starts from each successors, it sends a GVT-start in the way as LP_0 did. Every GVT-start in this phase defines $T_i^>$ for the traversed LP_i.

2. *MRG backward phase.* The arrival of GVT-start messages t the last node in MRG (LP_N) defines $T_N^< = T^*$. Now, starting from LP_N, GVT-1vt messages are propagated to LP_0 traversing MRG in the opposite direction; $T_i^<$ is defined for every LP_i. Note that LP_i propagates "back" as an estimate the minimum of LVT_i and the estimates received. When LP_0 receives GVT-1vts from its child LPs in the MRG, it can, with LVT_0, determine the new estimate $\widehat{GVT}(T^*)$ as the minimum overall received estimates and LVT_0.

3. *Broadcast GVT phase.* $\widehat{GVT}(T^*)$ is now propagated along the MRG.

Bellenot's algorithm sends less than $4N$ messages and uses overall $O(log(N))$ time per GVT prediction epoch after an $O(\log(N))$ time for the initial MRG embedding. It requires a FIFO, fault-free CS.

The *passive response* GVT (pGVT) algorithm [15] copes with faulty communication channels while relaxing (1) the FIFO requirement to CS and (2) the "centralized invocation" of the GVT computation. The latter is important since, if GVT advancement is made only upon the invocation by the GVT manager, GVT cycles due to message propagation delays can become unnecessarily long in real time. Moreover, frequent invocations can make GVT computations a severe performance bottleneck due to overwhelming communication load, whereas (argued in terms of simulated time) infrequent invocations causing lags in event commitment bears the danger of memory exhaustion due to delaying fossil collection overly long. An LP-initiated GVT estimation is proposed, that leaves it to individual LPs to determine when to report new GVT information to the GVT manager. Every LP in one GVT epoch holds the GVT estimate from the previous epoch as broadcast by the GVT master. Besides this, it locally maintains a GVT progress history, that allows each LP to individually determine *when* a new *local* GVT estimate (LGVT) should be reported to the manager. The algorithms executed by the GVT manager and the respective LP_is are described as follows:

GVT manager	LP_i, independently of all LP_l, $i \neq l$
1. Upon receipt of *LGVT*, determine new estimate $\widehat{GVT}'$. If $\widehat{GVT}' > \widehat{GVT}$, then	1. Recalculate the local GVT estimate: $LGVT = \min(LVT_i, ts(m_j) \in OQ_j)$ where $ts(m_j)$ is an *unacknowledged* output message, and
2. Recompute the k-sample average GVT increment as $$\overline{\Delta_{GVT}} = \frac{1}{k} \sum_{j=n-k}^{n} \Delta_{GVT_j}$$ where Δ_{GVT_j} is the jth GVT increment out of a history of k observations, and	2. Estimate K, the number of Δ_{GVT} cycles the reporting should be delayed, as the *real* time t_{s+ack} necessary to send a message to and have acknowledged from the manager, divided by the k-sample average *real* time in between two consecutive tuple arrivals from the manager as $$K = \frac{t_{s+ack}}{\frac{1}{k} \sum_{j=n-k}^{n} \Delta_{RT_j}}$$ and
3. Broadcast the tuple $\langle \widehat{GVT}', \overline{\Delta_{GVT}} \rangle$ to all LP_i.	3. Send the new *LGVT* information to the *GVT* manager whenever $\widehat{GVT} + K\overline{\Delta_{GVT}}$ exceeds the local *GVT* estimate $LGVT_i$.

It is clearly seen that a linear predictor of the GVT increment per unit of real time is used to trigger the reporting to the manager. The receipt of a straggler in LP_i with $ts(m) \leq LGVT$ naturally requires immediate reporting to the manager, even before the straggler is acknowledged itself. A key performance improvement of pGVT is that LPs simulating along the critical path will more frequently report GVT information than others (which do not have as great of a chance to improve $\widehat{GVT}$), i.e., communication resources are consumed for the targeted purpose rather than wasted for weak contributions to GVT progression.

Distributed GVT computation. A distributed GVT estimation procedure does not rely on the availability of common memory shared among LPs, and a centralized GVT manager

is not required. Although *distributed snapshot* algorithms [8] find a straightforward application, solutions more efficient than message acknowledgment (e.g., the delaying of sending event messages while awaiting control messages or piggybacking control information
onto event messages) are desired. Mattern [33] uses a "parallel" distributed snapshot
algorithm to approximate GVT, not related to any specific control topology such as a ring
or the MRG topology. Moreover, it does not rely on FIFO channels. To describe the basics
of Mattern's algorithm, distinguish external events $ee_i \in$ EE as either being *send events* se_i
$\in SE$ or *receive events* $re_i \in RE$. The set of events E in the distributed simulation is thus
the union of the set of internal events IE and the set of external events $EE = SE \cup RE$.
Both internal ($ie_i \in IE$) and external events ($ee_i \in EE$) can potentially change the state of
the CI in some LP (IQ, OQ, SS, etc.), but only events ee_i can change the state of CS, i.e.,
the number of messages in transit. Furthermore, let "→" be Lamport's *happens before*
relation [28] defining a partial ordering of $e \in E$ as follows:

1. If $e,e' \in E$ and e' is the next even if after e, then $e \rightarrow e'$,
2. If $e \in SE$ and $e' \in RE$ is the corresponding receive event, the $e \rightarrow e'$
3. If $e \rightarrow e'$ and $e' \rightarrow e''$ then $e \rightarrow e''$

A *consistent cut* is now defined as $C \subseteq E$ such that

$$(e' \in C) \wedge (e \rightarrow e') \Rightarrow (e \in C)$$

A consistent cut thus separates event occurrences in LPs to belong either to the simulations *past* or its *future*. Figure 35.11 illustrates a consistent cut C, whereas C' is inconsistent because of $e' \in C'$, $e \rightarrow e'$ but $e \notin C'$ (cut events are pseudoevents representing the
instants where a cut crosses the time line of an LP and have no correspondence in the simulation). A cut C' is *later* than a cut C if $C \subseteq C'$, i.e., the cut line of C' can be drawn right
to the one for C. The global state of a cut can now be seen as the local state of every LP_i,
i.e., all the event occurrences recorded in IQ, OQ, and SS up until the cut line, and the
state of the channels $ch_{i,j},(i \neq j)$ for which there exist messages in transit from the *past* of
LP_i into the future of LP_j at the time instant of the corresponding cut event (note that a
consistent cut can always be drawn as a vertical (straight) line after rearranging the events
without changing their relative positions).

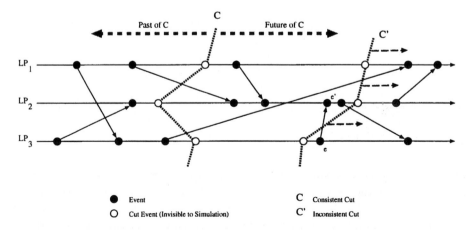

Figure 35.11 Mattern's GVT approximation using two contiguous cuts, C and C'

Mattern's GVT approximation is based on the computation of *two* cuts C and C', C' being later than C. For the computation of a single cut C' the following snapshot algorithm is proposed:

1. Every LP is colored *white,* initially, and one LP_{init} initiates the snapshot algorithm by broadcasting *red* control messages to all other LP_j ($i \neq j$). LP_{init} immediately turns to *red.* For all further steps, *white (red)* LPs can only send *white (red)* messages, and a *white (red)* message received by a *white (red)* LP does not change the LP's color.
2. Once a *white* LP_i receives a *red* message it takes a local snapshot $\Sigma_i(C')$ representing its state right *before* the receipt of that message, and turns to *red.*
3. Whenever a *red* LP_i receives a *white* message, it sends a copy of it, together with its local snapshot $\Sigma_i(C')$ (containing $LVT_i(C')$) to LP_{init} . (*White* messages received by a *red* LP are exactly the ones considered as "in transit.")
4. After LP_{init} has received all $\Sigma_i(C')$ (including the respective LVT_i's) and the last copy of all "in transit" messages, it can determine C' (i.e., the union of all $\Sigma_i(C')$). (Determinations of when the last copy of "in transit" messages has been received itself requires the use of a *distributed termination algorithm.*)

Note that the notion of a local snapshot $\Sigma_i(C')$ here is related to the cut C', as opposed to its relation to real time in Section 35.2.2.6. All $\Sigma_i(C')$s are drawn at different real times by the LPs, but are all related to the same cut. We can therefore also not follow the idea of constructing a global snapshot as $\Sigma(T) = \cup_{i=1}^{N} \Sigma_i(T) \cup CS(T)$ by combining all $\Sigma_i(T)$ and identifying $CS(T)$, which would then trivially let us compute $GVT(T)$. Nevertheless, Mattern's algorithm can be seen as an analogy: all local snapshots $\Sigma_i(C')$ are related to C', and the motivation is to determine a global snapshot $\Sigma(C')$ related to C'; however, the state of the communication system $CS(C')$ related to C' is not known. Some additional reasoning about the messages "in transit" at cut C' is necessary. The algorithm avoids an explicit computation of $CS(C')$, by assuming the availability of a previous cut C (C' is later than C) that isolates an epoch (of virtual time) between C and C' that guarantees certain conditions on the state of $CS(C')$.

Algorithmically, this means, that for the computation of a new GVT estimate *along* a "future" cut C' given the current cut C, C'. Determining the minimum of all local LVT_is from the $\Sigma_i(C')$s is trivial. To determine the minimum timestamp of all the message "in transit" copies at C' (i.e., messages crossing C' in forward direction; messages crossing C' in backward direction can simply be ignored since they do not harm GVT computation), C' is moved forward as far to the right of C as is necessary to guarantee that no message crossing C' originates before C; i.e., no message crosses C and C' (illustrated by dashed arrows in Fig. 35.11). A lower bound on the timestamp of all messages crossing C' can now be easily derived by the minimum of timestamps of all messages sent in between C and C'. Obviously, the closer C and C', the better the derived bound and the better the resulting GVT approximation. The "parallel" snapshot and GVT computation based on the ideas above (coloring messages and LPs, and establishing a GVT estimate based on the distributed computation of two snapshots) is sketched in Ref. [33].

35.2.2.9 Limiting the optimism to time buckets. Quite similar to the optimistic time windows approach, the *breathing time bucket* (BTB) protocol addresses the antimessage dilemma that exhibits instabilities in the performance of Time Warp. BTB is an optimistic windowing mechanism with a pessimistic message sendout policy to avoid the necessity of any antimessage by restricting potential rollback to affect only local history records (as in SRADS [14]). BTB basically processes events in time buckets of different size as determined by the event horizon (Fig. 35.12). Each bucket contains the maximum amount of causally independent events that can be executed concurrently. The local event horizon is the minimum timestamp of any new scheduled event as the consequence of the execution of events in the current bucket in some LP. The (global) event horizon EH then is the minimum over all local event horizons and defines the lower time edge of the next event bucket. Events are executed optimistically, but messages are sent out in a "risk-free" way, i.e., only if they conform to EH. Two methods have been proposed to determine when the last event in one bucket has been processed, and distribution/collection of event messages generated within that bucket can be started, but both lacking an efficient (pure) software implementation:

1. (Multiple) *asynchronous broadcast* can be employed to exchange local estimates of EH to locally determine the global EH. This operation can overlap the bucket computation, during which the CS is guaranteed to be free of event message traffic.
2. A system-wide *non-blocking sync* operation can be released by every LP as soon as it surpasses the *local* EH estimate, not hindering the LP to continue optimistically progressing computations. Once the last LP has issued the non-blocking sync, all the other LPs are interrupted and requested to send their event messages. Clearly, BTB can work efficiently only if a *sufficient* number of events are processed on average in one bucket.

The *breathing time warp* (BTW) [41] combines features of Time Warp with BTB aiming to eliminate shortcomings of the two protocols. The underlying idea again is the belief that the likelihood of an optimistically processed event being subject to a future correction decreases with the distance of its timestamp to GVT. The consequence for the protocol design is thus to release event messages with timestamps close to GVT, but delay the sendout of messages "distant" from GVT. The BTB protocol operates in two *modes*. Every bucket cycle starts in the *Time Warp mode*, sending up to M output messages aggressively with the hope that none of them will eventually be rolled back. M is the number of consecutive messages with timestamps right after GVT. If the LP has the chance to optimistically produce more than M output messages in the current bucket cycle, then BTW switches to the

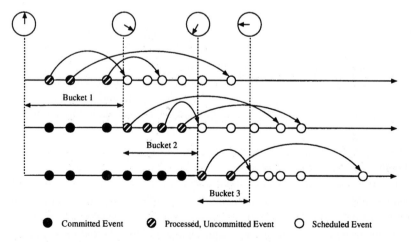

● Committed Event   ⊘ Processed, Uncommitted Event   ○ Scheduled Event

Figure 35.12 Event horizons in the breathing time buckets protocol

BTB mode, i.e. event processing continues according to BTB, but message sendout is suppressed. Should the EH be crossed in the BTB mode, then a GVT computation is triggered, followed by the invocation of fossil collection. If GVT can be improved, M is adjusted accordingly.

Depending on M (and the simulation model), BTW will perform somewhere between Time Warp and BTB; for simulation models with very small EH, BTW will mostly remain in Time Warp mode. Frequent GVT improvements will frequently adjust M and rarely allow it to be exceeded. This is indeed the desired behavior, since BTB would be overwhelmed by an increased synchronization and communication frequency in such a scenario. On the other hand, for large EH where Time Warp would overoptimistically progress due to the availability of many events in one bucket, BTW will adapt to a behavior similar to BTB, since M will often be exceeded and BTB will frequently be induced, thus "limiting" optimism.

35.2.2.10 Probabilistic optimism. A communication interface CI that considers the CMB protocol and Time Warp as two extremes in a spectrum of possibilities to synchronize LP's LVT advancements is the *probabilistic* distributed DES protocol [18a]. Let the event e with a scheduled occurrence time $t(e)$ in some LP_i be *causal* ($\rightarrow$) for a future event e' with $t(e')$ $= t(e) + \delta$, in LP_j; i.e., event e with some probability changes the state variables read by e'. Then, the CMB "block until safe-to-process" rule appears adequate for $P[e \rightarrow e'] = 1$, and is *overly pessimistic* for cases $P[e \rightarrow e'] < 1$, since awaiting the verification whether ($e \rightarrow e'$) or ($e \nrightarrow e'$) hinders the (probably correct) concurrent execution of events e and e' in different LPs. Clearly, if $P[e \rightarrow e'] \ll 1$, an optimistic strategy could have executed e' concurrently to e most of the time (in repeated simulations). This argument mainly motivates the use of an optimistic CI for simulation models with nondeterministic event causalities. On the other hand, as already seen with the discussion of rollback chains, an optimistic LP might tend towards *overoptimism*, i.e. optimism lacking rational justification. The *probabilistic* protocol aims to exploit information on the dynamic simulation behavior to be able to allow in every simulation step just that amount of optimism that can be justified.

As a more general argument, assume the timestamp of the *next* message m carrying the external event ee_{i+1} to arrive at LP_j to be in the time interval $[s, t]$ (t can be arbitrarily large), and that LVT_j has advanced to a certain value (see Fig. 35.13). External events have been received in the past and respected chronologically, some of the internal events are already scheduled for LP_j's future. Let the probability of the actual timestamp $t(ee_{i+1})$ of the forthcoming message be $P[t^* = x] = (1/(t - s)) \ \forall x \in [s, t]$, i.e. the next external event occurrence is equally likely for all points in $[s, t]$. Then the CMB protocol could have safely simulated all the internal events $t(ie_k) < t(ee_{i+1})$, but will block at LVT $= s$. Under the assumptions made above, however, by blocking, CMB fails to use a $(1 - \alpha)$ chance to produce useful simulation work by progressing over s and executing internal events in the time interval $[s, s + \alpha(t - s)]$. Again, an overpessimism is identified for CMB, at least for small values of α. Time Warp, on the other hand, would proceed simulating internal events even after surpassing t. This is overly optimistic in the sense that every internal event with $t(ie_k) > t$ will definitely (with probability of 1) have to be invalidated.

The *probabilistic* protocol appears to be a performance-efficient compromise between the two classical approaches. As opposed to CMB, it allows trespassing of s and progressing simulation up until $\hat{t} = t(ee_i) + \hat{\Delta}(\delta_1, \delta_2, ...\delta_n)$, $s \leq \hat{t} \leq t$, where $\hat{t}$ is an estimate based on the differences $\delta_k = t(ee_{i-n+k}) - t(ee_{i-n+k-1})$ of the observed arrival instants $A = (t(ee_{i-n+1}), t(ee_{i-n+2}), t(ee_i))$, n being the size of the (history) obser-

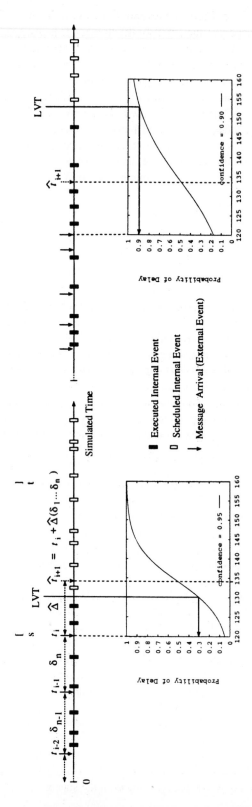

Figure 35.13 Probabilistic direct optimism control

vation window. Compared to Time Warp, it prevents from overoptimistically diffusing incorrect simulations too far ahead into the simulated future, and thus avoids unnecessary communication overhead due to rollbacks and consequential rollback chains.

The message arrival patterns observed are used to adapt the LP to a synchronization behavior that is the best trade-off between blocking and optimistically progressing with respect to the parallelism inherent to the simulation model, by conditioning the execution of *S3.2* through *S3.11* in the algorithm of Fig. 35.7 to a probabilistic "throttle." Assume that the arrival instants $A_i = (t(ee_{k-n+1}), t(ee_{k-n+2}), \dots t(ee_k))$ are collected in LP_j for every (input) channel $ch_{i,j}$, and that for $ch_{i,j}$ $t_{i \to j}$ is estimated with some confidence $0 \le \zeta(\hat{t}_{i \to j}) \le 1$. The probabilistic protocol is then obtained by replacing *S3.2–S3.11* in Fig. 35.7 with:

$$S3.2' = \begin{cases} \text{execute } S3.2\text{–}S3.11 & \text{with probability } 1 - \dfrac{1}{1 + e^{-(\gamma \zeta(\hat{t})\,(LVT - \hat{t}))}} \\[4mm] \text{skip } S3.2\text{–}S3.11, \text{delay } LP_j \text{ for } \bar{s} & \text{with probability } \dfrac{1}{1 + e^{-\gamma \zeta \hat{t}\,(LVT - \hat{t})}} \end{cases}$$

where $\hat{t} = \min_i(\hat{t}_{i \to j})$ and $\zeta(\hat{t})$ is the respective confidence level, $\bar{s}$ is the average step execution time (in real time), and LVT is the LP's current instant of simulated time. The confidence parameter $\zeta(\hat{t})$, with a scaling factor γ, describes a family of probability distribution functions for delaying the SE: should there be an estimate $\hat{t}$ provided by some statistical method characterizing the arrival process only at a low level of confidence ($\zeta(\hat{t})$ small), then the delay probability directly reflects the vagueness of information justifying optimism. The more confidence (evidence) there is in the forecast, the steeper the ascent of the delay probability as LVT progresses toward $\hat{t}$. (Steepness of the sigmoid function in Fig. 35.13 (left) with $\zeta(\hat{t}) = 0.95$ is higher than in Fig. 35.13 (right) $\zeta(\hat{t}) = 0.90$.) After simulation in LP_j has surpassed the estimate $\hat{t}$, delays become more and more probable, expressing the increasing rollback hazard LP_j encounters.

The choice of the size of the observation window n, as well as the selection of the forecast procedure is critical for the performance of the probabilistic protocol for two reasons: (1) the achievable prediction *accuracy* and (2) the computational and space complexity of the forecast method. Generally, the larger n is chosen, the more information on the arrival history is available in the statistical sense. Respecting much of the arrival history will at least theoretically give a higher prediction precision, but will in turn consume more memory space. Intuitively, complex forecast methods could give "better" predictions than trivial ones but are liable to intrude on the distributed simulation protocol with an unacceptable amount of computational resource consumption. Therefore, incremental forecast methods of low memory complexity are recommended, i.e., procedures where $\hat{t}_{i+2}$ based on $\hat{\Delta}(\delta_2, \delta_3, \dots \delta_{n+1})$ can be computed from $\hat{t}_{i+1}$ based on $\hat{\Delta}(\delta_1, \delta_2, \dots \delta_n)$ in $O(c)$ instead of $O(cn)$ time. Taking, for example, the observed mean

$$\hat{\Delta}_i = \frac{1}{n} \sum_{j=1}^{n} \delta_j$$

(without imposing any observation window) as the basis for an estimate of $\hat{t}$, then upon the availability of δ_{n+1}, $\hat{t}_{i+2}$ could be computed based on

$$\hat{\Delta}_{n+1} = \frac{n\hat{\Delta}_n + \delta_{n+1}}{n+1}$$

A possibility to weight recent history higher than past history could be an exponential smoothing of the observation vector by a smoothing factor $\alpha(|1 - \alpha| < 1)$:

$$\hat{\Delta}_{n+1} = \sum_{i=1}^{n} \alpha (1-\alpha)^{i-1} \delta_{n+1-i}$$

$\hat{\Delta}$ in this case has the incremental form

$$\hat{\Delta}_{n+1} = \alpha\delta_{n+1} + (1-\alpha)\hat{\Delta}_n$$

The arrival process of messages via a channel could also be considered to originate from an underlying but unknown stochastic process. *Autoregressive moving average* (ARMA) process models are a reasonable method to characterize that process by the relationship among a series of empirical *non-independent* observations $\{X_i\} = (X_1, X_2, \ldots X_n)$ (in our case $= (\delta_1, \delta_2, \ldots \delta_n)$). Assuming $\{X_i\}$ is already a centered series (i.e., transformed with respect to the series mean μ, $X_i = \delta_i - \mu$), then $X_t = \phi_1 X_{t-1} + \phi_2 X_{t-2} + \ldots + \phi_p X_{t-p} + \varepsilon_t$ is a pure autoregressive process of order $p(AR[p])$ with ε_t being a sequence of (independent, identically distributed) white noise random disturbances. X_t is usually called the centered response, and ϕ_i are the parameters that can be estimated from the realizations in various different ways, e.g., maximum likelihood or the Yule-Walker method. On the other hand, a process $X_t = \varepsilon_t \neq \theta_1 \varepsilon_{t-1} + \theta_2 \varepsilon_{t-2} + \ldots + \theta_q \varepsilon_{t-q}$ is a pure moving average process of order $q(MA[q])$ with $E(\varepsilon_i) = 0$, $\text{Var}(\varepsilon_i) = \sigma_\varepsilon^2$, and $E(X_t) = 0$. A (mixed) process ARMA$[p,q]$ is now defined as

$$X_t = \sum_{i=1}^{p} \phi_i X_{t-i} + \varepsilon_t + \sum_{i=1}^{q} \theta_i \varepsilon_{t-i}$$

Stationary ARMA$[p,q]$ processes are able to explain short term trends and even cycles in the observation pattern, thus characterizing the *transient* behavior of message arrivals (this is particularly desirable in cases where the simulation induces *phases* of message arrival patterns) to some extent. Several methods are available for ARMA$[p,q]$ processes to forecast X_{t+1} from $\{X_i\}$. An incremental way, for example, is the Durbin-Levinson method for one-step best linear predictions. Non-stationary series X_{t+1} can be treated with ARIMA$[p,d,q]$ processes, d in this case denoting the differencing order, i.e., the number of differencing transformations required to induce stationarity for the non-stationary ARIMA process (ARIMA$[p,0,q] \sim$ ARMA$[p,q]$). (See Refs. [17, 18a] for an application of ARIMA forecasts in the context of the simulation of Stochastic Petri nets.) Besides the basic statistical methods described above, also (nonlinear) mechanisms developed in control theory, machine learning and neural network simulations as well as Hidden Markov models could be used to control optimism in the probabilistic protocol.

The major strength of the *probabilistic* protocol is that the optimism of Time Warp can be automatically controlled by the model parallelism available as expressed by the likelihood of future messages, and can even adapt to a transient behavior of the simulated model. The asymptotic behavior of the protocol for simulation models with $P[e \rightarrow e'] \cong 1$

(for all pairs (e, e')) is close to CMB, while for models with $P[e \rightarrow e'] \cong 0$, it is arbitrarily close to (a throttled) Time Warp.

35.3   Conservative vs. Optimistic Protocols

The question of the relative qualities of conservative and optimistic protocols has often been raised. General rules of superiority cannot be formulated, since performance—due to a very high degree of interweaving of influencing factors—cannot be sufficiently characterized by models, although exceptions do exist [1]. Even full implementations often prohibit performance comparisons if different implementation strategies were followed or different target platforms were selected. Performance influences on behalf of the platform come from the hardware as such (communication/computation speed ratio), the communication model (FIFO, routing strategy, interconnection network topology, possibilities of broadcast/multicast operations, etc.) and the synchronization model (global control unit, asynchronous distributed memories, shared variables, etc.) making protocols widely not comparable across platforms. Protocol specific optimizations, i.e. optimizations in one protocol that do not have a counterpart in the other scheme (e.g. lazy cancellation) hinder even more a "fair" comparison. We therefore separate arguments in a more or less *rough* way, as shown in Table 35.1.

35.4   Conclusions and Outlook

The achievements attained in accelerating the simulation of the dynamics of complex discrete event systems using parallel or distributed multiprocessing environments have been presented. While parallel discrete event simulation (DES) governs the evolution of the system over simulated time in an iterative SIMD way, distributed DES tries to spatially decompose the event structure underlying the system, and executes event occurrences in spatial subregions by logical processes (LPs) usually assigned to different (physical) processing elements. Synchronization protocols are necessary in this approach to avoid timing inconsistencies and to guarantee the preservation of event causalities across LPs. Sources and levels of parallelism, synchronous vs. asynchronous simulation and principles of LP simulation have been outlined, ending up in the presentation of the "classical" LP simulation protocols and the various optimizations that appeared (comprehensive readings are Refs. [19] and [37]). The primary source for Time Warp is Ref. [27]; for conservative protocols it is Ref. [34]. A conclusive judgement on one of the protocols could not be drawn, since performance is always dependent on the respective simulation model and the technology (hardware, software) used for the implementation and execution.

Indeed, a protocol comparison is of less practical importance than indicated by much of the analysis literature. Issues with more practical relevance [20] are the design of simulation languages and the development of tools to support a simulation model description independently of the sequential, parallel or distributed DES algorithm or protocol to execute it. Further "silver bullets" [20], with the potential to make parallel simulation more accessible, are considered to be the development of easy to handle simulation modules and libraries (that avoid manual programming efforts) and automation of the parallelization process as far as possible, e.g., by the use of parallelizing compilers or parallelizing runtime environments. The automatic extraction of parallelism from sequential simulations is closely related to the problem of partitioning the simulation model into spatial or temporal [23] *regions* suitable for parallel execution. Spatial partitioning can be conducted at least semiautomatically if model specifications are made in a formalism abstract enough to sup-

TABLE 35.1   Separation of Arguments

Strategy	Conservative (CMB)	Optimistic (Time Warp)
Operational principle	*lcc* violation is strictly avoided; only *safe* ("good") events are processed	Lets *lcc* violation occur but recovers when detected (immediately or in the future); processes "good" and "bad" events, eventually commits good ones, cancels bad ones.
Synchronization	Synchronization mechanism is processor *blocking*; therefore prone to deadlock situations (deadlock is a protocol intrinsic, not a resource contention, problem); deadlock prevention protocols based on null messages are subject to severe communication overheads; deadlock detection and recovery protocols rely on a centralized deadlock manager.	Synchronization mechanism is rollback (of simulated time); consequential remote annihilation mechanisms are subject to severe communication overheads; cascades of rollbacks that will eventually terminate can burden execution performance and memory utilization.
Parallelism	Model parallelism cannot be fully exploited; if causalities are probable but seldom, protocol behaves overly pessimistically.	Model parallelism is fully exploitable; if causalities are probable and frequent, then Time Warp can gain most of the time.
Lookahead	Necessary to make CMB operable, essential for performance.	Time Warp does not rely on any model related lookahead information, but lookahead can be used to optimize the protocol.
Balance	CMB performs well as long as all static channels are equally utilized; large dispersion of events in space and time is not bothersome.	Time Warp performs well if average LVT progression is "balanced" among all LPs; space-time dispersion of events can degrade performance.
GVT	Implicitly executes along the GVT bound; no explicit GVT computation required.	Relies on explicit GVT, which is generally hard to compute; centralized GVT manager algorithms are subject to communication bottlenecks if no hardware support; distributed GVT algorithms impose high communication overhead and seem less effective.
States	Conservative memory utilization copes with simulation models having "arbitrarily" large state spaces.	Performs best when state space and storage requirement per state is small.
Memory	Conservative memory consumption (as a consequence of the scheme).	Aggressive memory consumption; state saving overhead; fossil collection requires efficient and frequent GVT computation to be effective; complex memory management schemes necessary to prevent memory exhaustion.
Messages and Communication	Timestamp order arrival of messages and event processing mandatory; strict separation of input channels required; static LP interconnection channel topology.	Messages can arrive out of chronological order, but must be executed in timestamp order; one single input queue; no static communication topology; no need to receive messages in sending order (FIFO), thus can be used on more general hardware platforms.
Implementation	Straightforward to implement; simple control and data structures.	Nontrivial to implement and debug; "tricky" implementations of control flow (interrupts) and memory organization essential; several performance influencing implementation optimizations possible.
Performance	Relies mainly on deadlock management strategy; computational and communication overhead per event is *small* on average; protocol in favor of "fine grain" simulation models; no general performance statement possible.	Relies mainly on optimism control and strategy to manage memory consumption; computational and communication overhead per event is high on average; protocol in favor of "large grain" simulation models; no general performance statement possible.

port a structural analysis [11, 17]. A traditional but still unsatisfied demand within the research community is the development of efficient performance prediction methods for distributed simulation protocols, such that the potential performance gain of a certain protocol or a specific simulation application can be studied before putting efforts into developing parallel code for it. The management and balancing of dynamic distributed simulation workloads [22, 38] is becoming more and more important with the shift from parallel processors and supercomputers to distributed computing environments (powerful workstations interconnected with high-speed networks and switches) as the preferred target architecture. In this context, protocols with the possibility of dynamic LP creation and migration (dynamic rescheduling) will have to be developed. In contrast to the approaches that "virtualize" time as presented in this work, the "virtualization" of space ambitiously studied at this time promises a shift of conventional parallel and distributed simulation paradigms. Further challenges are seen in the hierarchical combination or even the uniformization of protocols [2], the uniformization of continuous and discrete event simulation, the integration of real-time constraints into protocols (distributed *interactive* simulation), and so on.

Parallel and distributed simulation over the one and one-half decades of its existence has turned out to be more foundational than merely exercising on the duality of Lamport's logical clock problem. Today's availability of parallel and distributed computing and communication technology has given relevance to the field that could not have been foreseen in its early days. Indeed, parallel and distributed simulation protocol research has just started to ferment developments in computer science disciplines other than "simulation" in the classical sense. For example, simulated executions of SIMD programs in asynchronous environments can accelerate their execution [39], and parallel simulations executing parallel programs with message passing communication have already been shown to be possible [13]. Other work has shown that an intrusion-free monitoring and trace collection of distributed memory parallel program executions is possible by superimposing the execution with a distributed DES protocol. The difficult problem of debugging parallel programs finds a high likelihood of being tackled by similar ideas.

Remark. A more detailed presentation of the material, with supplementary examples, further references, and a set of tutorial slides (all in electronic format) is available directly from the author at http://www.ani.univie.ac.at/ or via ferscha@ani.univie.ac.at.

35.5 References

1. Akyildiz, I. F., L. Chen, R. Das, R. M. Fujimoto, and R. F. Serfozo. 1993. The effect of memory capacity on time warp performance. *Journal of Parallel and Distributed Computing,* 18(4):411–422.

2. Bagrodia, R., K. M. Chandy, and W. T. Liao. 1991. A unifying framework for distributed simulation. *ACM Transactions on Modeling and Computer Simulation,* 1(4):348–385.

3. Bain, W. L, and D. S. Scott. 1988. An algorithm for time synchronization in distributed discrete event simulation. *Proceedings of the SCS Multiconference on Distributed Simulation,* B. Unger and D. Jefferson, eds. 19 (3), 30–33.

4. Bal, D., and S. Hoyt. 1990. The adaptive time-warp concurrency control algorithm. In Distributed Simulation, *Proceedings of the SCS Multiconference on Distributed Simulation,* D. Nicol, ed. 174–177. Society for Computer Simulation Series, Volume 22, Number 1.

5. Bellenot, S. 1990. Global virtual time algorithms. *Proceedings of the Multiconference on Distributed Simulation,* 122–127.

6. Bryant, R. E. 1984. A switch-level model and simulator for mos digital systems. *IEEE Transactions on Computers,* C-33(2):160–177.

7. Cai, W., and St. J. Turner. 1990. An algorithm for distributed discrete-event simulation—the "carrier null message" approach. *Proceedings of the SCS Multiconference on Distributed Simulation,* Vol. 22 (1), 3–8.

8. Chandy, K. M., and J. Lamport. 1985. Distributed snapshots: determining global states of distributed systems. *ACM Transactions on Computer Systems,* 3(1):63–75.

9. Chandy, K. M., and J. Misra. 1979. Distributed simulation: a case study in design and verification of distributed programs. *IEEE Transactions on Software Engineering,* SE-5(5):440–452.

10. Chandy, K. M., and J. Misra. Asynchronous distributed simulation via a sequence of parallel computations. *Communications of the ACM,* 24(11):198–206.

11. Chiola, G., and A. Ferscha. 1993. Distributed simulation of Petri nets. *IEEE Parallel and Distributed Technology,* 1(3):33–50.

12. Das, S. R., and R. M. Fujimoto. 1994. An adaptive memory management protocol for time warp parallel simulation. *Proc. of the 1994 ACM Sigmetrics Conference on Measurement and Modeling of Computer Systems,* Nashville, 201–210.

13. Dickens, Ph. M., Ph. Eeidelberger, and D. M. Nicol. 1994. Paralleliz*ed Direct Execution Simulation of Message Passing Parallel Programs.* Technical report, ICASE. Hampton, Va.: NASA Langley Research Center.

14. Dickens, Ph. M., and P F. Reynolds. 1990. SRADS with local rollback. *Proceedings of the SCS Multiconference on Distributed Simulation,* Vol. 22 (1), 161–164.

15. D'Souza, L M., X. Fan, and P. A Wilsey. 1994. pGVT: an algorithm for accurate GVT estimation. *Proceedings of the 8th Workshop on Parallel and Distributed Simulation* (PADS '94), D. K. Arvind, R. Bagrodia, and J. Y.-B. Lin, eds., 102–109.

16. Felderman, R. E., and L. Kleinrock. 1990. An *Upper Bound on the Improvement of Asynchronous versus Synchronous Distributed Proce*ssing. Proc. of the SCS Multiconf. on Dist. Sim., D. Nicol, ed., Vol. 22, pages 131–136.

17. Ferscha, A. 1994. Concurrent execution of timed Petri nets. Proceedings of the 1994 Winter Simulation Conference, J. D. Tew, S. Manivannan, D. A. Sadowski, and A F. Seila, eds., 229–236.

18. Ferscha, A., and G. Chiola. Self-adaptive logical processes: the probabilistic distributed simulation protocol. *Proc. of the 27th Annual Simulation Symposium.*

18a. Ferscha, A. 1995. Probabilistic adaptive direct optimism control in Time Warp. *Proceedings of the 9th Workshop on Parallel and Distributed Simulation* (PADS '95), 120–129.

19. Fujimoto, R. M. 1990. Parallel discrete event simulation. *Communications of the ACM,* 33(10):30–53.

20. Fujimoto, R. M. 1993. Parallel discrete event simulation: will the field survive? *ORSA Journal of Computing,* 5(3):218–230.

21. Gafni, A. 1988. Rollback mechanisms for optimistic distributed simulation systems. *Proceedings of the SCS Multiconference on Distributed Simulation,* B. Unger and D. Jefferson, eds., 61–67.

22. Glazer, D. W., and C. Tropper. 1993. On process migration and load balancing in time warp. *IEEE Transactions on Parallel and Distributed Systems,* 4(3):318–327.

23. Greenberg, A. G., B. D. Lubachevsky, and I. Mitrani. 1991. Algorithms for unboundedly parallel simulations. *ACM Transactions on Computer Systems,* 9(3):201–221.

24. Groselj, B., and C. Tropper. 1988. The time-of-next-event algorithm. Proceedings of the SCS Multiconference on Distributed Simulation, B. Unger and D. Jefferson, eds., 19 (3), 25–29.

25. Jefferson, D. 1990. Virtual time II: the cancelback protocol for storage management in Time Warp. *Proc. of the 9th Annual ACM Symposium on Principles of Distributed Computing,* 75–90.

26. Jefferson, D., and P Reiher. 1991. Supercritical speedup. Proceedings of the 24th Annual Simulation Symposium, A. H. Rutan, ed. Los Alamitos, Calif.: IEEE Computer Society Press, 159–168.

27. Jefferson, D. A. 1985. Virtual time. *ACM Transactions on Programming Languages and Systems,* 7(3):404–425.

28. Lamport, L. 1978. Time, clocks, and the ordering of events in distributed systems. *Communications of the ACM,* 21(7):558–565.

29. Lin, Y.-B., B. Preiss, W. Loucks, and E. Lazowska. 1993. Selecting the checkpoint interval in time warp simulation. Proc. of the 7th Workshop on Parallel and Distributed Simulation, R. Bagrodia and D. Jefferson, eds. , Los Alamitos, Calif.: IEEE Computer Society Press, 3-10.

30. Lin, Y.-B., and B. R. Preiss. 1991. Optimal memory management for time warp parallel simulation. *ACM Transactions on Modeling and Computer Simulation,* 1(4):283–307.

31. Lubachevsky, B. D. 1988. Bounded lag distributed discrete event simulation. Proceedings of the SCS Multiconference on Distributed Simulation, B. Unger and D. Jefferson, eds., 19 (3), 183–191.

32. Lubachevsky, B. D., A. Weiss, and A. Shwartz. 1991. An analysis of rollback-based simulation. *ACM Transactions on Modeling and Computer Simulation,* 1(2):154–193.

33. Mattern, F. 1993. Efficient algorithms for distributed snapshots and global virtual time approximation. *Journal of Parallel and Distributed Computing,* 18(4):423–434.

34. Misra, J. 1986. Distributed discrete-event simulation. *ACM Computing Surveys,* 18(1):39–65.

35. Nicol, D. M. 1988. Parallel discrete-event simulation of fcfs stochastic queueing networks. *Proceedings of the ACM/SIGPLAN PPEALS 1988,* 124–137.

36. Nicol, D. M. 1991. Performance bounds on parallel self-initiating discrete event simulations. *ACM Transactions on Modeling and Computer Simulation,* 1(1):24–50.

37. Nicol, D. M., and R M. Fujimoto. 1994. Parallel simulation today. *Operations Research,* 1994.

38. Shanker, M. S., W. D. Kelton, and R. Padman. 1993. Measuring congestion for dynamic task allocation in distributed simulation. *ORSA Journal of Computing,* 5(1):54–68.

39. Shen, Sh., and L. Kleinrock. 1992. The virtual-time data-parallel machine. *Proc. of the 4th Symposium on the Frontiers of Massively Parallel Computation.* Los Alamitos, Calif.: IEEE Computer Society Press, 46–53.

40. Sokol, L. M., D. P. Briscoe, and A. P. Wieland. 1988. MTW: a strategy for scheduling discrete simulation events for concurrent execution. *Proc. of the SCS Multiconf. on Distributed Simulation,* 34–42.

41. J. S. Steinmann. Breathing Time Warp. *Proc. of the 7th Workshop on Parallel and Distributed Simulation,* R. Bagrodia and D. Jefferson, eds. Los Alamitos, Calif.: IEEE Computer Society Press, 109–118.

36

Parallelism for Image Understanding

Viktor K. Prasanna and Cho-Li Wang

Computer vision employs a broad spectrum of techniques from several areas such as image and signal processing, advanced mathematics, graph theory, relational algebra, and artificial intelligence. Even though vision has been classified as a Grand Challenge problem, the characteristics of the computations are significantly different from those of Grand Challenge problems in scientific and numerical computations.

Vision problems are generally classified into three levels: low-level, intermediate-level, and high-level. The low-level processing is mostly iconic, and the data communication is local to each pixel, but the higher levels use symbolic computations. The operations performed on each data item can be nonlocal, and the communication is also irregular compared with that used in low-level vision processing. This combination of computational needs provides a significant challenge to algorithm design as well as to implementing the algorithms on available parallel machines to obtain large speed-ups.

This chapter summarizes our work in using CM-5 for vision. The Connection Machine CM-5, a *synchronized* MIMD machine [15], is operated in SPMD (single-program, multiple data) mode, halfway between the highly synchronized SIMD model and the message-passing MIMD model. This provides the desired capabilities to solve vision tasks at different levels. We define a realistic abstract model of CM-5 in which explicit cost is associated with data routing and cooperative operations. We model the CM-5 as a set of SISD nodes connected by a high-bandwidth, low-latency data network and a control network. We assume that in a unit of time, a processing node (PN) can perform an arithmetic/logic operation on local data. We consider *start-up* times T_d and T_c as constant costs associated with every communication step performed using the data and control networks respectively. We assume τ_d and τ_c as the transmission rates (seconds per unit of data) for data communication using the data and control networks respectively. Combining (max, sum, prefix sum, etc.) a unit of data from each PN can be performed in τ_g time using the control network. Using this model, we develop scalable parallel algorithms for representative problems in vision computations at all three levels.

First, we study low-level vision processing on CM-5. Even though, low-level processing has been well understood and the real challenge lies in devising approaches to higher level processes, indeed, low-level operations can be very time consuming. For example, the well known Nevatia-Babu line finder implemented in LISP can take several hours to extract features in a 2048 × 2048 image on a state-of-the-art Sun Sparc station. Without a fast low-level system that provides the desired inputs to higher-level analyses, realizing a complete vision system remains a distant goal.

In this work, we discuss parallel implementation of a typical low-level vision task proposed in Ref. [21]. This is used to extract linear features from input images. The process of extracting linear features consists of two major procedures: *contour detection* and *linear approximation*. Contour detection involves detection of edges using convolution, removal of the "false" edges using a thinning operation, and removal of the "weak" edges using a thresholding operation. Upon detection of the edge pixels, a linking step and a linear approximation step are performed to group the detected edge pixels to form contours and approximate them by line segments.

Contour detection and linear approximation procedures exhibit different types of computational characteristics. In contour detection, the operations can be performed in a synchronized fashion with data communication between the pixels that is regular and local to each pixel. This type of operation maps well onto SIMD machines. However, in linear approximation, the communication can extend over large windows and is often irregular. This maps well onto MIMD machines.

Given an $n \times n$ image and a constant number of masks of size $m \times m$, the contour detection and the linear approximation procedures can be performed in $O(m^2n^2)$ time on a serial machine. This time complexity assumes a linear time heuristic for linear approximation. Based on our model of CM-5, we show that, using a partition of CM-5 having P PNs, contour detection can be performed in $O(n^2m^2/P)$ computation time and in

$$12T_d + \left(12 \left\lfloor \frac{m}{2} \right\rfloor \left\lfloor \frac{n}{\sqrt{P}} \right\rfloor + 24 \left\lfloor \frac{m}{2} \right\rfloor \right) \tau_d$$

communication time. Linear approximation can be performed in $O(n^2/P)$ computation time and in $(20P + 3 \log P)T_d + (24n + 80n^2/P)\tau_d$ communication time, in which $n \geq 0.5P^{3/2}$. This assumes that an $n \times n$ image has contours of maximum length $O(n)$.

We have performed implementations on CM-5 using C and CMMD message passing primitives. Our implementations show that, given a 2048 × 2048 grey level image as input, the extraction of linear features, which includes edge detection, thinning, linking, and linear approximation, can be performed in less than 1.2 seconds on a partition of CM-5 having 512 PNs. The serial execution time of our implementation in C (using an optimizing compiler) on a state-of-the-art Sun Sparc station takes more than 8 minutes on a 2048 × 2048 image.

In the second part of our work, we discuss an intermediate-level vision task, *perceptual grouping*. Perceptual grouping [18] can be defined as the process of imposing structural organization onto sensory data. The grouping process can be classified as an intermediate-level vision task directed toward closing the gap between what is produced by low-level processing (such as edge detectors) and what is desired as input (perfect contours, no noise, no fragmentation, etc.) to high-level analysis. Several computational frameworks have been proposed to realize perceptual grouping and have proven to be effective for extracting straight lines, curves, human-made structures, etc. We consider a typical application perceptual grouping described in Ref. [10] and design scalable parallel algorithms and implementations.

In Ref. [10], the grouping steps are used to detect and provide 3-D descriptions of buildings from monocular views of aerial scenes. The grouping process is composed of five group-grouping steps: *line grouping, junction grouping, parallel grouping, U-contour grouping, and rectangle grouping*. A feature hierarchy is established by grouping line segments to *lines* (a linear structure at a higher granularity level) *lines* to junctions, *lines* to parallels, parallels and *lines* to *U-contours* and *U-contours* to rectangles. The feature hierarchy including parallel relationships and portions of rectangles leads to the formation of building hypotheses.

Let *token* denote any of a line segment, a *line*, a pair of parallel lines, a *U-contour*, or a rectangle. The sequential algorithm [10] employs a search-based approach. The token data is first stored in the image plane before using it in line, junction, U-contour grouping steps. It is also stored as a sorted list in the parallel grouping step. Each token performs a search within a *window* in the image (or in the sorted list) to group the tokens in the window satisfying certain geometric constraints. Let S denote the set of input tokens to a grouping step and $W(S)$ denote the total area of the search windows generated by the grouping step. A grouping step can be performed in $O(W(S))$ time on a serial machine. We regard $W(S)$ as the total workload associated with the input tokens S.

The perceptual grouping process in Ref. [10] has been shown to produce good results in the building detection system and shows the promise of being extended to much more complex cases. While the users find the accuracy of the system useful, the computational speeds on current workstations such as a Sparc System 10 are simply not acceptable. For grouping 8k line segments (detected from an 1k × 1k image) to form building hypotheses, the serial implementation written in LISP takes nearly 10 minutes. To make these systems really usable, we need to bring the computation times down to a few seconds, and for large problem size to few tens of minutes at most.

Perceptual grouping is parallel and distributed in nature because the grouping operations can be performed independently at each token. However, while parallelizing the search-based grouping algorithm discussed above, the data communication patterns are generally global and irregular as a token may perform a search crossing over several subimage blocks and the distribution of tokens depends on the input image. In addition, it is possible that some tokens will produce a large number of search operations within a processor, which results in an unbalanced workload. These situations in the worst case can lead to severe performance deterioration.

Our parallel algorithms employ load balancing and data allocation strategies to distribute the computation evenly among the processors and reduce the communication cost. We show that, using a partition of CM-5 having P PNs, if $W(S) \geq P|S|$, a perceptual grouping step can be performed in $O(W(S)/P)$ computation time and $|S|r_c + PT_c$ communication time.

Our serial implementation, written in C on a Sun Sparc 400, shows that, using 8k line segments (detected from a 1k × 1k image) as input to perform the grouping process, our optimized serial code written in C takes nearly 14.2 seconds for line grouping and nearly 8.6 seconds for junction grouping on a Sun Sparc 400; our implementation on the CM-5 takes less than 1.5 seconds to complete the two steps using a partition of CM-5 having 32 PNs.

In the third part of our work, we consider a high-level vision task. Object recognition is a key step in an integrated vision system. Most model-based recognition systems work by hypothesizing matches between scene features and model features, predicting new matches, and verifying or changing the hypotheses through a search process [9]. Recently, geometric hashing [14] has been proposed as an alternate approach for object recognition.

In geometric hashing, a set of models is specified using their features points. For each model, all possible pairs of feature points are designated as a *basis set*. The coordinates of the features points of each model are computed relative to each of its basis. These coordinates are then used to hash into a hash table. The entries in the hash table are composed of *(model, basis)* pairs and are precomputed as follows: using a chosen basis, if a feature point in a model hashes into a bin, then the model and the basis are recorded in the bin. In the recognition phase, an arbitrary pair of feature points in the scene is chosen as a basis, and the coordinates of the feature points in the scene are computed relative to this basis. The new coordinates are used to hash into the hash table. Votes are accumulated for the *(model, basis)* pairs stored in the hashed locations. The pair winning the maximum number of votes is chosen as a candidate for matching. The execution of the recognition phase corresponding to a basis pair is termed a *probe*.

There have been two prior efforts in parallelizing the geometric hashing algorithm [3, 26]. Both implementations have been performed on SIMD hypercube-based machines. A major problem in both the implementations is the requirement of large number of processors. In Ref. [26], the number of processors used is same as the number of bins in the hash table. Thus, $O(Mn^3)$ processors are needed, where M is the number of models in the database and n is the number of feature points in each model.

We develop scalable parallel algorithms for a probe in the recognition phase. We employ a number of processors that depends on the size of the input, not on the size of the hash table. We assume each processor has a copy of the hash table. Given a scene S, we show that a probe of the recognition phase can be performed in $O(|V(S)|/P)$ time and $(40P + 1)T_d + (61|V(S)|)\tau_d/P + (P + 5)T_c + (P + 1)\tau_c + 4\tau_g$ time in network related operations on the model of CM-5 having P processors, where $V(S)$ denotes the votes cast by S.

An earlier implementation [26] results in 1.52 seconds for a probe of the recognition phase on an 8K processor CM-2 on an input scene consisting of 200 feature points. In this experiment, the model database has 1024 models, and each model is represented using 16 feature points. This performance is obtained under the following assumptions: the number of the processors $\approx$ number of hash bins, a distribution of hash table entries and the resulting votes such that no congestion occurs at a PN during hash bin access or during voting. We show that using the same hash function and the same number of hash bins as in Ref. [26], and using one copy of the hash table distributed over the entire processor array, a probe on a scene consisting of 256 feature points can be performed in less than 200 milliseconds on a 32-processor CM-5. In the implementation in Ref. [26], certain inputs result in hash bin access congestion wherein many processors access the data in a single PN. Such congestion does not occur in our implementation. In addition, our algorithm evenly distributes the generated votes. The performance of our algorithm depends on the total number of votes only. As in Ref. [26], synthesized model data and synthesized scenes were used.

Our implementations exploit the computation and communication characteristics of CM-5 architecture to lead to large speed-ups for the problem sizes typically used by the vision community. Our implementations have been developed in a modular fashion to permit various techniques to be employed for the individual steps of the processing. The implementations can be easily adapted to run on many state-of-the-art parallel machines such as Intel Paragon and IBM SP-2.

The organization of this chapter is as follows. Steps of the vision tasks considered here are outlined in Section 36.1. Section 36.2 discusses an abstract model of CM-5. Section 36.3 describes the key ideas of our algorithms and asymptotic analysis of these algorithms. Experimental results are shown and analyzed in Section 36.4. Finally, conclusions are presented in Section 36.5.

36.1   Vision Tasks

From a computational perspective, computer vision processing is usually organized as follows:

- *Low-level processing.* The input raw image is processed using image processing techniques. Most of these operations are performed on each pixel simultaneously using image data in the neighborhood of the pixel.
- *Intermediate-level processing.* This involves processing the incoming visual information from earlier levels to obtain information (such as geometric features) to be used for understanding images.
- *High-level processing.* Here, cognitive use of the acquired knowledge from the above processing is performed to infer semantic attributes of an image. Processing at this level can be classified as knowledge processing and/or symbolic processing. Search-based techniques are widely used at this level.

To illustrate our algorithms and implementations, we have considered parallelizing a task arising in each of low-level, intermediate-level, and high-level processing in a building detection system that has been developed at USC. The building detection system [10] is used to detect and provide 3-D descriptions of buildings from monocular views of aerial scenes. The input to the system is the aerial image of buildings in urban and suburban environments and the output is the buildings modeled as composition of rectangular blocks.

The system has been shown to produce good results and shows the promise of being extended to much more complex cases [10]. While the users find the accuracy of the system useful, the computational speeds on current workstations (such as a Sparc System 10) are simply not acceptable. For processing a 1k × 1k image, the serial implementation (written in LISP) takes nearly 30 minutes to complete the five grouping procedures, and processing a large image may take several hours [32]. To make these systems really usable, we need to bring computation times for small images down to a few seconds, and for large images to a few tens of minutes at most.

36.1.1   Overview of low-level vision processing

In this work, we have parallelized an image feature extraction task, the Nevatia-Babu line finder [21]. The objective of the task is to extract linear features from an input image. Input to the system is a 2-D image array of pixels (grey levels), and the output is a list of linear segments specified by end-point locations. The main processing steps are [21]:

1. *Edge detection.* Convolve the input image with masks corresponding to ideal step edges in a selected number of directions.
2. *Thinning and thresholding.* Compare each edge with its neighbors (in the direction orthogonal to the edge's orientation) and retain the edge if it is of greater magnitude and the magnitude is also greater than a fixed threshold.
3. *Edge linking.* Compare each edge with its neighbors in the direction of the edge's orientation and form a link to the neighbors if they are of similar orientation.
4. *Linear approximation.* Extract from the 2-D image array the edge locations (x and y coordinates) and approximate the resulting contours by piecewise linear segments.

The processing can be represented as two major tasks as shown in Fig. 36.1: *contour detection* followed by *linear approximation*. Given an $n \times n$ image and fixed number of masks of size $m \times m$, extraction of linear features, including contour detection and linear approximation, can be performed in $O(m^2 n^2)$ time on a serial machine.

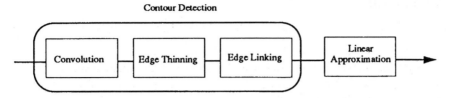

Figure 36.1 Overview of the Neva-Babu line finder

The above tasks have been implemented using C and LISP on several sequential machines, including Sparc workstations and Symbolics machine. This system is widely used to produce linear features for higher-level vision processing such as image matching, perceptual grouping, and object recognition.

36.1.2   Intermediate-level analysis

Perceptual grouping has proven to be effective for extracting straight lines 12], curves [6], man-made structures [10, 20], and so on. In general, the process groups the primitive features detected by low-level processing recursively to form structural hypotheses. We use the perceptual grouping steps performed in a building detection system [10] as an example of intermediate-level vision task to design the scalable parallel algorithms. Figure 36.2

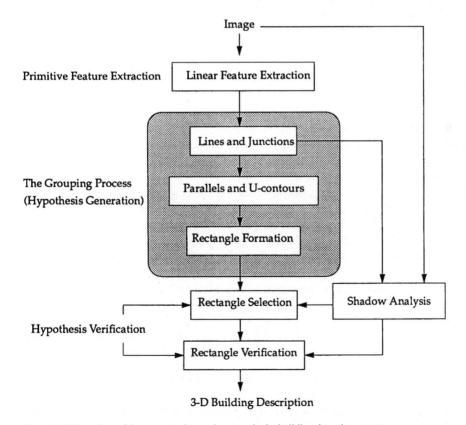

Figure 36.2 Overview of the perceptual grouping steps in the building detection system

shows the main steps in the system. The input to the system is the extracted line segments detected from aerial image of buildings in urban and suburban environments and the output is the buildings hypotheses modeled as composition of rectangular blocks. A series of grouping procedures are performed to group lines into junctions and parallels, parallels and lines into U-contours, and U-contours into rectangles. The grouping process makes explicit the geometric relationships among the image data and provides *focus of attention* for detection of the building structures in the image. For the sake of completeness, we outline the procedures in the system.

1. *Line Grouping.* This groups line segments that are closely bunched, overlapped, and parallel to each other to form a line (a linear structure at a higher granularity level). For each line segment, a search is performed within the region on both sides of the line segment to find other line segments that are parallel to it. The detected segments are grouped to form a line.

2. *Junction Grouping.* This groups two close right-angled lines to form a *junction.* For each *line*, a search is performed within the region on both sides of the *line* and near its end points to find *lines* that may jointly form right-angled corner(s).

3. *Parallel Grouping.* This groups two *lines* that are parallel to each other and have high percentage of overlap. For each *line*, a search is performed on a sorted *line* list (based on the slope of the *line*) to form a *parallel* by grouping with *line* having a difference of slope within a given threshold value and satisfying certain constraints with respect to overlap.

4. *U-contour Grouping.* This forms a *U-contour* if any *parallel* having its two lines aligned at one end. A search is performed within the window near the aligned end of each *parallel* to group with *lines* possibly connecting the end-points at the aligned end.

5. *Rectangle Grouping.* This forms a rectangle if any two *U-contours* share the same *parallel.*

Let $W(S)$ denote the total area of all the *search windows* generated by a set S of input *tokens* in a grouping step. The serial time complexity of a grouping step is $O(W(S))$. Note that the input set S to a grouping step is usually the output of an earlier grouping step.

36.1.3 Object recognition using geometric hashing

In a model-based recognition system, a set of objects is given, and the task is to find instances of these objects in a given scene. The objects are represented as sets of geometric features, such as points or edges, and their geometric relations are encoded using a minimal set of such features. The task becomes more complex if the objects overlap in the scene and/or other occluded unfamiliar objects exist in the scene.

Many model-based recognition systems are based on hypothesizing matches between scene features and model features, predicting new matches, and verifying or changing the hypotheses through a search process [9]. Geometric hashing, introduced by Lamdan and Wolfson [33], offers a different and more parallelizable paradigm. It can be used to recognize flat objects under weak perspective. The algorithm consists of two procedures, *preprocessing* and *recognition*. These are shown in Figs. 36.3 and 36.4, respectively. Additional details can be found in Ref. [33].

Preprocessing. The preprocessing procedure is executed off-line and only once. In this procedure the model features are encoded and are stored in a hash table data structure.

```
Preprocessing
For each model i such that 1 ≤ i ≤ M do
    Extract n feature points from the model;
        For j = 1 to n do
            For k = 1 to n do
                a. Compute the coordinates of all other features points in the
                   model by taking this pair as basis.
                b. Using the hash function on the transformed coordinates,
                   access a bin in the hash table. Add the (model,basis) pair
                   (i.e., (i, jk)) to the bin.
            end for
        end for
end for
end
```

Figure 36.3 Outline of steps in preprocessing

```
Recognition
1. Extract the set of feature points S from the input.
2. Selection:
       Select a pair of feature points as basis.
3. Probe:
       a. Compute the coordinates of all other features points in the
          scene relative to the selected basis.
       b. Using the given hash function on the transformed coordinates
          access the hash table obtained in the Preprocessing phase.
       c. Vote for the entries in the hash table.
       d. Select the (model, basis) pair receiving maximum number of votes
          as the matched model in the scene.
4. Verification:
       Verify the candidate model edges against the scene edges.
5. If a model wins the verification process, remove the corresponding
       feature points from the scene.
6. Repeat steps 2, 3, 4, and 5 (until some specified condition).
end
```

Figure 36.4 Outline of steps in recognition using geometric hashing

However, the information is stored in a highly redundant multiple-viewpoint way. Assume each model in the database has n feature points. In Step a in Fig. 36.3, for each ordered pair of feature points in the model chosen as a basis, the coordinates of all other points in the model are computed in the orthogonal coordinate frame defined by the basis pair. In Step b, (*model, basis*) pairs are entered into the hash table bins. The complexity of this preprocessing procedure is $O(n^3)$ for each model, hence $O(Mn^3)$ for M models.

Recognition. In the recognition procedure, a scene S of feature points is given as input. An arbitrary ordered pair of feature points in the scene is chosen. Taking this pair as a basis, the coordinates of the remaining feature points are computed. Using the hash function on the transformed coordinates, access a bin in the hash table (constructed in the preprocessing phase), and for every recorded (*model, basis*) pair in the bin, collect a vote for that pair.

The pair winning maximum number of votes is taken as the matching candidate. The execution of the recognition phase corresponding to one basis pair is termed as a probe. If no (*model, bases*) pair scores high enough, another basis from the scene is chosen, and a probe is performed.

There have been two prior efforts in parallelizing the geometric hashing algorithm [3, 26]. Both implementations have been performed on SIMD hypercube-based machines. A major problem in both the implementations is the requirement of large number of processors. In Ref. [26], the number of processors used is same as the number of bins in the hash table. Thus, $O(Mn^3)$ processors are needed, where M is the number of models in the database and n is the number of feature points in each model. their implementation results in 1.52 seconds for a probe of the recognition phase on an 8k processor CM-2 on an input scene consisting of 200 feature points. In this experiment, the model database has 1024 models, and each model is represented using 16 feature points.

We did not elaborate on parallelizing the preprocessing phase, since it is a one-time process and can be carried out offline. In the recognition phase, possible occurrence of the models in the scene is checked. The models are available in the hash table created during the preprocessing phase. An arbitrary ordered pair of feature points in the scene is chosen and broadcast to all the processors. Taking this pair as a basis, a probe of the model database is performed.

The time taken to perform a probe depends on the hash function employed to run. Assuming scene S results in $V(S)$ votes, a probe of the recognition phase can be implemented in $O(|V(S)|)$ time on a serial machine.

36.2   A Model of CM-5

A Connection Machine Model CM-5 system contains between 32 and 16,348 processing nodes (PNs). Each node is a 32 MHz Sparc processor with up to 32 MB of local memory. The peak performance of a node having 4 vector units is 128 MFLOPS. The PNs are interconnected by three networks: a data network, a control network, and a diagnostic network. The data network provides point-to-point data communication between any two PNs. Communication can be performed concurrently between pairs of PNs, and in both directions. The data network is a 4-ary *fat tree* [17]. The bandwidth continues to scale linearly up to 16,384 PNs [15]. The control network provides cooperative operations, including broadcast, synchronization, and scans (parallel prefix and suffix). The control network is a complete binary tree with all the PNs as leaves. Each partition consists of a control processor, a collection of processing nodes, and dedicated portions of data and control networks. Additional details can be found in [15]. Throughout this chapter, size of the machine refers to the number of PNs in a partition.

For our analysis, we will model the CM-5 as a set of high-performance SISD machines interacting through the data and control networks. We assume *cooperative message passing* [29]; the sending and receiving PNs must be synchronized before sending the message. Thus, the *start-up cost* including software overhead and synchronization overhead is associated with each message. We assume SPMD (*single program, multiple data*) mode execution in which each PN runs the same part of a program asynchronously until an synchronization point is reached. Synchronization points are inserted before a data communication step. The synchronization cost can be counted as part of the start-up time. In our analysis, we consider machine sizes that are not "large." The hardware latency (network interface overhead and network latency) in data network is small and is *hidden* by the software overheads.

Let T_d and T_c denote the startup time (seconds/message) for sending a message using data network and control network respectively. Let τ_d denote the transmission rate (seconds per unit of data) for data communication using the data network and τ_c denote the transmission rate for broadcasting data using control network. We make the following assumptions for our analysis:

1. In a unit of time, a PN can perform an arithmetic/logic operation on local data.
2. Sending a message containing m units of data from a PN to another PN or exchanging a message of size m between a pair of PNs takes $T_d + m\tau_d$ time using the data network.
3. Suppose each PN has m units of data to be routed to a single destination using the data network, and the set of all destinations is a permutation; then, the data can be routed in $T_d + m\,\tau_d$ time.
4. Broadcasting a message containing m units of data from a PN to all PNs can be performed in $T_c + m\,\tau_c$ time using the control network, where T_c is the start-up time for broadcasting and τ_c is the transmission rate (seconds per unit of data) for broadcasting using control network.
5. Combining (max, sum, prefix sum, etc.) a unit of data from each PN can be performed in τ_g time using the control network.

It has been measured [13] that the start-up time T_d is around 60 to 90 μs, and T_c is around 2 μs. These times are measured by sending a 0 byte message between two PNs using the data network or broadcasting a 0 byte message from a PN to all the PNs using the control network. The transmission rate τ_c has been observed to be 1.25 μs/B. The times for performing a combining operation (Υ_g) are 6 to 12 μs for integer value and 39 to 56 μs for double precision. For regular data communication pattern, τ_d is at the range of 0.100 to 0.123 μs/B. This model favors communicating long messages to communicating a large number of short messages. A unit of data is defined as a fixed size data structure to contain image data (a contour pixel, a label, etc.) in our analysis.

We assume SPMD mode execution in which each PN runs the same part of a program asynchronously until a synchronization point is reached. Synchronization points are inserted before starting a data permutation step. The synchronizing cost can be counted as part of the start-up time. Using this model, we can quantify the communication times and predict the running times of our implementations.

36.3   Scalable Parallel Algorithms

A parallel algorithm is considered scalable if the execution time of the algorithm on a machine with P processors varies as $1/P$[12]. Our goal is to design scalable algorithms that provide high-speed execution on available partition sizes of machines for problem sizes of interest to the vision community. In this section, we first discuss the design of scalable algorithms for low-level and intermediate-level vision tasks in the building detection system described in Section 36.1. Parallelization of object recognition using geometric hashing is also discussed.

36.3.1   Low-level processing—linear feature extraction

In the building detection system, the linear feature extraction (LFE) task consists of two main tasks: *contour detection* and *linear approximation*. The contour-detection task exhibits fine grained parallelism in a natural way. However, the linear approximation task does not have explicit parallelism; in the absence of efficient partitioning and routing techniques, the data communication overheads may dominate the overall execution time. We address these problems and present scalable algorithms for each component of the LFE system.

36.3.1.1 Contour detection. The contour-detection task consists of edge detection, thinning, and linking steps. These are *window operations,* in which the output at a pixel is based on the value of the input pixel and the value of its neighboring pixels. The neighborhood is defined by the window size. Several types of window operations with different window sizes are performed in the image feature extraction system [21].

The image array is divided into P blocks, where P is the number of processors available. Each block is of size

$$\frac{n}{\sqrt{P}} \times \frac{n}{\sqrt{P}}$$

The blocks are numbered in a shuffled row major order. PN_i receives the ith block, $0 \le i \le P - 1$. This mapping reduces the communication toward the root of the tree. The block i is mapped to PN_i (see Fig. 36.5).

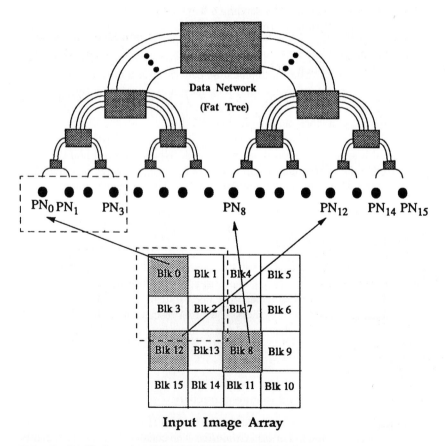

Input Image Array

Figure 36.5 Partitioning and mapping of the input image to CM-5 PN array

An important step of the implementation is the *image boundary padding*, which involves communication between PNs. In this step, the boundary data between neighboring image blocks is exchanged and stored in a local buffer (see Fig. 36.6). This step is required because the window operations performed on the boundary pixels of an image block may need pixel data stored in the neighboring PNs. The communication time depends on the size of the window.

Following the techniques described in Ref. [21], in edge detection, the input image is convolved with six masks. Each mask is of size 5×5. Boundary padding is needed to exchange the image data of boundary pixels before executing the convolution step. The outputs of the convolution step are the magnitude and the orientation of the edges. After the convolution step, boundary padding is needed. The edge magnitude and orientation information are then used to perform edge thinning that can be considered as a 3×3 window operation. Again, edge pixel information established by the thinning step needs to be updated. In edge linking, an edge pixel is linked with its neighbors, in the direction of its orientation, if they are of similar orientation. Some edge refining procedures are performed in this step to bridge the gap between disconnected edges pixels. This operation can be regarded as a 5×5 window operation.

The processing steps for the contour detection are outlined in Fig. 36.7. Given an $n \times n$ image and P PNs, we assume that each PN contains an image block of size n^2/P and the largest window is of size $m \times m$. The serial complexity of contour detection is $O(n^2 m^2)$, assuming a constant number of window operations are performed.

Theorem 1. Given an image of size $n \times n$, the contour detection procedure can be performed in $O(n^2 m^2/P)$ computation time and $12T_d + (12\lfloor m/2 \rfloor n/\sqrt{P} + 24\lfloor m/2 \rfloor)\tau_d$ communication time using a partition of CM-5 having P PNs.

The total computation time for Steps 2, 4, and 6 is $O(m^2 n^2/P)$. The communication time for each boundary data padding step is $4T_d + (4\lfloor m/2 \rfloor n/\sqrt{P} + 8\lfloor m/2 \rfloor)\tau_d$. This assumes that $\lfloor m/2 \rfloor < n/\sqrt{P}$. Note that, in practice, $n \geq 256$, $m \leq 10$, and $P \leq 1$k.

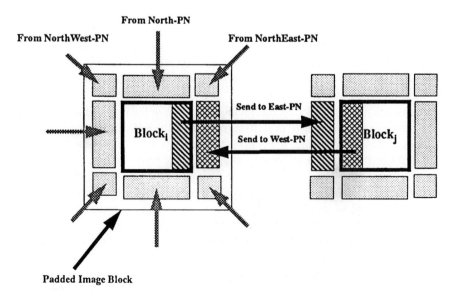

Figure 36.6 Data exchange of a boundary padding operation between PNs

```
Procedure: Contour-Detection
Input: n × n image and each PN has image block of size   n/√P × n/√P
Step 1: Perform boundary data padding
Step 2: Perform edge detection
Step 3: Perform boundary data padding
Step 4: Perform thinning and thresholding
Step 5: Perform boundary data padding
Step 6: Perform edge linking
end
```

Figure 36.7 A skeleton of the contour-detection procedure

36.3.1.2 Linear approximation. Several heuristics are available for approximating a contour by a set of line segments. A general discussion and evaluation of such techniques can be found in Ref. [7]. We employed a *strip-based* algorithm [27]. This algorithm has been shown to be competitive with respect to the quality of the output and the running time with other approximation algorithms. However, the analysis in this section is applicable to any linear time sequential approximation algorithm. For the sake of completeness, we briefly outline the algorithm. A detailed description can be found in Ref. [27].

The input to this algorithm is a set of (open) contours produced by the contour detection task. PN_i has the contours detected within the ith block where $0 \leq i \leq P - 1$. The contour can be represented as a linear linked list. The length of a contour denotes the number of contour pixels in the list. It is assumed that the maximum length of a contour in an $n \times n$ image is $O(n)$. Note that in the linking step, edges are linked based on orientation; sharp corners can be detected, and the contour can be disconnected at those pixels.

The algorithm proceeds as follows [27]: given a starting pixel p_a of a contour and an error bound e for controlling the quality of the approximation, it selects a pixel p_f on the contour that is at a distance $> e$ from p_a. If it can not find such a pixel, the algorithm stops and forms a line from p_a to the last pixel of the contour. Otherwise, draw a line using p_a and p_f. This line is referred to as the critical line. Next form two lines parallel to the critical line at a distance e. These lines are referred to as the *boundary lines* (see Fig. 36.8). Beginning with pixel p_{f+1}, examine the remaining pixels on the contour until a pixel p_b is found, where p_b is the first pixel lying outside the region formed by the boundary lines or p_b is the last pixel on the contour.

If p_b is outside the region formed by the boundary lines, then the line segment $\overline{p_a p_{b-1}}$ is the approximation to the contour from p_a to p_{b-1} and the same procedure is applied starting at p_{b-1}. If p_b is the last pixel, then $\overline{p_a p_b}$ is the approximation to the contour from p_a to p_b and the procedure stops. Assuming that the total number of pixels on the contour as l, it is easy to verify that the strip-based algorithm runs in $O(l)$ time on a serial machine.

Most of the known heuristics for linear approximation are inherently sequential in nature. One of the solutions to speed up the process is to let each PN concurrently approximate the contours whose starting pixels are located in its image block; if the contour crosses over the boundary, the approximation is continued by the PN containing the next part of the contour. The process continues until all contours are approximated. However, in this approach, some PNs can become bottlenecks if many contours pass through them and the approximation processes for each of these contours arrive at these PNs at the same time.

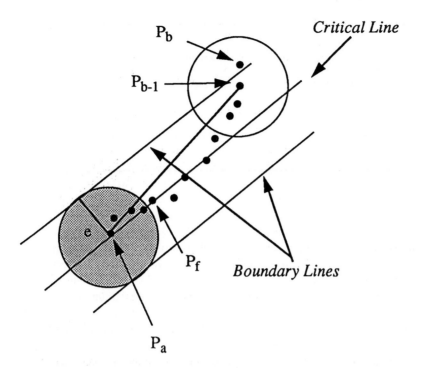

Figure 36.8 Strip-based linear approximation

Our approach is to assign the same label to each pixel of a contour and group all the pixels having the same label to a single PN to perform the linear approximation. Using this data allocation strategy, the contour redistribution is performed once to localize the contours into the PNs and the communication needed during the execution of the approximation procedure is minimal. An advantage of this approach is that any linear approximation heuristic (that has been proposed for serial machines) can be adapted without having to pay a communication penalty after the contours have been localized. We define *workload* on a PN as the total number of contour pixels to be processed by the PN. We employ a *load balancing* strategy that relocates the contour pixels such that all the connected contour pixels are moved to the same PN (or at most two PNs) and each PN has no more than $O(n^2/P)$ *workload*. The processing steps for the linear approximation on CM-5 are outlined in Fig. 36.9.

```
Procedure: Linear-Approximation
Input:  PNᵢ has contour data extracted from the i-th image block,
            0 ≤ i ≤ P - 1.
Step 1: Perform linear approximation
Step 2: Perform connected component labeling
Step 3: Group contour data having the same label to a PN
Step 4: Perform linear approximation
end
```

Figure 36.9 A skeleton of the linear approximation procedure

Theorem 2. Given an image of size $n \times n$, the linear approximation step can be performed in $O(n^2/P)$ computation time and $(20P + 3 \log P)T_d + (24n + 80n^2/P)\tau_d$ communication time on a partition of CM-5 having P PNs, where $n \geq 0.5P^{3/2}$.

Given an $n \times n$ image and P PNs, we assume that each PN contains an image block of size n^2/P, and all the contour pixels in the image block have been detected. An outline of our scalable algorithm is shown in Fig. 36.9.

In Step 1, any contour that does not cross an image boundary can be approximated within a PN. This approximation can be carried out for all the contours local to the PNs in $O(n^2/P)$ time. In the remainder of this discussion, we only consider contours that cross image boundaries. In Step 2, a divide-and-conquer strategy can be used to complete the component labeling in $O(n^2/P + n\sqrt{P})$ computation time and $24n\tau_d + 3 \log PT_d$ communication time. In Step 3, a modified *column sort* algorithm described in Ref. [30] can be used to move contours. The Leighton's *columnsort* [16] consists of a local sort followed by a data communication. These two steps are repeated four times. This algorithm is constrained by $r \geq 2(P-1)^2$, where r is the number of votes in each PN and P is the number of PNs. We use a modified *columnsort* to sort the votes in which a local sort step in *columnsort* is performed by a group of PNs by using *columnsort*. We partition the processor array into $P^{3/5}$ disjoint groups, each group having $P^{2/5}$ PNs. The local sort is performed on $r \times P^{2/5}$ elements in a group of $P^{2/5}$ PNs. The sorted elements are then shuffled among the groups. We repeat the local sort and the shuffle step four times. In this algorithm, each PN performs 16 sorting steps and 20 data communication steps. However, the constraint on n and P can be relaxed to $r \geq 2P^{4/5}$. We refer to this algorithm as a *two-level columnsort*. Note that we only sort the labels. Contour data having the same label are moved along with the label. Thus, the computation time for sorting is $O(n \log n/\sqrt{P})$. If $n \geq 0.5P^{3/2}$, the algorithm [30] requires 20 permutation steps. The total communication time is

$$\frac{80n^2}{P}\tau_d + 20PT_d$$

At the end of the sorting step, the contour pixels are redistributed such that each PN has at most n^2/P contour pixels and each contour is in, at most, two adjacent PNs and given two PNs, at most, one contour extends over them. The strip-based heuristic for approximating the contours can be applied within each PN. Step 4 can be completed in $O(n^2/P)$ time. On a partition of CM-5 having P PNs, the computation time is $O(n^2/P)$ and the communication time is

$$(20P + 3\log P)\,T_d + \left(24n + \frac{80n^2}{P}\right)\tau_d$$

where $n \geq 0.5P^{3/2}$. Note that if $n \geq 1.5\,P^{3/2}$, a modified *rotate sort* [19] described in Ref. [8] can be employed to reduce the communication time because it only requires eight data permutation steps.

36.3.2   Intermediate-level processing—perceptual grouping

Intermediate-level tasks operate on collection of image data extracted in low-level processing. In parallel implementations of intermediate-level tasks, the processors that hold these image data must coordinate their activity because the output depends on image data

Proof: The steps of the probe procedure are shown in Fig. 36.11. Step 1 takes $T_c + \tau_c$ time to broadcast a selected basis to all the PNs using control network. Step 2 consists of local computation only, which can be completed in $O(|S|/P)$ time.

The only step that needs careful implementation is the *Balance load* procedure. Let $|V(S)|$ denote the total number of votes cast by S. The *Balance load* procedure distributes the scene points such that each PN has $|V(S)|/P$ votes at the end of the *Vote* step. An outline of the *Balance load* procedure is shown in Fig. 36.12.

The *Balance load* procedure can be completed in $O(|V(S)|/P)$ time. As the key of each scene point is computed in Step 2, create a record for each scene point, i, having three fields: record$_i = (k_i, v_i, d_i)$, $0 \le i \le |S| - 1$ where k_i denotes the computed key, v_i is the number of votes cast by the scene point, and d_i is the index of the PN to which this scene point will be sent. Throughout the discussion we refer to this record as *scene point record*. Initially, v_i and d_i are undefined. Step A in Fig. 36.12 can be performed by finding the total number of votes cast within each PN and summing them over all PNs. This step can be completed in $O(|S|/P)$ computation time and $T_c + \tau_g$ time to sum up the votes. Step B computes the destination d_i, where $d_o = 0$ and

$$ d_i = \left\lfloor \frac{\sum\limits_{j=0}^{i-1} v_j}{\left\lceil \frac{|V(S)|}{P} \right\rceil} \right\rfloor $$

for $1 \le i \le |S| - 1$. This step can be completed in $O(|S|/P)$ computation time and $T_c + \tau_g$ time to perform a parallel prefix operation. Step C creates dummy records such that each PN contains $2|V(S)|/P$ records before sorting the records. Let s_j denote the set of all scene point records to be sent to PN$_j$ and $|s_j|$ denote the size of sj, $0 \le j \le P - 1$. We can compute $|s_j|$, $0 \le j \le P - 1$, using a segmented prefix sum operation where a *segment* denotes a set of records having the same destination. This can be completed in $O(|S|/P)$ computation time and $T_c + \tau_g$ time to perform a segmented prefix operation. Now copy $|s_j|$s to all PNs so that each PN has $|s_j|$, $0 \le j \le P - 1$. This can be completed in $PT_c + P\tau_c$ time using the control network. PN$_j$ creates $\lceil |V(S)|/P \rceil - |s_j|$ dummy records with the destination field having j and $\lceil |V(S)|/P \rceil - |S|/P + |s_j|$ dummy records with the destination field having P. At this time each PN has $\lceil (2|V(S)|)/P \rceil$ records. In step D (in Fig. 36.12), we sort the $\lceil (2|V(S)|)/P \rceil \times P$ records using the destination as the key. The column sort described in Section 36.3.1.2 can be employed. Note that the destination value is an integer in the range 0 to P. A local sort step can be performed in $O((|V(S)|)/P)$ time using radix sort. This step can be completed in $O((|V(S)|)/P)$ computation time and $20PT_d + (40(|V(S)|)/P)\tau_d$

```
Balance Load
    Step A: Compute |V(S)|.
    Step B: Compute the destination for each scene point record.
    Step C: Create dummy records.
    Step D: Sort the records.
    Step E: Send the records to their destination.
end
```

Figure 36.12 Steps in the *Balance load* procedure

communication time. After the sort, PN_i has s_{2i} and s_{2i+1}, for $0 \leq i \leq (P/2) - 1$, and all other PNs receive only dummy records. In step E, PN_i sends packets of size $\lceil |V(S)|/P \rceil$ to PN_{2i} and PN_{2i+1}. This can be completed in $T_d + (|V(S)|)/P)\tau_d$ communication time. Thus, the total time of the *Balance load* procedure is $O(|V(S)|)/P)$ computation time and

$$(20P + 1)\,T_d + \frac{41|V(S)|}{P}\tau_d + (P + 5)\,T_c + P\tau_c + 3\tau_g$$

time in network-related operations. Now, the scene point records are distributed such that if *record_i* generates v_i votes, it will be located in PN_j such that:

1. If $v_i \leq \lceil |V(S)|/P \rceil$, then PN_j will cast the votes generated by key k_i.
2. If $v_i > \lceil |V(S)|/P \rceil$, then PN_j, PN_{j+1} ... PN_{j+r} will cast at most $\lceil |V(S)|/P \rceil$ votes each on behalf of v_i, where PN_{j+r} is the least indexed PN with index $> j$ receiving a scene point record. Note that $v_i \leq (r + 1)\,\lceil |V(S)|/P \rceil$.

In Fig. 36.11, Step 4 is a table lookup that can be completed in $O((|V(S)|)/P)$ computation time. The hash bins are organized (in the preprocessing phase) such that they have multiple access points. The number of access points needs to be a function of P and can be precomputed. The last step that computes the winner can be performed by sorting the $|V(S)|$ votes using (*model,pair*) as the key. As the range of these keys is 0 to Mn^2, the sort of $|V(S)|$ items can be performed in $O(|V(S)|)/P)$ computation time and $20PT_d + 20((|V(S)|)/P)\tau_d$ communication time using column sort. At the end of the sort, the votes are distributed over the P PNs such that votes corresponding to a particular (*model, basis*) pair are in contiguous PNs. By comparing the votes within PNs and within adjacent PNs, and then computing a global max, Step 5 can be completed in $O((|V(S)|)/P)$ computation time and $T_c + \tau_g$ time to find the maximum vote count. The total execution time per probe is $O((|V(S)|)/P)$ computation time and

$$(40P + 1)T_d + 61((|V(S)|)/P)\tau_d + (P + 5)T_c + (P + 1)\tau_c + 4\tau_g$$

time in network-related operations.

36.4   Implementation Details and Experimental Results

We implemented our solutions on CM-5 partitions of 32, 64 (or 128), 256, and 512 PNs. The code was written using C and CMMD 3.0 message-passing library provided by the Thinking Machines Corporation. The program starts by reading the image data to the control processor and then distributing it to the PNs. Results are stored in the PNs and can be output through the control processor. In our implementations, parallel I/O and vector units were not employed. The times reported are the total CPU time measured by the CM-5 timer that has a resolution of 1 μs. The times reported on the serial machine are the total CPU time used by the user process. The resolution of the clock is 16.667 ms. All the reported times do not consider I/O time for initial loading of image data.

36.4.1   Low-level processing

At the time of this writing, the scalable linear approximation algorithm discussed in Section 36.3.2 was not implemented. We implemented an alternate approach that has two main steps:

1. Perform local approximation.
2. Exchange boundary information.

These two steps are repeated until there are no more contours to process. Indeed, to reduce frequent communications, Step (1) was performed until all the PNs had finished their local work. The rationale for choosing this approach follows. Note that the convolution step takes up a major part of the total execution time in the serial implementation. Thus, a reasonably good non-optimal parallel algorithm for implementation of a linear approximation task probably will not lead to severe degradation of the overall performance. The communication time of our scalable algorithm is $(20P + 3 \log P)T_d + (24n + 80n^2/P)\tau_d$. The serial time for linear approximation on a 2k × 2k image (the largest image we processed) is only 54 s. If $P = 512$ and we let $T_s = 60$ μs and $\tau_d = 0$ μs, the communication time is ≈ 600 ms. The speed-up achievable is less than 100. For the values of P (≤ 512) and n (≤ 2048) considered here, the alternate approach seems to lead to a reasonably fast implementation.

In Table 36.1, we show the total execution time for extracting linear features on various images using various partition sizes of CM-5. For the sake of comparison, the execution times of individual steps in contour detection and linear approximation procedures on Sun Sparc 400 are shown in Table 36.2. The largest image we have processed is of size 2048 × 2048. The total execution time for processing such an image (including contour detection and linear approximation) is 1.118 s. The same image processed by a Sun Sparc 400 station operating at 32 MHz using a optimized code written in C takes more than 8 min.

TABLE 36.1   Execution Times for Extracting Linear Features on Various Sizes of Images, Using Various Partitions of CM-5 (We have used e = 2.0 for performing linear approximation.)

	Total execution time (in seconds) on CM-5				
Machine size (no. of PNs)	Image size				
	128 × 128	256 × 256	512 × 512	1k × 1k	2k × 2k
32	0.097	0.347	1.001	3.536	14.958
64	0.065	0.239	0.537	1.853	7.602
256	0.053	0.170	0.086	0.599	2.010
512	0.052	0.179	0.123	0.374	1.118

TABLE 36.2   Execution Times on Various Image Sizes Using Sun Sparc 400 Operating at 33 MHz and Having 64 MB of Main Memory (The code was written in C and optimized using an optimizing compiler. We let e = 2.0 for performing linear approximation.)

	Execution time (in seconds) on Sun Sparc 400				
Processing steps	Image size				
	128 × 128	256 × 256	512 × 512	1k × 1k	2k × 2k
Convolution	1.23	4.90	19.73	80	314
Thinning	0.18	0.72	2.95	7.87	47
Linking	0.26	1.18	5.05	14.65	91
Approximation	0.18	0.70	2.95	8.41	54
Total time	1.86	7.50	30.68	112	506

Figure 36.13 shows the speed-up curves for performing the linear feature extraction task on various sizes of images. Speedup is calculated as the ratio of the execution time on Sparc 400 to the execution time on CM-5. For larger images, linear speed-up is observed.

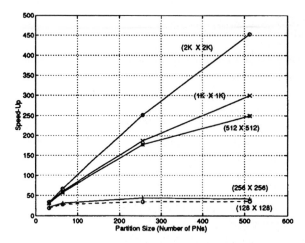

Figure 36.13 Speed-up curves for performing linear feature extraction on various sizes of images

Figures 36.14 through 36.19 show the raw images and the extracted line segments for various sizes of images on a partition of CM-5 having 32 PNs. These images were provided by the USC Vision Group.

36.4.2   Intermediate-level analysis

We have implemented the line grouping and junction grouping steps. The input was 8943 line segments detected from a 1k × 1k image. The serial implementation written in C results in 14.166 s for line grouping and 8.680 s for junction grouping using Sun Sparc 400. In Table 36.3, we show the execution times of line grouping and junction grouping steps. Given a 1k × 1k input image, the preprocessing phase, the line grouping, and the junction grouping can be performed in 5.0 s using a partition of CM-5 having 32 processing nodes.

TABLE 36.3   Execution Times (ms) for the Line Grouping and the Junction Grouping Steps on Various Partitions of CM-5 (Total number of junctions detected was approximately 3000.)

Step	Execution time (ms)			
	Line Grouping		Junction Grouping	
No. of tokens	8943		3736	
Partition size	32	64	32	64
Broadcast tokens	333	351	145	148
Load balancing	70	70	51	51
Grouping	339	179	535	326
Total time	742	600	731	525

36.4.3   Geometric hashing

The state of the art in image understanding techniques that employ geometric hashing results in the following scenario:

Figure 36.14 A 128×128 airport image

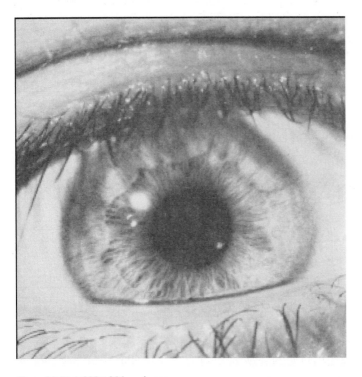

Figure 36.15 A 256×256 eye image

Figure 36.16 A 512×512 building image

Figure 36.17 Extracted line segments from contours of length ≥ 6 with $e = 3.0$ using CM-5 partition of 32 PNs; execution time ≈ 96 ms

Figure 36.18 Extracted line segments from contours of length ≥ 20 with $e = 5.0$ using CM-5 partition of 32 PNs; execution time ≈ 347 ms

Figure 36.20 Extracted line segments from contours of length ≥ 30 with $e = 5.0$ using CM-5 partition of 32 PNs; execution time ≈ 974ms

- number of models = few hundred up to a thousand
- number of feature points/model = 10 to 20
- number of scene points = few hundred up to a 1000

For a typical example, where 1024 models are used and each model has 16 points, the hash space has 4M points.

We have performed implementations on CM-5. These implementations exploit the computation and communication characteristics of the underlying architecture of the machine to lead to large speed-ups for the size of scenes and models typically used by the vision community. We show a simple implementation that performs well when the number of votes generated is small. However, the performance of this algorithm depends on the distribution of hash bin access, the distribution of votes generated, and the total number of votes generated. We refer to this as Algorithm A. Based on our scalable parallel algorithms, we develop two alternate implementations. The second implementation, Algorithm B, handles congestion in voting. The third implementation, Algorithm C, handles congestion in bin access as well as in voting. Algorithm C partitions the hash bins across the PNs and leads to superior performance in the worst case compared with the other two implementations. We briefly outline the procedures employed in our implementations.

1. The *Compute_Keys* procedure computes the transformed coordinates of the scene points and quantizes them according to a hash function $f()$ (details of the hash function are discussed later in this section).
2. The *Access_Bin* procedure routes the *keys* to their corresponding hash locations stored in $PN_{g(key)}$. The function $g()$ defines the mapping of the hash table locations onto the PN array. Let B denote the number of hash bins. If one copy of the hash table is used, then $g(i) = \lfloor i/\lceil (B/P) \rceil \rfloor, 0 \le i \le B - 1$.
3. For the *Collect_Vote* procedure, we designed a simple (asymptotically non-optimal) voting procedure in which each PN is responsible for collecting votes for Mn^2/P (*model, basis*) pairs. Thus, each PN has Mn^2/P counters. The *Collect-Vote* procedure accesses the hash bin entries locally, generates the votes and forwards them to the PN responsible for collecting the votes for that (*model, basis*) pair.
4. The *Compute_Winner1* procedure determines the (*model, basis*) pair receiving the maximum number of votes by scanning the vote counts recorded in each PN. This is performed in two steps: finding the local maximum of votes followed by finding the maximum over the entire PN array.
5. The *Generate_Vote* procedure is used by sort-based approach; it accesses the hash bin entries locally and copies the (*model, basis*) pairs into a list.
6. The *Compute_Winner2* procedure determines the (*model, basis*) pair receiving the maximum number of votes (*winner*) by sorting the (*model, basis*) list produced by *Generate Vote* procedure.

Based on the procedures described above, the following algorithms were developed.

Algorithm A—a straightforward approach. In this algorithm, we partition the hash table into *P blocks,* and each block has the same number of hash bins. PN_i receives $block_i, 0 \le i \le P - 1$. Note that all entries in a bin are stored in the same PN.

1. Broadcast a basis to all PNs.
2. Execute *Comp_Keys* procedure.

3. Execute *Access_Bin* procedure.
4. Execute *Collect_Vote* procedure.
5. Execute *Compute_Winnerl* procedure.

Algorithm B—a sort-based approach. In this algorithm, the distribution of the hash table is the same as in Algorithm A. However, the (*model, basis*) pair receiving the maximum number of votes is computed by sorting.

1. Broadcast a basis to all PNs.
2. Execute *Comp_Keys* procedure.
3. Execute *Access_Bin* procedure.
4. Execute *Generate_Vote* procedure.
5. Execute *Compute_Winner2* procedure.

Algorithm C—an alternate sort-based approach. In this algorithm, each bin is partitioned across the PNs. Thus, if a bin has k entries, then each PN receives $\lceil k/P \rceil$ entries.

1. Broadcast a basis to all PNs.
2. Execute *Comp_Keys* procedure.
3. Distribute the encoded scene points from each PN to all the other PNs. Thus, all PNs have the complete set of encoded scene points.
4. Execute *Generate_Vote* procedure.
5. Execute *Compute_Winner2* procedure.

In Algorithm C, we uniformly distribute the entries of each bin to all PNs such that each PN generates the same number of votes (for any given *S*) in Step 4. This results in an even distribution of workload over the PN array. Algorithm C leads to superior performance if the average bin length is at least equal to the size of the PN array.

Algorithm B and C were implemented using two sorting algorithms. The first algorithm is Leighton's *columnsort* [16] if $r \geq 2(P-1)^2$, where r is the number of votes in each PN and P is the number of PNs. In the second algorithm, we use 2-Level *columnsort* as described in Section 36.3 to sort the votes if $r \geq 2P^{4/5}$.

We have used a synthesized model database containing 1024 models. Each model consists of 16 randomly generated models in 2 dimensions. This results in a hash table having 4M entries. These model points were generated according to a Gaussian distribution with zero mean and unit standard deviation as in Ref. [26]. Similarly, scene points were synthesized using a normal distribution. The equalization technique in Ref. [26] was applied to quantify the transformed coordinates, i.e., for each transformed point (u,v), the following hash function (where σ denotes the standard deviation of the set of model points) is applied:

$$
f(u, v) = \left(1 - e^{-\frac{u^2 + v^2}{3\sigma^2}}, \; \mathrm{atan2}\,(u, v) \right)
$$

The above hash function uniformly distributes the data in the preprocessing phase over the hash space such that each hash bin has nearly the same length [26]. However, the performance of Algorithm C does not depend on the hash function employed as long as the average bin length ≥ number of PNs employed.

An earlier implementation [26] results in 1.52 s for a probe of the recognition phase on an 8k processor CM-2 on an input scene consisting of 200 feature points (see Table 36.4). In this experiment, the model database has 1024 models and each model is represented using 16 feature points. This performance is obtained under the following assumptions: the number of the processors ≈ number of hash bins, a distribution of hash table entries and the resulting votes such that no congestion occurs at a PN during hash bin access or during voting. We show that using the same hash function and the same number of hash bins as in Ref. [26] and using one copy of the hash table distributed over the entire PN array, a probe on a scene consisting of 256 feature points can be performed in less than 200 ms on a 32 processor CM-5. In the implementation in Ref. [26], certain inputs result in hash bin access congestion wherein many processors access the data in a single PN. Such congestion does not occur in our implementation. In addition, our algorithm evenly distributes the generated votes. The number of PNs employed in our algorithm does not depend on the number of hash bins. The performance of our algorithm depends on the total number of votes only. As in Ref. [26], synthesized model data and synthesized scenes were used.

TABLE 36.4 Comparison with Previous Implementations (The number of models is 1024; each model has 16 feature points.)

Methods	Machine size	No. of scene points	Average bin length	Total time
Hummel [26]	8k/CM-2	200	512	1.52 s
Algorithm A	32/CM-5	256	512	546 ms
Algorithm B	32/CM-5	256	512	364 ms
Algorithm C	32/CM-5	256	512	188 ms
Medioni [3]	8k/CM-2	x*	?	2.0–3.0 s

*Two real-life models (a key and a shaver having 15 feature points and 14 feature points, respectively) and a scene consisting of 28 feature points were used.

36.5 Concluding Remarks

We have presented scalable algorithms for low-level, intermediate-level, and high-level vision tasks on a model of CM-5. Our implementations exploit the features of CM-5. The experimental results are very encouraging and bring a promising future in using parallel machines to realize an integrated vision systems to support interactive execution mode. Indeed, vision computations have significantly different characteristics compared with other grand challenge problems in scientific and numerical computations. We believe modeling features of parallel machines, designing data partitioning techniques, and mapping the computations to balance the load using explicit message passing is a feasible approach to solve vision problems on coarse-grain message-passing parallel machines.

Acknowledgments

This research was supported in part by NSF under grant IRI-9217528 and in part by ARPA under grant F49620-93-1-0620. The implementations were performed on CM-5 at the Army High Performance Computing Center. This research was supported in part by the Army Research Office contract number DAALO3-89-C-0038 with the University of Minnesota Army High Performance Computing Research Center. We also thank Chung-An Lin of the University of Southern California for providing the images and for his assistance in visualizing our results.

36.6   References

1. Aho, A., J. Hopcroft, and J. Ullman. 1993. *Data Structures and Algorithms.* Reading, Mass.: Addison-Wesley, 1983.

2. Boldt, M., R. Weiss, and E. Riseman. 1989. Token-based extraction of straight lines. *IEEE Trans. Syst., Man, and Cybern.*, Vol. 19, No. 6, 1581–1594.

3. Bourdon, O., and G. Medioni. 1988. Object Recognition Using Geometric Hashing on the Connection Machine. *International Conference on Pattern Recognition*, 596–600.

4. Choudhary, A., and J. Patel. 1990. *Parallel Architectures and Algorithms for Integrated Vision Systems.* Kluwer Academic Publishers.

5. Lin, W., and V. Prasanna. 1994. Parallel algorithms and architectures for consistent labeling. In *Parallel Processing for Artificial Intelligence,* Vol. 1, L. Kanal, V. Kumar, H. Kitano, and C. Suttner, eds. Amsterdam: North-Holland.

6. Dolan, J., and E. Riseman. 1992. *Computing Curvilinear Structure by Token-Based Grouping.* CVPR, 264–270.

7. Dunham, J. G. 1986. Optimum uniform piecewise linear approximation of planar curves. *IEEE Transactions on Pattern Analysis and Machine Intelligence,* Vol. 8, No. 1, 67–75.

8. JáJá, J., and K. Ryu. 1994. The block distributed memory model for shared memory multiprocessors. *Proc. of International Parallel Processing Symposium,* 752–756.

9. Grimson, W. 1990. Object recognition by computer: the role of geometric constraints. Cambridge, Mass.: MIT Press.

10. Huertas, A., C. Lin, and R. Nevatia. 1993. Detection of buildings from monocular views of aerial scenes using perceptual grouping and shadows. *Image Understanding Work shop*, 253–260.

11. Khokhar, A. 1993. Scalable Data Parallel Algorithms and Implementations for Object Recognition. Ph.D. Thesis, Department of EE-Systems, University of Southern California.

12. Kumar, V., A. Grama, A. Gupta, and G. Karypis. 1994. *Introduction to Parallel Computing: Design and Analysis of Parallel Algorithms.* Benjamin/Cummings.

13. Kwan, T., B. Totty, and D. Reed. 1993. Communication and computation performance of the CM-5. *Proc. of Supercomputing '93,* 192–201.

14. Lamdan, Y., and H. Wolfson. 1988. Geometric hashing: a general and efficient model based recognition scheme. *Proc. International Conference on Computer Vision,* 218–249.

15. Leiserson, C., et al. 1992. *The Network Architecture of Connection Machine CM-5.* Technical Report, Thinking Machines Corporation.

16. Leighton, F. 1985. Tight bounds on the complexity of parallel sorting. *IEEE Transactions on Computers*, Vol. 34, No. 4, 344–354.

17. Leiserson, C. 1985. FAT-TREES: universal networks for hardware efficient supercomputing. *International Conference on Parallel Processing,* 393–402.

18. Lowe, D. 1985. *Perceptual Organization and Visual Recognition.* Kluwer Academic Press.

19. Marberg, J., and E. Gafni. 1988. Sorting in constant number of row and column phases on a mesh. *Algorithmica*, Vol. 3, 561–572.

20. Mohan, R., and R. Nevatia. 1989. Using Perceptual Organization to Extract 3-D Structures. *IEEE Trans. on Pattern Analysis and Machine Intelligence*, Vol. 11, No. 11, 1121–1139.

21. Nevatia, R., and K. Babu. 1980. Linear Feature Extraction and Description. *Computer Graphics and Image processing*, Vol. 13, 257–269.

22. Kumar, V. P. 1991. *Parallel Algorithms and Architectures for Image Understanding.* Boston, Mass.: Academic Press.

23. Prasanna, V., C. Wang, and A. Khokhar. 1993. Low level vision processing on connection machine CM-5. *Proc. IEEE Workshop on Computer Architectures for Machine Perception.*

24. Prasanna, V., and C. Wang. 1994. scalable parallel implementations of perceptual grouping on connection machine CM-5. To appear in *International Conference on Pattern Recognition*, 1994.

25. Prasanna, V., and C. Wang. 1994. Image feature extraction on connection machine CM-5. To appear in *Proc. Image Understanding Workshop.*

26. Rogoutsos, I., and R. Hummel. 1992. Massively parallel model matching: geometric hashing on the connection machine. *IEEE Computer*, 33–42.

27. Roberge, J. 1985. A data reduction algorithm for planar curves. *Computer Vision, Graphics, and Image Processing*, Vol. 29, 168–195.

28. Thinking Machines Corporation. 1991. CM-5: Technical Summary. Technical Report. Thinking Machines Corporation.

29. Thinking Machines Corporation 1992. CMMD Reference Guide Version 3.0. Thinking Machines Corporation.

30. Wang, C., V. K. Prasanna, H. Kim, and A. Khokhar. 1994. Scalable data parallel implementations of object recognition using geometric hashing. *Journal of Parallel and Distributed Computing* (March), 96–109.

31. Wang, C. Scalable Parallel Algorithms and Implementations for an Integrated Vision System. Ph.D. Thesis, in preparation, Department of EE-Systems, University of South ern California.

32. Lin, C. Personal communication, Institute for Robotics and Intelligent Systems, University of Southern California.

33. Wolfson, H. 1990. Model based object recognition by geometric hashing. *First Europe Conference on Computer Vision*, 526–536.

37

Parallel Computation in Biomedicine: Genetic and Protein Sequence Analysis

Tieng K. Yap, Ophir Frieder[*], and Robert L. Martino

Rapid progress in genetics and biochemistry research combined with the tools provided by modern biotechnology has generated massive volumes of genetic and protein sequence data. These sequence data are accessible to researchers in a variety of databases including the GenBank [6], EMBL (European Molecular Biology Laboratory) [10], SWISS-PROT (Swiss Proteins) [2], PIR (Protein Identification Resource) [26], and PDB (Protein Data Bank) [4] databases. They contain a large number of deoxyribose nucleic acid (*DNA*) and protein sequences with their associated biological and bibliographical information. The DNA sequences consist of a string of nucleotides identified by the base they contain. The protein sequences consist of a string of amino acids. For example, the current release of GenBank (rel. 79.0) contains a total of 143,492 sequences consisting of 157,152,442 bases. In the past, these sequences and their associated information were gathered from articles published in scientific journals. Presently, an increasing number of unpublished sequences are submitted electronically into these databases as they are discovered. As a result, the number of entries are growing almost exponentially with time.

As these databases grow in size, biomedical researchers need computational tools to retrieve biological information from them, analyze the sequence patterns they contain, predict the three-dimensional structure of the molecules the sequences represent, reconstruct evolutionary trees from the sequence data, and track the inheritance of chromosomes based on the likelihood of specific sequences occurring in different individuals [19]. These tools will be used to learn basic facts about biology such as which sequences of DNA are used to code proteins while other combinations of DNA are not used for protein synthesis. They will also be used to understand genes and how they influence diseases.

Many of the biomedical research activities mentioned in the previous paragraph contain computationally intensive tasks that can benefit from parallel computing methods. This chapter is concerned with developing a parallel computing tool for one such area, an im-

[*]This author was partially supported by the National Science Foundation under contract number IRI-93 5778S, by the Virginia Center for Innovative Technology under contract number INF-94-002, and by XPAND Corporation.

portant sequence pattern analysis activity, intersequence and intrasequence homology searching. When researchers discover new sequences, they are eager to search databases for sequences that are similar or relevant to their discoveries. They often search the databases at regular intervals since new sequences are continuously being added.

This chapter is organized as follows. The following section describes the types of biological sequences that are being discovered and stored in the various databases and the relationship between these types. We then describe the contents and storage format of a well known genetic sequence database, GenBank, that we used to evaluate various parallel computational approaches. Next, the residue substitution scoring matrices that define the similarity scores for all possible pairs of residues are discussed. After this section, we present some background on biological sequence comparison algorithms. We then discuss some previous parallel searching approaches and present a combined approach. We compare and evaluate the performance between the combined approach with two previous ones including a current state-of-the-art method. We continue, in the next section, by discussing some general aspects about the parallel computation approaches that we had evaluated and some issues specific to this application. We conclude the chapter by highlighting a problem that needs to be solved in the future.

37.1   The Origin of Genetic and Protein Sequence Data

A diagram giving the common transfers of sequence information in nature is shown in Fig. 37.1. DNA contains the complete genetic information that defines the structure, function, development, and reproduction of an organism. Genes are regions of DNA and proteins are the products of genes [11]. DNA *is* usually found in the form of a double helix with two chains wound in opposite directions around a central axis. Each chain consists of a sequence of nucleotide bases that can be one of the following four types: adenine (A), cytosine (C), guanine (G), and thymine (T). The two chains of a DNA molecule have complementary bases, with A-T and C-G being the only pairings that occur, so the sequence of a second chain can be determined if the sequence of the first chain is known. In replication, the chains unwind and each chain is used as a template to form a new companion chain.

The production of a protein from a segment of DNA occurs in two major steps, transcription and translation. A protein is a polymer consisting of a sequence of amino acids. Twenty different amino acids are commonly found in proteins. The DNA itself does not act as a template for protein synthesis. In transcription, a complementary RNA copy of one of the two DNA chains is formed with ribose nucleotides. RNA is a single stranded molecule similar to DNA except that the sugar backbone contains ribosome instead of deoxyribosome. During transcription, the base thymine (*T*) *is* encoded as uracil (*U*) while the three other bases remain the same. As a result, a sequence of RNA *is* composed of the bases *A, C, G,* and *U*. This RNA sequence is translated into a sequence of amino acids that combine to form a protein. During translation, three bases, referred to as a *codon*, are read at a time and translated into one amino acid.

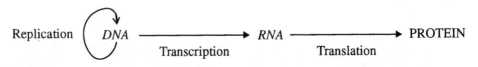

Figure 37.1 The transfer of sequence information in the synthesis of a protein

As shown in Table 37.1, there are 64 possible codons. Sixty-one of them specify the 20 possible amino acids, while the other 3 (*TAG, TGA,* and *TAA*) do not. These three codons are used to specify the end of the translation process of a polypeptide chain. Typically, methionine (*ATG*) is used to specify a start codon for the protein synthesis. To determine the amino acid that is specified by each genetic code, Table 37.1 can be used as follows. First, locate the first nucleic acid under the column labeled "First position." Then, locate the second nucleic acid under the column label "Second position." Finally, use the third nucleic acid to select one of the four lines in the square that the row of the first nucleic and the column of the second one meet. For example, the genetic code UUU specifies phenylalanine. The letter in parentheses after each full amino acid name is the single letter representation that will be used to refer to each amino acid in the remainder of this chapter.

TABLE 37.1   The Genetic Code

First position	Second position				Third position
(5' end)	U	C	A	G	(3' end)
U	Phenylalanine (F)	Serine (S)	Tyrosine (Y)	Cysteine (C)	U
	"	"	"	"	C
	Leucine (L)	"	STOP	STOP	A
	"	"	STOP	Tryptophan (W)	G
C	Leucine (L)	Proline (P)	Histidine (H)	Arginine (R)	U
	"	"	"	"	C
	"	"	Glutamine (Q)	"	A
	"	"	"	"	G
A	Isoleucine (I)	Threonine (T)	Asparagine (N)	Serine (S)	U
	"	"	"	"	C
	"	"	Lysine (K)	Arginine (R)	A
	Methionine (M)	"	"	"	G
G	Valine (V)	Alanine (A)	Aspartic Acid (D)	Glycine (G)	U
	"	"	"	"	C
	"	"	Glutamic Acid (E)	"	A
	"	"	"	"	G

Thousands of various types of proteins occur in biological organisms. They are responsible for catalyzing and regulating biochemical reactions and transporting molecules as well as being the basis for the structure of skin, hair, and muscle. One of the most important unsolved problems in biochemistry is predicting the three-dimensional structure of a protein molecule from its amino acid sequence. Table 37.2 gives some examples of proteins with their associated functions [27].

TABLE 37.2   Examples of Proteins

Protein	Function
Actin	Involved in muscle contraction
Casein	Stores ions in milk
Insulin	Controls blood glucose levels
Keratin	Contributes to the structure of hair

As can be seen from the previous discussion, two types of sequences are of interest. The first is the DNA sequence formed from the four-letter base alphabet A, C, G, and T. These sequences are found in the GenBank, *EMBL,* and *DDBJ* (DNA Database of Japan). The second type is the protein sequence that consists of the 20-letter amino acid alphabet. Protein sequences are collected in many databases such as *SWISS-PROT, PIR,* and *PDB.* The computational methods discussed in this chapter can be used for both types of sequences. RNA sequences will not be discussed as the computational methods used to perform RNA homology searching are the same as those used for DNA sequences.

37.2   An Example Database: GenBank

GenBank is an international database that contains a large number of DNA sequences and their associated biological and bibliographical information. Currently, GenBank is maintained and distributed by the National Center for Biotechnology Information (NCBI) at the National Institutes of Health [12]. The sequences in GenBank are obtained from scanning over 325,000 articles in 3,400 scientific journals annually. In addition, GenBank includes sequences from the European Molecular Biology Laboratory (EMBL) and the DNA Database of Japan (DDBJ). Recently, NCBI has accepted unpublished sequences via electronic submission directly from the scientific community for inclusion in the database. GenBank has grown exponentially over the last decade as can be seen in Fig. 37.2.

The GenBank distribution consists of the 22 flat files as listed in Fig. 37.3. To ensure portability and accessibility, these files are distributed in *ASCII* format. Portability is a very important feature as GenBank is distributed to scientists worldwide who work in a variety of computing environments. Accessibility is also important since most scientists do not want to develop and maintain special tools just to access the data in this database. This ASCII file format allows the scientists to use their existing text processing tools. These files, however, occupy a larger storage space and require longer access times than compressed binary files. For this reason, sophisticated users usually convert the distributed files to a special format that reduces the storage space and shortens the access time.

GenBank has four various file types (extensions), text (txt), index (idx), form (frm), and sequence (seq). The two text files contain the release notes and a short directory of the sequence entries. The release notes describe the file contents, formats, statistics, and other information. The short directory contains brief descriptions of all the sequences contained in GenBank. Each entry in this file contains the sequence name (*LOCUS*), its brief defini-

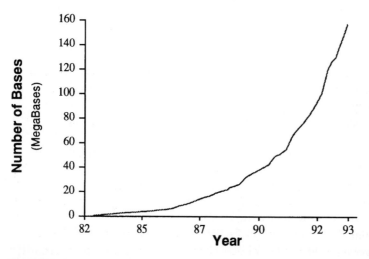

Figure 37.2 Growth of GenBank since 1982

	File	File Contents
1)	gbrel.txt	Release notes.
2)	gbsdr.txt	Short directory of the data bank.
3)	gbacc.idx	Index of the entries according to accession number.
4)	gbkey.idx	Index of the entries according to keyword phrase.
5)	gbaut.idx	Index of the entries according to author.
6)	gbjou.idx	Index of the entries according to journal citation.
7)	gbgen.idx	Index of the entries according to gene names.
8)	gbdat.frm	Forms for submitting sequences or corrections to GenBank
9)	gbpri.seq	Primate sequence entries.
10)	gbrod.seq	Rodent sequence entries.
11)	gbmam.seq	Other mammalian sequence entries.
12)	gbvrt.seq	Other vertebrate sequence entries.
13)	gbinv.seq	Invertebrate sequence entries.
14)	gbpln.seq	Plant sequence entries (including fungi and algae).
15)	gbbct.seq	Bacterial sequence entries.
16)	gbrna.seq	Structural RNA sequence entries.
17)	gbvrl.seq	Viral sequence entries.
18)	gbphg.seq	Phage sequence entries.
19)	gbsyn.seq	Synthetic and chimeric sequence entries.
20)	gbuna.seq	Unannotated sequence entries.
21)	gbest.seq	EST (expressed sequence tag) sequence entries.
22)	gbpat.seq	patent sequence entries.

Figure 37.3 GenBank distribution files

tion, and its length. Five index files that are used to index the sequence entries in the database. The five index keys are accession numbers, keywords, authors, journals, and gene symbols. These indexes are sorted alphabetically. File number 8 in Figure 37.3 is a sequence data submission form. This form can be filled out with a text editor and returned to the database via e-mail. The last file type is the sequence entry files that are used by sequence analysis software.

The GenBank database is the largest repository of genetic sequence information. In addition to human genetic sequences, it contains sequences from other species grouped into the fourteen divisions as shown Figure 37.3. Each sequence entry has the same format with a sequence of fields consisting of a label for the field and the field contents as shown in Figure 37.4.

Each line in these files has 80 characters. The field label occupies positions 1 to 10 while the content occupies positions 11 to 80 of the field's first and subsequent lines. The label may contain a keyword, a subkeyword, blank characters, a feature code, a number, or double slashes (//). Brief descriptions of the types of words that can occupy a label are as follows:

- Keyword is a record identifier that starts from position 1 (e.g, *Reference is* a keyword).
- Subkeyword starts from position 3 (e.g., *Authors* and *Title* are subkeywords of *Reference*). Blank characters in positions 1 to 10 indicate that this record is a continuation of the information under the keyword or subkeyword above it.
- Feature code, starting from position 5, indicates the nature of an entry in the Feature table. There are over 60 valid feature codes. The first few examples of these codes are allele, attenuator, and C_region.

```
LOCUS AAURRA  118 bp ss-rRNA RNA 16-JUN-1986
DEFINITION    A.auricula-judae (mushroom) 5S ribosomal RNA.
ACCESSION     K03160
KEYWORDS      5S ribosomal RNA; ribosomal RNA.
SOURCE        A.auricula-judae (mushroom) ribosomal RNA.
    ORGANISM  Auricularia auricula-judae
              Eukaryota;Fungi;Eumycota;Basidiomycotina;Phragmobasidiomycetes;
              Heterobasidiomycetidae; Auriculariales; Auriculariaceae.
REFERENCE     1 (bases 1 to 118)
    AUTHORS   Huysmans,E.,Dams,E.,Vandenberghe,A.and De Wachter,R.
    TITLE     The nucleotide sequences of the 5S rRNAs of four mushrooms and
              their use in studying the phylogenetic position of
              basidiomycetes among the eukaryotes
JOURNAL       Nucleic Acids Res. 11, 2871-2880 (1983)
    STANDARD  full automatic
FEATURES          Location/Qualifiers
    rRNA          1..118
                  /note="5S ribosomal RNA"
BASE COUNT    27 a 34 c 34 g 23 t
ORIGIN        5' end of mature rRNA.
         1    atccacggcc ataggactct gaaagcactg catcccgtcc gatctgcaaa gttaaccaga
        61    gtaccgccca gttagtacca cggtggggga ccacgcggga atcctgggtg ctgtggtt
```

Figure 37.4 An example GenBank sequence entry

- Number, ending at position 9, indicates the index of the first residue in the sequence on that line.
- Double slashes (//), in position 1 and 2, mark the end of the sequence entry.

A brief description of each record label (keyword or subkeyword) as presented in the GenBank release notes [12] is given below. Each sequence entry need not include all these keywords. The required keywords are indicated by the word "Mandatory" following the description.

- LOCUS—a short mnemonic name for the entry, chosen to suggest the sequence's definition. Mandatory keyword/exactly one record.
- Definition—a concise description of the sequence. Mandatory keyword/one or more records.
- Accession—the primary accession number is a unique, unchanging code assigned to each entry. Mandatory keyword/one or more records.
- Keywords—short phrases describing gene products and other information about an entry. Mandatory keyword in all annotated entries/one or more records.
- Segment—information on the order in which this entry appears in a series of discontinuous sequences from the same molecule. Optional keyword (only in segmented entries) / exactly one record.
- Source—common name of the organism or the name most frequently used in the literature. Mandatory keyword in all annotated entries/one or more records/includes one subkeyword.
- Organism—formal scientific name of the organism (first line) and taxonomic classification levels (second and subsequent lines). Mandatory subkeyword in all annotated entries/two or more records.
- Reference—citations for all articles containing data reported in this entry. Includes four subkeywords and may repeat. Mandatory keyword/one or more records.

- Authors—lists the authors of the citation. Mandatory subkeyword/one or more records.
- Title—full title of citation. Optional subkeyword (present in all but unpublished citations)/one or more records.
- Journal—lists the journal name, volume, year, and page numbers of the citation. Mandatory subkeyword/one or more records.
- Standard—lists information about the degree to which the entry has been annotated and the level of review to which it has been subjected. Mandatory subkeyword/exactly one record.
- Comment—cross-references to other sequence entries, comparisons to other collections, notes of changes in LOCUS names, and other remarks. Optional keyword/one or more records/may include blank records.
- Features—table containing information on portions of the sequence that code for proteins and RNA molecules and information on experimentally determined sites of biological significance. Optional keyword/one or more records.
- Base count—summary of the number of occurrences of each base code in the sequence. Mandatory keyword/exactly one record.
- Origin—specification of how the first base of the reported sequence is operationally located within the genome. Where possible, this includes its location within a larger genetic map. Mandatory keyword/ exactly one record. The ORIGIN line is followed by sequence data (multiple records).
- //—entry termination symbol. Mandatory at the end of an entry/exactly one record.

Table 37.3 highlights some statistics about GenBank. The last row summarizes the statistics of the entire GenBank. The length of the sequences are measured in base units, the number of bases in that sequence.

TABLE 37.3   GenBank (rel. 79.0) Statistics

Division	Shortest sequence	Longest sequence	No. of sequences	Total bases
bct	7	176196	13128	23424041
est	15	2916	25827	8115895
inv	15	80423	9638	15857331
mam	17	44594	4801	5488615
pat	1	9636	2730	1160837
phg	23	52297	919	1351364
pln	8	315338	13922	23524391
pri	1	180388	28427	2745501
rna	5	3029	3451	2056125
rod	5	94647	18641	20903697
syn	15	11454	1580	1882761
una	24	20500	1507	1415535
vrl	9	229354	13107	18017147
vrt	3	31111	5814	6499691
all	1	315338	143492	157152442

37.3   Residue Substitution Scoring Matrices

To compare two DNA or protein sequences, the similarities between the individual residues in these sequences must first be defined. A residue is a nucleic acid for DNA or an amino acid for a protein. A substitution scoring matrix that defines the similarities between all possible pairs of residues is needed. In this section, two such matrices are described. The first matrix is used with amino acid sequences, while the second is used with the nucleic acid sequences. These matrices contain the score for substituting or exchanging one residue by another one. In these matrices, the score for substituting or exchanging a residue by a similar one is always higher than a dissimilar one. A formal definition for the similarity between two residues will be given later.

A substitution scoring matrix for proteins must have at least 20×20 elements since there are 20 amino acids. One possible way of constructing such a matrix is to base it on the number of common bases between the two amino acids. Since each amino acid has three bases, a similarity score can be zero, one, two, or three units. That is, a score of zero is given if the two amino acids have no common bases, a score of one unit is given if there is one common base, and so on.

Dayhoff et al. [8] proved that simple scoring methods such as this one were not sophisticated enough to detect the similarity or relatedness between protein sequences. As a result, they presented a method to construct superior matrices based on combined effects of several biological factors. These factors include the nature of genetic code, the rates of mutation at the nucleotide level, and natural selection.

Table 37.4 shows a substitution scoring matrix that was constructed from Dayhoff's method. This table is referred to as the *PAM250* matrix [8]. The score M_{ij} in the matrix means that the amino acid i is expected to change to j $10^{M_{ij}/10}$ times after an evolutionary interval of 250 PAM units. One PAM (accepted point mutation) is equal to one accepted amino acid mutation per 100 links of protein. For example, amino acid C will change back to C about 16 times after an evolutionary interval of 250 PAM units. Table 37.4 is presented as a lower triangular matrix because the substitution score is symmetric. That is, the expected number of times that an amino acid X will change to Y is the same as that of Y will change to X. In addition to the 20 amino acids, this table has three additional residues (B, Z, X) and a dash. Sometimes, scientists do not wish to distinguish between amino acids D and N or E and Q. As a result, these letters are replaced by B and Z, respectively. For completeness, an unknown amino acid X is also included in the matrix, and it is assumed to be different from another unknown X. An unknown residue could be a result of a transmission error or storage media error where a valid residue was changed into an invalid one. A dash denotes a deletion. That is, a pair (A, –), or (–, A), means that A *is* deleted instead of substituted by a dash.

As pointed out by Dayhoff et al. [8], single mutations of nucleotides are seldom observed in the gene. These mutations may occur but they are probably rejected by natural selection. As a result, an identity substitution scoring matrix as shown in Table 37.5 is used for the nucleic acids (both DNA and RNA). That is, a score of one unit is given for substituting one nucleic acid by another identical one. Otherwise, a score of negative one is given. However, a score of one is also given for substituting T by U, or vice versa, since T of DNA *is* known to transcribe into U of RNA. For completeness, an unknown base X is also included in the matrix, and it is assumed to be different from another unknown base. The deletion score has been arbitrarily chosen to be –2 units so that a deletion of a base scores worse than its substitution.

Sometimes, researchers do not want to think in terms of negative scores. As a result, they add a constant positive number to all the scores in the matrix so that the largest negative number becomes zero. For example, to eliminate all negative numbers from Table 37.5, a positive number 2 must be added to all elements in this table.

TABLE 37.4   PAM250: A Substitution Scoring Matrix for Amino Acid Sequence Comparison

	C	S	T	P	A	G	N	D	E	Q	H	R	K	M	I	L	V	F	Y	W	B	Z	X	–
C	12																							
S	0	2																						
T	–2	1	3																					
P	–3	1	0	6																				
A	–2	1	1	1	2																			
G	–3	1	0	–1	1	5																		
N	–4	1	0	–1	0	0	2																	
D	–5	0	0	–1	0	1	2	4																
E	–5	0	0	–1	0	0	1	3	4															
Q	–5	–1	–1	0	0	–1	1	2	2	4														
H	–3	–1	–1	0	–1	–2	2	1	1	3	6													
R	–4	0	–1	0	–2	–3	0	–1	–1	1	2	6												
K	–5	0	0	–1	–1	–2	1	0	0	1	0	3	5											
M	–5	–2	–1	–2	–1	–3	–2	–3	–2	–1	–2	0	0	6										
I	–2	–1	0	–2	–1	–3	–2	–2	–2	–2	–2	–2	–2	2	5									
L	–6	–3	–1	–3	–1	–4	–3	–4	–3	–2	–2	–3	–3	4	2	6								
V	–2	–1	0	–1	0	–1	–2	–2	–2	–2	–2	–2	–2	2	4	2	4							
F	–4	–3	–3	–5	–4	–5	–4	–6	–5	–5	–2	–4	–5	0	1	2	–1	9						
Y	0	–3	–3	–5	–3	–5	–2	–4	–4	–4	0	–4	–4	–2	–1	–1	–2	7	10					
W	–8	–2	–5	–6	–6	–7	–4	–7	–7	–5	–3	2	–3	–4	–5	–2	–6	0	0	17				
B	–4	0	0	–1	0	0	2	3	3	1	1	–1	1	–2	–2	–3	–2	–4	–3	–5	3			
Z	–5	0	–1	0	0	0	2	3	3	1	1	–1	1	–2	–2	–3	–2	–5	–4	–6	2	3		
X	–3	0	0	–1	0	–1	–1	–1	–1	–1	–1	–1	–1	–1	–1	–1	–1	–2	–2	–4	–1	–1	–1	
–	–8	–8	–8	–8	–8	–8	–8	–8	–8	–8	–8	–8	–8	–8	–8	–8	–8	–8	–8	–8	–8	–8	–8	1

TABLE 37.5 Substitution Scoring Matrix for Nucleic Acid Sequence Comparison

	A	C	G	T	U	X	–
A	1	–1	–1	–1	–1	–1	–2
C	–1	1	–1	–1	–	–1	–2
G	–1	–1	1	–1	–1	–1	–2
T	–1	–1	–1	1	1	–1	–2
U	–1	–1	–1	1	1	–1	–2
X	–1	–1	–1	–1	–1	–1	–2
–	–2	–2	–2	–2	–2	–2	0

37.4   Sequence Comparison Algorithms

To compare two sequences, we need an algorithm that gives a measure of similarity between them. There are many simple methods for measuring the similarity between two sequences. One of these methods is to count the number of common residues between the two sequences. To take the positions of the residues into consideration, another method slides the shorter sequence along the longer one and counts the number of matches. The maximum number of matches can be used as a measure of similarity. These trivial methods do not provide good measures of similarity between two biological (genetic or protein) sequences. Biological sequences are known to mutate as they evolve. That is, some of their residues may be changed, deleted, or inserted. In addition, each pair of residues does not have the same measure of similarity as discussed in the previous section. As a result, useful methods take change (substitution), deletion, and insertion into account.

To define the terms substitution, deletion, and insertion precisely, consider an example. Let sequence A = CSTPGND and sequence B = CSDTND. One possible comparison consists of the following substitutions, deletions, and insertions.

$$CS - TPGND$$

$$CSDTN--D$$

When comparing two sequences by writing one sequence above the other, this process is referred to as *aligning the two sequences.* In the above alignment, we say that $B_3 = D$ is inserted into the first sequence, or it is deleted from the second one, depending on the view point. $A_4 = P$ is substituted by $B_5 = N$ or B_5 by A_4, again depending on the view point. The dash in the sequences represent a gap. Therefore, there is a gap of length one between A_2 and A_3 in the first sequence and a gap of length two between B_5 and B_6 in the second sequence.

Sequence comparison or alignment is a difficult problem when substitutions, deletions, and insertions are allowed. Waterman [36] proved that approximately $(1 + 2)^{2n+1} \sqrt{n}$ possible alignments exist between two sequences where n is the length of the two sequences. For example, if the length of each sequence is 1,000 residues, the number of possible alignments is approximately 10^{767}. As a result, it is hopeless to enumerate all possible alignments.

Needleman and Wunsch [25] introduced the first heuristic method that allowed substitution, deletion, and insertion for measuring the similarity between two protein sequences. Their method also gave the criterion for obtaining the optimal similarity score (measure). Basically, Needleman and Wunsch defined the similarity score between two sequences as the sum of all individual elementary similarities. The elementary similarities consist of substitutions, deletions, and insertions. This problem was later defined mathematically by Seller [30] in term of evolutionary distance between two sequences. Basically, the difference between the evolutionary distance and similarity score calculations is that the first one uses a minimization function while the other uses a maximization function. Seller's algorithm for computing the evolutionary distance between two sequences can be rewritten in term of similarity score as follows. The following notations are used in this algorithm.

$A = a_1 a_2 a_3 \ldots a_M.$
$B = b_1 b_2 b_3 \ldots b_N$
$S(A,B) =$ maximum similarity score between sequences A and B.
$s(a_i, b_j) =$ elementary score for substituting b_j for a_i.
$s(a_i, -) =$ elementary score for deleting a_i.

$s(-,b_j)$ = elementary score for inserting b_j.
$g = s(a_i,-) = s(-,b_j)$ = gap score.
$S_{i,j}$ = the accumulative similarity score between sequences A and B up to the ith and jth positions .

$$S_{i,j} = max \begin{cases} S_{i-1,j} + g \\ S_{i-1,j-1} + s(a_i, b_j) \\ S_{i,j-1} + g \end{cases} \tag{37.1}$$

Equation (37.1) is a recurrence relation that determines the similarity score between two sequences of length M and N by using the similarity scores of shorter sequences. That is, it starts from $i = 0$ and $j = 0$ and increments them until $i = M$ and $j = N$.

The initial conditions for this problem depend on the type of comparison that we wish to perform. There are three types of comparison: (1) sequence to sequence, (2) subsequence to sequence, and (3) subsequence to subsequence. If the two sequences have about the same length, we may want to use the comparison type 1 to find the similarity between the two sequences. If one sequence is significantly shorter than the other one, we may be interested in using the comparison type 2 to find segments of the longer sequence that are similar to the shorter sequence. If we are interested in finding subsequences that are common to both sequences, we use the type 3 comparison.

The three types of comparison algorithms have different initial conditions. They are defined as follows. All three types of comparisons have one condition in common that is

$$S_{0,0} = 0 \tag{37.2}$$

For the comparison type 1, a penalty is given for sliding either sequence along the other. In other words, the gaps at the end of the sequences are also penalized. These initial conditions are given as follows.

$$S_{i,0} = \sum_{k=0}^{i} g, \ (0 \le i \le M) \qquad S_{0,j} = \sum_{k=0}^{j} g, \ (0 \le j \le N) \tag{37.3}$$

For the comparison type 2, only the shorter sequence is allowed to slide along the longer one without penalty. If the shorter sequence is indexed by j, then the initial conditions are given as follows.

$$S_{i,0} = 0, \ (o \le i \le M) \qquad S_{0,j} = \sum_{k=0}^{i} g, \ (0 \le j \le N) \tag{37.4}$$

For the comparison type 3, either sequence can slide along the other one without penalty. The initial conditions are given as follows.

$$S_{i,0} = 0, \quad (0 \le i \le M) \quad S_{0,j} = 0 \quad (0 \le j \le N) \tag{37.5}$$

If the gap penalty is not constant, g must be replaced with the actual penalty function.

To see the complexity of this problem, consider the following example. Let us find the similarity score between sequences $A = aatcgatcct$ and $B = atccgcct$ using the comparison type 3. That is, we wish to find the longest segment that is common to both sequences. The substitution scoring matrix given previously in Table 37.5 is used where $g = -2$. The similarity matrix in Figure 37.5 is calculated as follows. We begin by initializing the first row $S_{0,j}$ and the first column $S_{i,0}$ to zeroes for $0 \leq i \leq 10$ and $0 \leq i \leq 8$. Then, the remaining elements can be calculated using Eq. (37.1), row by row or column by column. For example, $S_{1,1} = \text{MAX}\{0 + -2, 0 + 1, 0 + -2\} = 1$. Using this process, the entire matrix is calculated. The similarity score between these two sequences is the largest number in this matrix. For this example, the similarity score is 4.

The illustrated algorithm is a simple one. It has a constant gap function. That is, this algorithm does not distinguish between a sequence with one gap of length N from a sequence with N gaps of length one. In genetics, a mutated sequence with one gap of length N is more realistic than the one with N gaps of length one. As result, it is more useful to have an algorithm that assign a higher score to the sequence with one gap. Seller's algorithm was later generalized by Waterman et al. [37] and Smith et al. [33] to include this criterion. However, their algorithm requires $M^2 \times N$ computation steps, while the Seller's algorithm requires only $M \times M$ steps. Later, Gotoh [13] modified the Waterman et al. algorithm [37] to reduce the number of computational steps to $M \times N$. For this reason, we chose to implement the Gotoh algorithm. The Gotoh algorithm was also used by Sittig et al. [32] and Guan et al. [14] whose parallel computational approaches are discussed in detail later in this paper.

The Gotoh algorithm can be presented as follows. Let the two sequences be $A = a_1 a_2 a_3 \ldots a_M$ and $B = b_1 b_2 b_3 \ldots b_N$. Let $s(a_i, b_j)$ be a given similarity score for substituting residue a_i by b_j as given in Tables 37.4 and 37.5. The gap penalty function is defined as $w_k = -uk - v, (u,v \geq 0)$ where k is the gap length.

$$S_{i,j} = Max \begin{cases} P_{i,j} \\ S_{i-1,j-1} + s(a_i, b_i) \\ Q_{i,j} \end{cases} \qquad (37.6)$$

Sequence B

	−	a	t	c	c	g	c	c	t
−	0	0	0	0	0	0	0	0	0
a	0	1	−1	−1	−1	−1	−1	−1	−1
a	0	1	0	−2	−2	−2	−2	−2	−2
t	0	−1	−2	0	−2	−3	−3	−3	−1
c	0	−1	0	3	1	−1	−2	−2	−3
g	0	−1	−2	1	2	2	0	−2	−3
a	0	1	−1	−1	0	1	1	−1	−3
t	0	−1	2	0	−2	−1	0	0	0
c	0	−1	0	3	1	−1	0	1	−1
c	0	−1	−2	1	4	2	0	1	0
t	0	−1	0	−1	2	3	1	−1	2

Sequence A (row label along left side)

Figure 37.5 Similarity matrix for example sequences A and B

$$P_{i,j} = Max[S_{i-1,j} + w_1, P_{i-1,j} + u] \tag{37.7}$$

$$Q_{i,j} = Max[S_{i,j-1} + w_1, Q_{i,j-1} + u] \tag{37.8}$$

The initial conditions are defined in the same manner as we have done earlier. For the comparison type 3, the initial conditions are:

$$S_{i,0} = P_{i,0} = Q_{i,0} = 0 \text{ and } S_{0,j} = P_{0,j} = Q_{0,j} = 0, \text{ for } 0 \le i < M \text{ and } 0 \le j \le N$$

37.5   Parallel Techniques for Sequence Similarity Searching

One of the standard methods that researchers use to acquire new information from the genetic and protein sequence databases is to search the entire database for homologous sequences. When researchers discover new sequences, they are eager to search the databases for sequences that are similar or relevant to their discovered ones. If the newly discovered sequences contain homologies to some existing families, this provides an important clue to their possible function or structure. For example, recent studies illustrated that DNA sequence comparisons between widely varied organisms have been used to isolate or to confirm the identity of specific human genes with important functions [34].

In sequence similarity searching, we face one major difficulty that we do not encounter in business databases. As can be seen from the previous section, the comparison algorithms do not return a simple match or mismatch but instead a degree of similarity between the query and the target sequences (keys). For this reason, an index table, based on the sequences, cannot be constructed for the databases. As a result, the query sequence must be compared against all the sequences in the database for each search. On the other hand, the associated sequence information such as accession number, keyword phrases, authors, journal citations, and gene names can be indexed. For this reason, the search process is divided into two phases. In the first phase, we only retrieve the gene name, *LOCUS,* of the most similar sequence from the database. In the second phase, the retrieved gene name is used as an index to retrieve all of its associated information.

In the first phase, we need only to read the sequences and their identifications to do the comparisons. Thus, it is not efficient to use the original storage format. To achieve faster data access, we convert this format into a new one that contains only one record for each sequence entry. Each record, except the first one, contains three fields. The new format is as follows:

```
[<seq_len 1>]
[<seq_id 1> <seq_len 2> <seq 1>]...
[<seq_id n> <0> <seq n>].
```

There are two major objectives for using this format. The first objective is to be able to use an unformatted read function that is faster than the formatted one. The second objective is to perform only one read operation for each sequence entry. To achieve these objectives, all the records are stored as binary data instead of as *ASCII.* The first record contains a single number that is the length of the first sequence. The three fields of the remaining records, except the last one, are the sequence identification, the length of the next sequence, and the sequence. For the last record, the length of the next sequence is zero. The sequences have varying lengths but their identification length are fixed.

In this chapter, we only present computational approaches, for the first phase, that can be used to speed up the search for the most similar sequence in the database. Only the sequence identification, LOCUS, is retrieved. Once the LOCUS of the similar sequence has been retrieved, we can use a number of available software tools to retrieve its complete entry.

As the size of the databases increase, it is critical to find faster search techniques. To expedite the search, prior research efforts [9, 16–18, 22, 23, 31, 32] exploit the power of parallel computation. Two general methods to produce parallel operations can be used to speed up the search. One method is to make parallel the comparison operation in which all processors cooperate to determine each similarity score [9, 16–18]. The other method is to make parallel the entire search process in which each processor performs a number of comparisons independently [14, 22, 23, 31, 32]. In this second case, a processor computes the entire similarity score for the sequences it is comparing.

Later in this section, we will show that the first method has a very significant communication overhead, while the second one generally has much less. However, the second method must balance the workload among the processors. Achieving a uniform load is difficult since the sequences have varying length and require different amounts of time to compare. Thus, an effective technique must be developed to minimize the load imbalance. This partitioning is clearly an optimization problem that requires a large amount of computation to determine the optimal solution if an enumeration method is used. To obtain a satisfactory solution quickly, only heuristic approaches are considered.

The first method uses fine-grain parallelism techniques. The approaches used in this method involve a significant communication overhead. In this context, fine-grain parallelism refers to the parallelism within the comparison algorithm. To illustrate a fine-grain parallelism technique, consider an example of how each sequence comparison can be done by a parallel computer. Let $A = a_1 a_2 a_3 a_4$ and $B = b_1 b_2 b_3 b_4$. Let each processor calculate one row of the similarity matrix as shown Fig. 37.6b. The diagram in Fig. 37.6a shows the data dependency. To calculate an element in the similarity matrix, we need three previously calculated elements (top, left, and diagonal) as indicated by the arrows. In other words, the computation of an element $S_{i,j}$ depends on elements $S_{i-1,j}$, $S_{i,j-1}$, and $S_{i-1,j-1}$. Note that row 0 ($S_{0,j}$) and column 0 ($S_{i,0}$) are not shown but they are assumed to be initialized to zeroes.

Based on the dependency diagram, the elements on the same diagonal, where the sum of the i and j indexes $(i + j)$ are the same, can be computed in parallel. This is shown by the diagonal lines in the four processor implementation example represented by Figure 37.6b.

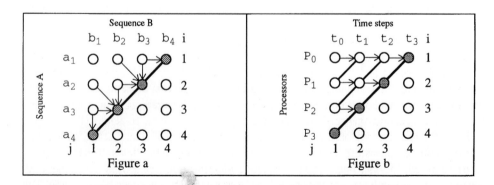

Figure 37.6 Data dependency and parallel computation diagrams for fine-grain parallel method

For example, $S_{4,1}$, $S_{3,2}$, $S_{2,3}$, and $S_{1,4}$ can be computed simultaneously at time step t_3. However, before $S_{4,1}$, $S_{3,2}$, and $S_{2,3}$, can be computed, processor P_j must obtain two elements from processor P_{j-1}. Before processor P_1 can compute $S_{2,3}$, it must obtain $S_{1,2}$ and $S_{1,3}$ from processor P_0. Only processor P_0 does not need to obtain any elements from any other processor since the elements in row 0 have been pre-initialized.

For this example, a serial computer takes $M \times N = 16$ steps to compute all elements of the similarity matrix. On a parallel computer, it takes only $M + N - 1 = 7$ steps using the decomposition strategy shown in Figure 37.6. The activity of each processor at each time step is shown in Table 37.6. Each entry $[i, j]$ in the table is an index of the similarity matrix. For instance, the entry $[1, 3]$ on row P_0 at column t_2 means that at time step 2, processor 0 calculates element $[1,3]$.

TABLE 37.6   Parallel Computation Time Steps for Fine-Grain Parallel Method

	t_0	t_1	t_2	t_3	t_4	t_5	t_6
P_0	[1, 1]	[1, 2]	[1, 3]	[1, 4]	idle	idle	idle
P_1	idle	[2, 1]	[2, 2]	[2, 3]	[2, 4]	idle	idle
P_2	idle	idle	[3, 1]	[3, 2]	[3, 3]	[3, 4]	idle
P_3	idle	idle	idle	[4, 1]	[4, 2]	[4, 3]	[4, 4]

All processors are not fully used until time step t_n in which $n + 1$ is the number of processors. At time step t_0, processor P_1 cannot begin its computation since the numbers that it needs from processor P_0 have not yet been computed and similarly for processors P_2 and P_3. Using this technique, each processor P_j must send two numbers to processor P_{j+1} for each computational step. For example, at time step t_3, processor P_0 must send $S_{1,2}$ and $S_{1,3}$ to processor P_1, processor P_1 must send $S_{2,1}$ and $S_{2,2}$ to processor P_2, and so on. As a result, there is significant communication within each comparison.

To search for the most similar sequence, we need to compare the query sequence against all the sequences in the database. The amount of communication and idle time is prohibitive for the fine-grain approach. To reduce the communication time for longer sequences where the number of processors is smaller than the length of the sequences, Edmiston et al. [9] decompose the similarity matrix into computational blocks. Each processor calculates all elements in each block before it communicates with its neighbors. This approach still has significant communication and processor idle time. Coulson et al. [7] had implemented a fine-grain approach on the *DAP*, Jones et al. [16, 17] and Lander et al. [18] on the Connection Machine, and Edmistron et al. [9] on the Intel Personal Supercomputer (*iPSC*).

The second method uses a coarser-grain parallelism approach that has lower communication overhead. However, the overall performance of the coarser-grain approach depends on its load balancing technique. The load balancing technique must be able to assign approximately about the same amount of work to each processor. We examine and later combine two previous load balancing techniques. The first technique was presented by Guan et al. [14] and the second one by Sittig et al. [31, 32].

The performance of any parallel computational approach depends on the balance of the workload among the processors. If the workload can be balanced well, better performance can be achieved. In other words, better performance can be achieved if the percentage of load imbalance is small. The percentage of load imbalance (*PLIB*) *is* defined by Eq. (37.9). One can also think of the percentage of load imbalance as approximately equaling the fraction of the overall processing time that the first finished processor has to wait for the last processor to finish. This number also indicates the degree of parallelism. For example, if *PLIB is* less than one, we achieve over 99 percent degree of parallelism. That is, over 99 percent of the total amount of processing is done in parallel. Therefore, the percentage of load imbalance can be used to compare different parallel computational approaches.

$$PLIB = \left(\frac{LargestLoad - SmallestLoad}{LargestLoad} \right) \times 100 \qquad (37.9)$$

Guan et al. [14] balanced the workload by partitioning the database into a number of portions according to the number of processors allocated. Ideally, each portion should be equal to the size of the database divided by the number of allocated processors. Typically, it is unlikely to obtain the ideal size for each portion since the database sequences have greatly varying length. That is, the last sequence assigned to each portion is either too short or too long. The last sequence is too short if it produces a portion that is smaller than the ideal one. It is too long if it causes the current portion to exceed the ideal size. If the last sequence assigned to portion P causes this portion to exceed the ideal size by more than X percent, it is reassigned to portion $P + 1$. This statement implies that X must be a non-negative number. If we want to minimize the percentage of load imbalance (*PLIB*), the value of X is should be set to zero. In a worst-case scenario, *PLIB* can be expressed as,

$$PLIB = \frac{(1 + X)\,Psize - (1 - (N - 1)\,X)\,Psize}{(1 + X)\,Psize} \times 100 \qquad (37.10)$$

where *Psize is* the ideal portion size and N is the number of nodes. In this equation, we assume that the size of every portion except the last one is equal to the allowable maximum: $(1 + X)Psize$. Equation (37.10) can be simplified into $PLIB = XN/(1 + X)$ that shows that, for any non-negative value of X, *PLIB* is minimal when X is zero.

In the Guan et al. approach, the advantage is low communication overhead. This advantage was achieved , a priori, by the search computation, assigning each processor to search its own portion independently without communicating with the other processors. The disadvantage is the potential of a high percentage of load imbalance because of the poor partitioning technique.

Sittig et al. [31, 32] also adopted a coarse-grain parallelism approach using a dynamic load balancing technique where the workload is assigned to each processor dynamically during query processing. This is in contrast to the static technique used by Guan et al. where the allocation of work is assigned prior to the commencement of the actual execution. This dynamic approach has the opposite advantage and disadvantage from the Guan et al. approach. The advantage is low *PLIB* while the disadvantage is higher communication overhead. In this approach, the sequences in the database were first sorted in decreasing length order. One processor is used as the master to distribute the sequences to the workers (the remaining processors) for comparison and to collect the similarity scores back from them. The main job of the master is to keep the workers busy as long as there is work to be done. That is, when a worker processor completes the processing of a sequence, it requests from the master an additional sequence. This form of dynamic load balancing continues until all the sequences have been compared. The sequences in the database were sorted to ensure that the last sequence compared is the shortest one.

This coarse-grain *master-worker* approach is significantly better than the fine-grain approach since it spends less time communicating among the processors. The workers need to communicate with the master only to acquire new tasks and to report new results. However, the communication overhead is higher than that of the Guan et al. approach. In addition, the master may not have enough work to do if there are too few workers. On the other hand, the master can become a system bottleneck if there are too many workers. If the master is overloaded, the workers are not kept busy.

To obtain low communication overhead, low *PLIB,* and to eliminate any processor bottlenecks, we combined the advantages of the Guan et al. and the Sittig et al. approaches. In this combined (*bucket*) approach, a greedy allocation approach (as found in Sittig et al.) *is* performed statically (as in Guan et al.). To achieve a near uniform allocation of sequences to *P* processors (buckets), the following greedy algorithm is used. First, the sequences are sorted by their length in decreasing order. Then, starting from the longest one, each sequence is placed into the bucket that has the smallest sum of sequence lengths. This, in fact, is the static equivalency of the dynamic approach of Sittig et al. One way to implement this sequence placement algorithm is given in Fig. 37.7.

Initially, the size of each bucket is set to zero by line 13. Before searching for the smallest one, we first initialize the temporary variable *size* to the largest integer. Then, to find the smallest bucket, we use a loop (lines 17 to 22) to examine all of the buckets. On line 23, the size of the chosen bucket is updated. On line 24, the sequence identification and its assigned bucket is written to the output file. When this procedure is completed, the differ-

```
1  INPUT:
2     inputfile = an input file which contains two
      field: a sequence identification and its length.
3     num_buck = the number of buckets.
4  OUTPUT:
5     outputfile = an output file which contains two field: a
      bucket number and sequence identification.
6  PROCEDURE
7     LET seq_id = a sequence identification.
8     LET seq_size = a sequence length.
9     LET bsize[i] = the sum of sequence lengths in bucket i.
10    LET size = a temporary variable.
11    LET MAXINT =the largest integer.
12 BEGIN
13    FOR (i=1 to num_buck) bsize[i]=0
14    WHILE (NOT END FILE)
15       READ seq_id and seq_size FROM inputfile
16       size=MAXINT
17       FOR (i=1 to num_buck)
18          IF(bsize[i]<size)
19             size=bsize[i]
20             buck=i
21          END_IF
22       END_FOR
23       bsize[buck]=bsize[buck]+seq_size
24       WRITE buck and seq_id TO outputfile
25    END_WHILE
26 END_PROCEDURE
```

Figure 37.7 Database sequence placement algorithm for the combined *bucket* method

ence between the smallest and the largest buckets can be found easily by examining all the sizes of buckets $bsize[1\ldots num\_buck]$.

This sequence placement problem is similar to bin packing. In the bin packing problem, the objective is to minimize the number of bins used and each bin has a limited capacity. The number of bins is minimized if the unused space in each bin is minimized. The slight distinction between the problems is that in the sequence placement problem, the number of buckets (bins) is fixed, and the capacity of each bucket is unlimited. Furthermore, the objective of this sequence placement problem is to minimize the difference between the largest and the smallest buckets not to minimize the number of buckets.

The Intel *iPSC/860* system at *NIH* was used in the following experiments to compare the different parallel approaches. It has 128 compute nodes, 8 I/O nodes, 15 disks, and 1 backup tape. To exploit the Intel *iPSC/860* concurrent I/O system, the bucket files are partitioned across the 15 disks in a manner to minimize the access time. The genetic sequences from GenBank are allocated into 128 buckets so that each compute node will not share an input file when all 128 compute nodes are used. If the number of nodes is fewer than 128, each node simply searches more than one file. Using this hypercube system, each node always searches the same number of files.

Each I/O node in this system controls two devices (disk or tape) as shown in Table 37.7. For example, I/O node 0 controls disks 0 and 8. Since there are eight I/O nodes, we want eight compute nodes to be able to read in the input data from the disks simultaneously. To achieve this goal, we distribute the 128 bucket files across the eight I/O nodes as shown in Table 37.7. Each file is stored across the two disks that are controlled by each I/O node except those files that are assigned to the last I/O node, namely I/O node 7 that has only one disk. Those files are stored on only one disk. For example, bucket file 0 is stored across disks 0 and 8 while bucket file 7 is stored on disk 7 only. To store a file across two disks, we first divide the file into 4 kB blocks and then distribute them between the two disks. Even numbered blocks are stored on the first disk and odd numbered blocks are stored on the second disk.

TABLE 37.7   Distribution of Bucket Files across Disks

I/O node	0		1		2		3		4		5		6		7
Disk	0	8	1	9	2	10	3	77	4	12	5	13	6	14	7
File	0	0	1	1	2	2	3	3	4	4	5	5	6	6	7
File	8	8	9	9	10	10	11	11	12	12	13	13	14	14	15
	⋮														
File	112	112	113	113	114	114	115	115	116	116	117	117	118	118	119
File	120	120	121	121	122	122	123	123	124	124	125	125	126	126	127

Using this parallel input/output method, the input data are accessed up to eight times faster than the serial method. This means that no additional reduction in access time is obtained for using more than eight compute nodes. If more than eight compute nodes are used, only eight of them can read in their data at the same time. Nodes 8 to 15 cannot read in their data until nodes 0 to 7 have read in theirs. The reason is that node i and $i + 8$ use the same I/O node to access their data. However, it is not a major problem that more than eight nodes cannot access the data simultaneously. First of all, not all nodes will access the data simultaneously, because the sequences in the database have varying length. Secondly, the compute node does not need its requested data right away. As long as the requested data arrives before it finished processing the previous sequence, no time is lost.

37.6   Performance

To compare the *bucket* approach with the Guan et al. *portion* approach, we calculated the *PLIB* for both approaches using the entire GenBank database. *PLIB* is calculated for a varying number of nodes based on our formula, Eq. (37.9). The variables, *LargestLoad* and *SmallestLoad,* in Eq. (37.9) are replaced by *LargestBucket* (*LargestPortion*) and *SmallestBucket* (*SmallestPorition*), respectively. The percentage of load imbalances for both approaches are shown in Table 37.8. As can be been in this table, the approach using the described greedy algorithm produced smaller *PLIB* than the *portion* approach for all cube sizes. For the *portion* approach, the *PLIB* for both a sorted and an unsorted database according to decreasing sequence lengths are presented. The sorted database was not always better than the unsorted one. The large *PLIB* found in the *portion* approach has demonstrated limited scalability [14]. For example, with 64 compute nodes, a scalability of only 58.76 was attained.

TABLE 37.8   Percentage of Load Imbalance (PLIB) Comparison between Portion and Bucket Approaches based on GenBank Database

	Portion		Bucket
No. of nodes	Unsorted	Sorted	Sorted
2	0.00135	0.00522	0.00000
4	0.01581	0.02233	0.00000
8	0.55388	0.09182	0.00002
16	1.10066	0.22202	0.00005
32	1.83130	2.77185	0.00006
64	10.10911	14.36256	0.00033
128	55.00905	35.53091	0.00033

To compare the *bucket* approach with the *master-worker* approach, we implemented both of them on the Intel *iPSCI860* hypercube. We then obtained run times for ten query sequences using the entire GenBank. Starting with the first sequence of 50 bases long, the lengths of these query sequences were increased by 50 bases. Incremental-length sequences were used as the query sequences to show the effect of sequence length on the *master-worker* approach. In this section, we identify these sequences by their lengths. For example, a query sequence with 50 bases is referred to as sequence 50.

The serial search time in seconds for the ten query sequences are shown in Table 37.9. These serial search times are obtained from executing a serial program on a single processor. For efficiency, the serial program (as well as the *master-worker* program) does not use the bucket files since only one processor is accessing the database. All sequences from the GenBank were stored across all fifteen disks as a single file. The advantage of storing the file this way is that it increases the speed of transferring the data by allowing the I/O nodes to use their caches. If the sequence is longer than 4k bases, only 4k bases are actually read from the disks. The remaining bases are read from the I/O node caches. That is, before the compute node reads the first 4 kB from the disks of the first I/O node, it informs the remaining I/O nodes to read 4 kB from their disks and place them in their caches. When the compute node finishes reading from the disk of I/O node 0, it simply reads the next 4 kB from the cache of I/O node 1 instead of from the disks and can continue on to read from the cache of I/O node 2 and so on. Reading from the cache is much faster than from the disk.

TABLE 37.9   Serial Search Times in Seconds for the Ten Query Sequences

Seq	50	100	150	200	250	300	350	400	450	500
Time	11347	22756	33796	45164	56308	67358	79611	93914	107935	122152

The parallel search times in seconds are presented in Table 37.10 for the *bucket* approach and in Table 37.11 for the *master-worker* approach. The numbers in the first row represent the sequence lengths and those in the first column represent the number of processors.

TABLE 37.10   The Search Times of the Bucket Approach in Seconds for the Ten Query Sequences and Different Numbers of Processors

	50	100	150	200	250	300	350	400	450	500
2	5687	11367	16979	22611	28339	33956	39783	47002	54038	61174
4	2890	5717	8513	11316	14158	16948	19881	23490	27006	30569
8	1411	2835	4239	5647	7072	8471	9942	11765	13537	15338
16	707	1418	2120	2824	3536	4235	4971	5873	6752	7643
32	359	711	1061	1413	1769	2118	2486	2937	3376	3822
64	191	361	534	708	886	1060	1244	1469	1689	1911
128	116	200	280	364	447	534	625	737	847	958

TABLE 37.11   The Search Times of the Master-Worker Approach in Seconds for the Ten Query Sequences and Different Numbers of Processors

	50	100	150	200	250	300	350	400	450	500
2	11336	22737	34198	45240	56350	67452	79743	93926	107955	122201
4	3821	7625	11432	15104	18819	22486	26525	31314	35974	40725
8	1633	3255	4887	6466	8061	9646	11375	13424	15437	17483
16	801	1533	2289	3023	3763	4504	5310	6267	7202	8150
32	470	783	1130	1478	1832	2187	2576	3037	3488	3947
64	469	469	617	770	933	1100	1287	1511	1730	1954
128	228	355	404	466	536	609	692	792	893	999

Using the run times in Tables 37.9, 37.10, and 37.11, we calculated the speedup factors for both approaches. We define the speedup factor as the ratio of the total run time of the serial version of the program to the total run time of the parallel version of the program. The speedup curves of the *master-worker* approach, based on the ten query sequences, are shown in Figure 37.8 and those of the *bucket* approach are shown in Figure 37.9. From these two figures, we make two observations. The first is the effect of the query sequence length on the performance of *master-worker* approach and the second is the difference between the speedup slopes of the two approaches. Relatively short sequences were used in our study since sequence fragments, rather than full length sequences, are commonly used in queries [21].

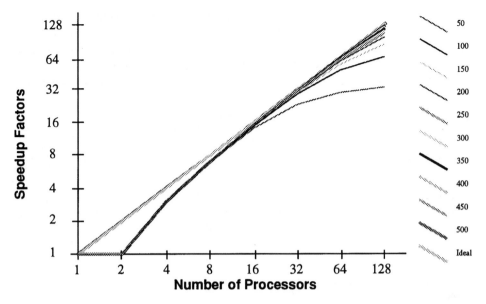

Figure 37.8 Speed-up curves of the master-worker approach for the ten incremental query sequences

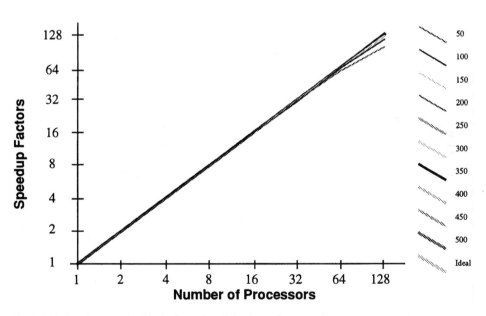

Figure 37.9 Speed-up curves of the bucket approach for the ten incremental query sequences

   In Fig. 37.8, we observed that the speedup curves deviate from the ideal with a larger deviation for smaller query sequences when the number of compute nodes was greater than sixteen. This effect is because of the fact that the master could not serve a large number of workers efficiently when a given query sequence was short because the workers finished the tasks faster than the master could distribute them. As a result, some workers had to wait idly to receive another task from the master while it was serving other workers. The shortest sequence had the greatest effect on the performance of this approach. The master becomes the bottleneck when greater than 16 nodes are used. The *master-worker* approach could achieve only a speedup factor of 34 for the shortest query sequence when 128 nodes were used. On the other hand, the length of the query sequence had only a very small effect on the *bucket* approach. That is, the speedup curves of the *bucket* approach have a very small degree of variation with respect to the sequence lengths.

Comparing the speedup curves of Fig. 37.8 with those of Fig. 37.9, we observed that the slopes of the two approaches are different. The *bucket* approach has steeper slopes. From one to two nodes, the *master-worker* approach did not achieve any speedup at all, since only one node performed the search. In addition, all the speedup factors in Fig. 37.8 are smaller than the ideal ones at all cubes smaller or larger than 16. On the other hand, all the speedup factors for the *bucket* approach are very close to the ideal ones. To quantify the improvement made by the *bucket* approach, we calculated the percentage of improvement for each combination of a query sequence length and a cube size as shown in Table 37.12. Again, the first row represents the sequence lengths and the first column represents the number of processors.

TABLE 37.12 Percentage of Improvement Made by the Bucket Approach over the Master-Worker Approach

	50	100	150	200	250	300	350	400	450	500
2	99.34	100.03	101.41	100.08	98.84	98.65	100.44	99.83	99.78	99.76
4	32.19	33.37	34.29	33.47	32.92	32.67	33.41	33.31	33.21	33.22
8	14.06	13.90	15.29	14.50	13.97	13.88	14.22	14.11	14.04	13.98
16	10.27	7.32	7.96	7.05	6.42	6.35	6.53	6.70	6.67	6.63
32	35.34	6.03	6.55	4.59	3.57	3.27	3.10	3.42	3.32	3.28
64	96.94	30.00	15.67	8.77	5.34	3.78	3.47	2.81	2.45	2.23
128	189.95	77.02	44.03	28.13	19.75	14.06	10.76	7.42	5.49	4.21

The percentage of improvement in Table 37.12 shows that the performance of the *bucket* approach is better than the *master-worker* approach in all cases. In particular, the *bucket* approach performed significantly better on both a small and a large cube. This was because the master was under-used in a smaller cube and was overloaded in a larger one. These effects can be seen more clearly in Fig. 37.10. Consider the shortest sequence as an example, starting from 16-node cube, and the improvement curves go up in either direction. That is, greater improvements were obtained when the cube size was decreased or increased from 16 nodes. When the cube size was decreased, the master was under-used. When the cube size was increased, the master was overloaded. The *master-worker* approach did not scale as well as the *bucket* approach when the query sequence was short and the number of nodes was large.

The computation time used to calculate the speedup covered the entire search including the I/O time, computation time, and communication time. That is, the entire program was timed. However, it did not include the preprocessing time involved in sorting and placing the database sequences into the 128 buckets since these tasks were not performed for every query but only once. For the current release of GenBank (rel 79.0), it took the host machine (SPARC station 2) only 70.8 seconds to sort the sequence indexes and 29.4 seconds to place these indexes into the 128 buckets.

We completed another comparison of the *master-worker* and *bucket* approaches using three quartile sequences from GenBank as the query sequences. The quartile sequences are defined as the 25th, 50th, and 75th percentile sequences. The *n*th percentile sequence is

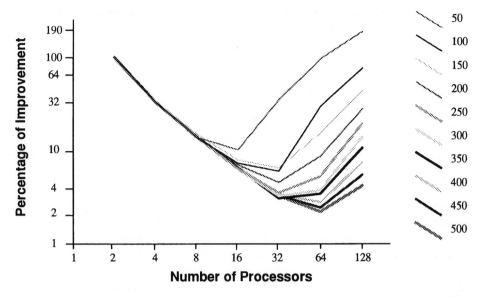

Figure 37.10 Graphic representation of the percentage of improvement made by the bucket approach over the master-worker approach

a sequence that is longer than or equal to at least *n* percent of the total number of sequences in GenBank. The 50th percentile sequence has a median length. The third quartile sequence used is 2.67 times longer than the previous longest query sequence tested. The information on the three query sequences is shown in Table 37.13.

TABLE 37.13 Quartile Query Sequences

Quartile	Locus	Length	Division
First	T00475	253 bp	EST
Second	HIV1MB503X	405 bp	VRL
Third	U01869	1335 bp	VRL

The speedup factors were calculated for all combinations of quartile query sequences and cube sizes as shown in Table 37.14. As expected from the previous experimentation, the *bucket* approach performed significantly better than the *master-worker* approach for the combination of a short query sequence and a large cube. The overall performance of the *bucket* approach is better than the *master-worker* approach in all cases. For the longest query sequence, the difference between the speedup factors of the *master-worker* and *bucket* approaches is smaller than that for the shortest query sequence, but the difference is still measurable.

37.7 Discussion and Conclusions

The results in Table 37.12 show that the *bucket* approach performed consistently better than the *master-worker* approach in all cases. In fact, the *bucket* approach achieved a near ideal speedup for all ten query sequences as shown in Fig. 37.9. On the other hand, the *master-worker* approach achieved significantly lower speedups than the ideal one when a

TABLE 37.14   Speed-Up on the Intel iPSC/860 for both the Master-Worker and the Bucket Approaches for the Three Quartile Query Sequences

No. of nodes	T00475		HIV1MB503X		U01869	
	Master...	Bucket	Master...	Bucket	Master...	Bucket
2	1.00	1.99	1.00	2.00	1.00	2.00
4	2.99	3.98	3.00	4.00	3.00	4.00
8	6.99	7.96	7.00	8.00	7.00	8.00
16	14.94	15.87	14.99	15.98	15.00	16.00
32	30.68	31.84	30.94	31.99	30.99	32.00
64	60.46	63.55	62.22	63.85	62.91	63.99
128	105.28	125.49	118.81	127.50	126.29	127.73

larger cube was used as can be in Fig. 37.8. This was because of the fact that the master was overloaded in a larger cube. As a result, we expect the *bucket* approach to perform significantly better than the *master-worker* approach on a system that has a larger number of nodes (more than 128). If we keep increasing the number of nodes, the master would eventually become a system bottleneck (heavily overloaded). On the other hand, this would not be a problem for the *bucket* approach providing the number of sequences also increases. Since the number of sequences in the GenBank is actually growing, the degree of scalability of the *bucket* approach will become of even greater importance.

In both approaches, the speedup factors of longer query sequences were better than the shorter ones. This was because of the fact that the amount of parallel computation of longer query sequences was greater than the shorter ones while the amount of communication and serial processing times remained the same. Furthermore, for shorter query sequences, there might be more contention among the compute nodes in getting access to the I/O nodes, especially when a larger number of nodes were used. That is, for short query sequences, the compute nodes process faster than the I/O nodes. Our system at *NIH* has only 8 I/O nodes. Without contention, data accessing would be completely overlapped with the computation. In other words, the computation on the current sequence does not stop while another sequence is being fetched. If more I/O nodes are available, the contention problem would become insignificant.

There are two drawbacks to our approach. The first drawback is the overhead for determining the workload for each bucket for each GenBank release. For the current release (rel. 79.0), it took the host machine (SPARC station 2) only 70.8 seconds to sort and 29.4 seconds to place all the sequences into 128 buckets. This overhead is very low relative to the overall processing time, which could take up to 34 hours, and this is not considered a significant limitation. Furthermore, GenBank is released only one to five times annually. The second drawback is a large variance in the bucket sizes that occurs if the daily updates are also placed into the buckets without rebalancing the workload. As a result, a larger percentage of load imbalance occurs. This problem is relatively limited in impact since, without significant timing penalty, we redetermine the workload in each bucket for every new GenBank release.

Both approaches—the *master-worker* and the *bucket*—do *not* work well if the database contains a sequence that is longer than the sum of all sequence lengths divided by the number of processors. With this situation, the workload cannot be balanced well among the processors. However, the GenBank database does not have this problem. As shown in Table 37.8, the percentage of load imbalance is very small.

In this chapter, we discussed only the parallel computational approaches that compare a query sequence with all the sequences in the database using the full dynamic programming algorithm. To reduce the execution time, some researchers use heuristic techniques to reduce the number of database sequences before they apply the full dynamic programming algorithm. Both Sittig et al. and Guan et al. showed that they could reduce the search time by applying some heuristic techniques. However, it is difficult to evaluate and compare these heuristic techniques because they do not produce the same set of answers.

37.8 Future Work

For the purpose of comparing our results with those of Sittig et al. and Guan et al., only the most similar sequence is retrieved from the database. In practice, a number of N similar sequences are often retrieved where N *is* usually between 5 and 50. These N sequences may belong to the same family. Therefore, future work in this area is to develop an efficient parallel computational approach for aligning these N sequences to examine their similarities in details. We want to know which parts of the sequences all the family members have in common and how the family members differ in other parts of the sequences. A multiple sequence alignment is a configuration that is obtained from stacking up sequences on top of each other. A minimal number of gaps are introduced in the sequences so that the number of matches is maximized according to a given optimization function. For some additional reading on multiple sequence alignment, see Refs. [1, 3, 20, 24].

37.9 References

1. Bacon, D. J., and W. F. Anderson. 1986. Multiple sequence alignment. *Journal of Molecular Biology,* Vol. 191, 153–162.
2. Bairoch, A., and B. Boeckmann. 1992. The SWISS-PROT protein sequence data bank. *Nucleic Acids Research,* Vol. 20, 2019–2022.
3. Berger, M. P., and P. J. Munson. 1991. A novel randomized iteration strategy for aligning multiple protein sequences. *CABIOS,* Vol. 7, No. 4, 479–484.
4. Bernstein, TF. C. F. Koetzle, G. J. B. Williams, E. F. Meyer Jr., M. D. Brice, J. R. Rodgers, O. Kennard, T. Shimanouchi, and M. Tasumi. 1977. The protein data bank: a computer-based archival file for macromolecular structures. *Journal of Molecular Biology,* Vol. 112, 535–542.
5. Bishop, M. J., and C. J. Rawlings, eds. *Nucleic Acid and Protein Sequence Analysis: A Practical Approach.* Oxford, U.K.: IRL Press Limited.
6. Burks, C., M. Cassidy, M. J. Cinkowsky, K. E. Cumella, P. Gilna, J. E. H. Hayden, G. M. Keen, T. A. Kelly, M. Kelly, D. Kristofferson, and J. Ryals. 1991. GenBank. *Nucleic Acids Research,* Vol. 19, suppl., 2221–2225.
7. Coulson, A. F. W. , J. F. Collins, and A. Lyall. 1987. Protein and nucleic acid sequence database searching: a suitable case for parallel processing. *The Computer Journal,* Vol. 30, No. 5, . 420–424.
8. Dayhoff, M. O., R. M. Schwartz, and B. C. Orcutt. 1978. A model of evolutionary change in proteins. In *Atlas of Protein Sequence and Structure,* M. O. Dayhoff, eds.,Vol. 5, Supplement 3. Silver Spring, Md.: The National Biomedical Research Foundation, 345–352.
9. Edmiston, E. W., N. G. Core, J. H. Saltz, and R. M. Smith. 1988. Parallel processing of biological sequence comparison algorithms. *International Journal of Parallel Programming,* Vol. 17, No. 3, 259–275.
10. European Molecular Biology Laboratory, Postfach 10.2209, D-6900 Heidelberg, Federal Republic of Germany, e-mail: DataLib@EMBL-Heidelberg.DE.
11. Friedland, P. and E. Kedes. 1985. Discovering the secrets of DNA. *Computer,* Vol. 18, No. 11, 49–69.
12. GenBank, National Center for Biotechnology Information, National Library of Medicine, 38A, 8N805, 8600 Rockville Pike, Bethesda, MD 20894, USA, Phone: (301) 496-2475, Fax: (301) 480-924, ftp: ncbi.nlm.nih.gov.
13. Gotoh, O. 1982. An improved algorithm for matching biological sequences. *Journal of Molecular Biology,* Vol. 162, 705–708.
14. Guan, X., R. Mural, R. Mann, and E. Uberbacher. 1991. On parallel search of *DNA* sequence databases. *Proceedings of the Fifth SIAM Conference on Parallel Processing for Scientific Computing,* 332–337.
15. G. von Heijne. 1987. *Sequence Analysis in Molecular Biology.* San Diego, Calif.: Academic Press.

16. Jones, r., W. Taylor, X. Zhang, J. P. Mesirov, and E. Lander. 1989. Protein sequence comparison the connection machine CM-2. In *Computers and DNA, SFI Studies in the Science of Complexity, G.* Bell and T. Marr, eds. Reading, Mass.: Addison-Wesley, Longman, Vol. VII, 1–9.

17. Jones, R. 1992. Sequence pattern matching on a massively parallel computer. *CABIOS,* Vol. 8, No. 4, 377–383.

18. Lander, E., J. P. Mesirov, and W. Taylor IV. 1989. Study of protein sequence comparison metrics on the Connection Machine CM-2. *Journal of Supercomputing,* Vol. 3, 255–269.

19. Lander, E., R. Langridge, and D. Saccocio. 1991. Computing in molecular biology: mapping and interpreting biological information. *Computer,* Vol. 24, No. 11, 6–13.

20. Martinez, H. M. 1988. A Flexible Multiple Sequence Alignment Program. *Nucleic Acids Research,* Vol. 16, No. 5, 1683–1691.

21. Michaels, G. S. personal communication.

22. Miller, P. L., P. M. Nadkarni, and N. M. Carriero. 1991. Parallel computation and FASTA: confronting the problem of parallel database search for a fast sequence comparison algorithm. *CABIOS,* Vol. 7., No. 1, 71–78.

23. Miller, P. L., P. M. Nadkarni, and W. R. Pearson. 1992. Comparing machine-independent versus machine-specific parallelization of a software platform for biological sequence comparison. *CABIOS,* Vol. 8. No. 2, 167–175.

24. Murata, M. 1990. Three-way Needleman-Wunsch algorithm. *Methods in Enzymology,* Vol. 183, 365–375.

25. Needleman, S. B., and C. D. Wunsch. 1970. A general method applicable to the search for similarities in the amino acid sequence of two sequences. *Journal of Molecular Biology,* Vol. 48, 443–453.

26. Protein Information Resource, National Biomedical Research Foundation, 3900 Reservoir Road, N.W., Washington, DC 20007, USA, e-mail: PIRMAIL@GUNBRF.bitnet.

27. Raven, P. H., and G. B. Johnson. 1992. *Biology,* third ed., St. Louis, Mo.: Mosby-Year Book.

28. Russell, P. J. 1992. *Genetics,* third ed. New York: Harper-Collins.

29. Sankoff, D., and J. B. Kruskal, eds. 1983. *Time warps, string edits, and macromolecules: the theory and practice of sequence comparison.* Reading, Mass.: Addison-Wesley.

30. Sellers, P. H. 1974. On the theory and computation of evolutionary distances. *SIAM J. Appl. Math.,* Vol. 26, No. 4, 787–793.

31. Sankoff, D., and J. B. Kruskal, eds. 1983. *Time Warps, String Edits, and Macromolecules: The theory and Practice of Sequence Comparison.* Reading, Mass.: Addison-Wesley.

32. Sittig, D., D. F., Foulser, N. Carriero, G. McCorkle, and P. L. Miller. 1991. A parallel computing approach to genetic sequence comparison: the master-worker paradigm with interworker communication. *Computers and Biomedical Research,* Vol. 24, 152–169.

33. Smith, T. F., and M. S. Waterman. 1981. Identification of common molecular subsequence. *Journal of Molecular Biology,* Vol. 147, 195–197.

34. U.S. Congress, Office of Technology Assessment. 1988. *Mapping Our Genes-The Genome Projects: How Big, How Fast?* OTA-BA-373 Washington, D.C.: U.S. Government Printing Office.

35. Waterman., M. S. 1989. Sequence alignments. In *Mathematical Methods for DNA Sequences,* M. S. Waterman, ed. Boca Raton, Fla.: CRC Press, 53–91.

36. Waterman., M. S. ed. 1989. *Mathematical Methods for DNA Sequences.* Boca Raton, Fla.: CRC Press.

37. Waterman, M. S., T. F. Smith, and W. A. Beyer. 1976. Some biological sequence metrics. *Advances in Mathematics,* Vol. 20, 367–387.

38. Ott, J. 1991. *Analysis of Human Genetic Linkage,* rev. ed. Baltimore: Johns Hopkins University Press.

38

Parallel Algorithms for Solving Stochastic Linear Programs

Amal De Silva and David Abramson

Parallel and distributed computing platforms are beginning to revolutionize the traditional engineering design process. Rather than performing direct experiments, researchers are now able to simulate extremely complex systems based on numerical models. When appropriately formulated, these models allow the designer to predict the behavior of real systems with great accuracy. However, the cost of such accuracy is a requirement for very large amounts of computing time. Until the advent of cheaply available parallel systems, very expensive supercomputers were required to use numerical simulation in this way. Because of the numerical techniques that are employed to solve the systems of equations in the models, it is often possible to achieve very high levels of performance from parallel systems.

Operations research techniques have been used for many years to allow engineers to optimize some aspect of their design. As in the simulation process discussed above, mathematical models are constructed that describe the problem, and then various algorithms can be applied to solve the system. However, to date, there have been very limited applications of parallel computers to these problems. The main reason for this is that many operations research algorithms are highly sequential, and thus it is not easy to achieve good speed-ups on large parallel systems. The problem with many operations research models is that the designer must make simplifications in the original real-world problem to produce a system that can be solved in a reasonable time. One area in which this is particularly relevant is in the process of incorporating uncertainty into the system. Traditionally, models are specified with exact values for parameters, and the sensitivity to those parameters is rarely explored. Stochastic programming is a technique that allows uncertainty to be incorporated in a problem's specification, and thus it is possible to produce a solution that takes account of the variance that is experienced in the real world. However, using conventional sequential computers, it is not possible to solve very large problems, because the model size becomes very large. Fortunately, some stochastic systems contain highly parallel data structures, and these can be exploited by parallel computers. Recently, there has been quite a high degree of research activity in this area. This chapter examines methods used for solving large linear stochastic optimization problems using parallel computers. Because this is a relatively new field of operations research, there are very few survey articles on the topic. Section 38.1 introduces the use of linear programming, and stochastic optimization. The methods are illustrated through a very simple example. The advanced reader may choose to skip Section 38.2. In Section 38.2 we present an overview of the various techniques that can be used to solve linear stochastic programs, and we look at their implementation on parallel systems. In Section 38.3, we present a comparison of the performance of the various methods.

38.1   Stochastic Linear Programming

38.1.1   Linear programming

Linear programming(LP) is a mathematical modeling technique used to optimize the allocation of limited resources. Typically, in such problems, one would make decisions regarding certain variables (for example the level of production in a factory). The process of selecting these variables is constrained by the availability of resources, for example the amount of machine capacity, storage space, or labor. Because the decisions interact, a change in one variable may require the change of another variable, and this means the process of selecting a set of optimal variables is non-trivial. Linear programming is a technique that allows the decision maker to compute the optimal values of the decision variables and evaluate a number of alternative designs quickly. Linear programming has been applied to many real world problems over the years. A few examples are given below.

- deciding on the production levels of manufacturing plants, constrained by factors such as machine capacity
- ensuring use of the cheapest combination of raw materials in the diet of farm animals while meeting the required nutritional specifications
- blending of crude oils to meet technical requirements and capacity constraints while maximizing profits
- scheduling crews of airline companies to minimize cost

For more details on applications of linear programming the reader is referred to Ref. [9].

To illustrate the use of linear programming, consider a manufacturing facility where the manager has to decide on how much of a certain product to produce (e.g., number of bars of chocolates) at the start of a given period, such that the amount of product that is unsold at the end of the period is minimized. To complicate the problem slightly, it may also be necessary to satisfy a known demand for the product over the time period. This problem can be formulated as a LP in the following manner:

$$
\begin{aligned}
\text{minimize} \quad & C \cdot i \\
\text{subject to} \quad & \\
x \quad -i \quad & = d \\
i \quad & \leq I \\
x \quad & \geq 0 \\
i \quad & \geq 0
\end{aligned}
$$

where x is a decision variable that represents how much to produce (e.g., number of chocolates) at the beginning of the period, and i represents the amount unsold at the end of the period. C is the cost incurred by each item that is unsold. Thus, linear program specifies that the total cost of unsold items is minimized. In this example there is no cost associated with the decision variable x and thus this variable does not make a contribution to the cost. d is the demand in this period. The first constraint expresses the relationship that the production is the demand of the product plus the amount unsold at the end of the period. The second constraint ensures that the amount unsold at the end of the period will always be less than a given maximum value I. The last two constraints ensure that the solution of the linear program will give non-negative values for the amount to be produced and the amount unsold. A key observation is that the cost and constraint equations are all linear, and therefore this modeling technique is known as *linear programming*. Problems that have a nonlinear relationship for either the cost function or the constraints are known as nonlinear programming problems. In this chapter we will only consider linear programming problems. A fairly simple analysis of the above problem yields a solution in that the number of unsold items equals zero and the production is equal to the demand. As a cost is incurred only for having unsold items at the end of the period, the optimum would be to produce just enough to satisfy the demand, and sell all the items manufactured. This can be done because the demand is known beforehand. Although the optimal cost of this simplified example can be evaluated by examination of the problem, for most practical problems the number of variables and constraints are large and a more scientific approach is required.

A general linear program can represented by the following matrix formulation:

$$(38.1)$$

$$
\begin{aligned}
\text{minimize} \quad & c^T v \\
\text{subject to} \quad & Av = b \\
& v \geq 0
\end{aligned}
$$

where v is a vector of n elements each corresponding to a decision variable of the linear program, and c is also a vector of n elements such that c_j for $j = 1 \ldots n$ is the cost incurred per item by variable i. A is an $m \times n$ matrix. Each row of A corresponds to a constraint in the linear program, where the number of constraints are m, and b is a vector of m elements that gives the right-hand side of the constraints. For the simple example described previously, m and n are both equal to two, since there are two variables and two constraints. The v, c, b vectors and A matrix are given by

$$
v = \begin{bmatrix} x \\ i \end{bmatrix} \quad c = \begin{bmatrix} 0 \\ C \end{bmatrix} \quad A = \begin{bmatrix} 1 & -1 \\ 0 & 1 \end{bmatrix} \quad b = \begin{bmatrix} d \\ 1 \end{bmatrix}
$$

Linear programs with inequality constraints can be formulated as equality constraints by adding extra variables that are referred to as slack variables. Maximization problems are converted to minimization problems by negating the cost function. There are two main approaches used to solve linear programs.

Simplex method. This method works by moving from one extreme point of the feasible region (the region of x that satisfies the constraints) to another while reducing the cost. Most introductory books on operations research or management science give a detailed description of the simplex method. (For example see Ref. [9].) It has been demonstrated with a contrived problem (Klee and Minty [31]) that the simplex method can trace a path through an exponential number of extreme points. Thus, the simplex method can have a worst case exponential time complexity. However, in practice it solves many problems in much less time than the theoretical worst case and is quite fast for reasonably sized problems.

38.1.1.1 Interior point methods. Karmarker [17] introduced the projective method to solve linear programs and demonstrate an algorithm that has a polynomial time complexity bound. This sparked a great deal of research, and a number of algorithms have been developed based on the work of Karmarker. These algorithms, which are broadly referred to as interior point methods, have the common property that all variables are kept away from the boundary of the non-negativity constraints of the linear program until the solution is obtained. While the number of iterations executed by an interior point code is generally far fewer than in the simplex method, the iterations are much slower. Thus, the polynomial time complexity bound does not guarantee that a solution is found in less time than with the simplex algorithm. Moreover, in the last few years there has been a great deal of competition between the simplex method and the interior point method, and the basic simplex algorithm has been improved enormously. Lustig, Marsten, and Shanno [19] have solved some extremely large linear programs with the their commercial interior point code OB1. While it is not clear whether the interior point codes are superior in general, they do seem to perform better on very large problems.

38.1.2   Stochastic linear programming

To model a real-world problem as a linear program, the model builder has to know the coefficients of the linear program with certainty [such as c, A, b of Eq. (38.1)], before it is solved. This is satisfactory when only one set of coefficients needs to be considered, and when the values are all know exactly. However, in many practical situations, the designer needs to include some values for the coefficients, because their exact values are not known. In many cases, a good approximation, or the expected value, is used for the uncertain coefficients. However, if the expected value is used in the LP, the resultant solution might not even be feasible (for an example see Ref. [15]) .

In the simple example given in the previous section, the demand is assumed to be deterministic (i.e., known beforehand). Therefore, it is possible to manufacture an amount that is exactly equal to the demand and to sell all the items produced. However, in practice, it is difficult to predict the actual demand for most goods that are manufactured, and thus there will always be some unsold items at the end of the period. This arises because the predicted demand is not equal to actual demand. Thus, a more realistic model would assume that the demand coefficient in the example is random.

Stochastic linear programming was proposed by George Dantzig in 1955 to overcome some of the limitations of deterministic linear programming. Stochastic programs allow the designer to incorporate variance in the model rather than using constants for the cost function and the constraint equations. However, despite of the need for stochastic coefficients, Stochastic programming has not been widely adopted. One of the main reasons for this is that stochastic programs are much larger than their deterministic equivalents and require much more computational resources to solve them. Until recently, computers have been unable to provide the performance required to solve the large real-world stochastic problems. However, recent advances in computer architecture have led to affordable vector and parallel supercomputers with the potential computational power to solve very large problems.

Stochastic programming allows the model designer to find an optimal solution under uncertainty. Rather than using exact, constant values, coefficients for uncertain events are represented as probability distributions. It is assumed that there is a known probability distribution for the random coefficients, and this may be either discrete or continuous. In this chapter, only discretely distributed stochastic linear programs will be considered, because these functions provide a convenient technique for modeling and solving real-world problems. If a continuous distribution is required, it is common to discretize the function using approximation techniques, and this is discussed elsewhere [11].

The stochasticity of the coefficients can be modeled by several methods. The reader is referred to Ref. [11] for various modeling techniques. In this chapter, a technique known as *stochastic linear programming with recourse* is considered. This modeling scheme provides a robust method to consider both the *anticipative* and *adaptive* nature of stochastic programs. The scheme is anticipative because the modeler can specify a number of potential scenarios that might occur; however, when one of them does occur in practice, it is possible to adapt the solution based on the choice.

Each realization of the discretely distributed stochastic linear program is called a *scenario*. We can illustrate the expressive power of stochastic linear programming with recourse by altering the simple example that was given earlier. Assume that the demand for a product can take two discrete values d_1 and d_2 with probability p_1 and p_2 rather than a fixed value. Because there are two possible choices, there are two scenarios and $p_1 + p_2 = 1$. If the two scenarios were considered separately, the following two LPs would be obtained. Thus, they must be considered together. However, in practice, only one demand value will occur, and thus the model must incorporate this.

$$
\begin{aligned}
minimize \quad & C \cdot i_1 \\
subject\ to \quad & \\
x_1 \quad -i_1 \quad & = d_1 \\
i_1 \quad & \leq I \\
x_1 \quad & \geq 0 \\
i_1 \quad & \geq 0
\end{aligned}
$$

LP1

$$
\begin{aligned}
minimize \quad & C \cdot i_2 \\
subject\ to \quad & \\
x_2 \quad -i_2 \quad & = d_2 \\
i_2 \quad & \leq I \\
x_2 \quad & \geq 0 \\
i_2 \quad & \geq 0
\end{aligned}
$$

LP2

The solution of LP1 and LP2 does not provide a practical method for accounting for the stochastic nature of coefficients, because the two linear programs produce two totally separate solutions. For LP1, the optimal solution is $x_1 = d_1$ and $i_1 = 0$ and, for LP2, it is $x_2 = d_2$ and $i_2 = 0$. Two different amounts are to be produced, depending on which scenario is going to occur. At the point where the decision has to be made, it is not possible tell which scenario is going to occur, and thus the decision maker would not know how much to produce. The expression "here and now" is used to describe the choice that must be made for x, because the value it takes must apply for both scenarios. On the other hand, the value for i, which is the amount unsold product held at the end of the period, depends on which scenario actually occurs. Thus, the variables in a stochastic linear program can be partitioned into two sets; one set represents decisions that have to be made "here and now," and the other set that represents the decisions that have to made after a particular scenario occurs. These two sets are called *first-stage* and *second-stage* variables, respectively. The first-stage variables should have the same value across all the scenarios. To force this to occur, some new constraints are added to specify that all first-stage variables must take the same value. These are referred to as *non-anticipatory constraints* because they avoid hindsight in selecting a scenario. In the example given before, x is a first-stage variable, while i is a second-stage variable. To form one new LP, the constraints of the two scenarios are merged to form a larger deterministic equivalent linear program. However, given that the two scenarios have two different objective functions, it is not clear what should be minimized. In the recourse models, the expected value of the objective functions is minimized. Thus, the cost function of each scenario is weighted by the probability of the scenario to form one new deterministic cost function. For scenarios with higher probability of occurrence the contribution to the objective is higher, and vice-versa. The equivalent deterministic linear program for the example is given by

$$
\begin{array}{l}
\text{minimize} \\
\quad p_1 C i_1 \qquad\qquad\qquad +p_2 C i_2 \\
\text{subject to} \\
\quad x_1 \quad -i_1 \qquad\qquad\qquad\qquad = d_1 \\
\quad\quad\ \ i_1 \qquad\qquad\qquad\qquad\quad \leq I \\
\qquad\qquad\qquad\quad x_1 \quad -i_2 \quad = d_1 \\
\qquad\qquad\qquad\qquad\ \ i_2 \quad \leq I \\
\quad x_1 \qquad\qquad -x_2 \qquad\qquad = 0 \\
\quad x_1 \quad \geq 0 \quad i_1 \qquad \geq 0 \\
\quad x_2 \quad \geq 0 \quad i_2 \qquad \geq 0
\end{array}
$$

There are two sets of constraints and variables for the two scenarios. By including a non-anticipativity constraint, $x_1 - x_2 = 0$, it is possible obtain a solution in which values for x_1 and x_2 are equal. The solution of this LP will provide a the same value for x_1 and x_2. i_1 and i_2 will have two different values. Thus the optimal amount to produce will be $x = x_1 = x_2$ and If scenario 1 occurs then i_1 is left at the end of the period while incurring a cost of $C \cdot i_1$. similarly if scenario 2 occurs, the amount leftover will be i_2. Thus, the second-stage variables provide a recourse action of adjustment after a scenario occurs.

Another possible formulation of the same problem is to use the same variable, x, in both constraints for the amount of product to be produced. This avoids the necessity for the non anticipativity constraint and also for the two variables x_1 and x_2. However, we will see later that having two separate sets of constraints is useful for parallelization of the models.

$$
\begin{array}{l}
\text{minimize} \\
\quad p_1 C i_1 \qquad\qquad\qquad +p_2 C i_2 \\
\text{subject to} \\
\quad x_1 \quad -i_1 \qquad\qquad = d_1 \\
\quad\quad\ \ i_1 \qquad\qquad\qquad \leq I \\
\quad x \qquad\qquad -i_2 \quad = d_1 \\
\qquad\qquad\qquad i_2 \quad \leq I \\
\quad x \geq 0 \quad i_1 \geq 0 \quad i_2 \geq 0
\end{array}
$$

For a general stochastic program with recourse to K scenarios, the equivalent linear program can be formulated as

$$
\begin{array}{llll}
\text{minimize} & & & \\
d^T x & +p_1 q_1{}^T y_1 & +p_2 q_2{}^T y_2 \quad \cdots & +pK q_K{}^T y_K \\
\text{subject to} & & & \\
Qx & & & = I \\
T_1 x & +W_1 y_1 & \cdots & = h_1 \\
T_2 x & & +W_2 y_2 \quad \cdots & = h_2 \\
\cdots & \cdots & \cdots \quad \cdots & = \cdots \\
T_K x & & +W_K y_K & = h_K \quad\quad 38.2
\end{array}
$$

where

K = number of scenarios

x = first-stage decision variables that have to be determined "here and now" before any scenario occurs,

d = cost vector of the first-stage decision variables,

y_j = second-stage decision (or recourse variable) of scenario j (This provides the recourse action required for each scenario to restore feasibility.)

p_j = probability of scenario $j \cdot \sum_j^K p_j = 1$

q_j = cost vector associated with second-stage variables of scenario j

I = right-hand side vector of the deterministic constraints

h_j = right-hand side vector of scenario j

Q = coefficient matrix of first-stage variables and deterministic constraints (Only the first-stage variables have nonzero coefficients in these rows.)

T_j = coefficient matrix of first-stage variables of scenario j.

W_j = coefficient matrix of second-stage variables of scenario j

If a problem has ξ random coefficients, and each random coefficients can take β independent realizations, then the total number of scenarios would be β^ζ. If ζ is equal to 10 and β is 10, then there would be 10^{10} scenarios. Thus, the size of the linear program would become too large to solve. If it were not for the non-anticipativity constraints, each scenario would be independent of the other and would be very easy to solve in a parallel, because all the scenarios are totally independent. Even with the non-anticipativity constraints, there is quite a high degree of data parallelism and, accordingly, researchers have been exploring the use of parallel computers to solve these problems. In the following sections, we will discuss some of the techniques that have been proposed and implemented, and also present some results of our own research in the area.

The two-stage techniques discussed so far have limited ability to model more complex systems in which a number of decisions must be made as scenarios unfold. To allow for this, it is possible to extend the scheme to specify a multistage problem . In this type of problem, observations are made in stages, and recourse actions are taken in each stage as the future unfolds. In general, the techniques that are used for two-stage models can be extended for multistage ones, and thus in this chapter, we will only consider two-stage systems.

38.2   Techniques for Solving Stochastic Linear Programs

In this section, we explore various algorithms solving for stochastic linear programs and the parallelization of these algorithms. Two main solution techniques have been proposed. One approach is to decompose the problem to a sequence of smaller subproblems (often one for each scenario) and combine these subproblems to solve a master problem. This technique is effective at reducing the time to find a solution on a wide range of computing platforms because it uses the solution of the smaller problems to solve the larger one. It also has value on parallel computers because the solution of the smaller systems can be performed concurrently. The second approach is to solve the large deterministic equivalent linear program using either the *simplex* or *interior point* method. Without any consideration of the structure of the matrices, this approach is not practical because, as was discussed earlier, the model grows exponentially with the number of scenarios. However, in the case of the interior point method, it is possible to exploit the block structure of the matrices to formulate a parallel or vector solution. Thus, unlike decomposition methods, this approach has value only on parallel and vector supercomputers.

Decomposition methods can be further divided into two basic types of algorithms: linear programming based and nonlinear programming based. (See Ref. [11].) The most popular linear programming based decomposition for stochastic linear programs is the *L-shape* method. The nonlinear programming based algorithms reformulate the stochastic linear programming problem to a nonlinear program.

There are numerous techniques for solving stochastic linear programs, and only the methods that have been made parallel will be discussed in this chapter. Some of the other methods may have potential for parallelization, although no work has been done in this direction to the authors' knowledge.

38.2.1   L-shaped method

There are number of techniques to decompose block structured linear programs such as stochastic linear programs. Some examples are Dantzig-Wolfe, the Benders [2], and the L-shaped method [29]. The latter has been used in practice to solve stochastic linear programs and will be discussed here. This method is applicable only for problems with a fixed T matrix. (i.e., the $T_1 = T_2 \ldots = T_K$ in the linear program given in Eq. (38.2)). The master problem is the following linear program.

$$
\begin{aligned}
&minimize &&d^T x \\
&subject\ to &&Qx = b \\
& &&x \geq 0
\end{aligned}
\tag{38.3}
$$

The jth subproblem is given by

$$
\begin{aligned}
&minimize &&q_j^T y_j \\
&subject\ to &&W_j y_j = h_j - Tx \\
& &&y_j \geq 0
\end{aligned}
\tag{38.4}
$$

The master problem consists of the constraints that only have coefficients from stage-one variables. Given a value for the x vector and substituting this in the right-hand side of the constraints yields a smaller linear program for each scenario (subproblem). It is possible to explain this approach intuitively. After solving the master problem the solution vector x is substituted in each subproblem and solved independently. The solutions of these subproblems are used to generate constraints that are then added to the master problem. The addition of these constraints makes the solution of the master problem equivalent to the solution of the stochastic linear program with respect to feasibility and optimality at some stage. This process is repeated until the condition for optimality is reached. The subproblem and the master problem are solved using the simplex method. The simplex method is used because it has the ability to obtain the optimal solution very quickly after a constraint has been added. It does this by using a method called *warm-starting*, and it uses the previously solved LP as a starting position in the vertex traversal process. The warm start capability of the interior point method is poor compared to the simplex algorithm and, for this reason, all implementations have used only the simplex method to solve both the master and subproblems of the L-shaped method. Developing a warm-start capability for the interior point method is active area of research at present. Birge ([5]) extended the L-shape method to multistage stochastic linear programs with recourse and referred to this algorithm as the *nested decomposition algorithm*.

38.2.1.1 Parallelization. A key observation about this method is that there are many subproblems, and they are independent. Consequently, it is fairly easy to solve them in parallel on a parallel computer. Only the right-hand side and cost vectors (h,q in (Equation 38.2)) differ in each subproblem and the constraint matrix is identical. Ariyawansa and Hudson [1] have bench marked a parallel version of this algorithm on some problems generated by the test problem generator GENSLP [16]. The results were produced on a Sequent/Balance shared memory multiprocessor. Good speed-ups where achieved for problems with 10,000 scenarios with 14 CPUs, while for the problems with 100 and 1000 scenarios, the speed-ups where more modest. The reason for the diminished speed-up on the smaller problems is that the master problem has to be solved sequentially, and this becomes a bottleneck in a parallel implementation. Another problem that can arise is that the convergence of this algorithm can be slow. Birge et al. [6] have implemented the parallel nested decomposition algorithm on network of RS6000 workstations.

38.2.2 Nonlinear programming-based techniques

Nonlinear programming based methods reformulate the stochastic linear program to an equivalent nonlinear program with the objective of decomposing it into independent subproblems. If each scenario is considered to be a subproblem, the requirement that the first-stage variables should take the same value in each subproblem makes the subproblems dependent on each other.

This class of algorithms work by decomposing the stochastic linear program into subproblems that are solved independently for each scenario. However, a subsequent computational step is undertaken either to aggregate the first-stage variable and/or to update the Lagrangian multipliers (explained later) in each iteration of the algorithm, rather than solving a different LP with new constraints as in the previous method.

A commonly used method in which stochastic linear programs can be decomposed into separate independent subproblems is the Augmented Lagrangian method [4]. The basic technique is to enforce the constraints by placing them in an Augmented cost function. When the augmented cost is minimized, then the original cost is minimized and the constraints are satisfied.

The Augmented Lagrangian method can be illustrated by the following example.

$$\begin{array}{ll} \textit{minimize} & f(x) \\ \textit{subject to } g_i(x) = 0 & \textit{for } i = 1 \dots m \end{array} \qquad (38.5)$$

The above problem can be used to form the following function

$$L(x, w) = f(x) + \sum_{i}^{m} w_i g_i(x) \qquad (38.6)$$

The function $L(x,w)$ is called the *Lagrangian function*, while, w_i variables s are called *Lagrangian multipliers*. There is one Lagrangian multiplier for each constraint $g_i(x) = 0$. Since a standard linear program has inequality constraints $x \geq 0$ as well as equality constraints, this method cannot be directly applied to a linear program.

The Lagrangian function can be used to find the optimal of problem given by equation (38.5). It can be shown that the optimal of equation (5) is a stationery point of the Lagrangian function. A stationery point can be either a minimum, maximum, or saddle point. To obtain the minimum of the original problem, the Lagrangian function is augmented by another penalty function to ensure that the Augmented Lagrangian can be minimized. The use of the penalty term excludes the possibility that the minimum of the original function $f(x)$, subject to the constraints, is a maximum or saddle point of the Lagrangian function. (See Ref. [12].)

This method can be used to reformulate the Stochastic linear program as Eq. (38.7), which is to be solved.

minimize

$$\phi(x,y,z,w) \equiv \sum_{j=1}^{K} p_j \{d_j^T x_j + q_j^T y_i + w_j(x_j - z) + \rho \| x_j - z \|^2\} \qquad (38.7)$$

subject to

$$\begin{array}{l} Qxj = b \\ Tx_j + W_1 y_1 = h_j \text{ for all } j = 1 \dots K \\ x_j \geq 0 \end{array} \qquad (38.8)$$

Solving this problem will give the optimum of the stochastic linear program given by Eq. (38.1). The objective function is the sum of K terms, one for each scenario weighted by its probability. Each of these terms consists of the objective function the linear program of each scenario $d_j x_j + q_j y_j$, the Lagrangian Multiplier term $w_j(x_j - z)$ for all $j = 1 \dots K$ and the Augmented Lagrangian penalty term $\rho \| x_j - z \|^2$. Here, the non-anticipativity constraints $x_j - z = 0$ are moved up to the objective function. The penalty term ensures that minimizing this function also gives the minimum of the stochastic linear program. Variable ρ is a positive penalty parameter. When the non-anticipativity constraints are not satisfied, the penalty term adds an extra cost to the objective function. Thus, minimizing this function minimizes the cost while satisfying the nonanticipativity constraints. It should be observed that the penalty term in the objective function is a quadratic, and therefore it is a quadratic program. Equation (38.7) cannot be directly decomposed into subproblems for each scenario, as there is a z term that links each subproblem.

Parallelization. There have been number of publications in which nonlinear programming based methods have been used to solve stochastic linear programs on parallel machines. Some of these techniques used are listed below

38.2.2.1 Progressive hedging algorithm. Rockafellar and Wets [25] introduced the progressive hedging algorithm that alternatively fixes z and minimizes Eq. (38.7) with respect to (x, y) and then fixes (x, y) and minimizes with respect to z. The first step can be performed in a parallel because equation (7) can be decomposed across all scenarios.

Mulvey and Vladimirou test this algorithm on a problem on stochastic portfolio management. Vladimirou and Mulvey [32] achieve near linear speed-up when a parallel implementation of the algorithm was tested on number of shared memory machines. While the algorithm scales well with the number of processors, it is inefficient compared to solving the large deterministic linear program using the interior point method. The reason for this is that this algorithm generally takes a longer time to solve than the interior point method and it provides a lower level of accuracy than the interior point algorithm because of convergence difficulties. Thus, in spite of good parallelization, it is not a particularly efficient sequential algorithm. Some results will be presented later.

38.2.2.2 Diagonal quadratic approximation (DQA) algorithm. In the DQA method the nonanticipativity constraints are reformulated as $x_j - x_j + 1 = 0$ for all $j = 1 \ldots K - 1$. Since this formulation does not have a variable z which links scenarios, it is possible to distribute scenarios to processors. But the penalty term in the Augmented Lagrangian function $\rho \| x_j -x_j+1 \|^2$ has cross-products, $x_j \cdot x_j + 1$, which makes it non-separable and thus not directly parallelizable. The DQA works by approximating the penalty term of the Augmented Lagrangian function with a local approximation. This approximation decomposes the problems into independent scenarios and thus is parallelizable. Each subproblem is a quadratic program that can be solved using a interior point based quadratic programming code. Mulvey and Ruszczynski [22] developed a parallel DQA code and used LOQO, a commercial interior point code, to solve each subproblem. The DQA method also requires a step in which Lagrangian multipliers are updated in each iteration of the algorithm. The scheme described by Mulvey and Ruszczynski does not solve each subproblem until optimality is achieved, but instead terminates after a small number of interior point iterations and sends the current values x_j to the next subproblem. This frequent updating increases the amount of communication required in a parallel implementation .

38.2.2.3 Regularized decomposition. The regularized decomposition algorithm [26] is a variant of the L-shaped method. A quadratic penalty term is added to the objective of subproblems to improve the stability and thus make it an nonlinear programming based algorithm.

38.2.2.4 Row action method. This method is based on the row action algorithm of Censor and Lent [10] and, unlike the previous two methods, is not based on the Augmented

Lagrangian. Neilsen and Zenios [24] implemented the row action algorithm on Connection Machine M-2 with up to 32,000 processing elements. This algorithm can directly solve only a certain class of nonlinear network programming problems (e.g., quadratic network programs). This method works by operating on one constraint at a time. The row action algorithm decomposes the problems by scenarios and also by rows that are data independent. Thus, a fine-grain parallelism can be achieved here, which suits the architecture of the CM-2.

38.2.3   Solving the deterministic equivalent linear program

As discussed earlier, it is possible to solve the deterministic equivalent linear program using either the simplex or interior point method. For small problems, it is practical to use commercially available simplex or interior point codes on conventional computing platforms. However, for most real-world problems, it is necessary to use many scenarios, and general linear programming packages are not efficient. Because the interior point method is generally considered to be superior to the simplex method for large problems, there have been a number of publications in which the interior point method has been directly applied to stochastic linear programs. Interior point methods are also considered more amenable for parallelization than the simplex method [33]. This because the matrices solved by the interior point method are, in general, denser than those used in the simplex method. In this section, we discuss the use of the interior point method to solve stochastic linear programs on parallel computers. The interior point method has a number of advantages over the decomposition schemes discussed. These include robustness, polynomial time complexity, and good convergence properties.

The major computational step of any interior point method in evaluating the optimal value of (1), is to solve the following equations in each iteration:

$$\begin{bmatrix} -D^{-2} & A^T \\ A & 0 \end{bmatrix} \begin{bmatrix} \Delta_v \\ \Delta_u \end{bmatrix} = \begin{bmatrix} \sigma \\ \rho \end{bmatrix} \tag{38.9}$$

in which A is the constraint matrix of the linear program. D is a matrix with only diagonal elements. The values of D change in each iteration. A^T is the transpose of the A matrix. Δv and $\Delta \mu$ are two direction vectors that are used to find the direction in each iteration to move from the current values of v towards the optimal values of v.

Most implementations solve equation (38.9) by first eliminating Δv to reduce it to the following equation

$$AD^2A^T\Delta\mu = AD^2\sigma + \rho \tag{38.10}$$

The matrix AD2AT is symmetric and positive definite provided that A is of full row rank (See Ref. [12]). Because of this property, the equation can be solved by the Cholesky factorization, which factors AD^2A^T into $LALT$. L is a lower triangular matrix were all the diagonal elements takes a value of 1.0 and A is a diagonal matrix with strictly positive elements. There are number of advantages in using the cholesky factorization. First, efficient sparse matrix codes are readily available to solve symmetric positive definite systems. Second, a factorization LAL^T is guaranteed to exist. Third, no pivoting is required for numerical stability, because A is guaranteed to have nonzero values. Because of this reason the creation of nonzeros can be predicted and the data structures required for factorization can be built before numerical factorization. This last point means that it is not necessary to manage an expanding data structure during each iteration of the factorization, reducing the cost of the method significantly. The disadvantage of this method is ADA^T matrix is a fully dense matrix if there is even one dense column in the A matrix. Thus, the advantage of using sparse cholesky factorization is lost. Consequently, another approach is to solve equation (38.9) by factoring the matrix

$$\begin{bmatrix} -D^{-2} & A^T \\ A & 0 \end{bmatrix} \tag{38.11}$$

This matrix is symmetric but not positive definite, and if factorization of this matrix to LAL^T is attempted as before, then the diagonal matrix is not guaranteed to have strictly positive values. This means that some values for A may be zero, and thus it is not possible to pivot these zero elements. Consequently, it is not feasible to use the cholesky factorization in these circumstances.

The Schur complement method is another technique [30] used to avoid the difficulty of dense columns. The constraint matrix is partitioned in the following manner:

$$[A] = [B \mid N]$$

in which N consists of the dense columns and B consists of sparse columns. Thus, the equation becomes

$$\begin{bmatrix} -D_B^{-2} & 0 & B^T \\ 0 & -D_N^{-2} & N^T \\ B & N & 0 \end{bmatrix} \begin{bmatrix} \Delta v_B \\ \Delta v_N \\ \Delta\mu \end{bmatrix} = \begin{bmatrix} \sigma_B \\ \sigma_N \\ \rho \end{bmatrix} \tag{38.12}$$

where v_B, D_B, s_B and v_N, D_N, s_N are v, D, s terms corresponding to columns in B and N respectively. Using the first set of equations to eliminate AVB results in

$$\begin{bmatrix} -D_N^{-2} & N^T \\ N & BD_B^2 B^T \end{bmatrix} \begin{bmatrix} \Delta v_N \\ \Delta\mu \end{bmatrix} = \begin{bmatrix} \sigma_N \\ \rho + BD_B^2 \sigma_B \end{bmatrix} \tag{38.13}$$

Permuting rows and columns we obtain the following system

$$\begin{bmatrix} BD_B^2 B^T & N \\ N^T & -D_N^{-2} \end{bmatrix} \begin{bmatrix} \Delta\mu \\ \Delta v_N \end{bmatrix} = \begin{bmatrix} \rho + BD_B^2 \sigma_B \\ \sigma_N \end{bmatrix} \tag{38.14}$$

Here the matrix $BD_B^2 B^T$ is symmetric positive definite (provided B is of full rank), and most importantly sparse since the dense columns have been removed. Assuming matrix

B is of full row rank after removal the dense columns, it is possible to factor it into LL^T efficiently.

$$\text{Let } P = BD_B^2B^T \; ; \; \hat{C} = -D_N^2$$

$$\varepsilon_1 = \rho + BD_B^2\sigma_B\varepsilon_2 = \sigma_N r_1 = \Delta\mu$$
$$r^2 = \Delta v_N$$

Then, equation (38.14) can be expressed as

$$\begin{bmatrix} P & N \\ N^T & \hat{C} \end{bmatrix} \begin{bmatrix} r_1 \\ r_2 \end{bmatrix} = \begin{bmatrix} \varepsilon_1 \\ \varepsilon_2 \end{bmatrix} \tag{38.15}$$

Vanderbei and Carpenter [30] have proven that the left-hand side matrix of Eq. (38.15) can be factored into $L\Lambda L^T$ where Λ is a non-singular diagonal matrix (i.e., none of the diagonal elements is zero but could take strictly negative of positive values). Thus, it is possible to factor the matrix such that

$$\begin{bmatrix} P & N \\ N^T & \hat{C} \end{bmatrix} = L\Lambda L^T \tag{38.16}$$

The matrix given by Eq. (32.16) consists of four submatrices $P, N, N^T, \hat{C}$. Thus, this matrix could be factored into $L\Lambda L^T$ as a block partitioned matrix by the following algorithm.

Step 1   Factor matrix P into $L_P L_P^T$
Step 2   Solve the triangular systems $L_P G = N$
Step 3   Solve the triangular systems $C = \hat{C} - G^T G$
Step 4   Factor matrix C into $L_C \Lambda_C L_C^T$

In this scheme, the Cholesky factorization is applied to submatrices instead of individual elements. Since P is positive definite and symmetric it is possible to use the standard cholesky factorization algorithm for Step 1 of the factorization. If the number of dense columns is small, then most the work is performed by step one, and the sparsity of P, can be exploited. Step 2 requires the solution of number of equations in the form $L_P\alpha = \beta$. The sparsity can be exploited in this step as well. Step 4 requires the factorization of a dense matrix and standard dense matrix algebra can be for this purpose. Using this technique it is possible to separate the sparse and dense parts of the constraint matrix and thereby exploit the sparsity of problem.

Although it is assumed that both the matrices A and B have full row rank in the theoretical development of the algorithm, no checks are made to see whether this condition is satisfied in practice. If these matrices do not have full row rank ADA^T and BDB^T would be positive semidefinite and zero diagonal elements would encountered while factoring. A commonly used method to overcome this difficulty is to add a small value to the diagonal elements to force ADA^T and BDB^T be positive definite [18].

As discussed previously, in a stochastic linear program, the linking between scenarios is through the first-stage variables. If the first-stage variables are considered as dense columns, then each block becomes independent, and therefore it is possible to implement Step 1 using a parallel algorithm.

38.2.4   Parallelization

The block structure of stochastic linear programs can be exploited for parallelization. If it is assumed that there are no first-stage constraints, (i.e., $Q = 0$) and the constraint matrix of the stochastic linear program is partitioned such that the first-stage variables are considered dense columns then

$$B = \begin{bmatrix} W_1 & & \cdots & 0 \\ & W_2 & & \\ & & \vdots & \vdots \\ & & \cdots & W_K \end{bmatrix} \quad N = \begin{bmatrix} T_1 \\ T_2 \\ \vdots \\ T_K \end{bmatrix} \tag{38.17}$$

and

$$P = BD_B^2 B^T = \begin{bmatrix} P_1 & 0 & \cdots & 0 \\ 0 & P_2 & \cdots & 0 \\ \vdots & \vdots & & \vdots \\ 0 & 0 & \cdots & P_K \end{bmatrix} \tag{38.18}$$

The diagonal block P_j is given by

$$P_j = W_j D_{yj}^2 W_j^T \tag{38.19}$$

As the matrix P consists only of diagonal blocks that are totally independent of each other, the factorization of P can be done in parallel. This is the basis of the parallelization of the interior point method. In most stochastic linear programming problems the number of blocks or scenarios is very large. Therefore, most of the time in solving equation (38.15) is taken by the factorization of the matrix P into $L_P L_P^T$ and for the triangular solution of either $L_{P\alpha} = \beta$ or $L_P^T \alpha = \beta$. Because of the independent diagonal blocks in P, all these computations can be done in parallel, and a substantial speed up can be achieved.

If the constraint matrix is partitioned in the presence of first-stage constraints $Qx = b$

$$B = \begin{bmatrix} 0 & & \cdots & 0 \\ W_1 & & \cdots & 0 \\ & W_2 & & \\ & & \vdots & \vdots \\ & & \cdots & W_K \end{bmatrix} \quad N = \begin{bmatrix} Q \\ T_1 \\ T_2 \\ \vdots \\ T_K \end{bmatrix} \tag{38.20}$$

The first set of constraints corresponding to $Qx = b$ constraints gives rise to an empty set of rows in the matrix B, and thus the matrix is rank deficient.

To avoid this problem, Birge and Qi [8] developed another factorization method (called BQ method) to overcome the rank deficiency of the B matrix when the first-stage columns are partitioned as dense columns. Although Birge and Qi derive this method using the Sherman-Morrison-Woodbury formula, it is computationally equivalent to reformulating $Qx = b$ by $Qx - I_{m1}x + I_{m1}z = b$, and partitioning

$$
B = \begin{bmatrix}
I_{m2} & 0 & \cdots & 0 \\
0 & W_1 & \cdots & 0 \\
0 & & W_2 & \\
\vdots & \vdots & & \vdots & \vdots \\
0 & 0 & \cdots & W_K
\end{bmatrix}
\qquad
N = \begin{bmatrix}
Q & -I_{m1} \\
T_1 & 0 \\
T_2 & 0 \\
\vdots \\
T_K & 0
\end{bmatrix}
\qquad (38.21)
$$

I_{ml} is an identity matrix with m_1 variables, where m_1 is the number of rows in the Q matrix.

Jessup, Yang, and Zenios [14] solve the system of equations in parallel that arise in each iteration of interior point method, Eq. (38.20), on a Intel iPSC/860 hypercube.

De Silva and Abramson [28] reformulated equation (38.1.2) to

minimize
$$p_1\{dx_1 + q_1y_1\} \qquad +p_2\{dx_2 + qy_2\} \qquad \cdots \qquad +pK\{dx_K + q_Ky_K\}$$
subject to
$$
\begin{aligned}
Qx_1 & & & & = 1 \\
T_1x_1 + W_1y_1 & & & & = h_1 \\
x_1 & & & & -z = 0 \\
& Qx_2 & \cdots & & = 1 \\
& T_2x_2 + W2y2 & \cdots & & -z = h_2 \\
& x_2 & & & -z = 0 \\
& & \cdots & \cdots & \cdots = \cdots \\
& & \cdots & Qx_K & = 1 \\
& & & T_Kx_K + W_Ky_K & = h_2 \\
& & & x_K & -z = 0
\end{aligned}
$$

$$(38.22)$$

The first-stage constraints are replicated in each scenario, and the non-anticipativity constraints are enforced by addition of a constraint $x_j - z = 0$. The variable, z, links the each block. The constraint matrix of this formulation is partitioned as

$$
B = \begin{bmatrix}
Q & 0 & \cdots & 0 \\
T_1W_1 & 0 & \cdots & 0 \\
I_{m_1} & 0 & \cdots & 0 \\
0 & Q & \cdots & 0 \\
0 & T_2W_2 & \cdots & 0 \\
0 & I_{m_1} & \cdots & 0 \\
\vdots & \vdots & \vdots & \vdots \\
0 & 0 & \cdots & Q \\
0 & 0 & \cdots & T_KW_K \\
0 & 0 & \cdots & I_{m_1}
\end{bmatrix}
\qquad
N = \begin{bmatrix}
0 \\
0 \\
I_{m_1} \\
0 \\
0 \\
-I_{m_1} \\
\vdots \\
0 \\
0 \\
-I_{m_1}
\end{bmatrix}
$$

$$(38.23)$$

The matrix, B, is also rank deficient because the rows due Q are linearly dependent on I_{ml}. Thus, BDB^T is positive semidefinite. De Silva and Abramson add a small multiple of the machine precision to the diagonal elements of BDB^T to force it to be positive definite. This technique was found to work very well in practice.

In this formulation, as in the previous case, the B matrix consists of independent blocks and the computations involving this matrix can be done in parallel. De Silva and Abramson [28] solve nine stochastic linear programming, problems that arise in portfolio management on a 128 processor Fujitsu AP1000 distributed memory computer.

38.3 Comparison of Methods

There are three significant issues that have to be considered in comparing the various methods available for solving stochastic linear programs. These are

- the accuracy of the solution obtained, and the stability and convergence of the algorithm
- the performance of the sequential algorithm
- the parallelization and efficiency of the parallel algorithm

Although there has been significant theoretical work on stochastic linear programming there are few comparative empirical evaluations of the schemes. Where real experimental data is available, the algorithms are often tested on different data sets and using different computer systems, and thus it is difficult to compare one scheme with another quantitatively. Furthermore, it is difficult to comment on the accuracy and convergence of all of the algorithms because this is not discussed by all researchers. Because of these difficulties, we now present a qualitative evaluation of the methods, and comment the aspects cited above where possible.

38.3.1 L-shaped method

Because there are no comparisons available between the L-shaped method and the other methods described in this chapter, it is difficult to comment on the performance of the sequential shaped method. Further, the parallelization properties seem to be poor compared to the other methods discussed in this chapter. The current folklore is that for some classes of problems the sequential L-shape method is significantly faster than the sequential interior point methods, however, we are unable to provide exact references.

The L-shaped method involves solving large number of similar subproblems and one master problem in each iteration. The advantage of the L-shape method is that the similarity between subproblems can be exploited in their solution by such techniques as shifting, bunching and bases updates [11]. Since this is a decomposition method it may experience convergence difficulties. Ariyawansa and Hudson tested a parallel implementation of the L-shaped method and achieved modest speed-ups [1]. As stated previously the master problem that is solved sequentially could become a significant bottleneck in this method. Birge et al. [6] have achieved a speed-up of 3.55 with a network of 4 RS6000 workstations on a parallel implementation of the Nested Decomposition Algorithm (An extension of the L-shape method to multistage problems). Some extremely large problems that were previously unsolvable have been solved using 8 workstations.

Ruszczynski [26] has simulated solving large multistage stochastic programming problems on a shared memory machine with seven processors. The simulations suggested near linear speed-ups on large problems.

38.3.2 Nonlinear programming

Mulvey and Vladimirou [32] tested the progressive hedging algorithm on portfolio management problems. They used a special solver to exploit the network structure of the subproblems. Lustig et al. [20] solved the same problem by solving the equivalent deterministic linear program by the interior point method with significantly lower execution times. The convergence of this algorithm was slow and the solutions obtained were not as accurate as with the use of the interior point method. Although the speed-ups achieved were good relative to the sequential progressive hedging algorithm, it was not attractive compared with solution of the deterministic equivalent linear program. De Silva and Abramson [27] implemented and tested this algorithm on two shared memory machines (4 CPU Silicon Graphics and 20 CPU Encore Multimax) using the same test problems of Mulvey and Vladimirou [23]. They used a quadratic interior point solver to solve the independent subproblems in the inner loop.

The DQA method seems to be more stable than the progressive hedging algorithm. Mulvey and Ruszczynski [21] compared the DQA method with a commercial interior point code, OB1, on the same test problems. The DQA method outperformed OB1 in all the problems tested although it took more iterations. The convergence criteria and the accuracy have not been published. A parallel implementation of this [22] method was tested on a cluster of workstations and the best speed up achieved was 5.27 with eight workstations. The speed-up decreased when the number of workstations exceeded eight. Because of the high cost of communication on a distributed network of workstations this algorithm should perform better on a tightly coupled parallel machine. Although more research is required into the DQA method, currently it is considered to be promising [19].

The row action algorithm is not compared here because it is only applicable to quadratic stochastic linear programs. Nielson and Zenios [24] have only published the timing and the MFLOPS rate on a CM-2 and no speed-ups are available. However, they report achieving 276 MFLOPS, which is quite low given the peak theoretical performance of the CM-2.

38.3.3 Interior point method

Solving the deterministic equivalent linear with the interior point method program instead of decomposing it into independent subproblems has the advantage that it inherits the convergence and proven complexity of this algorithm. To the authors' knowledge, there are no publications comparing the use of interior point method with the other methods that are available to solve stochastic linear programs.

De Silva and Abramson solved the nine portfolio management problems by Mulvey and Vladimirou on a Fujitsu AP1000, a 128-processor distributed memory machine. No numerical difficulties are reported, and 96.9 percent parallel efficiency is reported on an 800 scenario problem.

Jessup, Yang, and Zenios [14] have implemented the BQ method Intel iPSC/860 hypercube. They have only solved the system of equations that have to be solved in every iteration of the interior point method. Thus the numerical stability and accuracy of the method if solved until optimality is unknown. Almost linear speed-ups were reported when the number of first-stage variables were small compared to the number of second-stage variables for a sufficiently large number of scenarios. Birge and Holmes [7] also implement the BQ method on distributed network of workstations and tested these on problems with small numbers of scenarios. The speed-ups achieved by them was poor. (The best speed-up achieved was about 2.7 on four workstations, and it took more time to solve the same problem on five workstations than on one).

38.4   Conclusion and Future Directions

This chapter has presented and analyzed various parallel algorithms that have been developed for solving stochastic linear programs. Even though a quantitative comparison is difficult because of the variance in the reported work, we have provided a qualitative assessment of the methods.

Because of the variation in performance of optimization algorithms across various problem domains, it is unlikely that any method will be universally more efficient than others for solving stochastic problems. Thus, the efficiency of both the sequential and parallel forms of the algorithm is highly dependent on the problem. The progressive hedging method is generally considered inferior to the diagonal quadratic approximation (DQA) algorithm. The DQA method holds great promise, but further testing of this method is required. Direct solution by interior point method while exploiting the block structure of the problem is another promising direction. The current folklore is that for some classes of problems the sequential L-shape method is significantly faster than the sequential interior point methods. However, even for these problems the performance of the parallel algorithms vary greatly.

Our work has highlighted some problems in comparing the published results, because each project uses different test data and performs the work on varied computer architectures. In spite of this, a number of standard test sets have been used be researchers to test algorithms.

The three most popular test sets are

- Portfolio Management problems by Mulvey and Vladimirou [23]
- Ho and Loutes collection of staircase linear programs [13]
- U.S. Air force Scheduling problem called STORM [3]

To make a useful comparison of the available methods, it would be desirable to test all of the problems in the above test sets on the same parallel architecture.

In line with the increased competitiveness experienced by many industry sectors, many organizations are attempting to cut costs while improving productivity. Stochastic programming is a key technology in achieving these ends because it allows an organization to produce much more realistic models of their processes than has been possible to date. Moreover, the models make it possible to hedge against future uncertainty and achieve a reasonable outcome regardless of the eventual circumstances. Stochastic programming will only be viable for real-world problems if suitable algorithms can be developed that exploit the enormous computational power of current parallel computer systems. Current research has shown that this is possible for a limited class of problems. Future research must broaden the range of problems that can be solved, including non-linear and integer formulations.

38.5   References

1. Ariyawansa, K.A, and D.D Hudson. 1991. Performance of a benchmark parallel implementation of the Van Slyke and Wets algorithm for two-stage stochastic programs on a sequent/balance. *Concurrency: Practice and Experience.*, 3(2):109–128.
2. Bazaraa, M. S., and J.L. Jarvis. 1983. *Linear Programming and Network Flows.* New York: John Wiley & Sons.
3. Berger, A. J., J. M. Mulvey, and A. Ruszczynski. 1993. *Solving stochastic programs with convex objectives: extending the DQA algorithm.* Research Report SOR 93-16, Dept. of Civil Engineering and Operations Research, Princeton University.

4. Bertsekas, D. P. 1982. *Constrained Optimization and Lagrange Multiplier Methods.* San Francisco, Calif.: Academic Press.

5. Birge, J. R. 1985. Decomposition and partitioning methods for multistage stochastic linear programs. *Operations Research,* 33.

6. Birge, J. R., C.J. Donohue, D.F Holmes, and O.G. Svintsitski. 1994. *A parallel implementation of the nested decomposition algorithm for multistage stochastic linear programs.* Research Report 94-1, Department of Industrial and Operations Engineering, University of Michigan.

7. Birge, J. R., and D Holmes. 1992. Efficient solution of two-stage stochastic linear programs using interior point methods. *Computational Optimization and Applications,* 1:245–276.

8. Birge, J. R., and L. Qi. 1988. Computing block-angular karmarker projections with applications to stochastic programming. *Management Science,* 34:12:1472–1479.

9. Bradley, S. P., A. Hax, and T. L. Magnanti. 1988. *Applied Mathematical programming.* Reading, Mass.: Addison Wesley.

10. Censor, Y., and A. Lent. 1981. Iterative row action method for interval convex programming. *Journal of Optimization theory and Applications,* 34:321–353.

11. Ermoliev, Y., and R.G.B Wets. 1988. *Numerical Techniques for Stochastic Optimization.* Berlin: Springer.

12. Gill, P. E., W. Murray, and M.H. Wright. 1981. *Practical Optimization.* San Francisco: Academic Press.

13. Ho, J. K., and E. Loute. 1985. A set of staircase linear programming test problems. *Mathematical Programming,* 20:245–250.

14. Jessup, E. R., D. Yang, and S. A. Zenios. 1993. *Parallel factorization of structured matrices arising in stochastic programming.* Report 93-02, The Wharton School, University of Pennsylvania.

15. Kall, P. 1976. *Stochastic linear Programming.* Berlin: Springer-Verlag.

16. Kall, P., and E Keller. 1985. *Genslp: A program for generating input for stochastic linear programs with complete fixed recourse.* Manuscript, Institute for Operations Research, University of Zurich.

17. Karmarker, N. 1984. A new polynomial time algorithm for linear programming. *Proceedings of the 16th Annual ACM Symposium on Theory of Computing.*

18. Lustig, I. J., R. E. Marsten, and D. F Shanno. 1992. Computational experience with a primaldual interior point method for linear programming. *Linear Algebra and Its Applications,* 151:191–222.

19. Lustig, I. J., R. E. Marsten, and D. F Shanno. 1994. Interior point methods for linear programming: Computational state of the art. *ORSA Journal on Computing,* 64:1–14.

20. Lustig, I. J., J. M. Mulvey, and T.J. Carpenter. 1991. Formulating two-stage stochastic programs for interior point method. *Operations Research,* 39:5:757–770.

21. Mulvey, J. M., and A. Ruszczynski. 1990. *Diagonal quadratic approximation method for large scale linear programs.* Research Report SOR 90-8, Dept. of Civil Engineering and Operations Research, Princeton University.

22. Mulvey, J. M., and A. Ruszczynski. 1991. *A parallel interior point algorithm for large-scale stochastic optimization.* Research Report SOR 91-19, Dept. of Civil Engineering and Operations Research, Princeton University.

23. Mulvey, J. M., and H. Vladimirou. 1991. Applying the progressive hedging algorithm to stochastic generalized networks. *Annals of Operations Research,* 31:399–424.

24. Nielsen, S. S., and S. A. Zenios. A massively parallel algorithm for nonlinear stochastic network problems. *Operations Research,* 41(2):319–337.

25. Rockafellar, R. T., and R. J. B Wets. 1989. Scenarios and policy aggregation in optimization under uncertainty. *Mathematics of Operations Research,* 16(1):119–147.

26. Ruszczynski, A. 1993. A parallel decomposition of multistage of multistage stochastic programming problems. *Mathematical Programming,* 58:201–228.

27. De Silva, A., and D.A Abramson. 1993. Computational experience with parallel progressive hedging algorithm for stochastic linear programs. *Proceedings of the 6th Australian Transputer and OCCAM User Group Conference.*

28. De Silva, A., and D. A Abramson. 1994. *A parallel interior point method for stochastic linear program.* Research report CIT-94-4, Griffith University, Australia.

29. Van Slyke, R., and R. J. B. Wets. 1968. L-shaped linear programs with applications to optimal control and stochastic programming. *SIAM J. Appl. Math.,* 17:638–663.

30. Vanderbei, R. J., and T.J. Carpenter. 1993. Symmetric indefinite systems for interior point methods. *Mathematical Programming,* 58:1–32, 1993.

31. Minty, G.J., and V. L. Klee. 1970. *How good is the simplex method?* Mathematical note 643, Boeing Scientific Research Labs.

32. Vladimirou, H., and J. M. Mulvey. 1990. *Parallel and distributed computing for stochastic network programming.* Research Report SOR 90-11, Dept. of Civil Engineering and Operations Research, Princeton University.
33. Wright, M. H. 1991. Optimization and large scale computation. In *Very Large Scale Computation in the 21st Century.* J. Mesirov, ed. Thinking Machines Corporation.

39

Parallel Genetic Algorithms

Andrew Chipperfield and Peter Fleming

Stochastic search and optimization methods, based on the principles of natural biological evolution, have received considerable and increasing interest over the past decade. Introduced in the 1970s by Holland [1], genetic algorithms (GAs) are part of the larger class of evolutionary algorithms that also includes evolution strategies and genetic programming. Compared to traditional optimization methods, such as calculus-based and enumerative strategies, the GA is robust, global and generally more straightforward to apply. In recent years, GAs have been applied to a broad range of problems including ecosystem modeling, combinatorial and parametric optimization, machine intelligence, analysis of complex systems and financial prediction.

This Chapter provides a wide-ranging survey of the current trends and techniques in GAs and reviews a number of approaches that have been adopted to parallellize the GA. Starting with a tutorial level introduction, the basic operations in the GA are described in the form of a brief walk-through. Next, the major routines, such as fitness assignment and selection, are considered by referring to the model of a simple sequential GA and recent advances are described. Although the sequential model is used, most of the underlying mechanisms discussed are the same whether the GA implementation is sequential or parallel. A broad range of parallel implementations are described covering a representative selection of application areas. Finally, some conclusions are drawn about the benefits of parallel GAs over sequential ones.

39.1 What Are Genetic Algorithms?

The GA is a stochastic global search method that mimics the metaphor of natural biological evolution [1]. GAs operate on a population of potential solutions applying the principle of survival of the fittest to produce (hopefully) better and better approximations to a solution. At each generation, a new set of approximations is created by the process of selecting individuals according to their level of fitness in the problem domain and breeding them together using operators borrowed from natural genetics. This process leads to the evolution of populations of individuals that are better suited to their environment than the individuals that they were created from, just as in natural adaptation.

39.1.1   Overview of GAs

Individuals, or current approximations, are encoded as strings, *chromosomes*, composed over some alphabet(s), so that the *genotypes* (chromosome values) are uniquely mapped onto the decision variable (*phenotypic*) domain. The most commonly used representation in GAs is the binary alphabet {0, 1} although other representations can be used, e.g., ternary, integer, real-valued, etc. For example, a problem with two variables, x_1 and x_2, may be mapped onto the chromosome structure in the following way:

where x_1 is encoded with 10 bits and x_2 with 15 bits, possibly reflecting the level of accuracy or range of the individual decision variables. Examining the chromosome string in isolation yields no information about the problem we are trying to solve. It is only with the decoding of the chromosome into its phenotypic values that any meaning can be applied to the representation. However, as described below, the search process will operate on this encoding of the decision variables, rather than the decision variables themselves, except, of course, where real-valued genes are used.

Having decoded the chromosome representation into the decision variable domain, it is possible to assess the performance, or *fitness*, of individual members of a population. This is done through an objective function that characterizes an individual's performance in the problem domain. In the natural world, this would be an individual's ability to survive in its present environment. Thus, the objective function establishes the basis for selection of pairs of individuals that will be mated together during reproduction.

During the reproduction phase, each individual is assigned a fitness value derived from its raw performance measure given by the objective function. This value is used in the selection to bias towards more fit individuals. Highly fit individuals, relative to the whole population, have a high probability of being selected for mating whereas less fit individuals have a correspondingly low probability of being selected.

Once the individuals have been assigned a fitness value, they can be chosen from the population, with a probability according to their relative fitness, and recombined to produce the next generation. Genetic operators manipulate the characters (genes) of the chromosomes directly, using the assumption that certain individual's gene codes, on average, produce fitter individuals. The *recombination* operator is used to exchange genetic information between pairs, or larger groups, of individuals. The simplest recombination operator is that of single-point crossover.

Consider the two parent binary strings:

$$P_1 = 1\ 0\ 0\ 1\ 0\ 1\ 1\ 0, \text{ and}$$

$$P_2 = 1\ 0\ 1\ 1\ 1\ 0\ 0\ 0.$$

If an integer position, i, is selected uniformly at random from the range $[1, l - 1]$, where l is the string length, and the genetic information exchanged between the individuals about this point, then two new offspring strings are produced. The two offspring below are produced when the crossover point $i = 5$ is selected,

$$O_1 \; = \; 1 \;\; 0 \;\; 0 \;\; 1 \;\; 0 \;|\; 0 \;\; 0 \;\; 0, \text{ and}$$

$$O_2 \; = \; 1 \;\; 0 \;\; 1 \;\; 1 \;\; 1 \;|\; 1 \;\; 1 \;\; 0.$$

This crossover operation is not necessarily performed on all strings in the population. Instead, it is applied with a probability P_x when the pairs are chosen for breeding. A further genetic operator, called *mutation*, is then applied to the new chromosomes, again with a set probability, P_m. Mutation causes the individual genetic representation to be changed according to some probabilistic rule. In the binary string representation, mutation will cause a random bit to change its state, $0 \Rightarrow 1$ or $1 \Rightarrow 0$. So, for example, mutating the fourth bit of O_1 leads to the new string,

$$O_{1m} \; = \; 1 \;\; 0 \;\; 0 \;\; 0 \;\; 0 \;\; 0 \;\; 0 \;\; 0.$$

Mutation is generally considered to be a background operator that ensures the probability of searching a particular subspace of the problem space is never zero. This has the effect of tending to inhibit the possibility of converging to a local optimum, rather than the global optimum.

After recombination and mutation, the individual strings are then, if necessary, decoded, the objective function evaluated, a fitness value assigned to each individual and individuals selected for mating according to their fitness, and so the process continues through subsequent generations. In this way, the average performance of individuals in a population is expected to increase, as good individuals are preserved and bred with one another and the less fit individuals die out. The GA is terminated when some criteria are satisfied, e.g., a certain number of generations completed, a mean deviation in the performance of individuals in the population, or when a particular point in the search space is encountered.

39.1.2   GAs versus traditional methods

From the above discussion, it can be seen that the GA differs substantially from more traditional search and optimization methods. The four most significant differences are:

- GAs search a population of points in parallel, not a single point.
- GAs use probabilistic transition rules, not deterministic ones.
- GAs work on an encoding of the parameter set rather than the parameter set itself (except in where real-valued individuals are used).
- GAs do not require derivative information or other auxiliary knowledge; only the objective function and corresponding fitness levels influence the directions of search.

It is important to note that the GA provides a number of potential solutions to a given problem and the choice of final solution is left to the user. In cases where a particular problem does not have one individual solution, for example, a family of Pareto-optimal solutions, as is the case in multiobjective optimization problems, then the GA is potentially useful for identifying these alternative solutions simultaneously.

39.2   Major Elements of the Genetic Algorithm

The simple genetic algorithm (SGA) is described by Goldberg [2] and is used here to illustrate the basic components of the GA. A pseudocode outline of the SGA is shown in Fig. 39.1. The population at time t is represented by the time-dependent variable P, with the initial population of random estimates being $P(0)$. Using this outline of a GA, the remainder of this section describes the major elements of the GA.

39.2.1   Population representation and initialization

GAs operate simultaneously on a number of potential solutions, called a population, consisting of some encoding of the parameter set. Typically, a population is composed of between 30 and 100 individuals, although a variant called the micro GA uses very small populations, ~10 individuals, with a restrictive reproduction and replacement strategy in an attempt to satisfy real-time execution requirements [3].

The most commonly used representation of chromosomes in the GA is that of the single-level binary string. Here, each decision variable in the parameter set is encoded as a binary string and these are concatenated to form a chromosome (see the example in Section 39.1.1). The use of Gray coding has been advocated as a method of overcoming the hidden representational bias in conventional binary representation because the Hamming distance between adjacent values is constant [4]. Empirical evidence of Caruana and Schaffer [5] suggests that large Hamming distances in the representational mapping between adjacent values, as is the case in the standard binary representation, can result in the search process being deceived or unable to efficiently locate the global minimum. A further approach of Schmitendorgf, et al. [6], is the use of logarithmic scaling in the conversion of binary-coded chromosomes to their real phenotypic values. Although the precision of the parameter values is possibly less consistent over the desired range, in problems where the spread of feasible parameters is unknown a larger search space may be covered with the same number of bits than a linear mapping scheme, thus allowing the computational burden of exploring unknown search spaces to be reduced to a more manageable level.

While binary-coded GAs are most commonly used, there is an increasing interest in alternative encoding strategies, such as integer and real-valued representations. For some problem domains, it is argued that the binary representation is, in fact, deceptive in that it obscures the nature of the search [7]. In the subset selection problem [8], for example, the

```
procedure GA
begin
            t = 0;
            initialize P(t);
            evaluate P(t);
            while not finished do
            begin
                        t = t + 1;
                        select P(t) from P(t-1);
                        reproduce pairs in P(t);
                        evaluate P(t);
            end
end.
```

Figure 39.1 A simple genetic algorithm

use of an integer representation and look-up tables provides a convenient and natural way of expressing the mapping from representation to problem domain. Consider the traveling sales-person problem, the task is to find the shortest route visiting all the cities from a given set exactly once. By using integer labels, each candidate solution can be uniquely represented as a permutation of these elements. For example, in a seven-city tour, both {2, 7, 1, 3, 5, 6, 4} and {6, 4, 7, 1, 5, 3, 2} represent paths between the cities. Thus, the chromosomes used in a GA to solve this problem would contain seven integers, each integer corresponding to a city in the tour.

The use of real-valued genes in GAs is claimed by Wright [9] to offer a number of advantages in numerical function optimization over binary encodings. Efficiency of the GA is increased because there is no need to convert chromosomes to phenotypes before each function evaluation, less memory is required because efficient floating-point internal computer representations can be used directly, there is no loss in precision by discretization to binary or other values, and there is greater freedom to use different genetic operators. The use of real-valued encodings is described in detail by Michalewicz [10] and in the literature on Evolution Strategies. (See, for example, [11].)

Having decided on the representation, the first step in the SGA is to create an initial population. This is usually achieved by generating the required number of individuals using a random number generator that uniformly distributes numbers in the desired range. For example, with a binary population of N_{ind} individuals whose chromosomes are L_{ind} bits long, $N_{ind} \times L_{ind}$ random numbers uniformly distributed from the set {0, 1} would be produced.

A variation is the *extended random initialization* procedure of Bramlette [7] whereby a number of random initializations are tried for each individual and the one with the best performance is chosen for the initial population. Other users of GAs have seeded the initial population with some individuals that are known to be in the vicinity of the global minimum. (See, for example, [12] and [13].) This approach is, of course, only applicable if the nature of the problem is well understood beforehand or if the GA is used in conjunction with a knowledge based system.

39.2.2   The objective and fitness functions

The objective function is used to provide a measure of how individuals have performed in the problem domain. In the case of a minimization problem, the most fit individuals will have the lowest numerical value of the associated objective function. This raw measure of fitness is usually only used as an intermediate stage in determining the relative performance of individuals in a GA. Another function, the *fitness function*, is normally used to transform the objective function value into a measure of relative fitness [14], thus:

$$F(x) = g(f(x)) \tag{39.1}$$

where f is the objective function, g transforms the value of the objective function to a nonnegative number and F is the resulting relative fitness. This mapping is always necessary when the objective function is to be minimized as the lower objective function values correspond to fitter individuals. In many cases, the fitness function value corresponds to the number of offspring that an individual can expect to produce in the next generation. A commonly used transformation is that of proportional fitness assignment. (See, for example, [2].) The individual fitness, $F(x_i)$, of each individual is computed as the individual's raw performance, $f(x_i)$, relative to the whole population, i.e.,

$$F(x_i) = \frac{f(x_i)}{\displaystyle\sum_{i=1}^{N_{ind}} f(x_i)} \tag{39.2}$$

where N_{ind} is the population size and x_i is the phenotypic value of individual i. While this fitness assignment ensures that each individual has a probability of reproducing according to its relative fitness, it fails to account for negative objective function values.

A linear transformation that offsets the objective function [2] is often used prior to fitness assignment, such that,

$$F(x) = af(x) + b \tag{39.3}$$

where a is a positive scaling factor if the optimization is maximizing and negative if we are minimizing. The offset b is used to ensure that the resulting fitness values are nonnegative.

The use of linear scaling and offsetting outlined above is, however, a possible cause of rapid convergence. The *selection* algorithm (see below) selects individuals for reproduction on the basis of their relative fitness. Using linear scaling, the expected number of offspring is approximately proportional to that individual's performance. Because there is no constraint on an individual's performance in a given generation, highly fit individuals in early generations can dominate the reproduction causing rapid convergence to possibly suboptimal solutions. Similarly, if there is little deviation in the population, then scaling provides only a small bias towards the most fit individuals.

A further method of transforming the objective function values to fitness measures is power law scaling [2]. Here, the scaled fitness is taken as some specified power, k, of the raw fitness f:

$$F(x) = f(x)^k \tag{39.4}$$

The value of k is, in general, problem dependant and may be dynamically changed during the execution of the GA to shrink or stretch the range of fitness measures as required.

Baker [15] suggests that limiting the reproductive range, so that no individuals generate an excessive number of offspring, prevents premature convergence. Here, individuals are assigned a fitness according to their rank in the population rather than their raw performance. One variable, *SP*, is used to determine the bias, or *selective pressure*, towards the most fit individual and the fitness of the others is determined by,

$$F(x_i) = 2 - SP + 2(SP - 1)\frac{x_i - 1}{N_{ind} - 1} \tag{39.5}$$

where x_i is the position in the ordered population of individual i.

For example, for a population size of $N_{ind} = 40$ and selective pressure of $SP = 1.1$, individuals are given a fitness value in the range [0.9, 1.1]. The least fit individual has a fitness of 0.9 while the most fit is assigned a fitness of 1.1. The increment in the fitness value between adjacent ranks is thus 0.0051.

39.2.3   Selection

Selection is the process of determining the number of times, or *trials*, a particular individual is chosen for reproduction and, thus, the number of offspring that an individual will produce. The selection of individuals can be viewed as two separate processes:

1. determination of the number of trials an individual can expect to receive
2. conversion of the expected number of trials into a discrete number of offspring

The first part is concerned with the transformation of raw fitness values into a real-valued expectation of an individual's probability to reproduce and is dealt with in the previous subsection as fitness assignment. The second part is the probabilistic selection of individuals for reproduction based on the fitness of individuals relative to one another and is sometimes known as *sampling*. The remainder of this subsection will review some of the more popular selection methods in current usage.

Baker [16] presented three measures of performance for selection algorithms, *bias*, *spread* and *efficiency*. Bias is defined as the absolute difference between an individual's actual and expected selection probability. Optimal zero bias is therefore achieved when an individual's selection probability equals its expected number of trials.

Spread is the range in the possible number of trials that an individual may achieve. If $f(i)$ is the actual number of trials that individual i receives, then the "minimum spread" is the smallest spread that theoretically permits zero bias, i.e.,

$$f(i) \in \left\{ \lfloor et(i) \rfloor, \lceil et(i) \rceil \right\}$$
(39.6)

where $et(i)$ is the expected number of trials of individual i, $\lfloor et(i) \rfloor$ is the floor of $et(i)$ and $\lceil et(i) \rceil$ is the ceiling. Thus, while bias is an indication of accuracy, the spread of a selection method measures its consistency.

The desire for efficient selection methods is motivated by the need to maintain a GAs overall time complexity. It has been shown in the literature that the other phases of a GA (excluding the actual objective function evaluations) are $O(L_{ind} \cdot N_{ind})$ or better time complexity. The selection algorithm should thus achieve zero bias while maintaining a minimum spread and not contributing to an increased time complexity of the GA.

Roulette wheel selection methods. Many selection techniques employ a "roulette wheel" mechanism to select individuals probabilistically, based on some measure of their performance. A real-valued interval, *Sum*, is determined as either the sum of the individuals' expected selection probabilities or the sum of the raw fitness values over all the individuals in the current population. Individuals are then mapped one-to-one into contiguous intervals in the range [0, *Sum*]. The size of each individual interval corresponds to the fitness value of the associated individual. For example, in Fig. 39.2 the circumference of the roulette wheel is the sum of all six individual's fitness values. Individual 5 is the most fit individual and occupies the largest interval, whereas individuals 6 and 4 are the least fit and have correspondingly smaller intervals within the roulette wheel. To select an individual, a random number is generated in the interval [0, *Sum*] and the individual whose segment spans the random number is selected. This process is repeated until the desired number of individuals have been selected.

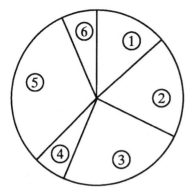

Figure 39.2 Roulette wheel selection

The basic roulette wheel selection method is stochastic sampling with replacement (SSR). Here, the segment size and selection probability remain the same throughout the selection phase and individuals are selected according to the procedure outlined above. SSR gives zero bias but a potentially unlimited spread. Any individual with a segment size > 0 could entirely fill the next population.

Stochastic sampling with partial replacement (SSPR) extends upon SSR by resizing an individual's segment if it is selected. Each time an individual is selected, the size of its segment is reduced by 1.0. If the segment size becomes negative, then it is set to 0.0. This provides an upper bound on the spread of $\lceil et(i) \rceil$. However, the lower bound is zero and the bias is higher than that of SSR.

Remainder sampling methods involve two distinct phases. In the integral phase, individuals are selected deterministically according to the integer part of their expected trials. The remaining individuals are then selected probabilistically from the fractional part of the individuals expected values. Remainder stochastic sampling with replacement (RSSR) uses roulette wheel selection to sample the individual not assigned deterministically. During the roulette wheel selection phase, individual's fractional parts remain unchanged and, thus, compete for selection between "spins." RSSR provides zero bias and the spread is lower bounded. The upper bound is limited only by the number of fractionally assigned samples and the size of the integral part of an individual. For example, any individual with a fractional part > 0 could win all the samples during the fractional phase. Remainder stochastic sampling without replacement (RSSWR) sets the fractional part of an individual's expected values to zero if it is sampled during the fractional phase. This gives RSSWR minimum spread, although this selection method is biased in favor of smaller fractions.

Stochastic universal sampling. Stochastic universal sampling (SUS) is a single-phase sampling algorithm with minimum spread and zero bias. Instead of the single selection pointer employed in roulette wheel methods, SUS uses N equally spaced pointers, where N is the number of selections required. The population is shuffled randomly and a single random number in the range [0 Sum/N] is generated, ptr. The N individuals are then chosen by generating the N pointers spaced by 1, [ptr, $ptr + 1$, ..., $ptr + N - 1$], and selecting the individuals whose fitnesses span the positions of the pointers. An individual is thus guaranteed to be selected a minimum of $\lfloor et(i) \rfloor$ times and no more than $\lceil et(i) \rceil$, thus achieving minimum spread. In addition, as individuals are selected entirely on their position in the population, SUS has zero bias. For these reasons, SUS has become one of the most widely used selection algorithm in current GAs.

39.2.4   Crossover (recombination)

The basic operator for producing new chromosomes in the GA is that of crossover. Like its counterpart in nature, crossover produces new individuals that have some parts of both parent's genetic material. The simplest form of crossover is that of single-point crossover, described in the Overview of GAs in Section 39.1.1. In this section, a number of variations on crossover are described and discussed and the relative merits of each reviewed.

Multipoint crossover. For multipoint crossover, m crossover positions,

$$k_i \in \{1, 2, ..., l-1\}$$

where k_i are the crossover points and l is the length of the chromosome, are chosen at random with no duplicates and sorted into ascending order. Then, the bits between successive crossover points are exchanged between the two parents to produce two new offspring. The section between the first allele position and the first crossover point is not exchanged between individuals. This process is illustrated in Fig. 39.3.

The idea behind multipoint, and indeed many of the variations on the crossover operator, is that the parts of the chromosome representation that contribute most to the performance of a particular individual may not necessarily be contained in adjacent substrings [17]. Further, the disruptive nature of multipoint crossover appears to encourage the exploration of the search space, rather than favoring convergence to highly fit individuals early in the search, thus making the search more robust [18].

Uniform crossover. Single and multipoint crossover define cross points as places between loci where a chromosome can be split. Uniform crossover [19] generalizes this scheme to make every locus a potential crossover point. A crossover mask, the same length as the chromosome structures is created at random and the parity of the bits in the mask indicates which parent will supply the offspring with which bits. Consider the following two parents, crossover mask and resulting offspring:

$$
\begin{aligned}
P_1 &= 1\,0\,1\,1\,0\,0\,0\,1\,1\,1 \\
P_2 &= 0\,0\,0\,1\,1\,1\,1\,0\,0\,0 \\
\text{Mask} &= 0\,0\,1\,1\,0\,0\,1\,1\,0\,0 \\
O_1 &= 0\,0\,1\,1\,1\,1\,0\,1\,0\,0 \\
O_2 &= 1\,0\,0\,1\,0\,0\,1\,0\,1\,1
\end{aligned}
$$

Here, the first offspring, O_1, is produced by taking the bit from P_1 if the corresponding mask bit is 1 or the bit from P_2 if the corresponding mask bit is 0. Offspring O_2 is created using the inverse of the mask or, equivalently, swapping P_1 and P_2.

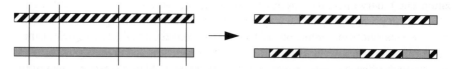

Figure 39.3 Multipoint crossover ($m = 5$)

Uniform crossover, like multipoint crossover, has been claimed to reduce the bias associated with the length of the binary representation used and the particular coding for a given parameter set. This helps to overcome the bias in single-point crossover towards short substrings without requiring precise understanding of the significance of individual bits in the chromosome representation. Spears and De Jong [20] have demonstrated how uniform crossover may be parameterized by applying a probability to the swapping of bits. This extra parameter can be used to control the amount of disruption during recombination without introducing a bias towards the length of the representation used. When uniform crossover is used with real-valued alleles, it is usually referred to as *discrete recombination*.

Other crossover operators. A related crossover operator is that of *shuffle* [21]. A single cross-point is selected, but before the bits are exchanged, they are randomly shuffled in both parents. After recombination, the bits in the offspring are unshuffled. This too removes positional bias as the bits are randomly reassigned each time crossover is performed.

The *reduced surrogate* operator [17] constrains crossover to always produce new individuals wherever possible. Usually, this is implemented by restricting the location of crossover points such that crossover points only occur where gene values differ.

Intermediate recombination. Given a real-valued encoding of the chromosome structure, intermediate recombination is a method of producing new phenotypes around and between the values of the parents phenotypes [22]. Offspring are produced according to the rule,

$$O_1 = P_1 \times \alpha (P_2 - P_1) \quad , \tag{39.7}$$

where α is a scaling factor chosen uniformly at random over some interval, typically $[-0.25, 1.25]$ and P_1 and P_2 are the parent chromosomes. (See, for example, Ref. [22].) Each variable in the offspring is the result of combining the variables in the parents according to the above expression with a new α chosen for each pair of parent genes. In geometric terms, intermediate recombination is capable of producing new variables within a slightly larger hypercube than that defined by the parents but constrained by the range of α as shown in Fig. 39.4.

Line recombination. Line recombination [22] is similar to intermediate recombination, except that only one value of α is used in the recombination. Fig. 39.5 shows how line recombination can generate any point on the line defined by the parents within the limits of the perturbation, α, for a recombination in two variables.

Discussion. The binary operators discussed in this section have all, to some extent, used disruption in the representation to help improve exploration during recombination. While these operators may be used with real-valued populations, the resulting changes in the genetic material after recombination would not extend to the actual values of the decision variables, although offspring may, of course, contain genes from either parent. The intermediate and line recombination operators overcome this limitation by acting on the deci-

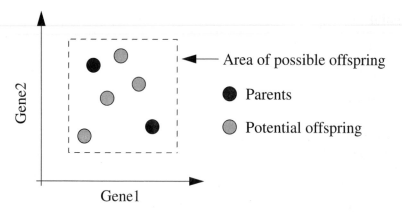

Figure 39.4 Geometric effect of intermediate recombination

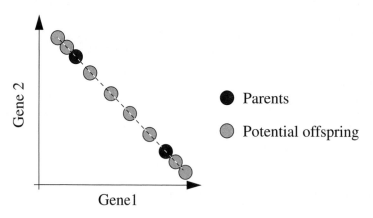

Figure 39.5 Geometric effect of line recombination

sion variables themselves. Like uniform crossover, the real-valued operators may also be parameterised to provide a control over the level of disruption introduced into offspring. For discrete-valued representations, variations on the recombination operators may be used that ensure that only valid values are produced as a result of crossover [23].

39.2.5 Mutation

In natural evolution, mutation is a random process where one allele of a gene is replaced by another to produce a new genetic structure. In GAs, mutation is randomly applied with low probability, typically in the range 0.001 and 0.01, and modifies elements in the chromosomes. Usually considered as a background operator, the role of mutation is often seen as providing a guarantee that the probability of searching any given string will never be zero and acting as a safety net to recover good genetic material that may be lost through the action of selection and crossover [2].

The effect of mutation on a binary string is illustrated in Fig. 39.6 for a 10-bit chromosome representing a real value decoded over the interval [0, 10] using both standard and Gray coding and a mutation point of 3 in the binary string. Here, binary mutation flips the

mutation point ⟶	binary	Gray
Original string - 0 0 ⌐0¬ 1 1 0 0 0 1 0	0.9659	0.6634
Mutated string - 0 0 ⌐1¬ 1 1 0 0 0 1 0	2.2146	1.8439

Figure 39.6 Binary mutation

value of the bit at the locus selected to be the mutation point. The effect of mutation on the decision variable, of course, depends on the encoding scheme used. Given that mutation is generally applied uniformly to an entire population of strings, it is possible that a given binary string may be mutated at more than one point.

With non-binary representations, mutation is achieved by either perturbing the gene values or random selection of new values within the allowed range. Wright [9] and Janikow and Michalewicz [24] demonstrate how real-coded GAs may take advantage of higher mutation rates than binary-coded GAs, increasing the level of possible exploration of the search space without adversely affecting the convergence characteristics. Indeed, Tate and Smith [25] argue that for codings more complex than binary, high mutation rates can be both desirable and necessary and show how, for a complex combinatorial optimization problem, high mutation rates and non-binary coding yielded significantly better solutions than the normal approach.

Many variations on the mutation operator have been proposed. For example, biasing the mutation towards individuals with lower fitness values to increase the exploration in the search without losing information from the fitter individuals [26] or parameterizing the mutation such that the mutation rate decreases with the population convergence [27]. Mühlenbein [22] has introduced a mutation operator for the real-coded GA that uses a nonlinear term for the distribution of the range of mutation applied to gene values. It is claimed that by biasing mutation towards smaller changes in gene values, mutation can be used in conjunction with recombination as a foreground search process. Other mutation operations include that of *trade mutation* [8], whereby the contribution of individual genes in a chromosome is used to direct mutation towards weaker terms, and *reorder mutation* [8], that swaps the positions of bits or genes to increase diversity in the decision variable space.

39.2.6   Reinsertion

Once a new population has been produced by selection and recombination of individuals from the old population, the fitness of the individuals in the new population may be determined. If fewer individuals are produced by recombination than the size of the original population, then the fractional difference between the new and old population sizes is termed a generation gap [28]. In the case where the number of new individuals produced at each generation is one or two, the GA is said to be steady-state [29] or incremental [30]. If one or more of the most fit individuals is deterministically allowed to propagate through successive generations then the GA is said to use an *elitist strategy*.

To maintain the size of the original population, the new individuals have to be reinserted into the old population. Similarly, if not all the new individuals are to be used at each generation or if more offspring are generated than the size of the old population then a reinsertion scheme must be used to determine which individuals are to exist in the new population. An important feature of not creating more offspring than the current population size at each generation is that the generational computational time is reduced, most dramatically in the case of the steady-state GA, and that the memory requirements are smaller as fewer new individuals need to be stored while offspring are produced.

When selecting which members of the old population should be replaced the most apparent strategy is to replace the least fit members deterministically. However, in studies, Fogarty [31] has shown that no significant difference in convergence characteristics was found when the individuals selected for replacement where chosen with inverse proportional selection or deterministically as the least fit. He further asserts that replacing the least fit members effectively implements an elitist strategy as the most fit will probabilistically survive through successive generations. Indeed, the most successful replacement scheme was one that selected the oldest members of a population for replacement. This is reported as being more in keeping with generational reproduction as every member of the population will, at some time, be replaced. Thus, for an individual to survive successive generations, it must be sufficiently fit to ensure propagation into future generations.

39.2.7 Termination of the GA

Because the GA is a stochastic search method, it is difficult to formally specify convergence criteria. As the fitness of a population may remain static for a number of generations before a superior individual is found, the application of conventional termination criteria becomes problematic. A common practice is to terminate the GA after a prespecified number of generations and then test the quality of the best members of the population against the problem definition. If no acceptable solutions are found, the GA may be restarted or a fresh search initiated.

39.3 Parallel GAs

Given the preceding description of the GA, it is clear that the GA may be made parallel in a number of ways. Indeed, there are many variations on parallel GAs, many of which are very different from the original GA presented by Holland [1]. Most of the major differences are encountered in the population structure and the method of selecting individuals for reproduction. The motivation for exploring parallel GAs is manifold. One may wish to improve speed and efficiency by employing a parallel computer, apply the GA to larger problems or try to follow the biological metaphor more closely by introducing structure and geographic location into the population. As this section will show, the benefits of using parallel GAs, even when run on a sequential machine, can be more than just a speed-up in the execution time.

In deciding whether a parallel GA is useful for a specific problem, the trade-off between population diversity, in terms of the population size, versus the execution time needs to be considered. For example, a small population size will yield a short execution time but may mean that some areas of the search space are not investigated because of the lack of genetic diversity in the population. This may mean that only suboptimal solutions are found, or, in the worst case, no satisfactory solutions are obtained. Large populations, on the other hand, may maintain diversity in the population but at the expense of execution time. Depending on the nature of the problem being addressed, excessive diversity may also mean that the GA is unable to find a satisfactory solution because of selective pressure driving reproduction towards suboptimal points. In many cases, because of the structure of the population and the use of local selection rules, parallel GAs offer an attractive mechanism for allowing diversity to exist within a population without unduly affecting the convergence characteristics of the GA. For some classes of problem, for example those characterized by multimodal search spaces or multiobjective formulations, the parallel GA can be shown to more effective that the sequential GA, allowing multiple, equally satisfactory, solution estimates to coexist in the global population simultaneously. The remainder of this chapter discusses these issues in detail.

In the remainder of this section, we describe a number of parallel GAs and use three broad categories to classify them: global, migration and diffusion. These categories reflect the various ways in which parallelism is exploited in the GA and the nature of the population structure and recombination mechanisms used. The global GA treats the entire population as a single breeding unit and aims to exploit the algorithmic parallelism inherent in the GA. Migration GAs divide the population into a number of subpopulations, each of which is treated as a separate breeding unit under the control of a conventional GA. To encourage the proliferation of good genetic material throughout the whole population, individuals migrate between the subpopulations from time to time. Generally, the migration GA is considered coarse-grained. The diffusion GA treats each individual as a separate breeding unit, the individuals it may mate with being selected from within a small local neighborhood. The use of local selection and reproduction rules leads to a continuous diffusion of individuals over the population. The diffusion GA is usually considered fine-grained.

39.3.1 Global GAs

Examination of the pseudocode outline of the sequential Simple GA given in Fig. 39.1 reveals that a significant proportion of the computation in a GA is composed of taking pairs of individuals, combining them to form new offspring, applying mutation and evaluating a cost function. Taking a population size of, say, 50 and assuming that reproduction of two individuals creates two new offspring, then the inner loop of Fig. 39.1 contains 25 discrete operations that may be performed concurrently at each generation. The worker/ farmer architecture in Fig. 39.7 demonstrates how this geometric parallelism may be exploited by a parallel computer.

The GA Farmer node initializes and holds the entire population, performs selection and assigns fitness to individuals. The Worker nodes recombine individuals, apply mutation and evaluate the objective function for the resulting offspring. Goldberg [2] describes a similar scheme, the *synchronous master-slave*, whereby a hybrid GA uses a local search

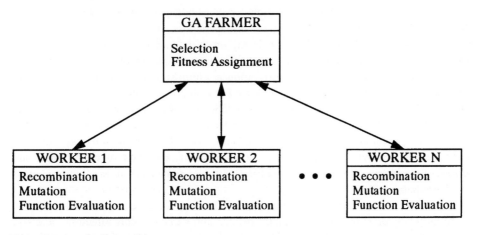

Figure 39.7 A worker/farmer GA

routine at each worker to further refine the estimates generated at each node. Others, notably Fogarty and Huang [30] and Dodd et al. [32] use the processor farm for the evaluation of objective functions only.

Although the farmed GA does not embrace all of the parallelism inherent in the GA, near linear speed-up has been reported in cases where the objective function is significantly more computationally expensive than the GA itself. In particular, when the objective function being minimized is of low computational cost, then there is potentially a bottleneck at the farmer while fitness assignment and selection are performed. The computational efficiency, of course, depends on the balance between the cost of the parallel parts of the GA and the sequential elements. Thus, the farmed GA may be inefficient if the objective function evaluation times vary greatly.

Goldberg [2] also describes a *semi-synchronous master-slave* GA that overcomes this potential bottleneck by relaxing the requirement for strict synchronous operation. Here, individuals are selected and inserted into the population as and when the worker nodes complete their tasks. However, both the synchronous and semi-synchronous models are potentially unreliable because of the dependence on the single farmer process.

A further issue is that of the message size used to pass individuals from the Farmer to the Workers. While smaller messages allow greater control over the load balance on the nodes in the parallel system, they may be less efficient in using available bandwidth. Thus, more than one pair of individuals may be sent from the Farmer to a Worker in a communication interval. If the objective function is sufficiently large, then the worker/farmer GA may prove a useful method of reducing the execution time of the GA when the nodes used are networked workstations.

A more robust extension to the worker/farmer implementations is the *asynchronous, concurrent* GA [2]. Using a number of identical processors, genetic operators and objective function evaluations are performed independently of one another on a population stored in a shared memory. This requires that no individual be accessed by more than one processor simultaneously. Although more complicated to implement than the conventional farmed GAs described above, this scheme is highly tolerant to processor and memory failure. Even if only one processor and some of the shared memory are functioning, it is still possible for useful processing to be performed.

39.3.2   Migration GAs

The GA as described thus far operates globally on a single population. That is, individuals are processed probabilistically on their performance in the population as a whole and any individual has the potential to mate with any other individual in the entire population. This treatment of the population as a single breeding unit is known as *panmixia*.

In natural evolution, species tend to reproduce within subgroups of the entire population, isolated to some extent from one another, but with the possibility of mating occurring across the boundaries of the subgroups. A population distributed among a number of semi-isolated breeding groups is known as *polytypic*. Humans, for example, are polytypic in that they consist of groups of the species isolated from one another geographically, culturally and economically. While breeding may occur between individuals from different subgroups of the species, it is much more likely that individuals from within the same group will reproduce together.

The *migration* or *island* model of the GA introduces geographic population distribution by dividing a large population into many smaller semi-isolated subpopulations or *demes*. Each subpopulation is a separate breeding unit using local selection and reproduction rules to locally evolve the species. From time to time, migration of individuals occurs between subpopulations such that individuals from one population are introduced into another subpopulation. The pattern of migration limits how much genetic diversity can occur in the global population. This pattern of migration is determined by the number of individuals migrated, the interval between migration and the migration paths between subpopulations. This movement of individuals between demes is often termed the *stepping stone model.*

The traditional sequential GA can readily be extended to encompass the migration model. A pseudocode outline of the modified algorithm for the migration GA is shown in Fig. 39.8. The population is divided into a number of subpopulations each of which is evolved by an independent GA. Additional routines are included to exchange individuals between subpopulations according to the communications topology employed and a global termination criteria introduced. Fig. 39.9 shows a possible implementation of the migration GA and some of the migration paths between the population islands.

```
-- Each node (GAi)
WHILE not finished
        SEQ
            ... Selection
            ... Reproduction
            ... Evaluation
            PAR
                ... Send emigrants
                ... Receive immigrants
```

Figure 39.8 Pseudo-code outline of the migration GA

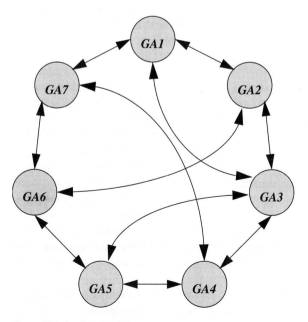

Figure 39.9 A migration GA

Grosso [33] first introduced a geographically isolated population structure in 1985 using an island model of the GA with five independent subpopulations. From his study, he found that semi-isolated populations improved the performance of the GA in terms of the quality of solution and the number of function evaluations required. He asserted that limited migration of individuals between subpopulations was more effective than either complete subpopulation interdependence or independence.

In 1987, Petty et al. [34] presented a migration GA influenced by earlier work on population isolation conducted by Wilson [35] in his ANIMAT classifier-based learning system and Schaffer's VEGA [36]. ANIMAT modelled an artificial creature seeking food and avoiding trees in a two-dimensional wood. Each individual specifies one action that this creature was to make, e.g,. in which direction to move. If an individual was selected for reproduction, then its mate was chosen from the subpopulation of individuals specifying the same action. Schaffer's VEGA, Vector Evaluated GA, was designed to solve problems with multiple objectives. The population was divided into a subpopulations, each subpopulation associated with a particular objective. Individuals were then selected according to their performance within a subpopulation, and mating was allowed to occur across the subpopulation boundaries. The idea behind this methodology is that highly-fit individuals in one objective domain mate with other fit individuals from other objective domains to produce new individuals well suited to all of the objectives.

Petty's migration GA was implemented on an Intel iPSC message-passing multiprocessor system with a binary n-cube interconnection network. One GA, called a nodal GA, was placed on each processor. At each generation, the best individual on each node is communicated to its neighbors. Each subpopulation therefore receives one individual from each neighboring node that it inserts into its current population. A number of insertion strategies were tested—replacing the worst individuals with immigrants, uniform replacement of individuals by immigrants and replacement of individuals with the smallest Hamming distance from the immigrants. However, the effect of the insertion methods is not reported. Petty et al. found a significant performance increase in terms of the convergence characteristics when the parallel GA was applied to numerical optimization problems selected from De Jong's test-bed functions , as well as the speed-up associated with running the GA on a parallel machine. Typically, the migration GA was able to converge to solutions not found by a global GA although the results showed a degree of problem dependence.

Tanse [37] also reported on a migration GA in 1987. He studied the migration model and compared its performance with a partitioned GA, i.e., a GA whose population is divided into subpopulations that evolve entirely independently with no migration between subpopulations. Again, the GA was implemented on a hypercube machine, although Tanse's parallel computer employed custom VAX-like CPUs. This implementation used two new parameters to specify the migration interval, at which generation migration should take place, and the migration rate, the number of individuals transferred between subpopulations. In early experiments , Tanse [37], selected individuals for migration probabilistically from the subset of the subpopulation whose fitness was at least equal to the average fitness of the subpopulation. Likewise, individuals were selected for replacement by immigrants probabilistically from the subset of individuals whose fitness was no greater than the average for that subpopulation. Later [38], it appeared that a new strategy was adopted. At a migration generation, each node produced more offspring than the current subpopulation size. The migrants were then uniformly selected from the offspring and removed from the subpopulation, thus maintaining the correct subpopulation size. The receiving subpopulation uniformly replaced individuals with immigrants. The philosophy behind this approach is that the most fit individuals are more likely to reproduce and are therefore most likely to migrate. The actual migration took place bidirectionally along one dimension of the hypercube, selected on the basis of the generation number. Thus, the neighbor selected to receive individuals from a node will also send its migrants to that node.

The results presented by Tanse showed a near-linear speed-up when compared against a sequential GA with a population size equal to the sum of the individual subpopulations. Comparing the migration GA with the partitioned GA, the migration GA consistently found superior individuals and had a higher average fitness over the entire population. However, because of the limited number of test functions, no conclusion can be drawn about the general effect of the migration rate and interval. The effect of the mutation and crossover operators used with the migration GA was also investigated. The results indicate that it is feasible to use different crossover and mutation rates on different nodes, allowing the balance between exploration and exploitation to be varied locally, but with migration ensuring that good individuals should survive in at least some subpopulations.

Similar results to Tanse are reported by Starkweather et al. [39] and Cohoon et al. [40]. Starkweathers et al.'s GA has a number of notable differences to the implementations described so far. Rather than using a generational GA in which most or all of the population is replaced at each generation, their parallel GA was based on Whitley's GENITOR program [29] that uses one-at-a-time reproduction replacing a single individual at each reproduction step. subpopulations are placed on a ring or a circle and migration of the fittest individuals takes place at regular predefined intervals. At migration step M, individuals on node X are migrated to the subpopulation given by $mod(M + X, N)$ where N is the number of subpopulations numbered 0 to $N - 1$. For example, at migration 1, subpopulation 0 individuals migrate to subpopulation 1 and at migration 2 individuals from subpopulation 0 migrate to subpopulation 2. The migration GA was applied to a wide range of problems including neural network optimization, a mapping problem and a 105-city traveling salesman problem. In all test cases, the migration GA produced better results than a comparable sequential one. When the migration GA was implemented on a sequential machine it was found that it would find better solutions and execute faster than a standard GA with the same population size on a number of test cases. In addition, the use of an adaptive mutation rate, initially high and reducing with generations, was found to improve the convergence characteristics of the GA.

In 1991, Mühlenbein et al. [41] described a real-valued parallel GA for use as a "black-box solver" in high-dimensional optimization problems. A conventional single-point crossover operator was employed that operated directly on the ANSI-IEEE floating-point representation of the decision variables. The mutation operator worked only on the fractional part of a variable's representation, thus mutation is exponentially biased towards producing a new individual in the region of the original rather than one a larger numerical distance away. In experiments, the parallel GA was able to find global solutions to problems of up to a dimension of 400.

The distributed breeder GA of Mühlenbein and Schlierkamp-Voosen [22], rather than modeling the natural and self-organized evolution of the earlier parallel GA described above, is based on a model of rational selection in human breeding groups. Whereas the parallel GA models natural selection, the breeder GA models artificial selection. Using influences from Evolution Strategies [11] and GAs, the breeder GA selects the best $T\%$ of a population, where T is a predefined parameter, and randomly mates them until sufficient offspring are produced. As well as ensuring that no individuals mate with themselves, the fittest individual also survives into the following generation. This selection and reproduction process is known as *truncation selection* as only a subset of each generation are used as potential parents.

The breeder GA operates on populations of real-valued individuals and has new genetic operators designed specifically for this representation, such as intermediate and line recombination, that have been described earlier in Sections 39.2.4 and 39.2.5. The parallel GA uses a local hill-climbing algorithm on certain individuals to improve a current local estimate. In the breeder GA, the mutation operator was found to be almost as effective as local hill-climbing but was much less complex and computationally demanding to implement. In all cases, the breeder GA was found to be more effective than the earlier parallel GA and managed to solve numerical optimization problems of dimension 1000.

A recent variation of the migration GA intended to overcome the problem of premature convergence is presented by Potts et al. [42]. This algorithm, called GAMAS (GA based on Migration and Artificial Selection), uses four interdependent binary populations labelled SPECIES I to SPECIES IV. At initialization, SPECIES II, SPECIES III and SPECIES IV are generated with a predefined bias. SPECIES II is created with a bias in favor of 0s in the chromosome representation, SPECIES IV biased towards 1s and SPECIES III with equal probability of 0s and 1s. Since the ratio of 0s to 1s contained in the optimal solution is not known *a priori*, this bias intends to encourage the exploration of different areas of the search space. SPECIES I is later filled as the result of an artificial selection process. The subpopulation size used in all SPECIES was set to 36 individuals.

To encourage both exploitation and exploration of the species, GAMAS uses various mutation rates in the three evolving subpopulations. SPECIES II is used for exploration and has a high mutation rate, SPECIES IV for exploitation with a low mutation rate and SPECIES III for both exploration and exploitation with a slightly higher mutation rate than SPECIES IV. At each generation, the most fit individuals from SPECIES II, II and IV are artificially selected and placed in SPECIES I, replacing less fit individuals where necessary. Thus, SPECIES I is used as a dominant incipient species and is held in isolation with no reproduction or crossover. To further enforce exploration, migration of randomly selected individuals takes place between SPECIES II, III and IV. This migration takes place at high frequency in early generations and is gradually reduced in subsequent generations. At predetermined generation intervals, the entire contents of SPECIES I are used to replace SPECIES IV, the exploitation species. The rationale being to further develop the good approximations already obtained. A further refinement to the migration GA is the concept of *recycling* whereby the three evolving species are reinitialized without affecting the individuals contained in SPECIES I. Recycling is intended to allow the GA to explore new regions of the search space and overcome the initial population bias reported by Goldberg [2].

GAMAS has been tested on a NP-hard combinatorial optimization problem—the file allocation problem, the XOR neural network that contains many optimal solutions and the multimodal, multivariate sine envelope sine wave function of Schaffer [43]. Again, in all these cases GAMAS found superior solutions to the sequential GA. In the XOR neural network problem, GAMAS also demonstrated an ability to find a number of different optimal solutions.

Clearly, the migration model of the GA is well suited to parallel implementation on MIMD machines. Given the range of possible population topologies and migration paths between them, efficient communications networks should be possible on most parallel architectures from small multiprocessor platforms to clusters of networked workstations. The semi-isolation of subpopulations and limited communication between them also encourages a high degree of fault tolerance. In a well designed migration GA, in the event of the loss of individual subpopulations or communications paths between them, the GA can still perform useful computation.

The migration GA has generally been reported as a more efficient search and optimization method than conventional sequential GAs. From the preceding text, it should be clear that this is the effect of local selection and migration rather than parallel implementation. However, the migration GA is slightly more complex to use as further parameters are introduced to control migration between subpopulations.

39.3.3 Diffusion GAs

An alternative model of a distributed population structure is provided by the diffusion GA. Whereas migration introduces discontinuities into the population structure with barriers between the borders of the islands containing the subpopulations, diffusion treats the population as a single continuous structure. Each individual is assigned a geographic location on the population surface and is allowed to breed with individuals contained in a small local neighborhood. This neighborhood is usually chosen from immediately adjacent individuals on the population surface and is motivated by the practical communication restrictions of parallel computers. The diffusion GA is also known as the *neighborhood, cellular* or *fine grained* GA.

Figure 39.10 shows a pseudocode outline of the diffusion GA. Consider the population distribution shown in Fig. 39.11a where each individual, $I_{j,k}$, is assigned a separate node on a toroidal-mesh parallel processing network. The Figure shows that there are no specific islands in the population structure, rather a contiguous geographic distribution of individuals, however, there is potential for a similar effect. Given that mating is restricted to adjacent processors, then individuals on distant processors may take as many generations to meet and mate as individuals in various subpopulations in the island model. Wright [44] refers to this form of isolation within a species as *isolation by distance*. From Fig. 39.10, each individual is first initialized, either randomly or using a heuristic, and its performance evaluated. Each node then sends its individual to its neighbors and receives individuals from those neighbors. For example, in Fig. 39.11a, individual $I_{3,1}$ sends a copy of itself to $I_{2,1}, I_{3,2}, I_{3,5}$ and $I_{4,1}$ and receives copies of the individuals on those nodes. The purpose of this communication is to provide a pool of potential mates from the incoming individuals. Thus, selection of a mate for individual $I_{3,1}$ is made on the basis of a neighborhood fitness over the individuals $I_{2,1}, I_{3,2}, I_{3,5}$ and $I_{4,1}$. Reproduction involves the usual crossover and mutation operators and is used to produce a single individual to replace the parent residing on the node. However, rules may be applied to retain the original parent if neither of the offspring is sufficiently fit to replace it.

```
-- Each node (Ii,j)

WHILE not finished
      SEQ
            ... Evaluate
            PAR
                  ... Send self to neighbours
                  ... Receive neighbours
            ... Select mate
            ... Reproduce
```

Figure 39.10 Pseudo-code outline of the diffusion GA

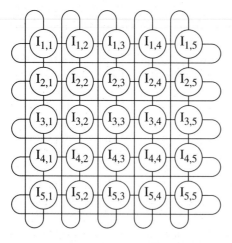

(a) Neighborhood Communications

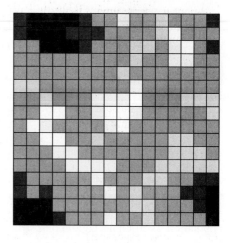

(b) Virtual Islands

Figure 39.11 A diffusion GA

At initialization, the distribution of genetic material over the population surface is random, assuming that the population has not been seeded heuristically. After a few generations, local clusters of individuals with similar genetic material and fitness may appear in the population giving rise to *virtual islands*. This phenomena is shown in Fig. 39.11b where the shading is used to represent individuals with similar genetic material. The drift in the population caused by local selection tends to reduce the number of clusters while increasing their size over generations as the most fit individuals diffuse over the population.

The first attempt at a massively-parallel fine-grained GA known to the authors was by Robertson [45] in 1987. Robertson used a SIMD Connection Machine and assigned one individual per processor. However, global selection and recombination was performed on the host machine and the individual processors were only used for function evaluation implementing a massive processor farm (Section 39.3.1). Even considering the large communications overhead of this scheme, the objective function evaluation was significantly large for a huge speed-up to be reported. By 1989 a more subtle and appropriate scheme for the Connection Machine was presented by Manderick and Spiessens [46] and later implemented on an AMT DAP [47]. Individuals were again placed on separate processors in a planar grid, but a local selection strategy based on a neighborhood fitness distribution was used. This first diffusion algorithm was not only motivated by the desire to use the Connection Machine's architecture more effectively, but also to align the GA more closely with natural biological evolution. Spiessens and Manderick argue that in nature there is no global selection or fitness-distribution. Rather, natural selection is a local phenomenon where individuals find a mate in their local environment. Their implementation is similar to that described here with the exception that one parent is chosen from the local neighborhood probabilistically on the basis of the neighborhood fitness function and recombined with a randomly selected mate within the same locality. This implementation was tested on the De Jong test-bed functions and compared with a conventional GA. The results indicate that the lower selective pressure, because of the local selection mechanism, encourages greater exploration of the search space and helps inhibit the early domination of the population by good individuals. The results also showed that the parallel GA was more effective when the objective function was multimodal.

About the same time, Mühlenbein [48] and Gorges-Schleuter [49] introduced an asynchronous parallel GA, ASPARAGOS. Using ideas from population genetics, the GA was implemented on a connected ladder network, or ring, topology using Transputers with one individual per processor as shown in Fig. 39.12. An individual's neighborhood is defined by its mobility, so, for example, given a "moving radius" of two yields a neighborhood size of eight.

ASPARAGOS also employs local hill-climbing when a new individual is created. The rationale behind this idea is that hill climbing can more quickly improve the fitness of an individual than genetic operators. When applied to the quadratic assignment problem, ASPARAGOS found a new optimum for the largest published problem [48]. In other areas, such as numerical optimization, ASPARAGOS has been found to perform well, and superlinear speed-up has been reported. Although, as Mühlenbein points out, measuring performance of parallel GAs is difficult because of the probabilistic nature of the algorithm and the effective change in the search strategy with varying numbers of processors. Furthermore, in some experiments, a single population GA was not able to locate the minimum at all.

Davidor [50], in his version of the diffusion GA, called ECO GA, used a 2-D grid with wrap-around to produce a population surface in which each individual had eight neighbors. He used a one-at-a-time reproduction strategy and allowed the offspring produced by a particular neighborhood to replace probabilistically an individual in the vicinity of its parents. Davidor also described the phenomena of niche and speciation where the virtual islands on the population surface represent near local optima.

An interesting variation on local selection is given by Collins and Jefferson [51]. Instead of selecting a mate from within a small local neighborhood, an individual takes a random walk and selects a mate from individuals encountered on the way. This was found to be highly efficient in the graph partitioning problem and demonstrated a capability of finding multiple optima in a single population. In addition, four metrics where used to measure the differences in evolutionary dynamics between polytypic and panmictic populations. They were the diversity of alleles and genotypes, an inbreeding coefficient measuring the similarity between parents and speed and robustness.

A hierarchically structured distributed GA is described by Voight et al. [52] that is suitable for fine and coarse grain multiprocessor implementations. A population model using varied diffusion rates corresponding to the hierarchical structure of the evolution surface is used. Interacting breeding units, or local environments, are organized hierarchically such that interaction within individual local environments is greater than the interactions between individuals in different local environments. Thus, the population structure allows

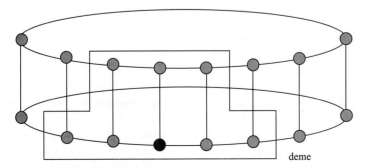

deme

Figure 39.12 ASPARAGOS population structure

the local interactions of the diffusion GA while also modeling the semiisolation of the migration GA. To accommodate different granularities, local environments or single individuals may reside on separate processors.

More recently, a number of researchers have focused on the nature of the population structure and its effect on the diffusion and convergence characteristics of the GA. For example, Baluja [53] reports on a comparative study of neighborhood topologies. Three topologies are considered—a linear neighborhood, a two dimensional toroidal array and a linear neighborhood with a rightward discontinuity. When tested on a wide range of testbed problems the toroidal array neighborhood consistently outperformed the other population structures. However, for some problems the linear neighborhoods were found to produce the best convergence pattern. Baluja argues that these results are because of a combination of the effect of genetic mobility and total population size. In particular, the parameters of the different implementations, such as crossover and mutation rates, where held constant and not tuned to a particular neighborhood structure. However, in an earlier study, Gorges-Schleuter [54] observed that as the borders in one-dimensional population structures are smaller than those in two dimensions, local niches once established tend to survive for longer periods.

The diffusion model provides a finer grain of parallelism than that of the worker/farmer and island models. It is suitable for implementation on a wide range of parallel architectures from single-bit digital array processors (DAPs) and massively parallel SIMD machines like the Connection Machine, to MIMD computers, such as Transputer networks. There are even reports in the literature of fine-grained GAs being implemented on clusters of networked workstations [55].

The basic operator to support the diffusion model is that of a local neighborhood selection mechanism. From the examples of diffusion models presented in this section, it is clear that this local selection results in performance superior to that of global and migration GAs with comparable population sizes. Good solutions are found faster, requiring fewer function evaluations, and different solution niches may be established in the same evolutionary cycle. Furthermore, diffusion appears to implement a more robust search in the presence of deceptive or GA-hard objective functions.

39.4   Conclusions and Future Trends

In this Chapter, a broad survey of the current trends in GAs has been presented. Many of the variations on the original GA have been discussed, such as different representation and selection strategies, and a number of parallel implementations have been described in detail. From the material presented in this text, it should be clear to the reader that the GA is a powerful and versatile search and optimization method applicable to a broad spectrum of activities. Further, it is apparent that parallel GAs employing population distribution with local selection and reproduction and some level of genetic mobility are superior to approaches where the population is treated globally.

Clearly, this is a particularly rewarding area for work in parallel algorithm design. Not only can we realize the algorithms efficiently on parallel architectures, but also the paradigm has revealed performance benefits that can even be realized on a sequential machine.

From the preceding discussion of migration and diffusion GAs, the following general points can be noted about parallel GAs when compared with global strategies:

- Parallel GAs typically require less function evaluations to find optimal solutions.
- They have the potential to find multiple optimal solutions.

- They may be synchronous or asynchronous.
- The implementation of parallel GAs may be adapted to efficiently exploit different parallel architectures.
- They have a higher degree of robustness.
- Parallel GAs may be made fault-tolerant.
- Parallel GAs come closer to the biological metaphor of evolution.

The use of varying crossover and mutation rates with different subpopulations is an attractive area for further study, particularly with regard to problems that have very large and/or deceptive search spaces. By encouraging both the exploration of the search space and the exploitation of the most promising areas, the parallel GA has much potential to yield a general approach to problems that have traditionally proved intractable using conventional approaches. That is not to say, however, that the GA can be seen as a universal panacea for all problems for which no adequate solution presently exists.

In this chapter we have described a range of population structures and selection methods that have been reported in the literature. Nature provides us with a rich source of inspiration for new paradigms that may be incorporated into the GA framework. Clearly, there is much scope for devising new selection strategies and population structures for specific problem domains. In addition, there is still much work to be done in laying the theoretical ground work for both sequential and parallel GAs.

We believe that the GA is a powerful and versatile search and optimization method. In this Chapter, we have concentrated on describing implementation methods, rather than focusing on research issues, in the hope that the reader will be encouraged to explore the potential of genetic algorithms in their own field.

39.5   References

1. Holland, J. 1975. *Adaptation in Natural and Artificial Systems*. Ann Arbor, Mich.: University of Michigan Press.
2. Goldberg, D. E. 1989. *Genetic Algorithms in Search: Optimization and Machine Learning*. Reading, Mass.: Addison-Wesley.
3. Karr, C. L. 1991. Design of an adaptive fuzzy logic controller using a genetic algorithm. *Proc. ICGA 4*, 450–457.
4. Holstien, R. B. 1971. *Artificial Genetic Adaptation in Computer Control Systems*. Ph.D. Thesis, Department of Computer and Communication Sciences. Ann Arbor, Mich.: University of Michigan.
5. Caruana, R. A., and J. D. Schaffer. 1988. Representation and hidden bias: gray vs. binary coding. *Proc. 6th Int. Conf. Machine Learning*, 153–161.
6. Schmitendorgf, W. E., O. Shaw, R. Benson and S. Forrest. 1992. Using genetic algorithms for controller design: simultaneous stabilization and eigenvalue placement in a region. *Technical Report No. CS92-9*, Dept. Computer Science, College of Engineering, University of New Mexico.
7. Bramlette, M. F. 1991. Initialization, mutation and selection methods in genetic algorithms for function optimization. *Proc ICGA 4*, 100–107.
8. Lucasius, C. B., and G. Kateman. 1992. Towards solving subset selection problems with the aid of the genetic algorithm. In *Parallel Problem Solving from Nature 2*, R. Männer and B. Manderick, eds. Amsterdam: North-Holland, 239–247.
9. Wright, A. H. 1991. Genetic algorithms for real parameter optimization. In *Foundations of Genetic Algorithms*, J. E. Rawlins, ed. San Mateo, Calif.: Morgan Kaufmann, 205–218.
10. Michalewicz, Z. 1992. *Genetic Algorithms + Data Structures = Evolution Programs*. New York: Springer Verlag.
11. Bäck, T., F. Hoffmeister and H.-P. Schwefel. 1991. A survey of evolution strategies. *Proc. ICGA 4*, 2–10.
12. Grefenstette, J. J. 1987. Incorporating problem specific knowledge into genetic algorithms. In *Genetic Algorithms and Simulated Annealing*, L. Davis, ed. San Mateo, Calif.: Morgan Kaufmann, 42–60.

13. Whitley, D., K. Mathias, and P. Fitzhorn. 1991. Delta coding: an iterative search strategy for genetic algorithms. *Proc. ICGA 4*, 77–84.

14. De Jong, K. A. 1975. *Analysis of the Behaviour of a Class of Genetic Adaptive Systems*. Ph.D. Thesis, Dept. of Computer and Communication Sciences, Ann Arbor, Mich.: University of Michigan.

15. Baker, J. E. 1985. Adaptive selection methods for genetic algorithms. *Proc. ICGA 1*, 101–111.

16. Baker, J. E. 1987. Reducing bias and inefficiency in the selection algorithm. *Proc. ICGA 2*, 14–21.

17. Booker, L. 1987. Improving search in genetic algorithms," In *Genetic Algorithms and Simulated Annealing*, L. Davis, ed. San Mateo, Calif.: Morgan Kaufmann, 61–73.

18. Spears, W. M., and K. A. De Jong. 1991. An analysis of multi-point crossover. In *Foundations of Genetic Algorithms*, J. E. Rawlins, ed. San Mateo, Calif.: Morgan Kaufmann, 301–315.

19. Syswerda, G. 1989. Uniform crossover in genetic algorithms. *Proc. ICGA 3*, 2–9.

20. Spears, W. M., and K. A. De Jong. 1991. On the virtues of parameterised uniform crossover. *Proc. ICGA 4*, 230–236.

21. Caruana, R. A., L. A. Eshelman, and J. D. Schaffer. 1989. Representation and hidden bias II: Eliminating defining length bias in genetic search via shuffle crossover. In *Eleventh International Joint Conference on Artificial Intelligence*, N. S. Sridharan, ed., Vol. 1. San Mateo, Calif.: Morgan Kaufmann, 750–755.

22. Mühlenbein, H., and D. Schlierkamp-Voosen. 1993. Predictive models for the breeder genetic algorithm. *Evolutionary Computation*, Vol. 1, No. 1, 25–49.

23. Furuya, H., and R. T. Haftka. 1993. Genetic algorithms for placing actuators on space structures. *Proc. ICGA 5*, 536–542.

24. Janikow, C. Z., and Z. Michalewicz. 1991. An experimental comparison of binary and floating point representations in genetic algorithms. *Proc. ICGA 4*, 31–36.

25. Tate, D. M., and A. E. Smith. 1993. Expected Allele Convergence and the Role of Mutation in Genetic Algorithms. *Proc. ICGA 5*, 31–37.

26. Davis, L. 1989. Adapting operator probabilities in genetic Algorithms. *Proc. ICGA 3*, 61–69.

27. Fogarty, T. C. 1989. Varying the probability of mutation in the genetic algorithm. *Proc. ICGA 3*, 104–109.

28. De Jong, K. A., and J. Sarma. 1993. Generation gaps revisited. In *Foundations of Genetic Algorithms 2*, L. D. Whitley, ed. M San Mateo, Calif.: Morgan Kaufmann.

29. Whitley, D. 1989. The GENITOR algorithm and selection pressure: Why rank-based allocations of reproductive trials is best. *Proc. ICGA 3*, 116–121.

30. Huang, R., and T. C. Fogarty. 1991. Adaptive classification and control-rule optimization via a learning algorithm for controlling a dynamic system. *Proc. 30th Conf. Decision and Control*, Brighton, England, 867–868.

31. Fogarty, T. C. 1989. An incremental genetic algorithm for real-time learning. *Proc. 6th Int. Workshop on Machine Learning*, 416–419.

32. Dodd, N., D. Macfarlane, and C. Marland. 1991. Optimization of artificial neural network structure using genetic techniques implemented on multiple transputers. *Transputing '91*, P. Welch, D. Stiles, T. L. Kunii, and A. Bakkers, eds. Vol. 2 687–700, IOS Press.

33. Grosso, P. B. 1985. *Computer Simulation of Genetic Adaptation: Parallel Subcomponent Interaction in a Multilocus Model*. Ph.D. Thesis, University of Michigan.

34. Petty, C. B., M. R. Leuze and J. J. Grefenstette. 1987. A parallel genetic algorithm. *Proc. ICGA 2*, 155–161.

35. Wilson, S. W. 1985. Knowledge growth in an artificial animal. *Proc. ICGA 1*, 16–23.

36. Schaffer, J. D. 1985. Multiple objective optimization with vector evaluated genetic algorithms. *Proc. ICGA 1*, 93–100.

37. Tanse, R. 1987. Parallel genetic algorithm for a hypercube. *Proc. ICGA 2*, 177–183.

38. Tanse, R. 1989. Distributed Genetic Algorithms. *Proc. ICGA 3*, 434–439.

39. Starkweather, T., D. Whitley and K. Mathias. 1990. Optimization using distributed genetic algorithms. *Proc. Parallel Problem Solving From Nature 1*, Lecture Notes in Computer Science No. 496, 176–185. New York: Springer-Verlag.

40. Cohoon, J. P., W. N. Martin, and D. S. Richards. 1991. A Multi-population genetic algorithm for solving the k-partition problem on hypercubes. *Proc. ICGA 4*, 244–248.

41. Mühlenbein, H., M. Schomisch, and J. Born. 1991. The parallel genetic algorithm as a function optimizer. *Parallel Computing*, No. 17, 619–632.

42. Potts, J. C., T. D. Giddens, and S. B. Yadav. 1994. The development and evaluation of an improved genetic algorithm based on migration and artificial selection. *IEEE Trans. Systems, Man and Cybernetics*, Vol. 24. No. 1, 73–86.

43. Schaffer, J. D., R. A. Caruana, L. J. Eshelman and R. Das. 1989. A study of control parameters affecting online performance of genetic algorithms for function optimization. *Proc. ICGA 3*, 36–40.

44. Wright, S. 1969. *Evolution and the Genetics of Populations*, Vol. 2. Chicago: University of Chicago Press.

45. Robertson, G. 1987. Parallel implementation of genetic algorithms in a classifier system. In *Genetic Algorithms and Simulated Annealing*, L. Davis, ed. London: Pitman, 129–140.

46. Manderick, B., and P. Spiessens. 1989. Fine-grained parallel genetic algorithms. *Proc. ICGA 3*, 428–433.

47. Spiessens, P., and B. Manderick. 1991. A massively parallel genetic algorithm: implementation and first analysis. *Proc. ICGA 4*, 279–286.

48. Mühlenbein, H. 1989. Parallel genetic algorithms, population genetics and combinatorial optimization. In *Parallelism, Learning, Evolution*; Workshop on Evolutionary Models and Strategies, J. D. Becker, I. Eisele, and F. W. Mundemann, eds. Lecture Notes in Artificial Intelligence, No. 565, 398–406. New York: Springer-Verlag.

49. Gorges-Schleuter, M. 1989. ASPARAGOS: an asynchronous parallel genetic optimization strategy. *Proc. ICGA 3*, 422–427.

50. Davidor, Y. 1991. A naturally occurring niche and species phenomenon: the model and first results. *Proc. ICGA 4*, 257–263.

51. Collins, R. J., and D. R. Jefferson. 1991. Selection in massively parallel genetic algorithms. *Proc. ICGA 4*, 249–256.

52. Voight, H. -M., I. Santibáñez-Koref, and J. Born. 1992. Hierarchically structured distributed genetic algorithms. In *Parallel Problem Solving from Nature 2*, R. Männer and B. Manderick, eds. Amsterdam: North-Holland, 145–154.

53. Baluja, S. 1993. Structure and performance of fine-grain parallelism in genetic search. *Proc. ICGA 5*, 155–162.

54. Georges-Schleuter, M. 1992. Comparison of local mating strategies in massively parallel genetic algorithms. In *Parallel Problem Solving from Nature 2*, R. Männer and B. Manderick, eds.Amsterdam: North-Holland, 553–562.

55. Maruyama, T, T. Hirose, and A. Konagaya. 1993. A fine-grained parallel genetic algorithm for distributed parallel systems. *Proc. ICGA 5*, 184–190.

40

Parallel Processing for Robotic Computations: A Review

Tarek M. Nabhan and Albert Y. Zomaya

This decade has witnessed the widespread use of robots and manipulators in the industry due to the increasing demand for flexible automation. Robotic systems belong to a class of real-time systems in which severe real-time constraints have to be met. Ideally, the system should provide adequate and sufficient computational power to enable the application of advanced algorithms.

The recent advances in VLSI technology and the need for more computing power have led to the current development of advanced computer architectures. While von Neumann machines have a single locus of control that determines the next instruction to be executed, advanced architectures are centered around the concept of concurrent computing. The central idea of parallelism is to have several processing units jointly executing a computational task to increase the performance.

This chapter reviews the application of parallel processing to robotic computations. The main aim is to cover the major contributions in this growing field. In the next few sections, the different components of a robotic system are introduced. The computational complexity of each component is then discussed, and the different solutions employing parallel processing techniques are studied. Finally, several directions for future research are identified.

40.1 Overview of Robotic Systems

Robotic systems are the by-product of numerically controlled machines. A robot is a software-controllable device that employs sensors to guide itself and (or) its end effectors through deterministic motions in order to manipulate physical objects.

Robots have rapidly evolved from theory to application over the last decade, primarily due to the need for improved productivity and quality. Robots are employed today primarily in the automotive industry. They are used for welding, machine loading, and assembly operations [1].

This work was supported by Australina Research Council grant 04/15/412/131.

To successfully complete a non trivial task in a moderately complicated environment, an integrated robotic system that consists of several components is employed as shown in Fig. 40.1.

The *task planner* converts a high-level language task specification to a robot-level specification in three steps. (1) Complex tasks are first decomposed into a series of necessary incremental actions called elementary operations, and each operation is represented by its initial and goal configurations. (2) A path is then generated that takes into account the objects in the workspace. The path is denoted by a series of via points that link the initial and final configurations of an elementary operation. (3) Finally, these points are converted to a temporal function of the robot configurations, which represents the continuous-time joint space (robot world) trajectory of the robot.

To perform its function properly, the task planner requires information about the robot and the objects in the environment. This is provided by environment sensors (which are sensory devices external to the robot itself) that detect the Cartesian positions and orientations (world information) of both the robot and the objects in the workspace.

The *controller* generates the forces and torques required to drive the robot such that it accurately follows the desired trajectory. The internal robot sensors determine the state of the robot in terms of the position and velocity of each joint. In addition, the forces and torques applied at the end-effector, due to its contact with the environment, are also sensed. The measurements are passed to the controller to provide feedback signals that maintain the integrity of the system.

The efficient utilization of robotic systems requires the execution of the different tasks at high sampling rate. A discussion of the computational complexity of robotic systems' components and a study of the computer architectures proposed to speedup their computations are detailed in the following text.

40.2   The Task Planner

The objective of the task planner is to transform high-level goals, such as the specification of a sophisticated mechanical assembly operation, into the sequence of commands neces-

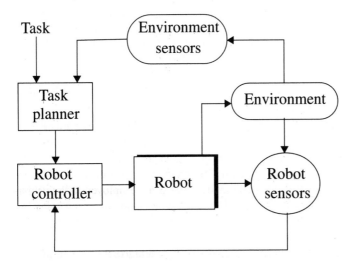

Figure 41.1 Main components of a robotic system

sary to achieve these goals. The task planner consists of three main components, namely: the *operations* planner, the *path* planner, and the *trajectory* planner. In the following text, the application of computer architectures to the different phases of this process is discussed.

40.2.1   Operations planning

The highest level of the planning process is concerned with the interfacing with a human operator, planning objectives for achieving the requested actions, and then coordinating the lower levels to meet these objectives. The latter level operates with less detailed information but with greater decision authority.

Artificial intelligence (AI) methods have been employed to improve the performance of this process. Intelligent search algorithms have been proposed in the literature for action planning [2]. Janabi-Sharifi et al. [3] proposed a learning model for grasping unknown objects. Moreover, an expert system for grasp mode selection was implemented. Bekey et al. [4] developed a knowledge-base grasp planner for multifingered robot hands. Brooks [5] proposed a planning checker algorithm that modifies existing plans to include sensing and adds constraints that ensures the success of the plan. In addition, Sicard and Levine [6] developed an expert system for arc welding applications.

To most efficiently process the information at this level, special VLSI-based chips have been designed to for executing declarative languages such as LISP and PROLOG [7–9]. These chips can be used as coprocessors in conventional computer architectures.

Advanced computer architectures that provide special features to support the execution of rule-based systems have been proposed [10]. Moldovan et al. [11] proposed a VLSI-based chip called the semantic network array processor (SNAP), which is a highly parallel architecture targeted to AI applications. The knowledge-base is distributed among the elements of the SNAP array, and the processing is performed locally where the knowledge is stored. Moreover, Moldovan [12] shows that sequential production systems can be transformed into equivalent parallel forms by performing an analysis of rule interdependence. A multiprocessor system called RUBIC (rule-based inference computer) was designed to implement the parallel processing model.

40.2.2   Path planning

The path planner receives the initial and goal configurations, developed by the operations planner, and generates a path (denoted by a series of via points) that links both configurations. Paths of the robot have to be short and must avoid unnecessary movements. In addition, robotic motions have to be executed without colliding with objects and other robots in the workspace.

The path planner is *static* if complete knowledge about the obstacles is known apriori. The difficulty of the problem increases if partial information about the obstacles in the environment is available at any given time. The problem becomes even more difficult if the obstacles in the environment are moving (*dynamic* environment).

Path planning is usually performed in the following steps. First, a representation of the robot's workspace and the obstacles is developed. A scheme is employed to represent the robot and the objects. An appropriate path planning techniques and a suitable search strategy are then used to generate a solution path, which is then optimized to yield a shorter and smoother path. For an excellent review of the different techniques employed in each step, the reader is referred to Hwang and Ahuja [13].

Path planning for robot manipulators is still in its infancy. Researchers are still endeavoring to find methods for implementing real-time path planners that run at practical speeds. With the advent of computer architectures and the increased speed of data processing technology, the level of applicability of path planners is increasing. Several researchers proposed path planners algorithms suitable for parallel processing. Cela et al. [14] proposed a general approach to the optimal path planning of industrial robots using nonlinear programming. The algorithm can be decomposed at the joints levels, which makes the technique implementable on parallel architectures. Cheng et al. [15] developed a novel parallel path searching method and applied the method to robot path planning problems. Lozano-Perez and O'Donnell [16] proposed a parallel path planning algorithm which employs a new technique for the concurrent computation of configuration space maps. The method was implemented on a Connection Machine with 8k processors. Prassler and Milios [17] presented a parallel distributed processing approach to robot navigation and path planning in unknown terrains. The method is based on massively parallel computation in a grid of simple processing elements, called cells.

Another approach to speed up the computation is by developing customized VLSI chips for path planning. Kameyama et al. [18] proposed a collision detection VLSI processor to achieve ultra high-performance processing with an ideal parallel processing scheme. Each processing element (PE) consists of an arithmetic unit for coordinate transformations and memory for the storage of manipulator and obstacle information.

40.2.3 Trajectory planning

The collision-free path generated by the path planner is passed to the trajectory planner to generate the temporal function representing the continuous-time joint space trajectory of the robot, which serves as an input to the robot controller. To ensure maximum performance, optimal trajectories are required. The problem is to generate a profile of positions, velocities, and accelerations for each joint, causing the end-effector of the robot manipulator to move in minimum time, while satisfying imposed constraints on maximum allowable forces, torques, velocities, accelerations, and 'jerk' values. (*Jerk* is a term used to describe the rate of change of acceleration. Jerk must be kept to low levels to avoid mechanical damage to the manipulator actuators.) Several methods have been proposed in the literature to handle this problem [19–22].

Zalzala and Morris [23] proposed an algorithm for concurrent computing the optimum trajectory using a multiprocessor architecture. The computational burden of the trajectory planning problem was reduced through a distributed realization of the formulation. The efficiency of the algorithm was demonstrated using a 45-transputer network employed for planning a trajectory for a PUMA 560 robot manipulator [24].

40.3 Sensing

For robots to "behave" intelligently, they must be equipped with an array of sensors to help them work efficiently in nonstructured and random environments. A robotic system may have several types of sensors. *Position control* is performed by means of optical encoders, tachometers, and acceleration gauges which provide the controller with the robot joints' positions, velocities, and accelerations, respectively. *Compliant controllers* employ data about the contact forces/torques received from a force/torque sensor. *Tactile* sensors allow the system to extract some features about the external world, such as hardness, elasticity, and roughness. *Proximity* sensors are devices that sense the distance from the end-effector, the robot hand, or any other part of the robot to other objects. *Range* sensors provide precise measurement of the distance between the sensor itself and an object. In addition, a *vision system* can also be employed to detect the locations and the shapes of the different objects in the environment. Some of the data received from robot sensors requires substantial computation to convert it to useful information. Techniques to speed up these computations are discussed in the following section.

40.3.1   Vision systems

Vision systems are considered the most powerful sensors in terms of the volume of data they provide [25]. A typical robotic vision system consists of a number of TV cameras capable of digitizing and processing the acquired information. The cameras are interfaced with a computer, where the images are analyzed, through a frame grabber and a visual information preprocessor. Computer processing of image data can range from simple template matching for identification of the objects in the workplace to sophisticated scene analysis and interpretation for autonomous robot navigation.

Image analysis and pattern recognition, as components of the computer vision process, are characterized by a hierarchy of processing levels. Problems typically move from low-level operations on pixels through parsing processes that infer intermediate objects such as edges, lines, regions, and surfaces, to high-level knowledge-based object recognition processes [26].

During real-time operations, the system is required to process numerous data sets every second. Due to its inherent parallelism, methods, techniques, and hardware components required for *low-level* processing of pixel images at video rates are well established [27, 28]. In addition, several researchers have proposed parallel algorithms for *medium-* and *high-level* image processing [29–32].

Robotic vision systems differ from many other vision applications by their critical time constraints since they must complete their processing in few seconds or less in order to meet industrial cycle times. Brady et al. [33] described a video-rate algorithm for detecting and tracking coplanar feature points on planar objects. The method provides important information for robot vehicle navigation, such as surface orientation. The algorithm was implemented on a heterogeneous parallel machine, PARADOX. Also, Trier and Fuchs [34] proposed a transputer-based image processing system for automated robot welding. The system supplies features such as contour tracking, process monitoring, and control of the automated welding process. Versions of the algorithm were tested using one to eight transputers, resulting in finishing times ranging from 215 ms to 50 ms per frame, respectively. Espinosa and Perkowski [35] developed a hierarchical Hough transform (HT) based on pyramidal architecture as a low-to-medium spatial vision subsystem for a mobile robot. The algorithm was implemented on a 386-based personal computer and proved to give results of high quality as compared with the standard HT implementation. Rygol et al. [36] implemented a 3D vision system upon a locally developed transputer-based hybrid parallel processing engine named MARVIN (multiprocessor architecture for vision), hosted by a SUN workstation. The competence of the system was shown by visually guiding a robot arm to pick up various objects in a clustered scene with a total processing time of 10 s.

Several customized VLSI architectures have also been proposed for the same purpose. Honda et al. [37] proposed a high-performance VLSI image processor based on a multiple-valued residue arithmetic circuit. Parallelism is hierarchically used to realize the high-performance processor. Graefe and Fleder [38] designed a programmable RISC (reduced instruction set computer) machine with a modified Harvard architecture, designed to operate as a coprocessor in combination with each parallel processor of a multiprocessor vision system. The performance of the processor has been shown using several real-world problems. Kent [39] developed the pipelined image processing engine (PIPE), which is a multi-stage, multi-pipelined unit. The chip can process sequences of images at video rates through a series of local neighborhood pixel processing operations. In addition, Anderson [40] proposed a customized VLSI chip for robotic vision processing. The chip employs eight specialized accumulators under the control of a programmable logic array for real-time processing of image data.

40.3.2   Tactile sensing

Tactile sensors are made to respond to contact forces that arise between the sensors and the surfaces of solid objects. These sensors are usually used to provide the computer with information about the topography and texture of contact surfaces. Tactile sensors usually employ force sensing over a small local area to acquire the surface conditions of external objects. It is important that these sensors be compact and light so that they do not interfere mechanically with operation of the robot.

Several papers addressed the computational complexity of the tactile sensing operation. Bhandarker and Sung [41] presented a parallel interpretation tree search algorithm for object recognition using sparse range or tactile data on the Intel iPSC/2 hypercube multicomputer. The objects are approximated as a piecewise combination of polyhedra. Three mapping strategies have been considered: breadth-first, depth-first, and depth-first with load sharing.

Raibert and Tanner [42] proposed a custom-designed VLSI device that performs transduction, tactile image processing, and communication. Forces are transduced using a conductive plastic technique in conjunction with metal electrodes on the surface of an integrated circuit. The VLSI chip contains an array of processors that perform two-dimensional convolution between programmable filtering masks and a binary tactile image. Goldwasser [43] described the Integrated Tactile Network Architecture (ITNA) which is a hierarchical system for managing the interaction of tactile sensing and motor control in the 3D active sensory environment. The advantage of this approach is that it keeps the processing electronics physically near to the sensor which aids to eliminate noise related problems.

40.4   Robot Control

The equations of motion for a robot manipulator can be written as a set of first-order partial differential equations:

$$\dot{x} = f(x(t), u(t)) , \ t \in R^+ \tag{40.1}$$

where x is a $2N \times 1$ state vector representing the joints' positions and velocities, i.e. $x(t) = [q_1(t), ..., q_N(t), \dot{q}_1, ..., \dot{q}_N]^T$, $f(x, u)$ is a differentiable vector-valued function of x and u, i.e. $f(x, u) = [f_1(x, u), ..., f_{2N}(x, u)]^T$, $u(t)$ is an $N \times 1$ vector representing the input joints' forces/torques, and N is the number of degrees of freedom. Equation represents a multi-input multi-output (MIMO) system in which the equations are tightly coupled and highly nonlinear.

The control problem is to design a controller which generates the desired control input $u(t)$ based on the information available at time t. This can be classified into two distinct categories: *gross motion control* and *compliant motion control*. In case of gross motion control, the robot is required to follow a certain prescribed trajectory $x_d(t)$ describing the initial and final positions (or orientations) of the end-effector. Define the tracking error vector as

$$\tilde{x}(t) = x(t) - x_d(t) \tag{40.2}$$

The problem is to design a control law $u(t)$ that ensures that $\tilde{x}(t) \to 0$ as $t \to \infty$.

If the motion of the end-effector is constrained, such as in deburring and assembling tasks, the robot must comply with the environmental constraints. This is usually referred to as *compliant motion control.*

Another classification is based on the trajectory coordinate space. If the desired trajectory is defined in terms of joints positions, velocities, and accelerations, the control scheme is called *joint based control* scheme. In many cases, the end-effector is required to follow straight lines or other geometrical shapes defined in Cartesian space. Control schemes based on forming errors in Cartesian space are referred to as *Cartesian based control* schemes [25].

The control of robot manipulators has represented a challenging problem for control engineers due to the highly nonlinear nature of robot dynamics. The increasingly sophisticated tasks required by robot manipulators have called for efficient control techniques to be used for high-speed tracking accuracy while operating in uncertain environments. Many control schemes were proposed in the literature to overcome this problem, ranging from simple joint servomechanisms assigned to the different joints of the arm to sophisticated adaptive control techniques [44–46].

Many advanced control schemes employ sophisticated computational models that render them, despite their superior performance, unimplementable on conventional uniprocessor machines due to their severe computational requirements. The advances in computer architectures have stimulated researchers to employ parallel processors to meet the hard real-time constraints of many control systems.

Several parallel architectures have been investigated for control systems applications. Several researchers proposed solutions using general purpose multiple instruction multiple data (MIMD) architectures [47, 48]. For example, Tahir and Virk [49] proposed a parallel algorithm suitable for MIMD systems for the optimal control of an aircraft longitudinal mode.

However, other architectures have also been used; Fleming et al. [50] investigated the suitability of a single-instruction multiple-data (SIMD) architecture to deal with a matrix-based controller structure. Gaston and Irwin [51] considered systolic arrays implementations of the Kalman filter, and Jacklin et al. [52] used an array processor chip to handle complex control calculations.

Most advanced robot controllers require the on-line computations of several analytic models of extreme complexity at high sampling rates, namely: robot *kinematics*, *Jacobian*, and *dynamics*. These models, due to their intensive computations, are some of the major obstacles that hinder the development of sophisticated robotic systems. The application of parallel processing and advanced computer architectures to speed up the computations of these models is the topic of the next section.

40.5   Applications of Advanced Architectures for Robot Kinematics and Dynamics

The six basic robotics computations in robot arm control are the forward and inverse kinematics, the forward and inverse dynamics, and the forward and inverse Jacobian equations. The on-line computations of these tasks are required at various stages of robot arm control and computer simulation of robot motion.

The kinematics is the study of motion of rigid objects without addressing the effects of the forces (or torques) involved. In order to analyze the robot motion, two coordinate systems are widely used: world coordinates and joint space coordinates. The world coordinates system is usually represented by a Cartesian system whose origin is fixed to the base of the manipulator. The joint space coordinates are a set of moving coordinates attached to each joint. The kinematic equations are the mathematical transformation model used to compute the position and orientation of the end-effector using measured data of angular (or linear) displacement of the joints. In addition, the mathematical relationship that relates the differential changes in joint space to the corresponding changes in the Cartesian space is called the manipulator's Jacobian [25].

The dynamic equations define the motion of a robot manipulator. The model relates forces (or torques) applied to drive the joint actuators to the positions, the velocities, and the accelerations of the joints. It can be classified into two distinct mathematical models:

1. the inverse dynamics, which compute the required forces (or torques) applied to the actuators, for a given trajectory
2. the forward dynamics, which compute the positions, the velocities, and the accelerations, for a given set of applied forces (or torques)

While the computation of the inverse dynamics is essential for many advanced control strategies, the computation of the forward dynamics is mainly used for analysis and simulation purposes.

Two main approaches for computing the inverse dynamics are discussed based on Lagrangian and Newtonian mechanics: the Lagrange-Euler (LE) and the Newton-Euler (NE). The LE technique expresses the dynamic model in the following compact matrix form:

$$\tau(t) = D(q)\ddot{q} + C(q,\dot{q}) + H(q) \tag{40.3}$$

where

D = an $N \times N$ matrix which represents the coupling and effective inertia terms
C = an $N \times 1$ vector which represents the centripetal and Coriolis terms
H = an $N \times 1$ vector representing the gravity loading effects
$q, \dot{q}, \ddot{q}$ = $N \times 1$ vectors representing positions, velocities, and accelerations, respectively
$\tau(t)$ = an $N \times 1$ vector of applied forces (or torques)
N = the number of degrees of freedom (dof)

The NE formulation is an alternative approach for computing the inverse dynamics using a recursive technique that treats each link as an isolated solid body. The model is based on d'Alembert principle, which states that for a mulibody system, the dynamic equilibrium is equivalent to the static equilibrium if the inertial forces are considered to be independent forces in equilibrium with all other forces acting on the system [53]. The NE formulation results from applying d'Alembert principle to each link of the manipulator.

The LE provides a better insight into the dynamic model than the NE. This is clear, since the NE is a recursive algorithm, whereas the LE provides a closed form solution that consists of three components of well defined characteristics. Moreover, the LE has been employed in many control schemes [44, 54–56]. However, the NE formulation is computationally more efficient than the LE formulation. While the computational complexity of this model is $O(N^3)$, the NE has a computational complexity of $O(N)$ [57].

The previous models are computationally expensive. Several techniques have been proposed that employ advanced computer architectures to speed up the computation of robot dynamics. Below, a review of the major solutions proposed for these tasks is presented.

Several researchers suggested the use of SIMD architectures for robotic tasks. Lin and Lee [58] showed that most robotic tasks possess highly regular properties and a common linear recursive structure. A reconfigurable, dual-network, SIMD machine has been designed to mach their characteristics. The different nodes are connected using a fault-tolerant cube network to maintain high reliability of the system. The forward dynamics problem was studied by Lee and Chang [59] who proposed two efficient parallel algorithms for computing the forward dynamics on SIMD architectures. The reconfigurable generalized-cube network was found to be the most suitable architecture for implementing the proposed parallel algorithms. Yeung and Lee [60] proposed generalized methods for the efficient computing of the Jacobian model on uniprocessor computers, SIMD architectures, and VLSI pipelines. Sadayappan et al. [61] designed a VLSI Robotics Vector Processor (RVP) chip which is tailored to exploit parallelism at the low-level matrix/vector operations of robot kinematics and dynamics models. The chip consists of a SIMD array of three floating point processor (FPP) units designed to provide parallelism both within and between the FPPs.

Pipelining is another approach for efficient computation by employing temporal parallelism. To exploit the nature of pipelined computer architectures, the analytic model must be expressed as a loop or series of loops of instruction sequences. Moreover, a simple addressing strategy for supplying the arguments to the pipeline must be employed. Robotic tasks are usually implemented on these architectures by reformulating the analytic model to match the above requirements.

Lathrop [62] proposed a systolic architecture for robot inverse dynamics. High speedup was achieved at the expense of increased cost. Wander and Tesar [63] developed a pipelined modeling software for the inertia matrix which is implemented on a medium-sized array processor to run in real-time. The algorithm is fourth-order in the number of links and can run over any array processor. Rahman and Meyer [64] described a parallel and pipelined bit-serial VLSI chip for robot inverse dynamics. In addition, a computational scheme on a pipelined vector processor for robot tasks was proposed by Cheng and Gupta [65]. A speedup of 1.82 was achieved by vector processing the inverse dynamics model of a general 6 dof robot manipulator.

Processor arrays have also been employed to speedup robotic computations [66, 67]. For example, Fijany and Scheid [67] developed fast parallel Preconditioned Conjugate Gradient (PCG) algorithms for robot manipulator forward dynamics. The authors discussed the implementation of the algorithms on two interconnected processor arrays and analyses the computation and communication complexities.

Data flow machines have been introduced in recent years to exploit maximum parallelism in programs. Geffin and Furht [68] proposed a dataflow multiprocessor for robot arm control. The proposed architecture computes the applied torques of a multi-link system using the Newton-Euler state space formulation. While maximum parallelism would require 1834 processors, a reasonable engineering solution would require only 42 processing elements.

Several researchers proposed customized VLSI chips capable of efficiently computing elementary mathematical functions such as Lee and Chang [69] and Walker and Cavallaro [70]. For example, Lee and Chang [69] developed a coordinate rotation digital computer

(CORDIC) architecture for the computation of closed-form inverse kinematics models. The chip is designed to efficiently compute a large set of elementary operations: multiplications, additions, divisions, square roots, trigonometric functions, and their inverses.

Among the different architectures used for robotic tasks, MIMD models have several attractive features. The architecture presents an inexpensive choice, because the system can be easily built using cheap and commercially available processors and connected to any personal computer. In addition, the model is suitable for a wide class of applications of varying nature. Moreover, multiprocessing improves the reliability of the system. A failure of one processor can be compensated, and the system as a whole can continue to function correctly with perhaps some loss in efficiency.

40.5.1 MIMD systems for robotic computations

Over the past decade, several techniques employing MIMD architectures have been proposed for the on-line computation of robotic tasks. Luh and Lin [71] assigned one CPU to each joint of a robot arm. They employed a scheduling strategy based on the *branch-and-bound* method to minimize the computation time. However, their work did not fully consider the recursive nature of the NE formulation. Kasahara and Narrita [72] developed a parallel processing scheme that employs two multiprocessor scheduling algorithms called, respectively, depth first/implicit heuristic search (DF/IHS) and critical path/most immediate successors first (CP/MISF). Binder and Herzog [73] distributed the recursive NE equations over multiple computing elements, one per joint. Concurrency was achieved by substituting the predicted values for the actual values of variables involved in the recursive equations. The method suffers from lack of accuracy, especially at high speed motions. Vukobratovic et al. [74] proposed a strategy that employs a modified branch-and-bound (BB) method combined with the largest-processing-time-first algorithm (LPTF). Barhen [75] developed a package called ROSES that employs a scheduling scheme based on heuristics combined with graph-theoretic impasse detection techniques. The efficiency of the work was demonstrated by computing the inverse dynamics problem on hypercube nodes. Hashimoto and Kimura [76] developed a new description of the NE formulation that exploits its potential parallelism. The scheme was then mapped onto a model of computation without any complex task scheduling. A real implementation of this work was later presented by Hashimoto et al. [77]. Zomaya [78] proposed a set of heuristic scheduling techniques for robotic computations. The algorithms were tested by an implementation on a network of transputers. Kircanski et al. [79] proposed two heuristic algorithms for the parallel computation of the NE model.

The previous techniques rely on heuristic methods to schedule the computational tasks. Hence, the performance of these schedulers is questionable. Chen et al. [80] proposed a scheduling technique that first finds a fast but suboptimal solution using a heuristic search algorithm called dynamic highest level first/most immediate successors first (DHLF/MISF). Then, the optimal solution is found using a state-space search method: the A* algorithm [81] combined with a heuristic function derived from the Fernandez and Bussell bound [82]. In retrospect, most of the aforementioned techniques have neglected inter-processor communication issues which have a major impact on the overall performance of any parallel implementation [83].

Toward this end, Ahmed and Li [84] developed a scheduling scheme called depth first/minimized overhead heuristic search (DF/MOHS). The method is an extension of the depth first/initial heuristic search (DF/IHS) method to include overhead and contention. The authors stated that there is no guarantee that the algorithm will always produce a near optimum schedule. Moreover, the technique is suitable only for shared memory MIMD systems. Lee and Chen [85] presented a scheduling technique that uses a weighted partite matching algorithm. However, the simulations and the experimental results showed that the algorithm is not appropriate when the average computation to average communication ratio (p/c) is less than 1. The authors also proposed another technique based on the simulated annealing algorithm. However, the cost function employed is heuristic in nature and computationally expensive.

Recently, Nabhan [86] proposed a Parallel Computing Engine (PCE) for analytic models. The analytic system is efficiently coded in its general form. The equations are then developed and simplified using the model parameters. A task graph presenting the interconnections among the different equations is generated. The graph can then be compressed to control the computation/communication requirements.

The task scheduler employs a graph-based iterative scheme, based on the simulated annealing algorithm, to map the vertices of the task graphs onto message passing MIMD type of architectures. The algorithm uses a non-analytic cost function that properly considers the computation capability of the processors, the network topology, the communication time, and congestion possibilities. Moreover, the proposed technique is simple, flexible, and computationally viable. In addition, the scheduler uses a simple strategy for the very fast rescheduling of tasks after each rearrangement. The nearly optimal schedules are automatically transformed to OCCAM code suitable for transputer architectures. In addition, a parallel simulated annealing algorithm is developed to speed up the computations of the task scheduler. The algorithm is based on the speculative computation method proposed by Witte [87] and Witte et al. [88]. The proposed strategy performs significantly better than Witte's method due to its much lower communication requirements. The main advantage of the algorithm is that its speedup increases with the size of the problem. The main components of the PCE toolset are shown in Fig. 40.2. The application of the PCE toolset to robotics tasks was reported by Nabhan and Zomaya [89].

40.6   Summary, Conclusions, and Future Directions

The design of advanced computer architectures has been constantly growing for the last two decades due to the use of high-density and high-speed processor and memory chips manufactured using improved VLSI technology. In this chapter, we have attempted to review the different applications of parallel processors in the field of robotics. The proposed solutions can be roughly divided into two main categories: implementations based on using off-the-shelf components and implementations based on using customized VLSI chips. The latter approach has recently attracted more attention due to advancements in the design of silicon compilers which have significantly reduced the design costs for custom VLSI circuits.

It is the opinion of the authors that the future application of computer architectures will follow one of two directions. The first approach is to employ a specialized computer architecture for efficient processing of each task. The research in this case is directed to provide efficient communication between these isolated architectures. The second approach is to employ a unified computer architecture suitable for all robotic tasks. REPLICA is an example of this type of architecture [90] which provides a flexible interconnection network and a large number of nodes that can be grouped into partitions. Moreover, each partition can be configured to operate in a SIMD or MIMD mode.

Another important research area is the development of programming paradigms that take advantage of the underlying sophisticated computer architecture. Cox and Gehani [91] dis-

model parameters

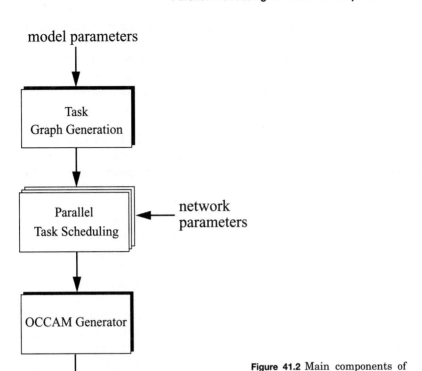

Figure 41.2 Main components of the PCE toolset

OCCAM code

cussed the use of Concurrent C for robot programming. In addition, Simon et al. [92] proposed a computer-aided design model for robot controllers. However, this area is still in its infancy, and future research in this direction is required.

40.7   References

1. Dorf, R. C. 1983. *Robotics and Automated Manufacturing*. Reston, Va.: Reston Publishing Company (Prentice Hall).
2. Lee, M. H. 1987. On the application of intelligent planning techniques in industrial robotics, in Wong, A. K. C. and Pugh, A., *Machine Intelligence and Knowledge Engineering for Robotic Applications*. New York: Springer-Verlag, 244–267.
3. Janabi-Sharifi, F., Wilson, W. J., and Pasng, G. K. H. 1993. A multi-layered learning model. *Journal of Intelligent and Robotic Systems*, vol. 8, 399–423.
4. Bekey, G. A., Liu, H., Tomovic, R., and Karplus, W. J. 1993. Knowledge-based control of grasping in robot hands using heuristics from human motor skills. *IEEE Transactions on Robotics and Automation*, vol. 9, no. 6, 709–722.
5. Brooks, R. A. 1982. Symbolic error analysis and robot planning. *International Journal of Robotics Research*, vol. 1, no. 4, 29–68.
6. Sicard, P. and Levine, M. D. 1987. MARS: An expert robot welding system. In Wong, A. K. C. and Pugh, A., *Machine Intelligence and Knowledge Engineering for Robotic Applications*. New York: Springer-Verlag, 355–385.
7. Pleszkun, A. R. and Thazhuthaveetil, M. J. 1987. The architecture of lisp machines. *IEEE Computer*, vol. 20, no. 3, 35–44.
8. Sussman, G. J., Holloway, J., Steel, G. L., Jr., and Bell, A. 1981. Scheme-79—list on a chip. *IEEE Computer*, vol. 14, no. 7, 10–21.

9. Wah, B. and Li, G. J. 1986. Survey on special purpose computer architectures for A.I. *SIGART Newsletter*, no. 96, 28–44.

10. Togai, M. and Watanabe, H. 1986. Expert system on a chip: an engine for real-time approximate reasoning. *IEEE Expert*, vol. 1, no. 3, 55–62.

11. Moldovan, D. I., Lee, W., and Lin, C. 1992. SNAP: A market-propagation architecture for knowledge processing. *IEEE Transactions on Parallel and Distributed Systems*, vol. 3, no. 4, 397–410.

12. Moldovan, D. I. 1989. RUPIC: A multiprocessor for rule-based systems. *IEEE Transactions on Systems, Man, and Cybernetics*, vol. 19, no. 4, 699–706.

13. Hwang, Y. K. and Ahuja, N. 1992. Gross motion planning—A survey. *ACM Computing Surveys*, vol. 24, no. 3, 219–291.

14. Cela, A., Hamam, Y., and Georges, D. 1991. Decomposition method for the constrained path planning of articulated systems. *Proceedings of the Fifth International Conference on Advanced Robotics: Robots in Unstructured Environments*, vol. 2, 994–999, Pisa, Italy.

15. Cheng, G. -X., Ikegami, M., and Tanaka, M. 1992. A resistive mesh analysis method for parallel path searching. *Proceedings of the 34th Midwest Symposium on Circuits and Systems*, vol. 2, 827–830, Monterey, CA, USA.

16. Lozano-Perez, T. and O'Donnell, P. A. 1991. Parallel robot motion planning. *Proceedings of the 1991 IEEE International Conference on Robotics and Automation*, vol. 2, 1000–1007, Sacramento, CA, USA.

17. Prassler, E. and Milios, E. 1990. Parallel distributed robot navigation in the presence of obstacles. *Proceedings of the Second IEEE Symposium on Parallel and Distributed Processing 1990*, 475–478, Dallas, TX, USA.

18. Kameyama, M., Amada, T., and Higuchi, T. 1992. Highly parallel collision detection processor for intelligent robots. *IEEE Journal of Solid-State Circuits*, vol. 27, no. 4, 500–506.

19. Bobrow, J. E., Dubowsky, S., and Gibson, J. S. 1983. On the optimal control of robotic manipulators with actuator constraints. *Proceedings of the 1983 American Control Conference*, 782–787.

20. Dissanayake, M. W. M. G., Goh, C. J., and Phan-Thien, N. 1990. Time-optimal trajectories for robot manipulators. *Robotica*, vol. 9, 131–138.

21. Khan, M. E. and Roth, B. 1971. The near-minimum-time control of open-loop articulated kinematic chains. *Transactions of ASME*, JDSMC, 164–172.

22. Singh, S. and Leu, M. C. 1987." Optimal trajectories generation for robotic manipulators using dynamic programming. *Transactions of the ASME, Journal of Dynamic Systems, Measurement & Control*, 88–96.

23. Zalzala, A. M. S. and Morris, A. S. 1988. An on-line distributed minimum-time trajectory generator for intelligent robot manipulators. *Research Report 358, Department of Control Engineering, University of Sheffield*, UK.

24. Morris, A. S., Zalazala, A. M. S., and Zomaya, A. Y. 1990. Transputer applications in robotics for dynamic modelling and path planning. *Microprocessors and Microsystems*, vol. 14, no. 9, 565–572.

25. Schilling, R. J. 1990. *Fundamentals of Robotics: Analysis & Control*. Englewood Cliffs, NJ: Prentice Hall.

26. Downton, A. C., Tregidgo, R. W. S., and Kabir, E. 1991. Recognition and verification of handwritten and hand-printed british postal addresses. *International Journal of Pattern Recognition and Artificial Intelligence*, vol. 5, no. 1 & 2, 265–291.

27. Fountain, T. J., Matthews, K. N., and Duff, M. J. B. 1988. The CLIP7 image processor. *IEEE Transactions on Pattern Analysis and Machine Intelligence*, no. 10, no. 3, 311–310.

28. Parkinston, D., Hunt, D. J., and MacQueen, K. S. 1988. The AMT DAP 500. *Proceedings of the 33rd IEEE Computer Society International Conference*, 196–1999.

29. Battiti, R. 1991. Real-time multi-scale vision on multi-computers. *Concurrency: Practice and Experience*, vol. 3, no. 2, 55–87.

30. Choudhary, A. N. and Ponnusamy, R. 1992. Run-time data decomposition for parallel implementation of image processing and computer vision tasks. *Concurrency: Practice and Experience*, vol. 4, no. 4, 313–334.

31. Rygol, M., Pollard, S., and Brown, C. 1991. MARVIN and TINA: A Multiprocessor 3-D Vision System. *Concurrency: Practice and Experience*, vol. 3, no. 4, 333–356.

32. Tucker, L., Feynman, C., and Fitzche, D. 1988. Object recognition using the connection machine. *Proceedings of IEEE Conference on Computer Vision and Pattern Recognition*, 871–877.

33. Brady, M., Han Wang, Shapiro, L. 1992. Video-rate detection and tracking of coplanar objects for visual navigation. *Proceedings of the Second International Conference on Automation, Robotics, and Computer Vision (ICARCV92)*, vol. 1, CV.11.1–CV.11.5.

34. Trier, W. and Fuchs, K. 1992. Image processing based on a parallel architecture for real-time control of industrial robots. *Proceedings of Parallel Computing and Transputer Applications*, vol. 2, 983–988.

35. Espinosa, C. and Perkowski, M. A. 1992. Hierarchical hough transform based on pyramidal architecture. *Proceedings of the Eleventh Annual International Phoenix Conference on Computers and Communications*, 743–750.

36. Rygol, M., Pollard, S., and Brown, C. 1991. Multiprocessor 3D vision system for pick and place. *Image and Vision Computing*, vol. 9, no. 1, 33–38.

37. Honda, M., Kameyama, M., and Higuchi, T. 1993. Multiple-valued VLSI image processor based on residue arithmetic and its evaluation. *IEICE Transactions on Electronics*, vol. E76-C, no. 3, 455–462.

38. Graefa, V. and Fleder, K. 1991. A powerful and flexible coprocessor for feature extraction in a robot vision system. *Proceedings of the 1991 International Conference on Industrial Electronics, Control, and Instrumentation (IECON91)*, vol. 3, 2019–2024.

39. Kent, E. W. 1987. PIPE: A specialized computer architecture for robot vision. In Graham, J. H., *Computer Architecture for Robotic and Automation*. New York: Gordon and Breach Science Publishers.

40. Andersson, R. L. 1985. Real-time gray-scale video processing using a moment-generating chip. *IEEE Journal of Robotics and Automation*, vol. RA-1, no. 2, 79–85.

41. Bhandarkar, S. M. and Sung, L. C. 1991. Object recognition on the Hypercube. *Proceedings of the 1991 IEEE International Conference on Systems, Man, and Cybernetics*, vol. 1, 625–630.

42. Raibert, M. H. and Tanner, J. E. 1982. Design and implementation of a VLSI tactile sensing computer. *International Journal of Robotics Research*, vol. 1, no. 3, 3–18.

43. Goldwasser, S. M. 1984. Computer architecture for grasping. *Proceedings of IEEE International Conference on Robotics and Automation*, 320–325, Atlanta.

44. Bejczy, A. K. 1974. Robot arm dynamics and control. *Technical Memorandum 33-669*. Pasadena, Calif.: Jet Propulsion Laboratory.

45. Fu, K. S., Gonzalez, R. C., and Lee, C. S. G. 1987. *Robotics: Control, Sensing, Vision, and Intelligence*. New York: McGraw-Hill.

46. Guo, L., and Angeles, J. 1989. Controller estimation for the adaptive control of robotic manipulator. *IEEE Transactions on Robotics and Automation*, vol. 5, no. 3, 315–323.

47. Fleming, P. J., Jones, D. I. 1988. Parallel processing in control. In Warwick, K. and Rees, D., *Industrial Digital Control Systems*. Stevenage, U. K.: Peter Peregrinus.

48. Gracia Nocetti, F., Thompson, H. A., De Oliveira, M. C. F., Jones, C. M., and Fleming, P. J. 1990. Implementation of a transputer-based flight controller. *Proceedings of IEE*, Pt D, vol. 137, 130–136.

49. Tahir, J. M. and Virk, G. S. 1990. A real-time distributed algorithm for an aircraft longitudinal optimal autopilot. *Concurrency: Practice and Experience*, vol. 2, no. 2, 109–121.

50. Fleming, P. J., Jones, D. I., and Jones, S. R. 1988. Parallel processing in real-time control: Applications, architectures, and implementations. In Jesshope, R. and Reinartz, K. D., eds. *CONPAR 88*, Cambridge University Press, U.K.

51. Gaston, M. F., and Irwin, G. W. 1989. Systolic approach to square root information Kalman filtering. *International Journal of Control*, vol. 50, 225–248.

52. Jacklin, S. A., Leyland, J. A., and Warmbrodt, W. 1986. Integrating computer architectures into the design of high-performance controllers. *IEEE Control Systems Magazine*, vol. 6, 3–8.

53. Luh, J. Y. S., Walker, M. W., and Paul, R. P. 1980. On-line computational scheme for mechanical manipulators. *Transactions of ASME Journal of Dynamic Systems Measurement & Control*, no. 102, 69–76.

54. Bayard, D. S. and Wen, J. T. 1988. A new class of stabilizing control laws for robotic manipulators, part II: adaptive case. *International Journal of Control*, vol. 47, no. 5, 1387–1406.

55. Khatib, O. 1987. A unified approach for motion and force control of robot manipulators: The operational space formulation. *IEEE Journal of Robotics and Automation*, vol. RA-3, no. 1, 43–53.

56. Wen, J. T. and Bayard, D. S. 1988. A new class of stabilizing control laws for robotic manipulators, part i: nonadaptive case. *International Journal of Control*, vol. 47, no. 5, 1361–1385.

57. Ramos, S. and Khosla, P. K. 1988. A parallel computational scheme for Lagrange-Euler dynamics. In Glasford, G., and Jabbour, K., *30th Midwest Symposium on Circuits and Systems*, 1024–1030.

58. Lin, C. T. and Lee, C. S. G. 1991. Fault-tolerant reconfigurable architecture for robot kinematics and dynamics computations. *IEEE Transactions on Systems, Man, and Cybernetics*, vol. 21, no. 5, 983–999.

59. Lee, C. S. G and Chang, P. R. 1988. Efficient parallel algorithms for robot forward dynamics computation. *IEEE Transactions on Systems, Man, and Cybernetics*, vol. 18, no. 2, 238–251.

60. Yeung, T. B. and Lee, C. S. G. 1989. Efficient parallel algorithms and VLSI architectures for manipulator Jacobian computation. *IEEE Transactions on Systems, Man, and Cybernetics*, vol. 19, no. 5, 1154–1166.

61. Sadayappan, P. Ling, Y. L. C., and Olson, K. W. 1989. A restructurable VLSI robotics vector processor architecture for real-time control. *IEEE Transactions on Robotics and Automation*, vol. 5, no. 5, 583–601.

62. Lathrop, R. H. 1985. Parallelism in manipulator dynamics. *International Journal of Robotics Research*, vol. 4, no. 2, 80–102.

63. Wander, J. P. and Tesar, D. 1987. Pipelined computation of manipulator modeling matrices. *IEEE Journal of Robotics and Automation*, vol. RA-3, no. 6, 556–566.

64. Rahman, M. and Meyer, D. G. 1987. A cost-efficient high-performance bit-serial architecture for robot inverse dynamics computation. *IEEE Transactions on Systems, Man, and Cybernetics*, vol. SMC-17, no. 6, 1050–1058.

65. Cheng, H. H. and Gupta, K. C. 1993. Vectorization of robot inverse dynamics on a pipelined vector processor. *IEEE Transactions on Robotics and Automation*, vol. 9, no. 6, 858–863.

66. Fijany, A. and Bejczy, A. K. 1989. A class of parallel algorithms for computation of the manipulator inertia matrix. *IEEE Transactions on Robotics and Automation*, vol. 5, no. 5, 600–615.

67. Fijany, A. and Scheid, R. E. 1994. Fast parallel preconditioned conjugate gradient algorithms for robot manipulator dynamics simulation. *Journal of Intelligent and Robotic Systems*, vol. 9, 73–99.

68. Geffin, S. and Furht, B. 1990. A dataflow multiprocessor system for robot arm control. *International Journal of Robotics Research*, vol.9, no. 3, 93–103.

69. Lee, C. S. G. and Chang, P. R. 1987. A maximum pipelined cordic architecture for inverse kinematic position computation. *IEEE Journal Of Robotics and Automation*, vol. RA-3, no. 5, 445–458.

70. Walker, I. D. and Cavallaro, J. R. 1994. Parallel VLSI architectures for real-time kinematics of redundant robots. *Journal of Intelligent and Robotic Systems*, vol. 9, 25–43.

71. Luh, J. Y. S., and Lin, C. S. 1982. Scheduling of parallel computation for a computer controlled mechanical manipulators. *IEEE Transactions on Systems, Man, and Cybernetics*, vol. 12, no. 2, 214–234.

72. Kasahara, H. and Narrita, S. 1985. Parallel processing of robot-arm control computation on a multimicroprocessor system. *IEEE Journal of Robotics and Automation*, vol. RA-1, no. 2, 104–113.

73. Binder, E., and Herzog, J. H. 1986. Distributed computer architecture and fast parallel algorithms in real time robot control. *IEEE Transactions on Systems, Man, and Cyber*netics, vol. 16, no. 4, 543–549.

74. Vukobratovic, M., Kircanski, N., and Li, S. G. 1988. An approach to parallel processing of dynamic robot models. *International Journal of Robotics Res*earch, vol. 7, no. 2, 64–71.

75. Barhen, J. 1987. Hypercube ensembles: An architecture for intelligent robots. In Graham, J. H., *Computer Architectures for Robotics and Automation*. New York: Gordon and Breach Science Publishers, 195–236.

76. Hashimoto, K. and Kimura, H. 1989. A new parallel algorithm for inverse dynamics. *International Journal of Robotics. Res.*, vol. 8, no. 1, 63–76.

77. Hashimoto, K., Ohashi, K., and Kimura, H. 1990. An implementation of a parallel algorithm for real-time model-based control on a network of microprocessors. *International Journal of Robotics Research*, vol. 9, no. 6, 37–47.

78. Zomaya, A. Y. 1992. *Modelling and Simulation of Robot Manipulators: A Parallel Processing Approach*. Singapore: World Scientific Publishing.

79. Kircanski, N., Petrovic, T., and Vukobratovic, M. 1993. Parallel computation of symbolic robot models and control laws: Theory and application to transputer networks. *Journal of Robotic Systems*, vol. 10, no. 3, 345–368.

80. Chen, C. L., Lee, C. S. G., and Hou, E. S. H. 1988. Efficient scheduling algorithms for robot inverse dynamics computation on a multiprocessor system. *IEEE Transactions on Systems, Man, and Cyber*netics, vol. 18, no. 5, 729–743.

81. Nilsson, N. J. 1980. *Principle of Artificial Intelligence*. Palo Alto, Calif.: Tioga Publishing Co.

82. Fernandez, E. B. and Bussell, B. 1973. Bound on the number of processors and time for multiprocessor optimal schedules. *IEEE Transactions on Computers*, vol. C-22, 745–751.

83. Zomaya, A. Y. and Morris, A. S. 1991. Dynamic simulation and modeling of robot manipulators using parallel architectures. *International Journal of Robotics and Automation*, vol. 6, no. 3, 129–139.

84. Ahmed, S., and Li, B. 1989. Robot control computation in microprocessor systems with multiple arithmetic processors using a modified DF/IHS scheduling algorithm. *IEEE Transactions on Systems, Man, and Cybernetics*, vol. 19, no. 5, 1167–1178.

85. Lee, C. S. G., and Chen, C. L. 1990. Efficient mapping algorithms for scheduling robot inverse dynamics computation on a multiprocessor system. *IEEE Transactions on Systems, Man, and Cybernetics*, vol. 20, no. 3, 582–595.

86. Nabhan, T. M. 1995. Parallel Computations for a Class of Time Critical Processes. Ph.D. dissertation, Electrical and Electronic Engineering Dept., The University of Western Australia, Perth, Western Australia.

87. Witte, E. E. 1990. Parallel simulated annealing using speculative computation. M.Sc. thesis, Dept. Comput. Sci., Washington University.

88. Witte, E. E., Chamberlain, R. D., and Franklin, M. A. 1991. Parallel simulated annealing using speculative computation. *IEEE Transactions on Parallel and Distributed systems*, vol. 2, no. 4, 483–494.

89. Nabhan, T. and Zomaya, A. 1995. Application of parallel processing to robotic computational tasks. *International Journal of Robotics Research*, vol. 14, no. 1, 76–86.

90. Ma, T. W. E. and Krishnamurti, R. 1984. REPLICA—A reconfigurable partitionable highly parallel computer architecture for active multi-sensory perception of 3-dimensional objects. *Proceedings of the IEEE International Conference on Robotics*, Atlanta, 176–182.

91. Cox, I. J. and Gehani, N. H. 1989. Concurrent programming and robotics. *International Journal of Robotics Research*, vol. 8, no. 2, 3–16.

92. Simon, D., Espiau, B., Castillo, E., and Kapellos, K. 1993. Computer-aided design of a generic robot controller handling of reactivity and real-time control issues. *IEEE Transactions on Control Systems Technology*, vol. 1, no. 4, 213–229.

41

Distributed Flight Simulation: A Challenge for Software Architecture

Rick Kazman

Flight simulation has always been an application that needed to be distributed to be computable at all. Although examples of single-processor flight simulators exist [1], they are typically simulators of low fidelity—often games. True flight simulation has extremely high fidelity demands: the virtual environment that the simulator creates must be as lifelike as possible to train the aircrew as effectively as possible.

Even in simulators created 30 years ago, distribution was utilized:

> Linear computing operations, such as summation, integration and sign changing, are performed by direct-current operational amplifiers, consisting of high gain drift correct amplifiers with appropriate resistive or capacitive feedback networks. There are 150 such operational amplifiers.... The non-linear part of the computing equipment consists of servo-multipliers, electronic time-division multipliers, and diode function generators. In addition, forty-eight high gain amplifiers are available for special computing circuits which may be built up for each simulation.... The individual computing amplifiers, potentiometers, multipliers, etc. are connected together to form the overall computation network by means of a central patching panel, having over 2300 terminations.[2]

Although the effects produced by such a simulator are primitive by today's standards, many of the principles behind the system's architecture have not changed. A single computer typically cannot provide the functionality and/or the raw computation power to serve a flight simulator (or if it could, it would be horrendously expensive). This situation is not likely to change: as computer power increases, so do our demands and so does the complexity of the air vehicles being simulated. Thus, a number of computers were "patched" together through a patch panel 30 years ago to create an early flight simulator; today, they typically would be connected through a fiber optic network.

41.1 The Challenge of Distributed Flight Simulation

Distributed flight simulation poses a serious software development challenge. This chapter discusses some of those challenges and what techniques have been applied to the software structure of flight simulators to address those challenges. Although these techniques were applied to flight simulation in the case study presented here, most of them are equally applicable to any complex distributed application.

41.1.1   Characteristics of flight simulation software

Flight simulation is characterized by the following properties:

Real-time performance constraints. Flight simulators must meet hard real-time deadlines. They must do so for a number of reasons:

1. *fidelity*: Flight simulators must execute at fixed frame rates (often called *harmonic frequencies*) that are high enough to ensure verisimilitude. Different senses require different frame rates to achieve verisimilitude. Within a frequency class—say 30 Hz or 60 Hz—all simulations must be able to execute to completion within the base time frame—1/30th or 1/60th of a second.
2. *coordination*: All portions of a simulator run at an integral multiple of the base rate. If the base rate is 60 Hz then slower portions of the simulation may run at 30 Hz, 20 Hz, 15 Hz, 12 Hz and so on. They may not run at a nonintegral multiple of the base rate, such as 25 Hz. One reason for this restriction is that the senses that a flight simulator provides for the crew being trained must be strictly coordinated. It would not do to have the pilot execute a turn but to not begin to see the change visually or feel the change for even a small period of time (say, 1/10th of a second). Even for delays that are too small to be consciously detectable, a lack of coordination may be a problem. Such delays may result in a phenomenon known as *simulator sickness*, a purely physiological reaction to imperfectly coordinated sensory inputs.

Continual development/modification. Simulators exist for only one purpose: to train users where the equivalent training on the actual vehicle would be much more expensive or dangerous. To provide a realistic training experience, a flight simulator must be faithful to the actual air vehicle. However, air vehicles, whether civilian or military, are continually being modified and updated. Therefore, the simulator software must be continually modified and updated to maintain verisimilitude.

Very large size and complexity. Flight simulators typically require tens of thousands of lines of code for the simplest training simulation, up to more than a million lines of code for complex, multi-user trainers. Furthermore, the complexity of flight simulators, mapped over a 30 year period, has shown exponential growth [3].

Development in geographically distributed areas. Military flight simulators are typically developed in a distributed fashion for two reasons, one technical and one political. The technical reason is that different portions of the development require different expertise, so it is common practice for the general contractor to subcontract portions of the work to specialists. The political reason is that high-technology jobs such as simulator development are political plums, and so many politicians fight to have a piece of the work in their jurisdiction. In either case, the integrability of the simulator—already a difficult consideration because of the size and complexity of the code—is increased because the paths of communication are long.

High expense of debugging/testing/modification. The complexity of flight simulation software, its real-time nature, and its tendency to be frequently modified all contribute to making the costs of testing, integrating, and modifying the software exceed the costs of development.

Muddy mapping between software structure and aircraft structure. Flight simulators have traditionally been built with run-time efficiency as their primary quality goal [4]. This is not surprising, given the above-mentioned performance and verisimilitude requirements of flight simulators, and given that simulators were initially built on platforms with extremely limited memory and processing power by today's standards. For example, a flight simulator for a Boeing 727 had 4 kB of "core" memory, and 40 kB of disk storage available. A state-of-the-art simulator in 1967 had "a 24-bit word, basic cycle time of [sic] the order of 2 microseconds, up to 64,000 words of core memory" [3].

However, because of the focus on run-time efficiency, the design of the flight simulator software has tended to bear little relationship to the design of the air vehicle it is modeling. This posed problems for simulator development (communication among the various stakeholders, i.e., software engineers, simulation engineers, programmers, managers, users and tool builders) was often nearly impossible. It is well accepted in the software engineering community that communication is of paramount importance in large, complex software development projects:

> The principal problems project teams face are communication, agreement about requirements, and managing change [5].

Communication was not facilitated by the models of flight simulation that had existed up until the mid-1980s, because these models were not concerned with qualities such as modifiability and integrability. Such an emphasis required a paradigm shift.

41.1.2 Managing the complexity

To manage the numerous challenges which flight simulation posed for its software designers, a new culture of software design for flight simulation needed to be created. This design framework, called "structural modeling," will be discussed for the remainder of this chapter. In brief, the design framework consists of object-oriented design to model the subsystems and components of the air vehicle, and real-time scheduling to control the execution order of the simulation's subsystems so that fidelity can be guaranteed. The main tenets of structural modeling are:

- simplicity and similarity of the system's substructures
- minimization of types of software objects
- small number of system-wide coordination strategies
- transparency of design

Each of these topics is addressed in detail in sections 41.4 and beyond.

41.2 A Generic Flight Simulator

Before discussing the structure and properties of flight simulator software any further, it is useful to look at the functionality which a flight simulator—any flight simulator—must support. A functional model of a flight simulator is shown in Fig. 41.1.

A flight simulator must provide a great number and great variety of services to its users. Its users consist of the crew being trained: pilot, co-pilot, navigator, and weapons personnel (if this is a military simulator), but also an "instructor-operator." The instructor-operator is in charge of the pedagogical aspects of flight simulation—it is this person who decides what mission will be run and under what conditions, hence the name "instructor." However, this person also controls the simulation in real time, hence the name "operator."

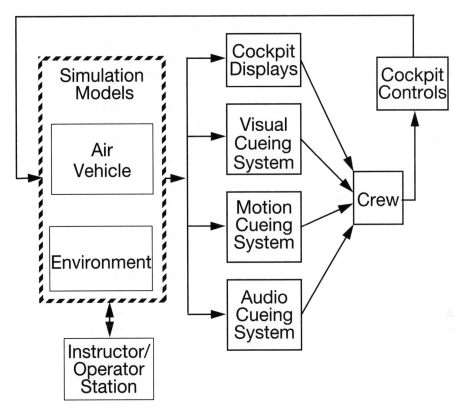

Figure 41.1 A generic flight simulation functional model

The instructor-operator will run the simulation, insert malfunctions (to test the flight crew's response to emergency situations), or freeze the simulation if some pedagogical point is to be made.

The flight simulator software must provide visual, audio and motion cues to the users, i.e., the aircrew. In addition, some simulators will have to provide force-feedback cues. These sensory cues are generated to provide the sights, sounds and feel of a normal air vehicle. The software must also simulate the air vehicle's normal set of instruments. These must all be simulated to a high degree of fidelity, and with a high degree of coordination to avoid simulator sickness.

The aircrew will be responding to their environment in real time. Some of these responses are continuous (the pilot's "stick," for example), and some will be events (changing the setting of a switch). These responses will need to be communicated back to the software simulation as they change the simulation state. In addition, as the air vehicle is being navigated, the environment through which it is passing changes which, in turn, will affect both the sensory inputs being fed to the aircrew and the state of the simulation.

Finally, the simulation software must attend to the instructor-operator who will initially construct the mission that the aircrew will be flying and who may, at any point, change the nature of the simulation.

It is important to keep this set of functions in mind, because the satisfaction of all aspects of this functionality, both in terms of the array of functions and their real-time performance, is required of any flight simulator.

41.3 Introduction to Software Architecture

Software architecture is a view of software that ignores the main concerns of computer science: data structures and algorithms. The main tenet of software architecture is that software has both functional and nonfunctional qualities that it must fulfill, and the fulfillment of nonfunctional qualities (for systems of nontrivial size) is determined by the system's software architecture rather than by code, algorithm, or data-structuring issues. Software architecture deals with systems at the level of computational components, their connections, their patterns of structure, their restrictions on usage, and their rationale [6, 7].

41.3.1 Nonfunctional qualities

Nonfunctional qualities, sometimes called software quality attributes, are things like portability, reliability, modifiability, maintainability, scalability and ease of construction. These attributes are orthogonal to a system's functionality, but have profound effects on a system's success over its lifetime.

The experience with flight simulator software suggests that the satisfaction of these qualities is the most important task for the designers of large, complex software systems. In fact, the Software Engineering Institute's experience with developing software architectures for flight simulators [4, 8, 9] suggests that system designers must first consider how their system will meet its nonfunctional quality goals (which will result in a high-level architecture) and then consider how they might implement the desired functionality within that architecture. This requires somewhat of a paradigm shift for software designers, who have traditionally concerned themselves with satisfying a system's functional requirements first and only considered nonfunctional qualities after the functional requirements have been satisfied.

41.3.2 Views of architecture

To properly discuss a system from the point of view of its software architecture, we must first define some terms. These terms and their definitions reflect a consensus in the software engineering literature [10]. We will describe a system's architecture in terms of its functionality, its structure, its allocation of functionality to structure, and its coordination model.

41.3.2.1 Functionality. A system's functionality is what the system does. It may be a single function or a bundle of related functions that together describe the system's overall behavior. For large systems, a partitioning divides the behavior into a collection of functions that together constitute the system's function but which are individually simple to describe or otherwise conceptualize.

Typically, a single system's functionality is decomposed through techniques such as structured analysis or object oriented analysis, but this is not always the case. In a mature domain (e.g., databases, user interfaces, flight simulators, and VLSI design), the partitioning of functionality has been exhaustively studied and is typically well understood, widely agreed upon, and canonized in implementation-independent terms.

The objective of understanding a system's functional partitioning is simple: to provide a principled way of breaking a system into manageable, codable, comprehensible parts. The consequences of analyzing the functional partitioning for the flight simulation domain (using object-oriented analysis) are as follows:

- One needs to analyze the problem domain. Although the functionality of flight simulators is well understood, any given flight simulator may deviate somewhat from the generic model.
- One needs to determine the chunks of functionality that will be allocated to the software architecture. For the sake of comprehensibility, there should be few classes of objects, and each class must have coherent functionality.

41.3.2.2 Structure. A system's software structure reveals how it is built from smaller connected pieces. The structure is described in terms of the following parts, as described in Fig. 41.2:

1. a collection of components that represent computational entities (e.g., a process) or persistent data repositories (e.g., a file)
2. a representation of the connections between the components; i.e., the communication and control relationships among the components

Square-cornered boxes with solid lines represent processes, or independent threads of control. Round-cornered boxes represent computational components that exist only *within* a process or within another computational component (e.g. procedures, modules). Square-cornered shaded boxes represent passive data repositories (files, databases). Solid, thin arrows represent data flow (uni- or bidirectional) and grey, wide arrows represent control flow (also uni- or bidirectional).

Note that there are primitives to describe computational entities, but also that data and control flow are distinguished from each other. This distinction is necessary to discuss "coordination models" as separate from the system's functionality or its data flow. These will be treated in Section 41.7.

41.3.2.3 Allocation. The allocation of function to structure identifies how the domain functionality is realized in the software structure. The purpose of making explicit this allocation is to understand the way in which the intended functionality is achieved by the developed system. There are many structural alternatives and many possible allocations from function into that structure. For example, Parnas [11], in his seminal paper investigating competing allocations of function to structure, gave as a canonical problem the cre-

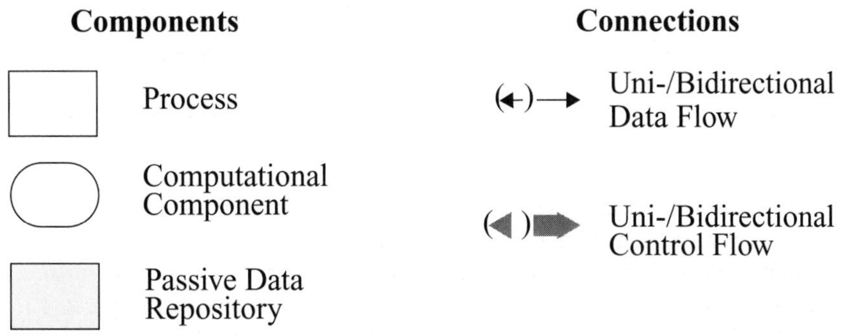

Components

☐ Process

◯ Computational Component

☐ Passive Data Repository

Connections

(◄)⟶ Uni-/Bidirectional Data Flow

(◄)⟹ Uni-/Bidirectional Control Flow

Figure 41.2 Architectural notation

ation of a *key word in context* (KWIC) index. Parnas presented two different software structures—one based upon functional decomposition and one based on abstract data types. Both of these systems contained exactly the same functionality but differed in the way that their functionality was allocated to their software structures.

Software designers choose structural alternatives on the basis of system requirements and constraints that are not directly implied by the system's functional description. The allocation of function to structure will be examined closely in this chapter, because these choices affect the achievement of a system's nonfunctional requirements.

41.3.2.4   Coordination. Coordination is the "glue" that binds separate computational activities into an ensemble that realizes the system's functionality. For example, the Linda language views coordination as separate from (in fact, orthogonal to) computation [12].

Coordination enables process:

- creation
- communication
- synchronization

It is process synchronization and communication that are of greatest importance to the domain of flight simulation.

41.4   Structural Modeling

A structural model is a reusable collection of software structures (objects) of differing levels of abstraction that provide the basis from which all flight simulator software is derived. Each of these software structures implements a coherent and consistent set of functions. The passing of data and control between the software structures is strictly limited by predefined mechanisms.

The objective of this chapter is to show how the practice of structural modeling—a characterization of software at the level of architecture—allowed software designers to take a problem of great scale and complexity and abstract it sufficiently to make it manageable, understandable, codable, modifiable and maintainable. The principles of structural modeling, developed for the flight simulator domain, can be applied to any large-scale distributed system.

41.4.1   Structural modeling principles

Structural modeling attempts, as its highest level objectives, to constrain programmers and designers to force-fit their design into a narrow set of structural and coordination primitives. The purpose of these constraints is to limit the unconstrained passing of data that object-oriented programming affords, and to limit the types of objects from which a system may be constructed and how these types may be composed. Structural modeling's basic principles may thus be enumerated as:

1. predefined partitioning of functionality among software elements
2. restriction of data and control flow
3. restriction to a small number of base types of elements
4. encapsulation of objects in their own internal state
5. removal of side effects
6. elimination of unnecessary levels of packaging

Creating a flight simulator that embodies these principles takes a great deal of discipline, planning, and analysis. The consequences and limitations of the principles of structural modeling are as follows.

41.4.1.1 Base types. Functionality is partitioned into objects analogous to aircraft components—things like pumps, actuators, valves, and so on. There are few classes of objects, and all objects within a class look the same and have the same entry points (or methods). For simplicity and consistency's sake, those parts of a simulation for which there is no real-world analogy (such as data gatherers) are modeled in the same way as the real-world entities, using the same software structures. The objective of minimizing the number of base types is to provide higher-level commonalities among groupings of objects. Thus, the defined types must encapsulate patterns of system behavior at a high level of abstraction.

Although this may at first appear to be a severe restriction, it in fact eases the development of large systems. Since every part of the system is composed of the same small set of base types, the system is more regular, easier to communicate and learn (and thus more reviewable), and easier to modify.

41.4.1.2 Fixed control patterns. Objects receive control in a pattern fixed by their parent. The coordination mechanisms of structural modeling are few, as we shall see. This allows temporal dependencies to be more easily understood and analyzed.

41.4.1.3 Coordination mechanisms. Coordination is analyzed separately from computation. This reduces the temporal dependencies that programmers can assume when creating their simulation models, thus making objects easier to distribute and control.

41.4.1.4 Data flow. Data is forced to flow through special data gatherers called "export areas." In standard object oriented programming, any object may invoke any other object's methods. Data thus flows around a system in arbitrary ways. In structural modeling, data may only flow through export areas. This has the effect of reducing one object's dependency on another, reducing the system's level of coupling. More precisely, it removes the requirement that one object know the *identity* of another object in order to exchange information with it. The objects are, in effect, decoupled.

Message traffic is channeled through well defined areas and therefore can be easily controlled, monitored, and measured.

41.4.1.5 Interaction. Integration is eased by limiting the ways in which components and subsystems can interact, which effectively limits the assumptions that are built into code. For example, each subsystem knows how to initialize, update and reconfigure itself, and does so without regard to the activities of any other subsystem.

41.5 Motivations for Structural Modeling

The motivations of structural modeling can be better understood if seen in a historical light, as we have indicated in earlier sections. The state of flight simulator software architecture up until the mid 1980s was characterized by the following properties and concerns [8]:

- Flight simulators were traditionally written in Fortran. Even a highly disciplined, structured, modularized version of Fortran is still a difficult vehicle for complex simulations.
- These simulations are typically multiprocess and multiprocessor.
- Communication between modules typically has been achieved through the use of global common (shared data) areas. For example, Marsman [13] indicates that "Over 50,000 data locations, with more than 4000 symbolic variable names, is 'common practice'."
- Flight simulations were typically optimized for efficiency, because for many years the hardware was so slow and so expensive that efficiency concerns had to be paramount to make a simulator run at all.
- Flight simulations emphasized "run" mode over other modes (such as "freeze" or "reposition" modes).
- Simulation functionality typically was fragmented for load balancing, and this fragmentation often was not along simple functional lines.
- Software was partitioned based on update rates rather than on conceptual or structural features.

The consequences of these characteristics were that the software making up a simulation was difficult to modify and difficult to update because there were too many intertwined relationships among the modules which were not controlled in a disciplined way, or even analyzed in a disciplined way.

A further consequence was that the system's functionality was difficult to distribute, and once a distribution had been decided upon, it was difficult to change.

These systems, because they were optimized for high efficiency in "run" mode, had low fidelity in non-run modes such as "freeze" or "reposition."

Finally, because the structure of the flight simulator's software did not resemble the structure of the air vehicle that was being modeled, it was difficult to for domain experts and simulation experts to communicate effectively, and it was difficult to insert malfunctions into the simulation code.

> Malfunctions in the physical aircraft, such as the failure of a pump in the hydraulics system, propagate their effects up to higher level systems of the aircraft; so the pump failure might lead to a failure in a hydraulically-controlled actuator for the air brake which then renders the air brake inoperable. The architecture of the simulation, however, did not explicitly represent the connection between the pump, the hydraulics system, the actuator, and the air brakes. Rather, the simulation would separately model the hydraulics system and the braking system, with no model for the actuator itself. An accurate portrayal of this malfunction would, therefore, require changes to both of these simulation models. Other malfunctions were even more complex, involving more simulation models whose logical connection in the physical aircraft were not represented explicitly in the simulation architecture.[8]

41.6   Flight Simulator Software Architecture: Overview

To address the limitations of the traditional ways of building flight simulators, a new software architecture was developed as part of the structural modeling discipline. The new software architecture can be divided into two main sets of concerns: *executive* and *application*.

The executive handles coordination issues: real-time scheduling of subsystems, synchronization between processors, event management from the instructor-operator station, data sharing, and data integrity. These functions are implemented via three structures:

- periodic sequencer
- event handler
- timeline synchronizer

The application handles computation, which consists solely of modeling the air vehicle. The application's functions are implemented by:

- components
- subsystems

These functional elements, which are the building blocks of the air vehicle structural model, will be discussed next. It is important to note two features of these building blocks:

1. *Their number.* Five building blocks serve to describe the significant functionality of the system.
2. *Their homogeneity.* Each of these building blocks has a small number of operations by which they can be invoked, encapsulates some state, and has a severely limited collection of external data upon which it relies.

41.7   Flight Simulator Software Architecture: Base Types

A structural model for an air vehicle consists of the 5 functional elements listed above, structured hierarchically into levels, as indicated in Fig. 41.3.

41.7.1   Timeline synchronizer

The executive level is itself hierarchically structured. Since a flight simulator is real-time, something must be responsible for the overall scheduling. The timeline synchronizer, as depicted in Fig. 41.4, fulfills this function. It controls data accesses, periodic processing

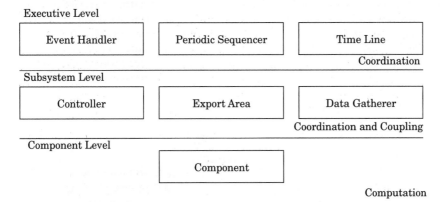

Figure 41.3 Levels in the structural model

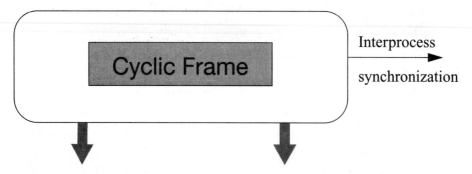

Figure 41.4 Timeline synchronizer

and event processing by scheduling the periodic sequencer and the event handler. The timeline synchronizer also maintains synchronization with other processes.

A typical processing frame is depicted in Fig. 41.5. Note that two portions of the processing frame are explicitly reserved for synchronization. The first synchronization ensures that all data is moved between the export areas of the various subsystems at the same time. This removes the possibility of contention for data and ensures that all subsystems receive the same data on a given frame. The second synchronization ensures that all periodic processing (the simulation modeling) is done simultaneously by all subsystems. This helps to alleviate one of the causes of simulator sickness, that of unsynchronized perceptual inputs. The data moves and periodic processing portions of the frame are handled by appropriate calls to the periodic sequencer, and the event portion is handled by the event handler.

41.7.2   Periodic sequencer

The periodic sequencer, depicted in Fig. 41.6, handles the regular scheduling of subsystems when it is invoked through its periodic processing operation. The various modes of flight simulator operation—run, initialize, freeze, etc.—can be described in terms of the subsystems which are invoked to achieve that mode, along with their operations (initialize, update, stabilize, and so on).

The periodic sequencer makes use of a predefined scheduling table (sometimes called a "piano roll") which describes a list of subsystem invocations required to achieve the desired mode. The scheduling table is empirically determined with the constraint that it represents worst-case behavior.

The periodic sequencer is also responsible for invoking subsystems to have them perform their "data moves" operation, transferring data from one export area to another.

Sync	Data moves	Event handling	Sync	Periodic sequencing	Time burning	End of frame

Figure 41.5 A typical processing frame

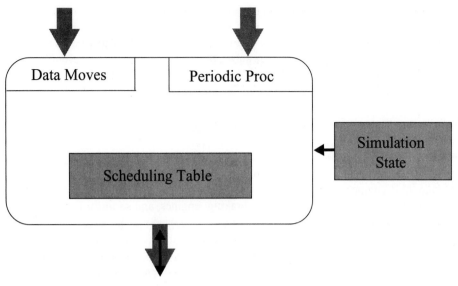

Figure 41.6 Periodic sequencer structure

In both cases, data moves and periodic scheduling, the periodic sequencer is limited to seeing only subsystems and their operations. It has no information as to how a subsystem is built from its components or how a subsystem's operations are achieved.

41.7.3   Event handler

Whereas most of a flight simulator's resources are spent in computing periodic operations, aperiodic events do occur, principally from the instructor-operator station. An event handler exists in a flight simulator architecture to handle such occurrences. The structure of an event handler is given in Fig. 41.7.

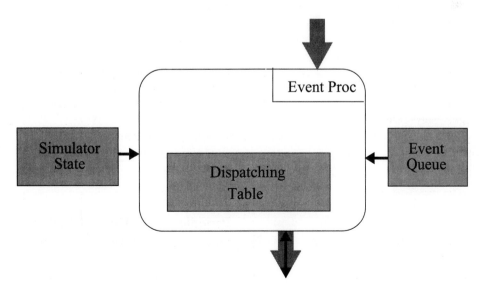

Figure 41.7 Event handler structure

It is these instructor-operator initiated events that change the mode of a simulation. Because of this, the event handler needs to know about the current state of the simulation and, in response to events which it receives from a queue, dispatches these events to subsystems that need to make a response. The subsystem, in turn, will relay the event to its affected components.

The event handler thus has a very limited view of the world. It knows only about events, subsystems, and the mapping between them. It specifically does not know anything about how those events are dealt with within the subsystems.

41.7.4   Subsystem controllers

Subsystem controllers coordinate the activities of the purely computational components and group these components into meaningful collections that represent subsystems within the air vehicle: hydraulics, air-frame, fuel, braking, engines, and so forth. To group components into meaningful collections of higher-level behaviors, the subsystem invokes each component in turn, passing it the relevant state data.

Subsystems can be invoked through either periodic operations such as "import," "update," and "stabilize," or aperiodic operations such as "initialize," "reconfigure," "set parameter," "process malfunction," or "hold parameter." The structure of a subsystem controller is given in Fig. 41.8.

Subsystems make use of export areas to exchange data. Each subsystem can write to exactly one export area but can read from any number of export areas. Only subsystem controllers can access export areas. Data is passed to or received from components by parameter passing. This relieves the individual components of the responsibility of accessing export areas, which effectively isolates them from their environment. This aids in the modifiability and integrability of the system software.

It is also interesting to note that components are coordinated by subsystem controllers in much the same way as subsystems are coordinated by the periodic sequencer. This is a direct manifestation of main emphasis of structural modeling: coordination and coupling, *not* computation.

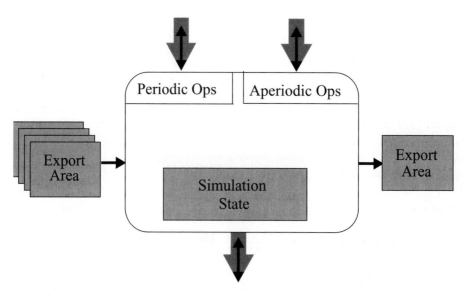

Figure 41.8 Subsystem controller structure

41.7.5   Components

Components—software models of things like reservoirs, valves, turbines, and actuators—do the actual simulation computation in a flight simulation. All of the rest of the structural model's parts exist only to control and to serve the components; i.e., to coordinate their executions, to ensure that they receive timely data, and to report their results to whatever other subsystems and components need to know them.

The structure of a typical component is given in Fig. 41.9. Components may be invoked via periodic or aperiodic operations, as with subsystem controllers. The periodic operation is simply "update"—given the current state, compute the component's new state. The aperiodic operations are "initialize," "set parameter," and "process malfunction." Note that whereas the components receive data and control from above (from subsystem controllers), they do not control or send data to any other objects. The hierarchy of components is a single level deep. There is no particular reason for this, except that, in practice, a single level of decomposition has sufficed to model flight simulator subsystems thus far.

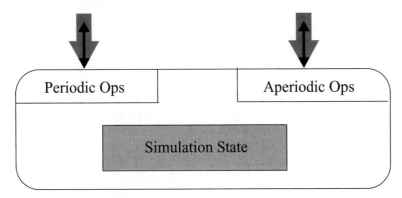

Figure 41.9 Component structure

41.8   A Simplified Software Structure

Now that we have defined the base types of the structural model, we are in a position to see what a complete software architecture for a flight simulator will look like. An example architecture has been given in Fig. 41.10.

This software structure represents flight simulators that have been built using over 1,000,000 lines of real-time Ada code. Its structure, however, is simpler to comprehend than many systems of 5,000 lines of code. This simplicity is the secret of success in large systems in general.

In large, distributed systems, simplicity and analyzability of the software architecture are of far greater importance in achieving a successful result than choosing the correct implementation language, data structures or algorithms.

41.9   Requirements of Flight Simulation

We discussed, in Section 41.3, that a system's requirements may be categorized into functional and nonfunctional portions. We will now discuss the requirements that are most germane to flight simulation, as well as the ways in which these requirements have

Executive

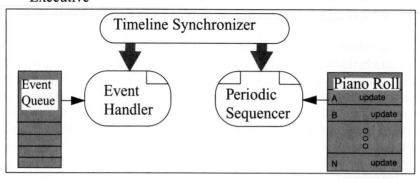

Application

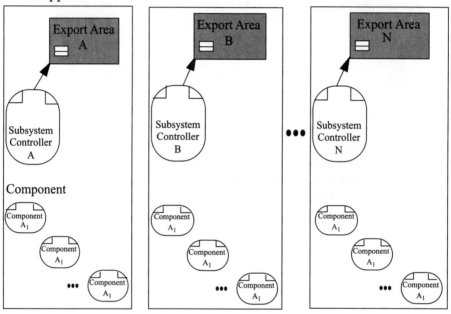

Figure 41.10 Example flight simulator architecture

affected the practice of structural modeling. In addition to functional and nonfunctional requirements, we will also discuss process requirements, since these have an enormous effect on the success of large-scale software systems.

41.9.1   Functional requirements

Systemic requirements—temporal coordination, managing the various subsystems which may operate at different update rates, and dealing with events—all occur because the simulation is typically implemented on tightly coupled multiprocessors, and because efficiency and scheduling concerns cannot be ignored.

Modeling fidelity is also of great concern to simulation developers, as has been discussed in Section 41.1.1. Simulators must mimic the real world. This implies two concerns for developers: choosing those characteristics of the real-world air vehicle that are to be modeled, and choosing the degree of modeling accuracy that is desired. These decisions will affect the software subsystems and components that are to be created and also affect their scheduling.

Supported capabilities must also be addressed when considering the functional requirements of a flight simulator. These comprise the various modes in which a flight simulator can operate—run mode, failure modes, freeze mode, record, playback, and reposition mode. In addition, a mission database must be provided.

A *prima facie* surprising observation is that systemic requirements and supported capabilities drive the form of the structural model, not modeling fidelity. This is because these functional requirements change little from simulator to simulator, whereas the modeling requirements will change drastically as different air vehicles are simulated.

41.9.2   Nonfunctional requirements

The nonfunctional requirements which motivated the structural model were maintainability (principally modifiability and extensibility), scalability and integrability.

Maintainability is promoted in the structural model by localizing changes (since subsystems are modeled after the air vehicle, and changes to the simulator are typically motivated by changes to the air vehicle itself), and by limiting the coupling between the subsystems so that changes to a subsystem do not have "rippling" effects throughout the entire simulation software.

Integrability is promoted by the structural model because of the localization of knowledge and the consistency of functional decomposition. Possibly more important than these factors, however, is the separation of computation from coordination.

Scalability is promoted by the structural model because of the predictability of its structure, the indirection of communication between subsystems (through the use of export areas), and the consistency of its composition mechanisms.

41.9.3   Process requirements

Process has been an increasingly important area of software engineering over the past decade [14]. It is claimed that without sufficient attention to software process, one can never achieve the reliable, repeatable creation of complex pieces of software. The success of any process relies on the software being understandable, reviewable, and repeatable.

For software to be *understandable*, the software architecture has to be simple enough, regular enough, and close enough to the air vehicle structure that simulation experts, software engineers, domain experts (pilots and instructors), and software coders could all discuss it in a mutually agreed-upon language. This means abstracting away unnecessary details and intentionally mandating and monitoring adherence to a simple software architecture.

The structural model makes systems more *reviewable* because of the simplicity of the structural model, the small number of structural elements, and the homogeneity of subsystem structure.

The process of creating flight simulators using the structural model is rendered *repeatable* because of the use of specification forms that software designers must use when specifying any structural element. This allows for consistent analysis and means that a designer can use only the provided building blocks. In addition, accompanying the structural model are sets of code templates that ensure that packaging is done consistently. Finally, exemplar subsystems are provided so that no one needs to begin the process of creation without having some examples of working subsystems to examine. The existence of exemplars has proven to aid consistency tremendously (far more than written guidelines).

41.10   Lessons Learned/Future Directions

Meeting the nonfunctional and process requirements is the greatest challenge in creating large, complex systems. However, few computer scientists are trained in techniques for meeting these requirements. They are, indeed, seldom mentioned in computer science curricula. This is because we are trained to analyze systems in terms of their functional requirements—this is the main assumption behind structured design and object-oriented design, for example. When nonfunctional or process requirements are mentioned, they are often an afterthought. Nonfunctional requirements are occasionally treated as language issues. (For example, we sometimes address maintainability or portability issues by discussing what programming language constructs to use or avoid.)

The contention of an increasing number of software engineers, however, is that process and nonfunctional requirements live in, and must be addressed by, the software architecture of a system. Thus, to address these requirements, we must define the software architecture

- first (before any software is designed)
- clearly (to facilitate communication)
- simply (to ease comprehension)

This work has been ongoing since 1986, and is still evolving. It has been successfully applied by the U.S. Air Force to the B-2 Weapon Systems Trainer, the C-17 Aircrew Training System and the Special Operations Forces Aircrew Training System (SOF-ATS). It has also been adopted for civilian use. For example, Boeing has based their DARTS architecture on structural modeling [15].

41.11   Summary

The results achieved from structural modeling have been positive: complex simulators have been easier to build, and easier to integrate and modify. In addition, these simulators have shown better fidelity in non-run modes—malfunction modes in particular.

The improvements in these simulators have principally accrued from a better understanding of, and adherence to, a well analyzed and well documented software architecture.

41.12   References

1. ASME. 1991. *PC-Based Instrument Flight Simulation: A First Collection of Papers*, New York, ASME.
2. Perry, D., L. Warton, and C. Welbourn. 1966. A Flight Simulator for Research into Aircraft Handling Characteristics. Aeronautical Research Council Reports and Memoranda No. 3566. London: Aeronautical Research Council.
3. Fogerty, L. 1967. Survey of Flight Simulation Computation Methods. *Third International Simulation and Training Conference*, April 1967, 36–40.
4. Rissman, M., R. D'Ippolito, K. Lee, and J. Stewart. 1990. Definition of Engineering Requirements for AFECO—Lessons from Flight Simulators. Pittsburgh, Pa.: Carnegie Mellon University Software Engineering Institute. Technical Report CMU/SEI-90-TR-25.
5. C. Potts, K. Takahashi, A. Anton. 1994. Inquiry-Based Requirements Analysis. *IEEE Software*, 11(2), 21–32.
6. Garlan, D., and M. Shaw. 1993. An Introduction to Software Architecture. In: *Advances in Software Engineering and Knowledge Engineering*, Volume I. World Scientific Publishing.

7. Perry, D., and A. Wolf. 1992. Foundations for the Study of Software Architecture. In: *SIGSOFT Software Engineering Notes*, 17(4), 40–52.

8. Abowd, G., L. Bass, L. Howard, and L. Northrop. 1993. Structural Modeling: an Application Framework and Development Process for Flight Simulators. Pittsburgh, Pa.: Carnegie Mellon University Software Engineering Institute. Technical Report CMU/SEI-93-TR-14.

9. Lee, K., M. Rissman, R. D'Ippolito, C. Plinta, and R. Van Scoy. 1988. An OOD Paradigm for Flight Simulators, 2nd Edition. Pittsburgh, Pa.: Carnegie Mellon University Software Engineering Institute. Technical Report CMU/SEI-88-TR-30.

10. Kazman, R., G. Abowd, L. Bass, and M. Webb. 1994. SAAM: A Method for Analyzing the Properties Software Architectures. In: *Proceedings of the 16th International Conference on Software Engineering* (Sorrento, Italy).

11. Parnas, D. 1972. On the Criteria to be Used in Decomposing Systems into Modules. *Communications of the ACM*, 15(12).

12. Gelernter, D., and N. Carriero. 1992. Coordination Languages and their Significance. *Communications of the ACM*, 55(2), 97–107.

13. Marsman, A. 1985. Flexible and High Quality Software on a Multi Processor Computer System Controlling a Research Flight Simulator. *AGARD Conference Proceedings No. 408: Flight Simulation*, 9-1–9-11.

14. Paulk, M., B. Curtis, M. B. Chrissis, et al. 1988. Capability Maturity Model for Software. Pittsburgh, Pa.: Carnegie Mellon University Software Engineering Institute. Technical Report CMU/SEI-91-TR-24.

15. Crispen, R., B. Freemon, K. King, and W. Tucker. 1993. DARTS: A Domain Architecture for Reuse in Training Systems. *15th I/ITSEC Proceedings*, 659–668.

Index

A

AAL, *see* ATM adaptation layer

AC^0_k, 119

AC^k, 118

Actus, 674, 675

ACU, 654

Acyclic scheduling, 638

Aggressive cancellation, 1019

Ahmed-Delosme-Morf array, 521

Algebraic complexity theory, 124

Algebraic nets, 82

Algorithm design techniques

    breadth-first, 179

    compression, 166

    depth-first, 179

    divide-and-conquer, 165

    Euler tour, 179

Algorithm scalability, 495

Algorithmic motif, 986

    loop splitting, 986

Algorithms

    asymptotic complexity, 174

    asynchronous, 144, 985

    back-substitution, 511

    bounded lag, 1020

    clustering, 260

    cographs, 394

    comparison based, 136

    computational complexity, 171, 174

    compute-aggregate-broadcast, 793

    diagonal quadratic approximation, 1107

    divide-and-conquer, 794

    embarassingly parallel, 986

    forward-substitution, 511

    fully-parallel, 172

    genetic, 1118

    GVT, 1027

    image smoothing, 673

    interval graphs, 384

    Las Vegas, 182

    loosely synchronous, 985

    Monte Carlo, 182

    one-sided error, 183

    parallel randomized, 169

    parallel, complexity of, 174

    parallel, efficiency of, 172

    parallel, speedup of, 172

    parallel, worst-case time of, 174

    pipeline, 794

    progressive hedging, 1107

    randomized, 152, 181

    reset, 155

    routing, 139

synchronous, 144, 985
systolic, 794
time bounded, 136
Tomasulo, 629
tree, 358
wait-free, 157
zero-error, 183
Alignment, 856
All nearest neighbors, 405
ALOGTIME-uniform family, 117
Alternation statement rule, 28
always, 39
Amdahl's law, 992
AND-parallelism, 845
Antidependence, 608, 626, 852
Antimessages, 1016
Application concurrency, 26
Application portability, 865
    criteria, 866-867, 892
    user priorities, 867, 892
Approximate agreement, 154
Architectures, *see also* Software
        architecture
    cache-only, 688
    dance-hall, 703
    distributed memory, 689, 703
    distributed-memory, 9
    equals, 645
    general-purpose, 9
    hybrid shared/distributed memory, 695
    less-than-or-equals, 645
    MIMD, 680
    MIN/tree topologies, 694
    RISC, 18, 595
    shared-memory, 9
    software, 1164
    special-purpose, 9
    tree topologies, 693
ARMA filter, 506
Arrangement, 562
Array control unit (ACU), 650
Array processors, 9, 1146
Artificial intelligence, 1146
Artificial rollback, 1025
Assignment statement axiom, 27
Associative processors, 653
Associative virtual shared memory, 775
Asymptotic measures, 91
    "omega", 91
    "theta", 91

Asynchronous communication, 26
Asynchronous model, 130
Asynchronous rings, 133
Asynchronous transfer mode, 813
    admission control, 823
    application programming interface,
        835
    bandwidth enforcement, 823
    cell loss priority, 824
    congestion control, 823
    management plane, 816
    physical medium dependent sublayer,
        817
    preventive control, 823
    protocol reference model, 816
    transmission convergence sublayer,
        817
    user plane, 816
    virtual channel, 815
    virtual channel identifier, 815
    virtual connections
    virtual path, 815
    virtual path identifier, 815
ATM adaptation layer, 819
    convergence sublayer, 819
    segmentation-and-reassembly
        sublayer, 819
    type 1, 819
    type 2, 819
    type 3/4, 820
    type 5, 821
ATM switches, 824
    Banyan-based fabrics, 829
    Batcher-Banyan network, 830
    knockout switch architecture, 828
    space-division architectures, 827
    time-division architecture, 825
ATM, *see* Asynchronous transfer mode
Atomic action, 33
Atomicity level (*see also* Granularity), 25
Augmented task graph, 251
Auralization, 914
Automatic parallelization, 851
Automatic resource sharing, 772
Automation, 1144
Autonomous robot navigation, 1148
Availability policy, 772
Await program, 51
await-then, 30, 33
Axiomatic semantic definitions, 24

B

Babbage's thesis, 101
Back-substitution algorithm, 511
Bandwidth, 10
Banyan-hypercube, 555
Basic blocks, 635
Basic engineering object, 318
Basic linear algebra subprograms (BLAS), 917
Batcher's Basic Lemma, 178
BBN Butterfly, 792
Benes network, 667
Bidiagonalization, 515
Bidirectional linear array, 506
Binary trees, 206
Binder object, 321
B-ISDN, *see* Integrated service data network
Bisection width, 569, 570
Bisimulation, 55
Bisimulation equivalent, 56
Bit-serial VLSI, 1152
Block algorithms, 925, 929
Block cyclic decomposition, 933
Block distributed memory model, 337
Block distribution, 856, 860
Block-based mode selection, *749*
Block-cyclic distribution, 857, 860
Boolean circuit model, 116
    depth, 117
    size, 117
Botzmann machine, 122
Boundary element methods, 930
Bounded lag, 1008
Branch graph, 264
Branch handling schemes, 624
    branch history buffer, 625
    branch target buffer, 625
    hardware branch prediction, 624
    history register table, 626
    pattern table, 626
    profiled branch prediction, 624
    software branch prediction, 624
    two-bit counter predictor, 625
Branch-and-bound, 1153
Breadth-first traversal, 445
Breathing time bucket, 1031
Breathing time warp, 1032
Brent's scheduling principle, 167

Brent's theorem, 415
Broadband integrated digital service network, 813
Broadcast elimination, 530
Broadcasting, 131, 145, 340, 555, 558, 562
    all-port communication model, 558
    one-port communication model, 558
Broadcasting with selective reduction (BSR), 193
Bulk synchronous PRAM, 114
Bulk-synchronous parallel model, 186, 189
Bus interconnection architectures, 685
Butterfly, 549
BW^0_k, 119
Bypassing, 599
Byzantine agreement, 149
Byzantine failures, 149, 152

C

Cache
    coherence, 706
    consistency, 682
    memories, 6
Cache-only architectures, 688
Calculus for communicating systems, 54
CAM, 652, 660, 661
Campbell's lenient unified model of parallel systems (CLUMPS), 115
Cancelback, 1024
Capability demand, 768
Capacity demand, 768
Capsule, 319
Carrier null-message protocol, 1014
Cartesian intersection, 51
Cartesian sum, 51
CC-NUMA, 706
CCS, 15
CGH, *see* Computer-generated holograms
Chain, 38
Chandy-Misra-Bryant protocols, 1009
Channel, 319
Checkpointing, 275
    adaptive, 298
    consistent, 288
    coordinated, 290
Checkpointing interval, 1022
Cholesky factorization, 917
Circuit complexity classes, 118
Circuit value problem (CVP), 105

Circular deadlock, 973
CKT, 118
Classic dynamic model, 228
Classic static dataflow computing model, 235
Client, 310
Cliques, 136
Clock frequency, RISC vs. CISC, 617
Clock sharing, 25
Clocks, 145
    logical, 143
    synchronization, 145
Closure, 92
Cluster, 139, 318
    home, 141, 145
    tree, 141, 145
Cluster computing, 762
    evolution, 763
Clustering, 10, 260
    benefits, 768
    infrastructure, 769
Clusters, dedicated, 772
CM-1, 791
CM-2, 656, 659, 662
Coarse-grain parallelism, 783
Coarse-grained, communication round model, 420
cobegin-coend, 30
Coherence problem, 706
Co-level, 257
Colored nets, 60
Coloring interval graphs, 384
COMA, 706
Combinational circuit, 195
Common decompositions
    block, 933
    cyclic, 933
Communicating sequential processes, 36, 848
Communication, 37
    BPC permutations, 574
    constant model, 574
    distance, 10
    geometry, 509
    interprocessor, 573
    latency, 494
    linear model, 574, 575
Communication
    bandwidth, 796

    displays, 903
    graphs, 128
Communication and synchronization, 25
Compatibility between generations, 644
Competition, 26
Compiler-based scheduling, 636
    global, 638
    local, 637
    sentinel, 639
Compilers, 1154
    complex instruction set, 596
    parallelizing, 886
Complexity, 188
    and neural networks, 121
    robotic system components, 1145
    theories of, 123
    VLSI computational, 121
Complexity theory of real functions, 124
Composite hypercube, 555
Compression, 166
Computation for the masses, era of, 786
Computational
    broadcast, 531
    interface, 311
    models, 307
    network, 167
    object, 313
    task, 1144
Computational geometry, 581
    convex hull, 582
    data structures, 584
    ECDF search, 582
    hypercube algorithms, 581
    hypercube cascading, 585
    multiple slanted range search, 587
    quadtrees, 587
    selection, 583
    smallest enclosing box, 582
    trapezoidal decomposition, 587
    triangulation, 582
Compute-aggregate-broadcast, 793
Computer architectures, 1146, 1150
Computer classifications, 649
Computer-generated holograms, 801
Computers
    configurable highly parallel, 788
    conventional, 5
    general-purpose parallel, 785
    IBM RP3, 788

NYU Ultracomputer, 788
personal, first, 786
uniprocessor, 5
Computing
a maximum clique, 387
a minimum clique cover, 391
a minimum dominating set, 392
a minimum spanning tree, 375
a vertex cover for cographs, 399
biconnected components, 378
connected components, 372
first "computing device", 784
high-performance, 5
models, demand-driven, 232
parallel and distributed (PDC), 5
real-time, 19
scientific, 981
CONCUR, 57
Concurrency, 37, 900
control, 509
implementation, 26
Concurrent computing, 1144
Concurrent programming, 24
Conditional star hooking, 372
Configurable highly parallel computer
(CHiP), 788
Configuration, 129
initial, 129
Conflict graph, 146
Conformance testing, 968, 971
Conjugate gradient method, 516
Co-NLOGSPACE (co-NL), 104
Connection Machine CM-2, 650, 664
Connection Machine CM-5, 650
co-NP, 95
Conrol unit (CU), 467
Consensus, 147, 151
agreement, 147
number, 157
validity, 147
Conservative time windows, 1013
Content-addressable memory, see CAM
Contour detection, 1051
Control dependence, 241
Control flow graphs, 635
Control parallelism, 496
Control-flow computing model, 223
Controller, 1145
Converge-cast, 131, 145

Convex hull, 404
Cooperation, 26
Coordination language, 846, 849
Coordination library, 774
Coordination problems, 147
CORDIC, 1153
Cover, 139
Covers, sparse, 138
Crash failures, 147
Cray X-MP and Y-MP, 792
Critical path, 260, 907
Crossbar switch, 686
Crossed cube architecture, 555
CSP, 15
CU, 468
Cube-connected cycles, 548, 561
Cycle scavenging, 771
Cycles per instruction, see CPI
Cyclic distribution, 856, 860
Cyclic scheduling, 641
cross-iteration dependencies, 643
polycyclic scheduling, 641
rotating registers, 644
software pipelining, 641

D
Dance-hall architectures, 703
Data
alignments, 858
broadcast, 531
broadcast elimination, 518
dependence, 242
dependency analysis, 852
flow graphs, 635
movement, 924
partitioning, 241
Data distribution, 469, 494, 859, 932
horizontal stripe method, 471, 472,
474, 476
square subimage, 476
square subimages, 469, 471, 474, 490
Data parallel algorithm, 496
Data parallel model, 335
Data parallelism, 466, 496, 728, 985
Data-conditional masking, 657
Data-driven computing models, dynamic,
233
Data-driven execution, 229, 233, 234, 235
Dataflow, 223

data-driven, 224, 225
demand-driven, 225
graphs, 224, 225, 231
machines, 9
Dataflow computers, 224
Dataflow computing models, 223
demand-driven, 230
dynamic, 228, 236
nondeterministic, 223
static, 226, 235
Data-race-free, 713
Datons, 232
de Bruijn network, 550
Deadlock
avoidance, 1011
circular, 973
detection/recovery, 1012
global, 973
Deadlock-doomed, 53
Deadlock-free, 53
Deadlock-prone, 53
Debuggers, 877
DECTOP, 883
NDBtool, 872
PDB, 883
Debugging, 955
Decider, 152
Declarative languages, 845
Decode unit, 622
Decomposition techniques, 512, 1104
Dedicated clusters, 772
Degree, 538
Delayed branches, 609, 611
Demand-driven computing models, 232
Demand-driven execution, 230, 231, 233, 234, 235, 236
Denelcor HEP, 790
Denotational semantic definitions, 24
Dense linear algebra libraries, 919
Depth-first execution, 229
Depth-first search, 380
DET, 119
Deterministic scheduling, 244
Deterministic testing, 965
DFS number, 141
DGEMM, 945
Diameter, 139
Diameter, network, 113
Diffractive-reflective optical interconnects, 799

Digital signal processing, 500
Direct naming, 956
Direct networks, 538
Directed acyclic graph, 227
Directory-based cache coherence, 706–707
Discrete event simulation
event-driven, 1004
time-driven, 1004
Discrete Fourier transform, 342
data parallel FFT, 343
Disjoint parallel programs, 32
Dispatch unit, 623
Distance, 139
Distributed heterogeneous supercomputing management system (DHSMS), 743
Distributed memory architectures, 689
Distributed shared memory, 717, 767
Distributed simulation, 1006
Distributed system complexity, 121
Distributed systems, 129
formal model, 129
Distribution, 856
block, 856
block-cyclic, 857
Divide and conquer, 165
cascading, 412
mesh, 417
rootish, 411
DLOGSPACE (L), 103
DLOGTIME-uniform family, 118
DMR-B-2, 279, 286
Dominant sequence, 260
Domino effect, 289
Double IPS cells, 506
Double pipelines, 506
DRAM, 700
Drinking philosophers, 146
DROI, *see* Diffractive-reflective optical interconnects
DSM, 717
DSP, *see* Digital signal processing
Dynamic data-driven computing model, 233
Dynamic data-driven dataflow computing models, 228, 236
Dynamic multistage interconnection networks, 659

Dynamic networks, 155
  eventual connectivity, 155
  eventual stability, 155
Dynamic parallelism, 227
Dynamic random access memory (DRAM), 700
Dynamic resource sharing, 772
Dynamic scheduling, 244, 263
Dynamics, 1151

E
Eager coherence, 715, 717
Eazyflow model, 234
Eduction, 234
Eduction model, 233
Effective concurrency, 26
Efficiency, 14, 173, 490
8-nearest-neighbor network, 659, 674
EISPACK, 917, 919
Elementary net systems, 59, 69
EM-4, 235
End-to-end, 155
Engineering model, 315
Enhanced hypercubes, 552
Ensemble computing, 762
Ensures, 35
Enterprise clusters, 771
Equals architecture, 645
Euler tour, 179, 408, 434
Event list (EVL), 1004
Events, 129
  concurrent, 143
  ordering, 142
eventually, 39
EXECUBE, 732
Execution, 129, 130
  admissible, 130
  data-driven, 229, 234, 235
  demand-driven, 230, 231, 234, 235, 236
  depth-first, 229
  instance, 264
  rule, 228
  time, 13, 285, 491
  unit, 623
Expandability, 10
Extended Church-Turing thesis, 92

F
Failures, 154

Byzantine, 149, 152
crash, 147
in synchronous systems, 146
link, 155
processor, 146, 151
False sharing, 707, 711, 717
Fan-in, 167
Fan-out, 167, 538
Fat tree, 1050
Fault tolerance, 274, 275
Fibonacci hypercubes, 554
Filtered rollback, 1020
Fine-grain parallelism, 783
Fine-structure complexity classes within P, 103
Firing rule, 230
Flight simulation, 1160
Flooding, 140
Flow dependencies, 626
Flynn's classes, 112
  MIMD, 112
  MISD, 112
  SIMD, 112
FNC (search NC), 110
Folded triangular array, 527
Forward-substitution algorithm, 511
Fossil collection, 1022
4-nearest-neighbor network, 659
Fractional cascading, 413
FRNC (randomized search NC), 110
Function partitioning, 241
Functional (control) parallelism, 728
Functional languages, 879
Functional programs, 231
Future function, 53

G
Gantt charts, 248
Gaussian elimination transformation, 513
gaxpy, 510
GenBank, 1074
General prefix computation (GPC), 196
General trees, 435
  binarization, 435
Genetic algorithms, 1118
  diffusion, 1137
  global, 1131
  migration, 1132
Genetic code, 1073
Gentleman-Kung triangular array, 522

Geometric hashing, 1044, 1058
Geometric hierarchies, 415
Geometric layering
  convex layers, 410
  layers of maxima, 410
Geometry of intervals, 204
GF11, 659, 662, 665, 666, 670
GFLOPS, 13
Givens rotations, 514
Global coin, 152
Global deadlock, 973
Global histogramming, 483, 485
Global masking, 657
Global PE masking, 658
Global router, 668
Global virtual time, 1022, 1027
Global virtual time algorithms, 1027
Goodyear MPP, 791
Grain, 241
  size, 881
Grand Challenge problem, 1042
Granularity, 12, 25, 509, 984
Graphic editors, 877
  Dagger, 882
Graphs
  concurrency, 969
  control flow, 635
  data flow, 635
  fan-in, 167
  fan-out, 167
  general algorithms, 372
  reachability, 969
  weighted control flow, 638
Graph-theoretic concepts and terminology,
  356
Gray code, 572
Grosch's law, 785
Guarded commands, 847
GVT, *see* Global virtual time

H

Hamming cube, 555
Heaps, 448
HeNCE, 739
Hennessy-Milner logic, 55
Heterogeneous computing, 725
  analytical benchmarking, 741
  conceptual model, 739
  CU/PE overlap, 728

matching, 747
mixed-component system, 732
mixed-machine system, 725
mixed-mode system, 725
scheduling, 747
selection theory, 742
task profiling, 741
Heterogeneous environments, 268
Heterogeneous networked environment,
  951
Heterogeneous parallel machine, 1148
Heuristic, 256
High Performance Fortran, 709, 847, 857,
  997
High-level net systems, 60
High-level processing, 1046
High-performance computing (HPC), 5
High-speed networks, 812
Holograms, computer-generated, 801
Holoswitch, 798
Host-to-network interfaces, 831
  direct memory access, 831
  programmed I/O, 831
Householder reflections, 514
HPF, *see* High Performance Fortran
Hybrid shared/distributed memory, 695
Hyper Petersen graph, 555
Hyperblocks, 640
Hypercomputing, 762
Hypercubes, 547, 558, 568, 664, 692, 1149
  (n, b, k)-cube, 554
  bisection width, 570
  bitonic sort, 579
  branch and bound, 587
  composite, 555
  dense algorithms, 581
  enhanced, 552
  Fibonacci, 554
  fine-grain, 577
  generalized, 554
  off-line routing, 574
  on-line permutation routing, 576
  random access concurrent read, 577
  r-ary k-cube, 554
  share sort, 580
  sparse algorithms, 581
  sparse enumeration sort, 580
  sparse share sort, 580
  wormhole routing, 576

Hyper-de Bruijn network, 555

I

I/O Automaton, 129
IBM GF11, 791
IBM RP3, 788
If-conversion, 639, 640
Illiac IV, 658, 662, 673
Image correlation, 477
Image processing engine, 1149
Image smoothing, 672
Image smoothing algorithm, 673
Imperative languages, 844
Implementation concurrency, 26
Impossibility, 151
Incomplete hypercubes, 559
Incremental state saving, 1021
Influences, 784
    direct, 143
In-forest, 250
Information-based numerical complexity
        theory, 124
Inherently sequential geometric problems,
        410
Inner product step, 504
In-order firing policy, 628
Instruction broadcast queue, 477
Instruction broadcast unit (IBU), 730
Instruction fetch unit, 622, 624, 655
Instruction-level parallelism, 621
Instructions retiring per cycle, 623
Integrated service data network, 812
Intensional data structures, 234
Interaction protocol, 26
Interactive environments, 950
Interceptor object, 321
Interconnection networks, 10, 537, 690
    hypercube topology, 692
    mesh, 473–474, 476
    mesh and torus topologies, 690
    MIN/tree topology, 694
    reconfigurable topology, 695
    ring, 473–474, 476, 690
    tree topologies, 693
Interconnection patterns, 793
Interface
    computational, 311
    stream, 313
Interference free, 33

Interior point methods, 1100, 1114
Interleaving semantics, 67
Interlocking property, 528
Intermediate-level processing, 1046
Interrupt (in CISC), 615
Intersection detection and enumeration, 405
Interval graphs, 447
Interval order, 252
Invariance thesis, 92
Invariant, 35
ISDN, *see* Integrated service data network
Isothetic line segments, 213
Issue rate, 626
I-structures, 228–229, 235
IVTRAN, 673

J

Jacobian equations, 1150

K

k-consensus, 154
Kinematics, 1150
knowing n, 135
Kolmogorov complexity theory, 123

L

Language-centered tools, 892
    Charm, 875
    platforms, 888
    summary, 886–887
Languages, 878, 879
    Actus, 675
    Charm, 882
    Charm++, 882
    Code 2, 884
    CODE2, 883
    coordination, 846, 849
    data flow, 879
    data parallel, 880
    data parallelism, 880
    dataflow, 883
    declarative, 845
    extensions, 880
    fortCharm, 882
    Fortran90, 880, 881, 886
    Fortran90D, 880, 881
    FortranD, 881
    FortranM, 884, 885, 886
    functional, 879

gather operations, 880
GLU, 883, 884
High-Performance Fortran, 880
HPF, 881, 886
imperative, 844
IVTRAN, 673
Jade, 880, 881, 886
Linda, 885
logical programming, 879
Lucid, 883
manipulate and construct arrays, 880
Mentat, 882
MPL, 674
Object-oriented Fortran, 882
PCN, 883, 884
PYRROS, 884
scatter operations, 880
side effects, 879
Sisal, 879
Split-C, 885, 886
TOPSYS, 883
Topsys, 884
user assertions, 880
virtual shared-memory extensions, 885
LAPACK, 917, 920
Lazy cancellation, 1019
Lazy coherence, 715, 717
Lazy re-evaluation, 1019
Lazy rollback, 1019
LCLV, *see* Liquid crystal light valve
Leader election problem, 132
Less-than-or-equals architecture, 645
Level, 257
Libraries, 871
    APPL, 871, 874
    Chameleon, 871, 874, 876
    Express, 871
    global operations, 872, 874, 875
    group operations, 873
    mapping, 873
    MPI, 868, 871, 873, 876, 892
    NX, 868, 876
    p4, 871, 874
    PARMACS, 871, 872, 874, 876
    PICL, 871, 872, 874, 876
    portable, 877
    portable, summary, 876
    process group, 872
    PVM, 868, 871, 872, 874, 875, 885
    safe message passing, 872, 873
    TCGMSG, 871, 874
    unsafe message passing, 876
    virtual topology, 872, 873
    XPVM, 872
Linda, 735, 994
Linear approximation, 1051
Linear array of processors, 539
Linear feature extraction, 1051
Linear pipelines, 506
Linear programming, 1098
Linear programming, stochastic, 1100
Linearizability, 156
Linked lists, 432
LINPACK, 917, 919
Liquid crystal light valve
List ranking, 176, 350, 408
List scheduling, 256
Load balance, 900
Load balancing, 986
Local causality constraint, 1008, 1015
Local virtual time, 1006–1007
Locality, 718
Locality, memory, 702
Lockup-free, 632
Log space uniform family, 117
Logical process simulation, 1006
    conservative, 1007
    optimistic, 1007
    synchronous, 1007
Logical processes, 1006
LogP, 114
LogP model, 337
Look-ahead, 1012
Loop bounding, 229, 230
Lower bound, 134, 136, 148
Lowest common ancestor computation, 361
Lowest common ancestors, 409
Low-level processing, 1046
LS problem, 518
L-shaped method, 1113
LU decomposition, 509
LU factorization, 917, 930
Luk triangular array, 528
LVT, *see* Local virtual time

M

Machine-independent complexity theory,
    123

Machine-independent semantics, 867, 875, 876, 879, 886, 890
    blocking receive, 868
    blocking send, 868
    buffered send, 868
    CMMD, 869
    non-blocking receive, 868
    non-blocking send, 868
    NX, 869
    ready send, 868
    standard send, 868
    synchronous send, 868
Mailbox naming, 956
Manchester Dataflow Machine, 229
Mapping problem, 796
Marked nets, 75
Markov reward model, 282
Masked and unmasked instructions, 657
Masking
    data-conditional, 657
    global, 657
    global PE, 658
    PE address, 658
MasPar MP-1, 792
MasPar MP-1, *see* MP-1
MasPar MP-2, *see* MP-2
MasPar Programming Language, *see* MPL
MasPar systems, 654
Massively Parallel Processor, *see* MPP
Massively parallel, defined, 782
Matching, 228
Mathematica, 950
Matlab, 950
Matrices, triangular, 512
Matrix
    similarity, 1082
    substitution scoring, 1077
Matrix matrix multiplication, 509
Matrix searching techniques, 413
Matrix transposition, 344, 487
Matrix vector multiplication algorithm, 507
Matrix-by-matrix multiplication, 488
Matrix-by-vector multiplication, 489
Maximal elements, 404
Maximum finishing time, 248
MCPR, 705
Memories, cache, 6
Memory consistency model, 711
Memory displays, 906

Memory interleaving, 701
Memory locality, 702
Memory sharing, 25
Mentat, 736
Merging, 407
Merging slopes technique, 205
Mesh, 659
Mesh divide and conquer, 417
Mesh of trees, 541
Mesh-connected computer, 539
Message
    complexity, 131
    logging, 296
    sendback, 1023
Message passing, 767
    asynchronous, 956
    interface (MPI), 943
    synchronous, 956
    model, 337
Message passing systems, 129
    asynchronous, 128
    synchronous, 128
Message-passing paradigm, 847
Method of moments, 930
Metric layering, 410
MFLOPS, 13
Microinstruction memory, 616
MIMD, 7, 112, 466, 482, 490, 496, 497, 680, 727, 982, 1042, 1150
    interconnection network, 468, 473, 476, 482
    model, 468
    PE, 468
Minimum coloring, 447
Minimum spanning tree, 136
MIPS, 13
MISD, 7, 112
Mixed-mode (hybrid SIMD/MIMD), 466
Mixed-mode machine, 496
Model interchangeability, 92
Module parallel computer, 113
Molecular dynamics, 990
Monsoon machine, 235
Moving time window, 1008, 1020
MP-1, 656, 662, 667, 668, 673, 674
MP-2, 656, 662, 667
MPL, 675
MPP, 13, 656
MPP system, 654

Multicomputers, 704
Multiobjective optimization, 1120
Multiple instruction, multiple data, *see* MIMD
Multiple quadratic forms, 493
Multiple-SIMD, 653
Multiple-SIMD machine, 490
Multiprocessor systems, 9, 1146
Multisearching, 418
Multistage interconnection network, 687
Multi-threaded models, 230
Multi-threaded processors, 707
Multi-writer coherence protocols, 717
Mutual admission, 25
Mutual exclusion, 25, 146

N
Naming
    direct, 956
    mailbox, 956
    port, 956
Nanoinstructions, 655
Narrowband integrated digital service network, 813
NC and SC, 110
NC reduction, 110
NC-class, 169, 180
NC^k, 118
n-dimensional grid, 664
Neighborhood, 134, 141, 142, 145
Net systems, high-level, 60
netlib, 923
Network diameter, 113
Network interface, 656
Network orchestration, 774
Network-based concurrent computing, 762
Networks
    4-nearest-neighbor, 659
    8-nearest-neighbor, 659, 674
    adaptive optical, 802
    AKS sorting, 546
    Bayan-hypercube, 555
    Benes, 667
    butterfly, 549
    composite hypercube, 555
    crossed cube, 555
    de Bruijn, 550
    direct, 538
    dynamic, 155

    dynamic multistage interconnection, 659
    Hamming Cube, 555
    hexagonal, 546
    hierarchical hypercube, 555
    high-speed, 812
    honeycomb, 546
    Hyper Petersen graph, 555
    hypercubic, 547
    hyper-de Bruijn, 555
    incomplete star graph, 546
    linear array of processors, 539
    mesh of trees, 541
    mesh-connected computer, 539
    meshes, 546
    Omega, 555
    optical, 780
    pancake, 542
    perfect shuttle, 554
    plus-minus interconnection, 554
    pyramid, 541
    shortest paths, 142
    star, 542
    Stirling, 546
    topology, 793
    trees, 540
    twisted k-cube, 555
Neural networks, 9
Neural networks and complexity issues, 121
Nick's class (NC), 103, 109
NLOG completeness, 105
NLOGSPACE, 104
Node, 319
Node degree, 113
Non-blocking cache, 632
Non-blocking implementation, 156
Nondeterministic, 847
Nondeterministic dataflow computing models, 223
Nondeterministic scheduling, 244
Nondeterministic testing, 964
Nonlinear programming, 1105, 1114
Non-strict functions, 230, 235
Non-strict operations, 233
Nonuniform families, 118
NORMA, 704
NP-complete, 96, 250
Nucleus, 319

Null messages, 1011
NUMA, 704
    memory management, 718
Numerical accuracy, 870, 876
NYU Ultracomputer, 788

O

Object-oriented programming, 998
Observation bisimilarity, 64
Observation equivalent, 56
Observation sequence, 55
OCCAM, 1154
ODP, 303
    computational viewpoint, 306
    engineering viewpoint, 306
    enterprise viewpoint, 305
    information viewpoint, 305
    technology viewpoint, 307
    viewpoints, 305
Onion peeling, 410
Open distributed processing, *see* ODP
Operation signature, 310
Operational net semantics, 77
Operational semantic definitions, 24
Operator nets, 231
Opposing forest, 251
OPSILA, 731
Optical communication
    free-space, 796
    parameters, 796
Optical interconnections, 780
    design, 797
Optical networks, 780
    3-D multistage, 803
    adaptive, 802
    dynamic holographic switch, 807
    two-wavelength photorefractive
        dynamic interconnect, 805
Optimal algorithms, 250
Optimality, 172
Optimization, 941, 1118, 1120
    multiobjective, 1120
Ordering of events, 142
OR-parallelism, 846
Outer product, 511
Out-forest, 250
Output dependence, 626, 852
Overhead, 490, 900

P

P and NP, 95
p4, 736, 871
Pancake, 542
Panel method, 930
Paracomputers, 113
Parallax, 266
Parallel and distributed computing (PDC), 5
Parallel
    architectures, 1147
    assessment window system (PAWS),
        743
    composition, 46
    computation, 1147
    computation thesis, 101
    data structure, 430
    disk arrays, 660
    efficiency, 490, 491
    heap, 449
        delete-think-insert, 448
    processor, defined, 783
    simulation, 1006
    time, 260
    virtual machine (PVM), 738
Parallel parentheses matching
    binary matching tree, 441
    encoding scheme, 441
    virtual copying/matching tree, 441
Parallel prefix, 407, 485
    scan operations, 423
Parallel processing, 1146
    economics, 784
    general-purpose parallel computers,
        785
    models, 783
    performance, 785
    technology, 784
Parallel random access machine, *see*
        PRAMs
Parallel techniques
    bucket, 1087
    master-worker, 1086
    portion, 1086
Parallelism, 924
    coarse-grain, 783
    coarser-grain, 1085
    dynamic, 227
    fine-grain, 783, 1084
    instruction-level, 621

loop, 227, 228
pipeline, 227
recursive, 227, 228
structural, 227
useful, 236
useless, 236
Parallelization
automatic, 851
tools, 877
Parallelizing compilers, 892
Parallelizing compilers and preprocessors, 886
automatic, 889
batch tools, 887
Forge 90, 890
Forge90, 880, 881, 888
interactive tools, 887, 889
KAP, 887, 888
ParaScope, 888
platforms, 891
summary, 890, 891
Parentheses matching, 202, 437, 440
Parenthesis string
edge associated, 436
embedded, 437
node associated, 436
PARLE, 57
Partially-ordered path, 969
Partitionable parallel machines, 489
Partitionable SIMD/MIMD (PASM) 650, 654, 730, 788
prototype, 478
PASM, see Partitionable SIMD/MIMD
Past function, 53
Path
collision-free, 1147
partially-ordered, 969
planning, 1146
Pattern of interaction, 25
Pattern of synchronization, 25
P-completeness, 105
P-completeness and NC, 111
PE address masking, 658
Percentage of load imbalance, 1085
Perceptual grouping, 1043, 1056
Performance, 867, 870, 874, 876, 879, 881, 882, 888, 890
choices of semantics, 871

code generation, 871
data locality, 880, 881
efficiency, 866
grain size, 881
impact of semantics, 874
monitoring, 898
overhead, 866, 874, 875, 882
scalability, 874, 890
speedup, 866
tuning, 900
visualization, 899
visualization tools, 898
Performance tools, 877
Ctool, 872
Etool, 872
Gauge, 883
Pablo, 884
ParaGraph, 872
PATOP, 883
Projections, 882
Upshot, 883
Vtool, 872
Xtool, 872
Persistent properties, 144
Petri nets, 15, 59
high-level, 60
Pipeline computations, 794
Pipelining, 6, 599, 641, 1149
Pixel, 469
Place/transition net, 73
Place/transition system, 73
Planar point location, 413
Plane-sweep, 406
Plural, 674
Plus-minus interconnection network, 554
Poisson solver, 671
Polygon triangulation, 405
POLYLOGSPACE, 103
Port naming, 956
Portability, 18
application, 865
Portable libraries
platforms, 878
summary, 876, 877
Postcondition, 27
Potential deadlock, 34
PPM, see Parallel parenthesis matching
PRAM-based complexity classes, 109
PRAMs, 11, 105, 168, 335, 409

arbitrary, 108
arithmetic, 112
basic algorithms, 175
BSP, 114
CRCW, 108
EREW, 107
LogP, 114
macro-operator, 112
nondeterministic, 109
priority, 108
probabalistic, 109
random, 109
variants, 107
XPRAM, 114
Precedence graph, 264
Precondition, 27
Preconditioned conjugate gradient, 509
Predicate/transition net systems, 60
Predicates, 639
Prefix sums, 201
Prefix sums problem, 175
Preprocessors, parallelizing, 886
Priority queues, 448
Probabilistic classes, 99
Probability, 876
Problems
    ear decomposition search, 179
    list ranking, 176
    list ranking by pointer jumping, 177
    merging, 177
    P-complete, 180
    prefix sums, 175
    recursive bitonic merge, 178
    sorting by comparison, 177
Process algebras, 59, 64
Processes, 36
Processing elements (PEs), 467
Processing-element-to-processing-element
        organization, 651
Processor array implementations, 500
Processor arrays, 650, 651
Processor consistency, 712
Processor displays, 900
Processor failures, in asynchronous
        systems, 151
Processor work, 286
Processors, 18, 129
    array, 9
    associative, 653

BBN Butterfly, 792
CM-1, 791
Cray X-MP and Y-MP, 792
Denelcor HEP, 790
faulty, 147, 151
Goodyear MPP, 791
IBM GF11, 791
MasPar MP-1, 792
nonfaulty, 147, 151
superscalar, 622
vector, 9
VLSI array, 650
Processor-to-memory organization, 651
Program dependence graph, 242
Programmable logic array (PLA), 118
Programming
    costs, 116
    coordination model, 983
    language semantics, 24
    linear, 1098
    model, 982
    nonlinear, 1105
    object-oriented, 998
    stochastic, 1098
Progress, 35
Proof outline, 33
Protocol object, 321
Protocols
    breathing time bucket, 1031
    breathing time warp, 1032
    Chandy-Misra-Bryant, 1009
    filtered rollback, 1020
    moving time window, 1020
Prune and search, 417
P-SPACE, 97
P-uniform family, 117
Pure-dependencies, 627
PVM, 738
Pyramid, 541
PYRROS, 266

Q
QR decomposition, 509
Queues, 445

R
RACTM (random access TM), 101
RAM, 11
RAM/ARRAY, 421
Ramsey's theorem, 136

Random access memory machine, 106
Randomized routing, 347
Raw temporal logic of actions, 40
Reachability testing, 967
Reactive servers, 950
Real-time
    computing, 19
    path planners, 1147
    processing, 495, 1152
Reconfigurable SIMD, 1152
Recursive doubling, 491, 670
Reduced instruction set computer, *see* RISC
Reducibility, 92
Reduction operations, 480
    generalized, 485
    image smoothing, 490
    min/max problem, 482
    recursive doubling, 480, 484, 490
    result-to-all-PEs, 483
    result-to-every-PE, 481
    result-to-one-PE, 481, 483
    transfer-op, 481
    tree summing, 480
Reference Model for Open Distributed
        Processing, 304
Regression testing, 955
Release consistency, 713
Reliability, 274
Renaming, 154
Rendezvous, 848, 957
    sequence, 960
Replay, 974
Reservation station, 623
Reset, 155
Resource allocation, 146
Resource-bounding functions, 90
    doubleexp, 91
    Exp, 90
    Explin, 90
    expoly, 91
    lin, 90
    log, 90
    poly, 90
    polylog, 91
    tripleexp, 91
Responsiveness, 15
RFCS, 279, 286
Rings, 132
    anonymous, 133

asynchronous, 133
    synchronous, 135
RISC, 18, 621, 1148
RM-ODP, 303
RNC (randomized decision NC), 110
Robot, 1144
    configurations, 1145
Robotic systems, 1144, 1145
Robotics vector processor, 1152
Rollback, 278, 1016
Rollback chains, 1018
Roll-forward, 279
Rotator graphs, 542, 562
Routing, 139, 555, 559, 560, 562
    deadlock, 556
    regional, 140, 141
    store-and-forward, 556
    wormhole, 556
RP problem class, 100
r-ruling set, 185
Ruzzo-uniform family, 117

S
Safe events, 1009
Safety, 35
Satisfiability problem, 97
saxpy, 510
Scalability, 913, 949
    of algorithms, 492, 495
ScaLAPACK, 917, 921
Schedule length, 248
Scheduling, 13, 130, 131, 1153
    acyclic, 638
    compiler-based, 636
    cyclic, 641
    heuristic, 1153
    polycyclic, 641
    unit, 623
Schwartz's parallel machine classes, 112
Scientific visualization, 909
Scoreboard, 623
Search, 1118
Search trees, 455
    2-3 trees, 459
    bottom-level balancing scheme, 457
    construction, 456
Self-stabilization, 155
Semantic definitions of programming
        languages, 24

Semantics
    axiomatic definition, 27
    denotational definition, 27
    machine-independent, 875
    operational definition, 27
Sensors, 1145
Sequence
    alignment, 1080
    comparison, 1080
    protein, 1072
    similarity score, 1080
Sequence statement rule, 28
Sequencing constraints, 972
Sequential consistency, 712
Server, 310
Set approach, 493
Shared memory, 682
    model, 336
    physically distributed, 688
    systems, 128, 130
    vs. message passing, 710–711
Shared objects, 156
Shared variables
    compare&swap, 128
    read/write, 128, 157
    read/write registers, 131
    read-modify-write, 128, 155
    read-write-modify, 131
    test&set, 128, 131, 157
Shared virtual memory, 717, 854
Shortcutting, 372
Shortest paths, 142
SIMD, 7, 112, 466, 482, 496, 649, 669, 727,
        856, 982, 1150
    CU, 467, 476, 478
    CU/PE overlap, 468, 476, 479
    interconnection network, 467, 473,
            476, 482
    masking, 471
    multiple, 490, 653
    PE, 467, 478
    PE-to-PE configuration, 467
    physically distributed memory
            organization, 467
    virtual machines, 653
SIMD/MIMD, partitionable, 730
Similarity, 148
Simple interlocking, 627
Simplex, 1099

Simulated annealing, 1154
Simulated concurrency, 26
Simulation
    distributed, 1006
    engines, 1004
    logical process, 1006
    parallel, 1006
Single instruction stream, single data
        stream, *see* SISD
Single instruction, multiple data, *see* SIMD,
        982
Single program, multiple data, *see* SPMD
Single program, multiple data, *see* SPMD
Singular value decomposition, 513
SISD, 7, 112, 982
Slackness, 170, 188
SLE, *see* System of linear equations
SmartNet, 755
Smoothing, 469, 491, 492
Snapshots
    atomic, 156
    distributed, 143
Snooping, 706
Software architecture, 1164
    allocation, 1165
    coordination, 1166
    functionality, 1164
    nonfunctional qualities, 1164
    structure, 1165
Solution of triangular systems, 509
SONET, *see* Synchronous optical network
Sorting, 201, 345, 407
    BDM sample sort, 349
    integers, 446
    quicksort, 345
    sample sort, 346
Spanning trees, 131
    minimum, 136
Specifying concurrency, 26
Speculation, 624
Speedup, 13, 173, 181, 470
    calculation, 470
Split-issue, 644
SPMD, 9, 466, 983, 1050
    model, 468
SRAM, 700
Stable, 35
Stacks, 440
Star, 542

Stars, 562
State functions, 39
State update, 632
    checkpoint repair, 634
    consistent state, 633
    exception, 633
    exception handler, 633
    fault, 633
    future file, 634
    imprecise fault, 632
    imprecise interrupts, 633
    instruction boundary, 633
    interrupt, 633
    messy register file, 633
    messy state, 633
    reorder buffer, 634
    trap, 633
    unit, 623
State variables, 1004
Static random-access memory (SRAM),
    700
Static scheduling, 244, 264
Steve's class (SC), 104
Stream interface, 313
Stretch, 140
Strict functions, 230
Strict operations, 233
Strongly-connected arc model, 230
Structural modeling, 1166
Structured operational semantics, 54
Stub object, 321
Subcompact process model, 230
Sublinear Turing complexity classes inside
    P, 103
Subtask parallelism, 491
SUM-OR tree, 655, 657, 664
Superblock formation, 638
Supercomputers, 6, 921
Supercritical speedup, 1019
Superpipelining, 18
Superscalar processors, 621, 622
Supporting object, 319
SVD, *see* Singular value decomposition
SVM, 717
Symmetry, 538
Synchronicity, 509
Synchronization, 11, 25
    events, 956
    fault, 962

Synchronization sequence, 956
    feasible, 961
    partially ordered, 959, 960
    simple, 974
    totally ordered, 959
    valid, 961
Synchronizers, 144
    α, 145
    η, 145
Synchronous communication, 26
Synchronous model, 130
Synchronous operation, 657
Synchronous optical networks, 817
    cell loss priority, 818
    generic flow control, 818
    header error check, 818
    payload type identifier, 818
    synchronous digital hierarchy, 817
    synchronous transport signal, 817
    virtual circuit identifier, 818
    virtual path identifier, 818
Synchronous rings, 135
Synchronous transfer mode, 813
Syntax directed translation, 29
System of linear equations, 510
Systolic arrays, 503, 1150
Systolic computations, 794

T
Tags, 228, 229, 232
Tail duplication, 639
Target machine, 247
Task
    allocation, 267
    computational, 1144
    displays, 905
    duplication, 276
    planner, 1145
    scheduling, 243
    specification, 1145
    trees, 242
Task graphs, 246, 1154
    augmented, 251
    in-forest, 250
    out-forest, 250
Temporal logic of actions, 39
Termination with high probability, 153
Testing, 955
    black-box, 963

conformance, 968, 971
deterministic, 965
nondeterministic, 964
reachability, 967
regression, 955
replay, 966
reproducible, 966
test generation, 968
white-box, 963
Texas Reconfigurable Array Computer
  (TRAC), 730
Three-dimensional views, 910
Three-path communication linear array,
  506
Throttling, 229
Throughput, 15
Time complexity, 132
Time stamps, logical, 143
Time steps, 1004
Time to market, 617
Time Warp, 1015
Time-stepped LP simulation, 1007
Timings for processor arrays, 508
TMR-F, 278, 286
Tomasulo algorithm, 629
  Common_Data_Bus, 630
  register renaming, 629
  reservation stations, 629
  result bus, 630
Topological sorting of n nonintersecting
  line segments, 413
Topologies
  clique, 136
  hypercube, 569, 692
  mesh, 690
  MIN/tree, 694
  reconfigurable, 695
  ring, 690
  torus, 690
Topology, 793
  degree, 538
  fan-out, 538
  interconnection network, 690
  symmetry, 538
Totally monotone matrix
  row minima, 414
  tube minima, 414
Totally-ordered path
TRAC, 730
Trace selection, 638

Trade-offs, 140, 145, 154, 941
Transactional model, 115
Transient faults, 275
Transition system, 61, 62
Transitive closure bottleneck, 180
Transparency system, 322
  concurrency, 323
  location, 323
  migration, 323
  Replication, 323
  resource, 323
Transputers, 1154
Trapezoidal decomposition, 413
Tree algorithms, 358
Tree contraction, 408
Trees, 131, 434, 540, 693
  binary, 206
  computing a postorder numbering, 358
  computing the number of descendents,
    359
  fat, 1050
  general, traversals of, 438
  level computation, 359
  path identification, 360
  reconstructing from traversals, 364
  spanning, 131
  SUM-OR, 655, 657, 664
Triangular matrices, 512
Triton, 731
True dependence, 852
Tuning, 941
Tuple space, 775, 850
Turing machine, 90, 93
  alternating, 98
  deterministic, 93
  generalized, 98
  nondeterministic, 93, 94
  probabilistic, 100
  random, 100
  stochastic, 101
Turing measures, 90
Twisted k-cube, 555

U
UCKT, 118
Ultracomputer, 113
Ultracomputing, 762
UMA, 704
Unconditional star hooking, 373

Unidirectional linear array, 505
UNITY, 34
Unless, 35
Useful parallelism, 236
Useless parallelism, 236
Utilization, 15

V
Validity, 151
Variable wavelength recording, 800
Variants of PRAMS, 107
Vector processors, 9, 1152
Vector summation, 483
Vectorization, 924
Very long instruction word, *see* VLIW
    processors
Viewpoints on ODP, 305
Virtual reality, 914
Virtual registers, 635
Virtual shared memory, 854, 994, 999
Virtual SIMD machines, 653
Virtual time, 1004
Visibility layers, 410
Visibility problems, 405
Vision systems, 1148
Visualization, 18, 897
    scientific, 909
VLIW processors, 634
VLSI, 7, 169, 186, 187, 500, 1144
    array processors, 650

bit-serial, 1152
complexity, 121
processor arrays, 501
Volume, 139
von Neumann machine, 169, 186
von Neumann model, 6, 223, 1144
Voronoi diagram, 405
VT, *see* Virtual time

W
Wait-free implementations, 156
Weighted control flow graph, 638
Window operations, 1052
Workstation
    array, 762
    cluster, 762
    farm, 762
Write buffers, 715
write-invalidate, 707
write-update, 707

X
Xab, 738
XNET, 668, 673, 674
XPRAM, 114

Z
Zig-zag grouping, 528
ZPNC(zero error randomized algorithm),
    110
ZPP class, 100

About the Editor

Albert Y. Zomaya is a senior lecturer in electrical and electronic engineering at the University of Western Australia, where he also leads the Parallel Computing Research Laboratory. Frequently published in technical journals, collaborative books, and conference papers, he is an associate editor of the *International Journal in Computer Simulation* and the *Control Engineering Practice Journal.* He is also the author of *Modelling and Simulation of Robot Manipulators: A Parallel-Processing Approach,* the coauthor of *Neuroadaptive Process Control: A Practical Approach,* and the editor of a volume titled *Parallel Computing: Paradigms and Applications.*

Dr. Zomaya is a member of the International Federation of Automatic Control Committees on Algorithms and Architectures for Real-Time Control and the Institute of Electrical and Electronic Engineers Technical Committee on Parallel Processing. He has served on the advisory boards and program committees of several national and international conferences. He is a chartered engineer and a member of the IEEE Computer Society, the Institute of Electrical Engineers (U.K.), and Sigma Xi. Dr. Zomaya's research interests lie in parallel processing, real-time systems, and adaptive computing systems.